ENCYCLOPEDIA OF
CRYPTOGRAPHY AND SECURITY

ENCYCLOPEDIA OF CRYPTOGRAPHY AND SECURITY

Editor-in-chief

Henk C.A. van Tilborg

Eindhoven University of Technology
The Netherlands

 Springer

Library of Congress Cataloging-in-Publication Data

A C.I.P. Catalogue record for this book is available from the Library of Congress.

Encyclopedia of Cryptography and Security, Edited by Henk C. A. van Tilborg

    p. cm.

ISBN-10: (HB) 0-387-23473-X
ISBN-13: (HB) 978-0387-23473-1
ISBN-10: (eBook) 0-387-23483-7
ISBN-13: (eBook) 978-0387-23483-0
Printed on acid-free paper.

Printed in the United States of America

9 8 7 6 5 4 3 2 1   SPIN 11327875 (HC) / 151464 (eBook)

springeronline.com

Dedicated to the ones I love

List of Advisory Board Members

List of Contributors

Carlisle Adams
Sacha Barg
Friedrich Bauer
Olivier Benoît
Eli Biham
Alex Biryukov
John Black
Robert Blakley
Gerrit Bleumer
Sharon Boeyen
Dan Boneh
Antoon Bosselaars
Gerald Brose
Marco Bucci
Mike Burmester
Christian Cachin
Tom Caddy
Ran Canetti
Anne Canteaut
Claude Carlet
Pascale Charpin
Hamid Choukri
Scott Contini
Claude Crépeau
Eric Cronin
Joan Daemen
Christophe De Canniere
Yvo Desmedt
Marijke de Soete
Yevgeniy Dodis
Glen Durfee
Cynthia Dwork
Carl Ellison
Toni Farley
Caroline Fontaine
Matthew Franklin
Martin Gagné
Daniel M. Gordon
Jorge Guajardo
Stuart Haber
Helena Handschuh
Darrel Hankerson
Clemens Heinrich
Tor Helleseth
Russ Housley

Hideki Imai
Anil Jain
Jill Joseph
Marc Joye
Mike Just
Gregory Kabatiansky
Burt Kaliski
Lars Knudsen
Çetin Kaya Koç
François Koeune
Hugo Krawczyk
Markus Kuhn
Peter Landrock
Kerstin Lemke
Arjen K. Lenstra
Paul Leyland
Benoît Libert
Moses Liskov
Steve Lloyd
Henri Massias
Patrick McDaniel
Alfred Menezes
Daniele Micciancio
Bodo Möller
François Morain
Dalit Naor
Kim Nguyen
Phong Q. Nguyen
Francis Olivier
Lukasz Opyrchal
Christof Paar
Pascal Paillier
Joe Pato
Sachar Paulus
Torben Pedersen
Benny Pinkas
David Pointcheval
Bart Preneel
Niels Provos
Jean-Jacques Quisquater
Vincent Rijmen
Ronald L. Rivest
Matt Robshaw
Arun Ross
Randy Sabett

Kazue Sako
David Samyde
Bruce Schneier
Berry Schoenmakers
Matthias Schunter
Nicolas Sendrier
Adi Shamir
Igor Shparlinski
Robert D. Silverman
Miles Smid
Jerome Solinas
Anton Stiglic
François-Xavier Standaert
Berk Sunar

Laurent Sustek
Henk van Tilborg
Assia Tria
Eran Tromer
Salil Vadhan
Pavan Verma
Colin Walter
Michael Ward
Andre Weimerskirch
William Whyte
Michael Wiener
Atsuhiro Yamagishi
Paul Zimmermann
Robert Zuccherato

Preface

The need to protect valuable information is as old as history. As far back as Roman times, Julius Caesar saw the need to encrypt messages by means of cryptographic tools. Even before then, people tried to hide their messages by making them "invisible." These hiding techniques, in an interesting twist of history, have resurfaced quite recently in the context of digital rights management. To control access or usage of digital contents like audio, video, or software, information is secretly embedded in the data!

Cryptology has developed over the centuries from an art, in which only few were skillful, into a science. Many people regard the "Communication Theory and Secrecy Systems" paper, by Claude Shannon in 1949, as the foundation of modern cryptology. However, at that time, cryptographic research was mostly restricted to government agencies and the military. That situation gradually changed with the expanding telecommunication industry. Communication systems that were completely controlled by computers demanded new techniques to protect the information flowing through the network.

In 1976, the paper "New Directions in Cryptography," by Whitfield Diffie and Martin Hellman, caused a shock in the academic community. This seminal paper showed that people who are communicating with each other over an insecure line can do so in a secure way with no need for a common secret key. In Shannon's world of secret key cryptography this was impossible, but in fact there was another cryptologic world of public-key cryptography, which turned out to have exciting applications in the real world. The 1976 paper and the subsequent paper on the RSA cryptosystem in 1978 also showed something else: mathematicians and computer scientists had found an extremely interesting new area of research, which was fueled by the ever-increasing social and scientific need for the tools that they were developing. From the notion of public-key cryptography, information security was born as a new discipline and it now affects almost every aspect of life.

As a consequence, information security, and even cryptology, is no longer the exclusive domain of research laboratories and the academic community. It first moved to specialized consultancy firms, and from there on to the many places in the world that deal with sensitive or valuable data; for example the financial world, the health care sector, public institutions, nongovernmental agencies, human rights groups, and the entertainment industry.

A rich stream of papers and many good books have been written on information security, but most of them assume a scholared reader who has the time to start at the beginning and work his way through the entire text. The time has come to make important notions of cryptography accessible to readers who have an interest in a particular keyword related to computer security or cryptology, but who lack the time to study one of the many books on computer and information security or cryptology. At the end of 2001, the idea to write an easily accessible encyclopedia on cryptography and information security was proposed. The goal was to make it possible to become familiar with a particular notion, but with minimal effort. Now, 4 years later, the project is finished, thanks to the help of many contributors, people who are all very busy in their professional life. On behalf of the Advisory Board, I would like to thank each of those contributors for their work. I would also like to acknowledge the feedback and help given by Mihir Bellare, Ran Canetti, Oded Goldreich, Bill Heelan, Carl Pomerance, and Samuel S. Wagstaff, Jr. A person who was truly instrumental for the success of this project is Jennifer Evans at Springer Verlag. Her ideas and constant support are greatly appreciated. Great help has been given locally by Anita Klooster and Wil Kortsmit. Thank you very much, all of you.

Henk van Tilborg

A

A5/1

A5/1 is the symmetric cipher used for encrypting over-the-air transmissions in the GSM standard. A5/1 is used in most European countries, whereas a weaker cipher, called A5/2, is used in other countries (a description of A5/2 and an attack can be found in [4]). The description of A5/1 was first kept secret but its design was reversed engineered in 1999 by Briceno, Golberg, and Wagner. A5/1 is a <u>synchronous stream cipher</u> based on <u>linear feedback shift registers (LFSRs)</u>. It has a 64-bit secret key.

A GSM conversation is transmitted as a sequence of 228-bit frames (114 bits in each direction) every 4.6 millisecond. Each frame is xored with a 228-bit sequence produced by the A5/1 running-key generator. The initial state of this generator depends on the 64-bit secret key, K, which is fixed during the conversation, and on a 22-bit public *frame number*, F.

The A5/1 running-key generator (see Figure 2) consists of three LFSRs of lengths 19, 22, and 23. Their characteristic polynomials are $X^{19} + X^5 + X^2 + X + 1$, $X^{22} + X + 1$, and $X^{23} + X^{15} + X^2 + X + 1$. For each frame transmission, the three LFSRs are first initialized (see Figure 1) to zero. Then, at time $t = 1, \ldots, 64$, the LFSRs are clocked, and the key bit K_t is xored to the feedback bit of each LFSR. For $t = 65, \ldots, 86$, the LFSRs are clocked in the same fashion, but the $(t - 64)$th bit of the frame number is now xored to the feedback bits.

After these 86 cycles, the generator runs as follows. Each LFSR has a clocking tap: tap 8 for the first LFSR, tap 10 for the second and the third ones (where the feedback tap corresponds to tap 0). At each unit of time, the majority value b of the three clocking bits is computed. A LFSR is clocked if and only if its clocking bit is equal to b. For instance, if the three clocking bits are equal to $(1, 0, 0)$, the majority value is 0. The second and third LFSRs are clocked, but not the first one. The output of the generator is then given by the xor of the outputs of the three LFSRs. After the 86 initialization cycles, 328 bits are generated with the previously described irregular clocking. The first 100 ones are discarded and the following 228 bits form the running-key.

Several time–memory trade-off attacks have been proposed on A5/1 [1, 2]. They require the knowledge of a few seconds of conversation plaintext and run very fast. But, they need a huge precomputation time and memory. Another attack due to Ekdahl and Johansson [3] exploits some weaknesses of the key initialization procedure. It requires a few minutes using 2–5 minutes of conversation plaintext without any notable precomputation and storage capacity.

Anne Canteaut

References

[1] Biham, E. and O. Dunkelman (2000). "Cryptanalysis of the A5/1 GSM stream cipher." *INDOCRYPT 2000*, Lecture Notes in Computer Science, vol. 1977, eds. B. Roy and E. Okamoto. Springer-Verlag, Berlin, 43–51.

[2] Biryukov, A., A. Shamir, and D. Wagner (2000). "Real time attack of A5/1 on a PC." *Fast Software Encryption 2000*, Lecture Notes in Computer Science, vol. 1978, ed. B. Schneier. Springer-Verlag, Berlin, 1–18.

[3] Ekdahl, P. and T. Johansson (2003). "Another attack on A5/1." *IEEE Transactions on Information Theory*, 49 (1), 284–289.

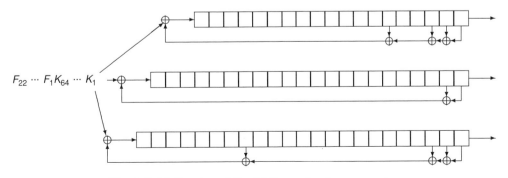

Fig. 1. Initialization of the A5/1 running-key generator

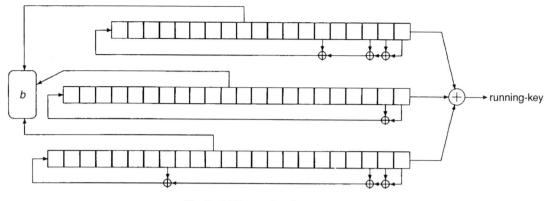

Fig. 2. A5/1 running-key generator

[4] Petrović, S. and A. Fúster-Sabater (2000). "Cryptanalysis of the A5/2 algorithm." Cryptology ePrint Archive, Report 2000/052. Available on http://eprint.iacr.org/

ABA DIGITAL SIGNATURE GUIDELINES

The American Bar Association provided a very elaborate, thorough, and detailed guideline on all the legal aspects of <u>digital signature schemes</u> and a <u>Public Key Infrastructure</u> (PKI) solution such as <u>X.509</u> at a time when PKI was still quite novel (1996). The stated purpose was to establish a safe harbor—a secure, computer-based signature equivalent—which will

1. minimize the incidence of electronic <u>forgeries</u>,
2. enable and foster the reliable <u>authentication</u> of documents in computer form,
3. facilitate commerce by means of computerized communications, and
4. give legal effect to the general import of the technical standards for authentication of computerized messages.

This laid the foundation for so-called *Certificate Policy Statements* (CPS) issued by <u>Certification Authorities</u> (CA), the purpose of which is to restrict the liability of the CA. It is fair to state that often these CPS are quite incomprehensible to ordinary users.

Peter Landrock

ACCESS CONTROL

Access control (also called *protection* or *authorization*) is a security function that protects shared resources against unauthorized accesses. The distinction between authorized and unauthorized accesses is made according to an *access control policy*. The resources which are protected by access control are usually referred to as *objects*, whereas the entities whose accesses are regulated are called *subjects*. A subject is an active system entity running on behalf of a human user, typically a process. It is not to be confused with the actual user.

Access control is employed to enforce security requirements such as *confidentiality* and *integrity* of data resources (e.g., files, database tables), to prevent the unauthorized use of resources (e.g., programs, processor time, expensive devices), or to prevent <u>denial of service</u> to legitimate users. Practical examples of security violations that can be prevented by enforcing access control policies are: a journalist reading a politician's medical record (confidentiality); a criminal performing fake bank account bookings (integrity); a student printing his essays on an expensive photo printer (unauthorized use); and a company overloading a competitor's computers with requests in order to prevent it from meeting a critical business deadline (denial of service).

ENFORCEMENT MECHANISM AND POLICY DECISION: Conceptually, all access control systems comprise two separate components: an *enforcement mechanism* and a *decision function*. The enforcement mechanism intercepts and inspects accesses, and then asks the decision function to determine if the access complies with the security policy or not. This is depicted in Figure 1.

An important property of any enforcement mechanism is the *complete mediation* property [17] (also called *reference monitor* property), which means that the mechanism must be able to intercept and potentially prevent *all* accesses to a resource. If it is possible to circumvent the enforcement mechanism no security can be guaranteed.

The complete mediation property is easier to achieve in centralized systems with a secure kernel than in distributed systems. General-purpose

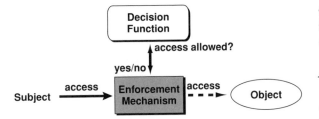

Fig. 1. Enforcement mechanism and decision function

operating systems, e.g., are capable of intercepting system calls and thus of regulating access to devices. An example for an enforcement mechanism in a distributed system is a packet filter <u>firewall</u>, which can either forward or drop packets sent to destinations within a protected domain. However, if any network destinations in the protected domain are reachable through routes that do not pass through the packet filter, then the filter is not a reference monitor and no protection can be guaranteed.

ACCESS CONTROL MODELS: An access control <u>policy</u> is a description of the allowed and denied accesses in a system. In more formal terms, it is a configuration of an *access control model*. In all practically relevant systems, policies can change over time to adapt to changes in the sets of objects, subjects, or to changes in the protection requirements. The model defines how objects, subjects, and accesses can be represented, and also the operations for changing configurations.

The model thus determines the flexibility and expressive power of its policies. Access control models can also be regarded as the languages for writing policies. The model determines how easy or difficult it is to express one's security requirements, e.g., if a rule like "all students except Eve may use this printer" can be conveniently expressed. Another aspect of the access model is which formal properties can be proven about policies, e.g., can a question like "Given this policy, is it possible that Eve can ever be granted this access?" be answered. Other aspects influenced by the choice of the access model are how difficult it is to manage policies, i.e., adapt them to changes (e.g., "can John propagate his permissions to others?"), and the efficiency of making access decisions, i.e. the complexity of the decision algorithm and thus the run-time performance of the access control system.

There is no single access model that is suitable for all conceivable policies that one might wish to express. Some access models make it easier than others to directly express confidentiality requirements in a policy ("military policies"), whereas others favor integrity ("commercial policies," [4]),

or allow to express history-based constraints ("Chinese Walls," [3]). Further detail on earlier security models can be found in [14].

Access Matrix Models

A straightforward representation of the allowed accesses of a subject on an object is to list them in a table or matrix. The classical *access matrix* model [12] represents subjects in rows, objects in columns, and permissions in entries. If an access mode *print* is listed in the matrix entry $M_{(Alice,LaserPrinter)}$, then the subject Alice may print-access the LaserPrinter object.

Matrix models typically define the sets of subjects, objects, and access modes ("rights") that they control directly. It is thus straightforward to express what a given subject may do with a given object, but it is not possible to directly express a statement like "all students except Eve may print." To represent the desired semantics, it is necessary to enter the access right *print* in the printer column for the rows of all subjects that are students, except in Eve's. Because this is a low-level representation of the policy statement, it is unlikely that administrators will later be able to infer the original policy statements by looking at the matrix, especially after a number of similar changes have been performed.

A property of the access matrix that would be interesting to prove is the *safety property*. The general meaning of *safety* in the context of protection is that no access rights can be leaked to an unauthorized subject, i.e. that there is no sequence of operations on the access matrix that, given some initial safe state, would result in an unsafe state. The proof by Harrison et al. [11] that safety is only decidable in very restricted cases is an important theoretical result of security research.

The access matrix model is simple, flexible, and widely used in practice. It is also still being extended and refined in various ways in the recent security literature, e.g., to represent both permissions and denials, to account for typed objects with specific rather than generic access modes, or for objects that are further grouped in domains.

Since the access matrix can become very large but is typically also very sparse, it is usually not stored as a whole, but either row-wise or column-wise. An individual matrix column contains different subjects' rights to access one object. It thus makes sense to store these rights per object as an *access control list* (ACL). A matrix row describes the access rights of a subject on all objects in the system. It is therefore appealing to store these rights per subject. From the subject's perspective, the row can be broken down to a list of access rights per object, or a *capability list*. The two approaches of implementing the matrix model using either ACLs or capabilities have different advantages and disadvantages.

Access Control Lists

An ACL for an object o is a list of tuples $(s, (r_1, \ldots, r_n))$, where s is a subject and the r_i are the rights of s on o. It is straightforward to associate an object's access control list with the object, e.g., a file, which makes it easy for an administrator to find out all allowed accesses to the object, or to revoke access rights.

It is not as easy, however, to determine a subject's allowed accesses because that requires searching all ACLs in the system. Using ACLs to represent access policies can also be difficult if the number of subjects in a system is very large. In this case, storing every single subject's rights results in long and unwieldy lists. Most practical systems therefore use additional aggregation concepts to reduce complexity, such as user groups or roles.

Another disadvantage of ACLs is that they do not support any kind of *discretionary access control* (DAC), i.e., ways to allow subjects to change the access matrix at their discretion. In the UNIX file system, e.g., every file object has a designated owner who may assign and remove access rights to the file to other subjects. If the recipient subject did not already possess this right, executing this command changes the state of the access matrix by entering a new right in a matrix entry. File ownership—which is not expressed in the basic access matrix—thus implies a limited form of administrative authority for subjects.

A second example of discretionary access control is the GRANT option that can be set in relational databases when a database administrator assigns a right to a user. If this option is set on a right that a subject possesses, this subject may itself use the GRANT command to propagate this right to another subject. This form of discretionary access control is also called *delegation*. Implementing controlled delegation of access rights is difficult, especially in distributed systems. In SQL, delegation is controlled by the GRANT option, but if this option is set by the original grantor of a right, the grantor cannot control which other subjects may eventually receive this right through the grantee. Delegation can only be prevented altogether.

In systems that support delegation there is typically also an operation to remove rights again. If the system's protection state after a revocation should be the same as before the delegation, removing a right from a subject which has delegated this right to other subjects requires transitively revoking the right from these grantees, too. This *cascading revocation* [9, 10] is necessary to prevent a subject from immediately receiving a revoked right back from one of its grantees.

Discretionary access control and delegation are powerful features of an access control system that make writing and managing policies easier when applications require or support cooperation between users. These concepts also support applications that need to express the delegation of some administrative authority to subjects. However, regular ACLs need to be extended to support DAC, e.g., by adding a meta-right GRANT and by tracing delegation chains. Delegation is more elegantly supported in systems that are based on capabilities or, more generally, credentials. A seminal paper proposing a general authorization theory and a logic that can express delegation is [13].

Capabilities and Credentials

An individual capability is a pair $(o, (r_1, \ldots, r_n))$, where o is the object and the $r_1, \ldots, r_n$ are access rights for o. Capabilities were first introduced as a way of protecting memory segments in operating systems [6, 8, 15, 16]. They were implemented as a combination of a reference to a resource (e.g., a file, a block of memory, a remote object) with the access rights to that resource. Capabilities were thus directly integrated with the memory addressing mechanism, as shown in Figure 2. Thus, the complete mediation property was guaranteed because there is no way of reaching an object without using a capability and going through the access enforcement mechanism.

The possession of a capability is sufficient to be granted access to the object identified by that capability. Typically, capability systems allow subjects to delegate access rights by passing on their capabilities, which makes delegation simple and flexible. However, determining who has access to a given object at a given time requires searching the capability lists of all subjects in the system. Consequently, blocking accesses to an object is more difficult to realize because access rights are not managed centrally.

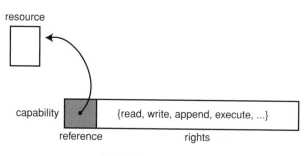

Fig. 2. A capability

Capabilities can be regarded as a form of <u>credentials</u>. A credential is a token issued by an authority that expresses a certain privilege of its bearer, e.g., that a subject has a certain access right, or is a member of an organization. A verifier inspecting a credential can determine three things: that the credential comes from a trusted authority, that it contains a valid privilege, and that the credential actually belongs to the presenter. A real-life analogy of a credential is registration badge, a driver's license, a bus ticket, or a membership card.

The main advantage of a credentials system is that verification of a privilege can be done, at least theoretically, off-line. In other words, the verifier does not need to perform additional communications with a decision function but can immediately determine if an access is allowed or denied. In addition, many credentials systems allow subjects some degree of freedom to delegate their credentials to other subjects. A bus ticket, e.g., may be freely passed on, or some organizations let members issue visitor badges to guests.

Depending on the environment, credentials may need to be authenticated and protected from theft. A bus ticket, e.g., could be reproduced on a photocopier, or a membership card stolen. Countermeasures against reproduction include holograms on expensive tickets, while the illegal use of a stolen driver's license can be prevented by comparing the photograph of the holder with the appearance of the bearer. Digital credentials that are created, managed, and stored by a trusted secure kernel do not require protection beyond standard memory protection. Credentials in a distributed system are more difficult to protect: <u>Digital signatures</u> may be required to authenticate the issuing authority, transport encryption to prevent eavesdropping or modification in transit, and binding the subject to the credential to prevent misuse by unauthorized subjects. Typically, credentials in distributed systems are represented in digital <u>certificates</u> such as <u>X.509</u> or <u>SPKI</u> [7], or stored in secure devices such as smart cards.

Role-Based Access Control (RBAC)

In the standard matrix model, access rights are directly assigned to subjects. This can be a manageability problem in systems with large numbers of subjects and objects that change frequently because the matrix will have to be updated in many different places. For example, if an employee in a company moves to another department, its subject will have to receive a large number of new access rights and lose another set of rights.

Aggregation concepts such as groups and roles were introduced specifically to make security

Fig. 3. The basic RBAC model

administration simpler. Because complex administrative tasks are inherently error-prone, reducing the potential for management errors also increases the overall security of a system. The most widely used role models are the family of models introduced in [19], which are called $\text{RBAC}_0, \ldots,$ RBAC_3. RBAC_0 is the base model that defines roles as a management indirection between users and permissions and is illustrated in Figure 3. Users are assigned to roles rather than directly to permissions, and permissions are assigned to roles.

The other role-based access control (RBAC) models introduce *role hierarchies* (RBAC_1) and *constraints* (RBAC_2). A role hierarchy is a partial order on roles that lets an administrator define that one role is senior to another role, which means that the more senior role inherits the junior role's permissions. For example, if a Manager role is defined to be senior to an Engineer role, any user assigned to the Manager role would also have the permissions assigned to the Engineer role.

Constraints are predicates over configurations of a role model that determine if the configuration is acceptable. Typically, role models permit the definition of mutual exclusion constraints to prevent the assignment of the same user to two conflicting roles, which can enforce separation of duty. Other constraints that are frequently mentioned include cardinality constraints to limit the maximum number of users in a role, or prerequisite role constraints, which express that, e.g., only someone already assigned to the role of an Engineer can be assigned to the Test-Engineer role. The most expressive model in the family is RBAC_3, which combines constraints with role hierarchies.

The role metaphor is easily accessible to most administrators, but it should be noted that the RBAC model family provides only an extensional definition of roles, so the meaning of the role concept is defined only in relation to users and permissions. Often, roles are interpreted in a task-oriented manner, i.e., in relation to a particular task or set of tasks, such as an Accountant role that is used to group the permissions for accounting. In principle, however, any concept that is perceived as useful for grouping users and permissions can be used as a role, even purely structural user groups such as IT-Department. Finding a suitable intensional definition is often an important prerequisite for modeling practical, real-life security policies in terms of roles.

Information Flow Models

The basic access matrix model can restrict the release of data, but it cannot enforce restrictions on the propagation of data after it has been read by a subject. Another approach to control the dissemination of information more tightly is based on specifying security not in terms of individual acess attempts, but rather in terms of the *information flow* between objects. The focus is thus not on protecting objects themselves, but the information contained *within* (and exchanged between) objects. An introduction to information flow models can be found in [18].

Since military security has traditionally been more concerned with controlling the release and propagation of information, i.e., confidentiality, than with protecting data against integrity violations, it is a good example for information flow security. The classic military security model defines four *sensitivity levels* for objects and four *clearance levels* for subjects. These levels are: unclassified, confidential, secret, and top secret. The classification of subjects and objects according to these levels is typically expressed in terms of *security labels* that are attached to subjects and objects.

In this model, security is enforced by controlling accesses so that any subject may only access objects that are classified at the same level for which the subject has clearance, or for a lower level. For example, a subject with a "secret" clearance is allowed access to objects classified as "unclassified," "confidential," and "secret," but not to those classified as "top secret." Information may thus only flow "upwards" in the sense that its sensitivity is not reduced. An object that contains information that is classified at multiple security levels at the same time is called a *multilevel object*.

This approach takes only the general sensitivity, but not the actual content of objects into account. It can be refined to respect the *need-to-know* principle. This principle, which is also called *principle of least privilege*, states that every subject should only have those permissions that are required for its specific tasks. In the military security model, this principle is enforced by designating *compartments* for objects according to subject areas, e.g., "nuclear." This results in a security classification that comprises both the sensitivity label and the compartment, e.g., "nuclear, secret." Subjects may have different clearance levels for different compartments.

The terms discretionary access control (DAC) and *mandatory access control* (MAC) originated in the military security model, where performing some kinds of controls was required to meet legal requirements ("mandatory"), viz. that classified information may only be seen by subjects with sufficient clearance. Other parts of the model, viz. determining whether a given subject with sufficient clearance also needs to know the information, involved some discretion ("discretionary").

The military security model (without compartmentalization) was formalized in [1]. This model defined two central security properties, the *simple security property* ("subjects may only read-access objects with a classification at or below their own clearance") and the *star-property* or *-property ("subjects may not write to objects with a classification below the subject's current security level"). The letter property ensures that a subject may not read information of a given sensitivity and write that information to another object at a lower sensitivity level, thus downgrading the original sensitivity level of the information. The model in [1] also included an ownership attribute for objects and the option to extend access to an object to another subject. The model was refined in [2] to address additional integrity requirements.

The permitted flow of information in a system can also more naturally be modeled as a *lattice* of security classes. These classes correspond to the security labels introduced above and are partially ordered by a flow relation "→" [5]. The set of security classes forms a lattice under "→" because a least upper bound and a greatest lower bound can be defined using a join operator on security classes. Objects are bound to these security classes. Information may flow from object a to b through any sequence of operations if and only if A "→" B, where A and B are the objects' security classes. In this model, a system is secure if no flow of information violates the flow relation.

Gerald Brose

References

[1] Bell, D.E. and L.J. LaPadula (1973). "Secure computer systems: A mathematical model." Mitre Technical Report 2547, vol. II.

[2] Biba, K.J. (1977). "Integrity considerations for secure computer systems." Mitre Technical Report 3153.

[3] Brewer, D. and M. Nash (1989). "The chinese wall security policy." *Proc. IEEE Symposium on Security and Privacy*, 206–214.

[4] Clark, D.D. and D.R. Wilson (1987). "A comparison of commercial and military computer security policies." *Proc. IEEE Symposium on Security and Privacy*, 184–194.

[5] Denning, D.E. (1976). "A lattice model of secure information flow." *Communications of the ACM*, 19 (5), 236–243.

[6] Dennis, J.B. and E.C. Van Horn (1966). "Programming semantics for multiprogrammed computations." *Communications of the ACM*, 9 (3), 143–155.

[7] Ellison, C.M., B. Frantz, B. Lampson, R. Rivest, B.M. Thomas, and T. Ylönen (1999). *SPKI Certificate Theory, RFC 2693*.

[8] Fabry, R.S. (1974). "Capability-based addressing." *Communications of the ACM*, 17 (7), 403–412.

[9] Fagin, R. (1978). "On an authorization mechanism." *ACM Transactions on Database Systems*, 3 (3), 310–319.

[10] Griffiths, P.P. and B.W. Wade (1976). "An authorization mechanism for a relational database system." *ACM Transactions on Database Systems*, 1 (3), 242–255.

[11] Harrison, M., W. Ruzzo, and J. Ullman (1976). "Protection in operating systems." *Communications of the ACM*, 19 (8), 461–471.

[12] Lampson, B.W. (1974). "Protection." *ACM Operating Systems Rev.*, 8 (1), 18–24.

[13] Lampson, B.W., M. Abadi, M. Burrows, and E. Wobber (1992). "Authentication in distributed systems: Theory and practice." *ACM Transactions on Computer Systems*, 10 (4), 265–310.

[14] Landwehr, C.E. (1981). "Formal models for computer security." *ACM Computing Surveys*, 13 (3), 247–278.

[15] Levy, H.M. (1984). *Capability-Based Computer Systems*. Butterworth-Heinemann, Newton, MA.

[16] Linden, T.A. (1976). "Operating system structures to support security and reliable software." *ACM Computing Surveys*, 8 (4), 409–445.

[17] Saltzer, J.H. and M.D. Schroeder (1975). "The protection of information in computer systems." *Proc. of the IEEE*, 9 (63), 1278–1308.

[18] Sandhu, R.S. (1993). "Lattice-based access control models." *IEEE Computer*, 26 (11), 9–19.

[19] Sandhu, R.S., E.J. Coyne, H.L. Feinstein, and C.E. Youman (1996). "Role-based access control models." *IEEE Computer*, 29 (2), 38–47.

ACCESS STRUCTURE

Let $\mathcal{P}$ be a set of parties. An access structure $\Gamma_{\mathcal{P}}$ is a subset of the powerset $2^{\mathcal{P}}$. Each element of $\Gamma_{\mathcal{P}}$ is considered *trusted*, e.g., has access to a shared secret (see *secret sharing scheme*). $\Gamma_{\mathcal{P}}$ is *monotone* if for each element of $\Gamma_{\mathcal{P}}$ each superset belongs to $\Gamma_{\mathcal{P}}$, formally: when $\mathcal{A} \subseteq \mathcal{B} \subseteq \mathcal{P}$ and $\mathcal{A} \in \Gamma_{\mathcal{P}}$, $\mathcal{B} \in \Gamma_{\mathcal{P}}$.

An *adversary structure* is the complement of an access structure; formally, if $\Gamma_{\mathcal{P}}$ is an access structure, then $2^{\mathcal{P}} \setminus \Gamma_{\mathcal{P}}$ is an adversary structure.

Yvo Desmedt

ACQUIRER

In retail payment schemes and electronic commerce, there are normally two parties involved, a customer and a shop. The Acquirer is the bank of the shop.

Peter Landrock

ADAPTIVE CHOSEN CIPHERTEXT ATTACK

An *adaptive* chosen ciphertext attack is a chosen ciphertext attack scenario in which the attacker has the ability to make his choice of the inputs to the decryption function based on the previous chosen ciphertext queries. The scenario is clearly more powerful than the basic chosen ciphertext attack and thus less realistic. However, the attack may be quite practical in the public-key setting. For example, plain RSA is vulnerable to chosen ciphertext attack (see RSA public-key encryption for more details) and some implementations of RSA may be vulnerable to *adaptive* chosen ciphertext attack, as shown by Bleichenbacher [1].

Alex Biryukov

Reference

[1] Bleichenbacher, D. (1998). "Chosen ciphertext attacks against protocols based on the RSA encryption standard PKCS#1." *Advances in Cryptology—CRYPTO'98*, Lecture Notes in Computer Science, vol. 1462, ed. H. Krawczyk. Springer-Verlag, Berlin, 1–12.

ADAPTIVE CHOSEN PLAINTEXT AND CHOSEN CIPHERTEXT ATTACK

In this attack the scenario allows the attacker to apply adaptive chosen plaintext and adaptive chosen ciphertext queries simultaneously. The attack is one of the most powerful in terms of the capabilities of the attacker. The only two examples of such attacks known to date are the boomerang attack [2] and the yoyo-game [1].

Alex Biryukov

References

[1] Biham, E., A. Biryukov, O. Dunkelman, E. Richardson, and A. Shamir (1999). "Initial observations on Skipjack: Cryptanalysis of Skipjack-3xor." *Selected Areas in Cryptography, SAC 1998*, Lecture Notes in Computer Science, vol. 1556, eds. S.E. Tavares and H. Meijer. Springer-Verlag, Berlin, 362–376.

[2] Wagner, D. (1999). "The boomerang attack." *Fast Software Encryption, FSE'99*, Lecture Notes in Computer Science, vol. 1636, ed. L.R. Knudsen. Springer-Verlag, Berlin, 156–170.

has the ability to make his choice of the inputs to the encryption function based on the previous chosen plaintext queries and their corresponding ciphertexts. The scenario is clearly more powerful than the basic chosen plaintext attack, but is probably less practical in real life since it requires interaction of the attacker with the encryption device.

Alex Biryukov

ADAPTIVE CHOSEN PLAINTEXT ATTACK

An *adaptive* chosen plaintext attack is a chosen plaintext attack scenario in which the attacker

plaintext	a	b	c	d	e	f	g	h	i	j	k	l	m	n	o	p	q	r	s	t	u	v	w	x	y	z
ciphertext	B	E	K	P	I	R	C	H	S	Y	T	M	O	N	F	U	A	G	J	D	X	Q	W	Z	L	V

plaintext	q	a	g	t	b	o	r	h	e	s	c	y	l	n	m	d	v	f	i	k	p	z	w	u	j	x
ciphertext	A	B	C	D	E	F	G	H	I	J	K	L	M	N	O	P	Q	R	S	T	U	V	W	X	Y	Z

ALBERTI ENCRYPTION

This is a polyalphabetic encryption with shifted, mixed alphabets.

As an example, let the mixed alphabet be given by:

	m	n	o	p	q	r	s	t	u	v	w	x	y	z
plaintext	a	b	c	d	e	f	g	h	i	j	k	l		
ciphertext	B	E	K	P	I	R	C	H	S	Y	T	M		
	O	N	F	U	A	G	J	D	X	Q	W	Z	L	V

or, reordered for decryption:

plaintext	q	a	g	t	b	o	r	h	e	s	c	y	l	n	m	d	v	f	i	k	p	z	w	u	j	x
ciphertext	A	B	C	D	E	F	G	H	I	J	K	L	M	N	O	P	Q	R	S	T	U	V	W	X	Y	Z

Modifying accordingly, the headline of a *Vigenère table* (see Vigenère cryptosystem) gives the *Alberti table*:

	q	a	g	t	b	o	r	h	e	s	c	y	l	n	m	d	v	f	i	k	p	z	w	u	j	x
A	A	B	C	D	E	F	G	H	I	J	K	L	M	N	O	P	Q	R	S	T	U	V	W	X	Y	Z
B	B	C	D	E	F	G	H	I	J	K	L	M	N	O	P	Q	R	S	T	U	V	W	X	Y	Z	A
C	C	D	E	F	G	H	I	J	K	L	M	N	O	P	Q	R	S	T	U	V	W	X	Y	Z	A	B
D	D	E	F	G	H	I	J	K	L	M	N	O	P	Q	R	S	T	U	V	W	X	Y	Z	A	B	C
E	E	F	G	H	I	J	K	L	M	N	O	P	Q	R	S	T	U	V	W	X	Y	Z	A	B	C	D
F	F	G	H	I	J	K	L	M	N	O	P	Q	R	S	T	U	V	W	X	Y	Z	A	B	C	D	E
G	G	H	I	J	K	L	M	N	O	P	Q	R	S	T	U	V	W	X	Y	Z	A	B	C	D	E	F
H	H	I	J	K	L	M	N	O	P	Q	R	S	T	U	V	W	X	Y	Z	A	B	C	D	E	F	G
I	I	J	K	L	M	N	O	P	Q	R	S	T	U	V	W	X	Y	Z	A	B	C	D	E	F	G	H
J	J	K	L	M	N	O	P	Q	R	S	T	U	V	W	X	Y	Z	A	B	C	D	E	F	G	H	I
K	K	L	M	N	O	P	Q	R	S	T	U	V	W	X	Y	Z	A	B	C	D	E	F	G	H	I	J
L	L	M	N	O	P	Q	R	S	T	U	V	W	X	Y	Z	A	B	C	D	E	F	G	H	I	J	K
M	M	N	O	P	Q	R	S	T	U	V	W	X	Y	Z	A	B	C	D	E	F	G	H	I	J	K	L
N	N	O	P	Q	R	S	T	U	V	W	X	Y	Z	A	B	C	D	E	F	G	H	I	J	K	L	M
O	O	P	Q	R	S	T	U	V	W	X	Y	Z	A	B	C	D	E	F	G	H	I	J	K	L	M	N
P	P	Q	R	S	T	U	V	W	X	Y	Z	A	B	C	D	E	F	G	H	I	J	K	L	M	N	O
Q	Q	R	S	T	U	V	W	X	Y	Z	A	B	C	D	E	F	G	H	I	J	K	L	M	N	O	P
R	R	S	T	U	V	W	X	Y	Z	A	B	C	D	E	F	G	H	I	J	K	L	M	N	O	P	Q
S	S	T	U	V	W	X	Y	Z	A	B	C	D	E	F	G	H	I	J	K	L	M	N	O	P	Q	R
T	T	U	V	W	X	Y	Z	A	B	C	D	E	F	G	H	I	J	K	L	M	N	O	P	Q	R	S
U	U	V	W	X	Y	Z	A	B	C	D	E	F	G	H	I	J	K	L	M	N	O	P	Q	R	S	T
V	V	W	X	Y	Z	A	B	C	D	E	F	G	H	I	J	K	L	M	N	O	P	Q	R	S	T	U
W	W	X	Y	Z	A	B	C	D	E	F	G	H	I	J	K	L	M	N	O	P	Q	R	S	T	U	V
X	X	Y	Z	A	B	C	D	E	F	G	H	I	J	K	L	M	N	O	P	Q	R	S	T	U	V	W
Y	Y	Z	A	B	C	D	E	F	G	H	I	J	K	L	M	N	O	P	Q	R	S	T	U	V	W	X
Z	Z	A	B	C	D	E	F	G	H	I	J	K	L	M	N	O	P	Q	R	S	T	U	V	W	X	Y

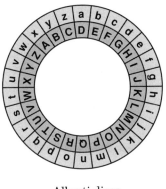

Alberti discs

An encryption example with the keytext "GOLD" of length 4 is:

plaintext	m	u	c	h	h	a	v	e	i	t	r	a	v	e	l	l	e	d
keytext	*G*	*O*	*L*	*D*	*G*	*O*	*L*	*D*	*G*	*O*	*L*	*D*	*G*	*O*	*L*	*D*	*G*	*O*
ciphertext	U	L	V	K	N	P	B	L	Y	R	R	E	W	W	X	P	O	D

Friedrich L. Bauer

Reference

[1] Bauer, F.L. (1997). "Decrypted secrets." *Methods and Maxims of Cryptology*. Springer-Verlag, Berlin.

ALPHABET

An alphabet is a set of characters (literals, figures, other symbols) together with a strict ordering (denoted by $<$) of this set. For good reasons it is usually required that a set of alphabetic characters has at least two elements and that it is finite. An alphabet Z of n elements is denoted Z_n, the order is usually the one of the listing.

$Z_{26} = \{a, b, c, \ldots, x, y, z\}$ is the common alphabet of *Latin* letters of present days. In former times and cultures, the Latin letter alphabet was smaller, so

$Z_{21} = Z_{26} \setminus \{j, k, w, x, y\}$ in Italian until about 1925,

$Z_{24} = Z_{26} \setminus \{k, w\}$ in Spanish until about 1950,

$Z_{25} = Z_{26} \setminus \{w\}$ in French and Swedish until about 1900.

In the Middle Ages, following the Latin tradition, 20 letters seem to have been enough for most writers (with v used for u),

$Z_{20} = Z_{26} \setminus \{j, k, u, w, x, y\}$.

Sometimes, mutated vowels and consonants like ä, ö, ü, ß (German), æ, œ (French), å, ø (Scandinavian), ł (Polish), č, ě, ř, š, ž (Czech) occur in literary texts, but in cryptography there is a tendency to suppress or transcribe them, i.e. to avoid diacritic marks.

The (present-day) *Cyrillic* alphabet has 32 letters (disregarding Ё):

$Z_{32} = \{$А, Б, В, Г, Д, Е, Ж, З, И, Й, К, Л, М, Н, О, П, Р, С, Т, У, Ф, Х, Ц, Ч, Ш, Щ, Ъ, Ы, Ь, Э, Ю, Я$\}$.

A set of m-tuples formed by elements of some set V is denoted V^m. If Z is an alphabet, Z^m has usually the lexicographic order based on the order of Z.

In mathematics and also in modern cryptography, the denotation $\mathbb{Z}_n$ is usually reserved for the set $\{0, 1, 2, \ldots, n-1\}$. It makes arithmetic modulo n possible (see modular arithmetic). Of course, $Z_{26} = \{a, b, c, \ldots, x, y, z\}$ can and often will be identified with $\mathbb{Z}_{26}$.

The following *number* alphabets are of particular historical interest:

$\mathbb{Z}_{10} = \{0, 1, 2, \ldots, 9\}$ (*denary* alphabet) with $0 < 1 < 2 < \cdots < 9$,

$\mathbb{Z}_4 = \{0, 1, 2, 3\}$ (*quaternary* alphabet) with $0 < 1 < 2 < 3$ (Alberti 1466),

$\mathbb{Z}_3 = \{0, 1, 2\}$ (*ternary alphabet*) with $0 < 1 < 2$ (Trithemius 1518),

$\mathbb{Z}_2 = \{0, 1\}$ (*binary alphabet*) with $0 < 1$ (Francis Bacon 1605). An element from $\mathbb{Z}_2$ is called *bit*, from bi(nary digi)t.

The technical utilization of the binary alphabet $\mathbb{Z}_2$ goes back to Jean Maurice Émile Baudot, 1874; at present one mainly uses quintuples and octuples of binary digits (called *bytes*).

The alphabet of m-tuples formed by elements of $\mathbb{Z}_n$ and ordered lexicographically is denoted $\mathbb{Z}_n^m$:

$Z_{32} = \mathbb{Z}_2^5$ (*teletype* alphabet or CCIT2 code), its cryptographic use goes back to Gilbert S. Vernam, 1917.

$Z_{256} = \mathbb{Z}_2^8$ (*bytes* alphabet), IBM ca. 1964 (cryptographic use by Horst Feistel, 1973).

Note that from a mathematical point of view, $\mathbb{Z}_{32} = \{0, 1, 2, \ldots, 31\}$ is not the same as $\mathbb{Z}_2^5 = \{(00000), (00001), (00010), (00011), (00100), \ldots, (11111)\}$. Of course, these two sets have the same cardinality, but arithmetically that does not make them the same. This can be seen from the way addition is defined for the elements of $\mathbb{Z}_{32}$ and $\mathbb{Z}_2^5$; while in $\mathbb{Z}_{32}$ arithmetic is done modulo 32, in $\mathbb{Z}_2^5$ every element added to itself gives (00000).

We mention the following alphabets:

standard alphabet: alphabet listed in its regular order.

mixed alphabet: standard alphabet listed in some permuted order.

reversed alphabet: standard alphabet listed in some backwards order.

shifted alphabet: standard alphabet listed with a cyclically shifted order.

A *vocabulary* is a set of characters (usually a standard alphabet), or of words, and/or phrases (usually alphabetically ordered), used to formulate the plaintext (*plaintext vocabulary*) or the ciphertext (*ciphertext vocabulary*) (see cryptosystem).

Friedrich L. Bauer

Reference

[1] Bauer, F.L. (1997). "Decrypted secrets." *Methods and Maxims of Cryptology*. Springer-Verlag, Berlin.

ANONYMITY

Anonymity of an individual is the property of being indistinguishable from other individuals in a certain respect. On the Internet, individuals may seek anonymity in sending certain messages, accessing certain chat rooms, publishing certain papers, etc. Consider a particular system, e.g., an electronic voting scheme, with participants P_1, $P_2, \ldots, P_n$ who seek anonymity with respect to a certain class A of action types $A_1, A_2, \ldots, A_m$, e.g., casting ballots B_1 (for candidate 1), B_2 for candidate 2, and so forth to B_m for candidate m, against an attacker who observes the system. In this system, anonymity with respect to the class A of action types means that for each i, the attacker cannot distinguish participant P_j ($1 \leq j \leq n$) executing action type A_i, denoted $[P_j : A_i]$, from any other participant P_k ($1 \leq k \leq n$) executing action type A_i. Expressed in terms of unlinkability, anonymity with respect to A means that for each action type A_i ($1 \leq i \leq m$) and each two participants P_j, P_k, the two events $[P_j : A_i]$ and $[P_k : A_i]$ are unlinkable (by the attacker). In this case, the *anonymity set* of the event $[P_j : A_i]$ is the set of all individuals $P_1, P_2, \ldots, P_n$, i.e., those who the attacker cannot distinguish from P_j when they execute action type A_i [3]. Sometimes, the anonymity set is more adequately defined in probabilistic terms as the set of all individuals who the attacker cannot distinguish with better than a small probability, which needs to be defined.

The anonymity set of an event is a volatile quantity that is beyond control of a single individual and typically changes significantly in size over time. For example, at the start of the voting period, only few participants may have reached the voting booths, while in the afternoon almost everyone may have cast his vote. Hence, soon after the start of the system, an attacker may not have a hard time guessing who has cast a particular vote he sees is cast in the system.

In order to apply this notion to a particular cryptographic scheme, the attacker model needs to be specified further. For example, is it a *passive attacker* such as an eavesdropper, or is it an *active attacker* (see cryptanalysis)? If passive, which communication lines can he observe and when. If active, how can he interact with the honest system participants (e.g., oracle access) and thereby stimulate certain behavior of the honest participants, or how many honest participants can he control entirely? (The number of honest participants an attacker can control without breaking a system is sometimes called the resilience of the system.) Is the attacker *computationally restricted* or *computationally unrestricted* (see computational security)? Based on a precise attacker model, anonymity can be defined with respect to specific classes of critical actions types, i.e., actions types of particular concern to the honest participants. Examples of critical actions are withdrawing and paying amounts in an electronic cash scheme, getting credentials issued and using them in an electronic credential scheme, casting ballots in electronic voting schemes, etc.

A measure of anonymity is the strength of the attacker model against which anonymity holds and the sizes of all anonymity sets. The stronger the attacker model is, the stricter the anonymity sets are defined, and the larger the sizes of all anonymity sets are, the stronger anonymity is achieved.

An important tool to achieve anonymity is pseudonyms [1, 2, 4]. Specific examples of anonymity are sender anonymity, recipient anonymity, and relationship anonymity. Sender anonymity can be achieved if senders use pseudonyms for sending messages, recipient anonymity can be achieved if recipients use pseudonyms for receiving messages, and relationship anonymity can be achieved if any two individuals use a joint pseudonym for sending and receiving messages to and from each other.

Anonymity can be regarded the opposite extreme of complete identifiability (accountability). Either extreme is often undesirable. The whole continuum between anonymity and complete identifiability is called *pseudonymity*. Pseudonymity is the use of pseudonyms as IDs for individuals. The use of pseudonyms may be rare, occasional, or frequent, and may be fully deliberate.

Gerrit Bleumer

References

[1] Chaum, David (1981). "Untraceable electronic mail, return addresses, and digital pseudonyms." *Communications of the ACM*, 24 (2), 84–88.

[2] Chaum, David (1986). "Showing credentials without identification—signatures transferred between unconditionally unlinkable pseudonyms." *Advances in Cryptology—EUROCRYPT'85*, Lecture Notes in Computer Science, vol. 219, ed. F. Pichler. Springer-Verlag, Berlin, 241–244.

[3] Pfitzmann, Andreas and Marit Köhntopp (2001). "Anonymity, unobservability, and pseudonymity—a proposal for terminology." *Designing Privacy Enhancing Technologies*, Lecture Notes in Computer Science, vol. 2009, ed. H. Frederrath. Springer-Verlag, Berlin, 1–9.

[4] Rao, Josyula R. and Pankaj Rohatgi (2000). "Can pseudonyms really guarantee privacy?" *9th Usenix Symposium, August 2000*.

ASYMMETRIC CRYPTOSYSTEM

The type of cryptography in which different keys are employed for the operations in the cryptosystem (e.g., encryption and decryption), and where one of the keys can be made public without compromising the secrecy of the other keys. See public-key encryption, digital signature scheme, key agreement, and (for the contrasting notion) symmetric cryptosystem.

Burt Kaliski

ATTRIBUTE CERTIFICATE

This is a certificate, i.e. a message digitally signed by some recognized Trusted Third Party, the content of which ties certain attributes to an ID, i.e. a user-ID. In the wake of the first PKI-euphoria (see Public Key Infrastructure), it was anticipated that there would be a great need for attribute certificates, and we may still come to see useful realizations of this concept. The original idea goes back to an early European project on PKI, where attribute certificates were introduced to represent e.g. power of attorney, executive rights etc., infor-

mation which currently is stored as official information on registered companies.

Peter Landrock

ATTRIBUTES MANAGEMENT

Attributes management is a subset of general "authorization data" management (see authorization architecture) in which the data being managed is *attributes* associated with entities in an environment. An *attribute* may be defined as follows [1]: "an inherent characteristic; an accidental quality; an object closely associated with or belonging to a specific person, thing, or office."

Carlisle Adams

Reference

[1] Merriam-Webster OnLine, http://www.m-w.com/cgi-bin/dictionary

AUTHENTICATED ENCRYPTION

INTRODUCTION: Often when two parties communicate over a network, they have two main security goals: privacy and authentication. In fact, there is compelling evidence that one should never use encryption without also providing authentication [8, 14]. Many solutions for the privacy and authentication problems have existed for decades, and the traditional approach to solving both simultaneously has been to combine them in a straightforward manner using so-called generic composition. However, recently there have been a number of new constructions which achieve both privacy and authenticity simultaneously, often much faster than any solution which uses generic composition. In this article we will explore the various approaches to achieving both privacy and authenticity, the so-called Authenticated Encryption problem. We will often abbreviate this as simply "AE." We will start with generic composition methods and then explore the newer combined methods.

Background

Throughout this article we will consider the AE problem in the "symmetric-key model." This means that we assume our two communicating parties, traditionally called "Alice" and "Bob," share a copy of some bit-string K, called the "key." This key is typically chosen at random and then distributed to Alice and Bob via one of various methods. This is the starting point for our work. We now wish to provide Alice and Bob with an AE algorithm such that Alice can select a message M from a predefined message-space, process it with the AE algorithm along with the key (and possibly a "nonce" N–a counter or random value), and then send the resulting output to Bob. The output will be the ciphertext C, the nonce N, and a short message authentication tag, σ. Bob should be able to recover M just given C, N, and his copy of the key K. He should also be able to certify that Alice was the originator by computing a verification algorithm using the above values along with the tag σ.

But what makes an AE algorithm "good?" We may have many requirements, and the relative importance of these requirements may vary according to the problem domain. Certainly one requirement is that the AE algorithm be "secure." We will speak more about what this means in a moment. But many other attributes of the algorithm may be important for us as well: performance, portability, simplicity/elegance, parallelizability, availability of reference implementations, or freedom from patents; we will pay attention to each of these concerns to varying levels as well.

Security

Certainly an AE scheme is not going to serve our needs unless it is secure. An AE scheme has two goals: privacy and authenticity. And each of these goals has a precise mathematical meaning [2, 3, 19]. In addition there is a precise definition for "authenticated encryption," the combination of both goals [5, 6, 26]. It would take us too far afield to carefully define each notion, but we will give a brief intuitive idea of what is meant. In our discussion we will use the term "adversary" to mean someone who is trying to subvert the security of the AE scheme, who knows the definition of the AE scheme, but who does not possess the key K.

Privacy means, intuitively, that a passive adversary who views the ciphertext C and the nonce N cannot "understand" the content of the message M. One way to achieve this is to make C indistinguishable from random bits, and indeed this is one definition of security for an encryption scheme that is sometimes used, although it is quite a strong one.

Authenticity means, intuitively, that an *active* adversary cannot successfully fabricate a cipher-

text C, a nonce N, and a tag σ in such a way that Bob will believe that Alice was the originator. In the formal security model we allow the adversary to generate tags for messages of his choice as if he were Alice for some period of time, and then he must attempt a forgery. We do not give him credit for simply "replaying" a previously generated message and tag, of course: he must construct a new value. If he does so with any significant probability of success, the authentication scheme is considered insecure.

Associated data

In many application settings we wish not only to encrypt and authenticate message M, but we wish also to include auxiliary data H which should be authenticated, but left unencrypted. An example might be a network packet where the payload should be encrypted (and authenticated) but the header should be unencrypted (and authenticated). The reason being that routers must be able to read the headers of packets in order to know how to properly route them.

This need spurred some designers of AE schemes to allow "associated data" to be included as input to their schemes. Such schemes have been termed AEAD (authenticated encryption with associated data) schemes, a notion which was first formalized by Rogaway [32]. As we will see, the AEAD problem is easily solved in the generic composition setting, but can become challenging when designing the more complex schemes. In his paper, Rogaway describes a few simple, but limited, ways to include associated data in any AE scheme, and then presents a specific method to efficiently add associated data to the OCB scheme, which we discuss below.

Provable security

One unfortunate aspect of most cryptographic schemes is that we cannot prove that any scheme meets the formal goals required of it. However, we can prove *some* things related to security, but it depends on the *type* of cryptographic object we are analyzing. If the object is a "primitive," such as a block cipher, no proof of security is possible, so instead we hope for security once we have shown that no known attacks (e.g., differential cryptanalysis) seem to work. However, for algorithms which are built on top of these primitives, called "modes," we *can* prove some things about their security; namely that they are as secure as the primitives which underlie them. Almost all of the AE schemes we will describe here are modes; only two of them are primitives.

Scheme	#Passes	Provably Secure	Assoc Data	Parallelizable	On-line	Patent-Free
IAPM	1	✓		✓	✓	
XECB	1	✓		✓	✓	
OCB	1	✓		✓	✓	
CCM	2	✓	✓			✓
EAX	2	✓	✓		✓	✓
CWC	2	✓	✓	✓	✓	✓
Helix	1		✓		✓	✓
SOBER-128	1		✓		✓	✓

Fig. 1. A comparison of the various AE schemes. Generic composition is omitted since answers would depend on the particular instantiation. For the schemes which do not support associated data, subsequent methods have been suggested to remedy this; for example, see [32]

AE schemes

The remainder of this article is devoted to the description and discussion of various AE algorithms. For convenience we list them in Figure 1. Note that we omit generic composition from the table since this approach comprises a class of schemes rather than a particular scheme.

Conventions

Let ϵ denote the empty string. Let Σ^n denote the set of all n-bit strings. In general, if S is a set we write S^+ to mean 1 or more repetitions of elements from S; that is, the set $\{s_1 s_2 \cdots s_m \mid m > 0, s_i \in S, 1 \le i \le m\}$. Thus $(\Sigma^n)^+$ is the set of all binary strings whose lengths are a positive multiple of n. If we write S^* we mean zero or more repetitions of elements from S. In other words, $S^* = S^+ \cup \{\epsilon\}$. We write $A \oplus B$ to mean the exclusive-or of strings A and B.

Many of our schemes use a block cipher. Throughout, n will be understood to be the block size of the underlying block cipher and k will be the size of its key. For block cipher E, we will write $E_K(P)$ to indicate invocation of block cipher E using the k-bit key K on the n-bit plaintext block P.

In order to process a message $M \in (\Sigma^n)^+$ we will often wish to break M into m strings, $M_1, \ldots, M_m$, each having n-bits such that $M = M_1 M_2 \cdots M_m$. For brevity, we will say "write $M = M_1 \cdots M_m$" and understand it to mean the above.

GENERIC COMPOSITION: Although AE did not get a formal definition until recently, the goal has certainly been implicit for decades. The traditional way of achieving both authenticity and privacy was to simply find an algorithm which yields each one and then use the combination of these two algorithms on our message. Intuitively it seems that this approach is obvious, straightforward, and completely safe. Unfortunately, there are many pitfalls accidentally "discovered" by well-meaning protocol designers.

One commonly made mistake is the assumption that AE can be achieved by using a non-cryptographic non-keyed hash function h and a good encryption scheme like CBC mode (Cipher Block Chaining mode; see modes of operation of a block cipher) with key K and initialization vector N. One produces $\mathrm{CBC}_{K,N}(M, h(M))$ and hopes this yields a secure AE scheme. However, these schemes are virtually always broken. Perhaps the best-known example is the Wired Equivalent Privacy (WEP) protocol used with 802.11 wireless networks. This protocol instantiates h as a Cyclic Redundancy Code (CRC) and then uses a stream cipher to encrypt. Borisov et al. showed, among other things, that it was easy to circumvent the authentication mechanism [15].

Another common pitfall is "key reuse." In other words, using some key K both for the encryption scheme and the MAC algorithm. This approach appliedly blindly almost always fails. We will later see that all of our "combined modes," listed after this section, do in fact use a single key, but they are carefully designed to retain security in spite of this.

It is now clear to researchers that one needs to use a *keyed* hash (i.e., a MAC) with some appropriate key $K1$ along with a secure encryption scheme with an independent key $K2$. However, it is unclear in what order these modes should be applied to a message M in order to achieve authenticated encryption. There are three obvious choices:

- *MtE:* MAC-then-Encrypt. We first MAC M under key $K1$ to yield tag σ and then encrypt the resulting pair (M, σ) under key $K2$.
- *EtM:* Encrypt-then-MAC. We first encrypt M under key $K2$ to yield ciphertext C and then compute $\sigma \leftarrow \mathrm{MAC}_{K1}(C)$ to yield the pair (C, σ).

- *E&M:* Encrypt-and-MAC. We first encrypt M under key $K2$ to yield ciphertext C and then compute $\sigma \leftarrow \text{MAC}_{K1}(M)$ to yield the pair (C, σ).

Also note that decryption and verification are straightforward for each approach above: for MtE decrypt first, then verify. For EtM and E&M verify first, then decrypt.

Security

In 2000, Bellare and Namprempre gave formal definitions for AE [5], and then systematically examined each of the three approaches described above in this formal setting. Their results show that if the MAC has a property called "strongly unforgeable," then it possible to achieve the strongest definition of security for AE only via the EtM approach. They further show that some known-good encryption schemes fail to provide privacy in the AE setting when using the E&M approach, and fail to provide a slightly stronger notion of privacy with the MtE approach.

These theoretical results generated a great deal of interest since three major pre-existing protocols, SSL/TLS (see Secure Socket Layer and Transport Layer Security), IPSec, and SSH, each used a different one of these three approaches: the SSL/TLS protocol uses MtE, IPSec uses EtM, and SSH uses E&M. One might think that perhaps security flaws exist in SSL/TLS and SSH because of the results of Bellare and Namprempre; however, concurrent with their work, Krawczyk showed that SSL/TLS was in fact secure because of the encoding used alongside the MtE mechanism [29]. And later Bellare, Kohno, and Namprempre showed that despite some identified security flaws in SSH, it could be made provably secure via a number of simple modifications despite its E&M approach.

The message here is that EtM with a provably secure encryption scheme and a provably secure MAC each with independent keys is the best approach for achieving AE. Although MtE and E&M can be secure, security will often depend on subtle details of how the data are encoded and on the particular MAC and encryption schemes used.

Performance

Simple methods for doing very fast encryption have been known for quite some time. For example, CBC mode encryption has very little overhead beyond the calls to the block cipher. Even more attractive is CTR mode (CounTeR mode; see modes of operation of a block cipher), which similarly has little overhead and in addition is paralleliz-

able. However, MACing quickly is not so simple. The CBC MAC (Cipher Block Chaining Message Authentication Code; see CBC MAC and variants) is quite simple and just as fast as CBC mode encryption, but there are well-known ways to go faster. The fastest software MAC in common use today is HMAC [1, 20]. HMAC uses a cryptographic hash function to process the message M and this is faster than processing M block-by-block with a block cipher. However even faster approaches have been invented using the Wegman–Carter construction [34]. This approach involves using a non-cryptographic hash function to process M, and then uses a cryptographic function to process the hash output. The non-cryptographic hash is randomly selected from a carefully designed family of hash functions, all with a common domain and range. The goal is to produce a family such that distinct messages are unlikely to hash to the same value when the hash function is randomly chosen from that family. This is the so-called universal hash family [16]. The fastest known MACs are based on the Wegman–Carter approach. The speed champions are UMAC [11] and hash127 [10], though neither of these are in common use yet.

Associated data

As we mentioned in the introduction, it is a common requirement in cryptographic protocols that we allow authenticated but non-encrypted data to be included in our message. Although the single-pass modes we describe next do not naturally allow for associated data, due to the fact that their encryption and authentication methods are intricately interwoven, we do not have this problem with generically composed schemes. Since the encryption and MAC schemes are entirely independent, we simply run the MAC on all the data and run the encryption scheme only on the data to be kept private.

Can we do better?

One obvious question when considering generically composed AE schemes is "can we do better?" In other words, might there be a way of achieving AE without using two different algorithms, with two different keys, and making two separate passes over the message. The answer is "yes," and a discussion of these results constitutes the remainder of this article.

SINGLE-PASS COMBINED MODES: It had long been a goal of cryptographers to find a mode of operation which achieved AE using only a single pass over the message M. Many attempts were made at such schemes, but all were broken. Therefore, until the year 2000, people still used generic composition to achieve AE, which as we have seen requires two passes over M.

IAPM

In 2000, Jutla at IBM invented two schemes which were the first correct single-pass AE modes [25]. He called these modes IACBC (Integrity-Aware Cipher Block Chaining) and IAPM (Integrity-Aware Parallelizable Mode). The first mode somewhat resembles CBC-mode encryption; however, offsets were added in before and after each block-cipher invocation, a technique known as "whitening." However, as we know, CBC-mode encryption is inherently serial: we cannot begin computation for the $(k+1)$th block-cipher invocation until we have the result of the kth invocation. Therefore, more interest has been generated around the second mode, IAPM, which does not have this disadvantage. Let's look at how IAPM works.

IAPM accepts a message $M \in (\Sigma^n)^+$, a nonce $N \in \Sigma^n$, and a key pair $K1, K2$ each selected from Σ^k for use with the underlying block cipher E. The key pair is set up and distributed in advance between the communicating parties; the keys are reused for a large number of messages. However, N and (usually) M vary with each transmission. First we break M into $M_1 \cdots M_{m-1}$ and proceed as follows.

There are two main steps: (1) offset generation and (2) encryption/tag generation. For offset generation we encipher N to get a seed value, and then encipher sequential seed values to get the remaining seed values. In other words, set $W_1 \leftarrow E_{K2}(N)$ and then set $W_i \leftarrow E_{K2}(W_1 + i - 2)$ for $2 \leq i \leq t$ where $t = \lceil \lg(m+2) \rceil$. Here lg means $\log_2$, so if we had a message M with 256 n-bit blocks, we would require $\lceil \lg(259) \rceil = 9$ block-cipher invocations to generate the W_i values. Finally, to derive our $m + 1$ offsets from the seed values, for i from 1 to $m + 1$, we compute $S_{i-1} \leftarrow \bigoplus_{j=1}^{t}(i[j] \cdot W_j)$ where $i[j]$ is the jth bit of i.

Armed with S_0 through S_m we are now ready to process M. First we encrypt each block of M by computing $C_i \leftarrow E_{K1}(M_i \oplus S_i) \oplus S_i$ for $1 \leq i \leq m - 1$. This xoring of S_i before and after the block-cipher invocation is the whitening we spoke of previously, and is the main idea in all schemes discussed in this section. Next we compute the authentication tag σ: set $\sigma \leftarrow E_{K1}(S_m \oplus \bigoplus_{i=1}^{m-1} M_i) \oplus S_0$. Notice that we are whitening the simple sum of the plaintext blocks with two *different* offset values, S_0 and S_m. Finally,

output $(N, C_1, \ldots, C_{m-1}, \sigma)$ as the authenticated ciphertext. Note that the output length is two n-bit blocks longer than M. This "ciphertext expansion," comparable to what we saw with generic composition, is quite minimal.

Given the $K1$, $K2$, and some output $(N, C_1, \ldots, C_{m-1}, \sigma)$, it is fairly straightforward to recover M and check the authenticity of the transmission. Notice that N is sent in the clear and so using $K2$ we can compute the W_i values and therefore the S_i values. We compute $M_i \leftarrow E_{K1}^{-1}(C_i \oplus S_i) \oplus S_i$ for $1 \leq i \leq m - 1$ to recover M. Then we check $E_{K1}(S_m \oplus \bigoplus_{i=1}^{m-1} M_i) \oplus S_0$ to ensure it matches σ. If we get a match, we accept the transmission as authentic, and if not we reject the transmission as an attempted forgery.

Comments on IAPM. Compared to generic composition, where we needed about $2m$ block-cipher invocations per message (assuming our encryption and authentication modes were block-cipher-based), we are now using only around $m \lg(m)$ invocations. Further refinements to IAPM reduce this even more, so the number of block-cipher invocations is nearly m in these optimized versions meaning that one can achieve AE at nearly the same cost of encryption alone.

Proving a scheme like IAPM secure is not a simple task, and indeed we cannot present such a proof here. The interested reader is encouraged to read Halevi's article which contains a rigorous proof that if the underlying block cipher is secure, then so are IACBC and IAPM [21].

XCBC and OCB

Quickly after announcement of IACBC and IAPM, other researchers went to work on finding similar single-pass AE schemes. Soon two other parties announced similar schemes: Gligor and Donescu produced a host of schemes, each with various advantages and disadvantages [18], and Rogaway, et al. announced their OCB scheme [33], which is similar to IAPM but with a long list of added optimizations.

Gligor and Donescu presented two classes of schemes: XCBC and XECB. XCBC is similar to CBC mode encryption just as IACBC was above, and XECB is similar to ECB mode encryption which allows parallelism to be exploited, much like the IAPM method presented above. Since many practitioners desire parallelizable modes, the largest share of attention has been paid to XECB. Similar to IAPM, XECB uses an offset to each message block, applied before and after a block cipher invocation. However, XECB generates these offsets in a very efficient manner, using arithmetic mod 2^n, which is very fast on most commodity processors. Once again, both schemes are highly optimized and provide AE at a cost very close to that of encryption alone. Proofs of security are included in the paper, using the reductionist approach we described above.

Rogaway, Bellare, Black, and Krovetz produced a single scheme called OCB (Offset CodeBook). This work was a follow-on to Jutla's IAPM scheme, designed to be fully parallelizable, along with a long list of other improvements. In comparison to IAPM, OCB uses a single block-cipher key, provides a message space of Σ^* so we never have to pad, and is nearly endian-neutral. Once again, a full detailed proof of security is included in the paper, demonstrating that the security of OCB is directly related to the security of the underlying block cipher.

OCB is no doubt the most aggressively optimized scheme of those discussed in this section. Performance tests indicate that OCB is about 6.4% slower than CBC mode encryption, and this is without exploiting the parallelism that OCB offers up. For more information, one can find an in-depth FAQ, all relevant publications, reference code, test vectors, and performance figures on the OCB Web page at http://www.cs.ucdavis.edu/~rogaway/ocb/.

Associated data. In many settings, the ability to handle associated data is crucial. Rogaway [32] suggests methods to handle associated data in all three of the single-pass schemes mentioned above, and for OCB gives an extension which uses PMAC [13] to give a particularly efficient variant of OCB which handles associated data.

Intellectual property. Given the importance of these new highly efficient AE algorithms, all of the authors decided to file for patents. Therefore, IBM and Gligor and Rogaway all have intellectual property claims for their algorithms and perhaps on some of the overriding ideas involved. To date, none of these patents have been tested in court, so the extent to which they are conflicting or interrelated is unclear. One effect, however, is that many would-be users of this new technology are worried that the possible legal entanglements are not worth the benefits offered by this technology. Despite this, OCB has appeared in the 802.11 draft standard as an alternate mode, and has been licensed several times. However, without IP claims it is possible all of these algorithms would be in common use today.

It was the complications engendered by the IP claims which spurred new teams of researchers to find further efficient AE algorithms which would not be covered by patents. Although not as fast as the single-pass modes described here, they still offer significant performance improvements over generic composition schemes. These schemes include CCM, CWC, and EAX, the latter invented in part by two researchers from the OCB team. We discuss these schemes next.

TWO-PASS COMBINED MODES: If we have highly efficient single-pass AE modes, why would researchers subsequently work to develop less efficient multi-pass AE schemes? Well, as we just discussed, this work was entirely motivated by the desire to provide patent-free AE schemes. The first such scheme proposed was CCM (CBC MAC with Counter Mode) by Ferguson, Housley, and Whiting. Citing several drawbacks to CCM, Bellare, Rogaway, and Wagner proposed EAX, another patent-free mode which addresses these drawbacks. And independently, Kohno, Viega, and Whiting proposed the CWC mode (Carter-Wegman with Counter mode encryption). CWC is also patent-free and, unlike the previous two modes, is fully parallelizable. We now discuss each of these modes in turn.

CCM Mode

CCM was designed with *AES* specifically in mind. It therefore is hard-coded to assume a 128-bit block size, though it could be recast for other block sizes. Giving all the details of the mode would be cumbersome, so we will just present the overriding ideas. For complete details, see the CCM specification [35].

CCM is parameterized. It requires that you specify a 128-bit block-cipher (eg, AES), a tag length (which must be one of 4, 6, 8, 10, 12, 14, or 16), and the message-length field's size (which induces an upperbound on the message length). Like all other schemes we mention, CCM uses a nonce N each time it is invoked, and the size of N depends on the the parameters chosen above; specifically, if we choose a longer maximum message length, we must accept a shorter nonce. It is left to the user to decide which parameters to use, but typical values might be to limit the maximum message length to 16 MBytes and then use a 96-bit nonce.

Once the parameters are decided, we invoke CCM by providing four inputs: the key K which will be used with AES, the nonce N of proper size, associated data H which will be authenticated but not encrypted, and the plaintext M which will be authenticated and encrypted. CCM operates in two passes: first we encode the above parameters into an initial block, prepend this block to H and M, and then run CBC MAC over this entire byte string using K. This yields the authentication tag σ. (The precise details of how the above concatenation is done *are* important for the security of CCM, but are omitted here.)

Next we form a counter-value using one of the scheme's parameters along with N and any necessary padding to reach 128 bits. This counter is then used with CTR mode encryption on $(\sigma \parallel M)$ under K to produce the ciphertext. The first 128 bits are the authentication tag, and we return the appropriate number of bytes according to the tag-length parameter. The subsequent bytes are the encryption of M and are always included in the output.

Decryption and verification are quite straightforward: N produces the counter-value and allows the recovery of M. Re-running CBC MAC on the same input used above allows verification of the tag.

Comments on CCM. It would seem that CCM is not much better than simple generic composition; after all, it uses a MAC scheme (the CBC MAC) and an encryption scheme (CTR mode encryption), which are both well-known and provably secure modes. But CCM *does* offer advantages over the straightforward use of these two primitives generically composed; in particular it uses the same key K for both the MAC and the encryption steps. Normally this practice would be very dangerous and unlikely to work, but the designers were careful to ensure the security of CCM despite this normally risky practice. The CCM specification does not include performance data or a proof of security. However, a rigorous proof was published by Jonsson [24]. CCM is currently the mandatory mode for the 802.11 wireless standard as well as currently being considered by NIST as a FIPS standard.

EAX Mode

Subsequent to the publication and subsequent popularity of CCM, three researchers decided to examine the shortcomings of CCM and see if they could be remedied. Their offering is called EAX [7] and addresses several perceived problems with CCM, including the following:

1. If the associated data field is fixed from message to message, CCM does not take advantage of this, but rather re-processes this data anew with each invocation.
2. Message lengths must be known in advance because the length is encoded into the first block before processing begins. This is not a problem in some settings, but in many applications we do not know the message length in advance.

3. The parameterization is awkward and, in particular, the trade-off between maximum message length and the size of the nonce seems unnatural.
4. The definition of CCM (especially the encodings of the parameters and length information in the message before it is processed) is complex and difficult to understand. Moreover, the correctness of CCM strongly depends on the details of this encoding.

Like CCM, EAX is a combination of a type of CBC MAC and CTR mode encryption. However, unlike CCM, the MAC used is not raw CBC MAC, but rather a variant. Two well-known problems exist with CBC MAC: (1) all messages must be of the same fixed length and (2) length must be a positive multiple of n. If we violate the first property, security is lost. Several variants to the CBC MAC have been proposed to address these problems: EMAC [9, 31] adds an extra block-cipher call to the end of CBC MAC to solve problem (1). Not to be confused with the AE mode of the same name above, XCBC [12] solves both problems (1) and (2) without any extra block-cipher invocations, but requires $k + 2n$ key bits. Finally, OMAC [23] improves XCBC so that only k bits of key are needed. The EAX designers chose to use OMAC with an extra input called a "tweak" which allows them to essentially get several different MACs by using distinct values for this tweak input. This is closely related to an idea of Liskov et al. who introduced tweakable block ciphers [30].

We now describe EAX at a high level. Unlike CCM, the only EAX parameters are the choice of block cipher, which may have any block size n, and the number of authentication tag bits to be output, τ. To invoke EAX, we pass in a nonce $N \in \Sigma^n$, a header $H \in \Sigma^*$ which will be authenticated but not encrypted, and the message $M \in \Sigma^*$ which will be authenticated and encrypted, and finally the key K, appropriate for the chosen block cipher. We will be using OMAC under key K three times, each time with a different tweak, written OMAC_K^0, OMAC_K^1, and OMAC_K^2; it's conceptually easiest to think of these three OMAC invocations as three separate MACs, although this is not strictly true. First, we compute $\mathrm{ctr} \leftarrow \mathrm{OMAC}_K^0(N)$ to obtain the counter value we will use with CTR mode encryption. Then we compute $\sigma_H \leftarrow \mathrm{OMAC}_K^1(H)$ to get an authentication tag for H. Then we encrypt and authenticate M with $C \leftarrow \mathrm{OMAC}_K^2(\mathrm{CTR}_K^{\mathrm{ctr}}(M))$. And finally we output the first τ bits of $\sigma = (\mathrm{ctr} \oplus C \oplus \sigma_H)$ as the authentication tag. We also output the nonce N, the associated data H, and the ciphertext C. The decryption and verification steps are quite straightforward.

Note that each of the problem areas cited above has been addressed by the EAX mode: no restriction on message length, no interdependence between the tag length and maximum message length, a performance savings when there is static header data, and no need for message length to be known up front. Also, EAX is arguably simpler to specify and implement. Once again, proving EAX secure is more difficult than just appealing to proofs of security for generically composed schemes since the key K is reused in several contexts which is normally not a safe practice.

CWC Mode

The CWC Mode [28] is also a two-pass mode: it uses a Wegman–Carter MAC along with CTR mode encryption under a common key K. Its main advantage over CCM and EAX is that it is parallelizable whereas the other two are not (due to their use of the inherently sequential CBC MAC type algorithms). Also, CWC strives to be very fast in hardware, a consideration which was not given nearly as much attention in the design of the other modes. In fact, the CWC designers claim that CWC should be able to encrypt and authenticate data at 10Gbps in hardware, whereas CCM and EAX will be limited to about 2Gbps because of their serial constraints.

As we discussed above in the section on generic composition, Wegman–Carter MACs require one specify a family of hash functions on a common domain and range. Typically we want these functions to (1) be fast to compute and (2) have a low collision probability. The CWC designers also looked for a family with additional properties: (3) parallelizability and (4) good performance in hardware. The function family they settled on is the well-known polynomial hash. Here a function from the family is named by choosing a value for x in some specified range, and then the polynomial

$$Y_1 x^\ell + Y_2 x^{\ell-1} + \cdots + Y_\ell x + Y_{\ell+1}$$

is computed modulo some integer (see modular arithmetic), typically a prime number. The specific family chosen by the CWC designers fixes $Y_1, \ldots, Y_\ell$ to be 96-bit integers, and $Y_{\ell+1}$ to be a 127-bit integer; their values are determined by the message being hashed. The modulus is set to the prime, $2^{127} - 1$.

Although it is possible to evaluate this polynomial quickly on a serial machine using Horner's method (and in fact, this may make sense in some cases), it is also possible to exploit parallelism in the computation of this polynomial. Assume n is odd and set $m = (n-1)/2$ and

$y = x^2 \bmod 2^{127} - 1$. Then we can rewrite the function above as

$$\left(Y_1 y^m + Y_3 y^{m-1} + \cdots + Y_\ell\right) x$$
$$+ \left(Y_2 y^m + Y_4 y^{m-1} + \cdots + Y_{\ell+1}\right) \bmod 2^{127} - 1.$$

This means that we can subdivide the work for evaluating this polynomial and then recombine the results using addition modulo $2^{127} - 1$. Building a MAC from this hash family is fairly straightforward, and therefore CWC yields a parallelizable scheme since CTR is clearly parallelizable.

The CWC designers go on to provide benchmark data to compare CCM, EAX, and CWC on a Pentium III, showing that the speed differences are not that significant. However, this is without exploiting any parallelism available with CWC. They do not compare the speed of CWC with that of OCB, where we would expect OCB to be faster even in parallel implementations.

CWC comes with a rigorous proof of security via a reduction to the underlying 128-bit block cipher (typically AES/Rijndael), and the paper includes a readable discussion of why the various design choices were made. In particular, it does not suffer from any of the above-mentioned problems with CCM.

AE PRIMITIVES: Every scheme discussed up to this point has been a mode of operation. In fact with the possible exception of some of the MAC schemes, every mode has used a block cipher as its underlying primitive. In this section we consider two recently developed modes which are stream ciphers which provide authentication in addition to privacy. That is to say, these are *primitives* which provide AE.

This immediately means there is no proof of their security, nor is there likely to ever be one. The security of primitives is usually a matter of opinion: does the object withstand all known attacks? Has it been in use for a long enough time? Have good cryptanalysts examined it?

With new objects, it is often hard to know how much trust to place in their security. Sometimes the schemes break, and sometimes they do not. We will discuss two schemes in this section: Helix and SOBER-128. Both were designed by teams of experienced cryptographers who paid close attention to their security as well as to their efficiency.

HELIX: Helix was designed by Ferguson et al. [17]. Their goal was to produce a fast, simple, patent-free stream cipher which also provided authentication. The team claims speeds of about 7 cycles per byte on a Pentium II, which is quite a bit faster than the fastest-known implementations of AES, which run at about 15 cycles per byte. At first glance this might be quite surprising: after all, AES does about 160 table look-ups and 160 32-bit XORs to encipher 16 bytes. This means AES uses about 10 look-ups and 10 XORs per byte. As we will see in a moment, Helix uses more operations than this per-byte! But a key difference is that AES does memory look-ups from large tables which perhaps are not in cache whereas Helix confines its work to the register file.

Helix takes a key K up to 32 bytes in length, and a 16-byte nonce N and a message $M \in (\Sigma^8)^*$. As usual, K will allow the encryption of a large amount of data before it needs to be changed, and N will be issued anew with each message encrypted, never to repeat throughout the life of K. Helix uses only a few simple operations: addition modulo 2^{32}, exclusive-or of 32-bit strings, and bitwise rotations. However, each iteration of Helix, called a "block," uses 11 XORs, 12 modular additions, and 20 bitwise rotations by fixed amounts on 32-bit words. So Helix is not simple to specify; instead we give a high-level description.

Helix keeps its "state" in five 32-bit registers (the designers were thinking of the Intel family of processors). The ith block of Helix emits one 32-bit word of key-stream S_i, requires two 32-bit words scheduled from K and N, and also requires the ith plaintext word M_i. It is highly unusual for a stream cipher to use the plaintext stream as part of its key-stream generation, but this feature is what allows Helix to achieve authentication as well as generating a key-stream.

As usual, the key-stream is used as a one-time pad to encrypt the plaintext. In other words, the ith ciphertext block C_i is simply $M_i \oplus S_i$. The five-word state resulting from block i is then fed into block $i + 1$ and the process continues until we have a long enough key-stream to encrypt M. At this point, a constant is XORed into one of the words of the resulting state, twelve more blocks are generated using a fixed plaintext word based on the length of M, with the key-stream of the four last blocks yielding the 128-bit authentication tag.

SOBER-128

A competitor to Helix is an offering from Hawkes and Rose called SOBER-128 [22]. This algorithm evolved from a family of simple stream ciphers (i.e., ciphers which did not attempt simultaneous authentication) called the SOBER family, the first of which was introduced in 1998 by Rose. SOBER-128 retains many of the characteristics of its ancestors, but introduces a method for authenti-

cating messages as well. We will not describe the internals of SOBER-128 but rather describe a few of its attributes at a higher level.

SOBER-128 uses a <u>linear-feedback shift register</u> in combination with several non-linear components, in particular a carefully-designed S-box which lies at its heart. To use SOBER-128 for AE one first generates a keystream used to XOR with the message M and then uses a separate API call "maconly" to process the associated data. The method of feeding back plaintext into the keystream generator is modeled after Helix, and the authors are still evaluating whether this change to SOBER-128 might introduce weaknesses.

Tests by Hawkes and Rose indicate that SOBER-128 is comparable in speed to Helix; however, both are quite new and are still undergoing cryptanalytic scrutiny—a crucial process when designing primitives. Time will help us determine their security.

BEYOND AE AND AEAD: Real protocols often require more than just an AE scheme or an AEAD scheme: perhaps they require something that more resembles a network transport protocol. Desirable properties might include resistance to replay and prevention against packet loss or packet reordering. In fact, protocols like SSH aim to achieve precisely this.

Work is currently underway to extend AE notions to encompass a broader range of such goals [27]. This is an extension to the SSH analysis referred to above [4], but considers the various EtM, MtE, and E&M approaches rather than focusing on just one. Such research is another step in closing the gap between what cryptographers produce and what consumers of cryptographic protocols require. The hope is that we will reach the point where methods will be available to practitioners which relieve them from inventing cryptography (which, as we have seen, is a subtle area with many insidious pitfalls) and yet allow them easy access to provably secure cryptographic protocols. We anticipate further work in this area.

NOTES ON REFERENCES: Note that AE and its extensions continue to be an active area of research. Therefore, many of the bibliographic references are currently to unpublished pre-prints of works in progress. It would be prudent for the reader to look for more mature versions of many of these research reports to obtain the latest revisions.

J. Black

References

[1] Bellare, M., R. Canetti, and H. Krawczyk (1996). "Keying hash functions for message authentication." *Advances in Cryptology—CRYPTO'96*, Lecture Notes in Computer Science, vol. 1109, ed. N. Koblitz. Springer-Verlag, Berlin, 1–15.

[2] Bellare, M., A. Desai, D. Pointcheval, and P. Rogaway (1998). "Relations among notions of security for public-key encryption schemes." *Advances in Cryptology—CRYPTO'98*, Lecture Notes in Computer Science, vol. 1462, ed. H. Krawczyk. Springer-Verlag, Berlin, 232–249.

[3] Bellare, M., J. Kilian, and P. Rogaway (2000). "The security of the cipher block chaining message authentication code." *Journal of Computer and System Sciences (JCSS)*, 61 (3) 362–399. Earlier version in CRYPTO'94. See www.cs.ucdavis.edu/~rogaway

[4] Bellare, M., T. Kohno, and C. Namprempre (2002). "Authenticated encryption in SSH: Provably fixing the SSH binary packet protocol." *ACM Conference on Computer and Communications Security (CCS-9)*. ACM Press, New York, 1–11.

[5] Bellare, M. and C. Namprempre (2000). "Authenticated encryption: Relations among notions and analysis of the generic composition paradigm." *Advances in Cryptology—ASIACRYPT 2000*, Lecture Notes in Computer Science, vol. 1976, ed. T. Okamoto. Springer-Verlag, Berlin.

[6] Bellare, M. and P. Rogaway (2000). "Encode-then-encipher encryption: How to exploit nonces or redundancy in plaintexts for efficient encryption." *Advances in Cryptology—ASIACRYPT 2000*, Lecture Notes in Computer Science, vol. 1976, ed. T. Okamoto. Springer-Verlag, Berlin, 317–330. See www.cs.ucdavis.edu/~rogaway

[7] Bellare, M., P. Rogaway, and D. Wagner (2003). "EAX: A conventional authenticated-encryption mode." Cryptology ePrint archive, reference number 2003/069, submitted April 13, 2003, revised September 9, 2003. See eprint.iacr.org

[8] Bellovin, S. (1996). "Problem areas for the IP security protocols." *Proceedings of the Sixth USENIX Security Symposium, July 1996*, 1–16.

[9] Berendschot, A., B. den Boer, J. Boly, A. Bosselaers, J. Brandt, D. Chaum, I. Damgård, M. Dichtl, W. Fumy, M. van der Ham, C. Jansen, P. Landrock, B. Preneel, G. Roelofsen, P. de Rooij, and J. Vandewalle (1995). *Final Report of Race Integrity Primitives*, Lecture Notes in Computer Science, vol. 1007, eds. A. Bosselaers and B. Preneel. Springer-Verlag, Berlin.

[10] Bernstein, D. (2000). "Floating-point arithmetic and message authentication." Available from http://cr.yp.to/hash127.html

[11] Black, J., S. Halevi, H. Krawczyk, T. Krovetz, and P. Rogaway (1999). "UMAC: Fast and secure message authentication." *Advances in Cryptology—CRYPTO'99*, Lecture Notes in Computer Science, vol. 1666, ed. J. Wiener. Springer-Verlag, Berlin.

[12] Black, J. and P. Rogaway (2000). "CBC MACs for arbitrary-length messages: The three-key constructions." *Advances in Cryptology—CRYPTO 2000*, Lecture Notes in Computer Science, vol. 1880, ed. M. Bellare. Springer-Verlag, Berlin.

[13] Black, J. and P. Rogaway (2002). "A block-cipher mode of operation for parallelizable message authentication." *Advances in Cryptology—EUROCRYPT 2002*, Lecture Notes in Computer Science, vol. 2332, ed. L. Knudsen. Springer-Verlag, Berlin, 384–397.

[14] Black, J. and H. Urtubia (2002). "Side-channel attacks on symmetric encryption schemes: The case for authenticated encryption." *Proceedings of the Eleventh USENIX Security Symposium, August 2002*, ed. D. Boneh, 327–338.

[15] Borisov, N., I. Goldberg, and D. Wagner (2001). "Intercepting mobile communications: The insecurity of 802.11." *MOBICOM*. ACM Press, New York, 180–189.

[16] Carter, L. and M. Wegman (1979). "Universal hash functions." *J. of Computer and System Sciences*, 18, 143–154.

[17] Ferguson, N., D. Whiting, B. Schneier, J. Kelsey, S. Lucks, and T. Kohno (2003). "Helix: Fast encryption and authentication in a single cryptographic primitive." *Fast Software Encryption, 10th International Workshop, FSE 2003*, Lecture Notes in Computer Science, vol. 2887, ed. T. Johansson. Springer-Verlag, Berlin.

[18] Gligor, V. and P. Donescu (2002). "Fast encryption and authentication: XCBC encryption and XECB authentication modes." *Fast Software Encryption, 8th International Workshop, FSE 2001*, Lecture Notes in Computer Science, vol. 2355, ed. M. Matsui. Springer-Verlag, Berlin, 92–108. See www.ece.umd.edu/~gligor/

[19] Goldwasser, S., S. Micali, and R. Rivest (1988). "A digital signature scheme secure against adaptive chosen-message attacks." *SIAM Journal of Computing*, 17 (2), 281–308.

[20] Krawczyk, H., M. Bellare, and R. Canetti (1997). "HMAC: Keyed hashing for message authentication." *IETF RFC-2104*.

[21] Halevi, S. (2001). "An observation regarding Jutla's modes of operation." Cryptology ePrint archive, reference number 2001/015, submitted February 22, 2001, revised April 2, 2001. See eprint.iacr.org

[22] Hawkes, P. and G. Rose (2003). "Primitive specification for SOBER-128." Available from http://www.qualcomm.com.au/Sober128.html

[23] Iwata, T. and K. Kurosawa (2003). "OMAC: One-key CBC MAC." *Fast Software Encryption*, Lecture Notes in Computer Science, vol. 2887, ed. T. Johansson. Springer-Verlag, Berlin.

[24] Jonsson, J. (2002). "On the security of CTR + CBC-MAC." *Selected Areas in Cryptography—SAC 2002*, Lecture Notes in Computer Science, vol. 2595, eds. K. Nyberg and H.M. Heys. Springer-Verlag, Berlin, 76–93.

[25] Jutla, C. (2001). "Encryption modes with almost free message integrity." *Advances in Cryptology—EUROCRYPT 2001*, Lecture Notes in Computer Science, vol. 2045, ed. B. Pfitzmann. Springer-Verlag, Berlin, 529–544.

[26] Katz, J. and M. Yung (2000). "Complete characterization of security notions for probabilistic private-key encryption." *Proceedings of the 32nd Annual Symposium on the Theory of Computing (STOC)*. ACM Press, New York.

[27] Kohno, T., A. Palacio, and J. Black (2003). "Building secure cryptographic transforms, or how to encrypt and MAC." Cryptology ePrint archive, reference number 2003/177, submitted August 28, 2003. See eprint.iacr.org

[28] Kohno, T., J. Viega, and D. Whiting (2003). "High-speed encryption and authentication: A patent-free solution for 10 Gbps network devices." Cryptology ePrint archive, reference number 2003/106, submitted May 27, 2003, revised September 1, 2003. See eprint.iacr.org

[29] Krawczyk, H. (2001). "The order of encryption and authentication for protecting communications (or: How secure is SSL?)." *Advances in Cryptology—CRYPTO 2001*, Lecture Notes in Computer Science, vol. 2139, ed. J. Kilian. Springer-Verlag, Berlin, 310–331.

[30] Liskov, M., R. Rivest, and D. Wagner (2002). "Tweakable block ciphers." *Advances in Cryptology—CRYPTO 2002*, Lecture Notes in Computer Science, vol. 2442, ed. M. Yung. Springer-Verlag, Berlin, 31–46.

[31] Petrank, E. and C. Rackoff (2000). "CBC MAC for real-time data sources." *Journal of Cryptology*, 13 (3), 315–338.

[32] Rogaway, P. (2002). "Authenticated-encryption with associated-data." *ACM Conference on Computer and Communications Security (CCS-9)*. ACM Press, New York, 196–205.

[33] Rogaway, P., M. Bellare, and J. Black (2003). "OCB: A block-cipher mode of operation for efficient authenticated encryption." *ACM Transactions on Information and System Security (TISSEC)*, 6 (3), 365–403.

[34] Wegman, M. and L. Carter (1981). "New hash functions and their use in authentication and set equality." *J. of Comp. and System Sciences*, 22, 265–279.

[35] Whiting, D., R. Housley, and N. Ferguson (2002). "Counter with CBC-MAC (CCM)." Available from csrc.nist.gov/encryption/modes/proposedmodes/

AUTHENTICATION

There is a rather common saying that cryptology has two faces. The first (and better known) face is cryptography in its narrow sense which should protect data (information) from being revealed to an opponent. The second face, known as authentication (also as information integrity), should guarantee with some confidence that a given information is authentic, i.e., has not been altered or substituted by the opponent. This confidence may depend on the computing power of the opponent (e.g., in underlined digital signature schemes this is the case). The latter is called unconditional authentication and makes use of symmetric cryptosystems.

The model of unconditional authentication schemes (or codes) consists of a sender, a receiver, and an opponent. The last one can observe all the information transmitted from the sender to the receiver; it is assumed (following Kerkhoff's maxim) that the opponent knows everything, even the original (plain) message (this is called *authentication without secrecy*), but he does not know the used key.

There are two kinds of possible attacks by the opponent. One speaks about an *impersonation attack* when the opponent sends a message in the hope that it will be accepted by the receiver as a valid one. In a *substitution attack* the opponent observes a transmitted message and then replaces it with another message. For authentication purposes it is enough to consider only so-called systematic *authentication codes* in which the transmitted message has the form $(m; z)$, where m is chosen from the set M of possible messages and $z = f(m)$ is its *tag* (a string of "parity-check symbols" in the language of coding theory). Let Z be the tag-set and let $F = \{f_1, \ldots, f_n\}$ be a set of n encoding maps $f_i : M \to Z$. To authenticate (or *code*) message m, the sender chooses randomly one of the encoding mappings f_i (the choice is in fact the secret key unknown to the opponent). One may assume without loss of generality that these encoding maps f_i are chosen uniformly. The corresponding probabilities of success for impersonation and substitution attacks are denoted by P_I and P_S respectively. The first examples of authentication codes were given in [3], among which is the following *optimal* scheme (known as *affine scheme*).

Let the set M of messages and the set Z of tags coincide with the finite field $\mathbf{F}_q$ of q elements (q should be a power of a prime number). The set F of encoding mappings consists of all possible affine functions, i.e. mappings of the form

$$f_{a,b}(m) = am + b.$$

For this scheme $P_I = P_S = q^{-1}$ and the scheme is optimal for both parameters—for P_I this is obvious and for P_S this follows from the *square-root bound* $P_S \geq 1/\sqrt{n}$ which is also derived in [3]. Although this scheme is optimal (meets this bound with equality), it has a serious drawback when being applied in practice since its key size (which is equal to $\log n = 2 \log q$) is two times larger than the message size.

For a long time (see [6, 10]), no known schemes (codes) had a key size that was much smaller than the message size. Schemes that did allow this were first constructed in [4]. They made use of a very important relationship between authentication codes and error-correcting codes (*ECC*, shortly) (see [8] and cyclic codes).

By definition (see [5]), an authentication code is a q-ary code V over the alphabet Z ($|Z| = q$) of length n consisting of $|M|$ codewords $(f_1(m), \ldots, f_n(m)) : m \in M$. Almost without loss of generality one can assume that all words in the *A-code* V have a uniform composition, i.e., all "characters" from the alphabet Z appear equally often in every codeword (more formally, $|\{i : v_i = z\}| = n/q$ for any $\underline{v} \in V$ and any $z \in Z$). This is equivalent to saying that P_I takes on its minimal possible value q^{-1}. The maximal probability of success of a substitution by the opponent is

$$P_S = 1 - n^{-1} d_A(V),$$

where $d_A(\underline{x}, \underline{y}) = n - q\gamma(\underline{x}, \underline{y})$, $\gamma(\underline{x}, \underline{y}) = \max\{|\{i : x_i = z, y_i = z'\}| : z, z' \in Z\}$ and $d_A(\overline{V})$ (the minimum A-distance of the code V) is defined as usual (see cyclic codes and McEliece public-key encryption scheme). The obvious inequality $d_A(V) \le d_H(V)$, with $d_H(V)$ being the *minimum Hamming distance* of V, allows one to apply known upper bounds for *ECC* to systematic A-codes and re-derive known nonexistence bounds for authentication codes as well as obtain new bounds (see [1, 5] for details).

On the other hand, the q-twisted construction proposed in [5] turns out to be a very effective tool to construct good authentication codes from *ECC* (in fact almost all known authentication schemes are implicitly or explicitly based on the q-twisted construction). Let C be an error-correcting code of length m over $\mathbf{F}_q$ with the minimal Hamming distance $d_H(C)$ and let U be its subcode of cardinality $q^{-1} |C|$ such that for all $U \in U$ and all $\lambda \in \mathbf{F}_q$ vectors $u + \lambda \mathbf{1}$ are distinct and belong to C, where $\mathbf{1}$ is the all-one vector. Then the following q-ary code $V_U := \{(u, u + \lambda_1 \mathbf{1}, \ldots, u + \lambda_q \mathbf{1}) : u \in U\}$ (where $\lambda_1, \ldots, \lambda_q$ are all different elements of the field $\mathbf{F}_q$) of length $n = mq$ is called q-twisted code and considered as A-code generates the authentication scheme [5] for protecting $|U|$ messages with the number of keys $n = mq$ providing probabilities

$$P_I = \frac{1}{q}, \qquad P_S = 1 - \frac{d_H(C)}{m}.$$

Application of the q-twisted construction to many optimal *ECC* (with enough large minimal code distance) produces optimal or near optimal authentication codes. For instance, *Reed–Solomon* codes generate authentication schemes which are the natural generalization of the aforementioned affine scheme (namely, $k = 1$) and have the following parameters ([2, 5]):

The number of messages is q^k, the number of keys is q^2, and the probabilities are $P_I = 1/q$, $P_S = k/q$, where $k + 1$ is the number of information symbols of the corresponding Reed–Solomon code.

Reed–Solomon codes are a particular case of *algebraic-geometry* (AG) codes and the corresponding application of q-twisted construction to AG codes leads to an asymptotically very efficient class of schemes with the important, additional property of being polynomial constructible (see [9]).

To conclude, we note that there is also another equivalent "language" to describe and investigate unconditional authentication schemes, namely, the notion of almost strongly two-universal hash functions (see [7] and also [10]).

Grigory Kabatiansky
Ben Smeets

References

[1] Bassalygo, L.A. and M.V. Burnashev (1996). "Authentication, identification and pairwise separated measures." *Problems of Information Transmission*, 32 (1), 41–47.

[2] den Boer, B. (1993). "A simple and key-economical unconditionally authentication scheme." *Journal on Computer Security*, 2 (1), 65–67.

[3] Gilbert, E.N., F.J. MacWilliams, and N.J.A. Sloane (1974). "Codes which detect deception." *Bell Syst. Tech. J.*, 33 (3), 405–424.

[4] Johansson, T., G.A. Kabatianskii, and B. Smeets (1994). "On the relation between A-codes and codes correcting independent errors." *Adavances in Cryptology—EUROCRYPT'93*, Lecture Notes in Computer Science, vol. 765, ed. T. Helleseth. Springer-Verlag, Berlin, 1–11.

[5] Kabatianskii, G.A., B. Smeets, and T. Johansson (1996). "On the cardinality of systematic authentication codes via error-correcting codes." *IEEE Transactions on Information Theory*, 42 (2), 566–578.

[6] Simmons, G.J. (1992). "A survey of information authentication. Contemporary cryptology." *The Science of Information Integrity*. IEEE Press, Piscataway, NJ.

[7] Stinson, D.R. (1994). "Universal hashing and authentication codes." *Designs, Codes and Cryptography*, 4, 369–380.

[8] van Tilborg, H.C.A. (1996). "Authentication codes: An area where coding and cryptology meet." *Cryptography and Coding V*, Lecture Notes in Computer Science, vol. 1025, ed. C. Boyd. Springer-Verlag, Berlin, 169–183.

[9] Vladuts, S.G. (1998). "A note on authentication codes from algebraic geometry." *IEEE Transactions on Information Theory*, 44, 1342–1345.

[10] Wegman, M.N. and J.L. Carter (1981). "New hash functions and their use in authentication and set equality." *J. Comput. Syst. Sci.*, 22, 265–279.

AUTHENTICATION TOKEN

The term "authentication token" can have at least three different definitions, but is generally used to refer to an object that is used to authenticate one entity to another (see authentication). The various definitions for "authentication token" include the credentials provided to an authenticating party as part of an identity verification protocol, a data structure provided by an authentication server for later use in authenticating to a different application server, and a physical device or computer file used to authenticate oneself. These definitions are expanded below.

CREDENTIALS PROVIDED TO AN AUTHENTICATING PARTY: In most identity verification or authentication protocols, the entity being authenticated must provide the authenticating entity with some proof of the claimed identity. This proof will allow the authenticating party to verify the identity that is being claimed and is sometimes called an "authentication token." Examples of these types of authentication tokens include functions of shared secret information, like passwords, known only to both the authenticating and authenticated parties and responses to challenges that are provided by the authenticating party but which could only be produced by the authenticated party.

DATA STRUCTURE PROVIDED BY AN AUTHENTICATION SERVER: In some security architectures end users are authenticated by a dedicated "authentication server" by means of an identity verification protocol. This server then provides the user with credentials, sometimes called an "authentication token," which can be provided to other application servers in order to authenticate to those servers. Thus, these credentials are not un-like those described above, which are provided directly by the end user to the authenticating party, except in that they originate with a third party, the authentication server.

Usually these tokens take the form of a data structure which has been digitally signed (see digital signature schemes) or MACed (see MAC algorithms) by the authentication server and thus vouch for the identity of the authenticated party. In other words, the authenticated party can assert his/her identity to the application server simply by presenting the token. These tokens must have a short lifetime since if they are stolen they can be used by an attacker to gain access to the application server.

DEVICE OR FILE USED FOR AUTHENTICATION: Quite often the credentials that must be provided to an authenticating party are such that they cannot be constructed using only data that can be remembered by a human user. In such situations it is necessary to provide a storage mechanism to maintain the user's private information, which can then be used when required in an identity verification protocol. This storage mechanism can be either a software file containing the private information and protected by a memorable password, or it can be a hardware device (e.g., a smart card and is sometimes called an "authentication token."

In addition to making many identity verification protocols usable by human end entities, these authentication tokens have another perhaps more important benefit. Since successful completion of the protocol now usually involves both something the end entity has (the file or device) and something the end entity knows (the password or PIN to access the smart card) instead of just something the end entity knows, the actual security of the authentication mechanism is increased. In particular, when the token is a hardware device, obtaining access to that device can often be quite difficult, thereby providing substantial protection from attack.

Robert Zuccherato

AUTHORIZATION ARCHITECTURE

Authentication and authorization are separate concepts (although authentication may be used in the service of authorization), and their respective architectures or infrastructures may be separately deployed and managed. Authentication allows

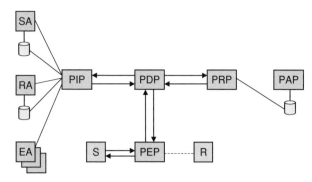

Fig. 1. Conceptual model of an authorization architecture

entity *A* to convince entity *B* of *A*'s identity with some degree of certainty (see <u>identification</u>, <u>identity verification protocol</u>, and <u>entity authentication</u>). Typically, however, this information is insufficient. Entity *A* may be trying to perform some task (e.g., execute an application, invoke a function, or access a file) and *B* needs to know not "who *A* is" as much as "whether *A* should be allowed to perform this task." Authorization allows *B* to make and enforce this decision. In some cases, *A*'s identity will be a critical input to the decision-making process ("is *A* allowed to read *A*'s medical record?"); in other cases, *A*'s identity may be almost irrelevant, useful for auditing purposes only ("the requester is an executive of the company and—regardless of who it is—all executives are allowed to see the quarterly results before they are announced"). Authentication answers the question "who is this entity?" and authorization answers the question "is this entity allowed to do what it is trying to do?"

AUTHORIZATION ARCHITECTURE: An authorization architecture is the set of components and data that allows authorization decisions to be made and enforced. The components of this architecture are shown in Figure 1 (note that this is a conceptual model; actual implementations will typically combine subsets of these components into single machines or even single processes).

COMPONENTS: The subject, S, sends a request to perform some action on a resource, R (e.g., read a file, POST to a Web site, execute an application, or invoke an object method). This request is intercepted by an entity called a policy enforcement point (PEP) whose job is to enforce a "PERMIT" or "DENY" decision with respect to this request. The decision itself is made by an entity called a policy decision point (PDP). The PDP makes this decision by gathering all the input data that is relevant to this request and evaluating it accord-

ing to an authorization policy that is applicable to this request. The relevant data includes the submitted request along with particular attributes about both the subject and the resource, and may also include attributes about the environment in which the request is submitted. Various authorities are responsible for creating and making available this attribute information: one or more subject authorities (SAs), a resource authority (RA), and one or more environmental authorities (EAs) package this information in a syntax that will be accessible by a policy information point (PIP), the entity that collects this data on behalf of the PDP. Similarly, a policy administration point (PAP) is responsible for creating authorization policies and making them accessible to a policy retrieval point (PRP), the entity that fetches policies for the PDP.

A given implementation may have variations on the basic architecture discussed above. For example, there may be multiple PDPs that work together to render an overall decision with respect to an authorization request.

INFORMATION FLOW: The flow of information in Figure 1 is as follows. The subject S submits a request to access a resource R. The PEP intercepts this access request and sends a request for an authorization decision to the PDP. The decision request will contain the information contained in the original access request, but may also contain additional information, such as some attributes of the subject, resource, or environment that are known to the PEP (e.g., the IP address of the machine from which the access request was made). The PDP will need to find an authorization policy that is relevant to this access request and so will supply the appropriate subject, resource, and action information to the PRP and ask it to retrieve the correct policy. Once the PDP has the authorization policy for this access request, it can examine the policy to see what subject, resource, or environment attributes are required in order for it to render a decision. If the PDP requires attributes that were not supplied by the PEP in the authorization decision request, the PDP will ask the PIP to retrieve these attributes. Once the PDP has all the data it requires (or has determined that some attribute data cannot be retrieved for some reason), it can evaluate the authorization policy and render a decision or produce a value of "indeterminate" (no decision possible due to missing attributes) or "error" (no decision possible due to network or processing difficulties). The PDP can then return its result to the PEP, which will enforce this result by granting access to the requested resource, or by returning an "access denied" or relevant error message to the subject.

ATTRIBUTES: An attribute is a piece of information that may be categorized as being associated with the subject, action, resource, or environment in an authorization architecture. Attributes may be static or dynamic. Static attributes of the subject are referred to by many names in various discussions and contexts, including privileges, permissions, rights, authorizations, properties, characteristics, entitlements, and grants. Static attributes can also be associated with resources and with actions. Groups, roles, and document labels are all examples of static attributes (even though a "role" is dynamic in another sense: that is, an entity may be able to step into or out of a role at will in the course of performing some aspects of its job).

Dynamic attributes are those whose values cannot be relied upon to remain unchanged between one time they are required (e.g., by the PDP) and the next time they are required. Example dynamic attributes of the subject include current account balance, amount of credit remaining, and IP address of requesting machine; dynamic attributes of the resource include the number of times it has been accessed; and dynamic attributes of the environment include current time of day, and time of receipt of the request.

Dynamic attributes are retrieved by the PDP/PIP in real time (i.e., at the time of access request evaluation) from the relevant authority. In order for this exchange to occur securely, it is necessary for the response to be authenticated so that the PDP/PIP can be confident that the intended authority created the response. In some cases, the request for these attributes may also need to be authenticated so that the authority can be confident that the legitimate PDP/PIP asked for this information. This authentication may take place independently on each message (e.g., using digital signatures), or may take place in the context of a secure session (such as an SSL (see Secure Socket Layer) session between the PDP/PIP and the relevant authority).

Static attributes need not be retrieved in real time from the authority; for example, they may be cached locally by the PDP or retrieved from an online repository such as a database or a directory. However, in such cases, the authenticity and integrity of the information must still be ensured. A method commonly employed is to put the attribute data into a data structure along with some representation of the entity to which it pertains (the identity of the subject, or the name of the resource, for example) and to have the relevant authority digitally sign this data structure. The signed data structure is the authority's "certificate" of the authenticity of the binding between the attribute data and the entity, which the entity may be able to use in a proof procedure with other parties to show ownership of the contained attributes.

When static attributes are available in an authorization architecture, the use of signed data structures binding such attributes to entities can have a number of attractive benefits. First, "offline" operation may be possible, in that relying parties such as the PDP and PIP do not need to access SAs or RAs in real time as access requests are being evaluated. Second, caching or other relatively local storage of this data at the PDP/PIP can significantly reduce network traffic when these attributes need to be retrieved. Third, extended trust and delegation of attribute granting authority are more readily achievable through the use of signed data structures. Finally, such an architecture can allow a simple mechanism to "turn off" all attributes for a given entity simultaneously (for example, if all attribute certificates are cryptographically linked to an entity's public-key certificate, then revoking that single public-key certificate will automatically revoke all associated attribute certificates—this can be a significant convenience when a company employee is fired or otherwise rendered inactive and access to many different networks and systems has to be cut off instantaneously).

POLICIES: An access control policy with respect to a specific resource or set of resources is the set of rules governing who can do what to those resources under what conditions. The term *authorization policy* includes access control policy, but has a broader definition, potentially including rules regarding the actual assignment of attributes to subjects or resources, the rules regarding the delegation of authority to assign such attributes, rules regarding the default behavior of various components in the absence of sufficient information, rules regarding the trusted system entities for each component in the architecture, and so on.

Terminology in this area is far from universally agreed, but the concepts are quite similar across many discussions. Typically a "rule" has an *effect* (indicating whether it is intended to contribute to a PERMIT decision or a DENY decision), a *scope* or a *target* of applicability (indicating the subject, resource, and action to which it applies), and a *condition* or *set of conditions* (indicating any restrictions, limitations, or qualifications to be imposed upon this subject being permitted or denied access to this resource). A "policy" is a collection of one or more rules along with an (implicit or explicit) algorithm for combining the rules that it contains or references. A well-known example combining algorithm is "deny overrides," in which any satisfied rule that has an effect of DENY takes precedence over all satisfied rules that have an effect of PERMIT. Another common example is "default deny," in which access is denied if for whatever reason an actual decision cannot be rendered by the PDP from the available data.

In many environments, policies will have what is referred to as "distributed authorship." That is, several different PAPs (policy administration points) may independently create policies that pertain to the same subject or to the same resource. For example, in a particular company or organization, there may be regulatory policies that govern access to certain types of data, legislative policies regarding the release of the same data, and corporate and even departmental policies regarding access to the same data. When a subject asks to read this data, all these policies must be taken into account by the PDP before it can render the appropriate decision. This means that the PDP must have some sort of *reconciliation* algorithm, determining the correct (i.e., intended) way in which to combine these various—potentially conflicting—policies. The reconciliation algorithm must be robust and comprehensive in order for the PDP to be able to deal in an automated fashion with all the possible ways in which independently created policies may interact. This aspect of authorization policy is still an area of much research.

ATTRIBUTE AND POLICY MANAGEMENT: Subject and resource attributes, as well as access control and authorization policies, need to be *managed* in an authorization architecture. Attributes and policies have life cycles: they may be created, used, versioned, audited, revoked, and archived. They may be "current" (i.e., active and valid) for a relatively short period of time or for a long period of time, and components in the architecture (especially the PDP) must readily be able to tell whether a particular attribute binding or policy statement can be relied upon or not. Various authorities in the architecture are responsible for managing the life cycle of this information, including SAs, RAs, and PAPs. Such authorities must be trusted to do this job in a reliable and timely fashion; thus, the establishment of a trust model (see trust models)

or trust infrastructure is critical to the success of the authorization architecture.

Another important aspect of management is attribute/policy storage and retrieval. How can this information be found by the components that need it (the PIP and PRP), when they need it? Attributes and policies must be indexed and stored in a manner that makes them easy to retrieve in real time, given only the information contained in the access request. Finding the best indexing mechanism, storage technology, and retrieval method for a given environment is an area of both theoretical and practical interest.

SYNTAX: The various pieces of information in the authorization architecture must be expressed and conveyed in a syntax that is understood by different components in the architecture. For example, the Subject Authority will bind attribute information to subject identifiers and express this binding in a data structure; the policy administration point will define an access control policy and express this policy in a data structure; the policy enforcement point will need a decision from a policy decision point regarding a particular access request and will package this decision request in a protocol message. In each case, the syntax and semantics of the data must be understood by multiple components in the architecture in order for proper enforcement of the intended authorization policies to take place.

Over the years, there have been many attempts to define a syntax to express attribute bindings and policy information, some based on Baccus-Nauer Form (BNF), some based on Abstract Syntax Notation One (ASN.1), and some more recent work based on Extensible Markup Language (XML). Examples include work in the Distributed Computing Environment (DCE), SESAME, and CORBA Security initiatives, PolicyMaker, PONDER, Distributed Management Task Force/Common Information Model (DMTF/CIM), IETF Simple Public Key Infrastructure (SPKI) s-expressions, ISO/ITU-T X.509 Attribute Certificate and PrivilegePolicy, OASIS XACML policy language, and OASIS SAML assertions and protocols.

It is unlikely that a single syntax for attribute binding information or for policy expression will meet the needs of all environments and architectures. However, the search for flexible, powerful syntaxes for these types of information continues throughout the academic and commercial communities. In the meantime, some of the efforts mentioned above have been found to be appropriate and useful in specific environments and communities of interest.

FURTHER READING: Further discussion on authorization models and architectures can be found in the references list.

Carlisle Adams

References

[1] Adams, C. and S. Lloyd (2003). *Understanding PKI: Concepts, Standards, and Deployment Considerations* (2nd ed.). Addison-Wesley, Reading, MA.

[2] CORBA Security Project, http://security.dstc.edu.au/projects/corba/

[3] Distributed Computing Environment (DCE), http://www.opengroup.org/dce/

[4] Godik, S. and T. Moses (2003). "eXtensible Access Control Markup Language (XACML) Version 1.0." OASIS Standard, 18 February 2003.

[5] Hallam-Baker, P. and E. Maler (2002). "Assertions and protocol for the OASIS security assertion markup language (SAML)." OASIS Standard, 5 November 2002.

AUTHORIZATIONS MANAGEMENT

Authorizations management is a subset of general "authorization data" management (see authorization architecture) in which the data being managed is *authorizations* associated with entities in an environment. An *authorization* may be defined as follows [1]: something (typically in writing) "empowering a person (or system entity) to perform an act or to execute an office."

Carlisle Adams

Reference

[1] Encyclopaedia Britannica, http://www.britannica.com/

AUTHORIZATION POLICY

Authorization policy is the policy used by a policy decision point (PDP), in conjunction with authorization data, to render authorization decisions. See authorization architecture for details.

Carlisle Adams

AUTOCORRELATION

Let $\{a_t\}$ be a sequence of *period* n (so $a_t = a_{t+n}$ for all values of t) with symbols being the integers mod q (see modular arithmetic). The periodic auto-correlation of the sequence $\{a_t\}$ at shift τ is defined as

$$A(\tau) = \sum_{t=0}^{n-1} \omega^{a_{t+\tau} - a_t},$$

where ω is a complex qth root of unity.

In most applications one considers binary sequences when $q = 2$ and $\omega = -1$. Then the auto-correlation at shift τ equals the number of agreements minus the number of disagreements between the sequence $\{a_t\}$ and its cyclic shift $\{a_{t+\tau}\}$. Note that in most applications one wants the autocorrelation for all nonzero shifts $\tau \neq 0 \pmod{n}$ (the out-of-phase autocorrelation) to be low in absolute value. For example, this property of a sequence is extremely useful for synchronization purposes.

Tor Helleseth

References

[1] Golomb, S.W. (1982). *Shift Register Sequences*. Aegean Park Press, Laguna Hills, CA.

[2] Helleseth, T. and P.V. Kumar (1998). "Sequences with low correlation." *Handbook of Coding Theory*, eds. V.S. Pless and W.C. Huffman. Elsevier, Amsterdam.

[3] Helleseth, T. and P.V. Kumar (1999). "Pseudonoise sequences." *The Mobile Communications Handbook*, ed. J.D. Gibson. CRC Press, Boca Raton, FL, Chapter 8.

AVAILABILITY

A service is of no practical use if no one is able to access it. *Availability* is the property that legitimate principals are able to access a service within a timely manner whenever they may need to do so. Availability is typically expressed numerically as the fraction of a total time period during which a service is available. Although one of the keystones of computer security, availability has historically not been emphasized as much as other properties of security such as *confidentiality* and *integrity*. This lack of emphasis on availability has changed recently with the rise of open Internet services.

Decreased availability can occur both inadvertently, through failure of hardware, software, or infrastructure, or intentionally, through attacks on the service or infrastructure. The first can be mitigated through redundancy, where the probability of all backups experiencing a failure simultaneously is (hopefully) very low. It is in regard to these random failures where "five-nines of availability" (available 99.999% of the time) are often used when describing systems. The second cause for loss of availability is of more interest from a security standpoint. When an attacker is able to degrade availability, it is known as a <u>Denial of Service</u> attack. Malicious attacks against availability can focus on the service itself (e.g., exploiting a common software bug to cause all backups to fail simultaneously), or on the infrastructure supporting the service (e.g., flooding network links between the service and the principal).

Eric Cronin

BEAUFORT ENCRYPTION

This is an encryption similar to the <u>Vigenère encryption</u> [1], but with shifted reversed standard alphabets. For encryption and decryption, one can use the *Beaufort table* below (Giovanni Sestri, 1710).

	a	b	c	d	e	f	g	h	i	j	k	l	m	n	o	p	q	r	s	t	u	v	w	x	y	z
A	Z	Y	X	W	V	U	T	S	R	Q	P	O	N	M	L	K	J	I	H	G	F	E	D	C	B	A
B	A	Z	Y	X	W	V	U	T	S	R	Q	P	O	N	M	L	K	J	I	H	G	F	E	D	C	B
C	B	A	Z	Y	X	W	V	U	T	S	R	Q	P	O	N	M	L	K	J	I	H	G	F	E	D	C
D	C	B	A	Z	Y	X	W	V	U	T	S	R	Q	P	O	N	M	L	K	J	I	H	G	F	E	D
E	D	C	B	A	Z	Y	X	W	V	U	T	S	R	Q	P	O	N	M	L	K	J	I	H	G	F	E
F	E	D	C	B	A	Z	Y	X	W	V	U	T	S	R	Q	P	O	N	M	L	K	J	I	H	G	F
G	F	E	D	C	B	A	Z	Y	X	W	V	U	T	S	R	Q	P	O	N	M	L	K	J	I	H	G
H	G	F	E	D	C	B	A	Z	Y	X	W	V	U	T	S	R	Q	P	O	N	M	L	K	J	I	H
I	H	G	F	E	D	C	B	A	Z	Y	X	W	V	U	T	S	R	Q	P	O	N	M	L	K	J	I
J	I	H	G	F	E	D	C	B	A	Z	Y	X	W	V	U	T	S	R	Q	P	O	N	M	L	K	J
K	J	I	H	G	F	E	D	C	B	A	Z	Y	X	W	V	U	T	S	R	Q	P	O	N	M	L	K
L	K	J	I	H	G	F	E	D	C	B	A	Z	Y	X	W	V	U	T	S	R	Q	P	O	N	M	L
M	L	K	J	I	H	G	F	E	D	C	B	A	Z	Y	X	W	V	U	T	S	R	Q	P	O	N	M
N	M	L	K	J	I	H	G	F	E	D	C	B	A	Z	Y	X	W	V	U	T	S	R	Q	P	O	N
O	N	M	L	K	J	I	H	G	F	E	D	C	B	A	Z	Y	X	W	V	U	T	S	R	Q	P	O
P	O	N	M	L	K	J	I	H	G	F	E	D	C	B	A	Z	Y	X	W	V	U	T	S	R	Q	P
Q	P	O	N	M	L	K	J	I	H	G	F	E	D	C	B	A	Z	Y	X	W	V	U	T	S	R	Q
R	Q	P	O	N	M	L	K	J	I	H	G	F	E	D	C	B	A	Z	Y	X	W	V	U	T	S	R
S	R	Q	P	O	N	M	L	K	J	I	H	G	F	E	D	C	B	A	Z	Y	X	W	V	U	T	S
T	S	R	Q	P	O	N	M	L	K	J	I	H	G	F	E	D	C	B	A	Z	Y	X	W	V	U	T
U	T	S	R	Q	P	O	N	M	L	K	J	I	H	G	F	E	D	C	B	A	Z	Y	X	W	V	U
V	U	T	S	R	Q	P	O	N	M	L	K	J	I	H	G	F	E	D	C	B	A	Z	Y	X	W	V
W	V	U	T	S	R	Q	P	O	N	M	L	K	J	I	H	G	F	E	D	C	B	A	Z	Y	X	W
X	W	V	U	T	S	R	Q	P	O	N	M	L	K	J	I	H	G	F	E	D	C	B	A	Z	Y	X
Y	X	W	V	U	T	S	R	Q	P	O	N	M	L	K	J	I	H	G	F	E	D	C	B	A	Z	Y
Z	Y	X	W	V	U	T	S	R	Q	P	O	N	M	L	K	J	I	H	G	F	E	D	C	B	A	Z

Friedrich L. Bauer

Reference

[1] Bauer, F.L. (1997). "Decrypted secrets". *Methods and Maxims of Cryptology*. Springer-Verlag, Berlin.

BERLEKAMP–MASSEY ALGORITHM

The *Berlekamp–Massey algorithm* is an algorithm for determining the <u>linear complexity</u> of a finite <u>sequence</u> and the feedback polynomial of a <u>linear feedback shift register (LFSR)</u> of minimal length which generates this sequence. This algorithm is due to Massey [3], who showed that the iterative algorithm proposed in 1967 by Berlekamp [1] for decoding BCH codes (see <u>cyclic codes</u>) can be used for finding the shortest LFSR that generates a given sequence.

For a given sequence $\mathbf{s}^n$ of length n, the Berlekamp–Massey algorithm performs n iterations. The tth iteration determines an LFSR of minimal length, which generates the first t digits of $\mathbf{s}^n$. The algorithm can be described as follows.

Input. $\mathbf{s^n} = s_0 s_1 \ldots s_{n-1}$, a sequence of n elements of $\mathbf{F}_q$.

Output. Λ, the linear complexity of $\mathbf{s^n}$ and P, the feedback polynomial of an LFSR of length Λ which generates $\mathbf{s^n}$.

Initialization.
$$P(X) \leftarrow 1, P'(X) \leftarrow 1, \Lambda \leftarrow 0, m \leftarrow -1, d' \leftarrow 1.$$

For t **from** 0 **to** $n - 1$ **do**
$$d \leftarrow s_t + \sum_{i=1}^{\Lambda} p_i s_{t-i}.$$

If $d \neq 0$ then
 $T(X) \leftarrow P(X)$.
 $P(X) \leftarrow P(X) - d(d')^{-1}P'(X)X^{t-m}$.
 if $2\Lambda \leq t$ then
  $\Lambda \leftarrow t + 1 - \Lambda$.
  $m \leftarrow t$.
  $P'(X) \leftarrow T(X)$.
  $d' \leftarrow d$.
Return Λ and P.

In the particular case of a binary sequence, the quantity d' does not need to be stored since it is always equal to 1. Moreover, the feedback polynomial is simply updated by

$$P(X) \leftarrow P(X) + P'(X)X^{t-m}.$$

The number of operations performed for computing the linear complexity of a sequence of length n is $\mathcal{O}(n^2)$.

It is worth noting that the LFSR of minimal length that generates a sequence $\mathbf{s^n}$ of length n is unique if and only if $n \geq 2\Lambda(\mathbf{s^n})$, where $\Lambda(\mathbf{s^n})$ is the linear complexity of $\mathbf{s^n}$.

EXAMPLE: The following table describes the successive steps of the Berlekamp–Massey algorithm applied to the binary sequence of length 7, $s_0 \cdots s_6 = 0111010$. The values of Λ and P obtained at the end of step t correspond to the linear complexity of the sequence $s_0 \cdots s_t$ and to the feedback polynomial of an LFSR of minimal length that generates it.

t	s_t	d	Λ	$P(X)$	m	$P'(X)$
			0	1	-1	1
0	0	0	0	1	-1	1
1	1	1	2	$1 + X^2$	1	1
2	1	1	2	$1 + X + X^2$	1	1
3	1	1	2	$1 + X$	1	1
4	1	0	2	$1 + X$	1	1
5	0	1	4	$1 + X + X^4$	5	$1 + X$
6	0	0	4	$1 + X + X^4$	5	$1 + X$

The linear complexity $\Lambda(\mathbf{s})$ of a linear recurring sequence $\mathbf{s} = (s_t)_{t \geq 0}$ is equal to the linear complexity of the finite sequence composed of the first n terms of $\mathbf{s}$ for any $n \geq \Lambda(\mathbf{s})$. Thus, the Berlekamp–Massey algorithm determines the shortest LFSR that generates an infinite linear recurring sequence $\mathbf{s}$ from the knowledge of any $2\Lambda(\mathbf{s})$ consecutive digits of $\mathbf{s}$.

It can be proved [2] that the Berlekamp–Massey algorithm and the Euclidean algorithm are essentially the same.

Anne Canteaut

References

[1] Berlekamp, E.R. (1967). *Algebraic Coding Theory*. McGraw-Hill, New York.
[2] Dornstetter, J.-L. (1987). "On the equivalence between Berlekamp's and Euclid's algorithms." *IEEE Transactions on Information Theory*, 33, 428–431.
[3] Massey, J.L. (1969). "Shift-register synthesis and BCH decoding." *IEEE Transactions on Information Theory*, 15, 122–127.

BERLEKAMP *Q*-MATRIX

The *Q-matrix* is the key component in Berlekamp's elegant algorithm [1] for factoring a polynomial over finite field.

Let $\mathbf{F}_q$ be a finite field and let $f(x)$ be a monic polynomial of degree d over $\mathbf{F}_q$:

$$f(x) = x^d + f_{d-1}x^{d-1} + \cdots + f_1 x + f_0,$$

where the coefficients $f_0, \ldots, f_{d-1}$ are elements of $\mathbf{F}_q$. The factorization of $f(x)$ has the form

$$f(x) = \prod_i h_i(x)^{e_i},$$

where each factor $h_i(x)$ is an irreducible polynomial and $e_i \geq 1$ is the *multiplicity* of the factor $h_i(x)$.

Berlekamp's algorithm exploits the fact that for any polynomial $g(x)$ over $\mathbf{F}_q$,

$$g(x)^q - g(x) = \prod_{c \in \mathbf{F}_q}(g(x) - c).$$

Accordingly, given a polynomial $g(x)$ such that

$$g(x)^q - g(x) \equiv 0 \bmod f(x),$$

one can find factors of $f(x)$ by computing the greatest common divisor (in terms of polynomials) of $f(x)$ and each $g(x) - c$ term. (This process may need to be repeated with other polynomials $g(x)$ until the irreducible factors $h_i(x)$ are found.) The Q-matrix is the key to obtaining the polynomial $g(x)$. In particular, Berlekamp shows how to transform the congruence above into a problem in linear algebra,

$$(Q - I)\mathbf{g} = \mathbf{0},$$

where Q is a $d \times d$ matrix over $\mathbf{F}_q$, and I is the $d \times d$ identity matrix. The elements of Q correspond to the coefficients of the polynomials $x^{q^i} \bmod f(x), 0 \leq i < d$. The elements of each solution $\mathbf{g}$, a vector over $\mathbf{F}_q$, are the coefficients of $g(x)$. The running time of the algorithm as described is polynomial time in d and q, but it can be improved to be polynomial in d and $\log q$, and more efficient algorithms are also available (e.g., [2]).

Burt Kaliski

References

[1] Berlekamp, E.R. (1970). "Factoring polynomials over large finite fields." *Mathematics of Computation*, 24, 713–735.

[2] Shoup, V. and E. Kaltofen (1998). "Subquadratic-time factorization of polynomials over finite fields." *Mathematics of Computation*, 67 (223), 1179–1197.

BINARY EUCLIDEAN ALGORITHM

The principles behind this algorithm were discovered by R. Silver and J. Tersian and independently by Stein [8]. The algorithm computes the greatest common divisor and is based on the following observations:

- If u and v are both even, then $\gcd(u, v) = 2\gcd(u/2, v/2)$;
- If u is even and v is odd, then $\gcd(u, v) = \gcd(u/2, v)$;
- Otherwise both are odd, and $\gcd(u, v) = \gcd(|u - v|/2, v)$.

The three conditions cover all possible cases for u and v. The algorithm systematically reduces u and v by repeatedly testing the conditions and accordingly applying the reductions. Note that the first condition, i.e., u and v both being even, applies only in the very beginning of the procedure. Thus, the algorithm first factors out the highest common power of 2 from u and v and stores it in g. In the remainder of the computation only the other two conditions are tested. The computation terminates when one of the operands becomes zero. The algorithm is given as follows.

The Binary GCD Algorithm
Input: positive integers x and y
Output: $g = \mathrm{GCD}(u, v)$
    $g \leftarrow 1$
    While u is even AND v is even do
      $u \leftarrow u/2$; $v \leftarrow v/2$; $g \leftarrow 2g$;
    End While
    While $u \neq 0$ do
      While u is even do $u \leftarrow u/2$;
      While v is even do $v \leftarrow v/2$;
      $t \leftarrow |u - v|/2$;
      If $u \geq v$ then
        $u \leftarrow t$;
      Else
        $v \leftarrow t$;
      End While
    End While
    Return (gv)

In the algorithm, only simple operations such as addition, subtraction, and divisions by two (shifts) are computed. Although the binary GCD algorithm requires more steps than the classical Euclidean algorithm, the operations are simpler. The number of iterations is known [6] to be bounded by $2(\log_2(u) + \log_2(v) + 2)$.

Similar to the extended Euclidean algorithm, the binary GCD algorithm was adapted to return two additional parameters s and t such that

$$su + tv = \gcd(u, v).$$

These parameters are essential for modular inverse computations. If $\gcd(u, v) = 1$ then it follows that $s = u^{-1} \bmod v$ and $t = v^{-1} \bmod u$. Knuth [5] attributes the extended version of the binary GCD algorithm to Penk. The algorithm given below is due to Bach and Shallit [1].

The Binary Euclidean Algorithm
Input: positive integers x and y
Output: integers s, t, g such that $su + tv = g$
where $g = \mathrm{GCD}(u, v)$
    $g \leftarrow 1$
    While u is even AND v is even do
      $u \leftarrow u/2$; $v \leftarrow v/2$; $g \leftarrow 2g$;
    End While
    $x \leftarrow u$; $y \leftarrow v$; $s'' \leftarrow 1$; $s' \leftarrow 0$; $t'' \leftarrow 0$;
    $t' \leftarrow 1$;
    L1 While x is even do
      $x \leftarrow x/2$;
      If s'' is even and t'' is even then
        $s'' \leftarrow s''/2$; $t'' \leftarrow t''/2$;
      Else
        $s'' \leftarrow (s'' + v)/2$; $t'' \leftarrow (t'' - u)/2$;
      End If
    End While
    While y is even do
      $y \leftarrow y/2$;
      If s' is even AND t' is even then
        $s' \leftarrow s'/2$; $t' \leftarrow t'/2$;
      Else
        $s' \leftarrow (s' + v)/2$; $t' \leftarrow (t' - u)/2$;
      End If
    End While
    If $x \geq y$ then
      $x \leftarrow x - y$; $s'' \leftarrow s'' - s'$; $t'' \leftarrow t'' - t'$;
    Else
      $y \leftarrow y - x$; $s' \leftarrow s' - s''$; $t' \leftarrow t' - t''$;
    End If

If $x = 0$ then
$$s \leftarrow s'; t \leftarrow t';$$
Else
    GoTo L1
End If
Return (s, t, gy)

The binary Euclidean algorithm may be used for computing inverses $a^{-1} \bmod m$ by setting $u = m$ and $v = a$. Upon termination of the execution, if $\gcd(u, v) = 1$ then the inverse is found and its value is stored in t. Otherwise, the inverse does not exist. In [6], it is noted that for computing multiplicative inverses the values of s'' and t'' do not need to be computed if m is odd. In this case, the evenness condition on s'' and t'' in the second while loop may be decided by examining the parity of s'. If m is odd and s' is even, then s'' must be even.

The run time complexity is $O((\log(n))^2)$ bit operations. Convergence of the algorithm, if not obvious, can be shown by induction. A complexity analysis of the binary euclidean algorithm was presented by Brent in [2]. Bach and Shallit give a detailed analysis and comparison to other GCD algorithms in [1].

Sorenson claims that the binary Euclidean algorithm is the most efficient algorithm for computing greatest common divisors [7]. In the same reference Sorenson also proposed a k-ary version of the binary GCD algorithm with worst case running time $O(n^2/\log(n))$.

In [3], Jebelean claims that Lehmer's Euclidean algorithm is more efficient than the binary GCD algorithm. The same author presents [4] a word-level generalization of the binary GCD algorithm with better performance than Lehmer's Euclidean algorithm.

See also Euclidean Algorithm.

Berk Sunar

References

[1] Bach, E. and J. Shallit (1996). *Algorithmic Number Theory, Volume I: Efficient Algorithms*. MIT Press, Cambridge, MA.

[2] Brent, R.P. (1976). "Analysis of the binary Euclidean algorithm." *Algorithms and Complexity*, ed. J.F. Traub. Academic Press, New York, 321–355.

[3] Jebelean, T. (1993). "Comparing several gcd algorithms." *11th IEEE Symposium on Computer Arithmetic*.

[4] Jebelean, T. (1993). "A generalization of the binary gcd algorithm." *Proceedings of the 1993 International Symposium on Symbolic and Algebraic Computation*. ACM Press, New York, 111–116.

[5] Knuth, D.E. (1997). *The Art of Computer Programming, Volume 2: Seminumerical Algorithms* (3rd ed.). Addison-Wesley, Longman Publishing Co., Inc., Reading, MA.

[6] Menezes, A.J., P.C. van Oorschot, and S.A. Vanstone (1997). *Handbook of Applied Cryptography*. CRC Press, Boca Raton, FL.

[7] Sorenson, J. (1994). "Two fast gcd algorithms." *Journal of Algorithms*, 16 (1), 110–144.

[8] Stein, J. (1967). "Computational problems associated with racah algebra." *Journal of Computational Physics*, 1, 397–405.

BINARY EXPONENTIATION

Most schemes for public-key cryptography involve exponentiation in some group (or, more generally, in some semigroup; a semigroup is an algebraic structure that is like a group except that elements need not have inverses, and that there may not even be a neutral element). The term *exponentiation* assumes that the group operation is written multiplicatively. If the group operation is written additively, one speaks of *scalar multiplication* instead, but this change in terminology does not affect the essence of the task.

Let $\circ$ denote the group operation and assume that the exponentiation to be performed is g^e where g is an element of the group (or semigroup) and e is a positive integer. Computing the result $g \circ \cdots \circ g$ in a straightforward way by applying the group operation $e - 1$ times is feasible only if e is very small; for $e \geq 4$, it is possible to compute g^e with fewer applications of the group operation. Determining the minimum number of group operations needed for the exponentiation, given some exponent e, is far from trivial; see fixed-exponent exponentiation. (Furthermore, the time needed for each single computation of the group operation is usually not constant: for example, it often is faster to compute a squaring $A \circ A$ than to compute a general multiplication $A \circ B$.) Practical implementations that have to work for arbitrary exponents need exponentiation algorithms that are reasonably simple and fast.

Assuming that for the exponentiation one can use no other operation on group elements than the group operation $\circ$ (and that one cannot make use of additional information such as the order of the group or the order of specific group elements), it can be seen that for l-bit exponents (i.e., $2^{l-1} \leq e < 2^l$), any exponentiation method will have to apply the group operation at least $l - 1$ times to arrive at the power g^e. The left-to-right binary exponentiation method is a very simple and memory-efficient technique for performing exponentiations in at most $2(l - 1)$ applications of the group operation for any l-bit exponent (i.e., within a factor of 2 from the lower bound). It is based on the binary representation of exponents e:

$$e = \sum_{i=0}^{l-1} e_i 2^i, \qquad e_i \in \{0, 1\}.$$

With l chosen minimal in such a representation, we have $e_{l-1} = 1$. Then g^e can be computed as follows:

```
A ← g
for i = l − 2 down to 0 do
    A ← A ∘ A
    if e_i = 1 then
        A ← A ∘ g
return A
```

If the group is considered multiplicative, then computing $A \circ A$ means squaring A, and computing $A \circ g$ means multiplying A by g; hence this algorithm is also known as the *square-and-multiply* method for exponentiation. If the group is considered additive, then computing $A \circ A$ means doubling A, and computing $A \circ g$ means adding g to A; hence this algorithm is also known as the *double-and-add* method for scalar multiplication.

The algorithm shown above performs a <u>left-to-right exponentiation</u>, i.e., it starts at the most significant digit of the exponent e (which, assuming big-endian notation, appears at the left) and goes toward the least significant digit (at the right). The binary exponentiation method also has a variant that performs a <u>right-to-left exponentiation</u>, i.e., starts at the least significant digit and goes toward the most significant digit:

```
flag ← false
B ← identity element
A ← g
for i = 0 to l − 1 do
    if e_i = 1 then
        if flag then
            B ← B ∘ A
        else
            B ← A    {Equiv. to B ← B ∘ A}
            flag ← true
    if i < l − 1 then
        A ← A ∘ A
return B
```

This algorithm again presumes that $e_{l-1} = 1$. The right-to-left method is essentially the traditional algorithm known as "Russian peasant multiplication," generalized to arbitrary groups.

For an l-bit exponent, the left-to-right and right-to-left binary exponentiation methods both need $l - 1$ squaring operations ($A \circ A$) and, assuming that all bits besides e_{l-1} are uniformly and independently random, $(l - 1)/2$ general group operations ($A \circ g$ or $B \circ A$) on average.

Various other algorithms are known that can be considered variants or generalizations of binary exponentiation: see <u>2^k-ary exponentiation</u> and <u>sliding window exponentiation</u> for other methods for computing powers (which can often be faster than binary exponentiation), and see <u>simultaneous exponentiation</u> for methods for computing power products. See also <u>signed digit exponentiation</u> for techniques that can improve efficiency in groups allowing fast inversion.

Bodo Möller

BINOMIAL DISTRIBUTION

If a two-sided coin is flipped n times, what is the probability that there are exactly k heads? This probability is given by the *binomial distribution*. If the coin is unbiased and the coin flips are independent of one another, then the probability is given by the equation

$$\Pr[k \text{ heads} \mid n \text{ coin flips}] = \binom{n}{k} 2^{-n}.$$

Here, the notation $\binom{n}{k}$, read "n choose k," is the number of ways of choosing k items from a set of n items, ignoring order. The value may be computed as

$$\binom{n}{k} = \frac{n!}{k!(n-k)!}.$$

For the first several values of n, the following probabilities are as follows for an unbiased coin (read k left to right from 0 to n):

$n = 0$:			1		
$n = 1$:		$\frac{1}{2}$		$\frac{1}{2}$	
$n = 2$:	$\frac{1}{4}$		$\frac{1}{2}$		$\frac{1}{4}$

$n = 3$: $\quad \frac{1}{8} \quad \frac{3}{8} \quad \frac{3}{8} \quad \frac{1}{8}$

$n = 4$: $\quad \frac{1}{16} \quad \frac{1}{4} \quad \frac{3}{8} \quad \frac{1}{4} \quad \frac{1}{16}$

More generally, if the coin flips are independent but the probability of heads is p, the binomial distribution is likewise biased:

$$\Pr[k \text{ heads} \mid n \text{ coin flips, probability } p \text{ of heads}]$$
$$= \binom{n}{k} p^k (1-p)^{n-k}.$$

The name "binomial" comes from the fact that there are two outcomes (heads and tails) and the probability distribution can be determined by computing powers of the two-term polynomial (binomial) $f(x) = px + (1 - p)$. The probability that there are exactly k heads after n coin flips is exactly the same as the x^k term of the polynomial $f(x)^n$.

As coin flips (either physical or their computational equivalent) are the basic building block of randomness in cryptography, the binomial distribution is likewise the foundation of probability analysis in this field.

Burt Kaliski

BIOMETRICS

A wide variety of systems require reliable authentication schemes to confirm the identity of an individual requesting their services (see identification). The purpose of such schemes is to ensure that the rendered services are accessed only by a legitimate user, and no one else. Examples of such applications include secure access to buildings, computer systems, laptops, cellular phones, and ATMs. In the absence of robust authentication schemes, these systems are vulnerable to the wiles of an impostor.

Traditionally, passwords (knowledge-based security) and ID cards (token-based security) have been used to restrict access to systems. However, security can be easily breached in these systems when a password is divulged to an unauthorized user or an ID card is stolen by an impostor. Further, simple passwords are easy to guess (by an impostor) and complex passwords may be hard to recall (by a legitimate user). The emergence of *biometrics* has addressed the problems that plague these traditional security methods. Biometrics refers to the automatic identification (or verification) of an individual (or a claimed identity) by using certain physiological or behavioral traits associated with the person. By using biometrics it is possible to establish an identity based on "who you are," rather than by "what you possess" (e.g., an ID card) or "what you remember" (e.g., a password). Current biometric systems make use of fingerprints, hand geometry, iris, retina, face, hand vein, facial thermograms, signature, voiceprint, etc. (Figure 1) to establish a person's identity [1,4]. While biometric systems have their limitations (e.g., additional cost, temporal changes in biometric traits, etc.), they have an edge over traditional security methods in that they cannot be easily stolen or shared.

Biometric systems also introduce an aspect of user convenience that may not be possible using

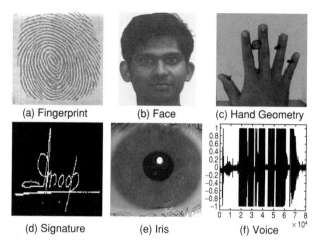

(a) Fingerprint   (b) Face   (c) Hand Geometry

(d) Signature   (e) Iris   (f) Voice

Fig. 1. Examples of some of the biometric traits used for authenticating an individual

traditional security techniques. For example, users maintaining different passwords for different applications may find it challenging to recollect the password associated with a specific application. In some instances, the user might even forget the password, requiring the system administrator to intervene and reset the password for that user. Maintaining, recollecting, and resetting passwords can, therefore, be a tedious and expensive task. Biometrics, on the other hand, addresses this problem effectively: a user can use the same biometric trait (e.g., right index finger) or different biometric traits (e.g., fingerprint, hand geometry, iris) for different applications, with "password" recollection not being an issue at all.

A typical biometric system operates by acquiring biometric data from an individual, extracting a feature set from the acquired data, and comparing this feature set against the template feature set stored in the database (Figure 2). In an *identification* scheme, where the goal is to recognize the individual, this comparison is done against templates corresponding to all the enrolled users (a one-to-many matching); in a *verification* scheme, where the goal is to verify a claimed identity, the comparison is done against only those templates corresponding to the claimed identity (a one-to-one matching). Thus, identification ("whose biometric data is this?") and verification ("does this biometric data belong to Bob?") are two different problems with different inherent complexities. The templates are typically created at the time of enrollment, and, depending on the application, may or may not require human personnel intervention.

Biometric systems are being increasingly deployed in large scale civilian applications. The Schiphol Privium scheme at the Amsterdam

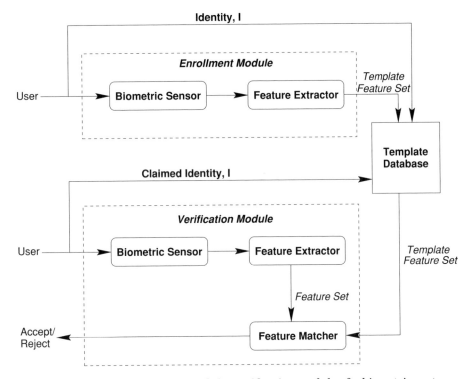

Fig. 2. The enrollment module and the verification module of a biometric system

airport, for example, employs iris scan cards to speed up the passport and visa control procedures. Passengers enrolled in this scheme insert their card at the gate and look into a camera; the camera acquires the image of the traveler's eye and processes it to locate the iris, and computes the Iriscode; the computed Iriscode is compared with the data residing in the card to complete user verification. A similar scheme is also being used to verify the identity of Schiphol airport employees working in high-security areas. Thus, biometric systems can be used to enhance user convenience while improving security.

A simple biometric system has four important modules: (i) *Sensor module* which acquires the biometric data of an individual. An example would be a fingerprint sensor that images the fingerprint ridges of a user. (ii) *Feature extraction module* in which the acquired biometric data is processed to extract a feature set that represents the data. For example, the position and orientation of ridge bifurcations and ridge endings (known as minutiae points) in a fingerprint image are extracted in the feature extraction module of a fingerprint system. (iii) *Matching module* in which the extracted feature set is compared against that of the template by generating a matching score. For example, in this module, the number of matching minutiae points between the acquired and template fingerprint images is determined, and a matching score reported. (iv) *Decision-making module* in which

the user's claimed identity is either accepted or rejected based on the matching score (verification). Alternately, the system may identify a user based on the matching scores (identification).

In order to analyze the performance of a biometric system, the probability distribution of genuine and impostor matching scores is examined. A genuine matching score is obtained when two feature sets corresponding to the *same* individual are compared, and an impostor matching score is obtained when feature sets from two *different* individuals are compared. When a matching score exceeds a certain threshold, the two feature sets are declared to be from the same individual; otherwise, they are assumed to be from different individuals. Thus, there are two fundamental types of errors associated with a biometric system: (i) a false accept, which occurs when an impostor matching score exceeds the threshold, and (ii) a false reject, which occurs when a genuine matching score does not exceed the threshold. The error rates of systems based on fingerprint and iris are usually lower compared to those based on voice, face, and hand geometry. A receiver operating characteristic (ROC) curve plots the false reject rate (FRR—the percentage of genuine scores that do not exceed the threshold) against the false accept rate (FAR—the percentage of impostor scores that exceed the threshold) at various thresholds. The operating threshold employed by a system depends on the nature of the application. In forensic

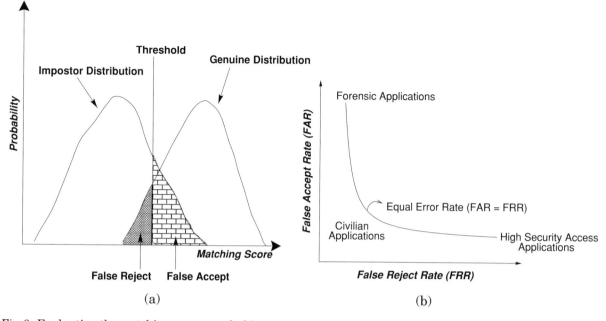

Fig. 3. Evaluating the matching accuracy of a biometric system: (a) histograms of genuine and impostor matching scores and the two types of errors that are possible in a biometric system; (b) a receiver operating characteristic curve indicating the operating point (threshold) for different kinds of applications

applications, for example, a low FRR is preferred, while in high security access facilities like nuclear labs, a low FAR is desired (Figure 3). Besides FAR and FRR, other types of errors are also possible in a biometric system. The failure to enroll (FTE) error refers to the inability of a biometric system to enroll an individual whose biometric trait may not be of good quality (e.g., poor quality fingerprint ridges). Similarly, a biometric system may be unable to procure good quality biometric data from an individual during authentication resulting in a failure to acquire (FTA) error.

A biometric system is susceptible to various types of attacks. For example, an impostor may attempt to present a fake finger or a face mask or even a recorded voice sample in order to circumvent the system. The problem of fake biometrics may be mitigated by employing challenge-response mechanisms or conducting liveness detection tests. Most biometric systems currently deployed are used for local authentication, i.e., seldom is the biometric data acquired from a user transmitted across a network channel. This avoids problems that would arise if a channel is compromised. Privacy concerns related to the use of biometrics and protection of biometric templates are issues that are currently being studied [3].

The increased demand for reliable and convenient authentication schemes, availability of inexpensive computing resources, development of cheap biometric sensors, and advancements in signal processing have all contributed to the rapid deployment of biometric systems in establishments ranging from grocery stores to airports. The emergence of multimodal biometrics has further enhanced the matching performance of biometric systems [2]. It is only a matter of time before biometrics integrates itself into the very fabric of society and impacts the way we conduct our daily business.

Anil K. Jain
Arun Ross

References

[1] Jain, Anil K., Ruud Bolle, and Sharath Pankanti, eds. (1999). *Biometrics: Personal Identification in Networked Society*. Kluwer Academic Publishers, Dordrecht.

[2] Jain, Anil K. and Arun Ross (2004). "Multibiometric systems." *Communications of the ACM*, 47 (1), 34–40.

[3] Prabhakar, Salil, Sharath Pankanti, and Anil K. Jain (2003). "Biometric recognition: Security and privacy concerns." *IEEE Security and Privacy Magazine*, 1 (2), 33–42.

[4] Wayman, James L. (2001). "Fundamentals of biometric authentication technologies." *International Journal of Image and Graphics*, 1 (1), 93–113.

BIRTHDAY PARADOX

The birthday paradox refers to the fact that there is a probability of more than 50% that among a group of at least 23 randomly selected people at least 2 have the same birthday. It follows from

$$\frac{365}{365} \cdot \frac{365-1}{365} \cdots \frac{365-22}{365} \approx 0.49 < 0.5;$$

it is called a paradox because the 23 is felt to be unreasonably small compared to 365. Further, in general, it follows from

$$\prod_{0 \le i \le 1.18\sqrt{p}} \frac{p-i}{p} \approx 0.5$$

that it is not unreasonable to expect a duplicate after about $\sqrt{p}$ elements have been picked at random (and with replacement) from a set of cardinality p. A good exposition of the probability analysis underlying the birthday paradox can be found in Corman et al. [1], Section 5.4.

Under reasonable assumptions about their inputs, common cryptographic k-bit hash functions may be assumed to produce random, uniformly distributed k-bit outputs. Thus one may expect that a set of the order of $2^{k/2}$ inputs contains two elements that hash to the same value. Such hash function collisions have important cryptanalytic applications. Another prominent cryptanalytic application of the birthday paradox is Pollard's rho factoring method (see integer factoring) where elements are drawn from $\mathbf{Z}/n\mathbf{Z}$ for some integer n to be factored. When taken modulo p for any unknown p dividing n, the elements are assumed to be uniformly distributed over $\mathbf{Z}/p\mathbf{Z}$. A collision modulo p, and therefore possibly a factor of n, may be expected after drawing approximately $\sqrt{p}$ elements.

Cryptanalytic applications of the birthday paradox where the underlying distributions are not uniform are the large prime variations of sieving based factoring methods. There, in the course of the data gathering step, data involving so-called large primes q is found with probability approximately inversely proportional to q. Data involving large primes is useless unless different data with a matching large prime is found. The fact that smaller large primes occur relatively frequently, combined with the birthday paradox, leads to a large number of matches and a considerable speed-up of the factoring method.

Arjen K. Lenstra

Reference

[1] Cormen, Thomas L., Charles E. Leiserson, Ronald L. Rivest, and Clifford Stein (2001). *Introduction to Algorithms* (2nd ed.). MIT Press, Cambridge, MA.

BLIND SIGNATURE

In a blind signature scheme signers have individual private signing keys and distribute their corresponding public verifying keys, just as in normal cryptographic digital signature schemes. Public verifying keys are distributed via authentication channels, for example, by means of public key infrastructures. There is also a publicly available verifying algorithm such that anyone who has retrieved a public verifying key y of a signer can verify whether a given signature s is valid for a given message m with respect to the signer's public verifying key y.

In a blind signature scheme, the signers neither learn the messages they sign, nor the signatures the recipients obtain for their messages. A verifier who seeks a signature for a message m' from a signer with verifying key y prepares some related message m and passes m to the signer. The signer provides a response s back to the recipient, such that the recipient can derive a signature s' from y, m, m', s such that s' is valid for m' with respect to y. The resulting signature s' is called a "blind signature," although it is not the signature that is blind, but the signer.

The first constructions of cryptographic blind signatures were proposed by David Chaum. These early blind signature schemes were based on RSA signatures. An example is the Chaum Blind Signature [7, 8].

The security of blind signature schemes is defined by a degree of unforgeability and a degree of blindness. Of the notions of unforgeability (see forgery) for normal cryptographic signature schemes defined by Goldwasser et al. [16], only unforgeability against *total break* and *universal break* apply to blind signature schemes. However, the notions of selective forgery and existential forgery are inappropriate for blind signature schemes, because they assume an active attack to be successful if after the attack the recipient has obtained a signature for a (new) message that the signer has not signed before. Obviously, this condition holds for every message a recipient gets signed in a blind signature scheme, and therefore the definition cannot discriminate attacks from normal use of the scheme. For blind signatures, one is interested in other notions of unforgeability, namely unforgeability against one-more forgery and restrictiveness (see forgery), both of which are mainly motivated by the use of blind signatures in untraceble electronic cash.

A one-more-forgery [19] is an attack that for some polynomially bounded integer n comes up with valid signatures for $n + 1$ pairwise different messages after the signer has provided signatures only for n messages. Blind signatures unforgeable against one-more forgery have attracted attention since Chaum et al. [10] and Chaum [9] used them to build practical offline and online <u>untraceable</u> <u>electronic cash</u> schemes. Most practical electronic cash schemes employ one-time blind signatures, where a customer can obtain only one signed message from each interaction with the bank during withdrawal of an electronic coin. This helps to avoid the problem of counterfeiting electronic coins [3–6, 22]. Formal definitions of one-time blind signatures have been proposed by Franklin and Yung [15] and by Pointcheval [19].

In a restrictive blind signature scheme, a recipient who passes a message m to a signer (using verifying key y) and receives information s in return can derive from y, m, m', s only valid signatures for those messages m' that observe the same structure as m. In offline electronic cash this is used to encode a customer's identity into the messages that are signed by the bank such that the messages obtained by the customer all have his identity encoded correctly. Important work in this direction was done by Chaum and Pedersen [7], Brands [1], Ferguson [12, 13], Frankel et al. [14] and Radu et al. [20, 21]. A formal definition of a special type of restrictive blind signatures has been given by Pfitzmann and Sadeghi [18].

Blindness is a property serving the privacy interests of honest recipients against cheating and collaborating signers and verifiers. The highest degree of unlinkability is unconditional unlinkability, where a dishonest signer and verifier, both with unconditional computing power, cannot distinguish the transcripts (m, s) seen by the signer in his interactions with the honest recipient from the recipient's outputs (m', s'), which are seen by the verifier, even if the signer and the verifier collaborate. More precisely, consider an honest recipient who first obtains n pairs of messages and respective valid signatures $(m_1, s_1), \ldots, (m_n, s_n)$ from a signer, then derives n pairs of blinded messages and signatures $(m_1', s_1'), \ldots, (m_n', s_n')$ from the former n pairs one by one, and later shows the latter n pairs to a verifier in random order. Then, the signer and the collaborating verifier should find each bijection of the former n pairs onto the latter n pairs to be equally likely to describe which of the latter pairs the honest recipient has derived from which of the former pairs. A weaker degree of blindness is defined as computational unlinkability, which is defined just as unconditional unlinkability except that the attacker is computationally

restricted (<u>computational complexity</u>). These are formalizations of the intended property that the signer does not learn "anything" about the message being signed.

On a spectrum between keeping individuals accountable and protecting their identities against unduly propagation or misuse, blind signature schemes tend toward the latter extreme. In many applications this strongly privacy oriented approach is not acceptable in all circumstances. While the identities of honest individuals are protected in a perfect way, criminal dealings of individuals who exploit such systems to their own advantage are protected just as perfectly. For example, Naccache and van Solms [17] have described "perfect crimes" where a criminal blackmails a customer to withdraw a certain amount of money from her account by using a blind signature scheme and then deposit the amount into the criminal's account.

Trustee based blind signature schemes have been proposed to strike a more acceptable balance between keeping individuals accountable and protecting their identities. Stadler et al. [22] have proposed *fair blind signatures*. Fair blind signatures employ a trustee who is involved in the key setup of the scheme and in an additional link-recovery operation between a signer and the trustee. The trustee can revoke the "blindness" of certain pairs of messages and signatures upon request. The link-recovery operation allows the signer or the judge to determine for each transcript (m, s) of the signing operation which message m' has resulted for the recipient, or to determine for a given recipient's message m' from which transcript (m, s) it has evolved. Similar approaches have been applied to constructions of electronic cash [3, 21].

Blind signatures have been employed extensively in cryptographic constructions of privacy oriented services such as untraceable <u>electronic cash</u>, anonymous <u>electronic voting schemes</u>, and <u>unlinkable</u> <u>credentials</u>.

Gerrit Bleumer

References

[1] Brands, Stefan (1993). "An efficient off-line electronic cash system based on the representation problem." Centrum voor Wiskunde en Informatica, Computer Science/Departement of Algorithmics and Architecture, Report CS-R9323, http://www.cwi.nl/

[2] Brands, Stefan (1994). "Untraceable off-line cash in wallet with observers." *Advances in*

Cryptology—CRYPTO'93, Lecture Notes in Computer Science, vol. 773, ed. D.R. Stinson. Springer-Verlag, Berlin, 302–318.

[3] Brickell, Ernie, Peter Gemmell, and David Kravitz (1995). "Trustee-based tracing extensions to anonymous cash and the making of anonymous change." *6th ACM-SIAM Symposium on Discrete Algorithms (SODA)*. ACM Press, New York, 457–466.

[4] Camenisch, Jan L., Jean-Marc Piveteau, and Markus A. Stadler (1994). "An efficient electronic payment system protecting privacy." *ESORICS'94 (Third European Symposium on Research in Computer Security), Brighton*, Lecture Notes in Computer Science, vol. 875, ed. D. Gollman. Springer-Verlag, Berlin, 207–215.

[5] Camenisch, Jan L., Jean-Marc Piveteau, and Markus A. Stadler (1995). "Blind signatures based on the discrete logarithm problem." *Advances in Cryptology—EUROCRYPT'94*, Lecture Notes in Computer Science, vol. 950, ed. A. De Santis. Springer-Verlag, Berlin, 428–432.

[6] Camenisch, Jan L., Jean-Marc Piveteau, and Markus A. Stadler (1996). "An efficient fair payment system." *3rd ACM Conference on Computer and Communications Security, New Delhi, India, March 1996*. ACM Press, New York, 88–94.

[7] Chaum, David (1983). "Blind signatures for untraceable payments." *Advances in Cryptology—CRYPTO'82*, Lecture Notes in Computer Science, eds. Plenum D. Chaum, R.L. Rivest, and A.T. Sherman. Plenum Press, New York, 199–203.

[8] Chaum, David (1990). "Showing credentials without identification: Transferring signatures between unconditionally unlinkable pseudonyms." *Advances in Cryptology—AUSCRYPT'90*, Lecture Notes in Computer Science, vol. 453. Springer-Verlag, Berlin, 246–264.

[9] Chaum, David (1990). "Online cash checks." *Advances in Cryptology—EUROCRYPT'89*, Lecture Notes in Computer Science, vol. 434, eds. J.-J. Quisquatere and J. Vandewalle. Springer-Verlag, Berlin, 288–293.

[10] Chaum, David, Amos Fiat, and Moni Naor (1990). "Untraceable electronic cash." *Advances in Cryptology—CRYPTO'88*, Lecture Notes in Computer Science, vol. 403, ed. S. Goldwasser. Springer-Verlag, Berlin, 319–327.

[11] Chaum, David and Torben P. Pedersen (1993). "Wallet databases with observers." *Advances in Cryptology—CRYPTO'92*, Lecture Notes in Computer Science, vol. 740, ed. E.F. Brickell. Springer-Verlag, Berlin, 89–105.

[12] Ferguson, Niels (1994). "Single term off-line coins." *Advances in Cryptology—EUROCRYPT'93*, Lecture Notes in Computer Science, vol. 765, ed. T. Helleseth. Springer-Verlag, Berlin, 318–328.

[13] Ferguson, Niels (1994). "Extensions of single-term coins." *Advances in Cryptology—CRYPTO'93*, Lecture Notes in Computer Science, vol. 773, ed. D.R. Stinson. Springer-Verlag, Berlin, 292–301.

[14] Frankel, Yair, Yiannis Tsiounis, and Moti Yung (1996). "Indirect discourse proofs: Achieving efficient fair off-line e-cash." *Advances in Cryptography—ASIACRYPT'96*, Lecture Notes in Computer Science, vol. 1163, eds. K. Kim and T. Matsumoto. Springer-Verlag, Berlin, 286–300.

[15] Franklin, Matthew and Moti Yung (1993). "Secure and efficient off-line digital money." *20th International Colloquium on Automata, Languages and Programming (ICALP)*, Lecture Notes in Computer Science, vol. 700, eds. A. Lingas, R.G. Karlsson, and S. Carlsson. Springer-Verlag, Heidelberg, 265–276.

[16] Goldwasser, Shafi, Silvio Micali, and Ronald L. Rivest (1988). "A digital signature scheme secure against adaptive chosen-message attacks." *SIAM Journal on Computing*, 17 (2), 281–308.

[17] Naccache, David and Sebastiaan von Solms (1992). "On blind signatures and perfect crimes." *Computers & Security*, 11 (6), 581–583.

[18] Birgit, Pfitzmann, and Ahmad-Reza Sadeghi (1999). "Coin-based anonymous fingerprinting." *Advances in Cryptology—EUROCRYPT'99*, Lecture Notes in Computer Science, vol. 1592, ed. J. Stern. Springer-Verlag, Berlin, 150–164.

[19] Pointcheval David (1998). "Strengthened security for blind signatures." *Advances in Cryptology—EUROCRYPT'98*, Lecture Notes in Computer Science, vol. 1403, ed. K. Nyberg. Springer-Verlag, Berlin, 391–405.

[20] Radu, Christian René Govaerts, and Joos Vandewalle (1996). "A restrictive blind signature scheme with applications to electronic cash." *Communications and Multimedia Security II*. Chapman & Hall, London, 196–207.

[21] Radu, Christian René Govaerts, and Joos Vandewalle (1997). "Efficient electronic cash with restricted privacy." *Financial Cryptography '97*. Springer-Verlag, Berlin, 57–69.

[22] Stadler, Markus, Jean-Marc Piveteau, and Jan L. Camenisch (1995). "Fair blind signatures." *Advances in Cryptology—EUROCRYPT'95*, Lecture Notes in Computer Science, vol. 921, eds. L.C. Guillou and J.-J. Quisquater. Springer-Verlag, Berlin, 209–219.

BLINDING TECHNIQUES

Blinding is a concept in cryptography that allows a client to have a provider compute a mathematical function $y = f(x)$, where the client provides an input x and retrieves the corresponding output y, but the provider would learn about neither x nor y. This concept is useful if the client cannot compute the mathematical function f all by himself, for example, because the provider uses an additional private input in order to compute f efficiently.

Blinding techniques can be used on the client side of client-server architectures in order to enhance the privacy of users in online transactions. This is the most effective way of dealing with server(s) that are not fully trusted.

Blinding techniques are also the most effective countermeasure against remote timing analysis of Web servers [4] and against power analysis and/or timing analysis of hardware security modules (see side-channel attacks and side-channel analysis).

In a typical setting, a provider offers to compute a function $f_x(m)$ using some private key x and some input m chosen by a client. A client can send an input m, have the provider compute the corresponding result $z = f_x(m)$, and retrieve z from the provider afterward. With a blinding technique, a client would send a transformed input m' to the provider, and would retrieve the corresponding result z' in return. From this result, the client could then derive the result $z' = f_x(m')$ that corresponds to the input m in which the client was interested in the first place. Some blinding techniques guarantee that the provider learns no information about the client's input m and the corresponding output z.

More precisely, blinding works as follows: consider a key generating algorithm *gen* that outputs pairs (x, y) of private and public keys (see public-key cryptography), two domains M, Z of messages, and a domain A of *blinding factors*. Assume a family of functions $z = f_x(m)$, where each member is indexed by a private key x, takes as input a value $m \in M$, and produces an output $z \in Z$. Let $\phi_{y,a} : M \to M$ and $\Phi_{y,a} : Z \to Z$ be two families of auxiliary functions, where each member is indexed by a public key y and a blinding factor a, such that the following two conditions hold for each key pair (x, y) that can be generated by *gen*, each blinding factor $a \in A$ and each input $m \in M$:

- the functions $\phi_{y,a}$ and $\Phi_{y,a}^{-1}$ are computable in polynomial time,
- $\Phi_{y,a}^{-1}(f_x(\phi_{y,a}(m))) = f_x(m)$ (as shown in the following diagram).

$$
\begin{array}{ccc}
M & \stackrel{f_x(m)}{\longrightarrow} & Z \\
\phi_{y,a}(m) \downarrow & & \uparrow \Phi_{y,a}^{-1}(z') \\
M & \stackrel{f_x(m')}{\longrightarrow} & Z
\end{array}
$$

In order to blind the computation of f_x by the provider, a client can use the auxiliary functions ϕ, Φ in a two-pass interactive protocol as follows:

1. The provider generates a pair (x, y) of a private key and a public key and publishes y.

2. The client chooses an input m, generates a blinding factor $a \in A$ at random, and transforms m into $m' = \phi_{y,a}(m)$.
3. The client sends m' to the provider and receives $z' = f_x(m')$ from the provider in return.
4. The client computes $z = \Phi_{y,a}^{-1}(z')$.

If both m and m' are equally meaningful or meaningless to the provider, then he has no way of distinguishing a client who sends the plain argument m from a client who sends a blinded argument m' in step 3.

The first blinding technique was proposed by Chaum as part of the Chaum Blind Signature [5,6]. It is based on a homomorphic property of the RSA signing function (see RSA digital signature scheme).

Let $n = pq$ be the product of two large safe primes, (x, y) being a pair of private and public RSA keys such that x is chosen randomly from $\mathbb{Z}_{(p-1)(q-1)}^*$ and $y = x^{-1} \pmod{(p-1)(q-1)}$ and $M = \mathbb{Z}_n^*$ be the domain of multiplicative inverses of the residues modulo n. The functions $f_x(m) = m^x \pmod{n}$ are the RSA signing functions. The families ϕ, Φ of auxiliary functions are chosen as follows:

$$\phi_{y,a}(m) = ma^y \pmod{n}$$
$$\Phi_{y,a}^{-1}(z') = z'a^{-1} \pmod{n}.$$

Other blinding techniques have been used in a variety of interactive protocols such as divertible proofs of knowledge [1,7,8], privacy-oriented electronic cash [2,3] and unlinkable credentials [6], and in anonymous electronic voting schemes.

Gerrit Bleumer

References

[1] Blaze, Matt, Gerrit Bleumer, and Martin Strauss (1998). "Divertible protocols and atomic proxy cryptography." *Advances in Cryptology—EUROCRYPT'98*, Lecture Notes in Computer Science, vol. 1403, ed. K. Nyberg. Springer-Verlag, Berlin, 127–144.

[2] Brands, Stefan (1994). "Untraceable off-line cash in wallet with observers." *Advances in Cryptology—CRYPTO'93*, Lecture Notes in Computer Science, vol. 773, ed. D.R. Stinson. Springer-Verlag, Berlin, 302–318.

[3] Brickell, Ernie, Peter Gemmell, and David Kravitz (1995). "Trustee-based tracing extensions to anonymous cash and the making of anonymous change." *6th ACM-SIAM Symposium on Discrete Algorithms (SODA)*. ACM Press, New York, 457–466.

[4] Brumley, David and Dan Boneh (2003). "Remote timing attacks are practical." *12th Usenix Security Symposium 2003*. http://www.usenix.org/publications/library/proceedings/sec03/

[5] Chaum, David (1993). "Blind signatures for untraceable payments." *Advances in Cryptology—CRYPTO'82*, Lecture Notes in Computer Science, eds. Plenum D. Chaum, R.L. Rivest, and A.T. Sherman. Plenum Press, New York, 199–203.

[6] Chaum, David (1990). "Showing credentials without identification: Transferring signatures between unconditionally unlinkable pseudonyms." *Advances in Cryptology—AUSCRYPT'90, Sydney, Australia, January 1990*, Lecture Notes in Computer Science, vol. 453. Springer-Verlag, Berlin, 246–264.

[7] Chaum, David and Torben P. Pedersen (1993). "Wallet databases with observers." *Advances in Cryptology—CRYPTO'92*, Lecture Notes in Computer Science, vol. 740, ed. E.F. Brickell. Springer-Verlag, Berlin, 89–105.

[8] Okamoto, Tatsuaki and Kazuo Ohta (1989). "Divertible zero-knowledge interactive proofs and commutative random self-reducibility." *Advances in Cryptology—EUROCRYPT'89*, Lecture Notes in Computer Science, vol. 434, eds. J.-J. Quisquater and J. Vandewalle. Springer-Verlag, Berlin, 134–149.

BLOCK CIPHERS

INTRODUCTION: In his milestone paper in 1949 [43] Shannon defines perfect secrecy for secret-key systems and shows that they exist. A secret-key cipher obtains *perfect secrecy* if for all plaintexts x and all ciphertexts y it holds that $\Pr(x) = \Pr(x|y)$ (see Information Theory and [43]). In other words, a ciphertext y gives no information about the plaintext. This definition leads to the following result.

COROLLARY 1. *A cipher with perfect secrecy is unconditionally secure against a ciphertext-only attack.*

As noted by Shannon the Vernam cipher, also called the *one-time pad*, obtains perfect secrecy. In the one-time pad the plaintext characters are added with independent key characters to produce the ciphertexts. However, the practical applications of perfect secret-key ciphers are limited, since it requires as many digits of secret key as there are digits to be enciphered. A more desirable situation would be if the same key could be used to encrypt texts of many more bits.

Two generally accepted design principles for practical ciphers are the principles of confusion and diffusion that were suggested by Shannon.

Confusion: "the ciphertext statistics should depend on the plaintext statistics in a manner too complicated to be exploited by the cryptanalyst."

Diffusion: "each digit of the plaintext and each digit of the secret key should influence many digits of the ciphertext" [29].

These two design principles are very general and informal.

Shannon also discusses two other more specific design principles. The first is to make the security of the system reducible to some known difficult problem. This principle has been used widely in the design of public-key systems, but not in secret-key ciphers. Shannon's second principle is to make the system secure against all known attacks, which is still the best known design principle for secret-key ciphers today.

A block cipher with n-bit blocks and a κ-bit key is a selection of 2^κ permutations (bijective mappings) of n bits. For any given key k, the block cipher specifies an encryption algorithm for computing the n-bit ciphertext for a given n-bit plaintext, together with a decryption algorithm for computing the n-bit plaintext corresponding to a given n-bit ciphertext.

The number of permutations of n-bit blocks is $2^n!$, which using Stirlings approximation is $\sqrt{2\pi 2^n}(\frac{2^n}{e})^{2^n}$ for large n. Since $\sqrt{2\pi 2^n}(\frac{2^n}{e})^{2^n} < 2^{(n-1)2^n}$ for $n \geq 3$, with $\kappa = (n-1)2^n$ one could cover all n-bit permutations, but typically κ is chosen much smaller for practical reasons. For example, for the AES (see Rijndael/AES and [38]) one option is the parameters $\kappa = n = 128$ in which case $(n-1)2^n \simeq 2^{135}$.

Most block ciphers are so-called *iterated ciphers* where the output is computed by applying in an iterative fashion a fixed key-dependent function r times to the input. We say that such a cipher is an r-round iterated (block) cipher. A *key-schedule algorithm* takes as input the user-selected κ-bit key and produces a set of subkeys.

Let g be a function which is invertible when the first of its two arguments is fixed. Define the sequence z_i recursively by

$$z_i = g(k_i, z_{i-1}), \qquad (1)$$

where z_0 is the plaintext, k_i is the ith subkey, and z_r is the ciphertext. The function g is called the *round function*.

$$z_0 \rightarrow \boxed{g}^{\overset{k_1}{\downarrow}} \rightarrow z_1 \rightarrow \boxed{g}^{\overset{k_2}{\downarrow}} \rightarrow z_2 \cdots \rightarrow z_{r-1} \rightarrow \boxed{g}^{\overset{k_r}{\downarrow}} \rightarrow z_r$$

In many block ciphers g consists of a layer of substitution boxes, or S-boxes, and a layer of bit permutations. Such ciphers are called *SP-networks* (see substitution-permutation (SP) network).

A special kind of iterated ciphers are the Feistel ciphers [10], which are defined as follows. Let n

(even) be the block size and assume the cipher runs in r rounds. Let z_0^L and z_0^R be the left and right halves of the plaintext, respectively, each of $n/2$ bits. The round function g operates as follows:

$$z_i^L = z_{i-1}^R$$
$$z_i^R = f(k_i, z_{i-1}^R) + z_{i-1}^L,$$

and the ciphertext is the concatenation of z_r^R and z_r^L. Here f can be any function taking as arguments an $n/2$-bit text and a round key k_i and producing $n/2$ bits. '+' is a commutative group operation on the set of $n/2$-bit blocks. If not specified otherwise, it will be assumed that '+' is bitwise addition modulo 2 (or in other terms, the exclusive-or operation denoted by $\oplus$). Also, variants where the texts are split into two parts not of equal lengths and variants where the texts are split into more than two parts have been suggested.

Two of the most important block ciphers are the Feistel cipher Data Encryption Standard (DES) [35] and the SP-network Advanced Encryption Standard (Rijndael/AES) [37].

In the following $e_k(\cdot)$ and $d_k(\cdot)$ denote, respectively the encryption operation and the decryption operation of a block cipher of block length n using the κ-bit key k.

We shall now describe Shannon's model which is standard in secret-key cryptology. The sender and the receiver share a common key k, which has been transmitted over a secure channel. The sender encrypts a plaintext x using the secret key k and sends the ciphertext y over an insecure channel to the receiver, who restores y into x using k. The attacker has access to the insecure channel and can intercept the ciphertexts (cryptograms) sent from the sender to the receiver. To avoid an attacker to speculate in how the legitimate parties have constructed their common key, the following assumption is often made.

ASSUMPTION 1. *All keys are equally likely and a key k is always chosen uniformly at random.*

Also it is often assumed that all details about the cryptographic algorithm used by the sender and receiver are known to the attacker, except for the value of the secret key. Assumption 2 is known as *Kerckhoffs's assumption* (see [17] or maxims).

ASSUMPTION 2. *The enemy cryptanalyst knows all details of the enciphering process and deciphering process except for the value of the secret key.*

The possible attacks against a block cipher are classified as follows, where A is the attacker.

Ciphertext-only attack. A intercepts a set of ciphertexts.

Known plaintext attack. A obtains $x_1, x_2, \ldots, x_s$ and $y_1, y_2, \ldots, y_s$, a set of s plaintexts and the corresponding ciphertexts.

Chosen plaintext attack. A chooses *a priori* a set of s plaintexts $x_1, x_2, \ldots, x_s$ and obtains in some way the corresponding ciphertexts $y_1, y_2, \ldots, y_s$.

Adaptively chosen plaintext attack. A chooses a set of plaintexts $x_1, x_2, \ldots, x_s$ interactively as he obtains the corresponding ciphertexts $y_1, y_2, \ldots, y_s$.

Chosen ciphertext attacks. These are similar to those of chosen plaintext attack and adaptively chosen plaintext attack, where the roles of plaintexts and ciphertexts are interchanged.

Also, one can consider any combination of the above attacks. The chosen text attacks are obviously the most powerful attacks. In many applications they are however also unrealistic attacks. If the plaintext space contains *redundancy* (see Information Theory),[1] it may be hard for an attacker to 'trick' a legitimate sender into encrypting nonmeaningful plaintexts and similarly hard to get ciphertexts decrypted, which do not yield meaningful plaintexts. But if a system is secure against an adaptively chosen plaintext/ciphertext attack, then it is also secure against all other attacks. An ideal situation for a designer would be to prove that his system is secure against an adaptively chosen text attack, although an attacker may never be able to mount more than a ciphertext-only attack.

The *unicity distance* of a block cipher is the smallest integer s such that essentially only one value of the secret key k could have encrypted a random selection of s plaintext blocks to the corresponding ciphertext blocks. The unicity distance depends on both the key size and on the redundancy in the plaintext space. However, the unicity distance gives no indication of the computational difficulty in breaking a cipher, it is merely a *lower* bound on the amount of ciphertext blocks needed in a ciphertext-only attack to be able to (at least in theory) identify a unique key. Let κ and n be the number of bits in the secret key respectively in the plaintexts and ciphertexts and assume that the keys are always chosen uniformly at random. In a ciphertext-only attack the unicity distance is defined as $n_u = \kappa/(n r_L)$, where r_L is the redundancy of the plaintexts, see e.g., [44]. The concept can be adapted also to the known or chosen plaintext scenarios. In these cases the redundancy of the plaintexts from the attacker's point of view is 100%. The unicity distance in a known or chosen plaintext attack is $n_v = \lceil \kappa/n \rceil$.

---

[1] Redundancy is an effect of the fact that certain sequences of plaintext characters appear more frequently than others.

The results of the cryptanalytic effort of the attacker A can be grouped as follows [21].

Total break. A finds the secret key k.

Global deduction. A finds an algorithm F, functionally equivalent to $e_k(\cdot)$ (or $d_k(\cdot)$) without knowing the key k.

Local deduction. A finds the plaintext (ciphertext) of an intercepted ciphertext (plaintext), which he did not obtain from the legitimate sender.

Distinguishing algorithm. A is given access to a black-box containing either the block cipher for a randomly chosen key or a randomly chosen permutation. He is able to distinguish between these two cases.

Clearly, this classification is hierarchical, that is, if a total break is possible, then a global deduction is possible and so on.

CRYPTANALYSIS: We begin by listing some attacks which apply to all block ciphers.

Exhaustive key search: this attack requires the computation of about 2^κ encryptions and requires n_u ciphertexts (ciphertext-only attack) or n_v plaintext/ciphertext pairs (known and chosen plaintext attack), where n_u and n_v are the unicity distances, cf. above.

Table attack: encrypt in a precomputation phase a fixed plaintext x under all possible keys, sort, and store all ciphertexts. Thereafter, a total break is possible requiring one chosen plaintext.

Dictionary attack: intercept and store all possible plaintext/ciphertext pairs. The running time of a deduction is the time of one table look-up.

Matching ciphertext attack: this attack applies to encryption using the (ecb), (cbc), and (cfb) modes of operation, see modes of operation for a block cipher or [40]. Collect s ciphertext blocks and check for collisions. For example, if y_i, y_j are n-bit blocks encrypted (using the same key) in the (cbc) mode, then if $y_i = y_j$, then $e_k(x_i \oplus y_{i-1}) = e_k(x_j \oplus y_{j-1}) \Rightarrow y_{i-1} \oplus y_{j-1} = x_i \oplus x_j$, thus information about the plaintexts is leaked. With $s \approx 2^{n/2}$ the probability of finding matching ciphertexts is about $1/2$, see birthday paradox.

Time-memory trade-off attack [14]: let us assume for the sake of exposition that the key space of the attacked cipher equals the ciphertext space, that is, $\kappa = n$. Fix some plaintext block x_0. Define the function $f(z) = e_z(x_0)$. Select m randomly chosen values $z^0, \ldots, z^{m-1}$. For each $j \in \{0, \ldots, m\}$ compute the values $z_i^j = f(z_{i-1}^j)$ for $i = 1, \ldots, t$, where $z^j = z_0^j$; store the pairs (start and end results) (z_0^j, z_t^j) for $j = 0, \ldots, m$ in a table T and sort the elements on the second components.

Subsequently, imagine that an attacker has intercepted the ciphertext $y = e_k(x_0)$. Let $w_0 = y$ and check if w_0 is a second component in T. If, say, $w_0 = z_t^\ell$, the attacker can find a candidate for the key k by computing forward from z_0^ℓ. If this does not lead to success, compute $w_i = f(w_{i-1})$ and repeat the above test for w_i for $i = 1, 2, \ldots, t$.

A close analysis [14] shows that if m and t are chosen such that $mt^2 \approx 2^\kappa$, there is a probability of about $mt/2^\kappa$ that in the above computations of $\{z_i^k\}$ the secret key has been used. If this is the case, the attack will find the secret key. If it is not the case, the attack fails. The probability of success can be increased by repeating the attack, e.g., with $2^{\kappa/3}$ iterations each time with $m = t = 2^{\kappa/3}$ one obtains a probability of success of more than $1/2$.

In summary, with $\kappa = n$ the attack finds the secret key with good probability after $2^{2\kappa/3}$ encryptions using $2^{2\kappa/3}$ words of memory. The $2^{2\kappa/3}$ words of memory are computed in a preprocessing phase, which takes the time of about 2^κ encryptions.

To estimate the complexity of a cryptanalytic attack, one must consider at least the time it takes, the amount of data that is needed, and the storage requirements. For an n-bit block cipher the following complexities should be considered.

Data complexity: the amount of data needed as input to an attack. Units are measured in blocks of length n.

Processing complexity: the time needed to perform an attack. Time units are measured as the number of encryptions an attacker has to do himself.

Storage complexity: the words of memory needed to do the attack. Units are measured in blocks of length n.

The complexity of an attack is often taken as the maximum of the three complexities above; however, in most scenarios the amount of data encrypted with the same secret key is often limited and for most attackers the available storage is small.

Iterated Attacks

Let x and y denote the plaintext and the ciphertext, respectively. In most modern attacks on iterated ciphers, the attacker repeats his attack for all possible values of (a subset of) the bits in the last-round key. The idea is that when he guesses the correct values of the key bits, he can compute the value of some bits of the ciphertexts before the last round, whereas when he guesses wrongly, these bits will correspond to ciphertext bits encrypted with a wrong key. If one can distinguish between these two cases, one might be able to extract bits of the last-round key. The *wrong key randomization* hypothesis, which is often used, says that when the attacker guesses a wrong value of the key, the resulting values are random and uniformly distributed. If an attacker succeeds in determining the value of the last-round key, he can peel off one round of the cipher and do a similar attack on a cipher one round shorter to find the second-last round key, etc. In some attacks it is advantageous to consider the first-round key instead of the last-round key or both at the same time, depending on the structure of the cipher, the number of key bits involved in each round, etc.

The two most general attacks on iterated ciphers are linear cryptanalysis and differential cryptanalysis.

Linear Cryptanalysis

Linear cryptanalysis [30, 34] is a known plaintext attack. Consider an iterated cipher, cf. (1). Then a linear approximation over s rounds (or an s-round linear hull) is

$$(z_i \cdot \alpha) \oplus (z_{i+s} \cdot \beta) = 0, \qquad (2)$$

which holds with a certain probability p, where $z_i, z_{i+s}, \alpha, \beta$ are n-bit strings and where '·' denotes the dot (or inner) product modulo 2. The strings α, β are also called *masks*. The quantity $|p - 1/2|$ is called the *bias* of the approximation. The expression with a '1' on the right side of (2) will have a probability of $1 - p$, but the biases of the two expressions are the same. The linear round approximations are usually found by combining several one-round approximations under the assumption that the individual rounds are mutually independent (for most ciphers this can be achieved by assuming that the round keys are independent). The complexity of a linear attack is approximately $|p - 1/2|^{-2}$. It was confirmed by computer experiments that the *wrong key randomization* hypothesis holds for the linear attack on the DES (see Data Encryption Standard). The attack on the DES was implemented in 1994, required a total of 2^{43} known plaintexts [31] and in 2002 was the fastest, known key-recovery attack on the DES. Linear cryptanalysis for block ciphers gives further details of the attack.

Differential Cryptanalysis

Differential cryptanalysis [3] is a chosen plaintext attack and was the first published attack which could (theoretically) recover DES keys in time less than that of an exhaustive search for the key. In a differential attack one exploits that for certain input differences the distribution of output differ-

ences of the nonlinear components is nonuniform. A difference between two bit strings, x and x' of equal length, is defined in general terms as $\Delta x = x \otimes (x')^{-1}$, where $\otimes$ is a group operation on bit strings and where the superscript $^{-1}$ denotes the inverse element. Consider an iterated cipher, cf. (1). The pair $(\Delta z_0, \Delta z_s)$ is called an s-round *differential* [27]. The probability of the differential is the conditional probability that given an input difference Δz_0 in the plaintexts, the difference in the ciphertexts after s rounds is Δz_s. Experiments have shown that the number of chosen plaintexts needed by the differential attack in general is approximately $1/p$, where p is the probability of the differential being used. For iterated ciphers, one often specifies the expected differences after each round of encryption. Such a structure over s rounds, i.e., $(\Delta z_0, \Delta z_1, \ldots, \Delta z_{s-1}, \Delta z_s)$, is called an s-round *characteristic*. The differential attack is explained in more details in differential cryptanalysis.

Extensions, Generalization, and Variations

The differential and linear attacks have spawned a lot of research in block cipher cryptanalysis and several extensions, generalizations, and variants of the differential and linear attacks have been developed. In [15] it was shown how to combine the techniques of differential and linear attacks. In particular, an attack on the DES reduced to eight rounds was devised, which on input only 512 chosen plaintexts finds the secret key. In [47] a generalization of both the differential and linear attacks, known as *statistical cryptanalysis*, was introduced. It was demonstrated that this statistical attack on the DES includes the linear attack by Matsui but without any significant improvement. In [18] an improved linear attack using multiple linear approximations was given. In [24] a linear attack is shown using nonlinear approximations in the outer rounds of an iterated cipher. In [12, 13] two generalizations of the linear attack were given.

A dth order differential [26] is the difference between two $(d - 1)$th order differentials and is a collection of 2^d texts, where a first-order differential is what is called a differential above. The main idea in the higher order differential attack is the fact that a dth order differential of a function of maximum algebraic degree d is a constant. Consequently, a $(d + 1)$th order differential of the function is zero. In [16, 22] these attacks were applied to various ciphers.

The boomerang attack [50] is a clever application of a second-order differential. Boomerangs are particularly effective when one can find a good differential covering the first half of the encryption operation and a good differential covering the first half of the decryption operation. More details of the attack can be found in <u>boomerang attack</u>.

Let $\{\alpha_0, \alpha_1, \ldots, \alpha_s\}$ be an s-round characteristic. Then $\{\alpha'_0, \alpha'_1, \ldots, \alpha'_s\}$ is called a *truncated characteristic*, if α'_i is a subsequence of α_i. Truncated characteristics were used to some extent in [3]. Note that a truncated characteristic is a collection of characteristics and therefore reminiscent of a differential. A truncated characteristic contains all characteristics $\{\alpha''_0, \alpha''_1, \ldots, \alpha''_s\}$ for which $\text{trunc}(\alpha''_i) = \alpha'_i$, where $\text{trunc}(x)$ is a truncated value of x not further specified here. The notion of truncated characteristics extends in a natural way to <u>truncated differentials</u> introduced in [22].

Other Attacks

Integral cryptanalysis [5, 25] can be seen as a dual to differential cryptanalysis and it is the best known attack on the advanced encryption standard. The attack is explained in more details in <u>multiset attacks</u>. In the interpolation attack [16] one expresses the ciphertext as a polynomial of the plaintext. If this polynomial has a sufficiently low degree, an attacker can reconstruct it from known (or chosen) plaintexts and the corresponding ciphertexts. In this way, he can encrypt any plaintext of his choice without knowing the (explicit) value of the secret key, see <u>interpolation attack</u> for more details. There has been a range of other correlation attacks most of which are relative to the attacked cipher, but which all exploit the nonuniformity of certain bits of plain- and ciphertexts [2, 8, 11, 19, 23, 47].

Key Schedule Attacks

One talks about <u>weak keys</u> for a block cipher, if there is a subspace of keys relative to which a certain attack can be mounted successfully, such that for all other keys the attack has little or zero probability of success. If there are only a small number of weak keys, they pose no problem for applications of encryption if the encryption keys are chosen uniformly at random. However, when block ciphers are used in other modes, e.g., for hashing, these attacks play an important role as demonstrated in [6, 42].

One talks about *related keys* for a block cipher, if for two (or more) keys k and k^* of a certain relation, there are certain (other) relations between the two (or more) encryption functions $e_k(\cdot)$ and $e_{k^*}(\cdot)$, which can be exploited in cryptanalytic attacks. There are several variants of this attack depending on how powerful the attacker A is assumed to be. One distinguishes between whether A gets encryptions under one or under several keys and whether there is a known or chosen relation between the keys (see <u>related key attack</u>).

The <u>slide attack</u> [4] applies to iterated ciphers where the list of round keys has a repeated pattern, e.g., if all round functions are identical, there are very efficient attacks.

BOUNDS OF ATTACKS: A motivation for the <u>Feistel cipher</u> design is the results by Luby and Rackoff (see <u>Luby-Rackoff cipher</u> or [28]). They showed how to build a $2n$-bit pseudorandom permutation from a pseudorandom n-bit function using the Feistel construction. For a three-round construction they showed that under a chosen plaintext attack, an attacker needs at least $2^{n/2}$ chosen texts to distinguish the Feistel construction from a random $2n$-bit function. Under a combined chosen plaintext and chosen ciphertext attack, this construction is however easily distinguished from random. For a four-round construction it was shown that even under this strong attack, an attacker needs at least $2^{n/2}$ chosen texts to distinguish the construction from a random $2n$-bit function.

In the *decorrelation theory* [48] one attempts to distinguish a given n-bit block cipher from a randomly chosen n-bit permutations. Using particular well-defined *metrics*, this approach is used to measure the distance between a block cipher and randomly chosen permutations. One speaks about decorrelation of certain orders depending on the type of attack one is considering. In [49] it was shown how this technique can be used to prove resistance against elementary versions of differential and linear cryptanalysis.

Resistance Against Differential and Linear Attacks

First it is noted that one can unify the complexity measures in differential and linear cryptanalysis. Let p_L be the probability of a linear approximation for an iterated block cipher, then define $q = (2p_L - 1)^2$ [32]. Let q denote the highest such quantity for a one-round linear approximation. Denote by p the highest probability of a one-round differential achievable by the cryptanalyst. It is possible to lower bound the probabilities of all differentials and all hulls in an r-round iterated cipher expressed in terms of p and q [21, 32, 40, 41]. The probabilities are taken as an average over all possible keys. It has further been shown that the round functions in iterated ciphers can be chosen in such a way that the probabilities of the differentials and the linear hulls are small [39,40]. In this way it is possible to construct iterated ciphers with a proof of security (as an average over all possible keys) against differential and linear cryptanalysis. This approach was used in the design of the block ciphers Misty1 [33] and *Kasumi* (see Kasumi/Misty1 or [1]).

ENHANCING EXISTING CONSTRUCTIONS

Multiple Encryption

In a *double encryption* with two keys k_1 and k_2, the ciphertext corresponding to x is $y = e_{k_2}(e_{k_1}(x))$. However, regardless of how k_1, k_2 are generated, there is a meet-in-the-middle attack that breaks this system with a few known plaintexts using about $2^{\kappa+1}$ encryptions and 2^{κ} blocks of memory, that is, roughly the same time complexity as key search in the original system. Assume some plaintext x and its corresponding ciphertext y encrypted as above are given. Compute $e_{k_1}(x)$ for all choices of the key $k_1 = i$ and store the results t_i in a table. Next compute $d_{k_2}(y)$ for all values of the key $k_2 = j$ and check whether the results s_j match a value in the table, that is, whether for some (i, j), $t_i = s_j$. Each such match gives a candidate $k_1 = i$ and $k_2 = j$ for the secret key. The attack is repeated on additional pairs of plaintext–ciphertext until only one pair of values for the secret key remains suggested. The number of known plaintexts needed is roughly $2\kappa - n$. There are variants of this attack with trade-offs between running time and the amount of storage needed [46]. In a *triple encryption* with three independent keys k_1, k_2, and k_3, the ciphertext corresponding to x is $y = e_{k_3}(e_{k_2}(e_{k_1}(x)))$. One variant of this idea is well known as *two-key triple encryption*, proposed in [45], where the ciphertext corresponding to x is $e_{k_1}(d_{k_2}(e_{k_1}(x)))$. Compatibility with a single encryption can be obtained by setting $k_1 = k_2$. However, whereas triple encryption is provably as secure as a single encryption, a similar result is not known for two-key triple encryption. A two-key triple encryption scheme with a proof of security appeared in [7].

Key-Whitening

Another method of increasing the key size is by key-whitening. One approach is the following: $y = e_k(x \oplus k_1) \oplus k_2$, where k is a κ-bit key, and k_1 and k_2 are n-bit keys. Alternatively, $k_1 = k_2$ may be used.

It was shown [20] that for attacks not exploiting the internal structure, the effective key size is $\kappa + n - \log_2 m$ bits, where m is the maximum number of plaintext/ciphertext pairs the attacker can obtain. (This method applied to the DES is named DES-X and attributed to Ron Rivest.)

Lars R. Knudsen

References

[1] (1999). V3.1.1 3GPP TS 35.202. Kasumi. Available at http://www.3gpp.org

[2] Biham, E. and A. Biryukov (1995). "An improvement of Davies' attack on DES." *Advances in Cryptology—EUROCRYPT'94*, Lecture Notes in Computer Science, vol. 950, ed. A. De Santis. Springer-Verlag, Berlin, 461–467.

[3] Biham, E. and A. Shamir (1993). *Differential Cryptanalysis of the Data Encryption Standard*. Springer-Verlag, Berlin.

[4] Biryukov, A. and D. Wagner (1999). "Slide attacks." *Fast Software Encryption, Sixth International Workshop, Rome, Italy, March 1999*, Lecture Notes in Computer Science, vol. 1636, ed. L.R. Knudsen. Springer-Verlag, Berlin, 245–259.

[5] Daemen, J., L. Knudsen, and V. Rijmen (1997). "The block cipher square." *Fast Software Encryption, Fourth International Workshop, Haifa, Israel, January 1997*, Lecture Notes in Computer Science, vol. 1267, ed. E. Biham. Springer-Verlag, Berlin, 149–165.

[6] Damgård, I.B. and L.R. Knudsen (1993). "The breaking of the AR hash function." *Advances in Cryptology—EUROCRYPT'93*, Lecture Notes in Computer Science, vol. 773, ed. T. Helleseth. Springer-Verlag, Berlin, 286–292.

[7] Damgård, I.B. and L.R. Knudsen (1998). "Two-key triple encryption." *The Journal of Cryptology*, 11 (3), 209–218.

[8] Davies, D. and S. Murphy (1995). "Pairs and triples of DES S-boxes." *The Journal of Cryptology*, 8 (1), 20–27.

[9] Davies, D.W. and W.L. Price (1989). *Security for Computer Networks*. John Wiley & Sons, New York.

[10] Feistel, H., W.A. Notz, and J.L. Smith (1975). "Some cryptographic techniques for machine-to-machine data communications." *Proceedings of IEEE*, 63 (11), 1545–1554.

[11] Gilbert, H., H. Handschuh, A. Joux, and S. Vaudenay (2001). "A statistical attack on RC6." *Fast Software Encryption, 7th International Workshop, FSE 2000, New York, USA, April 2000*, Lecture Notes in Computer Science, vol. 1978, ed. B. Schneier. Springer-Verlag, Berlin, 64–74.

[12] Harpes, C., G.G. Kramer, and J.L. Massey (1995). "A generalization of linear cryptanalysis and the applicability of Matsui's piling-up lemma." *Advances in Cryptology—EUROCRYPT'95*, Lecture Notes in Computer Science, vol. 921, eds. L. Guillou

and J.-J. Quisquater. Springer-Verlag, Berline, 24–38.

[13] Harpes, C. and J.L. Massey (1997). "Partitioning cryptanalysis." *Fast Software Encryption, Fourth International Workshop, Haifa, Israel, January 1997*, Lecture Notes in Computer Science, vol. 1267, ed. E. Biham. Springer-Verlag, Berlin, 13–27.

[14] Hellman, M. (1980). "A cryptanalytic time-memory trade-off." *IEEE Trans. on Information Theory*, IT-26 (4), 401–406.

[15] Hellman, M.E. and S.K. Langford (1994). "Differential–linear cryptanalysis." *Advances in Cryptology—CRYPTO'94*, Lecture Notes in Computer Science, vol. 839, ed. Y. Desmedt. Springer-Verlag, Berlin, 26–39.

[16] Jakobsen, T. and L. Knudsen (1997). "The interpolation attack on block ciphers." *Fast Software Encryption, Fourth International Workshop, Haifa, Israel, January 1997*, Lecture Notes in Computer Science, vol. 1267, ed. E. Biham. Springer-Verlag, Berlin, 28–40.

[17] Kahn, D. (1967). *The Codebreakers*. MacMillan, London.

[18] Kaliski, B.S. and M.J.B. Robshaw (1994). "Linear cryptanalysis using multiple approximations." *Advances in Cryptology—CRYPTO'94*, Lecture Notes in Computer Science, vol. 839, ed. Y. Desmedt. Springer-Verlag, Berlin, 26–39.

[19] Kelsey, J., B. Schneier, and D. Wagner (1999). "Mod n cryptanalysis, with applications against RC5P and M6." *Fast Software Encryption, Sixth International Workshop, Rome, Italy, March 1999*, Lecture Notes in Computer Science, vol. 1636, ed. L. Knudsen. Springer-Verlag, Berlin, 139–155.

[20] Kilian, J. and P. Rogaway (1996). "How to protect DES against exhaustive key search." *Advances in Cryptology—CRYPTO'96*, Lecture Notes in Computer Science, vol. 1109, ed. Neal Koblitz. Springer-Verlag, London, 252–267.

[21] Knudsen, L.R. (1994). "Block Ciphers—Analysis, Design and Applications." *PhD Thesis*, Aarhus University, Denmark.

[22] Knudsen, L.R. (1995). "Truncated and higher order differentials." *Fast Software Encryption—Second International Workshop, Leuven, Belgium*, Lecture Notes in Computer Science, vol. 1008, ed. B. Preneel. Springer-Verlag, Berlin, 196–211.

[23] Knudsen, L.R. and W. Meier (2001). "Correlations in RC6 with a reduced number of rounds." *Fast Software Encryption, 7th International Workshop, FSE 2000, New York, USA, April 2000*, Lecture Notes in Computer Science, vol. 1978, ed. B. Schneier. Springer-Verlag, Berlin, 94–108.

[24] Knudsen, L.R. and M.P.J. Robshaw (1996). "Nonlinear approximations in linear cryptanalysis." *Advances in Cryptology—EUROCRYPT'96*, Lecture Notes in Computer Science, vol. 1070, ed. U. Maurer. Springer-Verlag, Berlin, 224–236.

[25] Knudsen, L.R. and D. Wagner (2001). "Integral cryptanalysis." *FSE 2002*. To appear in proceedings from Springer-Verlag, Berlin.

[26] Lai., X. (1994). "Higher order derivatives and differential cryptanalysis." *Communication and Cryptography, Two Sides of One Tapestry*, ed. R. Blahut. Kluwer Academic Publishers, Dordrecht. ISBN 0-7923-9469-0.

[27] Lai, X., J.L. Massey, and S. Murphy (1992). "Markov ciphers and differential cryptanalysis." *Advances in Cryptology—EUROCRYPT'91*, Lecture Notes in Computer Science, vol. 547, ed. D.W. Davies. Springer-Verlag, Berlin, 17–38.

[28] Luby, M. and C. Rackoff (1988). "How to construct pseudorandom permutations from pseudorandom functions." *SIAM Journal of Computing*, 17 (2), 373–386.

[29] Massey, J.L. (1993). "Cryptography: Fundamentals and applications." *Copies of Transparencies, Advanced Technology Seminars*.

[30] Matsui, M. (1993). "Linear cryptanalysis method for DES cipher." *Advances in Cryptology—EUROCRYPT'93*, Lecture Notes in Computer Science, vol. 765, ed. T. Helleseth. Springer-Verlag, Berlin, 386–397.

[31] Matsui, M. (1994). "The first experimental cryptanalysis of the Data Encryption Standard." *Advances in Cryptology—CRYPTO'94*, Lecture Notes in Computer Science, vol. 839, ed. Y.G. Desmedt. Springer-Verlag, Berlin, 1–11.

[32] Matsui, M. (1996). "New structure of block ciphers with provable security against differential and linear cryptanalysis." *Fast Software Encryption, Third International Workshop, Cambridge, UK, February 1996*, Lecture Notes in Computer Science, vol. 1039, ed. D. Gollman. Springer-Verlag, Berlin, 205–218.

[33] Matsui, M. (1997). "New block encryption algorithm MISTY." *Fast Software Encryption, Fourth International Workshop, Haifa, Israel, January 1997*, Lecture Notes in Computer Science, vol. 1267, ed. E. Biham. Springer-Verlag, Berlin, 54–68.

[34] Matsui, M. and A. Yamagishi (1992). "A new method for known plaintext attack of FEAL cipher." *Advances in Cryptology—EUROCRYPT'92*, Lecture Notes in Computer Science, vol. 658, ed. R. Rueppel. Springer-Verlag, Berlin, 81–91.

[35] National Bureau of Standards (1977). "Data encryption standard." Federal Information Processing Standard (FIPS), Publication 46, National Bureau of Standards, U.S. Department of Commerce, Washington, DC.

[36] National Bureau of Standards (1980). "DES modes of operation." Federal Information Processing Standard (FIPS), Publication 81, National Bureau of Standards, U.S. Department of Commerce, Washington, DC.

[37] National Institute of Standards and Technology. Advanced encryption algorithm (AES) development effort. http://www.nist.gov/aes

[38] NIST (2001). "Advanced encryption standard." *FIPS 197*, US Department of Commerce, Washington, DC, November 2001.

[39] Nyberg, K. (1993). "Differentially uniform mappings for cryptography." *Advances in Cryptology—EUROCRYPT'93*, Lecture Notes in Computer Science, vol. 765, ed. T. Helleseth. Springer-Verlag, Berlin, 55–64.

[40] Nyberg, K. and L.R. Knudsen (1993). "Provable security against differential cryptanalysis." *Advances in Cryptology—CRYPTO'92*, Lecture Notes in Computer Science, vol. 740, ed. E.F. Brickell. Springer-Verlag, Berlin, 566–574.

[41] Nyberg, K. and L.R. Knudsen (1995). "Provable security against a differential attack." *The Journal of Cryptology*, 8 (1), 27–38.

[42] Preneel, B. (1993). "Analysis and Design of Cryptographic Hash Functions." *PhD Thesis*, Katholieke Universiteit Leuven.

[43] Shannon, C.E. (1949). "Communication theory of secrecy systems." *Bell System Technical Journal*, 28, 656–715.

[44] Stinson, D.R. (1995). *Cryptography—Theory and Practice*. CRC Press, Inc., Boca Raton, FL.

[45] Tuchman, W. (1979). "Hellman presents no shortcut solutions to DES." *IEEE Spectrum*, 16 (7), 40–41.

[46] van Oorschot, P.C. and M.J. Wiener (1996). "Improving implementable meet-in-the-middle attacks of orders of magnitude." *Advances in Cryptology—CRYPTO'96*, Lecture Notes in Computer Science, vol. 1109, ed. Neal Koblitz. Springer-Verlag, Berlin, 229–236.

[47] Vaudenay, S. (1995). "An experiment on DES—Statistical cryptanalysis." *Proceedings of the 3rd ACM Conferences on Computer Security, New Delhi, India*. ACM Press, New York, 139–147.

[48] Vaudenay, S. (1998). "Provable security for block ciphers by decorrelation." *STACS'98*, Lecture Notes in Computer Science, vol. 1373, eds. M. Morvan, C. Meinel, and D. Krob. Springer-Verlag, Berlin, 249–275.

[49] Vaudenay, S. (1999). "Resistance against general iterated attacks." *Advances in Cryptology—EUROCRYPT'99*, Lecture Notes in Computer Science, vol. 1592, ed. J. Stem. Springer-Verlag, Berlin.

[50] Wagner, D. (1999). "The boomerang attack." *Fast Software Encryption, Sixth International Workshop, Rome, Italy, March 1999*, Lecture Notes in Computer Science, vol. 1636, ed. L.R. Knudsen. Springer-Verlag, Berlin 156–170.

BLOWFISH

Blowfish [3] is a 64-bit <u>block cipher</u> designed by Bruce Schneier and published in 1994. It was intended to be an attractive alternative to DES (see <u>Data Encryption Standard</u>) or <u>IDEA</u>. Today, the Blowfish algorithm is widely used and included in many software products.

Blowfish consists of 16 <u>Feistel</u>-like iterations. Each iteration operates on a 64-bit datablock, split

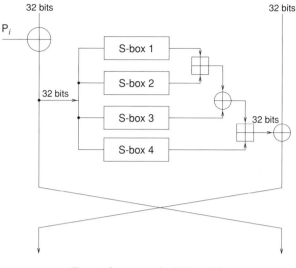

Fig. 1. One round of Blowfish

into two 32-bits words. First, a round key is XORed to the left word. The result is then input to four key-dependent 8×32-bit S-boxes, yielding a 32-bit output word which is XORed to the right word. Both words are swapped and then fed to the next iteration.

The use of key-dependent S-boxes distinguishes Blowfish from most other ciphers, and requires a rather complex key-scheduling algorithm. In a first pass, the lookup tables determining the S-boxes are filled with digits of π, XORed with bytes from a secret key which can consist of 32–448 bits. This preliminary cipher is then used to generate the actual S-boxes. Although Blowfish is one of the faster block ciphers for sufficiently long messages, the complicated initialization procedure results in a considerable efficiency degradation when the cipher is rekeyed too frequently. The need for a more flexible key schedule was one of the factors that influenced the design of <u>Twofish</u>, an *Advanced Encryption Standard* (see <u>Rijndael/AES</u>) finalist which was inspired by Blowfish.

Since the publication of Blowfish, only a few cryptanalytical results have been published. A first analysis was made by Vaudenay [4], who revealed classes of <u>weak keys</u> for up to 14 rounds of the cipher. Rijmen [2] proposed a second-order <u>differential attack</u> on a four-round variant of Blowfish. In a paper introducing <u>slide attacks</u> [1], Biryukov and Wagner highlighted the importance of XORing a different subkey in each round of Blowfish.

Christophe De Cannière

References

[1] Biryukov, A. and D. Wagner (1999). "Slide attacks." *Proceedings of Fast Software Encryption—FSE'99*,

Lecture Notes in Computer Science, vol. 1636, ed. L.R. Knudsen. Springer-Verlag, Berlin, 245–259.

[2] Rijmen, V. (1997). "Cryptanalysis and Design of Iterated Block Ciphers." *PhD Thesis*, Katholieke Universiteit Leuven.

[3] Schneier, B. (1994). "Description of a new variable-length key, 64-bit block cipher (Blowfish)." *Fast Software Encryption, FSE'93*, Lecture Notes in Computer Science, vol. 809, ed. R.J. Anderson. Springer-Verlag, Berlin, 191–204.

[4] Vaudenay, S. (1996). "On the weak keys of Blowfish." *Fast Software Encryption, FSE'96*, Lecture Notes in Computer Science, vol. 1039, ed. D. Gollmann. Springer-Verlag, Berlin, 27–32.

BLS SHORT DIGITAL SIGNATURES

It is well known that a digital signature scheme that produces signatures of length ℓ can have security at most 2^ℓ. In other words, it is possible to forge a signature on any message in time $O(2^\ell)$ just given the public key. It is natural to ask whether we can construct signatures with such security, i.e., signatures of length ℓ where the best algorithm for creating an existential forgery (with constant success probability) under a chosen message attack takes time $O(2^\ell)$. Concretely, is there a signature scheme producing 80-bit signatures where creating an existential forgery (with probability $1/2$) takes time approximately 2^{80}?

DSS signatures and Schnorr signatures provide security $O(2^\ell)$ with signatures that are 4ℓ-bits long. These signatures can be shortened [3] to about 3.5ℓ-bits without much affect on security. Hence, for concrete parameters, $\ell = 80$, shortened DSS signatures are 280-bits long.

Boneh et al. [2] describe a short signature scheme where 170-bit signatures provide approximately 2^{80} security, in the random oracle model. Hence, for $\ell = 80$, these signatures are approximately half the size of DSS signatures with comparable security. The system makes use of a group G where (i) the computational Diffie–Hellman problem is intractable, and (ii) there is an efficiently computable, nondegenerate, bilinear map $e : G \times G \to G_1$ for some group G_1. There are several examples of such groups from algebraic geometry where the bilinear map is implemented using the Weil pairing. Given such a group G of prime order q, the digital signature scheme works as follows:

Key Generation.
1. Pick an arbitrary generator $g \in G$.
2. Pick a random $\alpha \in \{1, \ldots, q\}$ and set $y = g^\alpha \in G$.
3. Let H be a hash function $H : \{0, 1\}^* \to G$.

Output (g, y, H) as the public key and (g, α, H) as the private key.

Signing. To sign a message $m \in \{0, 1\}^*$ using the private key (g, α, H) output $H(m)^\alpha \in G$ as the signature.

Verifying. To verify a message/signature pair $(m, s) \in \{0, 1\}^* \times G$ using the public key (g, y, H) test if $e(g, s) = e(y, H(m))$. If so, accept the signature. Otherwise, reject.

For a valid message/signature pair (m, s) we have that $s = H(m)^\alpha$ and therefore $e(g, s) = e(g, H(m)^\alpha) = e(g^\alpha, H(m)) = e(y, H(m))$. The second equality follows from the bilinearity of $e(,)$. Hence, a valid signature is always accepted. As mentioned above, the system is existentially unforgeable under a chosen message attack in the random oracle model, assuming the computational Diffie–Hellman assumption holds in G. Observe that a signature is a single element in G whereas DSS signatures are pairs of elements. This explains the reduction in signature length compared to DSS.

Recently, Boneh and Boyen [1] and Zhang et al. [4] describe a more efficient system producing signatures of the same length as BLS. However, security is based on a stronger assumption. Key generation is identical to the BLS system, except that the hash function used is $H : \{0, 1\}^* \to \mathbb{Z}_q$. A signature on a message $m \in \{0, 1\}^*$ is $s = g^{1/(\alpha+H(m))} \in G$. To verify a message/signature pair (m, s) test that $e(yg^{H(m)}, s) = e(g, g)$. We see that signature length is the same as in BLS signatures. However since $e(g, g)$ is fixed, signature verification requires only one computation of the bilinear map as opposed to two in BLS. Security of the system in the random oracle model is based on a nonstandard assumption called the t-Diffie–Hellman-inversion assumption. Loosely speaking, the assumption states that no efficient algorithm given $g, g^x, g^{(x^2)}, \ldots, g^{(x^t)}$ as input can compute $g^{1/x}$. Here t is the number of chosen message queries that the attacker can make. Surprisingly, a variant of this system can be shown to be existentially unforgeable under a chosen message attack *without the* random oracle model [1].

Dan Boneh

References

[1] Boneh, Dan and Xavier Boyen (2004). "Short signatures without random oracles." *Proceedings of Eurocrypt 2004*, Lecture Notes in Computer Science, vol. 3027, eds. C. Cachin and J. Camenisch. Springer-Verlag, Berlin, 56–73.

[2] Boneh, Dan, Ben Lynn, and Hovav Shacham (2004). "Short signatures from the Weil pairing." *J. of Cryptology*. Extended abstract in *Proceedings of ASIACRYPT 2001*, Lecture Notes in Computer

Science, vol. 2248, ed. C. Boyd. Springer-Verlag, Berlin.

[3] Naccache, David and Jacques Stern (2000). "Signing on a postcard." *Proceedings of Financial Cryptography 2000.*

[4] Zhang, Fangguo, Reihaneh Safavi-Naini, and Willy Susilo (2004). "An efficient signature scheme from bilinear pairings and its applications." *Proceedings of PKC 2004.*

BLUM INTEGER

A positive integer n is a *Blum integer* if it is the product of two distinct primes p, q where $p \equiv q \equiv 3 \pmod 4$. Blum integers are of interest in cryptography because the mapping

$$x \leftarrow x^2 \bmod n$$

is believed to be a trapdoor permutation (see trapdoor one-way function) on the quadratic residues modulo n. That is, exactly one of the four square roots of a quadratic residue modulo a Blum integer n is itself a quadratic residue. Inverting the permutation is equivalent to factoring n, but is easy given p and q. This fact is exploited in the Blum–Blum–Shub PRNG.

The permutation can be inverted when the prime factors p and q are known by computing both square roots modulo each factor, selecting the square root modulo each factor which itself is a square, then applying the Chinese remainder theorem. Conveniently, the square roots modulo the prime factors of a Blum integer can be computed with a simple formula: the solutions of $x^2 \equiv a \pmod p$ are given by $x \equiv \pm a^{(p+1)/4} \pmod p$ when $p \equiv 3 \pmod 4$. The appropriate square root can be selected by computing the Legendre symbol.

See also modular arithmetic, prime number.

Burt Kaliski

BLUM–BLUM–SHUB PSEUDORANDOM BIT GENERATOR

The Blum–Blum–Shub (BBS) pseudorandom bit generator [1] is one of the most efficient pseudorandom number generators known that is provably secure under the assumption that factoring large composites is intractable (see integer factoring). The generator makes use of modular arithmetic and works as follows:

Setup. Given a security parameter $\tau \in \mathbb{Z}$ as input, generate two random τ-bit primes p, q where $p = q = 3 \bmod 4$. Set $N = pq \in Z$. Integers N of this type (where both prime factors are distinct and are 3 mod 4) are called Blum integers. Next pick a random y in the group $\mathbb{Z}_N^*$ and set $s = y^2 \in \mathbb{Z}_N^*$. The secret seed is (N, s). As we will see below, there is no need to keep the number N secret.

Generate. Given an input $\ell \in \mathbb{Z}$ and a seed (N, s) we generate a pseudorandom sequence of length ℓ. First, set $x_1 = s$. Then, for $i = 1, \ldots, \ell$:
1. View x_i as an integer in $[0, N-1]$ and let $b_i \in \{0, 1\}$ be the least significant bit of x_i.
2. Set $x_{i+1} = x_i^2 \in \mathbb{Z}_N$.

The output sequence is $b_1 b_2 \cdots b_\ell \in \{0, 1\}^\ell$.

The generator can be viewed as a special case of the general *Blum–Micali generator* [2]. To see this, we show that the generator is based on a *one-way permutation* (see one-way function and substitutions and permutations) and a *hard-core predicate* of that permutation. For an integer N let $\mathrm{QR}_N = (\mathbb{Z}_N^*)^2$ denote the subgroup of quadratic residues in $\mathbb{Z}_N^*$ and let $F_N : \mathbb{Z}_N \to \mathbb{Z}_N$ denote the function $F_N(x) = x^2 \in \mathbb{Z}_N$. For Blum integers the function F_N is a permutation (a one-to-one map) of the subgroup of quadratic residues QR_N. In fact, it is not difficult to show that F_N is a *one-way* permutation of QR_N, unless factoring Blum integers is easy. Now that we have a one-way permutation we need a hard-core bit of the permutation to construct a Blum–Micali-type generator. Consider the predicate $B : \mathrm{QR}_N \to \{0, 1\}$ that on input $x \in \mathrm{QR}_N$ views x as an integer in $[1, N]$ and outputs the least significant bit of x. Blum, Blum, and Shub showed that $B(x)$ is a hard-core predicate of F_N assuming it is hard to distinguish quadratic residues in $\mathbb{Z}_N$ from nonresidues in $\mathbb{Z}_N$ with Jacobi symbol 1. Applying the Blum–Micali construction to the one-way permutation F_N and the hard-core predicate B produces the generator above. The general theorem of Blum and Micali now shows that the generator is secure assuming it is hard to distinguish quadratic residues in $\mathbb{Z}_N$ from nonresidues in $\mathbb{Z}_N$ with Jacobi symbol 1. Vazirani and Vazirani [5] improved the result by showing that $B(x)$ is a hard-core predicate under the weaker assumption that factoring random Blum integers is intractable.

One can construct many different hard-core predicates for the one-way permutation F_N defined above. Every such hard-core bit gives a slight variant of the BBS generator. For example, Hastad and Naslund [3] show that for most $1 \le j < \log_2 N$ the predicate $B_j(x) : \mathrm{QR}_N \to \{0, 1\}$ that returns the jth bit of x is a hard-core predicate of F_N assuming factoring Blum integers is intractable. Consequently, one can output bit j of x_i at every iteration and still obtain a secure generator, assuming factoring Blum integers is intractable.

One can improve the efficiency of the general Blum–Micali generator by outputting multiple simultaneously secure hard-core bits per iteration. For the function F_N it is known that the $O(\log \log N)$ least significant bits are simultaneously secure, assuming factoring Blum integers is intractable. Consequently, the simulator remains secure (asymptotically) if one outputs the $O(\log \log N)$ least significant bits of x_i per iteration.

Let I be the set of integers $I = \{1, \ldots, N\}$. We note that for a Blum integer N and a generator $g \in \mathbb{Z}_N^*$, Hastad et al. [4] considered the function $G_{N,g} : I \to \mathbb{Z}_N$ defined by $G_{N,g}(x) = g^x \in \mathbb{Z}_N$. They showed that *half* the bits of $x \in I$ are simultaneously secure for this function, assuming factoring Blum integers is intractable. Therefore, one can build a Blum–Micali generator from this function that outputs $(\log N)/2$ bits per iteration. The resulting pseudorandom generator is essentially as efficient as the BBS generator and is based on the same complexity assumption.

Dan Boneh

References

[1] Blum, L., M. Blum, and M. Shub (1983). "Comparison of two pseudo-random number generators." *Advances in Cryptology—CRYPTO'82*, eds. Plenum D. Chaum, R.L. Rivest, and A.T. Sherman. Springer-Verlag, Berlin, 61–78.

[2] Blum, M. and S. Micali (1982). "How to generate cryptographically strong sequences of pseudorandom bits." *Proceedings of FOCS'82*, 112–117.

[3] Hastad, J. and M. Naslund (2004). "The security of all RSA and discrete log bits." *Journal of the ACM*. Extended abstract in *Proc. of FOCS'98*, 510–521.

[4] Hastad, J., A. Schrift, and A. Shamir (1993). "The discrete logarithm modulo a composite hides $o(n)$ bits." *Journal of Computer and System Sciences (JCSS)*, 47, 376–404.

[5] Vazirani, U. and V. Vazirani (1984). "Efficient and secure pseudo-random number generation." *Proceedings of FOCS'84*, 458–463.

BLUM–GOLDWASSER PUBLIC KEY ENCRYPTION SYSTEM

The Blum–Goldwasser public key encryption system combines the general construction of Goldwasser–Micali [5] with the concrete Blum–Blum–Shub pseudorandom bit generator [2] to obtain an efficient semantically secure public key encryption whose security is based on the difficulty of factoring Blum integers. The system makes use of modular arithmetic and works as follows:

Key Generation. Given a security parameter $\tau \in \mathbb{Z}$ as input, generate two random τ-bit primes p, q where $p = q = 3 \bmod 4$. Set $N = pq \in Z$. The public key is N and private key is (p, q).

Encryption. To encrypt a message $m = m_1 \ldots m_\ell \in \{0, 1\}^\ell$:
1. Pick a random x in the group $\mathbb{Z}_N^*$ and set $x_1 = x^2 \in \mathbb{Z}_N^*$.
2. For $i = 1, \ldots, \ell$:
   (a) View x_i as an integer in $[0, N-1]$ and let $b_i \in \{0, 1\}$ be the least significant bit of x_i.
   (b) Set $c_i = m_i \oplus b_i \in \{0, 1\}$.
   (c) Set $x_{i+1} = x_i^2 \in \mathbb{Z}_N^*$.
3. Output $(c_1, \ldots, c_\ell, x_{\ell+1}) \in \{0, 1\}^\ell \times \mathbb{Z}_N$ as the ciphertext.

Decryption. Given a ciphertext $(c_1, \ldots, c_\ell, y) \in \{0, 1\}^\ell \times \mathbb{Z}_N$ and the private key (p, q) decrypt as follows:
1. Since N is a Blum integer, $\varphi(N)/4$ is odd (see Euler's totient function) and therefore $2^{\ell+1}$ has an inverse modulo $\varphi(N)/4$. Let $t = (2^{\ell+1})^{-1} \bmod (\varphi(N)/4)$.
2. Compute $x_1 = y^t \in \mathbb{Z}_N^*$. Observe that if $y \in (\mathbb{Z}_N^*)^2$ then $x_1^{(2^{\ell+1})} = y^{(t \cdot 2^{\ell+1})} = y \in \mathbb{Z}_N^*$.
3. Next, for $i = 1, \ldots, \ell$:
   (a) View x_i as an integer in $[0, N-1]$ and let $b_i \in \{0, 1\}$ be the least significant bit of x_i.
   (b) Set $m_i = c_i \oplus b_i \in \{0, 1\}$.
   (c) Set $x_{i+1} = x_i^2 \in \mathbb{Z}_N^*$.
4. Output $(m_1, \ldots, m_\ell) \in \{0, 1\}^\ell$ as the plaintext.

Semantic security of the system (against a passive adversary) follows from the proof of security of the Blum–Blum–Shub generator. The proof of security shows that an algorithm capable of mounting a successful semantic security attack is able to factor the Blum integer N in the public key.

We note that the system is XOR malleable: given the encryption $C = (c_1, \ldots, c_\ell, y)$ of a message $m \in \{0, 1\}^\ell$ it is easy to construct an encryption of $m \oplus b$ for any chosen $b \in \{0, 1\}^\ell$ (without knowing m). Let $b = b_1 \cdots b_\ell \in \{0, 1\}^\ell$. Simply set $C' = (c_1 \oplus b_1, \ldots, c_\ell \oplus b_\ell, y)$. Since the system is XOR malleable it cannot be semantically secure under a chosen ciphertext attack.

Interestingly, the same reduction given in the proof of semantic security shows that a chosen ciphertext attacker can factor the modulus N and therefore recover the private key from the public key. Consequently, as it is, the system completely falls apart under a chosen ciphertext attack. When using the system one must use a mechanism to defend against chosen ciphertext attacks. For example, Fujisaki and Okamoto [4] provide a general conversion from semantic security to chosen ciphertext security in the random oracle model. However, in the random oracle model one can construct more efficient chosen ciphertext secure systems that are also based on the difficulty of factoring [1, 3].

Dan Boneh

References

[1] Bellare, Mihir and Phillip Rogaway (1996). "The exact security of digital signatures: How to sign with RSA and Rabin." *Advances in Cryptology—EUROCRYPT'96*, Lecture Notes in Computer Science, vol. 1070, ed. U. Maurer. Springer-Verlag, Berlin, 399–416.

[2] Blum, L., M. Blum, and M. Shub (1983). "Comparison of two pseudo-random number generators." *Advances in Cryptology—CRYPTO'83*, ed. D. Chaum. Springer-Verlag, Berlin, 61–78.

[3] Boneh, Dan (2001). "Simplified OAEP for the RSA and Rabin functions." *Advances in Cryptology—CRYPTO 2001*, Lecture Notes in Computer Science, vol. 2139, ed. J. Kilian. Springer-Verlag, Berlin.

[4] Fujisaki, E. and T. Okamoto (1999). "Secure integration of asymmetric and symmetric encryption schemes." *Advances in Cryptology—CRYPTO'99*, Lecture Notes in Computer Science, vol. 1666, ed. J. Wiener. Springer-Verlag, Berlin, 537–554.

[5] Goldwasser, S. and S. Micali (1984). "Probabilistic encryption." *Journal of Computer and System Science (JCSS)*, 28 (2), 270–299.

BOLERO.NET

When exporters and importers wish to trade in goods onboard a ship, they use a document of title called a Bill of Lading (or B/L for short). It is issued by the ship operator as a receipt for the goods and, because he will only release the cargo against production of this document, the B/L has been used for centuries for trading and as financial security.

Making the functionality of this document available by electronic means is an undertaking similar to that of putting share trading online, and it is the job of bolero.net, a service operated since 1999 by the Through Transport Club, a mutual marine insurer, and S.W.I.F.T., the banks' cooperatively owned data network operator. The origins of the project stem from the mid-1980s and

it has been known as Bolero since 1994, when an early version was piloted in a project funded by the European Commission.

Bolero.net handles not just Bs/L but all other trade documentation too. However, it is the title function of the B/L which gives rise to the most interesting isuses.

In essence, the B/L is issued as a message from the shipowner to the original cargo owner, digitally signed and handled via bolero.net's secure message handling facility, the Core Messaging Platform (CMP). The CMP, when sent a new B/L, also passes a message to a bolero.net component called the Title Registry (TR). The TR sets up a new record in its database and from this point on, it is the information about ownership held in the TR which is ultimately authoritative in any dispute.

When the electronic B/L is being traded, the TR is updated through digitally signed messages from the users via the CMP. The TR will only accept instructions from the user currently recorded as the "holder," i.e., owner of the title being traded. This is a system similar to that operated by most dematerialized share trading schemes.

To enable electronic trading of negotiable documents such as the B/L, all you need is a database operated by a trusted third party, and digital signatures which can be verified by that party. Bolero.net is that trusted third party, and it operates its own Certification Authority, though there is a project under way to accept certificates from other issuers in the future.

As the traders would also like to keep their information confidential, the communications are handled as SSL protected exchanges (see Secure Socket Layer). The legal security of the transaction, i.e. the certainty that a trade carried out over bolero.net will be treated as legally binding by the courts, is provided via a multilateral user contract, known as the Rule Book. All users are bound by this contract and it creates a legal safety net which ensures that the traditional functionality of the ancient B/L can still be provided by electronic means.

For more information, see *www.bolero.net* and *www.bolerassociation.org*

Peter Landrock

BOOLEAN FUNCTIONS

Boolean functions play a central role in the design of most symmetric cryptosystems and in their security. In <u>stream ciphers</u>, they are usually used to combine the outputs to several <u>linear feedback shift registers</u> (see the corresponding entry and <u>Combination generator</u>), or to filter (and combine) the contents of a single one (see <u>Filter generators</u>). The sequence of their output, during a certain number of clock cycles, then produces the pseudorandom sequence which is used in a <u>Vernam cipher</u> (that is, which is bitwisely added to the plaintext to produce the ciphertext). In block ciphers (see <u>Block cipher</u>, <u>Data Encryption Standard (DES)</u>, *Advanced Encryption Standard* (<u>Rijndael/AES</u>)), the S-boxes are designed by appropriate composition of nonlinear Boolean functions.

An *n-variable* <u>Boolean function</u> f is a function from the set F_2^n of all binary vectors $x = (x_1, \ldots, x_n)$ of length n to the <u>field</u> $F_2 = \{0, 1\}$. The number n of variables is rarely large in practice. In the case of stream ciphers, it is most often less than 10; and the S-boxes used in most block ciphers are concatenations of sub S-boxes on at most eight variables. But determining and studying those Boolean functions which satisfy some conditions needed for cryptographic uses (see below) is not feasible through an exhaustive computer investigation, since the number of n-variable Boolean functions is too large when $n \geq 6$. However, clever computer investigations are useful for imagining or testing conjectures, and sometimes for generating interesting functions.

The *Hamming weight* $w_H(f)$ of a Boolean function f on F_2^n is the size of its *support* $\{x \in F_2^n / f(x) \neq 0\}$. The *Hamming distance* $d_H(f, g)$ between two functions f and g is the size of the set $\{x \in F_2^n / f(x) \neq g(x)\}$. Thus it equals the Hamming weight $w_H(f \oplus g)$ of the sum (modulo 2) of the functions.

Every n-variable Boolean function can be represented with its *truth table*. But the representation of Boolean functions, which is most usually used in cryptography, is the n-variable polynomial representation over F_2 of the form

$$f(x) = \bigoplus_{I \subseteq \{1, \ldots, n\}} a_I \left(\prod_{i \in I} x_i \right),$$

where $\oplus$ denotes the binary sum. The variables $x_1, \ldots, x_n$ appear in this polynomial with exponents smaller than or equal to 1 because, representing bits, they are equal to their own squares. This polynomial representation is called the *Algebraic Normal Form*, in brief, ANF (see also <u>Reed–Muller codes</u>).

EXAMPLE: The three-variable function f whose truth table equals

x_1	x_2	x_3	$f(x)$
0	0	0	0
0	0	1	1
0	1	0	0
0	1	1	0
1	0	0	0
1	0	1	1
1	1	0	0
1	1	1	1

has ANF: $(1 \oplus x_1)(1 \oplus x_2)x_3 \oplus x_1(1 \oplus x_2)x_3 \oplus x_1x_2x_3 = x_1x_2x_3 \oplus x_2x_3 \oplus x_3$. Indeed, the expression $(1 \oplus x_1)(1 \oplus x_2)x_3$, for instance, equals 1 if and only if $1 \oplus x_1 = 1 \oplus x_2 = x_3 = 1$, that is, $(x_1, x_2, x_3) = (0, 0, 1)$. ◇

A similar polynomial representation, called the Numerical Normal Form, in which the coefficients and the operation of summation take place in the group of integers instead of F_2, can also be used for studying Boolean functions.

The ANF of every Boolean function exists and is unique.

Two simple relations relate the truth table and the ANF:

$$\forall x \in F_2^n, \qquad f(x) = \bigoplus_{I \subseteq \mathrm{supp}(x)} a_I, \qquad (1)$$

$$\forall I \subseteq \{1, \ldots, n\}, \qquad a_I = \bigoplus_{x \in F_2^n / \mathrm{supp}(x) \subseteq I} f(x), \qquad (2)$$

where $\mathrm{supp}(x)$ denotes the support $\{i \in \{1, \ldots, n\}/ x_i = 1\}$ of x. Thus, the function is the image of its ANF by the *binary Möbius transform*, and *vice versa*.

The degree of the ANF is denoted by $d^\circ f$ and is called the *algebraic degree* of the function (some authors use also the term *nonlinearity order*). The algebraic degree is an *affine invariant* in the following sense: two functions f and g are called affinely (resp. linearly) *equivalent* if there exists an affine (resp. a linear) automorphism (i.e., invertible homomorphism) A of F_2^n such that $g = f \circ A$. A mapping is called affine invariant if it is invariant under affine equivalence.

The *affine* functions are those Boolean functions with degrees 0 or 1 (thus, with the simplest ANFs). Denoting by $a \cdot x$ the usual *inner product* $a \cdot x = a_1 x_1 \oplus \cdots \oplus a_n x_n$ in F_2^n, the general form of an n-variable affine function is $a \cdot x \oplus a_0$, with $a \in F_2^n$; $a_0 \in F_2$.

Another representation of Boolean functions can be used: the *trace representation*. The vector space F_2^n is endowed with the structure of the field F_{2^n}. Let us denote by tr the trace function from F_{2^n} to F_2: $\mathrm{tr}(x) = x + x^2 + x^{2^2} + \cdots + x^{2^{n-1}}$. Every

Boolean function on F_{2^n} can be represented in the form $\mathrm{tr}(P(x))$, where $x \in F_{2^n}$, and where $P(x)$ is a polynomial on one variable over F_{2^n}, of degree at most $2^n - 1$.

Almost all of the characteristics needed for Boolean functions in cryptography can be expressed by means of the *discrete Fourier transforms* of the functions. The discrete Fourier transform (also called Hadamard transform) of a Boolean function, or more generally of an integer-valued function φ on F_2^n, is the integer-valued function $\widehat{\varphi}$ defined on F_2^n by

$$\widehat{\varphi}(u) = \sum_{x \in F_2^n} \varphi(x)(-1)^{x \cdot u}. \tag{3}$$

There exists a simple divide-and-conquer butterfly algorithm to compute $\widehat{\varphi}$, whose complexity is $O(n2^n)$:

1. Write the table of the values of φ (its truth table if φ is Boolean), the binary vectors of length n being, say, in lexicographic order;
2. Let φ_0 be the restriction of φ to $\{0\} \times F_2^{n-1}$ and φ_1 its restriction to $\{1\} \times F_2^{n-1}$; the table of φ_0 (resp. φ_1) corresponds to the upper (resp. lower) half of the table of φ; replace the values of φ_0 by those of $\varphi_0 + \varphi_1$ and those of φ_1 by those of $\varphi_0 - \varphi_1$;
3. Apply recursively step 2 to φ_0 and to φ_1 (these $(n-1)$-variable functions taking the place of φ).

At each recursion, the number of variables of the functions decreases by 1. When the algorithm ends (i.e., when it arrives to functions on one variable each), the global table gives the values of $\widehat{\varphi}$.

EXAMPLE: This algorithm, applied for computing the Fourier transform of the three-variable function f already considered above, gives the following table.

x_1	x_2	x_3	$f(x)$			$\widehat{f}(x)$
0	0	0	0	0	0	3
0	0	1	1	2	3	-3
0	1	0	0	0	0	1
0	1	1	0	1	1	-1
1	0	0	0	0	0	-1
1	0	1	1	0	-1	1
1	1	0	0	0	0	1
1	1	1	1	-1	1	-1

For a given Boolean function f, the discrete Fourier transform can be applied to f itself (notice that $\widehat{f}(0)$ equals the Hamming weight of f). It can also be applied to the function $f(x) = (-1)^{f(x)}$ (often called the *sign function*), which

gives:

$$\widehat{f}(u) = \sum_{x \in F_2^n} (-1)^{f(x) \oplus x \cdot u}.$$

We call $\widehat{f}$ the *Walsh transform* of f. We shall use only this transform of Boolean functions in the sequel.

The discrete Fourier transform, as any other Fourier transform, has numerous properties. The two most important ones are the *inverse Fourier relation*: $\widehat{\widehat{\varphi}} = 2^n \varphi$, and *Parseval's relation*:

$$\sum_{u \in F_2^n} \widehat{\varphi}^2(u) = 2^n \sum_{x \in F_2^n} \varphi^2(x).$$

Parseval's relation applied to the Walsh transform of a Boolean function f gives:

$$\sum_{u \in F_2^n} \widehat{f}^2(u) = 2^{2n}.$$

The resistance of the diverse cryptosystems implementing Boolean functions to the known attacks can be quantified through some fundamental characteristics of the Boolean functions used in them. The design of cryptographic functions then needs to consider various characteristics (depending on the choice of the cryptosystem) simultaneously. Of course, these criteria are partially in conflict with each other, and trade-offs are necessary.

Criteria and Cryptographic Characteristics

1. Cryptographic functions must have *high algebraic degrees*, since all cryptosystems using Boolean functions can be attacked if the functions have low degrees.

   For instance, in the case of combining functions, if n LFSRs having lengths $L_1, \ldots, L_n$ are combined by the function $f(x) = \bigoplus_{I \subseteq \{1,\ldots,n\}} a_I \left(\prod_{i \in I} x_i\right)$, then the sequence produced by f can be obtained by a LFSR of length $L \leq \sum_{I \subseteq \{1,\ldots,n\}} a_I \left(\prod_{i \in I} L_i\right)$. The degree of f has therefore to be high so that L can have a high value (otherwise, the system does not resist the Berlekamp–Massey attack [2]). In the case of block ciphers, the use of Boolean functions of low degrees makes effective the "higher differential attack."

2. The output to any Boolean function f always has correlation to certain linear functions of its inputs. But this correlation should be small. In other words, the minimum Hamming distance between f and all affine functions must be high. Otherwise, an affine approximation of the Boolean function can be used to build attacks on

any kind of system implementing the function (see <u>Linear cryptanalysis for block ciphers</u> and <u>Linear cryptanalysis for stream ciphers</u>). The minimum distance between f and all affine functions is called the *nonlinearity* of f and denoted by $\mathcal{NL}(f)$ (see <u>Nonlinearity of Boolean functions</u> for more details). It can be quantified through the Walsh transform:

$$\mathcal{NL}(f) = 2^{n-1} - \tfrac{1}{2} \max_{u \in F_2^n} |\widehat{f}(u)|.$$

Parseval's relation then implies that for every n-variable Boolean function f:

$$\mathcal{NL}(f) \leq 2^{n-1} - 2^{n/2-1}.$$

3. Cryptographic functions must be *balanced* (their output must be uniformly distributed) for avoiding statistical dependence between the input and the output, which can be used in attacks. Note that f is balanced if and only if $\widehat{f}(0) = 0$.

   Moreover, any combining function $f(x)$ must stay balanced if we keep constant some coordinates x_i of x (at most m of them, where m is as large as possible). We say that f is then *m-resilient*. More generally, a (non necessarily balanced) Boolean function, whose output distribution probability is unaltered when any m of the input bits are kept constant, is called *mth order correlation-immune* (see <u>Correlation immune and resilient Boolean functions</u>).

4. The *propagation criterion* (*PC*), generalizing the *strict avalanche criterion* (*SAC*), quantifies the level of diffusion put in a cipher by a Boolean function. This criterion is more relevant to block ciphers. An n-variable Boolean function satisfies the propagation criterion $PC(l)$ of degree l if, for every vector x of Hamming weight at most l, the *derivative* $D_a f(x) = f(x) \oplus f(x + a)$ is balanced (see <u>Propagation characteristics of Boolean functions</u>).

   By definition, *SAC* is equivalent to $PC(1)$.

5. A vector $e \in F_2^n$ is called a *linear structure* of an n-variable Boolean function f if the derivative $D_e f$ is constant. Boolean functions used in block ciphers should avoid nonzero linear structures (see [1]). A Boolean function admits a nonzero linear structure if and only if it is linearly equivalent to a function of the form $f(x_1, \ldots, x_n) = g(x_1, \ldots, x_{n-1}) \oplus \varepsilon x_n$ where $\varepsilon \in F_2$.

6. Other characteristics of Boolean functions have been considered in the literature:
   - The *sum-of-squares indicator* $\mathcal{V}(f) = \sum_{a \in F_2^n} \left(\sum_{x \in F_2^n} (-1)^{D_a f(x)} \right)^2$ and the *absolute indicator* $\max_{a \in F_2^n, \, a \neq 0} \left| \sum_{x \in F_2^n} (-1)^{D_a f(x)} \right|$

quantify the global diffusion capability of the function (the lower they are, the better is the diffusion);

   - The *maximum correlation* between an n-variable Boolean function f and a subset I of $\{1, \ldots, n\}$ equals $C_f(I) = 2^{-n} \max_{g \in \mathcal{F}_I} \sum_{x \in F_2^n} (-1)^{f(x) \oplus g(x)}$, where $\mathcal{F}_I$ is the set of all n-variable Boolean functions whose values depend on $\{x_i, i \in I\}$ only. The maximum correlation $C_f(I)$ must be low for every nonempty set I of small size, to avoid nonlinear correlation attacks (note that mth order correlation immunity corresponds to an optimum maximum correlation to every subset I of size at most m, if we consider only affine approximations instead of all Boolean approximations).

Claude Carlet

References

[1] Evertse, J.H. (1988). "Linear structures in block ciphers." *Advances in Cryptology—EUROCRYPT'87*, Lecture Notes in Computer Science, vol. 304, eds. David Chaum and Wyn L. Price. Springer-Verlag, Berlin, 249–266.

[2] Massey, J.L. (1969). "Shift-register analysis and BCH decoding." *IEEE Trans. Inform. Theory*, 15, 122–127.

[3] Menezes, A., P. van Oorschot, and S. Vanstone (1996). *Handbook of Applied Cryptography*. CRC Press Series on Discrete Mathematics and its Applications. http://www.cacr.math.uwaterloo.ca/hac

BOOMERANG ATTACK

The boomerang attack is a <u>chosen plaintext</u> and <u>adaptive chosen ciphertext attack</u> discovered by Wagner [5]. It is an extension of <u>differential</u> attack to *two-stage* differential–differential attack which is closely related to <u>impossible differential</u> attack as well as to the <u>meet-in-the middle</u> approach. The attack may use *characteristics, differentials* as well as <u>truncated differentials</u>. The attack breaks constructions in which there are high-probability differential patterns propagating half-way through the cipher both from the top and from the bottom, but there are no good patterns that propagate through the full cipher.

The idea of the boomerang attack is to find good conventional (or truncated) differentials that cover half of the cipher but cannot necessarily be concatenated into a single differential covering the whole cipher. The attack starts with a pair of plaintexts P and P' with a difference Δ which goes to difference Δ^* through the upper half of the cipher. The attacker obtains the corresponding ciphertexts C and C', applies the difference ∇ to obtain ciphertexts $D = C + \nabla$ and $D' = C' + \nabla$, and decrypts them to plaintexts Q and Q'. The choice of ∇ is such that the difference propagates to the difference ∇^* in the decryption direction through the lower half of the cipher. For the *right quartet* of texts, difference Δ^* is created in the middle of the cipher between partial decryptions of D and D' which propagates to the difference Δ in the plaintexts Q and Q'. This can be detected by the attacker.

Moreover, working with quartets (pairs of pairs) provides boomerang attacks with additional filtration power. If one partially guesses the keys of the top round one has two pairs of the quartet to check whether the uncovered partial differences follow the propagation pattern, specified by the differential. This effectively doubles the attacker's filtration power.

The attack was demonstrated with a practical cryptanalysis of a cipher which was designed with provable security against conventional differential attack [4], as well as on round-reduced versions of several other ciphers. The related method of the *inside out* attack was given in the same paper. Further refinements of the boomerang technique have been found in papers on so-called *amplified boomerang* and *rectangle* attacks [1, 3]. In certain cases a free round in the middle may be gained due to a careful choice of the differences coming from the top and from the bottom [2, 5].

Alex Biryukov

References

[1] Biham, E., O. Dunkelman, and N. Keller (2002). "New results on boomerang and rectangle attacks." *Fast Software Encryption, FSE 2002*, Lecture Notes in Computer Science, vol. 2365, eds. J. Daemen and V. Rijmen. Springer-Verlag, Berlin, 1–16.

[2] Biryukov, A., C. De Cannire, and G. Dellkrantz (2003). "Cryptanalysis of SAFER++." *Advances in Cryptology—CRYPTO 2003*, Lecture Notes in Computer Science, vol. 2729, ed. D. Boneh. Springer-Verlag, Berlin. NES/DOC/KUL/WP5/028. Full version available at http://eprint.iacr.org/2003/109/

[3] Kelsey, J., T. Kohno, and B. Schneier (2001). "Amplified boomerang attacks against reduced-round MARS and Serpent." *Fast Software Encryption, FSE 2000*, Lecture Notes in Computer Science, vol. 1978, ed. B. Schneier. Springer-Verlag, Berlin, 75–93.

[4] Vaudenay, S. (1998). "Provable security for block ciphers by decorrelation." *STACS*, Lecture Notes in Computer Science, vol. 3404, eds. M. Morvan, C. Meinel, and D. Krob. Springer-Verlag, Berlin, 249–275.

[5] Wagner, D. (1999). "The boomerang attack." *Fast Software Encryption, FSE'99*, Lecture Notes in Computer Science, vol. 3404, ed. L.R. Knudsen. Springer-Verlag, Berlin, 156–170.

BROADCAST ENCRYPTION

CONCEPT DEFINITION AND APPLICATIONS: The concept of broadcast encryption deals with methods that allow to efficiently transmit information to a dynamically changing group of privileged users who are allowed to receive the data. It is often convenient to think of it as a revocation scheme, which addresses the case where some subset of the users are excluded from receiving the information.

The problem of a center transmitting data to a large group of receivers so that only a predefined subset is able to decrypt the data is at the heart of a growing number of applications. Among them are pay-TV applications, multicast (or secure group) communication, secure distribution of copyright-protected material (e.g., music), digital rights management, and audio streaming. Different applications impose different rates for updating the group of legitimate users. Users are excluded from receiving the information due to payments, subscription expiration, or since they have abused their rights in the past.

One special case is when the receivers are stateless. In such a scenario, a (legitimate) receiver is not capable of recording the past history of transmissions and change its state accordingly. Instead, its operation must be based on the current transmission and its initial configuration. Stateless receivers are important for the case where the receiver is a device that is not constantly on-line, such as a media player (e.g., a CD or DVD player where the "transmission" is the current disc [4, 10], a satellite receiver (GPS) and perhaps in multicast applications).

Broadcast encryption can be combined with tracing capabilities to yield trace-and-revoke schemes. A tracing mechanism enables the efficient tracing of leakage, specifically, the source of keys used by illegal devices, such as pirate decoders or clones. Trace-and-revoke schemes are of particular value in many scenarios: they allow to trace the identity of the user whose key was leaked; in turn, this user's key is revoked from the system for future uses.

What are the desired properties of a broadcast encryption scheme? A good scheme is characterized by

- Low bandwidth—we aim at a small message expansion, namely that the length of the encrypted content should not be much longer than the original message.
- Small amount of storage—we would like the amount of required storage (typically keys) at the user to be small, and as a secondary objective the amount of storage at the server to be manageable as well.
- Attentiveness—does the scheme require users to be on-line "all the time?" If such a requirement does not apply, then the scheme is called stateless.
- Resilience—we want the method to be resilient to large coalitions of users who collude and share their resources and keys.

In order to evaluate and compare broadcast encryption methods, we define a few parameters. Let N be the set of all users, $|\mathcal{N}| = N$, and $\mathcal{R} \subset N$ be a group of $|\mathcal{R}| = r$ users whose decryption privileges should be revoked. The goal of a broadcast encryption algorithm is to allow a center to transmit a message M to all users such that any user $u \in \mathcal{N} \setminus \mathcal{R}$ can decrypt the message correctly, while a coalition consisting of t or fewer members of $\mathcal{R}$ cannot decrypt it. The important parameters are therefore r, t, and N.

A system consists of three parts: (1) a key assignment scheme, which is an initialization method for assigning secret keys to receivers that will allow them to decrypt. (2) The broadcast algorithm—given a message M and the set $\mathcal{R}$ of users to revoke outputs a ciphertext message M' that is broadcast to all receivers. (3) A decryption algorithm—a (nonrevoked) user who receives ciphertext M' should produce the original message M using its secret information.

HISTORY OF THE PROBLEM: The issue of secure broadcasting to a group has been investigated earlier on, see for example [1]. The first formal definition of the area of *broadcast encryption*, including parameterization of the problem and its rigorous analysis (as well as coining the term) was done by Fiat and Naor in [5] and has received much attention since then; see for example [2, 6, 8, 9, 11, 13–15, 18, 19]. The original broadcast encryption method of [5] allows the removal of any number of users as long as at most t of them collude. There the message length is $O(t \log^2 t)$, a user must store a number of keys that is logarithmic in t and the amount of work required by the user is $\tilde{O}(r/t)$ decryptions. The scheme can be used in a stateless environment as it does not require

attentiveness. On the other hand, in the stateful case, gradual revocation of users is particularly efficient.

The logical-tree-hierarchy (LKH) scheme, suggested independently in the late 1990s by Wallner et al. [18] and Wong et al. [19], is designed to achieve secure group communication in a multicast environment. Useful mainly in the connected mode for multicast re-keying applications, it revokes or adds a single user at a time, and updates the keys of all remaining users. It requires a transmission of $2 \log N$ keys to revoke a single user, each user is required to store $\log N$ keys and the amount of work each user should do is $\log N$ encryptions (the expected number is $O(1)$ for an average user). These bounds are somewhat improved in [2, 3, 12, 16, 17], but unless the storage at the user is extremely high they still require a transmission of length $\Omega(r \log N)$. This algorithm may revoke any number of users, regardless of the coalition size.

Luby and Staddon [11] considered the information theoretic (see <u>computational complexity</u>, <u>information theory</u>, and <u>security</u>) setting and devised bounds for any revocation algorithms under this setting. Garay et al. [6] introduced the notion of *long-lived broadcast encryption*. In this scenario, keys of compromised decoders are no longer used for encryptions. The question they address is how to adapt the broadcast encryption scheme so as to maintain the security of the system for the good users.

CPRM, which stands for content protection for recordable media, [4] is a technology for protecting content on physical media such as recordable DVD, DVD Audio, Secure Digital Memory Card, and Secure CompactFlash. It is one of the methods that explicitly considers the stateless scenario. There, the message is composed of $r \log N$ encryptions, the storage at the receiver consists of $\log N$ keys, and the computation at the receiver requires a single decryption. It is a variant on the techniques of [5].

The subset difference method for broadcast encryption, proposed by Naor, Naor, and Lotspiech [13, 14], is most appropriate in the stateless scenario. It requires a message length of $2r - 1$ (in the worst case, or $1.38r$ in the average case) encryptions to revoke r users, and storage of $\frac{1}{2} \log^2 N$ keys at the receiver. The algorithm does not assume an upper bound of the number of revoked receivers, and works even if all r revoked users collude. The key assignment of this scheme is computational and not information theoretic, and as such it outperforms an information theoretic lower bound on the size of the message [11]. A rigorous security treatment of a family of schemes, including the subset difference method, is provided in [14]. Halevy and Shamir [8] have suggested a variant of subset difference called LSD (layered subset difference). The storage requirements are reduced to $O(\log^{1+\varepsilon} N)$ while the message length is $O(r/\varepsilon)$, providing a full spectrum between the complete subtree and subset difference methods. A reasonable choice is $\varepsilon = 2$.

Both LKH and the subset difference methods are hierarchical in nature and as such are particularly suitable to cases where related users must all be revoked at once, for instance, all users whose subscription expires on a certain day.

It is also important to realize that many implementations in this field remain proprietary and are not published both for security reasons (not to help the pirates) as well as for commercial reasons (not to help the competitors).

CONSTRUCTIONS: A high level overview of three fundamental broadcast encryption constructions is outlined below. Details are omitted and can be found in the relevant references. One technique that is commonly used in the key assignment of these constructions is the derivation of keys in a tree-like manner: a key is associated with the root of a tree and this induces a labeling of all the nodes of the tree. The derivation is done based on the technique first used by Goldreich, Goldwasser, and Micali (GGM) [7].

Fiat–Naor Construction

The idea of the construction in [5] is to start with the case where the coalition size (the number of users who collude and share their secret information) is t and reduce it to the case where the coalition size is 1, the basic construction. For this case, suppose that there is a key *associated* with each user; every user is given *all keys except the one associated with it*. (As an illustration, think of the key associated with a user as written on its forehead, so that all other users except for itself can see it.) To broadcast a message to the group $\mathcal{N} \setminus \mathcal{R}$, the center constructs a broadcast key by Xoring all keys associated with the *revoked users* $\mathcal{R}$. Note that any user $u \in \mathcal{N} \setminus \mathcal{R}$ can reconstruct this key, but a user $u \in \mathcal{R}$ cannot deduce the key from its own information. This naive key assignment requires every user to store $N - 1$ keys. Instead, by deriving the keys in a GGM tree-like process, the key assignment is made feasible by requiring every user to store $\log N$ keys only.

The construction is then extended to handle the case where up to t users may share their secret

information. The idea then is to obtain a scheme for larger t by various partitions of the user set, where for each such partition the basic scheme is used.

Logical Key Hierarchy

The LKH (logical key hierarchy) scheme [18–20] maintains a common encryption key for the active group members. It assumes that there is an initial set $\mathcal{N}$ of N users and that from time to time an active user leaves and a new value for the group key should be chosen and distributed to the remaining users. The operations are managed by a center which broadcasts all the maintenance messages and is also responsible for choosing the new key. When some user $u \in \mathcal{N}$ is revoked, a new group key K' should be chosen and all nonrevoked users in $\mathcal{N}$ should receive it, while no coalition of the revoked users should be able to obtain it; this is called a *leave* event. At every point a nonrevoked user knows the group key K as well as a set of secret "auxiliary" keys. These keys correspond to subsets of which the user is a member, and may change throughout the lifetime of the system.

Users are associated with the leaves of a full binary tree of height $\log N$. The center associates a key K_i with every node v_i of the tree. At initialization, each user u is sent (via a secret channel) the keys associated with all the nodes along the path connecting the leaf u to the root. Note that the root key K is known to all users and can be used to encrypt group communications.

In order to remove a user u from the group (a *leave* event), the center performs the following operations. For all nodes v_i along the path from u to the root, a new key K_i' is generated. The new keys are distributed to the remaining users as follows: let v_i be a node on the path and v_j be its child on the path and v_ℓ its child that is *not* on the path. Then K_i' is encrypted using K_j' and K_ℓ (the latter did not change), i.e., a pair of encryptions $\langle E_{K_j'}(K_i'), E_{K_\ell}(K_i') \rangle$. The exception is if v_i is the parent of the leaf u, in which case only a single encryption using the sibling of u is sent. All encryptions are sent to all the users.

Subset Difference

The subset difference construction defines a collection of subsets of users $S_1, \ldots, S_w$, $S_j \subseteq \mathcal{N}$. Each subset S_j is assigned a long-lived key L_j; a user u is assigned some secret information I_u so that every member of S_j should be able to deduce L_j from its secret information. Given a revoked set $\mathcal{R}$, the remaining users are partitioned into disjoint sets

$S_{i_1}, \ldots, S_{i_m}$ from the collection that entirely *cover* them (every user in the remaining set is in at least one subset in the cover) and a session key K is encrypted m times with $L_{i_1}, \ldots, L_{i_m}$. The message is then encrypted with the session key K.

Again, users are associated with the leaves of a full binary tree of height $\log N$. The collection of subsets $S_1, \ldots, S_w$ defined by this algorithm corresponds to subsets of the form "a group of receivers G_1 minus another group G_2," where $G_2 \subset G_1$. The two groups G_1, G_2 correspond to leaves in two full binary subtrees. Therefore, a valid subset S is represented by two nodes in the tree (v_i, v_j) such that v_i is an ancestor of v_j and is denoted as $S_{i,j}$. A leaf u is in $S_{i,j}$ iff it is in the subtree rooted at v_i but *not* in the subtree rooted at v_j, or in other words $u \in S_{i,j}$ iff v_i is an ancestor of u but v_j is not.

The observation is that for any subset $\mathcal{R}$ of revoked users, it is possible to find a set of at most $2r - 1$ subsets from the predefined collection that cover all other users $\mathcal{N} \setminus \mathcal{R}$.

A naive key assignment that assigns to each user all long-lived keys of the subsets it belongs to requires a user to store $O(N)$ key. Instead, this information (or rather a succinct representation of it) can be reduced to $\frac{1}{2} \log^2 N$ based on a GGM-like tree construction; for details see [14].

Dalit Naor

References

[1] Berkovits, S. (1991). "How to Broadcast a Secret." *Advances in Cryptology—EUROCRYPT'91*, Lecture Notes in Computer Science, vol. 547, ed. D.W. Davies. Springer, Berlin, 535–541.

[2] Canetti, R., J. Garay, G. Itkis, D. Micciancio, M. Naor, and B. Pinkas (1999). "Multicast security: A taxonomy and some efficient constructions." *Proc. of Proceedings IEEE INFOCOM'99*, vol. 2, 708–716.

[3] Canetti, R., T. Malkin, and K. Nissim (1999). "Efficient communication-storage tradeoffs for multicast encryption." *Advances in Cryptology—EUROCRYPT'99*, Lecture Notes in Computer Science, vol. 1592, ed. J. Stern. Springer, Berlin, 459–474.

[4] Content Protection for Recordable Media. Available: http://www.4centity.com/4centity/tech/cprm

[5] Fiat, A. and M. Naor (1994). "Broadcast encryption." *Advances in Cryptology—CRYPTO'93*, Lecture Notes in Computer Science, vol. 773, ed. D.R. Stinson. Springer, Berlin, 480–491.

[6] Garay, J.A., J. Staddon, and A. Wool (2000). "Long-lived broadcast encryption." *Advances in Cryptology—CRYPTO 2000*, Lecture Notes in Computer Science, vol. 1880, ed. M. Bellare. Springer-Verlag, Heidelberg, 333–352.

[7] Goldreich, O., S. Goldwasser, and S. Micali (1986). "How to construct random functions." *JACM*, 33 (4), 792–807.

[8] Halevy, D. and A. Shamir (2002). "The LSD Broadcast encryption scheme." *Advances in Cryptology—CRYPTO 2002*, Lecture Notes in Computer Science, vol. 2442, ed. M. Yung. Springer, Berlin.

[9] Kumar, R., R. Rajagopalan, and A. Sahai (1999). "Coding constructions for blacklisting problems without computational assumptions." *Advances in Cryptology—CRYPTO'99*, Lecture Notes in Computer Science, vol. 1666, ed. J. Wiener. Springer-Verlag, Heidelberg, 609–623.

[10] Lotspiech, J., S. Nusser, and F. Pestoni (2002). "Broadcast encryption's bright future." *Computer*, 35 (8), 57–63.

[11] Luby, M. and J. Staddon (1998). "Combinatorial bounds for broadcast encryption." *Advances in Cryptology—EUROCRYPT'98*, Lecture Notes in Computer Science, vol. 1403, ed. K. Nyberg. Springer-Verlag, Heidelberg, 512–526.

[12] McGrew, D. and A.T. Sherman (1998). Key establishment in large dynamic groups using one-way function trees. Available: www.csee.umbc.edu/sherman/

[13] Naor, D. and M. Naor (2003). "Protecting cryptographic keys: The trace and revoke approach." *The Computer Journal*, 36 (7), 47–53.

[14] Naor, D., M. Naor and J. Lotspiech (2001). "Revocation and tracing schemes for stateless receivers." *Advances in Cryptology—CRYPTO 2001*, Lecture Notes in Computer Science, vol. 2139, ed. J. Killian. Springer, Berlin, 41–62. Full version: ECCC Report 43, 2002. Available: http://www.eccc.uni-trier.de/eccc/

[15] Naor, M. and B. Pinkas (2001). "Efficient trace and revoke schemes." *Financial Cryptography, 4th International Conference, FC 2000 Proceedings*, Lecture Notes in Computer Science, vol. 1962, ed. Y. Frankel. Springer, Berlin, 1–20.

[16] Perrig, A., D. Song, and J.D. Tygar (2001). "ELK, a new protocol for efficient large-group key distribution." *2001 IEEE Symposium on Research in Security and Privacy*, 247–262.

[17] Waldvogel, M., G. Caronni, D. Sun, N. Weiler, and B. Plattner (1999). "The VersaKey framework: Versatile group key management." *IEEE Journal on Selected Areas in Communications*, 17 (9), 1614–1631.

[18] Wallner, D.M., E.J. Harder, and R.C. Agee (1999). "Key management for multicast: Issues and architectures." Internet Request for Comments 2627. Available: ftp.ietf.org/rfc/rfc2627.txt

[19] Wong, C.K., M. Gouda, and S. Lam (1998). "Secure group communications using key graphs." *Proc. ACM SIGCOMM'98*, 68–79.

[20] Wong, C.K. and S. Lam (2000). "Keystone: A group key management service." *International Conference on Telecommunications, Acapulco, Mexico, May 2000*.

C

CÆSAR CIPHER

Julius Cæsar is reported to have replaced each letter in the plaintext by the one standing three places further in the alphabet. For instance, when the key has the value 3, the plaintext word cleopatra will be encrypted by the ciphertext word fohrsdwud. Augustus allegedly found this too difficult and always took the next letter. Breaking the Cæsar cipher is almost trivial: there are only 26 possible keys to check (exhaustive key search) and after the first four or five letters are decrypted the solution is usually unique.

The Cæsar cipher is one of the most simple cryptosystems, with a monoalphabetic encryption: by counting down in the cyclically closed ordering of an alphabet, a specified number of steps.

Cæsar encryptions are special linear substitution (see substitutions and permutations) with $n = 1$ and the identity as homogeneous part φ. Interesting linear substitutions with $n \geq 2$ have been patented by Lester S. Hill in 1932.

Friedrich L. Bauer

Reference

[1] Bauer, F.L. (1997). "Decrypted secrets." *Methods and Maxims of Cryptology*. Springer-Verlag, Berlin.

CAMELLIA

Camellia [1] is a block cipher designed in 2000 by a team of cryptographers from NTT and Mitsubishi Electric Corporation. It was submitted to different standardization bodies and was included in the NESSIE Portfolio of recommended cryptographic primitives in 2003.

Camellia encrypts data in blocks of 128 bits and accepts 128-bit, 192-bit, and 256-bit secret keys. The algorithm is a byte-oriented Feistel cipher and has 18 or 24 rounds depending on the key length. The F-function used in the Feistel structure can be seen as a 1-round 8-byte substitution-permutation (SP) network. The substitution layer consists of eight 8×8-bit S-boxes applied in parallel, chosen from a set of four different affine equivalent transformations of the inversion function in $GF(2^8)$ (see Rijndael/AES). The permutation layer, called the P-function, is a network of byte-wise exclusive ORs and is designed to have a branch number of 5 (which is the maximum for such a network). An additional particularity of Camellia, which it shares with MISTY1 and KASUMI (see KASUMI/MISTY1), is the FL-layers. These layers of key-dependent linear transformations are inserted between every six rounds of the Feistel network, and thus break the regular round structure of the cipher.

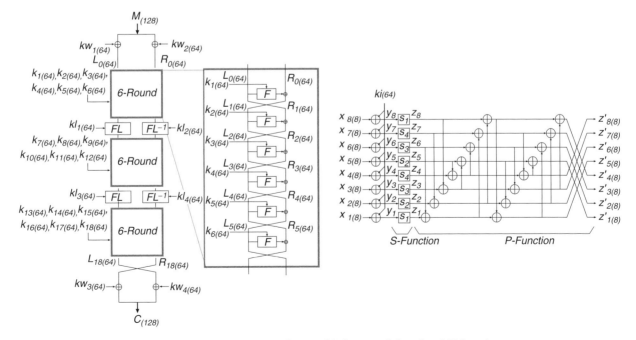

Fig. 1. Camellia: encryption for 128-bit keys and details of F-function

In order to generate the subkeys used in the *F*-functions, the secret key is first expanded to a 256-bit or 512-bit value by applying four or six rounds of the Feistel network. The key schedule (see block cipher) then constructs the necessary subkeys by extracting different pieces from this bit string.

The best attacks on reduced-round Camellia published so far are *square* and *rectangle* attacks (see integral attack and boomerang attack). The nine-round square attack presented by Yeom et al. [4] requires 2^{61} chosen plaintexts and an amount of work equivalent to 2^{202} encryptions. The rectangle attack proposed by Shirai [3] breaks ten rounds with 2^{127} chosen plaintexts and requires 2^{241} memory accesses. Hatano, Sekine, and Kaneko [2] also analyze an 11-round variant of Camellia using higher order differentials. The attack would require 2^{93} chosen ciphertexts, but is not likely to be much faster than an exhaustive search for the key, even for 256-bit keys.

Note that more rounds can be broken if the *FL*-layers are discarded. A linear attack on a 12-round variant of Camellia without *FL*-layers is presented in [3]. The attack requires 2^{119} known plaintexts and recovers the key after performing a computation equivalent to 2^{247} encryptions.

Christophe De Cannière

References

[1] Aoki, K., T. Ichikawa, M. Kanda, M. Matsui, S. Moriai, J. Nakajima, and T. Tokita (2001). "Camellia: A 128-bit block cipher suitable for multiple platforms—design and analysis." *Selected Areas in Cryptography, SAC 2000*, Lecture Notes in Computer Science, vol. 2012, eds. D.R. Stinson and S.E. Tavares. Springer-Verlag, Berlin, 39–56.

[2] Hatano, Y., H. Sekine, and T. Kaneko (2002). "Higher order differential attack of Camellia (II)." *Selected Areas in Cryptography, SAC 2002*. Lecture Notes in Computer Science, eds. H. Heys and K. Nyberg. Springer-Verlag, Berlin, 39–56.

[3] Shirai, T. (2002). "Differential, linear, boomerang and rectangle cryptanalysis of reduced-round Camellia." *Proceedings of the Third NESSIE Workshop*, NESSIE, November 2002.

[4] Yeom, Y., S. Park, and I. Kim (2002). "On the security of Camellia against the square attack." *Fast Software Encryption, FSE 2002*, Lecture Notes in Computer Science, vol. 2365, eds. J. Daemen and V. Rijmen. Springer-Verlag, Berlin, 89–99.

CAST

CAST is a design procedure for symmetric cryptosystems developed by C. Adams and S. Tavares

in 1993 [1, 2]. In accordance with this procedure, a series of DES-like block ciphers was produced (see Data Encryption Standard (DES)), the most widespread being the 64-bit block cipher CAST-128. The latest member of the family, the 128-bit block cipher CAST-256, was designed in 1998 and submitted as a candidate for the Advanced Encryption Standard (see Rijndael/AES).

All CAST algorithms are based on a Feistel cipher (a generalized Feistel network in the case of CAST-256). A distinguishing feature of the CAST ciphers is the particular construction of the *f*-function used in each Feistel round. The general structure of this function is depicted in Figure 1. The data entering the *f*-function is first combined with a subkey and then split into a number of pieces. Each piece is fed into a separate expanding S-box based on bent functions (see nonlinearity of Boolean functions). Finally, the output words of these S-boxes are recombined one by one to form the final output. Both CAST-128 and CAST-256 use three different 32-bit *f*-functions based on this construction. All three use the same four 8×32-bit S-boxes but differ in the operations used to combine the data or key words (the operations a, b, c, and d in Figure 1). The CAST ciphers are designed to support different key sizes and have a variable number of rounds. CAST-128 allows key sizes between 40 and 128 bits and uses

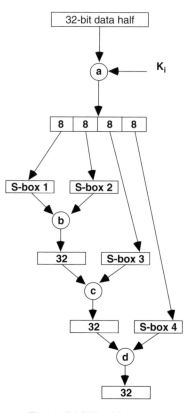

Fig. 1. CAST's *f*-function

12 or 16 rounds. CAST-256 has 48 rounds and supports key sizes up to 256 bits.

The first CAST ciphers were found to have some weaknesses. Rijmen et al. [5] presented attacks exploiting the nonsurjectivity of the *f*-function in combination with an undesirable property of the key schedule. Kelsey et al. [3] demonstrated that the early CAST ciphers were vulnerable to <u>related key attacks</u>. Moriai et al. [4] analyzed simplified versions of CAST-128 and presented a five-round attack using higher order differentials.

Christophe De Cannière

References

[1] Adams C.M. (1997). "Constructing symmetric ciphers using the CAST design procedure." *Designs, Codes and Cryptography*, 12 (3), 283–316.

[2] Adams, C.M. and Tavares S.E. (1993). "Designing S-boxes for ciphers resistant to differential cryptanalysis." *Proceedings of the 3rd Symposium on State and Progress of Research in Cryptography*, ed. W. Wolfowicz. Fondazione Ugo Bordoni, 181–190.

[3] Kelsey, J., B. Schneier, and D. Wagner (1997). "Related-key cryptanalysis of 3-WAY, Biham-DES, CAST, DES-X, NewDES, RC2, and TEA." *International Conference on Information and Communications Security, ICICS'97*, Lecture Notes in Computer Science, vol. 1334, eds. Y. Han, T. Okamoto, and S. Qing. Springer-Verlag, Berlin, 233–246.

[4] Moriai, S., T. Shimoyama, and T. Kaneko (1998). "Higher order differential attack of CAST cipher." *Fast Software Encryption, FSE'98*, Lecture Notes in Computer Science, vol. 1372, ed. S. Vaudenay. Springer-Verlag, Berlin, 17–31.

[5] Rijmen, V., B. Preneel, and E. De Win (1997). "On weaknesses of non-surjective round functions." *Designs, Codes, and Cryptography*, 12 (3), 253–266.

CBC-MAC AND VARIANTS

SIMPLE CBC-MAC: CBC-MAC is one of the oldest and most popular <u>MAC algorithms</u>. A MAC algorithm is a cryptographic algorithm that computes a complex function of a plaintext and a secret <u>key</u>; the resulting MAC value is typically appended to the plaintext to protect its authenticity. CBC-MAC is a MAC algorithm based on a <u>block cipher</u>; it is derived from the Cipher Block Chaining (CBC) <u>mode of operation</u>, which is a mode for encryption. CBC-MAC is very popular in financial applications and smart cards.

In the following, the block length and key length of the block cipher will be denoted by n and k respectively. The length (in bits) of the MAC value will be denoted by m. The encryption and decryp-

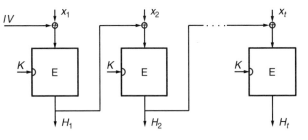

Fig. 1. CBC-MAC, where the MAC value is $g(H_t)$

tion with the block cipher E using the key K will be denoted by $E_K(\cdot)$ and $D_K(\cdot)$, respectively. An n-bit string consisting of zeroes will be denoted by 0^n.

CBC-MAC is an iterated MAC algorithm, which consists of the following steps (see also Figure 1):

- *Padding and splitting of the input.* The goal of this step is to divide the input into t blocks of length n; before this can be done, a padding algorithm needs to be applied. The most common padding method can be described as follows [12]. Let the message string before padding be $x = x_1, x_2, \ldots, x_{t'}$, with $|x_1| = |x_2| = \cdots = |x_{t'-1}| = t$ (here $|x_i|$ denotes the size of the string x_i in bits). If $|x_{t'}| = n$ append an extra block $x_{t'+1}$ consisting of one one-bit followed by $n - 1$ zero bits, such that $|x_{t'+1}| = n$ and set $t = t' + 1$; otherwise append a one-bit and $n - |x_{t'}| - 1$ zero bits, s.t. $|x_{t'}| = n$ and set $t' = t$. A simpler padding algorithm (also included in [12]) consists of appending $n - |x_{t'}|$ zero bits and setting $t' = t$. This padding method is not recommended, as it allows for trivial forgeries.

- *CBC-MAC computation*, which iterates the following operation:

$$H_i = E_K(H_{i-1} \oplus x_i), \qquad 1 \le i \le t.$$

The initial value is equal to the all zero string, or $H_0 = 0^n$ (note that for the *CBC* encryption mode, a random value H_0 is recommended).

- *Output transformation.* The MAC value is computed as $\mathrm{MAC}_K(x) = g(H_t)$, where g is the output transformation.

The simplest construction is obtained when the output transformation $g()$ is the identity function. Bellare et al. [4] have provided a security proof for this scheme. Their proof is based on the pseudorandomness of the block cipher and requires that the inputs are of fixed length. It shows a lower bound for the number of chosen texts that are required to distinguish the MAC algorithm from a random function, which demonstrates that CBC-MAC is a pseudorandom function. Note that this is a stronger requirement than being a secure MAC as this requires just unpredictability or computation resistance. An almost matching upper bound to this attack is provided by an internal <u>collision</u>

attack based on the birthday paradox (along the lines of Proposition 1 of MAC algorithms [6, 24]). The attack obtains a MAC forgery; it requires a single chosen text and about $2^{n/2}$ known texts; for a 64-bit block cipher such as Data Encryption Standard (DES), this corresponds to 2^{32} known texts.

If the input is not of fixed length, very simple forgery attacks apply to this scheme:

- given $\text{MAC}(x)$, one knows that $\text{MAC}_K(x\|(x \oplus \text{MAC}_K(x))) = \text{MAC}_K(x)$ (for a single block x);
- given $\text{MAC}(x)$ and $\text{MAC}(x')$, one knows that $\text{MAC}_K(x\|(x' \oplus \text{MAC}_K(x))) = \text{MAC}_K(x')$ (for a single block x').
- given $\text{MAC}(x)$, $\text{MAC}(x\|y)$, and $\text{MAC}(x')$, one knows that $\text{MAC}(x'\|y') = \text{MAC}(x\|y)$ if $y' = y \oplus \text{MAC}(x) \oplus \text{MAC}(x')$, where y and y' are single blocks.

A common way to preclude these simple forgery attacks is to replace the output transform g by a truncation to $m < n$ bits; $m = 32$ is a very popular choice for CBC-MAC based on DES ($n = 64$). However, Knudsen has shown that a forgery attack on this scheme requires $2 \cdot 2^{(n-m)/2}$ chosen texts and two known texts [16], which is only 2^{17} chosen texts for $n = 64$ and $m = 32$. Note that this is substantially better than an internal collision attack. The proof of security for fixed length inputs still applies however.

In order to describe attacks parameters in a compact way, an attack is quantified by the four-tuple $[a, b, c, d]$, where

- a is the number off-line block cipher encipherments,
- b is the number of known text-MAC pairs,
- c is the number of chosen text-MAC pairs, and
- d is the number of on-line MAC verifications.

Attacks are often probabilistic; in that case the parameters indicated result in a large success probability (typically at least 0.5). As an example, the complexity of exhaustive key search is $[2^k, \lceil k/m \rceil, 0, 0]$ and for a MAC guessing attack it is $[0, 0, 0, 2^m]$. The forgery attacks based on an internal collision for CBC-MAC as described above have attack parameters $[0, 2^{n/2}, 1, 0]$ if g is the identity function and $[0, 2, 2 \cdot 2^{(n-m)/2}, 0]$ if g is a truncation to m bits.

VARIANTS OF CBC-MAC: As a first comment, it should be pointed out that for most of these schemes, a forgery attack based on internal collisions applies with complexity $[0, 2^{n/2}, 1, 0]$ for $m = n$ and $[0, 2^{n/2}, \min(2^{n/2}, 2^{n-m}), 0]$ for $m < n$ (Propositions 1 and 2 in MAC algorithms).

The EMAC scheme uses as output transformation g the encryption of the last block with a different key. It was first proposed by the RIPE

Consortium in [27]; Petrank and Rackoff have provided a security proof in [23], which shows that this MAC algorithm is secure with inputs of arbitrary lengths:

$$g(H_t) = E_{K'}(H_t) = E_{K'}(E_K(x_t \oplus H_{t-1})),$$

where K' is a key derived from K.

A further optimization by Black and Rogaway [5] reduces the overhead due to padding; it is known as XCBC (or three-key MAC). XCBC uses a k-bit block cipher key K_1 and two n-bit whitening keys K_2 and K_3. It modifies the last encryption and padding such that the number of blocks before and after padding is equal or $t = t'$. If $|x_{t'}| = n$, then XOR the n-bit key K_2 to $x_{t'}$; otherwise append a one-bit and $n - |x_{t'}| - 1$ zero bits, s.t. $|x_{t'}| = n$, and then XOR the n-bit key K_3 to $x_{t'}$. The OMAC algorithm by Iwata and Kurosawa [13] reduces the number of keys to one by choosing $K_2 = `2' \cdot E_K(0^n)$ and $K_3 = `4' \cdot E_K(0^n)$ where '2' and '4' are two elements of the finite field $\text{GF}(2^n)$ (see [13] for the details of this representation) and "." represents multiplication in the finite field $\text{GF}(2^n)$. It is anticipated that NIST will standardize this algorithm under the name CMAC for use with AES.

RIPE-MAC [27] is a variant of EMAC with the following iteration:

$$H_i = E_K(H_{i-1} \oplus x_i) \oplus x_i, \qquad 1 \le i \le t.$$

This increases the complexity to find collisions even if one knows the key.

Because of the 56-bit key length, CBC-MAC with DES no longer offers adequate security. Several constructions exist to increase the key length of the MAC algorithm. No lower bounds on the security of these schemes against key recovery are known.

A widely used solution is the ANSI retail MAC, which first appeared in [3]. Rather than replacing DES by triple-DES, one processes only the last block with two-key triple-DES, which corresponds to an output transformation g consisting of a double-DES encryption:

$$g(H_t) = E_{K_1}(D_{K_2}(H_t)).$$

When used with DES, the key length of this MAC algorithm is 112 bits. However, Preneel and van Oorschot have shown that $2^{n/2}$ known texts allow for a key recovery in only $3 \cdot 2^k$ encryptions, compared to 2^{2k} encryptions for exhaustive key search [25] (note that for DES $n = 64$ and $k = 56$). If $m < n$, this attack requires an additional 2^{n-m} chosen texts. The complexity of these attacks is thus $[2^{k+1}, 2^{n/2}, 0, 0]$ and $[2^{k+1}, 2^{n/2}, 2^{n-m}, 0]$. Several key recovery attacks require mostly MAC verifications, with the following parameters:

$[2^k, 1, 0, 2^k]$ [20], $[2^{k+1}, \lceil (\max(k, n) + 1)/m \rceil, 0, \lceil (k - n - m + 1)/m \rceil \cdot 2^n)]$ [18] and for $m < n$: $[2^{k+1}, 0, 0, (\lceil n/m \rceil + 1) \cdot 2^{(n+m)/2-1}]$ [21].

The security of the ANSI retail MAC can be improved at no cost in performance by introducing a double DES encryption in the first and last iteration; this scheme is know as MacDES [20]:

$$H_1 = E_{K_2'}(E_{K_1}(X_1)) \quad \text{and} \quad g(H_t) = E_{K_2}(H_t).$$

Here K_2' is derived from K_2. The best known key recovery attack due to Coppersmith et al. [8] has complexity $[2^{k+3}, 2^{n/2+1}, 3s \cdot 2^{3n/4}, 0]$, for small $s \geq 1$; with truncation of the output to $m = n/2$ bits, this complexity increases to $[2^{k+s} + 2^{k+2p}, 0, 2^{n+3-p}, 2^{k+1}]$ with space complexity 2^{k-2s}. These attacks assume that a serial number is included in the first block; if this precaution is not taken, key recovery attacks have complexities similar to the ANSI retail MAC: $[2^{k+2}, 2^{n/2}, 2, 0]$ and $[2^{k+2}, 1, 1, 2^k]$ [9].

Several attempts have been made to increase the resistance against <u>forgery</u> attacks based on internal collisions. A first observation is that the use of serial numbers is not sufficient [7].

RMAC, proposed by Jaulmes et al. [14], introduces in the output transformation a derived key K' that is modified with a randomizer or "salt" R (which needs to be stored or sent with the MAC value):

$$g(H_t) = E_{K' \oplus R}(H_t).$$

The RMAC constructions offer increased resistance against forgery attacks based on internal collisions, but it has the disadvantage that its security proof requires resistance of the underlying block cipher against <u>related key attacks</u>. A security analysis of this scheme can be found in [17,19]. The best known attack strategy against RMAC is to recover K': once K' is known, the security of RMAC reduces to that of simple CBC-MAC. For $m = n$, the complexities are $[2^{k-s} + 2^{k-n}, 1, 2^s, 2^{n-1}]$ or $[2^{k-s} + 2^{k-n}, 1, 0, 2^{s+n-1} + 2^{n-1}]$, while for $m < n$ the complexities are $[2^{k-s} + 2^{k-m}, 0, 2^s, \lceil n/m + 1 \rceil \cdot 2^{(n+m)/2}]$ and $[2^{k-s} + 2^{k-m}, 0, 0, \lceil n/m + 1 \rceil \cdot 2^{(n+m)/2} + 2^{s+m-1}]$. A variant which exploits multiple collisions has complexity $[2^{k-1}/(u/t), 0, (t/e)2^{(t-1)nt}, 0]$ with $u = t + t(t - 1)/2$ (for $m = n$). These attacks show that the security level of RMAC is smaller than anticipated. However, when RMAC is used with three-key triple-DES, the simple key off-setting technique is insecure; [19] shows a full key recovery attack with complexity $[2^{64}, 2^8, 2^8, 2^{56}]$, which is much lower than anticipated. RMAC was included in NIST's 2002 draft special publication [22]; however, this draft has been withdrawn.

3GPP-MAC [1] uses a larger internal memory of $2n$ bits, as it also stores the sum of the intermediate values of the MAC computation. The MAC value is computed as follows:

$$\text{MAC} = g(E_{K_2}(H_1 \oplus H_2 \oplus \cdots H_t)).$$

Knudsen and Mitchell analyze this scheme in [18]. If g is the identity function, the extra computation and storage does not pay off: there exist forgery attacks that require only $2^{n/2}$ known texts, and the complexity of key recovery attacks is similar to that of the ANSI retail MAC. However, truncating the output increases the complexity of these attacks to an adequate level. For the 3GPP application, the 64-bit block cipher <u>KASUMI</u> is used with a 128-bit key and with $m = 32$. The best known forgery attack requires 2^{48} texts and the best known key recovery attacks have complexities $[2^{130}, 2^{48}, 2^{32}, 0]$ and $[2^{129}, 3, 0, 2^{64}]$.

STANDARDIZATION: CBC-MAC is standardized by several standardization bodies. The first standards included only simple CBC-MAC [2, 10]. In 1986, the ANSI retail MAC was added [3, 11]. The 1999 edition of ISO 9797-1 [12] includes simple CBC-MAC, EMAC, the ANSI retail MAC and MacDES (and two other schemes which are no longer recommended because of the attacks in [15]). 3GPP-MAC has been standardized by 3GPP [1]. It is anticipated that NIST will standardize OMAC.

B. Preneel

References

[1] 3GPP (2002). "*Specification of the 3GPP* confidentiality and integrity algorithms. Document 1: f8 and f9 specification." TS 35.201.

[2] ANSI X9.9 (revised) (1986). "Financial institution message authentication (wholesale)." American Bankers Association.

[3] ANSI X9.19 (1986). "Financial institution retail message authentication." American Bankers Association.

[4] Bellare, M., J. Kilian, and P. Rogaway (2000). "The security of cipher block chaining." *Journal of Computer and System Sciences*, 61 (3), 362–399. Earlier version in *Advances in Cryptology—CRYPTO'94*, Lecture Notes in Computer Science, vol. 839, ed. Y. Desmedt. Springer-Verlag, Berlin, 341–358.

[5] Black, J. and P. Rogaway (2000). "CBC-MACs for arbitrary length messages." *Advances in Cryptology—CRYPTO 2000*, Lecture Notes in Computer Science, vol. 1880, ed. M. Bellare. Springer-Verlag, Berlin, 197–215.

[6] Black, J. and P. Rogaway (2002). "A block-cipher mode of operation for parallelizable message

authentication." *Advances in Cryptology—EUROCRYPT 2002*, Lecture Notes in Computer Science, vol. 2332, ed. L. Knudsen. Springer-Verlag, Berlin, 384–397.

[7] Brincat, K. and C.J. Mitchell (2001). "New CBC-MAC forgery attacks." *Information Security and Privacy, ACISP 2001,* Lecture Notes in Computer Science, vol. 2119, eds. V. Varadharajan and Y. Mu. Springer-Verlag, Berlin, 3–14.

[8] Coppersmith, D., L.R. Knudsen, and C.J. Mitchell (2000). "Key recovery and forgery attacks on the MacDES MAC algorithm." *Advances in Cryptology—CRYPTO 2000*, Lecture Notes in Computer Science, vol. 1880, ed. M. Bellare. Springer-Verlag, Berlin, 184–196.

[9] Coppersmith, D. and C.J. Mitchell (1999). "Attacks on MacDES MAC algorithm." *Electronics Letters*, 35 (19), 1626–1627.

[10] FIPS 113 (1985). "Computer data authentication." NIST, US Department of Commerce, Washington, DC.

[11] ISO 8731:1987 (1987). "Banking—approved algorithms for message authentication." Part 1, DEA, Part 2, Message Authentication Algorithm (MAA).

[12] ISO/IEC 9797:1999 (1999). "Information technology—security techniques—message authentication codes (MACs)." Part 1: Mechanisms Using a Block Cipher.

[13] Iwata, T. and K. Kurosawa (2003). "OMAC: One key CBC MAC." *Fast Software Encryption*, Lecture Notes in Computer Science, vol. 2887, ed. T. Johansson. Springer-Verlag, Berlin, 129–153.

[14] Jaulmes, E., A. Joux, and F. Valette (2002). "On the security of randomized CBC-MAC beyond the birthday paradox limit: A new construction." *Fast Software Encryption*, Lecture Notes in Computer Science, vol. 2365, eds. J. Daemen and V. Rijmen. Springer-Verlag, Berlin, 237–251.

[15] Joux, A., G. Poupard, and J. Stern (2003). "New attacks against standardized MACs." *Fast Software Encryption*, Lecture Notes in Computer Science, vol. 2887, ed. T. Johansson. Springer-Verlag, Berlin, 170–181.

[16] Knudsen, L. (1997). "Chosen-text attack on CBC-MAC." *Electronics Letters*, 33 (1), 48–49.

[17] Knudsen, L. and T. Kohno (2003). "Analysis of RMAC." *Fast Software Encryption*, Lecture Notes in Computer Science, vol. 2887, ed. T. Johansson. Springer-Verlag, Berlin, 182–191.

[18] Knudsen, L.R. and C.J. Mitchell (2003). "Analysis of 3GPP-MAC and two-key 3GPP-MAC." *Discrete Applied Mathematics*, 128 (1), 181–191.

[19] Knudsen, L.R. and C.J. Mitchell (2003). "Partial key recovery attack against RMAC." Preprint.

[20] Knudsen, L. and B. Preneel (1998). "MacDES: MAC algorithm based on DES." *Electronics Letters*, 34 (9), 871–873.

[21] Mitchell, C.J. (2003). "Key recovery attack on ANSI retail MAC." *Electronics Letters*, 39, 361–362.

[22] NIST Special Publication 800-38B (2002). Draft Recommendation for Block Cipher Modes of Operation: The RMAC Authentication Mode.

[23] Petrank, E. and C. Rackoff (2000). "CBC MAC for real-time data sources." *Journal of Cryptology*, 13 (3), 315–338.

[24] Preneel, B. and P.C. van Oorschot (1995). "MDx-MAC and building fast MACs from hash functions." *Advances in Cryptology—CRYPTO'95*, Lecture Notes in Computer Science, vol. 963, ed. D. Coppersmith. Springer-Verlag, Berlin, 1–14.

[25] Preneel, B. and P.C. van Oorschot (1996). "A key recovery attack on the ANSI X9.19 retail MAC." *Electronics Letters*, 32 (17), 1568–1569.

[26] Preneel, B. and P.C. van Oorschot (1999). "On the security of iterated message authentication codes." *IEEE Trans. on Information Theory*, IT-45 (1), 188–199.

[27] RIPE (1995). "Integrity primitives for secure information systems." Final report of RACE integrity primitives evaluation (RIPE-RACE 1040). Lecture Notes in Computer Science, vol. 1007, eds. A. Bosselaers and B. Preneel. Springer-Verlag, Berlin.

CCIT2-CODE

This is a binary coding of the International Teletype Alphabet No. 2. The six control characters of the teletype machines are: 0: Void, 1: Letter Shift, 2: Word Space, 3: Figure Shift, 4: Carriage Return, 5: Line Feed.

```
0 t 4 o 2 h n m 5 l r g i p c v e z d b s y f x a w j 3 u q k 1
0 0 0 0 0 0 0 0 0 0 0 0 0 0 0 0 0 1 1 1 1 1 1 1 1 1 1 1 1 1 1 1 1  16
0 0 0 0 0 0 0 0 0 1 1 1 1 1 1 1 0 0 0 0 0 0 0 0 1 1 1 1 1 1 1 1  8
. . . . . . . . . . . . . . . . . . . . . . . . . . . . . . . .
0 0 0 0 1 1 1 1 0 0 0 0 1 1 1 1 0 0 0 0 1 1 1 1 0 0 0 0 1 1 1 1  4
0 0 1 1 0 0 1 1 0 0 1 1 0 0 1 1 0 0 1 1 0 0 1 1 0 0 1 1 0 0 1 1  2
0 1 0 1 0 1 0 1 0 1 0 1 0 1 0 1 0 1 0 1 0 1 0 1 0 1 0 1 0 1 0 1  1
0 1 2 3 4 5 6 7 8 9 10 11 12 13 14 15 16 17 18 19 20 21 22 23 24 25 26 27 28 29 30 31
```

Friedrich L. Bauer

Reference

[1] Bauer, F.L. (1997). "Decrypted secrets." *Methods and Maxims of Cryptology*. Springer-Verlag, Berlin.

CEPS STANDARD

The Common Electronic Purse Specifications (CEPS) define an electronic purse program built on the EMV specification. The CEPS scheme extends the EMV authentication architecture with a certification authority (CA) and issuer certificates to include the *Acquirer* side. The Acquirer is responsible for managing Point Of Sale (POS) transactions using a Purchase Secure Application Module (PSAM). The terminal (PSAM) authenticates itself to the smart card and does so using a method similar to the approach of EMV. On the card is stored an issuer-side CA index, an issuer certificate, and a card certificate which is transmitted to the terminal. Using a stored issuer CA certificate, the terminal verifies the issuer certificate and the card certificate. The terminal responds by generating a digital signature—using a terminal private key—which is encrypted by the card public key and transmitted to the card together with a corresponding terminal certificate on the public key, an acquirer certificate, and an acquirer-side CA index. As the card also has an acquirer-side CA certificate stored, the terminal can be properly authenticated.

In addition to the physical cards themselves, the majority of the cryptographic mechanisms and protocols comply with EMV (Personal Identification Number (PIN), MAC algorithms, card communication commands, public key cryptography, etc.). A few areas such as certain usages of symmetric cryptography are however at the (relative) discretion of the card issuer.

Peter Landrock

CERTIFICATE

A *certificate* is a data structure signed by an entity that is considered (by some other collection of entities) to be authoritative for its contents. The signature on the data structure binds the contained information together in such a way that this information cannot be altered without detection. Entities that retrieve and use certificates (often called "relying parties") can choose to rely upon the contained information because they can determine whether the signing authority is a source they trust and because they can ensure that the information has not been modified since it was certified by that authority.

The information contained in a certificate depends upon the purpose for which that certificate was created. The primary types of certificates are public-key certificates (see public-key infrastructure) and attribute certificate, although in principle an authority may certify any kind of information [1–3, 5]. Public-key certificates typically bind a public key pair[1] to some representation of an identity for an entity, although other relevant information may also be bound to these two pieces of data such as a validity period, an identifier for the algorithm for which the public key may be used, and any policies or constraints on the use of this certificate. Attribute certificates typically do not contain a public key, but bind other information (such as roles, rights, or privileges) to some representation of an identity for an entity. Public-key certificates are used in protocols or message exchanges involving authentication of the participating entities, whereas attribute certificates are used in protocols or message exchanges involving authorization decisions (see authorization architecture) regarding the participating entities.

Many formats and syntaxes have been defined for both public-key certificates and attribute certificates, including X.509 [4], SPKI [7] (see security standards activities), PGP [8], and SAML [6] (see privacy and also key management for a high-level overview of the X.509 certificate format). Management protocols have also been specified for the creation, use, and revocation of (typically X.509-based) public-key certificates.

Carlisle Adams

---

[1] The identity is bound explicitly to the public key, but implicitly to the private key as well. That is, only the public key is actually included in the certificate, but the underlying assumption is that the identified entity is the (sole) holder of the corresponding private key; otherwise, relying parties would have no reason to use the certificate to encrypt data for, or verify signatures from, that entity.

References

[1] Adams, C. and S. Farrell (1999). "Internet X.509 public key infrastructure: Certificate management protocols." Internet Request for Comments 2510.

[2] Adams, C. and S. Lloyd (2003). *Understanding PKI: Concepts, Standards, and Deployment Considerations* (2nd ed.). Addison-Wesley, Reading, MA.

[3] Housley, R. and T. Polk (2001). *Planning for PKI: Best Practices Guide for Deploying Public Key Infrastructure*. John Wiley & Sons, New York.

[4] ITU-T Recommendation X.509 (2000). "Information technology—open systems interconnection—the directory: public key and attribute certificate frameworks." (equivalent to ISO/IEC 9594-8: 2001).

[5] Myers, M., X. Liu, J. Schaad, and J. Weinstein (2000). "Certificate management messages over CMS." Internet Request for Comments 2797.

[6] OASIS Security Services Technical Committee (2002). "Security Assertion Markup Language (SAML) Version 1.0"; see http://www.oasis-open.org/committees/security/ for details.

[7] Simple Public Key Infrastructure (SPKI) Working Group charter (2003); see http://www.ietf.org/html.charters/spki-charter.html for details.

[8] Zimmermann, P. (1995). *The Official PGP User's Guide*. MIT Press, Cambridge, MA.

CERTIFICATE MANAGEMENT

Certificate management is the management of public-key underline{certificates}, covering the complete life cycle from the initialization phase, to the issued phase, to the cancellation phase. See key management for details.

Carlisle Adams

CERTIFICATE OF PRIMALITY

A *certificate of primality* (or *prime certificate*) is a small set of values associated with an integer that can be used to efficiently prove that the integer is a prime number. Certain primality proving algorithms, such as Elliptic Curves for Primality Proving, generate such a certificate. A certificate of primality can be independently verified by software other than the one that generated the certificate. This is useful in detecting any possible bugs in the implementation of a primality proving algorithm.

Anton Stiglic

CERTIFICATE REVOCATION

A certificate (see certificate and certification authority) is a binding between a name of an entity and that entity's public key (see public key cryptography). Normally, this binding is valid for the full lifetime of the issued certificate. However, circumstances may arise in which an issued certificate should no longer be considered valid, even though the certificate has not yet expired. In such cases, the certificate may need to be *revoked*. Reasons for revocation vary, but they may involve anything from a change in job status to a suspected private-key compromise. Therefore, an efficient and reliable method must be provided to revoke a public-key certificate before it might naturally expire.

Certificates must pass a well-established validation process before they can be used. Part of that validation process includes making sure that the certificate under evaluation has not been revoked. Certification Authorities (CAs) are responsible for making revocation information available in some form or another. Relying parties (users of a certificate for some express purpose) must have a mechanism to either retrieve the revocation information directly, or rely upon a trusted third party to resolve the question on their behalf.

Certificate revocation can be accomplished in a number of ways. One class of methods is to use periodic publication mechanisms; another class is to use online query mechanisms to a trusted authority. A number of examples of each class will be given in the sections below.

PERIODIC PUBLICATION MECHANISMS: A variety of periodic publication mechanisms exist. These are "prepublication" techniques, characterized by issuing the revocation information on a periodic basis in the form of a signed data structure. Most of these techniques are based on a data structure referred to as a Certificate Revocation List (CRL), defined in the ISO/ITU-T X.509 International Standard. These techniques include CRLs themselves, Certification Authority Revocation Lists (CARLs), End-entity Public-key certificate Revocation Lists (EPRLs), CRL Distribution Points (CDPs), Indirect CRLs, Delta CRLs and Indirect Delta CRLs, Redirect CRLs, and Certificate Revocation Trees (CRTs).

CRLs are signed data structures that contain a list of revoked certificates; the digital signature appended to the CRL provides the integrity and authenticity of the contained data. The signer of the CRL is typically the same entity that signed the issued certificates that are revoked by the CRL, but the CRL may instead be signed by an entity other than the certificate issuer.

Version 2 of the CRL data structure defined by ISO/ITU-T (the X.509v2 CRL) contains a powerful extension mechanism that allows additional information to be defined and placed in the CRL within the scope of the digital signature. Lacking this, the version 1 CRL has scalability concerns and functionality limitations in many environments. Some of the extensions that have been defined and standardized for the version 2 CRL enable great flexibility in the way certificate revocation is performed, making possible such techniques as CRL Distribution Points, Indirect CRLs, Delta CRLs, and some of the other methods listed above.

The CRL data structure contains a version number (almost universally version 2 in current practice), an identifier for the algorithm used to sign the structure, the name of the CRL issuer, a pair of fields indicating the validity period of the CRL ("this update" and "next update"), the list of revoked certificates, any included extensions, and the signature over all the contents just mentioned. At a minimum, CRL processing engines are to assume that certificates on the list have been revoked, even if some extensions are not understood, and take appropriate action (typically, not rely upon the use of such certificates in protocols or other transactions).

Extensions in the CRL may be used to modify the CRL scope or revocation semantic in some way. In particular, the following techniques have been defined in X.509.

- An Issuing Distribution Point extension and/or a CRL Scope extension may be used to limit the CRL to holding only CA certificates (creating a CARL) or only end-entity certificates (creating an EPRL).
- A CRL Distribution Point (CDP) extension partitions a CRL into separate pieces that together cover the entire scope of a single complete CRL. These partitions may be based upon size (so that CRLs do not get too large), upon revocation reason (this segment is for certificates that were revoked due to key compromise; that segment is for revocation due to privilege withdrawn; and so on), or upon a number of other criteria.
- The Indirect CRL component of the Issuing Distribution Point extension can identify a CRL as an Indirect CRL, which enables one CRL to contain revocation information normally supplied from multiple CAs in separate CRLs. This can reduce the number of overall CRLs that need to be retrieved by relying parties when performing the certificate validation process.
- The Delta CRL Indicator extension, or the Base Revocation Information component in the CRL Scope extension, can identify a CRL as a Delta CRL, which allows it to contain only incremental revocation information relative to some base CRL, or relative to a particular point in time. Thus, this (typically much smaller) CRL must be used in combination with some other CRL (which may have been previously cached) in order to convey the complete revocation information for a set of certificates. Delta CRLs allow more timely information with lower bandwidth costs than complete CRLs. Delta CRLs may also be Indirect, through the use of the extension specified above.
- The CRL Scope and Status Referral extensions may be used to create a Redirect CRL, which allows the flexibility of dynamic partitioning of a CRL (in contrast with the static partitioning offered by the CRL Distribution Point extension).

Finally, a Certificate Revocation Tree is a revocation technology designed to represent revocation information in a very efficient manner (using significantly fewer bits than a traditional CRL). It is based on the concept of a Merkle hash tree, which holds a collection of hash values in a tree structure up to a single root node, which is signed for integrity and authenticity purposes.

ONLINE QUERY MECHANISMS: Online query mechanisms differ from periodic publication mechanisms in that both the relying party and the authority with respect to revocation information (i.e., the CA or some designated alternative) must be online whenever a question regarding the revocation status of a given certificate needs to be resolved. With periodic publication mechanisms, revocation information can be cached in the relying party's local environment or stored in some central repository, such as an LDAP directory. Thus, the relying party may work offline (totally disconnected from the network) at the time of certificate validation, consulting only its local cache of revocation information, or may go online only for the purpose of downloading the latest revocation information from the central repository. As well, the authority may work offline when creating the latest revocation list and go online periodically only for the purpose of posting this list to a public location.

An online query mechanism is a protocol exchange—a pair of messages—between a relying party and an authority. The request message must indicate the certificate in question, along with any additional information that might be relevant. The response message answers the question (if it can be answered) and may provide supplementary data that could be of use to the relying party. In the simplest case, the requester asks the most basic question possible for this type of protocol: "has this certificate been revoked?" In other words, "if I was using a CRL instead of this online query mechanism, would this certificate appear on the CRL?" The response is essentially a yes or no answer, although an answer of "I don't know" (i.e., "unable to determine status") may also be returned. The IETF PKIX Online Certificate Status Protocol (OCSP) was created for exactly this purpose and has been successfully deployed in a number of environments worldwide.

However, the online protocol messages can be richer than the exchange described above. For example, the requester may ask not for a simple revocation status, but for a complete validation check on the certificate (i.e., is the entire certificate path "good," according to the rules of a well-defined path validation procedure). This is known as a Delegated Path Validation (DPV) exchange. Alternatively, the requester may ask the authority to find a complete path from the certificate in question to a specified trust anchor, but not necessarily to do the validation—the requester may prefer to do this part itself. This is known as a Delegated Path Discovery (DPD) exchange. The requirements for a general DPV/DPD exchange have been published by the IETF PKIX Working Group and a general, flexible protocol to satisfy these requirements (the Simple Certificate Validation Protocol, SCVP) is currently undergoing development in that group.

OTHER REVOCATION OPTIONS: It is important to note that there are circumstances in which the direct dissemination of revocation information to the relying party is unnecessary. For example, when certificates are "short-lived"—that is, have a validity period that is shorter than the associated need to revoke them—then revocation information need not be examined by relying parties. In such environments, certificates may have a lifetime of a few minutes or a few hours and the danger of a certificate needing to be revoked before it will naturally expire is considered to be minimal. Thus, revocation information need not be published at all.

Another example environment that can function without published revocation information is one in which relying parties use only brokered transactions. Many financial institutions operate in this way: online transactions are always brokered through the consumer's bank (the bank that issued the consumer's certificate). The bank maintains revocation information along with all the other data that pertains to its clients (account numbers, credit rating, and so on). When a transaction occurs, the merchant must always go to its bank to have the financial transaction authorized; this authorization process includes verification that the consumer's certificate had not been revoked, which is achieved through direct interaction between the merchant's bank and the consumer's bank. Thus, the merchant itself deals only with its own bank (and not with the consumer's bank) and never sees any explicit revocation information with respect to the consumer's certificate.

FURTHER READING: A survey of the various revocation techniques can be found in Chapter 8 of [1]. See also [2] for a good discussion of the many options in this area. The X.509 Standard [3] contains detailed specifications for most of the periodic publication mechanisms. For online query mechanisms, see the OCSP [4] and DPV/DPD Requirements [5] specifications.

Carlisle Adams

References

[1] Adams, C. and S. Lloyd (2003). *Understanding PKI: Concepts, Standards, and Deployment Considerations* (2nd ed.). Addison-Wesley, Reading, MA.

[2] Housley, R. and T. Polk (2001). *Planning for PKI: Best Practices Guide for Deploying Public Key Infrastructure*. Wiley, New York.

[3] ITU-T Recommendation X.509 (2000). "Information technology—open systems interconnection—the directory: Public key and attribute certificate frameworks." (equivalent to ISO/IEC 9594-8:2001).

[4] Myers, M., R. Ankney, A. Malpani, S. Galperin, and C. Adams (1999). "X.509 Internet public key infrastructure: Online certificate status protocol—OCSP." Internet Request for Comments 2569.

[5] Pinkas, D. and R. Housley (2002). "Delegated path validation and delegated path discovery protocol requirements." Internet Request for Comments 3379.

CERTIFICATION AUTHORITY

A Certification Authority[1] (CA) in a Public-Key Infrastructure (PKI) is an authority that is trusted by some segment of a population of entities—or perhaps by the entire population—to validly perform the task of binding public key pairs to identities. The CA certifies a key pair/identity binding by digitally signing (see digital signature scheme) a data structure that contains some representation of the identity of an entity (see identification) and the entity's corresponding public key. This data structure is called a "public-key certificate" (or simply a certificate, when this terminology will not be confused with other types of certificates, such as attribute certificates).

Although the primary and definitional duty of a CA is to certify key pair/identity bindings, it may also perform some other functions, depending upon the policies and procedures of the PKI in which it operates. For example, the CA may generate key pairs for entities upon request; it may store the key history for each entity in order to provide a key backup and recovery service; it may create identities for its subject community; and it may publicly disseminate revocation information for the certificates that it has issued. Alternatively, some or all these functions may be performed by other network entities that may or may not be under the explicit control of the CA, such as key generation servers, backup and recovery services, naming authorities, and on-line certificate status protocols (OCSP) responders.

---

[1] A CA is often called a "Certificate Authority" in the popular press and other literature, but this term is generally discouraged by PKI experts and practitioners because it is somewhat misleading: a CA is not an authority on *certificates* as much as it is an authority on the *process and act of certification*. Thus, the term "Certification Authority" is preferred.

The roles and duties of a CA have been specified in a number of contexts [1–5], along with protocols for various entities to communicate with the CA. As one example, the IETF PKIX Working Group (see security standards activities) has several standards-track specifications that are relevant to a CA operating in the context of an Internet PKI; see http://www.ietf.org/html.charters/pkix-charter.html for details.

Carlisle Adams

References

[1] Adams, C. and S. Farrell (1999). "Internet X.509 public key infrastructure: Certificate Management Protocols." Internet Request for Comments 2510.

[2] Adams, C. and S. Lloyd (2003). *Understanding PKI: Concepts, Standards, and Deployment Considerations* (2nd ed.). Addison-Wesley, Reading, MA.

[3] Housley, R. and T. Polk (2001). *Planning for PKI: Best Practices Guide for Deploying Public Key Infrastructure*. John Wiley & Sons, New York.

[4] ITU-T Recommendation X.509 (2000). "Information technology—open systems interconnection—the directory: public key and attribute certificate frameworks." (equivalent to ISO/IEC 9594-8:2001).

[5] Myers, M., X. Liu, J. Schaad, and J. Weinstein (2000). "Certificate management messages over CMS." Internet Request for Comments 2797.

CERTIFIED MAIL

Certified mail is the fair exchange of secret data for a receipt. It is the most mature instance of fair exchange that has been standardized in [4]: the players in a certified mail system are at least one sender S and one receiver R. Depending on the protocols used and the service provided, the protocol may involve one or more trusted third parties (TTPs) T. If reliable time stamping is desired, additional time-stamping authorities TS may be involved, too. For evaluating the evidence produced, a verifier V can be invoked after completion of the protocol. Sending a certified mail includes several actions [4]. Each of these actions may be disputable, i.e., may later be disputed at a verifier, such as a court (see Figure 1): a sender composes

Fig. 1. Framework for Certified Mail [4]: players and their actions

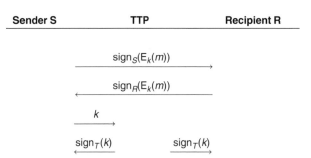

Fig. 2. Sketch of the protocol proposed in [7] (E_1 denotes symmetric encryption)

a signed message (nonrepudiation of origin) and sends it to the first TTP (nonrepudiation of submission). The first TTP may send, it to additional TTPs (nonrepudiation of transport) and finally to the recipient (nonrepudiation of delivery, which is a special case of nonrepudiation of transport). The recipient receives the message (nonrepudiation of receipt).

Like fair exchange and contract signing protocols, early research focused on two-party protocols [3, 5] fairly generating nonrepudiation of receipt tokens in exchange of the message. Like generic fair exchange, two-party protocols either have non-negligible failure probability or do not guarantee termination within a fixed time.

Early work on fair exchange with inline TTP was done in [8]. Optimistic protocols have been proposed in [1,2]. A later example of a protocol using an in-line TTP is the protocol proposed in [7]. The basic idea is that the parties first exchange signatures under the encrypted message. Then, the third party signs and distributes the key. The signature on the encrypted message together with the signatures on the key then forms the nonrepudiation of origin and receipt tokens. The protocol is sketched in Figure 2.

Matthias Schunter

References

[1] Asokan, N., Victor Shoup, and Michael Waidner (1998). "Asynchronous protocols for optimistic fair exchange." *1998 IEEE Symposium on Research in Security and Privacy*. IEEE Computer Society Press, Los Alamitos, 86–99.

[2] Bao, Feng, Robert Deng, and Wenbo Mao (1998). "Efficient and practical fair exchange protocols with off-line TTP." *1998 IEEE Symposium on Research in Security and Privacy*. IEEE Computer Society Press, Los Alamitos, 77–85.

[3] Blum, Manuel (1981). *Three Applications of the Oblivious Transfer*, Version 2. Department of Electrical Engineering and Computer Sciences, University of California at Berkeley, Berkley, CA.

[4] ISO/IEC (1997). *Information Technology—Security Techniques—Nonrepudiation*. Part 1: General; ISO/IEC International Standart 13888-1, (1st ed.).

[5] Rabin, Michael O. (1981). "Transaction protection by Beacons." Aiken Computation Laboratory, Harvard University, Cambridge, MA, Technical Report TR-29-81.

[6] Rabin, Michael O. (1983). "Transaction protection by beacons." *Journal of Computer and System Sciences* 27, 256–267.

[7] Zhou, Jianying and Dieter Gollmann (1996). "A fair non-repudiation protocol." *IEEE Symposium on Research in Security and Privacy Oakland*, 55–61.

CHAFFING AND WINNOWING

Chaffing and winnowing introduced by Ron Rivest [3] is a technique that keeps the contents of transmitted messages confidential against eavesdroppers without using encryption. Chaffing and winnowing was meant as a liberal statement in the debate about cryptographic policy in the 1990s as to whether law enforcement should be given authorized surreptitious access to the plaintext of encrypted messages. The usual approach proposed for such access was "key recovery," where law enforcement has a "back door" that enables them to recover the decryption key. Chaffing and winnowing was meant to obsolete this approach of key recovery because it reveals a technique of keeping messages confidential without using any decryption keys.

Here is how chaffing and winnowing works. A sender using the chaffing technique needs to agree with the intended recipient on an *authentication* mechanism, e.g., a message authentication code (see MAC algorithms) such as HMAC, and needs to establish an authentication key with the recipient. In order to send a message, the sender takes two steps:

Authentication: Breaks the message up into packets, numbers the packets consecutively, and authenticates each packet with the authentication key. The result is a sequence of "wheat" packets, i.e., those making up the intended message.

Chaffing: Fabricates additional dummy packets independent of the intended packets. Produces invalid MACs for the dummy packets, for example by choosing their MACs at random. These are the "chaff" packets, i.e., those used to hide the wheat packets in the stream of packets.

The sender sends all packets (wheat and chaff) intermingled in any order to the recipient. The recipient filters those packets containing a valid MAC (this is called *winnowing*), sorts them by packet number, and reassembles the message. An eavesdropper instead could not distinguish valid from invalid MACs because the required authentication key is only known to the sender and the recipient.

The problem of providing confidentiality by chaffing and winnowing is based on the eavesdropper's difficulty of distinguishing chaff packets from wheat packets. If the wheat packets each contain an English sentence, while the chaff packets contain random bits, then the eavesdropper will have no difficulty in detecting the wheat packets. On the other hand, if each wheat packet contains a single bit, and there is a chaff packet with the same serial number containing the complementary bit, then the eavesdropper will have a very difficult (essentially impossible) task. Being able to distinguish wheat from chaff would require him to break the MAC algorithm and/or know the secret authentication key used to compute the MACs. With a good MAC algorithm, the eavesdropper's ability to winnow is nonexistent, and the chaffing process provides perfect confidentiality of the message contents.

If the eavesdropper is as strong as some law enforcement agency that may monitor the main hubs of the Internet and may even have the power to force a sender to reveal the authentication key used, then senders could use alternative wheat messages instead of chaff. For an intended message the sender composes an innocuous looking cover message. The intended wheat message is broken into packets using the authentication key as described above. The cover wheat message is also broken into packets using a second authentication key that may or may not be known to the recipient. In this way, the sender could use several cover wheat messages for each intended wheat message. If the sender is forced to reveal the authentication key he used, he could reveal the authentication key of one of the cover wheat messages. Thus, he could deliberately "open" a transmitted message in several ways. This concept is similar to deniable encryption proposed by Canetti et al. [1].

In order to reduce the apparent overhead in transmission bandwidth, Rivest suggested that the chaffing could be done by an Internet Service Provider rather than by the sender himself. The ISP could then multiplex several messages, thus using the wheat packets of one message as chaff packets of another message and vice versa. He suggested other measures for long messages such that the relative number of chaff packets can be made quite small, and the extra bandwidth required for transmitting chaff packets might be insignificant in practice.

Instead of message authentication codes, senders could also use an underlined_signature scheme, which produces signatures that can only be verified by the intended recipients [2].

Gerrit Bleumer

References

[1] Canetti, Ran, Cynthia Dwork, Moni Naor, and Rafail Ostrovsky (1997). "Deniable encryption." *Advances in Cryptology—CRYPTO'97*, Lecture Notes in Computer Science, vol. 1294, ed. B.S. Kaliski. Springer-Verlag, Berlin, 90–104. ftp://theory.lcs.mit .edu/pub/tcryptol/96-02r.ps

[2] Jakobsson, Markus, Kazue Sako, and Russell Impagliazzo (1996). "Designated verifier proofs and their applications." *Advances in Cryptology— EUROCRYPT'96*, Lecture Notes in Computer Science, vol. 1070, ed. U. Maurer. Springer-Verlag, Berlin, 143–154.

[3] Rivest, R.L. (1998). Chaffing and Winnowing: Confidentiality without Encryption. http://theory. lcs.mit.edu/ rivest/chaffing.txt

CHALLENGE–RESPONSE IDENTIFICATION

In its simplest form, an identification protocol involves the presentation or submission of some information (a "secret value") from a claimant to a verifier (see Identification). Challenge–response identification is an extension in which the information submitted by the claimant is the function of both a secret value known to the claimant (sometimes called a "prover"), and a challenge value received from the verifier (or "challenger").

Such a challenge–response protocol proceeds as follows. A verifier V generates and sends a challenge value c to the claimant C. Using his/her secret value s and appropriate function $f()$, C computes the response value $v = f(c, s)$, and returns v to V. V verifies the response value v, and if successful, the claim is accepted. Choices for the challenge value c, and additionally options for the function $f()$ and secret s are discussed below.

Challenge–response identification is an improvement over simpler identification because it offers protection against replay attacks. This is achieved by using a challenge value that is time-varying. Referring to the above protocol, there are three general types of challenge values that might be used. The property of each is that the challenge value is not repeatedly sent to multiple claimants. Such a value is sometimes referred to as a *nonce*, since it is a value that is "*n*ot used more than once."

The challenge value could be a randomly generated value (see Random bit generation (hardware)), in which case V would send a random value c to C. Alternatively, the challenge value might be a sequence number, in which case the verifier V would maintain a sequence value corresponding to each challenger. At each challenge, the stored sequence number would be increased by (at least) 1 before sending to the claimant. Finally, the challenge value might be a function of the current time. In this case, a challenge value need not be sent from V to C, but could be sent by C, along with the computed verifier. As long as the time chosen was within an accepted threshold, V would accept.

There are three general classes of functions and secret values that might be used as part of a challenge–response protocol. The first is symmetric-key based in which the claimant C and verifier V *a priori* share a secret key K. The function $f()$ is a symmetric encryption function (see Symmetric Cryptosystem), a hash function, or a Message Authentication Code (see MAC algorithms). Both Kerberos (see Kerberos authentication protocol) and the Needham–Schroeder protocol are examples of symmetric-key based challenge–response identification. In addition, the protocols of ISO/IEC 9798-2 perform identification using symmetric key techniques.

Alternatively, a public key based solution may be used. In this case, the claimant C has the private key in a public key cryptosystem (see Public Key Cryptography). The verifier V possesses a public key that allows validation of the public key corresponding to C's private key. In general, C uses public key techniques (generally based on number-theoretic security problems) to produce a value v, using knowledge of his/her private key. For example, V might encrypt a challenge value and send the encrypted text. C would decrypt the encrypted text and return the value (i.e., the recovered plaintext) to V (note that in this case it would only be secure to use a random challenge, and not a sequence number or time-based value). Alternatively, V might send a challenge value to C and ask C to digitally sign and return the challenge (see Digital Signature Schemes). The Schnorr identification protocol is another example of public key based challenge–response identification.

Finally, a zero-knowledge protocol can be used (see Zero-Knowledge). In this case, the challenger demonstrates knowledge of his/her secret value without revealing any information (in an information theoretic sense—see "information theoretic security" in glossary) about this value. Such protocols typically require a number of "rounds" (each with its own challenge value) to be executed before a claimant may be successfully verified (see Zero-Knowledge and Identification).

Mike Just

References

[1] Menezes, A., P. van Oorschot, and S. Vanstone (1997). *Handbook of Applied Cryptography*. CRC Press, Boca Raton, FL.

[2] Stinson, D.R. (1995). *Cryptography: Theory and Practice*. CRC Press, Boca Raton, FL.

CHAUM BLIND SIGNATURE SCHEME

The Chaum Blind Signature Scheme [3,4] was the first blind signature scheme proposed in the publicly available literature. It was proposed by David Chaum and is based on the RSA signature scheme using the fact that RSA is an *automorphism* on $\mathbb{Z}_n^*$, the multiplicative group of units modulo an RSA integer $n = pq$, where n is the public modulus and p, q are safe RSA prime numbers. The tuple (n, e) is the public verifying key, where e is a prime between 2^{16} and $\phi(n) = (p-1)(q-1)$, and the tuple (p, q, d) is the corresponding private key of the signer, where $d = e^{-1} \bmod \phi(n)$ is the signing exponent. The signer computes signatures by raising the hash value $H(m)$ of a given message m to the dth power modulo n, where $H(\cdot)$ is a publicly known collision resistant hash function. A recipient verifies a signature s for message m with respect to the verifying key (n, e) by the following equation: $s^e = H(m) \pmod{n}$.

When a recipient wants to retrieve a blind signature for some message m', he chooses a blinding factor $b \in \mathbb{Z}_n$ and computes the auxiliary message $m = b^e H(m') \bmod n$. After passing m to the signer, the signer computes the response $s = m^d \bmod n$ and sends it back to the recipient. The recipient computes a signature s' for the intended message m' as follows: $s' = sb^{-1} \bmod n$. This signature s' is valid for m' with respect to the signer's public verifying key y because

$$
\begin{aligned}
s'^e &= (sb^{-1})^e \\
&= (m^d b^{-1})^e \\
&= m^{de} b^{-e} \\
&= mb^{-e} \\
&= b^e H(m') b^{-e} \\
&= H(m') \pmod{n}. \tag{1}
\end{aligned}
$$

(Note how the above-mentioned automorphism of RSA is used in the third rewriting.) It is conjec-tured that the Chaum Blind Signature Scheme is secure against a one-more-forgery, although this has not been proven under standard complexity theoretic assumptions, such as the assumption that the RSA verification function is one-way. The fact that the Chaum Blind Signature Scheme has resisted one-more-forgeries for more than 20 years led Bellare et al. [1] to isolate a nonstandard complexity theoretic assumption about RSA signatures that is sufficient to prove security of the Chaum Blind Signature in the random oracle model, i.e., by abstracting from the properties of any hash function $H(\cdot)$ chosen. They came up with a class of very strong complexity theoretic assumptions about RSA, which they called the one-more-RSA-inversion assumptions (or problems).

The Chaum Blind Signature Scheme achieves unconditional blindness [3] (see Blind Signature Scheme). That is if a signer produces signatures $s_1, \ldots, s_n$ for $n \in \mathbb{N}$ messages $m_1, \ldots, m_n$ chosen by a recipient, and the recipient later shows the resulting n pairs $(m'_1, s'_1), \ldots, (m'_n, s'_n)$ in random order to a verifier, then the collaborating signer and verifier cannot decide with better probability than pure guessing which message–signature pair (m_i, s_i) $(1 \le i \le n)$ resulted in which message–signature pair (m'_j, s'_j) $(1 \le j \le n)$.

Analogous to how Chaum leveraged the *automorphism* underlying the RSA signature scheme to construct a blind signature scheme, other digital signature schemes have been extended into blind signature schemes as well: Chaum and Pedersen [5] constructed blind Schnorr signatures [12]. Camenisch et al. [2] constructed blind Nyberg–Rueppel signatures [9] and blind signatures for a variant of DSA [8]. Horster et al. [7] constructed blind ElGamal digital signatures [6], and Pointcheval and Stern [10, 11] constructed blind versions of certain adaptations of Schnorr and Guillou–Quisquater signatures.

Gerrit Bleumer

References

[1] Bellare, Mihir, Chanathip Namprempre, David Pointcheval, and Michael Semanko (2001). "The one-more-RSA inversion problems and the security of Chaum's blind signature scheme." *Financial Cryptography 2001*, Lecture Notes in Computer Science, vol. 2339, ed. P.F. Syverson. Springer-Verlag, Berlin, 319–338.

[2] Camenisch, Jan, Jean-Marc Piveteau, and Markus Stadler (1995). "Blind signatures based on the discrete logarithm problem." *Advances in Cryptology—EUROCRYPT'94*, Lecture Notes in Computer Science, vol. 950, ed. A. De Santis. Springer-Verlag, Berlin, 428–432.

[3] Chaum, David (1993). "Blind signatures for untraceable payments." *Advances in Cryptology—CRYPTO'82*, eds. Plenum D. Chaum, R.L. Rivest, and A.T. Sherman. Plenum Press, New York, 199–203.

[4] Chaum, David (1990). "Showing credentials without identification: Transferring signatures between unconditionally unlinkable pseudonyms." *Advances in Cryptology—AUSCRYPT'90*, Lecture Notes in Computer Science, vol. 453, eds. J. Seberry and J. Pieprzyk. Springer-Verlag, Berlin, 246–264.

[5] Chaum, David and Torben P. Pedersen (1993). "Wallet databases with observers." *Advances in Cryptology—CRYPTO'92*, Lecture Notes in Computer Science, vol. 740, ed. E.F. Brickell. Springer-Verlag, Berlin, 89–105.

[6] ElGamal, Taher (1985). "A public key cryptosystem and a signature scheme based on discrete logarithms." *IEEE Transactions on Information Theory*, 31 (4), 469–472.

[7] Patrick, Horster, Markus Michels, and Holger Petersen (1994). "Meta-message recovery and meta-blind signature schemes based on the discrete logarithm problem and their applications." *Advances in Cryptography—ASIACRYPT'94*, Lecture Notes in Computer Science, vol. 917, eds. J. Pieprzyk and R. Safari-Naini. Springer-Verlag, Berlin, 224–237.

[8] National Institute of Standards and Technology (NIST) (1993). "Digital signature standard." Federal Information Processing Standards Publication (FIPS PUB 186).

[9] Nyberg, Kaisa and Rainer Rueppel (1993). "A new signature scheme based on the DSA giving message recovery." *1st ACM Conference on Computer and Communications Security, Proceedings, Fairfax, November 1993*. ACM Press, New York, 58–61.

[10] Pointcheval, David (1998). "Strengthened security for blind signatures." *Advances in Cryptology—EUROCRYPT'98*, Lecture Notes in Computer Science, vol. 1403, ed. K. Nyberg. Springer-Verlag, Berlin, 391–405.

[11] Pointcheval, David and Jacques Stern (1996). "Provably secure blind signature schemes." *Advances in Cryptography—ASIACRYPT'96*, Lecture Notes in Computer Science, vol. 1163, eds. K. Kim and T. Matsumoto. Springer-Verlag, Berlin, 252–265.

[12] Schnorr, Claus-Peter (1988). "Efficient signature generation by smart cards." *Journal of Cryptology*, 4 (3), 161–174.

CHINESE REMAINDER THEOREM

The Chinese remainder theorem (CRT) makes it possible to reduce <u>modular arithmetic</u> calculations with large moduli to similar calculations for each of the factors of the modulus. At the end, the outcomes of the subcalculations need to be pasted together to obtain the final answer. The big advantage is immediate: almost all these calculations involve much smaller numbers.

For instance, the multiplication 24×32 (mod 35) can be found from the same multiplication modulo 5 and modulo 7, since $5 \times 7 = 35$ and these numbers have no factor in common. So, the first step is to calculate:

$$24 \times 32 \equiv 4 \times 2 \equiv 8 \equiv 3 \pmod 5,$$
$$24 \times 32 \equiv 3 \times 4 \equiv 12 \equiv 5 \pmod 7.$$

The CRT, explained for this example, is based on a unique correspondence (Figure 1) between the

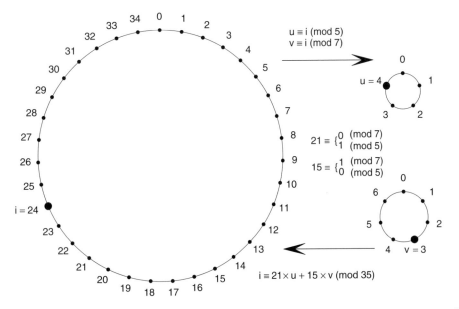

Fig. 1. The Chinese remainder theorem reduces a calculation modulo 35 to two calculations, one modulo 5 and the other modulo 7

integers $0, 1, \ldots, 34$ and the pairs (u, v) with $0 \leq u < 5$ and $0 \leq v < 7$. The mapping from i, $0 \leq i < 35$, to the pair (u, v) is given by the reduction of i modulo 5 and modulo 7, so $i = 24$ is mapped to $(u, v) = (4, 3)$. The mapping from (u, v) back to i is given by $i \equiv 21 \times u + 15 \times v$. The multiplier $a \equiv 21 \pmod{35}$ can be obtained from $a \equiv (v^{-1} \bmod u) \times v$, which is the solution of the two relations $a \equiv 1 \pmod{u}$ and $a \equiv 0 \pmod{v}$. The multiplier $b \equiv 15 \pmod{35}$ can be determined similarly. It follows that the answer of the multiplication above is given by $21 \times 3 + 15 \times 5 \equiv 33 \pmod{35}$.

The CRT finds applications in implementations of the RSA public-key encryption system, where one has to work with very large moduli that are the product of two prime numbers. Also Pohlig–Hellmann's method for taking discrete logarithms relies on the CRT (see discrete logarithm problem).

The CRT can be generalized to more than two factors and solves in general system of linear congruence relations of the form $a_i x \equiv b_i \pmod{m_i}$, $1 \leq i \leq k$, where the greatest common divisor of a_i and m_i should divide b_i for each $1 \leq i \leq k$.

Henk van Tilborg

Reference

[1] Shapiro, H.N. (1983). *Introduction to the Theory of Numbers*. John Wiley & Sons, New York.

CHOSEN CIPHERTEXT ATTACK

Chosen ciphertext attack is a scenario in which the attacker has the ability to choose ciphertexts C_i and to view their corresponding decryptions—plaintexts P_i. It is essentially the same scenario as a chosen plaintext attack but applied to a decryption function, instead of the encryption function. The attack is considered to be less practical in real life situations than chosen plaintext attacks. However, there is no direct correspondence between complexities of chosen plaintext and chosen ciphertext attacks. A cipher may be vulnerable to one attack but not to the other attack or the other way around. Chosen ciphertext attack is a very important scenario in public key cryptography, where known plaintext and even chosen plaintext scenarios are always available to the attacker due to publicly known encryption key. For example, the RSA public-key encryption system is not secure against adaptive chosen ciphertext attack [1].

Alex Biryukov

Reference

[1] Bleichenbacher, D. (1998). "Chosen ciphertext attacks against protocols based on the RSA encryption standard PKCS#1." *Advances in Cryptology—CRYPTO'98*, Lecture Notes in Computer Science, vol. 1462, ed. H. Krawczyk. Springer-Verlag, Berlin, 1–12.

CHOSEN PLAINTEXT ATTACK

Chosen plaintext attack is a scenario in which the attacker has the ability to choose plaintexts P_i and to view their corresponding encryptions—ciphertexts C_i. This attack is considered to be less practical than the known plaintext attack, but still a very dangerous attack. If the cipher is vulnerable to a known plaintext attack, it is automatically vulnerable to a chosen plaintext attack as well, but not necessarily the opposite. In modern cryptography differential cryptanalysis is a typical example of a chosen plaintext attack. It is also a rare technique for which conversion from chosen plaintext to known plaintext is possible (due to its work with pairs of texts). If a chosen plaintext differential attack uses m pairs of texts for an n bit block cipher, then it can be converted to a known-plaintext attack which will require $2^{n/2}\sqrt{2m}$ known plaintexts, due to birthday paradox-like arguments. Furthermore as shown in [1] the factor $2^{n/2}$ may be considerably reduced if the known plaintexts are redundant (for example for the case of ASCII encoded English text to about $2^{(n-r)/2}$ where r is redundancy of the text), which may even lead to a conversion of differential chosen-plaintext attack into a differential ciphertext-only attack.

Alex Biryukov

Reference

[1] Biryukov, A. and E. Kushilevitz (1998). "From differential cryptanalysis to ciphertext-only attacks." *Advances in Cryptology—CRYPTO'98*, Lecture Notes in Computer Science, vol. 1462, ed. H. Krawczyk. Springer-Verlag, Berlin, 72–88.

CHOSEN PLAINTEXT AND CHOSEN CIPHERTEXT ATTACK

In this attack the attacker is allowed to combine the underline{chosen plaintext attack} and *chosen* ciphertext attack together and to issue chosen queries both to the encryption and to the decryption functions.

Alex Biryukov

CIPHERTEXT-ONLY ATTACK

The ciphertext-only attack scenario assumes that the attacker has only passive capability to listen to the encrypted communication. The attacker thus only knows ciphertexts $C_i, i = 1, \ldots, N$, but not the corresponding plaintexts. He may however rely on certain redundancy assumptions about the plaintexts, for example that the plaintext is ASCII encoded English text. This scenario is the weakest in terms of capabilities of the attacker and thus it is the most practical in real life applications. In certain cases conversion of a known plaintext attack [2] or even chosen plaintext attack [1] into a ciphertext-only attack is possible.

Alex Biryukov

References

[1] Biryukov, A. and E. Kushilevitz (1998). "From differential cryptanalysis to ciphertext-only attacks." *Advances in Cryptology—CRYPTO'98*, Lecture Notes in Computer Science, vol. 1462, ed. H. Krawczyk. Springer-Verlag, Berlin, 72–88.

[2] Matsui, M. (1993). "Linear cryptanalysis method for DES cipher." *Advances in Cryptology—EUROCRYPT'93*, Lecture Notes in Computer Science, vol. 765, ed. T. Helleseth. Springer-Verlag, Berlin, 386–397.

CLAW-FREE

A pair of functions f and g is said to be *claw-free* or *claw-resistant* if it is difficult to find inputs x, y to the functions such that

$$f(x) = g(y).$$

Such a pair of inputs is called a *claw*, describing the two-pronged inverse.

The concept of claw-resistance was introduced in the digital signature scheme of Goldwasser et al. [3], which was based on claw-free trapdoor permutations (see trapdoor one-way function and substitutions and permutations). Damgård [1] showed that claw-free permutations (without the trapdoor) could be employed to construct collision-resistant hash functions (see also collision resistance).

Recently, Dodis and Reyzin have shown that the claw-free property is essential to obtaining good security proofs for certain signature schemes [2].

Burt Kaliski

References

[1] Damgård, Ivan Bjerre (1988). "Collision free hash functions and public key signature schemes." *Advances in Cryptology—EUROCRYPT'87*, Lecture Notes in Computer Science, vol. 304, eds. D. Chaum and W.L. Price. Springer, Berlin, 203–216.

[2] Dodis, Yevgeniy and Leonid Reyzin (2003). "On the power of claw-free permutations." *Security in Communications Networks (SCN 2002)*, Lecture Notes in Computer Science, vol. 2576, eds. S. Cimato, C. Galdi, and G. Persiano. Springer, Berlin, 55–73.

[3] Goldwasser, Shafi, Silvio Micali, and Ronald L. Rivest (1988). "A digital signature scheme secure against adaptive chosen-message attacks." *SIAM Journal on Computing*, 17 (2), 281–208.

CLIP SCHEME

Clip is MasterCard's instantiation of the Common Electronic Purse Specifications (see CEPS). As defined by CEPS standard, the physical infrastructure and logical methods such as smart cards, terminals, and cryptographic mechanisms are based upon the EMV standard with minor deviations. The amount involved in a purchase is debited directly on the card, and hence an additional layer of security is applied for the terminal to strongly authenticate itself to the card.

Peter Landrock

CLOCK-CONTROLLED GENERATOR

Let us consider a scheme that involves several registers and produces one output sequence. Based on some clocking mechanism, the registers go from one state to another, thereby producing an output bit. We can choose whether or not to synchronize these registers. In the second case, the output of the scheme will be more nonlinear than in the first case.

We will consider here registers whose clock is controlled by some events. The best studied case is the one of <u>Linear Feedback Shift Registers (LFSR)</u>, but this concept could be applied also to <u>Nonlinear Feedback Shift Registers (NLFSR)</u>.

So, the main idea here is that we have, for example, two LFSRs, say R_1 and R_2, and that the output of R_1 will determine the clocking of R_2. For example, if R_1 outputs a 1, then clock R_2 twice, and if R_1 outputs a 0, then clock R_2 three times. The output of the scheme could be the one of R_2.

EXAMPLE

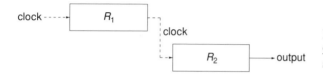

Some particular examples of such generators have been studied. We will mention here the *alternating step generator*, and the <u>shrinking generator</u>. We can also remark that a LFSR can manage its clocking by itself, since some of its internal bits can be chosen to determine its clocking; an example is the <u>self-shrinking generator</u>.

The alternating step generator consists of three LFSRs, say R, R_0, and R_1. The role of R is to determine the clocking of both R_0 and R_1. If R outputs a 0, then only R_0 is clocked, and if R outputs a 1, then only R_1 is clocked. At each step, a LFSR that is not clocked outputs the same bit as previously (a 0 if there is no previous step). So, at each step both R_0 and R_1 output one bit each, but only one of them has been clocked. The output sequence of the scheme is obtained by XORing those two bits.

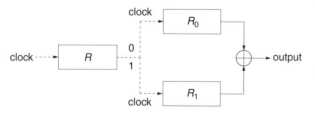

EXAMPLE. Let us suppose that R and R_1 are of length 2 and have period 3; the feedback relation for R is $s_{t+1} = s_t + s_{t-1}$. For R_1, let us consider $s_{t+1} = s_t + s_{t-1}$; R_0 has length 3, and its feedback relation is $s_{t+1} = s_t + s_{t-2}$. Then we have for

example (the first row corresponds to the initialization; the internal states are of the form $s_t s_{t-1}$ or $s_t s_{t-1} s_{t-2}$):

R		R_0		R_1		
State	Output	State	Output	State	Output	Output
11		010		01		
01	1	010	0	10	1	1
10	1	010	0	11	0	0
11	0	001	0	11	0	0
01	1	001	0	01	1	1
10	1	001	0	10	1	1
11	0	100	1	10	1	0
01	1	100	1	11	0	1
10	1	100	1	01	1	0
⋮	⋮	⋮	⋮	⋮	⋮	⋮

Some studies have been performed on the properties of the output sequence (period, <u>linear complexity</u>, etc.), according to the nature of the sequences associated with R, R_0, and R_1. A survey of techniques for attacking clock-controlled generators is given in [3], and more recent results are discussed in [1, 2, 4, 5].

Caroline Fontaine

References

[1] Golic, J. (1995). "Towards fast correlation attacks on irregularly clocked shift registers." *Advances in Cryptology—EUROCRYPT'95*, Lecture Notes in Computer Science, vol. 921, eds. L.C. Gvillou and J.-J. Quisquater. Springer-Verlag, Berlin, 248–262.

[2] Golic, J. and L. O'Connor (1995). "Embedding and probabilistic correlation attacks on clock-controlled shift registers." *Advances in Cryptology—EUROCRYPT'94*, Lecture Notes in Computer Science, vol. 950, ed. A. De Santis. Springer-Verlag, Berlin, 230–243.

[3] Gollmann, D. (1994). "Cryptanalysis of clock-controlled shift registers." *Fast Software Encryption*, Lecture Notes in Computer Science, vol. 809, ed. R.J. Anderson. Springer-Verlag, Berlin, 121–126.

[4] Johansson, T. (1998). "Reduced complexity correlation attacks on two clock-controlled generators." *Advances in Cryptography—ASIACRYPT'98*, Lecture Notes in Computer Science, vol. 1514, eds. K. Ohta and D. Pei. Springer-Verlag, Berlin, 342–356.

[5] Kholosha, A. (2001). "Clock-controlled shift registers and generalized Geffe key-stream generator." *Progress in Cryptology—INDOCRYPT'01*, Lecture Notes in Computer Science, vol. 2247, eds. C. Pandu Rangan and C. Ding. Springer-Verlag, Berlin, 287 ff.

CLOSEST VECTOR PROBLEM

The Closest Vector Problem (CVP) is a computational problem on lattices closely related to SVP (see Shortest Vector Problem). Given a lattice $\mathcal{L}$ and a target point $\vec{x}$, CVP asks to find the lattice point closest to the target. As for SVP, CVP can be defined with respect to any norm, but the Euclidean norm is the most common (see the entry lattice for a definition). A more relaxed version of the problem (used mostly in computational complexity) only asks to compute the distance of the target from the lattice, without actually finding the closest lattice vector.

CVP has been studied in mathematics (in the equivalent language of quadratic forms) since the nineteenth century. One of the first references to CVP (under the name "Nearest Vector Problem") in the computer science literature is [11], where the problem is shown to be NP-hard to solve exactly.

Many applications of the CVP only require finding a lattice vector that is not too far from the target, even if not necessarily the closest. A g-approximation algorithm for CVP finds a lattice vector within distance at most g times the distance of the optimal solution. The best known polynomial-time algorithms to solve CVP due to Babai [2] and Kannan [7] are based on lattice reduction, and achieve approximation factors that (in the worst case) are essentially exponential in the dimension of the lattice. In practice, heuristics approaches (e.g., the "embedding technique," see lattice reduction) seem to find relatively good approximations to CVP in a reasonable amount of time when the dimension of the lattice is sufficiently small.

CVP is widely regarded, both in theory and in practice, as a considerably harder problem than SVP. CVP is known to be NP-hard to solve approximately within any constant factor or even some slowly increasing subpolynomial function (see polynomial time) of the dimension n [1, 3]. However, CVP is unlikely to be NP-hard to approximate within small polynomial factors $g = O(\sqrt{n/\log n})$ [5]. Goldreich et al. [6] showed that any algorithm to efficiently approximate CVP can be used to efficiently approximate SVP within the same approximation factor and with essentially the same computational effort, formalizing the intuition that CVP is not an easier (and is a possibly harder) problem than SVP.

CVP is the basis of various cryptosystems (see lattice based cryptography) where the decryption process corresponds roughly to a CVP computation. These cryptosystems are based on the fact that any lattice admits many different representations (e.g., it can be represented by different bases), and some of them may have better geometric properties than others, so that they can be used as a decryption trapdoor. However, there are lattices [4, 8, 10] that admit no good representation, i.e., solving CVP (even approximately) is NP-hard no matter which basis (or other auxiliary information) is given. Therefore, the CVP instances used by lattice based cryptosystems (for which CVP can be efficiently solved using the decryption key) are conceivably easier than general CVP instances.

CVP has many applications in computer science, besides cryptography, and finding a good CVP approximation algorithm with approximation factors that grow as a polynomial in the dimension of the lattice is one of the major open problems in the area. For further information about CVP and other computational problems on lattices, the reader is referred to the book [9].

Daniele Micciancio

References

[1] Arora, S., L. Babai, J. Stern, and E.Z. Sweedyk (1997). "The hardness of approximate optima in lattices, codes, and systems of linear equations." *Journal of Computer and System Sciences*, 54 (2), 317–331. Preliminary version in *FOCS'93*.

[2] Babai, L. (1986). "On Lovasz' lattice reduction and the nearest lattice point problem." *Combinatorica* 6 (1), 1–13.

[3] Dinur, I., G. Kindler, and S. Safra (1998). "Approximating CVP to within almost-polynomial factors is NP-hard." *Proceedings of the 39th Annual Symposium on Foundations of Computer Science—FOCS'98, Palo Alto, CA, USA, November 1998*. IEEE, Piscataway, NJ, 99–111.

[4] Feige, U. and D. Micciancio (2004). "The inapproximability of lattice and coding problems with preprocessing." *Journal of Computer and System Sciences*, 69 (1), 45–46. To appear. Preliminary version in *CCC 2002*.

[5] Goldreich, O. and S. Goldwasser (2000). "On the limits of nonapproximability of lattice problems." *Journal of Computer and System Sciences*, 60 (3), 540–563. Preliminary version in *STOC'98*.

[6] Goldreich, O., D. Micciancio, S. Safra, and J.-P. Seifert (1999). "Approximating shortest lattice vectors is not harder than approximating closest lattice vectors." *Information Processing Letters*, 71 (2), 55–61.

[7] Kannan, R. (1987). *Annual Reviews of Computer Science, Volume 2, Chapter Algorithmic Geometry of Numbers*. Annual Review Inc., Palo Alto, CA, 231–267.

[8] Micciancio, D. (2001). "The hardness of the closest vector problem with preprocessing." *IEEE Transactions on Information Theory*, 47 (3), 1212–1215.

[9] Micciancio, D. and S. Goldwasser (2002). *Complexity of Lattice Problems: A Cryptographic Perspective*, The Kluwer International Series in Engineering and Computer Science, vol. 671, Kluwer Academic Publishers, Boston, MA.

[10] Regev, O. (2003). "Improved inapproximability of lattice and coding problems with preprocessing." *Proceedings of the 18th IEEE Annual Conference on Computational Complexity—CCC'03, Århus, Denmark, July 2003*. IEEE, Piscataway, NJ, 315–322.

[11] van Emde Boas, P. (1981). "Another NP-complete problem and the complexity of computing short vectors in a lattice." Technical Report 81-04, Mathematische Instituut, University of Amsterdam. Available on-line at URL: http://turing.wins.uva.nl/peter/

CODEBOOK ATTACK

A codebook attack is an example of a known plaintext attack scenario in which the attacker is given access to a set of plaintexts and their corresponding encryptions (for a fixed key): (P_i, C_i), $i = 1, \ldots, N$. These pairs constitute a codebook which someone could use to listen to further communication and which could help him to partially decrypt the future messages even without the knowledge of the secret key. He could also use this knowledge in a replay attack by replacing blocks in the communication or by constructing meaningful messages from the blocks of the codebook. Codebook attack may even be applied in a passive traffic analysis scenario, i.e., as a ciphertext-only attack, which would start with frequency analysis of the received blocks and attempts to guess their meaning. Ciphers with small block size are vulnerable to the Codebook attack, especially if used in the simplest Electronic Codebook mode of operation. Already with $N = 2^{n/2}$ known pairs, where n is the block size of the cipher, the attacker has good chances to observe familiar blocks in the future communications of size $O(2^{n/2})$, due to the birthday paradox. If communication is redundant, the size of the codebook required may be even smaller. Modern block ciphers use 128-bit block size to make such attacks harder to mount. A better way to combat such attacks is to use chaining modes of operation like Cipher-Block Chaining mode (which makes further blocks of ciphertext dependent on all the previous blocks) together with the authentication of the ciphertext.

Alex Biryukov

COLLISION ATTACK

A collision attack exploits repeating values that occur when a random variable is chosen with replacement from a finite set S. By the birthday paradox, repetitions will occur after approximately $\sqrt{|S|}$ attempts, where $|S|$ denotes the size of the set S. Many cryptographic attacks are based on collisions.

The most obvious application of a collision attack is to find collisions for a cryptographic hash function. For a hash function with an n-bit result, an efficient collision search based on the birthday paradox requires approximately $2^{n/2}$ hash function evaluations [10]. For this application, one can substantially reduce the memory requirements (and also the memory accesses) by translating the problem to the detection of a cycle in an iterated mapping [7]. Van Oorschot and Wiener propose an efficient parallel variant of this algorithm [9]. In order to make a collision search infeasible for the next 15–20 years, the hash result needs to be 180 bits or more. A collision attack can also play a role to find (second) preimages for a hash function: if one has $2^{n/2}$ values to invert, one expects to find at least one (second) preimage after $2^{n/2}$ hash function evaluations.

An *internal collision* attack on a MAC algorithm exploits collisions of the chaining variable of a MAC algorithm. It allows for a MAC forgery. As an example, a forgery attack for CBC-MAC and variants based on an n-bit block cipher requires at most $2^{n/2}$ known texts and a single chosen text [6]. For some MAC algorithms, such as MAA, internal collisions can lead to a key recovery attack [5].

A block cipher should be a one-way function from key to ciphertext (for a fixed plaintext). If the same plaintext is encrypted using $2^{k/2}$ keys (where k is the key length in bits), one expects to recover one of the keys after $2^{k/2}$ trial encryptions [1]. This attack can be precluded by the mode of operation; however, collision attacks also apply to these modes. In the Cipher Block Chaining (CBC) and Cipher FeedBack (CFB) mode of an n-bit block cipher, a repeated value of an n-bit ciphertext string leaks information on the plaintext [3,4] (see block ciphers for more details).

For synchronous stream ciphers that have a next state function that is a random function (rather than a permutation), one expects that the key stream will repeat after $2^{m/2}$ output symbols, with m being the size of the internal memory in bits. Such a repetition leaks the sum of the corresponding plaintexts, which is typically sufficient to recover them. This attack applies to a variant of the Output FeedBack (OFB) mode of a block cipher where less than n output bits are fed back to the input. If exactly n bits are fed back as specified by the OFB mode, one expects a repetition after the selection of $2^{n/2}$ initial values.

The best generic algorithm to solve the discrete logarithm problem in any group G requires time $O(\sqrt{p})$ where p is the largest prime dividing the order of G [8]; this attack is based on collisions.

In many cryptographic protocols, e.g., entity authentication protocols, the verifier submits a random challenge to the prover. If an n-bit challenge is chosen uniformly at random, one expects to find a repeating challenge after $2^{n/2}$ runs of the protocol. A repeating challenge leads to a break of the protocol.

A meet-in-the-middle attack is a specific variant of a collision attack which allows to cryptanalyze some hash functions and multiple encryption modes (see block ciphers).

A more sophisticated way to exploit collisions to recover a block cipher key or to find (second) preimages is a time-memory trade-off [2].

B. Preneel

References

[1] Biham, E. (2002). "How to decrypt or even substitute DES-encrypted messages in 2^{28} steps." *Inf. Process. Lett.*, 84 (3), 117–124.

[2] Hellman, M. (1980). "A cryptanalytic time-memory trade-off." *IEEE Trans. on Information Theory*, IT-26 (4), 401–406.

[3] Knudsen, L.R. (1994). "Block ciphers—analysis, design and applications." *PhD Thesis*, Aarhus University, Denmark.

[4] Maurer, U.M. (1991). "New approaches to the design of self-synchronizing stream ciphers." *Advances in Cryptology—EUROCRYPT'91*, Lecture Notes in Computer Science, vol. 547, ed. D.W. Davies. Springer-Verlag, Berlin, 458–471.

[5] Preneel, B., V. Rijmen, and P.C. van Oorschot (1997). "A security analysis of the message authenticator algorithm (MAA)." *European Transactions on Telecommunications*, 8 (5), 455–470.

[6] Preneel, B. and P.C. van Oorschot (1999). "On the security of iterated message authentication codes." *IEEE Trans. on Information Theory*, IT-45 (1), 188–199.

[7] Quisquater, J.-J. and J.-P. Delescaille (1990). "How easy is collision search? Application to DES." *Advances in Cryptology—EUROCRYPT'89*, Lecture Notes in Computer Science, vol. 434, eds. J.-J. Quisquater and J. Vandewalle. Springer-Verlag, Berlin, 429–434.

[8] Shoup, V. (1997). "Lower bounds for discrete logarithms and related problems." *Advances in Cryptology—EUROCRYPT'97*, Lecture Notes in Computer Science, vol. 1233, ed. W. Fumy. Springer-Verlag, Berlin, 256–266.

[9] Van Oorschot, P.C. and M. Wiener (1999). "Parallel collision search with cryptanalytic applications." *Journal of Cryptology*, 12 (1), 1–28.

[10] Yuval, G. (1979). "How to swindle Rabin." *Cryptologia*, 3, 187–189.

COLLISION RESISTANCE

Collision resistance is the property of a hash function that it is computationally infeasible to find two colliding inputs. This property is related to second preimage resistance, which is also known as weak collision resistance. A minimal requirement for a hash function to be collision resistant is that the length of its result should be 160 bits (in 2004). A hash function is said to be a collision resistant hash function (CRHF) if it is a collision resistant one-way hash function (OWHF) (see hash function). The exact relation between collision resistance, second preimage resistance, and preimage resistance is rather subtle, and depends on the formalization of the definition: it is shown in [8] that under certain conditions, *collision resistance* implies second preimage resistance and preimage resistance.

In order to formalize the definition of a collision resistant hash function (see [1]), one needs to introduce a class of functions indexed by a public parameter, which is called a key. Indeed, one cannot require that there does not exist an adversary who can produce a collision for a fixed hash function, since any simple adversary who stores two short colliding inputs for a function would be able to output a collision efficiently. Introducing a class of functions solves this problem, since an adversary cannot store a collision for each value of the key (provided that the key space is not too small).

For a hash function with an n-bit result, an efficient collision research based on the birthday paradox requires approximately $2^{n/2}$ hash function evaluations. One can substantially reduce the memory requirements (and also the memory accesses) by translating the problem to the detection of a cycle in an iterated mapping. This was first proposed by Quisquater and Delescaille [6]. Van Oorschot and Wiener propose an efficient parallel variant of this algorithm [10]; with a US\$ 10 million machine, collisions for MD5 (with $n = 128$) can be found in 21 days in 1994, which corresponds to 5 hours in 2004. In order to make a collision search infeasible for the next 15–20 years, the hash result needs to be 180 bits or more.

Second preimage resistance and collision resistance of hash functions have been introduced by Rabin in [7]; the attack based on the <u>birthday paradox</u> was first pointed out by Yuval [11]. Further work on collision resistance can be found in [1–5, 9, 12]. For an extensive discussion of the relation between collision resistance and (second) preimage resistance, the reader is referred to Rogaway and Shrimpton [8].

B. Preneel

References

[1] Damgård, I.B. (1990). "A design principle for hash functions." *Advances in Cryptology—CRYPTO'89*, Lecture Notes in Computer Science, vol. 435, ed. G. Brassard. Springer-Verlag, Berlin, 416–427.

[2] Gibson, J.K. (1990). "Some comments on Damgård's hashing principle." *Electronics Letters*, 26 (15), 1178–1179.

[3] Merkle, R. (1979). *Secrecy, Authentication, and Public Key Systems*. UMI Research Press, Ann Arbor, MI.

[4] Preneel, B. (1993). "Analysis and design of cryptographic hash functions." *Doctoral Dissertation*, Katholieke Universiteit Leuven.

[5] Preneel, B. (1999). "The state of cryptographic hash functions." *Lectures on Data Security*, Lecture Notes in Computer Science, vol. 1561, ed. I. Damgård. Springer-Verlag, Berlin, 158–182.

[6] Quisquater, J.-J. and J.-P. Delescaille (1990). "How easy is collision search? Application to DES." *Advances in Cryptology—EUROCRYPT'89*, Lecture Notes in Computer Science, vol. 434, eds. J.-J. Quisquater and J. Vandewalle. Springer-Verlag, Berlin, 429–434.

[7] Rabin, M.O. (1978). "Digitalized signatures." *Foundations of Secure Computation*, eds. R. Lipton and R. DeMillo. Academic Press, New York, 155–166.

[8] Rogaway, P. and T. Shrimpton (2004). "Cryptographic hash function basics: Definitions, implications, and separations for preimage resistance, second-preimage resistance, and collision resistance." *Fast Software Encryption*, Lecture Notes in Computer Science, Springer-Verlag, Berlin, to appear.

[9] Stinson, D. (2001). "Some observations on the theory of cryptographic hash functions." Technical Report 2001/020, University of Waterloo.

[10] van Oorschot, P.C. and M. Wiener (1999). "Parallel collision search with cryptanalytic applications." *Journal of Cryptology*, 12 (1), 1–28.

[11] Yuval, G. (1979). "How to swindle Rabin." *Cryptologia*, 3, 187–189.

[12] Zheng, Y., T. Matsumoto, and H. Imai (1990). "Connections between several versions of one-way hash functions." *Proc. SCIS90, The 1990 Symposium on Cryptography and Information Security, Nihondaira, Japan, January 31–February 2, 1990*.

COMBINATION GENERATOR

A *combination generator* is a <u>running-key</u> generator for <u>stream cipher</u> applications. It is composed of several <u>linear feedback shift registers (LFSRs)</u> whose outputs are combined by a <u>Boolean function</u> to produce the keystream. Then, the output sequence $(s_t)_{t \geq 0}$ of a combination generator composed of n LFSRs is given by

$$s_t = f(u_t^1, u_t^2, \ldots, u_t^n), \qquad \forall t \geq 0,$$

where $(u_t^i)_{t \geq 0}$ denotes the sequence generated by the ith constituent LFSR and f is a function of n variables. In the case of a combination generator composed of n LFSRs over $\mathbf{F}_q$, the combining function is a function from $\mathbf{F}_q^n$ into $\mathbf{F}_q$.

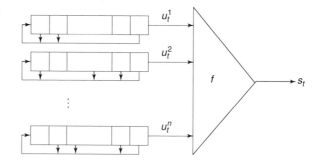

The combining function f should obviously be *balanced*, i.e., its output should be uniformly distributed. The constituent LFSRs should be chosen to have primitive feedback polynomials (see <u>primitive element</u>) for ensuring good statistical properties of their output sequences (see <u>Linear Feedback Shift Register</u> for more details).

The characteristics of the constituent LFSRs and the combining function are usually publicly known. The secret parameters are the initial states of the LFSRs, which are derived from the secret key of the cipher by a key-loading algorithm. Therefore, most attacks on combination generators consist in recovering the initial states of all LFSRs from the knowledge of some digits of the sequence produced by the generator (in a <u>known plaintext attack</u>), or of some digits of the ciphertext sequence (in a <u>ciphertext only attack</u>). When the feedback polynomials of the LFSR and the combining function are not known, the reconstruction attack presented in [2] enables to recover the complete description of the generator from the knowledge of a large segment of the ciphertext sequence.

STATISTICAL PROPERTIES OF THE OUTPUT SEQUENCE: The sequence produced by a combination generator is a linear recurring sequence.

Its period and its <u>linear complexity</u> can be derived from those of the sequences generated by the constituent LFSRs and from the *algebraic normal form* of the combining function (see <u>Boolean function</u>). Indeed, if we consider two linear recurring sequences **u** and **v** over $\mathbf{F}_q$ with linear complexities $\Lambda(\mathbf{u})$ and $\Lambda(\mathbf{v})$, we have the following properties:

- The linear complexity of the sequence $\mathbf{u} + \mathbf{v} = (u_t + v_t)_{t \geq 0}$ satisfies

$$\Lambda(\mathbf{u} + \mathbf{v}) \leq \Lambda(\mathbf{u}) + \Lambda(\mathbf{v}),$$

  with equality if and only if the <u>minimal polynomials</u> of **u** and **v** are relatively prime. Moreover, in the case of equality, the period of $\mathbf{u} + \mathbf{v}$ is the <u>least common multiple</u> of the periods of **u** and **v**.

- The linear complexity of the sequence $\mathbf{uv} = (u_t v_t)_{t \geq 0}$ satisfies

$$\Lambda(\mathbf{uv}) \leq \Lambda(\mathbf{u})\Lambda(\mathbf{v}),$$

  where equality holds if the minimal polynomials of **u** and **v** are primitive and if $\Lambda(\mathbf{u})$ and $\Lambda(\mathbf{v})$ are distinct and greater than 2. Other general sufficient conditions for $\Lambda(\mathbf{uv}) = \Lambda(\mathbf{u})\Lambda(\mathbf{v})$ can be found in [3–5].

Thus, the keystream sequence produced by a combination generator composed of n binary LFSRs with primitive feedback polynomials which are combined by a Boolean function f satisfies the following property proven in [5]. If all LFSR lengths $L_1, \ldots, L_n$ are distinct and greater than 2 (and if all LFSR initializations differ from the all-zero state), the linear complexity of the output sequence **s** is equal to

$$f(L_1, L_2, \ldots, L_n),$$

where the algebraic normal form of f is evaluated over integers. For instance, if four LFSRs of lengths $L_1, \ldots, L_4$ satisfying the previous conditions are combined by the Boolean function $x_1 x_2 + x_2 x_3 + x_4$, the linear complexity of the resulting sequence is $L_1 L_2 + L_2 L_3 + L_4$. Similar results concerning the combination of LFSRs over $\mathbf{F}_q$ can be found in [5] and [1]. A high linear complexity is a desirable property for a keystream sequence since it ensures that the <u>Berlekamp–Massey algorithm</u> becomes computationally infeasible. Thus, the combining function f should have a high *algebraic degree* (the algebraic degree of a Boolean function is the highest number of terms occurring in a monomial of its algebraic normal form).

KNOWN ATTACKS AND RELATED DESIGN CRITERIA: Combination generators are vulnerable to the <u>correlation attack</u> and its variants called <u>fast correlation attacks</u>. In order to make these attacks infeasible, the LFSR feedback polynomials should not be sparse. The combining function should have a high *correlation-immunity order*, also called *resiliency order* when the involved function is balanced (see <u>correlation-immune and resilient Boolean function</u>). But, there exists a trade-off between the correlation-immunity order and the algebraic degree of a Boolean function. Most notably, the correlation-immunity of a balanced Boolean function of n variables cannot exceed $n - 1 - \deg(f)$, when the algebraic degree of f, $\deg(f)$, is greater than 1. Moreover, the complexity of <u>correlation attacks</u> and of <u>fast correlation attacks</u> also increases with the <u>nonlinearity</u> of the combining function (see <u>correlation attack</u>). The trade-offs between high algebraic degree, high correlation-immunity order, and high nonlinearity can be circumvented by replacing the combining function by a finite state automaton with memory. Examples of such combination generators with memory are the summation generator and the stream cipher <u>E0</u> used in Bluetooth.

Anne Canteaut

References

[1] Brynielsson, L. (1986). "On the linear complexity of combined shift register sequences." *Advances in Cryptology—EUROCRYPT'85*, Lecture Notes in Computer Science, vol. 219, ed. F. Pichler. Springer-Verlag, Berlin, 156–160.

[2] Canteaut, A. and E. Filiol (2001). "Ciphertext only reconstruction of stream ciphers based on combination generators." *Fast Software Encryption 2000*, Lecture Notes in Computer Science, vol. 1978, ed. B. Schneier. Springer-Verlag, Berlin, 165–180.

[3] Herlestam, T. (1986). "On functions of linear shift register sequences." *Advances in Cryptology—EUROCRYPT'85*, Lecture Notes in Computer Science, vol. 219, ed. F. Pichler. Springer-Verlag, Berlin, 119–129.

[4] Göttfert, R. and H. Niederreiter (1995). "On the minimal polynomial of the product of linear recurring sequences." *Finite Fields and their Applications*, 1 (2), 204–218.

[5] Rueppel, R.A. and O.J. Staffelbach (1987). "Products of linear recurring sequences with maximum complexity." *IEEE Transactions on Information Theory*, 33 (1), 124–131.

COMMITMENT

COMMITMENT: A commitment scheme is a two-phase cryptographic <u>protocol</u> between two parties, a sender and a receiver, satisfying the following constraints. At the end of the Commit phase the

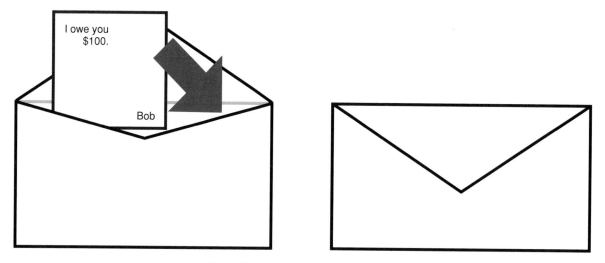

Fig. 1. Committing with an envelope

sender is committed to a specific value (often a single bit) that he cannot change later on (Commitments are binding) and the receiver should have no information about the committed value, other than what he already knew before the protocol (Commitments are concealing). In the Unveil phase, the sender sends extra information to the receiver that allows him to determine the value that was concealed by the commitment. Bit commitments are important components of zero-knowledge protocols [4, 16], and other more general two-party cryptographic protocols [19].

A natural intuitive implementation of a commitment is performed using an envelope (see Figure 1). Some information written on a piece of paper may be committed to by sealing it inside an envelope. The value inside the sealed envelope cannot be guessed (envelopes are concealing) without modifying the envelope (opening it) nor the content can be modified (envelopes are binding).

Unveiling the content of the envelope is achieved by opening it and extracting the piece of paper inside (see Figure 2).

The terminology of commitments, influenced by the legal vocabulary, first appeared in the contract signing protocols of Even [14], although it seems fair to attribute the concept to Blum [3] who implicitly uses it for coin flipping around the same time. In his Crypto 81 paper, Even refers to Blum's contribution saying: In the summer of 1980, in a conversation, M. Blum suggested the use of randomization for such protocols. So apparently Blum introduced the idea of using random hard problems to commit to something (coin, contract, etc.). However, one can also argue that the earlier work of Shamir et al. [26] on mental poker implicitly used commitments as well, since in order to generate a fair deal of cards, Alice encrypts the card names under her own encryption key, which is the basic idea for implementing commitments.

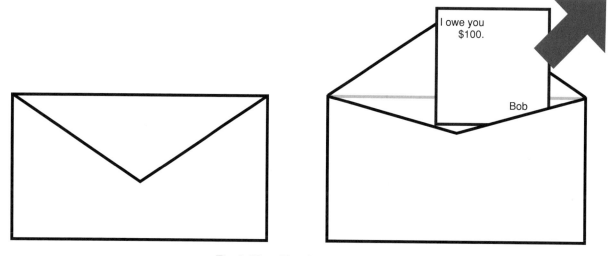

Fig. 2. Unveiling from an envelope

Under such computational assumptions, commitments come in two dual flavors: binding but computationally concealing commitments and concealing but computationally binding commitments.

Commitments of the first type may be achieved from any one-way function [18, 24] while those of the second type may be achieved from any one-way permutation (or at least regular one-way function) [25] or any collision-free hash function [17] (see also collision resistance and hash function). It is still an open problem to achieve commitments of the second type from one-way functions only.

A simple example of a bit commitment of the first type is obtained using the Goldwasser–Micali probabilistic encryption scheme with one's own pair of public keys (n, q) such that n is an RSA modulus (see RSA public key encryption) and q a random quadratic nonresidue modulo n with Jacobi symbol $+1$ (see also quadratic residue). Unveiling is achieved by providing a square root of each quadratic residue and of quadratic nonresidue multiplied by q. A similar example of a bit commitment of the second type is constructed from someone else's pair of public keys (n, r) such that n is an RSA modulus and r a random quadratic residue modulo n. A zero bit is committed using a random quadratic residue mod n while a one bit is committed using a random quadratic residue multiplied by r modulo n. Unveiling is achieved by providing a square root of quadratic residues committing to a zero and of quadratic residues multiplied by r used to commit to a one.

Unconditionally binding and concealing commitments can also be obtained under the assumption of the existence of a binary symmetric channel [10] and under the assumption that the receiver owns a bounded amount of memory [6]. In multiparty scenarios [2, 8, 16], commitments are usually achieved through Verifiable Secret Sharing Schemes [9]. However, the two-prover case [1] does not require the verifiable property because the provers are physically isolated from each other during the life span of the commitments.

In a quantum computation model (see quantum cryptography) it was first believed that commitment schemes could be implemented with unconditional security for both parties [5] but it was later demonstrated that if the sender is equipped with a quantum computer, then any unconditionally concealing commitment cannot be binding [22, 23].

Commitments exist with various extra properties: chameleon/trapdoor commitments [1, 15], commitments with equality (attributed to Bennett and Rudich in [11, 20]), nonmalleable commitments [13] (with respect to unveiling [12]), mutually independent commitments [21], and universally composable commitments [7].

Claude Crépeau

References

[1] Ben-Or, M., S. Goldwasser, J. Kilian, and A. Wigderson (1988). "Multi-prover interactive proofs: How to remove intractability assumptions." *Proceedings of 20th Annual AMC Symposium on Theory of Computing*, 113–122.

[2] Ben-Or, M., S. Goldwasser, and A. Wigderson (1988). "Completeness theorems for fault-tolerant distributed computing." *Proc. 20th ACM Symposium on Theory of Computing, Chicago, 1988.* ACM Press, New York, 1–10.

[3] Blum, M. (1982). "Coin flipping by telephone." *Advances in Cryptography, Santa Barbara, California, USA*, ed. Allen Gersho. University of California, Santa Barbara, 11–15.

[4] Brassard, G., D. Chaum, and C. Crépeau (1998). "Minimum disclosure proofs of knowledge." *JCSS*, 37, 156–189.

[5] Brassard, G., C. Crépeau, R. Jozsa, and D. Langlois (1993). "A quantum bit commitment scheme provably unbreakable by both parties." *29th Symp. on Found. of Computer Sci.* IEEE, Piscataway, NJ, 42–52.

[6] Cachin, C., C. Crépeau, and J. Marcil (1998). "Oblivious transfer with a memory-bounded receiver." *39th Annual Symposium on Foundations of Computer Science: Proceedings, November 8–11, 1998, Palo Alto, California.* IEEE Computer Society Press, Los Alamitos, CA, 493–502.

[7] Canetti Ran and Marc Fischlin (2001). "Universally composable commitments." *Advances in Cryptology—CRYPTO 2001, International Association for Cryptologic Research*, Lecture Notes in Computer Science, vol. 2139, ed. Joe Kilian. Springer-Verlag, Berlin, 19–40.

[8] Chaum, D., C. Crépeau, and I. Damgård (1988). "Multi-party unconditionally secure protocols." *Proc. 20th ACM Symposium on Theory of Computing, Chicago, 1988.* ACM Press, New York.

[9] Chor Benny, Shafi Goldwasser, Silvio Micali, and Baruch Awerbuch (1985). "Verifiable secret sharing and achieving simultaneity in the presence of faults (extended abstract)." *Proc. of 26th FOCS, Portland, OR, October 21–23, 1985.* IEEE, Piscataway, NJ, 383–395.

[10] Crépeau C. (1997). "Efficient cryptographic protocols based on noisy channels." *Advances in Cryptology—EUROCRYPT'97*, Lecture Notes in Computer Science, vol. 1233, ed. Walter Fumy. Springer-Verlag, Berlin, 306–317.

[11] Crépeau C., J. van de Graaf, and A. Tapp (1995). "Committed oblivious transfer and private multi-party computation." *Advances in Cryptology—CRYPTO'95*, Lecture Notes in Computer

Science, vol. 963, ed. Don Coppersmith. Springer-Verlag, Berlin, 110–123.

[12] Di Crescenzo, Giovanni, Yuval Ishai, and Rafail Ostrovsky (1998). "Non-interactive and non-malleable commitment." *30th Symposium on the Theory of Computing*, 141–150.

[13] Dolev, D., C. Dwork, and M. Naor (1991). "Non-malleable cryptography." *Proceedings of the Twenty Third Annual ACM Symposium on Theory of Computing, New Orleans, LA, May 6–8, 1991.* IEEE Computer Society Press, Los Alamitos, CA.

[14] Even, S. (1982). "Protocol for signing contracts." *Advances in Cryptography, Santa Barbara, CA, USA, 1982*, ed. Allen Gersho. University of California, Santa Barbara.

[15] Feige, U. and A. Shamir (1989). "Zero knowledge proofs of knowledge in two rounds." *Advances in Cryptology—CRYPTO'89*, Lecture Notes in Computer Science, vol. 435, ed. Gilles Brassard. Springer-Verlag, Berlin, 526–544.

[16] Goldreich Oded, Silvio Micali, and Avi Wigderson (1991). "Proofs that yield nothing but their validity or all languages in NP have zero-knowledge proof systems." *Journal of the Association for Computing Machinery*, 38 (3), 691–729.

[17] Halevi, S. and S. Micali (1996). "Practical and provably-secure commitment schemes from collision-free hashing." *Advances in Cryptology—CRYPTO'96*, Lecture Notes in Computer Science, vol. 1109, ed. Neal Koblitz. Springer-Verlag, Berlin, 201–215.

[18] Håstad, Johan, Russell Impagliazzo, Leonid A. Levin, and Michael Luby (1999). "A pseudorandom generator from any one-way function." *SICOMP. SIAM Journal on Computing*, 28 (4), 1364–1396.

[19] Kilian, J. (1988). "Founding cryptography on oblivious transfer." *Proc. 20th ACM Symposium on Theory of Computing, Chicago, 1988.* ACM Press, New York, 20–31.

[20] Kilian Joe (1992). "A note on efficient zero-knowledge proofs and arguments (extended abstract)." *Proceedings of the Twenty-Fourth Annual ACM Symposium on the Theory of Computing, Victoria, British Columbia, Canada, May 4–6, 1992*, 723–732.

[21] Liskov Moses, Anna Lysyanskaya, Silvio Micali, Leonid Reyzin, and Adam Smith (2001). "Mutually independent commitments." *Advances in Cryptology—ASIACRYPT 2001*, Lecture Notes in Computer Science, vol. 2248, ed. C. Boyd. Springer, Berlin, 385–401.

[22] Lo, H.-K. and F. Chau (1997). "Is quantum bit commitment really possible?" *Physical Review Letters*, 78 (17), 3410–3413.

[23] Mayers Dominic (1997). "Unconditionally secure quantum bit commitment is impossible." *Physical Review Letters*, 78 (17), 3414–3417.

[24] Naor Moni (1991). "Bit commitment using pseudorandomness." *Journal of Cryptology*, 4 (2), 151–158.

[25] Naor, M., R. Ostrovsky, R. Venkatesan, and M. Yung (1993). "Perfect zero-knowledge arguments for NP can be based on general complexity assumptions." *Advances in Cryptology—CRYPTO'92*, Lecture Notes on Computer Science, vol. 740, ed. E.F. Brickell. Springer-Verlag. This work was first presented at the DIMACS Workshop on Cryptography, October 1990.

[26] Shamir, A., R.L. Rivest, and L.M. Adleman (1981). "Mental poker." *The Mathematical Gardner*, ed. D. Klarner. Wadsworth, Belmont, CA.

COMMON CRITERIA

The Common Criteria (CC) is meant to be used as the basis for evaluation of security properties of IT products and systems. The objective desired is that by establishing a common base for criteria, the evaluation results of an IT product will be of more value to a wider audience.

The goal is for Common Criteria to permit comparability of products based on the results of independent security evaluations for various products evaluated by separate organizations in different countries. The vision is that by providing a common set of requirements for the security functions of IT products, and a common set of assurance measurements applied to them that the evaluation process will establish a level of confidence in the knowledge and trust of the evaluated products. The evaluation results may help consumers to determine whether an IT product or system is appropriate for their intended application and whether the security risks implicit in its use are acceptable.

Common Criteria is not a security specification that prescribes specific or necessary security functionality or assurance. Therefore, it is not intended to verify the quality or security of cryptographic implementations. In general, products that require cryptography are often required to attain a FIPS 140-2 validation for their cryptographic functionality before the common criteria evaluation can be completed. There are security products that are very important to security but may not incorporate cryptography as a part of their functionality. Examples include operating systems, firewalls, and IDS systems. Common Criteria is a methodology to gain assurance that a product is actually designed and subsequently performs according to the claims in the product's "Security Target" document. The level of assurance (EAL) can be specified to one of seven levels described later.

The Common Criteria specification has been published as International Standard ISO/IEC 15408:1999. It is sometimes also published in formats specific to a given country that facilities use in their individual test scheme. The content and requirements are intended to be identical.

Seven governmental organizations, which are collectively called "the Common Criteria Project Sponsoring Organizations," were formed to develop the standard and program. The countries and organizations are:

- **Canada:** Communications Security Establishment
- **France:** Service Central de la Scurit des Systmes dInformation
- **Germany:** Bundesamt fr Sicherheit in der Informationstechnik
- **Netherlands:** Netherlands National Communications Security Agency
- **United Kingdom:** Communications-Electronics Security Group
- **United States:** National Institute of Standards and Technology
- **United States:** National Security Agency

The Common Criteria Project Sponsoring Organizations approved the licensing and use of CC v2.1 to be the basis of ISO 15408. Because of its international basis, certifications under Common Criteria are under a "Mutual Recognition Agreement." This is an agreement that certificates issued by organizations under a specific scheme will be accepted in other countries as if they were evaluated under their own schemes. The list of countries that have signed up to the mutual recognition have grown beyond just the original sponsoring organizations.

Common Criteria incorporates a feature called a Protection Profile (PP). This is a document that specifies an implementation-independent set of security requirements for a category of products (i.e., Traffic Filters or smart cards) that meet the needs of specific consumer groups, communities of interest, or applications. Protection Profiles are considered a product in themselves, and are evaluated and tested for compliance to Common Criteria, just as a functional product would. Before a product can be validated using common criteria to a given protection profile (or a combination of them), the Protection Profiles have to be evaluated and issued certificates of compliance. Instead of the Security Target (a required document) referring to a protection profile for a set of security functionality and assurance criteria, it is acceptable for the product Security Target to independently state the security functionality and assurance level to which the product will be evaluated. The limitation is that this restricts the ability of product consumers or users to readily compare products of similar functionality.

EAL1. The objective for evaluation assurance level 1 (EAL1) is described as "functionally tested" is to confirm that the product functions in a manner consistent with its documentation, and that it provides useful protection against identified threats.

EAL1 is applicable where some confidence in correct operation is required, but the threats to security are not viewed as serious. The evaluation will be of value where independent assurance is required to support the contention that due care has been exercised with respect to the protection of personal or similar information.

EAL1 provides an evaluation of the product as made available to the customer, including independent testing against a specification, and an examination of the guidance documentation provided. It is intended that an EAL1 evaluation could be successfully conducted without assistance from the developer of the product, and for minimal cost and schedule impact.

EAL2. The objective for evaluation assurance level 2 (EAL2) is described as "structurally tested."

EAL2 requires the cooperation of the developer in terms of the delivery of design information and test results, but should not demand more effort on the part of the developer than is consistent with good commercial practice, and therefore, should not require a substantially increased investment of cost or time.

EAL2 is applicable in those circumstances where developers or users require a low to moderate level of independently assured security but does not require the submission of a complete development record by the vendor. Such a situation may arise when securing legacy systems, or where access to the developer may be limited.

EAL3. The objectives for evaluation assurance level 3 (EAL3) are described as "methodically tested and checked."

EAL3 permits a conscientious developer to gain maximum assurance from positive security engineering at the design stage without substantial alteration of existing sound development practices.

EAL3 is applicable in those circumstances where developers or users require a moderate level of independently assured security, and require a thorough investigation of the product and its development without substantial re-engineering.

EAL4. The objectives for evaluation assurance level 4 (EAL4) are described as "methodically designed, tested, and reviewed."

EAL4 permits a developer to gain maximum assurance from positive security engineering based on good commercial development practices, which, though rigorous, do not require substantial specialist knowledge, skills, and other resources.

EAL4 is therefore applicable in those circumstances where developers or users require a moderate to high level of independently assured security in conventional commodity products and are prepared to incur additional security-specific engineering costs.

EAL5. The objectives for evaluation assurance level 5 (EAL5) are described as "semiformally designed and tested."

EAL5 permits a developer to gain maximum assurance from security engineering based upon rigorous commercial development practices supported by moderate application of specialist security engineering techniques. Such a product will probably be designed and developed with the intent of achieving EAL5 assurance. It is likely that the additional costs attributable to the EAL5 requirements, relative to rigorous development without the application of specialized techniques, will not be large.

EAL5 is therefore applicable in those circumstances where developers or users require a high level of independently assured security in a planned development and require a rigorous development approach without incurring unreasonable costs attributable to specialist security engineering techniques.

EAL6. The objectives for evaluation assurance level 6 (EAL6) are described as "semiformally verified design and tested."

EAL6 permits developers to gain high assurance from application of security engineering techniques to a rigorous development environment in order to produce a premium product for protecting high value assets against significant risks.

EAL6 is therefore applicable to the development of security product for application in high-risk situations where the value of the protected assets justifies the additional costs.

EAL7. The objectives of evaluation assurance level 7 (EAL7) are described as "formally verified design and tested."

EAL7 is applicable to the development of security products for application in extremely high-risk situations and/or where the high value of the assets justifies the higher costs. Practical application of EAL7 is currently limited to products with tightly focused security functionality that is amenable to extensive formal analysis.

Common Criteria is documented in a family of three interrelated documents:
1. CC Part 1: Introduction and general model
2. CC Part 2: Security functional requirements
3. CC Part 3: Security assurance requirements.
The managed international homepage of the Common Criteria is available at www.commoncriteria .org. The homepage for US based vendors and customers is managed by NIST at http://csrc.nist .gov/cc.

Part 1, introduction and general model, is the introduction to the CC. It defines general concepts and principles of IT security evaluation and presents a general model of evaluation. Part 1 also presents constructs for expressing IT security objectives, for selecting and defining IT security requirements, and for writing high-level specifications for products and systems. In addition, the usefulness of each part of the CC is described in terms of each of the target audiences.

Part 2, security functional requirements, establishes a set of functional components as a standard way of expressing the functional requirements for Targets of Evaluation. Part 2 catalogs the set of functional components, families, and classes.

Part 3, security assurance requirements, establishes a set of assurance components as a standard way of expressing the assurance requirements for Targets of Evaluation. Part 3 catalogs the set of assurance components, families, and classes. Part 3 also defines evaluation criteria for Protection Profiles and Security Targets and presents evaluation assurance levels that define the predefined CC scale for rating assurance for Targets of Evaluation, which is called the Evaluation Assurance Levels.

Each country implements its own scheme of how it will implement the Common Evaluation Methodology for Information Technology Security.

Tom Caddy

COMMUNICATION CHANNEL ANONYMITY

Communication channel anonymity or *relationship anonymity* [4] is achieved in a messaging system if an eavesdropper who picks up messages from the communication line of a sender and the communication line of a recipient cannot tell with better probability than pure guess whether the sent message is the same as the received message. During the attack, the eavesdropper may also listen on all communication lines of the network, and he may also send and receive his own messages. It is clear that all messages in such a network must be encrypted to the same length in order to keep the attacker from distinguishing different messages by their content or length.

Communication channel anonymity implies either sender anonymity or recipient anonymity [4].

Communication channel anonymity can be achieved against computationally restricted eavesdroppers by MIX networks [1] and against computationally unrestricted eavesdroppers by DC networks [2,3].

Note that communication channel anonymity is weaker than communication link unobservability, where the attacker cannot even determine whether or not any message is exchanged between any particular pair of participants at any point of time. Communication link unobservability can be achieved with MIX networks and DC networks by adding dummy traffic.

Gerrit Bleumer

References

[1] Chaum, David (1981). "Untraceable electronic mail, return addresses, and digital pseudonyms." *Communications of the ACM*, 24 (2), 84–88.
[2] Chaum, David (1985). "Security without identification: Transaction systems to make Big Brother obsolete." *Communications of the ACM*, 28 (10), 1030–1044.
[3] Chaum, David (1988). "The dining cryptographers problem: Unconditional sender and recipient untraceability." *Journal of Cryptology*, 1 (1), 65–75.
[4] Pfitzmann, Andreas and Marit Köhntopp (2001). "Anonymity, unobservability, and pseudonymity— a proposal for terminology." *Designing Privacy Enhancing Technologies*, Lecture Notes in Computer Science, vol. 2009, ed. H. Frederrath. Springer-Verlag, Berlin, 1–9.

COMPROMISING EMANATIONS

Computer and communications devices emit numerous forms of energy. Many of these emissions are produced as unintended side effects of normal operation. For example, where these emissions take the form of radio waves, they can often be observed interfering with nearby radio receivers. Some of the unintentionally emitted energy carries information about processed data. Under good conditions, a sophisticated and well-equipped eavesdropper can intercept and analyze

such compromising emanations to steal information. Even where emissions are intended, as is the case with transmitters and displays, only a small fraction of the overall energy and information content emitted will ever reach the intended recipient. Eavesdroppers can use specialized more sensitive receiving equipment to tap into the rest and access confidential information, often in unexpected ways, as some of the following examples illustrate.

Much knowledge in this area is classified military research. Some types of compromising emanations that have been demonstrated in the open literature include:

- Radio-frequency waves radiated into free space
- Radio-frequency waves conducted along cables
- Power-supply current fluctuations
- Vibrations, acoustic and ultrasonic emissions
- High-frequency optical signals.

They can be picked up passively using directional antennas, microphones, high-frequency power-line taps, telescopes, radio receivers, oscilloscopes, and similar sensing and signal-processing equipment. In some situations, eavesdroppers can obtain additional information by actively directing radio waves or light beams toward a device and analyzing the reflected energy.

Some examples of compromising emanations are:

- Electromagnetic impact printers can produce low-frequency acoustic, magnetic, and power-supply signals that are characteristic for each printed character. In particular, this has been demonstrated with some historic dot-matrix and "golfball" printers. As a result, printed text could be reconstructed with the help of power-line taps, microphones, or radio antennas. The signal sources are the magnetic actuators in the printer and the electronic circuits that drive them.

- Cathode-ray tube (CRT) displays are fed with an analog video signal voltage, which they amplify by a factor of about 100 and apply it to a control grid that modulates the electron beam. This arrangement acts, together with the video cable, as a parasitic transmission antenna. As a result, CRT displays emit the video signal as electromagnetic waves, particularly in the VHF and UHF bands (30 MHz to 3 GHz). An AM radio receiver with a bandwidth comparable to the pixel-clock frequency of the video signal can be tuned to one of the harmonics of the emitted signal. The result is a high-pass filtered and rectified approximation of the original video signal. It lacks color information and each vertical edge appears merely as a line. Figure 1 demonstrates

Fig. 1. The top image shows a short test text displayed on a CRT monitor. In the bottom image, the compromising emanation from this text was picked up with the help of an AM radio receiver (tuned at 450 MHz, 50 MHz bandwidth) and a broadband UHF antenna. The output was then digitized, averaged over 256 frames to reduce noise, and finally presented as a reconstructed pixel raster

that text characters remain quite readable after this distortion. Where the display font and character spacing are predictable, automatic text recognition is particularly practical. For older, 1980s, video displays, even modified TV sets, with deflection frequencies adjusted to match those of the eavesdropped device, could be used to demonstrate the reconstruction of readable text at a distance [7]. In modern computers, pixel-clock frequencies exceed the bandwidth of TV receivers by an order of magnitude. Eavesdropping attacks on these require special receivers with large bandwidth (50 MHz or more) connected to a computer monitor or high-speed signal-processing system [3].

- CRT displays also leak the video signal as a high-frequency fluctuation of the emitted light. On this channel, the video signal is distorted by the afterglow of the screen phosphors and by the shot noise that background light contributes. It is possible to reconstruct readable text from screen light even after diffuse reflection, for example from a user's face or a wall. This can be done from nearby buildings using a telescope connected to a very fast photosensor (photomultiplier tube). The resulting signal needs to be digitally processed using periodic-averaging and deconvolution techniques to become readable. This attack is particularly feasible in dark environments, where light from the target CRT contributes a significant fraction of the overall illumination onto the observed surface. Flat-panel displays that update all pixels in a row simultaneously are immune from this attack [2].
- Some flat-panel displays can be eavesdropped via UHF radio, especially where a high-speed digital serial connection is used between the video controller and display. This is the case, for example, in many laptops and with modern graphics cards with a Digital Visual Interface

(DVI) connector. To a first approximation, the signal picked up by an eavesdropping receiver from a Gbit/s serial video interface cable indicates the number of bit transitions in the data words that represent each pixel color. For example, text that is shown in foreground and background colors encoded by the serial data words 10101010 and 00000000, respectively, will be particularly readable via radio emanations [3].

- Data has been eavesdropped successfully from shielded RS-232 cables several meters away with simple AM shortwave radios [5]. Such serial-interface links use unbalanced transmission. Where one end lacks an independent earth connection, the cable forms the inductive part of a resonant circuit that works in conjunction with the capacitance between the device and earth. Each edge in the data signal feeds into this oscillator energy that is then emitted as a decaying high-frequency radio wave.
- Line drivers for data cables have data-dependent power consumption, which can affect the supply voltage slightly. This in turn can cause small variations in the frequency of nearby oscillator circuits. As a result, the electromagnetic waves generated by these oscillators broadcast frequency-modulated data, which can be picked up with FM radios [5].
- Where several cables share the same conduit, capacitive and inductive coupling occurs. This can result in crosstalk from one cable to the other, even where the cables run parallel for just a few meters. With a suitable amplifier, an external eavesdropper might discover that the high-pass filtered version of a signal from an internal data cable is readable, for example, on a telephone line that leaves the building.
- Devices with low-speed serial ports, such as analog telephone modems with RS-232

interface, commonly feature light-emitting diodes (LEDs) that are connected as status indicators directly to data lines. These emit the processed data optically, which can be picked up remotely with a telescope and photo sensor [4]. Such optical compromising emanations are invisible to the human eye, which cannot perceive flicker above about 50 Hz. Therefore, all optical data rates above 1 kbit/s appear as constant light.

- The sound of a keystroke can identify which key on a keyboard was used. Just as guitar strings and drums sound very different depending on where they are hit, the mix of harmonic frequencies produced by a resonating circuit board on which keys are mounted varies with the location of the keystroke. Standard machine-learning algorithms can be trained to distinguish, for a specific keyboard model, keys based on spectrograms of acoustic keystroke recordings [1].

- Smart cards are used to protect secret keys and intermediate results of cryptographic computations from unauthorized access, especially from the cardholder. Particular care is necessary in their design with regard to compromising emanations. Due to the small package, eavesdropping sensors can be placed very close to the microprocessor, to record, for example, supply-current fluctuations or magnetic fields that leak information about executed instructions and processed data. The restricted space available in an only 0.8 mm thick plastic card makes careful shielding and filtering difficult. See also smartcard tamper resistance.

Video signals are a particularly dangerous type of compromising emanation due to their periodic nature. The refresh circuit in the video adapter transmits the display content continuously, repeated 60–90 times per second. Even though the leaked signal power measures typically only a few nanowatts, eavesdroppers can use digital signal-processing techniques to determine the exact repetition frequency, record a large number of frames, and average them to reduce noise from other radio sources. As frame and pixel frequencies differ by typically six orders of magnitude, the averaging process succeeds only if the frame rate has been determined correctly within at least seven digits precision. This is far more accurate than the manufacturing tolerances of the crystal oscillators used in graphics adapters. An eavesdropper can therefore use periodic averaging to separate the signals from several nearby video displays, even if they use the same nominal refresh frequency. Directional antennas are another tool for separating images from several computers in a building.

RF video-signal eavesdropping can be easily demonstrated with suitable equipment. Even in a noisy office environment and without directional antennas, reception across several rooms (5–20 meters) requires only moderate effort. Larger eavesdropping distances can be achieved in the quieter radio spectrum of a rural area or with the help of directional antennas. Eavesdropping of nonperiodic compromising signals from modern office equipment is usually only feasible where a sensor or accessible conductor (crosstalk) can be placed very close to the targeted device. Where an eavesdropper can arrange for special software to be installed on the targeted computer, this can be used to deliberately modulate many other emission sources with selected and periodically repeated data for easier reception, including system buses, transmission lines, and status indicators.

Compromising radio emanations are often broadband impulse signals that can be received at many different frequencies. Eavesdroppers tune their receivers to a quiet part of the spectrum, where the observed impulses can be detected with the best signal-to-noise ratio. The selected receiver bandwidth has to be small enough to suppress the powerful signals from broadcast stations on neighboring frequencies and large enough to keep the width of detected impulses short enough for the observed data rate.

Electromagnetic and acoustic compromising emanations have been a concern to military organizations since the 1960s. Secret protection standards (TEMPEST) have been developed. They define how equipment used to process critical secret information must be shielded, tested, and maintained. Civilian radio-emission limits for computers, such as the CISPR 22 and FCC Class B regulations, are only designed to help avoid interference with radio broadcast services at distances more than 10 meters. They do not forbid the emission of compromising signals that could be picked up at a quiet site by a determined receiver with directional antennas and careful signal processing several hundred meters away. Protection standards against compromising radio emanations therefore have to set limits for the allowed emission power about a million times (60 dB) lower than civilian radio-interference regulations. Jamming is an alternative form of eavesdropping protection, but this is not preferred in military applications where keeping the location of equipment secret is an additional requirement.

Markus Kuhn

References

[1] Asonov, Dmitri and Rakesh Agrawal (2004). "Keyboard acoustic emanations." *Proceedings 2004 IEEE Symposium on Security and Privacy, Oakland, California, May 9–12, 2004.* IEEE Computer Society Press, Los Alamitos, CA.

[2] Kuhn, Markus G. (2002). "Optical time-domain eavesdropping risks of CRT displays." *Proceedings 2002 IEEE Symposium on Security and Privacy, Oakland, California, May 12–15, 2002.* IEEE Computer Society Press, Los Alamitos, 3–18, ISBN 0-7695-1543-6.

[3] Kuhn, Markus G. (2003). "Compromising emanations: Eavesdropping risks of computer displays." Technical Report UCAM-CL-TR-577, University of Cambridge, Computer Laboratory.

[4] Loughry, Joe, and David A. Umphress (2002). "Information leakage from optical emanations." *ACM Transactions on Information Systems Security*, 5 (3), 262–289.

[5] Smulders, Peter (1990). "The threat of information theft by reception of electromagnetic radiation from RS-232 cables." *Computers & Security*, 9, 53–58.

[6] *1991 Symposium on Electromagnetic Security for Information Protection (SEPI'91), Proceedings, Rome, Italy, November 21–22, 1991.* Fondazione Ugo Bordoni.

[7] van Eck, Wim (1985). "Electromagnetic radiation from video display units: An eavesdropping risk?" *Computers & Security*, 4, 269–286.

COMPUTATIONAL COMPLEXITY

Computational complexity theory is the study of the minimal resources needed to solve computational problems. In particular, it aims to distinguish between those problems that possess efficient algorithms (the "easy" problems) and those that are inherently intractable (the "hard" problems). Thus computational complexity provides a foundation for most of modern cryptography, where the aim is to design cryptosystems that are "easy to use" but "hard to break." (See security.)

RUNNING TIME: The most basic resource studied in computational complexity is *running time*—the number of basic "steps" taken by an algorithm. (Other resources, such as *space* (i.e., memory usage), are also studied, but we will not discuss them here.) To make this precise, one needs to fix a model of computation (such as the Turing machine), but here we will informally think of it as the number of "bit operations" when the input is given as a string of 0's and 1's. Typically, the running time is measured as a function of the *input length*. For numerical problems, we assume the input is represented in binary, so the length of an integer N is roughly $\log_2 N$. For example, the elementary-school method for adding two n-bit numbers has running time proportional to n. (For each bit of the output, we add the corresponding input bits plus the carry.) More succinctly, we say that addition can be solved in time "order n," denoted $O(n)$ (see O-notation). The elementary-school multiplication algorithm, on the other hand, can be seen to have running time $O(n^2)$. In these examples (and in much of complexity theory), the running time is measured in the *worst case*. That is, we measure the maximum running time over all inputs of length n.

POLYNOMIAL TIME: Both the addition and multiplication algorithms are considered to be efficient, because their running time grows only mildly with the input length. More generally, polynomial time (running time $O(n^c)$ for a constant c) is typically adopted as the criterion of efficiency in computational complexity. The class of all computational problems possessing polynomial-time algorithms is denoted $\mathbf{P}$.[1] Thus ADDITION and MULTIPLICATION are in $\mathbf{P}$, and more generally we think of $\mathbf{P}$ as identifying the "easy" computational problems. Even though not all polynomial-time algorithms are fast in practice, this criterion has the advantage of robustness: the class $\mathbf{P}$ seems to be independent of changes in computing technology. $\mathbf{P}$ is an example of a *complexity class*—a class of computational problems defined via some algorithmic constraint, which in this case is "polynomial time."

In contrast, algorithms that do not run in polynomial time are considered infeasible. For example, consider the *trial division* algorithms for integer factoring or primality testing (see prime number). For an n-bit number, trial division can take time up to $2^{n/2}$, which is exponential time rather than polynomial time in n. Thus, even for moderate values of n (e.g., $n = 200$) trial division of n-bit numbers is completely infeasible for present-day computers, whereas addition and multiplication can be done in a fraction of a second. Computational complexity, however, is not concerned with the efficiency of a particular algorithm (such as trial division), but rather whether a problem has *any* efficient algorithm at all. Indeed,

---

[1] Typically, $\mathbf{P}$ is defined as a class of *decision problems* (i.e., problems with a yes/no answer), but here we make no such restriction.

for primality testing, there are polynomial-time algorithms known (see prime number), so PRIMALITY is in **P**. For integer factoring, on the other hand, the fastest known algorithm has running time greater than $2^{n^{1/3}}$, which is far from polynomial. Indeed, it is believed that FACTORING is not in **P**; the RSA and Rabin cryptosystems (see RSA public-key encryption, RSA digital signature scheme, Rabin cryptosystem, Rabin signature scheme) rely on this conjecture. One of the ultimate goals of computational complexity is to rigorously prove such *lower bounds*, i.e., establish theorems stating that there is no polynomial-time algorithm for a given problem. (Unfortunately, to date, such theorems have been elusive, so cryptography continues to rest on conjectures, albeit widely believed ones. More on this is given below.)

POLYNOMIAL SECURITY: Given the above association of "polynomial time" with feasible computation, the general goal of cryptography becomes to construct cryptographic protocols that have polynomial efficiency (i.e., can be executed in polynomial time) but super-polynomial security (i.e., cannot be broken in polynomial time). This guarantees that, for a sufficiently large setting of the *security parameter* (which roughly corresponds to the input length in complexity theory), "breaking" the protocol takes much more time than using the protocol. This is referred to as *asymptotic security*.

While polynomial time and asymptotic security are very useful for the theoretical development of the subject, more refined measures are needed to evaluate real-life implementations. Specifically, one needs to consider the complexity of using and breaking the system for fixed values of the input length, e.g., $n = 1000$, in terms of the actual time (e.g., in seconds) taken on current technology (as opposed to the "basic steps" taken on an abstract model of computation). Efforts in this direction are referred to as *concrete security*. Almost all results in computational complexity and cryptography, while usually stated asymptotically, can be interpreted in concrete terms. However, they are often not optimized for concrete security (where even constant factors hidden in O-notation are important).

Even with asymptotic security, it is sometimes preferable to demand that the growth of the gap between the efficiency and security of cryptographic protocols is faster than polynomial growth. For example, instead of asking simply for super-polynomial security, one may ask for *exponential security* (i.e., cannot be broken in time 2^{n^ϵ} for some $\epsilon > 0$). Based on the current best known algorithms, it seems that FACTORING may have

exponential hardness and hence the cryptographic protocols based on its hardness may have exponential security.[2]

COMPLEXITY-BASED CRYPTOGRAPHY: As described above, a major aim of complexity theory is to identify problems that cannot be solved in polynomial time and a major aim of cryptography is to construct protocols that cannot be broken in polynomial time. These two goals are clearly well-matched. However, since proving lower bounds (at least for the kinds of problems arising in cryptography) seems beyond the reach of current techniques in complexity theory, an alternative approach is needed.

Present-day complexity-based cryptography therefore takes a *reductionist approach*: it attempts to relate the wide variety of complicated and subtle computational problems arising in cryptography (forging a signature, computing partial information about an encrypted message, etc.) to a few, simply stated assumptions about the complexity of various computational problems. For example, under the assumption that there is no polynomial-time algorithm for FACTORING (that succeeds on a significant fraction of composites of the form $n = pq$), it has been demonstrated (through a large body of research) that it is possible to construct algorithms for almost all cryptographic tasks of interest (e.g., asymmetric cryptosystems, digital signature schemes, secure multiparty computation, etc.). However, since the assumption that FACTORING is not in **P** is only a conjecture and could very well turn out to be false, it is not desirable to have all of modern cryptography to rest on this single assumption. Thus another major goal of complexity-based cryptography is to abstract the properties of computational problems that enable us to build cryptographic protocols from them. This way, even if one problem turns out to be in **P**, any other problem satisfying those properties can be used without changing any of the theory. In other words, the aim is to base cryptography on assumptions that are as weak and general as possible.

Modern cryptography has had tremendous success with this reductionist approach. Indeed, it is now known how to base almost all basic cryptographic tasks on a few simple and general complexity assumptions (that do not rely on the

---

[2] In cryptography, a slightly different definition of exponential hardness is typically employed, with *exponential security* (compare *exponential time*) only referring to protocols that cannot be broken in time $2^{\epsilon n}$ for some $\epsilon > 0$. Accordingly, in cryptography, FACTORING is typically considered to provide *subexponential security* (compare subexponential time).

intractability of a single computational problem, but may be realized by any of several candidate problems). Among other things, the text below discusses the notion of a *reduction* from complexity theory that is central to this reductionist approach, and the types of general assumptions, such as the existence of *one-way functions*, on which cryptography can be based.

REDUCTIONS: One of the most important notions in computational complexity, which has been inherited by cryptography, is that of a *reduction* between computational problems. We say that problem Π reduces to problem Γ if Π can be solved in polynomial time given access to an "oracle" that solves Γ (i.e., a hypothetical black box that will solve Γ on instances of our choosing in a single time step). Intuitively, this captures the idea that problem Π is no harder than problem Γ. For a simple example, let us show that PRIMALITY reduces to FACTORING.[3] Suppose we have an oracle that, when fed any integer, returns its prime factorization in one time step. Then we could solve PRIMALITY in polynomial time as follows: on input N, feed the oracle with N, output "prime" if the only factor returned by the oracle is N itself, and output "composite" otherwise.

It is easy to see that if problem Π reduces to problem Γ, and Γ ∈ **P**, then Π ∈ **P**: if we substitute the oracle queries with the actual polynomial-algorithm for Γ, we obtain a polynomial-time algorithm for Π. Turning this around, Π ∉ **P** implies that Γ ∉ **P**. Thus, reductions provide a way to use an assumption that one problem is intractable to deduce that other problems are intractable. Much work in cryptography is based on this paradigm: for example, one may take a complexity assumption such as "there is no polynomial-time algorithm for FACTORING" and use reductions to deduce statements such as "there is no polynomial-time algorithm for breaking encryption scheme X." (As discussed later, for cryptography, the formalizations of such statements and the notions of reduction in cryptography are more involved than suggested here.)

NP: Another important complexity class is **NP**. Roughly speaking, this is the class of all computational problems for which solutions can be *verified* in polynomial time.[4] For example, given that PRIMALITY is in **P**, we can easily see that FACTORING is in **NP**: to verify that a supposed prime factorization of a number N is correct, we can simply test each of the factors for primality and check that their product equals N. **NP** can be thought of as the class of "well-posed" search problems: it is not reasonable to search for something unless you can recognize when you have found it. Given this natural definition, it is not surprising that the class **NP** has taken on a fundamental position in computer science.

It is evident that **P** ⊆ **NP**, but whether or not **P** = **NP** is considered to be one of the most important open problems in mathematics and computer science.[5] It is widely believed that **P** ≠ **NP**; indeed, we have seen that FACTORING is one candidate for a problem in **NP** \ **P**. In addition to FACTORING, **NP** contains many other computational problems of great importance, from many disciplines, for which no polynomial-time algorithms are known.

The significance of **NP** as a complexity class is due in part to the **NP**-*complete* problems. A computational problem Π is said to be **NP**-complete if Π ∈ **NP** and every problem in **NP** reduces to Π. Thus the **NP**-complete problems are the "hardest" problems in **NP**, and are the ones most likely to be intractable. (Indeed, if even a single problem in **NP** is not in **P**, then all the **NP**-complete problems are not in **P**.) Remarkably, thousands of natural computational problems have been shown to be **NP**-complete. (See [1].) Thus, it is an appealing possibility to build cryptosystems out of **NP**-complete problems, but unfortunately, **NP**-completeness does not seem sufficient for cryptographic purposes (as discussed later).

RANDOMIZED ALGORITHMS: Throughout cryptography, it is assumed that parties have the ability to make random choices; indeed this is how one models the notion of a secret key. Thus, it is natural to allow not just algorithms whose computation proceeds deterministically (as in the definition of **P**), but also consider *randomized algorithms*—ones that may make random choices in their computation. (Thus, such algorithms are designed to be implemented with a physical source of randomness. See <u>random bit generation (hardware)</u>.)

Such a randomized (*or probabilistic*) algorithm A is said to solve a given computational problem if on every input x, the algorithm outputs the correct answer with high probability (over its random

---

[3] Of course, this reduction is redundant given that PRIMALITY is in **P**, but suppose for a moment that we did not know this.

[4] **NP** stands for *nondeterministic polynomial time*. Like **P**, **NP** is typically defined as a class of decision problems, but again that constraint is not essential for our informal discussion.

[5] The significance of **P** versus **NP** in mathematics comes from the fact that it is equivalent to asking whether we can find short mathematical proofs efficiently.

choices). The error probability of such a random-ized algorithm can be made arbitrarily small by running the algorithm many times. For exam-ples of randomized algorithms, see the probabilis-tic primality tests in the entry on prime number. The class of computational problems having polynomial-time randomized algorithms is de-noted **BPP**.[6] A widely believed strengthening of the **P** $\neq$ **NP** conjecture is that **NP** $\not\subseteq$ **BPP**.

P VERSUS NP AND CRYPTOGRAPHY: The assumption **P** $\neq$ **NP** (and even **NP** $\not\subseteq$ **BPP**) is *nec-essary* for most of modern cryptography. For exam-ple, take any efficient encryption scheme and con-sider the following computational problem: given a ciphertext C, find the corresponding message M along with the key K and any randomization R used in the encryption process. This is an **NP** prob-lem: the solution (M, K, R) can be verified by re-encrypting the message M using the key K and the randomization R and checking whether the result equals C. Thus, if **P** = **NP**, this problem can be solved in polynomial time, i.e. there is an efficient algorithm for breaking the encryption scheme.[7]

However, the assumption **P** $\neq$ **NP** (or even **NP** $\not\subseteq$ **BPP**) does not appear *suffcient* for cryptography. The main reason for this is that **P** $\neq$ **NP** refers to *worst-case complexity*. That is, the fact that a computational problem Π is not in **P** only means that for every polynomial-time algorithm A, there *exist* inputs on which A fails to solve Π. However, these "hard inputs" could conceivably be very rare and very hard to find. Intuitively, to make use of intractability (for the security of cryptosystems), we need to be able to efficiently generate hard in-stances of an intractable computational problem.

ONE-WAY FUNCTIONS: The notion of a one-way function captures the kind of computational in-tractability needed in cryptography. Informally, a one-way function is a function f that is "easy to evaluate" but "hard to invert." That is, we require that the function f can be computed in polynomial time, but given $y = f(x)$, it is intractable to recover x. It is required that the difficulty of inversion holds even when the input x is chosen *at random*. Thus, we can efficiently generate hard instances

of the problem "find a preimage of y," by selecting x at random and setting $y = f(x)$. (Note that we actually generate a hard instance together with a solution; this is another aspect in which one-way functions are stronger than what follows from **P** $\neq$ **NP**.) To formalize the definition, we need the con-cept of a *negligible function*. A function $\epsilon : \mathbb{N} \to [0, 1]$ is negligible if for every constant c, there is an n_0 such that $\epsilon(n) \leq 1/n^c$ for all $n \geq n_0$. That is, ϵ vanishes faster than any polynomial. Then we have:

DEFINITION 1 (one-way function). *A one-to-one function f is* one-way *if it satisfies the following conditions.*
1. *(Easy to evaluate). f can be evaluated in polyno-mial time.*
2. *(Hard to invert). For every probabilistic polynomial-time algorithm A, there is a negli-gible function ϵ such that*

$$Pr[A(f(X)) = X] \leq \epsilon(n),$$

where the probability is taken over selecting an input X of length n uniformly at random and the random choices of the algorithm A.

For simplicity, we have only given the definition for one-to-one one-way functions. Without the one-to-one constraint, the definition should refer to the problem of finding *some* preimage of $f(X)$, i.e., re-quire the probability that $A(f(X)) \in f^{-1}(f(X))$ is negligible.[8]

The input length n can be thought of as corre-sponding to the *security parameter* (or *key length*) in a cryptographic protocol using f. If f is one-way, we are guaranteed that by making n sufficiently large, inverting f takes much more time than eval-uating f. However, to know how large to set n in an implementation requires a concrete security ana-logue of the above definition, where the maximum success probability ϵ is specified for A with a par-ticular running time on a particular input length n, and a particular model of computation.

The "inversion problem" is an **NP** problem (to verify that X is a preimage of Y, simply evaluate $f(X)$ and compare with Y.) Thus, if **NP** $\subseteq$ **BPP** then one-way functions do not exist. However, the converse is an open problem, and proving it would be a major breakthrough in complexity theory. For-tunately, even though the existence of one-way functions does not appear to follow from **NP** $\not\subseteq$ **BPP**, there are a number of natural candidates for one-way functions.

---

[6] **BPP** stands for "bounded-error probabilistic polynomial time."

[7] Technically, to conclude that the cryptosystem is broken re-quires that the message M is uniquely determined by cipher-text C. This will essentially be the case for most messages if the message length is greater than the key length. (If the message length is less than or equal to the key length, then there exist encryption schemes that achieve information-theoretic secu-rity (for a single encryption, e.g., the one-time pad), regardless of whether or not **P** = **NP**.)

[8] For technical reasons, we also need to require that f does not shrink its input too much, e.g. that the length of $|f(x)|$ and length of $|x|$ are polynomially related (in both directions.)

SOME CANDIDATE ONE-WAY FUNCTIONS: These examples are described informally, and may not all match up perfectly with the simplified definition above. In particular, some are actually *collections of one-way functions* $\mathcal{F} = \{f_i : \mathcal{D}_i \to \mathcal{R}_i\}$, in the functions f_i are parameterized by an index i that is generated by some randomized algorithm.[9]

1. (Multiplication) $f(p, q) = p \cdot q$, where p and q are primes of equal length. Inverting f is the FACTORING problem (see integer factoring), which indeed seems intractable even on random inputs of the form $p \cdot q$.

2. (Subset Sum) $f(x_1, \ldots, x_n, S) = (x_1, \ldots, x_n, \sum_{i \in S} x_i)$. Here each x_i is an n-bit integer and $S \subseteq [n]$. Inverting f is the SUBSET SUM problem (see knapsack cryptographic schemes). This problem is known to be **NP**-complete, but for the reasons discussed above, this does not provide convincing evidence that f is one-way (nevertheless it seems to be so).

3. (The Discrete Log Collection) $f_{G,g}(x) = g^x$, where G is a cyclic group (e.g., $G = \mathbb{Z}_p^*$ for prime p), g is a generator of G, and $x \in \{1, \ldots, |G| - 1\}$. Inverting $f_{G,g}$ is the DISCRETE LOG problem (see discrete logarithm problem), which seems intractable. This (like the next two examples) is actually a collection of one-way functions, parameterized by the group G and generator g.

4. (The RSA Collection) $f_{n,e}(x) = x^e \bmod n$, where n is the product of two equal-length primes, e satisfies $\gcd(e, \phi(n)) = 1$, and $x \in \mathbb{Z}_n^*$. Inverting $f_{n,e}$ is the RSA problem.

5. (Rabin's Collection) (see Rabin cryptosystem, Rabin digital signature scheme). $f_n(x) = x^2 \bmod n$, where n is a composite and $x \in \mathbb{Z}_n^*$. Inverting f_n is known to be as hard as factoring n.

6. (Hash Functions and Block Ciphers). Most cryptographic hash functions seem to be *finite* analogues of one-way functions with respect to *concrete* security. Similarly, one can obtain candidate one-way functions from block ciphers, say by defining $f(K)$ to be the block cipher applied to some fixed message using key K.

In a long sequence of works by many researchers, it has been shown that one-way functions are indeed the "right assumption" for complexity-based cryptography. On one hand, almost all tasks in cryptography imply the existence of one-way functions. Conversely (and more remarkably), many useful cryptographic tasks can be accomplished given any one-way function.

---

[9] Actually, one can convert a collection of one-way functions into a single one-way function, and conversely. See [3].

THEOREM 1. *The existence of one-way functions is necessary and sufficient for each of the following:*
- *The existence of* underlined{commitment} *schemes*
- *The existence of* pseudorandom number generators
- *The existence of* pseudorandom functions
- *The existence of* symmetric cryptosystems
- *The existence of* digital signature schemes.

These results are proven via the notion of reducibility mentioned above, albeit in much more sophisticated forms. For example, to show that the existence of one-way functions implies the existence of pseudorandom generators, one describes a general construction of a pseudorandom generator G from any one-way function f. To prove the correctness of this construction, one shows how to "reduce" the task of inverting the one-way function f to that of "distinguishing" the output of the pseudorandom generator G from a truly random sequence. That is, any polynomial-time algorithm that distinguishes the pseudorandom generator can be converted into a polynomial-time algorithm that inverts the one-way function. But if f is one-way, it cannot be inverted, so we conclude that the pseudorandom generator is secure. These reductions are much more delicate than those arising in, say, the **NP**-completeness, because they involve nontraditional computational tasks (e.g., inversion, distinguishing) that must be analyzed in the average case (i.e., with respect to non-negligible success probability).

The general constructions asserted in Theorem 1 are very involved and not efficient enough to be used in practice (though still polynomial time), so it should be interpreted only as a "plausibility result." However, from special cases of one-way functions, such as one-way permutations (see one-way function) or some of the specific candidate one-way functions mentioned earlier, much more efficient constructions are known.

TRAPDOOR FUNCTIONS: For some tasks in cryptography, most notably public-key encryption (see public-key cryptography), one-way functions do not seem to suffice, and additional properties are used. One such property is the *trapdoor* property, which requires that the function *can* be easily inverted given certain "trapdoor information." We do not give the full definition here, but just list the main properties. (See also trapdoor one-way function.)

DEFINITION 2 (trapdoor functions, informal). *A collection of one-to-one functions $\mathcal{F} = \{f_i : \mathcal{D}_i \to \mathcal{R}_i\}$ is a collection of* trapdoor functions *if*

1. *(Efficient generation). There is a probabilistic polynomial-time algorithm that, on input a security parameter n, generates a pair (i, t_i), where i is the index to a (random) function in the family and t_i is the associated "trapdoor information."*
2. *(Easy to evaluate). Given i and $x \in \mathcal{D}_i$, one can compute $f_i(x)$ in polynomial time.*
3. *(Hard to invert). There is no probabilistic polynomial-time algorithm that on input (i, $f_i(x)$) outputs x with non-negligible probability. (Here, the probability is taken over i, $x \in \mathcal{D}i$, and the coin tosses of the inverter.)*
4. *(Easy to invert with trapdoor). Given t_i and $f_i(x)$, one can compute x in polynomial time.*

Thus, trapdoor functions are *collections* of one-way functions with an additional trapdoor property (Item 4). The RSA and Rabin collections described earlier have the trapdoor property. Specifically, they can be inverted in polynomial time given the factorization of the modulus n.

One of the main applications of trapdoor functions is for the construction of public-key encryption schemes.

THEOREM 2. *If trapdoor functions exist, then public-key encryption schemes exist.*

There are a number of other useful strengthenings of the notion of a one-way function, discussed elsewhere in this volume: claw-free permutations, collision-resistant hash functions (see collision resistance), and universal one-way hash functions.

OTHER INTERACTIONS WITH CRYPTOGRAPHY: The interaction between computational complexity and cryptography has been very fertile. Above, we have described the role that computational complexity plays in cryptography. Conversely, several important concepts that originated in cryptography research have had a tremendous impact on computational complexity. Two notable examples are the notions of pseudorandom number generators and interactive proof systems. For more on these topics and the resulting developments in computational complexity, see [2].

FURTHER READING: Above, we have touched upon only a small portion of computational complexity, and even in the topics covered, many important issues were ignored (not to mention historical references). Thus, we refer the reader to the text [3] for more on computational complexity as

it relates to cryptography, and the texts [4, 5] for other aspects of complexity theory.

Salil Vadhan

References

[1] Garey, Michael R. and David S. Johnson (1979). *Computers and Intractability: A Guide to the Theory of NP-Completeness*. W.H. Freeman and Company, San Francisco.
[2] Goldreich, Oded (1999). *Modern Cryptography, Probabilistic Proofs, and Pseudorandomness*, Number 17 in Algorithms and Combinatorics. Springer-Verlag, Berlin.
[3] Goldreich, Oded (2001). *Foundations of Cryptography: Basic Tools*. Cambridge University Press, Cambridge, MA.
[4] Papadimitriou, Christos H. (1994). *Computational Complexity*. Addison-Wesley, Reading, MA.
[5] Sipser, Michael (1997). *Introduction to the Theory of Computation*. PWS Publishing, Boston, MA.

CONTRACT SIGNING

A contract is a nonrepudiable agreement on a given contract text, i.e., a contract can be used to prove agreement between the signatories to any verifier. A contract signing scheme [4] is used to fairly compute a contract such that, even if one of the signatories misbehaves, either both or none of the signatories obtain a contract. Contract signing generalizes fair exchange of signatures: a contract signing protocol does not need to output signatures but can define its own format instead. Contract signing can be categorized by the properties of fair exchange (like abuse-freeness) as well as the properties of the nonrepudiation tokens it produces (like third-party time stamping of the contract). Unlike agreement protocols, contract signing needs to provide a nonrepudiable proof that an agreement has been reached.

Early contract signing protocols were either based on an in-line Trusted Third Party [8], gradual exchange of secrets [5], or gradual increase of privilege [3]. Like fair exchange protocols, two-party contract signing protocols either do not guarantee termination or may else produce a partially signed contract. As a consequence, a trusted third party is needed for most practical applications. Optimistic contract signing [7] protocols optimize by involving this third-party only in case of exceptions. The first *optimistic* contract signing scheme has been described in [6]. An optimistic contract signing scheme for asynchronous networks has

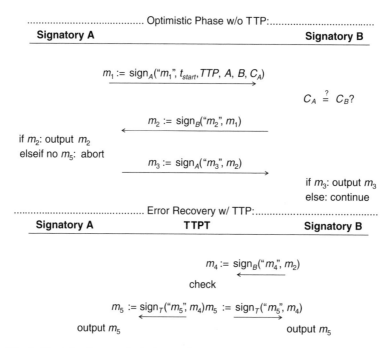

Fig. 1. Sketch of an optimistic synchronous contract signing protocol [7]

been described in [1]. An example for a multiparty abuse-free optimistic contract signing protocol has been published in [2]. A simple optimistic contract signing protocol for synchronous networks is sketched in Figure 1: party A sends a proposal, party B agrees, and party A confirms. If party A does not confirm, B obtains its contract from the TTP. (Note that the generic fair exchange protocol (see <u>fair exchange</u>) can be adapted for contract signing, too, by using a signature under C as $item_X$, using (C, X) as description $desc_X$, and using the signature verification function as the verification function.)

Matthias Schunter

References

[1] Asokan, N., Victor Shoup, and Michael Waidner (1988). "Optimistic fair exchange of digital signatures." *Advances in Cryptology—EUROCRYPT'98*, Lecture Notes in Computer Science, vol. 1403, ed. K. Nyberg. Springer-Verlag, Berlin, 591–606.

[2] Baum-Waidner, Birgit and Michael Waidner (2000). "Round-optimal and abuse-free optimistic multiparty contract signing." *27th International Colloquium on Automata, Languages and Programming (ICALP)*, Lecture Notes in Computer Science, vol. 1853, eds. U. Montanari, J.D.P Rolim, and E. Welzl. Springer-Verlag, Berlin, 524 ff.

[3] Ben-Or, M., O. Goldreich, S. Micali, and R.L. Rivest (1990). "A fair protocol for signing contracts." *IEEE Transactions on Information Theory*, 36 (1), 40–46.

[4] Blum, Manuel (1981). "Three applications of the oblivious transfer." Version 2. Department of Electrical Engineering and Computer Sciences, University of California at Berkeley, Berkley, CA.

[5] Blum, Manuel (1983). "Coin flipping by telephone, a protocol for solving impossible problems." *ACM SIGACT News*, 15 (1), 23–27.

[6] Even, Shimon (1983). "A protocol for signing contracts." *ACM SIGACT News*, 15 (1), 34–39.

[7] Pfitzmann, Birgit, Matthias Schunter, and Michael Waidner (1998). "Optimal efficiency of optimistic contract signing." *17th Symposium on Principles of Distributed Computing (PODC)*. ACM Press, New York, 113–122.

[8] Rabin, Michael O. (1983). "Transaction protection by beacons." *Journal of Computer and System Sciences*, 27, 256–267.

CONTROL VECTORS

A method introduced—and patented in a number of application scenarios—by IBM in the 1980s for the control of cryptographic key usage. The basic idea is that each cryptographic key has an associated control vector, which defines the permitted uses of the key within the system, and this is enforced by the use of tamper resistant hardware. At key generation, the control vector is cryptographically coupled to the key, e.g., by XOR-ring the key with the control vector before encryption and distribution. Each encrypted key and control vector is stored and distributed within the cryptographic system as a single <u>token</u>.

As an example, nonrepudiation may be achieved between two communicating hardware boxes by the use of a conventional MAC algorithms using symmetric methods. The communicating boxes would share the same key, but whereas one box would only be allowed to generate a MAC with that key, the other box would only be able to verify a MAC with the same key. The transform of the same key from, e.g., MAC-generation to MAC-verification is known as *key translation* and needs to be carried out in tamper resistant hardware as well. Similarly the same symmetric key may be used for encryption only enforced by one control vector in one device and for decryption only enforced by a different control vector in a different device.

Peter Landrock

COPY PROTECTION

Copy protection attempts to find ways which limit the access to copyrighted material and/or inhibit the copy process itself. Examples of copy protection include encrypted digital TV broadcast, access controls to copyrighted software through the use of license servers or through the use of special hardware add-ons (dongles), and technical copy protection mechanisms on the media.

Copy protection mechanisms can work proactively by aiming to prevent users from accessing copy protected content.

For content that is distributed on physical media such as floppy disks, *digital audio tape* (DAT), CD-ROM or *digital versatile disk* (DVD), copy protection can be achieved by *master copy control* and *copy generation control*:

Master copy control: If consumers are not allowed to even make backup copies of their master media, then one can mark the master media themselves in addition to or instead of marking the copyrighted content. This was an inexpensive and common way to protect software distributed for example on floppy disks. One of the sectors containing critical parts of the software was marked as bad such that data could be read from that sector, but could not be copied to another floppy disk.

Copy generation control: If consumers are allowed to make copies of their master copy, but not of copies of the master copy, then one needs to establish control over the vast majority of content recorders, which must be able to effectively prevent the making of unauthorized copies.

This approach is somewhat unrealistic because even a small number of remaining unregistered recorders can be used by organized hackers to produce large quantities of pirated copies.

Instead of protecting the distribution media of digital content, one can protect copyrighted digital content itself by *marking copyrighted content* and enforcing *playback control* by allowing only players that interpret these copyright marks according to certain access policies (access control). This approach works for digital content that is being distributed on physical media as well as being broadcast or distributed online. It is an example of a *digital rights management system* (*DRMS*).

Mark copyrighted content: If consumers are allowed to make a few backup copies for their personal use, then the copyrighted digital content itself can be marked as copy protected in order to be distinguishable from unprotected digital content. The more robust the marking, i.e., the harder it is to remove it without significantly degrading the quality of the digital content, the stronger copy protection mechanism can be achieved.

Playback control: Players for copyrighted content need to have a tamper resistant access circuitry that is aware of the copy protection marking, or players need to use online license servers to check the actual marks. Before converting digital content into audible or visible signals, the player compares the actual marking against the *licenses* or *tickets*, which are either built into their access circuitry or retrieved from a license server online, and stops if the former does not match the latter. The exact behavior of players is determined by access policies. There can be different kinds of tickets or licenses. Tickets of one kind may represent the right of ownership of a particular piece of content, i.e., the piece of content can be played or used as many times as the owner wishes. Tickets of another kind may represent the right of one-time play or use (*pay-per-view*). Other kinds of tickets can be defined. The more tamper resistant the access circuitry is or the more protected the communication with the license server and the license server itself, the stronger the copy protection mechanism that can be achieved.

Marking of copyrighted content can use anything from simple one-bit marks to XrML tags to sophisticated watermarking techniques. An example of the former is to define a new audio file format, in which the mark is a part of the header block but is not removable without destroying the original signal, because part of the definition of the file format requires the mark to be therein. In this case the signal would not really be literally "destroyed" but any application using this file format would not touch it without a valid mark. Some *electronic copyright management systems (ECMS)* propose mechanisms like this. Such schemes are weak as anyone with a computer or a digital editing workstation would be able to convert the information to another format and remove the mark at the same time. Finally this new audio format would be incompatible with the existing one. Thus the mark should really be embedded in the audio signal. This is very similar to S.C.M.S. (Serial Code Management System). When Phillips and Sony introduced the "S/PDIF" (Sony/Phillips Digital Interchange Format), they included the S.C.M.S. which provides a way copies of digital music are regulated in the consumer market. This information is added to the stream of data that contains the music when one makes a digital copy (a "clone"). This is in fact just a bit saying: digital copy prohibited or permitted. Some professional equipment are exempt from having S.C.M.S. With <u>watermarking</u> however, the copy control information is part of the audiovisual signal and aims at surviving file format conversion and other transformations.

An alternative to marking is *containing* copyrighted content. With this approach, the recording industry encrypts copyrighted digital content under certain encryption keys such that only players with appropriate decryption keys can access and playback the content.

Encrypt copyrighted content: The copyrighted digital content itself is encrypted in order to be accessible by authorized consumers only. The more robust the encryption, the stronger the copy protection mechanism that can be achieved.

Playback control: Players for copyrighted content need to have a <u>tamper resistant</u> access circuitry that is aware of certain decryption keys that are necessary to unlock the contents the consumer wants to be played. Before converting digital content into audible or visible signals, the player needs to look up the respective decryption keys, which are either built into the access circuitry of the player or are retrieved from a license server online. The exact behavior of players is determined by access policies. There can be different kinds of decrypting keys. Decrypting keys of one kind may represent the right of ownership of a particular piece of content, i.e., the piece of content can be played or used as many times as the owner wishes. Tickets of another kind may represent the right of one-time play or use (*pay-per-view*). Other

kinds of decryption keys can be defined. The more tamper resistant the access circuitry or the more protected the communication with the license server and the license server itself, the stronger the copy protection mechanism that can be achieved.

In order to effectively prevent consumers from copying digital content protected in this way, the players must not allow consumers to easily access the decrypted digital content. Otherwise, the containing approach would not prevent consumers from reselling, trading, or exchanging digital content at their discretion. As a first line of protection, players should not provide a high quality output interface for the digital content. A stronger level of protection is achieved if the decryption mechanism is integrated into the display, such that pirates would only obtain signals of degraded quality. The *content scrambling system* (CSS) used for *digital versatile disks* (DVDs) [2] is an example of the containing approach: in CSS, each of n manufacturers (n being several hundreds by 2002) has one or more manufacturer keys, and each player has one or more keys of its manufacturer built in. Each DVD has its own disk key dk, which is stored n times in encrypted form, once encrypted under each manufacturer key. The DVD content is encrypted under respective sector keys, which are all derived from the disk key dk.

Copy protection mechanisms can also work retroactively by deterring authorized users from leaking copies to unauthorized users. This approach requires solving the following two problems.

Mark copy protected content individually: Copy protected digital content carries information about its origin, i.e. the original source, author, distributor, etc. in order to allow to trace its distribution and spreading. It is like embedding a unique serial number in each authorized copy of protected content. The more robust the embedded marking, i.e., the harder it is to remove it without significantly degrading the quality of the digital content, the stronger the copy protection mechanism that can be achieved.

Deter from unauthorized access: Players need to have no <u>tamper resistant</u> access circuitry nor online access to license servers. Instead, each customer who receives an authorized copy is registered with the serial number of the copy provided. The marking serves as forensic evidence in investigations to figure out where and when unauthorized copies of original content have surfaced. This retroactive approach can be combined with the above mentioned proactive approach by using the embedded serial numbers as individual watermarks, which are recognized by players for the respective content.

This approach can use anything from hidden serial numbers to sophisticated fingerprinting techniques. Fingerprints are characteristics of an object that tend to distinguish it from other similar objects. They enable the owner to trace authorized users distributing them illegally. In the case of encrypted satellite television broadcasting, for instance, users could be issued a set of keys to decrypt the video streams and the television station could insert fingerprint bits into each packet of the traffic to detect unauthorized uses. If a group of users give their subset of keys to unauthorized people (so that they can also decrypt the traffic), at least one of the key donors can be traced when the unauthorized decoder is captured. In this respect, fingerprinting is usually discussed in the context of the traitor tracing *problem*.

Copy protection is inherently difficult to achieve in open systems for at least two reasons:

The requirements on watermarking are contradictory. In order to build an effective large-scale copy protection system, the vast majority of available players had to be equipped with some kind of tamper resistant circuitry or had online access to some license servers. Such circuitry had to be cheap and be integrated right into the players, and such online service had to be cheap and conveniently fast. Otherwise, the watermarking had no chance to gain any significant market share. However, tamper resistant hardware is expensive, so the cost per player limits the strength of the tamper resistance of its access circuitry. Online services incur communication costs on consumers and do not give them the independence and speed of offline access circuitry. The way how the CSS used for DVDs was "hacked" is just one more incident demonstrating the contradicting requirements: since the encryption mechanism was chosen to be a weak feedback shift register cipher, it was only a matter of time until a program called *DeCSS* surfaced, which can decipher any DVD. The access circuitry of players into which the deciphering algorithm is built was not well protected against reverse engineering the algorithm, and hence, the secret algorithm leaked, and with it the DVD keys one by one. The watermarking scheme of the *Secure Digital Music Initiative* (SDMI) [7], (a successor of MP3) was broken by Fabien Petitcolas [6]. Later, a public challenge of this watermarking scheme was broken by Felten et al. [3]. The SDMI consortium felt this piece of research might jeopardize the consortium's reputation and revenue so much that the SDMI consortium threatened to sue the authors if they would present their work at a public conference. Attacks on various other copy protection mechanisms have been described by Anderson in Section 20.3.30 of [1].

The requirements on fingerprinting are contradictory as well. On one hand the broadcaster or copyright holder may want to easily recognize the fingerprint, preferably visually. This allows easy tracing of a decoder that is used for illegal purposes. This approach is very similar to the commonly used watermarking by means of the logo of a TV station that is continuously visible in one of the corners of the screen. On the other hand, the fingerprint should be hidden, in order not to disturb paying viewers with program-unrelated messages on their screen, or to avoid any pirate detecting and erasing the fingerprint electronically. In the latter case, one may require specific equipment to detect and decode a fingerprint.

Despite the inherent technical difficulties to build effective large-scale copy protection systems, the content industries (TV producers, movie makers, audio makers, software companies, publishers) have and will continue to have a strong interest in protecting their revenues against pirates. They are trying to overcome the contradictory requirements mentioned above by two complementing approaches: they try to control the entire market of media players and recorders by contracting with the large suppliers. While they opt for relatively weak but inexpensive access circuitry for these players, they compensate for the weakness by promoting suitable laws that deter consumers from breaking this access circuitry or resorting to unauthorized players, or using unauthorized recorders. An example for trying to make secure access circuitry pervasive in the PC market is the *trusted computing platform alliance* (TCPA) [8]. An example of such legislative initiative is the *digital millenium copyright act* (DMCA) [4] in the United States. It prohibits the modification of any electronic copyright arrangement information (CMI) bundled with digital content, such as details of ownership and licensing, and outlaws the manufacture, importation, sale, or offering for sale of anything primarily designed to circumvent copyright protection technology.

It is clear that the issue of copy protection is a special case of the more general issue of *digital rights management* (DRM). Both issues bear the risk of a few companies defining the access policies, which are enforced by the players and thus determine what and how a vast majority of people would be able to read, watch, listen to, or work with. A moderate overview of the hot debate about content monopolies, pricing, free speech, democratic, privacy, and legislative issues, etc. is found at [5].

Gerrit Bleumer

References

[1] Anderson, Ross (2001). *Security Engineering*. Wiley & Sons, New York.

[2] Bloom, Jefrey A., Ingemar J. Cox, Ton Kalker, Jean-Paul, M.G. Linnartz, Matthew L. Miller, C. Brendan, and S. Traw (1999). "Copy protection for DVD video." *Proceedings of the IEEE*, 87 (7), 1267–1276.

[3] Craver, Scott A., Min Wu, Bede Liu, Adam Subblefield, Ben Swartzlander, Dan S. Wallach, Drew Dean, and Edward W. Felten (2001). "Reading between the lines: Lessons from the SDMI Challenge." *10th Usenix Security Symposium 2002, The USENIX Association 2001*, 353–363.

[4] Digital Millenium Copyright Act: http://www.loc.gov/copyright/legislation/dmca.pdf

[5] Digital Rights Management: http://www.epic.org/privacy/drm/

[6] Petitcolas, Fabien A., Ross Anderson, Markus G. Kuhn (1998). "Attacks on copyright marking systems." *Information Hiding*, Lecture Notes in Computer Science, vol. 1525, ed. D. Aucsmith. Springer-Verlag, Berlin, 218–238.

[7] Secure Digital Music Initiative http://www.sdmi.org

[8] Trusted Computing Platform Initiative: http://www.trustedcomputing.org

CORRECTING-BLOCK ATTACK

This attack can find collisions or (second) preimages for certain classes of hash functions. It consists of substituting all blocks of the input except for one or more blocks. This attack often applies to the last block and is then called a correcting-last-block attack, but it can also apply to the first block or to some blocks in the middle. For a preimage attack, one chooses an arbitrary message X and finds one or more correcting blocks Y such that $h(X\|Y)$ takes a certain value (here $\|$ denotes concatenation). For a second preimage attack on the target message $X\|Y$, one chooses X' and searches for one or more correcting blocks Y' such that $h(X'\|Y') = h(X\|Y)$ (note that one may select $X' = X$). For a collision attack, one chooses two arbitrary messages X and X' with $X' \neq X$; subsequently one searches for one or more correcting blocks denoted by Y and Y', such that $h(X'\|Y') = h(X\|Y)$.

The hash functions based on algebraic structures are particularly vulnerable to this attack, since it is often possible to invert the compression function using algebraic manipulations [9]. A typical countermeasure to a correcting-block attack consists of adding redundancy to the message blocks in such a way that it becomes computationally infeasible to find a correcting block with the necessary redundancy. The price paid for this solution is a degradation of the performance.

A first example is a multiplicative hash proposed by Bosset in 1977 [1], based on $GL_2(GF(p))$, the group of 2×2 invertible matrices over the finite field $GF(p)$, with $p = 10\,007$. Camion showed how to find a second preimage using a correcting-block attack that requires 48 correcting blocks of 6 bits each [2].

In 1984 Davies and Price [4] proposed a hash function with the following compression function f:

$$f = (H_{i-1} \oplus X_i)^2 \bmod N,$$

where X_i is the message block, H_{i-1} is the chaining variable, and N is an RSA modulus (see RSA digital signature scheme). In order to preclude a correcting block attack, the text input is encoded such that the most (or least) significant 64 bits of each block are equal to 0. However, Girault [5] has demonstrated that a second preimage can be found using the extended Euclidean algorithm (see Euclidean algorithm); improved results can be found in [6].

The 1988 scheme of CCITT X.509 Annex D [8] tried to remedy this approach by distributing the redundancy (one nibble in every byte). However, Coppersmith [3] applied a correcting-block attack to find two distinct messages X and X' such that

$$h(X') = 256 \cdot h(X).$$

This is a highly undesirable property, which a.o. implies that this hash function cannot be used in combination with a multiplicative digital signature scheme such as *RSA*. In 1998, ISO has adopted two improved schemes based on modular arithmetic (ISO/IEC 10118-4 *Modular Arithmetic Secure Hash*, MASH-1 and MASH-2 [7]), which so far have resisted correcting-block attacks.

B. Preneel

References

[1] Bosset, J. (1977). "Contre les risques d'altération, un système de certification des informations." *01 Informatique*, 107.

[2] Camion, P. (1986). "Can a fast signature scheme without secret be secure?" *Proc. 2nd International*

Conference on Applied Algebra, Algebraic Algorithms, and Error-Correcting Codes, Lecture Notes in Computer Science, vol. 228, ed. A. Poli. Springer-Verlag, Berlin, 215–241.

[3] Coppersmith, D. (1989). "Analysis of ISO/CCITT document X.509 annex D." IBM T.J. Watson Center, Yorktown Heights, NY. Internal Memo.

[4] Davies, D. and W.L. Price (1984). "Digital signatures, an update." *Proc. 5th International Conference on Computer Communication, October 1984*, 845–849.

[5] Girault, M. (1988). "Hash-functions using modulo-n operations." *Advances in Cryptology—EUROCRYPT'87*, Lecture Notes in Computer Science, vol. 304, eds. D. Chaum and W.L. Price. Springer-Verlag, Berlin, 217–226.

[6] Girault, M. and J.-F. Misarsky (1997). "Selective forgery of RSA signatures using redundancy." *Advances in Cryptology—EUROCRYPT'97*, Lecture Notes in Computer Science, vol. 1233, ed. W. Fumy. Springer-Verlag, Berlin, 495–507.

[7] ISO/IEC 10118 (1998). "Information technology—security techniques—hash-functions." *Part 4: Hash-Functions Using Modular Arithmetic*, 10118-4:1998.

[8] ITU-T X.500 (1988). "The directory—overview of concepts." ITU-T Recommendation X.500 (same as IS 9594-1, 1989).

[9] Preneel, B. (1993). "Analysis and design of cryptographic hash functions." *Doctoral Dissertation*, Katholieke Universiteit Leuven.

CORRELATION ATTACK FOR STREAM CIPHERS

The *correlation attack for stream ciphers* was proposed by Siegenthaler in 1985. It applies to any running-key generator composed of several linear feedback shift registers (LFSRs). The correlation attack is a divide-and-conquer technique: it aims at recovering the initial state of each constituent LFSRs separately from the knowledge of some keystream bits (in a known plaintext attack). A similar ciphertext only attack can also be mounted when there exists redundancy in the plaintext (see [3]).

The original correlation attack presented in [3] applies to some combination generators composed of n LFSRs of lengths $L_1, \ldots, L_n$. It enables to recover the complete initialization of the generator with only $\sum_{i=1}^{n} \left(2^{L_i} - 1\right)$ trials instead of the $\prod_{i=1}^{n} \left(2^{L_i} - 1\right)$ tests required by an exhaustive search. Some efficient variants of the original correlation attack can also be applied to other keystream generators based on LFSRs, like filter generators (see fast correlation attack for details).

ORIGINAL CORRELATION ATTACK ON COMBINATION GENERATORS: The correlation attack exploits the existence of a statistical dependence between the keystream and the output of a single constituent LFSR. In a binary combination generator, such a dependence exists if and only if the output of the combining function f is correlated to one of its inputs, i.e., if

$$p_i = \Pr[\, f(x_1, \ldots, x_n) \neq x_i \,] \neq \tfrac{1}{2}$$

for some i, $1 \leq i \leq n$. It equivalently means that the keystream sequence $\mathbf{s} = (s_t)_{t \geq 0}$ is correlated to the sequence $\mathbf{u} = (u_t)_{t \geq 0}$ generated by the ith constituent LFSR. Namely, the correlation between both sequences calculated on N bits

$$\sum_{t=0}^{N-1} (-1)^{s_t + u_t \bmod 2}$$

(where the sum is defined over real numbers) is a random variable which is binomially distributed with mean value $N(1 - 2p_i)$ and with variance $4Np_i(1 - p_i)$ (when N is large enough). It can be compared to the correlation between the keystream $\mathbf{s}$ and a sequence $\mathbf{r} = (r_t)_{t \geq 0}$ independent of $\mathbf{s}$ (i.e., such that $\Pr[s_t \neq r_t] = 1/2$). For such a sequence $\mathbf{r}$, the correlation between $\mathbf{s}$ and $\mathbf{r}$ is binomially distributed with mean value 0 and with variance N. Thus, an exhaustive search for the initialization of the ith LFSR can be performed. The value of the correlation enables to distinguish the correct initial state from a wrong one since the sequence generated by a wrong initial state is assumed to be statistically independent of the keystream. Table 1 gives a complete description of the attack.

In practice, an initial state is accepted if the magnitude of the correlation exceeds a certain

Table 1. Correlation attack

Input. $s_0 s_1 \ldots s_{N-1}$, N keystream bits,
$\quad p_i = Pr[\, f(x_1, \ldots, x_n) \neq x_i \,] \neq 1/2$.
Output. $u_0 \ldots u_{L_i - 1}$, the initial state of the i-th constituent LFSR.
For each possible initial state $u_0 \ldots u_{L_i - 1}$
$\quad$ Generate the first N bits of the sequence $\mathbf{u}$
$\quad\quad$ produced by the ith LFSR from the chosen
$\quad\quad$ initial state.
$\quad$ Compute the correlation between $s_0 s_1 \ldots s_{N-1}$
and
$\quad\quad u_0 u_1 \ldots u_{N-1}$:

$$\alpha \leftarrow \sum_{i=0}^{N-1} (-1)^{s_t + u_t \bmod 2}$$

$\quad$ If α is close to $N(1 - 2p_i)$
$\quad\quad$ return $u_0 u_1 \ldots u_{L_i - 1}$

---

decision threshold which is deduced from the expected false alarm probability P_f and the non-detection probability P_n (see [3]). The required keystream length N depends on the probability p_i and on the length L_i of the involved LFSR: for $P_n = 1.3 \times 10^{-3}$ and $P_f = 2^{-L_i}$, the attack requires

$$N \simeq \left(\frac{\sqrt{\ln(2^{L_i-1})} + 3\sqrt{2p_i(1-p_i)}}{\sqrt{2}(p_i - 0.5)} \right)^2$$

running-key bits. Clearly, the attack performs 2^{L_i-1} trials on average where L_i is the length of the target LFSR. The correlation attack only applies if the probability p_i differs from $1/2$.

Correlation Attack on other Keystream Generators:

More generally, the correlation attack applies to any keystream generator as soon as the keystream is correlated to the output sequence **u** of a finite state machine whose initial state depends on some key bits. These key bits can be determined by recovering the initialization of **u** as follows: an exhaustive search for the initialization of **u** is performed, and the correct one is detected by computing the correlation between the corresponding sequence **u** and the keystream.

Correlation Attack on Combination Generators Involving Several LFSRs:

For combination generators, the correlation attack can be prevented by using a combining function f whose output is not correlated to any of its inputs. Such functions are called first-order correlation-immune (or 1-resilient in the case of balanced functions). In this case, the running key is statistically independent of the output of each constituent LFSR; any correlation attack should then consider several LFSRs simultaneously. More generally, a correlation attack on a set of k constituent LFSRs, namely LFSR $i_1, \ldots,$ LFSR i_k, exploits the existence of a correlation between the running-key **s** and the output **u** of a smaller combination generator, which consists of the k involved LFSRs combined by a Boolean function g of k variables (see Figure 1). Since $\Pr[s_t \neq u_t] = \Pr[f(x_1, \ldots, x_n) \neq g(x_{i_1}, \ldots, x_{i_k})] = p_g$, this attack only succeeds when $p_g \neq 1/2$. The smallest number of LFSRs that can be attacked simultaneously is equal to $m + 1$ where m is the highest correlation-immunity order of the combining function. Moreover, the Boolean function g of $(m + 1)$ variables which provides the best approximation of f is the affine

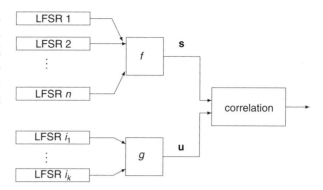

Fig. 1. Correlation attack involving several constituent LFSRs of a combination generator

function $\sum_{j=1}^{m+1} x_{i_j} + \varepsilon$ [1, 4]. Thus, the most efficient correlation attacks that can be mounted rely on the correlation between the keystream **s** and the sequence **u** obtained by adding the outputs of LFSRs $i_1, i_2, \ldots, i_{m+1}$. This correlation corresponds to

$$\Pr[s_t \neq u_t] = \frac{1}{2} - \frac{1}{2^{n+1}} |\widehat{f}(t)|,$$

where n is the number of variables of the combining function, t is the n-bit vector whose ith component equals 1 if and only if $i \in \{i_1, i_2, \ldots, i_{m+1}\}$ and $\widehat{f}$ denotes the Walsh transform of f (see Boolean functions). In order to increase the complexity of the correlation attack, the combining function used in a combination generator should have a high correlation-immunity order and a high nonlinearity (more precisely, its Walsh coefficients $\widehat{f}(t)$ should have a low magnitude for all vectors t with a small Hamming weight). For an m-resilient combining function, the complexity of the correlation attack is $2^{L_{i_1} + L_{i_2} + \cdots + L_{i_{m+1}}}$. It can be significantly reduced by using some improved algorithms, called fast correlation attacks.

Anne Canteaut

References

[1] Canteaut, A. and M. Trabbia (2000). "Improved fast correlation attacks using parity-check equations of weight 4 and 5." *Advances in Cryptology—EUROCRYPT 2000*, Lecture Notes in Computer Science number 1807, ed. B. Preneel. Springer-Verlag, Berlin, 573–588.

[2] Siegenthaler, T. (1984). "Correlation-immunity of nonlinear combining functions for cryptographic applications." *IEEE Trans. Inform. Theory*, IT-30 (5), 776–780.

[3] Siegenthaler, T. (1985). "Decrypting a class of stream ciphers using ciphertext only." *IEEE Trans. Computers*, C-34 (1), 81–84.

[4] Zhang, M. (2000). "Maximum correlation analysis of nonlinear combining functions in stream ciphers." *Journal of Cryptology*, 13 (3), 301–313.

CORRELATION IMMUNE AND RESILIENT BOOLEAN FUNCTIONS

Cryptographic Boolean functions must be *balanced* (i.e., their output must be uniformly distributed) for avoiding statistical dependence between their input and their output (such statistical dependence can be used in attacks).

Moreover, any combining function $f(x)$ (see combination generator), used for generating the pseudorandom sequence in a stream cipher, must stay balanced if we keep constant some coordinates x_i of x (at most m of them, where m is as large as possible). We say that f is then *m-resilient*. More generally, a (non necessarily balanced) Boolean function, whose output distribution probability is unaltered when any m of its input bits are kept constant, is called *mth order correlation-immune*. The notion of correlation-immune function is related to the notion of orthogonal array (see [1]). Only resilient functions are of practical interest as cryptographic functions.

The notion of correlation immunity was introduced by Siegenthaler in [5]; it is related to an attack on pseudorandom generators using combining functions: if such combining function f is not mth order correlation-immune, then there exists a correlation between the output of the function and (at most) m coordinates of its input; if m is small enough, a divide-and-conquer attack due to Siegenthaler (see correlation attack for stream ciphers) and later improved by several authors (see fast correlation attack) uses this weakness for attacking the system.

The maximum value of m such that f is m-resilient is called the *resiliency order* of f.

Correlation immunity and resiliency can be characterized through the Walsh transform $\widehat{f}(u) = \sum_{x \in F_2^n} (-1)^{f(x) \oplus x \cdot u}$, see [3]: f is mth order correlation-immune if and only if $\widehat{f}(u) = 0$ for all $u \in F_2^n$ such that $1 \leq w_H(u) \leq m$, where w_H denotes the Hamming weight (that is, the number of nonzero coordinates); and it is m-resilient if and only if $\widehat{f}(u) = 0$ for all $u \in F_2^n$ such that $w_H(u) \leq m$.

It is not sufficient for a combining function f, used in a stream cipher, to be m-resilient with large m. As any cryptographic function, it must also have high algebraic degree $d \circ f$ and high nonlinearity $\mathcal{NL}(f)$ (see Boolean functions). There are necessary trade-offs between the number of variables, the algebraic degree, the nonlinearity, and the resiliency order of a function.

- Siegenthaler's bound [5] states that any m-resilient function ($0 \leq m < n - 1$) has algebraic degree smaller than or equal to $n - m - 1$ and that any $(n - 1)$ resilient function is affine (Siegenthaler also proved that any n-variable mth order correlation-immune function has degree at most $n - m$).

- The values of the Walsh transform of an n-variable, m-resilient function are divisible by 2^{m+2} if $m \leq n - 2$, cf. [4] (and they are divisible by $2^{m+2+\left\lfloor \frac{n-m-2}{d} \right\rfloor}$ if f has degree d, see [2]). These divisibility bounds have provided nontrivial upper bounds on the nonlinearities of resilient functions [2, 4], also partially obtained in [6, 7]. The nonlinearity of any m-resilient function is upper bounded by $2^{n-1} - 2^{m+1}$. This bound is tight, at least when $m \geq 0.6n$, and any m-resilient function achieving it also achieves Siegenthaler's bound (see [6]).

High order resilient functions with high degrees and high nonlinearities are needed for applications in stream ciphers.

EXAMPLE [1]. Let r be a positive integer smaller than n; denote $n - r$ by s; let g be any Boolean function on F_2^s and let ϕ be a mapping from F_2^s to F_2^r. Define the function:

$$f_{\phi,g}(x, y) = x \cdot \phi(y) \oplus g(y) = \bigoplus_{i=1}^{r} x_i \phi_i(y) \oplus g(y),$$
$$x \in F_2^r, \qquad y \in F_2^s, \qquad (2)$$

where $\phi_i(y)$ is the ith coordinate of $\phi(y)$. Then, if every element in $\phi(F_2^s)$ has Hamming weight strictly greater than k, then $f_{\phi,g}$ is m-resilient with $m \geq k$. In particular, if $\phi(F_2^s)$ does not contain the null vector, then $f_{\phi,g}$ is balanced.

Examples of m-resilient functions achieving the best possible nonlinearity $2^{n-1} - 2^{m+1}$ (and thus the best degree) have been obtained for $n \leq 10$ and for every $m \geq 0.6n$ (n being then not limited, see [6]). They have been constructed by using the following methods permitting to construct resilient functions from known ones:

- [1, 5] Let g be a Boolean function on F_2^n. Consider the Boolean function on F_2^{n+1}: $f(x_1, \ldots, x_n, x_{n+1}) = g(x_1, \ldots, x_n) \oplus x_{n+1}$. Then, $\mathcal{NL}(f) = 2\mathcal{NL}(g)$ and $d \circ f = d \circ g$ if $d \circ g \geq 1$. If g is m-resilient, then f is $(m + 1)$-resilient.

- [5] Let g and h be two Boolean functions on F_2^n. Consider the function $f(x_1, \ldots, x_n, x_{n+1}) = x_{n+1}g(x_1, \ldots, x_n) \oplus '(x_{n+1} \oplus 1)h(x_1, \ldots, x_n)$ on F_2^{n+1}. Then, $\mathcal{NL}_f \geq \mathcal{NL}_g + \mathcal{NL}_h$ (moreover, if g and h are such that, for every word a, at least one of the numbers $\widehat{g}(a), \widehat{h}(a)$ is null, then $\mathcal{NL}(f)$ equals $2^{n-1} + \min(\mathcal{NL}(g), \mathcal{NL}(h)))$.

  If the algebraic normal forms of g and h do not have the same monomials of highest degree, then $d \circ f = 1 + \max(d \circ g, d \circ h)$.

  If g and h are m-resilient, then f is m-resilient (moreover, if for every $a \in F_2^n$ of weight $m + 1$, we have $\widehat{g}(a) + \widehat{h}(a) = 0$, then f is $(m + 1)$-resilient; this happens with $h(x) = g(x_1 \oplus 1, \ldots, x_n \oplus 1) \oplus \epsilon$, where $\epsilon = m \bmod 2$, see [1]).

- [6] Let g be any Boolean function on F_2^n. Define the Boolean function f on F_2^{n+1} by $f(x_1, \ldots, x_n, x_{n+1}) = x_{n+1} \oplus g(x_1, \ldots, x_{n-1}, x_n \oplus x_{n+1})$. Then, $\mathcal{NL}(f) = 2\mathcal{NL}(g)$ and $d \circ f = d \circ g$ if $d \circ g \geq 1$. If g is m-resilient, then f is m-resilient (and it is $(m + 1)$-resilient if $\widehat{g}(a_1, \ldots, a_{n-1}, 1)$ is null for every vector $(a_1, \ldots, a_{n-1})$ of weight at most m).

Claude Carlet

References

[1] Camion, P., C. Carlet, P. Charpin, and N. Sendrier (1992). "On correlation-immune functions." *Advances in Cryptology—CRYPTO'91*, Lecture Notes in Computer Science, vol. 576, ed. J. Feigenbaum. Springer, 86–100.

[2] Carlet, C. (2001). "On the coset weight divisibility and nonlinearity of resilient and correlation-immune functions." *Proceedings of SETA'01 (Sequences and their Applications 2001)*, Discrete Mathematics and Theoretical Computer Science. Springer, Berlin, 131–144.

[3] Xiao, Guo-Zhen and J.L. Massey (1988). "A spectral characterization of correlation-immune combining functions." *IEEE Trans. Inf. Theory*, IT-34 (3), 569–571.

[4] Sarkar, P. and S. Maitra (2000). "Nonlinearity bounds and constructions of resilient Boolean functions." *Advances in Cryptology—CRYPTO 2000*, Lecture Notes in Computer Science, vol. 1880, ed. Mihir Bellare. Springer, 515–532.

[5] Siegenthaler, T. (1984). "Correlation-immunity of nonlinear combining functions for cryptographic applications." *IEEE Transactions on Information theory*, IT-30 (5), 776–780.

[6] Tarannikov, Y.V. (2000). "On resilient Boolean functions with maximum possible nonlinearity." *Proceedings of INDOCRYPT 2000*, Lecture Notes in Computer Science, vol. 1977, eds. B.K. Roy and E. Okamoto. Springer, 19–30.

[7] Zheng, Y. and X.-M. Zhang (2001). "Improving upper bound on the nonlinearity of high order correlation

immune functions." *Proceedings of Selected Areas in Cryptography 2000*, Lecture Notes in Computer Science, vol. 2012, eds. D.R. Stinson and S.E. Tavares. Springer, 262–274.

COVERT CHANNELS

Lampson [8, p. 614] informally specified a special type of communication being:

> Covert channels, i.e. those not intended for information transfer at all, such as the service program's effect on the system load.

A more general definition can be found in [14, p. 110].

Covert channels often involve what is called timing channels and storage channels. An example of a timing channel is the start-time of a process. The modulation of disc space is an example of a storage channel. Methods that describe how covert channels can be fought can, e.g., be found in [9]. For more information about covert channels, see [1].

Simmons [10] introduced the research on covert channels to the cryptographic community by introducing a special type of channel, which he called a *subliminal channel*. (Simmons did not regard these channels as fitting the definition of a covert channel.) He observed that covert data could be hidden within the *authenticator* of an <u>authentication</u> code [10]. The capacity of this channel was not large (a few bits per authenticator), but as Simmons disclosed 10 years later [12, pp. 459–462] (see also [13]), the potential impact on treaty verification and on the national security of the USA could have been catastrophic.

In 1987 the concept of subliminal channel was extended to be hidden inside a <u>zero-knowledge interactive proof</u> [6]. The concept was generalized to a hidden channel inside any cryptographic system [3, 4]. Mechanisms to protect against subliminal channels were also presented [3, 4] and reinvented 5 years later [11]. Problems with some of these solutions were discussed in [5] (see [2]) for a more complete discussion.

A subliminal channel in a key-distribution scheme (see <u>key management</u>) could undermine <u>key escrow</u>, as discussed in [7]. *Kleptography* is the study of how a designer, making a black box cipher, can leak the user's secret <u>key</u> subliminally to the designer (see, e.g., [15]).

Yvo Desmedt

References

[1] Bishop, M. (2003). *Computer Security*. Addison-Wesley, Reading, MA.

[2] Burmester, M., Y.G. Desmedt, T. Itoh, K. Sakurai, H. Shizuya, and M. Yung (1996). "A progress report on subliminal-free channels." *Information Hiding, First International Workshop, Proceedings, May 30–June 1* (Lecture Notes in Computer Science, vol. 1174, ed. R. Anderson). Springer-Verlag, Cambridge, UK, 159–168.

[3] Desmedt, Y. (1990). "Abuses in cryptography and how to fight them." *Advances in Cryptology—CRYPTO'88*, Lecture Notes in Computer Science, vol. 403, ed. S. Goldwasser. Springer-Verlag, Santa Barbara, CA, 375–389.

[4] Desmedt, Y. (1990). "Making conditionally secure cryptosystems unconditionally abuse-free in a general context." *Advances in Cryptology—CRYPTO'89*, Lecture Notes in Computer Science, vol. 435, ed. G. Brassard. Springer-Verlag, Santa Barbara, CA, 6–16.

[5] Desmedt, Y. (1996). "Simmons' protocol is not free of subliminal channels." *Proceedings: 9th IEEE Computer Security Foundations Workshop, June 10–12*. Kenmare, Ireland, 170–175.

[6] Desmedt, Y., C. Goutier, and S. Bengio (1988). "Special uses and abuses of the Fiat–Shamir passport protocol." *Advances in Cryptology—CRYPTO'87*, Lecture Notes in Computer Science, vol. 293, ed. C. Pomerance. Springer-Verlag, Santa Barbara, CA, 21–39.

[7] Kilian, J. and T. Leighton (1995). "Failsafe key escrow, revisited." *Advances in Cryptology—CRYPTO'95*, Lecture Notes in Computer Science, vol. 963, ed. D. Coppersmith. Springer-Verlag, Santa Barbara, CA, 208–221.

[8] Lampson, B.W. (1973). "A note on the confinement problem." *Comm. ACM*, 16 (10), 613–615.

[9] Simmons, G.J. (1983). "Verification of treaty compliance-revisited." *Proc. of the 1983 IEEE Symposium on Security and Privacy, April 25–27, 1983*. IEEE Computer Society Press, Oakland, CA, 61–66.

[10] Simmons, G.J. (1983). "The prisoners' problem and the subliminal channel." *Advances in Cryptology—CRYPTO'83*, Lecture Notes in Computer Science, ed. D. Chaum. Plenum Press, New York, 51–67.

[11] Simmons, G.J. (1993). "An introduction to the mathematics of trust in security protocols." *Proceedings: Computer Security Foundations Workshop VI, June 15–17*. IEEE Computer Society Press, Franconia, NH, 121–127.

[12] Simmons, G.J. (1994). "Subliminal channels; past and present." *European Trans. on Telecommunications*, 5 (4), 459–473.

[13] Simmons, G.J. (1996). "The history of subliminal channels." *Information Hiding, First International Workshop, Proceedings, May 30–June 1*, (Lecture Notes in Computer Science, vol. 1174, ed. R. Anderson). Springer-Verlag, Cambridge, UK, 237–256.

[14] U.S. Department of Defense (1983). Department of Defense Trusted Computer System Evaluation Criteria. Also known as the Orange Book.

[15] Young, A. and M. Yung (1997). "Kleptography: Using cryptography against cryptography." *Advances in Cryptology—EUROCRYPT'97*, Lecture Notes in Computer Science, vol. 1233, ed. W. Fumy. Springer-Verlag, Konstanz, Germany, 62–74.

CPS, CERTIFICATE PRACTICE STATEMENT

A Certification Authority (CA) describes in a Certificate Practice Statement (CPS) the procedures and practices that it employs when managing certificates (issuing, revoking, renewing, and rekeying). The CPS describes manual processes for securely operating the CA and contains information on cryptographic aspects, including management of the keys used by the CA (see also key management). The certificate authority documents in its CPS that it manages certificates according to some *certificate policy* (see trust model). The certificate policy lists the requirements for the management of certificates by the CA and defines the applicability of a certificate issued under this policy. The policy might for example indicate that the certificate may be used for authenticating the subject (holder) in certain types of business transactions. The certificate policy under which a certificate is issued may be indicated in the certificate. For X.509 certificates a specific extension is defined for this purpose. This information allows relying parties to decide whether the certificate may be used in a given situation without knowing the CPS of the CA issuing the certificate.

Whereas the policy lists the requirements for the CA, the CPS describes how the CA meets these requirements. Thus the two documents will often have similar structures. The certificate policy may in particular define the rules for approving that a given CPS satisfies the policy and for validating that the CA continuously follows the CPS. It may, for example, be required that an independent auditor initially approves that the CPS complies with the certificate policy and yearly reviews the procedures actually followed by the CA against the CPS.

As a consequence, different certification authorities following different certificate practice statements may issue certificates under the same policy as long as their CPS satisfies the policy.

Torben Pedersen

Reference

[1] RfC2527: Internet X.509 Public Key Infrastructure Certificate Policy and Certification Practices Framework. See http://www.rfc-editor.org/rfc.html

CRAMER–SHOUP PUBLIC KEY SYSTEM

The Cramer–Shoup cryptosystem [6,8] is the first underlined:public-key cryptography system that is efficient and is proven to be chosen ciphertext secure without the random oracle model using a standard complexity assumption. Before describing the system we give a bit of history.

The standard notion of security for a public-key encryption system is known as semantic security under an adaptive chosen ciphertext attack and denoted by IND-CCA2. The concept is due to Rackoff and Simon [12] and the first proof that such systems exist is due to Dolev et al. [9]. Several efficient constructions for IND-CCA2 systems exist in the random oracle model. For example, OAEP is known to be IND-CCA2 when using the RSA trapdoor permutation [2, 10]. Until the discovery of the Cramer–Shoup system, there were no efficient IND-CCA2 systems that are provably secure under standard assumptions without random oracles.

The Cramer–Shoup system makes use of a group G of prime order q. It also uses a hash function $H : G^3 \to \mathbb{Z}_q$ (see also modular arithmetic). We assume that messages to be encrypted are elements of G. The most basic variant of the systems works as follows:

Key Generation. Pick an arbitrary generator g of G. Pick a random w in $\mathbb{Z}_q^*$ and random x_1, x_2, y_1, y_2, z in $\mathbb{Z}_q$. Set $\hat{g} = g^w$, $e = g^{x_1}\hat{g}^{x_2}$, $f = g^{y_1}\hat{g}^{y_2}$, $h = g^z$. The public key is $(g, \hat{g}, e, f, g, G, q, H)$ and the private key is $(x_1, x_2, y_1, y_2, z, G, q, H)$.

Encryption. Given the public key $(g, \hat{g}, e, f, h, G, q, H)$ and a message $m \in G$:
1. Pick a random u in $\mathbb{Z}_q$.
2. Set $a = g^u$, $\hat{a} = \hat{g}^u$, $c = h^u \cdot m$, $v = H(a, \hat{a}, c)$, $d = e^u f^{uv}$.
3. The ciphertext is $C = (a, \hat{a}, c, d) \in G^4$.

Decryption. To decrypt a ciphertext $C = (a, \hat{a}, c, d)$ using the private key $(x_1, x_2, y_1, y_2, z, G, q, H)$:
1. Test that $a, \hat{a}, c, d$ belong to G; output 'reject' and halt if not.
2. Compute $v = H(a, \hat{a}, c) \in \mathbb{Z}_q$. Test that $d = a^{x_1 + vy_1}\hat{a}^{x_2 + vy_2}$; output 'reject' and halt if not.
3. Compute $m = c/a^z \in G$ and output m as the decryption of C.

Cramer and Shoup prove that the system is IND-CCA2 if the DDH assumption [3] (see Decisional Diffie–Hellman problem) holds in G and the hash function H is collision resistant. They show that if a successful IND-CCA2 attacker exists, then (assuming H is collision resistant) one can construct an algorithm B, with approximately the same running as the attacker, that decides if a given 4-tuple $g, \hat{g}, a, \hat{a} \in G$ is a random DDH tuple or a random tuple in G^4. Very briefly, Algorithm B works as follows: it gives the attacker a public key for which B knows the private key. This enables B to respond to the attacker's decryption queries. The given 4-tuple is then used to construct the challenge ciphertext given to the attacker. Cramer and Shoup show that if the 4-tuple is a random DDH tuple, then the attacker will win the semantic security game with non-negligible advantage. However, if the input 4-tuple is a random tuple in G^4, then the attacker has zero advantage in winning the semantic security game. This behavioral difference enables B to decide whether the given input tuple is a random DDH tuple or not.

We briefly mention a number of variants of the system. Ideally, one would like an IND-CCA2 system that can be proven secure in two different ways: (i) without random oracles it can be proven secure using the decisional Diffie–Hellman assumption, and (ii) with random oracles it can be proven secure using the much weaker computational Diffie–Hellman assumption. For such a system, the random oracle model provides a hedge in case the DDH assumption is found to be false. A small variant of the Cramer–Shoup system above can be shown to have this property [8].

Occasionally, one only requires security against a weak chosen ciphertext attack in which the attacker can only issue decryption queries before being given the challenge ciphertext [1, 11]. A simpler version of the Cramer–Shoup system, called CS-Lite, can be shown to be secure against this weaker chosen ciphertext attack assuming DDH holds in G. This variant is obtained by computing d as $d = e^u$. There is no need for y_1, y_2, f or the hash function H. When decrypting we verify that $d = a^{x_1}\hat{a}^{x_2}$ in step 2.

Finally, one may wonder how to construct efficient IND-CCA2 systems using an assumption other than DDH. Cramer and Shoup [7] showed that their system is a special case of a more general paradigm. Using this generalization they construct a CCA2 system based on the Quadratic Residuosity assumption modulo a composite. They obtain a more efficient system using a stronger assumption known as the *Pallier assumption*. Other constructions for efficient

IND-CCA2 systems are given in [4, 5]. Finally, we note that Sahai and Elkind [13] show that the Cramer–Shoup system can be linked to the Naor–Yung double encryption paradigm [11].

Dan Boneh

References

[1] Bellare, M., A. Desai, D. Pointcheval, and P. Rogaway (1998). "Relations among notions of security for public-key encryption schemes." *Advances in Cryptology—CRYPTO'98*, Lecture Notes in Computer Science, vol. 1462, ed. H. Krawczyk. Springer, Berlin, 26–45.

[2] Bellare, M. and P. Rogaway (1994). "Optimal asymmetric encryption." *Advances in Cryptology—EUROCRYPT'94*, Lecture Notes in Computer Science, vol. 950, ed. A. De Santis. Springer, Berlin, 92–111.

[3] Boneh, Dan (1998). "The decision Diffie–Hellman problem." *Proceedings of the 3rd Algorithmic Number Theory Symposium*, Lecture Notes in Computer Science, vol. 1423, ed. J.P. Buhler. Springer-Verlag, Berlin, 48–63.

[4] Boneh, Dan and Xavier Boyen (2004). "Efficient selective-ID secure identity based encryption without random oracles." *Advances in Cryptology—EUROCRYPT 2004*, Lecture Notes in Computer Science, vol. 3027, eds. C. Cachin and J. Camenisch. 223–238.

[5] Canetti, Ran, Shai Halevi, and Jonathan Katz (2004). "Chosen-ciphertext security from identity-based encryption." *Advances in Cryptology—EUROCRYPT 2004*, Lecture Notes in Computer Science, vol. 3027, eds. C. Cachin and J. Camenisch. 207–222.

[6] Cramer, Ronald and Victor Shoup (1998). "A practical public key cryptosystem provably secure against adaptive chosen ciphertext attack." *Advances in Cryptology—CRYPTO'98*, Lecture Notes in Computer Science, vol. 1462, ed. Hugo Krawczyk. Springer-Verlag, Berlin, 13–25.

[7] Cramer, Ronald and Victor Shoup (2002). "Universal hash proofs and a paradigm for chosen ciphertext secure public key encryption." *Advances in Cryptology—EUROCRYPT 2002*, Lecture Notes in Computer Science, vol. 2332, ed. Lars Knudsen. Springer-Verlag, Berlin, 45–64.

[8] Cramer, Ronald and Victor Shoup (2004). "Design and analysis of practical public-key encryption schemes secure against adaptive chosen ciphertext attack." *SIAM Journal of Computing*, 33 (1), 167–226.

[9] Dolev, D., C. Dwork, and M. Naor (2000). "Non-malleable cryptography." *SIAM Journal of Computing*, 30 (2), 391–437.

[10] Fujisaki, Eiichiro, Tatsuaki Okamoto, David Pointcheval, and Jacques Stern (2004). "RSA-OAEP is secure under the RSA assumption." *J. Cryptology*, 17 (2), 81–104.

[11] Naor, Moni and Moti Yung (1990). "Public-key cryptosystems provably secure against chosen ciphertext attacks." *Proceedings of 22nd ACM Symposium on Theory of Computing, May 1990*, 427–437.

[12] Rackoff, C. and D. Simon (1991). "Noninteractive zero-knowledge proof of knowledge and chosen ciphertext attack." *Advances in Cryptology—CRYPTO'91*, Lecture Notes in Computer Science, vol. 576, ed. J. Feigenbaum. Springer-Verlag, Berlin, 433–444.

[13] Sahai, Amit and Edith Elkind (2002). "A unified methodology for constructing public-key encryption schemes secure against adaptive chosen-ciphertext attack." http://eprint.iacr.org/2002/042/

CREDENTIALS

In a general sense, credentials are something that gives a title to credit or confidence. In computer systems, credentials are descriptions of privileges that are issued by an authority to a subject. The privilege may be an access right, an eligibility, or membership (see also privilege management and access control). Examples from real life are driver's licenses, club membership cards, or passports. A credential can be *shown* to a verifier in order to prove one's eligibility or can be *used* toward a recipient in order to exercise the described privilege or receive the described service. The integrity of a credential scheme relies on the verifiers being able to effectively check the following three conditions before granting access or providing service:

1. The credential originates from a legitimate authority. For example, the alleged authority is known or listed as an approved provider of credentials for the requested service.
2. The credential is legitimately shown or used by the respective subject.
3. The privilege described is sufficient for the service requested.

In centralized systems, credentials are called *capabilities*, i.e., descriptions of the access rights to certain security critical objects (see access control). The centralized system manages the issuing of capabilities to subjects through a trusted issuing process, and all attempts of subjects to access objects through a trusted verifier, i.e., the access enforcement mechanism. If the subject has sufficient capabilities assigned, it is allowed to access the requested object, otherwise the access is denied. The capabilities and their assignment to subjects are stored in a central trusted repository, where they can be looked up by the access enforcement mechanism. Thus, in centralized systems, the integrity requirements 1, 2, 3 are enforced by trusted central processes.

In distributed systems there are autonomous entities acting as issuing authorities, as users who get credentials issued or show/use credentials, or as verifiers. Distributed credentials need to satisfy the above integrity requirements even in the presence of one or more cheating users, possibly collaborating. In addition, one can be interested in privacy requirements of users against cheating issuers and verifiers, possibly collaborating. David Chaum introduced credentials in this context of distributed systems in [8]. Distributed credentials have been proposed to represent such different privileges as electronic cash, passports, driver's licenses, diplomas, and many others. Depending on what privilege a credential represents, its legitimate use must be restricted appropriately (see integrity requirement 3 above). The following atomic *use restrictions* have been considered in the literature.

Nontransferable credentials cannot be (successfully) shown by subjects to whom they have not been issued in the first place. Such credentials could represent nontransferable privileges such as diplomas or passports.

Revocable credentials cannot be (successfully) shown after they have expired or have been revoked. Such credentials could represent revocable privileges such as driver's licenses or *public key certificates* commonly used in public key infrastructures (PKI).

Consumable credentials cannot be (successfully) shown after they have been used a specified number of times. Such credentials could represent privileges that get consumed when you use them, e.g., electronic cash.

More complex use restrictions can be defined by combining these atomic use restrictions in Boolean formulae. For example, revocable nontransferable credentials could be used as driver's licenses that expire after a specified period of, e.g., 2 years.

A credential scheme is called *online* if its credentials can be shown and used only by involving a central trusted authority that needs to clear the respective transactions. If the holder and verifier can do so without involving a third party, the credentials scheme is called *offline*. Online credential schemes are regarded as more secure for the issuers and verifiers, while offline credential schemes are regarded as more flexible and convenient to customers.

Credentials and their use could carry a lot of personal information about their holders. For example, consider an automated toll system that checks the driver's license of each car driver frequently but conveniently via wireless road check points.

Such a system would effectively ban drivers without a valid license, but it could also effectively monitor the moving of all honest drivers. Considerations like this led Chaum [8] to look for privacy in credentials:

Unlinkable credentials can be issued and shown/used in such a way that even a coalition of cheating issuers and verifiers has no chance to determine which issuing and showing/using or which two showings/usings originate from the same credential (see unlinkability).

Unlinkable credentials also leave the holders anonymous, because if transactions on the same credential cannot be linked, neither can such transactions be linked to the credential holder's identity. (Otherwise, they were no longer unlinkable.)

In the cryptographic literature, the term *credential* is most often used for nontransferable and unlinkable credentials, i.e., those that are irreversibly tied to human individuals, and protecting the privacy of users. Numerous cryptographic solutions have been proposed both for consumable credentials and for personal credentials alike. Chaum et al. [14] kicked off the development of consumable credentials. Improvements followed by Chaum et al. [10–12], Chaum and Pedersen [15], Cramer and Pedersen [17], Brands [2], Franklin and Yung [21], and others. Brands solution achieves overspending prevention by using a wallet-with-observer architecture (see [19] and electronic wallet), overspender detection without assuming tamper resistant hardware, and unconditional unlinkability of payments also without assuming tamper resistant hardware. Brands solution satisfied almost all requirements that had been stated by 1992 for efficient offline consumable credentials (e-cash) in a surprisingly efficient way.

Naccache and von Solms [23] pointed out later that unconditional unlinkability (which implies payer anonymity) might be undesirable in practice because it would allow blackmailing and money laundering. They suggested to strive for a better balance between the individuals' privacy and law enforcement. This work triggered a number of proposals for consumable credentials with anonymity revocation by Stadler et al. [24], Brickell et al. [4], Camenisch et al. [7], and Frankel et al. [20].

About the same amount of work has been done on developing personal credential schemes. Quite surprisingly, the problem of nontransferability between cheating collaborating individuals was neglected in many of the early papers by Chaum and Evertse [8, 9, 13] and Chen [16]. Chen's credentials are more limited than Chaum's and Evertse's because they can be shown only once. Damård [18] stated nontransferability as a security requirement but the proposed solution did not address nontransferability. Chaum and Pedersen [15] introduced the wallet-with-observer architecture and proposed personal credentials to be kept inside *wallet databases*, i.e., decentralized databases keeping all the privileges of their respective owners. Their proposal only partially addressed nontransferability by suggesting "distance bounding protocols" (Brands and Chaum [3]) in order to ensure the physical proximity of a wallet-with-observer during certain transactions. Distance bounding protocols can prevent Mafia frauds, where the individual present at an organization connects her device online to the wallet of another remote individual who holds the required credential and then diverts the whole communication with the organization to that remote wallet. Distance bounding cannot, however, discourage individuals from simply lending or trading their wallets. Lysyanskaya et al. [22] proposed a general scheme based on one-way functions and general zero-knowledge proofs, which is impractical, and a practical scheme that has the same limitations as Chen's: credentials can be shown only once.

The fundamental problem of enforcing nontransferability is simply this: the legitimate use of personal credentials (in contrast to consumable credentials) can neither be detected nor prevented by referring only to the digital activity of individuals. There must be some mechanism that can distinguish whether the individual who shows a personal credential is the same as the individual to whom that credential has been issued before. Since personal devices as well as personal access information such as PINs and passwords can easily be transferred from one individual to another, there is no other way to make this distinction but by referring to hardly transferable characteristics of the individuals themselves, for example, through some kind of (additional) biometric identification (see biometrics) of individuals. Then, illegitimate showing can be recognized during the attempt and thus can be prevented effectively, however, at the price of assuming tamper resistant biometric verification hardware. Bleumer proposed to enhance the wallet-with-observer architecture of Chaum and Pedersen [15] by a biometric recognition facility embedded into the tamper resistant observer in order to achieve transfer prevention [1].

Camenisch and Lysyanskaya [5] have proposed a personal credential scheme which enforces nontransferability by deterring individuals who are willing to transfer, pool, or share their credentials. Individuals must either transfer all their credentials or none (*all-or-nothing nontransferability*). They argue that even collaborating attackers would refrain from granting each other access to their credit card accounts, when they are collaborating only to share, e.g., a driver's license. Obviously, this deterrence from not transferring credentials is quickly neutralized if two or more participants mutually transfer credentials to each other. If any of them misuses a credit card account of the other, he may experience the same kind of misuse with his own credit card account as a matter of retaliation. It appears as if this concept promotes and protects closed groups of criminal credential sharers. In addition, it would be hard in practice to guarantee that for each individual the risk of sharing any particular credential is significantly higher than the respective benefits. Thus, for most real-world applications such as driver's licenses, membership cards, or passports, this deterrence approach to nontransferability would face severe acceptance problems from the issuers' and verifiers' perspective. Their scheme also supports anonymity revocation as an option, but at the cost of about a 10-fold increase of computational complexity. Subsequent work by Camenisch and Lysyanskaya [6] also shows how to revoke their anonymous credentials on demand. The price of this feature is an even higher computational complexity of the showing of credentials.

It appears that detecting a cheating individual who has lent his personal credentials to another individual, or who has borrowed a personal credential from another individual is technically possible, but is often unacceptable in practice. Unauthorized access may lead to disastrous or hard-to-quantify damage, which cannot be compensated after the access has been made regardless how individuals are persecuted and what measures of retaliation are applied.

The wisdom of more than 20 years of research on credentials is that in offline consumable credentials overspender detection can be achieved by digital means alone while overspending prevention can only be achieved by relying on tamper resistant hardware. In online consumable credentials, both overspender detection and overspending prevention can be achieved without relying on tamper resistant hardware.

In personal credentials, one is interested in transfer prevention, which we have called nontransferability. Considering a separate integrity requirement of transferer detection makes little sense in most applications because the potential damage caused by illegitimately transferring credentials is hard to compensate for.

Nontransferability can be achieved in a strict sense only by relying on <u>tamper resistant biometric</u> verification technology, regardless if it is an online or offline scheme. Nontransferability can be approximated by deterrence mechanisms integrated into the personal credential schemes, but it remains to be analyzed for each particular application how effective those deterrence mechanisms can be.

Gerrit Bleumer

References

[1] Bleumer, Gerrit (1998). "Biometric yet privacy protecting person authentication." *Information Hiding*, Lecture Notes in Computer Science, vol. 1525, ed. D. Aucsmith. Springer-Verlag, Berlin, 99–110.

[2] Brands, Stefan (1994). "Untraceable off-line cash in wallet with observers." *Advances in Cryptology—CRYPTO'93*, Lecture Notes in Computer Science, vol. 773, ed. D.R. Stinson. Springer-Verlag, Berlin, 302–318.

[3] Brands, Stefan and David Chaum (1994). "Distance-bounding protocols." *Advances in Cryptology—EUROCRYPT'93*, Lecture Notes in Computer Science, vol. 765, ed. T. Helleseth. Springer-Verlag, Berlin, 344–359.

[4] Brickell, Ernie, Peter Gemmell, and David Kravitz (1995). "Trustee-based tracing extensions to anonymous cash and the making of anonymous change." *6th ACM-SIAM Symposium on Discrete Algorithms (SODA) 1995*. ACM Press, New York, 457–466.

[5] Camenisch, Jan and Anna Lysyanskaya (2001). "Efficient non-transferable anonymous multishow credential system with optional anonymity revocation." *Advances in Cryptology—EUROCRYPT 2001*, Lecture Notes in Computer Science, vol. 2045, ed. B. Pfitzmann. Springer-Verlag, Berlin, 93–118.

[6] Camenisch, Jan and Anna Lysyanskaya (2002). "Dynamic accumulators and application to efficient revocation of anonymous credentials." *Advances in Cryptology—CRYPTO 2002*, Lecture Notes in Computer Science, vol. 2442, ed. M. Yung. Springer-Verlag, Berlin, 61–76.

[7] Camenisch, Jan, Ueli Maurer and Markus Stadler (1996). "Digital payment systems with passive anonymity-revoking trustees." *ESORICS'96*, Lecture Notes in Computer Science, vol. 1146, ed. V. Lotz. Springer-Verlag, Berlin, 33–43.

[8] Chaum, David (1985). "Security without identification: Transaction systems to make Big Brother obsolete." *Communications of the ACM* 28 (10), 1030–1044.

[9] Chaum, David (1990). "Online cash checks." *Advances in Cryptology—EUROCRYPT'89*, Lecture Notes in Computer Science, vol. 434, eds. J.-J. Quisquater and J. Vandewalle. Springer-Verlag, Berlin, 288–293.

[10] Chaum, David (1990). "Showing credentials without identification: Transferring signatures between unconditionally unlinkable pseudonyms." *Advances in Cryptology—AUSCRYPT'90, Sydney, Australia, January 1990*, Lecture Notes in Computer Science, vol. 453. Springer-Verlag, Berlin, 246–264.

[11] Chaum, David (1992). "Achieving electronic privacy." *Scientific American* 267 (2), 96–101.

[12] Chaum, David, Bert den Boer, Eugene van Heyst, Stig Mjlsnes, and Adri Steenbeek (1990). "Efficient offline electronic checks." *Advances in Cryptology—EUROCRYPT'89*, Lecture Notes in Computer Science, vol. 434, eds. J.-J. Quisquater and J. Vandewalle. Springer-Verlag, Berlin, 294–301.

[13] Chaum, David and Jan-Hendrik Evertse (1987). "A secure and privacy-protecting protocol for transmitting personal information between organizations." *Advances in Cryptology—CRYPTO'86*, Lecture Notes in Computer Science, vol. 263, ed. A. Odlyzko. Springer-Verlag, Berlin, 118–167.

[14] Chaum, David, Amos Fiat, and Moni Naor (1990). "Untraceable electronic cash." *Advances in Cryptology—CRYPTO'88*, Lecture Notes in Computer Science, vol. 403, ed. S. Goldwasser. Springer-Verlag, Berlin, 319–327.

[15] Chaum, David and Torben Pedersen (1993). "Wallet databases with observers." *Advances in Cryptology—CRYPTO'92*, Lecture Notes in Computer Science, vol. 740, ed. E.F. Brickell. Springer-Verlag, Berlin, 89–105.

[16] Chen, Lidong (1996). "Access with pseudonyms." *Cryptography: Policy and Algorithms*, Lecture Notes in Computer Science, vol. 1029, eds. E. Dawson and J. Golic. Springer-Verlag, Berlin, 232–243.

[17] Cramer, Ronald and Torben Pedersen (1994). "Improved privacy in wallets with observers (extended abstract)." *Advances in Cryptology—EUROCRYPT'93*, Lecture Notes in Computer Science, vol. 765, ed. T. Helleseth. Springer-Verlag, Berlin, 329–343.

[18] Damård, Ivan B. (1990). "Payment systems and credential mechanisms with provable security against abuse by individuals." *Advances in Cryptology—CRYPTO'88*, Lecture Notes in Computer Science, vol. 403, ed. S. Goldwasser. Springer-Verlag, Berlin, 328–335.

[19] Even, Shimon, Oded Goldreich, and Yacov Yacobi (1984). "Electronic wallet." *Advances in Cryptology—CRYPTO'83*, Lecture Notes in Computer Science, ed. D. Chaum. Plenum Press, New York, 383–386.

[20] Frankel, Yair, Yannis Tsiounis, and Moti Yung (1996). "Indirect discourse proofs: Achieving efficient fair off-line e-cash." *Advances in Cryptography—ASIACRYPT'96*, Lecture Notes in Computer Science, vol. 1163, eds. K. Kim and T. Matsumoto. Springer-Verlag, Berlin, 286–300.

[21] Franklin, Matthew and Moti Yung (1993). "Secure and efficient off-line digital money." *20th International Colloquium on Automata, Languages and Programming (ICALP)*, Lecture Notes in Computer Science, vol. 700, eds. A. Lingas, R.G. Karlsson, and S. Carlsson. Springer-Verlag, Berlin, 265–276.

[22] Lysyanskaya, Anna, Ron Rivest, A. Sahai, and Stefan Wolf (1999). "Pseudonym systems." *Selected Areas in Cryptography*, Lecture Notes in Computer Science, vol. 1758, eds. H.M. Heys and C.M. Adams. Springer-Verlag, Berlin, 302–318.

[23] Naccache, David and Sebastiaan von Solms (1992). "On blind signatures and perfect crimes." *Computers & Security*, 11 (6), 581–583.

[24] Stadler, Markus, Jean-Marc Piveteau, and Jan Camenisch (1995). "Fair blind signatures." *Advances in Cryptology—EUROCRYPT'95*, Lecture Notes in Computer Science, vol. 921, eds. L.C. Guillou and J.-J. Quisquater. Springer-Verlag, Berlin, 209–219.

CROSS-CORRELATION

Let $\{a_t\}$ and $\{b_t\}$ be two underline{sequences} of *period* n (so $a_t = a_{t+n}$ and $b_t = b_{t+n}$ for all values of t) over an alphabet being the integers mod q (see underline{modular arithmetic}). The cross-correlation between the sequences $\{a_t\}$ and $\{b_t\}$ at shift τ is defined as

$$C(\tau) = \sum_{t=0}^{n-1} \omega^{a_{t+\tau} - b_t},$$

where ω is a complex qth root of unity. Note that in the special case of binary sequences, $q = 2$ and $\omega = -1$.

In the special case when the two sequences are the same, the cross-correlation is the same as the underline{auto-correlation}. Many applications in underline{stream ciphers} and communication systems require large families of cyclically distinct sequences with a low maximum nontrivial value of the auto- and cross-correlation between any two sequences in the family.

Tor Helleseth

References

[1] Golomb, S.W. (1982). *Shift Register Sequences*. Aegean Park Press, Laguna Hills, CA.

[2] Helleseth, T. and P.V. Kumar (1999). "Pseudonoise sequences." *The Mobile Communications Handbook*, ed. J.D. Gibson. CRC Press, Boca Raton, FL, Chapter 8.

[3] Helleseth, T. and P.V. Kumar (1998). "Sequences with low correlation." *Handbook of Coding Theory*, eds. V.S. Pless and W.C. Huffman. Elsevier, Amsterdam, Chapter 21.

CRYPTANALYSIS

Cryptanalysis is the discipline of deciphering a ciphertext without having access to the keytext (see underline{cryptosystem}), usually by recovering more or less directly the plaintext or even the keytext used, in cases favorable for the attacker by reconstructing the whole cryptosystem used. This being the worst case possible for the attacked side, an acceptable level of security should rest completely in the key (see Kerckhoffs' and Shannon's underline{maximes}). "A systematic and exact reconstruction of the encryption method and the key used" (Hans Rohrbach, 1946) is mandatory if correctness of a cryptanalytic break is a to be proved, e.g., when a cryptanalyst is a witness to the prosecution.

TERMINOLOGY: Cryptanalysis can be *passive*, which is the classical case of intercepting the message without giving any hint that this was done, or *active*, which consists of altering the message or retransmitting it at a later time, or even of inserting own messages (some of these actions may be detected by the recipient).

A *compromise* is the loss (or partial loss) of secrecy of the key by its exposure due to cryptographic faults. We shall describe various kinds of key compromises.

A *plaintext-ciphertext compromise* is caused by a transmission of a message in ciphertext followed (e.g., because the transmission was garbled) by transmission of the same message in plaintext. If information on the encryption method is known or can be guessed, this results in exposure of the key. This attack may be successful for a plaintext of several hundred characters.

A *plaintext-plaintext compromise* is a transmission of two *isologs*, i.e., two different plaintexts, encrypted with the same keytext. If the encryption method is such that the encryption steps form a group (see underline{key group} and pure underline{crypto-system}), then a "difference" $p_1 - p_2$ of two plaintexts p_1, p_2 and a "difference" $c_1 - c_2$ of two plaintexts c_1, c_2 may be defined and the role of the keytext is cancelled out: $c_1 - c_2 = p_1 - p_2$. Thus, under suitable guesses on the plaintext language involved, e.g., on probable words and phrases, a "zig-zag" method (see below), decomposing $c_1 - c_2$, gives the plaintexts and then also the keytext. This compromise is not uncommon in the case of a shortage of keying material. It is even systemic if a periodic key is used.

A *ciphertext-ciphertext compromise* is a transmission of two *isomorphs*, i.e., the same plaintext, encrypted with two different keytexts.

Exchanging the role of plaintext and keytext, this case is reduced to and can be treated as a plaintext-plaintext compromise. This compromise is even systematic in message sets, where the same message is sent in different encryption to many places, such as it is common in public key cryptosystems.

One speaks of a *brute force attack* or <u>exhaustive key search</u> if all possible keytexts are tried out to decrypt a ciphertext (knowing or guessing the cryptosystem used). At present, with the still growing speed of supercomputers, every 10 years the number of trial and error steps that is feasible is increased by a factor of roughly 2^5.

Further commonly used terminology will be given now. In a <u>ciphertext-only attack</u>, only one or more ciphertexts under the same keytext are known. In a <u>known-plaintext attack</u> one knows one or more matching pairs of plaintext–ciphertext. Frequently, this attack is carried out with rather short fragments of the plaintext (e.g., probable words and phrases). In a <u>chosen-plaintext attack</u> one can choose plaintexts and obtain the corresponding ciphertexts. Sometimes this can be done with the proviso that the plaintexts may be chosen in a way that depends on the previous encryption outcomes. How to foist the plaintext on the adversary is not a cryptographer's problem, but is a problem of misleading the adversary and is to be executed by the secret services. Finally, in a <u>chosen-ciphertext attack</u> there is the possibility to choose different ciphertexts to be decrypted, with the cryptanalyst having access to the decrypted plaintext. An example may be the investigation of a tamperproof decryption box, with the hope of finding the key.

STATISTICAL APPROACHES TO CLASSICAL CRYPTOSYSTEMS: We shall now discuss some statistical methods that can be used by the cryptanalist.

an example: the frequency profile of the English language looks like

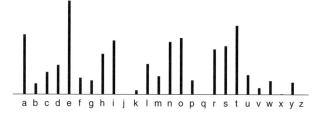

If a ciphertext of 349 characters has the following distribution:

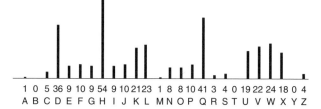

1	0	5	36	9	10	9	54	9	10	21	23	1	8	8	10	41	3	4	0	19	22	24	18	0	4
A	B	C	D	E	F	G	H	I	J	K	L	M	N	O	P	Q	R	S	T	U	V	W	X	Y	Z

it is easy to guess a <u>Cæsar</u> encryption that counts down three letters in the standard <u>alphabet</u>: a ≐ D, b ≐ E, c ≐ F, ..., z ≐ C. More difficult is the situation if a mixed alphabet is to be expected. Then the first step is to group the letters in cliques: the most frequent ones, the very rare ones, and some in between

{etaoin} {srh} {ld} {cumfpgwyb} {vk} {xjqz},

and to refine the decryption within these cliques by trial and error.

Depth is a notion used in connection with the cryptanalysis of polyalphabetic encryptions. It means the arrangement of a number of ciphertexts supposedly encrypted with the same keytext—for periodic polyalphabetic encryption broken down according to the assumed period.

Example: a depth of five lines:

T C C V L	E S K P T	X M P V W	H Y M V G	X B O R V	C W A R F
V L L B V	C K W F P	E H E C F	C G N Z E	K K K V I	H D D I D
M Y Y R D	M J W M C	U I G L O	K M X L R	E W H X M	T J H A S
B K Q T Z	B Z W K W	Z X G Z O	V T B A T	K W M G M	R J K L P
M Y Y V H	B W J D X	C P C Z O	H V T S I	V M E B S	O H R A U.

Frequency matching is a cryptanalytic method for breaking monoalphabetic (Cæsar type) encryptions. One determines the frequency of the characters in a ciphertext and compares them with the frequency of the characters in a language known or assumed to be used for the plaintext. To give

The lines of a depth are isologs: they are encrypted with the same key text and represent a plaintext–plaintext compromise.

By forming differences of the elements in selected columns, a reduction of depth to a monoalphabetic (Cæsar type) encryption is accomplished.

This makes it possible to derive the keytext

TRUTH ISSOP RECIO USTHA TITNE EDSAB

which decrypts the depth (by means of the Vigenère table) to

```
al i ce  wasbe  gi nni  ngtog  etver  yti re
cur i o  user a  ndcur  i ouse  rcri e  dal i c
the yw  e rei n  deeda  queer  l ooki  ngpar
i t was  t hewh  i te ra  bbi tt  rot t i  ngsl o
the ca  t erpi  l l ara  ndal i  cel oo  kedat.
```

Forming a depth is possible as soon as the value of the period of the periodic polyalphabetic encryption has been found, for instance by the Kasiski method below. Forming a depth is not possible, if the key is non-periodic. But even for periodic polyalphabetic encryptions, forming a depth of sufficiently many elements (usually more than six) is not possible if the keytext is short enough.

When the alphabets used in a polyalphabetic periodic substitution are a mixed alphabet and a shifted version of it, *symmetry of position* is the property that for any pair of letters their distance is the same in all rows of the encryption table. For a known period, it may allow, after forming a depth, the complete reconstruction of the substitution (Auguste Kerckhoffs, 1883).

Kasiski's method. If in a periodic polyalphabetic encryption the same plaintext sequence of characters happens to be encrypted with the same sequence of key characters, the same ciphertext sequence of characters will occur. Thus, in order to determine the period of a periodic polyalphabetic encryption, the distance between two "parallels" in the ciphertext (pairs, triples, quadruples etc. of characters) is to be determined; the distance of genuine parallels will be a multiple of the period. The greatest common divisor of these distances is certainly a period—it may, however, not be the smallest period. Moreover, the period analysis may be disturbed by faked parallels. Kasiski developed this fundamental test for key periodicity in 1863 and shattered the widespread belief that periodic polyalphabetic encryption is unbreakable.

The *Kappa test* is based on the relative frequency $\kappa(T, T')$ of pairs of text segments $T = (t_1, t_2, t_3, \ldots, t_M)$, $T' = (t'_1, t'_2, t'_3, \ldots, t'_M)$ of equal length, $M \geq 1$, with the same characters at the same positions (that is why this method is also called the *index of coincidence*, often abbreviated to I.C., William F. Friedman, 1925). The value of Kappa is rather typical for natural languages, since the expected value of $\kappa(T, T')$ is $\sum_{i=1}^{N} p_i^2$, where p_i is the probability of occurrence of the ith character of the vocabulary to which T and T'

belong. For sufficiently long texts, it is statistically roughly equal to $1/15 = 6.67\%$ for the English language and $1/12.5 = 8\%$ for the French language and the German language. Most importantly, it remains invariant if the two texts are polyalphabetically encrypted with the same keytext. If, however, they are encrypted with different keytexts or with the same key sequence, but with different starting positions, the character coincidence is rather random and the value of Kappa is statistically close to $1/N$, where N is the size of the vocabulary. The Kappa test applied to a ciphertext C and cyclically shifted versions $C^{(u)}$ of the ciphertext, where u denotes the number of shifts, yields the value $\kappa(C, C^{(u)})$. If the keytext is periodic with period d, then for $u = d$ and for all multiples of d, a value significantly higher than $1/N$ will occur, while in all other cases a value close to $1/N$ will be found. This is the use of the Kappa examination for finding the period; it turned out to be a more accurate instrument than the Kasiski method.

The Kappa test may also be used for adjusting two ciphertexts C, C' which are presumably encrypted with the same keytext, but with different starting positions (called *superimposition*). By calculating $\kappa(C^{(u)}, C')$, a shift d, determined as a value of u, for which $\kappa(C^{(u)}, C')$ is high, brings the two ciphertexts $C^{(d)}$ and C' "in phase", i.e., produces two isologs. In this way, a depth of n texts can be formed by superimposition from a ciphertext–ciphertext compromise of n ciphertexts.

The *De Viaris attack* is a cryptanalytic method invented by Gaëtan Henri Léon de Viaris in 1893 to defeat a polyalphabetic cryptosystem proposed by Étienne Bazeries, in which the alphabets did not form a Latin square. (A *Latin square* for a vocabulary of N characters is an $N \times N$ matrix over this alphabet such that each character occurs just once in every line and in every column.)

Pattern finding is a cryptanalytic method that can be applied to monoalphabetic encryptions. It makes use of patterns of repeated symbols. For example, the pattern *1211234322* with "signature" $4 + 3 + 2 + 1$ (four twos, three ones, two threes and one four) most likely allows in English nothing but peppertree, the pattern *1213143152* with the signature $4 + 2 + 2 + 1 + 1$ nothing but initiation (Andree 1982, based on Merriam-Webster's Dictionary).

Noncoincidence exhaustion. Some cryptosystems show peculiarities: genuine self-reciprocal permutations never encrypt a letter by itself. Porta encryptions even always encrypt a letter from the first half of the alphabet by a letter from the second half of the alphabet and vice versa. Such properties may serve to exclude many positions of a probable word (a *probable word* is a word or phrase that can be expected to be present in a message according to the context; it may form the basis for a known-plaintext attack).

Zig-zag exhaustion. For encryptions with a key group (see key), the difference of two plaintexts is invariant under encryption: it equals the difference of the corresponding ciphertexts. Thus in case of a plaintext–plaintext compromise (with a depth of 2), the difference of the ciphertexts can be decomposed into two plaintexts by starting with probable words or phrases in one of the plaintexts and determining the corresponding plaintext fragment in the other plaintext, and vice versa. This may lead in a zig-zag way ("cross-ruff") to complete decryption.

Theoretically, the decomposition is unique provided the sum of the relative redundancies of the two texts is at least 100%. For the English language, the redundancy (see information theory) is about 3.5 [bit/char] or 74.5% of the value $4.7 \approx \log_2 26$ [bit/char].

Multiple anagramming is one of the very few general methods for dealing with transposition ciphers, requiring nothing more than two plaintexts of the same length that have been encrypted with the same encryption step (so the encrypting transposition steps have been repeated at least once). Such a plaintext–plaintext compromise suggests a parallel to Kerkhoffs' method of superimposition. The method is based on the simple fact that equal encryption steps perform the same permutation of the plaintext letters. The ciphertexts are therefore written one below the other and the columns thus formed are kept together.

Friedrich L. Bauer

Reference

[1] Bauer, F.L. (1997). "Decrypted secrets." *Methods and Maxims of Cryptology*. Springer-Verlag, Berlin.

CRYPTO MACHINES

These are machines for automatic encryption using a composition of mixed alphabet substitutions often performed by means of rotors. Examples are: Enigma (Germany), Hebern Electric Code Machine (USA), Typex (Great Britain), SIGABA $\triangleq$ M-134-C (USA), and NEMA (Switzerland).

Rotor: wheel, sitting on an axle and having on both sides a ring of contacts that are internally wired in such a way that they implement a permutation.

The *Enigma machine* (Figures 1 and 2) was invented by the German Arthur Scherbius. In 1918, he filed a patent application for an automatic, keyboard-operated electric encryption machine performing a composition of a fixed number of polyalphabetic *substitution* (see substitutions and permutations) steps (four in the early commercial models) with shifted mixed alphabets performed by wired keying wheels (called *rotors*).

The key sequence was generated by the cyclometric, "counter-like" movement of the wheels. The fixed substitutions of the rotors were to be kept secret, the starting point of the key sequence was to be changed at short intervals. Later "improvements" were a reflector (Willi Korn, 1925) which made the encryption self-reciprocal (and opened a way of attack) and (by request from the German Armed Forces Staff) a plugboard performing a substitution that could be changed at

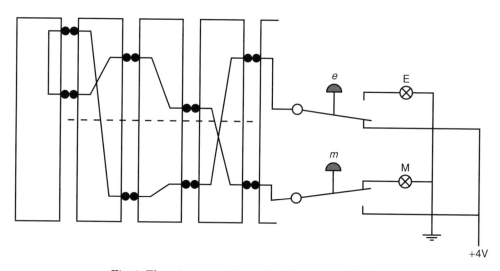

Fig. 1. Electric current through a three-rotor Enigma

Fig. 2. Cipher machine Enigma (four-rotor version)

short intervals (which in fact helped to avoid certain manifest attacks). The German *Wehrmacht* Enigma had three rotors (to be selected out of up to eight) and a reflector (to be selected out of two). In the Navy variant of 1942 the reflector was not fixed. Certain weaknesses of the Enigma encryption were not caused by the machine itself, but by cryptographic blunders in operating it.

An exceptional role was played by the German *Abwehr*, the Intelligence Service of the German Armed Forces, as far as ENIGMA goes: It used a variant of the old ENIGMA D with a pinion/tooth wheel movement of the rotors, with 11, 15, and 17 notches on the wheels ('11-15-17 ENIGMA'), and naturally without plugboard—following rather closely Willi Korn's German Patent No. 534 947, filed November 9, 1928, US Patent No. 1,938,028 of 1933.

A few specimens of a three-rotor-ENIGMA ('ENIGMA T'), likewise without plugboard, but with five-notched rotors, were destined for the Japanese Navy, but did not get out of the harbor and were captured by the Allies near Lorient, in Brittany.

In England, too, rotor machines were built during the Second World War: TYPEX was quite an improved ENIGMA (instead of the plugboard there was an entrance substitution performed by two fixed rotors which was not self-reciprocal).

In the USA, under the early influence of William Friedman (1891–1969) and on the basis of the Hebern development, there was in the early 1930s

a more independent line of rotor machines, leading in 1933 to the M-134-T2, then to the M-134-A (SIGMYC), and in 1936 to the M-134-C (SIGABA) of the Army, named CSP889 (ECM Mark II) by the Navy. The Germans obviously did not succeed in breaking SIGABA, which had five turning cipher rotors with irregular movement. It had been made watertight by Frank Rowlett (1908–1998), an aide of Friedman since 1930.

An interesting postwar variant of the ENIGMA with seven rotors and a fixed reflector was built and marketed by the Italian company Ottico Meccanica Italiana (OMI) in Rome.

The Swiss army and diplomacy used from 1947 on an ENIGMA variant called NEMA ('*Neue Maschine*') Modell 45. It was developed by Hugo Hadwiger (1908–1981), Heinrich Weber (1908–1997), and Paul Glur, and built by Zellweger A.G., Uster. It had ten rotors, six of which were active ones, while the others served for rotor movement only. The use of a reflector was unchanged.

Based on US-American experiences and similar to TYPEX was the rotor machine KL-7 of the NATO, in use until the 1960s.

The Swedish inventor Boris Hagelin created the *Hagelin machine* (Figure 3) in 1934. It was an

Fig. 3. The M-209 Hagelin machine

automatic mechanical encryption machine, with alphabet-wheel input and a printing device, performing by a "lug cage" adder a Beaufort substitution (see Beaufort encryption) controlled by five keying wheels with 17, 19, 21, 23, and 25 teeth according to a pre-chosen "step figure." The lug matrix and the step figure can be changed by activating suitable pins. This model, called C-35, was followed by a C-36, with six keying wheels and 17, 19, 21, 23, 25, and 26 teeth. Under the cover name M-209, C-36 was built by Smith-Corona for the US Army, Navy and Air Forces, altogether accounting for 140 000 machines. The Hagelin machines were less secure than the Enigma.

Friedrich L. Bauer

Reference

[1] Bauer, F.L. (1997). "Decrypted secrets." *Methods and Maxims of Cryptology*. Springer-Verlag, Berlin.

Fig. 1. Semagram. The message is in Morse code, formed by the short and long stalks of grass to the left of the bridge, along the river bank and on the garden wall. It reads: "compliments of CPSA MA to our chief Col. Harold R. Shaw on his visit to San Antonio May 11, 1945." (Shaw had been head of the Technical Operations Division of the US government's censorship division since 1943)

CRYPTOLOGY

Cryptology is the discipline of *cryptography* and cryptanalysis and of their interaction. Its aim is *secrecy* or *confidentiality*: the practice of keeping secrets, maintaining privacy, or concealing valuables. A further goal of cryptology is *integrity* and *authenticity*, usually given by a *message authentication code* (see MAC algorithms) or digital signature unique to the sender and serving for his identification.

Cryptography: the discipline of writing a message in *ciphertext* (see cryptosystem), usually by a translation from *plaintext* according to some (frequently changing) *keytext*, with the aim of protecting a secret from adversaries, interceptors, intruders, interlopers, eavesdroppers, opponents or simply attackers, opponents, enemies. Professional cryptography protects not only the plaintext, but also the key and more generally tries to protect the whole cryptosystem.

Steganography: counterpart of cryptography, comprising technical steganography (working with invisible inks and hollow heels, etc.) and linguistic steganography, the art of hiding information by linguistic means. Of the later, we mention semagrams, open code, comprising jargon code (masked secret writing, e.g., cues), and conceal-ment ciphers (veiled secret writings, like the null cipher and grilles (see substitutions and permutations)). Linguistic steganography has close connections with cryptography, e.g., the uses of grilles and transposition ciphers (see substitutions and permutations) are related.

Semagram: a picture or grapheme hiding a message behind innocent looking, frequently minute graphic details (Figure 1).

Open code: a class of linguistic steganography, selected words or phrases made to appear innocent in the surrounding text.

Cue: a short, prearranged jargon-code message, usually a word or phrase. The importance of the message is strongly linked to the time of transmission. Famous is the encrypted Japanese message HIGASHI NO KAZE AME ("east wind, rain") on December 7, 1941, a cue with the meaning "war with the USA."

Null cipher, where only certain letters are significant (the others being *nulls*):

- encryption rules of the type "the nth character after a particular character," specifically "the next character after a space" (*acrostics*),
- or rules for inserting a null between syllables (Tut Latin: TUT) or after phonetic consonants (Javanais: chaussure ↦ CHAVAUSSAVURAVE)
- or rules for suffixing nulls, e.g., un fou ↦ UNDREQUE FOUDREQUE (Metz 1670), also combined with shuffling of some letters (largonjem:

boucher $\mapsto$ LOUCHER<u>B</u>EM, pig Latin: third $\mapsto$ IRD<u>T̊HAY</u>);

- a borderline case is pure transposition: reversing letters of a word (back slang), e.g., tobacco $\mapsto$ OCCABOT).

Friedrich L. Bauer

Reference

[1] Bauer, F.L. (1997). "Decrypted secrets." *Methods and Maxims of Cryptology*. Springer-Verlag, Berlin.

CRYPTOSYSTEM

A cryptosystem (or *cipher system*) is a system consisting of an <u>encryption</u> *algorithm*, a *decryption algorithm*, and a well-defined triple of *text spaces*: *plaintexts*, *ciphertexts*, and *keytexts*. For a given keytext the encryption algorithm will map a plaintext to a (usually uniquely determined) ciphertext. For the corresponding keytext, the decryption algorithm will map the ciphertext to the (usually uniquely determined) plaintext. The cryptosystem may be performed by hand methods ("hand cipher"), by machine methods ("machine cipher"), or by software (see also <u>Shannon's model</u>).

Plaintext: text in an open language that is commonly understood among a larger group of people.

Ciphertext: text ("cryptogram") in a secret language that is understood only by few, authorized people, usually after decryption by hand or machine.

We mention two special properties: an *endomorphic cryptosystem* is a cryptosystem with identical plaintext and ciphertext space. Example: $\{a, b, c, \ldots, z\} \longrightarrow \{a, b, c, \ldots, z\}$. A *pure cryptosystem* is a cryptosystem that has the following property: whenever enciphering $\mathcal{E}_k$ with <u>key</u> k, followed by deciphering $\mathcal{D}_j$ with key j, followed by enciphering $\mathcal{E}_i$ with key i is performed, there is a key ℓ such that $\mathcal{E}_\ell$ has the same effect: $\mathcal{E}_i \mathcal{D}_j \mathcal{E}_k = \mathcal{E}_\ell$. In a pure cryptosystem, the mappings $\mathcal{D}_j \mathcal{E}_k$ ("$\mathcal{E}_k$ followed by $\mathcal{D}_j$") form a <u>group</u>, with $\mathcal{D}_k \mathcal{E}_k$ being the identity.

An endomorphic cryptosystem is pure if and only if its encipherings are closed under composition (Shannon), that is if its keys form a <u>group</u>, the *key group* (see <u>key</u>).

Friedrich L. Bauer

Reference

[1] Bauer, F.L. (1997). "Decrypted secrets." *Methods and Maxims of Cryptology*. Springer-Verlag, Berlin.

CRYPTREC

CRYPTREC was initially an abbreviation of Cryptography Research and Evaluation Committee, which was set up in 2000 by METI (Ministry of Economy, Trade and Industry, Japan) for the purpose of evaluating cryptographic techniques to assure their security for e-Government applications. However, since the **CRYPTREC Advisory Committee** was founded by MPHPT (Ministry of Public Management, Home Affairs, Posts and Telecommunications, Japan) and METI in 2001 to perform a policymaking study on the application of cryptographic techniques, it has been interpreted as the name of the organization of committees involved in the project for evaluation of cryptographic techniques available for the Japanese e-Government. The project itself is now referred to as CRYPTREC. From 2000 to 2002 the main goal of CRYPTREC was to publish a list of recommended ciphers for e-Government use. In March 2003 the list was published and established as the guiding principle in the usage of cryptographic techniques in the Japanese government ministries and agencies.

EVALUATION TARGETS: In the fiscal years 2000 and 2001, **CRYPTREC Evaluation Committee** called for the submission of the cryptographic techniques in order to compile a list of cryptographic techniques that could be employed for e-Government. In the public request for proposals, the CRYPTREC Evaluation Committee did not impose any restrictions on the national origin or the organization of the applicant in order to provide an opportunity for impartial evaluation for all applicants.

The CRYPTREC Evaluation Committee specified several cryptographic techniques as "indispensable cryptographic techniques." The CRYPTREC Evaluation Committee also evaluated several cryptographic techniques as "specific evaluation" target ciphers for special reasons such as requests from standardization organizations and the Law concerning Electronic Signatures and Certification.

Basically, the evaluation targets can be categorized into the following three types: "submitted cryptographic techniques," "cryptographic techniques for specific evaluation," and "indispensable cryptographic techniques."

Submitted Cryptographic Techniques

Cryptographic techniques in the categories of (1) digital signature, authentication, confidentiality, and key agreement for public-key cryptography, (2) 64-bit block ciphers, 128-bit block ciphers, and stream ciphers for symmetric-key cryptography, and (3) hash functions and pseudorandom number generators were sought for evaluation. The applicants who submitted techniques were asked to make their cryptographic techniques procurable by the end of the fiscal year 2002. CRYPTREC Evaluation Committee received a total of 63 applications in both fiscal years 2000 and 2001.

Indispensable Cryptographic Techniques

In addition to the cryptographic techniques submitted by the applicants, the CRYPTREC Evaluation Committee has selected techniques that were considered to be indispensable in the construction of e-Government systems. Such cryptographic techniques must have either comparatively long track records of use and evaluation, or have a long history of usage. These targets were selected for evaluation as "indispensable cryptographic techniques" whether or not an applicant submitted them.

Cryptographic Techniques for Specific Evaluation

"Cryptographic techniques for specific evaluation" are the cryptographic techniques that were evaluated by CRYPTREC on the basis of a special request, independent of their submission by an application and independent from a specification as "indispensable cryptographic techniques." Cryptographic techniques for specific evaluation in the fiscal years 2000 and 2001 are classified into the following three categories: (1) cryptographic techniques specified in Guidelines on the Law concerning Electronic Signatures and Certification Services, (2) cryptographic techniques used in SSL3.0/TLS1.0, and (3) contributions to **ISO/IEC JTC1/SC27**.

EVALUATION AND SELECTION OF CRYPTO-GRAPHIC TECHNIQUES: CRYPTREC has performed security evaluations in order to select cryptographic techniques that satisfy the level of security sufficient for e-Government usage. CRYPTREC has also performed software and hardware implementation evaluations to measure the processing speed and amount of system resources required.

In order to ensure that the technical evaluations are impartial and adequate, CRYPTREC has requested several specialists, besides its own members, to conduct evaluations (referred to as external evaluations).

In order to ensure that evaluations are fair for all the cryptographic techniques in the same category, CRYPTREC has applied the same evaluation methods as much as possible to allow relative comparisons.

Evaluation Items

The evaluation in CRYPTREC progressed gradually and in parallel to get a good understanding of algorithm properties and characteristics such as security and performance; the evaluation also assessed how easy it was to develop efficient implementations. There were four stages of the evaluation:

(1) *Screening evaluation.* Submitted documents were studied to investigate whether the target cryptographic technique had any problems in the design concept, design policies, security, or implementation.

(2) *Full evaluation.* The following items were investigated: (a) whether known attacks are applicable or not, (b) computational cost required for a known attack to succeed, (c) validity of provable security, (d) validity of parameter/key generating methods, (e) selection of auxiliary functions and methods used to implement them in the scheme, (f) anticipated problems of submitted cryptographic techniques in real systems, and (g) whether any attack can be mounted or not using the evaluators' expertise.

The techniques were also compared with other cryptographic techniques in order to assess relative strengths and weaknesses.

(3) *Software implementation evaluation.* The compatibility and portability with respect to computing resources and environments was verified by checking whether the software operated as described in the submitted documents in the following environments: (a) general PC environment, (b) most popular server environment, and (c) high-performance, high-end environment.

(4) *Hardware implementation evaluation.* It was investigated whether a third party could

design the hardware using the submitted documents only.

Evaluation Criteria

The following criteria were set for evaluation of cryptographic techniques according to the categories.

(1) Public-key cryptographic techniques. If a public-key cryptographic technique has a solid track record of operation and evaluation over a relatively long period of time and its specifications cannot be changed easily from the standpoint of interoperability, the following conditions must be satisfied:

   (a) The cryptographic techniques must have been evaluated and researched thoroughly by a number of researchers.

   (b) No security problem has been reported in a realistic system.

   Relatively new public-key cryptographic techniques were required to have at least "provable security." A comprehensive security evaluation was carried out in addition to checking the provable security, including issues such as the validity of number theoretic problems, the method of selecting recommended parameters, and the method of using auxiliary functions in a scheme.

(2) Symmetric cryptography techniques. Symmetric-key cryptographic techniques must satisfy either of the following conditions:

   (a) Even with the best attacking technique available to date, a computational cost of 2^{128} or more (i.e., exhaustive search for a secret key) is required to break symmetric-key cryptographic techniques. It is necessary to show at the techniques are secure against typical cryptanalytic attacks such as differential and linear cryptanalysis.

   (b) Widely used symmetric-key cryptographic techniques that have been evaluated in details and have no security problems in a realistic system are selected. In this case, a computational cost of 2^{100} or more is required to break them.

(3) Hash functions. Hash functions must satisfy either of the following conditions:

   (a) Even with the best attacking technique available to date, the computational cost to find the input value for a specific output value is not less than the computational cost required for an exhaustive search. Also, even if the best attacking technique is used, the computational cost to find a pair of different input values with the same output value is 2^{128} or more.

   (b) Widely used hash functions that have no security problems in realistic systems and with a hash length of 160 bits or more are selected.

(4) Pseudorandom number generators: Pseudorandom number generators must satisfy all the following conditions:

   (a) The statistical properties are close to that of a true random number. A past or future unknown output bit is hard to predict from the known output bit history.

   (b) The seed size must be large enough to be secure against an exhaustive key search of the system that uses a pseudorandom number generator.

   (c) The statistical properties of pseudorandom number generators must pass a typical statistical test suite for randomness such as SP800-22.

Requirements for the Draft of the e-Government Recommended Ciphers List

The CRYPTREC Advisory Committee has requested the CRYPTREC Evaluation Committee to evaluate the candidates for e-Government ciphers, cryptographic techniques that allow authentication, key agreement, confidentiality, and electronic signature functions in the e-Government system, and prepare an e-Government recommended ciphers list considering the following three points:

(1) Select several cryptographic techniques with sufficient security for use in the e-Government system (security guaranteed roughly 10 years).

(2) Select for each category at least one cryptographic technique that is being-incorporated or likely to be incorporated in commercial software (to be used by the general public).

(3) Confirm the specifications of cryptographic techniques recommended for e-Government to guarantee that ciphers satisfying these specifications can be procured.

E-GOVERNMENT RECOMMENDED CIPHERS LIST: As a three-year comprehensive project, the **"e-Government recommended ciphers list (draft)"** authored by the CRYPTREC Evaluation Committee was submitted to the CRYPTREC Advisory Committee for review. Then, MPHPT and METI invited comments from the general

Table 1. e-Government recommended ciphers list (draft) (prepared in November 2002)

Category		Name
Public-key ciphers	Signature	DSA
		ECDSA
		RSASSA-PKCS1-v1 5
		RSA-PSS
	Confidentiality	RSA-OAEP
		RSAES-PKCS1-v1 5[1]
	Key agreement	DH
		ECDH
		PSEC-KEM[2]
Symmetric-key ciphers	64-bit block ciphers[3]	CIPHERUNICORN-E
		Hierocrypt-L1
		MISTY1
		3-key Triple DES[4]
	128-bit block ciphers	AES
		Camellia
		CIPHERUNICORN-A
		Hierocrypt-3
		SC2000
	Stream ciphers	MUGI
		MULTI-S01
		128-bit RC4[5]
Others	Hash function	RIPEMD-160[6]
		SHA-1[6]
		SHA-256
		SHA-384
		SHA-512
	Pseudorandom number generator[7]	PRNG based on SHA-1 in ANSI X9.42-2001 Annex C.1
		PRNG based on SHA-1 for general purpose in FIPS 186-2 (+ change notice 1) Appendix 3.1
		PRNG based on SHA-1 for general purpose in FIPS 186-2 (+ change notice 1) revised Appendix 3.1

Notes:

[1] Use of this is permitted for the time being because it was used in SSL3.0/TLS1.0.

[2] On the assumption that this is used in the KEM (Key Encapsulation Mechanism)-DEM (Data Encapsulation Mechanism) construction.

[3] When constructing a new e-Government system, 128-bit block ciphers are preferable if possible.

[4] Using 3-key Triple DES is permitted for the time being under the following conditions:
  (1) It is specified as FIPS 46-3.
  (2) It is positioned as the de facto standard.

[5] It is assumed that the 128-bit RC4 will be used only in SSL3.0/TLS(1.0 or later). If any other cipher listed above is available, it should be used instead.

[6] If any hash functions with a longer hash value are available when constructing a new e-Government system, it is preferable that a 256-bit (or more) hash function be selected. However, this does not apply in cases where the hash function to be used has already been designated according to the public-key cryptographic specifications.

[7] Since pseudorandom number generators do not require interoperability due to their usage characteristics, no problems will be created by using a cryptographically secure pseudorandom number generating algorithm. Therefore, these algorithms are examples.

public. Finally, the draft was authorized as the "e-Government recommended ciphers list".

In Table 1, we show the e-Government recommended ciphers list. The notes are added to remind uses that some cryptographic techniques require in employing them for e-Government applications. For more details, the reader is referred to the CRYPTREC Web site [1].

OTHER ACTIVITIES

Revision of Guidelines on the Law Concerning Electronic Signatures and Certification Services

Guidelines on the Law concerning Electronic Signatures and Certification Services were revised corresponding to the CRYPTREC evaluation results in the fiscal year 2001.

SSL / TLS Evaluation Report

CRYPTREC evaluated the security of SSL/TLS (see Secure Socket Layer (SLS) and Transport Layer Security (TLS)) and reported as follows: SSL/TLS is secure against all known attacks. Using SSL/TLS, one needs to ensure that patches are applied and that parameters are properly selected. SSL/TLS is considered to offer an adequate security level for practical use. The functionality of TLS is still being extended. New security weaknesses can emerge as a result of these extensions. Therefore, it is necessary to monitor the status and progress of TLS and to keep investigating its security.

Publicizing External Evaluation Reports

CRYPTREC considers it important to publicize the cryptographic technique evaluation results in order to improve the reliability of security evaluations. All external evaluation reports that were compiled as a part of the evaluation activities of CRYPTREC are available on the CRYPTREC Web site [1].

Monitoring and Other Evaluations

After publishing the e-Government recommended ciphers list, the main responsibility of CRYPTREC has moved to monitoring the security of cryptographic techniques in the list. CRYPTREC also has started the evaluation of cryptographic modules and protocols.

Hideki Imai
Atsuhiro Yamagishi

Reference

[1] CRYPTREC Web site, http://www.shiba.tao.go.jp/ kenkyu/CRYPTREC/index.html and http://www. ipa.go.jp/security/enc/CRYPTREC/index.html

CUT-AND-CHOOSE PROTOCOL

CUT-AND-CHOOSE PROTOCOLS: A cut-and-choose protocol is a two-party protocol in which one party tries to convince another party that some data he sent to the former was honestly constructed according to an agreed upon method. Important examples of cut-and-choose protocols are interactive proofs [4], interactive arguments [1], zero-knowledge protocols [1, 3, 4] and *witness indistinguishable* and witness hiding protocols [2] for proving knowledge of a piece of information that is computationally hard to find. Such a protocol usually carries a small probability that it is successful despite the fact that the desired property is not satisfied.

The very first instance of such a cut-and-choose protocol is found in the protocol of Rabin [5] where the cut-and-choose concept is used to convince a party that the other party sent him an integer n that is a product of two primes p, q, each of which is congruent to 1 modulo 4. Note that this protocol was NOT zero-knowledge.

The expression cut-and-choose was later introduced by Chaum [1] in analogy to a popular cake sharing problem: given a complete cake to be shared among two parties distrusting each other (for reasons of serious appetite). A fair way for them to share the cake is to have one of them cut the cake in two equal shares, and let the other one choose his favourite share. This solution guarantees that it is in the former's best interest to cut the shares as evenly as possible.

Claude Crépeau

References

[1] Brassard, G., D. Chaum, and C. Crépeau (1988). "Minimum disclosure proofs of knowledge." *JCSS*, 37, 156–189.

[2] Feige, U. and A. Shamir (1990). "Witness indistinguishable and witness hiding protocols." *Proceedings of the 22nd Annual ACM Symposium on the Theory of Computing, Baltimore, MD, May*

1990, ed. Baruch Awerbuch. ACM Press, New York, 416–426.

[3] Goldreich, Oded, Silvio Micali, and Avi Wigderson (1991). "Proofs that yield nothing but their validity or all languages in NP have zero-knowledge proof systems." *Journal of the Association for Computing Machinery*, 38 (3), 691–729.

[4] Goldwasser, Shafi, Silvio Micali, and Charles Rackoff (1989). "The knowledge complexity of interactive proof systems." *SIAM Journal on Computing*, 18 (1), 186–208.

[5] Rabin, M.O. (1977). "Digitalized signatures." *Foundations of Secure Computation*. Papers presented at a 3 day workshop held at Georgia Institute of Technology, Atlanta, October 1977, eds. Richard A. DeMillo et al. Academic Press, New York, 155–166.

CYCLIC CODES

INTRODUCTION: For a general presentation of cyclic codes, our main reference is the *Handbook of Coding Theory*, especially the first chapter [4] (but also Chapters 11, 13, 14, and 19).

Cyclic codes were introduced as a particular practical class of *error-correcting codes* (ECC). Codes are devoted to the following fundamental problem: how to determine what message has been sent when only an approximation is received, due to a noisy communication channel. Cyclic codes belong to the class of *block codes* since here all messages have the same length k. Each of them is *encoded* into a *codeword* of length $n = k + r$. A t-error-correcting code is a well-chosen subset C of $\mathcal{A}^n$. Its elements are called *codewords* and have the property that each pair of them differs in at least $2t + 1$ coordinates. If the noisy channel generates not more than t errors during one transmission, the received vector will still lie closer to the originally transmitted codeword than any other codeword. This means that code C is able to correct t positions in each codeword.

A codeword will be denoted by

$$\mathbf{c} = (c_0, c_1, \ldots, c_{n-1}), \qquad c_i \in \mathcal{A}.$$

When the encoder is *systematic*, the first k symbols are called *information symbols* (they are the message) and the last r symbols are the redundancy symbols (added to help recover the message if errors occur). We consider here *linear* codes, meaning that $\mathcal{A}$ is a finite field and that C is a k-dimensional linear subspace of $\mathcal{A}^n$.

The *(Hamming) distance* between two codewords, $\mathbf{c}$ and $\mathbf{c}'$, is defined by:

$$d(\mathbf{c}, \mathbf{c}') = \text{card } \{i \in [0, n-1] \mid c_i \neq c_i'\}.$$

The *minimum distance* d of a code C is the smallest distance between different codewords; it determines the error correcting capabilities of C. Indeed, C can correct $t = \lfloor (d-1)/2 \rfloor$ errors. Since we focus on cyclic codes and on the most useful of them, the alphabet $\mathcal{A}$ will be a *finite field* $\mathbf{F}_q$ of characteristic 2, i.e., $q = 2^e$ for some integer $e \geq 1$. Moreover, the length of the codes will be generally $2^m - 1$, where e divides m; these codes are said *primitive*.

DEFINITION 1. *Consider the linear space $\mathbf{F}_q^n$ of all n-tuples over the finite field $\mathbf{F}_q$. An $[n, k, d]$ linear code C over $\mathbf{F}_q$ is a k-dimensional subspace of $\mathbf{F}_q^n$ with minimum distance d.*

By definition, a k-dimensional linear code C is fully determined by a basis over $\mathbf{F}_q$. When we put the k vectors of a basis as rows in a $k \times n$ matrix $\mathcal{G}$, we get a *generator matrix* of C. Indeed, C is given by $\{a v \mathcal{G} \mid a \in \mathbf{F}_q^k\}$.

The *Hamming weight* $wt(\mathbf{u})$ of any word $\mathbf{u}$ in $\mathbf{F}_q^n$ is the number of its nonzero coordinates. Note that $wt(\mathbf{u}) = d(\mathbf{u}, \mathbf{0})$. Obviously, for linear codes, the minimum distance is exactly the minimum weight of nonzero codewords.

PROPOSITION 1. *Let C be any $[n, k, d]$ linear code over $\mathbf{F}_q$. Then*

$$d = \min\{wt(\mathbf{c}) \mid \mathbf{c} \in C \setminus \{\mathbf{0}\}\}.$$

The *dual code* of C is the $[n, n-k]$ linear code:

$$C^\perp = \{\mathbf{y} \in \mathbf{F}_q^n \mid \mathbf{c} \cdot \mathbf{y} = 0, \text{ for all } \mathbf{c} \in C\},$$

where "$\cdot$" denotes the ordinary inner product of vectors: $\mathbf{c} \cdot \mathbf{y} = \sum_{i=0}^{n-1} c_i y_i$. An $(n-k) \times n$ generator matrix of $C^\perp$ is called a *parity check matrix* $\mathcal{H}$ for C. Note that $C = \{\mathbf{y} \in \mathbf{F}_q^n \mid \mathcal{H}\mathbf{y}^\mathrm{T} = \mathbf{0}^\mathrm{T}\}$.

When studying cyclic codes, it is convenient to view the labeling of the coordinate positions $0, 1, \ldots, n-1$ as integers modulo n (see modular arithmetic). In other words, viewing these coordinate positions as forming a cycle with $n-1$ being followed by 0.

DEFINITION 2. *A linear code C of length n over $\mathbf{F}_q$ is cyclic if and only if it satisfies for all $\mathbf{c} = c_0 \cdots c_{n-2} c_{n-1}$ in C:*

$$(c_0, \ldots, c_{n-2}, c_{n-1}) \in C \implies (c_{n-1}, c_0, \ldots, c_{n-2}) \in C.$$

The vector $c_{n-1} c_0 \cdots c_{n-2}$ is obtained from $\mathbf{c}$ by the cyclic shift of coordinates $i \mapsto i + 1$.

EXAMPLE 1. The generator matrix $\mathcal{G}$ below defines an $[8, 4, 2]$ binary cyclic code (length 8, dimension

4, and minimum weight 2):

$$\mathcal{G} = \begin{bmatrix} 1 & 0 & 0 & 0 & 1 & 0 & 0 & 0 \\ 0 & 1 & 0 & 0 & 0 & 1 & 0 & 0 \\ 0 & 0 & 1 & 0 & 0 & 0 & 1 & 0 \\ 0 & 0 & 0 & 1 & 0 & 0 & 0 & 1 \end{bmatrix}.$$

Cyclic codes are some of the most useful codes known. The involvement of Reed–Solomon (RS) codes and of Bose–Chaudhury–Hocquenghem (BCH) codes in a number of applications is well known. On the other hand, the Golay codes and the <u>Reed–Muller</u> (RM) codes, which are fundamental linear codes, can be represented as cyclic codes.

CONSTRUCTIVE DEFINITION: It seems difficult to construct a cyclic code C by means of Definition 2. So, useful definitions of cyclic codes are now considered.

An efficient definition is established by identifying each vector $\mathbf{c} = (c_0, c_1, \ldots, c_{n-1})$ with the polynomial $\mathbf{c}(x) = c_0 + c_1 x + \cdots + c_{n-1} x^{n-1}$. The fact that C is invariant under a cyclic shift is then expressed as follows:

$$\mathbf{c}(x) \in C \implies x\mathbf{c}(x) \pmod{x^n - 1} \in C.$$

Thus the proper context for studying cyclic codes of length n over $\mathbf{F}_q$ is the residue class ring

$$\mathcal{R}_n = \mathbf{F}_q[X]/(x^n - 1).$$

It is well known that $\mathcal{R}_n$ is a *principal ideal ring*. This means that any *ideal* I in $\mathcal{R}_n$ is *generated* by a single element g in I, i.e., $I = \{ag \mid a \in \mathcal{R}_n\}$. (An ideal in $\mathcal{R}_n$ is a subset I of $\mathcal{R}_n$ satisfying the properties: (1) for all i_1, i_2 in I also $i_1 - i_1 \in I$ and (2) for any $i \in I$ and $a \in \mathcal{R}_n$ also $ai \in I$.)

An alternative definition of a cyclic code can now be given.

DEFINITION 3. *A cyclic code C of length n over $\mathbf{F}_q$ is a principal ideal of the ring R_n. The codewords are polynomials in $\mathbf{F}_q[x]$ of degree less than n. Multiplication is carried out modulo $x^n - 1$.*

The next theorem, which is given in [4, Theorem 5.2], allows to determine the main parameters of any cyclic code of $\mathcal{R}_n$. We first recall some basic definitions.

DEFINITION 4. *Let α be a primitive nth root of unity in some* <u>extension field</u> *of $\mathbf{F}_q$. This means that $1, \alpha, \ldots, \alpha^{n-1}$ are all different and $\alpha^n = 1$. For each integer s with $0 \le s < n$, denote by $cl(s)$ the q-cyclotomic coset of s modulo n:*

$$cl(s) = \{s, qs, \ldots, q^{m-1}s \pmod{n}\},$$

where m is the smallest positive integer such that n divides $q^m - 1$ (so $\alpha \in \mathbf{F}_{q^m}$).

The <u>minimal polynomial</u> *of α^s over $\mathbf{F}_q$ is*

$$M_{\alpha^s}(x) = \prod_{i \in cl(s)} (x - \alpha^i),$$

where the $\alpha^i, i \in cl(s)$, are called the conjugates of α^s.

If ω is a <u>primitive element</u> of $\mathbf{F}_{q^m}$, then one can take $\alpha = \omega^{(q^m-1)/n}$. Note that $M_{\alpha^s}(x)$ is a polynomial over $\mathbf{F}_q$ while $\alpha \in \mathbf{F}_{q^m}$.

THEOREM 1. *Let C be a nonzero cyclic code of length n over $\mathbf{F}_q$. There exists a polynomial $g(x) \in C$, called the* generator polynomial *of C, with the following properties:*

(i) *$g(x)$ is the unique monic polynomial of minimum degree in C;*

(ii) *$g(x)$ is a generator of the ideal C in $\mathcal{R}_n$:*

$$C = \langle g(x) \rangle = \{a(x)g(x) \pmod{x^n - 1} \mid a(x) \in \mathbf{F}_q[x]\};$$

(iii) *$g(x)$ divides $x^n - 1$.*
Let $r = \deg(g)$, and let $g(x) = \sum_{i=0}^r g_i x^i$ where $g_r = 1$. Then

(iv) *the dimension of C is $k = n - r$; moreover the polynomials*

$$g(x), xg(x), \ldots, x^{k-1}g(x)$$

form a basis of C. The corresponding generator matrix is given by:

$$\mathcal{G} = \begin{bmatrix} g_0 & g_1 & \cdots & g_{n-k} & 0 & \cdots & 0 \\ 0 & g_0 & g_1 & \cdots & g_{n-k} & & 0 \\ \vdots & & \ddots & \ddots & & \ddots & \vdots \\ 0 & \cdots & 0 & g_0 & g_1 & \cdots & g_{n-k} \end{bmatrix}.$$

(v) *Let α be a primitive nth root of unity in some extension field of $\mathbf{F}_q$. Denote by M_{α^s} the minimal polynomial of α^s over $\mathbf{F}_q$; then*

$$g(x) = \prod_{s \in I} M_{\alpha^s}(x)$$

where I is a subset of representatives of the q-cyclotomic cosets modulo n.

The dual code of any cyclic code is cyclic too. The description of $C^\perp$ can be directly obtained from Theorem 1.

COROLLARY 1. *Let C be a cyclic code of length n over $\mathbf{F}_q$, with generator polynomial $g(x)$. Let $h(x)$ denote the* parity check polynomial *of C, defined by $x^n - 1 = h(x)g(x)$. Then the generator polynomial* *of $C^\perp$ is the polynomial*

$$\widetilde{h}(x) = \frac{x^k}{h_0} h(x^{-1}), \qquad where\ k = n - \deg(g).$$

EXAMPLE 2. Construction of the binary *Hamming code* of length $n = 15$.

$$x^{15} - 1 = (x^4 + x^3 + 1)(x^4 + x + 1)(x^4 + x^3 + x^2 + x + 1)(x + 1)(x^2 + x + 1).$$

Since $x^4 + x + 1$ is a *primitive polynomial* (meaning that its zeros are <u>primitive elements</u>), its root α is a <u>generator</u> of the cyclic group of the field $\mathbf{F}_{16}$. The polynomial $x^4 + x + 1$ is the minimal polynomial of α and has α and its conjugates as zeros. Consider the cyclic code C with generator polynomial:

$$g(x) = (x - \alpha)(x - \alpha^2)(x - \alpha^4)(x - \alpha^8)$$
$$= x^4 + x + 1.$$

The dimension of C is $15 - \deg(g) = 11$. Thus the code C is a $[15, 11, 3]$ cyclic code. The minimum distance of C is exactly 3 since $wt(g) = 3$ and no smaller weight can appear, as can be checked using a generator matrix $\mathcal{G}$ of C. According to Theorem 1, $\mathcal{G}$ is an 11×15 binary matrix whose lines are $g(x)$

x^0	x^1	x^2	x^3	x^4	x^5	x^6	x^7	x^8	x^9	x^{10}	x^{11}	x^{12}	x^{13}	x^{14}
$g(x)$	1	1	0	0	1	0	0	0	0	0	0	0	0	0	0

and the 10 shifts of $g(x)$. The parity check polynomial of C and the generator polynomial of $C^\perp$ are given by:

$$h(x) = \frac{x^{15} - 1}{x^4 + x + 1}$$
$$= x^{11} + x^8 + x^7 + x^5 + x^3 + x^2 + x + 1.$$

resp.

$$\widetilde{h}(x) = \sum_{i=0}^{11} h_{11-i} x^i$$
$$= x^{11} + x^{10} + x^9 + x^8 + x^6 + x^4 + x^3 + 1.$$

We then obtain a *parity check matrix* for C:

x^0	x^1	x^2	x^3	x^4	x^5	x^6	x^7	x^8	x^9	x^{10}	x^{11}	x^{12}	x^{13}	x^{14}
1	0	0	1	1	0	1	0	1	1	1	1	0	0	0
0	1	0	0	1	1	0	1	0	1	1	1	1	0	0
0	0	1	0	0	1	1	0	1	0	1	1	1	1	0
0	0	0	1	0	0	1	1	0	1	0	1	1	1	1

We explained here, by means of an example, the general definition of the binary Hamming code of length $2^m - 1$. It is a code whose parity check matrix has as columns all the nonzero vectors of $\mathbf{F}_2^m$. It is a $[2^m - 1, 2^m - m - 1, 3]$ code.

SOME CYCLIC CODES: From now on $q = 2^e$ and $n = q^m - 1$. Let C be any cyclic code of length n over $\mathbf{F}_q$ with generator polynomial $g(x)$. The roots of $g(x)$ are called the *zeros of the cyclic code* C. Thus, the code C is fully defined by means of its zero's set; this leads to the classical definition of the BCH codes and of other important families.

DEFINITION 5. *Let $q = 2$ and $n = 2^m - 1$; denote by α a primitive nth root of unity in $\mathbf{F}_{2^m}$. Let δ be an integer with $2 \le \delta \le n$.*

The binary BCH code of length n and designed distance δ is the cyclic code with zero's set: $\alpha, \alpha^2, \ldots, \alpha^{\delta-1}$ and their conjugates. In other words, the generator polynomial of this code is

$$g(x) = l\,cm\{M_\alpha(x), M_{\alpha^2}(x), \ldots, M_{\alpha^{\delta-1}}(x)\}.$$

EXAMPLE 3. Binary BCH codes of length 15. As in Example 2, we denote by α any root of the primitive polynomial $x^4 + x + 1$. Now the factorization of $x^{15} - 1$ into minimal polynomials is as follows:

$$x^{15} - 1 = (x - 1)\underbrace{(x^4 + x + 1)}_{M(\alpha)}\underbrace{(x^4 + x^3 + x^2 + x + 1)}_{M(\alpha^3)}$$
$$\underbrace{(x^2 + x + 1)}_{M(\alpha^5)}\underbrace{(x^4 + x^3 + 1)}_{M(\alpha^7)}.$$

There are three nontrivial BCH codes, whose zero's sets S_δ are as follows:

- $S_3 = \{\alpha^i \mid i = 1, 2, 4, 8\}$ for the $[15, 11, 3]$ BCH code;
- $S_5 = S_3 \cup \{\alpha^i \mid i = 3, 6, 9, 12\}$ for the $[15, 7, 5]$ BCH code;
- $S_7 = S_5 \cup \{\alpha^i \mid i = 5, 10\}$ for the $[15, 5, 7]$ BCH code.

For these codes, the designed distance δ is exactly the minimum distance; this property does not hold for any BCH code.

DEFINITION 6. *A Reed–Solomon code over F_q is a BCH code of length $n = q - 1$.*

RS codes appear in several cryptosystems. They determine, for instance, some <u>secret-sharing schemes</u> [3]. It is important to note that for RS codes, the designed distance is exactly the minimum distance. Moreover, the RS code with designed distance δ is an $[n, k, \delta]$ code with $k = n - \delta + 1$. Since this k attains the maximum value by the *Singleton bound* (see [4]), one says that RS codes are *maximum distance separable* (MDS) code.

DEFINITION 7. *Let α be a primitive root of $\mathbf{F}_{2^m}$. Any integer $s \in [0, 2^m - 1]$ can be identified by its* *binary expansion in* $\mathbf{F}_2^m$:

$$s = \sum_{i=0}^{m-1} s_i 2^i, \quad s_i \in \{0, 1\} \implies s = (s_0, \ldots, s_{m-1}).$$

The cyclic Reed–Muller *code of length* $2^m - 1$ *and order* r, *usually denoted by* $R^*(r, m)$, *is the binary cyclic code with zero set:*

$$S_r = \{\, \alpha^s \mid 1 \le wt(s) < m - r \,\},$$

where $wt(s)$ *is the Hamming weight of* s.

Note that if one extends all codewords in the cyclic Reed–Muller code above with an overall-parity check symbol, one obtains the regular Reed–Muller code.

Binary cyclic codes are related to the study and the construction of cryptographic primitives, mainly through Reed–Muller codes because of the large field of applications of <u>Boolean functions</u> and binary <u>sequences</u> in cryptography. They play a role in the study of cryptographic mapping on finite fields in general. One well-known application is the construction of *almost bent* (AB) mappings (see <u>nonlinearity of Boolean functions</u>), which resist both <u>differential</u> and <u>linear cryptanalysis</u> [1,2] (see next example). These connections are more explicit when using the trace representation of binary codewords of length $n = 2^m - 1$. We now label the coordinate positions by α^0, α^1, ..., α^{n-1}, where α is a primitive nth root of unity. Let $\mathbf{c}(x)$ be any codeword of some binary cyclic code C of length n. Define

$$T_{\mathbf{c}}(x) = \sum_{s=0}^{n-1} \mathbf{c}(\alpha^{n-s}) x^s. \tag{1}$$

This commonly known Fourier transform of $\mathbf{c}$ is called the *Mattson–Solomon polynomial* of $\mathbf{c}$ in algebraic coding theory. It follows that $T_{\mathbf{c}}(\alpha^j) = c_j$ and $T_{\mathbf{c}}(x)$ is a sum of traces from some subfields of $\mathbf{F}_{2^m}$ to $\mathbf{F}_2$.

The mapping $x \mapsto T_{\mathbf{c}}(x)$ is a Boolean function. On the other hand, any binary sequence of period n can be represented in this way (see entries <u>Boolean functions</u> and <u>Sequences</u>).

EXAMPLE 4. Consider any binary cyclic code of length $n = 2^m - 1$ whose generator polynomial is the product of two minimal polynomials, say $M_{\alpha^r}(x) M_{\alpha^s}(x)$. These codes are said to be *cyclic codes with two zeros* and usually denoted by $C_{r,s}$.

Now assume that $r = 1$ and $\gcd(s, 2^m - 1) = 1$. Then $C_{1,s}$ is an $[n, 2m, d]$ cyclic code; the dual of $C_{1,s}$ is a cyclic code which has two nonzeros only: α^{n-1} and α^{n-s} (apart from their conjugates). According to (1), $C_{1,s}^{\perp}$ is the set of binary codewords of length n defined as follows: each pair (a, b) of elements of $\mathbf{F}_{2^m}$ provides the ordered sequence of values

$$x \in \{1, \alpha, \ldots, \alpha^{n-1}\} \longmapsto \mathrm{Tr}(ax + bx^s),$$

where the *Trace* function Tr is defined by $\mathrm{Tr}(\beta) = \beta + \beta^2 + \cdots + \beta^{2^{m-1}}$. The *power function* $x \mapsto x^s$ is a permutation on $\mathbf{F}_{2^m}$. It is said to be *almost perfect nonlinear* [1] when $C_{1,s}$ has minimum distance 5. It is said to be an AB function when the nonzero weights of codewords of $C_{1,s}^{\perp}$ are either 2^{m-1} or $2^{m-1} \pm 2^{(m-1)/2}$; this is possible for odd m only.

Pascale Charpin

References

[1] Canteaut, A., P. Charpin, and H. Dobbertin (1999). "A new characterization of almost bent functions." *Fast Software Encryption (FSE6)*, Lecture Notes in Computer Science, vol. 1636, ed. L.R. Knudsen. Springer-Verlag, Berlin, 186–200.

[2] Carlet, C., P. Charpin, and V. Zinoviev (1998). "Codes, bent functions and permutations suitable for DES-like cryptosystems." *Designs Codes and Cryptography*, 15 (2), 125–156.

[3] McEliece, R.J. and D.V. Sarwarte (1981). "On sharing secrets and Reed–Solomon codes." *Comm. ACM*, 24, 583–584.

[4] Pless, V.S., W.C. Huffman, and R.A. Brualdi (1998). "An introduction to algebraic codes." *Handbook of Coding Theory, Part 1: Algebraic Coding*. Elsevier, Amsterdam, The Netherlands, Chapter 1.

D

DATA ENCRYPTION STANDARD (DES)

The Data Encryption Standard (DES) [31] has been around for more than 25 years. During this time the standard was revised three times: as FIPS-46-1 in 1988, as FIPS-46-2 in 1993 and as FIPS-46-3 in 1999. DES was an outcome of a call for primitives in 1974, which did not result in many serious candidates except for a predecessor of DES, Lucifer [15, 36] designed by IBM around 1971. It took another year for a joint IBM–NSA effort to turn *Lucifer* into DES. The structure of Lucifer was significantly altered: since the design rationale was never made public and the secret key size was reduced from 128-bit to 56-bits, this initially resulted in controversy, and some distrust among the public. After some delay, FIPS-46 was published by NBS (National Bureau of Standards)—now NIST (National Institute of Standards and Technology)—on January 15, 1977 [31] (see [35] for a discussion of the standardization process).

However, in spite of all the controversy it is hard to underestimate the role of DES [31]. DES was one of the first commercially developed (as opposed to government developed) ciphers whose structure was fully published. This effectively created a community of researchers who could analyse it and propose their own designs. This lead to a wave of public interest in cryptography, from which much of the cryptography as we know it today was born.

DESCRIPTION OF DES: The Data Encryption Standard, as specified in FIPS Publication 46-3 [31], is a <u>block cipher</u> operating on 64-bit data blocks. The <u>encryption</u> transformation depends on a 56-bit secret <u>key</u> and consists of sixteen <u>Feistel</u> iterations surrounded by two permutation layers: an initial bit permutation IP at the input, and its inverse IP^{-1} at the output. The structure of the cipher is depicted in Figure 1. The decryption process is the same as the encryption, except for the order of the round keys used in the Feistel iterations. As a result, most of the circuitry can be reused in hardware implementations of DES.

The 16-round Feistel network, which constitutes the cryptographic core of DES, splits the 64-bit data blocks into two 32-bit words (denoted by L_0 and R_0). In each iteration (or round), the second word R_i is fed to a function f and the result is added to the first word L_i. Then both words are swapped and the algorithm proceeds to the next iteration.

The function f is key-dependent and consists of four stages (see Figure 2). Their description is given below. Note that all bits in DES are numbered from left to right, i.e., the leftmost bit of a block (the most significant bit) is bit 1.

1. **Expansion (E).** The 32-bit input word is first expanded to 48 bits by duplicating and reordering half of the bits. The selection of bits is specified by Table 1. The first row in the table refers to the first 6 bits of the expanded word, the second row to bits 7–12, and so on. Thus bit 41 of the expanded word, for example, gets its value from bit 28 of the input word.

2. **Key mixing.** The expanded word is XORed with a round key constructed by selecting 48 bits from the 56-bit secret key. As explained below, a different selection is used in each round.

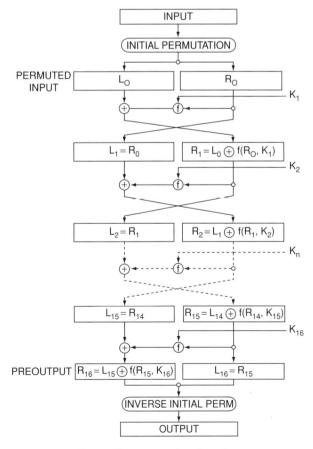

Fig. 1. The encryption function

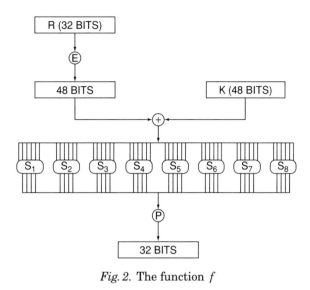

Fig. 2. The function f

3. **Substitution.** The 48-bit result is split into eight 6-bit words which are substituted in eight parallel 6×4-bit S-boxes. All eight S-boxes, called $S_1, S_2, \ldots, S_8$, are different but have the same special structure, as appears from their specifications in Table 2. Each row of the S-box tables consists of a permutation of the 4-bit values $0, \ldots, 15$. The 6-bit input word is substituted as follows: first a row is selected according to the value of the binary word formed by concatenating the first and the sixth input bit. The algorithm then picks the column given by the value of the four middle bits and outputs the corresponding 4-bit word.

4. **Permutation** (P). The resulting 32 bits are reordered according to a fixed permutation specified in Table 1 before being sent to the output. As before, the first row of the table refers to the first four bits of the output.

The selection of key bits in each round is determined by a simple key scheduling algorithm. The algorithm starts from a 64-bit secret key which includes 8 parity bits that are discarded after verification (the parity of each byte needs to be odd). The remaining 56 secret key bits are first permuted

Table 1. Expansion E and permutation P

		E						P	
32	1	2	3	4	5	16	7	20	21
4	5	6	7	8	9	29	12	28	17
8	9	10	11	12	13	1	15	23	26
12	13	14	15	16	17	5	18	31	10
16	17	18	19	20	21	2	8	24	14
20	21	22	23	24	25	32	27	3	9
24	25	26	27	28	29	19	13	30	6
28	29	30	31	32	1	22	11	4	25

according to a permutation PC_1 (see Table 4). The result is split into two 28-bit words C_0 and D_0, which are cyclically rotated over 1 position to the left after rounds 1, 2, 9, 16, and over 2 positions after all other rounds (the rotated words are denoted by C_i and D_i). The round keys are constructed by repeatedly extracting 48 bits from C_i and D_i at 48 fixed positions determined by a table PC_2 (see Table 4). A convenient feature of this key scheduling algorithm is that the 28-bit words C_0 and D_0 are rotated over exactly 28 positions after 16 rounds. This allows hardware implementations to efficiently compute the round keys on-the-fly, both for the encryption and the decryption.

CRYPTANALYSIS OF DES: DES has been subject to very intensive cryptanalysis. Initial attempts [16] did not identify any serious weaknesses except for the short key-size. It was noted that DES has a *complementation property*, i.e., given an encryption of the plaintext P into the ciphertext C under the secret key K: $E_K(P) = C$, one knows that the complement of the plaintext will be encrypted to the complement of the ciphertext under the complement of the key: $E_{\bar{K}}(\bar{P}) = \bar{C}$ (by complement we mean flipping of all the bits). Another feature was the existence of four weak keys, for which the cipher is an *involution*: $E_K(E_K(m)) = m$ (for these keys the contents of the key-schedule registers C and D is either all zeros or all ones), and six additional pairs of *semi-weak keys* for which $E_{K1}(E_{K2}(m)) = m$. The complementation and the weak-key properties are the result of interaction of the key-schedule, which splits the key-bits into two separate registers and the Feistel structure of the cipher. A careful study of the cycle structure of DES for weak and semi-weak keys has been given by Moore and Simmons [30]. See the book of Davies and Price [11] for a more detailed account on these and other features of DES identified prior to 1989. The properties of the group generated by DES permutations have also been studied intensively. Coppersmith and Grossman have shown [9] that in principle DES-like components can generate any permutation from the *alternating group* $A_{2^{64}}$ (all *even permutations*, i.e., those that can be represented with an even number of transpositions). However, DES implements only 2^{56} permutations, which is a tiny fraction of all the even permutations. If the set of 2^{56} DES permutations was closed under composition, then multiple encryption as used, for example in Triple-DES would be equivalent to single encryption and thus would not provide any additional strength. A similar weakness would be present if the size of the group generated by the DES permutations

Table 2. DES S-boxes

S_1 :	0	1	2	3	4	5	6	7	8	9	10	11	12	13	14	15
0 :	14	4	13	1	2	15	11	8	3	10	6	12	5	9	0	7
1 :	0	15	7	4	14	2	13	1	10	6	12	11	9	5	3	8
2 :	4	1	14	8	13	6	2	11	15	12	9	7	3	10	5	0
3 :	15	12	8	2	4	9	1	7	5	11	3	14	10	0	6	13

S_2 :	0	1	2	3	4	5	6	7	8	9	10	11	12	13	14	15
0 :	15	1	8	14	6	11	3	4	9	7	2	13	12	0	5	10
1 :	3	13	4	7	15	2	8	14	12	0	1	10	6	9	11	5
2 :	0	14	7	11	10	4	13	1	5	8	12	6	9	3	2	15
3 :	13	8	10	1	3	15	4	2	11	6	7	12	0	5	14	9

S_3 :	0	1	2	3	4	5	6	7	8	9	10	11	12	13	14	15
0 :	10	0	9	14	6	3	15	5	1	13	12	7	11	4	2	8
1 :	13	7	0	9	3	4	6	10	2	8	5	14	12	11	15	1
2 :	13	6	4	9	8	15	3	0	11	1	2	12	5	10	14	7
3 :	1	10	13	0	6	9	8	7	4	15	14	3	11	5	2	12

S_4 :	0	1	2	3	4	5	6	7	8	9	10	11	12	13	14	15
0 :	7	13	14	3	0	6	9	10	1	2	8	5	11	12	4	15
1 :	13	8	11	5	6	15	0	3	4	7	2	12	1	10	14	9
2 :	10	6	9	0	12	11	7	13	15	1	3	14	5	2	8	4
3 :	3	15	0	6	10	1	13	8	9	4	5	11	12	7	2	14

S_5 :	0	1	2	3	4	5	6	7	8	9	10	11	12	13	14	15
0 :	2	12	4	1	7	10	11	6	8	5	3	15	13	0	14	9
1 :	14	11	2	12	4	7	13	1	5	0	15	10	3	9	8	6
2 :	4	2	1	11	10	13	7	8	15	9	12	5	6	3	0	14
3 :	11	8	12	7	1	14	2	13	6	15	0	9	10	4	5	3

S_6 :	0	1	2	3	4	5	6	7	8	9	10	11	12	13	14	15
0 :	12	1	10	15	9	2	6	8	0	13	3	4	14	7	5	11
1 :	10	15	4	2	7	12	9	5	6	1	13	14	0	11	3	8
2 :	9	14	15	5	2	8	12	3	7	0	4	10	1	13	11	6
3 :	4	3	2	12	9	5	15	10	11	14	1	7	6	0	8	13

S_7 :	0	1	2	3	4	5	6	7	8	9	10	11	12	13	14	15
0 :	4	11	2	14	15	0	8	13	3	12	9	7	5	10	6	1
1 :	13	0	11	7	4	9	1	10	14	3	5	12	2	15	8	6
2 :	1	4	11	13	12	3	7	14	10	15	6	8	0	5	9	2
3 :	6	11	13	8	1	4	10	7	9	5	0	15	14	2	3	12

S_8 :	0	1	2	3	4	5	6	7	8	9	10	11	12	13	14	15
0 :	13	2	8	4	6	15	11	1	10	9	3	14	5	0	12	7
1 :	1	15	13	8	10	3	7	4	12	5	6	11	0	14	9	2
2 :	7	11	4	1	9	12	14	2	0	6	10	13	15	3	5	8
3 :	2	1	14	7	4	10	8	13	15	12	9	0	3	5	6	11

Table 3. Initial and final permutations

IP								IP^{-1}							
58	50	42	34	26	18	10	2	40	8	48	16	56	24	64	32
60	52	44	36	28	20	12	4	39	7	47	15	55	23	63	31
62	54	46	38	30	22	14	6	38	6	46	14	54	22	62	30
64	56	48	40	32	24	16	8	37	5	45	13	53	21	61	29
57	49	41	33	25	17	9	1	36	4	44	12	52	20	60	28
59	51	43	35	27	19	11	3	35	3	43	11	51	19	59	27
61	53	45	37	29	21	13	5	34	2	42	10	50	18	58	26
63	55	47	39	31	23	15	7	33	1	41	9	49	17	57	25

Table 4. DES key schedule bit selections

PC$_1$							PC$_2$					
57	49	41	33	25	17	9	14	17	11	24	1	5
1	58	50	42	34	26	18	3	28	15	6	21	10
10	2	59	51	43	35	27	23	19	12	4	26	8
19	11	3	60	52	44	36	16	7	27	20	13	2
63	55	47	39	31	23	15	41	52	31	37	47	55
7	62	54	46	38	30	22	30	40	51	45	33	48
14	6	61	53	45	37	29	44	49	39	56	34	53
21	13	5	28	20	12	4	46	42	50	36	29	32

were small. Using the special properties of the weak keys it has been shown that DES generates a very large group, with a lower-bound of 2^{2499} permutations [7, 8], which is more than enough to make the *closure attacks* [18] impractical.

In the two decades since its design three important theoretical attacks capable of breaking the cipher faster than exhaustive search have been discovered: differential cryptanalysis (1990) [5], linear cryptanalysis (1993) [22], and the *improved Davies' attack* [3, 12]. An interesting twist is that differential cryptanalysis was known to the designers of DES and DES was constructed in particular to withstand[1] this powerful attack [8]. This explains why the cipher's design criteria were kept secret. Many of these secrets became public with the development of differential cryptanalysis and were later confirmed by the designers [33]. Both differential and linear attacks as well as Davies' attack are not much of a threat to real-life applications since they require more than 2^{40} texts for the analysis. For example: a linear attack requires 2^{43} known plaintexts to be encrypted under the same secret key. If the user changes the key every 2^{35} blocks the success probability of the attack would

be negligible. Nevertheless, linear attacks were tested [23] in practice, run even slightly faster than theoretically predicted [17], and can potentially use twice less data in a chosen plaintext scenario [20]. In the case of the differential attack 2^{47} chosen plaintexts are required, though the attack would still work if the data is coming from up to 2^{33} different keys. However, the huge amount of *chosen* plaintext makes the attack impractical. In the case of Davies' attack the data requirement is 2^{50} *known plaintexts*, which is clearly impractical.

Although differential and linear attacks are hard to mount on DES, they proved to be very powerful tools for cryptanalysis; many ciphers which were not designed to withstand these attacks have been broken, some even with practical attacks. See for example the cipher FEAL [28, 29, 34]. In fact both attacks have been discovered while studying this cipher [4, 25], which was proposed as a more secure alternative to DES.

Exhaustive key *search* currently remains the biggest threat to the security of DES [31]. It was clear from the very beginning that a 56-bit key can be found in practical time by using a practical amount of resources. In 1977 a design for a key-search machine was proposed by Diffie and Hellman [13] with a cost of US\$ 20 million and the ability to find a solution in a single day. Later Hellman proposed a chosen plaintext time-memory tradeoff approach, which would allow to build an even cheaper machine, assuming that

---

[1] Note that DES is strong but not optimal against linear cryptanalysis or improved Davies' attack, for example simple reordering of the S-boxes would make the cipher less vulnerable to these attacks without spoiling its strength against the differential attack [24]. This could indicate that the designers of DES did not know about such attacks.

a precomputation of 2^{56} encryption steps is done once for a single chosen plaintext. An effective and complete ASIC design for a key-search machine has been proposed by Wiener in 1993 [38]. It was shown that the US$ 1 million machine would run through the full key-space in 7 hours. It became clear in 1993 that DES had to be upgraded to triple-DES or be replaced; however NIST decided to reconfirm the FIPS standard a second time in 1993 for another five years (as FIPS 46-2). In 1998 the Electronic Frontier Foundation (EFF) built a working dedicated hardware machine which cost less than US$ 250,000 and could run through the full key-space in four days [14]. In a parallel development it was shown that a network of tens of thousands of PCs (a computational power easily available to a computer virus, for example) could do the same work in several weeks. At that time the *AES competition* had been started. As a result of this effort DES has been replaced by a successor, AES, which is based on a 128-bit block 128/192/256-bit key cipher Rijndael/AES.

EXTENSIONS OF DES: So where is DES today? DES is not obsolete. Due to substantial cryptanalytic effort and the absence of any practical cryptanalytic attack, the structure of DES has gained public trust. There have been several proposals to remedy the short key size problem plaguing the cipher:

- **Triple-DES (Diffie–Hellman [13]).** The idea is to multiple encrypt the block using DES three times with two or three different keys. This method gains strength both against cryptanalytic attacks as well as against exhaustive search. It is weak against related key attacks, however, and the speed is three times slower than single DES [31]. A two-key variant in the form of *Encrypt-Decrypt-Encrypt (E-D-E)*, i.e., $E_{K_1}(D_{K_2}(E_{K_1}(m)))$ has been proposed by IBM (Tuchman, 1978) and is still in wide use by the banking community. The convenience of this option is that it is backward compatible with a single DES encryption, if one sets $K_1 = K_2$.
- **Independent subkeys (Berson [1]).** The idea is to use independently generated 48-bit subkeys in each round. The total key-size is 768 bits, which stops the exhaustive search attack. However, the cryptanalytic attacks like differential or linear do work almost as good as for DES [31]. The speed of this proposal is as for single DES, but it has a slower key-schedule.
- **Slow key-schedule (Quisquater et al. [32] or Knudsen [10]).** Exhaustive search is stopped by loosing key-agility of a cipher.

- **DES-X (Rivest, 1984).** The idea is to XOR additional 64-bits of secret key material at the input and at the output of the cipher. See the article on DES-X for more details. This is very effective against exhaustive search, but does not stop old cryptanalytic attacks on DES, and allows new related key attacks. This approach allows the reuse of old hardware. The speed is almost the same as that of a single DES.
- **Key-dependent S-boxes (Biham-Biryukov [2]).** The idea is similar to DES-X, but the secret key material is XORed before and after the S-boxes. S-boxes are reordered to gain additional strength. The result is secure against exhaustive search and improves the strength against cryptanalytic attacks (with the exception of related key attacks). This approach applies to software or to hardware which permits the loading of new S-boxes. The speed is the same as that of a single DES.

As of today two-key and three-key triple DES is still in wide use and is included in NIST (FIPS 46-3, the 1999 edition [31]) and ISO standards. However, two-key triple DES variants are not recommended for use due to dedicated meet-in-the-middle attack by Oorschot and Wiener [37] with complexity $2^{120-\log n}$ steps given $O(n)$ *known plaintexts* and memory. For example, if $n = 2^{40}$, complexity of attack is 2^{80} steps. This attack is based on an earlier attack by Merkle and Hellman [27] which required 2^{56} *chosen plaintexts*, steps, and memory. These attacks are hard to mount in practice, but they are an important certificational weakness.

The recommended usage mode for triple-DES is *Encrypt-Encrypt-Encrypt (E-E-E)* (or *Encrypt-Decrypt-Encrypt (E-D-E)*) with three independently generated keys (i.e. 168 key bits in total), for which the best attacks are the classical meet-in-the-middle attack with only three known plaintexts, 2^{56} words of memory and 2^{111} analysis steps; and the attack by Lucks [21] which requires 2^{108} time steps and 2^{45} known plaintexts. These attacks are clearly impractical.

The DES-X alternative is also in popular use due to its simplicity and almost no speed loss. Thorough analysis of a generic construction is given in [19] and the best currently known attack is a slide attack [6] with complexity of n known plaintexts and $2^{121-\log n}$ analysis steps (for example: 2^{33} known plaintexts and memory and 2^{87} analysis steps).

Alex Biryukov
Christophe De Cannière

References

[1] Berson, T.A. (1983). "Long key variants of DES." *Advances in Cryptology—CRYPTO'82*, Lecture Notes in Computer Science, eds. D. Chaum, R.L. Rivest, and A.T. Sherman. Plenum Press, New York, 311–313.

[2] Biham, E. and A. Biryukov (1995). "How to strengthen DES using existing hardware." *Advances in Cryptography—ASIACRYPT'94*, Lecture Notes in Computer Science, vol. 917, eds. J. Pieprzyk and R. Safavi-Naini. Springer-Verlag, Berlin, 395–412.

[3] Biham, E. and A. Biryukov (1997). "An improvement of Davies' attack on DES." *Journal of Cryptology*, 10 (3), 195–206.

[4] Biham, E. and A. Shamir. "Differential cryptanalysis of DES-like cryptosystems." In Menezes and Vanstone [26], 2–21.

[5] Biham, E. and A. Shamir (1993). "Differential cryptanalysis of the data encryption standard." *Advances in Cryptology—CRYPTO'90*, eds. A.J. Menezes and S.A. Vanstone. Lecture Notes in Computer Science, vol. 537. Springer-Verlag, Berlin, 2–21.

[6] Biryukov, A. and D. Wagner (2000). "Advanced slide attacks." *Advances in Cryptology—EUROCRYPT 2000*, Lecture Notes in Computer Science, vol. 1807, ed. B. Preneel. Springer-Verlag, Berlin, 589–606.

[7] Campbell, K.W. and M.J. Wiener (1993). "DES is not a group." *Advances in Cryptology—CRYPTO'92*, Lecture Notes in Computer Science, vol. 740, ed. E.F. Brickell. Springer-Verlag, Berlin, 512–520.

[8] Coppersmith, Don (1994). "The data encryption standard (DES) and its strength against attacks." *IBM Journal of Research and Development*, 38 (3), 243–250.

[9] Coppersmith, D. and E. Grossman (1975). "Generators for certain alternating groups with applications to cryptography." *SIAM Journal Applied Math*, 29 (4), 624–627.

[10] Damgard, I. and L.R. Knudsen (1998). "Two-key triple encryption." *Journal of Cryptology*, 11 (3), 209–218.

[11] Davies, D.W. and W.L. Price (1989). *Security for Computer Networks* (2nd ed.). John Wiley & Sons, New York.

[12] Davies, D.W. and S. Murphy (1995). "Pairs and triplets of DES S-Boxes." *Journal of Cryptology*, 8 (1), 1–25.

[13] Diffie, W. and M. Hellman (1997). "Exhaustive cryptanalysis of the NBS data encryption standard." *Computer*, 10 (6), 74–84.

[14] Electronic Frontier Foundation (EFF) (1998). "DES cracker." http://www.eff.org/DEScracker/

[15] Feistel, H. (1973). "Cryptography and computer privacy." *Scientific American*, 228, 15–23.

[16] Hellman, M.E., R. Merkle, R. Schroppel, L. Washington, W. Diffe, S. Pohlig, and P. Schweitzer (1976). "Results of an initial attempt to cryptanalyze the NBS Data Encryption Standard." Technical report, Stanford University, USA.

[17] Junod, P. (2001). "On the complexity of Matsui's attack." *Selected Areas in Cryptography, SAC 2001*, Lecture Notes in Computer Science, vol. 2259, eds. S. Vaudenay and A.M. Youssef. Springer-Verlag, Berlin, 199–211.

[18] Kaliski, B.S., R.L. Rivest, and A.T. Sherman (1988). "Is the data encryption standard a group?" *Journal of Cryptology*, 1 (1), 3–36.

[19] Kilian, J. and P. Rogaway (1996). "How to protect DES against exhaustive key search." *Advances in Cryptology—CRYPTO'96*, Lecture Notes in Computer Science, vol. 1109, ed. N. Koblitz. Springer-Verlag, Berlin, 252–267.

[20] Knudsen, L.R. and J.E. Mathiassen (2001). "A chosen-plaintext linear attack on DES." *Fast Software Encryption, FSE 2000*, Lecture Notes in Computer Science, vol. 1978, ed. B. Schneier. Springer-Verlag, Berlin, 262–272.

[21] Lucks, S. (1998). "Attacking triple encryption." *Fast Software Encryption, FSE'98*, Lecture Notes in Computer Science, vol. 1372, ed. S. Vaudenay. Springer-Verlag, Berlin, 239–257.

[22] Matsui, M. (1993). "Linear cryptanalysis method for DES cipher." *Advances in Cryptology—EUROCRYPT'93*, Lecture Notes in Computer Science, vol. 765, ed. T. Helleseth. Springer-Verlag, Berlin, 386–397.

[23] Matsui, M. (1994). "The first experimental cryptanalysis of the data encryption standard." *Advances in Cryptology—CRYPTO'94*, Lecture Notes in Computer Science, vol. 839, ed. Y. Desmedt. Springer-Verlag, Berlin, 1–11.

[24] Matsui, M. (1995). "On correlation between the order of S-boxes and the strength of DES." *Advances in Cryptology—EUROCRYPT'94*, Lecture Notes in Computer Science, vol. 950, ed. A. De Santis. Springer-Verlag, Berlin, 366–375.

[25] Matsui, M. and A. Yamagishi (1992). "A new method for known plaintext attack of FEAL cipher." *Advances in Cryptology—EUROCRYPT'92*, Lecture Notes in Computer Science, vol. 658, ed. R.A. Rueppel. Springer-Verlag, Berlin, 81–91.

[26] Menezes, A. and S.A. Vanstone (eds.) (1991). *Advances in Cryptology—CRYPTO'90*, Lecture Notes in Computer Science, vol. 537, eds. A.J. Menezes and S.A. Vanstone. Springer-Verlag, Berlin.

[27] Merkle, R.C. and M.E. Hellman (1981). "On the security of multiple encryption." *Communications of the ACM*, 14 (7), 465–467.

[28] Miyaguchi, S. (1990). "The FEAL-8 cryptosystem and a call for attack." *Advances in Cryptology—CRYPTO'89*, Lecture Notes in Computer Science, vol. 435, ed. G. Brassard. Springer-Verlag, Berlin, 624–627.

[29] Miyaguchi, S. "The FEAL cipher family." In Menezes and Vanstone (26), 627–638.

[30] Moore, J.H. and G.J. Simmons (1987). "Cycle structures of the DES with weak and semi-weak keys."

Advances in Cryptology—CRYPTO'86, Lecture Notes in Computer Science, vol. 263, ed. A.M. Odlyzko. Springer-Verlag, Berlin, 9–32.

[31] National Institute of Standards and Technology (1979). "FIPS-46: Data Encryption Standard (DES)." Revised as FIPS 46-1:1988, FIPS 46-2:1993, FIPS 46-3:1999, available at http://csrc.nist .gov/publications/fips/fips46-3/fips46-3.pdf

[32] Quisquater, J.-J., Y. Desmedt, and M. Davio (1986). "The importance of "good" key scheduling schemes (how to make a secure DES scheme with $\leq$ 48 bit keys)." *Advances in Cryptology—CRYPTO'85*, Lecture Notes in Computer Science, vol. 218, ed. H.C. Williams. Springer-Verlag, Berlin, 537–542.

[33] sci.crypt (1992). "Subject: DES and differential cryptanalysis." Unpublished, http://www.esat. kuleuven.ac.be/~abiryuko/coppersmith_letter.txt

[34] Shimizu, A. and S. Miyaguchi (1998). "Fast data encipherment algorithm FEAL." *Advances in Cryptology—EUROCRYPT'87*, Lecture Notes in Computer Science, vol. 304, eds. D. Chaum and W.L. Price. Springer-Verlag, Berlin, 267–278.

[35] Smid, M. and D. Branstad (1998). "The data encryption standard: past and future." *Proceedings of the IEEE*, 76 (5), 550–559.

[36] Smith, J.L. (1971). "The design of Lucifer: A cryptographic device for data communications." Technical Report, IBM T.J. Watson Research Center, Yorktown Heights, NY, USA.

[37] van Oorschot, P.C. and M.J. Wiener (1990). "A known plaintext attack on two-key triple encryption." *Advances in Cryptology—EUROCRYPT'90*, Lecture Notes in Computer Science, vol. 473, ed. I. Damgård. Springer-Verlag, Berlin, 318–325.

[38] Wiener, M. (1996). "Efficient des key search." *Practical Cryptography for Data Internetworks*, presented at the rump session of CRYPTO'93, 31–79.

DATA REMANENCE

Data remanence is the ability of computer memory to retain previously stored information beyond its intended lifetime. With many data storage techniques, information can be recovered using specialized techniques and equipment even after it has been overwritten. Examples:

- Write heads used on exchangeable media (e.g., floppy disks, magstripe cards) differ slightly in position and width due to manufacturing tolerances. As a result, one writer might not overwrite the entire area on a medium that had previously been written to by a different device. Normal read heads will only give access to the most recently written data, but special high-resolution read techniques (e.g., magnetic-force microscopy) can give access to older data that remains visible near the track edges.

- Even with a perfectly positioned write head, the hysteresis properties of ferromagnetic media can result in a weak form of previous data to remain recognizable in overwritten areas. This allows the partial recovery of overwritten data, in particular with older low-density recording techniques. Protection measures that have been suggested in the literature against such data remanence include encryption, multiple overwriting of sensitive data with alternating or random bit patterns, the use of special demagnetization ("degaussing") equipment, or even the physical destruction (e.g., shredding, burning) of media at the end of its life time.

- The CMOS flip-flop circuits used in static RAM have been observed to retain data for minutes, at low temperatures in some cases even for hours, after the supply voltage has been removed [4]. The data remanence of RAM can potentially be increased where memory cells are exposed to constant data for extended periods of time or in the presence of ionizing radiation ("burn in"). Protection measures that are used in some commercial security modules include sensors for low temperature and ionizing radiation. These have to be connected to battery-powered alarm circuits that purge security RAM instantly in unusual environments. Another protection technique inverts or rotates bit patterns every few seconds, to avoid long-term exposure of memory cells to a constant value ("RAM saver").

- File and database systems do not physically overwrite ("purge") data when it is deleted by the user, unless special data purging functions designed for security applications are used. When objects are deleted, normally their storage area is only marked as available for reallocation. This leaves deleted data available for recovery with special undelete software tools, until the time when the respective memory location is needed to store new data.

Markus Kuhn

References

[1] A guide to understanding data remanence in automated information systems. National Computer Security Center, NCSC-TG-025, United States Department of Defense, September 1991.

[2] Gutmann, Peter (2001). "Data remanence in semiconductor devices." *Proceedings of the 10th USENIX Security Symposium, Washington, DC, USA*, 13–17.

[3] Gutmann, Peter (1996). "Secure deletion of data from magnetic and solid-state memory." *Sixth*

USENIX Security Symposium Proceedings, San Jose, CA, 77–89.

[4] Skorobogatov, Sergei (2002). "Low temperature data remanence in static RAM." Technical Report UCAM-CL-TR-536, University of Cambridge, Computer Laboratory.

DAVIES–MEYER HASH FUNCTION

The Davies–Meyer hash function is a construction for a hash function based on a block cipher, where the length in bits of the hash result is equal to the block length of the block cipher. A hash function is a cryptographic algorithm that takes input strings of arbitrary (or very large) length and maps these to short fixed length output strings. The Davies–Meyer hash function is an unkeyed cryptographic hash function which may have the following properties: preimage resistance, second preimage resistance and collision resistance; these properties may or may not be achieved depending on the properties of the underlying block cipher.

In the following, the block length and key length of the block cipher will be denoted with n and k respectively. The encryption with the block cipher E using the key K will be denoted with $E_K(\cdot)$.

The Davies–Meyer scheme is an iterated hash function with a compression function that maps $k + n$ bits to n bits:

$$H_i = E_{X_i}(H_{i-1}) \oplus X_i. \tag{1}$$

By iterating this function in combination with MD-strengthening (see hash functions) one can construct a hash function based on this compression function; this hash function is known as the Davies–Meyer hash function. It has been shown by Black and Rogaway [1] that in the black-box cipher model, if $k \geq n$ finding a (second) preimage requires approximately 2^n encryptions and finding a collision requires approximately $2^{n/2}$ encryptions.

In order to achieve an acceptable security level against (2nd) preimage attacks, the block length n needs to be at least 80 bits (in 2004); for collision resistance, the block length should be at least 160 bits (in 2004). This means that this scheme should not be used with 64-bit block ciphers (e.g., CAST-128, Data Encryption Standard (DES), FEAL, GOST, IDEA, KASUMI/MISTY1); it should only be used for (2nd) preimage resistance with 128-bit block ciphers (e.g., Rijndael/AES, Camellia, CAST-256, MARS, RC6, TWOFISH, and SERPENT). Very few 256-bit block ciphers exist; one exception is the 256-bit version of RC6.

It is also important to note that a block cipher may have properties which pose no problem at all when they are used only for encryption, but which may result in the Davies–Meyer construction of the block cipher to be insecure [3, 4]. A typical example are the complementation property and weak keys of *DES*; it is also clear that the Davies–Meyer construction based on DES-X is highly insecure. The fact that the key is known to an opponent may also result in security weaknesses (e.g., differential attacks of Rijmen and Preneel [5]). Hirose defines a block cipher secure against a known plaintext attack for which the Davies–Meyer hash function is not 2nd preimage resistant [2].

Since there are very few block ciphers with a 256-bit block length, the Davies–Meyer construction is rarely used to obtain collision resistant hash functions. However, this construction is very popular in custom designed hash functions such as MD4, MD5, and the SHA family. Indeed, the compression functions of these hash functions are designed using an internal block cipher structure; the compression functions are made non-invertible by applying the Davies–Meyer construction to these internal block ciphers.

Bart Preneel

References

[1] Black, J., P. Rogaway, and T. Shrimpton (2002). "Black-box analysis of the block-cipher-based hash-function constructions from PGV." *Advances in Cryptology—CRYPTO 2002*, Lecture Notes in Computer Science, vol. 2442, ed. M. Yung. Springer-Verlag, Berlin, 320–355.

[2] Hirose, S. (2002). "Secure block ciphers are not sufficient for one-way hash functions in the Preneel-Govaerts-Vandewalle model." *Selected Areas in Cryptography*, Lecture Notes in Computer Science, vol. 2595, eds. K. Nyberg and H.M. Heys. Springer-Verlag, Berlin, 339–352.

[3] Preneel, B. (1993). "Analysis and design of cryptographic hash functions." *Doctoral Dissertation*, Katholieke Universiteit Leuven.

[4] Preneel, B., R. Govaerts, and J. Vandewalle (1994). "Hash functions based on block ciphers: A synthetic approach." *Advances in Cryptology—CRYPTO'93*, Lecture Notes in Computer Science, vol. 773, ed. D. Stinson. Springer-Verlag, Berlin, 368–378.

[5] Rijmen V. and B. Preneel (1995). "Improved characteristics for differential cryptanalysis of hash functions based on block ciphers." *Fast Software Encryption*, Lecture Notes in Computer Science, vol. 1008, ed. B. Preneel. Springer-Verlag, Berlin, 242–248.

DC NETWORK

The DC-Network is a synchronous network protocol by which the participants can broadcast messages anonymously and unobservably (see anonymity). A DC-Network can achieve sender and recipient anonymity even against computationally unrestricted attackers. The DC-Network protocol itself requires a network with a broadcast service. It was invented by David Chaum in 1984 [2–4] (hence the name DC-Network) and was somewhat re-discovered by Dolev and Ostrovsky in [5]. Messages can be addressed to one or more intended participants by encrypting them with their respective public encryption keys.

The basic DC-Network protocol allows one participant at a time to broadcast a message. In order to allow each participant to send at any time, the broadcast channel of the basic DC-Network needs to be allocated in a randomized fashion to all requesting senders. This can be achieved by well known contention protocols such as (slotted) ALOHA [8].

Consider the basic DC-Network protocol of n participants: $P_1, P_2, \ldots, P_n$. Messages are strings of k bits. As a preparation, all participants agree on pairwise symmetric keys, i.e., randomly chosen bitstrings of k bit length. Let us denote the key between P_i and P_j as $k_{i,j}$. Assume participant P_1 wants to send a message m anonymously to all other participants (*anonymous broadcast*). This can be achieved by the basic DC-Network protocol, which works as follows:

Compute partial sums: Each participant P_i ($1 \leq i \leq n$) computes the XOR sum s_i of all the keys $k_{i,j}$ ($1 \leq j \leq n$) it has exchanged with each other participant P_j ($j \neq i$), such that $s_i = \sum_{j \neq i} k_{i,j}$. Participant P_1 also adds his message m into his partial sum such that $s_1 = m + \sum_{j \neq 1} k_{j,1}$.

Broadcast partial sums: Each participant P_i broadcasts its partial sum s_i.

Compute global sum: Each participant P_i computes the global sum $s = \sum_{i=1}^{n} s_i = m + \sum_{i=1}^{n} \sum_{j \neq i} k_{i,j} = m$ in order to recover the message m. Note that because $k_{i,j} = k_{j,i}$, all the keys cancel out leaving only m standing out of the global sum.

The basic DC-Network protocol is computationally efficient but requires n reliable broadcasts for each message, and even more in case of resolving message collisions where two or more participants are sending their messages in the same round.

The basic DC-Network protocol runs on any network architecture. If all participants are honest, everyone obtains the message m. Chaum [4] has proved that the basic DC-Network protocol achieves <u>sender anonymity</u> and <u>recipient anonymity</u> even against computationally unrestricted attackers. However, the proof for recipient anonymity implicitly assumes that the partial sums are broadcast reliably, i.e., each message of an honest participant is broadcast to all participants without being modified [9].

DC-Network is the continued execution of the basic DC-Network <u>protocol</u>. In this case, unconditional sender anonymity can be maintained only by using fresh pairwise keys in each round, which is a similar situation as for the *one-time pad* (see <u>key</u>). Waidner has proposed to choose the pairwise keys for each round of the basic DC-Network protocol based on a pseudo-random number generator seeded with a selection of messages exchanged in previous rounds of the basic DC-Network protocol. This is more practical, but results in sender anonymity that holds only against computationally restricted attackers [9].

The core idea behind the DC-Network is to substantially involve more participants in each communication than just the intended sender and recipient in order to conceal their sending and receiving within the set of participants. This approach introduces an inevitable vulnerability in case not all of the participants honestly follow the protocol. In fact, the service of a DC-Network can be easily disrupted by one or more cheating participants, who either stop sending their partial sums or sending wrong partial sums or sending too many messages (denial-of-service attack). Disruptions of the DC-Network have been considered by Chaum [1], Bos and den Boer [4] and Waidner [9].

The *key graph* of a DC-Network is the graph where each participant is represented by a vertex and each pairwise key $k_{i,j}$ is represented by an edge connecting the vertices representing P_i and P_j. If the key graph is complete as in the example above, no coalition of non-senders except all of them together gains any information about who sent m. Less complete key graphs can be used in order to reduce the amount of pairwise keys. On the other hand, the less complete the key graph is chosen, the more vulnerable the basic DC-Network protocol is against cheating participants who may collude and exchange their views in and after the basic DC-Network protocol in order to strip away the honest participants' anonymity. Collusions of cheating participants can be represented in the key graph by eliminating their mutual pairwise keys. That is if P_i, P_j are cheating, then we remove the key $k_{i,j}$ from the key graph, which may lead to an unconnected graph. Any participant represented by an unconnected vertex is entirely stripped of its anonymity. Such a participant is fully observable by the collusion of cheating participants. It is worth noting that the key graph can be chosen independently of the underlying network topology (ring, star, bus, etc.).

Waidner points out in [9] that reliable broadcast is probably an unrealistic assumption because it cannot be achieved by cryptographic means alone as there is no byzantine agreement against computationally unrestricted active attackers who may arbitrarily control many participants [6]. Furthermore, Waidner has shown how to achieve recipient anonymity against computationally unrestricted active attackers by replacing the reliable broadcast by a fail-stop broadcast, where honest participants stop as soon as they receive inconsistent inputs. Fail-stop broadcast can be realized by $O(n)$ messages, each signed by an unconditionally secure authentication *code*, or more efficiently by a fail-stop signature [10].

Interestingly, no widely accepted formal definitions of sender and recipient anonymity in a network, i.e., continued transmission service, has come up yet. Thus, a fully formal treatment of DC-Network protocols is not possible to date. A new approach in this direction was proposed by Schneider and Sidiropoulos [7] based on the CSP process algebra (Communicating Sequential Processes).

Compared to MIX-Networks, DC-Networks achieve sender anonymity even against computationally unrestricted active attackers, while MIX networks only achieve sender anonymity against computationally restricted attackers.

Gerrit Bleumer

References

[1] Bos, Jurjen and Bert den Boer (1990). "Detection of disrupters in the DC protocol." *Advances in Cryptology—EUROCRYPT'89*, Lecture Notes in Computer Science, vol. 434, eds. J.-J. Quisquater and J. Vandewalle. Springer-Verlag, Berlin, 320–327.

[2] Chaum, David (1981). "Untraceable electronic mail, return addresses, and digital pseudonyms." *Communications of the ACM*, 24 (2), 84–88.

[3] Chaum, David (1986). "Showing credentials without identification—signatures transferred between unconditionally unlinkable pseudonyms." *Advances in Cryptology—EUROCRYPT'85*, Lecture Notes in Computer Science, vol. 219, ed. F. Pichler. Springer-Verlag, Berlin, 241–244.

[4] Chaum, David (1988). "The dining cryptographers problem: Unconditional sender and recipient untraceability." *Journal of Cryptology*, 1 (1), 65–75.

[5] Dolev, Shlomi and Rafail Ostrovsky (1997). "Efficient anonymous multicast and reception."

[6] Lamport, Leslie, Robert Shostak, and Marshall Pease (1982). "The Byzantine Generals problem." *ACM Transactions on Programming Languages and Systems*, 4 (3), 382–401.

[7] Schneider, Steve and Abraham Sidiropoulos, (1996). "CSP and anonymity." *ESORICS'96 (4th European Symposium on Research in Computer Security), Rome*, Lecture Notes in Computer Science, vol. 1146, ed. V. Lotz. Springer-Verlag, Berlin, 198–218.

[8] Tanenbaum, Andrew S. (1988). *Computer networks* (2nd ed.). Prentice-Hall, Englewood Cliffs.

[9] Waidner, Michael (1990). "Unconditional sender and recipient untraceability in spite of active attacks." *Advances in Cryptology—EUROCRYPT'89*, Lecture Notes in Computer Science, vol. 434, eds. J.-J. Quisquater and J. Vandewalle. Springer-Verlag, Berlin, 302–319.

[10] Waidner, Michael and Birgit Pfitzmann (1990). "The dining cryptographers in the disco: Unconditional sender and recipient untraceability with computationally secure serviceability." *Advances in Cryptology—EUROCRYPT'89*, Lecture Notes in Computer Science, vol. 434, eds. J.-J. Quisquater and J. Vandewalle. Springer-Verlag, Berlin, 690.

(continuation of reference [5])
"Efficient anonymous multicast and reception." *Advances in Cryptology—CRYPTO'97*, Lecture Notes in Computer Science, vol. 1294, ed. B.S. Kaliski. Springer-Verlag, Berlin, 395–409.

DEBRUIJN SEQUENCE

A k-ary deBruijn sequence of order n is a sequence of period k^n which contains each k-ary n-tuple exactly once during each period. DeBruijn sequences are named after the Dutch mathematician Nicholas deBruijn. In 1946 he discovered a formula giving the number of k-ary deBruijn sequences of order n, and proved that it is given by $((k-1)!)^{k^{n-1}} \cdot k^{k^{n-1}-n}$. The result was, however, first obtained more than 50 years earlier, in 1894, by the French mathematician C. Flye-Sainte Marie.

For most applications binary deBruijn sequences are the most important. The number of binary deBruijn sequences of period 2^n is $2^{2^{n-1}-n}$. An example of a binary deBruijn sequence of period $2^4 = 16$ is $\{s_t\} = 0000111101100101$. All binary 4-tuples occur exactly once during a period of the sequence. In general, binary deBruijn sequences are balanced, containing the same number of 0's and 1's in a period, and they satisfy many randomness criteria, although they may be generated using deterministic methods. They have been used as a source of pseudo-random numbers and in key-sequence generators of stream ciphers.

A deBruijn sequence can be generated by a <u>nonlinear feedback function</u> in n-variables. From the initial state $(s_0, s_1, \ldots, s_{n-1})$ and a nonlinear Boolean function $f(z_0, z_1, \ldots, z_{n-1})$ one can generate the sequence

$$s_{t+n} = f(s_t, s_{t+1}, \ldots, s_{t+n-1}), \quad \text{for } t = 0, 1, 2, \ldots$$

This can be implemented using an n-stage non-linear shift register. For example the binary deBruijn sequence above of period $16 = 2^4$ can be generated by $s_{t+4} = f(s_t, s_{t+1}, s_{t+2}, s_{t+3})$, using the initial state (0000) and the <u>Boolean function</u>

$$f(z_0, z_1, z_2, z_3) = 1 + z_0 + z_1 + z_1 z_2 z_3.$$

The binary *deBruijn graph* B_n of order n is a directed graph with 2^n nodes, each labeled with a unique binary n-tuple and having an edge from node $S = (s_0, s_1, \ldots, s_{n-1})$ to $T = (t_0, t_1, \ldots, t_{n-1})$ if and only if $(s_1, s_2, \ldots, s_{n-1}) = (t_0, t_1, \ldots, t_{n-2})$. The successive n-tuples in a deBruijn sequence therefore form a *Hamiltonian cycle* in the deBruijn graph, meaning that a full cycle visits each node exactly once.

There are many algorithms for constructing deBruijn sequences. The following is perhaps one of the easiest to describe. Start with n zeros and append a one whenever the n-tuple thus formed has not appeared in the sequence so far, otherwise append a zero. The sequence of length $2^4 = 16$ above is an example of a deBruijn sequence constructed in this way. It is known that the decision of which bit to select next can be based on local considerations and storage requirements can be reduced to only $3n$ bits.

Any Boolean function f such that the mapping

$$(z_0, z_1, \ldots, z_{n-1}) \to (z_1, z_2, \ldots, z_{n-1},$$
$$f(z_0, z_1, \ldots, z_{n-1}))$$

is a permutation of the set of binary n-tuples is called a *nonsingular* Boolean function. It can be written in the form,

$$f(z_0, z_1, \ldots, z_{n-1}) = z_0 + g(z_1, z_2, \ldots, z_{n-1})$$
$$(\text{mod } 2).$$

The truth table of a Boolean function $f(z_0, z_1, \ldots, z_{n-1})$ is a list of the values of $f(z_0, z_1, \ldots, z_{n-1})$ for all binary n-tuples. The weight of the truth table of f is the number of ones in this list.

Large classes of deBruijn sequences can be constructed by starting with a nonsingular Boolean function f that decomposes the deBruijn graph into several shorter disjoint cycles and then joining the cycles one by one until one arrives at a deBruijn sequence. To join two cycles one can find an n-tuple $(z_0, z_1, \ldots, z_{n-1})$ on a cycle (where we have

$(z_1, z_2, \ldots, z_{n-1}, f(z_0, z_1, \ldots, z_{n-1}))$ on the same cycle) and $(z_1, z_2, \ldots, z_{n-1}, 1 + f(z_0, z_1, \ldots, z_{n-1}))$ on a different cycle. Then the two cycles will be joined after changing(complementing) $g(z_1, z_2, \ldots, z_{n-1})$ (leading to two changes of the truth table of f).

One common starting function is the nonsingular function corresponding to $g = 0$, i.e., $f(z_0, z_1, \ldots, z_{n-1}) = z_0$, that is known to decompose B_n into the Pure Circulating Register(PCR), consisting of all cycles of period dividing n. This is known to contain $Z(n) = \frac{1}{n} \sum_{d|n} \phi(d) 2^{n/d}$ cycles. For $n = 4$ the PCR consists of the cycles $(0), (1), (01), (0001), (0011),$ and (0111). Another popular starting function is the Complementary Circulating Register(CCR) corresponding to $g = 1$, i.e., $f(z_0, z_1, \ldots, z_{n-1}) = z_0 + 1$ (mod 2). This is known to contain $Z^*(n) = \frac{1}{2} Z(n) - \frac{1}{2n} \sum_{2d|n} \phi(2d) 2^{n/2d}$ cycles. Another method to construct deBruijn sequences is to use recursive algorithms. There exist algorithms that take as input two deBruijn sequences of period 2^{n-1} and produce a deBruijn sequence of period 2^n.

The <u>linear complexity</u> of a deBruijn sequence is defined as the length of the shortest linear shift register that can be used to generate the sequence. The linear complexity L of a binary deBruijn sequence of period 2^n, $n \geq 3$, satisfies the double inequality,

$$2^{n-1} + n \leq L \leq 2^n - 1.$$

There exist deBruijn sequences that meet the upper and lower bounds with equality.

The *quadratic complexity* of a deBruijn sequence is the length of the shortest shift register that generates the sequence where the feedback function f is allowed to have quadratic terms. The quadratic complexity Q of a binary deBruijn sequence of period 2^n, $n \geq 3$, satisfies the double inequality

$$n + 2 \leq Q \leq 2^n - \binom{n}{2} - 1.$$

It is known that for any nonsingular Boolean function f, the number of cycles that it decomposes B_n into has the same parity as the weight of the truth table of g. Therefore for a deBruijn sequence the truth table of g has odd weight. It is further known that for a deBruijn sequence, the weight w of the truth table of g obeys,

$$Z(n) - 1 \leq w \leq 2^{n-1} - Z^*(n) + 1.$$

The lower bound can be achieved by starting with the PCR and joining cycles one at a time until we arrive at a deBruijn sequence. Each joining step will in this case increase the weight of the truth table of g by 1. Similarly we can construct deBruijn sequences of maximal weight by starting with the CCR and joining the cycles one by one, each joining step will in this case reduce the weight of the truth table of g by 1. For values $n < 7$ the number of deBruijn sequences of each possible weight of the truth table of g is known.

Tor Helleseth

References

[1] Fredricksen, H. (1982). "A survey of full length nonlinear shift register cycle algorithms." *SIAM Review*, 24 (2), 195–221.

[2] Golomb, S.W. (1982). *Shift Register Sequences*. Aegean Park Press, Laguna Hills, CA.

DECISIONAL DIFFIE–HELLMAN ASSUMPTION

The difficulty in computing discrete logarithms in some large finite groups has been the basis for many cryptographic schemes and protocols in the past decades, starting from the seminal Diffie–Hellman key agreement protocol [8], and continuing with encryption and digital signature schemes with a variety of security properties, as well as protocols for numerous other applications. Ideally, we would have liked to prove unconditional statements regarding the computational difficulty in computing discrete logarithms. However, since the current state of knowledge does not allow us to prove such claims, we formulate instead mathematical *assumptions* regarding the computational difficulty of this set of problems, and prove properties of the protocols we develop based on these assumptions.

A first assumption that is closely related to the Diffie–Hellman key exchange is the Computational Diffie–Hellman assumption (see Diffie–Hellman problem for more detail):

The Computational Diffie–Hellman (CDH) Problem: Given a group G, a generator g of G, and two elements $a = g^x, b = g^y \in G$, where x and y are unknown, compute the value $c = g^{xy} \in G$.

The Computational Diffie–Hellman (CDH) Assumption: Any probabilistic polynomial time algorithm solves the CDH problem only with negligible probability.

Notes:

(1) The probability is taken over the random choices of the algorithm. The probability is said to be negligible if it decreases faster than any inverse polynomial in the length of the input.

(2) As usual, the algorithm must run in time that is polynomial in the length of its input, namely in time that is polylogarithmic in the size of G. Also, a solution to the CDH problem is an algorithm that works for *all inputs*. Thus, the CDH assumption implies that there exists an infinite sequence of groups G for which no polytime algorithm can solve the CDH problem with probability that is not negligible. (Still, it is stressed that there exist infinite families of groups for which the CDH problem is in fact easy.)

(3) The assumption can be made with respect either to uniform-complexity or non-uniform complexity algorithms (i.e., circuit families.)

Indeed, the CDH assumption is very basic in cryptography. However, in many cases researchers were unable to prove the desired security properties of protocols based on the CDH assumption alone. (A quintessential example is the Diffie–Hellman key exchange protocol itself.) Furthermore, it appears that, at least in some groups, the CDH assumption captures only a mild flavor of the intractability of the Diffie–Hellman problem. Therefore the *Decisional* Diffie–Hellman assumption was formulated, as follows:

The Decisional Diffie–Hellman (DDH) Problem: Given a group G, a generator g of G, and three elements $a, b, c \in G$, decide whether there exist integers x, y such that $a = g^x$, $b = g^y$, and $c = g^{xy}$.

The Decisional Diffie–Hellman (DDH) Assumption (Version I): Any probabilistic polynomial time algorithm solves the DDH problem only with negligible probability.

The above formulation of the DDH assumption treats the problem as a worst-case computational problem (that is, an algorithm that solves the problem must work on *all* inputs. This formalization provides a useful comparison with the CDH problem. A much more useful alternative formulation of the DDH assumption only discusses the case where the inputs are taken from certain distributions. It is stated as follows:

The Decisional Diffie–Hellman (DDH) Assumption (Version II): The following two distributions are computationally indistinguishable:

- G, g, g^x, g^y, g^{xy}
- G, g, g^x, g^y, g^z

where g is a generator of group G and x, y, z are chosen at random from $\{1, \ldots, |G|\}$.

Note: More formally, the above two distributions are actually two distribution *ensembles,* namely two families of distributions where each distribution in a family is parameterized by the group G and the generator g. Recall that two distribution ensembles are computationally indistinguishable if, given a set of parameters (in our case, given G and g), no polytime algorithm can tell whether its input is drawn from the first ensemble or from the second. See more details in [10].

This version is useful since it asserts that, even when g^x and g^y are known, the value g^{xy} appears to be a "freshly chosen" random and independent number for any computationally bounded attacker. This holds in spite of the fact that the value g^{xy} is uniquely determined by g^x and g^y, thus its "entropy" (in the information-theoretic sense) is in fact zero. As shown in [12, 14], the two versions of the DDH assumption are equivalent. (Essentially, equivalence holds due to the *random self reducibility* property of the discrete logarithm problem.)

Clearly, the DDH assumption implies the CDH assumption. Furthermore, it appears to be considerably stronger. In fact, there are groups where DDH is clearly false, but CDH may still hold. Still, there exist groups where DDH is believed to hold, for instance multiplicative groups of large prime order. A quintessential example is the subgroup of size q of $\mathbf{Z}_p^*$ (see modular arithmetic) where $p = 2q + 1$ and p, q are primes. (In this case the larger prime p is called a safe prime, and the smaller prime q is called a Sophie-Germain prime.)

Note: To see an example of a family of groups where DDH does not hold but CDH may still hold, consider a group G where it is easy to check whether an element is a quadratic residue (e.g., let $G = \mathbf{Z}_p^*$ where p is prime and $|\mathbf{Z}_p^*| = p - 1$ is even). Here, the CDH assumption may hold, yet DDH is false: If the input is drawn from G, g, g^x, g^y, g^{xy} then it is never the case that the last element is a quadratic non-residue but the preceding two elements are quadratic residues. In contrast, if the input is taken from G, g, g^x, g^y, g^z then the above event happens with significant probability. Other examples of such groups also exist. Here let us mention in particular the case of bilinear and multilinear pairings in Elliptic-Curve groups, which have been recently shown to be useful in cryptography. See identity based cryptosystem and for example [3].

SOME APPLICATIONS OF DDH: The DDH assumption proves to be very useful in cryptographic analysis of protocols. It is immediate to show based on DDH that the Diffie–Hellman key exchange results in a "semantically secure" key, i.e., a key that is indistinguishable from random. (It is not known how to prove this statement based on CDH alone.) Similarly, it implies the semantic security of ElGamal public key encryption. In addition, it is used in proving the security of efficient pseudorandom functions [12], chosen-ciphertext-secure encryption [6], commitment and zero-knowledge protocols [7, 13], and many more.

VARIANTS OF DDH: The DDH assumption is only one of many assumptions that can be made on the intractability of the discrete logarithm problem. Several variants have been considered in the literature, some of which are *stronger* (allowing to prove stronger security properties of protocols), and some are *weaker* (and are believed to hold even in cases where DDH does not). Of the stronger ones, let us mention variants that allow the exponents x, y to be chosen from distributions other than uniform (or even in a semi-adversarial way) [5]. Other stronger variants are formalized in [3, 9, 11]. Of the weaker ones, we mention variants that give the distinguisher access only to a *hashed* version of the last element (either g^{xy} or g^z) e.g., [1].

BIBLIOGRAPHIC NOTE: The DDH assumption is implicit in many early works based on the Diffie–Hellman problem (starting with [8]). To the best of our knowledge, it was first formalized by Brands in [4] (in the context of undeniable signatures). It was further studied in [12, 14] and is widely used since. For further reading, see Boneh's survey [2].

Ran Canetti

References

[1] Abdalla, M., M. Bellare, and P. Rogaway (2001). "DHIES: An encryption scheme based on the Diffie–Hellman problem." *Topics in Cryptology— CT-RSA 2001*, Lecture Notes in Computer Science, vol. 2020, ed. D. Naccache. Springer-Verlag, Berlin, 143–158.

[2] Boneh, Dan (1998). "The decision Diffie–Hellman problem." *Proceedings of the Third Algorithmic Number Theory Symposium*, Lecture Notes in Computer Science, vol. 1423, ed. J.P. Buhler. Springer-Verlag, Berlin, 48–63.

[3] Boneh, Dan and Alice Silverberg (2002). "Applications of multilinear forms to cryptography." *Proceedings of the Conferences in memory of Ruth Michler*, Contemporary Mathematics, American Mathematical Society. Cryptology ePrint Archive, Report 2002/080. Available on http://eprint.iacr.org/

[4] Brands, S. (1993). "An efficient off-line electronic cash system based on the representation problem." CWI TR CS-R9323.

[5] Canetti, R. (1997). "Toward realizing random oracles: Hash functions that hide all partial information." *Advances in Cryptology—CRYPTO'97*, Lecture Notes in Computer Science, vol. 1294, ed. B.S. Kaliski Jr. Springer-Verlag, Berlin, 455–469.

[6] Cramer, R. and V. Shoup (1998). "A practical public-key cryptosystem provably secure against adaptive chosen ciphertext attack." *Advances in Cryptology—CRYPTO'98*. Lecture Notes in Computer Science, vol. 1462, ed. H. Krawczyk. Springer-Verlag, Berlin, 13–25.

[7] Damgård, I. (2000). "Efficient concurrent zero-knowledge in the auxiliary string model." *Advances in Cryptography—EUROCRYPT 2000*, Lecture Notes in Computer Science, vol. 1807, ed. B. Preneel. Springer-Verlag, Berlin, 418–430.

[8] Diffie, W. and M. Hellman (1976). "New directions in cryptography." *IEEE Trans. Info. Theory*, IT-22, 644–654.

[9] Dodis, Yevgeniy (2002). "Efficient construction of (Distributed) verifiable random functions." Cryptology ePrint Archive, Report 2002/133. Available on http://eprint.iacr.org/

[10] Goldreich, O. (2001). *Foundations of Cryptography: Volume 1—Basic Tools*. Cambridge University Press, Cambridge.

[11] Lysyanskaya, Anna (2002). "Unique signatures and verifiable random functions from the DH-DDH separation." *Advances in Cryptology—CRYPTO 2002*, Lecture Notes in Computer Science, vol. 2442, ed. M. Yung. Springer-Verlag, Berlin, 597–612.

[12] Naor, Moni and Omer Reingold (1997). "Number-theoretic constructions of efficient pseudo-random functions." Extended abstract in *Proc. 38th IEEE Symp. on Foundations of Computer Science*, 458–467.

[13] Pedersen, T.P. (1991). "Distributed provers with applications to undeniable signatures." *Advances in Cryptography—EUROCRYPT'91*, Lecture Notes in Computer Science, vol. 547, ed. D.W. Davis. Springer-Verlag, Berlin, 221–242.

[14] Stadler, M. (1996). "Publicly verifiable secret sharing." *Advances in Cryptography—EUROCRYPT'96*, Lecture Notes in Computer Science, vol. 1070, ed. U. Maurer. Springer-Verlag, Berlin, 190–199.

DECRYPTION EXPONENT

The exponent d in the RSA private key (n, d). See RSA public key encryption.

Burt Kaliski

DENIABLE ENCRYPTION

Suppose Alice sends a message to Bob in an informal chat conversation. If a typical encryption scheme as the ElGamal public key encryption scheme or Rijndael/AES is used, an authority can ask Alice to reveal what she sent Bob. Indeed, in the case of ElGamal, when Alice sends $(C_1, C_2) = (g^r, my^r)$ and is forced to reveal her randomness r used, anybody can obtain m. So, one can view the ciphertext as some commitment to the message. In the case of AES, when Alice is forced to reveal the key she shares with Bob, the authority again can obtain the message. (Using zero-knowledge, Alice is not required to reveal the key.)

The goal of deniable encryption [1] is that Alice can send a private message to Bob, without having the ciphertext result in a commitment. This can be viewed as allowing her to deny having sent a particular message. A scheme satisfying this condition is called a *sender-deniable* encryption scheme.

There is a similar concern from Bob's viewpoint. Can Bob be forced to open the received ciphertext? Again if the ElGamal public key encryption scheme is used, then using his secret key, Bob can help the authority to decipher the message. So, Bob "cannot deny" having received the message. A scheme that solves this issue is called a *receiver-deniable* encryption scheme.

An example of a sender-deniable scheme explained informally, works as follows. Suppose the sender (Alice) and the receiver (Bob) have agreed on some pseudorandomness, such that both can distinguish it from true randomness. When Alice wants to send a message bit 1, she will send some pseudorandom string, otherwise she sends true randomness. Since the authority cannot distinguish the pseudorandom from the real random, Alice can pretend she sent the opposite bit of what she did. For further details, see [1].

Canetti–Dwork–Naor–Ostrovsky demonstrated that a sender-deniable encryption scheme can be transformed into a receiver-deniable one, as follows:

Step 1. The receiver (Bob) sends the sender (Alice) a random r using a sender-deniable encryption scheme.

Step 2. The sender Alice sends Bob the ciphertext $r \oplus m$, where $\oplus$ is the exor.

A receiver-deniable scheme can also be transformed into a sender deniable one, as explained in [1].

Yvo Desmedt

Reference

[1] Canetti, R., C. Dwork, M. Naor, and R. Ostrovsky (1997). "Deniable encryption." *Advances in Cryptology—CRYPTO'97, Proceedings* Santa Barbara, CA, USA, August 17–21 (Lecture Notes in Computer Science vol. 1294), ed. B.S. Kaliski, Springer-Verlag, Berlin, 90–104.

DENIAL OF SERVICE

In the most literal sense, whenever a legitimate principal is unable to access a resource for any reason, it can be said that a denial of service has occurred. In common practice, however, the term *Denial of Service* (DoS) is reserved only to refer to those times when an interruption in <u>availability</u> is the intended result of a deliberate attack [3]. Often, especially in the press, DoS is used in an even more narrow sense, referring specifically to remote flooding attacks (defined below) against network services such as web servers. When attempting to prevent access to a target service, the target itself can be attacked, or, equally effectively, another service upon which the target depends can be attacked. For example, to cause a DoS of a web server, the server program could be attacked, or the network connection to the server could be attacked instead.

DoS attacks can be categorized as either *local Denial of Service* attacks or *remote Denial of Service* attacks. Local DoS attacks are a type of privilege escalation, where a principal with legitimate access to a service is able to deny others access to it. In many older UNIX-like operating systems, for example, when a user goes to change their password, the system first locks the global password file before asking the user for their new password; until the user enters their new password, the file remains locked and no other users are able to change passwords. Remote DoS attacks, on the other hand, often require no special rights for the attacker, or are against services which do not require any authentication at all. Flooding a web server with millions of requests is an example of a common remote DoS attack.

Some DoS attacks, referred to as *logic* attacks in [7], work by exploiting programming bugs in the service being attacked, causing it to immediately exit or otherwise stop responding. Examples of these types of attacks include the Windows 95 Ping-of-Death, BIND nameserver exit-on-error attacks, and countless buffer overflow attacks which crash, but do not compromise,[1] services. These kinds of DoS attacks are the easiest to prevent, since the attack is caused by invalid behavior that would not be expected from legitimate principals. By fixing bugs and more carefully filtering out bad input, these types of DoS attacks can be prevented. The attacks are also very *asymmetric* however, making them very dangerous until all vulnerable services have been upgraded. With these attacks, very little effort on the part of the attacker (a single malformed message typically) leads to a complete Denial of Service. An attacker with limited resources is able to quickly cause a great deal of damage with these attacks.

In contrast, *flooding* DoS attacks work by consuming limited resources on the server. Resources commonly targeted by these attacks include memory, disk space, CPU, and network bandwidth. Simple local DoS attacks such as acquiring and never releasing a shared lock also fall into this group. With these attacks, the problem lies in the rate the attacker does something, not in what they do. These attacks take a normal, acceptable activity such as forking a new process or requesting a web page, and raise it to an attack by performing the activity to excess.

Because these attacks involve behavior that would normally be perfectly acceptable, there is typically no way to tell with certainty that a request to a service is part of an attack. In some cases, particularly local DoS attacks, the consumption of resources can be limited by placing caps on how much of the resource any single user can consume. Limits can be placed on how much memory, disk space, or CPU a single user can use, and timeouts can be set whenever an exclusive lock is given out. The difficulty with using limits to prevent these attacks is that if a user needs to exceed one of these caps for legitimate purposes, they are unable to; the solution to the first DoS attack causes a Denial of Service of a different kind. Because most solutions to flooding attacks rely on some heuristic to determine when behavior is malicious, there are always some false positives which cause the prevention to be a DoS itself.

As with logic attacks, some flooding attacks are also highly asymmetric. In particular, many standard protocols (such as IP, TCP (see <u>firewall</u>) and SSL/TLS (see <u>Secure Socket Layer</u> and <u>Transport Layer Security</u>)) allow for asymmetric attacks because they require the service to keep state or perform expensive computations for the attacker.

---

[1] Technically, if an attack's primary purpose is to compromise a service, and, as a side effect, it crashes the service, this is not considered a DoS attack [3].

If an attacker begins many protocol sessions but never completes them, resources can quickly be exhausted. TCP SYN flood and incomplete IP fragment attacks both work by exhausting available buffers within the server's networking stack. When beginning SSL/TLS sessions the server must perform CPU-intensive public key cryptography operations which take orders of magnitude longer than it takes an attacker to send a request. To remove the asymmetry of these attacks, techniques that reduce or remove the state the server must keep [6] or force the client to perform a comparable amount of computation [1] have been proposed.

Flooding attacks which have similar resource requirements for the attacker and victim comprise the final group of common DoS attacks. While the previous attacks described have had an element of skill to them, finding and exploiting some programming error or imbalance in protocol design, these attacks are nothing more than a shoving match, with the participant with the most of the resource in question winning. Smurf attacks and DNS flooding are well known examples of these brute-force DoS attacks.

Often, the attacker does not have an excess of the resource (usually network bandwidth) themselves. Instead, to carry out their attack they first compromise a number of other hosts, turning them into *zombies*, and then have their zombies attack the service simultaneously. This type of attack is known as a *Distributed Denial of Service* (DDoS) attack, and has proven very effective in the past against a number of popular and very well connected Internet servers such as Yahoo! and eBay.

With all types of remote flooding attacks, if the source of the flood can be identified, it can be blocked with minimal disruption to non-attack traffic. With DDoS attacks, this identification is the main difficulty, since no single zombie produces an exceptionally large number of requests. Blocking the wrong source results in a DoS itself. Further complicating identification, many attackers mask themselves by forging, or *spoofing*, the source of requests. Traceback techniques [2,8] can be used to identify the true source, but their accuracy degrades as more sources are present. Egress filtering, which blocks packets from leaving edge networks if they claim to have not originated from that network, can prevent spoofing. Unfortunately, all networks must employ egress filtering before it is an adequate solution. Since most DoS attacks employ spoofing, Backscatter analysis [7] actually takes advantage of it, looking at replies from victims to the spoofed sources to determine world-wide DoS activity.

Once the true source of a flood has been identified, filters can be installed to block the attack. With bandwidth floods in particular, this blocking may need to occur close to the attacker in the network in order to fully block the DoS. This can either be arranged manually, through cooperation between network administrators, or automatically through systems like Pushback [5].

As seen above, many Denial of Service attacks have no simple solutions. The very nature of openly accessible services on the Internet leaves them vulnerable from these attacks. It is an interesting and rapidly evolving type of security attack. The list of resources at [4] is updated periodically with pointers to new attacks and tools for protecting services, and makes a good starting point for further exploring the causes and effects of DoS attacks, and the state of the art techniques in dealing with them.

Eric Cronin

References

[1] Aura, T., P. Nikander, and J. Leiwo (2000). "DoS resistant authentication with client puzzles." *Proc. of the Cambridge Security Protocols Workshop 2000.*

[2] Bellovin, S. (2002). "ICMP traceback messages." *Internet Draft.* draft-ietf-itrace-02.txt

[3] CERT Coordination Center (2001). "Denial of service attacks." http://www.cert.org/tech_tips/denial_of_service.html

[4] Dittrich, D. "Distributed denial of service (DDoS) attacks/tools." http://staff.washington.edu/dittrich/misc/ddos/

[5] Ioannidis, J. and S.M. Bellovin (2000). "Implementing pushback: Router-based defense against DDoS attacks." *Proc. of NDSS 2002.*

[6] Lemon, J. (2001). "Resisting SYN flood DoS attacks with a SYN cache." http://people.freebsd.org/~jlemon/papers/syncache.pdf

[7] Moore, D., G.M. Voelker, and S. Savage (2001). "Inferring internet denial-of-service activity." *Proc. of the 10th USENIX Security Symposium.*

[8] Savage, S., D. Wetherall, A. Karlin, and T. Anderson (2000). "Practical network support for IP traceback." *Proc. of ACM SIGCOMM 2000.*

DERIVED KEY

A derived key is a key, which may be calculated (derived) by a well-defined algorithm from a input consisting of public as well as secret data. As an example, the initial secret data might be a random seed, i.e., a string of random bits (see modular arithmetic), which is then exponentiated modulo, e.g., an *RSA*-modulus (say both of length 1024; see RSA public key encryption), after which the derived key may be the lower 128 bits of the result R (current seed), which is kept and exponentiated again for the derivation of the next key. The advantage is that if two parties share the same initial seed, they may independently of each other calculate identical derived keys by keeping track of the number of iterations.

Peter Landrock

DESIGNATED CONFIRMER SIGNATURE

Designated confirmer signatures (or sometimes simply 'confirmer signatures') are digital signatures that can be verified only by some help of a semi-trusted designated confirmer. They were introduced by Chaum in [3] as an improvement of convertible undeniable signatures. Unlike an ordinary digital signature that can be verified by anyone who has access to the public verifying key of the signer (*universal verifiability*), a designated confirmer signature can only be verified by engaging in a—usually interactive—protocol with the designated confirmer. The outcome of the protocol is an affirming or rejecting assertion telling the verifier whether the signature has originated from the alleged signer or not.

The main difference to (convertible) undeniable signatures is that the capabilities to produce signatures and to confirm signatures are laid into different hands, which has several advantages. Designated confirmer signatures improve the availability and reliability of the confirmation services for verifiers. Verifiers can rely on a designated confirmer instead of having to rely on the signers themselves. The designated confirmer can be organized as one or more authorities with a higher availability than each signer can afford to provide, and the designated confirmer can provide confirmation services according to a clearly stated confirmation policy, which can also be subject to independent audit on a regular basis. In practice, a designated confirmer would conceivably contract multiple signers and provide confirmation services to all their respective verifiers. Another way of increasing the availability of the confirmation services is by using an undeniable signature scheme with distributed provers as proposed by Pedersen [7]. Another advantage of designated confirmer signatures is that they alleviate the problem of coercable signers. In undeniable signature schemes, the signer may be blackmailed or bribed to confirm or disavow an alleged signature. This may be harder to accomplish with a designated confirmer organized as an authority with proper checks and balances.

Designated confirmer signatures are a useful tool to construct protocols for contract signing [1]. The trusted third party in contract signing takes the role of a designated confirmer. Each participant produces a designated confirmer signature of his statement and distributes it to all other participants and to the trusted third party. After the trusted third party has collected the statements and corresponding designated confirmer signatures from all participants, it converts them into ordinary digital signatures and circulates them to all participants according to a predefined policy. Designated confirmer signatures are also useful to construct *verifiable signature sharing* schemes [4].

A designated confirmer signature scheme has three operations: (i) An operation for generating double key pairs, one key pair of a private signing key with a public verifying key and another key pair of a private confirmer key with a public confirmer key, (ii) an operation for signing messages, and (iii) a *confirming operation* for proving signatures valid (confirmation) or invalid (disavowal). The private signing key is known only to the signer, the private confirmer key is known only to the confirmer, and the public verifying key as well as the public confirmer key are publicly accessible through authenticated channels, e.g., through a public key infrastructure (PKI). The signing operation is between a signer using the private signing key and a verifier using the public verifying key. The verifying operation is between the designated confirmer using its private confirmer key and a verifier using the public confirmer key. Furthermore, there is (iv) an *individual conversion operation* for converting individual designated confirmer signatures into ordinary digital signatures, and (v) a *universal verifying operation* to verify such converted signatures.

The characteristic security requirements of a designated confirmer signature scheme are similar to those of a convertible undeniable signature scheme [2]:

Unforgeability: Resistance against existential forgery under adaptive chosen message attacks by computationally restricted attackers.

Invisibility: A cheating verifier, given a signer's public verifying key, public confirmer key, a message, a designated confirmer signature and oracle access to the signer, cannot decide with probability better than pure guessing whether the signature is valid for the message with respect to the signer's verifying key or not. (This implies non-coercibility as described above.)

Soundness: A cheating designated confirmer cannot misuse the verifying operation in order to prove a valid signature to be invalid (non-repudiation), or an invalid signature to be valid (false claim of origin).

Non-transferability: A cheating verifier obtains no information from the confirming operation that allows him to convince a third party that the alleged signature is valid or invalid, regardless if the signature is valid or not.

Validity of Conversion: A cheating designated confirmer with oracle access to a signer cannot fabricate a converted signature valid for a message m with respect to the signer's public verifying key unless that signer has produced a designated confirmer signature for m before.

Practical constructions have been proposed by Chaum [3], Okamoto [6], Michels and Stadler [5], and by Camenisch and Michels [2]. All of them propose an *individual conversion operation*, but none of them discusses a *universal conversion operation* analogous to that of convertible undeniable signatures. Michels and Stadler [5] have discussed designated confirmer signatures that can be converted into well known ordinary signatures such as RSA digital signatures, Schnorr digital signatures, Fiat and Shamir signatures, or ElGamal digital signatures.

Designated confirmer signatures are a relatively young concept, which have not yet been blended with other interesting types of signature schemes such as threshold signatures, group signatures, or fail-stop signatures.

Gerrit Bleumer

References

[1] Asokan, N., Victor Shoup, and Michael Waidner (1998). "Optimistic fair exchange of digital signatures." *Advances in Cryptology—EUROCRYPT'98*, Lecture Notes in Computer Science, vol. 1403, ed. K. Nyberg. Springer-Verlag, Berlin, 591–606.

[2] Franklin, Matthew K. and Michael K. Reiter (1995). "Verifiable signature sharing." *Advances in Cryptology—EUROCRYPT'95*, Lecture Notes in Computer Science, vol. 921, eds. L.C. Guillou and J.-J. Quisquater. Springer-Verlag, Berlin, 50–63.

[3] Chaum, David (1995). "Designated confirmer signatures." *Advances in Cryptology—EUROCRYPT'94*, Lecture Notes in Computer Science, vol. 950, ed. A. De Santis. Springer-Verlag, Berlin, 86–91.

[4] Camenisch, Jan and Markus Michels (2000). "Confirmer signature schemes secure against adaptive adversaries." *Advances in Cryptography—*

[5] Michels, Markus and Markus Stadler (1998). "Generic constructions for secure and efficient confirmer signature schemes." *Advances in Cryptology—EUROCRYPT'98*, Lecture Notes in Computer Science, vol. 1403, ed. K. Nyberg. Springer-Verlag, Berlin, 406–421.

[6] Okamoto, Tatsuaki (1994). "Designated confirmer signatures and public-key encryption are equivalent." *Advances in Cryptology—CRYPTO'94*, Lecture Notes in Computer Science, vol. 839, ed. Y.G. Desmedt. Springer-Verlag, Berlin, 61–74.

[7] Pedersen, Torben Pryds (1991). "Distributed provers with applications to undeniable signatures (Extended abstract)." *Advances in Cryptology—EUROCRYPT'91*, Lecture Notes in Computer Science, vol. 547, ed. D.W. Davies. Springer-Verlag, Berlin, 221–242.

EUROCRYPT 2000, Lecture Notes in Computer Science, vol. 1807, ed. B. Preneel. Springer-Verlag, Berlin, 243–258.

DES-X (OR DESX)

DES-X is a 64-bit block cipher with a $2 \times 64 + 56 = 184$-bit key, which is a simple extension of DES (see Data Encryption Standard). The construction was suggested by Rivest in 1984 in order to overcome the problem of the short 56-bit key-size which made the cipher vulnerable to exhaustive key search attack. The idea is just to XOR a secret 64-bit key $K1$ to the input of DES and to XOR another 64-bit secret key $K2$ to the output of DES: $C = K2 \oplus DES_K(P \oplus K1)$. The keys $K1, K2$ are called *whitening keys* and are a popular element of modern cipher design. The construction itself goes back to the work of Shannon [6, p. 713], who suggested the use of a fixed mixing permutation whose input and output are masked by the secret keys. This construction has been shown to have provable security by Even–Mansour [2] if the underlying permutation is *pseudorandom* (i.e., computationally indistinguishable from a random permutation). A thorough study of DES-X was given in the work of Kilian–Rogaway [3], which builds on [2] and uses a blackbox model of security. Currently, the best attack on DES-X is a known-plaintext slide attack discovered by Biryukov–Wagner [1] which has complexity of $2^{32.5}$ known plaintexts and $2^{87.5}$ time of analysis. Moreover the attack is easily converted into a ciphertext-only attack with the same data complexity and 2^{95} offline time complexity. These attacks are mainly of theoretical interest due to their high time complexities. However, the attack is generic and would work for any cipher F used together with post- and pre-whitening with complexity $2^{(n+1)/2}$ known plaintexts and $2^{k+(n+1)/2}$ time steps (here n is the block size, and k is the key-size of the internal cipher F). A related key-attack on DES-X is given in [4]. Best conventional attack, which exploits the internal structure of DES, would be a <u>linear cryptanalysis attack</u>, using 2^{61} known plaintexts [3].

Alex Biryukov

References

[1] Biryukov, A. and D. Wagner (2000). "Advanced slide attacks." *Advances in Cryptology—EUROCRYPT 2000*, Lecture Notes in Computer Science, vol. 1807, ed. B. Preneel. Springer-Verlag, Berlin, 589–606.

[2] Even, S. and Y. Mansour (1997). "A construction of a cipher from a single pseudorandom permutation." *Journal of Cryptology*, 10 (3), 151–161. Springer-Verlag.

[3] Kaliski, B. and M. Robshaw (1996). "Multiple encryption: Weighing security and performance." *Dr. Dobb's Journal*, 243 (1), 123–127.

[4] Kelsey, J., B. Schneier, and D. Wagner (1997). "Related-key cryptanalysis of 3-WAY, Biham-DES, CAST, DES-X, NewDES, RC2, and TEA." *Proceedings of ICICS*, Lecture Notes in Computer Science, 1334, eds. Y. Han, T. Okamoto and S. Qing. Springer, Berlin, 233–246.

[5] Kilian, J. and P. Rogaway (1996). "How to protect against exhaustive key search." *Advances in Cryptology—CRYPTO'96*, Lecture Notes in Computer Science, vol. 1109, ed. N. Koblitz. Springer-Verlag, Berlin, 252–267.

[6] Shannon, C. (1949). "Communication theory of secrecy systems. A declassified report from 1945." *Bell Syst. Tech. J.* (28), 656–715.

DICTIONARY ATTACK (I)

Dictionary attack is an exhaustive <u>cryptanalysis</u> approach in which the attacker computes and stores a table of plaintext–ciphertext pairs $(P, C_i = E_{K_i}(P), K_i)$ sorted by the ciphertexts C_i. Here the plaintext P is chosen in advance among the most often encrypted texts like "login:", "Hello John", etc. and the key runs through all the possible keys K_i. If P is encrypted later by the user and the attacker observes its resulting ciphertext C_j, the attacker may search his table for the corresponding ciphertext and retrieve the secret <u>key</u> K_j.

The term <u>dictionary attack</u> is also used in the area of password guessing, but with a different meaning.

Alex Biryukov

DICTIONARY ATTACK (II)

A *dictionary attack* is a <u>password</u> [1] guessing technique in which the attacker attempts to determine a user's password by successively trying words from a *dictionary* (a compiled list of likely passwords) in the hope that one of these password guesses will be the user's actual password. In practice, the attacker's *dictionary* typically is not restricted to words from a traditional natural-language dictionary, but may include one or more of the following:

- variations on the user's first or last name, initials, account name, and other relevant personal information (such as address and telephone number, pet's name, and so on);
- words from various databases such as male and female names, places, cartoon characters, films, myths, and books;
- spelling variations and permutations of the above words, such as replacing the letter "o" with the number "0", using random capitalization, and so on;
- common word pairs.

Dictionary attacks can be quite successful in many environments because of the tendency of users to make poor password choices (unfortunately, passwords that are easily memorized by a legitimate user are also easily guessed by an attacker). These attacks can be performed in online mode (trying successive passwords until a login is successful) or offline mode (hashing or encrypting a dictionary of words and looking for any matches in a copied system file of hashed or encrypted user passwords). Server limits on the number of unsuccessful login attempts can help to thwart online attacks and the use of "salt" [see <u>salt</u>] can help to thwart offline attacks.

Carlisle Adams

References

[1] Schneier, B. (1996). *Applied Cryptography: Protocols, Algorithms, and Source Code in C* (2nd ed.). John Wiley & Sons, New York.

[2] Stallings, W. (1999). *Cryptography and Network Security: Principles and Practice* (2nd ed.). Prentice Hall.

DIFFERENTIAL CRYPTANALYSIS

Differential <u>cryptanalysis</u> is a general technique for the analysis of <u>symmetric cryptographic</u>

primitives, in particular of <u>block ciphers</u> and <u>hash functions</u>. It was first publicized in 1990 by Biham and Shamir [3, 4] with attacks against reduced-round variants of the <u>Data Encryption Standard (DES)</u> [14], and followed in 1991 by the first attack against DES which was faster than <u>exhaustive key search</u> [6].

Let P be a plaintext, and let C be the corresponding ciphertext encrypted under the (unknown) key K, such that $C = E_K(P)$. Let P^* be a second plaintext, and let C^* be the corresponding ciphertext under the same (unknown) key K, $C^* = E_K(P^*)$. We define the difference of the plaintexts as $P' = P \oplus P^*$, and the difference of the ciphertexts as $C' = C \oplus C^*$. Also for any intermediate data X during encryption (for example, the data after the third round, or the input to some operation in the fifth round), let the corresponding data during the encryption of P^* be denoted by X^*, and let the difference be $X' = X \oplus X^*$.

Differential cryptanalysis studies the differences, usually by means of exclusive-or (XOR), as they evolve in the various rounds and various operations of the cipher. Linear and affine operations do not affect the differences, or affect the differences in a predictable way: bit-permutation operations (that reorder the bits of the data X to $P(X)$) reorder the differences in the same way (i.e., to $P(X') = P(X) \oplus P(X^*)$); selections (that select some of the bits of the data) also select the bits of the differences; and XOR operations of two values $X \oplus Y$, also XOR the differences of the values to $X' \oplus Y' = (X \oplus Y) \oplus (X^* \oplus Y^*)$. An important observation is that mixing subkeys into the data may be discarded by means of differences: if the mixing of subkeys to the data is performed using an XOR operation by $Y = X \oplus K$, then in the second encryption it is $Y^* = X^* \oplus K$, and the output difference of the key mixing is $Y' = Y \oplus Y^* = (X \oplus K) \oplus (X^* \oplus K) = X'$, which is independent of the subkey. Key mixings may thus be ignored in the predictions of the differences.

For non-linear operations (such as S boxes) we can also study the evolvement of the differences. Certainly, when the difference of the input is 0, the two inputs are equal, and thus also the two outputs are equal, having a difference 0 as well. When the input difference is nonzero, we cannot predict the output difference, as it may have many different output differences for any input difference. However, it is possible to predict statistical information on the output difference given the input difference. Take for example S box S1 of DES. This S box has 6 input bits and 4 output bits. For each input difference X' there are 64 possible pairs of inputs with this difference (for any possible input X, the second input is computed by $X^* = X \oplus X'$). These 64 pairs may have various output differences. The main observation is that the output differences are not distributed uniformly. For example, for the input difference 34_x (the subscript x denotes that the number is in hexadecimal notation), no pair has output difference 0, nor 5 nor 6, and several other output differences; two pairs have output difference 4, eight pairs have output difference 1, and 16 of the 64 pairs with this input difference have output difference 2. For this input difference, a cryptanalyst can thus predict with probability 1/4 that the output difference is 2. A *difference distribution table* of an S box (or operation) is a table that lists the number of pairs which fulfill the input and output differences for each possible input and output differences, where the rows denote all the possible input differences, the columns all the output differences, and each entry contains the number of pairs with the corresponding differences. In the example above, the difference distribution table of S1 of DES has value 16 in row 34_x column 2.

Differential cryptanalysis defines characteristics that describe possible evolvements of the differences through the cipher. Each characteristic has a plaintext difference for which it predicts the differences in the following rounds. A pair of plaintexts for which the differences of the plaintexts and the intermediate data (when encrypted under the used key) are exactly as predicted by the characteristic are called *right pairs* (all other pairs are called *wrong pairs*). The probability that a characteristic succeeds to predict the differences (i.e., that a random pair is a right pair, given that the plaintext difference is as required by the characteristic) depends on the probabilities induced by the input and output differences for each S box (or each operation), where the total probability is the product of the probabilities of the various operations (assuming that the probabilities are independent, which is usually the case; otherwise the product is usually a good approximation for the probability).

Given the expected difference for the intermediate data before the last round (or more generally in some round near the end of the cipher), it may be possible to deduce the unknown key by a statistical analysis. The attack is a <u>chosen plaintext attack</u> that is performed in two phases: In the *data collection phase* the attacker requests encryption of a large number of pairs of plaintexts, where the differences of all the plaintext pairs are selected to have the plaintext difference of the characteristic. In the data analysis phase the attacker then recovers the key from the collected ciphertexts.

Assume that the probability of the characteristic is p (i.e., a fraction p of the pairs are expected to be right pairs). It is then expected that for a fraction p of the pairs, the difference of the data before the last round is as predicted by the characteristic. An (inefficient) method for deriving the subkey of the last round is then to try all the possibilities of the subkey of the last round. For each possible subkey partially decrypt all the ciphertexts by one round, and for each pair compute the differences of the data before the last round, by XORing the data resulting from the partial decryptions. For wrong guesses of the subkey it is expected that the difference predicted by the characteristic appears rarely, and for the correct value of the subkey it is expected that this difference appears for a fraction p or more of the pairs (as there is a fraction of about p of right pairs that are assured to suggest this difference, and as wrong pairs may also suggest this difference). In particular, if the probability p is not too low, it is expected that the correct subkey is the one which gives the expected difference most frequently. It should be noted that the derivation of the last subkey is usually much more efficient than (but equivalent in results to) this described algorithm, using the information of the input and output differences for each S box (or operation) in the last round. It should also be noted that in many cases characteristics shorter by more than one round than the cipher (usually up to three rounds shorter) can also be used for differential attacks.

Differential cryptanalysis usually requires a small multiple of $1/p$ pairs of chosen plaintexts, when using a characteristic with probability p, in order to ensure that sufficiently many right pairs appear in the data. This amount of encrypted data may be very large (about 2^{47} chosen plaintext blocks in the case of DES), making the complexity of the data collection phase larger than the complexity of the data analysis phase in most cases. The large number of chosen plaintexts may by itself make the attack impractical, as it transfers the responsibility of computing the major part of the attack from the attacker to the attacked party, who is required to encrypt a large number of chosen plaintexts for the attacker to be able to mount his attack. It is therefore common in such cases to quote the complexity of a differential attack to be the number of required chosen plaintexts.

After the publication of the differential cryptanalysis attack on DES, whose complexity is 2^{47} (it requires 2^{47} chosen plaintexts and the time of analysis is less than 2^{40}), IBM announced that they were aware of differential cryptanalysis when they designed DES, and actually designed it to withstand differential attacks. Moreover, differential attacks (to which they called the *T method*) were classified as top secret for purposes of US national security, and IBM were requested by the NSA not to publish any information on them.

There are various improvements of differential cryptanalysis aimed to reduce the complexity of differential attacks. One simple method is a combination of several characteristics in a single larger structure. In case two characteristics are used, such a structure is called a *quartet*. It contains four plaintexts of the form P, $P \oplus \Omega_p^1$, $P \oplus \Omega_p^2$, $P \oplus \Omega_p^1 \oplus \Omega_p^2$, for the plaintexts differences Ω_p^1 and Ω_p^2. It can easily be seen that in such a quartet each difference appears twice: the first difference appears as the difference of the first two plaintexts, and also as the difference of the other two plaintexts; the second difference appears as the difference of the first and third plaintexts, and also as the difference of the second and fourth plaintexts. Thus, a total of four pairs are contained in a quartet; without using quartets only two pairs are contained in the same number of plaintexts. Larger structures of eight plaintexts using three different characteristics contain 12 pairs. Such structures are useful when there are several high-probability characteristics that can be used for an attack, as if the second best characteristic has a relatively low probability, the benefit of getting pairs with such a difference is quite low.

Another improvement (which was also mentioned in the original publication on differential cryptanalysis) is using an extended form of differences, in which not all the bits of the difference are fixed. This type of differences was later called <u>truncated differences</u> [10]. An important type of truncated differences (in most cases truncated differences refer to this type) is the word-wise truncated differences. Word-wise truncated differences are differences in which the difference itself is not considered, but instead the differences are divided into two classes, namely zero differences and non-zero differences. In these cases the data blocks are divided to words (either 8-bit bytes, or 16-bit words, or words of a different size depending on the native structure of the cipher), and the analysis only considers whether the difference of a word is expected to be zero or not. Such consideration is useful when non-zero differences evolve to other (unknown in advance) non-zero differences, so that the information on the zero/non-zero difference evolve through many rounds of the cipher.

A third extension defines non-XOR differences, such as subtraction of integers (useful for cases where the native operation in the cipher is addition), or differences of division modulo a prime (useful for cases where the native operation in the cipher is multiplication modulo a prime, such as in IDEA [11]). Also a combination of different differences for different parts of the block, or for different rounds of the cipher is considered. For such cases, difference distribution tables where the input differences are defined with one operation and the output differences with another, are very useful (especially when the operation natively transforms one operation to another, such as in cases of exponentiation S boxes, or logarithm S boxes).

Higher-order differences [12] consider derivatives of a second or a higher order. Higher-order differences are shown successful in several cases where differential cryptanalysis is not applicable due to low probabilities of characteristics; in some of these cases higher-order differences prove the most successful attack. However, higher-order attacks are successful mainly against ciphers with a small number of rounds.

It was also observed that in most differential attacks, the intermediate differences predicted by the characteristics are not used, and thus can be ignored [11]. In such cases, the considered differences are only the plaintext differences and the difference after the final round of the characteristic. In most cases there are many different characteristics with the same plaintext difference and the same final difference; these characteristics sum up to one differential, whose probability is the sum of their probabilities.

The major method for protection against differential cryptanalysis is by bounding the probability of the best characteristic (or differential) to be very low. Whenever the designer wishes to prove that differential cryptanalysis is not applicable, he bounds the probability p of the best characteristic (or differential) such that $1/p$ is larger than the required complexity, or even larger than the size of the plaintext space (in which case even choosing the whole plaintext space is not sufficient for mounting an attack). These bounds were formalized into various theories of provable security against differential cryptanalysis.

A specially interesting theory for provable security against differential cryptanalysis (and also linear cryptanalysis) is the theory of decorrelation [16], which makes it possible to prove security of block ciphers against certain (restricted) kinds of attacks, including basic variants of differential and linear cryptanalysis.

Although the usual claims for security against differential cryptanalysis say that the probabilities of the highest-probability differentials are very low, and thus differential attacks require a huge amount of data and complexity, it was observed that even differentials with probability zero (i.e., that cannot occur—there are no right pairs under any key) can be used for attacks [1,9]. This kind of attacks is called *differential cryptanalysis using impossible differentials* (or shortly *impossible cryptanalysis*). The main idea is to select a large set of pairs with the plaintext difference of an impossible differential with $n-1$ (or slightly less) rounds, where n is the number of rounds of the block cipher, and to try all the possible subkeys of the extra round(s). If it appears that for some value of the subkey, decryption of the ciphertexts by one round (or the few rounds) leads to the impossible difference in any one of the pairs, then we are assured that the subkey is wrong, and thus can be discarded. After discarding sufficiently many subkeys, the attacker reduces his list of possible values of the subkeys to a short list (or even to one subkey), and he is assured that the correct subkey is in the list. Depending on the design of the cipher and the key schedule, for some ciphers it would be more efficient to try reducing the number of possible subkeys to 1 (i.e., only the correct subkey), while for others it would be more efficient to reduce the size of the short list to some larger size, and then perform an exhaustive search of the remaining possible keys.

There are also attacks that use differentials as their building blocks, while combining differentials in various ways. The most promising ones are boomerang [17], amplified boomerang [8], and rectangle [2] attacks. The main idea in all these attacks are the combination of four plaintexts, which for simplicity of description we assume are located on the corners of a square, where one short differential is used in both pairs for the first few rounds (the horizontal edges), while a second short differential is used for the rest of the rounds but on the orthogonal pairs (the vertical edges). Although the probabilities of the total structure are p^2q^2 where p and q are the probabilities of the two differentials, it appears that it is much easier in various cases to find good short differentials, than to find one full differential of a comparable probability.

Although differential cryptanalysis is basically a chosen plaintext attack (as the attacker needs to choose the plaintext differences), the attacker usually does not need to choose the exact values of the plaintexts. This observation allows conversion of chosen plaintext differential cryptanalysis attacks into known plaintext attacks [3], using the fact that in a sufficiently large set of random plaintexts there are many pairs whose difference is as required by the chosen plaintext attack. Once these pairs of plaintexts are identified, the original chosen plaintext attack may be performed on these pairs. This variant usually requires a huge number of known plaintexts, which is about $\sqrt{m}2^{n+1}$ where n is the size of the plaintext in bits and m is the number of chosen plaintext pairs required by the chosen plaintext attack. On some ciphers this is the best published known-plaintext attack.

In some cases it is also possible to convert differential cryptanalysis to ciphertext-only attacks. For more information on these conversions see [7].

Differential cryptanalysis was originally developed on FEAL-8 [13,15], a block cipher which was claimed to be faster and more secure than DES. It was then generalized and extended to DES and other schemes. Feal-8 was broken using a few hundred chosen plaintexts. Given the corresponding ciphertexts, it takes less than a minute on a personal computer to recover the key [5]. The first results on DES [4] showed that DES reduced to 15 rounds was vulnerable to a differential attack, while the full 16-round DES required 2^{58} chosen plaintexts for a successful attack, whose generation is slower than exhaustive search. In the following year an improvement of the technique was invented [6]. The main trick in the improved attack was the ability to receive the first round for free, using large specially designed structures, setting the characteristic from the second round on. This improvement made it possible to apply the 15-round attack on the full 16 rounds. Another improvement allowed to find the key when the first right pair is analyzed, rather than to wait till sufficiently many right pairs are found. This improvement is applicable when the attack considers all the key bits (or almost all the key bits) in a single counting phase. As a result, the improved attack could analyze the full 16-round DES given 2^{47} chosen plaintext and their corresponding ciphertexts, whose complexity of analysis was smaller than 2^{40}.

Eli Biham

References

[1] Biham, Eli, Alex Biryukov, and Adi Shamir (1999). "Cryptanalysis of skipjack reduced to 31 rounds using impossible differentials." *Advances in Cryptology—EUROCRYPT'99*, Lecture Notes in Computer Science, vol. 1592, ed. J. Stern. Springer, Berlin, 12–23.

[2] Biham, Eli, Orr Dunkelman, and Nathan Keller (2002). "New results on boomerang and rectangle attacks." *Proceedings of Fast Software Encryption*, Leuven, Lecture Notes in Computer Science, vol. 2365, eds. Daemen, J. and V. Rijmen. Springer, Berlin, 1–16.

[3] Biham, Eli and Adi Shamir (1993). *Differential Cryptanalysis of the Data Encryption Standard.* Springer-Verlag, Berlin, New York.

[4] Biham, Eli and Adi Shamir (1991). "Differential cryptanalysis of DES-like cryptosystems." *Journal of Cryptology*, 4 (1), 3–72.

[5] Biham, Eli and Adi Shamir (1991). "Differential cryptanalysis of FEAL and N-hash." Technical report CS91-17, Department of Applied Mathematics and Computer Science, The Weizmann Institute of Science, *Advances in Cryptology—EUROCRYPT'91*. The extended abstract appears in Lecture Notes in Computer Science, vol. 547, ed. D.W. Davies. Springer, Berlin, 1–16.

[6] Biham, Eli and Adi Shamir (1992). "Differential cryptanalysis of the full 16-round DES." *Advances in Cryptology—CRYPTO'92*, Lecture Notes in Computer Science, vol. 740, ed. E.F Brickel. Springer, Berlin, 487–496.

[7] Biryukov, Alex and Eyal Kushilevitz (1998). "From differential cryptanalysis to ciphertext-only attacks." *Advances in Cryptology—CRYPTO'98*, Lecture Notes in Computer Science, vol. 1462, ed. H. Krawczyk. Springer, Berlin, 72–88.

[8] Kelsey, John, Tadayoshi Kohno, and Bruce Schneier (2000). "Amplified boomerang attacks against reduced-round MARS and serpent." *Proceedings of Fast Software Encryption 7*, Lecture Notes in Computer Science, vol. 1978, ed. B. Schneier. Springer-Verlag, Berlin, 75–93.

[9] Knudsen, Lars Ramkilde (1998). "DEAL—a 128-bit block cipher." AES submission, available on http://www.ii.uib.no/~larsr/papers/deal.ps

[10] Knudsen, Lars (1995). "Truncated and higher order differentials." *Proceedings of Fast Software Encryption 2*, Lecture Notes in Computer Science, vol. 1008, ed. B. Preneel. Springer-Verlag, Berlin, 196–211.

[11] Lai, Xuejia, James L. Massey, and Sean Murphy (1991). "Markov ciphers and differential cryptanalysis." *Advances in Cryptology, Proceedings of EUROCRYPT'91*, Lecture Notes in Computer Science, vol. 547, ed. D.W. Davies. Springer, Berlin, 17–38.

[12] Lai, Xuejia (1994). "Higher order derivative and differential cryptanalysis." *Proceedings of Symposium on Communication, Coding and Cryptography*, in honor of J.L. Massey on the occasion of his 60th birthday.

[13] Miyaguchi, Shoji, Akira Shiraishi, and Akihiro Shimizu (1988). "Fast data encryption algorithm FEAL-8." *Review of Electrical Communications Laboratories*, 36 (4), 433–437.

[14] National Bureau of Standards (1977), Data Encryption Standard, U.S. Department of Commerce, FIPS pub. 46.

[15] Shimizu, Akihiro and Shoji Miyaguchi (1987). "Fast data encryption algorithm FEAL." *Advances in Cryptology—EUROCRYPT'87*, Lecture Notes in Computer Science, vol. 304, eds. David Chaum and Wyn L. Price. Springer, Berlin, 267–278.

[16] Vaudenay, Serge (1998). "Provable security for block ciphers by decorrelation." *Proceedings of STACS'98*, Lecture Notes in Computer Science, vol. 1373, eds. M. Morvan, C. Meinel, and D. Krob. Springer, Berlin, 249–275.

[17] Wagner, David (1999). "The boomerang attack." *Proceedings of Fast Software Encryption, FSE'99, Rome*, Lecture Notes in Computer Science, vol. 1636, ed. L. Knudsen. Springer, Berlin, 156–170.

DIFFERENTIAL–LINEAR ATTACK

Differential–Linear attack is a chosen plaintext *two-stage* technique of cryptanalysis (by analogy with two-stage rocket technology) in which the first stage is covered by differential cryptanalysis, which ensures propagation of useful properties midway through the block cipher. The second stage is then performed from the middle of the cipher and to the ciphertext using linear cryptanalysis. The technique was discovered and demonstrated on the example of 8-round *DES* (see Data Encryption Standard) by Langford and Hellman [4]. Given a *differential characteristic* with probability p for the rounds $1, \ldots, i$ and the *linear characteristic* with bias q for the rounds $i + 1, \ldots, R$, the bias of resulting linear approximation would be $1/2 + 2pq^2$ and the data complexity of the attack will be $O(p^{-2}q^{-4})$ [3, p. 65]. Thus the attack would be useful only in special cases when there are good characteristics or linear approximations half-way through the cipher, but no good patterns for the full cipher. Their attack enhanced with such refinements as packing data into *structures* and *key-ranking* (or *list decoding*) can recover 10-bits of the secret key for 8-round DES using 512 chosen plaintexts. In [1] the same technique is used to break 8-round FEAL with 12 chosen plaintexts and expensive analysis phase. Further applications and refinements of the technique are given in [2].

Alex Biryukov

References

[1] Aoki, K. and K. Ohta (1996). "Differential-linear cryptanalysis of FEAL-8." *IEICE Trans. on Fundamentals of Electronics, Communications and Computer Sciences*, E79A (1), 20–27.

[2] Eli Biham, Orr Dunkelman, and Nathan Keller, "Enhancing Differential-Linear Cryptanalysis", *Advances in Cryptology ASIACRYPT 2002*, Lecture Notes in Computer Science, vol. 2501, ed. Y. Zheng. Springer-Verlag, Berlin, p. 254–266.

[3] Langford, S.K. (1995). "Differential-linear cryptanalysis and threshold signatures." Technical report, *PhD Thesis*, Stanford University.

[4] Langford, S.K. and M.E. Hellman (1994). "Differential–linear cryptanalysis." *Advances in Cryptology—CRYPTO'94*. Lecture Notes in Computer Science, vol. 839, ed. Y. Desmedt. Springer-Verlag, Berlin, 17–25.

DIFFERENTIAL POWER ANALYSIS

Differential Power Analysis utilizes power consumption of a cryptographic device such as a smartcard as side-channel information. In *Simple Power Analysis* (SPA) an attacker directly observes a device's power consumption. It is known that the amount of power consumed by the device varies depending on the data operated on and the instructions performed during different parts of an algorithm's execution. Define a *power trace* as a set of power consumption measurements during a cryptographic operation. By simply examining power traces, it is possible to determine major characteristic details of a cryptographic device and the implementation of the cryptographic algorithm being used. SPA can therefore be used to discover implementation details, such as *DES* rounds (see Data Encryption Standard) and *RSA* operations (see RSA public key encryption). Moreover, SPA can reveal differences between multiplication and squaring operations, which can be used to recover the private key in RSA implementations. SPA can also reveal visible differences within permutations and shifts in DES implementations, which might lead to recovering the secret DES key.

While SPA attacks use primarily visual inspection to identify relevant power fluctuations, Differential Power Analysis (DPA) exploits characteristic behavior (e.g., power consumption behavior of transistors and logic gates) [2]. DPA uses an attacking model and statistical analysis to extract hidden information from a large sample of power traces obtained during "controlled" cryptographic computations. In case of SPA, direct observations of a device's power consumption would not allow identifying the effects of a single transistor switching. The use of statistical methods in a controlled DPA environment allows identifying small differences in power consumption, which can be used to recover specific information such as the individual bits in a secret key. This means secret key material can be recovered from tamper-resistant devices such as smartcards (smartcard tamper resistance). To execute an attack based on DPA, an attacker does not need to know as many details about how the algorithm is implemented.

The basis of a DPA attack is the use of an abstract model based on the power consumption characteristics of the logic that includes the noise components. When measuring the power consumption, various noise components are superimposed on the power traces. The main noise sources are external, intrinsic, quantization and algorithmic noise. Intrinsic and quantization noise are small compared to the power consumption. The external noise can be reduced by careful use of the measurement equipment. The algorithmic noise can be averaged out by the DPA strategy itself. To reduce the influence of noise in DPA one can increase the number of samples required to detect variations. Analysis can take place in the time and frequency domain.

The basis DPA technique is as follows. Assume that a sufficient number N of random power traces have been collected (e.g., N samples of ciphertexts obtained using the same encryption key). Each power trace is a collection of power samples $PS(n, t)$, which represent the power consumption at time t in trace n as the sum of the power dissipated by all circuitry. In practice, the number of measurements t in each power trace depends on the sampling rate and the memory capacity as well as the duration of the cryptographic operation. Next, partition the power samples $PS(n, t)$ into two sets S_0 and S_1 according to the outcome 0 or 1 of a partitioning or discrimination function D. The outcome value of the partitioning function D can be simply the value of a specific ciphertext bit. In general, the size of set S_0 will be roughly the same as the size of S_1. Next, compute the average power signal for each set S at time t. By subtracting the two averages, we obtain the DPA bias signal $B(t)$. Selecting an appropriate D-function will result in a DPA bias signal that an attacker can use to verify guesses of the secret key. The D-function is chosen such that at some point during implementation the device needs to calculate the value of this bit. When this occurs or any time data containing this bit is manipulated, there will be a slight difference in the amount of power dissipated depending on whether this bit is a zero or a one. Let ϵ denote this difference, and the instruction manipulating the D-bit occurs at time t', then the value ϵ is equal to the expectation difference

$$E[S \mid (D = 0)] - E[S \mid (D = 1)], \qquad \text{for } t = t'.$$

When $t \neq t'$ the device is manipulating bits other than the D-bit, and assuming that the power dissipation is independent of the D-bit, the difference in expectation of the two sets equals zero for sufficiently large N. Thus the bias function $B(t)$ will show power spikes of height ϵ at times t' and will appear flat at all other times. If the proper D-function was chosen, the bias signal will show spikes whenever the D-bit was manipulated and otherwise the resulting $B(t)$ will not show any bias. Using this approach an attacker can verify guesses for the hidden key bit information using the D-function. Repeating this approach for different D-bits, the secret key can be obtained bit by bit.

Variants or improvements of the classical DPA attack exist that use signals from multiple sources, use different measuring techniques, combine signals with different temporal offsets, use specific and more powerful differential functions, and apply more advanced signal processing functions and models. To enlarge the peak, a multiple-bit attack can be used.

A DPA attack involves hundreds to thousands of samples. After processing and statistical analysis, the DPA process can reconstruct the full secret or private key within several minutes. The whole process is easy to implement and requires only standard measurement equipment, which cost lies between a few hundred to a few thousand dollars. DPA attacks are non-invasive, which makes them difficult to detect. DPA requires little or no information about the target device and can be automated. DPA and SPA has successfully been applied to attack a large number of smartcards and PCMCIA cards [3]. See [1] for an approach how to counteract Power Analysis attacks.

Tom Caddy

References

[1] Chari, Suresh, Charanjit Jutla, Josyula R. Rao, and Pankaj Rohatgi (1999). "Towards sound approaches to counteracts power-analysis attacks." *Advances in Cryptology—CRYPTO'99*, Lecture Notes in Computer Science, vol. 1666, ed. M. Wiener. Springer-Verlag, Berlin, 389–412.

[2] Kocher, Paul, Joshua Jaffe, and Benjamin Jun (1999). "Differential power analysis." *Advances in Cryptology—CRYPTO'99*, Lecture Notes in Computer Science, vol. 1666, ed. M. Wiener. Springer-Verlag, Berlin, 388–397.

[3] Messerges, Thomas S., Ezzy A. Dabbish, and Robert H. Sloan (1999). "Investigations of power analysis attacks on smartcards." *Proceedings of USENIX Workshop on Smartcard Technology*, 151–161.

DIFFIE–HELLMAN KEY AGREEMENT

The Diffie–Hellman protocol is a type of key agreement protocol. It was originally described in Diffie and Hellman's seminal paper on public key cryptography.

This key agreement protocol allows Alice and Bob to exchange public key values, and from these values and knowledge of their own corresponding private keys, securely compute a shared key K, allowing for further secure communication. Knowing only the exchanged public key values, an eavesdropper is not able to compute the shared key.

As a preamble to the protocol, the following public parameters are assumed to exist (see Number Theory): a large prime number p such that discrete logarithms in the multiplicative group of integers from 1 to $p-1$ (Z_p^*) are intractable; and a generator g of Z_p^*. Alice randomly selects a value $0 < a < p-1$ and computes $r = g^a \bmod p$. Alice sends r to Bob. Similarly, Bob selects a value $0 < b < p-1$ and computes $s = g^b \bmod p$. Bob sends s to Alice. Given a and s, Alice computes $K = s^a \bmod p \equiv g^{ab} \pmod{p}$. Similarly, given b and r, Bob computes $K = r^b \bmod p \equiv g^{ab} \pmod{p}$. Thus, Alice and Bob are able to compute the same key value, K.

Now consider the information available to an eavesdropper. This includes g, p, r and s. Thus, the eavesdropper must attempt to compute $K \equiv g^{ab} \pmod{p}$ given $g^a \bmod p$ and $g^b \bmod p$. This is known as the decisional Diffie–Hellman problem and for appropriately chosen g and p, it is believed to be very difficult to solve.

Several variations to this simple protocol exist (see Key Agreement). Of particular note is the fact that the above protocol does not provide for the authentication of Alice and Bob. The Station-to-Station protocol provides one variation to this protocol that authenticates Alice and Bob.

Mike Just

References

[1] Menezes, Alfred, Paul van Oorschot, and Scott Vanstone (1997). *Handbook of Applied Cryptography*. CRC Press, Boca Raton.

[2] Stinson, Douglas R. (1995). *Cryptography: Theory and Practice*. CRC Press, Boca Raton.

DIFFIE–HELLMAN PROBLEM

In their pioneering paper Diffie and Hellman [15] proposed an elegant, reliable, and efficient way to establish a common key between two communicating parties. In the most general settings their idea can be described as follows (see Diffie–Hellman key agreement for further discussion). Given a cyclic group $\mathcal{G}$ and a generator g of $\mathcal{G}$, two communicating parties Alice and Bob execute the following protocol:

- Alice selects *secret x*, Bob selects *secret y*;
- Alice publishes $X = g^x$, Bob publishes $Y = g^y$;
- Alice computes $K = Y^x$, Bob computes $K = X^y$.

Thus at the end of the protocol the values $X = g^x$ and $Y = g^y$ have become *public*, while the value $K = Y^x = X^y = g^{xy}$ supposedly remains *private* and is known as the *Diffie–Hellman secret key*.

Thus the *Diffie–Hellman Problem*, DHP, with respect to the group $\mathcal{G}$ is to compute g^{xy} from the given values of g^x and g^y.

Certainly, only groups in which DHP is hard are of cryptographic interest. For example, if $\mathcal{G}$ is an *additive* group of the residue ring $\mathbb{Z}_m$ modulo m, see modular arithmetic, then DHP is trivial: using additive notations the attacker simply computes $x \equiv X/g \pmod{m}$ (because g is a generator of the additive group of $\mathbb{Z}_m$, we have $\gcd(g, m) = 1$) and then $K \equiv xY \pmod{m}$.

On the other hand, it is widely believed that using *multiplicative* subgroups of the group of units $\mathbb{Z}_m^*$ of the residue ring $\mathbb{Z}_m$ modulo m yields examples of groups for which DHP is hard, provided that the modulus m is carefully chosen. This belief also extends to subgroups of the multiplicative group $\mathbb{F}_q^*$ of a *finite field* $\mathbb{F}_q$ of q elements. In fact these groups are exactly the groups suggested by Diffie and Hellman [15]. Although, since that time the requirements on the suitable groups have been refined and better understood, unfortunately not too many other examples of "reliable" groups have been found. Probably the most intriguing and promising example, practically and theoretically, is given by subgroups of point groups on elliptic curves, which have been proposed for this kind of application by Koblitz [24] and Miller [36]. Since the original proposal, many very important theoretical and practical issues related to using elliptic curves in cryptography have been investigated, see [2, 17]. Even more surprisingly, elliptic curves have led to certain variants of the Diffie–Hellman schemes, which are not available in subgroups of $\mathbb{F}_q^*$ or $\mathbb{Z}_m^*$, see [5, 22, 23] and references therein.

DIFFIE–HELLMAN AND DISCRETE LOGA-RITHM PROBLEMS: It is immediate that if one can find x from the given value of $X = g^x$, that is, solve the underline{discrete logarithm problem}, DLP, then the whole scheme is broken. In fact, in our example of a "weak" group $\mathcal{G}$, this is exactly DLP which can easily be solved. Thus DHP is not harder than DLP. On the other hand, the only known (theoretical and practical) way to solve DHP is to solve the associated DLP. Thus a natural question arises whether DHP is equivalent to DLP or is strictly weaker. The answer can certainly depend on the specific group $\mathcal{G}$.

Despite a widespread assumption that this indeed is the case, that is, that in any cryptographically "interesting" group DHP and DLP are equivalent, very few theoretical results are known. In particular, it has been demonstrated in [6, 31, 32] that, under certain conditions, DHP and DLP are underline{polynomial time} equivalent. However, there are no unconditional results known in this direction.

Some quantitative relations between complexities of DHP and DLP are considered in [13].

CRYPTOGRAPHICALLY INTERESTING GROUPS: As we have mentioned, the choice of the group $\mathcal{G}$ is crucial for the hardness of DHP (while the choice of the generator g does not seem to be important at all). Probably the most immediate choice is $\mathcal{G} = \mathbb{F}_q^*$, thus g is a underline{primitive element} of $\mathbb{F}_q$. However, one can work in a subgroup of $\mathbb{F}_q^*$ of sufficiently large prime order ℓ (but still much smaller than q and thus more efficient) without sacrificing the security of the protocol. Indeed, we recall that based on our current knowledge we may conclude that the hardness of DLP in a subgroup $\mathcal{G} \subseteq \mathbb{F}_q^*$ (at least for some most commonly used types of fields; for further discussion see underline{discrete logarithm problem}) is majorised

1. by $\ell^{1/2}$ where ℓ is the largest prime divisor of #$\mathcal{G}$, see [35, 44];
2. by $L_q[1/2, 2^{1/2}]$ for a rigorous unconditional algorithm, see [37];
3. by $L_q[1/3, (64/9)^{1/3}]$ for the heuristic number field sieve algorithm, see [39, 40],

where as usual we denote by $L_x[t, \gamma]$ (see underline{L-notation}) any quantity of the form

$$L_x[t, \gamma] = \exp((\gamma + o(1))(\log x)^t (\log \log x)^{1-t}).$$

It has also been discovered that some special subgroups of some special underline{extension fields} are computationally more efficient and also allow one to reduce the information exchange without sacrificing the security of the protocol. The two most

practically and theoretically important examples are given by LUC, see [3, 43], and XTR, see [26–28], protocols (see, more generally, underline{subgroup cryptosystems}). Despite several substantial achievements in this area, these results are still to be better understood and put in a more systematic form [10].

One can also consider subgroups of the residue ring $\mathbb{Z}_m^*$ modulo a composite $m \geq 1$. Although they do not seem to give any practical advantages (at least in the original setting of the two party key exchange protocol), there are some theoretical results supporting this choice, for example, see [1].

The situation is more complicated with subgroups of the point groups of elliptic curves, and more generally of abelian varieties. For these groups not only the arithmetic structure of the cardinality $\mathcal{G}$ matters, but many other factors also play an essential role, see [2, 17, 19, 20, 25, 34, 38] and references therein.

BIT SECURITY OF THE DIFFIE–HELLMAN SECRET KEY: So far, while there are several examples of groups in which DHP (like DLP) is conjectured to be hard, as with other areas of cryptography, the security relies on unproven assumptions. Nevertheless, after decades of exploration, we have gained a reasonably high level of confidence in some groups, for example, in subgroups of $\mathbb{F}_p^*$. Of course, this assumes that p and #$\mathcal{G}$ are sufficiently large to thwart the discrete logarithm attack. Typically, nowadays, p is at least about 1024 bits, #$\mathcal{G}$ is at least about 160 bits. However, after the common key $K = g^{xy}$ is established, only a small portion of bits of K will be used as a common key for some pre-agreed underline{symmetric cryptosystem}.

Thus, a natural question arises: *Assume that finding all of K is infeasible, is it necessarilly infeasible to find certain bits of K?*

In practice, one often derives the secret key from K via a underline{hash function} but this requires an additional function, which generally must be modeled as a black box. Moreover, this approach requires a hash function satisfying some additional requirements which could be hard to prove unconditionally. Thus the security of the the obtained private key relies on the hardness of DHP and some assumptions about the hash function. Bit security results allow us to eliminate the usage of hash functions and thus to avoid the need to make any additional assumptions.

For $\mathcal{G} = \mathbb{F}_p^*$, Boneh and Venkatesan [8] have found a very elegant way, *using lattice basis reduction* (see underline{lattices}), to solve this question in the affirmative, see also [9]. Their result has been slightly improved and also extended to other groups in [21]. For the XTR version of DHP it has recently been done in [30]. The results of these papers can be summarized as follows: "error-free" recovery of even some small portion of information about the Diffie–Hellman secret key $K = g^{xy}$ is as hard as recovering the whole key (cf. underlined hard-core bit). Including the case where the recovering algorithm works with only some non-negligible positive probability of success is an extremely important open question. This would immediately imply that hashing K does not increase the security of the secret key over simply using a short substring of bits of K for the same purpose, at least in an asymptotic sense.

It is important to remark that these results do not assert that the knowledge of just a few bits of K for *particular* (g^x, g^y) translates into the knowledge of all the bits. Rather the statement is that given an efficient algorithm to determine certain bits of the common key corresponding to *arbitrary* g^x and g^y, one can determine all of the common key corresponding to *particular* g^x and g^y.

Another, somewhat dual problem involving some partial information about K is studied in [41]. It is shown in [41] that any polynomial time algorithm which for given x and y produces a list $\mathcal{L}$ of polynomially many elements of $\#\mathcal{G}$ where $K = g^{xy} \in \mathcal{L}$, can be used to design a polynomial time algorithm which finds K unambiguously.

NUMBER THEORETIC AND ALGEBRAIC PROPERTIES: As we have mentioned, getting rigorous results about the hardness of DHP is probably infeasible nowadays. One can however study some number theoretic and algebraic properties of the map $K : \mathcal{G} \times \mathcal{G} \to \mathcal{G}$ given by $K(g^x, g^y) = g^{xy}$. This is of independent intrinsic interest and may also shed some light on other properties of this map which are of direct cryptographic interest.

For example, many cryptographic protocols are based on the assumption of hardness of the decisional Diffie–Hellman problem, DDHP, rather than DHP itself. Roughly speaking, DDHP is the problem of deciding whether a triple $(u, v, w) \in \mathcal{G}^3$ of random elements of $\mathcal{G}$ is of the form (g^x, g^y, g^{xy}) for some x and y. Clearly, DDHP is no harder than DHP, and it is believed that in fact it is no easier, see [4]. Unfortunately there are no viable approaches to a proof of this conjecture. Motivated by this problem, in the series of works [11,12,18] several "statistical" results have been established, which show that if $\mathcal{G}$ is a sufficiently large subgroup of $\mathbb{F}_p^*$ then at least statistically the triples (g^x, g^y, g^{xy}) behave as triples of random elements.

One can also study algebraic properties of the set of points (g^x, g^y, g^{xy}) or even just (g^x, g^{x^2}) (which corresponds to the "diagonal" case $x = y$). In particular one can ask about the degree of polynomials F for which $F(g^x, g^y, g^{xy}) = 0$ or $F(g^x, g^y) = g^{xy}$ or $F(g^x, g^{x^2}) = 0$ or $F(g^x) = g^{x^2}$ for all or "many" $x, y \in \mathcal{G}$. Certainly it is intuitively obvious that such polynomials should be of very large degree and have a complicated structure. It is useful to recall the underlined interpolation attack on underlined block ciphers which is based on finding polynomial relations of similar spirit. It has been shown in [14] (as one would certainly expect) that such polynomials are of exponentially large degree, see also [42]. Several more results of this type can also be found in [16,33,45].

Igor E. Shparlinski

References

[1] Biham, E., D. Boneh, and O. Reingold (1999). "Breaking generalized Diffie–Hellman modulo a composite is no easier than factoring." *Inform. Proc. Letters*, 70, 83–87.

[2] Blake, I., G. Seroussi, and N. Smart (1999). *Elliptic Curves in Cryptography*. London Mathematical Society, Lecture Notes Series, vol. 265. Cambridge University Press, Cambridge.

[3] Bleichenbacher, D., W. Bosma, and A.K. Lenstra (1995). "Some remarks on Lucas-based cryptosystems." *Advances in Cryptology—CRYPTO'95*, Lecture Notes in Computer Science, vol. 963, ed. D. Coppersmith. Springer-Verlag, Berlin, 386–396.

[4] Boneh, D. (1998). "The decision Diffie–Hellman problem." *Proceedings of ANTS-III*, Lecture Notes in Computer Science, vol. 1423, ed. J.P. Buhler. Springer-Verlag, Berlin, 48–63.

[5] Boneh, D. and M. Franklin (2001). "Identity-based encryption from the Weil pairing." *Advances in Cryptology—CRYPTO 2001*, Lecture Notes in Computer Science, vol. 2139, ed. J. Kilian. Springer-Verlag, Berlin, 213–229.

[6] Boneh, D. and R. Lipton (1996). "Algorithms for black-box fields and their applications to cryptography." *Advances in Cryptology—CRYPTO'96*, Lecture Notes in Computer Science, vol. 1109, ed. N. Koblitz. Springer-Verlag, Berlin, 283–297.

[7] Boneh, D. and I.E. Shparlinski (2001). "On the unpredictability of bits of the elliptic curve Diffie–Hellman scheme." *Advances in Cryptology—CRYPTO 2001*, Lecture Notes in Computer Science, vol. 2139, ed. J. Kilian. Springer-Verlag, Berlin, 201–212.

[8] Boneh, D. and R. Venkatesan (1996). "Hardness of computing the most significant bits of secret keys in Diffie–Hellman and related schemes." *Advances in Cryptology—CRYPTO'96*, Lecture Notes

in Computer Science, vol. 1109, ed. N. Koblitz. Springer-Verlag, Berlin, 129–142.

[9] Boneh, D. and R. Venkatesan (1997). "Rounding in lattices and its cryptographic applications." *Proc. 8th Annual ACM-SIAM Symp. on Discr. Algorithms*. ACM, 675–681.

[10] Bosma, W., J. Hutton, and E. Verheul (2002). "Looking beyond XTR." *Advances in Cryptography— ASIACRYPT 2002*, Lecture Notes in Computer Science, vol. 2501, ed. Y. Zheng. Springer-Verlag, Berlin, 46–63.

[11] Canetti, R., J.B. Friedlander, S.V. Konyagin, M. Larsen, D. Lieman, and I.E. Shparlinski (2000). "On the statistical properties of Diffie–Hellman distributions." *Israel J. Math.*, 120, 23–46.

[12] Canetti, R., J.B. Friedlander, and I.E. Shparlinski (1999). "On certain exponential sums and the distribution of Diffie–Hellman triples." *J. London Math. Soc.*, 59, 799–812.

[13] Cherepnev, M.A. (1996). "On the connection between the discrete logarithms and the Diffie–Hellman problem." *Diskretnaja Matem.*, 6, 341–349 (in Russian).

[14] Coppersmith, D. and I.E. Shparlinski (2000). "On polynomial approximation of the discrete logarithm and the Diffie–Hellman mapping." *J. Cryptology*, 13, 339–360.

[15] Diffie, W. and M.E. Hellman (1976). "New directions in cryptography." *IEEE Trans. Inform. Theory*, 22, 109–112.

[16] El Mahassni, E. and I.E. Shparlinski (2001). "Polynomial representations of the Diffie–Hellman mapping." *Bull. Aust. Math. Soc.*, 63, 467–473.

[17] Enge, A. (1999). *Elliptic Curves and their Applications to Cryptography*. Kluwer Academic Publishers, Dordrecht.

[18] Friedlander, J.B. and I.E. Shparlinski (2001). "On the distribution of Diffie–Hellman triples with sparse exponents." *SIAM J. Discr. Math.*, 14, 162–169.

[19] Galbraith, S.D. (2001). "Supersingular curves in cryptography." *Advances in Cryptology— ASIACRYPT 2001*, Lecture Notes in Computer Science, vol. 2248, ed. C. Boyd. Springer-Verlag, Berlin, 495–513.

[20] Gaudry, P., F. Hess, and N.P. Smart (2002). "Constructive and destructive facets of Weil descent on elliptic curves." *J. Cryptology*, 15, 19–46.

[21] González Vasco, M.I. and I.E. Shparlinski (2001). "On the security of Diffie–Hellman bits." *Proc. Workshop on Cryptography and Computational Number Theory, Singapore, 1999*. Birkhäuser, 257–268.

[22] Joux, A. (2000). "A one round protocol for tripartite Diffie–Hellman." *Proc. of ANTS-IV*, Lecture Notes in Computer Science, vol. 1838, ed. W. Bosma. Springer-Verlag, Berlin, 385–393.

[23] Joux, A. (2002). "The Weil and Tate pairings as building blocks for public key cryptosystems." *Proc. of ANTS V*, Lecture Notes in Computer Science, vol. 2369, eds. D. Kohel and C. Fieker. Springer-Verlag, Berlin, 20–32.

[24] Koblitz, N. (1987). "Elliptic curve cryptosystems." *Math. Comp.*, 48, 203–209.

[25] Koblitz, N. "Good and bad uses of elliptic curves in cryptography." *Moscow Math. Journal*. To appear.

[26] Lenstra, A.K. and E.R. Verheul (2000). "The XTR public key system." *Advances in Cryptology— CRYPTO 2000*, Lecture Notes in Computer Science, vol. 1880, ed. M. Bellare. Springer-Verlag, Berlin, 1–19.

[27] Lenstra, A.K. and E.R. Verheul (2000). "Key improvements to XTR." *Advances in Cryptography— ASIACRYPT 2000*, Lecture Notes in Computer Science, vol. 1976, ed. T. Okamoto. Springer-Verlag, Berlin, 220–233.

[28] Lenstra, A.K. and E.R. Verheul (2001). "Fast irreducibility and subgroup membership testing in XTR." *PKC 2001*, Lecture Notes in Computer Science, vol. 1992, ed. K. Kim. Springer-Verlag, Berlin, 73–86.

[29] Lenstra, A.K. and E.R. Verheul (2001). "An overview of the XTR public key system." *Proc. the Conf. on Public Key Cryptography and Computational Number Theory, Warsaw 2000*. Walter de Gruyter, 151–180.

[30] Li, W.-C.W., M. Näslund, and I.E. Shparlinski (2002). "The hidden number problem with the trace and bit security of XTR and LUC." *Advances in Cryptology—CRYPTO 2002*, Lecture Notes in Computer Science, vol. 2442, ed. M. Yung. Springer-Verlag, Berlin, 433–448.

[31] Maurer, U.M. and S. Wolf (1999). "The relationship between breaking the Diffie–Hellman protocol and computing discrete logarithms." *SIAM J. Comp.*, 28, 1689–1721.

[32] Maurer, U.M. and S. Wolf (2000). "The Diffie–Hellman protocol." *Designs, Codes and Cryptography*, 19, 147–171.

[33] Meidl, W. and A. Winterhof (2002). "A polynomial representation of the Diffie–Hellman mapping." *Appl. Algebra in Engin., Commun. and Computing*, 13, 313–318.

[34] Menezes, A.J., N. Koblitz, and S.A. Vanstone (2000). "The state of elliptic curve cryptography." *Designs, Codes and Cryptography*, 19, 173–193.

[35] Menezes, A.J., P.C. van Oorschot, and S.A. Vanstone (1996). *Handbook of Applied Cryptography*. CRC Press, Boca Raton, FL.

[36] Miller, V.C. (1986). "Use of elliptic curves in cryptography." *Advances in Cryptology—CRYPTO'85* Lecture Notes in Computer Science, vol. 218, ed. H.C. Williams. Springer-Verlag, Berlin, 417–426.

[37] Pomerance, C. (1987). "Fast, rigorous factorization and discrete logarithm algorithms." *Discrete Algorithms and Complexity*. Academic Press, 119–143.

[38] Rubin, K. and A. Silverberg (2002). "Supersingular abelian varieties in cryptology." *Advances in Cryptology—CRYPTO 2002*, Lecture Notes in Computer Science, vol. 2442, ed. M. Yung. Springer-Verlag, Berlin, 336–353.

[39] Schirokauer, O. (1993). "Discrete logarithms and local units." *Philos. Trans. Roy. Soc. London, Ser. A*, 345, 409–423.

[40] Schirokauer, O., D. Weber, and T. Denny (1996). "Discrete logarithms: The effectiveness of the index calculus method." *Proceedings of ANTS-II*, Lecture Notes in Computer Science, vol. 1122, ed. H. Cohen. Springer-Verlag, Berlin, 337–362.

[41] Shoup, V. (1997). "Lower bounds for discrete logarithms and related problems." *Advances in Cryptology—EUROCRYPT'97*, Lecture Notes in Computer Science, vol. 1233, ed. W. Fumy. Springer-Verlag, Berlin, 256–266.

[42] Shparlinski, I.E. (2003). *Cryptographic Applications of Analytic Number Theory*. Birkhauser.

[43] Smith, P.J. and C.T. Skinner (1995). "A public-key cryptosystem and a digital signature system based on the Lucas function analogue to discrete logarithms." *Advances in Cryptography—ASIACRYPT'94*, Lecture Notes in Computer Science, vol. 917, eds. J. Pieprzyk and R. Safavi-Naini. Springer-Verlag, Berlin, 357–364.

[44] Stinson, D.R. (1995). *Cryptography: Theory and Practice*. CRC Press, Boca Raton, FL.

[45] Winterhof, A. (2001). "A note on the interpolation of the Diffie–Hellman mapping." *Bull. Aust. Math. Soc.*, 64, 475–477.

DIGITAL SIGNATURE SCHEMES

Digital signature schemes are techniques to assure an entity's acknowledgement of having sent a certain message. Typically, an entity has a private key and a corresponding public key which is tied to the entity's name (see also public key infrastructure). The entity generates a string called *signature* which depends on the message to sign and his private key.

The fact that the entity acknowledged, i.e. that he signed the message, can be verified by anyone using the entity's public key, the message, and the signature. *Data* authentication and signature schemes are sometimes distinguished in the sense that in the latter, verification can be done by anyone at any time after the generation of the signature. Due to this property, the digital signature scheme achieves non-repudiation property, that is, a signer cannot later deny the fact of signing.

Some examples of digital signature schemes are RSA digital signature scheme, ElGamal digital signature scheme, Rabin digital signature scheme, Schnorr digital signature scheme, Digital Signature Standard, and Nyberg-Rueppel signature scheme.

A digital signature scheme consists of three algorithms, namely the *key generation algorithm*, the *signing algorithm* and the *verification algorithm*. The security of digital signature is argued as follows: no adversary, without the knowledge of the private key, can generate a message and a signature that passes the verification algorithm. (See forgery for more discussions on the security of signatures.) There are two types of signature schemes, namely 'with *appendix*' and 'with *message recovery*'. In the former, the target message is the input of the verification algorithm; that is, the verifier must know the message in advance to verify the signature. In the latter, the target message is the output of the verification algorithm, so the message does not need to be sent with the signature. An example of the former is the ElGamal digital signature scheme and of the latter is the RSA digital signature scheme.

Kazue Sako

DIGITAL SIGNATURE STANDARD

The Digital Signature Standard (DSS), first proposed by Kravitz [2] in 1991, became a US federal standard in May 1994. It is published as Federal Information Processing Letters (FIPS) 186. The signature scheme is based on the ElGamal digital signature scheme and borrows ideas from Schnorr digital signatures for reducing signature size. We describe a slight generalization of the algorithm that allows for an arbitrary security parameter, whereas the standard only supports a fixed parameter. The signature scheme makes use of modular arithmetic and works as follows:

Key Generation. Given two security parameters $\tau, \lambda \in \mathbb{Z}$ $(\tau > \lambda)$ as input do the following:

1. Generate a random λ-bit prime q.
2. Generate a random τ-bit prime prime p such that q divides $p - 1$.
3. Pick an element $g \in \mathbb{Z}_p^*$ of order q.
4. Pick a random integer $\alpha \in [1, q]$ and compute $y = g^\alpha \in \mathbb{Z}_p^*$.
5. Let H be a hash function $H : \{0, 1\}^* \to \mathbb{Z}_q$. The FIPS 186 standard mandates that H be based on the SHA-1 cryptographic hash function.
6. Output the public key (p, q, g, y, H) and the private key (p, q, g, α, H).

Signing. To sign a message $m \in \{0, 1\}^*$ using the private key (p, q, g, α, H) do:

1. Pick a random integer $k \in [1, q-1]$.
2. Compute $r = (g^k \bmod p) \bmod q$. We view r as an integer $0 \leq r < q$.
3. Compute $s = k^{-1}(H(m) + \alpha r) \bmod q$.
4. Output the pair $(r, s) \in \mathbb{Z}_p^*$ as the signature on m.

Verifying. To verify a message/signature pair $(m, (r, s))$ using the public key (p, q, g, y, H) do:

1. Verify that $0 \leq r, s < q$, otherwise reject the signature.
2. Compute $u_1 = H(m)/s \bmod q$ and $u_2 = r/s \bmod q$.
3. Compute $v = (g^{u_1} y^{u_2} \bmod p) \bmod q$.
4. Accept the signature if $r = v \bmod q$. Otherwise, reject.

We first check that the verification algorithm accepts all valid message/signature pairs. For a valid message/signature pair we have

$$g^{u_1} y^{u_2} = g^{u_1 + \alpha u_2} = g^{(H(m)+\alpha r)/s} = g^k \pmod{p}.$$

It follows that $v = (g^{u_1} y^{u_2} \bmod p) \bmod q = r$ and therefore a valid message/signature is always accepted.

It is not clear how to analyze the security of this algorithm. Even the <u>random oracle model</u> does not seem to help since there is no hash function in the algorithm that can be modelled as a random oracle. It is believed that this is deliberate so that the algorithm does not infringe on existing patents. Security analysis for a generalization of DSS is given in [1].

To discuss signature length we fix concrete security parameters. At the present time the <u>discrete-logarithm problem</u> in the cyclic <u>group</u> $\mathbb{Z}_p^*$ where p is a 1024-bit prime and is considered intractable [3] except for a very well funded organization. DSS uses a subgroup of order q of $\mathbb{Z}_p^*$. When q is a 160-bit prime, the discrete log problem in this subgroup is believed to be as hard as discrete-log in all of $\mathbb{Z}_p^*$. Hence, for the present discussion we assume p is a 1024-bit prime and q is a 160-bit prime. Since a DSS signature contains two elements in $\mathbb{Z}_q$ we see that, with these parameters, a DSS signature is 320-bits long. This is the same length as a Schnorr signature. We note that <u>BLS short signatures</u> are half the size and provide comparable security.

Dan Boneh

References

[1] Brickell, Ernest, David Pointcheval, Serge Vaudenay, and Moti Yung (2000). "Design validations for discrete logarithm based signature schemes." *Proceedings of Public Key Cryptography 2000*, 276–292.

[2] Kravitz, D. (1993). "Digital signature algorithm." U.S. patent #5,231,668.

[3] Lenstra, Arjen and Eric Verheul (2001). "Selecting cryptographic key sizes." *Journal of Cryptology*, 14 (4), 255–293.

DIGITAL STEGANOGRAPHY

INTRODUCTION: Steganography is the art and science of hiding information by embedding messages within other, seemingly harmless messages. Steganography means "covered writing" in Greek. As the goal of steganography is to hide the *presence* of a message and to create a <u>covert channel</u>, it can be seen as the complement of cryptography, whose goal is to hide the *content* of a message.

A famous illustration of steganography is Simmons' "Prisoners' Problem" [10]: Alice and Bob are in jail, locked up in separate cells far apart from each other, and wish to devise an escape plan. They are allowed to communicate by means of sending messages via trusted couriers, provided they do not deal with escape plans. But the couriers are agents of the warden Eve (who plays the role of the adversary here) and will leak all communication to her. If Eve detects any sign of conspiracy, she will thwart the escape plans by transferring both prisoners to high-security cells from which nobody has ever escaped. Alice and Bob are well aware of these facts, so that before getting locked up, they have shared a secret codeword that they are now going to exploit for embedding a hidden information into their seemingly innocent messages. Alice and Bob succeed if they can exchange information allowing them to coordinate their escape and Eve does not become suspicious.

According to the standard terminology of information hiding [8], a legitimate communication among the prisoners is called *covertext*, and a message with embedded hidden information is called *stegotext*. The distributions of covertext and stegotext are known to the warden Eve because she knows what constitutes a legitimate communication among prisoners and which tricks they apply to add a hidden meaning to innocent looking messages.

The algorithms for creating stegotext with an embedded message by Alice and for decoding the message by Bob are collectively called a *stegosystem*. A stegosystem should hide the embedded message at least as well as an encryption scheme since it may be enough for the adversary to learn only a small amount of information about the embedded message to conclude that Alice and Bob are conspiring. But steganography requires more than that. The ciphertext generated by most encryption schemes resembles a sequence of random bits, and this is very likely to raise the suspicion of Eve. Instead, stegotext should "look" just like innocent covertext even though it contains a hidden message.

This intuition forms the basis of the recently developed formal approach to steganography [2, 3, 5, 6, 11]. It views a stegosystem as a cryptosystem with the additional property that its output, i.e., the stegotext, is not distinguishable from covertext to the adversary.

Formally, a stegosystem consists of a triple of algorithms for key generation, message encoding, and message decoding, respectively. In the symmetric-key setting considered here, the output of the key generation algorithm is given only to Alice and to Bob.

The covertext is modeled by a distribution C over a given set C. The covertext may be given explicitly as a list of values or implicitly as an oracle that returns a sample of C upon request. A stegosystem that does not require explicit knowledge of the covertext distribution is called *universal*.

A more general model of a covertext *channel* has also been proposed in the literature [5], which allows to model dependencies among repeated uses of the same covertext source. A channel consists of an unbounded sequence of values drawn from a set C whose distribution may depend in arbitrary ways on past outputs; access to the channel is given only by an oracle that samples from the channel. The assumption is that the channel oracle can be queried with an arbitrary prefix of a possible channel output, i.e., its past "history," and it will return the next symbol according to the channel distribution. In order to simplify the presentation, channels are not considered further here, but all definitions and constructions mentioned below can be readily extended to covertext channels.

We borrow the complexity-theoretic notions of probabilistic <u>polynomial-time</u> algorithms and *negligible functions*, in terms of a security parameter n, from modern cryptography [4].

DEFINITION 1 (Stegosystem). *Let C be a distribution on a set C of covertexts. A stegosystem is a triple of probabilistic polynomial-time algorithms* (*SK, SE, SD*) *with the following properties:*
- *The key generation algorithm SK takes as input the security parameter n and outputs a bit string sk, called the [stego] key.*

- *The steganographic encoding algorithm SE takes as inputs the security parameter n, the stego key sk and a message $m \in \{0, 1\}^l$ to be embedded and outputs an element c of the covertext space C, which is called stegotext. The algorithm may access the covertext distribution C.*
- *The steganographic decoding algorithm SD takes as inputs the security parameter n, the stego key sk, and an element c of the covertext space C and outputs either a message $m \in \{0, 1\}^l$ or a special symbol $\perp$. An output value of $\perp$ indicates a decoding error, for example, when SD has determined that no message is embedded in c.*

For all sk output by SK(1^n) and for all $m \in \{0, 1\}^l$, the probability that SD$(1^n, sk, SE(1^n, sk, m)) \neq m$ must be negligible in n.

Note that the syntax of a stegosystem as defined above is equivalent to that of a (symmetric-key) cryptosystem, except for the presence of the covertext distribution. The probability that the decoding algorithm outputs the correct embedded message is called the *reliability* of a stegosystem.

DEFINING SECURITY: The security of a stegosystem is defined in terms of an experiment that measures the capability of the adversary to detect the presence of an embedded message. In a secure stegosystem, Eve cannot distinguish whether Alice is sending legitimate covertext or stegotext.

The attack considered here is a *chosen-message attack*, where the adversary may influence the embedded message but has otherwise no access to the encoding and decoding functions. It parallels the notion of a chosen-plaintext attack against a cryptosystem.

Consider an adversary defined by a pair of algorithms (SA_1, SA_2). The experiment consists of four stages.

1. A key sk is generated by running the key generation algorithm SK.
2. Algorithm SA_1 is run with input the security parameter n; it outputs a tuple (m^*, s), where $m^* \in \{0, 1\}^l$ is a message and s is some additional information which the algorithm wants to preserve. SA_1 has access to the covertext distribution C.
3. A bit b is chosen at random and a *challenge covertext* c^* is determined depending on it: If $b = 0$ then $c^* \leftarrow SE(sk, m^*)$ (c^* becomes a steganographic encoding of m^*) otherwise $c^* \overset{R}{\leftarrow} C$ (c^* is chosen randomly according to C).

4. Algorithm SA_2 is run with inputs n, c^*, m^*, and s, and outputs a bit b'. The goal of SA_2 is to guess the value of b, i.e., to determine whether the message m^* has been embedded in c or whether c has simply been chosen according to $\mathcal{C}$.

The adversary succeeds to distinguish stegotext from covertext if $b' = b$ in the above experiment. Since it is trivial to achieve $\Pr[b' = b] = \frac{1}{2}$, what actually counts is the adversary's advantage above randomly guessing b. Formally, we define the *advantage* of adversary (SA_1, SA_2) to be

$$\Pr\Big[sk \leftarrow \mathsf{SK}; \ (m^*, s) \leftarrow SA_1(1^n); \ b \xleftarrow{R} \{0, 1\};$$
$$\textbf{if } b = 0 \textbf{ then } c^* \leftarrow \mathsf{SE}(1^n, sk, m^*)$$
$$\textbf{else } c^* \xleftarrow{R} \mathcal{C} : \ SA_2(1^n, c^*, m^*, s) = b \Big] - \frac{1}{2}.$$

Depending on the detection capabilities ascribed to the adversary in terms of its computational power, and depending on its advantage in distinguishing stegotext from covertext, one obtains the following security notions for stegosystems:

Perfectly secure steganography: The adversary is an arbitrary, unbounded algorithm and has advantage equal to 0.

Statistically secure steganography: The adversary is an arbitrary, unbounded algorithm and has only negligible advantage (in n).

Computationally secure steganography: The adversary is an arbitrary probabilistic, polynomial-time computable algorithm and has only negligible advantage in n.

The first two notions are collectively known as *information-theoretically secure steganography*. The third notion, computational security, is discussed further on (cf., security).

This discussion has so far assumed that the sender and the receiver share the same secret key before communicating (cf., symmetric cryptosystem). Departing from that assumption, it is also possible to define *public-key steganography* with computational security. This notion is analogous to public-key cryptography and is the subject of Section D (cf., asymmetric cryptosystem).

For the scope of this survey, the adversary is limited to *passive attacks* (see cryptanalysis). In the Prisoners Problem, this means that the couriers may not change the messages communicated between Alice and Bob and that Eve may not send a message generated by herself to Bob and observe his reaction to it. However, the adversary may influence the messages to be embedded; for example, the warden Eve may determine the details of Alice and Bob's escape plan by choosing to confine them in particular cells.

This survey is about the formal approach to steganography and about stegosystems that offer provable security. An overview of steganography with heuristic security and of the history of steganography is given by Anderson and Petitcolas [1].

What distinguishes steganography from other forms of information hiding is the focus on merely detecting the *presence* of a hidden message. Watermarking and fingerprinting are two different problems of information hiding, where the existence of a hidden message is public knowledge. The focus in these areas is on hiding the message in perceptual data from an observer that is typically a human, and on embedding the message robustly so that it cannot be removed without significantly distorting the data itself. The difference between watermarking and fingerprinting is that watermarking supplies digital objects with an identification of origin and all objects are marked in the same way; fingerprinting, conversely, attempts to identify individual copies of an object by means of embedding a unique marker in every copy that is distributed to a user.

INFORMATION-THEORETICALLY SECURE STEGANOGRAPHY

DEFINITION 2 (Perfect Security). *Given a covertext distribution $\mathcal{C}$, a stegosystem $(\mathsf{SK}, \mathsf{SE}, \mathsf{SD})$ is called* perfectly secure *with respect to $\mathcal{C}$ if for any adversary (SA_1, SA_2) with unbounded computational power, the advantage in the experiment above is equal to 0.*

Perfect security for a stegosystem parallels Shannon's notion of *perfect security* for a cryptosystem [9] (cf., Shannon's model). The requirement that every adversary has no advantage implies that the distributions of the challenge c^* are equal in the two cases where it was generated from SE (when $b = 0$) and sampled from $\mathcal{C}$ (when $b = 1$). Hence, the adversary obtains *no* information about b because she only observes the challenge c^* and the distribution of c^* is statistically independent of b. Perfectly secure stegosystems were defined by Cachin [3].

Perfectly secure stegosystems exist only for a very limited class of covertext distributions. For example, if the covertext distribution is uniform, the one-time pad is a perfectly secure stegosystem as follows.

Assume the covertext $\mathcal{C}$ is uniformly distributed over the set of n-bit strings for some positive n and let Alice and Bob share an n-bit key sk with uniform distribution. The encoding function computes the bitwise XOR of the n-bit message m and sk, i.e., $\mathsf{SE}(1^n, sk, m) = m \oplus sk$; Bob can decode this by computing $\mathsf{SD}(1^n, sk, c) = c \oplus sk$. The resulting stegotext is uniformly distributed in the set of n-bit strings. The one-time pad stegosystem is used like this in visual cryptography [7].

For covertext distributions that do not admit perfectly secure stegosystems, one may still achieve the following security notion.

DEFINITION 3 (Statistical Security). *Given a covertext distribution C, a stegosystem (SK, SE, SD) is called* statistically secure *with respect to C if for all adversaries (SA_1, SA_2) with unbounded computational power, there exists a negligible function ϵ such that the advantage in the experiment above is at most $\epsilon(n)$.*

Statistical security for stegosystems may equivalently be defined by requiring that for any sk and any m, the statistical distance between the probability distribution generated by $\mathsf{SE}(1^n, sk, m)$ and the covertext distribution is negligible.

Definition 3 was first proposed by Katzenbeisser and Petitcolas [6]. A very similar notion was defined by Cachin [3], using relative entropy between the stegotext and covertext distributions for quantifying the difference between them.

Here is a simple example of a statistically secure stegosystem, adopted from [3]. It is representative for a class of practical stegosystems that embed information in a digital image by modifying the least significant bit of every pixel representation [1]. Suppose that the cover space C is the set of n-bit strings with (C_0, C_1) being a partition of C and with distribution C such $|\mathrm{Pr}[c \overset{R}{\leftarrow} C : c \in C_0] - \mathrm{Pr}[c \overset{R}{\leftarrow} C : c \in C_1]| = \delta(n)$ for some negligible δ. Then there is a stegosystem for a one-bit message m using a one-bit secret key sk. The encoding algorithm SE computes $s \leftarrow m \oplus sk$ and outputs $c \overset{R}{\leftarrow} C_s$. Decoding works without error because $m = 0$ if and only if $c \in C_{sk}$. It is easy to see that the encoding provides perfect secrecy for m and that the stegosystem is statistically secure. Note, however, that finding the partition for a given distribution is an NP-hard combinatorial optimization problem.

There exist also statistically secure *universal* stegosystems, where the covertext distribution is only available as a sampling oracle. Information-theoretically secure stegosystems suffer from the same drawback as cryptosystems with unconditional security in the sense that the secret key may only be used once. This is not the case for computational security considered next.

COMPUTATIONALLY SECURE STEGANOGRAPHY

DEFINITION 4. (Computational Security). *Given a covertext distribution C, a stegosystem (SK, SE, SD) is called* computationally secure *with respect to C if for all probabilistic polynomial-time adversaries (SA_1, SA_2), there exists a negligible function ϵ such that the advantage in the experiment above is at most $\epsilon(n)$.*

The notion was formalized independently by Katzenbeisser and Petitcolas [6] and by Hopper, Langford, and von Ahn [5]. The latter work also presented the following construction of a computationally secure, universal stegosystem. It illustrates a popular encoding method that does not rely on knowledge of the covertext distribution, which is also used by some practical stegosystems.

The encoding method is based on an algorithm sample, which samples a covertext according to C such that a given bit string b of length $f = O(\log |C|)$ is embedded in it.

Algorithm sample
Input: security parameter n, a function $g : C \rightarrow \{0, 1\}^f$, and a value $b \in \{0, 1\}^f$
Output: a covertext x
  1: $j \leftarrow 0$
  2: **repeat**
  3:     $x \overset{R}{\leftarrow} C$
  4:     $j \leftarrow j + 1$
  5: **until** $g(x) = b$ or $j = n$
  6: **return** x

Intuitively, algorithm sample returns a covertext chosen from distribution C, but restricted to that subset of C which is mapped to the given b by g. sample may also fail and return a covertext c with $g(c) \neq b$, but this happens only with negligible probability in n.

Suppose $\{G_k\}$ is a pseudorandom function family indexed by k, with domain $\{0, 1\} \times C$ and range $\{0, 1\}^f$. (It can be thought of as a pair (G_0, G_1) of independent pseudorandom functions.) The secret key of the stegosystem consists of a randomly chosen k. The encoding algorithm $\mathsf{SE}(1^n, k, m)$ for an f-bit message m first "encrypts" m to $y \leftarrow G_k(0, c_0) \oplus m$ for a public constant $c_0 \in C$. Note that y is the ciphertext of a symmetric-key encryption of m and is computationally indistinguishable from a random f-bit string. This value y is then embedded by computing a stegotext $c \leftarrow \mathsf{sample}(n, G_k(1, \cdot), y)$. It can be shown that when

C is sufficiently random, as measured in terms of min-entropy, the output distribution of sample is statistically close to C [2,5].

The decoding algorithm $SD(1^n, k, c)$ outputs $m' \leftarrow G_k(1, c) \oplus G_k(0, c_0)$; it is easy to show that m' is equal to the message that was embedded using SE except with negligible probability.

This stegosystem is an extension of the example given above for statistical security. In fact, when G is a universal hash function and the encryption is realized using a one-time pad, this is a universal stegosystem with statistical security.

PUBLIC-KEY STEGANOGRAPHY: What if Alice and Bob did not have the time to agree on a secret key before being imprisoned? They cannot use any of the stegosystems presented so far because that would require them to share a common secret key. Fortunately, steganography is also possible without shared secrets, only with public keys, similar to public-key cryptography. The only requirement is that Bob's public key becomes known to Alice in a way that is not detectable by Eve.

Formally, a public-key stegosystem consists of a triple of algorithms for key generation, message encoding, and message decoding like a (secret-key) stegosystem, but the key generation algorithm now outputs a stego key pair (spk, ssk). The public key spk is made available to the adversary and is the only key needed by the encoding algorithm SE. The decoding algorithm SD needs the secret key ssk as an additional input.

DEFINITION 5 (Public-key Stegosystem). *Let C be a distribution on a set C of covertexts. A public-key stegosystem is a triple of probabilistic polynomial-time algorithms (SK, SE, SD) with the following properties:*

- *The key generation algorithm SK takes as input the security parameter n and outputs a pair of bit strings (spk, ssk), called the [stego] public key and the [stego] secret key.*
- *The steganographic encoding algorithm SE takes as inputs the security parameter n, the stego public key spk and a message $m \in \{0, 1\}^l$ and outputs a covertext $c \in C$.*
- *The steganographic decoding algorithm SD takes as inputs the security parameter n, the stego secret key ssk, and a covertext $c \in C$, and outputs either a message $m \in \{0, 1\}^l$ or a special symbol $\perp$.*

For all (spk, ssk) output by the key generation algorithm and for all $m \in \{0, 1\}^l$, the probability that $SD(1^n, ssk, SE(1^n, spk, m)) \neq m$ must be negligible in n.

Security is defined analogously to the experiment of Section 2 with the difference that the public key spk is additionally given to the adversary algorithms SA_1 and SA_2 and that the challenge covertext is computed using spk only. With these modifications, a public-key stegosystem (SK, SE, SD) is called *secure against chosen-plaintext attacks* if it is computationally secure according to Definition 4.

Secure public-key stegosystems can be constructed using the method of Section D, but with the pseudorandom function G_0 (which is used for "encryption") replaced by a public-key cryptosystem that has almost uniform ciphertexts. This property means that the output of the encryption algorithm is computationally indistinguishable from a uniform bit string of the same length.

The definition and several constructions of public-key stegosystems have been introduced by von Ahn and Hopper [11] and by Backes and Cachin [2]. The latter work also goes beyond the case of passive adversaries considered here and models adaptive chosen-covertext attacks, which are similar to adaptive chosen-ciphertext attacks against public-key cryptosystems. Achieving security against such attacks results in the strongest security notion known today for public-key cryptosystems and for public-key stegosystems.

As this brief survey of steganography shows, the evolution of the formal approach to stegosystems has gone through the same steps as the development of formal models for cryptosystems. The models and the formulation of corresponding stegosystems that offer provable security have greatly enhanced our understanding of this important area of information security.

Christian Cachin

References

[1] Anderson, R.J. and F.A. Petitcolas (1998). "On the limits of steganography." *IEEE Journal on Selected Areas in Communications*, 16.

[2] Backes, M. and C. Cachin (2005). "Public-key steganography with active attacks." *Proceedings 2nd Theory of Cryptography Conference (TCC 2005)*, Lecture Notes in Computer Science, vol. 3378, ed. J. Kilian. Springer, Berlin, 210–226.

[3] Cachin, C. (2004). "An information-theoretic model for steganography." Information and Computation, vol. 192, pp. 41–56. (Preliminary version appeared in Proc. 2nd Workshop on Information Hiding, Lecture Notes in Computer Science, vol. 1525, Springer, Berlin, 1998).

[4] Goldreich, O. (2001). *Foundations of Cryptography: Basic Tools*. Cambridge University Press, Cambridge.

[5] Hopper, N.J., J. Langford, and L. von Ahn (2002). "Provably secure steganography." *Advances in Cryptology—CRYPTO 2002*, Lecture Notes in Computer Science, vol. 2442, ed. M. Yung. Springer, Berlin.

[6] Katzenbeisser, S. and F.A.P. Petitcolas (2002). "Defining security in steganographic systems." *Security and Watermarking of Multimedia Contents IV, Proceedings of SPIE*, International Society for Optical Engineering, vol. 4675, eds. E.J. Delp and P.W. Won, 260–268.

[7] Naor, M. and A. Shamir (1995). "Visual cryptography." *Advances in Cryptology—EUROCRYPT'94*, Lecture Notes in Computer Science, vol. 950, ed. A. De Santis. Springer, Berlin, 1–12.

[8] Pfitzmann, B. (1996). "Information hiding terminology." *Information Hiding, First International Workshop*, Lecture Notes in Computer Science, vol. 1174, ed. R. Anderson. Springer, Berlin, 347–350.

[9] Shannon, C.E. (1949). "Communication theory of secrecy systems." *Bell System Technical Journal*, (28), 656–715.

[10] Simmons, G.J. (1984). "The prisoners' problem and the subliminal channel." *Advances in Cryptology—CRYPTO'83*, Lecture Notes in Computer Science, ed. D. Chaum. Plenum, New York, 51–67.

[11] von Ahn, L. and N.J. Hopper (2004). "Public-key steganography." *Advances in Cryptology–EUROCRYPT 2004*, Lecture Notes in Computer Science, vol. 3027, eds. C. Cachin and J. Camenisch. Springer, Berlin, 322–339.

DISCRETE LOGARITHM PROBLEM

Let G be a cyclic group of order n, and g be a generator for G. Given an element $y \in G$, the *discrete logarithm problem* is to find an integer x such that

$$g^x = y.$$

The discrete logarithm problem has been of particular interest since Diffie and Hellman (see Diffie–Hellman key agreement) invented a cryptographic system based on the difficulty of finding discrete logarithms (a similar system was created around the same time by Malcolm Williamson at the Government Communications Headquarters (GCHQ) in the UK, but not revealed until years later). Given two people Alice and Bob who wish to communicate over an insecure channel, each decides on a private key x_A and x_B. Alice sends g^{x_A} to Bob, and Bob sends g^{x_B} to Alice. Each of them can then raise the received message to their private

key to compute

$$(g^{x_A})^{x_B} = (g^{x_B})^{x_A} = g^{x_A x_B}.$$

An eavesdropper Eve who only knows g^{x_A} and g^{x_B} must figure out $g^{x_A x_B}$. This is widely believed to be difficult. Clearly if Eve can solve the discrete logarithm problem, she can compute x_A and x_B and so break the system.

Other systems, such as the ElGamal digital signature scheme and the Digital Signature Standard, also depend on the difficulty of solving the discrete logarithm problem.

Pohlig and Hellman [9], and independently Silver, observed that if G has a subgroup of order l, then by raising g and y to the (n/l)th power we may solve for x modulo l. Thus, the difficulty of the discrete logarithm problem depends on the largest prime factor of n. For the rest of this article we will assume that n is prime.

THE DISCRETE LOGARITHM PROBLEM IN DIFFERENT GROUPS: Any finite group may be used for a Diffie–Hellman system, but some are more secure than others. The main groups used are:

- The multiplicative subgroup of a finite field $GF(q)$, with q a prime or a power of 2.
- The points on an elliptic curve E over a finite field (see elliptic curves).
- The class group of a quadratic number field.

Finite fields $GF(2^n)$ were popular into the 1980s, but attacks by Blake, Fuji-Hara, Mullin and Vanstone, and Coppersmith showed that the fields were easier to attack than similarly-sized prime fields. Index calculus attacks may also be applied to prime fields.

Hafner and McCurley [6] gave a subexponential attack for class groups of imaginary quadratic number fields, and Buchmann [2] extended this to real quadratic and, conjecturally, higher-degree number fields. Most elliptic curves, on the other hand, have no known subexponential attacks. See the entry on elliptic curve cryptography for more details.

GENERIC ALGORITHMS FOR DISCRETE LOGARITHMS: We will first consider *generic* algorithms, which do not use any special information about the group G, but only compose elements and check for equality. Nechaev [7] and Shoup [15] showed that generic algorithms must take $\Omega(\sqrt{n})$ time (see O-notation). Shor [14] showed that a quantum computer can solve a discrete logarithm problem in *any* group in polynomial time, but whether a sufficiently large quantum computer can be built is still an open problem.

Shanks' Baby Step–Giant Step Method

Shanks [13] gave the first algorithm better than a brute-force search. Let $m = \lceil \sqrt{n} \rceil$. We construct two tables, one starting at 1 and taking "giant steps" of length m:

$$1, g^m, g^{2m}, \ldots, g^{(m-1)m}$$

and one of "baby steps" of length one from y:

$$y, yg, yg^2, \ldots, yg^{m-1}.$$

Sort these lists and look for a match. If we find $g^{im} = yg^j$, then $y = g^{im-j}$, and so $x = im - j$. Any $x \in [0, n-1]$ may be written in this form for $i, j \le m$, so we are certain to find such a match.

The time for this algorithm is $O(\sqrt{n})$ group operations, plus the time to find collisions in the two lists. This may be done either by sorting the lists or using hash tables.

Pollard's ρ Method

The drawback to Shanks's algorithm is that it requires $O(\sqrt{n})$ space as well as time. Pollard [10] gave two methods that use negligible space and still run in $O(\sqrt{n})$ time: the ρ method and the kangaroo method, which are discussed below. They are not deterministic, but depend on taking pseudo-random walks in G.

Divide the elements of G into three subsets, S_1, S_2 and S_3, say by the value of a hash of the elements modulo three. We define a walk by $h_0 = 1$ and

$$h_{i+1} = \begin{cases} h_i y, & \text{if } h_i \in S_1 \\ h_i^2, & \text{if } h_i \in S_2 \\ h_i g, & \text{if } h_i \in S_3. \end{cases}$$

At each step we know

$$h_i = g^{a_i} y^{b_i} = g^{a_i + x b_i}$$

for some a_i, b_i. (In particular, we have $(a_0, b_0) = (0, 0)$ initially, and $(a_{i+1}, b_{i+1}) = (a_i, b_i + 1)$, $(2a_i, 2b_i)$, or $(a_i + 1, b_i)$, depending on the hash value.) Eventually, this walk must repeat. If $h_i = h_j$, we have

$$x \equiv \frac{a_j - a_i}{b_i - b_j} \pmod{n}.$$

If $b_i - b_j$ is relatively prime to n (which is very likely if n is prime), this gives us x.

Figure 1 illustrates the ρ method walk.

Rather than store all of the steps to detect a collision, we may simultaneously compute h_i and h_{2i}, and continue around the cycle until they agree. Assuming that this map behaves as a random walk, we will need $O(\sqrt{n})$ steps to find a repeat.

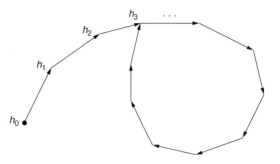

Fig. 1. ρ method walk, with a collision at h_3

Parallelized Collision Search

The ρ method has two main drawbacks. One is that it is difficult to parallelize. Having k processors do random walks only results in an $O(\sqrt{k})$ speedup, since the different walks are independent, and the probability of one of k cycles of length l having a collision is much less than one cycle of length kl (see [8] for details). Another is that after the collision occurs, many more steps around the cycle are needed before the collision is detected. Parallelized collision search [8] is a variant of the ρ method which fixes both problems.

We designate a small fraction of elements of G *distinguished points*, say if the last several bits of the element are all zero. Then a walk will begin at a random point, proceed as for the ρ method, and end when we hit a distinguished point. We save that point along with the starting point of the path, and then begin a walk at a new random point. When a distinguished point is hit for the second time, we have a collision and with high probability can determine x.

By picking the right fraction of elements of G to be distinguished points, we may ensure that not too much memory is needed to store the paths, and not much time is wasted after a collision occurs. Also, this algorithm may be trivially parallelized, with a linear speedup.

Figure 2 shows this method.

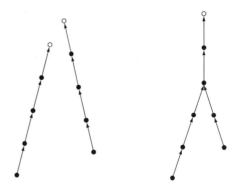

Fig. 2. Parallelized collision search paths, with three distinguished points and one collision

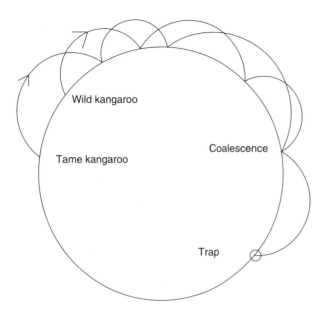

Fig. 3. Kangaroo method paths, with one distinguished point and collision

Pollard's Kangaroo Method

Another method due to Pollard also uses a random walk in G. In this algorithm the steps are limited: $h \longrightarrow hg^{s(h)}$, where the hop length $s(h)$ is a pseudorandom function of h with values between 1 and $\sqrt{n}$.

The idea is to start from two points, say g (the "tame" kangaroo, since we know its discrete logarithm at all times) and y (the "wild" kangaroo), and alternately take hops with length determined by $s(h)$. We will set "traps" when a kangaroo hits a distinguished point. If the wild kangaroo and tame kangaroo paths meet or "coalesce" at any point, they will take the same hops from then on, and any traps encountered by one after they coalesce will also be encountered by the other. When one reaches a trap that the other one hit, we have a collision and can determine x.

The main advantage of the kangaroo method is when x is known to be in a certain range, say $[0, L]$ for some $L \ll |G|$. In that case we may start the tame kangaroo from $g^{L/2}$, and the wild kangaroo from y. We expect to find a collision before we get far out of $[0, L]$, and so this will take $O(\sqrt{L})$ time. See Figure 3 for an illustration.

SUBEXPONENTIAL METHODS: The lower bound for generic algorithms means that to find a faster algorithm we must use information about the group. The main method for doing this is called *index calculus*, and is described in this section.

Index Calculus Methods

Let

$$L_x[t, \gamma] = e^{(\gamma + o(1))(\log x)^t (\log \log x)^{1-t}},$$

for $x \to \infty$ (see L-notation for further discussion). This function interpolates between slow and fast algorithms; $L_x[1, \gamma] \approx x^\gamma$ is exponential in $\log x$ (see exponential time), while $L_x[0, \gamma] \approx (\log n)^\gamma$ is polynomial (see polynomial time). All the algorithms of the previous section are $O(\sqrt{n}) = L_n[1, 1/2]$. With early index calculus methods we may reduce this to $L_n[1/2, c]$, and the number field sieve further improves this to $L_n[1/3, c]$ for appropriate constants c.

All index calculus algorithms for discrete logarithms have three main parts:

1. Gather equations relating the discrete logarithms of a factor base of "small" elements.
2. Solve a linear system to find the factor base discrete logarithms.
3. To find the discrete logarithm of an element y, reduce y to a product of elements in the factor base.

The first step is the same as in integer factoring. The second step is also done in factoring, but modulo 2 instead of modulo n. The third step is only done for discrete logarithms, typically by multiplying y by random powers of g, and attempting to express the result as a product of smaller numbers, possibly recursively breaking those numbers into smaller ones until everything is in the factor base.

The factor base is a set of elements such as small primes or low-degree polynomials, such that other elements have a reasonable chance of being "smooth": expressible as a product of these small elements (see the entry on smoothness). To optimize the algorithm we need to know the probability of this happening; see the section on number theory for more information.

Typically the first two steps require large computations, and finding individual logarithms is much quicker.

For additional technical details on these methods, please see the entry *index calculus*.

Discrete Logarithms in Prime Fields

Coppersmith, Odlyzko, and Schroppel gave an $L_p[1/2, c]$ algorithm for prime fields GF(p), which turned out to be special case of the Number Field Sieve (using imaginary quadratic fields). In their method there are two factor bases, one of small rational primes and another of small primes in the imaginary quadratic field.

The Number Field Sieve, which is the fastest known algorithm for factoring integers, may also be applied to finding discrete logarithms [5, 12]. The factor base used consists of small rational primes and representatives of small prime ideals in a number field. The asymptotic complexity is the same as for factoring. The sieving phase is the same, but solving the linear system modulo $p - 1$ instead of modulo 2 makes discrete logarithms harder than factoring problems of the same size.

Because the number field sieve works better for special numbers (such as primes $p = r^e + s$ for r and s small), it has been suggested that the Digital Signature Standard could be given a "trapdoor" by using a prime for which the Number Field Sieve runs faster than on a typical prime of that size. However, in [4], it is shown that such trapdoors may be detected, and that it is easy to specify primes which were clearly not chosen with a trapdoor.

Discrete Logarithms in Fields of Characteristic 2

Until the 1980s, fields GF(q) with $q = 2^n$ received the most attention, because of their applications to shift registers and ease of implementation in hardware. However, it turned out that attacks on these fields ran much faster than prime fields, and so few cryptosystems today depend on discrete logarithms in these fields.

Blake et al. [1] gave an attack which ran in time $L_q[1/2, c]$. Their factor base consists of polynomials in GF(2)[x] of low degree. This was improved by Coppersmith [3], who gave the first index calculus algorithm which runs in time $L_q[1/3, c]$. It was not realized until much later, but Coppersmith's method was a special case of the function field sieve (see the entry sieving in function fields).

Other Fields

Schirokauer [11] has looked at GF(q) for $q = p^m$ with $p > 2$ and $m > 1$. By combining features of the number field sieve and function field sieve, he gives an algorithm which is conjectured to run in $L_q[1/3, c]$ for some c for fields with $q \longrightarrow \infty$ and $m > (\log p)^2$ or $m < (\log p)^{1/2-\epsilon}$. In the "gap" between these constraints the algorithm is conjectured to run in time $L_q[2/5, c']$.

Recently Lenstra and Verheul invented a cryptosystem called XTR, which depends on the security of discrete logarithms in GF(p^6), for $p^6 \approx 1024$ bits. Weber [17] has computed discrete logarithms in fields GF(p^2) for small p.

ATTACKS ON ELLIPTIC CURVE DISCRETE LOGARITHMS: The elliptic curve discrete logarithm problem (ECDLP) was suggested as a basis for cryptosystems in 1985 by Neal Koblitz and Victor Miller. Because no subexponential attack was known for them, much shorter key sizes could be used.

Since then, several attacks on special elliptic curves have been developed, but no index calculus attack for general curves are known.

CHALLENGES AND ATTACKS: In 1989, Kevin McCurley gave a challenge problem. Let $q = (7^{149} - 1)/6$, and $p = 2 \times 739q + 1$. McCurley gave two numbers modulo p which equal 7^x and 7^y for some x and y, and issued a challenge to find 7^{xy}.

The form of p was intended to make it easy to show that p is prime and that 7 is a primitive root modulo p. Unfortunately, soon afterwards the number field sieve was discovered, which showed that the special form of this p made the system much less secure. The challenge was broken in 1998 by Weber and Denny [18] using the special number field sieve.

Joux and Lercier found discrete logarithms modulo a nonspecial 120-digit prime in 2001. For fields of characteristic 2, the record is GF(2^{607}), which was done in 2001 by Thomé [16].

In 1997 Certicom issued a series of ECDLP challenges. The problems ranged from easy (curves over 79-bit fields), to very difficult (359-bit fields). The largest challenge problem solved to date is a curve over GF(p) for a 109-bit prime p by a group at Notre Dame in 2002, using parallelized collision search.

Daniel M. Gordon

References

[1] Blake, I.F., R. Fuji-Hara, R.C. Mullin, and S.A. Vanstone (1984). "Computing logarithms in fields of characteristic two." *SIAM Journal of Algebraic and Discrete Methods*, 5, 276–285.

[2] Buchmann, Johannes (1990). "A subexponential algorithm for the determination of class groups and regulators of algebraic number fields." *Séminaire de Théorie des Nombres, Paris 1988–1989, Progr. Math.*, vol. 91, Birkhäuser, Boston, 27–41.

[3] Coppersmith, D. (1984). "Fast evaluation of discrete logarithms in fields of characteristic two." *IEEE Transactions on Information Theory*, 30, 587–594.

[4] Gordon, D.M. (1992). "Designing and detecting trapdoors in discrete log cryptosystems." *Advances in Cryptology—CRYPTO'92*, Lecture Notes

in Computer Science, vol. 740, ed. E.F. Brickell. Springer, Berlin, 66–75.

[5] Gordon, D.M. (1993). "Discrete logarithms in GF(p) using the number field sieve." *SIAM J. Discrete Math.*, 6, 124–138.

[6] Hafner, J. and K. McCurley (1989). "A rigorous subexponential algorithm for computation of class groups." *J. Amer. Math. Soc.*, 2 (4), 837–850.

[7] Nechaev, V.I. (1994). "On the complexity of a deterministic algorithm for a discrete logarithm." *Math. Zametki*, 55, 91–101.

[8] van Oorschot, P.C. and M.J. Wiener (1999). "Parallel collision search with cryptanalytic applications." *J. Cryptology*, 12, 1–28.

[9] Pohlig, S.C. and M.E. Hellman (1978). "An improved algorithm for computing logarithms over $GF(p)$ and its cryptographic significance." *IEEE Trans. Info. Theory*, IT-24, 106–110.

[10] Pollard, J.M. (1978). "Monte Carlo methods for index computation (mod p)." *Mathematics of Computation*, 32, 918–924.

[11] Schirokauer, O. "The impact of the number field sieve on the discrete logarithm problem in finite fields." *Proceedings of the 2002 Algorithmic Number Theory workshop at MSRI.*

[12] Schirokauer, O. (1993). "Discrete logarithms and local units." *Philos. Trans. Roy. Soc. London Ser. A*, 345, 409–423.

[13] Shanks, D. (1971). "Class number, a theory of factorization, and genera." In *1969 Number Theory Institute (Proc. Sympos. Pure Math., Vol. XX, State Univ. New York, Stony Brook, NY, 1969)*, Amer. Math. Soc., Providence, RI, 415–440.

[14] Shor, P.W. (1997). "Polynomial-time algorithms for prime factorization and discrete logarrithms on a quantum computer." *SIAM J. Comput.*, 26, 1484–1509.

[15] Shoup, V. (1997). "Lower bounds for discrete logarithms and related problems." *Advances in Cryptolog—EUROCRYPT'97*, Lecture Notes in Computer Science, vol. 1233, ed. W. Furny. Springer, Berlin, 256–266.

[16] Thomé, E. (2001). "Computation of discrete logarithms in GF(2^{607})." *Advances in Cryptography—ASIACRYPT 2001*, Lecture Notes in Computer Science, vol. 2248, ed. C. Boyd. Springer, Berlin, 107–124.

[17] Weber, D. (1998). "Computing discrete logarithms with quadratic number rings." *Advances in Cryptology—EUROCRYPT'98*, Lecture Notes in Computer Science, vol. 1403, ed. K. Nyberg. Springer, Berlin, 171–183.

[18] Weber, D. and T.F. Denny (1986). "The solution of McCurley's discrete log challenge." *Advances in Cryptology—CRYPTO'98*, Lecture Notes in Computer Science, vol. 1462, ed. H. Krawczyk. Springer, Berlin, 458–471.

E

E0 (BLUETOOTH)

E0 is a stream cipher, designed especially for Bluetooth communications (Bluetooth is a standard for wireless short-range connectivity, see [1]). As usual for stream ciphers, the main point is the keystream generation. For E0, it is derived from the summation generator, with four input LFSRs (see linear feedback shift register). The four LFSRs have lengths respectively, 25, 31, 33 and 39; their feedback polynomials are all primitive, with five nonzero terms. The global system looks like this:

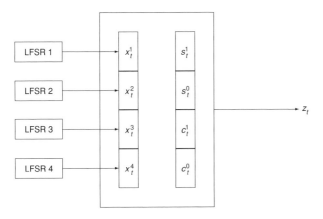

where t denotes the time, and the internal bits satisfy (in the three first equations, the addition is taken modulo 2; in the last one, it is the usual integer addition, followed by a rounding off downwards):

$$z_t = x_t^1 \oplus x_t^2 \oplus x_t^3 \oplus x_t^4 \oplus c_t^0$$
$$c_{t+1}^0 = s_{t+1}^0 \oplus c_t^0 \oplus c_{t-1}^0 \oplus c_{t-1}^1$$
$$c_{t+1}^1 = s_{t+1}^1 \oplus c_t^1 \oplus c_{t-1}^0$$
$$\left(s_{t+1}^0, s_{t+1}^1\right) = \left\lfloor \frac{x_t^1 + x_t^2 + x_t^3 + x_t^4 + c_t^0 + 2c_t^1}{2} \right\rfloor.$$

More precise details can be found in [1].

Several studies and attacks have been proposed [2, 3], but the more powerful is the one of Golic et al. [5]; they proposed a linear cryptanalysis for this cipher, based on the fact that the produced keystream sequence is short and that the system is frequently reinitialized. This attack is going with a work factor of $\mathcal{O}(2^{70})$, with a precomputing stage of complexity $\mathcal{O}(2^{80})$. The authors suggest to reinforce E0 according to the improvement techniques presented in [6].

A more recent paper, only available at the moment as a preprint, deals with an improvement of this attack [4].

Caroline Fontaine

References

[1] Bluetooth™ (2001). *Bluetooth Specifications*, Version 1.1.
[2] Ekdahl, P. and T. Johansson (2000). "Some results on correlations in the Bluetooth stream cipher." *Proceedings of the 10th Joint Conference on Communications and Coding*.
[3] Fluhrer, S. and S. Lucks (2001). "Analysis of the E_0 encryption scheme." *Selected Areas in Cryptography*. Springer-Verlag, Berlin.
[4] Fluhrer, S. (2002). "Improved key recovery of level 1 of the Bluetooth encryption system." IACR ePrint Archive 2002/068.
[5] Golic, J., V. Bagini and G. Morgari (2002). "Linear cryptanalysis of Bluetooth stream cipher." *Advances in Cryptology—EUROCRYPT 2002*, Lectures Notes in Computer Science, vol. 2332, ed. L. Knudsen. Springer-Verlag, Berlin, 238–255.
[6] Hermelin, M. and K. Nyberg (1999). "Correlation properties of the Bluetooth combiner." *Information Security and Cryptology—ICISC'99*, Lecture Notes in Computer Science, vol. 1787, ed. J. Song. Springer-Verlag, Berlin, 17–29.

EAVESDROPPER

An eavesdropper (see also Shannon's model) is a person or party who tries to get unauthorized access to data, e.g. by breaking into a computer system or tapping into a communication channel. The use of a proper cryptosystem should make it impossible for the eavesdropper to determine the meaning of an intercepted message. Meaningful plaintext has been replaced by unintelligible ciphertext.

A common distinction is between *passive* eavesdroppers, who only read or listen to the ciphertext, and *active* eavesdroppers who may replace a ciphertext by another one, retransmit a ciphertext at a different moment, insert their own texts, etc.

Jean-Jacques Quisquater

ECC CHALLENGES

In 1997 Certicom [1] issued a series of underline{elliptic curve discrete logarithm problem} (ECDLP) challenges. Each challenge asks for the solution of an ECDLP instance in $\langle P \rangle$, where P is a point of prime order n on an underline{elliptic curve} E defined over a underline{finite field} $\mathbb{F}_q$. The difficulty of the challenge is measured by the bitlength of the order n.

The challenges are of three kinds. In the following, ECCp-k denotes a randomly selected elliptic curve over a prime field, ECC2-k denotes a randomly selected elliptic curve over a characteristic two finite field $\mathbb{F}_{2^m}$ where m is underline{prime}, and ECC2K-k denotes a *Koblitz curve* (that is, an elliptic curve whose defining equation has coefficients in the binary field $\mathbb{F}_2$); k is the bitlength of n. In all cases, the bitlength of the order of the underlying field is equal to or slightly greater than k (so the elliptic curves have prime order or almost prime order). An underlined entry denotes that the challenge has been solved as of April 2004.

1. Randomly generated curves over prime fields: underline{ECCp-79}, underline{ECCp-89}, underline{ECCp-97}, underline{ECCp-109}, ECCp-131, ECCp-163, ECCp-191, ECCp-239, and ECCp-359.
2. Randomly generated curves over characteristic two finite fields: underline{ECC2-79}, underline{ECC2-89}, underline{ECC2-97}, underline{ECC2-109}, ECC2-131, ECC2-163, ECC2-191, ECC2-238, and ECC2-353.
3. Koblitz curves over $\mathbb{F}_2$: underline{ECC2K-95}, underline{ECC2K-108}, ECC2K-130, ECC2K-163, ECC2K-238, and ECC2K-358.

The challenges were solved using the parallel collision search variant of Pollard's ρ algorithm as explained in the underline{discrete logarithm problem} entry. The computation was performed on workstations distributed over the Internet. The hardest challenges solved to date were ECCp-109 and ECC2-109. Of the challenges that remain unsolved, the one that is expected to be the easiest to solve is ECC2K-130.

Darrel Hankerson
Alfred Menezes

Reference

[1] Certicom ECC Challenge. Available on http://www .certicom.com

ELECTROMAGNETIC ATTACK

INTRODUCTION: Kerchoff's laws (see underline{maxim's}) recommend basing cryptographic security solely on the secrecy of the underline{key} and not on the concealment of the underline{encryption} algorithm. A cryptosystem that uses some specific encryption method may, however, be imperfect as to its physical implementation. One or several leakages of all possible kinds may in that case provide an attacker with relevant information. Physical signals can often be used as a leakage source to conduct underline{side channel cryptanalysis} [9] (see also underline{side-channel attacks}) Time, power consumption or electromagnetic radiations can, for instance, be used. Electromagnetic radiation leakage has been known for a long time now, [6] and it also constitutes the subject of very recent research [11]. When analysing cryptographic implementations, the near and far field of cryptographic processors may offer a leakage source that should be seriously taken into account.

HISTORY: It is quite difficult to fix with precision the advent of side channel cryptanalysis. It even seems that this date is rather to be situated at the end of the XIXth century or at the very beginning of XXth. J. Maxwell has established its theory on electromagnetic waves in 1873. Some cross talk problems in telephone links were mentioned at the end of the XIXth century. The obtained information was only copied on another media in order to listen to it afterwards. In 1918, H. Yardley and its team discovered that classified information could leak from electric materials. This knowledge enabled them to rediscover the handled secrets. The data contained in a cryptographic device modulated a signal on the tape of a nearby recording source. The study of the IBM typewriter in the middle thirties indicated that the leakage of information was important and had to be seriously taken into account. Many years and some interception cases later, militaries seriously worried about this new threat and initiated the underline{TEMPEST} program. It is amusing to notice that the first analyses were based on electromagnetic radiation, rather than directly on the analysis of the consumed current. This is due to the ease of measuring the radiated field, which needs no physical access to the device, unlike consumption measurements.

In the early seventies, people began mentioning cases of interference between some of their electronic equipment. Electromagnetic radiation has even been described by standard, thus allowing the peaceful coexistence of various devices at the same time in the same place. All electronic devices are sensitive to outside disturbances and may in some cases, themselves, be disturbing elements. So an office computer can interfere with a radio receiver and this is the idea on which the study of electromagnetic fields emitted by processors is based upon [13, 20]. The idea has been applied to smart cards, and allowed, realizing that their radiation could easily be measured [8].

PRINCIPLES: It is thus possible to investigate radiations coming from electronic components while they are executing a sensitive computation involving some secret. A solution therefore is to measure the electromagnetic radiation of the chip during computation. The principles of Simple Power Analysis (SPA) and Differential Power Analysis (DPA) [10] (see differential power analysis) are based on consumption differences generated during the computation in function of the value of some key bit, which can be used to recover the key. Similarly, Simple ElectroMagnetic Analysis (SEMA) and Differential ElectroMagnetic Analysis (DEMA) allow retrieving the key as well, based on the same concept [14]. It is important to notice that refinements such as Automatic Template Analysis, Dictionary attacks, High Order DPA or Multiple bits DPA [10, 12] are also useful in the case of electromagnetic analyses.

size of the used amplifiers, their power and the quantity of changing bits. But it is possible to carry out the same measure in the case of leakage by consumption measurements. It is possible to decouple the power supply by judiciously placing a small capacitor between the power supply line and the ground of the device under measure. This is a principle that is well-known to electronics engineers, and all disruptions that are present on the power supply line will in this way go through the capacitor to end up at the ground. As previously, only the commutations are visible. By measuring the currents that transit through this capacitor, it is possible to highly reduce the number of samples that are required for a differential measure. This current measurement can be carried out by an electromagnetic measurement.

The figure depicted below shows two curves, representing the initialisation of a smart card, the execution of a DES (see Data Encryption Standard) and the stopping of the card. Both curves have signal to noise ratios that are very close, but the green curve, which details the current through the capacitor, requires twenty-five times less traces than the one that represents the current measurement.

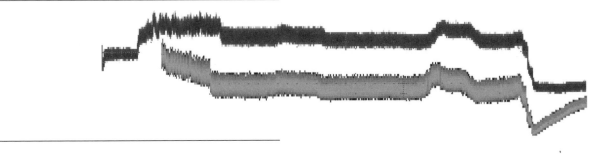

In the case of the current analysis, however, the only possible measure contains the sum of the contributions of all actions of the processor at a given moment. Computer architecture and the massive use of commutation electronics generate interesting properties for electromagnetic analysis. The modifications linked to the evolution of the clock characterize the system for the case of a synchronous processor and the bus transfers have a consumption that is proportional to the number of bit transitions between two cycles. It is indeed possible to easily use sensors that are only sensitive to commutations inside the chip. When the value of a logic gate is established and does not vary anymore, the only existing currents are continuous currents. During the commutation from one value to another, on the contrary, the involved currents contain very different frequency components. Every continuous current will indeed not provide any contribution to the sensor. The oscillators and commutation lines, on the contrary, provide contributions that are directly linked to the

As for power consumption measurements, the analysis of the currents, obtained with an electromagnetic field sensor can be performed in the time domain as well as in the frequency domain. The internal clocks of the components, as well as the oscillator of the charge pump required for the writing into some non-volatile memories, can in this way easily be found. In certain cases (charge pumping oscillator . . .), the sensitive information modulates the radiated signals and a simple demodulation suffices to recover the data [1, 14, 17].

The cards are well protected against a wide variety of attacks (see tamper detection, tamper response and tamper resistance) in order to avoid fault insertion [3, 4, 19] (see fault attack). As a consequence, they recreate their clock internally and by electromagnetic analysis it is possible to recover this clock, to re-amplify it and to multiply it in order to be able to synchronize the acquisition frequency of the device as well as possible.

One of the advantages of EMA is the locality principle. Using an adapted sensor it is possible to locally measure the field radiated by a chip [7, 15]. But the equations are much more complex than announced in [7], and the near field approximation does not require the same equations at all. Using a coil as the electromagnetic sensor and considering it as an adapted sensor is only a first order approximation.

ADVANTAGES: An advantage of Electromagnetic Analysis is that it allows obtaining at least the same result as power analysis [17]. But the most accurate information leakage model (see side-channel attacks) is based on bit commutations between two states. Moreover, for some so-called classes of 'bad' instructions, it allows to deduce results, where power analysis fails [5, 18]. So, EMA could be used to reduce the effectiveness of existing countermeasures against Power Analysis. Buses and registers constitute the major leakage sources. In addition, it is possible to use the possible leakage sources jointly. Rao and his team have shown that attacks that are based on signals which, taken together, do not necessarily present the best signal to noise ratio, can lead to satisfactory results [2]. Smart cards are particularly vulnerable, as they can hardly detect a listening material (see tamper detection, tamper response and tamper resistance). In certain cases electromagnetic analysis allows to recover PIN codes (see personal identification number) or, by applying an a little bit more complex approach (active sensors), to insert faults [16].

CONTEXT: Maxwell's equations indicate that it is possible to theoretically predict the radiation of a cryptoprocessor. The complexity of the computation is, however, often prohibitive and inhibits from using such a procedure. From this observation on, only an empirical approach, based on practice and experimentation, allows to rapidly obtain useful results. Once this step is taken, one can turn back to a numerical simulation and provide more reliable numerical values to the used model.

The practical approach is simple, but the principle of analysis by measuring the electromagnetic radiation is, however, more complex to put in place than the one that uses power consumption measurements. As for power analysis, a sensor and an acquisition system are sufficient to recover the sampled data. But sometimes the obtained signal has to be amplified before it can be correctly measured. Different sensors can therefore be used, but they do not all offer the same information. The measured spectrum also vary in the function of the implementations and the packaging of the components.

COUNTERMEASURES: There exist multiple countermeasures at the hardware level, but they are however not all well suited to all cases and have sometimes to be locally adjusted. The first of all these countermeasures is of course the use of a Faraday cage, in order to stop all kinds of radiation leakage. This countermeasure, although being ideally the most perfect one, is also the most difficult to put in place. There are multiple and heavy constraints when using a Faraday cage and they cannot always all be relaxed. In order to reduce radiation, a thin metal layer (ideally a ferromagnetic one) may sometimes suffice to render measurements more difficult. In cases where a Faraday cage cannot be used because of bounding wires, power supply lines, or simply because of mechanical constraints (the thickness of a smart card is fixed at 0.76 mm), one should define a protection zone around the device to be measured. But once again, this cannot always be done.

Electronic designs call more and more upon low consumption techniques. As a consequence, this reduces the commuted currents and thus reduces their radiation. These techniques are, however, not sufficient. One is then forced to use asynchronism techniques and classical DPA countermeasures (Dual rail logic and precharged logic ...) (see side-channel attacks). Some of the new architectures seem, however, to be able to break the locality principle and scatter around the computation over the processor.

CONCLUSION: Analysis by electromagnetic radiation has to be taken into account seriously, especially when it enables discovering cryptographic keys. Practical examples that could be threatened by such an attacks/are numerous (see hardware security module and EMV-standard). Therefore, there exist some evaluation criterions (see security evaluation criteria and countermeasures to stop this type of attack. Even if the signals are noisier, electromagnetic analysis has some serious advantages compared to power consumption analysis. The combination of both often allows reducing the required number of samples to recover the cryptographic keys, because of the improvement of the signal to noise ratio. The statistical analyses that can be performed afterwards are numerous and also allow improving the efficiency of the attack compared to a classical differential analysis. The use of side channels to recover a cryptographic key is primordial when a physical access to the device is at disposal. But electromagnetic analysis strongly depends on the architecture of the chip and some knowledge of the internal circuitry of the processor also highly facilitates the work.

Jean-Jacques Quisquater
Samyde David

References

[1] Agrawal, D., B. Archambeault, J.R. Rao, and P. Rohatgi (2002). "The EM side-channel(s)." *Proc. of the Cryptographic Hardware and Embedded Systems, CHES 2002, Redwood City*, Lecture Notes in Computer Science, vol. 2523, ed. B. Kaliski. Springer-Verlag, Berlin, 29–45. Also available on http://ece.gmu.edu/crypto/ches02/talks.htm

[2] Agrawal, D., J.R. Rao, and P. Rohatgi (2003). "Multi-channel attacks." *Proc. of the Cryptographic Hardware and Embedded Systems, CHES 2003, Cologne, Germany*, Lecture Notes in Computer Science, vol. 2779, ed. C. Walter. Springer-Verlag, Berlin, 2–16.

[3] Biham, E. and A. Shamir (1997). "Differential fault analysis of secret key cryptosystems." *Proc. of Advances in Cryptology—CRYPTO'97*, Lecture Notes in Computer Science, vol. 1294, ed. B.S. Kaliski Jr. Springer-Verlag, Berlin, 513–525. Also available on http://citeseer.nj.nec.com/biham97differential.html

[4] Boneh, D., R.A. Demillo, and R.J. Lipton (1997). "On the importance of checking cryptographic protocols for faults." *Advances in Cryptology—EUROCRYPT'97*, Lecture Notes in Computer Science, vol. 1294, ed. B.S. Kaliski Jr. Springer-Verlag, Berlin, 37–51. Also available on http://citeseer.nj.nec.com/boneh97importance.html

[5] Chari, S., J.R. Rao, and P. Rohatgi (2002). "Template Attacks." *Proc. of the Cryptographic Hardware and Embedded Systems, CHES 2002, Redwood City, USA*, Lecture Notes in Computer Science, vol. 2523, ed. B. Kaliski. Springer-Verlag, Berlin, 13–28. Also available on http://ece.gmu.edu/crypto/ches02/talks.htm

[6] http://www.cryptome.org

[7] Gandolfi, K., C. Mourtel, and F. Olivier (2001). "Electromagnetic attacks: Concrete results." *Proc. of the Cryptographic Hardware and Embedded Systems, CHES 2001, Paris, France*, Lecture Notes in Computer Science, vol. 2162, ed. D. Naccache. Springer-Verlag, Berlin, 251–256. Also available on http://www.gemplus.com/smart/r_d/publications/pdf/GMO01ema.pdf

[8] Hess, E., N. Jansen, B. Meyer, and T. Schutze (2000). "Information leakage attacks against smart card implementations of cryptographic algorithms and countermeasures." *Eurosmart, Proc. of the Eurosmart Conference, Nice, France*, 55–63.

[9] Kelsey, J., B. Schneier, D. Wagner, and C. Hall (1998). "Side channel cryptanalysis of product ciphers." *Proc. of ESORICS'98*, Lecture Notes in Computer Science, vol. 1485, eds. Quisquater, Deswarte, Meadows, and Gollmann. Springer-Verlag, Louvain la Neuve, Belgium, 97–110. Also available on http://www.schneier.com/paper-side-channel.html

[10] Kocher, P., J. Jaffe, and B. Jun (1999). "Differential power analysis." *Advances in Cryptology—CRYPTO'99*, Lecture Notes in Computer Science, vol. 1666, ed. M. Wiener. Springer-Verlag, Berlin, 388–397. Also available on http://www.cryptography.com/resources/whitepapers/DPA.html

[11] Kuhn, M., G. and Ross J. Anderson (1998). "Soft tempest: Hidden data transmission using electromagnetic emanations." *Proc. of Information Hiding, Second International Workshop, IH'98, Portland, Oregon, USA*, 124–142. Also available on http://www.cl.cam.ac.uk/~mgk25/ih98-tempest.pdf

[12] Messerges, T.S., E.A. Dabbish, and R.H. Sloan (1999). "Investigations of power analysis attacks on smartcards." *USENIX Workshop on Smartcard Technology*, 151–162. Also available on http://www.usenix.org/publications/library/proceedings/smartcard99/full_papers/messerges/messerges.pdf

[13] Muccioli, J.P. and M. Catherwood (1993). "Characteristics of near-field magnetic radiated emissions from VLSI microcontroller devices." *EMC Test and Design*.

[14] Quisquater, J.-J. and D. Samyde (2000). "A new tool for non-intrusive analysis of smart cards based on electro-magnetic emissions: The SEMA and DEMA methods." *Eurocrypt Rump Session, Bruges, Belgium*.

[15] Quisquater J.-J. and D. Samyde (2001). "Electromagnetic analysis (EMA): Measures and countermeasures for smart cards." *Proc. of the International Conference on Research in Smart Cards E-Smart 2001, Cannes, France*, Lecture Notes in Computer Science, vol. 2140, ed. I. Attali and T. Jensen. Springer-Verlag, Berlin, 200–210.

[16] Quisquater, J.-J. and D. Samyde (2002). "Eddy currents for magnetic analysis with active sensor." *Eurosmart, Proc. of the ESmart Conference, Cannes, France*, 185–194.

[17] Rao J.R. and P. Rohatgi (2001). {EMpowering} Side-Channel Attacks, preliminary technical report. Available on http://citeseer.nj.nec.com/cache/papers/cs/22094/http:zSzzSzeprint.iacr.orgzSz2001zSz037.pdf/rao01empowering.pdf

[18] Rao, J.R., P. Rohatgi, H. Scherzer, and S. Tinguely (2002). "Partitioning attacks or how to rapidly clone some GSM cards." *IEEE Symposium on Security and Privacy, Berkeley, CA*. Available on http://www.research.ibm.com/intsec/gsm.ps

[19] Skorobogatov, S. and R. Anderson (2002). "Optical fault induction attacks." *Proc. of the Cryptographic Hardware and Embedded Systems, CHES 2002, Redwood City, USA*, Lecture Notes in Computer Science, vol. 2523, ed. B. Kaliski. Springer-Verlag, Berlin, 2–12. Also available on http://ece.gmu.edu/crypto/ches02/talks.htm

[20] Slattery, K.P., J.P. Muccioli, and T. North (2000). "Modeling the radiated emissions from microprocessors and other VLSI devices." *IEEE 2000 International Symposium on Electromagnetic Compatibility.*

ELECTRONIC CASH

Electronic Cash is a self-authenticating digital payment instrument that can be stored in an electronic wallet (or *electronic purse*) just like traditional cash is stored in a traditional wallet. Electronic cash is a sort of pre-paid electronic payment, i.e., payers withdraw electronic cash from their bank accounts prior to making a purchase and payment. To make a payment the payer simply passes the required amount of electronic cash to the payee. The payee is not referred to any bank account of the payer.

Like electronic payment schemes in general, electronic cash schemes shall satisfy the following security requirements:

Payment authorization: Electronic cash typically comes in the form of *electronic coins* of various face values, which have attached a digital signature by the issuing bank. Any payee can immediately verify the validity of such elecronic coins by checking them against the public verifying key of the respective issuing bank.

No counterfeiting: The overall value of all payment instruments shall not be increased without further action by an authorized minting bank. This is partly achieved by the digital signatures of electronic coins, which ensure that electronic coins cannot be forged. However, the digital signatures alone cannot prevent cheating payers from overspending. In some systems, such attacks can be detected after the fact, but they cannot be prevented unless electronic wallets employ a piece of tamper resistant hardware that controls the spending of coins effectively. A moderate level of tamper resistance can be achieved by smartcards. This approach is taken, e.g., by [4]. Stronger levels of tamper resistance can only be achieved by more tamper responsive electronic wallets.

Confidentiality: Certain payment information may be required to be kept confidential from prying eyes (privacy). The purchase content, the payment amount, or the time of payment shall not be disclosed to individuals not involved in the transaction. This is usually achieved by using a point-to-point connection between the payer's electronic wallet and the payee's merchant device (e.g., a point of sale terminal), or by using an SSL/TLS tunnel over the Internet (see Secure Socket Layer and Transport Layer Security).

Payer anonymity and payment unlinkability can be achieved by electronic coins that are statistically independent of the payers who use them. According to work by Chaum and Brands [1, 3] this can be achieved by blind signatures as follows: When a payer opens an account with her issuing bank, she identifies herself to the issuer and establishes a role pseudonym to be used for her withdrawels of electronic coins from the issuer. When the payer withdraws an electronic coin from her bank account, the issuer provides a blind signature for the payer's role pseudonym. Different public verifying keys can be used to encode different face values of electronic coins, such that the face value cannot be changed by the blinding of signatures. The payer then transforms the blind signature into a signature for a one-time pseudonym of the payer. The resulting electronic coin is statistically independent of any other electronic coin of this and every other payer, thus achieving payer anonymity and payment unlinkability even against computationally unlimited payee's who collaborate with the issuer to figure out origins of electronic coins. The remarkable property of Brands proposal is that he shows how to construct the payment protocol such that the payer automatically loses her anonymity once she spends any of her electronic coins twice. The payment protocol ensures that such cheating will immediately reveal to the payee the role pseudonym the payer uses with her issuer. Thus, Brands electronic cash scheme achieves overspender detection even if implemented without tamper resistant electronic wallets.

Reliability: The payment transaction must be atomic in the sense that it is either processed completely or not at all. Even if the network or system crashes, there must be recovery mechanisms in place that either allow to re-synchronize the devices of all participants automatically or at least enable all participants to make their just claims. This is usually not addressed in the cryptographic literature, and for many electronic cash schemes actually in use it is not described in great detail.

Nacchache and van Solms [5] pointed out that anonymous electronic cash can be misused if there is no way of revoking the payer's anonymity in case of suspected money laundering and other kinds of financial abuse. Their work sparked more sophisticated proposals of anonymous electronic cash, for example by Brickell, Gemmell and Kravitz [2] where centralized or decentralized trustees are capable of revoking the anonymity from electronic coins.

Electronic cash provides customers a way of offline underline{electronic payment}, i.e., no bank or other underline{trusted third party} is involved in the payments. This may be appealing to certain groups of customers, but it is not favored by banks, and banks have argued against offline electronic payments by saying it is less secure than online electronic payments.

Another practical issue in any electronic cash product is how customers are protected against loss of electronic coins in critical cases such as when their electronic wallets fail or if disaster or bankruptcy strikes their issuer.

It is thus conceivable that electronic cash will remain a method of payment for smaller amounts, while online electronic payments methods will remain to be used for larger amounts.

Gerrit Bleumer

References

[1] Brands, S. (1994). "Untraceable off-line cash in wallet with observers." *Advances in Cryptology—CRYPTO'93*, vol. 773, ed. D.R. Stinson. Springer-Verlag, Berlin, 302–318.

[2] Brickell, E., P. Gemmell, and D. Kravitz (1995). "Trustee-based tracing extensions to anonymous cash and the making of anonymous change." *6th ACM–SIAM Symposium on Discrete Algorithms (SODA), 1995*. ACM Press, New York, 457–466.

[3] Chaum, D. (1983). "Blind signatures for untraceable payments." *Advances in Cryptology—CRYPTO'82*, Lecture Notes in Computer Science, ed. D. Chaum. Plenum, New York, 199–203.

[4] http://www.mondex.com

[5] Naccache, D. and Sebastiaan von Solms (1992). "On blind signatures and perfect crimes." *Computers & Security*, 11 (6), 581–583.

ELECTRONIC CHEQUE

This term is used quite freely and could mean anything from an electronic payment instruction of some kind to a digitally signed electronic counterpart of a paper based cheque, and may even be considered as the so-called negotiable instrument (as opposed to a crossed cheque, which may not be forwarded as a payment but needs to be cashed).

Peter Landrock

ELECTRONIC NEGOTIABLE INSTRUMENTS

A negotiable instrument is a document which, according to law, can be traded freely, such as cash, endorsable cheques or Bills of Lading (which actually is only quasi-negotiable). All of these types of documents may appear in electronic form as well. The only challenge is to prevent what is known as double-spending (e.g., for cash). This can be achieved in two ways: either by having an on-line underline{Trusted Third Party} to keep track of ownership, or by using tamper resistant devices to prevent double-spending.

Peter Landrock

ELECTRONIC PAYMENT

Since the Internet spread beyond the research communities and made significant inroads into the commercial world, more and more customers became connected to the Internet. Customers first got equipped with personal computers, then with palm pilots, and more recently with cell phones. By the end of the 1990s most customers in the developed countries were hooked to the Internet by one device or another. The wide availability of customer devices and the Internet itself sparked the development of electronic payment instruments throughout the 1980s and 1990s and many of them have been put to trial.

In traditional payment systems as well as in electronic payment systems, payers and payees keep and manage their money in bank accounts. The payer's bank is sometimes called the *issuer*, while the payee's bank is called the *aquirer*. A payment system is a way to move a specified amount of money from the payer's bank account into the payee's bank account in a secure fashion. In order to transfer money from the payer's account at the issuer to the payee's account at the acquirer, the payer and payee can use various electronic payment instruments. All electronic payment instruments are electronic representations of cash, a payment order, funds transfer order, or the like that authorizes the transfer of a specified amount of money. Electronic payments can be initiated by the payer or by the payee.

In *indirect payment systems*, the payer initiates a payment with the acquirer into the payees bank account, or the payee initiates a payment with the issuer from the payer's account. In either case, the payer and payee have no online interaction during the payment. Respective examples are *electronic funds transfer* and *automatic clearing house (ACH)*.

In *direct payment schemes*, the payer and payee interact online during the payment while connecting their devices, either directly, e.g., by inserting the payer's card into the payee's card reader and terminal or by a point-to-point IR connection, or by connecting the two devices through a wired or wireless network. A direct payment scheme is called *online* if the payment protocol requires the issuer or the acquirer to participate in the payment protocol online. Otherwise, it is called *offline*. Online payment schemes are perceived as more secure because each payment transaction is overseen by an issuer or by an acquirer, who are regarded as trusted participants. Offline (direct) payment schemes require payers to use electronic wallets, i.e., secure hardware devices, in order to prevent overspending (overspending prevention). Direct payment schemes can be classified as follows:

Pre-pay: At the time of payment, the payee's bank account is credited, but the payer has to have withdrawn a sufficient amount of money from her or his accounts BEFORE making the payment. This is usually called electronic cash.

Pay-now: At the time of payment, the bank account of the payers is debited and the bank account of the payee is credited. Examples are electronic checks and debit cards.

Pay-later: At the time of payment, the payee's bank account is credited, but the payer's bank account is debited some time later. Typical examples are electronic credit cards.

Payment schemes must satisfy a number of security requirements:

Payment authorization: Payers shall not find money deducted from their accounts without their consent. Thus, all payments shall be authorized at least by the payer. This will not necessarily imply that payers have to authenticate their identity to payees. In pre-pay systems, i.e., e-cash systems, the payment instruments are self-authenticated, and payers may remain anonymous.

A payer can authorize a payment by out-of-band means such as by phone or regular mail. This is common with credit cards payments for phone orders or mail orders. In lasting business relationships, the payer and payee can agree on a shared secret such as a password, passphrase or PIN. The payer then needs to type the shared secret in order to authorize a payment to the payee sharing the secret. The highest degree of security can be achieved if the payer uses a digital signature to authorize payments. Distributing the respective public verifying keys requires a public key infrastructure (PKI), but ensures non-repudiation, i.e., only the intended payer is capable to produce a signature for the payment with respect to the public verifying key certified to the payer's name.

No counterfeiting: The overall value of all payment instruments cannot increase without further action by an authorized minting bank. In other words, payees shall not find their accounts credited without anyone actually paying for this amount.

Confidentiality: Certain payment information may be required to be kept confidential from prying eyes (privacy). The purchase content, the payment amount, or the time of payment shall not be disclosed to individuals not involved in the transaction. If anonymity of payer or payee, unlinkability of payments or untraceability of payments are an issue, then the identities of the payer and/or payee must be disclosed neither to outsiders nor to certain participants of the payment transaction.

Reliability: The payment transaction must be atomic in the sense that it is either processed completely or not at all. Even if the network or system crashes, there must be recovery mechanisms in place that either allow to re-synchronize the devices of all participants automatically or at least enable all participants to make their just claims.

In order to support frequent payments of small amounts, typically less than 1\$ each, special micro payment schemes have been proposed. They involve no complex cryptographic computations for the payment itself, but require some overhead between the payer and payee in order to set up the micropayment option. Typical applications are pay-per-view or pay-per-click or pay-per-phone tick. If micropayments between a payer and a payee are so rare that even the small overhead to set up the micropayment option is not justified, then they can still use a micro payment scheme, such as μ-iKP [2], by employing a broker who frequently receives micro payments from the payer and makes micropayments to the payee. This way, one leverages on existing business relationships spanning from the payer over the broker to the payee.

The predominant standard for on-line pay-now schemes (electronic checks) is SET, the Secure Electronic Transactions [3] standard, a merger of VISA's Secure Transaction Technology (STT) and

Master Card's Secure Electronic Payment Protocol (SEPP). Marketing, branding and compliance testing is organized by SetCo, Inc., a joint subsidiary of VISA and MasterCard.

There is a large and quickly changing variety of proposals for electronic payment schemes; some more directed to research activities, others striving for market share. A good overview is given by Asokan, Janson, Steiner and Waidner in [1].

Gerrit Bleumer

References

[1] Asokan, N., P.A. Janson, M. Steiner, and M. Waidner (1997). "The state of the art in electronic payment systems." *Computer*, 30 (9), 28–35.

[2] Bellare, M., J. Garay, R. hauser, A. Herzberg, H. Krawczyk, M. Steiner, G. Tsudik, E. Van Herreweghen, and M. Waidner (2000). "Design, implementation and deployment of the iKP secure electronic payment system." *IEEE Journal on Selected Areas in Communications*, 18 (4), 611–627.

[3] http://www.setco.org/set_specifications.html

ELECTRONIC POSTAGE

Electronic postage is a way to pay for postal transportation services in an electronic way. Customers who have less than five or ten mailpieces to send per day on an average will use stamps, and customers who have several hundreds of mailpieces of equal weight and size will use rebated bulk mail options. For many other customers, electronic postage is a convenient option. Electronic postage comes in two form factors, as a software application running on a regular personal computer, or built right into a desktop printer, or integrated into a postage metering device [5].

Large Postal Services such as the US Postal Services, Deutsche Post AG, and Canada Post Corporation have started initiatives that will replace mechanical postage metering devices in their respective Postal markets by electronic metering devices within 3 to 5 years. Other Postal Services are likely to follow these initiatives because electronic metering devices reduce the risk of fraud significantly, and they support the integration of value added features such as track and trace services.

The first specification of electronic postage was published by the US Postal Services in 1996 [7]. The first publication specified *closed systems*, i.e., postage metering devices that couple the electronic postage vault together with the printing mechanism. It was later accompanied by a spec-

ification of *open systems*, which means systems based on a regular personal computer connected to a desktop printer. Both specifications enforced that electronic postage would only be stored inside certified hardware security modules, which were called *postal security devices*. In closed systems, the postal security devices would be integrated within the postage metering devices in order to faciliate high throughputs of mailpieces. For open systems, there were two options. Either the postal security device was held inside a server at the postage provider, such that customers could use some application software in order to download postage every time they had to produce an indicium for a mail piece. This approach is called *online PC postage*. The other option was to build postal security devices into the customers personal computers, which would then be used more or less like a postage meter. This approach is called *offline PC postage*. In 1999, a third type of system was specified, called *centralized systems*, where customers employ one postal security device in a central location of a network, and connect several printers or postage metering devices (without built-in postal security devices) to the network, for example, one for each department. Each printer or postage metering device would then receive its indicia from the central postal security device. In practice, profitable business models have only been developed for electronic postage metering devices and for online PC postage (see listings at [3, 7]).

The main idea behind the Information Based Indicia Program (IBIP) is this: each postal security device serves as a secure storage for pre-paid electronic postage, and produces a digital signature for each mail piece a customer is going to send. Typically, the digital signature is produced in real time, such that all actual mail piece parameters such as weight, size, mail category, date of mailing, etc. can be taken into account by the digital signature. All parameters and the digital signature of a mail piece are encoded into a two-dimensional barcode, which is printed in the upper right corner of an envelope. A similar kind of bar code can also be used for parcels. Typically, those barcodes are printed to a label, which the customer affixes to the respective parcel. IBIP specifies which signature algorithms are permitted (the RSA digital signature scheme, the Digital Signature Standard (DSS), and Elliptic curve signature schemes (ECDSA)), which minimum resolution to be used for printing indicia, and which barcode symbologies are permitted (PDF417 and Datamatrix [1]). IBIP also specifies the length of the keys to be used for the digital signature scheme, and in case of

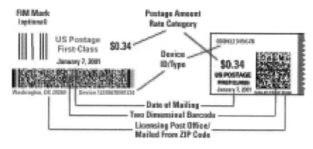

Fig. 1. Sample indicia

ECDSA specifies a set of permitted elliptic curves based on recommendations by NIST [4]. The size of the footprint of the resulting barcodes is between 3×1 inches (PDF417), and 1 by 1 inch (Datamatrix). Sample indicia with either bar code symbology are displayed in Figure 1.

The postal mail sorting centers have to be equipped with CCD cameras in order to read the barcodes. Furthermore, the postal infrastructure must be enabled to decode the mail piece parameters and the respective digital signatures from each barcode and verify these digital signatures. In order to do this, the postage providers are required by IBIP to submit the public verifying keys of each postal security device under their or one of their customers operation to the US Postal Services. The US Postal Services have set up a Public Key Infrastructure (PKI) to which each postage provider must get registered before it can be approved to provide electronic postage to the US postal market.

The postal security devices must be certified according to FIPS 140-2 Level 3 with additional requirements on physical security [4]. They must have an active tamper response mechanism that permanently shuts off the signing functionality of the postal security device as soon as specified attempts of tampering are detected. The operating software of postal security devices must be designed to implement a finite state machine specified by IBIP. An important part of the internal state of a PSD is its set of postal registers that keeps track of its electronic postage. Mainly, there is a descending register, an ascending register and a control total register, which are initially set to zero. The descending register is increased by a respective amount every time the postal security device downloads electronic postage from the postage provider, and it is decreased by the face value of a requested indicium every time the postal security device produces a signature. The ascending register is increased by the face value of a requested indicium every time the postal security device produces a signature. The total setting register is increased by a respective amount every time the postal security device downloads electronic postage from the postage provider. At any point in time during the life cycle of a postal security device, these three postal registers shall observe the following relation:

$$\text{descending register} + \text{ascending register}$$
$$= \text{total settings register} \qquad (1)$$

Other Postal Services such as Deutsche Post AG [3] or Canada Post Corporation have specifications that rely on the same principles as the Information Based Indicia Program including the use of FIPS 140-2 certified hardware security modules with the above mentioned postal registers, but may specify other types of digital signatures, key lengths, elliptic curves, or integrity check codes based upon message authentication codes in place of digital signatures.

The deployment of CCD cameras, high speed cryptographic equipment for verifying digital signatures and a countrywide distributed database for detecting and rejecting replays of indicia are the major investments for any Postal Service making the transition towards electronic postage. Since the early 1990s there is also increasing activity into innovation and patenting by the postage providers. Because of these commitments, electronic postage can be expected to become an industrial application area of security and cryptography mechanisms.

All of these specifications value traceability of customers higher than their privacy. For example, there is no option in any of these specifications to send mail anonymously. The likely reasons are firstly that the by far largest group of customers are companies, who naturally have an interest to be recognized, not to remain anonymous, and secondly, that the main motivation of the Postal Services to retire the mechanical postage meters in their markets was to reduce fraud by meter manipulation. Nevertheless, if electronic postage should take over significant market share from conventional stamps then privacy probably becomes a customer requirement to be addressed [3]. Sending mail anonymously can be misused to cover up attacks against mail recipients by using mail bombs or contaminated mail pieces. But making products available that allow anonymous electronic postage would neither encourage such attacks, nor would prohibiting these products prevent such attacks.

Electronic postage for postcards has been addressed by Pintsov and Vanstone in [6].

Gerrit Bleumer

References

[1] ANSI/AIM BC11. International Symbology Specification—Data Matrix; 11/96. Available on http://www.rvsi.com/cimatrix/DataMatrix.html

[2] Bleumer, G. (2000). "Secure PC-franking for everyone." *EC-Web 2000*, Lecture Notes in Computer Science, vol. 1875, eds. K. Bauknecht, S. K. Madria, and G. Pernul. Springer-Verlag, Berlin, 94–109.

[3] Deutsche Post AG. http://www.deutschepost.de/stampit

[4] National Institute of Standards and Technology (2002). "Security requirements for cryptographic modules." FIPS PUB 140-2. Available on http://csrc.nist.gov/cryptval/140-2.htm

[5] Pastor, J. (1991) "CRYPTOPOST. A cryptographic application to mail processing." *Journal of Cryptology*, 3 (2), 137–146.

[6] Pintsov, L.A. and Scott A. Vanstone (2002). "Postal revenue collection in the digital age." *Financial Cryptography 2000*, Lecture Notes in Computer Science, vol. 1962, ed. Y. Frankel. Springer-Verlag, Berlin, 105–120.

[7] United States Postal Services. "Information based indicia program." Available on http://www.usps.gov/postagesolutions

ELECTRONIC VOTING SCHEMES

A system is called an electronic voting system when ballots are directly recorded electronically. Standards for such systems have been set by the Federal Election Commission. In most such systems used today, voters go to designated polling places, and cast their votes electronically after being identified and authorized by conventional, non-electronic means. In the near future, we envision systems in which voters securely send their ballots to the authorities over a network. Current research on electronic voting schemes is concerned with maintaining the privacy of the ballots while ensuring their validity, and with reliably verifying the final tally.

The general technique of secure multiparty computation can be applied to solve this problem in theory. However, such a solution places a heavy computational load on the voter's computer, and requires transaction with all other voters, making it impractical. A more realistic approach is to establish voting authorities and develop protocols specific to voting so that a voter only needs to communicate with the authorities, with a moderate computational and communication cost. The goal of such protocols is to prevent the authorities from learning who voted for whom (or what), while enabling them to check the validity of the votes and compute the correct tally.

Two approaches for maintaining the privacy of votes are: (1) hiding the voter's identity and (2) hiding the vote data. In the former approach, we assume that there is an anonymous channel by which authorities can receive a ballot, but cannot determine who sent it. For this approach it is important to check the validity of such anonymous ballots, and to verify that no one has voted multiple times or has voted when they are not authorized to vote. In the latter approach, the authorities can identify the voter but the ballots are encrypted. Therefore, it is necessary to employ techniques enabling the voting authorities to tally the ballots without decrypting each ballot individually. Below we outline mechanisms to solve the problems in each strategy.

A key technique in the former approach is blind signatures. A blind signature scheme enables a signer to sign a message without seeing it. This may sound paradoxical, but is a powerful tool in achieving privacy in voting and payments systems (see electronic cash). Prior to voting, the authorities identify the voter and, using the blind signature scheme, issue a digital signature to his ballot. Thus the authorities cannot learn the vote. During the voting phase, the voter anonymously sends the signed ballot.

The authorities can check the validity of the ballot by verifying their own signature on it. The use of blind signature scheme prevents the authorities from learning to which voter the signature was issued. Along with blind signatures, we need a mechanism for distinguishing between two valid ballots from two different voters with the same choice and two or more ballots from a single voter. This can be done by including random sequences in the ballot format. By making a long enough sequence, the probability that two distinct voters choose the same sequence can be made negligibly small. If two ballots are ever submitted with the same sequence, they are assumed to be copies of a single legitimate ballot, and only one is counted.

This random sequence can also serve as a key to verify the authority's activity. By searching for this sequence in a published list of the accepted ballots, a voter can confirm that his vote was indeed counted in the tally. The ballot format should be designed to prevent a voter from receiving multiple valid ballots in a single blind signature issuing procedure.

For the second approach, we use a special encryption property that enables the tallying of votes without decrypting each vote. The property is called homomorphism. If we represent yes-votes by 1 and no-votes by 0, the sum of votes gives the number of yes-votes. If the votes are encrypted using a homomorphic encryption function, the encrypted votes may be combined to create an encryption of the sum of these votes, without decrypting any of the votes. Decrypting the encrypted total gives the final tally. In order to use this idea, we must use <u>probabilistic public-key encryptions</u>; otherwise, all encrypted yes-votes would look exactly the same, and it would be straightforward to determine how people voted.

Another method of tallying encrypted votes while preserving privacy is by using a cryptographic *shuffling procedure*. In this method, the authorities do decrypt each ballot but only when the ballots are shuffled so that the voter's identity cannot be matched to any voter list. In cryptographic shuffling, the list of encrypted ballots is transformed to another list of encrypted ballots, where the order of the entries is shuffled, so that no one can determine the correspondence between entries in the new list and entries in the old list. To do so, the appearance of each encrypted entry must be changed without altering its decrypted result. We use an encryption scheme that is *malleable* in this regard: that is, this procedure can be performed by a party without knowing the decryption key. An example of such a scheme is <u>ElGamal public key encryption</u>.

Each of the two strategies described above has its own advantages and disadvantages; each is based on different assumptions. The former strategy requires an anonymous channel. The latter requires that all voters trust the authorities holding the decryption key (in more sophisticated schemes, it is only necessary to trust that one of the authorities is honest). Verifiability of the tally is also different. In the former, one can verify only his or her own vote, while in the latter, anyone can verify the integrity of voter signatures on the accepted vote list and the correctness of the sum based on the list.

Neither strategy has a good solution for the *receipt-free* problem. That is, a malicious voter (or a coerced innocent voter) generates 'a receipt' or a proof of how he or she voted to trade for money (or to satisfy the coercer). Current solutions require physical assumptions or work on a limited coercer model.

Kazue Sako

References

[1] Chaum, D. (1981). "Untraceable electronic mail, return addresses, and digital pseudonyms." *Communications of the ACM*, 24 (2), 84–88.

[2] Chaum, D. (1985). "Security without identification: Transaction systems to make big brother obsolete." *Communications of the ACM*, 28 (10), 1030–1044.

[3] Cramer, R., R. Gennaro, and B. Schoenmakers (1997). "A secure and optimally efficient multi-authority election scheme." *European Transactions on Telecommunications*, 8, 481–489. Preliminary version in *Advances in Cryptology—EUROCRYPT'97*.

[4] Fujioka, A., T. Okamoto, and K. Ohta (1992). "A practical secret voting scheme for large scale elections." *Advances in Cryptology—AUSCRYPT'92*, Lecture Notes in Computer Science, vol. 718, ed. T. Okamoto. Springer-Verlag, Berlin, 244–251.

[5] Hirt, M. and K. Sako (2000). "Efficient receipt-free voting based on homomorphic encryption." *Advances in Cryptology—EUROCRYPT 2000*, Lecture Notes in Computer Science, vol. 1807, ed. B. Preneel. Springer-Verlag, Berlin, 539–556.

ELECTRONIC WALLET

In a general business sense, an electronic wallet (or electronic purse) is a consumer device providing some additional security compared to a mere credit card solution. An electronic wallet could be as simple as an encrypted storage of credit card information that saves consumers to re-enter their credit card data manually each time they make a payment. There is a considerable variety of products and services, each called "electronic wallet" or "electronic purse", that turn up in the marketplace and in investors' press conferences, while the technical specifications and the life time of these products and services are left unclear. We will thus concentrate on the more specific use of the term "electronic wallet" in the cryptographic literature.

In a more specific, cryptographic sense, an electronic wallet is a consumer device designed to store and manage electronic funds or <u>electronic cash</u>. In particular, an electronic wallet is used to download funds from a bank account, to store those funds inside the electronic wallet and transfer deliberate amounts to other electronic wallets or point of sale terminals in order to make purchases. Even, Goldreich and Yacobi [6] were one of the first to consider electronic wallets as consumer devices with keypad and display allowing their holders to also inspect the current amount of stored electronic cash and configure their own devices.

If such electronic wallets have an offline electronic cash scheme installed, consumers experience increased autonomy because they can make payments offline and manage the amounts of electronic cash remaining in their e-wallets offline, i.e., without a bank or any other trusted third party being involved in the payments or management actions. Such electronic wallets differ from ATM cards, debit cards or credit cards, which are merely tokens that identify their holders to access the respective bank accounts or allow payees to deduct certain payment amounts from the respective payers' bank accounts.

An electronic wallet shall satisfy security requirements of their holders and of the banking industry operating the financial infrastructure behind those electronic wallets. Pfitzmann, Pfitzmann, Schunter and Waidner [8] distinguish the following three security requirements on mobile user devices in general and on electronic wallets in particular:

Personal agent trust: In normal operation, consumers demand their electronic cash to be stored reliably by their electronic wallets. All sorts of mishaps, e.g., hitting unintended commands at the keypad, interrupting communication lines, or experiencing downtimes of the banking servers, shall all be smoothly tolerated by the electronic wallets without losing a penny of electronic cash.

Consumers may have additional privacy requirements such as anonymity or unlinkability of transactions, or untraceability of their payments. As a minimum privacy requirement they may want to prevent payment providers, payees and others to implant cookies on their electronic wallets, which could be used to trace their purchase behavior.

Captured agent trust: In exceptional cases where an electronic wallet is lost or stolen, attackers shall find it infeasible to misuse a captured electronic wallet in order to pay for their own dealings. In general, a captured electronic wallet shall stop all its services and render itself totally useless to a potential attacker. In situations where a consumer is blackmailed and forced to authorize a certain payment on behalf of an attacker, for example by entering a password, there should be mechanisms in place by which the victim could purposely lock up his electronic wallet and optionally set off some form of alarm, preferably some wireless alarm that goes unnoticed by the attacker [2].

Undercover agent trust: The banking industry demands that legitimate holders cannot misuse their electronic wallets, for example, by manipulating electronic wallets such that the amount of electronic cash is increased without being paid for, or by spending electronic cash more than once, or by using premium services although they should not be available to a consumer because they have not been paid for. In a certain sense, an electronic wallet is a pocket branch of the issuing bank in the hands of a consumer.

Like with any real branch, it should be infeasible to break in and steal the assets.

Most commercially available electronic wallets are based on smart cards in order to keep the price tag low. Prominent examples are Mondex [7] and [4] CEPS (Common Electronic Purse Specifications), which is based on the EMV specification by Europay, MasterCard, and Visa. Smart cards, however, can hardly meet all of these security requirements with a high level of assurance (see Security Evaluation Criteria). For example, in order to achieve undercover agent trust, an electronic wallet had to have some active tamper response mechanism, which would lock up the electronic wallet in case it detects an attempt of tampering with the secure housing of the circuitry. Security assurance can be evaluated and certified according to the industry standard FIPS 140-2.

In a landmark paper, Chaum and Pedersen [3] have proposed a hardware architecture for electronic wallets that can support all of the above security requirement at the same time. The main idea is to embed a tamper resistant piece of hardware, which is called an *observer*, into a larger piece of hardware, called a *wallet*, which is totally under the holder's control. Chaum and Pedersen devised protocols for the:

- **holder withdrawing** e-cash from her account at an issuer into her wallet with observer,
- **holder paying** chunks of e-cash from her wallet with observer to a payee (merchant) to make a purchase, and
- **merchant depositing** paid e-cash into his account at an acquirer.

Each of these protocols includes actions of the respective issuer or acquirer, the wallet and the observer. The observers are all certified by the banking infrastructure behind the issuer and the acquirer and the observers' task in all the protocols is to authenticate the holder's wallet to the respective issuer or acquirer and to authorize the requested transaction. For example, the observer keeps track of the amount of funds remaining inside its wallet. If the holder requests to spend an amount that exceeds the available funds in her wallet, then the observer will not authorize the payment and the payee will not acknowledge the payment. However, the observer has no way of communicating directly with the issuer or the acquirer. All communication between the observer and an issuer or acquirer is relayed through the wallet, which transforms all messages in such a way that it preserves the authentication and authorization by the observer, but keeps the payment transactions of all holders statistically unlinkable, even if the issuers and the payees all collaborate against the payers.

Cramer and Pedersen [5] improved this proposal by achieving the same strong unlinkability of payments even if the attacker coalition of issuers and payees captures the wallets with the observer and manages to extract all information stored by the observer.

Brands [1] proposed an independent suite of protocols for withdrawal, payment, and deposit that achieved the same level of security as Cramer and Pedersen [5] and the additional feature that a holder who spends any piece of electronic cash more than once would automatically revoke its own anonymity, such that the acquirer who collects two or more deposits originating from the same piece of electronic cash would efficiently recover the cheating holder's identity.

Gerrit Bleumer

References

[1] Brands, S. (1994). "Untraceable off-line cash in wallet with observers." *Advances in Cryptology—CRYPTO'93*, Lecture Notes in Computer Science, vol. 773, ed. D.R. Stinson. Springer-Verlag, Berlin, 302–318.

[2] Chaum, D. (1992). "Achieving electronic privacy." *Scientific American*, 267, 96–101.

[3] http://www.cepsco.com

[4] Chaum, D. and Torben Pedersen (1993). "Wallet databases with observers." *Advances in Cryptology—CRYPTO'92*, Lecture Notes in Computer Science, vol. 740, ed. E.F. Brickell. Springer-Verlag, Berlin, 89–105.

[5] Cramer, R. and Torben Pedersen (1994). "Improved privacy in wallets with observers" (Extended abstract). *Advances in Cryptology—EUROCRYPT'93*, Lecture Notes in Computer Science, vol. 765. ed. T. Helleseth. Springer-Verlag, Berlin, 329–343.

[6] Even, S., O. Goldreich, and Y. Yacobi (1984). "Electronic wallet." *Advances in Cryptology—CRYPTO'83*, ed. D. Chaum. Plenum, New York, 383–386.

[7] http://www.mondex.com

[8] Pfitzmann, A., B. Pfitzmann, M. Schunter, and M. Waidner (1997). "Trusting mobile user devices and security modules." *Computer*, 30 (2), 61–68.

ELGAMAL DIGITAL SIGNATURE SCHEME

The ElGamal signature scheme [1] is one of the first digital signature scheme based on an arithmetic modulo a prime (see modular arithmetic). It can be viewed as an ancestor of the Digital Signature Standard and Schnorr signature scheme.

ElGamal signatures are much longer than DSS and Schnorr signatures. As a result, this signature scheme is not used often and is mostly of interest for historical reasons. We present a small modification of the original scheme that includes a hash function needed for the security analysis:

Key Generation. Given a security parameter $\tau \in \mathbb{Z}$ as input do the following:
1. Generate a random τ-bit prime p. Pick a random generator $g \in \mathbb{Z}_p^*$ of the group $\mathbb{Z}_p^*$.
2. Pick a random integer $\alpha \in [1, p-2]$ and compute $y = g^\alpha \in \mathbb{Z}_p^*$.
3. Let H be a hash function $H : \{0,1\}^* \to \{1, \ldots, p-2\}$.
4. Output the public key (p, g, y, H) and the private key (p, g, α, H).

Signing. To sign a message $m \in \{0,1\}^*$ using the private key (p, g, α, H) do:
1. Pick a random integer $k \in [1, p-2]$ with $\gcd(k, p-1) = 1$.
2. Compute $r = g^k \bmod p$. We view r as an integer $1 \le r < p$.
3. Compute $s = k^{-1}(H(m\|r) - \alpha r) \bmod p - 1$. Here $m\|r$ denotes concatenation of m and r.
4. Output the pair $(r, s) \in \mathbb{Z}_p^*$ as the signature on m.

Verifying. To verify a message/signature pair $(m, (r, s))$ using the public key (p, g, y, H) do:
1. Verify that $1 \le r < p$, otherwise reject the signature.
2. Compute $v = y^r r^s \in \mathbb{Z}_p$.
3. Accept the signature if $v = g^{H(m\|r)} \in \mathbb{Z}_p$. Otherwise, reject.

We first check that the verification algorithm accepts all valid message/signature pairs. For a valid message/signature pair we have:

$$v = y^r r^s = g^{\alpha r} g^{ks} = g^{\alpha r + ks} = g^{\alpha r + H(m\|r) - \alpha r}$$
$$= g^{H(m\|r)} \pmod{p}$$

and therefore a valid message/signature is always accepted.

The signature can be shown to be existentially unforgeable (see existential forgery) under a chosen message attack in the random oracle model, assuming the discrete logarithm problem in the group generated by g is intractable [4]. In the proof of security, the function H is assumed to be a random oracle. In practice, one derives H from some cryptographic hash function such as SHA-1. We note that the check in Step 1 of the verification algorithm is required. Without this check, there is a simple forging algorithm that, just given the public key, is able to forge a signature on any message m. In these forged signatures the value r is an integer in the range $1 \ldots p(p-1)$. Similarly, we note that signing the message m directly without first hashing it with H results in a system for which a simple existential forgery is possible, just given the public key.

To discuss signature length we fix concrete security parameters. At the present time the discrete-log problem in the cyclic group $\mathbb{Z}_p^*$ where p is a 1024-bit prime is considered intractable [2] except for a very well funded organization. With these parameters an ElGamal signature is 2048-bit long—much longer than the related DSS or Schnorr signatures.

There are many variants of the ElGamal signature scheme. We refer to [3] for a comparison of some six variants. The signature scheme can be made to work in any finite cyclic group in which the discrete log problem is intractable. In particular, there is an analogous scheme that works on elliptic curves instead of finite fields. A variant that supports message recovery was proposed in [3].

Dan Boneh

References

[1] ElGamal, Taher (1985). "A public key cryptosystem and signature scheme based on the discrete logarithms." *IEEE Transactions on Information Theory*, 31, 469–472.

[2] Lenstra, Arjen and E. Verheul (2001). "Selecting cryptographic key sizes." *Journal of Cryptology*, 14 (4), 255–293.

[3] Nyberg, K. and R. Rueppel (1996). "Message recovery for signature schemes based on the discrete logarithm problem." *Design Codes and Cryptography*, 7, 61–81.

[4] Pointcheval, D. and Jacques Stern (2000). "Security arguments for digital signatures and blind signatures." *Journal of Cryptology*, 13 (3), 361–96.

ELGAMAL PUBLIC KEY ENCRYPTION

In the ElGamal public key encryption scheme [1] $\langle g \rangle$ is a finite cyclic group of large enough order. A value q (a multiple of) the *order* of g, denoted as $\mathrm{ord}(g)$ (not necessarily a prime), is public. In the original ElGamal scheme, $\langle g \rangle = \mathbb{Z}_p^*$, p a prime and $q = p - 1$.

If Alice wants to make a public key, she chooses a random element a in Z_q and she computes $y_A := g^a$ in the group $\langle g \rangle$. Her public key will be (g, q, y_A). If a group of users uses the same g and q, the public key could be shorter. Her secret key is a.

If Bob, knowing Alice's public key (g, q, y_A), wants to encrypt a message $m \in \langle g \rangle$ to be sent to Alice, he chooses a random k in Z_q and computes $(c_1, c_2) := (g^k, m \cdot y_A^k)$ in the group and sends $c = (c_1, c_2)$. To decrypt Alice (using her secret key a) computes $m' := c_2 \cdot (c_1^a)^{-1}$ in this group.

The security of this scheme is related to the Diffie–Hellman problem. A non-malleable variant of this scheme was proposed independently by Tsiounis and Yung [4] and Jakobsson [2], by combining Schnorr's signature scheme [3] with the ElGamal encryption. The proof of non-malleability uses some cryptographic assumptions and the random oracle model.

Yvo Desmedt

References

[1] ElGamal, T. (1985). "A public key cryptosystem and a signature scheme based on discrete logarithms." *IEEE Trans. Inform. Theory*, 31, 469–472.

[2] Jakobsson, M. (1998). "A practical mix." *Advances in Cryptology—EUROCRYPT'98*, Lecture Notes in Computer Science, vol. 1403, ed. K. Nyberg. Springer-Verlag, Berlin, 448–461.

[3] Schnorr, C.P. (1990). "Efficient identification and signatures for smart cards." *Advances in Cryptology—CRYPTO'89*, Lecture Notes in Computer Science, vol. 435, ed. G. Brassard. Springer-Verlag, Berlin, 239–252.

[4] Tsiounis, Y. and M. Yung (1998). "The security of ElGamal based encryption." *Pub lic Key Cryptography, First International Workshop on Practice and Theory in Public Key Cryptography, PKC'98, Pacifico Yokohama, Japan*, Lecture Notes in Computer Science, eds. H. Imai and Y. Zheng. Springer-Verlag, Berlin, 117–134.

ELLIPTIC CURVE

Elliptic curves have been used in integer factoring algorithms and in primality proving algorithms, and also for designing public-key cryptosystems. This section introduces elliptic curves and associated group operations, along with basic structural properties of particular interest in cryptography.

DEFINING EQUATION: An *elliptic curve* E over a field F is defined by a Weierstrass equation

$$E/F : y^2 + a_1 xy + a_3 y = x^3 + a_2 x^2 + a_4 x + a_6 \tag{1}$$

with $a_1, a_2, a_3, a_4, a_6 \in F$ and $\Delta \neq 0$, where Δ is the *discriminant* of E and is defined as follows:

$$
\left.
\begin{aligned}
\Delta &= -d_2^2 d_8 - 8d_4^3 - 27d_6^2 + 9d_2 d_4 d_6 \\
d_2 &= a_1^2 + 4a_2 \\
d_4 &= 2a_4 + a_1 a_3 \\
d_6 &= a_3^2 + 4a_6 \\
d_8 &= a_1^2 a_6 + 4a_2 a_6 - a_1 a_3 a_4 + a_2 a_3^2 - a_4^2.
\end{aligned}
\right\} \tag{2}
$$

If L is any <u>extension field</u> of F, then the set of *L-rational points* on E is

$$
\begin{aligned}
E(L) = \{(x, y) \in L \times L : y^2 + a_1 xy + a_3 y - x^3 \\
- a_2 x^2 - a_4 x - a_6 = 0\} \cup \{\infty\},
\end{aligned}
$$

where ∞ is the *point at infinity*.

Two elliptic curves E_1 and E_2 defined over F and given by Weierstrass equations (1) are said to be *isomorphic* over F if there exist $u, r, s, t \in F, u \neq 0$, such that the change of variables

$$
(x, y) \rightarrow (u^2 x + r, u^3 y + u^2 s x + t) \tag{3}
$$

transforms equation E_1 into equation E_2. The transformation (3) is called an *admissible change of variables*.

SIMPLIFIED WEIERSTRASS EQUATION: A Weierstrass Equation (1) defined over F can be simplified considerably by applying admissible changes of variables.

1. If the characteristic of F is not equal to 2 or 3, then the admissible change of variables

$$
(x, y) \rightarrow \left(\frac{x - 3a_1^2 - 12a_2}{36}, \frac{y - 3a_1 x}{216} \right.
$$
$$
\left. - \frac{a_1^3 + 4a_1 a_2 - 12a_3}{24} \right)
$$

transforms E to the curve

$$
y^2 = x^3 + ax + b, \tag{4}
$$

where $a, b \in F$. The discriminant of this curve is $\Delta = -16(4a^3 + 27b^2)$.

2. If the characteristic of F is 2, then there are two cases to consider. If $a_1 \neq 0$, then the admissible change of variables

$$
(x, y) \rightarrow \left(a_1^2 x + \frac{a_3}{a_1}, a_1^3 y + \frac{a_1^2 a_4 + a_3^2}{a_1^3} \right)
$$

transforms E to the curve

$$
y^2 + xy = x^3 + ax^2 + b, \tag{5}
$$

where $a, b \in F$. The discriminant is $\Delta = b$. If $a_1 = 0$, then the admissible change of variables

$$
(x, y) \rightarrow (x + a_2, y)
$$

transforms E to the curve

$$
y^2 + cy = x^3 + ax + b, \tag{6}
$$

where $a, b, c \in F$. The discriminant is $\Delta = c^4$.

3. If the characteristic of F is 3, then there are two cases to consider. If $a_1^2 \neq -a_2$, then the admissible change of variables

$$
(x, y) \rightarrow \left(x + \frac{d_4}{d_2}, y + a_1 x + a_1 \frac{d_4}{d_2} + a_3 \right),
$$

where $d_2 = a_1^2 + a_2$ and $d_4 = a_4 - a_1 a_3$ transforms E to the curve

$$
y^2 = x^3 + ax^2 + b, \tag{7}
$$

where $a, b \in F$. The discriminant is $\Delta = -a^3 b$. If $a_1^2 = -a_2$, then the admissible change of variables

$$
(x, y) \rightarrow (x, y + a_1 x + a_3)
$$

transforms E to the curve

$$
y^2 = x^3 + ax + b, \tag{8}
$$

where $a, b \in F$. The discriminant is $\Delta = -a^3$.

GROUP LAW: Let E be an elliptic curve defined over the field F. There is a *chord-and-tangent rule* for adding two points in $E(F)$ to give a third point in $E(F)$. Together with this addition operation, the set of points $E(F)$ forms an abelian <u>group</u> with ∞ serving as its identity. It is this group that is used in the construction of elliptic curve cryptographic schemes.

The addition rule is best explained geometrically. Figure 1 illustrates the addition and doubling rule for the curve $y^2 = x^3 - x$ over the real numbers $\mathbb{R}$. Let $P = (x_1, y_1)$ and $Q = (x_2, y_2)$ be two distinct points on an elliptic curve E. Then the *sum* R of P and Q is obtained by drawing a line through P and Q; this line intersects the elliptic curve at a third point. Then R is the reflection of this point about the x-axis. The *double* R of P, also denoted $2P$, is obtained by drawing the tangent line to the elliptic curve at P. This line intersects the elliptic curve at a second point. Then R is the reflection of this point about the x-axis.

Algebraic formulas for the group law can be easily derived from the geometric description. These formulas are presented next for elliptic curves E of the simplified Weierstrass form (4) in affine coordinates when the characteristic of the underlying field F is not 2 or 3, and for elliptic curves E of the form (5) over characteristic 2 finite fields.

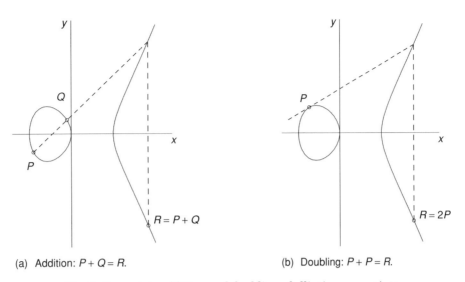

(a) Addition: $P + Q = R$.   (b) Doubling: $P + P = R$.

Fig. 1. Geometric addition and doubling of elliptic curve points

Group law for $E/F : y^2 = x^3 + ax + b$, $\mathrm{char}(F) \neq 2, 3$

1. *Identity.* $P + \infty = \infty + P = P$ for all $P \in E(F)$.
2. *Negatives.* If $P = (x, y) \in E(F)$, then $(x, y) + (x, -y) = \infty$. The point $(x, -y)$ is denoted by $-P$ and is called the *negative* of P; note that $-P$ is indeed a point in $E(F)$. Also, $-\infty = \infty$.
3. *Point addition.* Let $P = (x_1, y_1) \in E(F)$ and $Q = (x_2, y_2) \in E(F)$, where $P \neq \pm Q$. Then $P + Q = (x_3, y_3)$, where

$$x_3 = \left(\frac{y_2 - y_1}{x_2 - x_1}\right)^2 - x_1 - x_2$$

and

$$y_3 = \left(\frac{y_2 - y_1}{x_2 - x_1}\right)(x_1 - x_3) - y_1.$$

4. *Point doubling.* Let $P = (x_1, y_1)' \in E(F)$, where $P \neq -P$. Then $2P = (x_3, y_3)$, where

$$x_3 = \left(\frac{3x_1^2 + a}{2y_1}\right)^2 - 2x_1$$

and

$$y_3 = \left(\frac{3x_1^2 + a}{2y_1}\right)(x_1 - x_3) - y_1.$$

Group law for $E/F : y^2 + xy = x^3 + ax^2 + b$, $\mathrm{char}(F) = 2$

1. *Identity.* $P + \infty = \infty + P = P$ for all $P \in E(F)$.
2. *Negatives.* If $P = (x, y) \in E(F)$, then $(x, y) + (x, x + y) = \infty$. The point $(x, x + y)$ is denoted by $-P$ and is called the *negative* of P; note that $-P$ is indeed a point in $E(F)$. Also, $-\infty = \infty$.

3. *Point addition.* Let $P = (x_1, y_1) \in E(F)$ and $Q = (x_2, y_2) \in E(F)$, where $P \neq \pm Q$. Then $P + Q = (x_3, y_3)$, where

$$\lambda = \frac{y_1 + y_2}{x_1 + x_2}, \quad x_3 = \lambda^2 + \lambda + x_1 + x_2 + a,$$

and

$$y_3 = \lambda(x_1 + x_3) + x_3 + y_1.$$

4. *Point doubling.* Let $P = (x_1, y_1) \in E(F)$, where $P \neq -P$. Then $2P = (x_3, y_3)$, where

$$x_3 = \left(x_1 + \frac{y_1}{x_1}\right)^2 + \left(x_1 + \frac{y_1}{x_1}\right) + a$$

and

$$y_3 = x_1^2 + \left(x_1 + \frac{y_1}{x_1}\right)x_3 + x_3.$$

GROUP ORDER: Let E be an elliptic curve defined over a <u>finite field</u> $F = \mathbb{F}_q$. The number of points in $E(\mathbb{F}_q)$, denoted $\#E(\mathbb{F}_q)$, is called the *order* of E over $\mathbb{F}_q$. Since the Weierstrass equation (1) has at most two solutions for each $x \in \mathbb{F}_q$, we know that $\#E(\mathbb{F}_q) \in [1, 2q + 1]$. Hasse's theorem provides tighter bounds for $\#E(\mathbb{F}_q)$:

$$q + 1 - 2\sqrt{q} \leq \#E(\mathbb{F}_q) \leq q + 1 + 2\sqrt{q}.$$

$\#E(\mathbb{F}_q)$ can be computed in <u>polynomial time</u> by Schoof's algorithm or one of its many derivatives.

Let p be the characteristic of $\mathbb{F}_q$. An elliptic curve E defined over $\mathbb{F}_q$ is *supersingular* if p divides t, where $t = q + 1 - \#E(\mathbb{F}_q)$. If p does not divide t, then E is *non-supersingular*. The elliptic curves (6) and (8) are supersingular.

If E is an elliptic curve defined over $\mathbb{F}_q$, then E is also defined over any extension $\mathbb{F}_{q^n}$ of $\mathbb{F}_q$. The group $E(\mathbb{F}_q)$ of $\mathbb{F}_q$-rational points is a subgroup of the group $E(\mathbb{F}_{q^n})$ of $\mathbb{F}_{q^n}$-rational points and hence $\#E(\mathbb{F}_q)$ divides $\#E(\mathbb{F}_{q^n})$. If $\#E(\mathbb{F}_q)$ is known, then $\#E(\mathbb{F}_{q^n})$ can be efficiently determined by the following result due to Weil. Let $\#E(\mathbb{F}_q) = q + 1 - t$. Then $\#E(\mathbb{F}_{q^n}) = q^n + 1 - V_n$ for all $n \geq 2$, where $\{V_n\}$ is the sequence defined recursively by $V_0 = 2$, $V_1 = t$, and $V_n = V_1 V_{n-1} - q V_{n-2}$ for $n \geq 2$.

GROUP STRUCTURE: Let E be an elliptic curve defined over $\mathbb{F}_q$. Then $E(\mathbb{F}_q)$ is isomorphic to $\mathbb{Z}_{n_1} \oplus \mathbb{Z}_{n_2}$ where n_1 and n_2 are uniquely determined positive integers such that n_2 divides both n_1 and $q - 1$.

Note that $\#E(\mathbb{F}_q) = n_1 n_2$. If $n_2 = 1$, then $E(\mathbb{F}_q)$ is a cyclic group. If $n_2 > 1$, then $E(\mathbb{F}_q)$ is said to have *rank 2*. If n_2 is a small integer (e.g., $n = 2, 3$ or 4), we sometimes say that $E(\mathbb{F}_q)$ is *almost cyclic*. Since n_2 divides both n_1 and $q - 1$, one expects that $E(\mathbb{F}_q)$ is cyclic or almost cyclic for most elliptic curves E over $\mathbb{F}_q$.

EXAMPLE. Consider the elliptic curve

$$E : y^2 = x^3 + 2x + 4$$

defined over $\mathbb{F}_{13}$, the integers modulo 13. The points in $E(\mathbb{F}_{13})$ are:

$(0, 2)$ $(2, 4)$ $(5, 3)$ $(7, 6)$ $(8, 5)$ $(9, 6)$ $(10, 6)$ $(12, 1)$ ∞
$(0, 11)$ $(2, 9)$ $(5, 10)$ $(7, 7)$ $(8, 8)$ $(9, 7)$ $(10, 7)$ $(12, 12)$

so $\#E(\mathbb{F}_{13}) = 17$. Examples of the group law are $(8, 5) + (2, 4) = (7, 6)$, and $2(8, 5) = (0, 2)$. Since the group order 17 is prime, $E(\mathbb{F}_{13})$ is a cyclic group.

Darrel Hankerson
Alfred Menezes

References

[1] Koblitz, N. (1994). *A Course in Number Theory and Cryptography* (2nd ed.). Springer-Verlag, Berlin.
[2] Silverman, J. (1986). *The Arithmetic of Elliptic Curves*. Springer-Verlag, Berlin.
[3] Silverman, J. (1994). *Advanced Topics in the Arithmetic of Elliptic Curves*. Springer-Verlag, Berlin.
[4] Washington, L. (2003). *Elliptic Curves: Number Theory and Cryptography*. CRC Press, Boca Raton, FL.

ELLIPTIC CURVE CRYPTOGRAPHY

Elliptic curve cryptographic schemes were proposed independently in 1985 by Neal Koblitz [3] and Victor Miller [5]. They are the elliptic curve analogues of schemes based on the discrete log-arithm problem where the underlying group is the group of points on an elliptic curve defined over a finite field. The security of all elliptic curve signature schemes, elliptic curve key agreement schemes and elliptic curve public-key encryption schemes is based on the apparent intractability of the elliptic curve discrete logarithm problem (ECDLP). Unlike the case of the ordinary discrete logarithm problem in the multiplicative group of a finite field, or with the integer factoring problem, there is no subexponential-time algorithm known for the ECDLP. Consequently, significantly smaller parameters can be selected for elliptic curve schemes than for ordinary discrete logarithm schemes or for RSA, and achieve the same level of security. Smaller parameters can potentially result in significant performance benefits, especially for higher levels of security.

Darrel Hankerson
Alfred Menezes

References

[1] Blake, I., G. Seroussi, and N. Smart (1999). *Elliptic Curves in Cryptography*. Cambridge University Press.
[2] Hankerson, D., A. Menezes, and S. Vanstone (2003). *Guide to Elliptic Curve Cryptography*. Springer-Verlag, Berlin.
[3] Koblitz, N. (1987). "Elliptic curve cryptosystems." *Mathematics of Computation*, 48, 203–209.
[4] Menezes, A. (1993). *Elliptic Curve Public Key Cryptosystems*. Kluwer Academic Publishers, Boston, MA.
[5] Miller, V. (1986). "Use of elliptic curves in cryptography." *Advances in Cryptology—CRYPTO'85*, Lecture Notes in Computer Science, vol. 218, ed. H.C. Williams. Springer-Verlag, Berlin, 417–426.

ELLIPTIC CURVE DISCRETE LOGARITHM PROBLEM

Let E be an elliptic curve defined over a finite field $\mathbb{F}_q$, and let $P \in E(\mathbb{F}_q)$ be a point of order n. Given $Q \in \langle P \rangle$, the elliptic curve discrete logarithm problem (ECDLP) is to find the integer l, $0 \leq l \leq n - 1$, such that $Q = lP$.

The ECDLP is a special case of the discrete logarithm problem in which the cyclic group G is represented by the group $\langle P \rangle$ of points on an elliptic curve. It is of cryptographic interest because its apparent intractability is the basis for the security of elliptic curve cryptography.

If the order n of the base point P is composite and its factorization is known, then the Pohlig-Hellman algorithm [14] (see the <u>discrete logarithm problem</u> entry) can be used to efficiently reduce the ECDLP in $\langle P \rangle$ to instances of the ECDLP in proper subgroups of $\langle P \rangle$. Thus, the difficulty of the original ECDLP instance depends on the size of the largest prime factor of n. In order to maximize resistance to the Pohlig-Hellman algorithm, n should be prime, as we will henceforth assume.

POLLARD'S ρ METHOD: Pollard's ρ method [15] (see <u>discrete logarithm problem</u>) is the best general-purpose algorithm known for solving the ECDLP. The algorithm, as improved by Teske [21], has an expected running time of $\sqrt{\pi n/2}$ elliptic curve operations and has negligible storage requirements. Van Oorschot and Wiener [22] showed how Pollard's ρ method can be effectively parallelized so that r processors could jointly work on solving one ECDLP instance with a net speedup by a factor of r. The processors do not have to communicate with each other, and only occasionally transmit data to a central processor. This method is also called parallel collision search (see <u>discrete logarithm problem</u>).

Gallant, Lambert and Vanstone [4] and Wiener and Zuccherato [23] observed that Pollard's ρ method can be modified to operate on equivalence classes determined by the negation map acting on points (which maps a point P to $-P$), rather than on points themselves. The running time of this modified version is

$$(\sqrt{\pi n})/2,$$

a speedup by a factor of $\sqrt{2}$.

Gallant, Lambert and Vanstone [4] and Wiener and Zuccherato [23] also observed that Pollard's ρ method can be accelerated by exploiting other efficiently-computable endomorphisms of an elliptic curve. For the two *Koblitz elliptic curves* (also known as *anomalous binary curves*) which are defined over $\mathbb{F}_2$ by the equations $y^2 + xy = x^3 + 1$ and $y^2 + xy = x^3 + x^2 + 1$, the *Frobenius map* $\phi : (x, y) \mapsto (x^2, y^2)$ is an endomorphism on $E(\mathbb{F}_{2^m})$ and can be efficiently computed. By operating on the equivalence classes of points determined by the negation and Frobenius maps, Pollard's ρ method for Koblitz curves can be accelerated further by a factor of $\sqrt{m}$ for a resulting running time of

$$(\sqrt{\pi n})/2\sqrt{m}.$$

The parallelized version of Pollard's ρ algorithm has been used in practice to solve several <u>ECC challenges</u>.

INDEX-CALCULUS METHODS: Unlike the case of the discrete logarithm problem in the multiplicative group of a finite field, there is no <u>index-calculus method</u> known for solving the ECDLP that has a subexponential (or better) running time. No appropriate choice is known for the elements of the <u>factor base</u> required in the index-calculus method. In the case of elliptic curves over prime fields, the most natural choice for factor base elements is obtained by regarding an elliptic curve point as having coordinates in the field of the rational numbers, and selecting those points that have small *height*. Miller [13] and Silverman and Suzuki [19] presented convincing arguments why this approach is doomed to fail.

Silverman [18] proposed a different idea for attacking the ECDLP, which he termed *xedni calculus*. Shortly after, Jacobson et al. [7] gave compelling theoretical and experimental evidence why xedni calculus would be ineffective for solving the ECDLP.

SPECIAL-PURPOSE ALGORITHMS: Algorithms that are faster than Pollard's ρ method for solving the ECDLP are known for some special classes of elliptic curves. When selecting an elliptic curve for use in a cryptographic scheme, one should verify that the elliptic curve chosen is not vulnerable to these special-purpose attacks.

Attack on Prime-Field Anomalous Curves

An elliptic curve E over a prime field $\mathbb{F}_p$ is said to be *prime-field anomalous* if the number of points in $E(\mathbb{F}_p)$ is equal to p. Satoh and Araki [16], Semaev [17], and Smart [20] showed that for such curves, the ECDLP in $E(\mathbb{F}_p)$ can be efficiently solved. Hence, when selecting an elliptic curve E over a prime field $\mathbb{F}_p$ for cryptographic use, it is important to verify that $\#E(\mathbb{F}_p) \neq p$.

Weil and Tate Pairing Attacks

Let E be an elliptic curve defined over a finite field $\mathbb{F}_q$. Menezes, Okamoto and Vanstone [10] and Frey and Rück [3] showed how the *Weil* and *Tate pairings* can be used to efficiently reduce the ECDLP in $E(\mathbb{F}_q)$ to the <u>discrete logarithm problem</u> in the multiplicative group of an <u>extension field</u> $\mathbb{F}_{q^k}$, where <u>subexponential-time</u> <u>index calculus methods</u> are known. The reduction is only useful for solving the ECDLP instance if the discrete logarithm problem in $\mathbb{F}_{q^k}$ is tractable—this imposes the restriction that k not be too large. Now, the smallest permissible value for the extension degree k is the smallest integer k such that n divides $q^k - 1$. Hence by verifying that n does not divide $q^k - 1$ for all integers $k \in [1, c]$ (where c is chosen so that the discrete logarithm problem in $\mathbb{F}_{q^c}$ is deemed to be intractable), the Weil and Tate pairing attacks can be circumvented.

Weil Descent

Frey [2] proposed a general methodology using *Weil descent* for reducing the ECDLP in an elliptic curve E over a characteristic two finite field $\mathbb{F}_{2^m}$ to the discrete logarithm problem in the jacobian $J_C(\mathbb{F}_{2^n})$ of an algebraic curve C defined over a subfield $\mathbb{F}_{2^n}$ of $\mathbb{F}_{2^m}$. Gaudry, Hess and Smart [6] gave an explicit algorithm for the case where C is a hyperelliptic curve of genus g defined over $\mathbb{F}_{2^n}$; their method is called the *GHS attack*. Since subexponential-time algorithms are known for the discrete logarithm problem in high genus hyperelliptic curves (see [1, 5]), the GHS attack can potentially solve the ECDLP faster than Pollard's ρ method.

Menezes and Qu [11] showed that the GHS attack is slower than Pollard's ρ method for all elliptic curves defined over finite fields $\mathbb{F}_{2^m}$ where m is prime and $m \in [160, 600]$. Thus the GHS attack is ineffective for elliptic curves over these fields.

The GHS attack for elliptic curves over $\mathbb{F}_{2^m}$ where m is composite has been extensively analyzed (see [8, 9, 12]). This body of work shows that the GHS attack is indeed effective in solving the ECDLP for some elliptic curves over some fields $\mathbb{F}_{2^m}$ with m composite. In view of these attacks, it seems prudent to avoid use of elliptic curves over finite fields $\mathbb{F}_{2^m}$ where m is composite.

Darrel Hankerson
Alfred Menezes

References

[1] Adleman, L., J. DeMarrais, and M. Huang (1994). "A subexponential algorithm for discrete logarithms over the rational subgroup of the jacobians of large genus hyperelliptic curves over finite fields." *Algorithmic Number Theory—ANTS-I*, Lecture Notes in Computer Science, vol. 877, eds. L. Adleman and M.-D. Huang. Springer-Verlag, Berlin, 28–40.

[2] Frey, G. (2001). "Applications of arithmetical geometry to cryptographic constructions." *Proceedings of the Fifth International Conference on Finite Fields and Applications.* Springer-Verlag, Berlin, 128–161.

[3] Frey, G. and H. Rück (1994). "A remark concerning m-divisibility and the discrete logarithm in the divisor class group of curves." *Mathematics of Computation*, 62, 865–874.

[4] Gallant, R., R. Lambert, and S. Vanstone (2000). "Improving the parallelized Pollard lambda search on anomalous binary curves." *Mathematics of Computation*, 69, 1699–1705.

[5] Gaudry, P. (2000). "An algorithm for solving the discrete log problem in hyperelliptic curves." *Advances in Cryptology—EUROCRYPT 2000*, Lecture Notes in Computer Science, vol. 1807, ed. B. Preneel. Springer-Verlag, Berlin, 19–34.

[6] Gaudry, P., F. Hess, and N. Smart (2002). "Constructive and destructive facets of Weil descent on elliptic curves." *Journal of Cryptology*, 15, 19–46.

[7] Jacobson, M., N. Koblitz, J. Silverman, A. Stein, and E. Teske (2000). "Analysis of the xedni calculus attack." *Designs, Codes and Cryptography*, 20, 41–64.

[8] Jacobson, M., A. Menezes, and A. Stein (2001). "Solving elliptic curve discrete logarithm problems using Weil descent." *Journal of the Ramanujan Mathematical Society*, 16, 231–260.

[9] Maurer, M., A. Menezes, and E. Teske (2002). "Analysis of the GHS Weil descent attack on the ECDLP over characteristic two finite fields of composite degree." *LMS Journal of Computation and Mathematics*, 5, 127–174.

[10] Menezes, A., T. Okamoto, and S. Vanstone (1993). "Reducing elliptic curve logarithms to logarithms in a finite field." *IEEE Transactions on Information Theory*, 39, 1639–1646.

[11] Menezes A. and M. Qu (2001). "Analysis of the Weil descent attack of Gaudry, Hess and Smart." *Topics in Cryptology—CT-RSA 2001*, Lecture Notes in Computer Science, vol. 2020, ed. D. Naccache. Springer-Verlag, Berlin, 308–318.

[12] Menezes, A., E. Teske, and A. Weng (2004). "Weak fields for ECC." *Topics in Cryptology—CT-RSA 2004*, Lecture Notes in Computer Science, vol. 2964, ed. T. Okamoto. Springer-Verlag, Berlin, 366–386.

[13] Miller, V. (1986). "Use of elliptic curves in cryptography." *Advances in Cryptology—CRYPTO'85*, Lecture Notes in Computer Science, vol. 218, ed. H.C. Williams. Springer-Verlag, Berlin, 417–426.

[14] Pohlig, S. and M. Hellman (1978). "An improved algorithm for computing logarithms over $GF(p)$ and its cryptographic significance." *IEEE Transactions on Information Theory*, 24, 106–110.

[15] Pollard, J. (1978). "Monte Carlo methods for index computation (mod p)." *Mathematics of Computation*, 32, 918–924.

[16] Satoh, T. and K. Araki (1998). "Fermat quotients and the polynomial time discrete log algorithm for anomalous elliptic curves." *Commentarii Mathematici Universitatis Sancti Pauli*, 47, 81–92.

[17] Semaev, I. (1998). "Evaluation of discrete logarithms in a group of p-torsion points of an elliptic curve in characteristic p." *Mathematics of Computation*, 67, 353–356.

[18] Silverman, J. (2000). "The xedni calculus and the elliptic curve discrete logarithm problem." *Designs, Codes and Cryptography*, 20, 5–40.

[19] Silverman, J. and J. Suzuki (1998). "Elliptic curve discrete logarithms and the index calculus." *Advances in Cryptography—ASIACRYPT'98*, Lecture Notes in Computer Science, vol. 1514, eds. K. Ohta and D. Pei. Springer-Verlag, Berlin, 110–125.

[20] Smart, N. (1999). "The discrete logarithm problem on elliptic curves of trace one." *Journal of Cryptology*, 12, 193–196.

[21] Teske, E. (1998). "Speeding up Pollard's rho method for computing discrete logarithms." *Algorithmic Number Theory—ANTS-III*, Lecture Notes in Computer Science, vol. 1423, ed. J.P. Buhler. Springer-Verlag, Berlin, 541–554.

[22] van Oorschot, P. and M. Wiener (1999). "Parallel collision search with cryptanalytic applications." *Journal of Cryptology*, 12, 1–28.

[23] Wiener, M. and R. Zuccherato (1999). "Faster attacks on elliptic curve cryptosystems." *Selected Areas in Cryptography—SAC'98*, Lecture Notes in Computer Science, vol. 1556, eds. S. Tavares and H. Meijer. Springer-Verlag, Berlin, 190–200.

ELLIPTIC CURVE KEY AGREEMENT SCHEMES

In the elliptic curve analogue of the basic Diffie–Hellman key agreement scheme [4], two users A and B share domain parameters $D = (q, \mathrm{FR}, S, a, b, P, n, h)$ (see elliptic curve keys). A selects an integer $d_A \in_R [1, n-1]$ and sends $Q_A = d_A P$ to B. Similarly, B selects an integer $d_B \in_R [1, n-1]$ and sends $Q_B = d_B P$ to A. A computes $K = d_A Q_B = d_A d_B P$, and B similarly computes $K = d_B Q_A$. The shared secret point K is used to derive a secret key which can then be used to encrypt or authenticate messages using symmetric-key schemes. Security against passive adversaries is based on the hardness of the elliptic curve analogue of the Diffie–Hellman problem: given the domain parameters D and points Q_A and Q_B, compute K.

In order to provide security against active adversaries, the exchanged points have to be authenticated. Many variants of the basic Diffie-Hellman scheme that provide authentication have been proposed including the station-to-station protocol [5] and the MQV key agreement scheme [8]. For a survey of authenticated Diffie–Hellman protocols, see [2,3]. Elliptic curve analogues of these protocols have been standardized in ANSI X9.63 [1], IEEE 1363-2000 [6], and ISO 15946-3 [7].

Darrel Hankerson
Alfred Menezes

References

[1] ANSI X9.63 (2001). *Public Key Cryptography for the Financial Services Industry: Key Agreement and Key Transport Using Elliptic Curve Cryptography*. American National Standards Institute.

[2] Blake-Wilson, S. and A. Menezes (1999). "Authenticated Diffie–Hellman key agreement protocols." *Selected Areas in Cryptography—SAC'98*, Lecture Notes in Computer Science, vol. 1556, eds. S.E. Tavares and H. Meijer. Springer-Verlag, Berlin, 339–361.

[3] Boyd, C. and A. Mathuria (2003). *Protocols for Key Establishment and Authentication*. Springer-Verlag, Berlin.

[4] Diffie, W. and M. Hellman (1976). "New directions in cryptography." *IEEE Transactions on Information Theory*, 22, 644–654.

[5] Diffie, W., P. van Oorschot, and M. Wiener (1992). "Authentication and authenticated key exchanges." *Designs, Codes and Cryptography*, 2, 107–125.

[6] IEEE Std 1363-2000 (2000). *IEEE Standard Specifications for Public-Key Cryptography*.

[7] ISO/IEC 15946-3 (2002). *Information Technology–Security Techniques—Cryptographic Techniques Based on Elliptic Curves—Part 3: Key Establishment*.

[8] Law, L., A. Menezes, M. Qu, J. Solinas, and S. Vanstone (2003). "An efficient protocol for authenticated key agreement." *Designs, Codes and Cryptography*, 28, 119–134.

ELLIPTIC CURVE KEYS

Key pairs for elliptic curve cryptography are associated with a set of *domain parameters* $D = (q, \mathrm{FR}, S, a, b, P, n, h)$ which consist of:

1. The order q of the underlying field $\mathbb{F}_q$.
2. An indication FR of the representation used for the elements of $\mathbb{F}_q$.
3. A seed S if the elliptic curve was generated verifiably at random using a method such as those described in FIPS 186-2 [1].
4. Two field elements a and b that define the equation of the elliptic curve: $y^2 = x^3 + ax + b$ in the case that the characteristic of $\mathbb{F}_q$ is not 2 or 3, and $y^2 + xy = x^3 + ax^2 + b$ if $\mathbb{F}_q$ has characteristic 2.
5. A point $P \in E(\mathbb{F}_q)$ of prime order.
6. The order n of P.
7. The cofactor $h = \#E(\mathbb{F}_q)/n$.

Domain parameters may either be shared by a group of users, or they may be specific to each user.

Typically the cofactor h is small (e.g., $h = 1, 2, 3$ or 4). A suitable elliptic curve can be found by randomly selecting elliptic curves E over $\mathbb{F}_q$ until

$\#E(\mathbb{F}_q)$ is a prime or almost prime. The number of points $\#E(\mathbb{F}_q)$ can be determined using Schoof's algorithm [5] and its derivatives. For the case where the characteristic of $\mathbb{F}_q$ is 2, $\#E(\mathbb{F}_q)$ can be computed extremely rapidly using Satoh's algorithm or the AGM method and their variants (see [2–4, 6]).

Given a set of domain parameters $D = (q, \mathrm{FR}, S, a, b, P, n, h)$, an elliptic curve *key pair* is (d, Q), where $d \in_R [1, n - 1]$ is the *private key*, and $Q = dP$ is the corresponding *public key*. Computing the private key from the public key is an instance of the underlined elliptic curve discrete logarithm problem.

Darrel Hankerson
Alfred Menezes

References

[1] FIPS 186-2 (2000). *Digital Signature Standard (DSS)*, Federal Information Processing Standards Publication 186-2. National Institute of Standards and Technology, Gaithersburg, MD.

[2] Fouquet, M., P. Gaudry, and R. Harley (2000). "An extension of Satoh's algorithm and its implementation." *Journal of the Ramanujan Mathematical Society*, 15, 281–318.

[3] Gaudry, P. (2002). "A comparison and a combination of SST and AGM algorithms for counting points of elliptic curves in characteristic 2." *Advances in Cryptography—ASIACRYPT 2002*, Lecture Notes in Computer Science, vol. 2501, ed. Y. Zheng. Springer-Verlag, Berlin, 311–327.

[4] Satoh, T. (2000). "The canonical lift of an ordinary elliptic curve over a prime field and its point counting." *Journal of the Ramanujan Mathematical Society*, 15, 247–270.

[5] Schoof, R. (1985). "Elliptic curves over finite fields and the computation of square roots mod p." *Mathematics of Computation*, 44, 483–494.

[6] Skjernaa, B. (2003). "Satoh's algorithm in characteristic 2." *Mathematics of Computation*, 72, 477–487.

ELLIPTIC CURVE METHOD

The Elliptic Curve Method (ECM for short) was invented in 1985 by H.W. Lenstra, Jr. [5]. It is suited to find small—say 9–30 digits—prime factors of large numbers. Among the different factorization algorithms whose complexity mainly depends on the size of the factor searched for (trial division, Pollard rho, Pollard $p - 1$, Williams $p + 1$—see underlined integer factoring), it is asymptotically the best method known. ECM can be viewed as a generalization of Pollard's $p - 1$ method, just like underlined elliptic curves for primality proving (ECPP)

generalizes the $n - 1$ primality test. ECM relies on Hasse's theorem: if p is prime, then an elliptic curve over $\mathbb{Z}/p\mathbb{Z}$ (see underlined modular arithmetic) has underlined group order $p + 1 - t$ with $|t| \leq 2\sqrt{p}$, where t depends on the curve. If $p + 1 - t$ is a smooth number (see underlined smoothness), then ECM will—most probably—succeed and reveal the unknown factor p.

Since 1985, many improvements have been proposed to ECM. Lenstra's original algorithm had no second phase. Brent proposes in [2] a "underlined birthday paradox" second phase, and further more technical refinements. In [7], Montgomery presents different variants of phase two of ECM and Pollard $p - 1$, and introduces a parameterization with homogeneous coordinates, which avoids inversions modulo n, with only 6 and 5 modular multiplications per addition and duplication on E, respectively. It is also possible to choose elliptic curves with a group order divisible by 12 or 16 [1, 7, 8].

Phase one of ECM works as follows. Let n be the number to factor. An elliptic curve is $E(\mathbb{Z}/n\mathbb{Z}) = \{(x : y : z) \in \mathbb{P}^2(\mathbb{Z}/n\mathbb{Z}), y^2z \equiv x^3 + axz^2 + bz^3 \bmod n\}$, where a, b are two parameters from $\mathbb{Z}/n\mathbb{Z}$, and $\mathbb{P}^2(\mathbb{Z}/n\mathbb{Z})$ is the projective plane over $\mathbb{Z}/n\mathbb{Z}$. The neutral element is $\mathcal{O} = (0 : 1 : 0)$, also called point at infinity. The key idea is that computations in $E(\mathbb{Z}/n\mathbb{Z})$ project to $E(\mathbb{Z}/p\mathbb{Z})$ for any prime divisor p of n, with the important particular case of quantities which are zero in $E(\mathbb{Z}/p\mathbb{Z})$ but not in $E(\mathbb{Z}/n\mathbb{Z})$. Pick at random a curve E and a point P on it. Then compute $Q = k \cdot P$ where k is the product of all prime powers less than a bound B_1 (see underlined elliptic curves). Let p be a prime divisor of n: if the order of E over $\mathbb{Z}/p\mathbb{Z}$ divides k, then Q will be the neutral element of $E(\mathbb{Z}/p\mathbb{Z})$, thus its z-coordinate will be zero modulo p, hence $\gcd(z, n)$ will reveal the factor p (unless z is zero modulo another factor of n, which is unlikely).

Phase one succeeds when all prime factors of $g = \#E(\mathbb{Z}/p\mathbb{Z})$ are less than B_1; phase two allows one prime factor g_1 of g to be as large as another bound B_2. The idea is to consider two families $(a_i Q)$ and $(b_j Q)$ of points on E, and check whether two such points are equal over $E(\mathbb{Z}/p\mathbb{Z})$. If $a_i Q = (x_i : y_i : z_i)$ and $b_j Q = (x'_j : y'_j : z'_j)$, then $\gcd(x_i z'_j - x'_j z_i, n)$ will be non-trivial. This will succeed when g_1 divides a non-trivial $a_i - b_j$. Two variants of phase two exist: the *birthday paradox continuation* chooses the a_i's and b_j's randomly, expecting that the differences $a_i - b_j$ will cover most primes up to B_2, while the *standard continuation* chooses the a_i's and b_j's so that every prime up to B_2 divides at least one $a_i - b_j$. Both continuations may benefit from the use of fast polynomial arithmetic, and are then called "FFT extensions" [8].

The expected running time of ECM is conjectured to be $\mathcal{O}(L(p)^{\sqrt{2}+o(1)}M(\log n))$ to find *one* factor of n (see O-notation), where p is the (unknown) smallest prime divisor of n, $L(x) = e^{\sqrt{\log x \log\log x}}$ (cf. L-notation), $M(\log n)$ represents the complexity of arithmetic modulo n, and the $o(1)$ in the exponent is for p tending to infinity. The second phase decreases the expected running time by a factor $\log p$. Optimal bounds B_1 and B_2 may be estimated from the (usually unknown) size of the smallest factor of n, using Dickman's function [9]. For RSA moduli (see RSA public key encryption), where n is the product of two primes of roughly the same size, the running time of ECM is comparable to that of the Quadratic Sieve.

ECM has been used to find factors of Cunningham numbers ($a^n \pm 1$ for $a = 2, 3, 5, 6, 7, 10, 11, 12$). In particular Fermat numbers $F_n = 2^{2^n} + 1$ are very good candidates for $n \geq 10$, since they are too large for general purpose factorization methods. Brent completed the factorization of F_{10} and F_{11} using ECM, after finding a 40-digit factor of F_{10} in 1995, and two factors of 21 and 22 digits of F_{11} in 1988 [3]. Brent, Crandall, Dilcher and Van Halewyn found a 27-digit factor of F_{13} in 1995, a (different) 27-digit factor of F_{16} in 1996, and a 33-digit factor of F_{15} in 1997.

Some applications of ECM are less obvious. The factors found by the Cunningham project [4] help to find primitive polynomials over the finite field GF(q). They are also used in the Jacobi sum and cyclotomy tests for primality proving [6].

Brent maintains a list of the ten largest factors found by ECM (ftp://ftp.comlab.ox.ac.uk/pub/Documents/techpapers/Richard.Brent/champs.txt); his extrapolation from previous data would give an ECM record of 70 digits in year 2010, 85 digits in year 2018, and 100 digits in year 2025. As of October 2003, the ECM record is a factor of 57 digits.

Paul Zimmermann

References

[1] Atkin, A.O.L. and F. Morain (1993). "Finding suitable curves for the elliptic curve method of factorization." *Mathematics of Computation*, 60 (201), 399–405.

[2] Brent, R.P. (1986). "Some integer factorization algorithms using elliptic curves." *Australian Computer Science Communications*, 8, 149–163. Also available on http://web.comlab.ox.ac.uk/oucl/work/richard.brent/pub/pub102.html

[3] Brent, R.P. (1999). "Factorization of the tenth Fermat number." *Mathematics of Computation*, 68 (225), 429–451.

[4] Brillhart, J., D.H., Lehmer, J.L., Selfridge, B. Tuckerman, and S.S. Wagstaff (2002). *Factorizations of $b^n \pm 1$ for $b = 2, 3, 5, 6, 7, 10, 11, 12$ up to high powers* (3rd ed.). *Contemporary Mathematics*, vol. 22. American Math. Society, Providence, RI. Also available on http://www.cerias.purdue.edu/homes/ssw/cun/third/

[5] Lenstra, H.W. (1987). "Factoring integers with elliptic curves." *Annals of Mathematics*, 126, 649–673.

[6] Mihailescu, P. (1998). "Cyclotomy primality proving—recent developments." *Proc. of ANTS III, Portland, OR*, Lecture Notes in Computer Science, vol. 1423. ed. J.P. Buhler. Springer-Verlag, Berlin, 95–110.

[7] Montgomery, P.L. (1987). "Speeding the Pollard and elliptic curve methods of factorization." *Mathematics of Computation*, 48 (177), 243–264.

[8] Montgomery, P.L. (1992). "An FFT extension of the elliptic curve method of factorization." *PhD Thesis*, University of California, Los Angeles. Also available on ftp.cwi.nl:/pub/pmontgom/ucladissertation.psl.gz

[9] Van de Lune, J. and E. Wattel (1969). "On the numerical solution of a differential-difference equation arising in analytic number theory." *Mathematics of Computation*, 23 (106), 417–421.

ELLIPTIC CURVE POINT MULTIPLICATION USING HALVING

Elliptic curve cryptographic schemes require calculations of the type

$$kP = \underbrace{P + \cdots + P}_{k},$$

where k is a large integer and the addition is over the elliptic curve (see elliptic curves). The operation is known as *scalar* or *point multiplication*, and dominates the execution time of signature and encryption schemes based on elliptic curves. Double-and-add variations of familiar square-and-multiply methods (see binary exponentiation) for modular exponentiation are commonly used to find kP. Windowing methods can significantly reduce the number of point additions required, but the number of point doubles remains essentially unchanged.

Among techniques to reduce the cost of the point doubles in point multiplication, perhaps the best known is illustrated in the case of Koblitz curves (elliptic curves over the field $\mathbb{F}_{2^m}$ with coefficients in $\mathbb{F}_2$; see [8]), where point doubling is replaced by inexpensive field squarings. Knudsen [4] and Schroeppel [6,7] proposed a point halving operation which shares strategy with τ-adic methods on Koblitz curves in the sense that most point doublings are replaced with less-expensive operations. The improvement is not as dramatic as that obtained on Koblitz curves; however, halving applies to a wider class of curves.

We restrict our attention to elliptic curves E over binary fields $\mathbb{F}_{2^m}$ defined by the equation

$$y^2 + xy = x^3 + ax^2 + b,$$

where $a, b \in \mathbb{F}_{2^m}$, $b \neq 0$. To simplify the exposition, we consider only the case that the trace function Tr (see <u>Boolean functions</u>) satisfies $\text{Tr}(a) = 1$; see [4] (where "minimal two-torsion" corresponds to $\text{Tr}(a) = 1$) for the necessary adjustments and computational costs for $\text{Tr}(a) = 0$ curves. We further assume that m is <u>prime</u>. These properties are satisfied by the five random curves over binary fields recommended by NIST in the FIPS 186-2 standard [2].

Let $P = (x, y)$ be a point on E with $P \neq -P$. The (affine) coordinates of $Q = 2P = (u, v)$ can be computed as follows:

$$\lambda = x + y/x \tag{1}$$
$$u = \lambda^2 + \lambda + a \tag{2}$$
$$v = x^2 + u(\lambda + 1) \tag{3}$$

(see <u>elliptic curves</u>). *Point halving* is the following operation: given $Q = (u, v)$, compute $P = (x, y)$ such that $Q = 2P$. The basic idea for halving is to solve (2) for λ, (3) for x, and finally (1) for y.

When $\mathcal{G}$ is a subgroup of odd order n in E, point doubling and point halving are automorphisms of $\mathcal{G}$. Therefore, given a point $Q \in \mathcal{G}$, there is a unique point $P \in \mathcal{G}$ such that $Q = 2P$. An efficient algorithm for point halving in $\mathcal{G}$, along with a point multiplication algorithm based on halving, are outlined in the following sections.

POINT HALVING: The notion of *trace* plays a central role in deriving an efficient algorithm for point halving. The *trace function* $\text{Tr} : \mathbb{F}_{2^m} \to \mathbb{F}_{2^m}$ is defined by $\text{Tr}(c) = c + c^2 + c^{2^2} + \cdots + c^{2^{m-1}}$. The map is linear, $\text{Tr}(c) \in \{0, 1\}$, and $\text{Tr}(u) = \text{Tr}(a)$ for $(u, v) \in \mathcal{G}$ (from an application of Tr to (2)).

Given $Q = (u, v) \in \mathcal{G}$, point halving seeks the unique point $P = (x, y) \in \mathcal{G}$ such that $Q = 2P$. The first step of halving is to find $\lambda = x + y/x$ by solving the equation

$$\widehat{\lambda}^2 + \widehat{\lambda} = u + a \tag{4}$$

for $\widehat{\lambda}$. It is easily verified that $\lambda \in \{\widehat{\lambda}, \widehat{\lambda} + 1\}$. If $\text{Tr}(a) = 1$, then it follows from (3) that $\widehat{\lambda} = \lambda$ if and

only if $\text{Tr}(v + u\widehat{\lambda}) = 0$. Hence λ can be identified, and then (3) is solved for the unique root x. Finally, if needed, $y = \lambda x + x^2$ may be recovered with one field multiplication.

Let the λ-*representation* of a point $Q = (u, v)$ be (u, λ_Q), where $\lambda_Q = u + v/u$. Given the λ-representation of Q as the input to point halving, we may compute $t = v + u\widehat{\lambda}$ without converting to affine coordinates, since

$$t = v + u\widehat{\lambda} = u\left(u + u + \frac{v}{u}\right) + u\widehat{\lambda} = u(u + \lambda_Q + \widehat{\lambda}).$$

In point multiplication, repeated halvings may be performed directly on the λ-representation of a point, with conversion to affine only when a point addition is required.

ALGORITHM 1. *Point halving*

Input: λ-representation (u, λ_Q) or affine representation (u, v) of $Q \in \mathcal{G}$.
Output: λ-representation (x, λ_P) of $P = (x, y) \in \mathcal{G}$, where $\lambda_P = x + y/x$ and $Q = 2P$.
1. *Find a solution $\widehat{\lambda}$ of $\widehat{\lambda}^2 + \widehat{\lambda} = u + a$.*
2. *If the input is in λ-representation, then compute $t = u(u + \lambda_Q + \widehat{\lambda})$; else compute $t = v + u\widehat{\lambda}$.*
3. *If $\text{Tr}(t) = 0$, then $\lambda_P \leftarrow \widehat{\lambda}$, $x \leftarrow \sqrt{t + u}$; else $\lambda_P \leftarrow \widehat{\lambda} + 1$, $x \leftarrow \sqrt{t}$.*
4. *Return (x, λ_P).*

The point halving algorithm requires a field multiplication and three main steps: computing the trace of t, solving the quadratic equation (4), and computing a square root. In a normal basis, field elements are represented in terms of a basis of the form $\{\beta, \beta^2, \ldots, \beta^{2^{m-1}}\}$. The trace of an element $c = \sum c_i \beta^{2^i} = (c_{m-1}, \ldots, c_0)$ is given by $\text{Tr}(c) = \sum c_i$. The square root computation is a right rotation: $\sqrt{c} = (c_0, c_{m-1}, \ldots, c_1)$. Squaring is a left rotation, and $x^2 + x = c$ can be solved bitwise. These operations are expected to be inexpensive relative to field multiplication. However, field multiplication in software for normal basis representations tends to be slow in comparison to multiplication with a polynomial basis. We shall restrict our discussion to computations in a polynomial basis representation, where $c \in \mathbb{F}_{2^m}$ is expressed as $c = \sum_{i=0}^{m-1} c_i z^i$ with $c_i \in \{0, 1\}$.

Trace Computations

The trace of c may be calculated as $\text{Tr}(c) = \sum_{i=0}^{m-1} c_i \text{Tr}(z^i)$, where the values $\text{Tr}(z^i)$ are precomputed.

As an example, $\mathbb{F}_{2^{163}}$ with reduction polynomial $f(z) = z^{163} + z^7 + z^6 + z^3 + 1$ has $\text{Tr}(z^i) = 1$ if and only if $i \in \{0, 157\}$, and finding $\text{Tr}(c)$ is an essentially free operation.

Solving the Quadratic Equation

For an odd integer m, define the *half-trace* $H : \mathbb{F}_{2^m} \to \mathbb{F}_{2^m}$ by $H(c) = \sum_{i=0}^{(m-1)/2} c^{2^{2i}}$. Then

$$H(c) = H\left(\sum_{i=0}^{m-1} c_i z^i\right) = \sum_{i=0}^{m-1} c_i H(z^i)$$

is a solution of the equation $x^2 + x = c + \text{Tr}(c)$. For elements c with $\text{Tr}(c) = 0$, the calculation produces a solution $H(c)$ of $x^2 + x = c$, and requires an expected $m/2$ field additions and storage for m field elements $H(z^i)$.

The storage and time required to solve the quadratic equation can be reduced. The basic strategy is to write $H(c) = H(c') + s$ where c' has fewer nonzero coefficients than c. The property $H(c) = H(c^2) + c + \text{Tr}(c)$ for $c \in \mathbb{F}_{2^m}$ may be applied directly to eliminate storage of $H(z^i)$ for even i. Repeated applications may yield further improvements. For example, if the reduction polynomial $f(z) = z^m + r(z)$ has $\deg r < m/2$, then the strategy can be applied in an especially straightforward fashion to eliminate storage of $H(z^i)$ for odd i, $m/2 < i < m - \deg r$. If $\deg r$ is small, the storage requirement is reduced to approximately $m/4$ elements. Such strategies are outlined in [4, Appendix A]; see also [3] for details.

Computing Square Roots

The square root of c may be expressed as

$$\sqrt{c} = \sum_{i \text{ even}} c_i z^{\frac{i}{2}} + \sqrt{z} \sum_{i \text{ odd}} c_i z^{\frac{i-1}{2}}.$$

The value $\sqrt{z}$ may be precomputed, and finding $\sqrt{c}$ is expected to be significantly less expensive than a field multiplication. Note that if the reduction polynomial f is a trinomial, then substantial improvements are possible based on the observation that $\sqrt{z}$ may be obtained directly from f; see [3].

POINT MULTIPLICATION: Halve-and-add variants of point multiplication methods replace most point doublings with halvings. However, point halving is performed on affine (or λ) representations, and hence some modifications may be required if projective coordinates are used. The algorithm presented in this section illustrates the use of projective coordinates (and a windowing method) with halving.

Let $w \geq 2$ be an integer. A *width-w NAF* of a positive integer k is an expression $k = \sum_{i=0}^{l-1} k_i 2^i$ where each nonzero coefficient k_i is odd, $|k_i| < 2^{w-1}$, $k_{l-1} \neq 0$, and at most one of any w consecutive digits is nonzero [8]. A positive integer k has a unique width-w NAF denoted $\text{NAF}_w(k)$, with length at most one more than the length of the binary representation. As an illustration, coefficients in the binary representation of $k = 29$ and the width-2 and width-3 NAFs are given by:

$$
\begin{array}{rrrrrr}
k = & 1 & 1 & 1 & 0 & 1 \\
\text{NAF}_2(k) = 1 & 0 & 0 & -1 & 0 & 1 \\
\text{NAF}_3(k) = 1 & 0 & 0 & 0 & 0 & -3.
\end{array}
$$

The average density of nonzero digits among all width-w NAFs of length l is approximately $1/(w + 1)$. The signed digit representation $\text{NAF}_2(k)$ is known as the non-adjacent form. In point multiplication, the use of signed digit representations is motivated by the property that point subtraction is as efficient as addition.

If kP is to be found for a given scalar k, then a conversion is required for halving-based methods. If k' is defined by

$$k \equiv k'_{t-1}/2^{t-1} + \cdots + k'_2/2^2 + k'_1/2 + k'_0 \pmod{n},$$

where $k'_i \in \{0, 1\}$ and n is the order of $\mathcal{G}$, then $kP = \sum_{i=0}^{t-1} k'_i/2^i P$; i.e., $(k'_{t-1}, \ldots, k'_0)$ may be used by halving-based methods. This can be generalized to width-w NAF: if $\sum_{i=0}^{l-1} k'_i 2^i$ is the w-NAF representation of $2^{t-1}k \bmod n$, then

$$k \equiv \sum_{i=0}^{t-1} \frac{k'_{t-1-i}}{2^i} + 2k'_t \pmod{n},$$

where it is understood that $k_i = 0$ if $i \geq l$.

Algorithm 2 presents a right-to-left version of a halve-and-add method with the input $2^{t-1}k \bmod n$ represented in w-NAF. Point halving occurs on the input P rather than on accumulators (which may be in projective form). The expected running time is approximately

$$(\text{step 4 cost}) + (t/(w + 1) - 2^{w-2})A' + tH$$

where H denotes a point halving and A' is the cost of a point addition when one of the inputs is in λ-representation. If projective coordinates are used for Q_i, then the additions in step 3.1 and 3.2 are mixed-coordinate. Step 4 may be performed by calculating $Q_i \leftarrow Q_i + Q_{i+2}$ for odd i from $2^{w-1}-3$ to 1, and then the result is given by $Q_1 + 2\sum_{i \in I \setminus \{1\}} Q_i$ [5, Exercise 4.6.3-9].

Table 1. Field operation costs for point and curve operations, where M, I, and V denote field multiplication, inversion, and division, respectively. The cost H of halving is an estimate. A' denotes the cost of a point addition when one of the inputs is in λ-representation

		Field operations	
Calculation	Point operations	Affine	Projective[a]
Point operation			
Addition	A	$M+V$	$8M$
Addition[b]	A'	$2M+V$	$9M$
Double	D	$M+V$	$4M$
Halve[c]	H	$2M$	
Curve operation (with width-2 NAF)			
kP via doubling	$(1/3)tA + tD$	$(4/3)t(M+V)$	$(20/3)tM + (2M+I)$
kP via Alg 2	$(1/3)tA' + tH$	$(8/3)tM + (1/3)tV$	$5tM + (2M+I)$

[a] Mixed-coordinate additions and $a \in \{0, 1\}$.
[b] A field multiplication converts λ-representation to affine.
[c] Estimated.

ALGORITHM 2. *Halve-and-add w-NAF (right-to-left) point multiplication*

Input: Window width w, $\mathrm{NAF}_w(2^{t-1}k \bmod n) = \sum_{i=0}^{t} k_i' 2^i$, $P \in \mathcal{G}$.
Output: kP. (Note: $k = k_0'/2^{t-1} + \cdots + k_{t-2}'/2 + k_{t-1}' + 2k_t' \bmod n$.)
1. *Set $Q_i \leftarrow \infty$ for $i \in I = \{1, 3, \ldots, 2^{w-1} - 1\}$.*
2. *If $k_t' = 1$ then $Q_1 = 2P$.*
3. *For i from $t-1$ down to 0 do:*
   3.1 If $k_i' > 0$ then $Q_{k_i'} \leftarrow Q_{k_i'} + P$.
   3.2 If $k_i' < 0$ then $Q_{-k_i'} \leftarrow Q_{-k_i'} - P$.
   3.3 $P \leftarrow P/2$.
4. *$Q \leftarrow \sum_{i \in I} i Q_i$.*
5. *Return(Q).*

The computational costs in terms of field operations are summarized in Table 1 for the case that Algorithm 2 is used with $w = 2$. Only field multiplications and inversions are considered, under the assumption that field addition is relatively inexpensive. The choice between affine coordinates and projective coordinates is driven primarily by the cost of inversion (I) relative to multiplication (M). If division has approximate cost $I + M$, then the estimates in the table show that projective coordinates will be preferred in Algorithm 2 with $w = 2$ whenever an inversion costs more than six multiplications.

Summary

The performance advantage of halving methods is clearest in the case of point multiplication kP where P is not known in advance, and smaller field inversion to multiplication ratios generally favor halving. However, significant storage (e.g., $m/4$ field elements) for the solve routine appears to be essential for performance. It should be noted, however, that the precomputation for the solve and square root routines is per field.

Darrel Hankerson
Alfred Menezes

References

[1] Bellare, M. (2000). *Advances in Cryptology— CRYPTO 2000*, Lecture Notes in Computer Science, vol. 1880, ed. M. Bellare. *20th Annual International Cryptology Conference, Santa Barbara, CA.* Springer-Verlag, Berlin.

[2] FIPS 186-2 (2000). *Digital Signature Standard (DSS)*. Federal Information Processing Standards Publication 186-2. National Institute of Standards and Technology, Gaithersburg, MD.

[3] Fong, K., D. Hankerson, J. López, and A. Menezes (2004). "Field inversion and point halving revisited." *IEEE Transactions on Computers*, 53 (8), 1047–1059.

[4] Knudsen, E. (1999). "Elliptic scalar multiplication using point halving." *Advances in Cryptography— ASIACRYPT'99*, Lecture Notes in Computer Science, vol. 1716, eds. K.Y. Lam, E. Okamoto, and C. Xing. Springer-Verlag, Berlin, 135–149.

[5] Knuth, D. (1998). *The Art of Computer Programming—Seminumerical Algorithms* (3rd ed.). Addison-Wesley, Reading, MA.

[6] Schroeppel, R. (2000). "Elliptic curves: Twice as fast!" *Advances in Cryptology—CRYPTO 2000*, Lecture Notes in Computer Science, vol. 1880, ed. M. Bellare. Presentation at the *CRYPTO 2000* [1] Rump Session.

[7] Schroeppel, R. (2000). "Elliptic curve point ambiguity resolution apparatus and method." International Application Number PCT/US00/31014.

[8] Solinas, J. (2000). "Efficient arithmetic on Koblitz curves." *Designs, Codes and Cryptography*, 19, 195–249.

ELLIPTIC CURVE PUBLIC-KEY ENCRYPTION SCHEMES

It is possible to describe elliptic curve analogues of all the variants of the ElGamal public-key encryption scheme [3]. We describe one such variant, the *Elliptic Curve Integrated Encryption Scheme* (ECIES), proposed by Abdalla, Bellare and Rogaway [1].

The elliptic curve domain parameters are $D = (q, \mathrm{FR}, S, a, b, P, n, h)$, and an entity A's key pair is (d, Q) (see elliptic curve keys). E denotes a symmetric cryptosystem such as the Rijndael/AES, and MAC (see MAC algorithms) denotes a message authentication code algorithm such as HMAC. In order to encrypt a message m to A, an entity B does the following:

1. Select $k \in_R [1, n-1]$.
2. Compute $R = kP$ and $Z = kQ$.
3. Derive two keys k_1 and k_2 from Z and R.
4. Compute $c = E_{k_1}(m)$ and $t = \mathrm{MAC}_{k_2}(c)$.
5. Send (R, c, t) to A.

A decrypts using her private key d as follows:

1. Compute $Z = dR$.
2. Derive two keys k_1 and k_2 from Z and R.
3. Compute $t' = \mathrm{MAC}_{k_2}(c)$; reject the ciphertext if $t \neq t'$.
4. Compute $m = E_{k_1}^{-1}(c)$.

ECIES has been proven to be semantically secure against adaptive chosen-ciphertext attacks [4] under the assumptions that the encryption scheme E is secure, that the MAC algorithm is secure, and that certain non-standard variants of the Diffie–Hellman problem are intractable [1].

Another noteworthy elliptic curve public-key encryption scheme is the elliptic curve analogue of the Cramer–Shoup scheme [2]. This scheme has the advantage that its security has been proven under standard assumptions only; however, encryption and decryption are slower than in schemes such as ECIES.

Darrel Hankerson
Alfred Menezes

References

[1] Abdalla, M., M. Bellare, and P. Rogaway (2001). "The oracle Diffie–Hellman assumptions and an analysis of DHIES." *Topics in Cryptology—CT-RSA 2001*, Lecture Notes in Computer Science, vol. 2020, ed. D. Naccache. Springer-Verlag, Berlin, 143–158.

[2] Cramer, R. and V. Shoup (1998). "A practical public key cryptosystem provably secure against adaptive chosen ciphertext attack." *Advances in Cryptology—CRYPTO'98*, Lecture Notes in Computer Science, vol. 1462, ed. H. Krawczyk. Springer-Verlag, Berlin, 13–25.

[3] ElGamal, T. (1985). "A public key cryptosystem and a signature scheme based on discrete logarithms." *IEEE Transactions on Information Theory*, 31, 469–472.

[4] Rackoff, C. and D. Simon (1992). "Non-interactive zero-knowledge proof of knowledge and chosen ciphertext attack." *Advances in Cryptology—CRYPTO'91*, Lecture Notes in Computer Science, vol. 576, ed. J. Feigenbaum. Springer-Verlag, Berlin, 433–444.

ELLIPTIC CURVE SIGNATURE SCHEMES

Many variants of the ElGamal digital signature scheme [1] have been proposed including the Digital Signature Algorithm (DSA; see Digital Signature Standard) [2]; Schnorr digital signature scheme [8]; the Nyberg-Rueppel signature scheme [6]; and the Korean certificate-based digital signature algorithm (KCDSA) [5]. Some of these variants have been proven to be existentially unforgeable by adaptive chosen-message attacks [3] under certain assumptions including intractability of the elliptic curve discrete logarithm problem (ECDLP) (see [7]). We outline *ECDSA*, an elliptic curve analogue of the DSA; for further details, see [4].

The elliptic curve domain parameters are $D = (q, \mathrm{FR}, S, a, b, P, n, h)$, and an entity A's key pair is (d, Q) (see elliptic curve keys). In order to sign a message m, A does the following:

1. Select $k \in_R [1, n-1]$.
2. Compute $R = kP$.
3. Compute $r = x \bmod n$ (see modular arithmetic), where x is the x-coordinate of R. If $r = 0$ then go to step 1.
4. Compute $e = H(m)$, where H is a cryptographic hash function.
5. Compute $s = k^{-1}(e + dr) \bmod n$. If $s = 0$ then go to step 1.
6. The digital signature on m is (r, s).

To verify A's signature (r, s) on m, and entity B does the following:

1. Verify that r and s are integers in the interval $[1, n-1]$.
2. Compute $e = H(m)$.
3. Compute $w = s^{-1} \bmod n$, $u_1 = ew \bmod n$ and $u_2 = rw \bmod n$.

4. Compute $X = u_1 P + u_2 Q$, and verify that $X \neq \infty$.

5. Let $v = x \bmod n$, where x is the x-coordinate of X.

6. Accept the signature if and only if $v = r$.

Signature verficiation works because if a signature (r, s) on a message m was indeed generated by the legitimate signer, then $s \equiv k^{-1}(e + dr)$ (mod n). Rearranging gives

$$k \equiv s^{-1}(e + dr) \equiv s^{-1}e + s^{-1}rd \equiv we + wrd$$

$$\equiv u_1 + u_2 d \pmod{n}.$$

Thus $X = u_1 P + u_2 Q = (u_1 + u_2 d)P = kP$, and so $v = r$ as required.

Darrel Hankerson
Alfred Menezes

References

[1] ElGamal, T. (1985). "A public key cryptosystem and a signature scheme based on discrete logarithms." *IEEE Transactions on Information Theory*, 31, 469–472.

[2] FIPS 186-2 (2000). *Digital Signature Standard (DSS)*, Federal Information Processing Standards Publication 186-2. National Institute of Standards and Technology, Gaithersburg, MD.

[3] Goldwasser, S., S. Micali, and R. Rivest (1988). "A digital signature scheme secure against adaptive chosen-message attacks." *SIAM Journal on Computing*, 17, 281–308.

[4] Johnson, D., A. Menezes, and S. Vanstone (2001). "The elliptic curve digital signature algorithm (ECDSA)." *International Journal of Information Security*, 1, 36–63.

[5] Lim, C. and P. Lee (1998). "A study on the proposed Korean digital signature algorithm." *Advances in Cryptography—ASIACRYPT'98*, Lecture Notes in Computer Science, vol. 1514, eds. K. Ohta, and D. Pei. Springer-Verlag, Berlin, 175–186.

[6] Nyberg, K. and R. Rueppel (1996). "Message recovery for signature schemes based on the discrete logarithm problem." *Designs, Codes and Cryptography*, 7, 61–81.

[7] Pointcheval, D. and J. Stern (2000). "Security arguments for digital signatures and blind signatures." *Journal of Cryptology*, 13, 361–396.

[8] Schnorr, C. (1991). "Efficient signature generation by smart cards." *Journal of Cryptology*, 4, 161–174.

ELLIPTIC CURVES FOR PRIMALITY PROVING

Proving the primality of an integer N (see primality proving algorithm) is easy if $N - 1$ can be factored: N is a prime number if and only if the multiplicative group of invertible elements $(\mathbb{Z}/N\mathbb{Z})^*$ is cyclic of order $N - 1$ (see modular arithmetic). To prove that an integer g is a generator of $(\mathbb{Z}/N\mathbb{Z})^*$ and hence that the group is cyclic, it suffices to check that $g^{N-1} \equiv 1 \bmod N$ and $g^{(N-1)/q} \not\equiv 1 \bmod N$ for all prime factors q of $N - 1$. (It is quite easy to find a generator, or to prove that none exists, given the prime factors.)

The above method is the converse of Fermat's Little Theorem. However, it is rare that $N - 1$ is easy to factor. Less rare is the case where $N - 1$ has a large prime cofactor C, in which case the primality of $N - 1$ can be proven in the same way, modulo the assumption that C can be proven prime in turn. This approach of primality cannot succeed to prove the primality of all numbers in reasonable time. Other approaches have therefore been pursued for proving that a number is prime. For instance, the first deterministic algorithm for this task was designed by Adleman et al. in the early 1980s [2] and improved by Cohen and Lenstra [7] among others [5, 6, 12].

The basic idea of primality proving with elliptic curves is to enlarge the set of groups that can be used for proving the primality of N. We add to $(\mathbb{Z}/N\mathbb{Z})^*$ the groups formed by elliptic curves over $\mathbb{Z}/N\mathbb{Z}$, that is sets $E(\mathbb{Z}/N\mathbb{Z}) = \{(x, y) \in (\mathbb{Z}/N\mathbb{Z})^2, y^2 \equiv x^3 + ax + b \bmod N\}$, where $\gcd(4a^3 + 27b^2, N) = 1$. If N is indeed prime, then $E(\mathbb{Z}/N\mathbb{Z})$ is a group of order $m = N + 1 - t$ for some integer $t \in [-2\sqrt{N}, 2\sqrt{N}]$, by Hasse's theorem. A primality testing theorem analogous to the converse of Fermat's theorem can be proven, so that it is enough to find a generator P of the curve satisfying equalities in the group $E(\mathbb{Z}/N\mathbb{Z})$. To each curve corresponds a distinct m. A lot of them should be factorizable numbers. This will give us a lot of candidates to try for primality, and not just $N - 1$.

For use in primality, two problems must be solved: find algorithms to compute the group law on $E(\mathbb{Z}/N\mathbb{Z})$ and compute the cardinality of the group. The first task is easy, many algorithms being known. The second task is not so easy, but there exists a deterministic point counting algorithm due to R. Schoof for this, which runs in polynomial time.

In 1986, two primality proving algorithms using elliptic curves were proposed, somewhat anticipated in 1985 by Bosma, Chudnovsky and Chudnovsky. One is due to Goldwasser and Kilian [9, 10], the other one to Atkin [3]. The Goldwasser–Kilian algorithm uses random curves whose cardinality has to be computed with Schoof's algorithm. The analysis shows that proving primality can be done in random polynomial time for almost all primes, a result eventually proven for all primes [1] by Adleman and Huang using curves of genus 2, shortly after the work of Goldwasser and Kilian.

The drawback of the Goldwasser–Kilian algorithm is that Schoof's algorithm is too slow to be interesting in practical primality proving, despite the improvements made by Elkies and Atkin.

Atkin's approach replaces random curves by curves which are reductions of curves with complex multiplication and for which the cardinality is known in advance. The analysis of this algorithm, also called ECPP [4], is heuristic and yields a running time $O((\log N)^6)$ (see [11] and O-notation) for deciding the primality of N using classical arithmetic. In practice, due to many improvements of several people including the present author, primes of cryptographic size (say 512 or 1024 bits) can be proved in a few seconds, and large numbers tackled as well, the record being around 5000 decimal digits. An interesting feature of this algorithm is the fact that it gives a certificate of primality that can be checked in time $O((\log N)^4)$.

As a matter of fact, a faster version of ECPP was outlined by J. O. Shallit, with a complexity of $O((\log N)^5)$ (see [11]). With fast multiplication techniques, this is a $\tilde{O}((\log N)^4)$ method and its practicality has been recently demonstrated with the primality of several numbers with more than 15000 decimal digits [8, 13].

François Morain

References

[1] Adleman, L.M. and M.-D.A. Huang (1996). "Primality testing and Abelian varieties over finite fields." *Proc. of International Workshop TYPES'96*, Lecture Notes in Math, vol. 1512, eds. E. Gimenez and C. Paulin-Mohring. Springer-Verlag, Berlin.

[2] Adleman, L.M., C. Pomerance, and R.S. Rumely (1983). "On distinguishing prime numbers from composite numbers." *Ann. Math.* (2), 117 (1), 173–206.

[3] Atkin, A.O.L. (1986). Manuscript. Lecture Notes of a Conference, Boulder CO.

[4] Atkin, A.O.L. and F. Morain (1993). "Elliptic curves and primality proving." *Math. Comp.*, 61 (203), 29–68.

[5] Bosma, W. and M.-P. van der Hulst (1990). "Primality proving with cyclotomy." *PhD Thesis*, Universiteit van Amsterdam.

[6] Cohen, H. and A.K. Lenstra (1987). "Implementation of a new primality test." *Math. Comp.*, 48 (177), 103–121.

[7] Cohen, H. and H.W. Lenstra, Jr. (1984). "Primality testing and Jacobi sums." *Math. Comp.*, 42 (165), 297–330.

[8] Franke, J., T. Kleinjung, F. Morain, and T. Wirth. Proving the primality of very large numbers with fastecpp. In D. Buell, editor, Algorithmic Number Theory, volume 3076 of Lecture Notes in Computer Science, pages 194–207. Springer, Berlin, 2004. 6th

International Symposium, ANTS-VI, Burlington, VT, USA, June 2004.

[9] Goldwasser, S. and J. Kilian (1986). "Almost all primes can be quickly certified." *Proc. 18th STOC, Berkeley, CA.* ACM, New York, 316–329.

[10] Goldwasser, S. and J. Kilian (1990). "Primality testing using elliptic curves." *J. ACM*, 46 (4), 450–472.

[11] Lenstra, A.K. and H.W. Lenstra, Jr. (1996). "Algorithms in number theory." *Handbook of Theoretical Computer Science, Volume A: Algorithms and Complexity*, ed. J. van Leeuwen. North Holland, Amsterdam, Chapter 12, 674–715.

[12] Mihăilescu, P. (1997). "Cyclotomy of rings and primality testing." Diss. ETH no. 12278, Swiss Federal Institute of Technology Zürich.

[13] Morain, F. (2005). Implementing the asymptotically fast version of the elliptic curve primality proving algorithm. Available at http://www.lix.polytechnique.fr/Labo/Francois.Morain/

EMV[1]

OVERVIEW: Traditional credit and debit payment cards carry a magnetic stripe, a hologram, a specimen signature of the cardholder, and one or more payment brands, and they may be embossed with the cardholder's name, their account number and the expiry date of the card. Magnetic stripe technology provides little in the way of card authentication. Point-of-Sale terminals cannot authenticate magnetic stripe cards and, even if sent "online" for authorization by the card issuer, because of the static nature of the magnetic stripe, the issuer is not able to distinguish card data originating from a genuine card from replayed card data or card data read from a copied (cloned) card. However with the advent of EMV chip cards, card authentication can be performed by the terminal or issuer using dynamic techniques that distinguish genuine cards from clones. Both the card and terminal implement offline risk management processes that control whether a transaction is approved or declined offline, or whether online authorization should be sought.

EMV defines different mechanisms for chip card authentication. In the so-called on-line authentication method (on-line CAM) the authentication of the payment card and transaction is done through an on-line communication to the card issuer during the transaction while the off-line authentication method (off-line CAM) enables the payment terminal to authenticate the card without this online communication.

---

[1] EMV2000 Integrated Circuit Card Specification for Payment Systems, Version 4.0, December 2000. A joint specification, originally developed by Europay, MasterCard and Visa, and now administered by EMVCo, LLC. © 2000 EMVCo, LLC. EMV is a trademark of EMVCo, LLC.

THE EMV ARCHITECTURE

The Transaction Flow

The figure below provides an overview on the transaction flow for EMV.

Off-line CAM

The off-line card authentication mechanism is either "static", wherein the same authentication data is provided by the card to the terminal for every transaction, or "dynamic", wherein the authentication data provided by the card will be different for each transaction. The EMV2000 [1] specifications define one static offline CAM (SDA) and two dynamic offline CAMs: Dynamic Data Authentication (DDA) and Combined Data Authentication (CDA).

For SDA the issuer pre-signs unique static card data to protect against alteration of the data after personalization. During a transaction the terminal can retrieve this signed static data from the chip card and verify the correctness of this data (Figure 2).

For both DDA and CDA the issuer personalizes the chip card with a certified private RSA key unique to the chip card and then during a transaction the card produces a dynamic signature on a random challenge received from the terminal. By verifying this dynamic signature, the terminal can authenticate the chip card itself (under the assumption that the chip card's private key is known only to the chip card), and confirm the legitimacy of static chip card data. Additionally with CDA the dynamic signature of the chip card covers all the transaction data necessary for the terminal to confirm the integrity of the chip card's response for the current transaction (Figure 3).

The EMV specifications have specified CDA as a result of a study on the use of "wedge" devices for offline transactions. Wedge devices alter

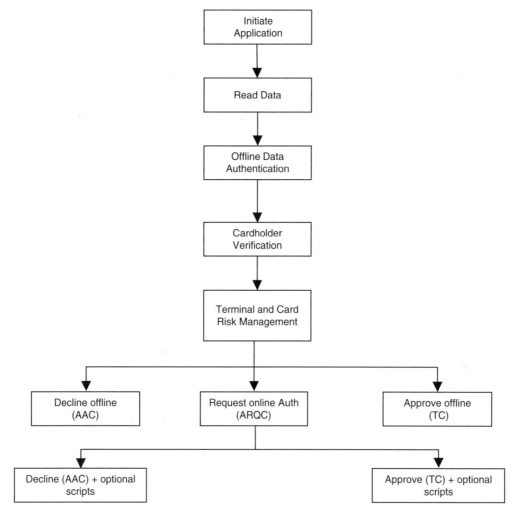

Fig. 1. EMV transaction flow

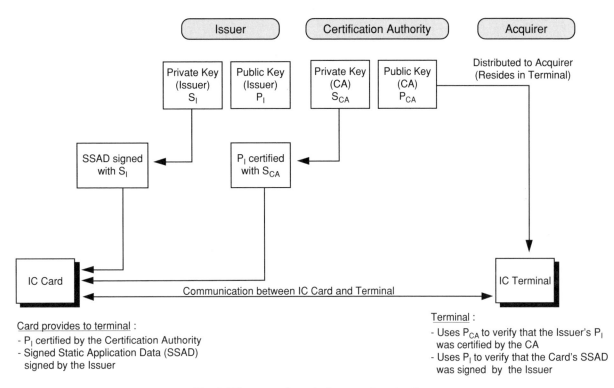

Fig. 2. Diagram of static data authentication

data exchanged between the genuine card and the terminal and such alterations may not be detected when using SDA or DDA. Thus whereas DDA authenticates the card but not the transaction data, with CDA the card digitally signs all the important transaction data including the value, and so any modifications to this data can be detected.

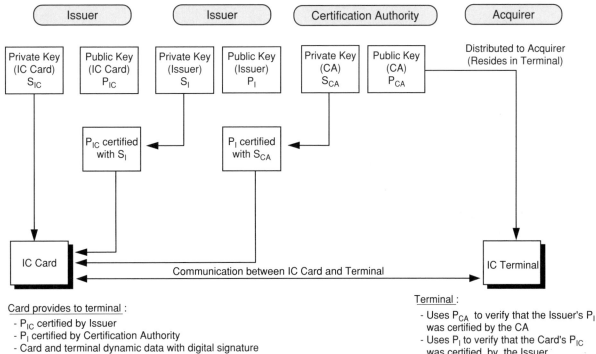

Fig. 3. Diagram of dynamic data authentication

On-line CAM and Application Cryptograms

In order to understand the significance of the offline CAMs other aspects of the EMV transaction must be introduced, namely application cryptograms. During an EMV transaction the terminal will obtain from the card at least one Application Cryptograms (AC) that can be verified by the card issuer. These cryptograms, which are dynamically generated, can be one of the following:

ARQC   generated when the chip card requests online authorization

TC   generated when the chip card approves the transaction

AAC   generated when the chip card declines the transaction

AAR   generated for authorization referrals

If the card generates an ARQC then the terminal sends this to the issuer who responds with a cryptogram called the ARPC which can be verified by the card. This process is known as online dynamic data authentication or On-line Mutual Authentication (OMA). TCs are retained by the terminal for inclusion in the clearing records.

These transaction cryptograms are dynamically generated by the chip card and use symmetric cryptography, employing a unique key derived from a master key shared between the chip card and the issuer.

Cardholder Verification Method

The introduction of chip cards allows Issuers to choose whether their cards support online PIN and/or offline PIN verification (see Personal Identification Number). With online PIN the PIN entered by the cardholder is encrypted and sent over the network to the issuer for verification in the usual way (as per magnetic stripe). With offline PIN verification the PIN entered by the cardholder is sent from the PIN pad to the card for verification. In order to perform offline PIN verification the card must securely store a reference copy of the cardholder's PIN.

EMV specifies two methods for offline PIN verification: plaintext and enciphered.

- With offline plaintext PIN verification the card receives the PIN in clear from the terminal.
- For offline enciphered PIN verification the card must be a DDA-capable card. The card will receive the PIN encrypted under a card public key and using a randomized RSA encryption method specified in EMV. The card will decrypt the PIN using its corresponding private RSA key.

For simple Lost and Stolen cards the use of offline PIN verification will vastly reduce fraud as compared to the use of handwritten signatures.

Risk Management

The decision for off-line or on-line authorization for a given chip card transaction is based on the outcome of the card and terminal risk management, consisting of

- Floor limit checking. Merchant terminals contain a floor limit value. If the transaction value exceeds this number then the terminal should request online authorization.
- Random transaction selection by the terminal, ensuring that transactions go on-line periodically.
- Velocity Checking, where the terminal can choose to go on-line depending on card risk management, aiming to limit the number of consecutive offline transactions performed by the card.

Moreover chip card technology allows the issuer to block or unblock the card, change the PIN and some of the card risk management parameters, using issuer to card script processing. These scripts are protected using symmetric cryptography ensuring integrity and/or confidentiality, as appropriate.

CRYPTOGRAPHIC MECHANISMS AND PROTOCOLS

Message Authentication Code

The computation of an s-byte MAC ($4 \leq s \leq 8$; see MAC algorithms) is according to ISO/IEC 9797-1 [2] using a 64-bit block cipher ALG in CBC mode as specified in ISO/IEC 10116. More precisely, the computation of a MAC S over a message MSG consisting of an arbitrary number of bytes with a MAC Session Key K_S takes place in the following steps:

1. *Padding and Blocking*

    Pad the message M according to ISO/IEC 7816-4 (which is equivalent to method 2 of ISO/IEC 9797-1), hence add a mandatory '80' byte to the right of MSG, and then add the smallest number of '00' bytes to the right such that the length of resulting message **MSG** := (MSG || '80' || '00' || '00' || . . . || '00') is a multiple of 8 bytes.

    MSG is then divided into 8-byte blocks X_1, $X_2, \ldots, X_k$.

2. *MAC Session Key*

The MAC Session Key K_S either consists of only a leftmost key block $K_S = K_{SL}$ or the concatenation of a leftmost and a rightmost key block $K_S = (K_{SL} || K_{SR})$.

3. *Cryptogram Computation*

Process the 8-byte blocks $X_1, X_2, \ldots, X_k$ with the block cipher in CBC mode (see modes of operation) using the leftmost MAC Session Key block K_{SL}:

$$H_i := \mathrm{ALG}(K_{SL})[X_i \oplus H_{i-1}], \text{ for } i = 1, 2, \ldots, k$$

with initial value $H_0 :=$
('00' || '00' || '00' || '00' || '00' || '00' || '00' || '00').
Compute the 8-byte block H_{k+1} in one of the following two ways:

- According to ISO/IEC 9797-1 Algorithm 1:

$$H_{k+1} := H_k.$$

- According to Optional Process 1 of ISO/IEC 9797-1 Algorithm 3:

$$H_{k+1} := \mathrm{ALG}(K_{SL})[\mathrm{ALG}^{-1}(K_{SR})[H_k]].$$

The MAC S is then equal to the s most significant bytes of H_{k+1}.

Digital Signature Scheme Giving Message Recovery

This section describes the special case of the digital signature scheme giving message recovery using a hash function according to ISO/IEC 9796-2 [3], which is used in the EMV specification for both static and dynamic data authentication.

The digital signature scheme uses the following two types of algorithms.

- A reversible asymmetric algorithm consisting of a signing function $\mathrm{Sign}(S_K)[\ ]$ depending on a Private Key S_K and a recovery function $\mathrm{Recover}(P_K)[\ ]$ depending on a Public Key P_K. Both functions map N-byte numbers onto N-byte numbers and have the property that

$$\mathrm{Recover}(P_K)[\mathrm{Sign}(S_K)[X]] = X,$$

for any N-byte number X.

- A hash algorithm $\mathrm{Hash}[\ ]$ that maps a message of arbitrary length onto an 20-byte hash code.

Signature Generation

The computation of a signature S on a message MSG consisting of an arbitrary number L of at least $N - 21$ bytes takes place in the following way.

1. Compute the 20-byte hash value $H :=$ $\mathrm{Hash}[\mathrm{MSG}]$ of the message M.
2. Split MSG into two parts $\mathrm{MSG} = (\mathrm{MSG}_1 \ || \ \mathrm{MSG}_2)$, where MSG_1 consists of the $N - 22$ leftmost (most significant bytes) of MSG and MSG_2 of the remaining (least significant) $L - N + 22$ bytes of MSG.
3. Define the byte $B := $ '6A'.
4. Define the byte $E := $ 'BC'.
5. Define the N-byte block X as the concatenation of the blocks B, MSG_1, H and E, hence

$$X := (B || \mathrm{MSG}_1 || H || E).$$

6. The digital signature S is then defined as the N-byte number.

$$S := \mathrm{Sign}(S_K)[X].$$

Signature Verification

The corresponding signature verification takes place in the following way:

1. Check whether the digital signature S consists of N bytes.
2. Retrieve the N-byte number X from the digital signature S:

$$X = \mathrm{Recover}(P_K)[S].$$

3. Partition X as $X = (B \ || \ \mathrm{MSG}_1 \ || \ H \ || \ E)$, where
   - B is one byte long,
   - H is 20 bytes long,
   - E is one byte long,
   - MSG_1 consists of the remaining $N - 22$ bytes.
4. Check whether the byte B is equal to '6A'.
5. Check whether the byte E is equal to 'BC'.
6. Compute $\mathrm{MSG} = (\mathrm{MSG}_1 \ || \ \mathrm{MSG}_2)$ and check whether $H = \mathrm{Hash}[\mathrm{MSG}]$.

If and only if these checks are correct is the message accepted as genuine.

Approved Algorithms

The double-length key triple DES encipherment algorithm is the approved cryptographic algorithm to be used in the encipherment (using ECB or CBC as specified in ISO/IEC 10116 [4]) and MAC mechanisms specified above (see ISO/IEC CD 18033-3 [7]).

Single DES (see Data Encryption Standard) is only approved for use with the version of the MAC mechanism specified in Algorithm 3 of ISO 9797-1 (triple DES applied to the last block).

The digital signature scheme and offline PIN encryption use the RSA transform (see RSA digital signature scheme) as defined in ISO/IEC CD 18033-2 [6].

The approved algorithm for hashing is SHA-1 as specified in ISO/IEC 10118-3 [5].

Marijke de Soete
Michael Ward

References

[1] EMV2000 (2002). *Integrated Circuit card Specification for Payment Systems*, version 4.0.
[2] ISO/IEC 9797-1 (2002). "Information technology—security techniques, message authentication codes—Part 1: Mechansims using a block cipher algorithm."
[3] ISO/IEC 9796-2 (2002). "Information technology—security techniques, digital signature schemes giving message recovery—Part 2: Integer factorization based mechanisms."
[4] ISO/IEC 10116 (1997). "Information technology—security techniques, modes of operations for an n-bit block cipher."
[5] ISO/IEC 10118-3 (1998). "Information technology—security techniques, hash-functions—Part 3: Dedicated hash-functions."
[6] ISO/IEC 18033-2: FCD 2004 (2004). "Information technology—security techniques, encryption algorithms—Part 2: Asymmetric ciphers."
[7] ISO/IEC 18033-3: FCD 2004 (2004). "Information technology—security techniques, encryption algorithms—Part 3: Block ciphers."

ENCRYPTION

An encryption (also called *enciphering*), is a mapping of plaintext to ciphertext, based on some chosen keytext. It is performed by a stepwise application of a (more or less formalized) encryption algorithm (see cryptosystem).

An *encryption step* is an encryption applied to a particular sequence of plaintext characters, using, in a way that depends on the *encryption key*, a particular *encryption rule* of an encryption algorithm.

Often *padding* is necessary to give the message the proper length as required by the encryption algorithm. With padding, one also means the filling of gaps between meaningful messages, frequently by special padding characters (called nulls). Both meaningful messages and padding characters are encrypted, thus masking the occurrence of idle times. Careless padding may corrupt some encryption systems.

A *product cipher* or *superencryption* consists of an encryption A applied to the result of encryption B; it is denoted by encryption AB (product of B and A).

Homophones are two or more ciphertext elements belonging to the same plaintext element.

EXAMPLE. The bipartite, monographic encryption (see substitutions and permutations and alphabet) $Z_{26} \longrightarrow [\mathbf{Z}_{10} \backslash \{0\}]^2$ given by:

			1	2	3	4	5	6	7	8	9
9,	6,	3	a	b	c	d	e	f	g	h	i
8,	5,	2	j	k	l	m	n	o	p	q	r
7,	4,	1	s	t	u	v	w	x	y	z	

The plaintext letter e has the three homophones $95, 65, 35$. The table also defines the inverse, a unipartite, digraphic substitution $[\mathbf{Z}_{10} \backslash \{0\}]^2 \longrightarrow Z_{26}$, which turns out to be polyphonic.

Polyphony is an encryption that assigns to each plaintext element one out of several ciphertext elements (to be chosen arbitrarily, randomly in the best case).

Decryption or *deciphering* is the inverse operation of an encryption; it maps ciphertext to plaintext, based on the (authorized) knowledge of the keytext. It may consist of several *decryption steps*.

One speaks of an *unauthorized* decryption if it takes place without authorized access to the key.

Examples of classical encryption schemes are often named after people (see Cæsar, Alberti, Vigenère, Porta, Beaufort, Vernam, Polybios, and Playfair).

Friedrich L. Bauer

Reference

[1] Bauer, F.L. (1997). "Decrypted secrets." *Methods and Maxims of Cryptology*. Springer-Verlag, Berlin.

ENCRYPTION EXPONENT

The exponent e in the RSA public key (n, e). See RSA public key encryption and RSA digital signature scheme.

Burt Kaliski

ENTITLEMENTS MANAGEMENT

Entitlements management is a subset of general "authorization data" management (see authorization architecture) in which the data being managed are *entitlements* granted or bestowed upon entities in an environment. An *entitlement* may be defined as follows [1]: "a right to benefits specified especially by law or contract".

Carlisle Adams

Reference

[1] Merriam-Webster OnLine. Available on http://www.m-w.com/cgi-bin/dictionary

ENTITY AUTHENTICATION

Entity authentication is the process by which one entity (the *verifier*) is assured of the identity of a second entity (the *claimant*) that is participating in a protocol (see identity verification protocol). This assurance is usually obtained by requiring the claimant to provide corroborating evidence of the claimed identity to the verifier. The claimed identity can either be presented to the verifier as part of the protocol or can be presumed by context.

The term identification is sometimes used as a synonym for entity authentication, however it is also sometimes used to simply refer to the process of claiming or stating an identity without providing the corroborating evidence required for entity authentication. Care must be taken to ensure, when using this term, that the correct interpretation is used.

The corroborating evidence required in order to obtain entity authentication is sometimes also called credentials or an authentication code and is usually calculated using one of the following three types of input:

1. *Something known*: Typically this involves the claimant providing a password (see password) or PIN (see Personal Identification Code) that is then either presented directly to the verifier or is used to compute credentials that are presented to the verifier and are then validated.
2. *Something possessed*: This type of input includes physical devices that are used to compute the credentials presented to the verifier as well as software files that must be possessed by the claimant in order to compute the credentials. Examples of physical devices include smart cards, magnetic stripe cards and one-time password generators (see one-time password). Software files will typically contain secret or private keys that are used to compute credentials, however physical devices may contain these keys as well. Note that these physical devices are sometimes called authentication codes.
3. *Something inherent*: This category includes that class of inputs known as biometrics. In this category human physical characteristics, like fingerprints, retinal scans or handwriting, are used to produce the credentials.

The main reason why entity authentication is usually required is to restrict access to protected resources. Examples include remote access to computer accounts, access to web sites, and bank account access at automated teller machines. If entity authentication is being used for this purpose then it necessarily must be combined with an access control method or privilege management technique; see also authorization architecture) in order to restrict access. The claimant will usually first be authenticated and then the access control or privilege management technique will be used to determine the level of access allowed.

Important standards that describe methods of entity authentication include:

- ISO/IEC 9798 [1]: This is a five part international standard dealing with entity authentication that includes a general introduction and parts focusing on entity authentication based upon symmetric encryption, public-key signatures, cryptographic check functions, and zero-knowledge techniques.
- ANSI X9.26 [2]: This is a US banking standard that specifies techniques for entity authentication using passwords and DES-based challenge response protocols.
- ISO 11131 [3]: This is the international version of ANSI X9.26.
- FIPS 196 [4]: This is a US government standard that specifies techniques for entity authentication using public-key techniques. It is based upon techniques described in ISO/IEC 9798-3.

Robert Zuccherato

References

[1] ISO/IEC 9798-1 (1997). Information technology—security techniques—entity authentication.
[2] ANSI X9.26 (1991). American National Standard—financial institution sign-on authentication for wholesale financial transactions, ASC X9 Secretariat—American Bankers Association.
[3] ISO 11131 (1992). Banking—financial institution sign-on authentication.
[4] FIPS 196 (1997). *Entity Authentication using Public Key Cryptography*, Federal Information Processing Standards Publication 196. U.S. Department of Commerce/NIST, National Technical Information Service, Springfield, VA.

EUCLIDEAN ALGORITHM

The Euclidean algorithm was reported by Euclid in his Elements [3]. For a brief historical account the reader is referred to Knuth [4]. The Euclidean algorithm provides a simple and efficient means for computing the greatest common divisor (GCD) denoted $\gcd(u, v)$ of two positive integers u and v without finding their factorizations. The gcd operation exhibits the following properties:

1. $\gcd(g, 0) = |g|$;
2. $\gcd(u, v) = \gcd(v, u - v)$;
3. $\gcd(u, v) = \gcd(v, u \bmod v) = \gcd(v, u - qv)$, where $q = \lfloor u/v \rfloor$.

The execution of the Euclidean algorithm is based on the repetitive application of the property $\gcd(u, v) = \gcd(v, u \bmod v)$. Assume $u \geq v$ and let $r_0 = u$ and $r_1 = v$, then the computation proceeds as $\gcd(r_0, r_1) = \gcd(r_1, r_2) = \ldots = \gcd(r_k, 0) = r_k$, where the residues r_i are related as follows:

$$r_{i+2} = r_i - q_i r_{i+1} \quad \text{with} \quad q_i = \lfloor r_i / r_{i+1} \rfloor.$$

The sequence r_i is strictly decreasing and by induction it may be shown to quickly converge to zero. When $r_{k+1} = 0$ then r_k holds the desired result and the algorithm terminates. The Euclidean Algorithm is given below:

The Euclidean Algorithm
Input: positive integers u and v, with $u \geq v$
Output: $g = \gcd(u, v)$
    While $v > 0$ do
        $q \leftarrow \lfloor u/v \rfloor$; $r \leftarrow u - qv$;
        $u \leftarrow v$; $v \leftarrow r$;
    End While
    $g \leftarrow u$;
    Return (g)

In many cryprographic applications the "extended" version of the Euclidean algorithm plays an important role. In addition to the greatest common divisor, the Extended Euclidean Algorithm (EEA) returns two unique integers s and t. Using these integers the greatest common divisor may be expressed as a linear combination of u and v:

$$us + vt = \gcd(u, v).$$

If u and v are relatively prime, it immediately follows that

$$us = 1 \; (\bmod \; v).$$

Hence, the Extended Euclidean Algorithm provides an efficient method to compute modular inverse $u^{-1} \bmod v = s$. The original Euclidean algorithm described above is modified to compute the parameters s and t. We define new iteration parameters s_i and t_i such that

$$r_i = s_i u + t_i v.$$

Consider the $i + 1$-st iteration $r_{i+1} = r_{i-1} - q_{i-1} r_i$. By substituting r_i and r_{i-1} in this identity we obtain the following relation:

$$\begin{aligned} r_{i+1} &= (s_{i-1}u + t_{i-1}v) - q_{i-1}(s_i u + t_i v) \\ &= (s_{i-1} - q_{i-1}s_i)u + (t_{i-1} - q_{i-1}t_i)v. \end{aligned}$$

Also since $r_{i+1} = s_{i+1}u + t_{i+1}v$, the values of the s and t parameters in iteration $i + 1$ are found as follows:

$$s_{i+1} = s_{i-1} - q_{i-1}s_i$$
$$t_{i+1} = t_{i-1} - q_{i-1}t_i.$$

As seen from these equations for the computation of the sequences s_i and t_i only the last two values of s and t need to be available. In the first two iterations of the algorithm, $r_0 = s_0 u + t_0 v$ and $r_1 = s_1 u + t_1 v$ since $r_0 = u$ and $r_1 = v$ the parameters need to be initialized as $s_0 = 1$, $s_1 = 0$, $t_0 = 0$, and $t_1 = 1$. A generic description of the Extended Euclidean algorithm is given below:

The Extended Euclidean Algorithm
Input: positive integers u and v, with $u \geq v$
Output: $g = \gcd(u, v)$ and integers s, t
    satisfying $us + vt = g$
    $s'' \leftarrow 1$; $s' \leftarrow 0$; $t'' \leftarrow 0$; $t' \leftarrow 1$
    While $v > 0$ do
        $q \leftarrow \lfloor u/v \rfloor$; $r \leftarrow u - qv$;
        $s \leftarrow s'' - qs'$; $t \leftarrow t'' - qt'$;
        $u \leftarrow v$; $v \leftarrow r$;
        $s'' \leftarrow s'$; $s' \leftarrow s$; $t'' \leftarrow t'$; $t' \leftarrow t$;
    End While
    $g \leftarrow u$;
    $s \leftarrow s''$; $t \leftarrow t''$;
    Return (g, s, t)

The algorithm recycles the temporary variables s', s'', t', and t'' by shifting their values in each iteration of the loop.

The most complex operation used in the EEA is the integer division operation. If the goal of the EEA is to compute a modular inverse, i.e., $u^{-1} \bmod v$, then the operations required for computing t may be entirely omitted.

The exact number of iterations required for the algorithm to converge is more difficult to analyze.

According to [2], Lamé, Dixon, and Heilbronn developed an upper bound for the number of iterations in the Euclidean algorithm as

$$\left\lceil \frac{\log(n\sqrt{5})}{\log((1+\sqrt{5})/2)} \right\rceil - 2 \approx 2.078 \log(n) + 1.672$$

and came up with the following approximation for the average number of iterations:

$$\frac{12 \log(2)}{\pi^2} \log(n) + 0.14 \approx 0.843 \log(n) + 0.14,$$

where $u, v \leq n$. Hence, the number of iterations grows logarithmically with the size of the inputs. In each iteration of the Euclidean algorithm a costly division operation is computed which takes time $O((\log(n))^2)$ (see <u>O-notation</u>). This gives a running time of $O((\log(n))^3)$. However, by careful implementation it is possible to systematically reduce the sizes of u and v in each step, and achieve the overall computation in time $O((\log(n))^2)$.

For multiprecision integers, there is a useful variant due to Lehmer [5] with time complexity $O((\log(n))^2)$. Lehmer's algorithm is based on the observation that the quotient in the Euclidean algorithm is in general dependant only on the leading digits of u and v. More clearly, if $\bar{u}$ and $\bar{v}$ denote the leading digits of u and v, respectively. Then the following inequality always holds:

$$\frac{\bar{u}}{\bar{v}+1} \leq \frac{u}{v} \leq \frac{\bar{u}+1}{\bar{v}}.$$

Hence, to determine the quotient in the Euclidean iteration one may compute $q = \bar{u}/(\bar{v}+1)$ and $q' = (\bar{u}+1)/\bar{v}$. If $q = q'$ then it must be that $q' = u/v = q$, and the quotient is determined via two simple single precision divisions. Otherwise, a multiprecision division is performed as in the original Euclidean iteration. The advantage of this method is that multiprecision division is only performed when absolutely essential. Lehmer's algorithm is presented below:

Lehmer's Euclidean Algorithm
Input: positive integers u and v with $u \geq v$
Output: $g = \gcd(u, v)$
    While $v > B$ do
      $\bar{u} \leftarrow \lfloor u/2^k \rfloor$; $\bar{v} \leftarrow \lfloor v/2^k \rfloor$;
      $s'' \leftarrow 1$; $s' \leftarrow 0$; $t'' \leftarrow 0$; $t' \leftarrow 1$;
      While $\bar{v} + s' \neq 0$ AND $\bar{v} + t' \neq 0$ do
        $q \leftarrow \lfloor (\bar{u}+s'')/(\bar{v}+s') \rfloor$;
        $q' \leftarrow \lfloor (\bar{u}+t'')/(\bar{v}+t') \rfloor$;
        If $q \neq q'$ then Goto L
        $z \leftarrow s'' - qs'$; $s'' \leftarrow s'$; $s' \leftarrow z$;
        $z \leftarrow t'' - qt'$; $t'' \leftarrow t'$; $t' \leftarrow z$;
        $z \leftarrow \bar{u} - q\bar{v}$; $\bar{u} \leftarrow \bar{v}$; $\bar{v} \leftarrow z$;

End While
  L If $t'' = 0$ then
    $q \leftarrow \lfloor u/v \rfloor$;
    $w \leftarrow u - qv$; $u \leftarrow v$; $v \leftarrow w$;
  Else
    $w \leftarrow s''u + t''v$; $r \leftarrow s'u + t'v$;
    $u \leftarrow w$; $v \leftarrow r$;
  End If
End While
Compute $g = \gcd(u, v)$ via the
Euclidean Algorithm and Return(g)

Note that in the algorithm $\bar{u}, \bar{v}, s', s'', t', t'', q, q', z$ are single precision variables, whereas w and r are multiprecision. In the first step B denotes the largest value a single precision integer (word) may hold. In the next step k is made as large as to truncate all bits except the leading word of u and v. In the innermost loop the two approximations of the quotient are compared and accordingly either the single precision operations in the next three steps or the multiprecision operations in the following if branch are computed. The While loop repeats until v is sufficiently reduced to fit into a single precision variable. Finally, in the last step, the execution continues with the classical Euclidean algortihm which computes and returns the GCD of the latest contents of u and v. For a more detailed treatment of the subject see [4]. An extended version of Lehmer's algorithm is given in [1]. Sorenson presents a complexity analysis of Lehmer's algorithm in [6].

The Euclidean Algorithm may be adapted to work on polynomials as well. Since there is no natural ordering among polynomials, the greatest common divisor of two polynomials is defined as the largest degree monic polynomial that divides both polynomials. Here we restrict our attention to polynomials defined over a field. It should be noted however that it is possible to develop Euclidean algorithms for polynomials defined over an arbitrary domain. The Extended Euclidean algorithm for polynomials is given below.

The Extended Euclidean Algorithm for Polynomials
Input: polynomials $u(x)$ and $v(x)$, with
      $\deg(u) \geq \deg(v)$
Output: $g(x) = \gcd(u(x), v(x))$ and polynomials
      $s(x), t(x)$
      satisfying $u(x)s(x) + v(x)t(x) = g(x)$
      $s''(x) \leftarrow 1$; $s'(x) \leftarrow 0$; $t''(x) \leftarrow 0$;
      $t'(x) \leftarrow 1$
      While $v(x) > 0$ do
        $r(x) \leftarrow u(x) \bmod v(x)$;
        $q(x) \leftarrow (u(x) - r(x))/v(x)$;

$$s(x) \leftarrow s''(x) - q(x)s'(x) ;$$
$$t(x) \leftarrow t''(x) - q(x)t'(x) ;$$
$$u(x) \leftarrow v(x) ; v(x) \leftarrow r(x) ;$$
$$s''(x) \leftarrow s'(x) ; s'(x) \leftarrow s(x) ;$$
$$t''(x) \leftarrow t'(x) ; t'(x) \leftarrow t(x) ;$$

End While
$a \leftarrow$ leading nonzero coefficient
of $u(x)$;
$$g(x) \leftarrow a^{-1}u(x) ;$$
$$s(x) \leftarrow a^{-1}s''(x) ; t(x) \leftarrow a^{-1}t''(x) ;$$
Return $(g(x), s(x), t(x))$

The modular reduction and the division operations within the loop may be achieved by a single polynomial division. The most significant difference in the polynomial version is the final scaling of $g(x), s(x)$ and $t(x)$ with a^{-1} to obtain a monic greatest common divisor polynomial. For this the inverse of a needs to be computed in the field over which the polynomials are defined.

See also <u>Binary Euclidean Algorithm</u>.

Berk Sunar

References

[1] Cohen, H. (1993). *A Course in Computational Algebraic Number Theory*. Springer-Verlag, Berlin.

[2] Crandall, R. and C. Pomerance (2001). *Prime Numbers: A Computational Perspective*. Springer-Verlag, New York.

[3] Euclid (1956). *Thirteen Books of Euclid's Elements* (2nd ed.). Dover, New York.

[4] Knuth, D.E. (1997). *The Art of Computer Programming. Volume 2: Seminumerical Algorithms* (3rd ed.). Addison-Wesley, Reading, MA.

[5] Lehmer, D.H. (1938). "Euclid's algorithm for large numbers." *American Mathematical Monthly*, 45, 227–233.

[6] Sorenson, J. (1995). "An analysis of Lehmer's Euclidean GCD algorithm." *Proceedings of the 1995 International Symposium on Symbolic and Algebraic Computation*. ACM Press, New York, 254–258.

EULER'S TOTIENT FUNCTION

The multiplicative <u>group</u> of the <u>ring</u> $\mathbf{Z}_n$ consists of the integers (or more formally, residue classes) between 0 and $n - 1$ that are relatively prime to the modulus n. The number of elements in this group, denoted $\phi(n)$, is called *Euler's totient function* of n.

If n is prime, then $\phi(n)$ is $n - 1$, since 0 is the only integer in the set that is not relatively prime to n.

If n is a prime power, $n = p^k$, $k > 1$, then $\phi(n)$ is $(p-1)p^{k-1}$, since any integer in the set that is divisible by p is not relatively prime to n.

If n is a general composite of the form

$$n = \prod_{i=1}^{d} p_i^{k_i},$$

where $d > 2$, $p_1, \ldots, p_d$ are distinct primes, and each $k_i \geq 1$, then

$$\phi(n) = \prod_{i=1}^{d} \phi(p_i^{k_i}) = \prod_{i=1}^{d} (p_i - 1)p_i^{k_i-1}.$$

In particular, if n is a composite of the form $n = pq$ where p and q are distinct primes, which is a common case in cryptography, then $\phi(n) = (p - 1)(q - 1)$.

By a basic result of group theory, since $\phi(n)$ is the <u>order</u> of the multiplicative <u>group</u> $\mathbf{Z}_n^*$, it follows that every element $x \in \mathbf{Z}_n^*$ has multiplicative order dividing $\phi(n)$, i.e.,

$$x^{\phi(n)} \equiv 1 \,(\text{mod } n).$$

This is known as *Euler's Theorem*. (<u>Fermat's Little Theorem</u> covers the special case where n is prime.) A related function is the λ function, which is the smallest positive integer such that

$$x^{\lambda(n)} \equiv 1 \,(\text{mod } n)$$

for every element $x \in \mathbf{Z}_n^*$. The λ function is the same as the ϕ function if n is a prime or prime power; if n is composite it is defined as

$$\lambda(n) = \text{lcm}(\phi(p_1^{k_1}), \ldots, \phi(p_d^{k_d})),$$

which is a divisor of $\phi(n)$.

Burt Kaliski

EXHAUSTIVE KEY SEARCH

INTRODUCTION: The simplest approach to cryptanalyzing a <u>block cipher</u> is exhaustive key search. The <u>cryptanalyst</u> wishes to find the <u>key</u> k that was used with block cipher E to encrypt some plaintext P to produce ciphertext C, $C = E_k(P)$ (see Figure 1 and <u>Shannon's model</u>).

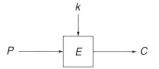

Fig. 1. Block cipher encryption

Once k is known, the cryptanalyst can find the plaintext by decrypting the ciphertext, $P = E_k^{-1}(C)$. A very simple approach to finding k is known as exhaustive search: the cryptanalyst tries decrypting the known ciphertext C with each possible key in turn until the correct key k is found.

Implicit in the definition of exhaustive search is the assumption that the cryptanalyst can tell whether a given guess of the key is correct. This requires that the cryptanalyst has some information about the plaintext. For example, the plaintext may start with a known header (e.g., "Dear Sir"), or the bytes of the plaintext may be limited to ASCII character values. In general, the cryptanalyst requires as much information about the plaintext as he has uncertainty about the key. Otherwise, there are likely to be many keys corresponding to plausible plaintexts.

The time required to complete an exhaustive key search depends on the number (K) of possible keys, the time (t) it takes to test a candidate key, and the number (p) of processors performing the search. Each processor is responsible for approximately K/p keys and would take time Kt/p to test them all. On average we expect to find the key about half way through the search making the expected run time approximately $Kt/(2p)$.

Although the focus here is on block ciphers, exhaustive search methods can be applied to <u>stream ciphers</u> and <u>MAC algorithms</u> as well.

In the following sections, we discuss some of the history of exhaustive search including the level of effort required for hardware and software based attacks (Section "History"), the invulnerability of modern block ciphers to exhaustive search (Section "Modern block ciphers"), countermeasures for preventing exhaustive search attacks (Section "Countermeasures"), and other related attacks (Section "Related attacks").

HISTORY: Much of the history of exhaustive key search concerns the U.S. <u>Data Encryption Standard</u> (DES) [4]. DES is a block cipher with 64-bit plaintext and ciphertext blocks, and 56-bit keys ($K = 2^{56}$). When DES was first proposed by the US government, there was considerable controversy over its short keys. In 1977, Diffie and Hellman [2] estimated that a machine capable of recovering DES keys in a day could be built for \$20 million. Some argued that this estimate was too low and that even if it were correct, it provides an adequate barrier to cryptanalysis. On the other hand, advancing technology would reduce costs quickly, and the cost of using larger keys is quite low. For

example, increasing the DES key length by just 8 bits to 64 bits would increase the cost of exhaustive search by a factor of $2^8 = 256$.

The debate about the DES key size raged on for many years (see Diffie's foreword to "Cracking DES" [3] for more details). At Crypto '93, a paper was presented that gave a detailed design for a \$1 million DES key search machine consisting of $p = 57{,}600$ custom chips, each capable of testing a DES key every $t = 20$ ns [10]. Using the expression for expected run time from the Introduction, this machine takes, on average, $Kt/(2p) = 3.5$ hours to recover a DES key. Although general purpose computers are poorly suited to performing DES key search, by the late 1990s various group efforts succeeded in searching the DES key space using months of computer time.

By 1998, the Electronic Frontier Foundation had actually built a hardware-based DES key search machine called "Deep Crack" [3]. Because their budget was limited, they used custom gate array chips that are slower and more expensive than fully custom chips, but have considerably lower design and fabrication start-up costs. Deep Crack cost \$200,000 to build, consisted of 1536 chips each capable of searching through 60 million keys per second, and required, on average, 4.5 days to recover a DES key.

As the evidence mounted over the years, it became inceasingly difficult to pretend that DES provides adequate security, although a great deal of legacy DES-based equipment is still in use today because of a desire to avoid the cost of replacing it.

MODERN BLOCK CIPHERS: The reign of DES is finally giving way to new ciphers with longer keys. The most significant of these new ciphers is <u>Rijndael/AES</u> [1], which was selected by the U.S. National Institute of Standards and Technology (NIST) to be the Advanced Encryption Standard (AES) [6]. The AES plaintext and ciphertext block size is 128 bits, and its keys can be 128, 192, or 256 bits long.

To assess the difficulty of AES exhaustive key search, we use the 1993 DES key search design as a starting point. By Moore's law, we expect the speed of a key search machine to double every 18 months. From 1993 to 2003, this is about a factor of 100. A \$1 million DES key search machine in 2003 would then take about 2 minutes to recover a key rather than 3.5 hours. If we ignore the fact that AES is more complex than DES to implement on a chip, even the smallest key size of AES is 72 bits longer than a DES key meaning that AES requires 2^{72} times as much effort to recover a key. Even if we increase the budget to \$1 trillion for a machine, the time increases to billions of years. AES is essentially invulnerable to key search.

Note that these arguments apply only to key search. It is conceivable that a clever cryptanalyst will find some short cut to recovering AES keys. AES has withstood considerable scrutiny so far, and time will tell whether the current confidence in its security is justified.

COUNTERMEASURES: The most effective way to prevent exhaustive key search attacks is to use sufficiently long keys. By making the number of keys K large, the key search time $Kt/(2p)$ becomes large. In some cases system designers are constrained to some key size limit (e.g., interoperating with a legacy system, export controls). In the following subsections, we examine some of the means that have been tried to prevent key search attacks.

Frequent Key Changes

A common suggestion for avoiding key search attacks is to change keys frequently. If it takes an hour to find a key, then change keys every half hour. Implicit in this suggestion is the assumption that the key will not be useful after the half hour is up, which is true in only certain types of systems.

It turns out that this technique does not work well, because key search does not take exactly an hour, but takes a time that is uniformly distributed between zero and two hours. The probability that the key will be found in the first half hour is 25%. If the cryptanalyst does not succeed before the key is changed, then he abandons the search and starts to work on the next key. On average, one quarter of the keys will be found while they are still in use. Even if the keys are changed very frequently, say every minute, the cryptanalyst expects to wait about 2 hours to recover a key quickly enough that it is still in use. Frequent key changes only slow the cryptanalyst down by at most a factor of two.

Eliminate Known Plaintext

In most analyses of key search time, it is assumed that the cryptanalyst has some known plaintext P corresponding to ciphertext C. The cryptanalyst can then take a candidate key, use it to decrypt C and compare the result to P. If we could encrypt only the unpredictable parts of messages or find

some other way to eliminate known plaintext, we could slow down the cryptanalyst. This is not as easy as it sounds.

Suppose that the cryptanalyst has no known plaintext, but knows that the plaintext is coded with one ASCII character per byte. This means that each byte has a leading zero bit. In the case of DES with a 64-bit block size, each plaintext block contains 8 zero bits at known positions. When the cryptanalyst tries decrypting a ciphertext block with the wrong key, the random result has only a 1 in 2^8 chance of having the correct form. Thus the cryptanalyst can eliminate most keys in just a single decryption. The remaining wrong keys can be eliminated with one or more additional decryptions of other ciphertext blocks. Overall, the impact on run time is negligible. Even if the cryptanalyst knows only one bit of redundancy per plaintext block, half the keys will be eliminated on the first decryption, a quarter of the keys will be eliminated on the second decryption, etc. Overall, this only slows the search by a factor of two.

Different Modes of Encryption

There are several standardized modes of operation of a block cipher [5]. The simplest is called electronic codebook (ECB) mode where each plaintext block is encrypted in turn and the resulting ciphertext blocks are concatenated. However, there are other modes where the previous ciphertext block is involved in encrypting the next plaintext block. The 1993 DES key search paper [10] shows that a key search design can be adapted to cipher-block chaining (CBC), cipher-feedback (CFB), and output feedback (OFB) modes for a run-time penalty of less than 2.5.

Extensive Key Setup

In most block ciphers, there is a key setup process to take a key and produce a set of subkeys for direct use in the algorithm. Collectively, these subkeys are usually much longer than the original key. When encrypting, the key setup can be performed once per key, and the resulting subkeys can be used for the life of the key. However, for key search, it is necessary to compute new subkeys for each key that is tried. This leads to the idea of making the key setup process expensive to slow down key search, while having (hopefully) minimal impact on encryption.

For example, if key setup is 100 times more effort than encrypting a block, then key search is slowed down by a factor of approximately 100. But from the cryptographer's point of view, if a key is used to encrypt significantly more than 100 blocks, the impact is minimal.

There are a number of potential problems with this approach. A system may be designed to encrypt short messages making the key setup overhead significant. Subkeys would have to be stored rather than recomputed for each encryption, which may be a problem in small devices that handle many keys. If the key space is small enough, the cryptanalyst might precompute and store the subkeys for each key and use the resulting database to perform key search quickly. The cryptanalyst may have many keys to attack at once and can compute subkeys from each key once and try the subkeys on each of the problem instances, thereby spreading the cost of subkey generation across multiple problem instances.

All-Or-Nothing Encryption

Rivest had an interesting idea for slowing down key search called all-or-nothing encryption [8]. This encryption mode is designed so that it is not possible to recover a single plaintext block (even knowing the key) until the entire message is decrypted. This means that the cryptanalyst cannot eliminate most keys by decrypting a single block, and the key search effort increases by a factor of the number of blocks in the message.

RELATED ATTACKS: Exhaustive key search belongs to a larger class of attacks sometimes called black-box cryptanalysis. These attacks do not depend on the exact details of the scheme being attacked, but only on its external characteristics. With exhaustive search, the only details about the block cipher that matter are the key size, block size, and complexity of performing an encryption or decryption in hardware or software. Some view these attacks as uninteresting because they do not make use of details of the scheme being attacked, but on the other hand, the broad applicability of black-box attacks makes them quite powerful techniques. Other examples of black-box cryptanalysis are attacks on double- and triple-encryption with a block cipher, generic methods of finding hash function collisions, and generic methods of finding discrete logarithms in cyclic groups [7, 9, 11].

CONCLUSION: Exhaustive key search is a simple approach to attacking a block cipher and has had an interesting history in connection with DES. However, modern ciphers with keys of 128 bits and longer have made exhaustive cryptanalysis infeasible with current technology.

Michael J. Wiener

References

[1] Daemen, J. and V. Rijmen (2002). *The Design of Rijndael: AES—The Advanced Encryption Standard*. Springer-Verlag, Berlin.

[2] Diffie, W. and M. Hellman (1977). "Exhaustive cryptanalysis of the NBS data encryption standard." *Computer*, 10 (6), 74–84.

[3] Electronic Frontier Foundation (1998). *Cracking DES: Secrets of Encryption Research, Wiretap Politics, and Chip Design*. O'Reilly and Associates, Sebastopol, CA.

[4] FIPS 46 (1977). *Data Encryption Standard*. Federal Information Processing Standards Publication 46. U.S. Department of Commerce/National Bureau of Standards, National Technical Information Service, Springfield, VA (revised as FIPS 46-1 in 1988, FIPS 46-2 in 1993).

[5] FIPS 81 (1980). *DES Modes of Operation*. Federal Information Processing Standards Publication 81. U.S. Department of Commerce/National Bureau of Standards, National Technical Information Service, Springfield, VA.

[6] FIPS 197 (2001). *Advanced Encryption Standard (AES)*. Federal Information Processing Standards Publication 197. U.S. National Institute of Standards and Technology (NIST). Available on http://csrc.nist.gov/CryptoToolkit/aes/

[7] Pollard, J.M. (1978). "Monte Carlo methods for index computation (mod *p*)." *Mathematics of Computation*, 32 (143), 918–924.

[8] Rivest, R.L. (1997). "All-or-nothing encryption and the package transform." *Fast Software Encryption '97, 4th International Workshop*, Lecture Notes in Computer Science, vol. 1267, ed. E. Biham. Springer-Verlag, Berlin, 210–218.

[9] van Oorschot, P.C. and M.J. Wiener (1999). "Parallel collision search with cryptanalytic applications." *Journal of Cryptology*, 12 (1), 1–28.

[10] Wiener, M.J. (1996). "Efficient DES key search." *The Rump Session of CRYPTO'93*. Reprinted in *Practical Cryptography for Data Internetworks*, ed. W. Stallings. IEEE Computer Society Press, 31–79.

[11] Wiener, M.J. (2004). "The full cost of cryptanalytic attacks." *Journal of Cryptology*, 17 (2), 105–124.

EXISTENTIAL FORGERY

Existential forgery is a weak, message related forgery against a cryptographic digital signature schemes. Given a victim's verifying key, an existential forgery is successful, if the attacker finds a signature s for any new message m, such that the signature s is valid for m with respect to the victim's verifying key. The message m need not be sensical or useful in any obvious sense.

Gerrit Bleumer

EXPONENTIAL TIME

An *exponential-time* algorithm is one whose running time grows as an exponential function of the size of its input. Let x denote the length of the input to the algorithm (typically in bits, but other measures are sometimes used). Let $T(x)$ denote the running time of the algorithm on inputs of length x. Then the algorithm is exponential-time if the running time satisfies

$$T(x) \leq cb^x$$

for all sufficiently large x, where the coefficient and the base $b > 1$ are constants. The running time will also satisfy the bound

$$T(x) \leq (b')^{c'x}$$

for another base base $b' > 1$, and an appopriate constant c'. The term "exponential" comes from the fact that the size x is in the exponent. In O-notation, this would be written $T(x) = O(b^x)$, or equivalently $T(x) = (b')^{O(x)}$. The notations $2^{O(x)}$ and $e^{O(x)}$ are common.

Exhaustive key search is one example of an algorithm that takes exponential time; if x is the key size in bits, then key search takes time $O(2^x)$.

For further discussion, see polynomial time.

Burt Kaliski

EXPONENTIATION ALGORITHMS

The problem of computing an exponentiation occurs frequently in modern cryptography. In particular, it is the core operation in the three most popular families of public-key algorithms: integer factorization based schemes (e.g., RSA public key encryption); discrete logarithm problem based schemes (e.g., Digital Signature Standard or Diffie–Hellman key agreement); and elliptic curve cryptography. Exponentiation is defined as the repeated application of the group operation to a single group element. If we assume a multiplicative group, i.e., the group operation is called "multiplication", and we denote the group element by g, we write an exponen-

tation as

$$\underbrace{g \cdot g \cdot \ldots \cdot g}_{e \text{ times}} = g^e.$$

This case is in particular relevant for RSA and discrete logarithm schemes in finite fields. It should be kept in mind that the corresponding notation for additive groups, i.e., groups where the group operation is an addition, looks as follows:

$$\underbrace{g + g + \ldots + g}_{e \text{ times}} = e \cdot g.$$

This case, repeated addition of the same group element, is the core operation in elliptic curve cryptosystems. Strictly speaking, this operation is not called "exponentiation" but multiplication. Accordingly, in elliptic curve schemes the corresponding operation is referred to as "scalar point multiplication."

There is a wealth of different exponentiation methods. The general goal of these algorithms is to perform an exponentiation with a minimum of arithmetic operations combined with optimized storage requirements. For groups where inversion of a field element is not a trivial operation, e.g., in finite fields or integer rings, we can distinguish between three main families of exponentiation algorithms. For clarity, we assume multiplicative groups:

General Exponentiation. This case is given if neither the base element g nor the exponent e is known ahead of time. Hence, no precomputations can be be performed. This is the most general case. The basic algorithm in this situation is the binary exponentiation method, also known as square-and-multiply algorithm. For random exponents e, this method takes on average $\lceil \log_2 e \rceil - 1$ squarings and $0.5 (\lceil \log_2 e \rceil - 1)$ multiplications in the group. The binary method has negligible storage requirements. Generalizations of the binary method, referred to as the 2^k-ary exponentiation and sliding window exponentiation, lead to reductions of the number of group multiplications, while increasing the storage requirements. However, both of the latter algorithms essentially do not reduce the number of squarings required.

Fixed-base Exponentiation. In certain cryptographic schemes the basis g of an exponentiation algorithm is known à priori and, thus, precomputations can be performed. This leads to much faster exponentiation times compared to the general case. These methods are described in the entry fixed-base exponentiation.

Fixed-exponent Exponentiation. The other special case is given in applications where the exponent e is known ahead of time. Again, precomputations can be performed. The gain compared to the general exponentiation case is relatively moderate, however. Corresponding techniques are described in <u>fixed-exponent exponentiation</u>. These techniques are closely related to the theory of addition chains.

In situations where taking the inverse of a group element is trivial (as it is the case in elliptic curve groups, where inversion is often a single subtraction of two field elements), there are additional exponentiation methods. Those methods are based on the idea that e can sometimes be represented more efficiently when negative digits are allowed as well as positive digits. For instance, the decimal number 31 has the binary representation $(1, 1, 1, 1, 1)$ but the signed-digit representation $(1, 0, 0, 0, 0, -1)$. Since the complexity of exponentiation algorithms often depends on the number of non-zero digits, the latter representation can lead to faster exponentiations. Corresponding algorithms are described in <u>signed-digit exponentiation</u>.

Another special case is given when exponentiations of the following form are required:

$$g_1^{e_1} \cdot g_2^{e_2}.$$

Such exponentiations are referred to as <u>simultaneous exponentiation</u>. They can be computed together considerably more efficiently than the two individual exponentiations $g_1^{e_1}$ and $g_2^{e_2}$. Efficient methods exist both for multiplicative groups and for additive groups. In the latter case the notation

$$e_1 \cdot g_1 + e_2 \cdot g_2$$

is being used. This is the core operation in the verification step of the elliptic curve digital signature algorithm (or ECDSA), which is described in the entry <u>Elliptic Curve signature schemes</u>.

Christof Paar

EXTENSION FIELD

Let $F = (S, +, \times)$ be a <u>field</u> and let $f(x)$ be a monic <u>irreducible polynomial</u> of degree d over F. That is,

let $f(x)$ be a polynomial

$$f(x) = x^d + f_{d-1}x^{d-1} + \cdots + f_1 x + f_0.$$

Let α denote a root of this polynomial. Then the set of elements generated from field operations on elements of F and α is itself a field, denoted $F(\alpha)$. The field $F(\alpha)$ is called an *extension field* of F of *extension degree* d. F is a *subfield* of $F(\alpha)$, since F is a subset of $F(\alpha)$ and is itself a field.

The definition given here is for a *simple* extension field. In general, an extension field of a field F is any field that contains F as a subfield; an example of a non-simple extension of F is $F(\alpha, \beta)$ where β is another root. Simple extension fields are the ones usually encountered in cryptography.

The elements of a simple extension field may be viewed as polynomials of degree at most $d - 1$ in α, where the coefficients are elements of F, so that an element $A \in F(\alpha)$ corresponds to the polynomial

$$a_{d-1}\alpha^{d-1} + \cdots + a_1\alpha + a_0.$$

Field addition in the extension field corresponds to coefficient-wise addition of the d coefficients, while field multiplication corresponds to polynomial multiplication modulo the <u>field polynomial</u> $f(x)$.

If F is a <u>finite field</u> with q elements, i.e., $\mathbf{F}_q$, then $F(\alpha)$ is a finite field with q^d elements, i.e., $\mathbf{F}_{q^d}$.

For instance, a binary finite field $\mathbf{F}_{2^k}$ is an extension field of the finite field $\mathbf{F}_2$, and likewise an odd-characteristic extension field $\mathbf{F}_{p^k}$ with $p \geq 3$ is an extension field of $\mathbf{F}_p$.

A <u>number field</u> $\mathbf{Q}(\alpha)$ is an extension field of the rational numbers $\mathbf{Q}$.

The representation of elements given above is a *polynomial basis representation*, since the polynomial terms $\alpha^{d-1}, \ldots, \alpha, 1$ form a *basis* for the extension field when viewed as *vector space* over F (i.e., linear combinations of the basis elements generate the vector space). However, various other representations have been developed and can be applied in cryptography.

See also <u>primitive element</u>.

Burt Kaliski

F

FACTOR BASE

The term *factor base* refers to the set of small prime numbers (and more generally, prime powers and irreducible polynomials) among which relations are constructed in various integer factoring algorithms (see also sieving) as well as in certain algorithms for solving the discrete logarithm problem (see also index calculus).

The choice of factor base—and specifically the maximum value in the factor base, called the *smoothness bound*—plays a crucial role in the time and hardware requirements for these algorithms. For further discussion, see the entry on smoothness.

Burt Kaliski

FACTORING CIRCUITS

In [1] Bernstein proposed a new approach to the relation combination step of the number field sieve (see integer factoring), based on sorting in a large mesh of small processors. A variant based on routing in a similar mesh was later described in [2]. A hardware approach to relation collection based on more traditional methods was also proposed in [1]; a more detailed hardware design is described in [4] (see TWIRL). As of the summer of 2003, it seems realistic that a US$ 10 million device, not counting research and development costs, should be able to factor 1024-bit RSA moduli at the rate of one modulus per year (see RSA problem). See also [3]. Combined with Moore's law this does not bode well for long term prospects of the security offered by such moduli.

An alternative way to measure the cost of integer factoring was presented in [1] as well, namely as the product of the runtime and the cost of the hardware required, as opposed to just the runtime (see also [5]). Refer to [1, 2] for a discussion of the effect this has on the parameter choices and asymptotic cost of the Number Field Sieve.

Arjen K. Lenstra

References

[1] Bernstein, D.J. (2001). Circuits for integer factorization: A proposal. Available at cr.yp.to/papers. html#nfscircuit

[2] Lenstra, A.K., A. Shamir, J. Tomlinson, and E. Tromer (2002). "Analysis of Bernstein's factorization circuit." *Advances in Cryptography—ASIACRYPT 2002*, Lecture Notes in Computer Science, vol. 2501, ed. Y. Zheng. Springer-Verlag, Berlin, 1–26.

[3] Lenstra, A.K., E. Tromer, A. Shamir, W. Kortsmit, B. Dodson, J. Hughes, and P. Leyland (2003). "Factoring estimates for a 1024-bit RSA modulus." *Advances in Cryptography—ASIACRYPT 2003*, Lecture Notes in Computer Science, vol. 2894, ed. C.S. Laih. Springer-Verlag, Berlin.

[4] Shamir, A. and E. Tromer (2003). "Factoring large numbers with the TWIRL device." *Advances in Cryptology—CRYPTO 2003*, Lecture Notes in Computer Science, vol. 2729, ed. D. Bonch. Springer-Verlag, Berlin.

[5] Wiener, M.J. (2004). "The full cost of cryptanalytic attacks." *J. of Cryptology*, 17 (2): 105–124.

FAIL-STOP SIGNATURE

Fail-stop signatures are digital signatures where signers enjoy unconditional unforgeability (with an unavoidable but negligible error probability), while the verifiers bear the risk of forged signatures, and therefore enjoy computational security only. If a signer is confronted with an alleged signature that she has not produced, then the signer can with overwhelming probability prove that the alleged signature is in fact forged. Afterwards, the signer can revoke her verifying key, thus the name fail-stop signature scheme. Fail-stop signatures were introduced by Pfitzmann [4] who gives an in-depth introduction in [6]. The security for the signer is strictly stronger than the strongest security defined by Goldwasser, Micali, and Rivest (see GMR signatures [2]), where signers enjoy computational security while verifiers are unconditionally secure against forgery. In other words, the signer is secure even against counterfeiting by a computationally unrestricted attacker. In the same sense, fail-stop signatures provide strictly stronger security for the signer than all the standard digital signature schemes such as RSA digital signatures, Rabin digital signatures, ElGamal digital signatures, Schnorr digital signatures, Digital Signature Algorithm (DSA), elliptic curve signature schemes (ECDSA), etc. A comparative overview is found in [5]. Fail-stop signatures are more fundamental than ordinary digital signatures because the latter can be constructed from the former, but not vice versa [3].

In a fail-stop signature scheme, each signing member has an individual signing key pair of a private signing key and a public verifying key (see public key cryptography). The verifying algorithm takes the signer's public verifying key, a message, and a corresponding signature and returns a Boolean result indicating whether the signature is valid for the message with respect to the public verifying key. If a signer is confronted with a signature that she did not produce, but which holds against the verifying algorithm, then the signer can use an additional algorithm to produce a proof of forgery. Any third party verifier can then verify by means of an additional algorithm that the proof of forgery is valid. The first efficient fail-stop signature scheme was constructed by van Heijst, Pedersen and Pfitzmann [3, 10, 11].

Pedersen and Pfitzmann have shown that fail-stop signatures can be produced and verified about as efficiently as ordinary digital signatures, but that the length of the signer's private signing key is proportional to the number of messages that can be signed with it [3]. However, signers need not store complete long signing keys at any one time, but can generate random bits for them on demand.

Fail-stop signatures have been proposed to be used in electronic cash schemes such that customers need not bear the risk of very powerful banks forging some of their customer's signatures. Although this may appear far fetched for the case of general customers, it may be a necessary condition to open the door for processing high value transactions (millions of dollars) fully electronically.

A fail-stop signature scheme has the following operations: (i) an operation for generating pairs of a private signing key and a public verifying key for an individual, (ii) an operation to produce fail-stop signatures, (iii) an operation to verify fail-stop signatures, (iv) an operation to produce proofs of forgery, and (v) an operation to verify proofs of forgery.

The characteristic security requirements of a fail-stop signature scheme are:

Unforgeability (Recipient's Security): resistance against existential forgery under adaptive chosen message attacks by computationally restricted attackers.

Nonrepudiation: a computationally restricted signer cannot produce a fail-stop signature that he can later prove to be a forged signature.

Signer's Security: if a computationally unlimited attacker fabricates a signature, the alleged

signer can produce a valid proof of forgery with overwhelming probability.

Constructions have been based on groups, in which the discrete logarithm problem is hard [3, 10, 11], on the RSA digital signature scheme [7], and on authentication codes [9].

Fail-stop signatures schemes can be equipped with additional features: Chaum et al. have proposed undeniable fail-stop signatures [1]. Susilo et al. have proposed threshold fail-stop signatures [8].

Gerrit Bleumer

References

[1] Chaum, David, Eugéne van Heijst, and Birgit Pfitzmann (1992). "Cryptographically strong undeniable signatures, unconditionally secure for the signer." *Advances in Cryptology—CRYPTO'91*, Lecture Notes in Computer Science, vol. 576, ed. J. Feigenbaum. Springer-Verlag, Berlin, 470–484.

[2] Goldwasser, Shafi, Silvio Micali, and Ronald L. Rivest (1988). "A digital signature scheme secure against adaptive chosen-message attacks." *SIAM Journal on Computing*, 17 (2), 281–308.

[3] Pedersen, Torben P. and Birgit Pfitzmann (1997). "Fail-stop signatures." *SIAM Journal on Computing*, 26 (2), 291–330.

[4] Pfitzmann, Birgit (1991). "Fail-stop signatures; principles and applications." *Proc. COMPSEC'91, 8th World Conference on Computer Security, Audit and Control*. Elsevier, Oxford, 125–134.

[5] Pfitzmann, Birgit (1993). "Sorting out signature schemes." *1st ACM Conference on Computer and Communications Security, Fairfax, November 1993*. ACM Press, New York, 74–85.

[6] Pfitzmann, Birgit (1996). "Digital signature schemes general framework and fail-stop signatures." Lecture Notes in Computer Science, vol. 1100, ed. B. Pfitzmann. Springer-Verlag, Berlin.

[7] Susilo, Willy, Reihaneh Safavi-Naini, and Josef Pieprzyk (1999). "RSA-based fail-stop signature schemes." *International Workshop on Security (IWSEC'99)*. IEEE Computer Society Press, Los Alamitos, CA, 161–166.

[8] Susilo, Willy, Rei Safavi-Naini, and Josef Pieprzyk (1999). "Fail-stop threshold signature schemes based on elliptic curves." *Information Security and Privacy*, Lecture Notes in Computer Science, vol. 1587, eds. J. Pieprzyk, R. Safari-Naini, and J. Seberry. Springer-Verlag, Berlin, 103–116.

[9] Susilo, Willy, Reihaneh Safavi-Naini, Marc Gysin, and Jennifer Seberry (2000). "A new and efficient fail-stop signature scheme." *The Computer Journal*, 43 (5), 430–437.

[10] van Heijst, Eugéne and Torben P. Pedersen (1993). "How to make efficient fail-stop signatures." *Advances in Cryptology—EUROCRYPT'92*, Lecture

Notes in Computer Science, vol. 658, ed. R.A. Rueppel. Springer-Verlag, Berlin, 366–377.

[11] van Heijst, Eugéne, Torben P. Pedersen, and Birgit Pfitzmann (1993). "New constructions of fail-stop signatures and lower bounds." *Advances in Cryptology—CRYPTO'92*, Lecture Notes in Computer Science, vol. 740, ed. E.F. Brickell. Springer-Verlag, Berlin, 15–30.

FAIR EXCHANGE

An exchange protocol transmits a digital token from each participant to each other participant in a group. An exchange protocol is called fair, if and only if

1. The protocol either ensures that all honest participants receive all items as expected or releases no useful information about any item of any honest party.
2. The protocol terminates after a fixed time.

In the two-party case, this means that a honest party releases its item if it receives the expected item from the peer. Otherwise, no additional information is released. A practical example is Ebay's escrow service that works as follows: the buyer deposits a payment at Ebay, the seller sends the item to the buyer, and if the buyer approves, Ebay forwards the payment to the seller. Note that the exact formalization of "fixed time" depends on the network model: in synchronous networks with a global notion of rounds this is formalized by "fixed number of rounds." In asynchronous networks without global time, a maximum logical time [6] is used.

Early research focused on two-party protocols (see multiparty protocols) solving particular instances of fair exchange such as contract signing [2] and [3] or certified mail [7]. However, these protocols are either unfair with a probability that is linear in the number of rounds (e.g., 1/10 for 10 rounds) or cannot guarantee termination in a fixed time [4].

Fairness with a negligible error probability within a limited time can only be guaranteed by involving a Trusted Third Party (TTP) or by assuming a trusted majority. A TTP ensures that the outcome is fair even if one party cheats. The TTP solves the fundamental problem that a party sending the complete item while not having received a complete item is always at disadvantage. Third parties can either be on-line, i.e., involved in each protocol run or optimistic, i.e., involved only if an exception occurs.

Generic fair exchange protocols [1] can exchange many types of items. Examples are data (described by a publicly known property), receipts, signatures on well-known data, or payments. Besides generic protocols, there exist a variety of protocols that optimize or extend exchanges of particular items: payment for receipt, payment for data ("fair purchase"), data for receipt ("certified mail"), or signature for signature ("contract signing").

A nonoptimistic generic fair exchange protocol similar to the contract signing protocol in [8] is depicted in Figure 1: each party, say A, sends its item $item_A$ and a description $desc_B$ of the item expected in return to the TTP. The TTP exchanges the items if the expectations are met. This is determined by a function verify() that verifies that an item matches its description. Otherwise, the exchange is aborted and no additional information about the items is released. Fair exchange protocols can be further classified by the number of participants, the network model they assume, the properties of the items they produce, and whether they are abuse-free or not. In synchronous networks there exists a global notion of rounds that limits the time needed to transmit messages between correct parties. In asynchronous networks messages between honest participants are guaranteed to be delivered eventually; but the transmission time is unbounded. Besides these two well-known network models, some protocols assume special network models that assume trusted hosts such as reliable messaging boards or reliable file-transfer

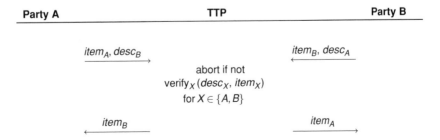

Fig. 1. A generic fair exchange protocol with on-line trusted third party

servers. Abuse-free protocols [5] provide the additional guarantee that if a failed protocol runs it never produces any evidence. This is of some importance for contract signing where a dishonest party may gain an advantage by being able to prove that a particular honest party was willing to sign a particular contract.

Matthias Schunter

References

[1] Asokan, N., Matthias Schunter, and Michael Waidner (1997). "Optimistic protocols for fair exchange." *4th ACM Conference on Computer and Communications Security, Zürich, April 1997*, 6–17.

[2] Blum, Manuel (1981). *Three Applications of the Oblivious Transfer*, Version 2. Department of Electrical Engineering and Computer Sciences, University of California at Berkeley.

[3] Even, Shimon (1983). "A protocol for signing contracts." *ACM SIGACT News*, 15 (1), 34–39.

[4] Even, Shimon and Yacov Yacobi (1980). "Relations among public key signature systems." Technical Report Nr. 175, Computer Science Department, Technion, Haifa, Israel.

[5] Garay, Juan A., Markus Jakobsson, and Philip MacKenzie (1999). "Abuse-free optimistic contract signing." *Advances in Cryptology—CRYPTO'99*, Lecture Notes in Computer Science, vol. 1666, ed. J. Wiener. Springer-Verlag, Berlin, 449–466.

[6] Lynch, Nancy A. (1966). *Distributed Algorithms*. Morgan Kaufmann, San Francisco.

[7] Rabin, Michael O. (1981). *Transaction Protection by Beacons*. Aiken Computation Laboratory, Harvard University, Cambridge, MA, Technical Report TR-29-81.

[8] Rabin, Michael O. (1983). "Transaction protection by beacons." *Journal of Computer and System Sciences*, 27, 256–267.

FAST CORRELATION ATTACK

Fast correlation attacks were first proposed by W. Meier and O. Staffelbach [12] in 1988. They apply to running-key generators based on linear feedback shift registers (LFSRs), exactly in the same context as the *correlation attack*, but they are significantly faster. They rely on the same principle as the correlation attack: they exploit the existence of a correlation between the keystream and the output of a single LFSR, called the *target LFSR*, whose initial state depends on some bits of the secret key. In the original correlation attack, the initial state of the target LFSR is recovered by an exhaustive search. Fast correlation attacks avoid examining all possible initializations of the target LFSR by using some efficient error-correcting techniques (see cyclic codes). But, they require the knowledge of a longer segment of the keystream (in the context of a known-plaintext attack). As for the correlation attack, similar algorithms can be used for mounting a ciphertext only attack when there exists redundancy in the plaintext.

FAST CORRELATION ATTACKS AS A DECODING PROBLEM: The key idea of fast correlation attacks consists in viewing the correlation attack as a decoding problem. If there exists a correlation between the keystream $\mathbf{s}$ and the output $\mathbf{u}$ of the target LFSR, then the running-key subsequence $(s_t)_{t<N}$ can be seen as the result of the transmission of $(u_t)_{t<N}$ through the binary symmetric channel with error probability $p = \Pr[s_t \neq u_t] < 1/2$ (if $\Pr[s_t \neq u_t] > 1/2$, the bitwise complement of $\mathbf{s}$ is considered). Moreover, all bits of the LFSR sequence $\mathbf{u}$ depend linearly on the LFSR initial state, $u_0 \cdots u_{L-1}$. Therefore, $(u_t)_{t<N}$ is a codeword of a linear code of length N and dimension L defined by the LFSR feedback polynomial. Thus, recovering the LFSR initial state consists in decoding the running-key subsequence relatively to the LFSR code.

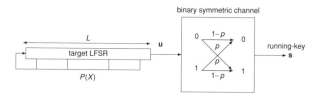

With this formulation, the original correlation attack proposed by Siegenthaler consists in applying a maximum-likelihood decoding algorithm to the linear code defined by the LFSR feedback polynomial. It has complexity 2^L, where L is the length of the target LFSR. The complexity can be reduced by using faster decoding algorithms. But, they usually require a larger number of running-key bits.

DECODING TECHNIQUES FOR FAST CORRELATION ATTACKS: Several algorithms can be used for decoding the LFSR code, based on the following ideas:

- Find many sparse linear recurrence relations satisfied by the LFSR sequence (these relations correspond to sparse multiples of the feedback polynomial), and use them in an iterative decoding procedure dedicated to low-density parity-check codes [3, 13, 14]. The complexity of this attack may significantly decrease when the feedback polynomial of the target LFSR is sparse. Thus, the use of sparse LFSR feedback polynomials should be avoided in LFSR-based running-key generators;

- Construct a convolutional code [6] (or a turbo code [7]) from the LFSR code, and use an appropriate decoding algorithm for this new code (Viterbi algorithm or turbo-decoding);
- Construct a new linear block code with a lower dimension from the LFSR code and apply to this smaller linear code a maximum-likelihood decoding algorithm [4], or a polynomial reconstruction technique [8].

A survey of all these techniques and their computational complexities can be found in [9] and [10]. In practice, the most efficient fast correlation attacks enable to recover the initial state of a target LFSR of length 60 for an error-probability $p = 0.4$ in a few hours on a PC from the knowledge of 10^6 running-key bits.

FAST CORRELATION ATTACKS ON COMBINATION GENERATORS: In the particular case of a combination generator, the target sequence **u** is the sequence obtained by adding the outputs of $(m + 1)$ constituent LFSRs, where m is the correlation-immunity order of the combining function (see correlation attack). Thus, this sequence **u** corresponds to the output of a unique LFSR whose feedback polynomial is the greatest common divisor of the feedback polynomials of the $(m + 1)$ involved LFSRs. Since the feedback polynomials are usually chosen to be primitive, the length of the target LFSR is the sum of the lengths of the $(m + 1)$ LFSRs. The keystream corresponds to the received word as output of the binary symmetric channel with error probability

$$p = \Pr[s_t \neq u_t] = \frac{1}{2} - \frac{1}{2^{n+1}} |\hat{f}(t)|,$$

where n is the number of variables of the combining function, t is the n-bit vector whose ith component equals 1 if and only if $i \in \{i_1, i_2, \ldots, i_{m+1}\}$ and $\hat{f}$ denotes the Walsh transform of f (see correlation attack and Boolean functions).

FAST CORRELATION ATTACKS ON FILTER GENERATORS: In the case of a filter generator, the target LFSR has the same feedback polynomial as the constituent LFSR, but a different initial state. Actually, if the keystream is given by

$$s_t = f(v_{t+\gamma_1}, v_{t+\gamma_2}, \ldots, v_{t+\gamma_n}),$$

where **v** is the sequence generated by the constituent LFSR, the optimal target sequence **u** then corresponds to

$$u_t = \sum_{i=1}^{n} \alpha_i v_{t+\gamma_i},$$

where $\alpha = (\alpha_1, \ldots, \alpha_n)$ is the vector which maximizes the magnitude of the Walsh transform of the filtering function. Thus, the keystream corresponds to the received word as output of the binary symmetric channel with error probability

$$p = \Pr[s_t \neq u_t] = \frac{\mathcal{NL}(f)}{2^n},$$

where $\mathcal{NL}(f)$ is the nonlinearity of the filtering function. The fast correlation attacks on filter generators can be improved by using several target LFSRs together [2, 10].

Other particular fast correlation attacks apply to filter generators, like conditional block-oriented correlation attacks [1, 5, 11] (see filter generator).

Anne Canteaut

References

[1] Anderson, R.J. (1995). "Searching for the optimum correlation attack." *Fast Software Encryption 1994*, Lecture Notes in Computer Science, vol. 1008, ed. B. Preneel. Springer-Verlag, Berlin, 137–143.

[2] Canteaut, A. and E. Filiol (2002). "On the influence of the filtering function on the performance of fast correlation attacks on filter generators." *Symposium on Information Theory in the Benelux, May 2002*.

[3] Canteaut, A. and M. Trabbia (2000). "Improved fast correlation attacks using parity-check equations of weight 4 and 5." *Advances in Cryptology—EUROCRYPT 2000*, Lecture Notes in Computer Science, vol. 1807, ed. B. Preneel. Springer-Verlag, Berlin, 573–588.

[4] Chepyshov, V., T. Johansson, and B. Smeets (2000). "A simple algorithm for fast correlation attacks on stream ciphers." *Fast Software Encryption 2000*, Lecture Notes in Computer Science, vol. 1978, ed. B. Schneier. Springer-Verlag, Berlin, 181–195.

[5] Golić, J.Dj. (1996). "On the security of nonlinear filter generators." *Fast Software Encryption 1996*, Lecture Notes in Computer Science, vol. 1039, ed. D. Gollman. Springer-Verlag, Berlin, 173–188.

[6] Johansson, T. and F. Jönsson (1999). "Improved fast correlation attack on stream ciphers via convolutional codes." *Advances in Cryptology—EUROCRYPT'99*, Lecture Notes in Computer Science, vol. 1592, ed. J. Stern. Springer-Verlag, Berlin, 347–362.

[7] Johansson, T. and F. Jönsson (1999). "Fast correlation attacks based on turbo code techniques." *Advances in Cryptology—CRYPTO'99*, Lecture Notes

in Computer Science, vol. 1666, ed. J. Wiener. Springer-Verlag, Berlin, 181–197.

[8] Johansson, T. and F. Jönsson (2000). "Fast correlation attacks through reconstruction of linear polynomials." *Advances in Cryptology—CRYPTO 2000*, Lecture Notes in Computer Science, vol. 1880, ed. M. Bellare. Springer-Verlag, Berlin, 300–315.

[9] Jönsson, F. (2002). "Some results on fast correlation attacks." *PhD Thesis*, University of Lund, Sweden.

[10] Jönsson, F. and T. Johansson (2002). "A fast correlation attack on LILI-128." *Information Processing Letters*, 81 (3), 127–132.

[11] Lee, S., S. Chee, S. Park, and S. Park (1996). "Conditional correlation attack on nonlinear filter generators." *Advances in Cryptography—ASIACRYPT'96*, Lecture Notes in Computer Science, vol. 1163, eds. K. Kim and T. Matsumoto. Springer-Verlag, Berlin, 360–367.

[12] Meier, W. and O. Staffelbach (1988). "Fast correlation attacks on stream ciphers." *Advances in Cryptology—EUROCRYPT'88*, Lecture Notes in Computer Science, vol. 330, ed. C.G. Günther. Springer-Verlag, Berlin, 301–314.

[13] Meier, W. and O. Staffelbach (1989). "Fast correlation attack on certain stream ciphers." *J. Cryptology*, 1 (3), 159–176.

[14] Mihaljevic, M.J., M.P.C. Fossorier, and H. Imai (2000). "A low-complexity and high performance algorithm for the fast correlation attack." *Fast Software Encryption 2000*, Lecture Notes in Computer Science, vol. 1978, ed. B. Schneier. Springer-Verlag, Berlin, 196–212.

FAULT ATTACK

A successful fault attack on an Integrated Circuit Card (ICC) or smartcard requires two steps: fault injection and the fault exploitation. The first step consists in injecting a fault at the appropriate time during the process. Fault injection is very dependent on the hardware and therefore the ICC. The

second step consists in exploiting the erroneous result or unexpected behavior. Fault exploitation depends on the software design and implementation. In the case of an algorithm it will also depend on its specification since the fault exploitation will be combined with cryptanalysis most of the time. Depending on the type of analysis performed, the fault injection will have to be done at a precise instant or roughly in a given period of time.

There are many ways to generate a fault in an ICC. We can already distinguish three major means of fault injection:

Electrical perturbation on the standard ISO contact of the smartcard

- Vcc glitch (see below)
- Clock duty cycle and/or frequency alteration

Light-beam perturbation (contact-less)

- Global light-beam (wide spectrum)
- Focused light-beam (wide spectrum)
- Laser-beam (single wavelength)

Electromagnetic field perturbation (contact-less).

The effectiveness of each fault injection method strongly depends on the hardware design, manufacturing process, and technology. The chip behavior under a fault injection can be of four types:

- No effect
- Wrong results or unexpected behavior (exploitable fault)
- No response from the card (hardware reset required)
- Card is dead (physically damaged).

Of course, only one out of the four listed behavior might be exploitable. Moreover, the perturbation can have a transient effect, permanent effect, or an intermediary state in between:

- Transient effect (only during fault injection)
- Semipermanent effect (for a variable period of time from a few minutes to a few days)
- Permanent effect.

In practice, transient faults are easier to exploit as we will see in the fault exploitation section.

The main difficulty of the fault injection step is to find the appropriate parameters of the perturbation for the ICC in consideration. Inappropriate parameters will not lead to an exploitable fault such as a wrong result or an unexpected behavior. Therefore, there is a risk to irremediably kill the chip. Besides, if the hardware implements some security sensors and/or protection mechanism, it will be even more difficult to inject an exploitable fault without triggering a security mechanism.

A Vcc glitch is defined by many parameters among which are its shape, the falling and rising slope, the low and high level, and the low level duration. Very good results have been obtained on some IC with a limited control over these parameters. In fact, the three main parameters

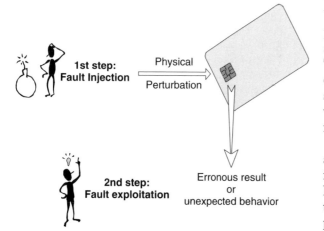

1st step: Fault Injection — Physical Perturbation

2nd step: Fault exploitation — Erronous result or unexpected behavior that needed to be tuned are the low and high voltage values and the glitch duration. Besides, the combination of all parameters quickly gives a huge search space and therefore it is not practical to have too many parameters to play with.

A white light perturbation can be very effective on some ICC. The hardware required for such an attack can be easy to set up by using a photo camera bulk and the associated electronic. Then, the major parameters to be tuned are the light emission intensity, the light emission duration and, eventually, the selection of a specific area of the chip to be exposed (focused light attack) rather than the complete die.

Using a laser such as a pulsed Nd:YAG laser is even more powerful because the energy level, the wavelength, the beam duration, and the beam size are much more under control. It is important to control the energy level in order to avoid destruction of the chip (too much energy) or an ineffective beam (not enough energy).

Fault exploitation is the mandatory second step of a successful fault attack. The fault exploitation directly depends on the fault effect, the fault localization in time, and the target of the attack. We distinguish two major types of target for a fault attack:

- The operating system and application sensitive process
- The cryptographic algorithm.

A sensitive process is defined as a piece of code processing data which is known to the external world but should not be modified. The fault injection will precisely modify such data. By definition, if the data should not be modified, it means that modifying it will compromise the system security to some extent.

Differential fault analysis (DFA) mainly consists in analyzing an algorithm result (ciphertext) under regular condition and under abnormal condition for the same input (plaintext). The abnormal condition is usually obtained by fault injection during the process (transient fault) or before the process (permanent fault). Differential fault analysis has been widely studied from a theoretical point of view. In September 1996, three researchers from Bellcore identified a new attack against *plain* RSA (see RSA digital signature scheme), see Boneh et al. [1], when performed with the Chinese Remainder Theorem. This attack was reported in a Bellcore press release entitled *New Threat Model Breaks Crypto Codes*, but no technical details were provided. Later, another researcher wrote a short memo that seemed to describe a more realistic attack. In the case of a computation error, the Bellcore researchers showed how to recover the secret factors

p and q of the public RSA modulus n from *two* signatures of the same message: a correct one and a faulty one. Lenstra remarked that only the faulty signature was required. At the same time, Joye and Quisquater noted that Lenstra's observation could actually be applied to all RSA-type cryptosystems, including variants based on Lucas sequences (LUC) and elliptic curves (KMOV, Demytko). See Joye et al. [2].

In conclusion, a fault attack is a threat for any secure token (whatever the form factor) and must be taken into consideration at all steps of the product design and specification. Countermeasures and protection can be designed both in hardware and in software to thwart such attacks.

Olivier Benoît

References

[1] Boneh, Dan, Richard A. DeMillo, and Richard J. Lipton (1997). "On the importance of checking cryptographic protocols for faults." *Advances in Cryptology—EUROCRYPT'97*, Lecture Notes in Computer Science, vol. 1233, ed. W. Fumy. Springer-Verlag, Berlin, 37–51.

[2] Joye, Marc, Arjen K. Lenstra, and Jean-Jacques Quisquater (1999). "Chinese remaindering cryptosystems in the presence of faults." *Journal of Cryptology*, 12 (4), 241–245.

FEAL

The Fast Data Encipherment Algorithm (FEAL) [9] is a family of block ciphers developed by Shimizu and Miyaguchi. Since the introduction of its first version (FEAL-4, presented in 1987), the block cipher has stimulated the development of some of the most useful cryptanalytical techniques available today.

FEAL was designed to be a more efficient alternative to the Data Encryption Standard (DES), in particular in software. The block cipher encrypts data in blocks of 64 bits and uses a 64-bit key. It is a Feistel cipher, just as DES, but the components have been modified in order to be more suitable for word-oriented processors. The complete cipher can be implemented using only additions modulo 2^8, rotations over 2 bits, and XORs. The 32-bit f-function used in the Feistel network takes a 16-bit subkey as a parameter and mixes it with the data using the functions S_0 and S_1. Both functions take two bytes as input and return a single output byte. They are defined as:

$$S_i(x, y) = [x + y + i \pmod{2^8}] \lll 2, \qquad i \in \{0, 1\}.$$

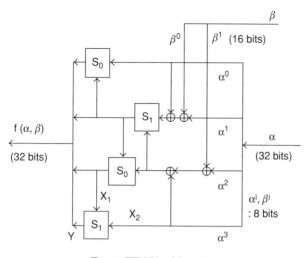

Fig. 1. FEAL's *f*-function

Another structural difference with DES is the additional layers inserted before the first and after the last round of FEAL. In these layers, an extra 64-bit subkey is mixed with the data and the left half of the data is XORed with the right half.

While most of its components are far simpler, FEAL has a more complex *key schedule* (see block cipher) than DES. The secret key is expanded in a Feistel-like ladder network using a nonlinear f_k-function similar to the *f*-function mentioned above.

FEAL has been studied by many researchers, and various interesting techniques were developed to analyze its security. The next paragraph gives a short overview of the most important developments.

The first vulnerabilities in the original four-round version, FEAL-4, were discovered by den Boer [4] in 1988. He developed an adaptively chosen plaintext attack which was later improved by Murphy [8]. The improved attack required only 20 chosen plaintexts and at that time recovered the key in less than 4 hours. At the Securicom conference in 1989, Biham and Shamir demonstrated a chosen plaintext attack on FEAL-8, the eight-round version of the cipher. The attack, which would turn out to be a direct application of differential cryptanalysis, is mentioned in different papers [1, 5, 7], but its details were only published in 1991 [2]. In 1990, Gilbert and Chassé [5] proposed a statistical chosen plaintext attack on FEAL-8. A year later, Tardy-Corfdir and Gilbert [10] presented a *known* plaintext attack on FEAL-4 and FEAL-6. The first attack had many ideas in common with differential cryptanalysis, which was being developed around the same time; the second attack contained elements which would later be used in linear cryptanalysis. In 1992, a first variant of this linear cryptanalysis was in-

troduced by Matsui and Yamagishi [6]. The new attack could recover FEAL-4 keys in 6 minutes using only five known plaintexts. In 1993, Biham and Shamir [3] developed a more efficient differential attack against FEAL-8 deriving the secret key in 2 minutes on a PC given 128 chosen plaintexts.

The development of these new techniques forced the designers of FEAL to considerably increase the number of rounds, sacrificing its original efficiency.

Christophe De Cannière

References

[1] Biham, E. and A. Shamir (1991). "Differential cryptanalysis of DES-like cryptosystems." *Advances in Cryptology—CRYPTO'90*, Lecture Notes in Computer Science, vol. 537, eds. A. Menezes and S.A. Vanstone. Springer-Verlag, Berlin, 2–21.

[2] Biham, E. and A. Shamir (1991). "Differential cryptanalysis of Feal and N-Hash." *Advances in Cryptology—EUROCRYPT'91*, Lecture Notes in Computer Science, vol. 547, ed. D.W. Davies. Springer-Verlag, Berlin, 1–16.

[3] Biham, E. and A. Shamir (1993). *Differential Cryptanalysis of the Data Encryption Standard*. Springer-Verlag, Berlin.

[4] den Boer, B. (1988). "Cryptanalysis of FEAL." *Advances in Cryptology—EUROCRYPT'88*, Lecture Notes in Computer Science, vol. 330, ed. C.G. Günther. Springer-Verlag, Berlin, 293–299.

[5] Gilbert, H. and G. Chassé (1991). "A statistical attack of the FEAL-8 cryptosystem." *Advances in Cryptology—CRYPTO'90*, Lecture Notes in Computer Science, vol. 537, eds. A. Menezes and S.A. Vanstone. Springer-Verlag, Berlin, 22–33.

[6] Matsui, M. and A. Yamagishi (1993). "A new method for known plaintext attack of FEAL cipher." *Advances in Cryptology—EUROCRYPT'92*, Lecture Notes in Computer Science, vol. 658, ed. R.A. Rueppel. Springer-Verlag, Berlin, 81–91.

[7] Miyaguchi, S. (1990). "The FEAL-8 cryptosystem and a call for attack." *Advances in Cryptology—CRYPTO'89*, Lecture Notes in Computer Science, vol. 435, ed. G. Brassard. Springer-Verlag, Berlin, 624–627.

[8] Murphy, S. (1990). "The cryptanalysis of FEAL-4 with 20 chosen plaintexts." *Journal of Cryptology*, 2 (3), 145–154.

[9] Shimizu, A. and S. Miyaguchi (1988). "Fast data enciphery algorithm FEAL." *Advances in Cryptology—EUROCRYPT'87*, Lecture Notes in Computer Science, vol. 304, eds. D. Chaum and W.L. Price. Springer-Verlag, Berlin, 267–278.

[10] Tardy-Corfdir, A. and H. Gilbert (1992). "A known plaintext attack of FEAL-4 and FEAL-6." *Advances in Cryptology—CRYPTO'91*, Lecture Notes in Computer Science, vol. 576, ed. J. Feigenbaum. Springer-Verlag, Berlin, 172–181.

FEISTEL CIPHER

One popular class of the modern iterative block-ciphers is the *Feistel* ciphers (named so after Horst Feistel—cryptanalyst who worked with the IBM crypto group in the early 1970s). The round of a Feistel cipher uses the product of two involutions (a function G is called an involution if it is its own inverse: $G(G(x)) = x$) in order to achieve the very comfortable similarity of encryption and decryption processes. Given an n-bit block, a Feistel round function divides it into two halves L (left) and R (right). Then some function $F(R, k)$ is applied to the right half and the result is XORed with the left half (this is the first involution):

$$(L, R) \to (R, L \oplus F(R, k)).$$

Here k is the round subkey produced by the key scheduling algorithm; it may vary from round to round. Then the halves are swapped (the second involution) and the process is repeated. Another convenience in this construction is that it is always a permutation, and thus is invertible no matter what function F is used and thus a designer may now concentrate on the security properties of this function. Many modern ciphers are designed as Feistel ciphers. One prominent example of a Feistel cipher is the Data Encryption Standard.

Note that the division into halves in a Feistel cipher can be replaced by division into quarters, octets, etc. Such ciphers are called *generalized Feistel ciphers*. Several modern ciphers are of this type, for example the CAST family of ciphers or the cipher Skipjack.

Alex Biryukov

FERMAT PRIMALITY TEST

The Fermat primality test is a primality test, giving a way to test if a number is a prime number, using Fermat's little theorem and modular exponentiation (see modular arithmetic).

Fermat's Little Theorem states that if a is relatively prime to a prime number p, then $a^{p-1} \equiv 1 \bmod p$. Fermat's little theorem is not true for composite numbers generally, and so it is an excellent tool to use to test for the primality of a number. Basically, to test whether p is prime, we can see if a randomly chosen a satisfies Fermat's little theorem. This is called the *Fermat primality test*. If a and p do not satisfy Fermat's little theorem, we can be sure that p is not prime, and thus the test is completed. However, if a and p do satisfy Fermat's little theorem, we

cannot necessarily be convinced that p is prime, as Fermat's little theorem sometimes holds when p is not prime.

Unhappily for primality testers, there are some composite numbers, called Carmichael numbers, which pass the Fermat test for *every* base a. The smallest one is 561. Carmichael numbers are relatively rare; asymptotically, if $C(n)$ is the number of Carmichael numbers less than n, then

$$n^{2/7} < C(n) < n^{1 - \frac{\ln \ln \ln n}{\ln \ln n}}.$$

As such, the Fermat primality test has a limit: it cannot guarantee that a composite number won't slip through undetected. However, any number that fails the Fermat test is certainly composite.

Moses Liskov

FERMAT'S LITTLE THEOREM

Pierre de Fermat (1601–1665) was one of the most reknowned mathematicians in history. He focused much of his work on Number Theory, though he made great contributions to many other areas of mathematics. Fermat's "last theorem" was a remark Fermat made in a margin of a book, for which he claimed to have a proof but the margin was too small to write it down.

Fermat's little theorem concerns modular arithmetic. The statement of the theorem is that if p is a prime number and a is any number not divisible by p, then $a^{p-1} \equiv 1 \bmod p$.

The theorem follows easily from the following observations. If we consider the product $(a)(2a)(3a) \cdots ((p-1)a)$, then on the one hand we can write this as $(p-1)!a^{p-1}$. On the other hand, the list of terms in the product is a complete list between 1 and $p-1$, since no two terms in the list are equivalent modulo p (if $na \equiv ma \bmod p$ where $n > m$, then p divides $n - m$, but $n - m$ cannot be as large as p, so it must be 0). Thus, we deduce that this product is $(p-1)! \bmod p$. Thus $(p-1)!a^{p-1} \equiv (p-1)! \pmod{p}$ or in other words, $a^{p-1} \equiv 1 \pmod{p}$.

Euler's theorem (see Euler's totient function) generalizes Fermat's little theorem to handle composite moduli as well as primes.

Fermat's little theorem is important to cryptography in that it gives rise to techniques for testing prime numbers. See also Fermat primality test.

Moses Liskov

FIAT–SHAMIR IDENTIFICATION PROTOCOL AND THE FIAT–SHAMIR SIGNATURE SCHEME

INTRODUCTION: There are several variants of the Fiat–Shamir identification protocol. One way to classify these is based on the number of secrets. In the basic one [4] each prover knows only one secret. Another is to distinguish between identity based and public key based ones. In both, a trusted center made public $n = p \cdot q$ such that p and q are secret a prime numbers only known to the center.

In the identity based system [3, p. 152] (see also [2, 4, 5]) a trusted center gives each user a secret key, partially based on biometrics. In particular, to receive an *identity* from the trusted center, Alice goes to the center. There her fingerprints and, other biometrics information is collected and her identity verified. I is the string which contains: Alice's identity (name), Alice's biometrics, and other information to identify Alice uniquely. The center chooses k small j_i, $1 \leq i \leq k$, such that $x_i := f(I, j_i)$ are quadratic residues modulo n, where f is a public function (see also modular arithmetic). The center calculates as secrets the smallest $s_i := (\sqrt{x_i})^{-1} \bmod n$ and gives these secrets to Alice (write these in Alice's smart card).

In the public key based system, Alice chooses her secrets and publishes a public key. In particular, Alice chooses uniformly random s_i in Z_n and computes $x_i := s_i^{-2} \bmod n$. This x_i with Alice's identity is published in an authentic public directory.

PRELIMINARY: In the *identity based* scheme, when Bob wants to verify Alice's identity, Bob asks Alice's identity I together with the j_i and Bob checks Alice's biometrics. If correct, Bob then calculates $x_i := f_i(I, j_i)$, or else Bob halts.

In the *public key* based scheme, Alice reveals her identity and Bob finds her public key in the public key directory.

THE PROTOCOL: Repeat the protocol t times:
Step 1. Alice chooses uniformly random r and computes $z := r^2 \bmod n$ and sends t to Bob.
Step 2. Bob sends Alice bits e_i ($1 \leq i \leq k$).
Step 3. Alice sends $\alpha := r \cdot \prod_{e_i=1} s_i^{e_i} \bmod n$.
Step 4. Bob verifies that $\alpha^2 \cdot \prod_{e_i=1} x_i = z$.
If all the verifications are correct, Bob accepts.

PARAMETERS: The size of t and k as functions of n is chosen depending whether one uses Fiat–Shamir as an identification protocol, or as a zero-knowledge interactive proof. For more details see [1].

FIAT–SHAMIR SIGNATURE SCHEME: The Fiat–Shamir scheme can be used to digitally sign (see also digital signature schemes). This is called the Fiat–Shamir signature. To sign a message m, just use a secure hash function h to compute e_i in the above protocol as follows:

$$e_i := h(z, m), \tag{1}$$

where z is as above. Use the Fiat–Shamir protocol and publish as signature: $(\mathbf{e}, \alpha)$ where $\mathbf{e} = (e_1, \ldots, e_k)$ and α is as above.

To verify the signature the verifier computes z' as in Step 4. Then the verifier uses this z' to compute $\mathbf{e}'$ using Equation 1. If $\mathbf{e}' = \mathbf{e}$, the verifier accepts the signature of m.

Yvo Desmedt

References

[1] Burmester, M.V.D. and Y.G. Desmedt (1989). "Remarks on the soundness of proofs." *Electronics Letters*, 25 (22), 1509–1511.

[2] Fiat, A. and A. Shamir (1987). "How to prove yourself: Practical solutions to identification and signature problems." *Advances in Cryptology—CRYPTO'86*, Lecture Notes in Computer Science, vol. 263, ed. A. Odlyzko. Springer-Verlag, Santa Barbara, CA.

[3] Fiat, A. and A. Shamir (1987). "Unforgeable proofs of identity." *Securicom 87, March 4–6, 1987, Paris, France*, 147–153.

[4] Shamir, A. (1986). "Interactive identification." Presented at the *Workshop on Algorithms, Randomness and Complexity, Centre International de Rencontres Mathématiques (CIRM), Luminy (Marseille), France*.

[5] Shamir, A. (1987). "The search for provably secure identification schemes." *Proceedings of the International Congress of Mathematicians, August 3–11, 1987, Berkeley, CA, USA*, 1488–1495.

FIELD

A *field* $F = (S, +, \times)$ is a ring that has a multiplicative identity (denoted 1) and satisfies two additional properties:

- **Commutativity of $\times$:** For all $x, y \in S$, $x \times y = y \times x$.
- **Multiplicative inverse:** For all $x \in S$ such that $x \neq 0$, there exists a *multiplicative inverse* z such that $x \times z = z \times x = 1$.

Thus, a field is a commutative ring with a multiplicative identity where every element except the additive identity has a multiplicative inverse, so that division is defined in the usual way (i.e., the quotient y/x is defined for all $x \neq 0$ and y as $y \times z$ where z is the multiplicative inverse of x).

The *characteristic* of a field is the least positive integer k such that for all $x \in S$, $kx = 0$, if such a k exists and is defined as 0 otherwise. A <u>finite field</u> is field with a finite number of elements; the characteristic of a finite field is always a <u>prime number</u>. The rational numbers $\mathbf{Q}$ are an example of an infinite field under ordinary addition and multiplication.

Burt Kaliski

FIELD POLYNOMIAL

The <u>irreducible polynomial</u> with respect to which field operations are computed in an <u>extension field</u>.

Burt Kaliski

FILTER GENERATOR

A *filter generator* is a <u>running-key</u> generator for <u>stream cipher</u> applications. It consists of a single <u>linear feedback shift register (LFSR)</u> which is filtered by a nonlinear function. More precisely, the output <u>sequence</u> of a filter generator corresponds to the output of a nonlinear function whose inputs are taken from some stages of the LFSR. If $(u_t)_{t \geq 0}$ denotes the sequence generated by the LFSR, the output sequence $(s_t)_{t \geq 0}$ of the filter generator is given by

$$s_t = f(u_{t+\gamma_1}, u_{t+\gamma_2}, \ldots, u_{t+\gamma_n}), \qquad \forall t \geq 0,$$

where f is a function of n variables, n is less than or equal to the LFSR length, and $(\gamma_i)_{1 \leq i \leq n}$ is a decreasing sequence of non-negative integers called the *tapping sequence*.

In order to obtain a keystream sequence having good statistical properties, the filtering function f should be *balanced* (i.e., its output should be uniformly distributed), and the feedback polynomial of the LFSR should be chosen to be a *primitive polynomial* (see <u>linear feedback shift register</u> for more details).

In a filter generator, the LFSR feedback polynomial, the filtering function, and the tapping sequence are usually publicly known. The secret parameter is the initial state of the LFSR which is derived from the secret <u>key</u> of the cipher by a key-loading algorithm. Therefore, most attacks on filter generators consist in recovering the LFSR initial state from the knowledge of some digits of the sequence produced by the generator (in a <u>known plaintext attack</u>), or of some digits of the ciphertext sequence (in a <u>ciphertext only attack</u>). The attack presented in [8] enables to construct an equivalent keystream generator from the knowledge of a large segment of the ciphertext sequence when the LFSR feedback polynomial is the only known parameter (i.e., when the filtering function, the tapping sequence and the initial state are kept secret).

Any filter generator is equivalent to a particular <u>combination generator</u>, in the sense that both generators produce the same output sequence. An equivalent combination generator consists of n copies of the LFSR used in the filter generator with shifted initial states; the combining function corresponds to the filtering function.

Statistical Properties of the Output Sequence

The output <u>sequence</u> **s** of a filter generator is a linear recurring sequence. Its <u>linear complexity</u>, $\Lambda(\mathbf{s})$, is related to the LFSR length and to the *algebraic degree* of the filtering function f (the algebraic degree of a <u>Boolean function</u> is the highest number of terms occurring in a monomial of its algebraic normal form). For a binary LFSR with a primitive feedback polynomial, we have

$$\Lambda(\mathbf{s}) \leq \sum_{i=0}^{d} \binom{L}{i},$$

where L denotes the LFSR length and d denotes the algebraic degree of f [6, 8]. The period of **s** divides $2^L - 1$. Moreover, if L is a large prime, $\Lambda(\mathbf{s})$ is at least $\binom{L}{d}$ for most filtering functions with algebraic degree d (see [9]).

Thus, to achieve a high linear complexity, the LFSR length L and the algebraic degree of the

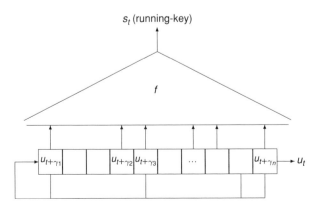

s_t (running-key)

f

filtering function should be large enough. More precisely, the keystream length available to an attacker should always be much smaller than $\binom{L}{\deg(f)}$.

Known Attacks and Related Design Criteria

Filter generators are vulnerable to <u>fast correlation attacks</u> because their output sequence is correlated to some linear function of the stages of the constituent LFSR. Efficient fast correlation attacks on filter generators are described in [JJ02] and [CF02]. In order to make the fast correlation attacks computationally infeasible, the filtering function should have a high <u>nonlinearity</u>. Moreover, it should have many nonzero Walsh coefficients. Another design criterion is that the LFSR feedback polynomial should not be sparse.

Another attack on any filter generator is the generalized <u>inversion attack</u>. It depends highly on the *memory size* of the generator, which corresponds to the largest spacing between two taps, i.e., $M = \gamma_1 - \gamma_n$. To make this attack infeasible, the tapping sequence should be such that the memory size is large and preferably close to its maximum possible value $L - 1$, where L is the LFSR length. Moreover, when the <u>greatest common divisor</u> of all spacing between the taps, $\gcd(\gamma_i - \gamma_{i+1})$, is large, the effective memory size can be reduced by a decimation technique (see <u>inversion attack</u>). Then, the greatest common divisor of all $(\gamma_i - \gamma_{i+1})$ should be equal to 1.

The choice of the tapping sequence also conditions the resistance to the so-called *conditional correlation attacks* [1, 4, 7]. The basic idea of these particular correlation attacks is that some information on the LFSR sequence may leak when some patterns appear in the keystream sequence. Actually, the keystream bits s_t and $s_{t+\tau}$, with $\tau \geq 1$, respectively depend on the LFSR-output bits $u_{t+\gamma_1}, \ldots, u_{t+\gamma_n}$ and $u_{t+\gamma_1+\tau}, \ldots, u_{t+\gamma_n+\tau}$. Therefore, the pair $(s_t, s_{t+\tau})$ only depends on $M - I(\tau)$ bits of the LFSR sequence, where M is the memory size and $I(\tau)$ is the size of the intersection between $\{\gamma_i, 1 \leq i \leq n\}$ and $\{\gamma_i + \tau, 1 \leq i \leq n\}$, i.e., the number of pairs (i, j) with $i < j$ such that $\gamma_i - \gamma_j = \tau$. It is then clear that a given observation of $(s_t, s_{t+\tau})$ may provide some information on the $(M - I(\tau))$ involved bits of the LFSR sequence when $I(\tau)$ is large. Thus, $I(\tau)$ should be as small as possible for all values of $\tau \geq 1$. It can be proved that the lowest possible value of $\max_{\tau \geq 1} I(\tau)$ is 1 and it is achieved when the tapping sequence is a *full positive difference set*, i.e., when all differences $\gamma_i - \gamma_j$, $i < j$, are distinct (a full positive difference set is also called a *Golomb ruler*). Such a tap-

ping sequence of n integers only exists if the LFSR length exceeds $n(n-1)/2$ (see [4] for details).

Advanced algebraic techniques, like Gröbner bases, also provide powerful <u>known plaintext attacks</u> on filter generator, called *algebraic attacks* [3]. Any keystream bit can be expressed as a function of the L initial bits of the LFSR. Thus, the knowledge of any N keystream bits leads to an algebraic system of N equations of L variables. The degree of these equations corresponds to the algebraic degree of the filtering function. But, efficient Gröbner bases techniques enable to substantially lower the degree of the equations by multiplying them by well-chosen multivariate polynomials. Then, it may be possible to recover the LFSR initial state by solving the algebraic system even if the filtering function has a high degree.

Anne Canteaut

References

[1] Anderson, R.J. (1995). "Searching for the optimum correlation attack." *Fast Software Encryption 1994*, Lecture Notes in Computer Science, vol. 1008, ed. B. Preneel. Springer-Verlag, Berlin, 137–143.

[2] Canteaut, A. and E. Filiol (2002). "On the Influence of the Filtering Function on the Performance of Fast Correlation Attacks on Filter Generators." *Symposium* on *Information theory* in the Benelux, *May 2002*.

[3] Courtois, N.T and W. Meier (2003). "Algebraic attacks on stream ciphers with linear feedback." *Advances in Cryptology—EUROCRYPT 2003*, Lecture Notes in Computer Science, vol. 2656, ed. E. Biham. Springer-Verlag, Berlin, 345–359.

[4] Golić, J.Dj. (1996). "On the security of nonlinear filter generators." *Fast Software Encryption 1996*, Lecture Notes in Computer Science, vol. 1039, ed. D. Gollman. Springer-Verlag, Berlin, 173–188.

[5] Jönsson, F. and T. Johansson (2002). "A fast correlation attack on LILI-128." *Information Processing Letters*, 81 (3), 127–132.

[6] Key, E.L. (1976). "An analysis of the structure and complexity of nonlinear binary sequence generators." *IEEE Transactions on Information Theory*, 22, 732–736.

[7] Lee, S., S. Chee, S. Park, and S. Park (1996). "Conditional correlation attack on nonlinear filter generators." *Advances in Cryptography—ASIACRYPT'96*, Lecture Notes in Computer Science, vol. 1163, eds. K. Kim and T. Matsumoto. Springer-Verlag, Berlin, 360–367.

[8] Massey, J.L. (2001). "The ubiquity of Reed–Muller codes." *Applied Algebra, Algebraic Algorithms and Error-Correcting Codes—AAECC-14*, Lecture Notes in Computer Science, vol. 2227, eds. S. Boztas and I. Shparlinski. Springer-Verlag, Berlin, 1–12.

[9] Rueppel, R.A. (1986). *Analysis and Design of Stream Ciphers*. Springer-Verlag, Berlin.

FINGERPRINTING

Fingerprinting is a technique that aims at preventing unauthorized redistribution of digital content. This assumes a multitude of scenarios under which identical or very close copies of a document v (software, images or other digital media) are made available to a large number of users of the system by paid subscription. Fingerprinting consists in embedding a mark x (*a fingerprint*) in the document with the purpose of encoding the identity of the user. Fingerprints should not be easily detectable or removable, they must be also designed in a way that makes forgery difficult or expensive. In the area of content distribution, the functionality associated with the fingerprinting system is its resistance to a *collusion attack*. A group or *coalition* of users of the system is said to perform a collusion attack if they produce a copy of the document v with either an obfuscated (for instance, totally removed) fingerprint or a fingerprint that does not enable the distributor D to trace it back to any of the members of the coalition. A closely related concept of underline{watermarking} is aimed at tracing the owner of the contents and is not intended to fight collusion attacks.

The first mention of fingerprinting in the literature dates back to 1983 [15]. Presently there are two main trends in fingerprinting depending on the acceptance of the so-called *marking assumption*. Under this assumption, the document v and the fingerprint x are strings over some finite alphabet Σ (a typical length of v is in the range of Mbytes or greater, while the length of x is on the order of several kbytes). The location of x in v is not known to the users nor is it assumed that the bits of x form a consecutive substring of v. However, for practical reasons, fingerprints occupy the same positions in all the copies of v. A pirate user or a coalition of t such users can detect some positions that belong to x. It is assumed that *changing any undetectable position of the user's copy of the document makes it useless* and obliterates its market value. This scenario applies for instance to fingerprinting licensed software. Fingerprinting under the marking assumption emerged in the defining paper [8] which studied distribution of decryption keys used to access pay-per-view TV programs and similar applications. The main tools of analysis of fingerprinting with the marking assumption are combinatorial, information- and coding-theoretic.

In the absence of the marking assumption, both the distributor and the users are allowed to add relatively small distortion to the original document. The main application of this scenario is *fingerprinting of digital media* such as image, video, audio, and speech content. In such applications, slight differences from the original version will not affect the quality of users' copies. It is assumed that both the signal and the fingerprint are real random variables that obey some probability distribution. Establishing resilience of the system against collusion attacks relies on probabilistic or information-theoretic analysis.

MEDIA FINGERPRINTING: It is natural to assume that the original document v is represented by a vector in $\mathbb{R}^n$. Let M be the total number of users of the system. The marked copy of the ith user is obtained by computing $x_i = v + w_i$, where w_i is the ith fingerprint (another real vector).

The most common assumption in the literature is that the document is a random vector $(v_1, \ldots, v_n)$ whose coordinates are real random $N(0, 1)$-variables (Gaussian random variables with zero mean and unit variance). Similarly, fingerprints $w_i, i = 1, \ldots, M$, are chosen as Gaussian vectors or taken as linear combinations of orthogonal pseudonoise vectors. By assumption, the deviation of x_i from v has to be bounded above uniformly for all i: $\|x_i - v\| \le \delta_1 \sqrt{n}$, where δ_1 is an appropriately chosen threshold. In the case of Gaussian fingerprints it is assumed that the variance α^2 of their coordinates satisfies the inequality $\alpha < \delta_1$ (otherwise with large probability, marked copies will violate the bounded deviation assumption).

Collusion attacks in this context proceed by averaging the copies of the document belonging to the users in the coalition or by taking their convex combination and adding some Gaussian noise or by some similar general method. Let y be the fingerprint formed by a pirate coalition. Detection is performed by correlation analysis: the ith user is assumed guilty if $(y, x_i) \ge \delta_2 \|y\|$, where δ_2 is another threshold. It is proved in [10] that for Gaussian fingerprints, the minimum size of the coalition that succeeds in removing any fingerprint with probability bounded away from zero must be proportional to $\sqrt{n / \log M}$.

Another class of problems in the same context, called *asymmetric fingerprinting* [13], arises when it is assumed that not only coalitions of users but also the distributor may create unauthorized copies of the content. The goal of asymmetric fingerprinting is to bar the distributor from issuing an unregistered copy, while at the same time allowing it to trace the source of such copy to the user whose mark it contains.

FINGERPRINTING UNDER MARKING ASSUMPTION: This group of problems was introduced in [8] as a part of a broader area known as underline{traitor tracing}. Let $C = (c_1, \ldots, c_M)$, where for all

$i, c_i \in \Sigma^n$, be a collection of their fingerprints, also called a *fingerprinting code*. Suppose that the maximum size of a pirate coalition U is t. Properties and constructions of fingerprinting codes depend crucially on the exact rule that the coalition follows to produce an unregistered fingerprint y.

Let $E(U) = \{y\}$ be the set of all fingerprints y that can be created by the coalition U, called the *envelope* of the coalition. The current art distinguishes between narrow-sense and wide-sense envelopes. The *narrow-sense* envelope $E(U)$ of the coalition U of users consists of all the n-vectors $y = (y^1, \ldots, y^n) \in \Sigma^n$ that satisfy $y^i \in \{u^i : u \in U\}, i = 1, \ldots, n$. Let

$$E(C) = \bigcup_{U \in \binom{C}{t}} E(U)$$

be the union of the t-envelopes of all coalitions of size t or less. The code C that satisfies

$$\bigcap_{U : y \in E(U)} U \neq \emptyset \quad \text{for every } y \in E(C)$$

is said to possess a *t-identifiable parent property*. Intuitively, for every unregistered fingerprint y, such a code enables the content provider to identify at least one of the members of the pirate coalition. The study of such codes was undertaken in [1, 2, 4, 6, 9]. It is easy to show that if $|\Sigma| \leq t$, the maximum number of users of the system is $M \leq t$. For $|\Sigma| \geq t + 1$ it is possible [1, 4] to construct codes with t-identifiable parent property codes of size M such that $\log M = \Omega(n)$. Upper bounds on the size of such codes were obtained in [1, 2, 6].

The wide-sense envelope reflects a more general problem statement under which the members of the coalition have more amplitude in creating an unregistered copy of the document, Namely, once the position is detected by the members of the coalition as belonging to the fingerprint, they can substitute this position in y with any letter from the alphabet. Accordingly the set of fingerprints that can be generated by the coalition U under this strategy is called the *wide-sense* envelope of U. This envelope is larger than the narrow-sense one for all alphabets of size $|\Sigma| \geq 3$.

Properties of fingerprinting codes resistant to the collusion attack in the wide sense were established in [3, 7] for any alphabet size $|\Sigma| \geq 2$. Both papers assume that the collusion attack is adversarial in the sense that in creating a forged fingerprint, the coalition pursue the strategy that minimizes distributor's chance for correct identification. One of the central facts is that any fingerprinting scheme whose probability of wrong identification under a collusion attack is

small (for instance, approaches zero as the code length n grows) must rely on a family of codes $\mathcal{C} = (C_1, \ldots, C_K)$ rather than an individual code. The family $\mathcal{C}$ is publicly known while the choice of a specific code C_i forms the secret key of the distributor. The error probability of identification (defined in a suitable way) is bounded below as follows [3]:

$$p_e(\mathcal{C}) \geq \frac{t-1}{(2t-1)K}.$$

This shows that no individual code (i.e., $K = 1$) can provide reliable protection against collusion attack in the wide sense.

Turning to the practically important case of the binary alphabet, we note that under this restriction, the narrow- and wide-sense envelope conditions give rise to the same problem. As mentioned above, finger-printing codes that provide *exact identification* in this case do not exist. Relaxing the requirement of exact identification of pirate users enables the content provider to accommodate a large number of subscribers even if the fingerprint alphabet is binary. Suppose that with some small probability ϵ, the distributor implicates an innocent user (false positive) or misses all of the pirates (false negative). For this problem, paper [7] suggested fingerprinting schemes with $n = O(t^4 \log(M/\epsilon) \log(1/\epsilon))$ (see O-notation). This implies that the error probability ϵ cannot decrease faster than $\exp(-\omega(\sqrt{n}))$ and then $\log M = O(\sqrt{n})$. For a fixed t, paper [3] improved this result to $\log M = O(n), \epsilon = \exp(-\Omega(n))$. Existence of codes of length $n = O(t^2 \log(M/\epsilon))$ was proved in [14] by random choice.

Algorithmic aspects of identification with fingerprinting codes were addressed in [3, 5, 11]. In particular, [3, 5] provide large-size codes with polynomial identification of pirate users.

Finally, a general analysis of information hiding based on an information-theoretic model of fingerprinting as coding with side information is performed in [12]. The analysis in this paper also subsumes such contexts as watermarking and digital steganography.

Alexander Barg
Gregory Kabatiansky

References

[1] Alon, N., G. Cohen, M. Krivelevich, and S. Litsyn (2003). "Generalized hashing and parent-identifying codes." *J. Combinatorial Theory, Ser. A*, 104, 207–215.

[2] Alon, N. and U. Stav (2004). "New bounds on parent-identifying codes: The case of multiple parents."

Combinatorics, Probability, and Computing, 13 (6), 795–807.

[3] Barg, A., G.R. Blakley, and G. Kabatiansky (2003). "Digital fingerprinting codes: Problem statements, constructions, identification of traitors." *IEEE Trans. Inform. Theory*, 49 (4), 852–865.

[4] Barg, A., G. Cohen, S. Encheva, G. Kabatiansky, and G. Zémor (2001). "A hypergraph approach to the identifying parent property: The case of multiple parents." *SIAM J. Discrete Math.*, 14, 423–431.

[5] Barg, A. and G. Kabatiansky (2004). "A class of i.p.p. codes with efficient identification." *J. Complexity*, 20 (2–3), 137–147.

[6] Blackburn, S.R. (2003). "An upper bound on the size of a code with the k-identifiable parent property." *J. Comb. Theory Series A*, 102, 179–185.

[7] Boneh, D. and J. Shaw (1998). "Collusion-secure fingerprinting for digital data." *IEEE Trans. Inform. Theory*, 44 (5), 1897–1905.

[8] Chor, B., A. Fiat, and M. Naor (1994). "Traitor tracing." *Advances in Cryptology—CRYPTO'94*, Lecture Notes Computer Science, vol. 839, ed. Y.G. Desmedt. Springer-Verlag, Berlin, 480–491.

[9] Hollmann, H.D.L., J.H. van Lint, J.-P. Linnartz, and L.M.G.M. Tolhuizen (1998). "On codes with the identifiable parent property." *J. Combin. Theory, Ser. A*, 82 (2), 121–133.

[10] Kilian, J., F.T. Leighton, L.R. Matheson, T.G. Shamoon, R.E. Tarjan, and F. Zane (1998). "Resistance of digital watermarks to collusive attacks." Princeton Computer Science Technical Report, 585–98.

[11] Silverberg, A., J. Staddon, and J.L. Walker (2001). "Efficient traitor tracing algorithm using list decoding." *Advances in Cryptography—ASIACRYPT 2001*, Lecture Notes Computer Science, vol. 2248, ed. C. Boyd. Springer-Verlag, Berlin, 175–192.

[12] Moulin, P. and J. O'Sullivan (2003). "Information-theoretic analysis of information hiding." *IEEE Trans. Inform. Theory*, 49 (3), 563–593.

[13] Pfitzmann, B. and M. Schunter (1996). "Asymmetric fingerprinting." *Advances in Cryptology—EUROCRYPT'96*, Lecture Notes Computer Science, vol. 1070, ed. U. Maurer. Springer-Verlag, Berlin, 84–95.

[14] Tardos, G. (2003). "Optimal probabilistic fingerprint codes." *Proceedings of the 35th Annual ACM Symposium on Theory of Computing*, 116–125.

[15] Wagner, N. (1983). "Fingerprinting." *Proc. 1983 IEEE Symp. Security and Privacy, April 1983*, 18–22.

FINITE FIELD

A *finite field* is a field with a finite number of elements. The number of elements or the order of a finite field is a power $q = p^k$ of a prime number $p \geq 2$, where $k \geq 1$. The finite field with q elements is denoted $\mathbf{F}_q$ or $GF(q)$ (the latter meaning *Galois Field*). The prime p is the *characteristic* of the field, i.e., for all $x \in \mathbf{F}_q$, $px = 0$.

(One generally refers to *the* finite field with q elements in the sense that all finite fields with a given order have the same structure, i.e., they are *isomorphic* to one another.)

Finite fields are commonly organized into three types in cryptography:

- *Characteristic-2* or *binary fields*, where $p = 2$.
- *Prime-order fields*, where $p \geq 3$ and $k = 1$.
- *Odd-characteristic extension fields*, where $p \geq 3$ and $k > 1$.

Finite fields are widely employed in cryptography. The IDEA and Rijndael/AES algorithms, for instance, both involve operations over relatively small finite fields. Public-key cryptography generally involves much larger finite fields. For example, the Digital Signature Algorithm (see Digital Signature Standard) operates in a subgroup of the multiplicative group of a prime-order finite field. Elliptic curve cryptography can be defined over a variety of different finite fields.

The *multiplicative group* of a finite field, denoted $\mathbf{F}_q^*$, is the group consisting of the elements of $\mathbf{F}_q$ with multiplicative inverses, i.e., the elements except for 0, under the multiplication operation. This is a cylic group of order $q - 1$; all the elements of the group can be obtained as powers of a single generator.

Efficient implementation of finite field arithmetic has been a major area of research in cryptography. For a discussion, see inversion in finite fields and rings and optimal extension fields.

A classic volume in this "field" is Lidl and Niederreiter's text [1]; a book by Menezes et al. [2] gives further treatment of the cryptographic applications.

Burt Kaliski

References

[1] Lidl, R. and H. Niederreiter (1986). *Introduction to Finite Fields and Their Applications*. Cambridge University Press, Cambridge.

[2] Menezes, A.J., I.F. Blake, X. Gao, R.C. Mullin, S.A. Vanstone, and T. Yaghoobian (1992). *Applications of Finite Fields*. Kluwer Academic Publishers, Dordrecht.

FIPS 140-2

The full name is Federal Information Processing Standard (FIPS) 140-2, titled: Security Requirements For Cryptographic Modules [2]. This document was issued May 25, 2001 and supercedes FIPS 140-1, which was published January 11, 1994 [1]. FIPS 140-1 and FIPS 140-2 were developed not only as documents to communicate requirements, but also as complete programs with the objective to provide the end customer with cryptographic products that can be trusted and used with confidence. The Program is called the cryptographic module validation program (CMVP). The program's intent is to balance several objectives to maximize the effectiveness for end users of the cryptographic products including security functionality, assurance, cost, and schedule. The program implementation includes testing requirements, lab accreditations, thorough report and validation result review by the regulators, and certificates of validation issued by the NIST/CSE, including an actively maintained Web site of currently approved products http://csrc.nist.gov/cryptval/.

The FIPS 140-1 and -2 program was established in response to the Information Technology Management Reform Act of 1996 and the Computer Security Act of 1987. The program is an equal partnership between U.S. NIST and CSE of the Canadian Government with both the USA and Canadian organizations being integrally involved with all aspects of the program.

FIPS 140-2 is published and maintained by the U.S. Department of Commerce; National Institute of Standards and Technology; Information Technology Laboratory; and Cryptographic Module Validation Program. The standard is designed to be flexible enough to adapt to advancements and innovations in science and technology. The standard will be reviewed every 5 years in order to consider new or revised requirements that may be needed to meet technological and economic changes.

FIPS 140-2 is applicable to cryptographic-based security systems that may be used in a wide variety of computer and telecommunication applications (e.g., data storage, access control and personal identification, network communications, radio, facsimile, and video) and in various environments (e.g., centralized computer facilities, office environments, and hostile environments). The cryptographic services (e.g., encryption, authentication, digital signature, and key management) provided by a cryptographic module are based on many factors that are specific to the application and environment. The security level to which a cryptographic module is validated must be chosen to provide a level of security appropriate for the security requirements of the application and environment in which the module will be utilized and the security services that the module will provide.

The security requirements for a particular security level include both the security requirements specific to that level and the security requirements that apply to all modules regardless of the level.

It is important to note the difference between FIPS 140-2 and verification of correct cryptographic algorithm testing. Algorithm testing is limited only to verifying that a design correctly performs the logical and mathematical processes to comply with the specified algorithm functionality, such as, but not limited to:

- FIPS PUB 46-3, Data Encryption Standard (DES).
- FIPS PUB 81, DES Modes of Operation.
- FIPS PUB 113, Computer Data Authentication.
- FIPS PUB 180-1, Secure Hash Standard SHA family.
- FIPS PUB 186-2, Digital Signature Standard (DSS).

In contrast FIPS 140-2 is a comprehensive security standard that integrates module access control, key management, interface control, design assurance, operational assurance, and utilizes algorithm testing to verify that each specific algorithm used is correct.

FIPS 140-2 provides a standard that is used by government and commercial organizations when they specify that cryptographic-based security systems are to be used for providing protection of sensitive or valuable data. The protection of data, processes, and critical security parameters that is provided by a cryptographic module embedded within a security system is necessary to maintain the access control, confidentiality, and integrity of the information entrusted to that system.

This standard specifies the security requirements that will be satisfied by a cryptographic module. These requirements have been developed over a number of years by cryptographic and information security experts from government organizations, international representatives, the Department of Defense, users, and vendors with the objective being something that is meaningful, but also reasonable with regard to technology, usability, and affordability. The requirements have been compiled to address a broad range of threats and vulnerabilities, which trusted security focal points (modules) are subject to.

The security requirements cover areas related to the secure design and implementation of a cryptographic module. These areas include:

1. Cryptographic module specification
2. Cryptographic module ports and interfaces
3. Roles, services, and authentication
4. Finite state model
5. Physical security

6. Operational environment
7. Cryptographic key management
8. Electromagnetic interference/electromagnetic compatibility (EMI/EMC)
9. Self-tests
10. Design assurance
11. Mitigation of other attacks.

The standard provides four increasing qualitative levels of security intended to cover a wide range of potential applications and environments.

Security Level 1 provides the lowest level of security, the basic security requirements are specified for a cryptographic module. No specific physical security mechanisms are required in a Security Level 1 cryptographic module beyond the basic requirement for production-grade components. Such implementations may be appropriate for some low-level security applications when other controls, such as physical security, network security, and administrative procedures, are limited or nonexistent.

Security Level 2 enhances the physical security mechanisms of a Security Level 1 cryptographic module by adding the requirement for physical security in the form of tamper-evidence, tamper-evident seals or pick-resistant locks are placed on covers or doors to protect against unauthorized physical access. Security Level 2 also requires the module operators to authenticate to the module to assume a specific role and perform a corresponding set of services. Level 2 allows the software and firmware components of a cryptographic module to be executed on a general purpose computing system using an operating system that meets the functional requirements specified in specific Common Criteria (CC) Protection Profiles and has been evaluated to a CC assurance level 2 (EAL2 or higher).

Security Level 3 increases the physical security of the cryptographic module to inhibit the intruder from gaining access to *critical security parameters* (CSPs) held within the cryptographic module. Physical security mechanisms required at Security Level 3 are intended to have a high probability of detecting and responding to attempts at physical access, use or modification of the cryptographic module through the implementation of strong enclosures and tamper detection/response that includes circuitry that zeroizes all plaintext CSPs when the removable covers/doors of the cryptographic module are opened.

Level 3 requires stronger identity-based authentication mechanisms, enhancing the security provided by the role-based authentication mechanisms specified for Security Level 2. Level 3 specifies more robust key management processes including the entry or output of plaintext CSPs using split knowledge procedures or to be performed using ports that are physically separated from other ports, or interfaces that are logically separated using a trusted path from other interfaces. Plaintext CSPs may be entered into or output from the cryptographic module in encrypted form (in which case they may travel through enclosing or intervening systems).

Level 3 allows the software and firmware components of a cryptographic module to be executed on a general purpose computing system using an operating system that meets the functional requirements specified in specified *protection profiles* (PPs) if they also meet the additional functional requirement of a Trusted Path (FTP_TRP.1) and has been evaluated at the CC EAL3 (or higher) with the additional assurance requirement of an Informal *Target of Evaluation* (TOE) Security Policy Model (ADV_SPM.1).

Security Level 4 provides the highest level of security defined in this standard. At this security level, the physical security mechanisms provide a complete envelope of protection around the cryptographic module with the intent of detecting and responding to all unauthorized attempts at physical access. Penetration of the cryptographic module enclosure from any direction has a very high probability of being detected, resulting in the immediate zeroization of all plaintext CSPs. Security Level 4 cryptographic modules are useful for operation in physically unprotected and potentially hostile environments.

Level 4 also protects a cryptographic module against a security compromise due to environmental conditions or fluctuations outside of the module's normal operating ranges for voltage and temperature.

Level 4 allows the software and firmware components of a cryptographic module to be executed on a general purpose computing system using an operating system that meets the functional requirements specified for Security Level 3 and is evaluated at the CC evaluation assurance level EAL4 (or higher).

One of the important concepts that is fundamental to the application of FIPS 140-2 is that of a cryptographic boundary. The Cryptographic boundary shall consist of an explicitly defined perimeter that establishes the physical bounds of a cryptographic module. If a cryptographic module consists of software or firmware components, the cryptographic boundary shall contain the processor(s) and other hardware components that store and protect the software and firmware components. Hardware, software, and firmware components of a cryptographic module can be excluded from the requirements of this standard if shown that these components do not affect the security of the module. This concept allows for the evaluation of products to fixed and established confines, thereby making the process feasible. It also provides for a totally self-contained crypto module that contains enough functionality to protect itself and be able to be trusted.

For purposes of specifying physical security mechanisms, FIPS 140-2 defines three possible physical embodiments for a module and the associated security mechanisms for each.

– Single-chip cryptographic modules are physical embodiments in which a single integrated circuit is employed
– Multiple-chip embedded cryptographic modules are physical embodiments in which two or more IC chips are interconnected and are embedded within an enclosure or a product that may not be physically protected.
– Multiple-chip standalone cryptographic modules are physical embodiments in which two or more IC chips are interconnected and the entire enclosure is physically protected.

For a module to receive a certificate of validation, it must follow a process defined by the CMVP program and be tested by an accredited laboratory. Laboratories are accredited to perform FIPS 140-2 testing based on proving to the National Voluntary Laboratory Accreditation Program (NVLAP) that they have the appropriate quality system, quality process, cryptographic skills, and FIPS 140-2 knowledge to warrant accreditation. Laboratory reaccreditation occurs on an annual basis.

Testing by the laboratory is based on a document related to FIPS 140-2 termed the Derived Test Requirements (DTR) document [3]. This document outlines the responsibilities of the test laboratory and the analysis and testing that is performed. It also describes the information and material that the vendor must provide the laboratory for the purpose of evaluating the compliance of the module to FIPS 140-2 requirements.

Tom Caddy

References

[1] National Institute of Standards and Technology (NIST) (1994). *Security Requirements for Cryptographic Modules*. Federal Information Processing Standards Publication 140-1 (FIPS PUB 140-1). http://csrc.nist.gov/publications/fips/fips140-1/fips1401.pdf
[2] National Institute of Standards and Technology (NIST) (2001). *Security Requirements for Cryptographic Modules*. Federal Information Processing Standards Publication 140-2 (FIPS PUB 140-2). http://csrc.nist.gov/publications/fips/fips140-2/fips1402.pdf
[3] National Institute of Standards and Technology (NIST) (2003). Derived Test Requirements for FIPS PUB 140-2. http://csrc.nist.gov/cryptval/140-1/fips1402DTR.pdf

FIREWALL

A *firewall* is a network device that enforces security policy for network traffic. The term originates from *fire wall*, a fireproof wall used as a barrier to prevent the spread of fire. An Internet firewall creates a barrier between separate networks by imposing a point of control that traffic needs to pass before it can reach a different network [2]. A firewall may limit the exposure of hosts to malicious network traffic, e.g., remote adversaries attempting to exploit security holes in vulnerable applications, by preventing certain packets from entering networks protected by the firewall.

When inspecting a network packet, a firewall decides if it should drop or forward the packet. The decision is based on a firewall's security policy and its internal state. Before forwarding a packet, a firewall may modify the packet's content. Packet inspection may occur at several different layers:

- The *link layer* provides physical addressing of devices on the same network. Firewalls operating on the link layer usually drop packets based on the *media access control* (MAC) addresses of communicating hosts.
- The *network layer* contains the *Internet protocol* (IP) headers that support addressing across networks so that hosts not on the same physical network can communicate with each other.
- The *transport layer* provides data flows between hosts. On the Internet, the *transmission control protocol* (TCP) and the *user datagram protocol* (UDP) are used for this purpose. Most firewalls operate at the network and transport layer. TCP provides reliable data flow between hosts. UDP is a much simpler but unreliable transport protocol.
- The *application layer* contains application specific protocols like the *hypertext transfer protocol* (HTTP). Inspection of application specific protocols can be computationally expensive because more data needs to be inspected and more states are required.

The IP header [7] contains information required by routers to forward packet between hosts. For an IPv4 header, firewalls usually inspect the following fields: header length, total length, protocol, source IP address, and destination IP address. The protocol field states which transport protocol follow after the IP header. The most frequently used transport protocols are TCP, UDP, and ICMP. While the IP protocol provides addressing between different hosts on the Internet, TCP and UDP allow the addressing of services by employing source and destination port numbers. For example, a mail client sends email by contacting a mail server on destination port 25 which is used for the *simple mail transport protocol* (SMTP) and Web servers usually listen for HTTP requests on port 80. A data flow between hosts is uniquely identified by its 4-tuple:

(source IP address, source port number, destination IP address, destination port number).

A firewall's policy is described by a list of filter rules that specify which packets may pass and which packets are blocked. A rule contains an action and several specifiers. The specifiers describe the packets for which a rule should be applied. The action of the last matching rule determines if the firewall forwards or drops the inspected packet. Many firewalls use a format to describe filter rules that is similar to the following:

action dir interface protocol srcip srcport destip dstport flags,

The first word *action* can be either *pass* or *block* and determines if a packet that matches this rule should be forwarded or dropped. These two actions are supported by all firewalls, but many modern firewall systems also support logging of packets that match specific rules, network address translation, and generating return packets. The remaining words are specifiers that restrict which packets a rule applies to. The different elements in a filter rule are explained below:

- *dir* refers to the direction a packet is taking. It may be either *incoming* or *outgoing*. Depending on the firewall solution, the direction may be specific to a network interface or to a network boundary.
- *interface* specifies the physical or virtual network device upon which the packet was received. This is often used to classify the network

from which the packet arrived as internal or external.

- *protocol* allows a filer to specify for which transport protocols the rule should match. Using a protocol specifier allows us to differentiate, for example between UDP and TCP packets.
- *srcip* specifies which source IP addresses should match the rule. This can either be a single address or IP network ranges.
- *srcport* applies only to TCP or UDP packets and determines the range of port numbers a packet should have to match the rule.
- *dstip* and *dstport* specify which destination IP addresses and destination ports to match and are very similar to the source specifiers described above.
- *flags* may contain firewall specific features. For example, the *SYN* flag indicates that rule matches only where the packet contains a TCP connection setup message (see below).

In the following example, we show a simple firewall configuration that blocks all traffic except SMTP connections to a central mail server:

Figure 1 shows the rules of a firewall that uses only the information contained in the inspected packet to decide if the packet should be dropped or forwarded. Any incoming packet that is not sent to port 25 of the mail server matches only the first firewall rule and will be dropped. A firewall that reaches policy decisions based only on the content of the inspected packet is called *stateless*. Filtering packets that enter a network is called *ingress* filtering and filtering packets that leave a network is called *egress* filtering.

If we want to enforce a policy that allows only outgoing HTTP connections, the rules for stateless firewalls are more difficult to configure. When a client initiates an HTTP connection to port 80 of a Web server, the server sends back data with source port 80. A naive firewall configuration might allow all TCP packets with source port 80 to enter the network. As a result, an adversary is allowed to initiate any kind of connection as long as the packets arrive with the correct source port. We can improve the firewall configuration by inspecting the control flags used by TCP. A packet to initiate a connection has only the *SYN* flag set. The receiver acknowledges the connection attempt by setting

action	dir	interface	protocol	srcip	srcport	dstip	dstport	flags
block	in	external		all				
block	out	external		all				
pass	in	external	tcp	any	any	mail-server	25	
pass	out	external	tcp	mail-server	25	any	any	

Fig. 1. A simple configuration that allows only traffic to and from the SMTP port of a specific mail server

action	dir	interface	protocol	srcip	srcport	dstip	dstport	flags
block	in	external		all				
block	out	external		all				
pass	in	external	tcp	any	80	any	any	
block	in	external	tcp	any	80	any	any	S/SA
pass	out	external	tcp	any	any	any	80	

Fig. 2. A configuration that allows only outgoing HTTP traffic. The rules use TCP control flags to prevent outsiders from initiating connections to our internal network

both the *SYN* and *ACK* flag. After a connection has been established, the *SYN* flag is not used again. A better firewall configuration is shown in Figure 2. The third rule permits all external tcp traffic with source port 80 but the subsequent rule blocks all Web server packets that contain a SYN flag but no ACK flag. In other words, external connection attempts are denied.

A more flexible configuration is permitted by *stateful* firewalls [8] which track connection states for ICMP, TCP, and UDP packets. While UDP and ICMP are connectionless protocols, a stateful firewall may create short-lived states for them. The connection state allows a firewall to determine if a packet is part of an ongoing connection and drop packets that are outside the valid range of parameters. A stateful firewall is easier to configure because packets that match connection state are forwarded without consulting the firewall rules. The HTTP client example from above can be realized with a stateful firewall using only one rule; see Figure 3.

Because IP is independent of the physical network layer, it is possible that an IP packet needs to be fragmented into several smaller packets that fit within the frame size of the physical medium. The first IP fragment contains all addressing information including ports. The firewall may enforce its policy only for the first fragment. Even if all subsequent fragments are forwarded by the firewall, without the first fragment it is not possible for the end host to completely reassemble the IP packet.

The situation is more complicated if the firewall decides to forward the first fragment. As it is possible for fragments to overlap, an adversary may decide to send an IP fragment that overlaps the transport header data of the first fragment. In that case, an adversary may be able to rewrite the port numbers and effectively bypass the firewall's security policy. Overlapping IP fragments may also confuse *network intrusion detection systems* (NIDS) [5].

There are several methods to send packets that may bypass application level firewalls and evade intrusion detection systems [6]. Because a NIDS does not have complete knowledge of network topology and end host state, an adversary may evade security policy by exploiting ambiguities in the traffic stream monitored by the NIDS. For example, a packet may be seen by the monitoring device but never reach the end host. As a result, the traffic seen by the monitor differs from the traffic received by an end host and the monitor may make an incorrect decision.

One possible solution to remove ambiguities from the network traffic is called *traffic normalization* [4]. It rewrites ambiguous traffic so that the interpretation of traffic at the end host is known to the NIDS. As a result, an adversary loses direct control over the specific layout of traffic, making it more difficult to evade intrusion detection systems. Recently, firewalls have started to support traffic normalization. Handley et al. enumerate methods to normalize traffic. For example, a firewall may reassemble all IP fragments into complete IP packets before forwarding them.

Firewalls may also help to limit the problem of *address spoofing* and *denial of service* attacks [3]. Address spoofing refers to IP packets that carry an incorrect IP source address. An adversary may be able to exploit trust relationships by sending packets via an external network that claim to originate from the internal network [1]. Denial of service is aimed at disrupting network services and often uses thousands of machines to generate network traffic that saturates the available bandwidth at

action	dir	interface	protocol	srcip	srcport	dstip	dstport	flags
block	in	external		all				
block	out	external		all				
pass	**out**	**external**	**tcp**	**any**	**any**	**any**	**80**	**keep state**

Fig. 3. A stateful firewall is easier to configure because it keeps track of established connections. Packets that are part of an ongoing connection do not need to match any rules

the target host. To prevent tracing the attack back to the originators, denial of service attacks often use IP address spoofing.

Organizations that employ firewalls should enforce a policy that does not allow spoofed traffic to either enter or leave the network by employing both ingress and egress filtering. Denying packets with spoofed source addresses from leaving the network prevents denial of service attacks to hide their origin.

ADDITIONAL DEFINITIONS

- Ingress filtering: filtering packets that enter a network.
- Egress filtering: filtering packets that leave a network.
- Stateful packet inspection: decisions to drop or forward a packet are based not only on the content of the inspected packet but also on previous network traffic.
- Denial of service: disrupting a computer service by exhausting resources like network bandwidth, memory, or computational power.
- Network intrusion detection system: a system that monitors network traffic to identify abnormal behavior that indicates network attacks.

Niels Provos

References

[1] Bellovin, S.M. (1989). "Security problems in the TCP/IP protocol suite." ACM *Computer Communications Review* 2:19, pp. 32–48, April 1989.

[2] Cheswick, William R. and Steven M. Bellovin (1994). *Firewalls and Internet Security Repelling the Willy Hacker*. Addison-Wesley, Reading, MA.

[3] Ferguson, P. and D. Senie (2000). "Network ingress filtering: Defeating denial of service attacks which employ IP source address spoofing." RFC 2827.

[4] Handley, M., C. Kreibich, and V. Paxson (2001). "Network intrusion detection: Evasion, traffic normalization and end-to-end protocol semantics." *Proceedings of the 10th USENIX Security Symposium, August 2001*.

[5] Paxson, V. (1988). "Bro: A system for detecting network intruders in real-time." *Proceedings of the 7th USENIX Security Symposium, January 1998*.

[6] Ptacek, Thomas and Timothy Newsham (1998). "Insertion, evasion, and denial of service: Eluding network intrusion detection." Secure Networks Whitepaper, August 1998.

[7] Stevens, W.R. (1994). *TCP/IP Illustrated*, vol. 1. Addison-Wesley, Reading, MA.

[8] van Rooij, Guido (2000). "Real stateful TCP packet filtering in IP filter." *Proceedings of the 2nd International SANE Conference, May 2000*.

FIXED-BASE EXPONENTIATION

There are many situations where an exponentiation of a fixed base element $g \in G$, with G being some group, by an arbitrary positive integer exponent e is performed. For instance, such cases occur at the Diffie–Hellman key agreement. Fixed-base exponentiation aims to decrease the number of multiplications compared to general exponentiation algorithms such as the binary exponentiation algorithm. With a fixed base, precomputation can be done once and then used for many exponentiations. Thus the time for the precomputation phase is virtually irrelevant. Using precomputations with a fixed base was first introduced by Brickell et al. (and thus it is also referred to as the BGMW method) [1]. In the basic version, values $g_0 = g$, $g_1 = g^2$, $g_2 = g^{2^2}, \ldots, g_t = g^{2^t}$ are precomputed, and then the binary exponentiation algorithm is used without performing any squarings. Having an exponent e of bit-length $n + 1$, such a strategy requires on average $n/2$ multiplications whereas the standard binary exponentiation algorithm requires $(3/2)n$ multiplications. However, there is quite some storage required for all precomputed values, namely storage for $t + 1$ values. The problem of finding an efficient algorithm for fixed-base exponentiation can be rephrased as finding a short vector-addition chain for given base elements $g_0, g_1, \ldots, g_t$ (see fixed-exponent exponentiation). Note that there is always a trade-off between the execution time of an exponentiation and the number t of precomputed group elements.

In order to reduce the computational complexity, one might use a precomputed version of the k-ary exponentiation by making the precomputation phase only once. However, time is saved by multiplying together powers with identical coefficients, and then raising the intermediate products to powers step by step. The main idea of the *fixed-base windowing method* is that $g^e = \prod_{i=0}^{t} g_i^{e_i} = \prod_{j=1}^{h-1} (\prod_{e_i=j} g_i)^j$ where $0 \le e_i < h$ [1]. Assume that g^e is to be computed where g is fixed. Furthermore, there is a set of integers $\{b_0, b_1, \ldots, b_t\}$ such that any appropriate positive exponent e can be written as $e = \sum_{i=0}^{t} e_i b_i$, where $0 \le e_i < h$ for some fixed positive integer h. For instance, choosing $b_i = 2^i$ is equivalent to the basic BGMW method described above. Algorithm 1 takes as input the set of precomputed values $g_i = g^{b_i}$ for $0 \le i \le t$, as well as h and e.

ALGORITHM 1. Fixed-base windowing method
INPUT: *a set of precomputed values* $\{g^{b_0}, g^{b_1}, \ldots, g^{b_t}\}$, *the exponent* $e = \sum_{i=0}^{t} e_i b_i$, *and the positive integer* h
OUTPUT: g^e
1. $A \leftarrow 1$, $B \leftarrow 1$
2. For j from $(h-1)$ down to 1 do
   2.1 For each i for which $e_i = j$ do $B \leftarrow B \cdot g^{b_i}$
   2.2 $A \leftarrow A \cdot B$
3. Return A

This algorithm requires at most $t + h - 2$ multiplications, and there is storage required for $t + 1$ precomputed group elements. The most common version of this method is the case where the exponent e is represented in radix b, where b is a power of 2, i.e., $b = 2^w$. The parameter w is also called the window size. The exponent is written as $e = \sum_{i=0}^{t} e_i (2^w)^i$ or $(e_t \ldots e_1 e_0)_{2^w}$ where $t + 1 = \lceil n/w \rceil$ and $b_i = (2^w)^i$ for $0 \le i \le t$, and $h = 2^w$. Then on average there are $(t+1)(2^w - 1)/2^w + 2^w - 3$ multiplications required. Consider the following example [5] for $e = 862$ and $w = 2$, i.e., $b = 4$. Then $b_i = 4^i$, $0 \le i \le 4$, such that the values g^1, g^4, g^{16}, g^{64}, and g^{256} are precomputed. Furthermore, this gives $t = 4$ and $h = 4$. Table 1 displays the values of A and B at the end of each iteration of step 2:

A method to reduce the required memory storage for precomputation even further was proposed by Lim and Lee [5] which is called the *fixed-base comb method*. Here, the binary representation of the exponent e is written in h rows such that there is a matrix EA (exponent array) established. Then v columns of the matrix are processed one at a time. Assume that the exponent is written as $e = (e_n \cdots e_1 e_0)_2$. Then select an integer h (the number of rows of EA) with $1 \le h \le n + 1$ and compute $a = \lceil (n+1)/h \rceil$ (the number of columns of EA). Furthermore, select an integer v (the number of columns of EA that are processed at once) with $1 \le v \le a$, and compute $b = \lceil a/v \rceil$ (the number of processing steps). Let $X = (R_{h-1} \| R_{h-2} \| \cdots \| R_0)$ be a bit-string formed from e by padding e on the left with 0's such that X has bit-length ah and such that each R_i is a bit-string of length a. Form the $h \times a$ array EA where row i of EA is the bit-string R_i. The fixed-base comb method algorithm has two phases. First, there is a precomputation phase that is done only once for a fixed base element g, and then there is the exponentiation phase that is done for each exponentiation. Algorithm 2 [6] describes the precomputation phase.

ALGORITHM 2. Fixed-base comb method—precomputation phase
INPUT: *a group element* g *and parameters* h, v, a, *and* b
OUTPUT: $\{G[j][i] : 1 \le i < 2^h, 0 \le j < v\}$
1. For i from 0 to $(h-1)$ do
   1.1 $g_i \leftarrow g^{2^{ia}}$
2. For i from 1 to $(2^h - 1)$ (where $i = (i_{h-1} \ldots i_0)_2$), do
   2.1 $G[0][i] \leftarrow \prod_{j=0}^{h-1} g_j^{i_j}$
   2.2 For j from 1 to $(v-1)$ do
      2.2.1 $G[j][i] \leftarrow (G[0][i])^{2^{jb}}$
3. Return $G[j][i]$ for $1 \le i < 2^h, 0 \le j < v$

Now let $I_{j,k}, 0 \le k < b, 0 \le j < v$, be the integer whose binary representation is column $(jb + k)$ of EA, where column 0 is on the right and the least significant bit of a column is at the top. Algorithm 3 displays the fixed-base comb method exponentiation phase.

ALGORITHM 3. Fixed-base comb method—exponentiation phase
INPUT: *a group element* g *and an exponent* e *as well as the precomputed values* $G[i][j]$
OUTPUT: g^e
1. $A \leftarrow 1$
2. For k from $(b-1)$ down to 0 do
   2.1 $A \leftarrow A \cdot A$
   2.2 For j from $(v-1)$ down to 0 do
      2.2.1 $A \leftarrow G[j][I_{j,k}] \cdot A$
3. Return A

The number of multiplications required in the computation phase is at most $a + b - 2$ of which there are at least $b - 1$ squarings. Furthermore, there is space required for $v(2^h - 1)$ precomputed group elements. Note that the required computational complexity depends on h and v, i.e., on the available memory capacity for storing precomputed elements. In practice, values $h = 4$ or 8 and $v = 1$ or 2 offer a good trade-off between running time and memory requirements. Again, assume $e = 862 = (1101011110)_2$, i.e., $t = 9$. Choose $h = 3$ and thus $a = 4$, and choose $v = 2$ such that $b = 2$. In the first phase algorithm 2 is applied. Table 2 displays the precomputed values. Here, all possible values that might occur in a column of the EA matrix are precomputed. Note that the values

Table 1. Example for the windowing method

j		3	2	1
B	1	$g^4 g^{256} = g^{260}$	$g^{260} g = g^{261}$	$g^{261} g^{16} g^{64} = g^{341}$
A	1	g^{260}	$g^{260} g^{261} = g^{521}$	$g^{521} g^{341} = g^{862}$

Table 2. Example for fixed-base comb method precomputation

i	1	2	3	4	5	6	7
$G[0][i]$	g_0	g_1	g_1g_0	g_2	g_2g_0	g_2g_1	$g_2g_1g_0$
$G[1][i]$	g_0^4	g_1^4	$g_1^4g_0^4$	g_2^4	$g_2^4g_0^4$	$g_2^4g_1^4$	$g_2^4g_1^4g_0^4$

of the second row are the values of the first one to the power of $a = 4$ such that later on two columns can be processed at a time. Recall that $g_i = g^{2^{ia}}$.

Now form the bit-string $X = (001101011110)$ with two padded zeros. Table 3 displays the exponent array EA. Note that the least significant bit of e is displayed in the upper right corner of EA and the most significant bit in the lower left corner.

Finally, Algorithm 3 is performed. Table 4 displays the steps of each iteration. Note that only the powers of the three base values g_i are displayed.

The last row of the table corresponds to $g^e = g_0^{l_0} g_1^{l_1} g_2^{l_2} = g^{14} g^{16 \cdot 5} g^{256 \cdot 3} = g^{862}$. The fixed-base comb method is often used for implementations as it promises the shortest running times at given memory constraints for the precomputed values. A compact description of the algorithm can be found in [4].

Further examples and explanations can be found in [3, 5]. An improvement of the fixed-base windowing method, which is called *fixed-base*

Euclidean method, was proposed by de Rooij [2]. However, in most situations the fixed-base comb method is more efficient.

André Weimerskirch

References

[1] Brickell, E.F., D.M. Gordon, K.S. McCurley, and D.B. Wilson (1992). "Fast exponentiation with precomputations." *Advances in Cryptology—EUROCRYPT'92*, Lecture Notes in Computer Science, vol. 658, ed. R.A. Rueppel. Springer-Verlag, Berlin.

[2] de Rooij, P. (1994). "Efficient exponentiation using precomputation and vector addition chains." *Advances in Cryptology—EUROCRYPT'94*, Lecture Notes in Computer Science, vol. 950, ed. A. De Santis. Springer-Verlag, Berlin.

[3] Gordon, D.M. (1998). "A survey of fast exponentiation methods." *J. Algorithms*, 27, 129–146.

[4] Hankerson, D., J.L. Hernandez, and A. Menezes (2000). "Software implementation of elliptic curve cryptography over binary fields." *Proceedings of Cryptographic Hardware and Embedded Systems, CHES 2000*, Lecture Notes in Computer Science, vol. 1965, eds. Ç.K. Koç and C. Paar. Springer-Verlag, Berlin.

[5] Lim, C. and P. Lee (1994). "More flexible exponentiation with precomputation." *Advances in Cryptology—CRYPTO'94*, Lecture Notes in Computer Science, vol. 839, ed. Y.G. Desmedt. Springer-Verlag, Berlin.

[6] Menezes, A.J., P.C. van Oorschot, and S.A. Vanstone (1996). *Handbook of Applied Cryptography*. CRC Press. Boca Raton, FL.

Table 3. Example for exponent array EA

$$a = 4$$

$I_{1,1}$	$I_{1,0}$	$I_{0,1}$	$I_{0,0}$
$e_3 = 1$	$e_2 = 1$	$e_1 = 1$	$e_0 = 0$
$e_7 = 0$	$e_6 = 1$	$e_5 = 0$	$e_4 = 1$
0	0	$e_9 = 1$	$e_8 = 1$

$h = 3$

$$v = 2$$
$$b = \lceil a/v \rceil = 2$$

FIXED-EXPONENT EXPONENTIATION

There are several situations where an exponentiation g^e of an arbitrary element $g \in G$, with G being some group, by a fixed exponent e needs to be performed. For instance, such cases occur in RSA public key encryption and decryption as well as in ElGamal decryption. Fixed-exponent exponentiation algorithms aim to decrease the number of multiplications compared to general exponentiation algorithms such as the binary exponentiation algorithm. They are based on the fact that certain precomputations can be performed once for a fixed exponent. The problem of finding the smallest number of multiplications to compute g^e is equivalent to finding the shortest *addition chain* of e. An addition chain is a sequence of numbers such that each number is the sum of two

Table 4. Example for fixed-base comb method exponentiation

$$A = g_0^{l_0} g_1^{l_1} g_2^{l_2}$$

k	j	l_0	l_1	l_2
1	–	0	0	0
1	1	4	0	0
1	0	5	0	1
0	–	10	0	2
0	1	14	4	2
0	0	14	5	3

previous ones and such that the sequence starts with 1. For instance, an addition chain of 19 is $V = (1, 2, 4, 5, 9, 10, 19)$.

DEFINITION 1. *An addition chain V of length s for a positive integer e is a sequence $u_0, u_1, \ldots, u_s$ of positive integers, and an associated sequence $w_1, \ldots, w_s$ of pairs $w_i = (i_1, i_2)$, $0 \le i_1, i_2 < i$, having the following properties: (i) $u_0 = 1$ and $u_l = e$; and (ii) for each $u_i, 1 \le i \le s, u_i = u_{i_1} + u_{i_2}$.*

Since we are considering fixed exponents, we do not take into account the time for precomputations. However, determining the shortest addition chain for e is believed to be computationally infeasible in most cases for chains of relevent length. Thus in practice there are heuristics used to obtain nearly optimal addition chains. Knuth [4] describes several such heuristics. It is wise to implement multiple heuristics in order to compare the results, and finally to choose the shortest addition chain. Such precomputation can be computationally very demanding, though. After an addition chain for an exponent e had been obtained, exponentiation can be performed as described in algorithm 1.

ALGORITHM 1. Addition chain exponentiation
INPUT: *a group element g, an addition chain $V = (u_0, u_1, \ldots, u_s)$ of length s for a positive integer e, and the associated sequence $(w_1, \ldots, w_s)$, where $w_i = (i_1, i_2)$*
OUTPUT: g^e
1. $g_0 \leftarrow g$
2. For i from 1 to s do
   2.1 $g_i \leftarrow g_{i_1} \cdot g_{i_2}$
3. Return (g_s)

Algorithm 1 computes g^e for a precomputed addition chain for e of length s with s multiplications. For instance, for above example of $e = 19$ it is $V = (1, 2, 4, 5, 9, 10, 19)$. Algorithm 1 then works as follows:

i	0	1	2	3	4	5	6
w_i	–	$(0,0)$	$(1,1)$	$(0,2)$	$(2,3)$	$(3,3)$	$(4,5)$
g_i	g	g^2	g^4	g^5	g^9	g^{10}	g^{19}

In some cases *addition–subtraction chains* might be used to shorten the length of a chain. In such cases a number of the chain is the sum or subtraction of two previous elements in the chain. For instance, the shortest addition chain for 31 is $V = (1, 2, 3, 5, 10, 11, 21, 31)$ whereas there exists a shorter addition–subtraction chain $C' = (1, 2, 4, 8, 16, 32, 31)$ [3]. However, addition–subtraction chains only make sense in cases where an inversion in the underlying group is computationally cheap. Thus, addition–subtraction chains are not used for exponentiation in an RSA modulus but might be applied to elliptic curve operations.

There are two generalized versions of addition chains. An *addition sequence* is an addition chain $V = (u_0, \ldots, u_s)$ such that it contains a specified set of values $r_1, \ldots, r_t$. They are used when an element g needs to be raised to multiple powers $r_i, 1 \le i \le t$. Especially when the exponents $r_1, r_2, \ldots, r_t$ are far apart, an addition sequence might be faster in such a case. Finding the shortest addition sequence is known to be NP-complete [2] and thus heuristics are used to find short sequences.

In cases of simultaneous exponentiations such as in the digital signature standard (DSS), a generalized addition chain called *vector-addition chain* can be applied. These are used to compute $g_0^{e_0} g_1^{e_1} \cdots g_{k-1}^{e_{k-1}}$ where the g_is are arbitrary elements in G and the e_is are fixed positive integers. In a vector addition chain, each vector is the sum of two previous ones. For instance, a vector-addition chain C of the vector $[15, 5, 12]$ is $([1, 0, 0], [0, 1, 0], [0, 0, 1], [1, 0, 1], [2, 0, 2], [2, 1, 2], [3, 1, 2], [5, 2, 4], [6, 2, 5], [12, 4, 10], [15, 5, 12])$.

DEFINITION 2. *Let s and k be positive integers and let v_i denote a k-dimensional vector of non-negative integers. An ordered set $V = \{v_i : -k + 1 \le i \le s\}$ is called a* vector-addition chain *of length s and dimension k if V satisfies the following: (i) Each $v_i, -k + 1 \le i \le 0$, has a 0 in each coordinate position, except for coordinate position $i + k - 1$, which is a 1. (Coordinate positions are labeled 0 through $k - 1$.). (ii) For each $v_i, 1 \le i \le s$, there exists an associated pair of integers $w_i = (i_1, i_2)$ such that $-k + 1 \le i_1, i_2 < i$ and $v_i = v_{i_1} + v_{i_2}$ ($i_1 = i_2$ is allowed).*

Again, the time for precomputations is irrelevant. There is a 1-1 correspondence between vector-addition chains and addition sequences [6]. Hence determining the shortest vector-addition chain is NP-complete and thus heuristics are used to obtain short vector-addition chains [1]. Having a vector-addition chain, exponentiation can be performed as shown in algorithm 2. The algorithm needs s multiplications for a vector-addition chain of length s to compute $g_0^{e_0} g_1^{e_1} \cdots g_{k-1}^{e_{k-1}}$ for arbitrary base and fixed exponents.

ALGORITHM 2. Vector-addition chain exponentiation

INPUT: *group elements* $g_0, g_1, \ldots, g_{k-1}$ *and a vector-addition chain* V *of length* s *and dimension* k *with associated sequence* $w_1, \ldots, w_s$, *where* $w_i = (i_1, i_2)$.

OUTPUT: $g_0^{e_0} g_1^{e_1} \ldots g_{k-1}^{e_{k-1}}$ *where* $v_s = (e_0, \ldots, e_{k-1})$.

1. For i from $(-k+1)$ to 0 do
   1.1 $a_i \leftarrow g_{i+k-1}$
2. For i from 1 to s do
   2.1 $a_i \leftarrow a_{i_1} \cdot a_{i_2}$
3. Return (a_s)

An overview of addition chains was given by Knuth [4]. Further examples of fixed-exponent exponentiation can be found in [3, 5]. A lower bound for the shortest length of addition chains was proven by Schönhage [7], an upper bound is obtained by constructing an addition chain of e from its binary representation, i.e., by the binary exponentiation algorithm. Yao proved bounds for addition sequences [8].

André Weimerskirch

References

[1] Bos, J. and M. Coster (1990). "Addition chain heuristic." *Advances in Cryptology—CRYPTO'89*, Lecture Notes in Computer Science, vol. 435, ed. G. Brassard. Springer-Verlag, Berlin.

[2] Downey, P., B. Leong, and R. Sethi (1981). "Computing sequences with addition chains." *SIAM Journal on Computing*, 10, 638–646.

[3] Gordon, D.M. (1998). "A survey of fast exponentiation methods." *Journal of Algorithms*, 27, 129–146.

[4] Knuth, D.E. (1997). *The Art of Computer Programming, Volume 2: Seminumerical Algorithms* (3rd ed.). Addison-Wesley, Reading, MA.

[5] Menezes, A.J., P.C. van Oorschot, and S.A. Vanstone (1996). *Handbook of Applied Cryptography*. CRC Press, Boca Raton, FL.

[6] Olivos, J. (1981). "On vectorial addition chains." *Journal of Algorithms*, 2, 13–21.

[7] Schönhage, A. (1975). "A lower bound for the length of addition chains." *Theoretical Computer Science*, 1, 1–12.

[8] Yao, A.C. (1976). "On the evaluation of powers." *SIAM Journal of Computing*, 5, 100–103.

FORGERY

The term forgery usually describes a message related attack against a cryptographic digital signature scheme. That is an attack trying to fabricate a digital signature for a message without having access to the respective signer's private signing key. The security requirement of *unforgeability* of digital signatures is also called nonrepudiation.

A generally accepted formal framework for ordinary digital signatures has been given by Goldwasser et al. [2]. According to their definition a digital signature scheme consists of three algorithms: one for generating key pairs, one for signing messages, and one for verifying signatures. The key generator produces a key pair (see public key cryptography) of a signing key (private key) and a verifying key (public key). The signing algorithm takes as input a signing key and a message, and returns a signature. The verifying algorithm takes as input a verifying key, a message, and a signature, and returns a Boolean value. A signature is usually called *valid* for a message with respect to a verifying key if the verifying algorithm returns TRUE on input this verifying key, message, and signature.

The GMR definition identifies four types of forgery against digital signature schemes. They are in order of decreasing strength: (i) to figure out the signing key (*total break*), (ii) an equivalent algorithm that also produces signatures valid with respect to the victim's verifying key (*universal forgery*), (iii) to find a signature for a new message selected by the attacker (selective forgery), or (iv) to find a signature for any one new message (existential forgery).

They further define *passive attacks* and *active attacks* (see cryptanalysis) for cryptographic signature schemes. An attack is called passive if it is restricted to the access of the verifying key and a number of signed messages (*known message attack*). An attack is called active if it also accesses the signer to ask for signatures on messages chosen by the attacker (*chosen message attack*). The attack is successful if the attacker can come up with a signature for a new message, i.e., one for which the signer has not provided a signature before. A stronger active attack is where each message may be chosen by taking into account the signer's responses to all previously chosen messages (*adaptive chosen message attack*). Orthogonal to the classification of active attacks into whether they are adaptive or not is the following classification based on how the attacker may interleave his queries to the signer [5]. In a *sequential attack*, the attacker may ask the signer to sign messages one by one. In a *concurrent attack*, the attacker may ask the signer to sign more than one message at the same time, while the attacker may interleave his queries arbitrarily.

Security requirements are then defined as security against a certain type of forgery under a certain type of attack. For example, the GMR signature scheme has been proven to be secure against existential forgery under an adaptive chosen message attack.

There are several different types of cryptographic signatures schemes to which other or additional security requirements apply. The better known types of cryptographic signature schemes are in alphabetical order (i) blind signatures, (ii) designated confirmer signatures, (iii) fail-stop signatures, (iv) group signatures, (v) threshold signatures, (vi) undeniable signatures, (vii) *transitive signatures* [3], and (viii) *monotone signatures* [4]. All of them impose security requirements alternatively or in addition to the security requirements of ordinary cryptographic signature schemes described above. Most of the alternative/additional security requirements are defined by means of alternative/additional types of forgery.

In a blind signature scheme, the verifier does not tell the signer which message to sign, but instead "blinds" the intended message and requests a signature for the blinded message from the signer. Then the signer provides information to the verifier, from which the verifier can efficiently derive a valid signature for the intended message. The GMR framework does not apply to blind signatures because in a blind signature scheme the notion of a successful attack is undefined. If an attacker comes up with a signature for a message, it makes no sense to ask whether that message has been signed by the signer before or not, simply because the signer has no idea which messages he has signed before. Instead, two other types of active attacks have been defined for blind signature schemes: *one-more forgery* [6] and *forgery of restrictiveness* [1, 5].

A one-more-forgery [6] is an attack that for some polynomially bounded integer n comes up with valid signatures for $n + 1$ pairwise different messages after the signer has provided signatures only for n messages.

A forgery of restrictiveness is an attack that comes up with a signature for a message that does not observe a predefined internal structure.

Gerrit Bleumer

References

[1] Brands, Stefan (1994). "Untraceable off-line cash in wallet with observers." *Advances in Cryptology—CRYPTO'93*, Lecture Notes in Computer Science, vol. 773, ed. D.R. Stinson. Springer-Verlag, Berlin, 302–318.

[2] Goldwasser, Shafi, Silvio Micali, and Ronald L. Rivest (1988). "A digital signature scheme secure against adaptive chosen-message attacks." *SIAM Journal on Computing*, 17 (2), 281–308.

[3] Micali, Silvio and Ronald Rivest (2002). "Transitive signature schemes." *Topics in Cryptology—CT-RSA 2002*, Lecture Notes in Computer Science, vol. 2271, ed. B. Preneel. Springer-Verlag, Berlin, 236–243.

[4] Naccache, David, David Pointcheval, and Christophe Tymen (2001). "Monotone signatures." *Financial Cryptography 2001*, Lecture Notes in Computer Science, vol. 2339, ed. P.F. Syverson. Springer-Verlag, Berlin, 305–318.

[5] Pfitzmann, Birgit and Ahmad-Reza Sadeghi (1999). "Coin-based anonymous fingerprinting." *Advances in Cryptology—EUROCRYPT'99*, Lecture Notes in Computer Science, vol. 1592, ed. J. Stern. Springer-Verlag, Berlin, 150–164.

[6] Pointcheval, David (1998). "Strengthened security for blind signatures." *Advances in Cryptology—EUROCRYPT'98*, Lecture Notes in Computer Science, vol. 1403, ed. K. Nyberg. Springer-Verlag, Berlin, 391–405.

G

GAP

A gap of length k in a binary <u>sequence</u> is a set of k consecutive 0s flanked by 1s See also <u>run</u> or [1].

Tor Helleseth

Reference

[1] Golomb, S.W. (1982). *Shift Register Sequences.* Aegean Park Press, Laguna Hills, CA.

GENERALIZED MERSENNE PRIME

A <u>prime number</u> of the form

$$p = f(2^m),$$

where $f(t)$ is a low-degree polynomial with small integer coefficients. Generalized Mersenne primes are useful in <u>public-key cryptography</u> because reduction modulo these primes can be very fast using a generalization of the technique used for <u>Mersenne primes</u>. In particular, the integer division by p is replaced by a small number of additions and subtractions and some bit shifts.

In practice, m is taken to be a multiple of the word size of the machine in order to eliminate the need for shifting bits within words. The precise rule for modular reduction (see <u>modular arithmetic</u>) depends on $f(t)$.

The simplest example is

$$p = 2^{192} - 2^{64} - 1.$$

In this case, $f(t) = t^3 - t - 1$, and one has

$$\sum_{i=0}^{5} c_i t^i \equiv b_2 t^2 + b_1 t + b_0 \pmod{f(t)},$$

where

$$b_0 = a_0 + a_3 + a_5$$
$$b_1 = a_1 + a_3 + a_4 + a_5$$
$$b_2 = a_2 + a_4 + a_5.$$

This yields the following reduction algorithm for $p = f(2^{64})$. A positive integer modulo p^2 is represented as the concatenation of six 64-bit words

$$(c_5 \quad c_4 \quad c_3 \quad c_2 \quad c_1 \quad c_0),$$

which represents

$$c = \sum_{i=0}^{5} c_i \cdot 2^{64i}.$$

Then

$$c \equiv s_0 + s_1 + s_2 + s_3 \pmod{p},$$

where the s_is are the integers represented by the following concatentations of the six words:

$$s_0 = (c_2 \quad c_1 \quad c_0)$$
$$s_1 = (0 \quad c_3 \quad c_3)$$
$$s_2 = (c_4 \quad c_4 \quad 0)$$
$$s_3 = (c_5 \quad c_5 \quad c_5).$$

In the general case, one derives the reduction formulae by writing down the reductions (mod $f(t)$) of t^j for $d \leq j < 2d$, where d is the degree of f. One then rearranges the terms to minimize the numbers of additions and subtractions. Details can be found in [2].

To find a generalized Mersenne prime of a given bit size, one lists the small-coefficient integer polynomials $f(t)$ of appropriate degrees and chooses the one for which $f(2^m)$ is prime and whose reduction rules require the smallest number of additions and subtractions.

The U.S. National Institute for Standards and Technology (NIST) has standardized four generalized Mersenne primes, including the above example, in its <u>Digital Signature Standard</u> [1].

A useful generalization of generalized Mersenne primes is generalized Mersenne numbers. A near-prime (i.e., a small multiple of a large prime) of the generalized Mersenne form can be used in <u>elliptic curve cryptography</u> almost as easily as a prime modulus. If m is a generalized Mersenne number divisible by a large prime p, then one implements the cryptography over an elliptic curve over the <u>field</u> $\mathbb{F}_p$. The arithmetic is carried out modulo m (using the appropriate reduction rules) and is additionally reduced modulo p at the end of the calculation. This can sometimes be advantageous if a generalized Mersenne near-prime exists requiring significantly fewer additions and subtractions than any available generalized Mersenne prime.

Jerome Solinas

References

[1] Federal Information Processing Standard (FIPS) 186-2 (February 2000). "Digital Signature Standard." http://csrc.nist.gov/encryption/tkdigsigs.html

[2] Solinas, Jerome A. (1999). "Generalized Mersenne Numbers." Centre for Applied Cryptographic Research (CACR) Tech. Report 99-39. http://www.cacr.math.uwaterloo.ca/

GENERATOR

An element g of a group is said to *generate* that group if the set of elements

$$g, g^2, g^3, \ldots$$

(where the group law is multiplication) traverses all of the elements in the group. Such an element is a *generator*. In other words, g is a generator of a group if and only if every element y in the group can be expressed as

$$y = g^x$$

for some x. Every cyclic group has at least one generator, and a group that is generated by a single element is cyclic by definition.

When the group is a multiplicative group modulo an integer n, a generator is also called a *primitive root* of n.

A generator (sometimes also called the *base*) is one of the parameters in several cryptosystems. See Diffie–Hellman key agreement, Digital Signature Standard.

Burt Kaliski

GMR SIGNATURE

The GMR Signature Scheme was invented by Goldwasser et al. [2, 3]. In their landmark publication [2], they presented a formal framework of security definitions of cryptographic signature schemes (see digital signature schemes and public key cryptography) and they proposed the GMR Signature Scheme, which, under the assumption that integer factoring is hard, is provably secure by their strongest security definition, i.e., it resists existential forgery under an adaptive chosen message attack.

The value of the GMR Signature Scheme is not in its practical use, but in showing the existence of a provably secure signature scheme. Other more efficient signature schemes such as

RSA digital signature scheme, DSA or ECDSA (see Digital Signature Standard) are memoryless, i.e., in order to produce a signature, one does not need to memorize any previously produced signatures (in whole or in part). Also, the computational effort of signing and verifying and the length of signatures are independent of the number of signatures a signer has produced before. However, the GMR Signature Scheme is not memoryless, i.e., each signature a signer produces depends on every other signature the signer has produced before. Even worse, the time to produce and to verify a signature and the length of a signature increase (logarithmically) with the number of signatures that a signer has produced before. At the cost of some memory, the average signing performance can be made independent of the number of signatures that a signer has produced before. In essence, the overall performance of the GMR Signature Scheme (in terms of time and memory) decreases steadily over the lifetime of each signing key pair.

In a nutshell, the GMR Signature Scheme works as follows: a signer who anticipates to produce 2^b signatures constructs a key pair for security parameter k as follows: The public verifying key consists of four components:

1. The maximum number $B = 2^b$ ($b \in \mathbb{N}_0$) of messages to be signed.
2. Two randomly chosen Blum integers $n_1 = p_1 q_1$ and $n_2 = p_2 q_2$, i.e., each the product of two prime numbers both congruent to 3 modulo 4, but not congruent to each other modulo 8, such that n_1 and n_2 are each of length k-bit.
3. The two infinite families of pairwise claw-free trap-door permutations (also see trap-door one-way functions) $f_{i,n}(x) = \pm 4^{\text{rev}(i)} x^{2^{\overline{\text{len}(i)}}} \bmod n$ for $n = n_1$ and $n = n_2$, where the function $\text{rev}(\cdot)$ takes a bit string $i \in \{0, 1\}^+$ and returns the integer represented by the bits of i in reversed order, and the function $\text{len}(\cdot)$ takes a bit string $i \in \{0, 1\}^+$ and returns its number of bits. The sign plus or minus is selected such that the image of $f_{i,n}$ is positive and smaller than $n/2$.
4. A randomly chosen integer r in the domain of $f_{i,n_1}(\cdot)$.

The private signing key consists of the primes p_1, q_1, p_2, q_2, which allow to efficiently compute the inverse permutations $f_{i,n_1}^{-1}(y)$ and $f_{i,n_2}^{-1}(y)$.

Let us first consider the simple case of signing only one message m, i.e., $B = 1$ and $b = 0$. The signer chooses a random element r_0 from the domain of $f_{i,n_1}(\cdot)$, and computes the signature (t, r_0, s) for m such that:

$$t = f_{r_0,n_1}^{-1}(r) \quad \text{and} \quad s = f_{m,n_2}^{-1}(r_0). \quad (1)$$

The signature (t, r_0, s) is displayed in Figure 1 as a

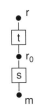

Fig. 1. Signing one message

chain connecting the signer's root element r with the actual message m. A verifier checks the signature by computing from the message m up to the signer's root element r as follows: First check that $f_{m,n_2}(s) = r_0$ and then check that $f_{r_0,n_1}(t) = r$.

If there are $B = 2^b > 1$ messages to be signed, the signer expands the root element r from the case $B = 1$ above into a binary authentication tree (see Merkle [1]) with B leaves $r_0, r_1, \ldots, r_{B-1}$. All nodes of the authentication tree except for the root r are chosen uniformly at random from the domain of $f_{i,n_1}(\cdot)$ analogously to r_0 in Equation (1) above. Each node R is unforgeably tied to both its children R_0 and R_1 by computing a tag $f^{-1}_{R_0\|R_1,n_1}(R)$ analogously to how the tag t was computed in Equation (1) above. Finally, the message m_j ($0 \le j \le B - 1$) is unforgeably tied to r_j, which hangs off of the jth leaf of the authentication tree by computing a tag $s_j = f^{-1}_{m_j,n_2}(r_j)$ analogously to how the tag s was computed in Equation (1) above. The signature for message m_j then consists of (i) the sequence of b nodes from the root r down to item r_j, (ii) the b tags associated to these nodes (analogously to t above), and (iii) the tag s_j (analogously to s above). The authentication tree and associated tags for the case $B = 8, b = 3$ is depicted in Figure 2 as a binary tree connecting the signer's root element r with the B messages $m_0, m_1, \ldots, m_{B-1}$. A verifier can check the signature in reverse order beginning from the message m_j and working back up to the root element r of the signer.

Obviously, the authentication tree need not be precomputed in the first place, but can be built step-by-step as the need arises. For example, signing the first message requires one to compute the complete path from the root r down to the item r_0, but signing the second message only requires one to build one more leaf of the authentication tree and the next item r_1, while all previously built nodes of the authentication tree can be reused. So, on average, each signature requires one to build two new nodes of the authentication tree and one additional item. A complete security analysis is given in [3].

Gerrit Bleumer

References

[1] Merkle, Ralph (1982). "Secrecy authentication, and public-key systems." *PhD Dissertation*, Electrical Engineering Department, ISL SEL 79-017k, Stanford University, Stanford, CA.

[2] Goldwasser, Shafi, Silvio Micali, and Ronald L. Rivest (1984). "A "Paradoxical" solution to the signature problem." *25th Symposium on Foundations of Computer Science (FOCS) 1984*. IEEE Computer Society, 441–448.

[3] Goldwasser, Shafi, Silvio Micali, and Ronald L. Rivest (1988). "A digital signature scheme secure against adaptive chosen-message attacks." *SIAM Journal on Computing*, 17 (2), 281–308.

GOLDWASSER–MICALI ENCRYPTION SCHEME

The Goldwasser–Micali encryption scheme (see public key cryptography) is the first encryption scheme that achieved semantic security against a passive adversary under the assumption that solving the quadratic residuosity problem is hard. The scheme encrypts 1 bit of information, and the resulting ciphertext is typically 1024 bits long.

In the Goldwasser–Micali encryption scheme, a public key is a number n, that is a product of two primes numbers, say p and q. Let Y be a quadratic nonresidue modulo n (see quadratic residue and modular arithmetic), whose Jacobi Symbol is 1. The decryption key is formed by the prime factors of n.

The Goldwasser–Micali encryption scheme encrypts a bit b as follows. One picks an integer r ($1 < r < n - 1$) and outputs $c = Y^b r^2 \bmod n$ as ciphertext. That is, c is quadratic residue if and only if $b = 0$. Therefore a person knowing the prime factors of n can compute the quadratic residuosity of the ciphertext c, thus obtaining the value of b.

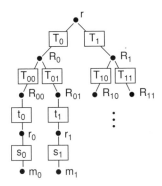

Fig. 2. Signing more than one message

If the quadratic residuosity problem is hard, then guessing the message from the ciphertext is equivalently hard.

Kazue Sako

Reference

[1] Goldwasser, S. and S. Micali (1984). "Probabilistic encryption." *Journal of Computer and System Sciences*, 28, 270–299.

GOLOMB'S RANDOMNESS POSTULATES

No finite sequence constructed by a linear feedback shift register is a truly random sequence. Golomb [1] introduced the notion of a *pseudorandom* sequence for a periodic binary sequence that satisfies three randomness postulates. These postulates reflect properties one would expect to find in a random sequence.

A run in a binary sequence is a set of consecutive 0s or 1s. A run of 0s is denoted a gap and a run of 1s is denoted a *block*. A gap of length k is a set of k consecutive 0s flanked by 1s. A block of length k is a set of k consecutive 1s flanked by 0s. A run of length k is is a gap or length k or a block of length k. In this terminology, the three randomness postulates of a periodic sequence are as follows:

R-1: In a period of the sequence, the number of 1s and the number of 0s differ by at most 1.

R-2: In every period, half the runs have length 1, one-fourth have length 2, one-eighth have length 3, etc., as long as the number of runs so indicated exceeds 1. Moreover, for each of these lengths, there are equally many gaps and blocks.

R-3: The out-of-phase auto-correlation of the sequence always has the same value.

The R-2 postulate implies the R-1 postulate, but otherwise the postulates are independent, which follows from the observation that there exist sequences that satisfy some but not all of the postulates. Some examples are:

Example 1: The m-sequence 00001001011001111 10001101110101 of period 31 is an R-1, R-2 and R-3 sequence.

Example 2: The sequence 00100011101 of period 11 is an R-1 sequence. It is also an R-3 sequence since the out-of-phase auto-correlation is equal to −1. However, the sequence is not an R-2 sequence since there are two blocks of length 1 but only one gap of length 1.

Example 3: The sequence 0000010001101 of period 13 is an R-3 sequence with a constant out-of-phase auto-correlation being 1, but the sequence violates the R-1 and R-2 conditions.

Example 4: The sequence 01001101 of period 8 is an R-1 and R-2 sequence but not an R-3 sequence.

The three randomness postulates are inspired by the properties obeyed by maximum-length linear sequences (m-sequences). Also they can be interpreted as properties of flipping a perfect coin. R-1 says that heads and tails occur about equally often; R-2 says that after a run of n heads (tails) there is a probability $1/2$ that the run will end with the next coin-flip. R-3 is the notion of independent trials. Knowing the outcome of a previous coin-flip gives no information of the current coin-flip.

Any sequence obeying both R-1 and R-3 can be shown to have a value of the out-of-phase auto-correlation of −1, and therefore the period must be odd. Sequences which obey the randomness postulates of Golomb are sometimes also called pseudonoise sequences.

Tor Helleseth

Reference

[1] Golomb, S.W. (1982). *Shift Register Sequences*. Aegean Park Press, Laguna Hills, CA.

GOST

GOST is an encryption algorithm adopted as a standard by the former Soviet Union in 1989 [5]. The specifications, translated from Russian in 1993, describe a DES-like 64-bits block cipher (see Data Encryption Standard) and specify four modes of operation.

The GOST encryption algorithm is a very simple 32-round Feistel cipher. It encrypts data in blocks of 64 bits and uses a 256-bit secret key. The 32-bit F-function used in the Feistel construction consists of three transformations. First, a 32-bit subkey is mixed with the data using an addition modulo 2^{32}. The result is then split into 4-bit segments, fed in parallel to eight 4×4-bit S-boxes. Finally, the output values are merged again and rotated over 11 bits. The *key schedule* (see block cipher) of GOST is also particularly simple: the 256-bit secret key is divided into eight 32-bit words and directly used as subkeys in rounds 1–8, 9–16, and 17–24. The same eight subkeys are reused one more time in rounds 25–32, but in reverse order Figure 1.

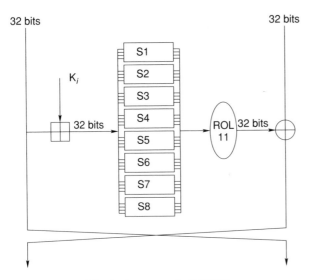

32 bits

32 bits

K_i

32 bits

ROL 11

32 bits

S1
S2
S3
S4
S5
S6
S7
S8

Fig. 1. One round of GOST

A remarkable property of the GOST standard is that the eight S-boxes are left unspecified. The content of these lookup tables is considered to be a secondary long-term secret key, common to a network of computers, and selected by some central authority. The set of S-boxes can be strong or weak, depending on the authority's intention. The S-boxes can even be hidden in an encryption chip, thus keeping them secret to the users of the device. As explained in a short note by Saarinen [3], recovering the secret S-boxes would not be very hard, however. If a user is allowed to select the 256-bit key of the cipher, he can easily mount a "chosen key attack" and efficiently derive the secret contents of the lookup tables.

The best attacks on GOST exploit the simplicity of its key schedule. The first attack in open literature is a related key attack by Kelsey et al. [2]. Biryukov and Wagner [1] have shown that the cipher is vulnerable to slide attacks. Their cryptanalysis breaks 20 rounds out of 32, but also reveals weak key classes for the full cipher. Finally, Seki and Kaneko [4] have applied differential cryptanalysis to reduced-round versions of GOST. Combined with a related-key approach, the attack breaks 21 rounds.

Christophe De Cannière

References

[1] Biryukov, A. and D. Wagner (2000). "Advanced slide attacks." *Advances in Cryptology—EUROCRYPT 2000*, Lecture Notes in Computer Science, vol. 1807, ed. B. Preneel. Springer-Verlag, Berlin, 589–606.

[2] Kelsey, J., B. Schneier, and D. Wagner (1996). "Key-schedule cryptanalysis of IDEA, G-DES, GOST, SAFER, and Triple-DES." *Advances in Cryptology—CRYPTO'96*, Lecture Notes in Computer Science, vol. 1109, ed. N. Koblitz. Springer-Verlag, Berlin, 237–251.

[3] Saarinen, M.-J. (1998). *A Chosen Key Attack Against the Secret S-Boxes of GOST*, unpublished.

[4] Seki, H. and T. Kaneko (2001). "Differential cryptanalysis of reduced rounds of GOST." *Selected Areas in Cryptography, SAC 2000*, Lecture Notes in Computer Science, vol. 2012, eds. D.R. Stinson and S.E. Tavares. Springer-Verlag, Berlin, 315–323.

[5] Zabotin, I.A., G.P. Glazkov, and V.B. Isaeva (1989). "Cryptographic protection for information processing systems: Cryptographic transformation algorithm." Technical Report, Government Standard of the USSR, GOST 28147-89 (translated by A. Malchik, with editorial and typographic assistance of W. Diffie).

GREATEST COMMON DIVISOR

The greatest common divisor (gcd) of a set of positive integers $\{a_1, \ldots, a_k\}$ is the largest integer that divides every element of the set. This is denoted by $\gcd(a_1, \ldots, a_k)$ or sometimes just $(a_1, \ldots, a_k)$. For example, $\gcd(21, 91) = 7$ because 7 divides both 21 and 91, and no integer larger than 7 divides both of these values. If the gcd is 1, then the integers are said to be relatively prime.

An important property of the gcd is that it can always be written as an integer linear combination of the elements of the set. In other words, there exist integers $x_1, \ldots, x_k$ such that $\sum_{i=1}^{k} a_i \cdot x_i = \gcd(a_1, \ldots, a_k)$. In the example above, one such linear combination is given by $x_1 = -4$ and $x_2 = 1$, since $21 \cdot (-4) + 91 \cdot 1 = 7$.

The gcd of a set of integers can be quickly computed using a method due to Euclid, known as the Euclidean algorithm. The integer linear combination that gives the gcd can also be computed quickly, using what is known as the *extended Euclidean algorithm*.

Scott Contini

GROUP

A *group* $G = (S, \circ)$ is defined by a set of elements S and a group operation $\circ$ that satisfy the following *group axioms*:

- **Closure:** For all $x, y \in S$, $x \circ y \in S$.
- **Associativity:** For all $x, y, z \in S$, $(x \circ y) \circ z = x \circ (y \circ z)$.

- **Identity:** There exists an *identity element*, denoted I, such that for all $x \in S$, $x \circ I = I \circ x = x$.
- **Inverse:** For all $x \in S$, there exists an *inverse* y such that $x \circ y = y \circ x = I$.

A group is *commutative* (also called *Abelian*) if the group operation does not depend on the ordering of the elements, i.e., if for all $x, y \in S$, $x \circ y = y \circ x$. A group is *cyclic* if it has a single generator, i.e., an element g such that every element of the group can be obtained by repeated composition with g, starting with the identity element. The *order* of a group G, denoted $\#G$, is the number of elements in G.

Groups commonly employed in cryptography include the following:

- A *multiplicative group modulo a prime*, where S consists of the set of integers (i.e., residue classes) modulo a prime p, excluding 0, and the group operation is multiplication. This group is typically denoted by $\mathbf{Z}_p^*$, where $\mathbf{Z}_p$ denotes the integers modulo p, and * denotes that the group operation is multiplication and 0 is excluded. The order of $\mathbf{Z}_p^*$ is $p - 1$.
  (This group is the same as the multiplicative group of the finite field $\mathbf{F}_p$.)
- A *subgroup* of a multiplicative group modulo a prime, i.e., a subset of elements in $\mathbf{Z}_p^*$ that is closed under multiplication. Typically, the subgroup is selected so that its order is a prime number. For instance, the Digital Signature Algorithm (see Digital Signature Standard) operates in a subgroup of order q of $\mathbf{Z}_p^*$, where p and q are large primes and q divides $p - 1$.
- A subgroup of the multiplicative group of an extension field $\mathbf{F}_{q^d}$.
- A subgroup of an elliptic curve group. If the elliptic curve group has a prime order, then the subgroup is the same as the elliptic curve group.

All these examples are cyclic and commutative; an elliptic curve group itself may be noncyclic but the subgroup of interest itself is typically cyclic.

A group is formally denoted by both the set and the group operation, i.e., $(S, \circ)$, but in cryptography sometimes only the set S is denoted and the group operation is implied. In cryptography, the group operation is typically either denoted by multiplication or addition. In the former case, repeated application of the group operation is denoted by *exponentiation* (e.g., g^a); in the latter, it is denoted by *scalar multiplication* (e.g., aP where P is the group element).

Groups are primarily associated with public-key cryptography, but they also have some applications to symmetric cryptography. For instance, several researchers investigated whether the set of keys in the Data Encryption Standard forms a group [2,3], which would significantly weaken the standard; the set of keys does not.

Braid groups are a notable recent example of a noncommutative group in public-key cryptography [1,4,5].

See also modular arithmetic and prime number.

Burt Kaliski

References

[1] Anshel, Iris, Michael Anshel, and Dorian Goldfeld (1999). "An algebraic method for public-key cryptography." *Mathematical Research Letters*, 6, 287–291.

[2] Campbell, K.W. and M.J. Wiener (1992). "DES is not a group." *Advances in Cryptology—CRYPTO'92*, Lecture Notes in Computer Science, vol. 740, ed. E.F. Brickell. Springer, Berlin, 512–520.

[3] Kaliski Jr., Burton S., Ronald L. Rivest, and Alan T. Sherman (1988). "Is the Data Encryption Standard a group?" *Journal of Cryptology*, 1, 3–36.

[4] Ko, Ki Hyoung, Sang Jin Lee, Jung Hee Cheon, Jae Woo Han, Ju-sung Kang, and Choonsik Park (2000). "New public-key cryptosystem using braid groups." *Advances in Cryptology—CRYPTO 2000*, Lecture Notes in Computer Science, vol. 1880, ed. M. Bellare. Springer, Berlin, 166–183.

[5] Lee, Eonkyung and Je Hong Park (2003). "Cryptanalysis of the public-key encryption based on braid groups." *Advances in Cryptology—EUROCRYPT 2003*, Lecture Notes in Computer Science, vol. 2656, ed. E. Biham. Springer, Berlin, 477–490.

GROUP KEY AGREEMENT

INTRODUCTION

Definition. Group Key Agreement (GKE) or *conference keying* (or *group key distribution*) is an extension of two-party key agreement to groups of $n \geq 2$ parties: it allows a group of parties to share a common *session key* (see key) or *conference key*.

Applications. Many computer applications involve dynamic peer groups. Examples include: teleconferencing, multi-user computations, multicast messaging, pay-per-view, distributed interactive simulations, real-time information systems, replication services and generally, distributed applications. These require a secure communication channel (see also Shannon's model) that links all the parties in a group. Such a channel can be established by using a *symmetric encryption scheme* (see also symmetric cryptosystem) with key, a *group session key*.

Requirements. Given the openness of most networking systems, it is important that key agreement is achieved efficiently and securely. Other requirements include, scalability, freshness of the session keys, forward secrecy, re-keying and of course, reliability.

Efficiency: The efficiency of a GKE protocol is measured in terms of its computational complexity (usually the number of modular exponentiations), its *communication* complexity (the number of communicated messages) and its *rounds* complexity (the number of rounds of communication exchanges). The complexity of a multiparty protocol involving n parties is *scalable* if it is bounded by $\mathcal{O}(\log n)$ (see $\mathcal{O}$-notation). Scalability is particularly important when the number of parties in the group is large.

Security: Security involves both the privacy of the session keys and authentication. Privacy (*key secrecy*) requires that in a *passive attack* it is hard to *distinguish* the session key from a random key. Passive attacks are eavesdropping attacks in which the adversary has access only to traffic communicated in public (prior to key agreement). There is a stronger version of privacy that requires that it is hard to *compute* the session key in a passive attack. *Freshness* of the session keys is required to prevent *known key attacks* (see also related key attack) in which the adversary succeeds in getting hold of session keys by analyzing large amounts of communication traffic encrypted with the same group key, or by noncryptographic means. Session keys must therefore be regularly updated.

Forward secrecy: A compromised session key should only affect its session, and not jeopardize earlier sessions.

Re-keying: Groups are dynamic, with new members joining the group or old members leaving the group. Whenever a new group is formed, a new session key must be agreed: otherwise information regarding previous sessions (or future sessions) would become available to those joining (or leaving) the group.

Reliability: The communication channel must be reliable.

GROUP KEY AGREEMENT PROTOCOLS: Several GKE protocols have been proposed in the literature (see e.g., [5–11,14,16,17,20–23,26,27,29,30]. Most of these are based on the Diffie–Hellman key agreement protocol [13] or variants of it. We shall describe some of these, starting with the simplest ones and point out their strengths and weaknesses. We first consider GKE protocols for small groups which are secure against passive attacks. We will then consider authenticated GKE protocols. Finally we shall describe a GKE protocols for large groups.

Private Group Key Agreement for Small Groups

We consider GKE protocols that are secure against *passive attacks* (eavesdropping). These involve a group $G = \langle g \rangle$ generated by an element g whose order is a k-bit prime q. The group could be a subgroup of the multiplicative group Z_m^* (which consists of all positive integers less than m and relative prime to m; see modular arithmetic), the group of an *elliptic curve* (see elliptic curves), or more generally, any finite group whose operation can be efficiently computed and for which one can establish membership and select random elements efficiently. For convenience we shall call the elements of G, numbers.

The parameters of the group G are selected by a *Trusted Center* (see also Trusted Third Party); k is the security parameter. In this section we shall assume that the number of parties n involved in the key agreement protocol is small. That is, $n \ll q$ ($n = poly(k)$). The protocols we shall describe involve random selections of elements from G. We use the notation $x \in_R X$ to indicate that x is selected at random (independently) from the set X with uniform distribution.

1. A basic centralized GKE protocol [11]. Let $\mathcal{U} = \{U_1, U_2, \ldots, U_n\}$, $n \ll q$, be a group of parties. A designated member of the group called the *chair*, say U_1, selects the session key $sk \in_R G$.

Round 1. The chair exchanges a key $K_i \in G$ with every member $U_i \in \mathcal{U} \backslash U_1$ by using the Diffie–Hellman key agreement protocol.

Round 2. The chair sends to each member $U_i \in \mathcal{U} \backslash U_1$ the "number": $X_i = sk/K_i \in G$.

Key Computation. Each member $U_i \in \mathcal{U} \backslash U_1$ computes the key: $sk = X_{1i} \cdot K_{1i}$.

Security and Efficiency. We clearly have forward secrecy. *Privacy* reduces to the Diffie–Hellman problem [18]. More specifically, *distinguishing* the session key from a random key in a passive attack (*key secrecy*) is as hard as the Decisional Diffie–Hellman problem; *computing* the session key is as hard as the *Computational Diffie–Hellman problem* (see Decisional Diffie–Hellman problem). The complexity of this protocol is unbalanced: whereas the chair has to exchange Diffie–Hellman keys with each one of the other members of the group and then send the X_i's, the other members need only exchange a Diffie–Hellman key with the chair. It is possible [12] to share the burden evenly among the members by arranging the group in a graph tree, with root the chair, and then exchanging Diffie–Hellman keys only along root-to-leaf paths. In this case, all members of the group (including the chair) exchange a Diffie–Hellman key with their adjacent nodes in the tree (their parent and children). Then the per-user complexity is shared evenly, however there is now a time-delay which is proportional to the height of the tree [12]. The per-user complexity of the tree-based protocol is $\mathcal{O}(\log n)$.

2. The Burmester–Desmedt distributed GKE protocol [10, 11, 18].

Let $\mathcal{U} = \{U_1, U_2, \ldots, U_n\}$ be the group of parties. The protocol uses a broadcast channel (this could be replaced by $\binom{n}{2}$ point-to-point channels). There are two rounds. In the first round each member $U_i \in \mathcal{U}$ broadcasts a random number (element) $z_i = g^{r_i} \in G$. In the second round U_i broadcasts the number $X_i = (z_{i+1}/z_{i-1})^{r_i} \in G$. The session key is: $sk = g^{r_1 r_2 + r_2 r_3 + \cdots + r_n r_1} \in G$. More specifically:

Round 1. Each member $U_i \in \mathcal{U}$ selects an $r_i \in_R Z_q$ and broadcasts: $z_i = g^{r_i} \in G$.

Round 2. Each member $U_i \in \mathcal{U}$ broadcasts $X_i = (z_{i+1}/z_{i-1})^{r_i} \in G$, where the indices are taken in a cycle.

Key Computation. Each member $U_i \in \mathcal{U}$ computes the session key:

$$sk_i = (z_{i-1})^{n r_i} \cdot X_i^{n-1} \cdot X_{i+1}^{n-2} \cdots X_{i-2} \in G.$$

Remark. If all members adhere to the protocol then each will compute the same key: $sk = g^{r_1 r_2 + r_2 r_3 + \cdots + r_n r_1}$. Indeed, let $A_{i-1} = (z_{i-1})^{r_i} = g^{r_{i-1} r_i}$, $A_i = (z_{i-1})^{r_i} \cdot X_i = g^{r_i r_{i+1}}$, $A_{i+1} = (z_{i-1})^{r_i} \cdot X_i \cdot X_{i+1} = g^{r_{i+1} r_{i+2}}$, and so on. Then $sk_i = A_{i-1} \cdot A_i \cdot A_{i+1} \cdots A_{i-2} = sk$.

In the special case when there are only two members, we have $X_1 = X_2 = 1$ and $sk = g^{r_1 r_2 + r_2 r_1} = g^{2 r_1 r_2}$. This is essentially the <u>Diffie–Hellman key agreement</u> (in this case there is no need to broadcast X_1, X_2).

Security and Efficiency. We have forward secrecy, as in the previous protocol. Privacy is reduced to the Decisional Diffie–Hellman problem [15]. (Computing the session key, in a passive attack, is as hard as the Computational Diffie–Hellman problem [10] when n is even; this extends to the odd case if one of the parties behaves as two virtual parties.) The complexity per user is $\mathcal{O}(1)$ (constant). This is roughly double that of the Diffie–Hellman key agreement protocol: there are two (broadcast) rounds,

X_1	g	g^{r_1}		
X_2	g^{r_2}	g^{r_1}	$g^{r_1 r_2}$	
X_3	$g^{r_2 r_3}$	$g^{r_1 r_3}$	$g^{r_1 r_2}$	$g^{r_1 r_2 r_3}$
X_4	$g^{r_2 r_3 r_4}$	$g^{r_1 r_3 r_4}$	$g^{r_1 r_2 r_4}$	$g^{r_1 r_2 r_3}$

Fig. 1. The sets X_i for a group with four members

three exponentiations, and a few multiplications. Because of its efficiency, for re-keying, we just re-run the protocol.

Variants. By arranging the connectivity of group $\mathcal{U}$ in different ways to route messages, e.g., in a star, a cycle, or a tree, we get several GKE variants which can be used to meet specific requirements [11, 12].

3. The Ingemarsson–Tang–Wong distributed GKE protocol [7, 8, 14].

For this protocol the group of parties $\mathcal{U}$ is arranged in a cycle or ring: $(U_1, U_2, \ldots, U_n)$. The protocol uses sets of numbers $X_i \subset G$, $i = 1, 2, \ldots, n$ that are defined as follows. Select $r_1, r_2, \ldots, r_n \in_R Z_q$. Define $X_1 = \{x_{1,0} = g, \ x_{1,1} = g^{r_1}\}$, and recursively $X_i = \{x_{i,0} = (x_{i-1,0})^{r_i}, \ \ldots, \ x_{i,i-2} = (x_{i-1,i-2})^{r_i}; \ x_{i,i-1} = x_{i-1,i-1}, x_{i,i} = (x_{i-1,i-1})^{r_i}\}$, for $1 < i \le n-1$. Finally $X_n = \{x_{n,0} = (x_{n-1,0})^{r_n}, \ \ldots, \ x_{n,i-2} = (x_{n-1,n-2})^{r_n}; \ x_{n,n-1} = x_{n-1,n-1}\}$. In Figure 1 we illustrate these sets for the case when $n = 4$.

The protocol has two phases: *up-flow* and *down-flow*. In the up-flow phase, each member U_i sends to the next member U_{i+1} in the ring the set X_i. In the down-flow phase, U_n broadcasts X_n. The session key is $sk = g^{x_1 x_2 \cdots x_n} \in G$. More specifically:

Phase 1

For $i = 1$ to $n-1$

U_i sends to U_{i+1}: $X_i \subset G$, where $r_i \in_R Z_q$ is selected by U_i.

Phase 2

U_n broadcasts: $X_n \subset G$, where $r_n \in_R Z_q$ is selected by U_n.

Key Computation. Each member $U_i \in \mathcal{U}$ computes the key:

$$sk_i = (g^{x_1 x_2 \cdots x_{i-1} x_{i+1} \cdots x_n})^{x_i} = g^{x_1 x_2 \cdots x_n} \in G,$$

where $g^{x_1 x_2 \cdots x_{i-1} x_{i+1} \cdots x_n}$ is obtained from X_n.

Remark. If all members adhere to the protocol then each will compute the same key.

Security and Efficiency. We have forward secrecy, as in the previous protocols. Privacy is guaranteed under the *Group Computational Diffie–Hellman* (GCDH) assumption, which we describe below.

GCDH Assumption. Let $G = \langle g \rangle$ be a group of order a k-bit prime q, $n = poly(k)$, $I_n = \{1, 2, \ldots, n\}$, 2^{I_n} be the powerset of I_n, and $\Gamma \subset 2^{I_n} \setminus I_n$.

Given:

$$GCDH_\Gamma = \{\cup_{J \in \Gamma}(J, g^{\prod_{j \in J} x_j}) \mid (x_1, \ldots, x_n) \in_R (Z_q)^n\},$$

it is hard to find $g^{x_1 x_2 \cdots x_n}$.

The complexity of this protocol is $\mathcal{O}(\log n)$. There is a time-delay for the $n + 1$ steps, and the computation of the sets X_i requires on average $\frac{1}{2}n$ exponentiations.

Authenticated Group Key Agreement

Katz and Yung [15] recently described a compiler $\mathcal{C}$ which will transform any GKE protocol P that is secure against *passive attacks* into a protocol $P' = \mathcal{C}(P)$ that is secure against *active adversaries* (see cryptanalysis) that control all communication in the network. The compiler adds only one communication round to protocol P and authenticates (see <u>authentication</u>) all communication traffic.

The security model used is that of Bresson et al. [8], which is based on the Random Oracle model of Bellare and Rogaway [3, 4] (see also [2]). In this model, each member $U \in \mathcal{U}$ is allowed to execute the protocol P many times with several groups of partners (and to interleave the rounds of the executions). Each such execution of P is called an *instance*. The ith instance executed by U is denoted by Π_U^i.

The following assumptions regarding the inputs P to the compiler $\mathcal{C}$ are made: (i) P is a GKE protocol that is secure against passive attacks (key secrecy); (ii) each message sent by instance Π_U^i includes the *identity* of the sender U and a *sequence number* j which starts at 0 and is incremented by 1 each time Π_U^i sends a new message m (i.e., Π_U^i sends $U||j||m$, the concatenation of U, j and m); and (iii) all messages are broadcasted to the entire group $\mathcal{U}$ taking part in the execution of P.

It is easy to see that any GKE protocol $\tilde{P}$ can readily be converted to a protocol P that satisfies these three assumptions for compiler $\mathcal{C}$. Moreover, if $\tilde{P}$ is secure against passive attacks then P will also be. Finally, the round complexity of P is the same as that of $\tilde{P}$.

The compiler $\mathcal{C}$ uses a digital signature scheme that is secure against <u>adaptive chosen message</u> attacks. Let $\mathcal{P}$ be the set of potential users of protocol P. Each party $U \in \mathcal{P}$ should have a public/secret signature key pair (PK_U, SK_U). Given protocol P, the compiler now constructs protocol P' as follows:

1. Let $\mathcal{U} = U_1, \ldots, U_n$ be a group of parties that wish to establish a session key and let $\mathcal{U}' = U_1 || \cdots || U_n$ (concatenation). Each party U_i selects a random nonce $t_i \in \{0, 1\}^k$ and broadcasts $U_i||0||t_i$, where 0 is the (initial) sequence number for this message. After receiving all the broadcast messages, each instance Π_U^i stores $\mathcal{U}'$ and $t' = t_1 || \cdots || t_n$ as part of its state information.

2. The parties in group $\mathcal{U}$ now execute protocol P with the following changes:
   - Whenever instance Π_U^i has to broadcast $U||j||m$ in protocol P, instead the instance broadcasts $U||j||m||\sigma$, where σ is the signature of party U on $j||m||\mathcal{U}'||t'$ with the key SK_U.
   - Whenever instance Π_U^i receives message $V||j||m||\sigma$, it checks that: (i) party $V \in \mathcal{U}$ and message m is valid (for the protocols described earlier this means that $z_i, X_i \in G$, or $x_{ij} \in G$); (ii) j is the next expected sequence number for V, and (iii) σ is a valid signature of V on $j||m||\mathcal{U}'||t'$. If any of these fails, Π_U^i aborts the protocol.

Compiler $\mathcal{C}$ can thus be used to convert any GKE protocol that is secure against a passive adversary into a GKE protocol that is secure against active adversaries in the <u>random oracle model</u> as extended in [8].

We finally consider the case when the number of parties n in the group G is large.

Group Key Agreement for Large Groups

Many emerging applications, such as teleconferencing, pay-per-view and real-time information systems, are based on a group communication model in which data is broadcasted to large dynamically evolving groups (see also <u>broadcast encryption</u>). Securing such communication is particularly challenging because of the requirement for frequent re-keying as a result of members leaving the group or new members joining the group. Several GKE protocols for large groups have been proposed in the literature (see e.g., [1, 17, 19, 22, 23, 28, 32], also [27, 31]). These are centralized algorithms which offer scalable solutions to the group key management problem and employ appropriate key graphs that securely distribute re-keying information after members leave or join the group. In this section, we shall briefly describe the OFT protocol which employs a key graph tree for scalable re-keying.

4. The McGrew–Sherman one-way function tree (OFT) protocol [17, 25].

Let k be the security parameter and n the number of potential users of the system. We shall assume $n \gg k$ (n superpolynomial in k). A trusted manager is responsible for the selection and management of all the *symmetric* user keys (see symmetric cryptosystem), as well as the maintenance of user-key relation. Each member U of the group $\mathcal{G}$ is given a set of keys $X_U \subset 2^{\{0,1\}^k}$, which includes the group session key. The sets X_U are constructed in a special way by using a one-way function [18] $g\{0,1\}^k \to \{0,1\}^k$ and a pseudo-random function [18] $f: \{0,1\}^{2k} \to \{0,1\}^k$.

Structure of the OFT Tree. This is a binary tree maintained by the manager. Nodes are assigned keys as follows. First each leaf u of the tree is assigned a random key $k_u \in \{0,1\}^k$. Then starting from the bottom of the tree each internal node x is assigned a key which is computed by using the rule:

$$k_x = f(g(k_{left(x)}), g(k_{right(x)})), \qquad (1)$$

where $left(x)$ is the left child of x and $right(x)$ is the right child of x. The key k_{root} of the root is the group session key.

Each party U who wishes to join the group is issued with a set of keys X_U which consists of: the key k_u of a leaf u and all the "blinded" node keys $k_x' = g(k_x)$ that are siblings to nodes on its path to the root. From these keys U can compute all the node keys on its path to the root by using rule (1). An illustration is given in Figure 2.

Removing or Adding Members. Whenever a member U associated with leaf u is to be removed, the key k_u of U is invalidated and the group member V associated with the sibling of u is assigned to the parent of u and given a new leaf key k_v^{new} which is selected at random from $\{0,1\}^k$ by the manager. All internal node keys along the path of this node to the root are then assigned new values using rule (1).

The new key values must be communicated securely to all the members of the group who store the corresponding old values. For this purpose, the manager broadcasts the new values encrypted in such a way that only the appropriate members can decrypt them. For the member V associated with the sibling v of u, the manager encrypts the new leaf key k_v^{new} with the old key of V and broadcasts $E_{k_v^{old}}[k_v^{new}]$, where E is a *symmetric encryption* function [18] (see also symmetric cryptosystem). For each of the internal nodes x on the path from u to the root, only the new blinded values $(k_x^{new})' = g(k_x^{new})$ are needed (members store one leaf key and the blinded keys of the siblings of the nodes on their paths to the root). Observe that group members that possess the old blind value $(k_x^{old})'$ know the unblinded key value k_s of the sibling s of x. The manager therefore only needs to broadcast the encryption $E_{k_s}[(k_x^{new})']$.

When a new member V joins the group, an existing leaf node u is split: the member U associated with u is now associated with $left(u)$ and the new member is associated with $right(u)$. Members U, V are given keys $k_u^{new} = k_{left(u)}, k_v = k_{right(u)}$, respectively, selected at random from $\{0,1\}^k$. All the keys of the internal nodes x along the path from u to the root are then assigned new values k_x^{new} using rule (1). The new value k_u^{new} and the blinded values of the new keys $(k_x^{new})'$ are then encrypted with the appropriate keys and broadcasted as in the previous case. The key k_v is given privately to V.

Security and Efficiency. The security of the OFT protocol reduces to that of the one-way function g, the pseudo-random function f, and the symmetric encryption function E. We have *forward* and *backward* security (see also group signature): members who leave the group cannot read *future* messages and members who join the group cannot read *previous* messages.

The communication complexity of adding or removing a member is bounded by $hk + h$ bits, where h is the height of the tree and k the length of the keys. This is because up to h keys must be broadcasted, and h bits are needed to describe the member that joins or leaves the group. If the tree is balanced, then the operation of adding or removing a member with the OFT protocol is scalable (with complexity $O(\log n)$).

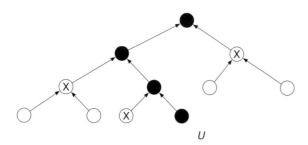

Fig. 2. An OFT tree: member U possesses the keys of the black nodes and the blinded keys of their siblings "$\otimes$"

Mike Burmester

References

[1] Balenson, D.A., D.A. McGrew, and A.T. Sherman (1998). "Key management for large dynamic groups: One-way function trees and amortized initialization." *Advanced Security Research Journal—NAI Labs*, 1 (1), 27–46.

[2] Bellare, M., D. Pointcheval, and P. Rogaway (2000). "Authenticated key exchange secure against dictionary attacks." *Advances in Cryptology—EUROCRYPT 2000*, Lecture Notes in Computer Science, vol. 1807, ed. B. Preneel. Springer-Verlag, Berlin, 139–155.

[3] Bellare, M. and P. Rogaway (1994). "Entity authentication and key distribution." *Advances in Cryptology—CRYPTO'93*, Lecture Notes in Computer Science, vol. 773, ed. D.R. Stinson. Springer-Verlag, Berlin, 232–249.

[4] Bellare, M. and P. Rogaway (1995). "Provably secure session key distribution: The three party case." *Proc. 27th ACM STOC*, 1995, 57–64.

[5] Blundo, C., A. De Santis, A. Herzberg, S. Kutten, U. Vaccaro, and M. Yung (1993). "Perfectly-secure key distribution for dynamic conferences." *Advances in Cryptology—CRYPTO'92*, Lecture Notes in Computer Science, vol. 740, ed. E.F. Brickell. Springer-Verlag, Berlin, 471–486.

[6] Boyd, C. and J.M.G. Nieto (2003). "Round-optimal contributory conference key agreement." *PKC 2003*, Lecture Notes in Computer Science, vol. 2567, ed. Y.G. Desmedt. Springer-Verlag, Berlin, 161–174.

[7] Bresson, E., O. Chevassut, and D. Pointcheval (2001). "Provably authenticated group Diffie–Hellman key exchange—The dynamic case." *Advances in Cryptography—ASIACRYPT 2001*, Lecture Notes in Computer Science, vol. 2248, ed. C. Boyd. Springer-Verlag, Berlin, 290–300.

[8] Bresson, E., O. Chevassut, D. Pointcheval and J.-J. Quisquater (2001). "Provably authenticated group Diffie–Hellman key exchange." *ACM Conference on Computer and Communications Security*, 255–264.

[9] Bresson, E., O. Chevassut, and D. Pointcheval (2002). "Group Diffie–Hellman key exchange under standard assumptions." *Advances in Cryptology—EUROCRYPT 2001*, Lecture Notes in Computer Science, vol. 2045, ed. B. Pfitzmann. Springer-Verlag, Berlin, 321–336.

[10] Burmester M. and Y. Desmedt (1994). "A secure and efficient conference key distribution system." *Pre-proceedings of EUROCRYPT'94, Scuola Superiore Guglielmo Reiss Romoli (SSGRR), Perugia, Italy, May 9–12*, 279–290.

[11] Burmester M. and Y. Desmedt (1995). "A secure and efficient conference key distribution system." *Advances in Cryptology—EUROCRYPT'94*, Lecture Notes in Computer Science, vol. 950, ed. A. De Santis. Springer-Verlag, Berlin, 275–286.

[12] Burmester, M. and Y. Desmedt (1996). "Efficient and secure conference key distribution." *Security Protocols, International Workshop, Cambridge, UK, 1996*, Lecture Notes in Computer Science, vol. 1189, ed. M. Lomas. Springer-Verlag, Berlin, 119–129.

[13] Diffie, W. and M.E. Hellman (1976). "New directions in cryptography." *IEEE Transactions on Information Theory*, IT-22 (6), 644–654.

[14] Ingemarsson, I., D.T. Tang, and C.K. Wong (1982). "A conference key distribution system." *IEEE Transactions on Information Theory*, 28 (5), 714–720.

[15] Katz, J. and M. Yung (2003). "Authenticated group key exchange in constant rounds." *Advances in Cryptology—CRYPTO 2003*, Lecture Notes in Computer Science, vol. 2729, ed. D. Boneh. Springer-Verlag, Berlin, 110–125.

[16] Koyama, K. and K. Ohta (1988). "Identity-based conference key distribution systems." *Advances in Cryptology—CRYPTO'87*, Lecture Notes in Computer Science, vol. 293, ed. C. Pomerance. Springer-Verlag, Berlin, 175–185.

[17] McGrew, D. and A.T. Sherman (1998). "Key establishment in large dynamic groups using one-way function trees." Manuscript. The journal version is [25].

[18] Menezes, A., P. van Oorschot, and A.A. Vanstone. (1997). *Handbook of Applied Cryptography*. CRC Press, Boca Raton, FL.

[19] Mittra, S. "Iolus: A framework for scalable secure multicasting." *Proceedings of the ACM SIGCOMM'97*, 11–19.

[20] Okamoto, E. (1988). "Key distribution systems based on identification information." *Advances in Cryptology—CRYPTO'87*, Lecture Notes in Computer Science, vol. 293, ed. C. Pomerance. Springer-Verlag, Berlin, 224–314.

[21] Okamoto, E. and K. Tanaka (1989). "Key distribution system based on identification information." *IEEE Journal of Selected Areas in Communication*, 7 (4), 481–485.

[22] Perrig, A., D. Song, and J. Tygar (May 2001). "ELK, a new protocol for efficient large-group key distribution." In *Proceedings of the 2001 IEEE Symposium on Security and Privacy, May 2001*, 247–262.

[23] Setia, S., S. Koussih, and S. Jajodia Kronos (2000). "A scalable group re-keying approach for secure multicast." *IEEE Symposium on Security and Privacy*, 215–228.

[24] Schneier, B. (1996). *Applied Cryptography*. John Wiley & Sons, New York.

[25] Sherman, A.T. and D.A. McGrew (2003). "Key establishment in large dynamic groups using one-way function trees." *IEEE Software Engineering*, 444–448.

[26] Steiner, M., G. Tsudik, and M. Waidner (1996). "Diffie–Hellman key distribution extended to group communication." *ACM Conference on*

Computer and Communications Security, 31–37.

[27] Steiner, M., G. Tsudik, and M. Waidner (May 1998). "CLIQUES: A new approach to group key agreement." *Proceedings of the 18th International Conference on Distributed Computing Systems (ICDCS'98), Amsterdam.*

[28] Steiner, M., G. Tsudik, and M. Waidner (2000). "Key agreement in dynamic peer groups." *IEEE Transactions on Parallel and Distributed Systems*, 11 (8), 769–780.

[29] Tsujii, S. and T. Itoh (1989). "An ID-based cryptosystem based on the discrete logarithm." *IEEE J. Selected Areas in Communication*, 7, 467–473.

[30] Yacobi, Y. and Z. Shmuely (1990). "On key distribution systems." *Advances in Cryptology—CRYPTO'89*, Lecture Notes in Computer Science, vol. 435, ed. G Brassard. Springer-Verlag, Berlin, 344–355.

[31] Wallner, D.M., E.J. Harder, and R.C. Agee (1997). "Key management for multiast: Issues and architectures." *Internet Engineering Task Force*, July 1, ftp://ftp.ietf.org/internet-drafts/draft-wallner-key-arch-01.txt

[32] Wong, C.K., Chung Kei, M. Gouda, and S.S. Lam. (1998). "Secure group communications using key graphs." *Proceedings of SIGCOMM'98*. ACM Press, New York, 68–79.

GROUP SIGNATURES

Group signatures are digital signatures where signers can establish groups such that each member of the group can produce signatures anonymously on behalf of the group. Each group can be managed by a *trusted group authority*, which oversees joining and leaving the group and can re-identify individual signers in case of disputes according to a clearly stated policy. Obviously, many groups can choose to be managed by the same trusted group authority, or a group can choose to fully distribute the group management among its members such that every member is involved in all management transactions. The concept of group signatures and first practical constructions were introduced by Chaum and van Heijst [4].

The term 'group signatures' or 'group-oriented signatures' is sometimes also used for another type of signature scheme where signers also form groups such that, for example, any t-out-of-n members of a group can together produce a signature. In this case, the capability of producing signatures is granted only to large enough coalitions within a group. This was first introduced by Boyd as *multisignatures*, but according to Desmedt [9] there is growing consensus to call them threshold signatures.

In a group signature scheme, each signing member of a group has an individual signing key pair. If individuals generate their key pairs without having to agree on common domain parameters, the group sigature scheme is called *separable* [6]. An individual is registered for a group by presenting a suitable ID certificate to the respective trusted group authority and submitting her or his public verifying key. The trusted group authority constructs a group key pair, which consists of a private group key and a public group key, and publishes the public group key through one or more authentic channels such as a public key infrastructure (PKI). A member leaves a group by revoking her or his public verifying key from the trusted group authority. It is the responsibility of the trusted group authority to keep track of who belongs to the group at any point of time. A group signature scheme is called *static* if the public group key needs to be updated after members join or leave the group, or update their individual key pairs. Constructions were proposed in [2, 4, 7]. All of them suffer from the drawback that the size of a public group key and that of the signatures are proportional to the size of a group. If the public group key remains unchanged after members join or leave the group, or update their individual key pairs, the group signature scheme is called *dynamic* [13]. The first construction of a dynamic group signature scheme was proposed by Camenisch and Stadler [8]. Their construction has the additional advantage that the size of the public group key and that of the signatures are independent of the size of the group. Their paper sparked more work on dynamic group signature schemes [1, 5, 6, 13].

Everyone who has access to the public group key of group G can verify every signature produced by every member of group G. When a signature is disputed, the respective verifier can request the signer's identity from the respective trusted group authority. The trusted group authority then uses the private group key in order to recover the signer's identity. If and when the signer's identity is recovered and released is controlled by the group management policy that governs the operations of the trusted group authority.

Robust key management is a particularly important issue in group signatures. For example, if the public group key changes whenever members join or leave the group or update their private signing keys, then it becomes a burden for the trusted group authority to publish the public group keys for all recent time intervals in a timely fashion. Moreover, if a private signing key of a group member is compromised, then the attacker can freely produce signatures on behalf of the group until the respective public verifying key is revoked. Therefore, all the signatures of the victimized group member must be regarded invalid if there is no way of distinguishing the signatures produced by the honest group member from those produced by the attacker. These problems are addressed by an approach called *forward security*. Forward secure group signature schemes allow individual group members to join or leave a group or update their private signing keys without affecting the respective public group key. By dividing the lifetime of all individual private signing keys into discrete time intervals, and by tying all signatures to the time interval when they are produced, group members who are revoked in time interval i have their signing capability effectively stripped away in time interval $i + 1$, while all their signatures produced in time interval i or before (and, of course, the signatures of all other group members) remain verifiable and anonymous [13]. Forward security in group signature schemes is similar to *forward security* in threshold signature schemes.

Group signatures are useful, for example, to build secure auction systems [7], where all the bidders form a group and authorize their tenders by group signatures. After the winning tender has been determined, the trusted group authority can re-identify the lucky bidder, while the other bidders remain anonymous. Group signature schemes are dually related to *identity escrow schemes* [10], i.e., group-member identification schemes with revocable anonymity.

A group signature scheme has the following operations: (i) an operation for generating pairs of a private signing key and a public verifying key for an individual; (ii) an operation for generating pairs of a private group key and a public group key for a trusted group authority; (iii) operations for group management such as joining and revoking group members; (iv) an operation for signing messages; (v) an operation for verifying signatures against a public group key; and (vi) an operation to identify (de-anonymize) a group member by one of her or his signatures.

The characteristic security requirements of a group signature scheme are:

Unforgeability: Resistance against existential forgery under adaptive chosen message attacks by computationally restricted attackers.

Unlinkability: Any cheating verifier except the trusted group authority, given any two messages m_1, m_2 and respective signatures s_1, s_2 that are valid for these messages with respect to the public group key of group G cannot decide with non-negligible probability better than pure guessing whether the two signatures have been produced by the same group member or two different group members. Note that unlinkability implies signer anonymity.

Exculpability: A cheating trusted group authority (or any coalition of signers) cannot produce a valid signature that identifies an originating group member who has in fact not produced the signature (false claim of origin).

Traceability: Any coalition of cheating signers cannot produce valid signatures for which not at least one of them is held responsible.

Forward Security: Signers who leave the group can no longer sign messages in behalf of the group. Other additional security requirement are proposed by Song [13].

Constructions have been based on groups, in which the discrete logarithm problem is hard [8], and on the RSA signature scheme [1, 5, 13].

Group signature schemes can be equipped with additional features: Cramer et al. [3] proposed to add threshold properties into a group signature scheme such that any subset of group members can be authorized to produce signatures on behalf of the group. Lysyanskaya and Ramzan [11] proposed blind group signatures based on the dynamic group signature scheme by Camenisch and Stadler [8]. Another blind group signature scheme based on [8] was proposed by Nguyen et al. [12].

Gerrit Bleumer

References

[1] Ateniese, Giuseppe, Jan Camenisch, Marc Joye, and Gene Tsudik (2000). "A practical and provably secure coalition-resistant group signature scheme." *Advances in Cryptology—CRYPTO 2000*, Lecture Notes in Computer Science, vol. 1880, ed. M. Bellare. Springer-Verlag, Berlin, 255–270.

[2] Camenisch, Jan (1997). "Efficient and generalized group signatures." *Advances in Cryptology—EUROCRYPT'97*, Lecture Notes in Computer Science, vol. 1233, ed. W. Fumy. Springer-Verlag, Berlin, 465–479.

[3] Cramer, Ronald, Ivan Damgård, Berry Schoenmakers (1994). "Proofs of partial knowledge and simplified design of witness hiding Protocols." *Advances in Cryptology—CRYPTO'94*, Lecture Notes in Computer Science, vol. 839, ed. Y.G. Desmedt. Springer-Verlag, Berlin, 174–187. *CWI Quarterly Journal*, 8 (2), (1995), 111–128.

[4] Chaum, David, Eugene van Heijst (1991). "Group signatures." *Advances in Cryptology—EUROCRYPT'91*, Lecture Notes in Computer Science, vol. 547, ed. D.W. Davies. Springer-Verlag, Berlin, 257–265.

[5] Camenisch, Jan, Markus Michels (1998). "A group signature scheme with improved efficiency." *Advances in Cryptography—ASIACRYPT'98*, Lecture

Notes in Computer Science, vol. 1514, eds. K. Ohta and D. Pei. Springer-Verlag, Berlin, 160–174.

[6] Camenisch, Jan, Markus Michels (1999). "Separability and efficiency of generic group signatures." *Advances in Cryptology—CRYPTO'99*, Lecture Notes in Computer Science, vol. 1666, ed. J. Wiener. Springer-Verlag, Berlin, 413–430.

[7] Chen, Lidong, Torben P. Pedersen (1995). "New group signature schemes." *Advances in Cryptology—EUROCRYPT'94*, Lecture Notes in Computer Science, vol. 950, ed. A. De Santis. Springer-Verlag, Berlin, 171–181.

[8] Camenisch, Jan, Markus Stadler (1997). "Efficient group signature schemes for large groups." *Advances in Cryptology—CRYPTO'97*, Lecture Notes in Computer Science, vol. 1294, ed. B.S. Kaliski. Springer-Verlag, Berlin, 410–424.

[9] Desmedt, Yvo (1993). "Threshold cryptography." *Advances in Cryptology—AUSCRYPT'92*, Lecture Notes in Computer Science, vol. 718, eds. J. Seberry and Y. Zheng. Springer-Verlag, Berlin, 3–14.

[10] Kilian, Joe and Erez Petrank (1998). "Identity Escrow." *Advances in Cryptology—CRYPTO'98*, Lecture Notes in Computer Science, vol. 1642, ed. H. Krawczyk. Springer-Verlag, Berlin, 169–185.

[11] Lysyanskaya, Anna, Zulfikar Ramzan (1998). "Group blind digital signatures: A scalable solution to electronic cash." *Financial Cryptography '98*, Lecture Notes in Computer Science, vol. 1465, eds. D.M. Goldschlag and S.G. Stubblebine. Springer-Verlag, Berlin, 184–197.

[12] Nguyen, Khanh Quoc, Yi Mu, Vijay Varadharajan (1999). "Divertible zero-knowledge proof of polynomial relations and blind group signature." *Australasian Conference on Information Security and Privacy, ACISP'99*, Lecture Notes in Computer Science, vol. 1587, eds. J. Pieprzyk, R. Safari-Naini, and J. Seberry. Springer-Verlag, Berlin, 117–128.

[13] Song, Dawn Xiaodong (2001). "Practical forward secure group signature schemes." *8th ACM Conference on Computer and Communications Security (CCS-8)*. ACM Press, New York.

H

HARD-CORE BIT

Let f be a one-way function. According to the definition of such a function, it is difficult, given $y = f(x)$ where x is random, to recover x. However, it may be easy to determine certain information about x. For instance, the RSA function $f(x) = x^e \bmod n$ (see RSA public-key encryption) is believed to be one-way, yet it is easy to compute the Jacobi symbol of x, given $f(x)$:

$$\left(\frac{x^e \bmod n}{n} \right) = \left(\frac{x}{n} \right)^e = \left(\frac{x}{n} \right).$$

Another example is found in the discrete exponentiation function $f(x) = g^x \bmod p$ (see discrete logarithm problem), where the least-significant bit of x is revealed from the Legendre symbol of $f(x)$, i.e., $f(x)^{(p-1)/2}$, which indicates whether $f(x)$ is a square and hence whether x is even.

It has therefore been of considerable interest in cryptography to understand which parts of the inverse of certain one-way functions are hardest to compute. This has led to the notion of a *hard-core bit*. Informally, a function B from inputs to $\{0, 1\}$ is *hard-core* with respect to f if it is infeasible to approximate $B(x)$, given $f(x)$. Here, "approximating" means predicting with probability significantly better than $1/2$.

Hard-core bits have been identified for the main hard problems in public-key cryptography, including the discrete logarithm problem and the RSA problem. Blum and Micali [2] gave the first unapproximability results for the former; Alexi et al. [1], the first complete results for the latter. (See also [4] for recent enhancements.)

In addition, Goldreich and Levin [3] have given a very elegant method for constructing a hard-core bit from *any* one-way function:

1. Define $g(x, r) = (f(x), r)$, where the length of r is the same as the length of x.
2. Define $B(x, r) = x_1 r_1 \oplus \cdots \oplus x_k r_k$, where x_i, r_i are the bits of x and r, and $\oplus$ is the exclusive-or operation.

If f is one-way, then g is clearly one-way; Goldreich and Levin's key result is a proof that the predicate B so constructed is hard-core with respect to g. Intuitively, this can be viewed as saying that the XOR of a random subset of bits of the inverse is hard to predict.

Researchers have also studied the related problem of *simultaneous security* of multiple bits, that is, whether an individual hard-core bit remains difficult to approximate even if the values of other hard-core bits are known. Typically, the order of $\log k$ bits can be shown to be simultaneously secure for the functions mentioned above, where k is the size of the input to the function; up to $k/2$ have been proven simultaneously secure for related functions [5]. It has been conjectured that half of the bits of the the RSA/Rabin functions are simultaneously secure.

The primary application of these results is in constructing a pseudo-random number generator. Following early work by Yao [6] and Blum and Micali [2], if B is a hard-core bit and f is a one-way permutation, then the sequence

$$B(x_0), B(x_1), B(x_2), \ldots,$$

where $x_i = f(x_{i-1})$ and x_0 is a random seed, is indistinguishable from a truly random sequence of the same length. The proof proceeds by showing that any efficient algorithm to distinguish the sequence from a truly random sequence can also be used to distinguish the supposed hard-core bit from a random value, and hence to approximate the bit.

Burt Kaliski

References

[1] Alexi, W.B., B. Chor, O. Goldreich, and C.-P. Schnorr (1988). "RSA and Rabin functions: Certain parts are as hard as the whole." *SIAM Journal on Computing*, 17 (2), 194–209.

[2] Blum, M. and S. Micali (1984). "How to generate cryptographically strong sequences of pseudo-random bits." *SIAM Journal on Computing*, 13 (4), 850–863.

[3] Goldreich, O. and L. Levin (1989). "A hard-core predicate for all one-way functions." *Proceedings of the 21st Annual ACM Symposium on Theory of Computing*, 25–32.

[4] Håstad, J. and M. Näslund (2004). "The security of all RSA and discrete log bits." *Journal of the ACM*, 51 (2), 187–230.

[5] Håstad, J., A.W. Schrift, and A. Shamir (1993). "The discrete logarithm modulo a composite hides $O(n)$ bits." *Journal of Computer and System Sciences*, 47 (3), 376–404.

[6] Yao. A. (1982). "Theory and applications of trapdoor functions." *Proceedings of the 23rd Annual IEEE Symposium on Foundations of Computer Science (FOCS)*, 80–91.

HARDWARE SECURITY MODULE

Security devices are playing an essential role in our everyday life, ensuring security for the financial transactions we could directly or indirectly make. These devices provide the ability to make transactions within a distributed and virtual environment, satisfying the request of Trust and Privacy.

One of the commonly used trusted token is the Smart Card.

When you want to pay for some gift, the vendor wants to be assured that you are the real owner—not a hacker collecting credit card numbers—of the banking account. On the other side, you need to know that the vendor is really who he claims to be before paying. Both parties in this transaction need to authenticate each other's identity.

In addition, depending of the transaction's value, both parties may want the exchanges protected against hacker spying, disclosing and intercepting the exchanges of the transaction. Both parties need confidentiality and integrity.

Security technology can be divided into two types: software defined and hardware achieved.

Security Processors are computational devices that are used to execute security functions in a short execution time. In the context of Information Technology, these Security Processors have to execute trustfully these operations whether the environment is hostile or not, so they are usually integrated into a tamper resistant device or package. These devices are known by a variety of names, including:
- Tamper Resistant Security Module (TRSM)
- Network Security Processor (NSP)
- Host/Hardware Security Module (HSM).

An HSM is a physically secure, tamper-resistant security server that provides cryptographic functions to secure transactions in retail and financial applications. This includes:
- PIN (see Personal Identification Number) encryption and verification
- Debit card validation
- Stored value card issuing and processing
- Chip card issuing and processing
- Message authentication (see MAC algorithms)
- Symmetric key management.

With a DSP-RSA Module, the HSM can also support public key cryptographic operations including digital signatures, certificates, and asymmetric key management.

Acting as a peripheral to a host computer, the HSM provides the cryptographic facilities needed to implement a wide range of data security tasks.

Banks, corporations, and probably some branches of the military are using HSMs as part of their security chain.

Hardware Security Modules offer a higher level of security than software. They are normally evaluated by third parties, such as the USA's "National Institute of Standards and Technology" (NIST), through the Federal Information Processing Standards Publication (FIPS PUB 140) or French "Direction Centrale de la Sécurité des Systmès d'Information" (DCSSI). This level of security is required by some highly secured web applications, Public Key Infrastructures and Certification Authorities.

FIPS140-1 level	Description
1	Software-only implementation performs correctly algorithms.
2	Both software and hardware implementations can meet this level. Hardware must incorporate a limited degree of tamper-evident design or employ locks to secure sensitive informations. Role-based authentication is required to authorize a defined set of services.
3	This level requires a hardware implementation, designed to prevent an intruder from obtaining secrets from the device. Detected tampering must result in logical-destruction of sensitive informations.
4	Hardware certified at this level must resist the most sophisticated attacks. Cryptographic functions are performed within an 'envelope' protected by the security parameters. The intent of Level 4 protection is to detect a penetration of the device from any direction. Such devices are suitable for operation in physically unprotected environments.

Hardware Security Modules perform cryptographic operations, protected by hardware. These operations may include:
- Random number generation
- Key generation (asymmetric and symmetric) (see asymmetric cryptosystem and symmetric cryptosystem)
- Asymmetric private key storage while providing protection (security) from attack (i.e., no unencrypted private keys in software or memory):
  - Private keys used for signing and decryption
  - Private keys used in PKI for storing Root Keys.

The diagram below describes the internal elements of an HSM.

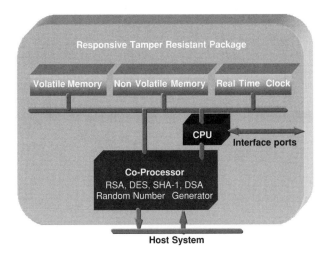

The diagram below illustrates various hardware encryption technologies and their respective positions of cost relative to security.

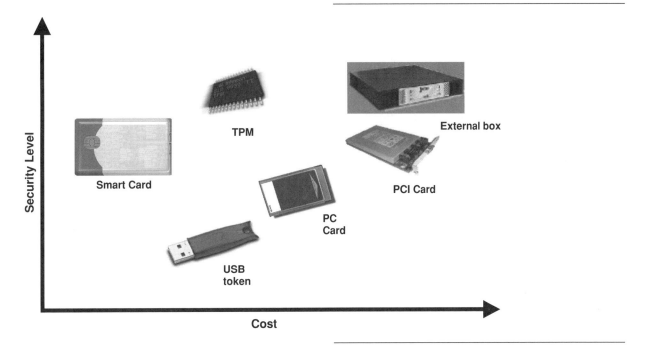

An HSM is a key element in the security chain of a system. It provides all necessary cryptographic functions in association to a secured key management for generation, storage and handling. Its cost may mean that only companies can afford to use such devices. Overall benefit is the third party evaluation of this piece of hardware and software, through governmental certification schemes.

USEFUL LINKS:

- More on HSM: *http://www.cren.net/crenca/onepagers/additionalhsm.html*
- "Building a High-Performance, Programmable Secure CoProcessor" by Sean W. Smith and Steve Weingart.
- Cryptographic equipments list: *http://www.ssi.gouv.fr/fr/reglementation/liste_entr/index.html*
- FIPS PUB 140-2 Security Requirements for Cryptographic Modules:
  - *http://csrc.nist.gov/cryptval/*
  - *http://csrc.nist.gov/cryptval/140-2.htm*
- PKCS #11—Cryptographic Token Interface Standard: *http://www.rsasecurity.com/rsalabs/pkcs/pkcs-11/index.html*

DEFINITIONS (EXTRACTED FROM ISO 15408)

Tamper Evidence Requirement

A device that claims Tamper Evidence characteristics *shall* be designed and constructed as follows:

- *Substitution*: To protect against substitution with a forged or compromised device, a device is designed so that it is not practical for an attacker to construct a duplicate from commercially available components that can reasonably be mistaken for a genuine device.
- *Penetration*: To ensure that penetration of an SCD is detected, the device *shall* be so designed

and constructed that any successful penetration *shall* require that the device be subject to physical damage or prolonged absence from its authorized location such that the device cannot be placed back into service without a high probability of detection by a knowledgeable observer.

Tamper Resistance Requirements

- *Penetration*: An SCD *shall* be protected against penetration by being Tamper Resistant to such a degree that its passive resistance is sufficient to make penetration infeasible both in its intended environment and when taken to a specialized facility where it would be subjected to penetration attempts by specialized equipment.
- *Modification*: The unauthorized modification of any key or other sensitive data stored within an SCD, or the placing within the device of a tap, e.g., active, passive, radio, etc. to record such sensitive data, *shall* not be possible unless the device be taken to a specialized facility and this facility be subjected to damage such that the device is rendered inoperable.
- *Monitoring*: Monitoring *shall* be countered by using tamper resistant device characteristics. The passive physical barriers *shall* include the following:
  - shielding against electromagnetic emissions in all frequencies in which sensitive information could be feasibly disclosed by monitoring the device;
  - privacy shielding such that during normal operation, keys pressed will not be easily observable to other persons. (For example, the device could be designed and installed so that the device can be picked up and shielded from monitoring by the user's own body.)

Where parts of the device cannot be appropriately protected from monitoring; these parts of the device *shall* not store, transmit or process sensitive data. The device *shall* be designed and constructed in such a way that any unauthorized additions to the device, intended to monitor it for sensitive data, *shall* have a high probability of being detected before such monitoring can occur.

- Removal: If protection against removal is required, the device *shall* be secured in such a manner that it is not economically feasible to remove the device from its intended place of operation.

Laurent Sustek

HASH FUNCTIONS

INTRODUCTION: Cryptographic hash functions take input strings of arbitrary (or very large) length and map these to short fixed length output strings. The term *hash functions* originates from computer science, where it denotes a function that compresses a string of arbitrary length to a string of fixed length. Hash functions are used to allocate as uniformly as possible storage for the records of a file. For cryptographic applications, we distinguish between unkeyed and keyed hash functions. We consider here only cryptographic hash functions without a secret parameter or secret key; these are also known as *Manipulation Detection Codes* (or *MDCs*). An important class of keyed hash functions are MAC Algorithms, which are used for information authentication. They can reduce the authenticity of a large quantity of information to the secrecy and authenticity of a short secret key.

Unkeyed cryptographic hash functions or MDCs will in the remainder of this article be called hash functions. These functions can also provide information authentication in a natural way: one replaces the protection of the authenticity of the original input string by the protection of the authenticity of the short hash result. A simple example of such a process is the communication of a large data file over an insecure channel. One can protect the authenticity of this data file by sending the hash result of the file over an authenticated channel, e.g., by reading it over the phone or by sending it by regular mail or telefax.

The most common application of hash functions is in digital signatures: one will apply the signing algorithm to the hash result rather than to the original message; this brings both performance and security benefits. With hash functions one can also compare two values without revealing them or without storing the reference value in the clear. The typical examples are passwords and passphrases: the verifier will store the image of the passphrase under a hash function; on receipt of the passphrase, it will apply the hash function and check whether the result is equal to this image. Hash functions can also be used to commit to a value without revealing it. The availability of efficient hash functions has resulted in their use as pseudo-random functions (for key derivation), and as building blocks for MAC algorithms (e.g., HMAC), block ciphers (e.g., Bear and Lion [5] and Shacal [41]) and stream ciphers (e.g., SEAL). Finally, many results in cryptology rely on the random oracle model: informally one assumes the existence of a random function that can be queried on arbitrary inputs to yield a random output. If one needs to instantiate a random oracle in practice, one typically uses a hash function.

DEFINITIONS: It will be assumed that the description of the hash function h is publicly known; one also requires that given the inputs, the computation of the hash result must be *efficient*.

One-Way Hash Function (OWHF)

The concept of one-way hash functions was introduced by Diffie and Hellman in [25]. The first informal definition was given by Merkle [52,53] and Rabin [68].

DEFINITION 1. *A one-way hash function (OWHF) is a function h satisfying the following conditions:*

1. *The argument X can be of arbitrary length and the result h(X) has a fixed length of n bits.*
2. *The hash function must be one-way in the sense that given a Y in the image of h, it is computationally infeasible to find a message X such that h(X) = Y (preimage resistant) and given X and h(X) it is computationally infeasible to find a message X' ≠ X such that h(X') = h(X) (second preimage resistant).*

Typical values for the length n of the hash result are $64 \ldots 128$. A function that is preimage resistant is known as a one-way function (but preimage resistance is typically used for hash functions). It is clear that for permutations or injective functions, second preimage resistance is not relevant. Note that some authors call second preimage resistance as *weak collision resistance*. For some applications (e.g., pseudo-random functions and MAC algorithms based on hash functions), a large part of the input of the hash function may be known, yet one requires that it is hard to recover the unknown part of the input. This property is known as *partial preimage resistance*.

Collision Resistant Hash Function (CRHF)

The importance of collision resistance for hash functions used in digital signature schemes was first pointed out by Yuval [83]. The first formal definition of a CRHF was given by Damgård [19,20]. An informal definition was given by Merkle [53].

DEFINITION 2. *A collision resistant hash function (CRHF) is a function h satisfying the following conditions:*

1. *The argument X can be of arbitrary length and the result h(X) has a fixed length of n bits.*
2. *The hash function must be an OWHF, i.e., preimage resistance and second preimage resistant.*
3. *The hash function must be collision resistant: it is computationally infeasible to find two distinct messages that hash to the same result.*

Note that one finds in the literature also the terms *collision freeness* and *collision intractible*.

Relation between Definitions

It is clear that finding a second preimage cannot be easier than finding a collision: therefore the second preimage condition in the definition of a CRHF seems redundant. However, establishing the exact relation between these conditions requires formal definitions. Under certain conditions, detailed in [74], collision resistance implies both second preimage resistance and preimage resistance. A formalization of collision resistance requires a public parameter (also called a key), even if in practice one uses a fixed function. If such a parameter is also introduced for preimage and second preimage resistance, one has then the choice between randomizing the challenge (for the second preimage or preimage attack), the key, or both. The relationship between these definitions is studied by Rogaway and Shrimpton [74]. Earlier work on this topic can be found in [21, 37, 62, 78, 85].

A universal one-way hash function (UOWHF) was defined by Naor and Yung [59]. It is a class of functions indexed by a public parameter (called a key), for which finding a second preimage is hard; the function instance (or parameter) is chosen after the challenge input, which means that finding collisions for a particular instance does not help an attacker. It corresponds to one of the cases considered by Rogaway and Shrimpton.

Simon [77] provides a motivation to treat collision resistant hash functions as independent cryptographic primitives. He shows hat no provable construction of a CRHF can exist based on a black-box one-way permutation, i.e., a one-way permutation treated as an oracle.

One may need other properties of hash functions: for example, when hash functions are combined with a multiplicative digital signature scheme (e.g., the plain RSA digital signature standard [73]), one requires that the hash function is not multiplicative, that is, it should be hard to find inputs x, x', x'' such that $h(x) \cdot h(x') = h(x'')$ for some group operation "·". See Section "Attacks dependent... Scheme" for an example and [3] for a more extensive discussion on these aspects.

If one digitally signs a hash value of a message rather than the message itself, or if one uses a hash function to commit to a value, a CRHF hash function is necessary. For other applications of hash functions, such as protecting passphrases, preimage resistance is sufficient.

ITERATED HASH FUNCTIONS: Most practical hash functions are based on a compression function with fixed size input; they process every message block in a similar way. Lai and Massey call this an *iterated* hash function [50]. The input is first padded such that the length of the input is a multiple of the block length. Next it is divided into t blocks X_1 through X_t. The hash result is then computed as follows:

$$H_0 = IV$$
$$H_i = f(X_i, H_{i-1}), \qquad i = 1, 2, \ldots, t$$
$$h(X) = g(H_t).$$

Here IV is the abbreviation of *Initial Value*, H_i is called the *chaining variable*, the function f is called the *compression* function or *round* function, and the function g is called the *output transformation*. Most constructions used for g is the identity function. Two elements in this definition have an important influence on the security of a hash function: the choice of the padding rule and the choice of the IV. It is essential that the padding rule is unambiguous (i.e., there do not exist two messages that can be padded to the same padded message); at the end one should append the length of the message; and the IV should be defined as part of the description of the hash function (this is called MD-strengthening after Merkle [53] and Damgård [21]).

A natural question can now be formulated: which properties should be imposed on f to guarantee that h satisfies certain properties? Two partial answers have been found. The first result is by Lai and Massey [50] and gives necessary and sufficient conditions for f in order to obtain an *ideally secure* hash function h.

THEOREM 1 (Lai–Massey). *Assume that the padding contains the length of the input string, and that the message X (without padding) contains at least 2 blocks. Then finding a second preimage for h with a fixed IV requires 2^n operations if and only if finding a second preimage for f with arbitrarily chosen H_{i-1} requires 2^n operations.*

The fact that the condition is necessary follows from the following argument: if one can find a second preimage for f in 2^s operations (with $s < n$),

one can find a second preimage for h in $2^{(n+s)/2+1}$ operations with a meet-in-the-middle attack (cf. Section "Meet-in-the-middle attack").

A second result by Damgård [21] and independently by Merkle [53] states that for h to be a CRHF, it is sufficient that f is a collision resistant function.

THEOREM 2 (Damgård–Merkle). *Let f be a collision resistant function mapping l to n bits (with $l - n > 1$). If an unambiguous padding rule is used, the following construction yields a CRHF:*

$$H_1 = f(0^{n+1} \| X_1)$$
$$H_i = f(H_{i-1}X_i), \qquad for\ i = 2, 3, \ldots, t.$$

The construction can be improved slightly, and extended to the case where $l = n + 1$, at the cost of an additional assumption on f (see [21] for details and Gibson's comment [37]). This variant, which is used in practice, avoids the prefixing of a zero or one bit. This construction can also extended to a tree, which allows for increased parallelism [21, 60].

METHODS OF ATTACK ON HASH FUNCTIONS: A distinction is made between attacks that only depend on the size n in bits of the hash result attacks that depend on the black-box properties of the compression function and cryptanalytic attacks that exploit the detailed structure of the compression function.

The security of hash functions is heuristic; only in a few slow constructions, it can be reduced to a number theoretic problem. Therefore, it is recommended to be conservative in selecting a hash function: one should not use hash functions for which one can find 'near' collisions or (second) preimages, or hash functions for which one can only find 'randomly looking' collisions or (second) preimages. Such a property may not be a problem for most applications, but one can expect that attacks can be refined to create collisions or (second) preimages that satisfy additional constraints. It is also important to note that many attacks only impose constraints on the last one or two blocks of the input.

Black-Box Attacks on Hash Functions

These attacks depend only on the size n in bits of the hash result; they are independent of the specific details of the algorithm. It is assumed that the hash function behaves as a random function: if this is not the case this class of attacks will typically be more effective.

For attacks that require a limited memory size and not too many memory accesses, 2^{70} operations is considered to be on the edge of feasibility (in 2004). In view of the fact that the speed of computers is multiplied by 4 every 3 years (this is one of the formulations of Moore's law), 2^{80} operations is sufficient for the next 5–10 years, but it will be only marginally secure within 15 years. For applications that require protection for 20 years, one should try to design the scheme such that an attack requires at least 2^{90} operations. For a more detailed discussion of the cost of brute force attacks, see the entry on <u>exhaustive key search</u>.

Random (Second) Preimage Attack. One selects a random message and hopes that a given hash result will be obtained. If the hash function has the required 'random' behavior, the success probability equals $1/2^n$. This attack can be carried out off-line and in parallel. If t hash results can be attacked simultaneously, the work factor is divided by a factor t, but the storage requirement becomes t n-bit blocks. For example, if $t = 2^{n/2}$, on average $2^{n/2}$ attempts are required to hit one of the values. This linear degradation of security can be mitigated by parameterizing the hash function and changing the parameter (or <u>salt</u>) for every instance [52] (see also <u>preimage resistance</u> and <u>second preimage resistance</u>).

Birthday Attack. The <u>birthday paradox</u> states that for a group of 23 people, the probability that at least two people have a common birthday exceeds $1/2$. Intuitively one expects that the probability is much lower. However, the number of pairs of people in such a group equals $23 \times 22/2 = 253$. This can be exploited to find collisions for a hash function in the following way: one generates r_1 variations on a bogus message and r_2 variations on a genuine message. The expected number of collisions equals $r_1 \cdot r_2/n$. The probability of finding a bogus message and a genuine message that hash to the same result is given by $1 - \exp(-r_1 \cdot r_2/2^n)$, which is about 63% when $r = r_1 = r_2 = 2^{n/2}$. Finding the collision does not require r^2 operations: after sorting the data, which requires $O(r \log r)$ operations, the comparison is easy. This attack was first pointed out by Yuval [83].

One can substantially reduce the memory requirements (and also the memory accesses) for collision search by translating the problem to the detection of a cycle in an iterated mapping. This was first proposed by Quisquater and Delescaille [66]. Van Oorschot and Wiener propose an efficient parallel variant of this algorithm [80]; with a 10 million US\$ machine, collisions for <u>MD5</u> (with $n =$ 128) can be found in 21 days in 1994, which corresponds to 5 hours in 2004. In order to make a collision search infeasible, n should be at least 160 bits; security for 15–20 years or more requires at least 180 bits.

Black-Box Attacks on the Compression Function

This class of attacks depends on some high level properties of the compression function f. They are also known as chaining attacks, since they exploit the way in which multiple compression functions are combined.

Meet-in-the-Middle Attack. This attack applies to hash functions for which the compression function f is easy to invert (see also Theorem 1). It is a variant of the birthday attack that allows to find a (second) preimage in time $2^{n/2}$ rather than 2^n. The opponent generates r_1 variations on the first part of a bogus message and r_2 variations on the last part. Starting from the initial value and going backwards from the hash result, the probability for a matching intermediate variable is given by $1 - \exp(-r_1 \cdot r_2/2^n)$. The only restriction that applies to the meeting point is that it cannot be the first or last value of the chaining variable. The memory cost can be made negligible using a cycle finding algorithm [67]. For more details, see the entry <u>meet-in-the-middle attack</u>.

A generalized meet-in-the-middle attack has been proposed by Coppersmith [15] and Girault et al. [39] to break p-fold iterated schemes, i.e., weak schemes with more than one pass over the message as proposed by Davies and Price [22].

Correcting-Block Attack. This attack consists of substituting all blocks of the message except for one or more blocks. This attack often applies to the last block and is then called a correcting-last-block attack, but it can also apply to the first block or to some blocks in the middle. For a collision attack, one chooses two arbitrary messages X and X' with $X' \neq X$; subsequently one searches for one or more correcting blocks denoted with Y and Y', such that $h(X' \| Y') = h(X \| Y)$. A similar approach is taken for a (second) preimage attack. Hash functions based on algebraic structures are particularly vulnerable to this attack. See <u>correcting-block attack</u> for more details.

Fixed Point Attack. A fixed point for a compression f is a pair (H_{i-1}, X_i) that satisfies $f(X_i, H_{i-1}) = H_{i-1}$. If the chaining variable is equal to H_{i-1}, one can insert an arbitrary number of blocks equal to

X_i without modifying the hash result. Producing collisions or a second preimage with this attack is only possible if the chaining variable can be made equal to H_{i-1}: this is the case if IV can be chosen equal to a specific value, or if a large number of fixed points can be constructed (e.g., if one can find an X_i for a significant fraction of H_{i-1}'s). This attack can be made more difficult by appending a block count or bit count and by fixing IV (MD-strengthening, see Section "Iterated hash functions").

Attacks Dependent on the Internal Details of the Compression Function

Most cryptanalytical techniques that have been applied to block ciphers have a counterpart for hash functions. As an example, differential cryptanalysis has been shown to be a very effective attack tool on hash functions [9]. Differential attacks of hash functions based on block ciphers have been studied in [62,69]. Special cryptanalytic techniques have been invented by Dobbertin to cryptanalyze MD4, MD5 and the RIPEMD family [26, 27]. For hash functions based on block ciphers, weaknesses of block ciphers that may not be a problem for confidentiality protection can be problematic for hashing applications. For example, one needs to take into account the complementation property and fixed points [58] of DES, as well as the existence of weak keys (see also the weak hash keys defined by Knudsen [47]).

Attacks Dependent on the Interaction with the Signature Scheme

Signature schemes can be made more efficient and secure by compressing the information to be signed with a hash function and to sign the hash result. Even if the hash function is collision resistant, it might be possible to break the resulting signature scheme. This attack is then the consequence of an interaction between both schemes. In the known examples of such an interaction, both the hash function and the signature scheme have some multiplicative structure (Coppersmith's attack on X.509 Annex D [16], see correcting block attack).

A more subtle point is that problems can arise if the hash function and the digital signature scheme are not coupled. For example, given $h(X)$, with h a strong CRHF, one could try to find a value X' such that $h'(X') = h(X)$, where h' is a weak hash function, and then claim that the signer has signed X' with h' instead of X with h. This problem can be addressed to some extent by signing together

with the hash result a unique hash identifier (e.g., as defined in [42]). However, Kaliski has pointed out that this approach has some limitations [46]. A simpler approach is to allow only one hash function for a given signature scheme (DSA [34] and SHA-1 [32]).

AN OVERVIEW OF PRACTICAL HASH FUNCTIONS: This section presents three types of hash functions: custom designed hash functions, hash functions based on a block cipher, and hash functions based on algebraic structures (modular arithmetic, knapsack, and lattice problems). It is important to be aware of the fact that many proposals have been broken; due to space limitations it is not possible to include all proposals found in the literature and the attacks on them. For a more detailed discussion, the reader is referred to [65].

Custom Designed Hash Functions

This section discusses a selection of custom designed hash functions, i.e., algorithms that have been designed for hashing operations. Most of these algorithms use the Davies–Meyer approach (cf. Section "Hash functions . . . cipher"): the compression function is a block cipher, keyed by the text input X_i; the plaintext is the value H_{i-1}, which is also added to the ciphertext (feedforward).

MD2 [45] is a hash function with a 128-bit result that was published by Rivest in 1990. The algorithm is software oriented, but due to the byte structure it is not very fast on 32-bit machines. It inserts at the end a checksum byte (a weak hash function) of all the inputs. Rogier and Chauvaud have found collisions for a variant of MD2, that omits the checksum byte at the end [75].

A much faster algorithm by the same designer is MD4 [71]; it also dates back to 1990 and has a 128-bit result. It is defined for messages of shorter than 2^{64} bits. An important contribution of MD4 is that it has introduced innovative design principles; it was the first published cryptographic algorithm that made optimal use of logic operations and integer arithmetic on 32-bit processors. The compression function hashes a 128-bit chaining variable and a 512-bit message block to a 128-bit chaining variable. The algorithms derived from it (with improved strength) are often called the *MDx-family*. This family contains the most popular hash functions used today. Dobbertin has found collisions for MD4; his attack combines algebraic techniques and optimization techniques such as genetic algorithms [26, 27]. It can be extended to result in 'meaningful' collisions: the complete message (except for a few dozen bytes) is under complete control of the attacker. His attack also applies to the compression function of *extended MD4* [71], which consist of two loosely coupled instances of MD4. Later Dobbertin et al. showed that a reduced version of MD4 (2 rounds out of 3) is not preimage resistant [29].

Following early attacks on MD4 by den Boer and Bosselaers [23] and by Merkle, Rivest quickly proposed a strengthened version, namely MD5 [72]. It was however shown by den Boer and Bosselaers [24] that the compression function of MD5 is not collision resistant (but their collisions are of the special form $f(H_{i-1}, X_i) = f(H'_{i-1}, X_i)$, which implies they have no direct impact on applications). Dobbertin has extended his attack on MD4 to yield collisions for the compression function of MD5, i.e., $f(H_{i-1}, X_i) = f(H_{i-1}, X'_i)$, where he has some control over H_{i-1} [28]. It is believed that it is feasible to extend this attack to collisions for MD5 itself (i.e., to take into account the *IV*).

HAVAL was proposed by Zheng et al. at Auscrypt'92 [86]; it consists of several variants (outputs length between 128 and 256 bits and 3, 4 or 5 rounds). The 3-round version was broken by Van Rompay et al. [81] for all output lengths.

NIST has published a series of variants on MD4 as FIPS standards under the name Secure Hash Algorithm family or SHA family. The first Secure Hash Algorithm was published by NIST [31] in 1993 (it is now referred to as SHA-0). The size of the hash result is 160 bits. In 1995, NIST discovered a certificational weakness in SHA-0, which resulted in a new release of the standard published under the name SHA-1 [32]. In 1998, Chabaud and Joux have published an attack that finds collisions for SHA-0 in 2^{61} operations [13]; it is probably similar to the (classified) attack developed earlier that prompted the upgrade to SHA-1. In 2002, three new hash functions have been published with longer hash results: SHA-256, SHA-384, and SHA-512 [33]. In December 2003, SHA-224 has been added in a change notice to [33]. SHA-256 and SHA-224 have eight 32-bit chaining variables, and their compression function takes message blocks of 512 bits and chaining variables of 256 bits. SHA-384 and SHA-512 operate on 64-bit words; their compression function processes messages in blocks of 1024 bits and chaining variables of 512 bits. They are defined for messages shorter than 2^{128} bits.

Yet another improved version of MD4, called RIPEMD, was developed in the framework of the EEC-RACE project RIPE [70]. RIPEMD has two independent paths with strong interaction at the end of the compression function. It resulted later in the RIPEMD family. Due to partial attacks by Dobbertin [26], RIPEMD was upgraded by Dobbertin et al. to RIPEMD-128 and RIPEMD-160, which have a 128-bit and a 160-bit result, respectively [30]. Variants with a 256 and 320-bit result have been introduced as well.

Whirlpool is a design by Barreto and V. Rijmen [6]; it consists of a 512-bit block cipher with a 512-bit key in the Miyaguchi–Preneel mode (see Section "Hash functions ... cipher"); it offers a result of 512 bits. The design principles of Whirlpool are closely related to those of Rijndael/AES.

Together with SHA-256, SHA-384, and SHA-512, Whirlpool has been recommended by the NESSIE project. The ISO standard on design hash functions (ISO/IEC 10118-3) contains RIPEMD-128, RIPEMD-160, SHA-1, SHA-256, SHA-384, SHA-512, and Whirlpool [42]. Other custom designed hash functions include FFT-Hash III [76], N-hash [57], Snefru [54], Subhash [18] and Tiger [4].

Hash Functions Based on a Block Cipher

Hash functions based on a block cipher have been popular as this construction limits the number of cryptographic algorithms which need to be evaluated and implemented. The historic importance of DES, which was the first standard commercial cryptographic primitive that was widely available, also plays a role. The main disadvantage of this approach is that custom designed hash functions are likely to be more efficient. This is particularly true because hash functions based on block ciphers require a key change after every encryption.

The encryption operation E will be written as $Y = E_K(X)$. Here X denotes the plaintext, Y the ciphertext, and K the key. The size of the plaintext and ciphertext or the block length (in bits) will be denoted with r, while the key size (in bits) will be denoted with k. For DES, $r = 64$ and $k = 56$. The *hash rate* of a hash function based on a block cipher is defined as the number of r-bit input blocks that can be processed with a single encryption.

The discussion in this section will be limited to the case $k \approx r$. For $k \approx r$, a distinction will be made between the cases $n = r$, $n = 2r$, and $n > 2r$. Next the case $k \approx 2r$ will be discussed. A more extensive treatment can be found in [65].

Size of Hash Result Equal to the Block Length. All known schemes of this type have rate 1. The first secure construction for such a hash function was the 1985 scheme by Matyas et al. [51]:

$$H_i = E^{\oplus}_{s(H_{i-1})}(X_i).$$

Here $s()$ is a mapping from the ciphertext space to the key space and $E^{\oplus}_K(X)$ denotes $E_K(X) \oplus X$. This scheme has been included in ISO/IEC 10118–2 [42]. The dual of this scheme is the Davies–Meyer scheme:

$$H_i = E^{\oplus}_{X_i}(H_{i-1}). \tag{1}$$

It has the advantage that it extends more easily to block ciphers for which key size and block size are different. A variant on these schemes was proposed in 1989 independently by Miyaguchi et al. [43,57] and Preneel et al. [63] (it is known as the Miyaguchi–Preneel scheme):

$$H_i = E^{\oplus}_{s(H_{i-1})}(X_i) \oplus H_{i-1}.$$

In 1993, Preneel et al. identify 12 secure variants [64]; Black et al. [10] offer a concrete security proof for these schemes in the black-box cipher model (after earlier work in [82]): this implies that in this model, finding a collision requires $\approx 2^{r/2}$ encryptions and finding a (second) preimage takes $\approx 2^r$ encryptions. This shows that these hash functions can only be collision resistant if the block length is 160 bits or more. Most block ciphers have a smaller block length, which motivates the constructions in the next section.

Size of Hash Result Equal to Twice the Block Length. The goal of *double block length* hash functions is to achieve a higher security level against collision attacks. Ideally a collision attack on such a hash function should require 2^r encryptions, and a (second) preimage attack 2^{2r} encryptions.

The few proposals that survive till today have rate less than 1. Two important examples are MDC-2 and MDC-4 with hash rate 1/2 and 1/4, respectively. They have been designed by Brachtl et al. [12], and are also known as the Meyer–Schilling hash functions after the authors of the first paper described these schemes [55].

$$T^1_i = E^{\oplus}_{u(H^1_{i-1})}(X_i) = LT^1_i \parallel RT^1_i$$
$$T^2_i = E^{\oplus}_{v(H^2_{i-1})}(X_i) = LT^2_i \parallel RT^2_i$$
$$H^1_i = LT^1_i \parallel RT^2_i$$
$$H^2_i = LT^2_i \parallel RT^1_i.$$

The variables H^1_0 and H^2_0 are initialized with the values IV_1 and IV_2 respectively, and the hash result is equal to the concatenation of H^1_t and H^2_t. The functions u, v map the ciphertext space to the key space and need to satisfy the condition $u(IV^1) \neq v(IV^2)$. For $k = r$, the best known preimage and collision attacks on MDC-2 require $2^{3r/2}$ and 2^r operations, respectively [50]. A collision attack on MDC-2 based on the Data Encryption Standard (DES) ($r = 64$, $k = 56$) requires at most 2^{55} encryptions. Note that the compression function of MDC-2 is rather weak, hence Theorems 1 and 2 cannot be applied: preimage and collision attacks on the compression function require at most 2^r and $2^{r/2}$ operations. MDC-2 has been included in ISO/IEC 10118-2 [42].

One iteration of MDC-4 consists of the concatenation of two MDC-2 steps, where the plaintexts in the second step are equal to $H2_{i-1}$ and $H1_{i-1}$. The rate of MDC-4 is equal to 1/4. For $k = r$, the best known preimage attack for MDC-4 requires $2^{7r/4}$ operations. This shows that MDC-4 is probably more secure than MDC-2 against preimage attacks. However, finding a collision for MDC-4 itself requires only 2^{r+2} encryptions, while finding a collision for its compression function requires $2^{3r/4}$ encryptions [49,62]. The best known (2nd) preimage attack on the compression function of MDC-4 requires $2^{3r/2}$ encryptions.

A series of proposals attempted to achieve rate 1 with constructions of the following form:

$$H^1_i = E_{A^1_i}(B^1_i) \oplus C^1_i$$
$$H^2_i = E_{A^2_i}(B^2_i) \oplus C^2_i,$$

where A^1_i, B^1_i, and C^1_i are binary linear combinations of H^1_{i-1}, H^2_{i-1}, X^1_i, and X^2_i and where A^2_i, B^2_i, and C^2_i are binary linear combinations of H^1_{i-1}, H^2_{i-1}, X^1_i, X^2_i, and H^1_i. The hash result is equal to the concatenation of H^1_t and H^2_t. However, Knudsen *et al.* showed that for all hash functions in this class, a preimage attack requires at most 2^r operations, and a collision attack requires at most $2^{3r/4}$ operations (for most schemes this can be reduced to $2^{r/2}$) [48].

Merkle describes an interesting class of proposals in [53], for which he proves in the black-box cipher model that the compression function is collision resistant. The most efficient scheme has rate 1/4 and offers a security level of 2^{56} encryptions when used with DES. Other results on output length doubling have been obtained by Aiello and Venkatesan [1].

Size of Hash Result Larger than Twice the Block Length. Knudsen and Preneel propose a collision resistant compression function, but with parallel encryptions only [49]. They show how a class of efficient constructions for hash functions can be obtained based on non-binary error-correcting codes. Their schemes can achieve a provable security level against collisions equal to 2^r, $2^{3r/2}$ (or more) and this with rates larger than 1/2 (based on a rather strong assumption). The internal memory of the scheme is larger than two or three blocks, which implies that an output transformation is required. Two of their schemes have been included in ISO/IEC 10118-2 [42].

Size of the Key Equal to Twice the Block Length. In this case making efficient constructions is a little easier. A scheme in this class was proposed by Merkle [52], who observed that a block cipher with key length larger than block length is a natural compression function:

$$H_i = E_{H_{i-1} \| X_i}(C),$$

with C a constant string. Another construction can be found in [50].

Two double length hash functions with rate 1/2 have been proposed by Lai and Massey [50]; they form extensions of the Davies–Meyer scheme. One scheme is called *Tandem Davies–Meyer*, and has the following description:

$$H_i^1 = E^{\oplus}_{H_{i-1}^2 \| X_i}(H_{i-1}^1)$$
$$H_i^2 = E^{\oplus}_{X_i \| (H_i^1 \oplus H_{i-1}^1)}(H_{i-1}^2).$$

The second scheme is called *Abreast Davies–Meyer*:

$$H_i^1 = E^{\oplus}_{H_{i-1}^2 \| X_i}(H_{i-1}^1)$$
$$H_i^2 = E^{\oplus}_{X_i \| H_{i-1}^1}(\overline{H}_{i-1}^2).$$

Here $\overline{H}$ denotes the bitwise complement of H. The best known attacks for a preimage and a collision require 2^{2r} and 2^r encryptions, respectively. Faster schemes in this class have been developed in [49].

Hash Functions Based on Algebraic Structures

It should be pointed out that several of these hash functions are vulnerable to the insertion of trapdoors, which allow the person who chooses the design parameters to construct collisions. Therefore one needs to be careful with the generation of the instance. For an RSA public key encryption modulus, one could use a distributed generation as developed by Boneh and Franklin [11] and Frankel et al. [35].

Hash Functions Based on Modular Arithmetic. For several schemes there exists a security re-

duction to a number theoretic problem that is believed to be difficult. However, they are very slow: typically they hash $\log_2 \log_2 N$ bits per modular squaring (or even per modular exponentiation). Damgård provides two hash functions for which finding a collision is provably equivalent to factoring an RSA modulus [19]. Gibson proposes a construction based on the discrete logarithm problem modulo a composite [38]. A third approach by Bellare et al. [7] uses the discrete logarithm problem in a group of prime order p denoted with G_p. Every non-trivial element of G_p is a generator. The hash function uses t random elements α_i from G_p ($\alpha_i \neq 1$). The hash result is then computed as

$$H_{t+1} = \prod_{i=1}^{t} \alpha_i^{\tilde{X}_i}.$$

Here $\tilde{X}_i$ is obtained by considering the string X_i as the binary expansion of a number and prepending 1 to it. This avoids trivial collisions when X_i consists of all zeroes.

There exists a large number of *ad hoc* schemes for which there is no security reduction; many of these have been broken. The most efficient schemes are based on modular squaring. The best schemes seem to be of the form:

$$H_i = ((X_i \oplus H_{i-1})^2 \bmod N) \oplus H_{i-1}.$$

In order to preclude a correcting block attack, it is essential to add redundancy to the message blocks X_i and to perform some additional operations. Two constructions, MASH-1 and MASH-2 (for Modular Arithmetic Secure Hash) have been standardized in ISO/IEC 10118-4 [42]. MASH-1 has the following compression function:

$$H_i = \big(((X_i \oplus H_{i-1}) \vee A)^2 (\bmod N)\big) \oplus H_{i-1},$$

Here $A = \texttt{0xF00...00}$, the four most significant bits in every byte of X_i are set to $\texttt{1111}$, and the output of the squaring operation is chopped to n bits. A complex output transformation is added, which consists of a number of applications of the compression function; its goal is to destroy all the remaining mathematical structure. The final result is at most $n/2$ bits. The best known preimage and collision attacks on MASH-1 require $2^{n/2}$ and $2^{n/4}$ operations [17]; they are thus not better than brute force attacks. MASH-2 is a variant of MASH-1 which uses exponent $2^8 + 1$ [42]. This provides for an additional security margin.

Hash Functions Based on Knapsack and Lattice Problems. The knapsack problem (see knapsack cryptographic schemes) of dimensions n and $\ell(n)$ can be defined as follows: given a set of n l-bit integers $\{a_1, a_2, \ldots, a_n\}$, and an integer S, find a vector X with components x_i equal to 0 or 1 such that

$$\sum_{i=1}^{n} a_i \cdot x_i = S \bmod 2^{\ell(n)}.$$

For application to hashing, one needs $n > \ell(n)$; knapsack problems become more difficult when $n \approx \ell(n)$; however, the performance of the hash function decreases with the value $n - \ell(n)$. The best known attacks are those based on underlined lattice reduction (LLL) [44] and an algebraic technique which becomes more effective if $n \gg \ell(n)$ [61]. It remains an open problem whether for practical instances a random knapsack is sufficiently hard.

Ajtai introduced a function that is one-way (or preimage resistant) if the problem of approximating the shortest vector in a lattice to polynomial factors is hard [2]. Goldreich et al. have proved that the function is in fact collision resistant [40]. Micciancio has proposed a CRHF for which the security is based on the worst case hardness of approximating the covering radius of a lattice [56].

Several multiplicative knapsacks have also been proposed, such as the schemes by Zémor [84] and Tillich and Zémor [79]. Their security is based on the hardness of finding short factorizations in certain groups. In some cases one can even prove a lower bound on the Hamming distance between colliding messages. Attacks on these proposals (for certain parameters) can be found in [14,36].

Incremental Hash Functions. A hash function (or any cryptographic primitive) is called *incremental* if it has the following property: if the hash function has been evaluated for an input x, and a small modification is made to x, resulting in x', then one can update $h(x)$ in time proportional to the amount of modification between x and x', rather than having to recompute $h(x')$ from scratch. If a function is incremental, it is automatically parallelizable as well.

This concept was first introduced by Bellare et al. [7]. They also made a first proposal based on exponentiation in a group of prime order. Improved constructions were proposed by Bellare and Micciancio [8].

Bart Preneel

References

[1] Aiello, W., and R. Venkatesan (1996). "Foiling birthday attacks in length-doubling transformations. Benes: A non-reversible alternative to Feistel." *Advances in Cryptology—EUROCRYPT'96*, Lecture Notes in Computer Science, vol. 1070, ed. U. Maurer. Springer-Verlag, Berlin, 307–320.

[2] Ajtai, M. (1996). "Generating hard instances of lattice problems." *Proceedings of 28th ACM Symposium on the Theory of Computing*, 99–108.

[3] Anderson, R. (1995). "The classification of hash functions." *Codes and Cyphers: Cryptography and Coding IV*, ed. P.G. Farrell. Institute of Mathematics & Its Applications (IMA), 83–93.

[4] Anderson, R. and E. Biham (1996). "Tiger: A new fast hash function." *Fast Software Encryption*, Lecture Notes in Computer Science, vol. 1039, ed. D. Gollmann. Springer-Verlag, Berlin, 89–97.

[5] Anderson, R. and E. Biham (1996). "Two practical and provably secure block ciphers: BEAR and LION." *Fast Software Encryption*, Lecture Notes in Computer Science, vol. 1039, ed. D. Gollmann. Springer-Verlag, Berlin, 113–120.

[6] Barreto, P.S.L.M. and V. Rijmen (2000). "The Whirlpool hashing function." NESSIE submission.

[7] Bellare, M., O. Goldreich, and S. Goldwasser (1994). "Incremental cryptography: The case of hashing and signing." *Advances in Cryptology—CRYPTO'94*, Lecture Notes in Computer Science, vol. 839, ed. Y. Desmedt. Springer-Verlag, Berlin, 216–233.

[8] Bellare, M. and D. Micciancio (1997). "A new paradigm for collision-free hashing: Incrementality at reduced cost." *Advances in Cryptology—EUROCRYPT'97*, Lecture Notes in Computer Science, vol. 1233, ed. W. Fumy. Springer-Verlag, Berlin, 163–192.

[9] Biham, E. and A. Shamir (1993). *Differential Cryptanalysis of the Data Encryption Standard*. Springer-Verlag, Berlin.

[10] Black, J., P. Rogaway, and T. Shrimpton (2002). "Black-box analysis of the block-cipherbased hash-function constructions from PGV." *Advances in Cryptology—CRYPTO 2002*, Lecture Notes in Computer Science, vol. 2442, ed. M. Yung. Springer-Verlag, Berlin, 320–355.

[11] Boneh, D. and M. Franklin (1997). "Efficient generation of shared RSA keys." *Advances in Cryptology—CRYPTO'97*, Lecture Notes in Computer Science, vol. 1294, ed. B. Kaliski. Springer-Verlag, Berlin, 425–439.

[12] Brachtl, B.O., D. Coppersmith, M.M. Hyden, S.M. Matyas, C.H. Meyer, J. Oseas, S. Pilpel, and M. Schilling (1990). *Data Authentication Using Modification Detection Codes Based on a Public One Way Encryption Function*, U.S. Patent Number 4,908,861, March 13, 1990.

[13] Chabaud, F. and A. Joux (1998). "Differential collisions: An explanation for SHA-1." *Advances in Cryptology—CRYPTO'98*, Lecture Notes in Computer Science, vol. 1462, ed. H. Krawczyk. Springer-Verlag, Berlin, 56–71.

[14] Charnes, C. and J.Pieprzyk (1995). "Attacking the SL_2 hashing scheme." *Advances in Cryptography—ASIACRYPT'94*, Lecture Notes in Computer Science, vol. 917, eds. J. Pieprzyk and R. Safavi-Naini. Springer-Verlag, Berlin, 322–330.

[15] Coppersmith, D. (1985). "Another birthday attack." *Advances in Cryptology—CRYPTO'85,* Lecture Notes in Computer Science, vol. 218, ed. H.C. Williams. Springer-Verlag, Berlin, 14–17.

[16] Coppersmith, D. (1989). "Analysis of ISO/CCITT Document X.509 Annex D." IBM T.J. Watson Center, Yorktown Heights, NY, Internal Memo, June 11, 1989 (also ISO/IEC JTC1/SC20/WG2/N160).

[17] Coppersmith, D. and B. Preneel (1995). "Comments on MASH-1 and MASH-2." February 21, ISO/IEC JTC1/SC27/N1055.

[18] Daemen, J. (1995). "Cipher and Hash Function Design. Strategies Based on Linear and Differential Cryptanalysis." *Doctoral Dissertation*, Katholieke Universiteit Leuven.

[19] Damgård, I.B. (1988). "Collision free hash functions and public key signature schemes." *Advances in Cryptology—EUROCRYPT'87,* Lecture Notes in Computer Science, vol. 304, eds. D. Chaum and W.L. Price. Springer-Verlag, Berlin, 203–216.

[20] Damgård, I.B. (1988). "The Application of claw Free Functions in Cryptography." *PhD Thesis*, Aarhus University, Mathematical Institute.

[21] Damgård, I.B. (1990). "A design principle for hash functions." *Advances in Cryptology—CRYPTO'89,* Lecture Notes in Computer Science, vol. 435, ed. G. Brassard. Springer-Verlag, Berlin, 416–427.

[22] Davies, D. and W.L. Price (1980). "The application of digital signatures based on public key cryptosystems." NPL Report, DNACS 39/80, December 1980.

[23] den Boer, B. and A. Bosselaers (1992). "An attack on the last two rounds of MD4." *Advances in Cryptology—CRYPTO'91,* Lecture Notes in Computer Science, vol. 576, ed. J. Feigenbaum. Springer-Verlag, Berlin, 194–203.

[24] den Boer, B. and A. Bosselaers (1994). "Collisions for the compression function of MD5." *Advances in Cryptology—EUROCRYPT'93,* Lecture Notes in Computer Science, vol. 765, ed. T. Helleseth. Springer-Verlag, Berlin, 293–304.

[25] Diffie, W. and M.E. Hellman (1976). "New directions in cryptography." *IEEE Transactions on Information Theory*, IT-22 (6), 644–654.

[26] Dobbertin, H. (1997). "RIPEMD with two-round compress function is not collisionfree." *Journal of Cryptology*, 10 (1), 51–69.

[27] Dobbertin, H. (1996). "Cryptanalysis of MD4." *Journal of Cryptology*, 11 (4), 1998, 253–271. See also *Fast Software Encryption*, Lecture Notes in Computer Science, vol. 1039, ed. D. Gollmann. Springer-Verlag, Berlin, 53–69.

[28] Dobbertin, H. (1996). "The status of MD5 after a recent attack." *CryptoBytes*, 2 (2), 1–6.

[29] Dobbertin, H. (1998). "The first two rounds of MD4 are not one-way." *Fast Software Encryption*, Lecture Notes in Computer Science, vol. 1372, ed. S. Vaudenay. Springer-Verlag, Berlin, 284–292.

[30] Dobbertin, H., A. Bosselaers, and B. Preneel (1996). "RIPEMD-160: A strengthened version of RIPEMD." *Fast Software Encryption*, Lecture Notes in Computer Science, vol. 1039, ed. D. Gollmann. Springer-Verlag, Berlin, 71–82. See also http://www.esat.kuleuven.ac.be/~bosselae/ripemd160

[31] FIPS 180 (1993). *Secure Hash Standard.* Federal Information Processing Standard (FIPS), Publication 180, National Institute of Standards and Technology, US Department of Commerce, Washington, DC, May 11.

[32] FIPS 180-1 (1995). *Secure Hash Standard.* Federal Information Processing Standard (FIPS), Publication 180-1, National Institute of Standards and Technology, US Department of Commerce, Washington, DC, April 17.

[33] FIPS 180-2 (2003). *Secure Hash Standard.* Federal Information Processing Standard (FIPS), Publication 180-2, National Institute of Standards and Technology, US Department of Commerce, Washington, DC, August 26, 2002 (Change notice 1 published on December 1).

[34] FIPS 186 (1994). *Digital Signature Standard.* Federal Information Processing Standard (FIPS), Publication 186, National Institute of Standards and Technology, US Department of Commerce, Washington, DC, May 19.

[35] Frankel, Y., P.D. MacKenzie, and M. Yung (1998). "Robust efficient distributed RSA-key generation." *Proceedings of 30th ACM Symposium on the Theory of Computing*, 663–672.

[36] Geiselmann, W. (1995). "A note on the hash function of Tillich and Zémor." *Cryptography and Coding. 5th IMA Conference*, ed. C. Boyd. Springer-Verlag, Berlin, 257–263.

[37] Gibson, J.K. (1990). "Some comments on Damgård's hashing principle." *Electronic Letters*, 26 (15), 1178–1179.

[38] Gibson, J.K. (1991). "Discrete logarithm hash function that is collision free and one way." *IEEE Proceedings*, 138 (6), 407–410.

[39] Girault, M., R. Cohen, and M. Campana (1988). "A generalized birthday attack." *Advances in Cryptology—EUROCRYPT'88,* Lecture Notes in Computer Science, vol. 330, ed. C.G. Günther. Springer-Verlag, Berlin, 129–156.

[40] Goldreich, O., S. Goldwasser, and S. Halevi (1996). "Collision-free hashing from lattice problems." *Theory of Cryptography Library*, http://philby.ucsd.edu/cryptolib.html

[41] Handschuh, H., L.R. Knudsen, and M.J.B. Robshaw (2001). "Analysis of SHA-1 in encryption mode." *Topics in Cryptology—CT-RSA 2001,* Lecture Notes in Computer Science, vol. 2020, ed. D. Naccache. Springer-Verlag, Berlin, 70–83.

[42] ISO/IEC 10118, *Information technology—security techniques—hash-functions, Part 1: General, 2000, Part 2: Hash-functions using an n-bit block cipher algorithm, 2000. Part 3: Dedicated hash-functions, 2003. Part 4: Hash-functions using modular arithmetic, 1998.*

[43] ISO-IEC/JTC1/SC27/WG2 N98, *Hash functions using a pseudo random algorithm.* Japanese contribution, 1991.

[44] Joux, A. and L. Granboulan (1995). "A practical attack against knapsack based hash functions." *Advances in Cryptology—EUROCRYPT'94,* Lecture Notes in Computer Science, vol. 950, ed. A. De Santis. Springer-Verlag, Berlin, 58–66.

[45] Kaliski Jr., B.S. (1992). "The MD2 Message-Digest algorithm." *Request for Comments (RFC) 1319,* Internet Activities Board, Internet Privacy Task Force.

[46] Kaliski Jr., B.S. (2002). "On hash function firewalls in signature schemes." *Topics in Cryptology—CT-RSA 2002,* Lecture Notes in Computer Science, vol. 2271, ed. B. Preneel. Springer-Verlag, Berlin, 1–16.

[47] Knudsen, L.R. (1995). "New potentially 'weak' keys for DES and LOKI." *Advances in Cryptology—EUROCRYPT'94,* Lecture Notes in Computer Science, vol. 950, ed. A. De Santis. Springer-Verlag, Berlin, 419–424.

[48] Knudsen, L.R., X. Lai, and B. Preneel (1998). "Attacks on fast double block length hash functions." *Journal of Cryptology,* 11 (1), 59–72.

[49] Knudsen, L. and B. Preneel (2002). "Enhancing the security of hash functions using non-binary error correcting codes." *IEEE Transactions on Information Theory,* 48 (9), 2524–2539.

[50] Lai, X. and J.L. Massey (1993). "Hash functions based on block ciphers." *Advances in Cryptology—EUROCRYPT'92,* Lecture Notes in Computer Science, vol. 658, ed. R.A. Rueppel. Springer-Verlag, Berlin, 55–70.

[51] Matyas, S.M., C.H. Meyer, and J. Oseas (1985). "Generating strong one-way functions with cryptographic algorithm." *IBM Technology Disclosure Bulletein,* 27 (10A), 5658–5659.

[52] Merkle, R. (1979). *Secrecy, Authentication, and Public Key Systems.* UMI Research Press.

[53] Merkle, R. (1990). "One way hash functions and DES." *Advances in Cryptology—CRYPTO'89,* Lecture Notes in Computer Science, vol. 435, ed. G. Brassard. Springer-Verlag, Berlin, 428–446.

[54] Merkle, R. (1990). "A fast software one-way hash function." *Journal of Cryptology,* 3 (1), 43–58.

[55] Meyer, C.H. and M. Schilling (1998). "Secure program load with Manipulation Detection Code." *Proceedings of Securicom,* 111–130.

[56] Micciancio, D. (2002). "Improved cryptographic hash functions with worst-case/average case connection." *Proceedings of 34th Annual ACM Symposium on Theory of Computing,* 609–618.

[57] Miyaguchi, S., M. Iwata, and K. Ohta (1989). "New 128-bit hash function." *Proceeding of 4th International Joint Workshop on Computer Communications, Tokyo, Japan, July 13–15,* 279–288.

[58] Moore, J.H. and G.J. Simmons (1987). "Cycle structure of the DES for keys having palindromic (or antipalindromic) sequences of round keys." *IEEE Transactions on Software Engineering,* 13, 262–273.

[59] Naor, M. and M. Yung (1990). "Universal one-way hash functions and their cryptographic applications." *Proceedings of 21st ACM Symposium on the Theory of Computing,* 387–394.

[60] Pal, P. and P. Sarkar (2003). "PARSHA-256—A new parallelizable hash function and a multithreaded implementation." *Fast Software Encryption,* Lecture Notes in Computer Science, vol. 2887, ed. T. Johansson. Springer-Verlag, Berlin, 347–361.

[61] Patarin, J. (1995). "Collisions and inversions for Damgård's whole hash function." *Advances in Cryptography—ASIACRYPT'94,* Lecture Notes in Computer Science, vol. 917, eds. J. Pieprzyk and R. Safavi-Naini. Springer-Verlag, Berlin, 307–321.

[62] Preneel, B. (1993). "Analysis and Design of Cryptographic Hash Functions." *Doctoral Dissertation,* Katholieke Universiteit Leuven.

[63] Preneel, B., R. Govaerts, and J. Vandewalle (1989). "Cryptographically secure hash functions: an overview." ESAT Internal Report, K.U. Leuven.

[64] Preneel, B., R. Govaerts, and J. Vandewalle (1994). "Hash functions based on block ciphers: A synthetic approach." *Advances in Cryptology—CRYPTO'93,* Lecture Notes in Computer Science, vol. 773, ed. D. Stinson. Springer-Verlag, Berlin, 368–378.

[65] Preneel, B. (2004). "Hash functions and MAC algorithms: State of the art." Lecture Notes in Computer Science, ed. B. Preneel. Springer-Verlag, Berlin, in print.

[66] Quisquater, J.-J. and J.-P. Delescaille (1990). "How easy is collision search? Application to DES." *Advances in Cryptology—EUROCRYPT'89,* Lecture Notes in Computer Science, vol. 434, eds. J.-J. Quisquater and J. Vandewalle. Springer-Verlag, Berlin, 429–434.

[67] Quisquater, J.-J. and J.-P. Delescaille (1990). "How easy is collision search. New results and applications to DES." *Advances in Cryptology—CRYPTO'89,* Lecture Notes in Computer Science, vol. 435, ed. G. Brassard. Springer-Verlag, Berlin, 408–413.

[68] Rabin, M.O. (1978). "Digitalized signatures." *Foundations of Secure Computation,* eds. R. Lipton and R. DeMillo. Academic Press, New York, 155–166.

[69] Rijmen, V. and B. Preneel (1995). "Improved characteristics for differential cryptanalysis of hash functions based on block ciphers." *Fast Software Encryption,* Lecture Notes in Computer Science, vol. 1008, ed. B. Preneel. Springer-Verlag, Berlin, 242–248.

[70] RIPE (1995). "Integrity primitives for secure information systems." Final Report of RACE Integrity Primitives Evaluation (RIPE-RACE 1040). Lecture Notes in Computer Science, vol. 1007, eds. A. Bosselaers and B. Preneel. Springer-Verlag, Berlin.

[71] Rivest, R.L. (1991). "The MD4 message digest algorithm." *Advances in Cryptology—CRYPTO'90,* Lecture Notes in Computer Science, vol. 537,

ed. S. Vanstone. Springer-Verlag, Berlin, 303–311.

[72] Rivest, R.L. (1992). "The MD5 message-digest algorithm." *Request for Comments (RFC) 1321*, Internet Activities Board, Internet Privacy Task Force.

[73] Rivest, R.L., A. Shamir, and L. Adleman (1978). "A method for obtaining digital signatures and public-key cryptosystems." *Communications ACM*, 21, 120–126.

[74] Rogaway, P. and T. Shrimpton (2004). "Cryptographic hash function basics: Definitions, implications, and separations for preimage resistance, second-preimage resistance, and collision resistance." *Fast Software Encryption*, Lecture Notes in Computer Science, vol. 3017, eds. W. Meier, and B.K. Roy. Springer-Verlag, Berlin, 371–388.

[75] Rogier, N. and P. Chauvaud (1997). "MD2 is not secure without the checksum byte." *Designs, Codes, and Cryptography*, 12 (3), 245–251.

[76] Schnorr, C.P. and S. Vaudenay (1994). "Parallel FFT-Hashing." *Fast Software Encryption,* Lecture Notes in Computer Science, vol. 809, ed. R. Anderson. Springer-Verlag, Berlin, 149–156.

[77] Simon, D. (1998). "Finding collisions on a one-way street: Can secure hash functions be based on general assumptions?" *Advances in Cryptology—EUROCRYPT'98,* Lecture Notes in Computer Science, vol. 1403, ed. K. Nyberg. Springer-Verlag, Berlin, 334–345.

[78] Stinson, D. (2001). "Some observations on the theory of cryptographic hash functions." Technical Report 2001/020, University of Waterloo.

[79] Tillich, J.-P. and G. Zémor (1994). "Hashing with SL_2." *Advances in Cryptology—CRYPTO'94,* Lecture Notes in Computer Science, vol. 839, ed. Y. Desmedt. Springer-Verlag, Berlin, 40–49.

[80] van Oorschot, P.C. and M. Wiener, (1999). "Parallel collision search with cryptanalytic applications." *Journal of Cryptology*, 12 (1), 1–28.

[81] Van Rompay, B. A. Biryukov, B. Preneel, and J. Vandewalle (2003). "Cryptanalysis of 3-pass HAVAL." *Advances in Cryptography—ASIACRYPT 2003,* Lecture Notes in Computer Science, vol. 2894, ed. C.S. Lai. Springer-Verlag, Berlin, 228–245.

[82] Winternitz, R. (1984). "A secure one-way hash function built from DES." *Proceedings of the IEEE Symposium on Information Security and Privacy*. IEEE Press, 88–90.

[83] Yuval, G. (1979). "How to swindle Rabin." *Cryptologia*, 3, 187–189.

[84] Zémor, G. (1994). "Hash functions and Cayley graphs." *Designs, Codes, and Cryptography*, 4 (4), 381–394.

[85] Zheng, Y., T. Matsumoto, and H. Imai (1990). "Connections between several versions of one-way hash functions." *Transactions on IEICE E*, E73 (7), 1092–1099.

[86] Zheng, Y., J. Pieprzyk, and J. Seberry (1993). "HAVAL—a one-way hashing algorithm with variable length output." *Advances in Cryptology—AUSCRYPT'92*, Lecture Notes in Computer Science, vol. 718, eds. J. Seberry and Y. Zheng. Springer-Verlag, Berlin, 83–104

HMAC

HMAC is a MAC algorithm designed by Bellare et al. [1] in 1996. A MAC algorithm is a cryptographic algorithm that computes a complex function of a data string and a secret key; the resulting MAC value is typically appended to the string to protect its authenticity. HMAC is a MAC algorithm based on a hash function. This type of construction became popular, because in the mid to late 1990's no secure and efficient custom designed MAC algorithms were available and hash functions (such as MD5) offered a much better performance than block ciphers; this implies that HMAC is faster than CBC-MAC. HMAC offers the advantage that it can be implemented without making any modification to the code of the hash function itself.

We present a description of HMAC for use with hash functions such as MD5, RIPEMD-160, and SHA-1, that process inputs in blocks of 512 bits; it is straightforward to extend this description to other hash functions:

$$\mathrm{MAC}_K(x) = h((K \oplus opad)\|h((K \oplus ipad)\|x)).$$

Here K is the key of the MAC algorithm (padded with zeroes to a block of 512 bits) and *opad* and *ipad* are constant 512-bit strings (equal to 64 times the hexadecimal values '36_x' and '$5c_x$', respectively). The resulting MAC value can (optionally) be truncated to 80 bits. In practice, one can apply the compression function f of the hash function (see Hash functions) to the strings $K \oplus ipad$ and $K \oplus opad$, respectively, and store the resulting values as initial values for the inner and outer hashing operations. This will reduce the number of operations of the compression function by two.

The designers of HMAC have proved that HMAC is a secure MAC algorithm if the following conditions hold: (i) the hash function is collision resistant for a secret value and random value $IV = H_0$ (the initial value of the iteration); (ii) its compression function is a secure MAC (with the secret key in the H_i input and the text in the X_i input); (iii) the compression function is 'weakly pseudo-random' (for more details see [1]). Note that the first condition is weaker than normal collision resistance with known IV. The other two conditions are different from standard conditions on hash functions, but some other hash function applications assume that they hold.

The best attack known on HMAC is a forgery attack based on internal collisions [5]. For HMAC based on MD5, this attack requires 2^{64} known text-MAC pairs (and a similar number of chosen texts if truncation is applied). For HMAC based on SHA-1 and RIPEMD-160, this increases to 2^{80} known text-MAC pairs.

HMAC has been included in ISO/IEC 9797-2 [3] and in FIPS 198 [2]. It is also widely used on the Internet [4]: HMAC-SHA-1 with a 160-bit key and a truncation to 96 bits is the mandatory algorithm for providing message *authentication* at the network layer (IPsec), and HMAC-MD5 is optional. Transport Layer Security (TLS) also uses both algorithms.

B. Preneel

References

[1] Bellare, M., R. Canetti, and H. Krawczyk (1996). "Keying hash functions for message authentication." *Advances in Cryptology—CRYPTO'96*, Lecture Notes in Computer Science, vol. 1109, ed. N. Koblitz. Springer-Verlag, Berlin, 1–15. Full version http://www.cs.ucsd.edu/users/mihir/papers/hmac.html

[2] FIPS 198 (2002). *The Keyed-Hash Message Authentication Code (HMAC)*. NIST, US Department of Commerce, Washington, DC.

[3] ISO/IEC 9797 (2002). "Information technology—Security techniques—Message Authentication Codes (MACs), Part 2: Mechanisms using a dedicated hash-function."

[4] Krawczyk, H., M. Bellare, and R. Canetti (1997). *HMAC: Keyed-Hashing for Message Authentication*. RFC 2104.

[5] Preneel, B. and P.C. van Oorschot (1995). "MDx-MAC and building fast MACs from hash functions." *Advances in Cryptology—CRYPTO'95*, Lecture Notes in Computer Science, vol. 963, ed. D. Coppersmith. Springer-Verlag, Berlin, 1–14.

HOMOMORPHISM

A mapping between two groups that preserves the group structure is a *homomorphism*. Let $(S, \circ)$ and $(T, \bullet)$ be two groups, and let f be a mapping from S to T. The mapping f is a homomorphism if, for all $x, y \in S$,

$$f(x \circ y) = f(x) \bullet f(y).$$

As an example, the mapping $f(x) = x^e \bmod n$ in RSA public-key encryption is a homomorphism

since

$$f(xy) \equiv (xy)^e \equiv f(x)f(y) \pmod{n}.$$

In this case, the sets S and T are the operations $\circ$ and $\bullet$ are the same. Another homomorphism is the mapping $f(x) = g^x \bmod p$ related to the discrete logarithm problem:

$$f(x + y) \equiv f(x)f(y) \pmod{p}.$$

Here, the sets and the operations are different between the two sides.

Ring homomorphisms that preserve group structure with respect to two operations may be defined similarly.

Homomorphisms are important in cryptography since they preserve relationships between elements across a transformation such as encryption. Sometimes the structure enables additional security features (as in blind signatures), and other times it enables additional attacks (e.g., Bleichenbacher's attack on an unprotected form of RSA encryption [1]).

Burt Kaliski

Reference

[1] Bleichenbacher, Daniel (1998). "Chosen ciphertext attacks against protocols based on RSA encryption standard PKCS #1. In *Advances in Cryptology—CRYPTO '98*, ed. H. Krawczyk, vol. 1462 of Lecture Notes in Computer Science, Springer, Berlin, 1–12.

HTTPS, SECURE HTTPS

HTTPS is a variant of http for handling secure transactions. A secure http request is made using an URL of the type "https://..." instead of the "http://..." request used for ordinary http. The default "https" port number is 443, as assigned by the Internet Assigned Numbers Authority.

In a secure http transaction, data sent to and received from an https server are protected using Secure Socket Layer (SSL) or Transaction Layer Security (TLS). Thus https is a two-step process in which security mechanisms and the necessary session keys are agreed initially. These session keys establish a secure tunnel during which the actual messages can be subsequently transmitted. The secure http server must have a certified public key (see certificate and public key cryptography), which is used when exchanging the session keys (e.g., the client generates and encrypts a common secret under the public key of the server). Only a server having the private key corresponding to the public key in the certificate is able to recover the exchanged secret and obtain a common key with the requesting entity. By looking up the information in the serverís certificate, the client can therefore be confident that it communicates with the right server and that only this server can see the contents of the messages transmitted over https.

HTTPS optionally supports identification of the client during key exchange. In this case, the client must have a certified key pair as well. During the initial key agreement, this key pair is used to authenticate the client (e.g., by making a digital signature). Thus the server may base it decision to proceed on the identity of the client. In particular, this allows the server to control access to its services.

Torben Pedersen

I

IDEA

IDEA (previous name *IPES*) is a 64-bit, 8.5-round non-Feistel block cipher with 128-bit keys, proposed by Lai and Massey in 1991 [12]. It is a modified version of a previous design called PES (Proposed Encryption Standard) by the same authors [11], with added strength against differential cryptanalysis. The *key-schedule* of the cipher is completely linear. The main idea behind the design is the mix of non-commuting group operations: addition mod 2^{16} (denoted by $\boxplus$), XOR (denoted by $\oplus$), multiplication mod $(2^{16}-1)$ (denoted by $\odot$, with $0 \equiv 2^{16}$). These operations work with 16-bit words. One round of IDEA is split into two different half-round operations: key mixing (denoted by T) and M-mixing denoted by $M = s \circ MA$, where MA denotes a multiplication–addition structure and s denotes a swap of two middle words.[1] T divides the 64-bit block into four 16-bit words X_1, X_2, X_3, X_4 and mixes the key words Z_1, Z_2, Z_3, Z_4 with the data using $\odot$ and $\boxplus$:

$$(X_1, X_2, X_3, X_4) \xrightarrow{T} (X_1 \odot Z_1, X_2 \boxplus Z_2,$$
$$X_3 \boxplus Z_3, X_4 \odot Z_4).$$

The transform MA provides *diffusion* between different words and mixes in two more key words Z_5, Z_6:

$$Y_1 = ((X_1 \oplus X_3) \odot Z_5 \boxplus (X_2 \oplus X_4)) \odot Z_6,$$
$$Y_2 = Y_1 \boxplus ((X_1 \oplus X_3) \odot Z_5),$$
$$(X_1, X_2, X_3, X_4) \xrightarrow{MA} (X_1 \oplus C_2, X_2 \oplus C_1,$$
$$X_3 \oplus C_2, X_4 \oplus C_1).$$

Both MA and s are involutions. The full 8.5-round IDEA can be written as

$$\text{IDEA} = T \circ s \circ (s \circ MA \circ T)^8 = T \circ s \circ (M \circ T)^8.$$

The only changes between IDEA and its predecessor PES, are in the order of operations in the key mixing subround T: PES uses the order $(\odot, \odot, \boxplus, \boxplus)$, while IDEA uses the order $(\odot, \boxplus, \boxplus, \odot)$, and in the swap of the words after the MA subround. In IDEA the outer words X_1, X_4 are not swapped. These changes were motivated by a differential attack on PES given in [12].

Since its publication, IDEA resisted intensive cryptanalytic efforts [1–8, 13]. In [10, p. 79] IDEA reduced to four rounds was claimed to be secure against differential attacks. Progress in cryptanalyzing round-reduced variants was very slow, starting with an attack on a two-round variant of IDEA in 1993 [13] by Meier, an improvement to 2.5 rounds by Daemen et al. [4], then an attack on 3.5 rounds published in 1997 [3] by Borst et al. Impossible differential attack significantly improved previous results for 3 and 3.5 rounds and could break up to 4.5 rounds [1] using the full codebook and 2^{112} steps. Finally the current best attack marginally breaks five rounds with a new variant of meet-in-the-middle attack due to Demirci et al. [6] and uses 2^{24} chosen plaintexts, 2^{58} memory and 2^{126} steps of analysis. This approach has higher analysis complexity than the impossible differential attack for 4 and 4.5 rounds, but requires less data.

In addition to these attacks three relatively large easily detectable classes of weak keys were found: in [5] 2^{51} weak keys out of the 2^{128} keys were found to be detectable with 16 chosen plaintexts and 2^{17} steps using differential *membership tests*, and in [7] 2^{63} weak keys were found to be detectable given 20 chosen plaintexts with a negligible complexity under differential–linear membership tests. Recently a boomerang membership test allowed to find a class of 2^{64} keys [2] with 2^{16} steps and queries for the membership test. Still the chance of choosing a weak key at random is about 2^{-64} which is extremely low. Using this approach for a 5-round reduced IDEA one key out of 2^{31} can be recovered with just 2^{10} adaptive chosen plaintext and chosen ciphertext queries, which compares favourably to the currently best attacks. Also related key attacks on 3.5 rounds [9] and on 4 rounds [7] of IDEA were developed.

Alex Biryukov

References

[1] Biham, E., A. Biryukov, and A. Shamir (1999). "Miss in the middle attacks on IDEA and Khufu." *Fast Software Encryption, FSE'99*, Lecture Notes in Computer Science, vol. 1636, ed. L.R. Knudsen. Springer-Verlag, Berlin, 124–138.

[2] Biryukov, A., J.N. Jr., B. Preneel, and J. Vandewalle (2002). "New weak-key classes of IDEA."

---

[1] As usual the composition of transformations is applied from right to left, i.e., MA is applied first, and the swap s is applied to the result.

International Conference on Information and Communications Security, ICICS 2002, Lecture Notes in Computer Science, vol. 2513, eds. R.H. Deng, S. Qing, F. Bao, and J. Zhou. Springer-Verlag, Berlin, 315–326.

[3] Borst, J., L.R. Knudsen, and V. Rijmen (1997). "Two attacks on reduced IDEA (extended abstract)." *Advances in Cryptology—EUROCRYPT'97*, Lecture Notes in Computer Science, vol. 1233, ed. W. Fumy. Springer-Verlag, Berlin, 1–13.

[4] Daemen, J., R. Govaerts, and J. Vandewalle (1993). "Cryptanalysis of 2.5 rounds of IDEA (extended abstract)." Technical Report 93/1, Department of Electrical Engineering, ESAT-COSIC.

[5] Daemen, J., R. Govaerts, and J. Vandewalle (1994). "Weak keys for IDEA." *Advances in Cryptology—CRYPTO'93*, Lecture Notes in Computer Science, vol. 773, ed. D.R. Stinson. Springer-Verlag, Berlin, 224–231.

[6] Demirci, H., A. Selçuk, and E. Türe (2004). "A new meet-in-the-middle attack on the IDEA block cipher." *Selected Areas in Cryptography, SAC 2003*, Lecture Notes in Computer Science, vol. 3006, eds. M. Matsui and R. Zuccherato. Springer-Verlag, Berlin.

[7] Hawkes, P. (1998). "Differential–linear weak key classes of IDEA." *Advances in Cryptology—EUROCRYPT'98*, Lecture Notes in Computer Science, vol. 1403, ed. K. Nyberg. Springer-Verlag, Berlin, 112–126.

[8] Hawkes, P. and L. O'Connor (1996). "On applying linear cryptanalysis to IDEA." *Advances in Cryptography—ASIACRYPT'96*, Lecture Notes in Computer Science, vol. 1163, eds. K. Kim and T. Matsumoto. Springer-Verlag, Berlin, 105–115.

[9] Kelsey, J., B. Schneier, and D. Wagner (1996). "Key-schedule cryptanalysis of IDEA, G-DES, GOST, SAFER, and Triple-DES." *Advances in Cryptology—CRYPTO'96*, Lecture Notes in Computer Science, vol. 1109, ed. N. Koblitz. Springer-Verlag, Berlin, 237–251.

[10] Lai, X. (1992). "On the Design and Security of Block Ciphers." *Doctoral Dissertation*, Swiss Federal Institute of Technology, Zurich.

[11] Lai, X. and J.L. Massey (1990). "A proposal for a new block encryption standard." *Advances in Cryptology—EUROCRYPT'90*, Lecture Notes in Computer Science, vol. 473, ed. I.B. Damgard. Springer-Verlag, Berlin, 389–404.

[12] Lai, X., J.L. Massey, and S. Murphy (1991). "Markov ciphers and differential cryptanalysis." *Advances in Cryptology—EUROCRYPT'91*, Lecture Notes in Computer Science, vol. 547, ed. D.W. Davies. Springer-Verlag, Berlin, 17–38.

[13] Meier, W. (1993). "On the security of the IDEA block cipher." *Advances in Cryptology—EUROCRYPT'93*, Lecture Notes in Computer Science, vol. 765, ed. T. Helleseth. Springer-Verlag, Berlin, 371–385.

IDENTIFICATION

A "name", or *identity*, is a set of information that distinguishes a specific entity from every other within a particular environment. In some environments, the "name" may just be a given name; in other environments, it will be a given name and a family name; in still others, it may include additional data such as a street address, or may be some other form entirely (for example, an employee number). In all cases, however, the identity depends upon the environment: the size and characteristics of the environment determine the amount of information required for uniqueness.

Identification is the claim of an identity. Each of two entities is involved in this process: the *claimant* claims an identity either explicitly or implicitly ("I am x"), and the *verifier* makes a corresponding claim with respect to the same identity ("The entity with whom I am dealing is X", where X is either x or a mapping from x to some other namespace that is meaningful to the verifier). In order for the verifier to believe its own claim enough to rely upon it for some purpose, there must be corroborating evidence of the claimant's claim. This evidence may come from the claimant directly or may come from some third party that is trusted by the verifier. In either case, the process of obtaining and verifying this evidence is known as authentication (see also entity authentication) and may make use of a protocol exchange between the verifier and the claimant (see identity verification protocol).

Depending upon the purpose for which the verifier needs the identity of the claimant (that is, depending upon how much the verifier must rely upon this identity), the authentication process associated with an identification step may be relatively weak, relatively strong, or somewhere in between. The verifier needs to assess and manage the risk that this identity may have been stolen by another entity who is now trying to impersonate the true holder of this identity (see impersonation attack). Impersonation can potentially lead to unauthorized access to personal or corporate data, networks, applications, or functions; the verifier will typically use stronger authentication mechanisms if a successful impersonation attack leads to the release of sensitive data.

IDENTITY UNIQUENESS: Identification is only possible within a domain when all identities in that domain are unique (i.e., no two entities in the domain have the same "name"). There are two general schemes for achieving identity uniqueness within a domain: *hierarchical namespaces* and *flat namespaces*. In a flat namespace, the Naming Authority (the authority that officially assigns identities to, or associates identities with, entities) is responsible for binding a unique identity to every entity in the domain and uses a fixed, nonextensible syntax to express this identity. Within a company, an *n*-digit employee number is an example of a flat namespace scheme.

In a hierarchical namespace, the Naming Authority is responsible for binding a unique identity to only a subset of the entities; these entities in turn are responsible for binding unique identities to other groups of entities, and so on, until every entity has a unique identity. The identity syntax is typically flexible and extensible. Examples of hierarchical namespace schemes include e-mail address, IP address, and X.500 Distinguished Name.

In general, there is a tradeoff between uniqueness and usability (user-friendliness) in a flat namespace. Furthermore, for many large domains—in particular, the domain of the entire world—there is no single recognized naming authority. Consequently, hierarchical schemes are typically used in practice in large-scale environments.

AUTHORITIES FOR NAMING AND AUTHENTICATION: Although they may in theory be the same, typically the Naming Authority and the Authentication Authority in a domain are distinct entities. The Naming Authority associates a "name" with an entity for the purposes of identification, while the Authentication Authority associates that same "name" with corroborating evidence for the purposes of authentication, and/or with an authentication mechanism for the purposes of identity verification. A Certification Authority in a Public Key Infrastructure (PKI) is an example of the latter type of Authentication Authority in that it binds a "name" (such as an e-mail address) to a public key pair that will be used as an authentication mechanism by the entity associated with that "name". A government department issuing driver's licenses is an example of the former type of Authentication Authority in that it writes a "name" (in this case, a given and family name, along with address and other information) into an official document (the driver's license) that may be used as corroborating evidence to validate a claim of identity.

Carlisle Adams

IDENTITY-BASED CRYPTOSYSTEMS

INTRODUCTION: Identity-based public key cryptography is a paradigm (see also identity-based encryption) introduced by Shamir in 1984 [29]. His motivation was to simplify key management and remove the need for public key certificates as much as possible by letting the user's public key be the binary sequence corresponding to an information identifying him in a nonambiguous way (e-mail address, IP address combined to a user name, telephone number...). The removal of certificates allows avoiding the trust problems encountered in current public key infrastructures (PKIs): it is no longer necessary to bind a public key to its owner's name since it is one single thing and it also simplifies key management since public keys are human-memorizable. These systems involve trusted authorities called private key generators (PKGs) that have to deliver private keys to users after having computed them from their identity information (users do not generate their key pairs themselves) and from a master secret key. End users do not have to enquire for a certificate for their public key. The only things that still must be certified are the public keys of trusted authorities (PKGs). This does not completely remove the need for certificates but, since many users depend on the same authority, this need is drastically reduced. Several practical solutions for identity-based signatures (IBS) have been devised since 1984 [13, 17, 28] but finding a practical identity-based encryption scheme (IBE) remained an open challenge until 2001 when Boneh and Franklin [5] proposed to use bilinear maps (the Weil or Tate pairing) over supersingular elliptic curves to achieve an elegant identity-based encryption method. Other identity-based signature and key agreement schemes based on pairings were proposed after 2001 [11, 18, 30].

Basically, an identity-based cryptosystem is made of four algorithms. First, a *Setup* algorithm that is run by a PKG and takes as input a security parameter to output a public/private key pair (P_{pub}, mk) for the PKG (P_{pub} is its public key and mk is its master key that is kept secret). Second, a key generation algorithm *Keygen* that is run by a PKG: it takes as input the PKG's master key mk and a user's identity ID to return the user's private key d_{ID}. In the case of identity-based encryption, the third algorithm is an encryption algorithm *Encrypt* that is run by anyone and takes as input a plaintext M, the recipient's identity and the PKG's public key P_{pub} to output a ciphertext C. The last algorithm is then the decryption algorithm *Decrypt* that takes as input the ciphertext C and the private decryption key d_{ID} to return a plaintext M. In the case of identity-based signatures, the last two algorithms are the signature generation algorithm *Sign* that, given a message M, the PKG's public key and a private key d_{ID} generates a signature on M that can be verified by anyone thanks to the signature verification algorithm *Verify*. The latter takes as input the PKG's key P_{pub} and the alleged signer's identity ID to return 1 or 0 depending on whether the signature is acceptable or not.

In this chapter we give a survey of the main advancements achieved in the field of identity-based cryptography since Shamir's call for proposals in 1984. We first recall two schemes obtained from simple modular arithmetic before showing examples obtained from bilinear maps over elliptic curves.

ID-Based Cryptosystems from Modular Arithmetic:
This section presents two simple identity-based cryptosystems: the Guillou–Quisquater [17] digital signature and Cocks's public key encryption scheme [12]. Both are obtained from modular arithmetic and their security relies on the intractability of factoring large integers. The first one uses the RSA trapdoor permutation while the second one is based on quadratic residues.

The Guillou–Quisquater Signature Scheme

This scheme is derived from a three round identification scheme. It was proposed in 1988 and is made of the four following algorithms:

Setup: Given a security parameter k_0, the private key generator (PKG) picks two $k_0/2$-bit primes p and q and computes $n = pq$. It also picks a prime number $e \in \mathbb{Z}_{\varphi(n)}$ such that $\gcd(e, \varphi(n)) = 1$ and chooses a cryptographic hash function $h :\rightarrow \mathbb{Z}_e$ and a redundancy function $R : \{0, 1\}^* \rightarrow \mathbb{Z}_n$. The pair (n, e) is its public key while the pair (p, q) is kept secret and is its master key. The functions R and h are also made public.

Keygen: Given a user's identity ID, the PKG computes $I = R(ID) \in \mathbb{Z}_n^*$ and $a \in \mathbb{Z}_n^*$ such that $Ia^e \equiv 1 \pmod{n}$. The obtained a is returned to the user as a private key.

Sign: Given a message m, the signer does the following:
1. Pick a random $k \leftarrow_R \mathbb{Z}_n^*$ and compute $r = k^e \bmod n$
2. Compute $\ell = h(m \| r) \in \mathbb{Z}_e$
3. Calculate $s = ka^\ell \bmod n$

The signature on m is the pair (s, ℓ).

Verify: To verify a signature $(s\ell)$ on m,
1. Compute $I = R(ID)$ from the signer's identity ID.
2. Compute $u = s^e I^\ell \bmod n$.
3. Accept the signature if $\ell = h(m \| u)$.

To verifiy the consistency of the scheme, we note that

$$u \equiv s^e I^\ell \equiv (ka^\ell)^e I^\ell \equiv k^e(a^e I)^\ell \equiv k^e \equiv r \pmod{n}.$$

Hence $u = r$ and then $h(m\|u) = h(m\|r)$. This signature scheme is derived from the Guillou–Quisquater identification protocol (GQ) using the Fiat–Shamir heuristic [13] (that allows the conversion of any identification scheme into a digital signature by the replacement of the verifier's challenge with the hash value of the message to sign concatenated to the commitment). That is why the output of the hash function h must be smaller than e (the set of challenges is $\mathbb{Z}_e$ in the underlying identification protocol). The GQ signature scheme can be proved to be existentially unforgeable provided it is hard to invert the RSA function (see RSA public-key encryption) by using the proof technique of Pointcheval and Stern [26, 27]. The public exponent is taken as a prime for provable security purposes. The redundancy function R aims at preventing attacks that could take advantage of multiplicative relations between identities. In order to avoid birthday paradox attacks on the hash function, it is recommended to use public exponents e of at least 160 bits (in the corresponding identification scheme, shorter exponents are allowed). The security parameters should be at least 1024 or 2048 to avoid attacks trying to factor the modulus.

Cocks's Identity-Based Cryptosystem

This encryption scheme, due to Cocks in 2001, is based on quadratic residues and on the properties of the Legendre and Jacobi symbols for Blum integers (i.e., composite numbers n that are a product of two primes p and q such that $p \equiv q \equiv 3 \pmod 4$). It is made of the four algorithms depicted below:

Setup: The PKG picks prime numbers p and q such that $p \equiv q \equiv 3 \pmod 4$, computes their product $n = pq$ that is made public together with a hash function $H : \{0, 1\}^* \rightarrow \mathbb{Z}_n^*$. The PKG's master key is (p, q).

Keygen: Given an identity ID, the PKG computes a chain of hash values starting from ID until obtaining $a = H(H(H \ldots (ID))) \in \mathbb{Z}_n^*$ such that $(\frac{a}{n}) = 1$. For such a $a \in \mathbb{Z}_n^*$, either a or $-a$ is a square in $\mathbb{Z}_n^*$. It is easy to verify that $r = a^{\frac{n+5-(p+q)}{8}} \bmod n$ satisfies $a = r^2 \bmod n$ or $a = -r^2 \bmod n$ depending on whether $(\frac{a}{p}) = (\frac{a}{q}) = 1$ or

$(\frac{a}{p}) = (\frac{a}{q}) = -1$. The obtained r is returned to the user as a private key.

Encrypt: The sender A ignores which of a or $-a$ is a square in $\mathbb{Z}_n^*$. We first assume we are in the case $a = r^2 \bmod n$. A generates a symmetric transport key K and encrypts the plaintext M with it. Each bit x of that symmetric key is then encrypted before being sent to the receiver B. To do this, A encodes x in $\{-1,1\}$ rather than in $\{0,1\}$ and does the following.

1. Pick a random $t \in \mathbb{Z}_n^*$ such that $(\frac{t}{n}) = x$.
2. Compute $s = (t + \frac{a}{t}) \bmod n$ (since $(\frac{t}{n}) \neq 0$, t is coprime with p and q and thus invertible in $\mathbb{Z}_n$) and send it to B.

Since A does not know which of a or $-a$ is the square of B's decryption key, A has to repeat the above process for a new t and, this time, send $s = (t - a/t) \bmod n$. Hence, $2|n|$ bits, where $|x|$ denotes the bitlength of x, have to be transmitted for each bit of the symmetric key.

Decrypt: B recovers x as follows. Given that

$$t\left(1 + \frac{r}{t}\right)^2 \equiv t + 2r + \frac{r^2}{t} \equiv t + 2r + \frac{a}{t}$$
$$\equiv s + 2r \pmod{n},$$

B can compute $(\frac{s+2r}{n}) = (\frac{t}{n}) = x$ and recover x using his/her private key r thanks to the multiplicative properties of the Jacobi symbol. Once the symmetric key K is obtained in clear, the ciphertext can be decrypted.

For 128-bit symmetric keys, the scheme is reasonably computationally cheap: the sender's computing time is dominated by 2×128 Jacobi symbol evaluations and 2×128 modular inversions (see inversion in finite fields and rings). The receiver just has to compute 128 Jacobi symbols since he/she knows which of a or $-a$ is the square of his/her private key. The drawback of the scheme is its bandwidth overhead: for a 1024-bit modulus n and a 128-bit symmetric transport key, at least 2×16 KB need to be transmitted if all the integers s are sent together. This remains reasonable for applications such as secure e-mail.

Cocks provided no formal security proof for his scheme. He informally demonstrated that his construction is secure against chosen-plaintext attacks under the Quadratic Residuosity Assumption (i.e., the hardness of deciding whether or not a random integer a such that $(\frac{a}{n}) = 1$ is a square or not). He argued that several generic transformations can be applied to the scheme to turn it into a chosen-ciphertext secure one.

ID-BASED CRYPTOSYSTEMS FROM PAIRINGS OVER ELLIPTIC CURVES: Pairings are bilinear maps that are efficiently computable for a particular kind of elliptic curves. These curves have been thought to be unsuitable for cryptographic purposes since the MOV reduction [22] from the elliptic curve discrete logarithm problem (ECDLP) to the discrete logarithm in a finite field. In 2000, Joux [20] showed how to use pairings over these curves in constructive cryptographic applications. Namely, he showed how to devise a one-round tripartite Diffie–Hellman protocol using pairings. Since that seminal paper, pairings provided a couple of other applications [5, 7–11, 15, 18, 25, 28]. One of the most important was Boneh and Franklin's identity-based encryption protocol [5] that is described further.

Overview of Pairings

Let us consider groups $\mathbb{G}_1$ and $\mathbb{G}_2$ of the same prime order q. We need a bilinear map $\hat{e} : \mathbb{G}_1 \times \mathbb{G}_2 \to \mathbb{G}_2$ satisfying the following properties:
1. Bilinearity: $\forall P, Q \in \mathbb{G}_1, \forall a, b \in \mathbb{Z}_q^*$, we have $\hat{e}(aP, bQ) = \hat{e}(P, Q)^{ab}$.
2. Non-degeneracy: for any point $P \in \mathbb{G}_1, \hat{e}(P, Q) = 1$ for all $Q \in \mathbb{G}_1$ iff $P = \mathcal{O}$.
3. Computability: there exists an efficient algorithm to compute $\hat{e}(P, Q) \quad \forall P, Q \in \mathbb{G}_1$.

The modified Weil pairing [5] and the Tate pairing are admissible applications. $\mathbb{G}_1$ is a cyclic subgroup of the additive group of points of a supersingular elliptic curve (i.e., a curve for which the number of points is a multiple of the underlying field's characteristic) over a finite field. $\mathbb{G}_2$ is a cyclic subgroup of the multiplicative group associated to a finite extension of $\mathbb{F}_p$. The security of the schemes described in this section relies on the hardness of the following problems.

DEFINITION 1.
- Given a group $\mathbb{G}_1$ with generator P, the **Computational Diffie–Hellman problem** (CDH) is, given (aP, bP), to compute $abP \in \mathbb{G}_1$
- In a group $\mathbb{G}_1$, the **Decisional Diffie–Hellman problem** (DDH) is, given (aP, bP, cP), to decide whether $c \equiv ab \pmod{q}$ or not. Tuples (aP, bP, cP) satisfying this equality are called "valid Diffie–Hellman tuple".
- Given groups $\mathbb{G}_1$ and $\mathbb{G}_2$ of prime order q, a bilinear map $\hat{e} : \mathbb{G}_1 \times \mathbb{G}_2 \to \mathbb{G}_2$ and a generator P of $\mathbb{G}_1$, the **bilinear Diffie–Hellman problem** (BDH) in $(\mathbb{G}_1, \mathbb{G}_2, \hat{e})$ is to compute $\hat{e}(P, P)^{abc}$ given (P, aP, bP, cP).

So far, it is not known whether the BDH problem is strictly easier than the CDH problem or not. However, no algorithm is known to solve it in a polynomial time. It was shown by Joux and

Nguyen [21] that elliptic curves for which pairings can be efficiently computed give rise to groups in which the CDH problem is still considered intractable while the DDH problem is easy. Indeed, given an instance (aP, bP, cP) of the DDH problem, it sufices to compare $\hat{e}(aP, bP)$ and $\hat{e}(P, cP)$ to decide whether $c \equiv ab \pmod{p}$ or not. This kind of algebraic group is called "Gap Diffie–Hellman groups" according to the terminology used in [24]. The separation between computational and decisional problems is used in recent new cryptographic constructions like [8, 11] and many others.

The Boneh–Franklin Identity-Based Encryption Scheme [6]

We describe the basic version of the scheme [6]. This version is only provably secure against chosen-plaintext attacks and has some similarities with El Gamal's cryptosystem. Boneh and Franklin showed that applying the Fujisaki-Okamoto generic transformation [14] allows turning this basic scheme into a chosen-ciphertext secure one. In fact, they introduced an extended security model for chosen-ciphertext security in the identity-based setting. That model considers the fact that, before trying to find information on the plaintext corresponding to a ciphertext produced using a given public key, an attacker might be in possession of private keys associated to other identities. Their scheme, proposed in 2001 (a short time before Cocks' one) turns out to be the first one to be really secure and practical.

In the notation below, the symbol $\oplus$ denotes the exclusive bitwise OR operation.

Setup: Given a security parameter k, the PKG chooses groups $\mathbb{G}_1$ and $\mathbb{G}_2$ of prime order $q > 2^k$. A generator P of $\mathbb{G}_1$, a randomly chosen master key $s \in \mathbb{Z}_q^*$ and the associated public key $P_{pub} = sP$ are selected. The space of plaintexts is $\mathcal{M} = \{0, 1\}^n$ for a fixed n while $\mathcal{C} = \mathbb{G}_1^* \times \{0, 1\}^n$ is the ciphertext space. Cryptographic hash functions

$$H_1 : \{0, 1\}^* \to \mathbb{G}_1^* \quad \text{et} \quad H_2 : \mathbb{G}_2 \to \{0, 1\}^n.$$

are also chosen. The system's public parameters are params $= (\mathbb{G}_1, \mathbb{G}_2, \hat{e}, n, P, P_{pub}, H_1, H_2)$.

Keygen: Given identity $ID \in \{0, 1\}^*$, the PKG computes $Q_{ID} = H_1(ID) \in \mathbb{G}_1$ and $d_{ID} = sQ_{ID} \in \mathbb{G}_1$ that is given to the user as a private key.

Encrypt: To encrypt a message $M \in \mathcal{M}$, the sender follows the steps below:
1. Compute $Q_{ID} = H_1(ID) \in \mathbb{G}_1$.
2. Pick a random $r \in \mathbb{Z}_q^*$.
3. Compute $g_{ID} = \hat{e}(Q_{ID}, P_{pub}) \in \mathbb{G}_2$.
   The ciphertext $C = (rP, M \oplus H_2(g_{ID}^r))$ is sent to the receiver.

Decrypt: Given a ciphertext $C = (U, V)$, the receiver uses his/her private key d_{ID} to decrypt by computing $M = V \oplus H_2(\hat{e}(d_{ID}, U))$.

The protocol's consistency is easy to check. Indeed, if the sender correctly encrypted the message, we have $U = rP$ and

$$\hat{e}(d_{ID}, U) = \hat{e}(sQ_{ID}, rP) = e(Q_{ID}, P_{pub})^r = g_{ID}^r.$$

We note that all the above algorithms make use of a hash function that has the group $\mathbb{G}_1$ as a range. It is explained in [5] how to efficiently implement such a hash function.

It is shown in [5] that, when padded with the Fujisaki-Okamoto transform, the above basic scheme is provably secure against adaptive chosen-ciphertext attacks provided the Bilinear Diffie–Hellman problem is hard.

In the version described above, the crucial information is the PKG's master key: all the system's privacy is compromised if that master key is ever stolen by an attacker. In order to avoid having a single target for attacks (and remove key escrow), it is possible to split the PKG into several partial PKGs in such a way that these partial PKGs jointly generate a discrete logarithm key pair and each of these eventually holds a piece of the master key. Users then have to visit all the different PKGs to obtain their partial private decryption keys that can then be recombined into a full decryption key (see also secret sharing scheme and threshold scheme).

An Identity-Based Signature from Pairings

Guillou and Quisquater [17] showed an example of identity-based signature that was based on simple arithmetic and obtained from an identification scheme using the Fiat–Shamir heuristic. Their scheme is provably secure under the RSA assumption using Pointcheval and Stern's forking lemma [26, 27]. In 2000, Sakai et al. proposed another identity-based signature based on pairings over elliptic curves [28]. Unfortunately, they did not provide any security proof for that scheme. Paterson also proposed another ID-based signature [25] that is still not provably secure. In 2002 appeared a first example of provably secure identity-based signature obtained from these fashionable bilinear maps. That scheme was proven secure under the Diffie–Hellman assumption using the Pointcheval–Stern proof technique. It can be depicted as follows.

Setup: The PKG generates the same public parameters as in the Boneh–Franklin scheme. The only diference is that it uses hash functions $H_1 : \{0, 1\}^* \times \mathbb{G}_1 \to \mathbb{Z}_q$ and $H_2 : \{0, 1\}^* \to \mathbb{G}_1$.

The pubic parameters are $(\mathbb{G}_1, \mathbb{G}_2, \hat{e}, P, P_{pub}, H_1, H_2)$. The master key is $s \in \mathbb{Z}_q$.

Keygen: As in the Boneh–Franklin scheme, for an identity ID, the PKG computes $Q_{ID} = H_2(ID)$ and the private key $d_{ID} = s\,Q_{ID}$ is transmitted to the user.

Sign: Given his/her private key d_{ID} and a message $M \in \mathcal{M}$ to sign, the signer randomly picks $r \leftarrow_R \mathbb{Z}_q$ and computes

$$U = r\,Q_{ID}, \ h = H_1(M, U), \ \text{et } V = (r + h)d_{ID}.$$

The signature is $\sigma = (U, V)$.

Verify: To verify a signature $\sigma = (U, V)$ on a message M and an identity ID, one first computes $h = H_1(M, U)$ and then verifies if $(P, P_{pub}, U + hQ_{ID}, V)$ is a valid Diffie–Hellman tuple (by checking if $\hat{e}(P, V) = \hat{e}(P_{pub}, U + hQ_{ID})$ as explained in Section "Overview of pairings"). If this condition holds, the verifier accepts the signature. It rejects it otherwise.

To check the consistency of the scheme, one can note that, if $\sigma = (U, V)$ is a valid signature on a message M and an identity ID, then we have

$$\begin{aligned}
&(P, P_{pub}, U + hQ_{ID}, V) \\
&= (P, P_{pub}, (r + h)Q_{ID}, (r + h)d_{ID}) \\
&= (P, sP, (r + h)Q_{ID}, s(r + h)Q_{ID}) \\
&= (P, sP, x(r + h)P, xs(r + h)P),
\end{aligned}$$

because $Q_{ID} = xP$ for some $x \in \mathbb{Z}_q$ since $\mathbb{G}_1$ is a cyclic group. We thus actually have a valid Diffie–Hellman tuple.

The signature generation algorithm is quite efficient since it requires two multiplications in $\mathbb{G}_1$ as most expensive operations. The verification algorithm requires two pairing evaluations to solve an instance of the DDH problem and is thus more expensive than the signature generation operation. Globally, the performances of the complete protocol are comparable to those of the Boneh–Franklin encryption scheme. From a bandwidth point of view, if the scheme is implemented using a suitable elliptic curve, one can obtain 320-bit signatures.

Hierarchical Identity-Based Cryptography

A shortcoming of the Boneh–Franklin identity-based encryption scheme is that in a large network, the PKG's key generation task rapidly becomes a bottleneck since many private keys have to be computed and secure channels have to be established to transmit them to their legitimate owner. To overcome this problem, a solution is to set up a hierarchy of PKGs in which each PKG only computes private keys for entities (other PKGs or end-users) that are immediately below them in the hierarchy. In such hierarchies, entities are represented by a tuple of identifying strings IDs (i.e., a concatenation of their identifier to those of all their ancestors'one: for example a child of $\langle ID_1, \ldots, ID_i \rangle$ has an address $\langle ID_1, \ldots, ID_i, ID_{i+1} \rangle$) rather than a single identifier as in the Boneh–Franklin scheme.

In this section, we give an example, proposed by Gentry and Silverberg [16], of such a hierarchical scheme that can be viewed as a scalable extension of Boneh and Franklin's proposal (both schemes are identical if the hierarchy has a single level). Unlike another hierarchical scheme proposed by Horwitz and Lynn [19], this one supports an arbitrary number of levels ℓ. The end-users are always located at the leafs of the hierarchy. Lower-level PKGs (i.e., PKGs other than Root PKG located at the top of the hierarchy) generate private keys for their children by using some public information coming from their ancestors together with a private information that is only known to them. Each of them then adds some information to the parameters that are known to their children.

In our notation, we call $Level_i$ the set of entities at level i, $Level_0$ being the sole Root PKG. The simplified version of the scheme is made of the following algorithms:

Root Setup: Given a security parameter k, the root PKG

1. generates groups $\mathbb{G}_1$ and $\mathbb{G}_2$ of prime order q and a symmetric bilinear map $\hat{e} : \mathbb{G}_1 \times \mathbb{G}_1 \to \mathbb{G}_2$.
2. picks a generator $P_0 \in \mathbb{G}_1$.
3. chooses $s_0 \in \mathbb{Z}_q$ and sets $Q_0 = s_0 P_0$.
4. chooses hash functions $H_1 : \{0, 1\}^* \to \mathbb{G}_1$ and $H_2 : \mathbb{G}_2 \to \{0, 1\}^n$ for some n denoting the size of plaintexts.

The space of plaintexts $\mathcal{M} = \{0, 1\}^n$ is that of ciphertexts $\mathcal{C} = \mathbb{G}_1^t \times \{0, 1\}^n$ where t is the level of the ciphertext's recipient in the hierarchy. The public paramers are $\texttt{params} := (\mathbb{G}_1, \mathbb{G}_2, \hat{e}, P_0, Q_0, H_1, H_2)$. The master key is $s_0 \in \mathbb{Z}_q$.

Lower Level Setup: An entity E_t at level $Level_t$ randomly picks $s_t \in \mathbb{Z}_q$ and keeps it secret.

Keygen$_{t-1}$: At level $Level_t$, we consider an entity E_t of address $(ID_1, \ldots, ID_t)$, where $(ID_1, \ldots, ID_i)$, for $1 \leq i \leq t - 1$, is the address of its ancestor at $Level_i$. Let S_0 be the unit element of $\mathbb{G}_1$. At $Level_{t-1}$, the father E_{t-1} of E_t generates E_t's private key as follows:

1. It computes $P_t = H_1(ID_1, \ldots, ID_t) \in \mathbb{G}_1$.
2. It computes E_t secret point $S_t = S_{t-1} + s_{t-1}P_t$ and transmits it to E_t. We thus have

$$S_t = s_0 P_1 + s_1 P_2 + \cdots + s_{t-1}P_t = \sum_{i=1}^{t} s_{i-1}P_i$$

S_t then depends on secret elements $S_{t-1} \in \mathbb{G}_1$ and $s_{t-1} \in \mathbb{Z}_q$ of E_{t-1}.

3. E_{t-1} also transmits $Q_i = s_i P_0 \in \mathbb{G}_1$ to E_t for $1 \leq i \leq t-1$ (it must have computed $Q_{t-1} = s_{t-1} P_0$ itself from its secret s_{t-1} and have received $Q_0, \ldots, Q_{t-2}$ from its ancestors).

E_t thus obtains its private key $(S_t, Q_1, \ldots, Q_t, s_t)$. The part $(S_t, Q_1, \ldots, Q_{t-1})$ is received from its father E_{t-1} and it generates the components s_t and $Q_t = s_t P_0$ itself.

Encrypt: To encrypt a message $M \in \mathcal{M}$ for an entity E_t of address $(ID_1, \ldots, ID_t)$,

1. Alice computes $P_i = H_1(ID_1, \ldots, ID_i) \in \mathbb{G}_1$ for $1 \leq i \leq t$.
2. She randomly picks $r \in \mathbb{Z}_q$.
3. The ciphertext is $C = [r P_0, r P_2, \ldots, r P_t, M \oplus H_2(g^r)]$ avec $g = \hat{e}(Q_0, P_1) \in \mathbb{G}_2$.

Decrypt: E_t receives $C = [U_0, U_2, \ldots, U_t, V] \in \mathcal{C}$. To decrypt it, he/she computes

$$V \oplus H_2 \left(\frac{\hat{e}(U_0, S_t)}{\prod_{i=2}^{t} \hat{e}(Q_{i-1}, U_i)} \right) = M.$$

The bilinearity of the map $\hat{e}$ allows verifying the consistency of the scheme:

$$\hat{e}(U_0, S_t) = \hat{e}(r P_0, \sum_{i-1}^{t} s_{i=1} P_i)$$
$$= \hat{e}(P_0, P_1)^{r s_0} \hat{e}(P_0, P_2)^{r s_1} \cdots \hat{e}(P_0, P_t)^{r s_{t-1}}$$
$$= \hat{e}(Q_0, P_1)^r \hat{e}(Q_1, P_2)^r \cdots \hat{e}(Q_{t-1}, P_t)^r$$
$$= g^r \hat{e}(Q_1, U_2) \cdots \hat{e}(Q_{t-1}, U_t)$$

and

$$\frac{\hat{e}(U_0, S_t)}{\prod_{i=2}^{t} \hat{e}(Q_{i-1}, U_i)} = g^r$$

for the g computed by Alice at the encryption.

The above version of the scheme is a simplified one satisfying only the security notion of one-wayness against chosen-plaintexts attacks. To convert it into a chosen-ciphertext secure one, the Fujisaki-Okamoto generic transformation [14] is simply applied to it. Unlike the 2-level solution proposed by Horwitz and Lynn in 2002, the resulting scheme provably resists a collusion between any number of dishonest users (i.e., a set of users combining their private information in an attempt to threaten the confidentiality of messages sent to a honest user) provided the Bilinear Diffie–Hellman problem is hard. It is shown in [16] how to turn the above encryption scheme into a hierarchical identity-based signature and how to shorten the ciphertexts produced by the scheme (as well as the signatures outputted by the derived hierarchical signature).

CONCLUSIONS AND OPEN PROBLEMS: This chapter gave a basic overview of the major advancements in the field of identity-based cryptography since 1984. Although many really interesting breakthroughs have been achieved, there still remain problems that might hamper this kind of cryptosystem to replace today's certificate based public key infrastructures. For example, ID-based cryptography only provides automatic revocation of a user's public key privileges by concatenating a validity period to users' identifying information (in such a way that a key is never used outside its validity period and different private keys have to be given to a user for each time period). Except a method proposed in [4] (which requires the use of an online server holding a piece of each user's private key), no easy method to provide fine-grain revocation (which is mandatory in case of a user's key compromise) has been found so far. Another problem is the key escrow that might be undesirable for some applications: except the case of hierarchical encryption where it is somewhat restricted (entities are only able to decrypt messages intended for their children), PKGs are able to decrypt ciphertexts intended for any user depending on them. Solutions to this problem have been proposed [2, 15] but they imply the loss of the easier key management advantage that is provided by human-memorizable public keys in ID-based cryptography.

Other problems are still unsolved in this area: for example no really practical identity-based encryption scheme avoiding the use of pairings has been proposed so far (Cocks's one is not really practical from a bandwidth point of view since a ciphertext's size is many times that of the plaintext). Another open challenge is to devise an identity-based cryptosystem that is provably secure in the security model described in [5] in the standard model rather than the random oracle model [3] (such a result was obtained in [10] for a security model weaker than the one presented in [5]).

We also mention the existence of many other proposals for ID-based cryptographic protocols from pairings. They are not discussed here but are referenced in [1].

Benoît Libert
Jean-Jacques Quisquater

References

[1] Barreto, P.-S.-L.M. (2002). *The Pairing Based Crypto Lounge*. Web page located at http://planeta.terra.com.br/informatica/paulobarreto/pblounge.html

[2] Al-Riyami, S. and K.G. Paterson (2003). "Certificateless public key cryptography." Eprint available at http://eprint.iacr.org/2003/126/

[3] Bellare, M. and P. Rogaway (1993). "Random oracles are practical: A paradigm for designing efficient protocols." *Proc. of the 1st ACM Conference on Computer and Communications Security*, 62–73.

[4] Boneh, D., X. Ding, G. Tsudik, and M. Wong (2001). "A method for fast revocation of public key certificates and security capabilities." *Proceedings of the 10th USENIX Security Symposium*, 297–308.

[5] Boneh, D. and M. Franklin (2001). "Identity based encryption from the Weil pairing." *Advances in Cryptology—CRYPTO 2001*, Lecture Notes in Computer Science, vol. 2139, ed. J. Kilian. Springer-Verlag, Berlin, 213–229.

[6] Boneh, D. and M. Franklin (2003). "Identity based encryption from the Weil pairing." *SIAM J. of Computing*, 32 (3), 586–615, extended version of [5].

[7] Boneh, D., C. Gentry, B. Lynn, and H. Shacham (2003). "Aggregate and verifiably encrypted signatures from bilinear maps." *Advances in Cryptology—EUROCRYPT'03*, Lecture Notes in Computer Science, vol. 2656, ed. E. Biham. Springer, Berlin, 416–432.

[8] Boneh, D., B. Lynn, and H. Shacham (2001). "Short signatures from the Weil pairing." *Advances in Cryptography—ASIACRYPT 2001*, Lecture Notes in Computer Science, vol. 2248, ed. C. Boyd. Springer, Berlin, 514–532.

[9] Boyen, X. (2003). "Multipurpose identity-based signcryption: A swiss army knife for identity-based cryptography." *Advances in Cryptology—CRYPTO 2003*, Lecture Notes in Computer Science, vol. 2729, ed. D. Boneh. Springer-Verlag, Berlin, 382–398.

[10] Canetti, R., S. Halevi, and J. Katz (2003). "A forward-secure public-key encryption scheme." *Advances in Cryptology—EUROCRYPT 2003*, Lecture Notes in Computer Science, vol. 2656, ed. E. Biham. Springer, Berlin, 255–271.

[11] Cha, J.C. and J.H. Cheon (2003). "An identity-based signature from gap Diffie–Hellman groups." *Proceedings of PKC 2003*, Lecture Notes in Computer Science, vol. 2567, ed. Y.G. Desmedt. Springer-Verlag, Berlin, 18–30.

[12] Cocks, C. (2001). "An identity based encryption scheme based on quadratic residues." *Proc. of Cryptography and Coding*, Lecture Notes in Computer Science, vol. 2260, ed. B. Honary. Springer, Berlin, 360–363.

[13] Fiat, A. and A. Shamir (1986). "How to prove yourself: practical solutions to identification and signature problems." *Advances in Cryptology—CRYPTO'86*, Lecture Notes in Computer Science, vol. 263, ed. A. Odlyzko. Springer-Verlag, Berlin, 186–194.

[14] Fujisaki, E. and T. Okamoto (1999). "Secure integration of asymmetric and symmetric encryption schemes." *Advances in Cryptology—CRYPTO'99*, Lecture Notes in Computer Science, vol. 1666, ed. J. Wiener. Springer-Verlag, Berlin, 537–554.

[15] Gentry, C. (2003). "Certificate-based encryption and the certificate revocation problem." *Advances in Cryptology—EUROCRYPT 2003*, Lecture Notes in Computer Science, vol. 2656, ed. E. Biham. Springer-Verlag, Berlin, 272–293.

[16] Gentry, C. and A. Silverberg (2002). "Hierarchical ID-based cryptography." *Advances in Cryptography—ASIACRYPT 2002*, Lecture Notes in Computer Science, vol. 2501, ed. Y. Zheng. Springer-Verlag, Berlin, 548–566.

[17] Guillou, L. and J.-J. Quisquater (1988). "A paradoxical identity-based signature scheme resulting from zero-knowledge." *Advances in Cryptology—CRYPTO'88*, Lecture Notes in Computer Science, vol. 403, ed. S. Goldwasser. Springer-Verlag, Berlin, 216–231.

[18] Hess, F. (2003). "Efficient identity based signature schemes based on pairings." *Proceedings of SAC 2002*, Lecture Notes in Computer Science, vol. 2595, eds. K. Nyberg and H.M. Heys. Springer-Verlag, Berlin, 310–324.

[19] Horwitz, J. and B. Lynn (2002). "Toward hierarchical identity-based encryption." *Advances in Cryptology—EUROCRYPT 2002*, Lecture Notes in Computer Science, vol. 2332, ed. L. Knudsen. Springer-Verlag, Berlin, 466–481.

[20] Joux, A. (2000). "A one round protocol for tripartite Diffie–Hellman." *Proc. of ANTS-IV*, Lecture Notes in Computer Science, vol. 1838, ed. W. Bosma. Springer-Verlag, Berlin, 385–394.

[21] Joux, A. and K. Nguyen (2001). "Separating decision Diffie–Hellman from Diffie–Hellman in cryptographic groups." Eprint available at http://eprint.iacr.org/2001/003/

[22] Menezes, A.J., T. Okamoro, and S. Vanstone (1993). "Reducing elliptic curve logarithms to logarithms in a finite field." *IEEE Trans. on Inf. Theory*, 39, 1639–1646.

[23] Menezes, A.J. (1995). *Elliptic Curve Public Key Cryptosystems* (2nd ed.). Kluwer Academic Publishers, Boston, MA.

[24] Okamoto, T. and D. Pointcheval (2001). "The gap-problems: a new class of problems for the security of cryptographic schemes." *Proc. of of PKC 2001*, Lecture Notes in Computer Science, vol. 1992, ed. K. Kim. Springer-Verlag, Berlin, 104–118.

[25] Paterson, K.G. (2002). "ID-based signatures from pairings on elliptic curves." Available on http://eprint.iacr.org/2002/004/

[26] Pointcheval, D. and J. Stern (1996). "Security proofs for signature schemes." *Advances in Cryptology—EUROCRYPT'96*, Lecture Notes in Computer Science, vol. 1070, ed. U. Maurer. Springer-Verlag, Berlin, 387–398.

[27] Pointcheval, D. and J. Stern (2000). "Security arguments for digital signatures and blind signatures." *Journal of Cryptology*, 13 (3), 361–396.

[28] Sakai, R., K. Ohgishi, and M. Kasahara (2000). "Cryptosystems based on pairing." *The 2000*

Sympsium on Cryptography and Information Security, Okinawa, Japan.

[29] Shamir, A. (1984). "Identity based cryptosystems and signature schemes." *Advances in Cryptology—CRYPTO'84*, Lecture Notes in Computer Science, vol. 196, eds. G.R. Blakley and D. Chaum. Springer-Verlag, Berlin.

[30] Smart, N.P. (2002). "An identity based authenticated key agreement protocol based on the Weil pairing." *Electronic Letters*, 38 (13), 630–632.

IDENTITY-BASED ENCRYPTION

An identity-based encryption (IBE) scheme (see also identity-based cryptography) is a public key encryption scheme in which the public key can be an arbitrary string. For example, if Alice wants to send a message to Bob at bob@yahoo.com, she simply encrypts the message using the string "bob@yahoo.com" as the public key.

HISTORY: The concept of identity-based encryption was first introduced by Shamir in 1984 [12]. His original motivation was to eliminate the need for directories and certificates by using the identity of the receiver as the public key. Efficient solutions for the related notion of identity-based signature were quickly found [5, 6], but identity-based encryption proved to be much more challenging. Most schemes proposed since 1984 [4, 9, 11, 13, 14] were unsatisfactory because they were too computationally intensive, they required tamper resistant hardware, or they were not secure if users colluded. The first usable IBE scheme was proposed in 2001 by Boneh and Franklin [1]. Their scheme was soon adapted to provide additional functionalities, such as authentication [10], non-repudiation [2] and the ability to support a hierarchical structure [7, 8]. Cocks also proposed a relatively efficient scheme [3], but his scheme has not been proven secure against chosen ciphertext attack.

DEFINITION: More formally, an **identity-based encryption scheme** consists of four randomized algorithms: **Setup**, **Extract**, **Encrypt** and **Decrypt**.

Setup: takes as input a security parameter and outputs params (system parameters) and master-key. The system parameters must include the description of the message space $\mathcal{M}$ and the ciphertext space $\mathcal{C}$. The system parameters will be publicly known while the master-key is known only to the private key generator.

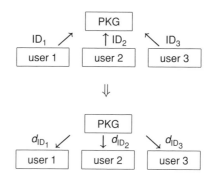

Fig. 1. Private key request in an IBE scheme

Extract: takes as input the system parameters params, the master-key and an arbitrary string ID $\in \{0, 1\}^*$ and outputs the private key d_{ID} corresponding to the public key ID.

Encrypt: takes as input the system parameters params, a public key ID and a plaintext $M \in \mathcal{M}$ and outputs a corresponding ciphertext.

Decrypt: takes as input the system parameters params, a private key d_{ID} and a ciphertext $C \in \mathcal{C}$ and outputs the corresponding plaintext.

The algorithm **Setup** is run by a trusted third party, the private key generator (PKG). The PKG also runs the algorithm **Extract** at the request of a user who wishes to obtain the private key corresponding to some string (see Figure 1). Note that the user needs to prove to the PKG that he is the legitimate "owner" of this string (for example, to obtain the private key corresponding to "bob@yahoo.com", the user must prove that bob@yahoo.com is truly his email address), and the private key must be returned to the user on a secure channel in order to keep the private key secret. The algorithms **Encrypt** and **Decrypt** are run by the users to encrypt and decrypt messages. They must satisfy the standard consistency constraints, namely if all the algorithms are applied correctly, then any message in the plaintext space encrypted with the algorithm **Encrypt** should be correctly decrypted by the algorithm **Decrypt**.

APPLICATIONS OF IDENTITY-BASED ENCRYPTION: We already mentioned that the original motivation for identity-based encryption was to simplify certificate management. Here are a few other applications.

Revocation of Public Keys

Public key certificates contain a preset expiration date. In an identity-based encryption scheme, we can make the keys expire by encrypting the messages using the public key "receiver-address ‖ current-date" where current-date can be the day, week, month, or year depending on the frequency at which we want the users to renew their private key. Note that unlike traditional public key infrastructure, the senders do not need to obtain new certificates every time the private keys are renewed, however, the receiver must query the PKG each time to obtain the new private key. So identity-based encryption is a very efficient way of implementing ephemeral public keys. This is also useful if, for example, the private key is kept on a laptop: if the laptop is stolen, only the private key corresponding to that period of time is compromised, the master-key is unharmed. This approach can also be used to send messages into the future since the receiver will not be able to decrypt the message until he gets the private key for the date specified by the sender from the PKG.

Managing User Credentials

By encrypting the messages using the address "receiver-address ‖ current-date ‖ clearance-level", the receiver will be able to decrypt the message only if he has the required clearance. This way, the private key generator can be used to grant user credentials. To revoke a credential, the PKG simply stops providing the private key in the next time period.

Delegations of Decryption Keys

Suppose a manager has several assistants each responsible for a different task. Then the manager can act as the private key generator and give his assistants the private keys corresponding to their responsibilities (so the public key would be 'Duty'). So each assistant can decrypt the messages whose subject fall within his or her responsibilities, but cannot decrypt messages intended for other assistants. The manager can decrypt all the messages using his or her master-key.

Example

We present the scheme proposed by Boneh and Franklin [1]. It requires a bilinear map, i.e., a map $\hat{e} : \mathbb{G}_1 \times \mathbb{G}_1 \to \mathbb{G}_2$, where G_1 and G_2 are groups, such that for any $g, h \in \mathbb{G}_1$ and $a, b \in \mathbb{Z}$,

$$\hat{e}(g^a, h^b) = \hat{e}(g, h^b)^a = \hat{e}(g^a, h)^b = \hat{e}(g, h)^{ab}.$$

The security of this scheme is related to the difficulty of computing $\hat{e}(g, g)^{abc}$ when given g, g^a, g^b, g^c for $g \in \mathbb{G}_1$ and $a, b, c \in \mathbb{Z}$ (see discrete logarithm problem and for the sequel also modular arithmetic).

Setup: Given a security parameter k,

(1) pick two cyclic groups $\mathbb{G}_1$, $\mathbb{G}_2$ of prime order p and a bilinear map $\hat{e} : \mathbb{G}_1 \times \mathbb{G}_1 \to \mathbb{G}_2$ corresponding to the security parameter (say p could be a k-bit prime),
(2) pick a random generator $g \in \mathbb{G}_1$,
(3) pick a random $s \in \mathbb{Z}_p$ and compute $g_{pub} = g^s$,
(4) pick cryptographic hash functions

$$H_1 : \{0, 1\}^* \to \mathbb{G}_1^*, H_2 : \mathbb{G}_2 \to \{0, 1\}^n,$$
$$H_3 : \{0, 1\}^n \times \{0, 1\}^n \to \mathbb{Z}_p^*,$$
$$H_4 : \{0, 1\}^n \to \{0, 1\}^n \text{ for some integer } n > 0.$$

The plaintext space is $\mathcal{M} = \{0, 1\}^n$ and the ciphertext space is $\mathcal{C} = \mathbb{G}_1^* \times \{0, 1\}^n \times \{0, 1\}^n$. The public system parameters are params $= \langle \mathbb{G}_1, \mathbb{G}_2, \hat{e}, \quad p, n, g, g_{pub}, H_1, H_2, H_3, H_4 \rangle$. The master-key is s.

Extract: Given a string ID $\in \{0, 1\}^*$, the master-key s and system parameters params, compute $h_{ID} = H_1(ID) \in \mathbb{G}_1$ and $d_{ID} = h_{ID}^s$, and return d_{ID}.

Encrypt: Given a plaintext $M \in \mathcal{M}$, a public key ID and public parameters params,
(1) compute $h_{ID} = H_1(ID)$,
(2) pick a random $\sigma \in \{0, 1\}^n$ and compute $r = H_3(\sigma, M)$,
(3) compute $\gamma = \hat{e}(g_{pub}, h_{ID})$,
(4) set the ciphertext to $C = \langle g^r, \sigma \oplus H_2(\gamma^r), M \oplus H_4(\sigma) \rangle$.

Decrypt: Given a ciphertext $\langle U, V, W \rangle \in \mathcal{C}$, a private key d_{ID} and system parameters params,
(1) compute $\gamma' = \hat{e}(U, d_{ID})$,
(2) compute $\sigma = V \oplus H_2(\gamma')$,
(3) compute $M = W \oplus H_4(\sigma)$,
(4) compute $r = H_3(\sigma, M)$. If $U \neq g^r$, reject the ciphertext, else return M.

Martin Gagné

References

[1] Boneh, D. and M. Franklin (2001). "Identity based encryption scheme from the Weil pairing." *Advances in Cryptology—CRYPTO 2001*, Lecture Notes in Computer Science, vol. 2139, ed. J. Kilian. Springer, Berlin, 213–229.

[2] Boyen, X. (2003). "Multipurpose identity-based signcryption: A Swiss army knife for identity-based cryptography." *Advances in Cryptology—CRYPTO 2003*, Lecture Notes in Computer Science, vol. 2729, ed. D. Boneh. Springer, Berlin, 383–399.

[3] Cocks, C. (2001). "An identity based encryption scheme based on quadratic residues." *Cryptography and Coding*, Lecture Notes in Computer Science, vol. 2260, ed. B. Honary. Springer, Berlin, 360–363.

[4] Desmedt, Y. and J.-J. Quisquater (1986). "Public-key systems based on the difficulty of tampering." *Advances in Cryptology—CRYPTO'86*, Lecture

Notes in Computer Science, vol. 263, ed. A. Odlyzko. Springer, Berlin, 111–117.

[5] Feige, U., A. Fiat, and A. Shamir (1988). "Zero-knowledge proofs of identity." *Journal of Cryptology*, 1, 77–94.

[6] Fiat, A. and A. Shamir (1987). "How to prove yourself: Practical solutions to identification and signature problems." *Advances in Cryptology—CRYPTO'86*, Lecture Notes in Computer Science, vol. 263, ed. A. Odlyzko. Springer, Berlin, 186–194.

[7] Gentry, C. and A. Silverberg (2002). "Hierarchical ID-based cryptography." Cryptology ePrint Archive, Report 2002/056.

[8] Horwitz, J. and Ben Lynn (2002). "Toward hierarchical identity-based encryption." *Advances in Cryptology—EUROCRYPT 2002*, Lecture Notes in Computer Science, vol. 2332, ed. L. Knudsen. Springer, Berlin, 466–481.

[9] Hühnlein, D., M. Jacobson, and D. Weber (2000). "Towards practical non-interactive public key cryptosystems using non-maximal imaginary quadratic orders." *Proceedings of SAC 2000*, Lecture Notes in Computer Science, vol. 2021, eds. D. R. Stinson and S.E. Tavares, Springer, Berlin, 275–287.

[10] Lynn, B. (2001). "Authenticated identity-based encryption." Cryptology ePrint Archive, Report 2002/072.

[11] Maurer, U. and Y. Yacobi (1991). "Non-interactive public-key cryptosystem." *Advances in Cryptology—EUROCRYPT'91*, Lecture Notes in Computer Science, vol. 547, ed. D.W. Davies. Springer, Berlin, 498–507.

[12] Shamir, A. (1984). "Identity-based cryptosystems and signature schemes." *Advances in Cryptology—CRYPTO'84*, Lecture Notes in Computer Science, vol. 196, eds. G.R. Blakley and D. Chaum. Springer, Berlin, 47–53.

[13] Tsuji, S. and T. Itoh (1989). "An ID-based cryptosystem based on the discrete logarithm problem." *IEEE Journal on Selected Areas in Communication*, 7 (4), 467–473.

[14] Tanaka, H. (1987). "A realization scheme for the identity-based cryptosystem." *Advances in Cryptology—CRYPTO'87*, Lecture Notes in Computer Science, vol. 293, ed. C. Pomerance, 341–349.

IDENTITY MANAGEMENT

Identity Management is the set of processes, tools, and social contracts surrounding the creation, maintenance, and termination of a digital identity for people or, more generally, for systems and services, to enable secure access to an expanding set of systems and applications.

Traditionally, identity management has been a core component of system security environments where it has been used for the maintenance of account information for login access to a system or a limited set of applications. An administrator issues accounts so that resource access can be restricted and monitored. Control has been the primary focus for identity management. More recently, however, identity management has exploded out of the sole purview of information security professionals and has become a key enabler for electronic business.

As the richness of our electronic lives mirrors our physical world experience, as activities such as shopping, discussion, entertainment and business collaboration are conducted as readily in the cyber world as in person, we begin to expect more convenience from our electronic systems. We expect our personal preferences and profile to be readily available so that, for example, when we visit an electronic merchant we need not tediously enter home delivery information; when participating in a discussion, we can check the reputation of other participants; when accessing music or videos, we first see the work of our favorite artists; and when conducting business, we know that our partners are authorized to make decisions. Today, identity management systems are fundamental to underpinning accountability in business relationships; providing customization to user experience; protecting privacy; and adhering to regulatory controls.

WHAT IS DIGITAL IDENTITY: Identity is a complicated concept having many nuances ranging from philosophical to practical. For the purposes of this discussion, we define the identity of an individual as the set of information known about that person. For example, a person's identity in the real world can be a set of a name, an address, a driver's license, a birth certificate, a field of employment, etc. This set of information includes items such as a name which is used as an *identifier*—it allows us to refer to the identity without enumerating all of the items; a driver's license or birth certificate which are used as an *authenticator*—they are issued by a relevant authority and allow us to determine the legitimacy of someone's claim to the identity; a driver's license which is used as a *privilege*—it establishes the permission to operate a motor vehicle.

Digital identity is the corresponding concept in the digital world. As people engage in more activities in the cyber world, the trend has been to link the real world attributes of identity with an individual's cyber world identity, giving rise to privacy concerns.

ELEMENTS OF AN IDENTITY MANAGEMENT SYSTEM: Identity management solutions are

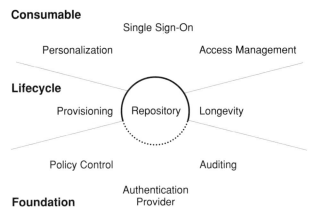

Consumable

Single Sign-On

Personalization   Access Management

Lifecycle

Provisioning (Repository) Longevity

Policy Control   Auditing

Authentication
Foundation   Provider

Fig. 1. Identity management system components

modular and composed of multiple service and system components. This section outlines components of an example identity management architecture illustrated in Figure 1.

Identity Management Foundation Components

The following are some foundational components of an identity management system.

- **Repository**—At the core of the system is the logical data storage facility and identity data model that is often implemented as an LDAP (Lightweight Directory Access Protocol) accessible directory or meta-directory. Policy information governing access to and use of information in the repository is generally stored here as well.
- **Authentication Provider**—The <u>authentication</u> provider, sometimes referred to as the *identity provider*, is responsible for performing primary authentication of an individual that will link them to a given identity. The authentication provider produces an *authenticator*—a token that allows other components to recognize that primary authentication has been performed. Primary authentication techniques include mechanisms such as password verification, proximity token verification, smartcard verification, biometric scans, or even X.509 PKI certificate verification. Each identity may be associated with more than one authentication provider. The mechanisms employed by each provider may be of different strengths and some application contexts may require a minimum strength to accept the claim to a given identity.
- **Policy Control**—Access to and use of identity information is governed by <u>policy</u> controls. Authorization policies determine how information is manipulated; privacy policies govern how identity information may be disclosed. Policy controls may cause events to be audited or even

for the subject of an identity to be notified when information is accessed.

- **Auditing**—Secure auditing provides the mechanism to track how information in the repository is created, modified and used. This is an essential enabler for forensic analysis—which is used to determine how and by whom policy controls were circumvented.

Identity Management Lifecycle Components

The following are two lifecycle components of an identity management system.

- **Provisioning**—Provisioning is the automation of all the procedures and tools to manage the lifecycle of an identity: creation of the *identifier* for the identity; linkage to the authentication providers; setting and changing attributes and *privileges*; and decommissioning the identity. In large-scale systems, these tools generally allow some form of self-service for the creation and on-going maintenance of an identity and frequently use a workflow or transactional system for verification of data from an appropriate authority and to propagate data to affiliated systems that may not directly consume the repository.
- **Longevity**—Longevity tools create the historical record of an identity. These tools allow the examination of the evolution of an identity over time.

Identity Management Consumable Value Components

The following are some consumable value components of an identity management system.

- **Single Sign-on**—Single sign-on allows a user to perform primary authentication once and then access the set of applications and systems that are part of the identity management environment.
- **Personalization**—Personalization and preference management tools allow application specific as well as generic information to be associated with an identity. These tools allow applications to tailor the user experience for a given individual leading to a streamlined interface for the user and the ability to target information dissemination for a business.
- **Access Management**—Similar to the policy controls within the identity management system foundation components, <u>access control</u> components allow applications to make authorization and other policy decisions based on privilege and policy information stored in the repository.

TRENDS DRIVING IDENTITY MANAGEMENT:
Several trends have combined to drive the need
for identity management systems. Consumers,
e-businesses, enterprises, and governments all
see value in the emergence of mature identity
management systems. Often the requirements
of these communities are complementary, but in
some cases conflicting needs raise new issues.

Consumer Trends

With each new website a user discovers, consumers find themselves creating a new digital
identity. This proliferation of accounts is tedious
both in the work needed to keep information correct and in the need to remember unique account
name password combinations. Often this leads to
security vulnerabilities such as when consumers
choose poor, easy-to-remember passwords, or use
the same password at a collection of independent
sites. Consumers are looking for web-based single
sign-on that allows easy access to a variety of sites.

The emergence of information aggregators for
financial services in the late 1990s is evidence
that consumers are driven to the convenience of
easy access—even at the expense of disclosing
some sensitive information to a third party. These
aggregators provided a portal that extracted information from the consumer's financial service
providers. To access this information, consumers
needed to disclose account information and access passwords to the independent aggregator
service.

Consumers, however, have demonstrated resistance to the notion of a single universally usable
digital identity. The selective disclosure inherent
in managing independent identities allows users
to maintain different personas for different interaction environments. This is consistent with how
people interact in the physical world and is illustrated in Figure 2. As a result, consumers are look-

Views of Identity

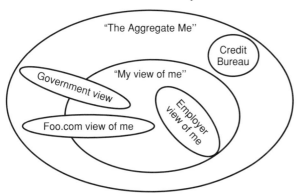

Fig. 2. Multiple views of identity

ing for identity management systems that support
some degree of anonymity or pseudonymity.

e-Business Trends

Electronic businesses are motivated to please
their customers and therefore to deploy the ease
of use aspects enabled by identity management
systems. Perhaps more importantly, they are also
looking to extract direct value from the system.
For large conglomerates, an identity management
system allows e-businesses to consolidate their
relationship with customers—it allows the organization to present a single face to the consumer. Personalization systems allow the business to learn about the consumer and then target
advertisement and special offers based on individual history and stated preferences.

Enterprise Trends

User account and password management has long
been a major expense for enterprise IT organizations. Network operating systems and environments have provided some relief, by allowing a single account and password to work on a collection
of machines, but this has failed to provide true
single sign-on for heterogeneous environments. As
enterprises are driven to greater degrees of collaboration with business partners, as they integrate
supply chains the number and diversity of systems
and applications increases. Enterprises are driven
toward identity management solutions that will
address heterogeneity issues and allow them to
integrate with their business partners. They need
systems that will provide for independent administration and that will provide strong accountability for business transactions.

Government Trends

With the evolution of e-government initiatives,
governments share many of the concerns motivating e-businesses. Scale, however, is more of a
concern for government organizations—few businesses have a customer base the size of a government's citizenry.

Governments, however, do have some other concerns. Privacy regulations such as the EU privacy directive or US sector specific legislation such
as the Gramm-Leach-Bliley act of 1999 or the
Health Insurance Portability And Accountability
act of 1996 create specific controls on how person-
ally identifiable information can be processed in
IT systems. These regulations establish requirements for the privacy policy control component
of an identity management system, and impose

constraints on how businesses exploit identity information.

MODELS FOR DEPLOYING IDENTITY MANAGEMENT:
Identity management systems are primarily being deployed in one of three models: as silos, as walled gardens, and as federations.

Silo

This is the predominant model on the Internet today. In this model the identity management environment is put in place and operated by a single entity for a fixed user community.

Walled Garden

Walled gardens represent a closed community of organizations. A single identity management system is deployed to serve the common user community of a collection of businesses. Most frequently this occurs in business-to-business exchanges and specific operating rules govern the entity operating the identity management system.

Federation

Federated identity management environments are now emerging. These include systems like Microsoft's .Net Passport and .Net TrustBridge and the Liberty Alliance Project: Liberty Architecture. The central difference between federated identity systems and walled gardens is that there is no single entity that operates the identity management system. Federated systems support multiple identity providers and a distributed and partitioned store for identity information. Clear operating rules govern the various participants in a federation—both the operators of components and the operators of services who rely on the information provided by the identity management system. Most systems exhibit strong end-user controls over how identity information is disseminated amongst members of the federation.

IDENTITY MANAGEMENT ISSUES:
Identity management systems bring great value to the digital world and federated identity environments, in particular, hold great promise for widespread deployment. As the distinction between real world identity and digital identity becomes more blurred, however, a number of issues remain to be considered[1]:

- **Authenticity of identity.** How is the accuracy and validity of identity information measured and determined? What are the trust services that must be in place to generate confidence in information in the identity management service?
- **Longevity of information.** Do identity management systems provide adequate care to track changes to identity information over time? Do they maintain the necessary artifacts to support historical investigations?
- **Privacy.** Do identity management systems provide adequate controls to preserve individual privacy? Does the system provide adequate support for anonymity and multiple user controlled personas?
- **Identity theft.** Do widespread identity management systems make it easier to perpetrate identity theft or identity fraud?
- **Legal structures.** What protections are in place for the holder of the identity or for the relying party? Do these protections go beyond contractual obligations when digital identity systems are used for interactions that today are limited to the physical world?

Joe Pato

IDENTITY VERIFICATION PROTOCOL

An identity verification protocol is a protocol used to obtain entity authentication of one entity to another entity. The authentication provided by the protocol can be either *unilateral* (i.e., authenticates just one of the entities to the other entity) or *mutual* (i.e., authenticates both entities). In addition to the two entities directly involved in the authentication (the claimant and verifier), some protocols require the participation of a trusted third party in order to achieve authentication.

There are a number of different identity verification protocols that exist, but most of them fall into three main types. These are: password-based schemes, challenge–response protocol schemes and zero-knowledge identification techniques.

Password-based: These schemes typically involve the claimant providing a password or PIN that is either sent directly to the verifier or used to generate a token (see authentication token), or credentials, that can be validated by the verifier, who also knows the password. Most of these schemes are susceptible to replay attacks,

[1] For a more detailed examination of issues with large scale identity systems, see the National Research Council's Computer Science and Telecommunications Board report

IDs—Not That Easy: Questions About Nationwide Identity Systems (2002) available at *http://www.cstb.org/web/project_authentication*.

password guessing and dictionary attacks and compromise of the password at either the claimant or verifier. Thus, these schemes usually provide limited security, but have the advantage that they are easy to implement and deploy.

Challenge–response: These protocols require the claimant to verify its identity to the verifier by proving knowledge of a secret value that is known only to the claimant (and possibly the verifier) and will not be revealed as part of the protocol. Proving knowledge of this secret is accomplished by responding to a particular challenge provided by the verifier. Typically the verifier produces a time-variant parameter (e.g., a nonce) and the claimant is required to apply some cryptographic transformation to that parameter using a key that only it (or possibly also the verifier) knows. Examples of challenge–response identity verification protocols can be found in ISO/IEC 9798-2 [1], ISO/IEC 9798-3 [2], FIPS 196 [3] and TLS (see Transport Layer Security) [4].

Zero-knowledge: These protocols use asymmetric techniques in order to prove the claimant's identity, but are based upon interactive proof systems and zero-knowledge techniques and thus differ from asymmetric-based challenge–response protocols. These protocols are designed to reveal no information whatsoever beyond whether or not the claimant knows a secret (and thus has the claimed identity) and then only to the verifier. Examples of zero-knowledge identification protocols include the Fiat–Shamir identification protocol, the GQ identification protocol [5], and the Schnorr identification protocol.

It should be noted that identity verification protocols only provide entity authentication at the instant of protocol execution and do not necessarily prove that the authenticated entity was participating in an entire session. If entity authentication is required for the entire lifetime of a given session then either the identity verification protocol can be repeatedly performed throughout the lifetime of the session, or it can be combined with a key establishment mechanism and an ongoing integrity service, as is done in TLS.

Robert Zuccherato

References

[1] ISO/IEC 9798-2 (1999). "Information technology—security techniques—entity authentication—Part 2: Mechanisms using symmetric encipherment algorithms."

[2] ISO/IEC 9798-3 (1997). "Information technology—security techniques—entity authentication—Part 3: Entity authentication using a public-key algorithm."

[3] FIPS 196 (1997). "Entity authentication using public key cryptography." Federal Information Processing Standards Publication 196, U.S. Department of Commerce/NIST, National Technical Information Service, Springfield, Virginia.

[4] Dierks, T. and Allen, C. (1999). "The TLS protocol—version 1.0". *RFC 2246.*

[5] Guillou, L.C. and J.-J. Quisquater (1988). "A practical zero-knowledge protocol fitted to security microprocessor minimizing both transmission and memory". *Advances in Cryptology—EUROCRYPT'88* Lecture Notes in Computer Science, vol. 330, ed. C.G. Günther. Springer, Berlin, 123–128.

IMPERSONATION ATTACK

An *impersonation attack* is an attack in which an adversary successfully assumes the identity of one of the legitimate parties in a system or in a communications protocol. The goal of a strong identification or entity authentication protocol is to make negligible the probability that, for a given party A, any party C distinct from A, carrying out the protocol and playing the role of A, can cause another party B to complete and accept A's identity.

Carlisle Adams

Reference

[1] Menezes, A., P. van Oorschot, and S. Vanstone (1997). *Handbook of Applied Cryptography*. CRC Press, Boca Raton, FL.

IMPOSSIBLE DIFFERENTIAL ATTACK

Impossible differential attack is a chosen plaintext attack and is an extension of differential cryptanalysis. Impossible differential attack [1] was defined in 1998 and has been shown to break 31 out of 32 rounds of the cipher Skipjack [5], designed by the NSA and declassified in 1998. Independently, an attack based on similar principles was used by Knudsen in 1998 to cryptanalyse 6-rounds of the cipher *DEAL* [3] which was one of his proposals for the AES (see Rijndael/AES). The attack, using impossible differentials, was shown to be a generic tool for cryptanalysis [2] and was applied to improve on the best known attacks for such strong and long standing <u>block ciphers</u> as <u>IDEA</u> and Khufu [4], breaking round-reduced versions of these ciphers. The two main ideas were the <u>miss-in-the-middle</u> technique for construction of impossible events inside ciphers and the *sieving technique* for filtering wrong key-guesses.

Once the existence of impossible events in a cipher is proven, it can be used directly as a *distinguisher* from a *random permutation* (see <u>substitutions and permutations</u>). Furthermore one can find the keys of a cipher by analyzing the rounds surrounding the impossible event, and guessing the subkeys of these rounds. All the keys that lead to a contradiction are obviously wrong. The impossible event in this case plays the role of a *sieve*, methodically rejecting the wrong key guesses and leaving the correct key. It is important to note that the miss-in-the-middle technique is only one of the ways to construct impossible events and that the sieving technique is only one of the possible ways to exploit them.

In order to get a feel of the attack, consider a cipher $E(\cdot)$ with n-bit blocks, a set of input differences $\mathcal{P}$ of cardinality 2^p and a corresponding set of output differences $\mathcal{Q}$ of cardinality 2^q. Suppose that no difference from $\mathcal{P}$ can cause an output difference from $\mathcal{Q}$. One may ask how many chosen texts should be requested in order to distinguish $E(\cdot)$ from a random permutation? In general about 2^{n-q} pairs with differences from $\mathcal{P}$ are required. This number can be reduced by using *structures* (a standard technique for saving chosen plaintexts in <u>differential attacks</u>). In the optimal case one may use structures of 2^p texts which contain about 2^{2p-1} pairs with differences from $\mathcal{P}$. In this case $2^{n-q}/2^{2p-1}$ structures are required, and the number of chosen texts used by this distinguishing attack is about $2^{n-p-q+1}$ (assuming that $2p < n - q + 1$). Thus the higher $p+q$ is, the better is the distinguisher based on the impossible event.

Alex Biryukov

References

[1] Biham, E., A. Biryukov, and A. Shamir (1999). "Cryptanalysis of Skipjack reduced to 31 rounds using impossible differentials." *Advances in Cryptology—EUROCRYPT'99*, Lecture Notes in Computer Science, vol. 1592, ed. J. Stern. Springer-Verlag, Berlin, 12–23.

[2] Biham, E., A. Biryukov, and A. Shamir (1999). "Miss in the middle attacks on IDEA and Khufu." *Fast Software Encryption, FSE'99*, Lecture Notes in Computer Science, vol. 1636, ed. L. R. Knudsen. Springer-Verlag, Berlin, 124–138.

[3] Knudsen, L.R. (1998). "DEAL—a 128-bit block cipher." Technical Report 151, Department of Informatics, University of Bergen, Norway.

[4] Merkle, R.C. (1991). "Fast software encryption functions." *Advances in Cryptology—CRYPTO'90*, Lecture Notes in Computer Science, vol. 537, eds. A. Menezes and S.A. Vanstone. Springer-Verlag, Berlin, 476–501.

[5] NIST (1998). "Skipjack and KEA algorithm specification." Technical Report, http://csrc.nist.gov/CryptoToolkit/skipjack/skipjack-kea.htm. Version 2.0.

INDEX CALCULUS

Index calculus refers to a method for computing discrete logarithms in a <u>finite field</u> in <u>subexponential time</u> complexity (see also <u>Discrete Logarithm Problem</u>). The basic ideas appeared first in the work of Western and Miller [9]. The original algorithm was invented independently by Adleman [1], Merkle [5] and Pollard [7] according to Odlyzko [6]. The first partial analysis of the complexity of the algorithm is due to Adleman [1].

Consider a finite field k with multiplicative <u>group</u> G <u>generated</u> by g. The main and trivial observation on which index calculus relies is that once the discrete logarithms $\log_g(g_i)$ are known, then the discrete logarithm for an element defined by $\prod_i g_i^{n_i}$ is given by the sum

$$\sum_i n_i \log_g(g_i) \bmod |G|.$$

Given the element h of G, index calculus computes the discrete logarithm $\log_g(h)$ in two main steps:
1. In a precomputation step, $\log_g(g_i)$ for sufficiently many elements g_i of G is computed.
2. Find an element g^y such that hg^y has the form $hg^y = \prod_i g_i^{n_i}$. Then by the above remark the discrete logarithm $\log_g(h)$ can be computed easily.

The way the discrete logarithms in step 1 are obtained is as follows:

Assume we have a surjective map $\phi: R \to G$ from a <u>ring</u> R, in which we have unique factorization in prime elements, to G. In the case that the surjection ϕ can be efficiently inverted we can lift elements from G to R. Assume we have a factorization

$$\phi^{-1}(g^x) = \prod_{p \in S} p^{n_{p,x}},$$

where S, the <u>factor base</u>, denotes a subset of the set P of <u>prime</u> elements of R. This implies, via application of ϕ, that the equality $g^x = \prod_{p \in S} \phi(p)^{n_{p,x}}$ in G holds. Hence taking discrete logarithms to the base g we obtain

$$x = \sum_{p \in S} n_{p,x} \log_g(\phi(p)) \bmod |G|.$$

In order to recover the values $\log_g(\phi(p))$ for $p \in S$ it is now sufficient to run through two steps:

- In the first step (often referred to as <u>sieving</u> step), we collect more than $|S|$ distinct relations, these then form an overdefined system of linear equations modulo the order of the group $|G|$.
- Solve the resulting system of linear equations $\bmod |G|$ using linear algebra and obtain the values $\log_g(\phi(p))$ for $p \in S$.

Given any element $h \in G$, we then search for an element g^y such that $\phi^{-1}(hg^y)$ has the form

$$\phi^{-1}(hg^y) = \prod_{p \in S} p^{n_{p,y}}.$$

$\log_g(h)$ can then be computed by

$$\log_g(h) \equiv -y + \sum_{p \in S} n_{p,y} \log_g(\phi(p)) \bmod |G|.$$

IMPLEMENTATION CHOICE AND COMPLEXITY ESTIMATE: The most obvious choices for the ring R are:

1. The natural numbers $\mathbb{N}$ in the case of prime fields $G = \mathbb{F}_p^*$.
2. The ring $\mathbb{F}_p[x]$ of polynomials over $\mathbb{F}_p$ in the case of extension fields $G = \mathbb{F}_{p^n}$ (see the entry <u>extension field</u>).

For the set S we choose smooth prime elements of the ring R (see <u>smoothness</u>):

1. The prime numbers less than or equal to a smoothness bound B:

$$S = \{ p \in \mathbb{N} \text{ prime} \mid p \leq B \}$$

in the case of prime fields.

2. The <u>irreducible polynomials</u> of norm less than or equal to a smoothness bound B:

$$S = \{ f \in \mathbb{F}_p[x] \text{ irreducible} \mid p^{\deg(f)} \leq B \}$$

in the case of extension fields.

Using results on the distribution of smooth elements one sees that using a smoothness bound $B = L_{p^n}(1/2, \sqrt{1/2})$, the first step of collecting relations by computing random powers g^x takes expected time $L_{p^n}(1/2, \sqrt{2})$ (see the entry <u>L-notation</u> for a definition of the preceding notation). In the case of prime fields, we have $n = 1$ in this and all subsequent expressions. Here one has to be careful to also take into account the time needed for the smoothness test of the lifted elements $\phi^{-1}(g^x)$. However, using the <u>Elliptic Curve Method</u> for factoring in the prime field case one has an expected running time of $L_p(1/2, 1)$ for factorization (in the worst case), while in the case of extension fields,

for a polynomial f over $\mathbb{F}_p$ of degree n factorization using the Berlekamp algorithm has complexity $O(n^3 + tpn^2)$ where t denotes the number of irreducible factors of the polynomial f (see <u>O-notation</u>).

For the linear algebra part one notes that the resulting system of around $L_{p^n}(1/2, \sqrt{1/2})$ linear equations is sparse, hence special methods can be applied for the solution of this system which have an expected running time $L_{p^n}(1/2, \sqrt{2})$.

Finally for the last step of finding a smooth lift of the form $\phi^{-1}(g^x)$ the running time estimation of $L_{p^n}(1/2, \sqrt{2})$ of the first step applies as well, resulting in an overall expected running time of $L_{p^n}(1/2, \sqrt{2})$ for the complete index calculus algorithm.

VARIANTS OF THE ALGORITHM: By replacing $\mathbb{N}$ or $\mathbb{F}_p[x]$ with different rings also allowing unique factorisation one obtains variants of the classical index calculus algorithm.

One of the most important variants in the case of prime fields uses the ring of Gaussian integers $\mathbb{Z}[i]$ of the imaginary quadratic field $\mathbb{Q}(i)$ and is therefore called the *Gaussian integer method* (refer to [4] for more details). It has expected running times $L_p(1/2, 1)$ for the precomputation part of the algorithm and $L_p(1/2, 1/2)$ for the computation of individual discrete logarithms.

In the case of extension fields Coppersmith's method is of special importance. It is applicable only in the case of small characteristics and was the first method to compute discrete logarithms in subexponential complexity better than $L_{2^n}(1/2, c)$, its expected running time is $L_{2^n}(1/3, 1.588)$ (see [2]). It was realized later that this method in fact is a special case of the function field sieve (see <u>sieving in function fields</u>).

INDEX CALCULUS USING THE NUMBER FIELD SIEVE: A different idea to solve the discrete logarithm problem in a finite field $k = \mathbb{F}_p$ is to find integers s and t such that the equation $g^s \cdot h^t = w^q$ holds for some $w \in k$, where q is a divisor of the order of the multiplicative group $k^\times$. If t is coprime to q we have then computed $\log_g(h) \bmod q$ since we have

$$\log_g(h) \equiv -st^{-1} \bmod q.$$

Having computed $\log_g(h)$ modulo the different primes dividing the order of $k^\times$ we can then recover $\log_g(h)$ using the <u>Chinese Remainder Theorem</u>.

We are thus led to the problem of constructing q–th powers in k, i.e., relations of the form $g^s \cdot h^t = w^q$. Index calculus techniques can be applied to this problem. If we choose R to be the ring of integers of a more general, non trivial number field, we are led to techniques related to the Number Field Sieve.

Contrary to the techniques described above, the Number Field Sieve approach uses two factor bases: one consisting of small rational primes, the other consisting of algebraic primes of small norm.

This is due to the fact that the Number Field Sieve approach uses two different maps ϕ: one is the natural projection $\phi_1 : \mathbb{N} \to \mathbb{F}_p$, while the second one ϕ_2 is a certain projection from the ring of algebraic integers R to $\mathbb{F}_p$.

In the sieving stage, pairs of smooth elements s_1, s_2 are collected which have the additional property that the equality $\phi_1(s_1) = \phi_2(s_2)$ holds. Again we are led to a system of linear equations, solving this system will lead to the construction of a qth power in $\mathbb{F}_p$ and will thus yield the solution of the discrete logarithm problem. Schirokauer presented an algorithm based on this approach that has complexity $L_p(1/3, (64/9)^{1/3})$ (see [8]).

Details regarding the application of the Number Field Sieve in this setting can be found in [3] and [8]. See also the entry on the Discrete Logarithm Problem for more details on recent challenges and attacks.

INDEX CALCULUS AND ELLIPTIC CURVES: The elliptic curve discrete logarithm problem (ECDLP) has attracted great interest since its introduction to cryptography in 1985. One the most interesting features of this problem is that it has up till now resisted all attempts to apply index calculus techniques to it.

Consider the ECDLP on a curve over the finite field $\mathbb{F}_{p^n}$. The most naive idea would be to take a surjection $\phi : K \to \mathbb{F}_{p^n}$, choose an elliptic curve E' over K reducing to E via ϕ and thus obtain a map between elliptic curves $\phi : E'(K) \to E(\mathbb{F}_{p^n})$, the most obvious one being $\phi : \mathbb{Q} \to \mathbb{F}_p$ in the case of prime fields. However, properties of elliptic curves over global fields imply that there is no chance to lift sufficiently many points from $E(\mathbb{F}_p)$ to $E(\mathbb{Q})$ with reasonably sized coefficients in order to be able to apply index calculus techniques. The main ingredient here is the existence of a quadratic form called canonical height on the Mordell-Weil group (modulo torsion) of the global elliptic curve.

A different approach (known as XEDNI calculus) suggested to first lift sufficiently many points from $E(\mathbb{F}_p)$ to $\mathbb{Q}$ and then fit a globally defined elliptic curve through these global points. However this idea was proven to have expected running time $O(p)$, far worse then the square root complexity of the exhaustive search approach.

Kim Nguyen

References

[1] Adleman, L.M. (1979). "A subexponential algorithm for the discrete logarithm problem with applications to cryptography." *Proc. 20th IEEE Found. Comp. Sci. Symp.*, 55–60.

[2] Coppersmith, D. (1984). "Fast evaluation of discrete logarithms in fields of characteristic two." *IEEE Transactions on Information Theory*, 30, 587–594.

[3] Gordon, Daniel M. (1979). "Secrecy, Authentication and Public Key Systems." *PhD Dissertation*, Dept. of Electrical Engineering, Stanford University.

[4] Lamacchia, B.A. and A.M. Odlyzko (1991). "Computation of discrete logarithms in prime fields." *Designs, Codes and Cryptography*, 1, 47–62.

[5] Merkle, R. (1993). "Discrete logarithms in $\mathbb{F}_p$ using the number field sieve." *SIAM J. Discrete Math.*, 6, 124–138.

[6] Odlyzko, A. (1985). "Discrete logarithms in finite fields and their cryptographic significance." *Advances in Cryptology—EUROCRYPT'84*, Lecture Notes in Computer Science, vol. 209, eds. T. Beth, N. Cot, and I. Ingemarsson. Springer-Verlag, Berlin, 224–314.

[7] Pollard, J. (1978). "Monte Carlo methods for index computations (mod p)." *Math. Comp.*, 32, 918–924.

[8] Schirokauer, O. (1993). "Discrete logarithms and local units." *Philos. Trans. Roy. Soc. London Ser. A*, 345, 409–423,

[9] Wester, A.E. and J.C.P. Miller. (1968). "Tables of indices and primitive roots." *Royal Society Mathematical Tables*, vol. 9. Cambridge University Press, Cambridge.

INFORMATION THEORY

The *entropy* function $H(X)$ is a measure of the uncertainty of X, in formula

$$H(X) = - \sum_{a \,:\, p_X(a) > 0} p_X(a) \cdot \log_2 p_X(a),$$

where $p_X(a) = \Pr[X = a]$ denotes the probability that random variable X takes on value a. The interpretation is that with probability $p_X(a)$, X can be described by $\log_2 p_X(a)$ bits of information.

The *conditional entropy* or *equivocation* (Shannon 1949) $H(X|Y)$ denotes the uncertainty of X provided Y is known:

$$H(X|Y) = - \sum_{a,b \,:\, p_{X|Y}(a|b) > 0} p_{X,Y}(a, b) \cdot \log_2 p_{X|Y}(a|b)$$

where $p_{X,Y}(a, b) =_{\text{def}} \Pr[(X = a) \wedge (Y = b)]$ and $p_{X|Y}(a|b)$ obeys Bayes' rule for conditional probabilities:

$$p_{X,Y}(a, b) = p_Y(b) \cdot p_{X|Y}(a|b), \text{ thus}$$
$$-\log_2 p_{X,Y}(a, b) = -\log_2 p_Y(b) - \log_2 p_{X|Y}(a|b).$$

The basic relation on conditional entropy follows from this:

$$H(X, Y) = H(X \mid Y) + H(Y).$$

In particular, we note that the entropy is additive if and only if X and Y are independent:

$$H(X, Y) = H(X) + H(Y),$$

in analogy to the additive entropy of thermodynamical systems.

The *redundancy* of a text is that part (expressed in bits) that does not carry information. In common English, the redundancy is roughly 3.5 [bit/char], the information is roughly 1.2 [bit/char], redundancy and information sum up to $4.7 = \log_2 26$ [bit/char].

We shall now use the terminology above to describe three possible properties of a cryptosystem. A cryptosystem is *of Vernam type* if $H(K) = H(C)$, where $H(K)$ is the entropy of the key K and $H(C)$ is the entropy of the ciphertext C. A cryptosystem has *independent key* if the plaintext P and keytext K are mutually independent: $H(P) = H(P \mid K)$ and $H(K) = H(K \mid P)$ ("knowledge of the keytext does not change the uncertainty of the plaintext, and knowledge of the plaintext does not change the uncertainty of the keytext").

A cryptosystem is called *perfect* if plaintext and ciphertext are mutually independent: $H(P) = H(P \mid C)$ and $H(C) = H(C \mid P)$ ("knowledge of the ciphertext does not change the uncertainty of the plaintext, and knowledge of the plaintext does not change the uncertainty of the ciphertext"). This means that the security of the system depends entirely on the key; perfect cryptosystems correspond to holocryptic keytexts (see key), which are extremely difficult to achieve in practice.

SHANNON'S MAIN THEOREM: In a cryptosystem, where the key character is uniquely determined by the plaintext character and the ciphertext character ("ciphertext and plaintext together allow no uncertainty on the keytext"), any two of the following three properties of the cryptosystem imply the third one:

Vernam type, independent key, perfect.

The *unicity distance* for a given language, a given cryptosystem and a given cryptanalytic procedure of attack is the minimal length of the plaintext such that decryption is unique. Example: let Z be the cardinality of keytext space, assume simple *substitution* (see substitutions and permutations) and an attack by letter frequency. Then for English with an alphabet of 26 letters the unicity distance U is given by

(1) $U \approx \frac{1}{0.53} \log_2 Z$ for decryption with single-letter frequencies,

(2) $U \approx \frac{1}{1.2} \log_2 Z$ for decryption with bigram frequencies,

(3) $U \approx \frac{1}{1.5} \log_2 Z$ for decryption with trigram frequencies,

(w) $U \approx \frac{1}{2.1} \log_2 Z$ for decryption with word frequencies,

(∗) $U \approx \frac{1}{3.5} \log_2 Z$ for decryption using all grammatical and semantical rules.

For simple substitution with $Z = 26!$, one has $\log_2 Z \approx 88.38$. This leads to the values 167, 74, 59, 42, and 25 for the unicity distance, which are confirmed by practical experience.

For bigram substitution with $Z = 676!$, there is $\log_2 Z \approx 5385.76$ and $U \approx 1530$ for decryption using all grammatical and semantical rules.

The situation is rather similar for German, French, Italian, Russian, and related Indo-European languages.

For holocryptic sequences of key elements, the unicity distance is infinite.

Friedrich L. Bauer

References

[1] Bauer, F.L. (1997). "Decrypted secrets." *Methods and Maxims of Cryptology*. Springer-Verlag, Berlin.

[2] McEliece, R.J. (1977). "The theory of information and coding." *Encyclopedia of Mathematics and its Applications*, vol. 3. Addison-Wesley Publishing Company, Reading, MA.

INTEGER FACTORING

DEFINITION: Integer factoring is the following problem: given a positive composite integer n, find positive integers v and w, both greater than 1, such that $n = v \cdot w$.

RELATION TO INFORMATION SECURITY: Integer factoring is widely assumed to be a hard problem. Obviously, it is not hard for all composites, but composites for which it is believed to be difficult can easily be generated. This belief underlies the security of RSA public-key encryption and the RSA digital signature scheme. To the present day, no proof of the difficulty of factoring has been published. This is quite unlike the discrete logarithm problem, where the difficulty is provable for a generic group [19, 27]. However, this result does not have much practical relevance. In particular it does not say anything about the hardness of computing discrete logarithms in multiplicative groups of finite fields, a problem that is widely regarded as being as hard (or as easy) as integer factoring. On a quantum computer both problems are easy in the sense that they allow polynomial-time solutions. Given the current state of the art in quantum computer manufacturing, this is not yet considered to be a threat affecting factoring or discrete logarithm based cryptosystems. Quantum computer factoring is not discussed here.

METHODS FOR INTEGER FACTORIZATION: RSA cryptosystems are faster when smaller composites are used, but believed to be more secure for larger ones. Finding the right middle-ground between efficiency and security requirements requires the study of theoretical and practical aspects of the integer factorization methods. Often, two types of integer factoring methods are distinguished: general purpose and special purpose methods. For general purpose methods the factoring effort depends solely on the size of the composite n to be factored. For special purpose methods properties of n (mostly but not always of one of the factors of n) come into play as well. RSA composites are generally chosen in such a way that special purpose methods would be less efficient than general purpose ones. Special purpose methods are therefore hardly relevant for RSA composites. For randomly selected composites, however, special purpose methods are on average very effective. For example, almost 92% of all positive integers have a factor <1000; if such a factor exists it will be found very quickly using trial division, the simplest of the special purpose methods (see below).

Here the following factoring methods are sketched:

Special purpose—trial division; Pollard's rho method; Pollard's $p - 1$ method and generalizations thereof; Elliptic Curve Method.

General purpose—Fermat's method and congruence of squares; Dixon's random squares method; continued fraction method (CFRAC); linear sieve; Quadratic Sieve; Number Field Sieve.

For a more complete survey refer to [6] and the references therein.

ESTABLISHING COMPOSITENESS: Fermat's little theorem says that $a^{p-1} \equiv 1 \bmod p$ if p is a prime number and a is a positive integer $<p$ (see modular arithmetic). Thus, an $a \in \{1, 2, \ldots, n-1\}$ for which $a^{n-1} \not\equiv 1 \bmod n$ would establish the compositeness of n at the cost of a single exponentiation modulo n. The proof of compositeness does not provide any information that may be useful to find a nontrivial factor of n. Also, this type of compositeness proof does not work for all composites, because for some composites $a^{n-1} \equiv 1 \bmod n$ for all a that are coprime to n. There are infinitely many of such composites, the so-called Carmichael numbers [1].

Fermat's little theorem allows an alternative formulation for which the converse is always useful for compositeness testing. Let $n - 1 = 2^t \cdot u$ for integers t and u with u odd. If $n > 2$ were prime, then any integer $a \in \{2, 3, \ldots, n-1\}$ satisfies the condition that either $a^u \equiv 1 \bmod n$ or $a^{2^i u} \equiv -1 \bmod n$ for some $i \in \{0, 1, \ldots, t-1\}$. An integer $a \in \{2, 3, \ldots, n-1\}$ for which this condition does not hold is called a 'witness to the compositeness of n.' For odd composite n at least 75% of the numbers in $\{2, 3, \ldots, n-1\}$ are witnesses to their compositeness [24]. Therefore, it can in general be expected that n's compositeness can be proved at the cost of at most a few exponentiations modulo n, simply be trying elements of $\{2, 3, \ldots, n-1\}$ at random until a witness has been found. This probabilistic compositeness test is often referred to as the Miller-Rabin probabilistic primality test. If n itself is randomly selected too (as may happen during the search for a prime number), it is usually faster to establish its compositeness using trial division (see below).

DISTINCT FACTORS: Let a be a witness to the compositeness of n, as above. This witness can be used to check that n is not a prime power at negligible additional cost. By squaring the number $a^{2^{t-1}u} \bmod n$ that was last calculated, one calculates $(a^n - a) \bmod n$. If it is zero, then n is not a prime power because the odd parts of the $t + 2$ factors

$$a \cdot (a^u - 1) \cdot \prod_{i=0}^{t-1}(a^{2^i u} + 1)$$

of $a^n - a$ are pairwise relatively prime; actually, in that case one of those $t + 2$ factors has a nontrivial factor in common with n, which can easily be found. If $(a^n - a) \bmod n \neq 0$, one verifies that $\gcd(a^n - a, n) = 1$, which shows that n is not a prime power: if n were p^k for a prime p, then $a^p \equiv a \bmod p$ and thus also $a^n = a^{p^k} \equiv a \bmod p$, so that p would divide $a^n - a$.

REPEATED FACTORS: If n is an odd composite and not a prime power, it may still be a proper power of a composite (i.e., $n = m^\ell$ for $m, \ell \in \mathbf{Z}_{>1}$ with m composite) or it may properly contain a square (i.e., $n = m^\ell \cdot w$ for $m, \ell, w \in \mathbf{Z}_{>1}$ with

$\gcd(m, w) = 1$). Proper powers can be recognized by approximating ℓth roots of n for $1 \le \ell \le \lceil \frac{\log n}{\log 3} \rceil$ using a standard numerical method such as Newton's method. At present there is in general no better way to find out if n properly contains a square than factoring n.

Trial division up to bound B is the process of checking for all primes $\le B$ in succession if they divide n, until the smallest prime factor p of n is found or until it is shown that $p > B$. This takes time proportional to $\log n \cdot \min(p, B)$. For randomly selected n trial division can be expected to be very effective. It cannot be recommended to use B larger than, say, 10^6 (with the precise value depending on the relative speeds of implementations) because larger p can be found more efficiently using one of the methods described below.

Pollard's rho method [21] is based on the birthday paradox: if $x_0, x_1, x_2, \ldots$ is a random walk on $\mathbf{Z}/p\mathbf{Z}$, then for any p there is a fair probability that $x_i = x_j$ for some indices $i \ne j$ up to about $\sqrt{p}$. Similarly, if $x_0, x_1, x_2, \ldots$ is a random walk on $\mathbf{Z}/n\mathbf{Z}$, then for any $p < n$ there is a fair probability for a collision $x_i \equiv x_j \bmod p$ for $i \ne j$ up to about $\sqrt{p}$; if p is an unknown divisor of n, such a collision can be recognized because it implies that p divides $\gcd(n, x_i - x_j)$.

In Pollard's rho method a walk on $\mathbf{Z}/n\mathbf{Z}$ is defined by selecting $x_0 \in \mathbf{Z}/n\mathbf{Z}$ at random and by defining $x_{i+1} = (x_i^2 + 1) \bmod n$. There is no a priori reason why this would define a random walk on $\mathbf{Z}/n\mathbf{Z}$, but if it does it may reveal the smallest factor p of n after only about $\sqrt{p}$ iterations. At the i-th iteration, this would require $i - 1$ gcd-computations $\gcd(n, x_i - x_j)$ for $j < i$, making the method slower than trial division. This problem is overcome by means of Floyd's cycle-finding method: at the i-th iteration compute just $\gcd(n, x_i - x_{2i})$ (thus requiring computation of not only x_i but x_{2i} as well). As a result, and under the assumption that the walk is random, the expected time to find p becomes proportional to $(\log n)^2 \cdot \sqrt{p}$; this closely matches practical observations. The name of the method is based on the shape of the Greek character rho ('ρ') depicting a sequence that bites in its own tails. The method is related to Pollard's rho method for solving the discrete logarithm problem.

In practice the gcd-computation per iteration is replaced by a single gcd-computation of n and the product modulo n of, say, 100 consecutive $(x_i - x_{2i})$'s. In the unlikely event that the gcd turns out to be equal to n, one backs up and computes the gcd's more frequently. See also [16].

POLLARD'S $p - 1$ METHOD [20]: It follows from Fermat's little theorem that if a is coprime to a prime p and k is an integer multiple of $p - 1$, then $a^k \equiv 1 \bmod p$. Thus, if p is a prime factor of n, then p divides either $\gcd(a, n)$ or $\gcd(a^k - 1, n)$ where a is randomly selected from $\{2, 3, \ldots, n - 2\}$. This means that primes p dividing n for which $p - 1$ is B-smooth (smoothness), may be found by selecting an integer $a \in \{2, 3, \ldots, n - 2\}$ at random, checking that $\gcd(a, n) = 1$, and computing $\gcd(a^k - 1, n)$ where k is the product of the primes $\le B$ and appropriately chosen small powers thereof. This takes time proportional to $(\log n)^2 \cdot B$. In a 'second stage' one may successively try $k \cdot q$ as well for the primes q between B and B', thereby finding p for which $p - 1$ is the product of a B-smooth part and a single larger prime factor up to B'; the additional cost is proportional to $B' - B$.

For n with unknown factorization, the best values B and B' are unknown too and, in general, too large to make the method practical. However, one may try values B, B' depending on the amount of computing time one finds reasonable and turn out to be lucky; if not one gives up as far as Pollard's $p - 1$ method is concerned. Despite its low probability of success, the method is quite popular, and has led to some surprising factorizations.

GENERALIZATIONS OF POLLARD'S $p - 1$ METHOD: Pollard's $p - 1$ method is the special case $d = 1$ of a more general method that finds a prime factor p of n for which the order $p^d - 1$ of the multiplicative group $\mathbf{F}_{p^d}^*$ of $\mathbf{F}_{p^d}$ is smooth. Because

$$X^d - 1 = \prod_{t \text{ dividing } d} \Phi_t(X),$$

where $\Phi_t(X)$ is the tth cyclotomic polynomial (thus, $\Phi_1(X) = X - 1$, $\Phi_2(X) = \frac{X^2-1}{X-1} = X + 1$, $\Phi_3(X) = \frac{X^3-1}{X-1} = X^2 + X + 1$, $\Phi_4(X) = \frac{X^4-1}{(X-1)(X+1)} = X^2 + 1$, etc.), the order of $\mathbf{F}_{p^d}^*$ is smooth if and only if $\Phi_t(p)$ is smooth for all integers t dividing d. For each t the smoothness test (possibly leading to a factorization of n) consists of an exponentiation in a ring modulo n that contains the order $\Phi_t(p)$ subgroup of the multiplicative group of the subfield $\mathbf{F}_{p^t}$ of $\mathbf{F}_{p^d}$. For $t = d = 2$ the method is known as Williams' $p + 1$ method [31]; for general d it is due to Bach and Shallit [3].

USAGE OF 'STRONG PRIMES' IN RSA: It is not uncommon, and even prescribed in some standards, to use so-called strong primes as factors of RSA moduli. These are primes for which both $p - 1$

and $p + 1$ have a very large prime factor, rendering ineffective a $p - 1$ or $p + 1$ attack against the modulus. This approach overlooks other $\Phi_t(p)$ attacks (which, for random moduli, have an even smaller probability of success). More importantly it overlooks the fact that the resulting RSA modulus is just as likely to be vulnerable to a single elliptic curve when using the Elliptic Curve Method for Factoring. It follows that usage of strong primes does in general not make RSA moduli more resistant against factoring attacks. See also [26].

CYCLING ATTACKS AGAINST RSA:
These attacks, also called 'superencryption attacks' work by repeatedly re-encrypting an RSA ciphertext, in the hope that after k re-encryptions (for some reasonable k) the original ciphertext appears. They are used as an additional reason why strong primes should be used in RSA. However, it is shown in [26] that a generalized and more efficient version of cycling attacks can be regarded as a special purpose factoring method that is successful only if all prime factors of $p - 1$ are contained in $e^k - 1$ for one of the primes p dividing n, where e is the RSA public exponent. The success probability of this attack is therefore small, even when compared to the success probability of Pollard's $p - 1$ method.

ELLIPTIC CURVE METHOD FOR FACTORING [13]:
The success of Pollard's $p - 1$ method (or its generalizations) depends on the smoothness of the order of one of the groups $\mathbf{F}_p^*$ (or $\mathbf{F}_{p^d}^*$) with p ranging over the prime factors of n. Given n, the group orders are fixed, so the method works efficiently for some n but for most n it would require too much computing time. In the elliptic curve method each fixed group $\mathbf{F}_p^*$ of fixed order (given n) is replaced by the group E_p of points modulo p of an elliptic curve modulo n. For randomly selected elliptic curves modulo n, the order $\#E_p$ of E_p behaves as a random number close to p. If $\#E_p$ is smooth, then p can efficiently be found using arithmetic in the group of points of the elliptic curve modulo n. It is conjectured that the smoothness behavior of $\#E_p$ is similar to that of ordinary integers of that size (smoothness probability), which implies that the method works efficiently for all n. It also implies that the method can be expected to find smaller factors faster than larger ones. In the worst case where n is the product of two primes of about the same size the heuristic expected runtime is $L_n[1/2, 1]$, with L_n as in L-notation; this is subexponential in $\log(n)$. See Elliptic Curve Method for Factoring for a more complete description and more detailed expected runtimes.

FERMAT'S METHOD AND CONGRUENCE OF SQUARES:
Fermat's method attempts to factor n by writing it as the difference of two integer squares. Let $n = p \cdot q$ for odd p and q with $p < q$, so that $q - p = 2y$ for an integer y. With $x = p + y$ it follows that $n = (x - y)(x + y) = x^2 - y^2$. Thus, if one tries $x = [\sqrt{n}] + 1, [\sqrt{n}] + 2, [\sqrt{n}] + 3, \ldots$ in succession until $x^2 - n$ is a perfect square, one ultimately finds $x^2 - n = y^2$. This is efficient only if the resulting y, the difference between the factors, is small; if it is large the method is inferior even to trial division.

Integers x and y that satisfy the similar but weaker condition

$$x^2 \equiv y^2 \bmod n$$

may also lead to a factorization of n: from the fact that n divides $x^2 - y^2 = (x - y)(x + y)$ it follows that

$$n = \gcd(n, x - y) \gcd(n, x + y).$$

If x and y are random solutions to $x^2 \equiv y^2 \bmod n$, then there is a probability of at least 50% that this yields a non-trivial factorization of n. All general purpose factoring methods described below work by finding 'random' solutions to this equation.

THE MORRISON–BRILLHART APPROACH:
To construct solutions to $x^2 \equiv y^2 \bmod n$ that may be assumed to be sufficiently random, Kraïtchik in the 1920s proposed to piece together solutions to $x^2 \equiv a \bmod n$. In the Morrison–Brillhart approach this is achieved using the following two steps [18]:

Relation collection. Fix a set P of primes (often called the factor base), and collect a set V of more than $\#P$ integers v such that

$$v^2 \equiv \left(\prod_{p \in P} p^{e_{v,p}} \right) \bmod n, \quad \text{with } e_{v,p} \in \mathbf{Z}.$$

These identities are often called 'relations' modulo n. If P is the set of primes $\leq B$, then v's such that $v^2 \bmod n$ is B-smooth lead to relations. For each v the exponents $e_{v,p}$ are regarded as a $\#P$-dimensional vector, denoted $(e_{v,p})_{p \in P}$.

Relation combination. Because $\#V > \#P$, the $\#P$-dimensional vectors $(e_{v,p})_{p \in P}$ are linearly dependent and there exist at least $\#V - \#P$ linearly independent subsets S of V such that

$$\sum_{v \in S} e_{v,p} = 2(s_p)_{p \in P}, \quad \text{with } (s_p)_{p \in P} \in \mathbf{Z}^{\#P}.$$

These subsets S with corresponding vectors $(s_p)_{p \in P}$ give rise to at least $\#V - \#P$ independent solutions to $x^2 \equiv y^2 \bmod n$, namely

$$x = \left(\prod_{v \in S} v\right) \bmod n, \quad y = \left(\prod_{p \in P} p^{s_p}\right) \bmod n,$$

and thereby at least $\#V - \#P$ independent chances to factor n.

All current general purpose factoring methods are based on the Morrison–Brillhart approach. They differ in the way the relations are collected, but they are all based on, more or less, the same relation combination step.

MATRIX STEP: Because S and $(s_p)_{p \in P}$ as above can be found by looking for linear dependencies modulo 2 among the rows of the $(\#V \times \#P)$-matrix $(e_{v,p})_{v \in V, p \in P}$, the relation combination step is often referred to as the 'matrix step.' With Gaussian elimination the matrix step can be done in $(\#P)^3$ steps (since $\#V \approx \#P$). Faster methods, such as conjugate gradient, Lanczos, or Wiedemann's coordinate recurrence method, require $O(w \cdot \#P)$ steps (see O-notation), where w is the number of non-zero entries of the matrix $(e_{v,p} \bmod 2)_{v \in V, p \in P}$. See [5, 11, 17, 23, 29, 30] for details.

In the various runtime analyses below, $\#P$ is measured using the L-notation and w turns out to be $c \cdot \#P$ for a c that disappears in the $o(1)$ of the L-notation, so that the runtime $O(w \cdot \#P)$ simplifies to $(\#P)^2$.

DIXON'S RANDOM SQUARES METHOD [8]: The simplest relation collection method is to define P as the set of primes $\leq B$ for some bound B and to select different v's at random from $\mathbf{Z}/n\mathbf{Z}$ until more than $\pi(B)$ ones have been found for which $v^2 \bmod n$ is B-smooth. The choice of B, and the resulting expected runtime, depends on the way the values $v^2 \bmod n$ are tested for B-smoothness. If smoothness is tested using trial division, then $B = L_n[1/2, 1/2]$ (with L_n as in L-notation). For each candidate v, the number $v^2 \bmod n$ is assumed to behave as a random number $\leq n = L_n[1, 1]$, and therefore, according to smoothness probability, B-smooth with probability $L_n[1/2, -1]$. Testing each candidate for B-smoothness using trial division takes time $\#P = \pi(B) = L_n[1/2, 1/2]$ (using the properties of L_n as set forth in L-notation), so collecting somewhat more than $\#P$ relations can be expected to take time

number of relations to be collected	trial division	inverse of smoothness probability
$\overbrace{L_n[1/2, 1/2]}$	$\cdot \overbrace{L_n[1/2, 1/2]}$	$\cdot \overbrace{(L_n[1/2, -1])^{-1}}$

$$= L_n[1/2, 2].$$

Gaussian elimination on the $\#V \times \#P$ matrix takes time

$$L_n[1/2, 1/2]^3 = L_n[1/2, 3/2].$$

Because at most $\log_2(n)$ entries are non-zero for each vector $(e_{v,p})_{p \in P}$, the total number of non-zero entries of the matrix is $\#V \cdot \log_2(n) = L_n[1/2, 1/2]$ and the matrix step can be done in

$$L_n[1/2, 1/2]^2 = L_n[1/2, 1]$$

steps using Lanczos or Wiedemann algorithms. In either case the runtime is dominated by relation collection and the total expected time required for Dixon's method with trial division is $L_n[1/2, 2]$. Unlike most methods described below, the expected runtime of the trial division variant of Dixon's method can rigorously be proved, i.e., it does not depend on any heuristic arguments or conjectures.

If B-smoothness is tested using the elliptic curve method, the time to test each $v^2 \bmod n$ is reduced to $L_n[1/2, 0]$: the entire cost of the smoothness tests disappears in the $o(1)$. As a result the two stages can be seen to require time $L_n[1/2, 3/2]$ each when Gaussian elimination is used for the matrix step. In this case, i.e., when using the elliptic curve method for smoothness testing, the runtime can be further reduced by using Lanczos or Wiedemann methods and a different value for B. Redefine B as $L_n[1/2, \sqrt{1/2}]$ so that relation collection takes time

$$L_n[1/2, \sqrt{1/2}] \cdot L_n[1/2, 0] \cdot (L_n[1/2, -\sqrt{1/2}])^{-1}$$
$$= L_n[1/2, \sqrt{2}]$$

and the matrix step requires $L_n[1/2, \sqrt{1/2}]^2 = L_n[1/2, \sqrt{2}]$ steps. The overall runtime of Dixon's method becomes

$$L_n[1/2, \sqrt{2}] + L_n[1/2, \sqrt{2}] = L_n[1/2, \sqrt{2}] :$$

asymptotically relation collection and combination are equally expensive. As described here, the expected runtime of this elliptic curve based variant of Dixon's method depends on the conjecture involved in the expected runtime of the elliptic curve method. It is shown in [22], however, that the expected runtime of a variant of the elliptic curve smoothness test can rigorously be proved. That leads to a rigorous $L_n[1/2, \sqrt{2}]$ expected runtime for Dixon's method.

CONTINUED FRACTION METHOD (CFRAC) [18]: The quadratic residues $v^2 \bmod n$ in Dixon's method provably behave with respect to smoothness probabilities as random non-negative integers less than n. That allows the rigorous proof of the expected runtime of Dixon's method. However, this theoretical advantage is not a practical concern. It would be preferable to generate smaller quadratic residues, thereby improving the smoothness chances and thus speeding up relation collection, even though it may no longer be possible to rigorously prove the expected runtime of the resulting method. The earliest relation collection method where quadratic residues were generated that are substantially smaller than n was due to Morrison and Brillhart and is based on the use of continued fractions; actually, this method (dubbed 'CFRAC') predates Dixon's method.

If a_i/b_i is the ith continued fraction convergent to $\sqrt{n}$, then $|a_i^2 - nb_i^2| < 2\sqrt{n}$. Thus, if v is chosen as a_i for $i = 1, 2, \ldots$ in succession, then $v^2 \bmod n = a_i^2 - nb_i^2$ is a quadratic residue modulo n that is $< 2\sqrt{n}$ and thus much smaller than n. In practice this leads to a substantially larger smoothness probability than in Dixon's method, despite the fact that if prime p divides $v^2 \bmod n$, then $(a_i/b_i)^2 \equiv n \bmod p$ so that n is a quadratic residue modulo p. With $B = L_n[1/2, 1/2]$, P the set of primes $p \leq B$ with $(\frac{n}{p}) = 1$, and elliptic curve smoothness testing, the heuristic expected runtime becomes $L_n[1/2, 1]$. The heuristic is based on the assumption that the residues $v^2 \bmod n$ behave, with respect to smoothness properties, as ordinary random integers $\leq n$ and that the set of primes $p \leq B$ for which $(\frac{n}{p}) \neq 1$ does not behave unexpectedly. In that case, when the <u>L-notation</u> is used to express smoothness probabilities, the difference with truly random integers disappears in the $o(1)$.

In [14] it is shown how this same expected runtime can be achieved rigorously (by a method that is based on the use of class groups). If elliptic curve smoothness testing is replaced by trial division, $B = L_n[1/2, \sqrt{1/8}]$ is optimal and the heuristic expected runtime becomes $L_n[1/2, \sqrt{2}]$.

NOTE ON THE SIZE OF RSA MODULI: In the mid 1970s CFRAC (with trial division based smoothness testing) was the factoring method of choice. Strangely, at that time, no one seemed to be aware of its (heuristic) subexponential expected runtime $L_n[1/2, \sqrt{2}]$. Had this been known by the time the RSA challenge [9] was posed, Ron Rivest may have based his runtime estimates on CFRAC instead of Pollard's rho (with its exponential expected runtime) [25], come up with more realistic estimates for the difficulty of factoring a 129-digit modulus, and could have decided that 129 digits were too close for comfort (as shown in [2]). As a result, 512-bit RSA moduli may have become less popular.

Linear Sieve

It was quickly realized that the practical performance of CFRAC was marred by the trial division based smoothness test. In the late 1970s Richard Schroeppel therefore developed a new way to generate relatively small residues modulo n that can be tested for smoothness very quickly: look for small integers i, j such that

$$f(i, j) = (i + [\sqrt{n}])(j + [\sqrt{n}]) - n \approx (i + j)\sqrt{n}$$

is smooth. Compared to CFRAC the residues are somewhat bigger, namely $(i + j)\sqrt{n}$ as opposed to $2\sqrt{n}$. But the advantage is that smoothness can be tested for many i, j simultaneously using a sieve (see <u>sieving</u>): if p divides $f(i, j)$ then p divides $f(i + kp, j + \ell p)$ for any $k, \ell \in \mathbf{Z}$. This means that if $f(i, j)$ is tested for B-smoothness for $0 \leq i < I$ and $0 \leq j < J$, the smoothness tests no longer take time $I \cdot J \cdot \pi(B) \approx I \cdot J \cdot B/\log B$, but

$$\sum_{p \leq B} \sum_{0 \leq i < I} \sum_{0 \leq j < J} \frac{1}{p} = O(I \cdot J \cdot \log\log(B)).$$

This leads to a heuristic expected runtime $L_n[1/2, 1]$. Inconveniently, $(i + [\sqrt{n}])(j + [\sqrt{n}])$ is not automatically a square, which means that for all values $i + [\sqrt{n}]$ and $j + [\sqrt{n}]$ that occur in smooth $f(i, j)$'s columns have to be included in the matrix. The effect this has on the expected runtime disappears in the $o(1)$ in L_n.

This method, dubbed 'linear sieve,' was the first factoring method that was heuristically shown (by Schroeppel) to have subexponential expected runtime. (That the earlier CFRAC also had subexponential expected runtime was realized only later; see also [10].) Its main historical significance is, however, that it led to the <u>Quadratic Sieve</u>, for many years the world's most practical factoring method.

Quadratic Sieve

The first crude version of the quadratic sieve was due to Carl Pomerance who realized that it may be profitable to take $i = j$ in Schroeppel's linear sieve. Although smoothness could still be tested quickly using a sieve and the heuristic expected runtime (with sieving) turned out to be a low $L_n[1/2, 1]$, in practice the method suffered from deteriorating smoothness probabilities (due to the linear growth of the quadratic residue $f(i, i)$). This problem was, however, quickly overcome by Jim Davis and Diane Holdridge which led to the first factorization of a number of more than 70 decimal digits [7]. Since then the method has been embellished in various ways (most importantly by

Peter Montgomery's multiple polynomial version, as described in [28]) to make it even more practical. See <u>Quadratic Sieve</u> for details. At this point the largest factorization obtained using quadratic sieve is the 135-digit factorization reported in [15].

Number Field Sieve

Until the late 1980s the best factoring methods, including the most practical one (quadratic sieve), shared the same expected runtime $L_n[1/2, 1]$ despite the fact that the underlying mathematics varied considerably: heuristically for quadratic and linear sieve, CFRAC, and the worst case of the elliptic curve method, and rigorously for the class group method from [14]. This remarkable coincidence fostered the hope among users of the RSA cryptosystem that $L_n[1/2, 1]$, halfway between linear time $\log n$ and exponential time n (<u>L-notation</u>), is the 'true' complexity of factoring.

The situation changed, slowly, when in late 1988 John Pollard distributed a letter to a handful of colleagues. In it he described a novel method, still based on the Morrison–Brillhart approach, to factor integers close to a cube and expressed his hope that, one day, the method may be used to factor the ninth Fermat number $F_9 = 2^{2^9} + 1$, back then the world's 'most wanted' composite. It was quickly established that for certain 'nice' n Pollard's new method should work in heuristic expected runtime $L_n[1/3, (\frac{32}{9})^{1/3}] \approx L_n[1/3, 1.526]$. This was the first indication that, conceivably, the complexity of factoring would not be stuck at $L_n[1/2, \ldots]$. The initial work was soon followed by the factorization of several large 'nice' integers, culminating in 1990 in the factorization of F_9 [12]. Further theoretical work removed the 'niceness' restriction and led to the method that is now referred to as the 'number field sieve': a general purpose factoring method with heuristic expected runtime $L_n[1/3, (\frac{64}{9})^{1/3}] \approx L_n[1/3, 1.923]$. The method as it applies to 'nice' numbers is now called the 'special number field sieve.' See <u>Number Field Sieve</u> for details. The first time a 512-bit RSA modulus was factored, using the number field sieve, was in 1999 [4].

With hindsight the property that all $L_n[1/2, 1]$ factoring methods have in common is their dependence, in one way or another, on smoothness of numbers of order $n^{O(1)}$. The number field sieve breaks through the $n^{O(1)}$ barrier and depends on smoothness of numbers of order $n^{o(1)}$.

Arjen K. Lenstra

References

[1] Alford, W.R., A. Granville, and C. Pomerance (1994). "There are infinitely many Carmichael numbers." *Ann. of Math*, 140, 703–722.

[2] Atkins, D., M. Graff, A.K. Lenstra, and P.C. Leyland (1995). "The magic words are squeamish ossifrage." *Advances in Cryptography—ASIACRYPT'94*, Lecture Notes in Computer Science, vol. 917, eds. J. Pieprzyk and R. Safavi-Nami. Springer-Verlag, Berlin, 265–277.

[3] Bach, E. and J. Shallit (1989). "Cyclotomic polynomials and factoring." *Math. Comp.*, 52, 201–219.

[4] Cavallar, S., B. Dodson, A.K. Lenstra, W. Lioen, P.L. Montgomery, B. Murphy, H. te Riele et al. (2000). "Factorization of a 512-bit RSA modulus." *Advances in Cryptology—EUROCRYPT 2000*, Lecture Notes in Computer Science, vol. 1807, ed. B. Preneel. Springer-Verlag, Berlin, 1–18.

[5] Coppersmith, D. (1994). "Solving homogeneous linear equations over GF(2) via block Wiedemann algorithm." *Math. Comp.*, 62, 333–350.

[6] Crandall, R.E., and C. Pomerance (2001). *Prime Numbers: A Computational Perspective*. Springer-Verlag, Berlin.

[7] Davis, J.A. and D.B. Holdridge (1983). "Factorization using the quadratic sieve algorithm." Tech Report SAND 8301346, Sandia National Laboratories, Albuquerque, NM.

[8] Dixon, J.D. (1981). "Asymptotically fast factorization of integers." *Math. Comp.*, 36, 255–260.

[9] Gardner, M. (1977). "Mathematical games: A new kind of cipher that would take millions of years to break." *Scientific American*, 120–124.

[10] Knuth, D.E. (1998). *The Art of Computer Programming, Volume 2, Seminumerical Algorithms* (3rd ed.). Addison-Wesley, Reading, MA.

[11] LaMacchia, B.A. and A.M. Odlyzko (1990). "Solving large sparse linear systems over finite fields." *Advances in Cryptology—CRYPTO'90*, Lecture Notes in Computer Science, vol. 537, eds. A.J. Menezes and S.A. Vanstone. Springer-Verlag, Berlin, 109–133.

[12] Lenstra, A.K., H.W. Lenstra, Jr., M.S. Manasse, and J.M. Pollard (1993). "The factorization of the ninth Fermat number." *Math. Comp.*, 61, 319–349.

[13] Lenstra, Jr., H.W. (1987). "Factoring integers with elliptic curves." *Ann. of Math.*, 126, 649–673.

[14] Lenstra, Jr., H.W. and C. Pomerance (1982). "A rigorous time bound for factoring integers." *J. Amer. Math. Soc.*, 5, 483–516.

[15] Leyland, P., A.K. Lenstra, B. Dodson, A. Muffett, and S. Wagstaff (2002). "MPQS with three large primes." *Proceedings ANTS V*, Lecture Notes in Computer Science, vol. 2369, eds. C. Fieker and D.R. Kohel. Springer, Berlin, 448–462.

[16] Montgomery, P.L. (1987). "Speeding the Pollard and elliptic curve methods of factorization." *Math. Comp.*, 48, 243–264.

[17] Montgomery, P.L. (1995). "A block Lanczos algorithm for finding dependencies over GF(2)."

Advances in Cryptology—EUROCRYPT'95, Lecture Notes in Computer Science, vol. 921, ed. L.C. Guillou. Springer-Verlag, Berlin, 106–120.

[18] Morrison, M.A. and J. Brillhart (1975). "A method of factorization and the factorization of F_7." *Math. Comp.*, 29, 183–205.

[19] Nechaev, V.I. (1994). "Complexity of a determinate algorithm for the discrete logarithm." *Mathematical Notes*, 55 (2), 155–172. Translated from *Matematicheskie Zametki*, 55 (2), (1994) 91–101. This result dates from 1968.

[20] Pollard, J.M. (1974). "Theorems on factorization and primality testing." *Proceedings of the Cambridge Philosophical Society*, 76, 521–528.

[21] Pollard, J.M. (1978). "Monte Carlo methods for index computation (mod p)." *Math. Comp.*, 32, 918–924.

[22] Pomerance, C. (1987). "Fast, rigorous factorization and discrete logarithm algorithms." *Discrete Algorithms and Complexity*, eds. D.S. Johnson, T. Hishizeki, A. Nozaki, and H.S. Wilf. Academic Press, Orlando, FL, 119–143.

[23] Pomerance, C. and J.W. Smith (1992). "Reduction of huge, sparse matrices over finite fields via created catastrophes." *Experimental Math.*, 1, 89–94.

[24] Rabin, M.O. (1980). "Probabilistic algorithms for primality testing." *J. Number Theory*, 12, 128–138.

[25] Rivest, R.L. (1979). Letter to M. Gardner containing an estimate of the difficulty of factoring a 129-digit modulus using Pollard's rho method?

[26] Rivest, R.L. and R.D. Silverman (1997). "Are 'strong' primes needed for RSA?" Manuscript, available on www.iacr.org

[27] Shoup, V. (1997). "Lower bounds for discrete logarithms and related problems." *Advances in Cryptology—EUROCRYPT'97*, Lecture Notes in Computer Science, vol. 1233, ed. W. Fumy. Springer-Verlag, Berlin, 256–266.

[28] Silverman, R.D. (1987). "The multiple polynomial quadratic sieve." *Math. Comp.*, 46, 327–339.

[29] Villard, G. (1997). "Further analysis of Coppersmith's block Wiedemann algorithm for the solution of sparse linear systems (extended abstract)." *Proceedings of 1997 International Symposium on Symbolic and Algebraic Computation*. ACM Press, New York, 32–39.

[30] Wiedemann, D. (1986). "Solving sparse linear equations over finite fields." *IEEE Transactions on Information Theory*, IT-32, 54–62.

[31] Williams, H.C. (1982). "A $p+1$ method of factoring." *Math. Comp.*, 39, 225–234.

INTERACTIVE ARGUMENT

An **interactive argument** (or *computationally sound proof system*) is a relaxation of an interactive proof, introduced in [1]. The difference is that the prover is restricted to be a polynomial time algorithm for an interactive argument, whereas no such restrictions on the prover apply for an interactive proof. The prover's advantage over the verifier is that the prover gets a private input, which allows the prover to perform his or her task in polynomial time (completeness).

The soundness condition for an interactive argument, referred to as *computational soundness*, states as before that executions of the protocol between the prover and the verifier should result in the verifier rejecting the proof, if $x \notin L$ holds; here, the prover is not required to follow the protocol, that is, the prover may behave arbitrarily, but the prover is limited to be a (*probabilistic*) *polynomial-time* algorithm.

Hence, cheating by the prover is not required to be impossible; rather, cheating is required to be *infeasible*. Therefore, interactive arguments are easier to achieve than interactive proofs; in particular, while perfect zero-knowledge arguments are known to exist for every language in NP, it is considered unlikely that perfect zero-knowledge proofs exist for every language in NP.

Berry Schoenmakers

References

[1] Brassard, G., D. Chaum, and C. Crépeau (1988). "Minimum disclosure proofs of knowledge." *Journal of Computer and System Sciences*, 37 (2), 156–189.

[2] Goldreich, O. (2001). *Foundations of Cryptography—Basic Tools*. Cambridge University Press, Cambridge.

[3] Naor, M., R. Ostrovsky, R. Venkatesan, and M. Yung (1998). "Zero-knowledge arguments for NP can be based on general assumptions." *Journal of Cryptology*, 11 (2), 87–108.

INTERACTIVE PROOF

The notion of an *interactive proof* plays an important role in complexity theory. An interactive proof is a protocol between two parties, called the *prover* and the *verifier*. The crucial point is that the verifier is restricted to be a (probabilistic) polynomial time algorithm, whereas no such restriction applies to the prover. By means of an interactive proof the prover convinces the verifier of the validity of a given statement. A statement is of the form $x \in L$, where L is a formal language. The interesting languages are those for which no polynomial time membership tests (are known to) exist. It follows that the verifier cannot determine on its own whether $x \in L$ holds.

An interactive proof is required to satisfy two conditions. The first condition is *completeness*, which means that executions of the protocol between the prover and the verifier should result in the verifier accepting the proof, if $x \in L$ holds. The second condition is *soundness*, which means that executions of the protocol between the prover and the verifier should result in the verifier rejecting the proof, if $x \notin L$ holds; here, the prover is not required to follow the protocol, that is, the prover may behave arbitrarily.

A simple example of an interactive proof runs as follows. Consider the language L_H consisting of graphs containing a Hamiltonian cycle. It is well-known that the problem of determining membership for L_H is NP-complete. Hence, it is supposedly hard to determine whether a given graph contains a Hamiltonian cycle. However, given a purported Hamiltonian cycle for a graph, it is easy to check whether this is indeed the case. An interactive proof for L_H is obtained if the prover simply sends a Hamiltonian cycle for the graph under consideration to the verifier. The conditions of completeness and soundness are clearly satisfied.

In the context of cryptography, interactive proofs are usually required to satisfy some additional conditions. Many interactive proofs are in fact proofs of knowledge. Also, zero-knowledge proofs are a main example of interactive proofs used for cryptographic purposes, noting that zero-knowledge interactive arguments and witness hiding protocols are possible alternatives. The above interactive proof for L_H is not zero-knowledge nor witness hiding, as the prover simply gives away a Hamiltonian cycle.

Berry Schoenmakers

References

[1] Goldwasser, S., S. Micali, and C. Rackoff (1989). "The knowledge complexity of interactive proof systems." *SIAM Journal on Computing*, 18, 186–208. Preliminary version in *17th ACM Symposium on the Theory of Computing*, 1982.

[2] Goldreich, O. (2001). *Foundations of Cryptography—Basic Tools*. Cambridge University Press, Cambridge.

[3] Shamir, A. (1992). "IP = SPACE." *Journal of the ACM*, 39 (4), 869–877.

INTERPOLATION ATTACK

The *interpolation attack* is a technique for attacking block ciphers built from simple algebraic functions. It was introduced by Jakobsen and Knudsen [2, 3] in 1997 and applied to variants of SHARK, a predecessor of Rijndael/AES.

The attack is based on a well-known principle: given an unknown polynomial $y = f(x)$, if the degree of $f(x)$ does not exceed $n - 1$, then its coefficients can efficiently be recovered by taking n distinct samples (x_i, y_i) with $y_i = f(x_i)$. The *Lagrange interpolation formula* provides a general expression for the polynomial reconstructed this way:

$$f(x) = \sum_i y_i \prod_{j \neq i} \frac{x - x_j}{x_i - x_j}.$$

This mathematical property has interesting implications when considering a block cipher with a fixed but unknown secret key. If the ciphertext is written as a polynomial (with unknown coefficients) of the plaintext, and if the degree of this polynomial is sufficiently low, then a limited number of plaintext–ciphertext pairs suffice to completely determine the encryption function. This allows the attacker to encrypt and decrypt data blocks for the unknown key without doing any key-recovery.

An interesting property of the basic interpolation attack is that it is not affected by the internal structure of the cipher, apart from the degree of the polynomial representing the encryption function. In fact, a low degree is not strictly necessary for an efficient attack; it suffices that the number of unknown coefficients is sufficiently small, and this happens to be the case for a number of ciphers which were optimized against linear and differential attacks (for example, the $\mathcal{KN}$ cipher by Knudsen and Nyberg).

The attack outlined above can be extended and generalized in many ways. It can, for example, also be applied by only expressing a part of the ciphertext as a function of the plaintext, or by constructing an implicit polynomial expression involving parts of the plaintext and the ciphertext. The latter could be derived from a rational expression, or obtained by applying a meet-in-the-middle attack. Furthermore, in a subsequent paper [1], Jakobsen demonstrated that the interpolation ideas can still be applied when the polynomials are probabilistic. The method is based on Sudan's algorithm, designed to decode Reed-Solomon codes (see cyclic codes). Finally note that all attacks described above can easily be turned into key-recovery attacks by guessing the last round key of the cipher and checking the correctness of the guess by applying the interpolation attack on the remaining rounds.

Christophe De Cannière

References

[1] Jakobsen, T. (1998). "Cryptanalysis of block ciphers with probabilistic non-linear relations of low degree." *Advances in Cryptology—CRYPTO'98*, Lecture Notes in Computer Science, vol. 1462, ed. H. Krawczyk. Springer-Verlag, Berlin, 212–223.

[2] Jakobsen, T. and L.R. Knudsen (1997). "The interpolation attack on block ciphers." *Proceedings of Fast Software Encryption—FSE'97*, Lecture Notes in Computer Science, vol. 1267, ed. E. Biham. Springer-Verlag, Berlin, 28–40.

[3] Jakobsen T. and L.R. Knudsen (2001). "Attacks on block ciphers of low algebraic degree." *Journal of Cryptology*, 14, 197–210.

INTRUSION DETECTION

Intrusion Detection (ID) systems attempt to recognize unauthorized or anomalous use of a computer system or network. ID solutions use sensors to ascertain the state of the environment. Intrusions are detected by recognizing some characteristic in the sensed data. These characteristics may reflect overt illegal access or use of system resources, a match with a known pattern of intrusive activity, or deviate from some model of normal activity. The intruder may or may not commit harmful activities as part of the sensed attack, and may not even be successful in completing an attempted intrusion. ID is concerned with detecting the existence of an intrusion, regardless of its nature or affect, and initiating an appropriate response.

ID techniques are often categorized as either *misuse* or *anomaly* detection. In both of these techniques, collected data is compared against a model of system behavior (see Fig. 1). Such data often includes computer audit/log data and network traffic data [6]. The difference between the two techniques lies in the underlying model: misuse detection uses a model of "bad" system behavior, and anomaly detection uses a model of "normal" system behavior. Loosely speaking, an attack is detected if the observed behavior matches anomalous or diverges from normal behavior, respectively.

Misuse Detection compares current system data with models of known intrusive activities. The model for this technique is designed using specifically known patterns of unauthorized behavior to predict and detect subsequent similar attempts. These *signatures* rely on information about known intrusions. The signature consists of a specific behavioral pattern or profile. The ID system attempts to match a sequence of observed events with known pattern of events in the signature (e.g., password guessing, buffer overflow attacks [2]). The events a particular misuse system will measure is dependent on the environment and the types of attacks it is attempting to detect. For example, one might gage the number and frequency of login failures when trying to detect password guessing attacks. While active, any sequence of events that do not match an attack signature is deemed normal or at least harmless. While misuse detection is highly effective at detecting known attacks, it cannot detect attacks for which there is no available signature.

Anomaly detection systems attempt to detect abnormal behavior. An anomaly detection system establishes a model of normalcy based on known or expected system behavior (e.g., as measured from past behavior). The model crucially relies on the accuracy of this *normal profile*. The normal profile is a data set or approximation that defines the baseline state of the system under normal operating conditions. The model may also contain information gained from investigating user patterns, such as profiling the programs a user executes. Armed with this model, any significant deviation from the norm profile is considered anomalous, and is thus a potential indicator of intrusive activity. Unlike signature-based schemes, anomaly-based techniques present the possibility of detecting previously unknown attacks. While this is certainly desirable, anomaly-based techniques may generate *false positives*. In this case, a normal activity is incorrectly flagged as intrusive, resulting in the initiation of unnecessary counter-measures. High false positive rates have limited the effectiveness of past anomaly detection solutions.

To simplify, depending on the environments and target attacks, intrusion detection techniques can be deployed within hosts or networks. Host-based ID techniques detect intrusions based on data

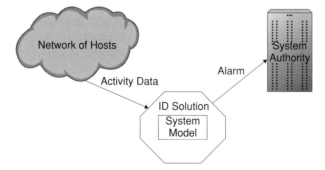

Fig. 1. Intrusion detection techniques compare environmental state (e.g., activity data) with a model of normal or anomalous system behavior

collected from the operating system or user applications. Network-based techniques use observed network state or protocol interactions to detect attacks. Neither of these solutions make any assumption about the source of the attack: an intrusion may emanate from a legitimate (insider) user or an anonymous external adversary.

Host-based intrusions often involve adversaries operating as authorized users of a computer system, but conducting activities that violate a site's security policies (see policy). These types of intrusions are most often detected using operating system audit and log data. These logs commonly include actions that are restricted to access by system administrators (such as adding or deleting user accounts), as well those permitted by non-administrative users (e.g., adding or deleting files, executing processes or programs, logging in and out of a system, etc.). Host-based ID is effective for detecting attacks mounted by otherwise authorized users because of the sensor's visibility into user behavior [3].

Network intrusions often involve adversaries who are not authorized to use any particular part of the environment, but abuse open network protocols and services. These types of intrusions may be detected using network traffic data. Network-based ID techniques often obtain the raw network packets used to perform an intrusion by passively tapping all traffic traversing a network segment [3]. Some ID solutions examine packet headers for signs of suspicious activity. For example, many IP-based denial-of-service (DoS) and fragmented packet (Teardrop) attacks can only be identified by looking at the packet headers as they travel across a network. An ID technique can also investigate the packet payload, looking for commands of syntax used in specific attacks to recognize embedded payload attacks. An ID technique may also model traffic as *flows*. Flows are virtual or real network connections that represent aggregated related and concurrent communication. Flows represent higher layer communication sessions, and can be used to infer malicious intent, where the individual packets may appear harmless (e.g., in a *distributed* Dos (DDdoS) attack). Network-based ID is effective at detecting foreign attacks because of their visibility into the communication media (i.e., network).

Network-based ID techniques are often strategically deployed at critical access points [5] (see Fig. 2). These systems can alleviate much of the need for hosts to monitor for network-based attacks when such configurations are used. A network ID solution may also be placed outside of a network (or in the network demilitarized zone

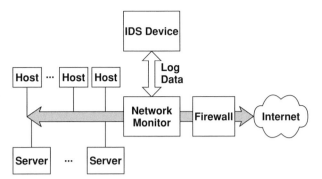

Fig. 2. Deploying IDS—a network-based IDS can be strategically deployed between the Internet and sensitive data hosts and servers. In this way, all traffic coming from the open Internet is monitored by the IDS

(DMZ)) to detect attacks intended for resources behind the network's perimeter defenses. Because network-based ID often operates in real time and on separate devices, it is difficult or impossible for an adversary to later remove evidence of an attack. Note that a network-based ID solution need not only indicate the method of attack, but also infer or extract other information that may help lead to identification and prosecution of the adversary.

Host-based ID solutions frequently offer better forensic tools than their network-based counterparts. Because a host-based ID solution has access to the operating system internals, it has many more opportunities to collect sensor data. For example, while an attack on a critical server made via a keyboard cannot be detected by an ID network-based system, a host-based ID system that captures keystrokes can easily detect the attack.

Ultimately, the kinds of attacks that host and network-based systems most effectively detect are quite different. The different approaches (host vs. network) complement each other well. The combination of these two technologies is a foundation upon which many environments build their security infrastructure.

The response to a detected attack varies from system to system. More frequently than not, the ID solution forwards an alarm event to a network management system or authority for display and resolution. In a host-based system, this may be the administrator for the host or some local network administrator. Many sensors may be deployed throughout a network in a large network-based ID solution. The sensors and ID components communicate with each other to generate and correlate alarms at, for example, a network management console.

ID solutions are packaged as Intrusion Detection Systems (IDS). These software solutions come in a variety of forms due to the unconstrained design space for detecting intrusions [1] (e.g., Generic Intrusion Detection Model, Network Security Monitor Model, Autonomous Agents Model, Behavior-based Intrusion Detection Model, Predictive Pattern Generation Model, and Knowledge-based Intrusion Detection Model [4]). An IDS may be deployed as the example of a network IDS displays in Figure 2. One popular open source network IDS available for free downloading is called SNORT, and is available at www.snort.org. This system sniffs and logs network packets, and employs a signature-based technique for network-based ID.

Toni Farley
Jill Joseph

References

[1] Galvin, P.B., A. Silberschatz, and G. Gagne (2003). *Operating Systems Concepts* (6th ed.). John Wiley & Sons, Inc., New York.

[2] Mohay, George (2003). *Computer and Intrusion Forensics* (1st ed.). Artech House, Norwood, MA.

[3] Proctor, Paul E. (2001). *Practical Intrusion Detection HandBook* (3rd ed.). Prentice-Hall, Englewood Cliffs, NJ.

[4] Sherif, J.S. and T.G. Dearmond (2002). "Intrusion detection: systems and models." *Enabling Technologies: Infrastructure for Collaborative Enterprises, 2002 11th IEEE International Workshops*, 115–133.

[5] Tannenbaum, Andrew S. (2001). *Modern Operating Systems* (2nd ed.). Prentice-Hall, Englewood Cliffs, NJ.

[6] Ye, N. (2003). *The Handbook of Data Mining* (1st ed.). Lawrence Erlbaum Associates, Mahwah, NJ.

INVASIVE ATTACKS

INTRODUCTION: A smart card contains an embedded microchip with metallic contacts on the front. The smartcard does not usually have its own power supply, yet it operates as a very small computer. The embedded operating system (OS) controls application execution, access condition, cryptographic routines and communication protocols with the outside world (usually a terminal). Some smartcards have several applications embedded at a time; these are called "multi-application" smart cards.

The largest application is the GSM. The subscriber identity module (SIM) is found in all handsets that use the GSM wireless communication standard (Europe, Asia, Latin America and increasingly North America). Smart cards are also widely used in banking, pay TV, access control, health insurance, public transportation, government and online services. Broadly speaking, the smartcard is used to control access to specific devices, networks or services. When smart cards are combined with a Public Key Infrastructure (PKI), they efficiently implement secure spaces within a broader IT environment, which are useful for applications such as online payment and identification.

SMART CARD SECURITY: The purpose of a smartcard is to ensure the secure storage of sensitive data, and also the integrity and tamper-resistant execution of cryptographic applications. Highly sensitive data is never released outside the card; all operations are carried out by the operating system inside the card. The operating system also handles security and data access for each of these applications. This section describes the mechanisms that protect on-card data and applications.

Smartcards have an integrated CMOS circuit commonly refered to as a chip comprising:
- Logical functions.
- CPU from 8 bits up to 32 bits.
- Different kinds of memories like ROM, EEPROM, RAM and more recently Flash and Feram (Ferroelectrics RAM). Here the ROM is a permanent memory storing all or part of the operating system. The EEPROM is a nonvolatile erasable memory that stores application data, mainly keys, PIN codes or personal data and, in some cases, parts of the operating system and it's applications. The RAM is a volatile memory that stores temporary data such as intermediate internal application data or session keys or access rights.
- I/O interface for communication with the external world.
- Peripherals such as crypto-coprocessors, Random-Number Generators.

OVERVIEW OF ATTACKS ON SMART CARDS: Basically, smart card attacks can be classified into three main categories: social, logical and physical attacks.

Social Attacks

These are the oldest. The idea behind these attacks is to obtain information directly from the manufacturer using classical social engineering techniques. Countermeasures in this case range

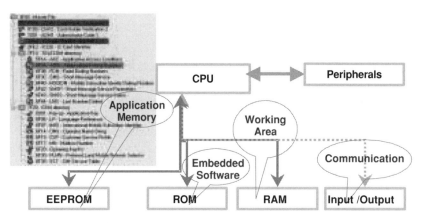

Fig. 1. Microprocessor card architecture

from physical security such as access control to the sites, physically isolated labs, and the education of employees to the application of strict security procedures. This kind of attack falls outside the scope of this document.

Logical Attacks

These attacks are used to recover secret data from secure devices without actually damaging the device. By monitoring execution time, power consumption or electromagnetic radiation of a chip, it is frequently possible to infer information about the processed data. Performing <u>Side-channel analysis</u> on a secure device requires advanced knowledge in electronics, cryptography, signal processing and statistics. A well-known class of attacks in this group is based on the analysis of smart card *power consumption*. This class includes <u>Differential Power Analysis</u> (DPA), Simple Power Analysis (SPA), and timing analysis.

- **SPA** uses variations in the global power consumption of the chip and infers information that is normally held within the chip. For example, an increase in power consumption might indicate that a modular exponentiation (an important cryptographic function) is being performed. In general, SPA will give better results if the attacker has extensive knowledge of the hardware architecture.
- **DPA** is more sophisticated than the SPA. Statistical analysis of power consumption curves is carried out for several executions of the same algorithm. The input data is changed in such a way that sensitive information can be deduced.
- <u>**Timing attacks**</u> have posed serious problems in the past, because in older designs the execution time would vary according to the data and/or the cryptographic keys that were being processed. Current smart card chips have been

designed with constant timing; or at least timings that do not depend on data or secret keys.

- **Electromagnetic analysis** is a newer type of attack. It is based on the same techniques as those used for DPA and SPA, but the physical quantities that are measured are not the same. In this case, the RF signals provide the essential information. Such attacks fall into the category of side-channel attacks, but they differ in a number of crucial points from power attacks.
- <u>**Fault attacks**</u> are conducted using a combination of several environmental conditions that cause the chip to produce computational errors that can leak protected information. Sensors that detect abnormal operating conditions are thus used to preclude the need for costly software and hardware countermeasures (it is always better to anticipate rather than to correct errors).
- **Software attacks** target software flaws using a normal communication channel to interface with the card. These flaws may weaken the security features of the card or allow them to be bypassed, leaving the system open to frauds. There is a wide range of software attacks, some of which are not specific to the smart card. Incorrect file access conditions, malicious code, flaws in cryptographic protocols, design and implementation errors are common flaws in computing systems.

Invasive Attacks

This refers to attacks of physical systems where the physical properties of the chip are irreversibly modified. Different kinds of attacks are possible using "standard" reverse engineering techniques with optical or scanning electron microscopes (SEM). The aim is to capture information stored in memory areas, or data flowing through the data bus, registers, etc. Such techniques are also used to disconnect circuits, to override sensors, or to defeat blown fuse links (using probe stations or Focused Ion Beam [FIB]). These attacks are not specific to Smartcards. At first, tools used for invasive attacks were dedicated to failure analysis and debugging by semiconductor manufacturers.

Such attacks are detailed in the following section.

INVASIVE ATTACKS: The goal of these attacks is to unearth information stored in memories, or data flowing through the data bus, registers, etc. These attacks are not specific to smart cards but are generic to CMOS components. Modifying an electronic component requires considerable time and resources, sophisticated and expensive tools and extensive hardware expertise. This cost can be considered as a first barrier. The more you know about the internal architecture of the chip the easier these attacks will be.

Access to Silicon

Delayering. The first step consists in doing some reverse engineering. Reverse engineering allows block localization, like memories, buses, random number generation, inputs etc. and also an understanding of the chip architecture.

The chip is covered by a globe-top made of epoxy resin that can be removed by using hot fuming nitric acid. At this stage, the chip's surface is not yet accessible for probing or modification; only optical or electrical analysis could be feasible depending on the chip manufacturer's process.

CMOS chip structure is made of multiplayer stacking going from three to five metal layers. The upper layer is called the passivation layer (silicon oxide), which protects the chip against environmental hazards and ionic contaminations.

As a large amount of stress can be generated on the die during the assembly process, a thick film of polyimide over the passivation layer is deposited. These stresses may lead to cracking of the protective passivation layer. Before getting access to the silicon, the polyimide and passivation layers must be removed using ethylendiamine and fluoridric acid respectively. For the passivation layer the most convenient depassivation technique is to use a laser cutter, but this technique can only be used for create small windows used for probing.

There is no particular way to prevent optical reverse engineering except by increasing the design complexity and sharing specific parts of the circuit among different layers.

Fig. 2. Example of window opened in the passivation layer using laser cutter

Block Localization. Once the passivation layer has been removed, the upper metal layer becomes accessible. Each metal layer can be selectively removed by chemical attacks. Depending on the layer structure, it can be dry-etched using plasma or wet etching. This step can be destructive: several chips are usually required for these attacks.

Memory blocks like ROM, EEPROM, RAM and parts of analogue blocks such as charge pumps and capacitors are easily recognizable, an example is given in Figure 7.

Memory Content Extraction
- **ROM**

    The ROM is a critical part of the circuit because it is where the code (operating system, JAVA virtual machine, API) is stored. In new generations of products (0.18 μm) it is preferable to use a metal or ionic implantation ROM process. The difficulty of reverse-engineering the ROM content in these two cases is identical. In the case of a metal ROM a selective delayering of the chip metal layers is necessary in order to reach the appropriate metal layer (usually M1) and the metal connections. In the case of an ionic implantation ROM a complete delayering is required in order to reach the silicon level. A chemical attack is further performed to reveal doping[1] differences and make the content readable by optical observations, a process that can be automated, and then ROM mapping of zero and one is extracted.

    In order to protect the ROM contents from illegal extraction, all chip manufacturers encrypt the ROM. The ciphering algorithm depends on both the data value and the address. So a complete reverse engineering of the decoder's logic

---

[1] In order to create active region in the semiconductor, impurities like Phosphor or boron are introduced.

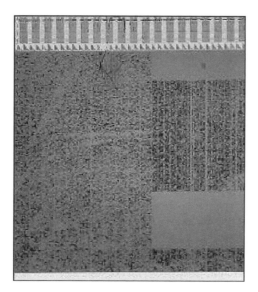

Fig. 3. Metal ROM showing unused area

is necessary. In addition, ROM code will be systematically filled with random data when the code size is not equal to the complete ROM size.

- **RAM**
  The RAM is a temporary working area. The RAM data is, for example, used for cryptographic calculations and session key transfers. Using a SEM specialized in voltage contrast, it is possible to observe RAM activity in operating mode. These microscopes have the ability of detecting variations in voltage. Applying power to the chip and observing the chip in image mode reveals the DC (direct current) conditions on the surface layers of the chip.

  Like for the ROM, chip manufacturers encrypt or scramble the RAM. Scrambling depends only on the address, while ciphering depends on both the address and the data: these mechanisms can be static or dynamic. Frequency sensors, used for restricting the operating range of the chip, could protect against the use of SEM in voltage contrast.

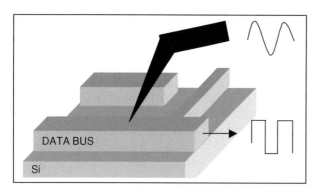

Fig. 4. Mechanical probing

- **EEPROM or FLASH**
  This write/erase nonvolatile memory is used to store application data, part of executable code and sensitive data like PIN code or session keys. Reversing the EEPROM or Flash memory consists in finding the state of the floating gate. At this time, no method has been found to extract the complete memory content. As before, non-volatile memory can be encrypted or scrambled to increase security level. There is an emergence of new nonvolatile memory point like Flash and Feram in the smartcard chip market. These new memory maps have the particularity of being very compact, of taking less space, and are probably more difficult to attack. We will soon have Smartcards where the executable code will be loaded in flash. With a strong memory management and a shrinking technology, one would think that these products will be more secure. However, the resulting level of security depends on other factors such as the time available before the product needs to be finished.

Bus Localization. Using Voltage contrast techniques, bus activity can be observed. Besides, in a slow scan image mode, it is possible to visualize different clock rates across the bus lines. Observing the changing contrast on the signal path can be correlated to the changing logic state. However manufacturers forecast specific countermeasures comprising:
- Static ciphering
- Dynamic ciphering
- Complemented logic
- Buried Buses
- Dummy activity on buses.

Chip Probing

Once the first step of reverse engineering is completed, chip probing can start. As buses are connected to the CPU and also to the memories, there is a great interest in taping data passing through these buses. Hence, it could be possible to retrieve the full running program. In this case a small reverse engineering must have been done previously and passivation removal or circuit modification is sometimes required.

Probing could be mechanical using a microprobe on a micromanipulator with a probe station. This technique is very difficult for sub-Micronics technology smaller than 0.35 μm.

The smaller the line is, the more difficult it will be to access them by mechanical probing. Instead, e-beam probing which is a Contactless method

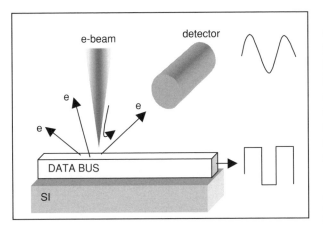

Fig. 5. e-Beam probing

based on secondary electron analysis giving voltage values, will be used.

Chip probing could be prevented by adding countermeasures like:

- An active shield that is a metal mesh where data passes continuously on lines. If there is a disconnection or modification of this mesh then the chip does not operate anymore.
- A passive shield where a full metal mesh covers some sensitive parts of the circuit.
- Bus static or dynamic scrambling.
- Buried lines.

Chip Modification. Modifying or disconnecting part of a circuit in order can constitute an interesting attack method. Using this method it is possible to connect or disconnect hardware security mechanisms.

The Focused Ion Beam (FIB) is a convenient and powerful tool. A FIB allows material deposited for the creation of metal lines or metal cross-pads allowing access to the bus, as illustrated in the following figure:

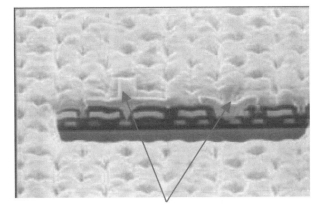

Fig. 6. Probe pads added by FIB to reach M1 through a shield

In a similar manner, removing materials using a FIB allows the track to be cut, the disconnection of security sensors, or the opening of a window through the passivation layer to get access to buried levels. As mentioned previously, countermeasures can be added in order to prevent such attacks:

- Active shield
- Complexity of the design
- Glue logic.

PROTECTION AGAINST PHYSICAL ATTACKS: Conducting physical attacks against smart cards at the semiconductor level requires expensive equipment and considerable technical expertize. Therefore, the threat is limited to few organizations and specialists. However, smart card manufacturers cannot afford to release cards without effective countermeasures against such attacks. Some of the principles used to physically secure cards are described below.

Modern smart cards use semiconductor technology for the chip, making reverse engineering by observation difficult. The size and density of the transistors on the chip surface has drastically shrunk (0.18 μm), and chips are now considered secure against visual analytical reverse engineering.

Functional blocks are mixed, producing what is called a glue logic design. This makes it much more difficult for an attacker to analyze the structure of the chip and to localize functional blocks (CPU, RAM, ROM, EEPROM, buses, registers, etc.).

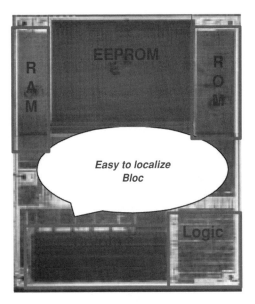

Fig. 7. Old chip design

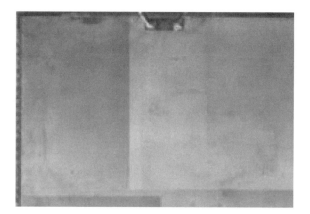

Fig. 8. New chip design

Moreover, the buses are scrambled or ciphered and are thus inaccessible from outside the chip, so connections cannot be easily made to recover memory contents. Memories are also scrambled or ciphered in order to protect the chip from selective access/erasure of individual data bytes.

supply. If this layer is removed, the chip will no longer operate. This layer prevents analysis of electrical voltage on the chip to infer information.

Moreover, a set of sensors is activated to detect abnormal variations of environmental variables (see also <u>smartcard tamper resistance</u>). It guarantees that the chip will not be able to operate in abnormal conditions of use. These sensors measure values such as voltage, temperature, clock frequency, and light. Such sensors offer protections against fault attacks (among others).

CONCLUSION: Smartcards turn out to be the strongest components in the system. In practice, it tends to be much easier to exploit weaknesses in protocols or in the implementation, than to physically penetrate the card and extract its secrets. In the last five years the level of hardware security found in smartcards has increased enormously. In 1998 the first publications on potential smartcard vulnerabilities were written. Afterwards chip manufacturers have made tremendous efforts to increase the security levels.

This is illustrated in the following figure:

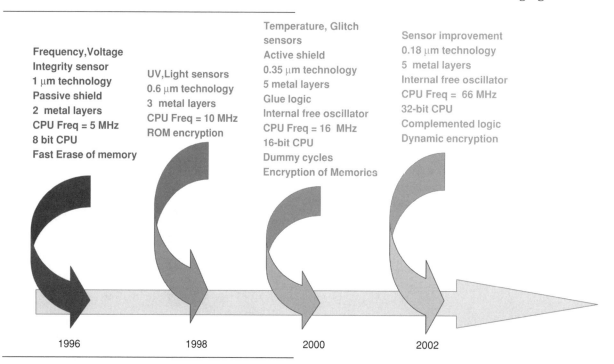

Chips are made of multiple layers, allowing manufacturers to hide sensitive components (e.g., data lines, connections) in between different layers that contain less sensitive components. For instance, the ROM is located in the lower (least accessible) layers of the chip.

A current-carrying protective layer or active shield is added at the top of the chip for power

However security by the use of countermeasures must be added at each smartcard process level such as component, software layer, applicative layer etc. Although, a smartcard device with a good use security countermeasure is the most secure token, soon:

• Chip design technologies will be smaller and smaller (0.07 μm),

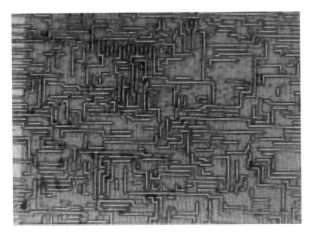

Fig. 9. Part of Glue logic design

- CPU frequencies will reach 100 MHz
- New nonvolatile memories will be introduced (Feram, MRam, etc.)
- More complex design and countermeasures will be developed.

Hence, invasive attacks will be increasingly difficult and will require powerful tools and expert knowledge in chip architecture and electronics.

Assia Tria
Hamid Choukri

INVERSION ATTACK

The *inversion attack* is a known plaintext attack on some particular filter generators. It was proposed by Golić in 1996 [1]. A generalization [2] to any filter generator, called *generalized inversion attack*, was presented by Golić, Clark and Dawson in 2000. Both inversion attack and generalized inversion attack aim at recovering the initial state of the linear feedback shift register (LFSR) from a segment of the running-key when the LFSR feedback polynomial, the tapping sequence, and the filtering function are known.

ORIGINAL INVERSION ATTACK: The original inversion attack only applies when the filtering function f is linear in its first input variable (forward attack) or in its last input variable (backward attack), i.e., when

$$f(x_1, x_2, \ldots, x_n) = x_1 + g(x_2, \ldots, x_n)$$

or

$$f(x_1, x_2, \ldots, x_n) = g(x_1, \ldots, x_{n-1}) + x_n,$$

where g is a Boolean function of $n - 1$ variables. In the first case, the keystream **s** is defined by

$$\begin{aligned} s_t &= f(u_{t+\gamma_1}, u_{t+\gamma_2}, \ldots, u_{t+\gamma_n}) \\ &= u_{t+\gamma_1} + g(u_{t+\gamma_2}, \ldots, u_{t+\gamma_n}), \end{aligned}$$

where $(u_t)_{t \geq 0}$ is the sequence generated by the LFSR and $(\gamma_i)_{1 \leq i \leq n}$ is a decreasing sequence of non-negative integers. The attack relies on the fact that bit $u_{t+\gamma_1}$ can be deduced from the $(\gamma_1 - \gamma_n)$ previous terms, $(u_{t+\gamma_1+1}, \ldots, u_{t+\gamma_n})$ if the running-key bit s_t is known. The relevant parameter of the attack is then the *memory size* of the filter generator, defined by $M = \gamma_1 - \gamma_n$. Indeed, the complete initialization of the LFSR can be recovered by an exhaustive search on only M bits as described in Table 1.

The backward attack, which applies when the filter function is linear in its last variable, is similar. The complexity of both forward and backward attacks is $\mathcal{O}(L2^M)$. It follows that the memory size of a filter generator should be large and preferably close to its maximum possible value $L - 1$.

Moreover, the complexity of the attack dramatically decreases when the greatest common divisor of all spacings between the taps, $d = \gcd(\gamma_i - \gamma_{i+1})$, is large. Indeed, the inversion attack can be applied to the *d-decimation* of the LFSR sequence, i.e., to the sequence obtained by sampling the LFSR sequence at intervals of d clock cycles (see maximum length linear sequence or [1]). Therefore, the effective memory size of the filter generator corresponds to

$$M' = \frac{\gamma_1 - \gamma_n}{\gcd(\gamma_i - \gamma_{i+1})}.$$

Table 1. Inversion attack

Input. $s_0 s_1 \ldots s_{N-1}$, N keystream bits.
Output. $u_{\gamma_n} \ldots u_{L+\gamma_n-1}$, L consecutive bits of the LFSR sequence, where L is the LFSR length.
For each choice of the M-bit vector $u_{\gamma_n} \ldots u_{\gamma_1-1}$
   Compute the next $(L - M)$ bits of the LFSR sequence by

   $u_{t+\gamma_1} \leftarrow s_t + g(u_{t+\gamma_2}, \ldots, u_{t+\gamma_n}), \quad 0 \leq t \leq L - M$.

   Compute $(N - L)$ additional bits of the LFSR sequence with the LFSR recurrence relation, and the
      corresponding running-key bits, $\hat{s}_t$, for $L - M \leq t < N - M$.
   If the $N - L$ bits $\hat{s}_t$ are equal to the observed keystream sequence, then return $(u_{\gamma_n} \ldots u_{L+\gamma_n-1})$.

The related design criterion is then that the greatest common divisor of all spacings between the taps should be equal to 1.

GENERALIZED INVERSION ATTACK: A similar attack can be mounted even if the filtering function is not linear in its first or last variable. In the general case, the keystream is given by

$$s_t = f(u_{t+\gamma_1}, u_{t+\gamma_2}, \ldots, u_{t+\gamma_n}).$$

Exactly as in the original inversion attack, the basic step of the attack consists in deducing bit $u_{t+\gamma_1}$ from the knowledge of the keystream bit s_t and of the M previous terms of the LFSR sequence, $(u_{t+\gamma_1+1}, \ldots, u_{t+\gamma_n})$. For fixed values of s_t and of $(u_{t+\gamma_1+1}, \ldots, u_{t+\gamma_n})$, the unknown bit $u_{t+\gamma_1}$ may take 0, 1 or 2 possible values. Then, an exhaustive search on the M bits $u_{\gamma_n}, \ldots, u_{\gamma_1-1}$ of the LFSR sequence, can still be performed. For a given value of the M-bit vector $u_{\gamma_n}, \ldots, u_{\gamma_1-1}$, a binary tree of depth $L - M$ representing all the solutions for the next $(L - M)$ bits of **u** is formed. Each node at level t corresponds to a guessed value of $(u_{t+\gamma_n}, \ldots, u_{t+\gamma_1-1})$. Then, the number of edges out of this node is 0, 1 or 2 according to the number of solutions x of the equation $s_t = f(x, u_{t+\gamma_2}, \ldots, u_{t+\gamma_n})$. If a tree of depth $L - M$ can be constructed from a given M-bit root, some additional bits of the LFSR sequence are computed and their consistency with the observed keystream is checked. It is shown that the typical number of surviving nodes at level $L - M$ is linear in L. Then, the typical complexity of the attack is $\mathcal{O}(L2^M)$. Exactly as in the inversion attack, the parameter involved in the attack is the effective memory size, i.e.,

$$M' = \frac{\gamma_1 - \gamma_n}{\gcd(\gamma_i - \gamma_{i+1})}.$$

Another technique based on a trellis representation and on the Viterbi algorithm is described in [3]. Its efficiency is comparable to the generalized inversion attack.

Anne Canteaut

References

[1] Golić, J.Dj. (1996). "On the security of nonlinear filter generators." *Fast Software Encryption 1996*, Lecture Notes in Computer Science, vol. 1039, ed. D. Gollman. Springer-Verlag, Berlin, 173–188.

[2] Golić, J.Dj., A. Clark, and E. Dawson (2000). "Generalized inversion attack on nonlinear filter generators." *IEEE Transactions on Computers*, 49 (10), 1100–1108.

[3] Leveiller, S., J. Boutros, P. Guillot, and G. Zémor (2001). "Cryptanalysis of nonlinear filter generators with {0, 1}-metric Viterbi de-coding." *Cryptography and Coding—8th IMA International Conference*, Lecture Notes in Computer Science, vol. 2260. Springer-Verlag, Berlin, 402–414.

INVERSION IN FINITE FIELDS AND RINGS

The need to compute the multiplicative inverse of an element of a finite field (or Galois field) or of a finite ring occurs frequently in cryptography. The main application domains are asymmetric cryptosystems, for instance in the computation of the private-public key pair in RSA (see RSA public key encryption schems) or in the group operation of elliptic curve cryptosystems. The finite structures in asymmetric algorithms are typically and relatively large. A second application domains are inversions in small finite fields which occur in the context of block ciphers, e.g., within the S-box of the *Advanced Encryption Standard (Rijndael/AES)*.

In the case of inversion in a finite integer ring or polynomial ring, the extended Euclidean algorithm can be used. Let u be the element whose inverse is to be computed and v the modulus. Note that u and v must be relatively prime in order for the inverse to exist. The extended Euclidean algorithm computes the coefficients s and t such that: $us + vt = \gcd(u, v) = 1$. The parameter s is the inverse of u modulo v. In the case of integer rings, using the binary Euclidean algorithm often leads to faster executions on digital computers. The binary Euclidean algorithm does not require integer divisions but only simple operations such as shifts and additions.

There are several approaches to computing multiplicative inverses of non-zero elements in finite fields:

Extended Euclidean algorithm. This is the most general and in many cases most efficient method. The application is completely analogous to the case of finite rings as discussed above. In the case of prime fields, the standard extended Euclidean algorithm applies. The binary Euclidean algorithm is often an advantage, too, in this case. If the inverse in an extension field is to be computed, the Euclidean algorithm with polynomials has to be used.

Fermat's little theorem. This method has a higher computational complexity than the extended Euclidean algorithm but can nevertheless be relevant in certain situations, e.g., if a fast exponentiation unit is available or if an algorithm with a simple control structure is desired. From <u>Fermat's little theorem</u> it follows immediately that for any element $A \in GF(q^m)$, $A \neq 0$, the inverse can be computed as $A^{-1} = A^{(q^m - 2)}$. For fields of characteristic two, i.e., fields $GF(2^m)$, the use of addition chains allows to dramatically reduce the number of multiplications (though not the number of squarings) required for computing the exponentiation to the $(2^m - 2)$th power. This method is referred to as <u>Itoh–Tsujii Inversion</u>.

Look-up tables. A conceptually simple method is based on look-up tables. In this case, the inverses of all field elements are pre-computed once with one of the methods mentioned above, and stored in a table. Assuming the table entries can be accessed by an appropriate method, e.g., by the field elements themselves in a binary representation, the inverses are available quickly. The drawback of this method are the storage requirements, since k memory locations are needed for <u>fields</u> $GF(k)$. Since the storage requirements are too large for the finite fields commonly needed in public-key cryptography, inversion based on look-up tables is mainly useful in cases of small finite fields, e.g., $GF(256)$, which have applications in block ciphers or which are subfields of larger extension fields.

Reduction to subfield inversion. In the case of <u>extension fields</u> $GF(q^m)$, $m \geq 2$, inversion in the field $GF(q^m)$ can be reduced to inversion in the field $GF(q)$. This reduction comes at the cost of extra operations (multiplications and additions) in the field $GF(q^m)$. If the inversion in the subfield $GF(q)$ is sufficiently inexpensive computationally compared to extension field inversion, the method described in Theorem 1 of <u>Itoh–Tsujii Inversion</u> can have a low over-all complexity. The method was introduced for fields in normal basis representation in [2] and generalized to fields in polynomial basis representation in [1]. The method can be applied iteratively in fields with multiple field extensions, sometimes referred to as tower fields. In the case of fields $GF(2^m)$, m a prime, the Itoh–Tsujii algorithm degenerates into inversion based on Fermat's little theorem. It should be stressed that this method is not a complete inversion algorithm since it is still necessary to eventually perform an inversion in the subfield. However, inversion in a (small) subfield can often be done fast with one of the methods described above.

Direct inversion. This method is applicable to extension fields $GF(q^m)$, and mainly relevant for fields where m is small, e.g., $m = 2, 3, 4$. Similar to <u>Itoh–Tsujii Inversion</u>, direct inversion also reduces extension field inversion to subfield inversion. As an example, we demonstrate the method for fields $GF(q^2)$, introduced in [3]. Let us consider a non-zero element $A = a_0 + a_1 x$ from $GF(q^m)$, where $a_0, a_1 \in GF(q)$. Let us assume the irreducible <u>field polynomial</u> has the form $P(x) = x^2 + x + p_0$, where $p_0 \in GF(q)$. If the inverse is denoted as $B = A^{-1} = b_0 + b_1 x$, the equation

$$A \cdot B = [a_0 b_0 + p_0 a_1 b_1] + [a_0 b_1 + a_1 b_0 + a_1 b_1] x$$
$$= 1$$

must be satisfied, which is equivalent to a set of two linear equations in b_0, b_1 over $GF(q)$ with the solution:

$$\left. \begin{array}{l} b_0 = \dfrac{a_0 + a_1}{\Delta} \\[2mm] b_1 = \dfrac{a_1}{\Delta} \end{array} \right\},$$
$$\text{where } \Delta = a_0(a_0 + a_1) + p_0 a_1^2. \tag{1}$$

The advantage of this algorithm is that all operations are performed in $GF(q)$. Note that there is one inversion in the subfield $GF(q)$ of the parameter Δ required. The algorithm can be applied recursively. The relationship between direction inversion and the Itoh–Tsujii method is sketched in [4].

Christof Paar

References

[1] Guajardo, J. and C. Paar (2002). "Itoh–Tsujii Inversion in standard basis and its application in cryptography and codes." *Designs, Codes and Cryptography*, 25, 207–216.

[2] Itoh, T. and S. Tsujii (1988). "A fast algorithm for computing multiplicative inverses in $GF(2^m)$ using normal bases." *Information and Computation*, 78, 171–177.

[3] Morii, M. and M. Kasahara (1989). "Efficient construction of gate circuit for computing multiplicative inverses over $GF(2^m)$." *Trans. of the IEICE*, E 72, 37–42.

[4] Paar, C. (1995). "Some remarks on efficient inversion in finite fields." *1995 IEEE International Symposium on Information Theory*, Whistler, B.C. Canada 58.

IPES

IPES is an alternative name for the <u>IDEA</u> cipher. IPES stands for "improved PES", where

PES [1] is a cipher predecessor of IDEA which was cryptanalysed by differential cryptanalysis in [2]. The only changes between IDEA and PES are in the order of operations in the key-mixing sub-round: PES uses the order $(\odot, \odot, \boxplus, \boxplus)$, while IDEA uses the order $(\odot, \boxplus, \boxplus, \odot)$, and in the swap of the words after the *MA* subround. In IDEA the the outer words X_1, X_4 are not swapped.

Alex Biryukov

References

[1] Lai, X. and J.L. Massey (1990). "A proposal for a new block encryption standard." *Advances in Cryptogoly—Proceedings of Eurocrypt '90*, Lecture Notes in Computer Science, vol. 473, ed. I.B. Damgard, Springer-Verlag, Berlin, 389–404.

[2] Lai, X., J.L. Massey, and S. Murphy (1991). "Markov ciphers and differential cryptanalysis." *Advances in Cryptology—Proceedings of Eurocrypt '91*. Lecture Notes in Computer Science, vol. 547, ed. D.W. Davies, Springer-Verlag, Berlin, 17–38.

IPsec

Prior to the explosion of computer networks in the late 1980s, enterprize environments were largely isolated collections of hosts. The protocols used to connect those computers did not require much security. Indeed, few security issues were considered by original designers of the *Internet Protocol* (IP) suite upon which those and subsequent networks are based. While the openness of these protocols is a key ingredient to the Internet's success, the lack of security has led to many troublesome problems. For example, many otherwise safe systems have been compromised by an adversary who forges IP addresses. Such address "spoofing" is trivial on the current Internet. These and other security problems continue to confound the users and administrators of the Internet.

IPsec is a protocol suite that adds security to the existing IP protocols [4]. Standardized by the Internet Engineering Task Force [3], IPsec defines new IP message formats and the infrastructure used to define and manage security relevant state. IPsec is a general purpose architecture. Hosts, networks, and gateways define policies for each class of traffic they wish to secure. These policies define what security services they desire to apply to the traffic (e.g., authenticity, confidentiality).

IPsec provides security properties that are specific to the network medium. For example, the IPsec authentication service ensures that a host

receiving a packet is able to determine that a packet was transmitted by the host or network that claims to have sent it. A related property, *integrity*, allows that same host to assert that the packet data, called payload, was not modified in transit. Authenticity and integrity are implemented in IPsec using the *authentication header* (AH) transform and optionally by the *encapsulating security payload*.

Confidentiality is also defined by IPsec with respect to the hosts and networks implementing it. Where confidentiality is configured, an IPsec host is guaranteed that the data transmitted between hosts (e.g., payload) is only visible to the intended recipient. To put it in another way, the payload and optionally the IP header cannot be viewed by any intermediate node or external entity on the intermediate network. Confidentiality is implemented in IPsec using the *encapsulating security payload* (ESP) transform.

IPsec defines the requirements of end-point authentication (e.g., credentials and methods) and the protocols and procedures used for the creation and management of state. These services define not only how key agreement is reached, but also how a concrete set of services and parameters is negotiated. These procedures are implemented by the abstract *Internet Security Association and Key Management Protocol* (ISAKMP) [5] and concrete *Internet Key Exchange* (IKE) [2] protocols.

IPsec provides a general architecture for secure networks. It has been particularly useful in supporting interesting network services. For example, IPsec is an ideal technology for implementing virtual private networks (VPNs). This is principally because of its ability to operate in tunnel mode, where intermediate gateways can securely transport traffic between private networks over untrusted networks like the Internet. IPsec has also been extremely useful in providing security in environments where the physical media is exceptionally vulnerable (e.g., wireless networks).

The remainder of this entry will explore the architecture and operation of the IPsec protocol suite. We begin in the next section by describing the entities and services comprising the IPsec infrastructure.

THE IPSEC ARCHITECTURE: The IPsec architecture [4] coordinates hosts and network elements to ensure that the information flowing between hosts is secured. How the security is defined for a particular environment is determined by policy. Policies are configured manually or obtained from the emerging IPsec policy system [1] and applied to packet traffic.

The IPsec the *Security Policy Database* (SPD) states processing rules for network traffic. The selector is a unique (to the SPD) collection of address, protocol, and type of service bits (TOS). The SPD maps selectors onto IPsec policies. These policies define the processing discipline applied to matching packets. The processing policies are similar to those found in <u>firewalls</u>, where a packet can be processed by IPsec, passed (without modification), or dropped. It is up to the host or network administration to determine which policy is most appropriate for particular network traffic.

A central and essential artifact of IPsec state is the *security association* (SA). An SA is a data structure used to define and track communication state and configuration between IPsec end-points. Each SA stores the packet-processing transform (e.g., AH or ESP), parameters (e.g., cryptographic algorithms) and other security-relevant data (e.g., keys, sequence numbers). The SA is created by the key management service as defined below. An SA is uniquely identified by its security parameter index (SPI). The SPI is negotiated at the start of an IPsec session and is later placed in the header field of all relevant IPsec packets. The SAs established by a host are held in the *Security Association Database* (SAD). Unlike most security protocols, IPsec defines uni-directional security associations. That is, the keys and policy are applied in a single direction. The advantage of this approach is that network administrators are free to establish different security policies for each direction or even to restrict the flow of data to one direction.

The following example illustrates the processing of IP traffic by the elements of the IPsec architecture. Assume initially that a local host A (in Figure 1) has not communicated with B recently.[1] An arbitrary application on A attempts to send data to some external host B via the User Data Protocol (UDP) [6] (**1**). Upon reception at the IPsec implementation on A in the host operating system, the protocol (UDP), target address (B), and other information is mapped in the SPD to an IPsec policy. In this example, the policy mandates that ESP in confidentiality only mode and automated key management be used. In response, the IKE protocol is executed between the two hosts and session keys are established (**2**). The successful completion of the IKE protocol results in the creation of an SA defining the keying material, ESP policy, and SA lifetime. The original packet is transformed per the ESP policy specification (**3**) and transmitted

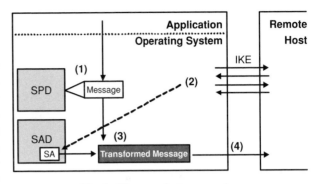

Fig. 1. IPsec packet processing

to the remote host (**4**). Note that each subsequent packet sent during the SA's lifetime and matching the original selector will use the same SA (and hence, will not require further involvement of the key management protocol, IKE).

TRANSFORMS: An IPsec *transform* defines a packet format and a set of associated processing rules that address a particular set of security guarantees. IPsec defines two transforms, the authentication header (AH) and the encapsulating security payload (ESP). These transforms operate in either transport or tunnel mode. A transform operating *transport mode* secures the payload but not the IP header. This is useful when you need to apply security to the upper layer protocols only (e.g., authenticate TCP header and payload data only). *Tunnel mode* provides the guarantees over the entire IP packet. This is useful when you need to provide guarantees over the IP header fields (e.g., confidentiality of source and target addresses), and is very useful when dealing with operational issues (e.g., simplifies private addressing).

In transport mode, AH includes a *message authentication code* (<u>MAC algorithms</u>) in the IPsec header. The keyed MAC is calculated using the keys defined in the SA and over the payload data. ESP ensures confidentiality by encrypting the payload with the key defined in the SA. Where enabled, ESP includes an authenticating MAC similar to AH. Conversely, AH and ESP tunnel modes completely encapsulate the IP header and payload. The entirety of the IP packet is treated as payload, and a new IP packet is formed around it.

The original IP header data is protected (and not visible in the case of ESP) in tunnel mode. This has the advantage that an observer will (and again in ESP can only) see the packet as originating from the device that tunneled it. This allows network architects to use IPsec tunneling devices as *security gateways*. These gateways serve to separate and conceal sensitive traffic traveling across

---

[1] IPsec SAs have an explicit lifetime, after which the state is discarded. This example assumes the lifetime of any such previous communication between the hosts has been exceeded.

untrusted networks. Such gateways are a central element of contemporary virtual private networks (VPNs).

Another security guarantee supported by both transforms is replay protection (e.g., also known as anti-replay). Replay prevention ensures that packets are processed by the receiving host at most once, and hence are not "replayed" maliciously or accidentally. This feature is implemented by authenticating packet sequence numbers. Every SA has a sequence number that is initially set to 0 and incremented by one after a packet transmission. A transmission window set by policy to some agreed size (typically 64 or 128). Every time a packet is received, the sequence number is checked. Any correctly authenticated packet who's sequence number falls in the window and has not been seen before is accepted. If the packet it newer (has a sequence larger than the the right side of the window), the window is moved to the right to accept the packet. If the packet is older (to the left of the window), it is dropped.

KEY MANAGEMENT: IPsec would not be very useful without key management. The purpose of key management facility in IPsec is to determine and distribute the keys used by the payload-processing transforms, and to secondarily negotiate the <u>policy</u> defining the SAs. All of these functions are implemented in the abstract ISAKMP architecture by IKE. Strictly speaking, ISAKMP, IKE, and IPsec are separate standards, but in recent years have become largely inseparable.

IPsec provides for two kinds of key management, static and automatic. Environments with static key management simply identify the keys to be used to secure the communication between the end-points (e.g., the IPsec SAs). While avoiding the complexity and cost of implementing a key management protocol, this approach can be difficult to manage. Each end-point must be configured with an SA used to communicate with every other end-point. This is time-consuming and error prone where many end-points must be managed. Moreover, because it potentially exposes a large amount cipher-text (because the same key may be used for an indeterminate time), manual keying is not frequently viewed as good security practice.

ISAKMP is an abstract key management architecture. It defines the possible states, transitions, and (abstract) exchanges one uses to establish a shared key. Based on an authenticated <u>Diffie–Hellman key agreement</u> and built on the ISAKMP architecture, IKE is a concrete protocol. IKE serves three purposes: (a) to establish the policy

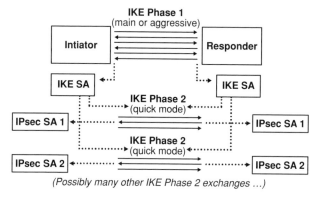

Fig. 2. IKE Phase 1 and 2 protocol flow

for an IKE session, (b) to establish a key for the IKE SA (see below), and (c) to establish particular SAs for the IPsec processing of IP and upper layer data.

As illustrated in Figure 2, IKE works in two phases. The first phase (creatively called *phase 1*) allows the two end-points to establish an *IKE SA*. An IKE SA defines the keys and policy used to establish regular payload processing SAs, called *IPsec SAs*. Phase one can operate in two modes. In *main-mode*, IKE implements a six message protocol which provides identity protection. Identity protection ensures the initiator's (the host that first tries to initiate communication) identity is not exposed to an attacker who is actively attacking the system (e.g., by hijacking the address of the intended host). Where such protection is not needed, the protocol can use a simpler, but less secure, three message *aggressive mode* protocol. The end result of either the main or aggressive mode protocol is the same, an IKE SA.

The simpler IKE Phase two exchange is appropriately named *quick mode*. This phase uses a previously established IKE SA to create an IPsec SA. The hosts perform the quick mode operation in three messages: an initial request with keying material, a response to the request, and an acknowledgment of the response. To simplify, the keys and configuration used to define the IPsec SA are generated from the values passed between the parties. If such a thing is deemed desireable, the parties may optionally engage in another Diffie–Hellman exchange. Note that a single IKE SA can be used to establish many different IPsec SA, even simultaneously.

CONSIDERATIONS: If all these modes and transforms seem complicated, they are. A central criticism of IPsec has been its complexity, mostly due to its attempt to be a suite of protocols that addresses the requirements of many constituents. The reasons for its broad task are obvious: because the world uses IP, IPsec must address the problems of the world. However, addressing such problems have evidently led to extremely difficult to implement (and often to manage) protocols. Recently, draft proposals have surfaced within the Internet Engineering Task Force (IETF) standards organization that attempt to simplify IKE, so there may be some relief on the horizon.

Patrick McDaniel

References

[1] Blaze, M., A. Keromytis, M. Richardson, and L. Sanchez (2003). "IP Security Policy (IPSP) requirements." *Internet Engineering Task Force*, RFC 3586.

[2] Harkins, D. and D. Carrel (1998). "The Internet key exchange." *Internet Engineering Task Force*, RFC 2409.

[3] Internet Engineering Task Force (IETF), 2004.

[4] Kent, S. and R. Atkinson (1998). "Security architecture for the Internet Protocol." *Internet Engineering Task Force*, RFC 2401.

[5] Maughan, D., M. Schertler, M. Schneider, and J. Turner (1998). "Internet Security Association and Key Management Protocol (ISAKMP)." *Internet Engineering Task Force*, RFC 2408.

[6] Postel, J. (1980). "User datagram protocol." *Internet Engineering Task Force*, RFC 768.

IRREDUCIBLE POLYNOMIAL

A polynomial that is not divisible by any smaller polynomials other than trivial ones is an *irreducible polynomial*. Let $f(x)$ be a polynomial

$$f(x) = f_d x^d + f_{d-1} x^{d-1} + \cdots + f_1 x + f_0,$$

where the coefficients $f_0, \ldots, f_d$ are elements of a field **F**. If there is another polynomial $g(x)$ over **F** with degree between 1 and $d-1$ such that $g(x)$ divides $f(x)$, then $f(x)$ is *reducible*. Otherwise, $f(x)$ is irreducible. (Nonzero polynomials of degree 0, i.e., nonzero elements of **F**, divide every polynomial so are not considered.)

As an example, the polynomial $x^2 + 1$ over the finite field $\mathbf{F}_2$ is reducible since $x^2 + 1 = (x+1)^2$, whereas the polynomial $x^2 + x + 1$ is irreducible.

A representation of the finite field $\mathbf{F}_{q^d}$ can be constructed from a representation of the finite field $\mathbf{F}_q$ together with an irreducible polynomial of degree d, for any d; the polynomial $f(x)$ is called the *field polynomial* for this field. In the example just given, $x^2 + x + 1$ would be a field polynomial for $\mathbf{F}_4$ over $\mathbf{F}_2$. See extension field.

Burt Kaliski

ISSUER

In retail payment schemes and electronic commerce, there are normally two parties involved: a customer and a shop. The Issuer is the bank of the customer.

Peter Landrock

ITOH–TSUJII INVERSION ALGORITHM

Originally introduced in [5], the Itoh and Tsujii algorithm (ITA) is an exponentiation-based algorithm for inversion in finite fields which reduces the complexity of computing the inverse of a nonzero element in $GF(2^n)$, when using a normal basis representation, from $n-2$ multiplications in $GF(2^n)$ and $n-1$ cyclic shifts using the binary exponentiation method to at most $2\lfloor \log_2(n-1) \rfloor$ multiplications in $GF(2^n)$ and $n-1$ cyclic shifts. As shown in [4], the method is also applicable to finite fields with a polynomial basis representation.

For the discussion that follows, it is important to point out that there are several possibilities to represent elements of a finite field. Thus, in general, given an irreducible polynomial $P(x)$ of degree m over $GF(q)$ and a root α of $P(x)$ (i.e., $P(\alpha) = 0$), one can represent an element $A \in GF(q^m)$, $q = p^n$ and p prime, as a polynomial in α, i.e., as $A = a_{m-1}\alpha^{m-1} + a_{m-2}\alpha^{m-2} + \cdots + a_1\alpha + a_0$ with $a_i \in GF(q)$. The set $\{1, \alpha, \alpha^2, \ldots, \alpha^{m-1}\}$ is then said to be a polynomial basis (or standard basis) for the finite field $GF(q^m)$ over $GF(q)$ (see also extension field). Another type of basis is called a normal basis. Normal bases are of the form $\{\beta, \beta^q, \beta^{q^2}, \ldots, \beta^{q^{m-1}}\}$ for an appropriate element $\beta \in GF(q^m)$. Then, an element $B \in GF(q^m)$ can be represented as $B = b_{m-1}\beta^{q^{m-1}} + b_{m-2}\beta^{q^{m-2}} + \cdots + b_1\beta^q + b_0\beta$ where $b_i \in GF(q)$. It can be shown that for any field $GF(q)$ and any extension field $GF(q^m)$, there exists always a normal basis of $GF(q^m)$ over $GF(q)$ (see [6, Theorem 2.35]). Notice that $(\beta^{q^i})^{q^k} = \beta^{q^{i+k}} = \beta^{q^{i+k \bmod m}}$ which follows from the fact that $\beta^{q^m} \equiv \beta$ (see also Fermat's little theorem). Thus, raising an element $B \in GF(q^m)$ to the qth power can be easily accomplished through a cyclic shift of its coordinates, i.e., $B^q = (b_{m-1}\beta^{q^{m-1}} + b_{m-2}\beta^{q^{m-2}} + \cdots + b_1\beta^q + b_0\beta)^q = b_{m-2}\beta^{q^{m-1}} + b_{m-3}\beta^{q^{m-2}} + \cdots + b_0\beta^q + b_{m-1}\beta$, where we have used the fact that in any field of characteristic p, $(x+y)^q = x^q + y^q$, where $q = p^n$.

Now, we can show how to compute the multiplicative inverse of $A \in GF(2^n)$, $A \neq 0$, according to the binary method for exponentiation. From Fermat's little theorem we know that $A^{-1} \equiv A^{2^n - 2}$ can be computed as

$$A^{2^n - 2} = A^2 \cdot A^{2^2} \cdots A^{2^{n-1}}.$$

This requires $n - 2$ multiplications and $n - 1$ cyclic shifts. Notice that because we are working over a field of characteristic two (see <u>finite field</u>), squaring is a linear operation. In addition, if a normal basis is being used to represent the elements of the field, we can compute A^2 for any $A \in GF(2^n)$ with one cyclic shift.

Itoh and Tsujii proposed in [5] three algorithms. The first two algorithms describe *addition chain* (see <u>fixed-exponent exponentiation</u>) for exponentiation-based inversion in fields $GF(2^n)$ while the third one describes a method based on subfield inversion. The first algorithm is only applicable to values of n such that $n = 2^r + 1$, for some positive r, and it is based on the observation that the exponent $2^n - 2$ can be re-written as $(2^{n-1} - 1) \cdot 2$. Thus if $n = 2^r + 1$, we can compute $A^{-1} \equiv (A^{2^{2^r} - 1})^2$. Furthermore, we can rewrite $2^{2^r} - 1$ as

$$2^{2^r} - 1 = (2^{2^{r-1}} - 1) \, 2^{2^{r-1}} + (2^{2^{r-1}} - 1). \tag{1}$$

Equation (1) and the previous discussion lead to Algorithm 1.

ALGORITHM 1. *Multiplicative inverse computation in $GF(2^n)$ with $n = 2^r + 1$ [5, Theorem 1]*

Input: $A \in GF(2^n)$, $A \neq 0$, $n = 2^r + 1$
Output: $C = A^{-1}$
  $C \leftarrow A$
  for $i = 0$ to $r - 1$ **do**
    $D \leftarrow C^{2^{2^i}}$ {NOTE: 2^i cyclic shifts}
    $C \leftarrow C \cdot D$
  end for
  $C \leftarrow C^2$
  Return (C)

Notice that Algorithm 1 performs $r = \log_2(n - 1)$ iterations. In every iteration, one multiplication and i cyclic shifts, for $0 \leq i < r$, are performed which leads to an overall complexity of $\log_2(n - 1)$ multiplications and $n - 1$ cyclic shifts.

EXAMPLE 1. Let $A \in GF(2^{17})$, $A \neq 0$. Then according to Algorithm 1 we can compute A^{-1} with the

following addition chain:

$$A^2 \cdot A = A^3$$

$$\left(A^3\right)^{2^{2^1}} \cdot A^3 = A^{15}$$

$$\left(A^{15}\right)^{2^{2^2}} \cdot A^{15} = A^{255}$$

$$\left(A^{255}\right)^{2^{2^3}} \cdot A^{255} = A^{65535}$$

$$\left(A^{65535}\right)^2 = A^{131070}.$$

A quick calculation verifies that $2^{17} - 2 = 131070$. Notice that in accordance with Algorithm 1 we have performed four multiplications in $GF(2^{17})$ and, if using a normal basis, we would also require $2^4 = 16$ cyclic shifts.

Algorithm 1 can be generalized to any value of n [5]. First, we write $n - 1$ as

$$n - 1 = \sum_{i=1}^{t} 2^{k_i}, \quad \text{where } k_1 > k_2 > \cdots > k_t. \tag{2}$$

Using the fact that $A^{-1} \equiv (A^{2^{n-1} - 1})^2$ and (2), it can be shown that the inverse of A can be written as:

$$(A^{2^{n-1}-1})^2 = \Bigg[(A^{2^{2^{k_t}} - 1}) \left(\left(A^{2^{2^{k_{t-1}}} - 1}\right) \cdots \right.$$

$$\left. \left[(A^{2^{2^{k_2}} - 1})(A^{2^{2^{k_1}} - 1})^{2^{2^{k_2}}} \right]^{2^{2^{k_3}}} \cdots \right)^{2^{2^{k_t}}} \Bigg]^2. \tag{3}$$

An important feature of (3) is that in computing $A^{2^{2^{k_1}} - 1}$ all other quantities of the form $A^{2^{2^{k_i}} - 1}$ for $k_i < k_1$ have been computed. Thus, the overall complexity of (3) can be shown to be:

$$\#\text{MUL} = \lfloor \log_2(n - 1) \rfloor + HW(n - 1) - 1$$
$$\#\text{CSH} = n - 1, \tag{4}$$

where $HW(\cdot)$ denotes the Hamming weight of the operand, i.e., the number of ones in the binary representation of the operand (see also <u>cyclic codes</u>), MUL refers to multiplications in $GF(2^n)$, and CSH refers to cyclic shifts over $GF(2)$ when using a normal basis.

EXAMPLE 2. Let $A \in GF(2^{23})$, $A \neq 0$. Then according to (2) we can write $n - 1 = 22 = 2^4 + 2^2 + 2$ where $k_1 = 4$, $k_2 = 2$, and $k_3 = 1$. It follows that we can compute $A^{-1} \equiv A^{2^{23} - 2}$ with the following addition chain:

$$A^{2^2-1} = A^2 \cdot A$$

$$A^{2^4-1} = \left(A^3\right)^{2^2} \cdot A^3$$

$$A^{2^8-1} = \left(A^{15}\right)^{2^4} \cdot A^{15}$$

$$A^{2^{16}-1} = \left(A^{255}\right)^{2^8} \cdot A^{255}$$

$$A^{2^{23}-2} = \left(A^{2^2-1} \cdot \left(A^{2^4-1} \cdot \left(A^{2^{16}-1}\right)^{2^4}\right)^{2^2}\right)^2.$$

The above addition chain requires 6 multiplications and 22 cyclic shifts which agrees with the complexity of (4).

In [5], the authors also notice that the previous ideas can be applied to extension fields $GF(q^m)$, $q = 2^n$. Although this inversion method does not perform a complete inversion, it reduces inversion in $GF(q^m)$ to inversion in $GF(q)$. It is assumed that subfield inversion can be done relatively easily, e.g., through table look-up or with the extended <u>Euclidean algorithm</u>. These ideas are summarized in Theorem 1. The presentation here follows [4] and it is slightly more general than [5] as a subfield of the form $GF(2^n)$ is not required, rather we allow for general subfields $GF(q)$.

THEOREM 1. *[5, Theorem 3] Let $A \in GF(q^m)$, $A \neq 0$, and $r = (q^m - 1)/(q - 1)$. Then, the multiplicative inverse of an element A can be computed as*

$$A^{-1} = (A^r)^{-1} A^{r-1}. \tag{5}$$

Computing the inverse through Theorem 1 requires four steps:
Step 1. Exponentiation in $GF(q^m)$, yielding A^{r-1}.
Step 2. Multiplication of A and A^{r-1}, yielding $A^r \in GF(q)$.
Step 3. Inversion in $GF(q)$, yielding $(A^r)^{-1}$.
Step 4. Multiplication of $(A^r)^{-1} A^{r-1}$.
Steps 2 and 4 are trivial since both A^r, in Step 2, and $(A^r)^{-1}$, in Step 4, are elements of $GF(q)$ [6]. Both operations can, in most cases, be done with a complexity that is well below that of one single extension field multiplication. The complexity of Step 3, subfield inversion, depends heavily on the subfield $GF(q)$. However, in many cryptographic applications the subfield can be small enough to perform inversion very efficiently, for example, through table look-up [2,3], or by using the Euclidean algorithm (see also <u>inversion in finite fields</u>). What remains is Step 1, exponentiation to the $(r - 1)$th power in the extension field $GF(q^m)$.

First, we notice that the exponent can be expressed in q-adic representation as

$$r - 1 = q^{m-1} + \cdots + q^2 + q = (1 \cdots 110)_q. \tag{6}$$

This exponentiation can be computed through repeated raising of intermediate results to the q-th power and multiplications. The number of multiplications in $GF(q^m)$ can be minimized by using the addition chain in (3). Thus, computing A^{r-1} requires [5]:

$$\#\text{MUL} = \lfloor \log_2(m - 1) \rfloor + HW(m - 1) - 1$$

$$\#q\text{-EXP} = m - 1, \tag{7}$$

where q-EXP refers to the number of exponentiations to the qth power in $GF(q)$.

EXAMPLE 3. Let $A \in GF(q^{19})$, $A \neq 0$, $q = p^n$ for some prime p. Then, using the q-adic representation of $r - 1$ from (6) and the addition chain from (3), we can find an addition chain to compute $A^{r-1} = A^{q^{18}+q^{17}+\cdots+q^2+q}$ as follows. First, we write $m - 1 = 18 = 2^4 + 2$ where $k_1 = 4$, and $k_2 = 1$. Then, $A^{r-1} = (A^{q^{16}+q^{15}+\cdots+q^2+q})^{q^2} \cdot (A^{q^2+q})$ and we can compute $A^{q^{16}+q^{15}+\cdots+q^2+q}$ as

$$A^{q^2} = \left(A^q\right)^q$$

$$A^{q^2+q} = A^q \cdot A^{q^2}$$

$$A^{\sum_{i=1}^4 q^i} = \left(A^{q^2+q}\right)^{q^2} \cdot A^{q^2+q}$$

$$A^{\sum_{i=1}^8 q^i} = \left(A^{\sum_{i=1}^4 q^i}\right)^{q^4} \cdot A^{\sum_{i=1}^4 q^i}$$

$$A^{\sum_{i=1}^{16} q^i} = \left(A^{\sum_{i=1}^8 q^i}\right)^{q^8} \cdot \left(A^{\sum_{i=1}^8 q^i}\right).$$

Notice that in computing $A^{q^{16}+q^{15}+\cdots+q^2+q}$, we have computed A^{q^2+q}. The complexity to compute A^{r-1} (and, thus, the complexity to compute A^{-1} if the complexity of multiplication and inversion in $GF(q)$ can be neglected) in $GF(q^{19})$ is found to be 5 multiplications in $GF(q^{19})$ and 18 exponentiations to the qth power in agreement with (7).

We notice that [5] assumes a normal basis representation of the field elements of $GF(q^m)$, $q = 2^n$, in which the exponentiations to the qth power are simply cyclic shifts of the m coefficients that represent an individual field element. In polynomial (or standard) basis, however, these exponentiations are, in general, considerably more expensive.

Reference [4] takes advantage of finite field properties and of the algorithm characteristics to improve on the overall complexity of the ITA in polynomial basis. The authors make use of two facts: (i) the algorithm performs alternating multiplications and several exponentiations to the qth power in a row and (ii) raising an element $A \in GF(q)$, $q = p^n$, to the q^eth power is a linear operation in $GF(q^m)$, since q is a power of the field characteristic.

In general, computing A^{q^e} has a complexity of $m(m-1)$ multiplications and $m(m-2)+1 = (m-1)^2$ additions in $GF(q)$ [4]. This complexity is roughly the same as one $GF(q^m)$ multiplication, which requires m^2 subfield multiplications if we do not assume fast convolution techniques (e.g., the <u>Karatsuba algorithm</u> for multiplication). However, in polynomial basis representation computing A^{q^e}, where $e > 1$, can be shown to be as costly as a single exponentiation to the qth power. Thus, [4] performs as many subsequent exponentiations to the qth power in one step between multiplications as possible, yielding the same multiplication complexity as in (7), but a reduced number of q^e-exponentiations. This is summarized in Theorem 2.

THEOREM 2. *[4, Theorem 2]. Let $A \in GF(q^m)$. One can compute A^{r-1} where $r - 1 = q + q^2 + \cdots + q^{(m-1)}$ with no more than*

$$\#MUL = \lfloor \log_2(m-1) \rfloor + HW(m-1) - 1$$
$$\#q^e\text{-}EXP = \lfloor \log_2(m-1) \rfloor + HW(m-1)$$

operations, where $\#MUL$ and $\#q^e$-EXP refer to multiplications and exponentiations to the q^eth power in $GF(q^m)$, respectively.

We would like to stress that Theorem 2 is just an upper bound on the complexity of this exponentiation. Thus, it is possible to find addition chains which yield better complexity as shown in [1]. In addition, we see from Theorem 2 that Step 1 of the ITA algorithm requires about as many exponentiations to the q^eth power as multiplications in $GF(q^m)$ if a polynomial basis representation is being used. In the discussion earlier in this section it was established that raising an element $A \in GF(q^m)$ to the q^eth power is roughly as costly as performing one multiplication in $GF(q^m)$. Hence, if it is possible to make exponentiations to

the q^eth power more efficient, considerable speedups of the algorithm can be expected. Three classes of finite fields are introduced in [4] for which the complexity of these exponentiations is in fact substantially lower than that of a general multiplication in $GF(q^m)$. These are:

- Fields $GF((2^n)^m)$ with binary field polynomials.
- Fields $GF(q^m)$, $q = p^n$ and p an odd prime, with binomials as field polynomials.
- Fields $GF(q^m)$, $q = p^n$ and p an odd prime, with binary equally spaced field polynomials (ESP), where a binary ESP is a polynomial of the form $x^{sm} + x^{s(m-1)} + x^{s(m-2)} + \cdots + x^{2s} + x^s + 1$.

Jorge Guajardo

References

[1] Chung, Jae Wook, Sang Gyoo Sim, and Pil Joong Lee (2000). "Fast Implementation of Elliptic Curve Defined over $GF(p^m)$ on CalmRISC with MAC2424 Coprocessor." *Workshop on Cryptographic Hardware and Embedded Systems—CHES 2000*, Lecture Notes in Computer Science, vol. 1965, eds. Ç.K. Koç and C. Paar. Springer-Verlag, Berlin, 57–70.

[2] De Win, E., A. Bosselaers, S. Vandenberghe, P. De Gersem, and J. Vandewalle (1996). "A fast software implementation for arithmetic operations in $GF(2^n)$." *Advances in Cryptography—ASIACRYPT'96*, Lecture Notes in Computer Science, vol. 1233, eds. K. Kim and T. Matsumoto. Springer-Verlag, Berlin, 65–76.

[3] Guajardo, J. and C. Paar (1997). "Efficient algorithms for elliptic curve cryptosystems." *Advances in Cryptology—CRYPTO'97*, Lecture Notes in Computer Science, vol. 1294, ed. B. Kaliski. Springer-Verlag, Berlin, 342–356.

[4] Guajardo, J. and C. Paar (2002). "Itoh–Tsujii inversion in standard basis and its application in cryptography and codes." *Design, Codes, and Cryptography*, 25 (2), 207–216.

[5] Itoh, T. and S. Tsujii (1988). "A fast algorithm for computing multiplicative inverses in $GF(2^m)$ using normal bases." *Information and Computation*, 78, 171–177.

[6] Lidl, R. and H. Niederreiter (1997). *Finite Fields, Encyclopedia of Mathematics and its Applications*, vol. 20 (2nd ed.). Cambridge University Press, Cambridge.

J

JACOBI SYMBOL

The *Jacobi symbol* generalizes the Legendre symbol to all odd integers. Let n be an odd, positive integer with prime factorization

$$n = p_1^{a_1} p_2^{a_2} \cdots p_k^{a_k},$$

where the prime numbers $p_1, \ldots, p_k$ are distinct, and let x be an integer. The Jacobi symbol of x modulo n equals the product of Legendre symbols of x with respect to each of the primes:

$$\left(\frac{x}{n}\right) = \left(\frac{x}{p_1}\right)^{a_1} \left(\frac{x}{p_2}\right)^{a_2} \cdots \left(\frac{x}{p_k}\right)^{a_k}.$$

If n is prime then the Jacobi symbol is the same as the Legendre symbol.

The Jacobi symbol may be computed efficiently even when the prime factorization of n is unknown by the *Quadratic Reciprocity Theorem*, which was proved by Gauss (see Chapter 5 of [1]). See also Quadratic Residuosity Problem.

Burt Kaliski

Reference

[1] Ireland, Kenneth F. and M. Rosen (1991). *A Classical Introduction to Modern Number Theory, vol. 84. Graduate Texts in Mathematics*. Springer-Verlag, Berlin.

K

KARATSUBA ALGORITHM

The Karatsuba algorithm (KA) for multiplying two polynomials was introduced in 1962 [3]. It saves coefficient multiplications at the cost of extra additions compared to the schoolbook or ordinary multiplication method. The basic KA is performed as follows. Consider two degree-1 polynomials $A(x)$ and $B(x)$ with $n = 2$ coefficients:

$$A(x) = a_1 x + a_0$$
$$B(x) = b_1 x + b_0.$$

Let $D_0, D_1, D_{0,1}$ be auxiliary variables with

$$D_0 = a_0 b_0$$
$$D_1 = a_1 b_1$$
$$D_{0,1} = (a_0 + a_1)(b_0 + b_1).$$

Then the polynomial $C(x) = A(x)B(x)$ can be calculated in the following way:

$$C(x) = D_1 x^2 + (D_{0,1} - D_0 - D_1)x + D_0.$$

This method requires three multiplications and four additions. The schoolbook method requires n^2 multiplications and $(n-1)^2$ additions, i.e., four multiplications and one addition. Clearly, the KA can also be used to multiply integer numbers.

The KA can be generalized for polynomials of arbitrary degree [6]. The following algorithm describes a method to multiply two arbitrary polynomials with n coefficients using the *one-iteration KA*.

ALGORITHM 1. Generalized One-Iteration KA
Consider two degree-d polynomials with $n = d + 1$ coefficients

$$A(x) = \sum_{i=0}^{d} a_i x^i, \qquad B(x) = \sum_{i=0}^{d} b_i x^i.$$

Compute for each $i = 0, \ldots, n - 1$

$$D_i := a_i b_i.$$

Calculate for each $i = 1, \ldots, 2n - 3$ and for all s and t with $s + t = i$ and $t > s \geq 0$

$$D_{s,t} := (a_s + a_t)(b_s + b_t).$$

Then $C(x) = A(x)B(x) = \sum_{i=0}^{2n-2} c_i x^i$ *can be computed as*

$$c_0 = D_0$$
$$c_{2n-2} = D_{n-1}$$

$$c_i = \begin{cases} \sum_{s+t=i;t>s\geq0} D_{s,t} \\ \quad - \sum_{s+t=i;n-1\geq t>s\geq0} (D_s + D_t) \\ \quad \text{for odd } i, \, 0 < i < 2n - 2 \\ \sum_{s+t=i;t>s\geq0} D_{s,t} \\ \quad - \sum_{s+t=i;n-1\geq t>s\geq0} (D_s + D_t) \\ \quad + D_{i/2} \\ \quad \text{for even } i, \, 0 < i < 2n - 2. \end{cases}$$

The number of auxiliary variables is given as:

$$\#D_i = n$$
$$\#D_{s,t} = n^2/2 - n/2$$
$$\#D = \#D_i + \#D_{s,t} = n^2/2 + n/2.$$

The operational complexity is as follows:

$$\#\text{MUL} = n^2/2 + n/2$$
$$\#\text{ADD} = 5/2\, n^2 - 7/2\, n + 1.$$

For example, consider the KA for three coefficients. Let $A(x)$ and $B(x)$ be two degree-2 polynomials:

$$A(x) = a_2 x^2 + a_1 x + a_0$$
$$B(x) = b_2 x^2 + b_1 x + b_0$$

with auxiliary variables

$$D_0 = a_0 b_0$$
$$D_1 = a_1 b_1$$
$$D_2 = a_2 b_2$$
$$D_{0,1} = (a_0 + a_1)(b_0 + b_1)$$
$$D_{0,2} = (a_0 + a_2)(b_0 + b_2)$$
$$D_{1,2} = (a_1 + a_2)(b_1 + b_2).$$

Then $C(x) = A(x)B(x)$ is computed by an extended version of the KA using 6 multiplications and 13 additions. The schoolbook method requires in this case nine multiplications and four additions:

$$C(x) = D_2 x^4 + (D_{1,2} - D_1 - D_2)x^3$$
$$+ (D_{0,2} - D_2 - D_0 + D_1)x^2$$
$$+ (D_{0,1} - D_1 - D_0)x + D_0.$$

The KA can be performed recursively to multiply two polynomials as shown in algorithm 2. Let the number of coefficients be a power of 2, $n = 2^i$. To apply the algorithm both polynomials are split into a lower and an upper half:

$$A(x) = A_u(x)x^{n/2} + A_l(x)$$
$$B(x) = B_u(x)x^{n/2} + B_l(x).$$

These halves are used as before, i.e., as if they were coefficients. The polynomials A_u, A_l, B_u, and B_l are split again in half in the next iteration step. In the final step of the recursion, the polynomials degenerate into single coefficients. Since every step exactly halves the number of coefficients, the algorithm terminates after $t = \log_2 n$ steps. Note that there are overlaps of the coefficient positions when combining the result that lead to further additions. For example, consider algorithm 2 where N is the number of coefficients of $A(x)$ and $B(x)$. The polynomial D_0 has $N - 1$ coefficients such that the upper $N/2 - 1$ coefficients of D_0 are added to the lower $N/2 - 1$ coefficients of $(D_{0,1} - D_0 - D_1)$.

Let #MUL and #ADD be the number of multiplications and additions, respectively, in the underlying coefficient ring. Then the complexity to multiply two polynomials with $n = 2^i$ coefficients is as follows [5]:

$$\text{\#MUL} = n^{\log_2 3}$$
$$\text{\#ADD} \le 6n^{\log_2 3} - 8n + 2.$$

ALGORITHM 2. Recursive KA, $C = KA(A, B)$

INPUT: *Polynomials $A(x)$ and $B(x)$*
OUTPUT: *$C(x) = A(x) B(x)$*
$N \leftarrow max(degree(A),\ degree(B)) + 1$
if $N = 1$ return $A \cdot B$
Let $A(x) = A_u(x)\,x^{N/2} + A_l(x)$
and $B(x) = B_u(x)\,x^{N/2} + B_l(x)$
$D_0 \leftarrow KA(A_l, B_l)$
$D_1 \leftarrow KA(A_u, B_u)$
$D_{0,1} \leftarrow KA(A_l + A_u, B_l + B_u)$
return $D_1 x^N + (D_{0,1} - D_0 - D_1)x^{N/2} + D_0$

The recursive KA can be generalized in various ways. If the number of coefficients n is not a power of 2, algorithm 2 is slightly altered by splitting the polynomials into a lower part of $\lceil N/2 \rceil$ coefficients and an upper part of $\lfloor N/2 \rfloor$ coefficients. We call this variant the *simple recursive KA*. In this case the KA is less efficient than for powers of 2. A lower bound for the number of multiplications is given by $\text{\#MUL}_{low} = n^{\log_2 3}$ whereas an upper bound is $\text{\#MUL}_{up} = 1.39n^{\log_2 3}$. When applying the one-iteration KA for two and three coefficients as basis of the recursion, i.e., when applying the KA for two and three coefficients as final recursion

step, the upper bound improves to $\text{\#MUL}_{up} = 1.24n^{\log_2 3}$. When applying the one-iteration KA for two, three, and nine coefficients as basis of the recursion, the upper bound further improves to $\text{\#MUL}_{up} = 1.20n^{\log_2 3}$.

In the same manner we obtain bounds for the number of additions $\text{\#ADD}_{low} = 6n^{\log_2 3} - 8n + 2$ and $\text{\#ADD}_{up} = 7.30n^{\log_2 3}$ which improves for a basis of two and three coefficients to $\text{\#ADD}_{up} = 6.85n^{\log_2 3}$. For a basis of two, three, and nine coefficients, it further improves to $\text{\#ADD}_{up} = 6.74n^{\log_2 3}$.

When using the recursive KA for $n = p^j$ coefficients, i.e., applying the KA for p coefficients for j recursion levels, the number of multiplications is given as $\text{\#MUL} = n^{\log_p(1/2p^2 + 1/2p)}$. For $p = 2$ the number of multiplications is given by $\text{\#MUL} = n^{\log_2 3}$ whereas for large p this converges to $(1/2)^j n^2$.

Now consider two polynomials $A(x)$ and $B(x)$ with $n = \prod_{i=1}^{j} n_i$ coefficients that we multiply with a variant of the KA, that we call *generic recursive KA*. First we write $A(x) = \sum_{s=0}^{n_j - 1} A_s x^{s \cdot \prod_{i=1}^{j-1} n_i}$ as polynomial with n_j coefficients A_i and do the same for $B(x)$. Each of these "coefficients" is itself a polynomial with $\prod_{i=1}^{j-1} n_i$ coefficients. Then we use the recursive KA for n_j coefficients, and so on until the recursion eventually terminates. The number of needed multiplications is as follows:

$$\text{\#MUL}_{\prod_{i=1}^{j} n_i} = \left(\frac{1}{2}\right)^j \prod_{i=1}^{j} n_i(n_i + 1).$$

The number of additions is a complex expression [6]. There is a simple rule for the generic recursive KA to be most efficient: use the factorization of a number n with multiple prime factors combined with an increasing sequence of steps, i.e., KA for $n = \prod_{i=1}^{j} n_i$ with $n_i \geq n_{i+1}$, e.g., $2 \cdot 2 \cdot 3 \cdot 5$ for polynomials with 60 coefficients. In this case the polynomials with 60 coefficients are first split into an upper and lower half of 30 coefficients each, and the KA for two coefficients is applied to these halves. These polynomials of 30 coefficients are again split in half, and the KA for two coefficients is applied to the two halves. In the next recursion step the polynomials of 15 coefficients are split into three parts of five coefficients each, and so on. Note that in most cases the simple recursive KA that splits the operands in two halves of size $\lfloor n/2 \rfloor$ and $\lceil n/2 \rceil$ is more efficient than the general recursive KA, especially when the number of coefficients n has large prime factors.

The recursive KA versions are more efficient than the one-iteration KA. For example, instead of using the one-iteration KA for $n = 31$ coefficients one can use the recursive KA and split the 31-coefficient polynomial into two polynomials of 15 and 16 coefficients, respectively. Alternatively you could split the 31-coefficient polynomial into three parts of 10, 10, and 11 coefficients, respectively. In the next recursion step the polynomials are again split in two or three parts, and so on. However, a number of intermediate results have to be stored due to the recursive nature. This might reduce the efficiency of the recursive KA variants for small-sized polynomials.

The simple recursive KA is the easiest and most efficient way to implement the KA when the number of coefficients is unknown at implementation time or if it often changes. It is especially efficient if special cases are implemented as basis of the recursion. Providing special cases for $n = 2$ and $n = 3$ coefficients using the one-iteration KA as well as for $n = 9$ using the KA that splits the operands into three parts recursively yield efficient running times.

Further improvements are possible due to the use of dummy coefficients. For example, to multiply two polynomials of $n = 15$ coefficients it might be useful to append a zero coefficient and use a recursive KA for $k = 16$ coefficients. Some operations can be saved whenever the leading zero coefficient is involved. However, this gains only little improvement to the simple recursive KA without using dummy coefficients.

We now introduce the time ratio between a multiplication and an addition on a given platform $r' = t_m/t_a$ where t_m and t_a denote the cost for a multiplication and an addition, respectively. Let #MUL and #ADD be the number of multiplications and additions that the KA needs to multiply two polynomials of n coefficients, and $r = \frac{\#ADD - (n-1)^2}{n^2 - \#MUL}$. If the actual ratio r' on a given hardware platform is larger than r then it is more efficient to use the KA instead of the ordinary method. One can show that the KA always outperforms the ordinary multiplication method if $r' > 3$, i.e., if one multiplication takes longer than three additions the KA performs faster than the ordinary schoolbook method. For some cases the KA is always faster than the schoolbook method, i.e., it needs less multiplications and additions. However, there are also cases where it is more efficient to use a combination of KA and the schoolbook method. For example, consider the case $n = 8$ and a platform with $r' = 2$. When applying the elementary recursive KA it needs 27 multiplications and 100 additions, resulting in $r = 1.38$. Since $r' > r$ it is more efficient to use the KA. However, for $n = 4$ we obtain the ratio $r = 2.14$. Thus it is more efficient to use the ordinary method to multiply polynomials with

four coefficients. Therefore it is efficient to apply one recursive KA step, and then use the schoolbook method to multiply the polynomial halves of degree 3.

For some applications it is wise to use efficient underlying macros to multiply two polynomials with w coefficients. For example, there might be a macro to multiply two polynomials with four coefficients. Then two polynomials with $n = 20$ coefficients can be multiplied by using the KA for $20/4 = 5$ coefficients. In most cases it is efficient to use a mixture of different recursive steps combined with different underlying macros. Note that there might be optimized versions of the KA for special underlying coefficient rings like binary or prime fields. Also note that the KA can be applied to squaring polynomials by simply replacing all the coefficient multiplications by coefficient squarings. Although there is no special form of a squaring KA there might still be a performance gain compared to the ordinary squaring method which requires n squarings, $n(n-1)/2$ multiplications and $(n-1)^2$ additions. However, this varies for different platforms and depends on the ratio in time between a squaring and a multiplication. In most cases, i.e., if n is not very small, the squaring KA outperforms the ordinary squaring method [6].

Further information about the Karatsuba and similar algorithms can be found in [1, 4]. Further aspects of efficient implementations are presented in [2, 6].

André Weimerskirch

References

[1] Bernstein, D.J. (1998). "Multidigit multiplication for mathematicians." to appear in Advances in Applied Mathematics.

[2] Erdem, S.S. (2001). "Improving the Karatsuba–Ofman multiplication algorithm for special applications." *PhD Thesis*, Department of Electrical & Computer Engineering, Oregon State University.

[3] Karatsuba, A. and Y. Ofman (1963). "Multiplication of multidigit numbers on automata." *Sov. Phys.—Dokl.*, 7, 595–596.

[4] Knuth, D.E. (1997). *The Art of Computer Programming. Volume 2: Seminumerical Algorithms* (3rd ed.). Addison-Wesley, Reading, MA.

[5] Paar, C. (1994). "Efficient VLSI architecture for bit parallel computation in Galois fields." *PhD Thesis*, Institute for Experimental Mathematics, University of Essen, Germany.

[6] Weimerskirch, A. and C. Paar (2002). "Generalizations of the Karatsuba algorithm for efficient Implementations." Technical Report, Ruhr-University Bochum, 2002. Available at http://www.crypto.rub.de

KASUMI/MISTY1

MISTY1 [3] is a 64-bit <u>block cipher</u> designed by Mitsuru Matsui and first published in 1996. A variant of this cipher, called KASUMI [4], was adopted in 1999 as part of the confidentiality and integrity system for mobile communication specified by the Third Generation Partnership Project (3GPP). MISTY1 itself was submitted to the <u>NESSIE</u> project and included in its portfolio of recommended cryptographic primitives in 2003.

MISTY1 and KASUMI operate on 64-bit blocks and require a 128-bit secret key. Both have a similar recursive structure. The top level consists of an 8-round <u>Feistel cipher</u> built around a 32-bit non-linear <u>Boolean function</u> FO (Figure 1). The function FO itself is a 3-round Feistel-like "ladder" network containing a 16-bit non-linear function FI. The function FI, on its turn, consists of a similar 3-round (MISTY1) or 4-round (KASUMI) ladder network, using 7×7-bit and 9×9-bit S-boxes called S_7 and S_9. Note that these S-boxes differ slightly in both ciphers.

The key material is mixed with the data at different stages in the cipher, both in the FO and the FI functions. An additional component present in MISTY1 and KASUMI is the FL-function (see also <u>Camellia</u>). These functions are key dependent linear transformations, used in different ways in both ciphers: in MISTY1, FL-layers separate every two rounds of the cipher; in KASUMI, they are inserted before and after the FO-functions for odd and even rounds respectively. Another noteworthy difference between both ciphers is the expansion of the key. The *key schedule* (see <u>block cipher</u>) consists of a cyclical network of non-linear FI-functions in the case of MISTY1, but is completely linear in KASUMI. More details and motivations for the differences between MISTY1 and KASUMI can be found in [5].

MISTY1 has been widely studied since its publication, but no serious flaws have been found. Currently, the best attack on reduced-round variants of MISTY1 is the 5-round <u>integral attack</u> by Knudsen and Wagner [1]. The attack requires 2^{34} chosen plaintexts and has a time complexity of 2^{48}. Many other attacks have been proposed for variants without the FL-layers. The best attack on KASUMI was suggested by Kühn [2] and breaks 6 rounds using <u>impossible differentials</u>.

Christophe De Cannière

References

[1] Knudsen, L.R. and D. Wagner (2002). "Integral cryptanalysis (extended abstract)." *Fast Software Encryption, FSE 2002*, Lecture Notes in Computer Science, vol. 2365, eds. J. Daemen and V. Rijmen. Springer-Verlag, Berlin, 112–127.

[2] Kühn, U. (2001). "Cryptanalysis of reduced-round MISTY." *Advances in Cryptology—EUROCRYPT 2001*, Lecture Notes in Computer Science, vol. 2045, ed. B. Pfitzmann. Springer-Verlag, Berlin, 325–339.

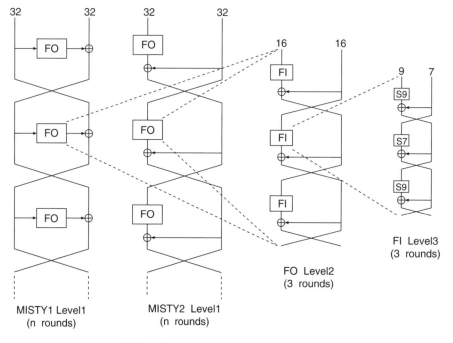

Fig. 1. Recursive structure of MISTY

[3] Matsui, M. (1997). "Block encryption algorithm MISTY." *Fast Software Encryption, FSE'97*, Lecture Notes in Computer Science, vol. 1267, ed. E. Biham. Springer-Verlag, Berlin, 64–74.

[4] Third Generation Partnership Project (1999). "KASUMI specification." Technical Report, Security Algorithms Group of Experts (SAGE), Version 1.0.

[5] Third Generation Partnership Project (2001). "3GPP KASUMI evaluation report." Technical Report, Security Algorithms Group of Experts (SAGE), Version 2.0.

KERBEROS AUTHENTICATION PROTOCOL

Kerberos is a network underline{authentication} protocol that uses underline{symmetric cryptography} (i.e., cryptography based upon a common secret underline{key} such as the underline{Data Encryption Standard DES)} to provide strong authentication for client/server applications. A client and a server prove their identity to each other through the mediation of a trusted third party called a *Key Distribution Center* (KDC). The Kerberos protocol also establishes a session key that may be used to provide confidentiality and integrity for subsequent communications between the authenticated participants.

Kerberos was created at the Massachusetts Institute of Technology (MIT) and is based upon the underline{Needham–Schroeder protocol}. A high-level description of the Kerberos protocol is provided in the underline{key management} entry.

Carlisle Adams

References

[1] Kerberos information (including papers and documentation) may be found at the following site: http://web.mit.edu/kerberos/www/

[2] Kohl, J. and B.C. Neuman (1993). "The Kerberos network authentication service (V5)." Internet Request for Comments 1510.

[3] Kohl, J., B.C. Neuman, and T.Y. Ts'o (1994). "The evolution of the kerberos authentication system." *Distributed Open Systems*. IEEE Computer Society Press, Los Alamitos, CA, 78–94.

[4] Neuman, B.C. and T. Ts'o (1994). "Kerberos: An authentication service for computer networks." *IEEE Communications*, 32 (9), 33–38.

KEY

A *key* is an element from an alphabet (the *key alphabet*) that selects, resp. defines a particular encryption step. A *keytext* is a sequence of *key elements* from a key alphabet that select, resp. define a sequence of particular encryption steps.

A *polyalphabetic encryption*, which is also called a *polyalphabetic substitution cipher*, is a substitution (see underline{substitutions and permutations}) with more than one alphabet, each one designated by a key element.

A *double key* is a polyalphabetic encryption with shifted mixed underline{alphabets} (see underline{Alberti encryption}). It is cryptologically equivalent to a polyalphabetic encryption with a underline{Vigenère} table ("tabula recta") whose plaintext standard alphabet is replaced by a mixed alphabet—the mixed alphabet being the 'second key'. Moreover, a *treble key* is a double key with the additional proviso that the standard alphabet for the keys of a Vigenère table is replaced by a mixed alphabet—this mixed alphabet being the 'third key'.

A *periodic key* or *repeated key* is a key sequence which repeats itself after d steps ($d > 1$), while a *nonperiodic key* is a key sequence that cannot be viewed as the repetition of a shorter sequence. A *running key* is an infinitely long nonperiodic key.

With *autokey* (French: *autochiffrant*, German: Selbstchiffrierung) one means the method to derive a nonperiodic key from the plaintext itself, using a *priming key character* or *key phrase* Δ to begin with. A simple example (Blaise de Vigenère, 1586) with priming key character $\Delta = \underline{D}$, is the following encryption, that makes use of the underline{Porta} encryption table:

plain	a	u	n	o	m	d	e	l	e	t	e	r	n	e	l
key	$\underline{D}$	*A*	*U*	*N*	*O*	*M*	*D*	*E*	*L*	*E*	*T*	*E*	*R*	*N*	*E*
cipher	X	I	A	H	G	U	P	T	M	L	S	H	I	X	T

A disadvantage is the spreading of encryption errors—a general weakness of all autokeying methods.

Arvid Damm (1919) proposed an autokeying variant of encryption ("influence letter") that was used in the German WW II teletype cipher machines SZ 40 and T 52. Claude Shannon gave (1949) the warning that autokeying Vigenère encryption with a priming key Δ of length d is equivalent to Vigenère encryption of d-grams with period $2 \cdot d$, i.e. by successively adding and subtracting Δ; thus it offers little security.

Chaitin defines a *random sequence* as an infinite sequence such that no finite subsequence has a shorter algorithmic characterization than the listing of the subsequence—no subsequence can be condensed into a shorter algorithmic description. A keytext, i.e., a sequence of key elements, with this property is called *holocryptic*. No sequence

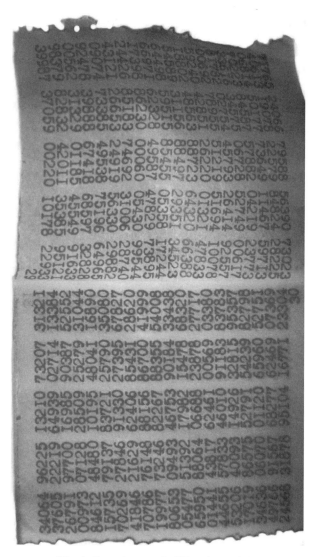

Fig. 1. One-time pad of Russian origin

generated by a deterministic, finite-state machine, i.e. by a deterministic algorithm, even if it does not terminate, has this property.

An *individual key* is a keytext (German 'i-Wurm') that is not copied from any source whatsoever. A *one-time key* is a keytext that is used just one time. Any written version of it (*one-time pad*, OTP) should be destroyed after use.

A *random key* is a random sequence used as a key sequence. To be cryptographically useful, it should by necessity be an individual key and a one-time key. Randomness should be achieved during the generation of the key sequence. Tests for randomness can only disprove it, but cannot prove it. Chaitin's definition has only theoretical value—it is mainly used as an instrument to show that a particular key sequence is nonrandom.

In an endomorphic cryptosystem, the encryption steps (governed by the key characters) may

form a group under composition: the *key group*. Such a cryptosystem is a pure cryptosystem. Examples are the Vigenère encryption and the Beaufort encryption, where the key group is a cyclic group, and the Vernam encryption by addition mod. 2, where the key group is $(C_2)^n$, the n-fold direct group of the cyclic group of order 2. A trivial example is encryption by a monoalphabetic selfreciprocal permutation π, where the key group is C_2, the cyclic group of order 2 consisting of π and the identity. Encryption with a key group, although it offers technical convenience, should be avoided since it opens particular ways of cryptanalytic attack.

We conclude this entry with some words on various roles that keys can play and how they are communicated in cryptosystems. (This is called *key negotiation*.)

In a classical setting, if a communication from participant A to participant B should be protected by encryption, A has to tell B what key to use for decryption, or B has to tell A what key to use for encryption. In command structures, there is also the possibility that the command tells both participants which keys to use. To this end, they make use of *key directives*: directories containing all relevant keys. Of course, transmission of all the information concerning the key should be done after encryption with a different cryptosystem. Violating this maxime for their Enigma traffic was a serious cryptographic blunder of the German *Wehrmacht* staff. (We note that modern cryptographic protocols (e.g., the Diffie–Hellman key agreement) scheme) may generate the same key for A and B without encrypted communication.

A *session key* or *message-encrypting key* is a keytext used during one communication session. In the Enigma traffic a session key was called '*Spruchschlüssel*' (text setting). A *base key* is a key used for encrypting keys ('key-encrypting key'). In the Enigma traffic, they were called '*Grundstellung*' (basic wheel setting) and formed part of the '*Tagesschlüssel*' (daily key).

A symmetric or *conventional* or *classical* cryptosystem is a communication line with two partners who are at different times both sender and receiver and use the same cryptosystem, each one having a private key for encryption and one for decryption—altogether four keys. If in an endomorphic cryptosystem selfreciprocal permutations (see substitutions and permutations) are used as key elements, keys for encryption and for decryption coincide.

A *private key* cryptosystem is a cryptosystem where sender and recipient share encryption and decryption keys (to be kept secret). In a *secret key* cryptosystem sender and recipient share one common key (to be kept secret).

Key symmetric cryptosystem: If two operations defined by keys $\mathcal{A}, \mathcal{B}$ commute and are mutually reciprocals: $\mathcal{A} \cdot \mathcal{B} = \mathcal{B} \cdot \mathcal{A} =$ identity, $\mathcal{A}$ may be used by partner A for encryption and by partner B for decryption, while $\mathcal{B}$ may be used by partner A for decryption and by partner B for encryption—altogether only two keys. If in an endomorphic cryptosystem selfreciprocal permutations are used as key elements, only one key is needed.

In many cryptosystems, actually in all classical ones, knowledge of the encryption key allows for an easy determination of the decryption key. This, however, is not necessarily so: As James H. Ellis pointed out in 1970, there may exist encryption methods where the knowledge of an encryption key does not suffice to derive the decryption key efficiently—in 1973 Clifford Cocks found in the multiplication of sufficiently large prime numbers the wanted, practically non-invertible operation. This led to the idea of public key cryptography, which is also called *asymmetric* cryptography. It was published in this form for the first time in 1976 by Whitfield Diffie and Martin E. Hellman. In this system, a key $\mathcal{A}$ is publicly announced by participant A with the proviso that he possesses a key $\mathcal{B}$ such that he can decrypt with $\mathcal{B}$ any message sent to him by anybody as long as it is encrypted with $\mathcal{A}$. This allows a star-like communication system. The advantage that no key negotiation is necessary and the key directory is open to the public is burdened by the fact that secrecy is only guaranteed to the extent that reconstruction of the (secret) decryption key (private key) from the public key needs exponential time and therefore is intractable.

Friedrich L. Bauer

Reference

[1] Bauer, F.L. (1997). "Decrypted secrets." *Methods and Maxims of Cryptology*. Springer-Verlag, Berlin.

KEY AGREEMENT

Key agreement refers to one form of *key exchange* (see also key encryption key) in which two or more users execute a protocol to securely share a resultant key value. As an alternative to key agreement, a *key transport protocol* may be used. The distinguishing feature of a key agreement protocol is that participating users each contribute an equal portion toward the computation of the resultant shared key value (as opposed to one user computing and distributing a key value to other users).

The original, and still most famous, protocol for key agreement was proposed by Diffie and Hellman (see Diffie–Hellman key agreement) along with their concept for public-key cryptography. Basically, users Alice and Bob send public-key values to one another over an insecure channel. Based on the knowledge of their corresponding private keys, they are able to correctly and securely compute a shared key value. An eavesdropper, however, is unable to similarly compute this key using only knowledge of the public key values.

There are numerous variations to the basic Diffie–Hellman key agreement protocol. One classification is based upon the longevity of the public keys shared between Alice and Bob. For example, the public keys may be long-term, or *static*, in which case each public key would likely be contained in a public-key certificate. Alternatively, the public keys may be short-term, or *ephemeral*, in which case the public keys would be for one-time use during the protocol session. Hybrid protocols combine both uses; for example, Alice may use an ephemeral public key while Bob might use a static public key.

The protocol instantiation in which both Alice and Bob use ephemeral public keys is vulnerable to a man-in-the-middle attack (see man-in-the-middle attack), unless additional precautions are taken. Use of static public keys helps to ensure that exchanged values are properly authenticated. In addition, the exchanged values may be further protected against attack. The station-to-station protocol is such a protocol in which exchanged values are encrypted and signed.

Although the original Diffie–Hellman key agreement protocol is presented as a communication between two users, the protocol has been extended to allow more than two users to agree upon a key. Several variations for such a protocol have been described in the literature, and vary based upon the number of protocol rounds, the amount of information exchanged, the number of broadcast messages, and other parameters.

Mike Just

References

[1] Menezes, A., P. van Oorschot, and S. Vanstone (1997). *Handbook of Applied Cryptography*. CRC Press, Boca Raton, FL.

[2] Stinson, D.R. (1995). *Cryptography: Theory and Practice*. CRC Press, Boca Raton, FL.

KEY AUTHENTICATION

Key authentication is the property obtained when performing a key establishment protocol (see also key agreement and key management) and one entity has the assurance that only a particularly identified other party may possibly know the negotiated key. This property may be unilateral if only one party participating in the protocol has the assurance, or it may be mutual if both parties have the assurance. Key authentication is sometimes referred to as "implicit key authentication" to distinguish it from "explicit key authentication", which is discussed below.

(Implicit) Key authentication can be obtained within a key establishment protocol in a number of ways. One possible method of obtaining this property is to encrypt the key to be established, k, for the other party using his (symmetric or asymmetric) key. In this case, since the only other party that could possibly decrypt the encrypted key is the intended recipient, the appropriate assurance is obtained. Many of the variants of the Diffie–Hellman protocol (see Diffie–Hellman key exchange protocol) also provide key authentication. For example, consider the case where both parties A and B have static authenticated (i.e., certified) Diffie–Hellman public keys α^a and α^b, respectively. If the agreed-upon key is simply $k = \alpha^{ab}$, then both parties have assurance that only the other party could possibly compute this key.

A property of key establishment protocols that is similar to key authentication is the property known as "key confirmation". Key confirmation is the property obtained when one party has the assurance that some other party actually has possession of the negotiated key. Notice that this property is distinct from key authentication in that the assurance is obtained relative to "some other party" instead of "a particularly identified other party". Thus, with key confirmation the other party need not be identified or even known at all. Also note that key confirmation provides assurance that the key is actually known by the other party whereas key authentication only provides assurance that the other party could possibly know the key. As with key authentication, key confirmation may be mutual or unilateral.

Typically key confirmation is obtained in one of three ways. First, a (one-way) hash of the negotiated key could be sent from one party to the other. Second, the key (or a key derived from the negotiated key) could be used in a MAC (see message authentication code) to authenticate a message. Finally, the key (or a key derived from the negotiated key) could be used to encrypt an agreed upon message. Any of these mechanisms will prove to the legitimate recipient that someone has possession of the key and used it to create the received values.

In many environments both (implicit) key authentication and key confirmation are required properties. In such circumstances, when both properties are obtained, it is said that "explicit key authentication" has been achieved. Explicit key authentication is the property obtained when one party has assurance that only a particularly identified other party actually has knowledge or possession of the negotiated key. Again, this property may be either mutual or unilateral.

A popular, typical example of a key establishment protocol that provides mutual explicit key authentication is the station-to-station protocol/STS protocol. In fact most of the protocols in use today that provide explicit key authentication are based upon the STS protocol. Examples include the SSL protocol (see secure socket layer) and the protocols used in IP$_{\text{SEC}}$.

Entity authentication is the assurance that the identified party is actually alive and participating in the protocol at that time. Quite often protocols that provide explicit key authentication will also provide entity authentication since the identified party must prove knowledge of the negotiated key. However, it is not always the case that any key negotiation protocol that includes entity authentication will also provide explicit (or implicit) key authentication. Care must be taken to ensure that the entity whose identity has been authenticated is the same entity as the one establishing the key. Otherwise, subtle attacks may allow one entity to have its identity authenticated and another entity to establish the key.

Robert Zuccherato

KEY ENCRYPTION KEY

Most cryptographic systems require some supporting Key Management, e.g., to enable *key exchange* or *key transport*. In order to ensure that keys are not used for different purposes, as otherwise lack of duality indirectly might thwart the system, one introduces key labels and key usage as well as key layers. Whereas *data keys* (used to encrypt data) at the bottom layer are exchanged frequently as so-called *session keys*, key encryption keys are used to exchange session keys or other key exchange keys belonging to layers just below, and are typically rarely or even never exchanged, and if so, then either by key custodians or public key techniques.

Peter Landrock

KEY ESCROW

Key Escrow: *"Something (e.g., a document, an encryption key) is delivered to a third person to be given to the grantee only upon the fulfillment of a condition."*
Escrowed Encryption Standard (EES),
FIPS 185 [1]

On April 16, 1993, the U.S. Government announced a new encryption initiative aimed at providing a high level of communications security and privacy without jeopardizing effective law enforcement, public safety, and national security. This initiative involved the development of tamper resistant cryptographic chips (Clipper and Capstone) that implemented an encryption/decryption algorithm (SKIPJACK) for the protection of sensitive information transmitted between two parties. What was special about these chips was that each one contained a device unique key that would give a third party, in possession of the key, the capability to decrypt all data encrypted using the chip. The purpose of this feature was to provide a means by which properly authorized law enforcement officials could decrypt encrypted communications. Authorization involved procedures modeled on those required for the authorization of a wiretap [2].

SKIPJACK was designed by the National Security Agency. A review group of four experts reviewed the originally classified Skipjack algorithm, and reported that there was no significant risk that the algorithm had "trapdoors" or could be broken by any known method of attack [3]. The National Institute of Standards and Technology (NIST) specified some details of the escrow parameters in FIPS 185. NIST also worked with representatives of the Justice Department, the Treasury Department, the National Security Agency, and the Federal Bureau of Investigation to develop and implement a Key Escrow System for the protection and controlled release of the information necessary to reconstruct the device unique keys [4]. The system was designed so that no single person or organization could compromise the device unique key.

Although the use of escrow cryptography was not mandatory, it raised concerns from the civil libertarian, product vendor, and academic communities. Civil libertarians feared that escrow might someday be made mandatory; product vendors wondered whether the marked would support cryptographic systems that provided the U.S. Government access to the protected information; and academics worried about whether the risks were worth the benefits. An ad hoc group of cryptographers and computer scientists argued that key escrow systems "are inherently less secure, more costly, and more difficult to use" [5].

Nevertheless, many data storage system owners wished to recover data encrypted by the users in the event that the user loses, destroys, or is unable to produce the encryption key. Researchers and encryption product vendors began to design and implement systems that provided for the recovery of user keys (often by the system administrator) [6]. This process is commonly referred to as key recovery. Today, many encryption product manufacturers provide a key recovery capability in their products or in the systems that make use of their products. Key recovery in these systems is primarily for the benefit of the user or the system owner. Authorized law enforcement officials would have to present their authorization to the system administrator/owner before obtaining access to any keys or information. The system administrator/owner would then be able to decide on the appropriate action. This is a well-understood and accepted process that has been used for many years.

While the initial concept of Key Escrow was not successful, it led to a greater appreciation for the need of users and system owners to have backup capabilities for the recovery of encryption keys or the data that they protect.

Miles E. Smid

References

[1] National Institute of Standards and Technology (1994). "Escrowed Encryption Standard (EES)." Federal Information Processing Standard (FIPS PUB 185).
[2] Delaney, D.P. et al. (1993). "Wiretap laws and procedures: What happens when the government taps a line." Available from Georgetown University, Department of Computer Science, Washington, DC, from cpsr.org, or by e-mail from denning @cs.georgetown.edu.
[3] Brickell, E.F. et al. (1993). "The SKIPJACK review, Interim Report: The SKIPJACK Algorithm." Available from Georgetown University, Office of Public Affairs, Washington, DC, from cpsr.org, or by e-mail from denning@cs.georgetown.edu.

[4] Denning, D.E. and M. Smid (1994). "Key escrowing today." *IEEE Communications Magazine*, 32 (9), 58–68.

[5] Abelson, H. et al. "The risks of key recovery, key escrow, & trusted third party encryption." Available on cdt.org/crypto/risks98 or epic.org/crypto/key_escrow.

[6] Denning, D.E. (1996). "Descriptions of key escrow systems." *Communications of the ACM*, February 26, 1997 version available on cosc.Georgetown.edu/~denning/crypto/appendix.html.

KEY MANAGEMENT

INTRODUCTION: Cryptographic keys are used to encrypt/decrypt data or to create/verify underline digital signatures (see key). One of the biggest issues associated with cryptography is the secure distribution of these keys to the appropriate communicating parties. This is referred to as key distribution or key establishment. The life cycle associated with this keying material (i.e., the initialization, distribution, and cancellation of the keys) is referred to as *key management*. The purpose of this section is to discuss key management, with particular emphasis on key distribution.

Before we discuss key management, it is important to understand that there are two basic types of cryptography: (1) *symmetric* or *secret key* and (2) *asymmetric* or *public key*.

Symmetric cryptography is characterized by the fact that the same key is used to perform both the encryption and decryption. This means that the communicating parties must have copies of the same cryptographic key, and a method to securely convey these keys to the appropriate parties must be available. Compromise of the secret key naturally leads to the compromise of any data that was encrypted using that key.

Public key cryptography is characterized by the fact that the key used to perform a cryptographic operation (e.g., digital signature creation) is not the key used to perform the inverse cryptographic operation (e.g., digital signature verification). Public key cryptography is based on the notion of key pairs. One key is referred to as the *public key* and can be revealed to anyone. The other key is referred to as the *private key* and is not revealed to anyone other than the end-entity associated with that key (although there are exceptions such as private key backup with a trusted third party when required). These keys are mathematically related; however, knowledge of the public key does not divulge enough information to allow an attacker to determine the private key

efficiently. The concept of asymmetric cryptography was first introduced to the general public in 1976 (see [3]), but much of the technology necessary to support public key cryptography was not available until the mid-1990s.

As illustrated below, symmetric cryptography and asymmetric cryptography are not necessarily mutually exclusive. In fact, these techniques can be used together in order to offer a complementary set of services. For example, symmetric cryptography can be used to encrypt a message and asymmetric cryptography can be used to securely transfer the secret key used to encrypt the file to the intended recipient(s). However, this is not always possible and other distribution mechanisms may be required.

To illustrate these concepts in more detail, we will first discuss key management associated with a secret key only system. This will be followed by a discussion of public key cryptography and how public key and secret key cryptography can be used together.

SYMMETRIC OR SECRET KEY CRYPTOGRAPHY

Background

When the first electronic symmetric cryptosystems were deployed, key management was physical in nature and it required a significant amount of human involvement. Keys were centrally generated and recorded on media such as paper or magnetic tape and the keying material was physically distributed to the appropriate locations. This was sometimes accomplished through the use of couriers (sometimes humorously referred to as "sneaker net"). The keying material was physically destroyed when no longer needed.

However, modern symmetric cryptosystems are more advanced and typically use some form of electronic key distribution. One possible model for electronic distribution of symmetric keys is based on a trusted third party component known as a *Key Distribution Center* (KDC) (e.g., see [5]). Before an end-entity (e.g., an end user) can access a target resource (e.g., a server), the end-entity makes a request to the KDC to establish a session key that can be used to secure the communication between the end-entity and the target resource. This model is illustrated in Figure 1.

The outbound arrows between the KDC and the communicating parties are logical representations of the key distribution process. In practice, The KDC may distribute one copy of the symmetric key directly to the end-entity and another copy to the target resource, or both copies of the symmetric

Key Distribution Center Model

KDC

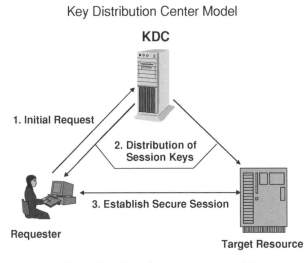

Fig. 1. Key distribution center model

Kerberos to illustrate the concepts associated with symmetric key management.

Initialization

In the Kerberos scheme, end-entities and target resources are referred to as Principals. Kerberos maintains a database of all Principals and their associated symmetric keys. This allows the session keys for each principal to be protected when they are in transit. These symmetric keys are initialized separately and must be established before an end-entity can send a request to the AS.

Distribution

When an end-entity needs to communicate with a target resource for the first time, the end-entity makes a request to the AS. The request contains the end-entity's identifier as well as the identifier of the target resource. Typically the initial target resource is the TGS and for the purposes of this example, we will assume that the request is for a session with the TGS. However, this may not always the case (see [4] for more information).

Assuming that the end-entity and target resource (i.e., TGS) are in the Kerberos database, the AS will generate a symmetric key (referred to as a *session key*) and encrypt one copy of the session key using the symmetric key of the end-entity and another copy of the session key using the symmetric key of the target resource (this is referred to as a *ticket*). These encrypted copies of the session key are returned to the requesting end-entity.

The end-entity decrypts its copy of the session key and uses it to encrypt the end-entity's identity information and a time stamp (this is referred to as the *authenticator*). The time stamp is necessary to prevent replay attacks. The session key encrypted by the AS for the target resource (i.e., the ticket for the TGS) and the encrypted authenticator are then sent to the target resource. The two communicating parties now have copies of the session key and are able to communicate securely.

From that point forward the end-entity can make additional requests to either the AS or the TGS (usually the TGS, but see [4] for more details) in order to establish a session key between the end-entity and any other target resource. The difference between requesting this information from the AS and from the TGS is that the end-entity's copy of the session key is encrypted with the end-entity's symmetric key obtained from the Kerberos database when dealing with the AS, but the end-entity's copy of the session key is

key may be distributed back to the end-entity and the end-entity would then pass the symmetric key to the target resource. In both cases two copies are needed since the symmetric key is encrypted for the intended recipients (i.e., one copy of the key is encrypted for the end-entity and another copy of the key is encrypted for the target resource). This is necessary to prevent someone in a position to intercept the session key from being able to use the key to eavesdrop on the subsequent communication. This implies that the KDC and the communicating parties must have been pre-initialized with keys that can be used to protect the distribution of the session keys. These are sometimes referred to as Key Encrypting Keys (KEKs).[1] Note that this pre-initialization step should not be considered unusual since some form of initial bootstrap process is typically required in any cryptographic system.

A classic example of an electronic secret key distribution based on this model is Kerberos V5 (see [4]). Kerberos enables electronic key distribution in a client/server network environment. Kerberos is comprised primarily of two logical components: (1) the Authentication Server (AS) and (2) the Ticket Granting Server (TGS). The AS and TGS may be physically separate, or they may reside on the same platform, or they may even be part of the same process. Collectively, these two logical components can be thought of as the KDC as described above. Since Kerberos is a well known, publicly available symmetric cryptosystem, we will use

---

[1] It is possible to have all communicating parties pre-initialized with the same KEK; however, this practice is generally considered to be less secure since compromise of the single KEK would impact the entire community rather than a single entity.

encrypted using the TGS session key when dealing with the TGS.

Cancellation

In terms of key cancellation, we need to consider that there are actually two types of keys being used. Each principal (i.e., end user or server) has a shared secret key used to protect the distribution of the session keys. The lifetime of these shared secret keys is typically very long. Cancellation of a given key is usually facilitated through replacement (i.e., keys can be changed in accordance with local policy) or deletion of an entry (e.g., when a principal no longer belongs within the Kerberos realm). The second type of key is the session key. The lifetime of the session keys is directly coupled with the lifetime of the session itself. Once the session between the client and server is terminated, the session key is no longer used.

Summary and Observations

In summary, the Kerberos database is initialized with entries for each principal. Each entry includes the shared secret key associated with that principal which is used to protect the session keys. The session keys are generated by the KDC in response to requests from clients and are securely distributed to the appropriate principals. The shared secret keys generally have long lifetimes, but the session keys are ephemeral (i.e., short-lived).

One of the criticisms associated with symmetric only key management is that it does not scale well. It has also been criticized for having a single point of failure (i.e., what happens when the KDC goes down?) as well as a single point of attack (all of the pre-initialized symmetric keys are stored in the KDC database). However, alternative key management schemes exist that help to alleviate some of these problems. In particular, asymmetric cryptography can be used to exchange secret keys as discussed in the next section.

ASYMMETRIC OR PUBLIC KEY CRYPTOGRAPHY: Asymmetric cryptography is often implemented in association with a supporting infrastructure referred to as Public Key Infrastructure (PKI). PKI refers to the policies, procedures, personnel, and components necessary to support the key management process. The primary components of the PKI include the Certification Authority (CA) and *Local Registration Authority (LRA)*. (Other components may also be present, but these are not relevant for the purposes of this discussion.) The consumers of

the PKI-related services are referred to as *end-entities* and may be end users, devices, processes, or servers.

Initialization

The generation of the public/private key pairs associated with the end-entities can occur within the CA, LRA, or the end-entity's system. If necessary (depending on where key generation occurred), the private component of the public/private key pair must be securely distributed to the appropriate end-entity. Several protocols have been defined to accomplish this (e.g., see [6] or [1]). The private key is stored securely in standard formats such as PKCS #5 and #8. The public key component is populated in a signed data structure issued by a CA. This data structure is referred to as a public key certificate. The latest version of the public key certificate (version 3) is defined in [7] and a high-level representation of a version 3 public key certificate is provided in Figure 2.

The digital signature appended to the public key certificate provides two things. First, the integrity of the certificate can be verified so any modifications to the data contained within the certificate after it was issued can be detected. Second, the identity of the issuing CA can be verified. This allows the users of the certificate to determine if the certificate originated from a trustworthy source. Since both the content and source of the certificate can be verified, the certificate can be distributed via potentially nonsecure channels. For example, the public key certificate can be stored "in the clear" in a public repository (e.g., an X.500 directory) which allows end-entities to retrieve these certificates easily when required.

When end-entities enroll with the PKI, they typically use one or more shared secrets to demonstrate they are the end-entity that they claim to be. The shared secrets may have been established at some point in the past, or they may be distributed to the end-entity as part of a formal registration process. This latter method is typically required when the certificate(s) are used in conjunction with high assurance and/or sensitive transactions. This often requires that the end-entity present himself or herself to an LRA along with acceptable forms of identification (e.g., a driver's license or employee ID). In any case, the shared secrets facilitate the initial bootstrap process in which end-entities are first initialized with their keys.

Distribution

Once the end-entities are initialized with the PKI, they can engage in secure communication

Version 3 Public Key Certificate

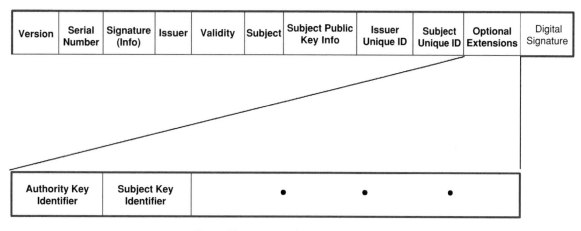

Version	Serial Number	Signature (Info)	Issuer	Validity	Subject	Subject Public Key Info	Issuer Unique ID	Subject Unique ID	Optional Extensions	Digital Signature

Authority Key Identifier	Subject Key Identifier	●	●	●

Fig. 2. Version 3 public key certificate

(e.g., secure e-mail) with their peers. Theoretically, it would be possible for end-entities to use public key cryptography to encrypt data for their peers by using the public key of each peer (assuming that the asymmetric algorithm supports encryption/decryption). However, there are a few practical issues that make this an unattractive approach. First, asymmetric cryptography is notoriously slow when compared to symmetric cryptography. Asymmetric cryptography is therefore suitable only for the encryption of small amounts of data. Second, it would be extremely wasteful to encrypt the data N times, once for each intended recipient—especially when dealing with large amounts of data (consider an e-mail with file attachments). (Note that this is also true even when symmetric algorithms are used.) Third, not all asymmetric algorithms support encryption/decryption (e.g., RSA does, but DSA does not). Thus, we would like to take advantage of the speed of symmetric cryptography, but we want to avoid the key distribution problems mentioned in the previous section. We also want to avoid encryption of the data multiple times (once for each recipient). This is why PKI supports both asymmetric and symmetric cryptography. Symmetric cryptography is used for data encryption (but the data is only encrypted once regardless of the number of recipients), and asymmetric cryptography is used for the distribution of the secret key to the intended recipients.

We can illustrate how this works using the example illustrated in Figure 3. In this example, we are using a symmetric algorithm to encrypt data (e.g., an e-mail message) and an asymmetric algorithm such as RSA (see RSA public key encryption) to enable the secure distribution of the secret key that was used to encrypt the e-mail.

Asymmetric Key Distribution

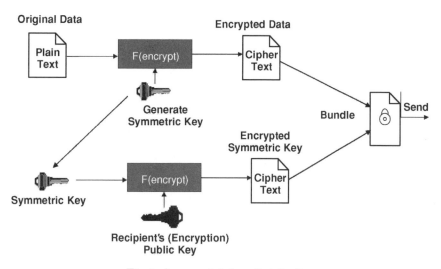

Fig. 3. Asymmetric key distribution

Essentially, the system generates a secret key that is then used to encrypt the message. Any number of symmetric algorithms could be used for this purpose (e.g., CAST-128 or Rijndael/AES). The secret key is then encrypted using the intended recipient's public (encryption) key. If multiple recipients were involved, the original data would still be encrypted once using the generated secret key, and the secret key would be encrypted *N* times, once for each recipient. The public key certificate for each recipient can be retrieved from a repository, or perhaps the certificate may have been conveyed to the originator in a previous exchange. On receipt, each recipient can use his/her corresponding private decryption key to decrypt the symmetric key necessary to decrypt the original data. This provides an efficient and secure key distribution mechanism that does not suffer from the drawbacks discussed in the previous section.

Asymmetric cryptography can also be used to support a process known as key agreement. In this case, the communicating parties negotiate an ephemeral secret key. (See Diffie–Hellman key exchange protocol for additional information.)

Cancellation

In terms of key cancellation, there are two things to consider: (1) the secret key used to encrypt the data and (2) the public/private key pair. In terms of the secret key, this survives as long as the data is encrypted, which could be indefinitely. However, the secret key is deleted/destroyed when the associated file is deleted or (permanently) decrypted. There may also be cases when the protection is no longer considered adequate (e.g., the symmetric algorithm has been compromised or the key length used no longer provides adequate protection). In this event, the file is decrypted and re-encrypted using a new algorithm and/or key. The original key would be deleted/destroyed.

In terms of the public/private key pairs, public key certificates are issued with a fixed lifetime, typically on the order of 2–5 years depending on the purpose of the certificate and the associated local policy. In some PKIs, the certificates (and associated private key) are renewed automatically before the existing certificate(s) expire. In other PKIs, end-entities must request new certificate(s) when their existing certificate(s) expire.

It is possible to establish a different lifetime for the private component of the public/private key pair when that key pair is used in conjunction with digital signatures (see digital signature schemes). This is done in comprehensive PKIs where it is desirable to have a grace period between the time the private signing key can no longer be used and the time that the associated public key certificate expires so that digital signatures created before the corresponding private key expired can still be verified without exposing the end user to needless warning messages. These comprehensive PKIs generally update the public/private key pairs (and public key certificate) automatically.

Finally, it is possible to revoke certificates before they naturally expire. This might be done for a variety of reasons, including suspected private key compromise. (See certificate revocation for additional information related to certificate revocation.)

The interested reader can find a more comprehensive discussion of public key life cycle management in Chapter 7 of [2].

Summary and Observations

PKI provides comprehensive key management through a combination of asymmetric and symmetric cryptography. Symmetric cryptography is used for bulk data encryption/decryption and asymmetric cryptography is used for key distribution. The use of asymmetric cryptography to facilitate key distribution is an extremely powerful tool that serves to eliminate many of the problems associated with symmetric-only cryptosystems.

Steve Lloyd

References

[1] Adams, C. and S. Farrell (1999). "Internet X.509 public key infrastructure: Certificate management protocols." Internet Request for Comments 2510.

[2] Adams, C. and S. Lloyd (2003). *Understanding PKI: Concepts, Standards, and Deployment Considerations* (2nd ed.). Addison-Wesley, Reading, MA. ISBN 0-672-32391-5.

[3] Diffie, W. and M. Hellman (1976). "New directions in cryptography." *IEEE Transactions on Information Theory*, 22, 644–654.

[4] Kohl, J. and C. Neuman (1993). "The Kerberos network authentication service (V5)." Internet Request for Comments 1510.

[5] Needham, R. and M. Schroeder (1978). "Using encryption for authentication in large networks of computers." *Communications of the ACM*, 21 (12). 993–999.

[6] Ramsdell, B. (1999). "S/MIME Version 3 certificate handling." Internet Request for Comments 2632.

[7] ITU-T Recommendation X.509 (2000). "Information technology—open systems interconnection—the directory: Public key and attribute certificate frameworks." (Equivalent to ISO/IEC 9594-8:2001).

KNAPSACK CRYPTOGRAPHIC SCHEMES

INTRODUCTION: The knapsack problem originates from operational research. Suppose one wants to transport some goods which have a given economical value and a given size (e.g., volume). The transportation medium, for example a truck, is however limited in size. The question then is to maximize the total economical value to transport, given the size limitations of the transportation medium.

The above mentioned knapsack problem is not the one that was proposed for cryptographic purposes. The one used is only a special case namely, the one in which the economical value of each good is equal to its size. This special problem is known as the *subset sum problem* [24]. Merkle and Hellman initiated the use of the subset problem—which they called knapsack—for cryptographic purposes.

DEFINITION 1. *In the* **subset sum problem** *n integers a_i are given (the size of n goods). Given a certain integer (the size of the transportation medium) S, the problem is to decide whether a subset of the n numbers exist such that by adding them together one obtains S, formally to decide/find (whether there are) bits x_i such that:*

$$S = \sum_{i=1}^{n} x_i \cdot a_i. \tag{1}$$

The problem to decide whether such a subset exists is NP-complete [24].

For now on, when the "the knapsack problem" is mentioned, it is used as a synonym for "the subset sum problem". Note that there is also a subset product problem, which was used in the so-called multiplicative public key knapsack cryptographic systems [38]. The *multiplicative knapsack* and its security will be surveyed in Section "The multiplicative knapsack scheme and its history".

Most research on cryptographic knapsack schemes was related to public key encryption/ decryption, i.e., to protect privacy. Cryptographic knapsack schemes which protect the authenticity are briefly discussed in Section "The trapdoor knapsack schemes to protect signatures and authenticity."

THE CRYPTOGRAPHIC KNAPSACK SCHEME: AN INTRODUCTION: Except for x which is usually binary and except when explicitly mentioned, all numbers used are natural numbers or integers (depending on the context).

The Encryption in Additive Knapsack Schemes

In most *additive knapsack* systems the encryption operation works as follows: Suppose Bob wants to send a binary $\mathbf{x} = (x_1, x_2, \ldots, x_n)$ message to Alice, and Alice's public key is $\mathbf{a} = (a_1, a_2, \ldots, a_n)$. To encrypt the message Bob computes the ciphertext (see cryptosystem):

$$S = \sum_{i=1}^{n} x_i \cdot a_i \tag{2}$$

which he sends to Alice. So this defines an encryption function $E_{\mathbf{a}}(\cdot)$ that maps $\mathbf{x}$ into S.

Since the encryption key is public and S can be eavesdropped, it must be "difficult" to find $\mathbf{x}$ from S and $\mathbf{a}$. This problem is the subset sum problem, which is an NP-complete problem. So, no efficient polynomial time algorithm exist to find $\mathbf{x}$ in the worst case (over all S and $\mathbf{a}$). So it seemed that breaking the cryptographic knapsack was hard. It is important to notice the term "worst case." Indeed if $\mathbf{a} = (1, 2, 4, 8, \ldots, 2^{n-1})$ it is trivial to find for all S the corresponding $\mathbf{x}$, by writing S in binary form. Sequences $\mathbf{a}$ for which it is easy to find, for all S, its corresponding $\mathbf{x}$, have been called *easy*.

To allow unique decryption, the encryption function $E_{\mathbf{a}}(\cdot)$ has to be *one-to-one*. Shamir [47] called a sequence $\mathbf{a}$ that leads to such a one-to-one encryption function a *one-to-one system*. It is co-NP-complete to decide whether a given sequence is a one-to-one system [47].

The Decryption

If $\mathbf{a}$ is chosen randomly by Alice, there is no known method for her to decrypt S and find the plaintext (see cryptosystem). To allow this, Merkle and Hellman [38] introduced some trapdoor (see trapdoor one-way function). The secret information used to make the trapdoor is called the *decryption key. It is now the trapdoor technique which turns out to allow the breaking of the cryptographic public knapsacks.*

Most knapsack schemes differ only in the use of other trapdoor techniques. However, some knapsack schemes allow the x_i to have more values than just binary. Others add some kind of noise to the plaintext.

THE MERKLE–HELLMAN TRAPDOOR: In the Merkle–Hellman case, when Alice constructs her public encryption key $\mathbf{a}$ she will first generate a *superincreasing* sequence $\mathbf{a}^1$ of natural numbers $(a_1^1, a_2^1, \ldots, a_n^1)$, which is defined as follows:

DEFINITION 2. *A vector* $\mathbf{a}^1 = (a_1^1, a_2^1, \ldots, a_n^1)$ *is said to be a superincreasing sequence if:*

$$\textit{for each } i \ (2 \leq i \leq n): \quad a_i^1 > \sum_{j=1}^{i-1} a_j^1.$$

$(1, 2, 4, 8, \ldots, 2^{n-1})$ *is a superincreasing sequence. As will be explained further on, a superincreasing sequence is an "easy" sequence.*

To hide the superincreasing structure Alice transforms $\mathbf{a}^j$ into $\mathbf{a}^{j+1}$ starting with $j = 1$. For each transformation j, she chooses an (w_j, m_j) such that Conditions 3 and 4 are satisfied

$$\sum_{i=1}^n a_i^j < m_j \tag{3}$$

$$\gcd(w_j, m_j) = 1 \tag{4}$$

and then computes:

$$a_i^{j+1} \equiv a_i^j \cdot w_j \ (\text{mod } m_j), \quad \text{where } 0 < a_i^{j+1} < m \tag{5}$$

(see underline{modular arithmetic}). When Alice used k transformations, her public key is given by: $\mathbf{a} = \mathbf{a}^{k+1}$.

We will refer to the transformation defined in Equations (3)–(5) as the *Merkle–Hellman transformation*. We call the condition in (3) the *Merkle–Hellman dominance* condition. In the case one uses this transformation in the direction from $\mathbf{a}^{j+1}$ to $\mathbf{a}^j$, we call it the *reverse Merkle–Hellman transformation*:

$$a_i^j \equiv a_i^{j+1} \cdot w_j^{-1} \ (\text{mod } m_j), \quad \text{where } 0 < a_i^j < m. \tag{6}$$

When $\mathbf{a}^1$ is superincreasing and only one transformation is used, the resulting public key scheme is called the *basic Merkle–Hellman scheme*, or sometimes the *single iterated Merkle–Hellman scheme*. The case that two transformations are used instead of one is called the *doubly iterated* one.

Let us now explain the decryption for the Merkle–Hellman scheme. When Alice receives the ciphertext S, she iteratively computes S^j from S^{j+1} starting from $j = k$ and $S^{k+1} = S$ as follows:

$$S^j = S^{j+1} \cdot w_j^{-1} \ (\text{mod } m), \quad \text{where } 0 \leq S^j < m_j. \tag{7}$$

It is trivial to understand that $S^j \equiv \sum_{i=1}^n x_i a_i^j \bmod m$ and, as a consequence of the inequality in Equation (7) and the Merkle–Hellman dominance condition (see Equation (3)), $S^j = \sum_{i=1}^n x_i a_i^j$. So finally, Alice ends up with S^1. Finally, it is "easy" to find the message $\mathbf{x}$ from S^1. Indeed, start with $h = n$. If $S^1 > \sum_{i=1}^{h-1} a_i^1$ then x_h has to be 1, else 0.

Continue iteratively by subtracting $x_h a_h$ from S^1, with h decrementing from n to 1 during the iterations. In fact a rather equivalent process is used to write numbers in binary notation. Indeed the sequence $(1, 2, 4, 8, \ldots, 2^{n-1})$ is superincreasing.

In Section "The decryption" we have seen that an important condition for the public key is that it has to form a one-to-one system. This is the case for the Merkle–Hellman knapsack scheme by applying the following Lemma as many times as transformations were used, and by observing that a superincreasing sequence forms a one-to-one system.

LEMMA 1. *Suppose that* $(a_1^1, a_2^1, \ldots, a_n^1)$ *is a one-to-one knapsack. If* $m > \sum a_i^1$ *and* $\gcd(w, m) = 1$, *then any set* $(a_1, a_2, \ldots, a_n)$, *such that* $a_i \equiv a_i^1 \cdot w \bmod m$, *is a one-to-one system.*

PROOF. By contradiction: Suppose that $(a_1, a_2, \ldots, a_n)$ does not form a one-to-one system. Then there exist $\mathbf{x}$ and $\mathbf{y}$ such that $\mathbf{x} \neq \mathbf{y}$ and $\sum x_i a_i = \sum y_i a_i$. Thus evidently, $\sum x_i a_i \equiv \sum y_i a_i \bmod m$, and also $(\sum x_i a_i) \cdot w^{-1} \equiv (\sum y_i a_i) \cdot w^{-1} \bmod m$, because w^{-1} exists $(\gcd(w, m) = 1)$. So $\sum x_i a_i^1 \equiv \sum y_i a_i^1 \bmod m$. Since $0 \leq \sum x_i a_i^1 \leq \sum a_i^1 < m$ and analogously $0 \leq \sum y_i a_i^1 \leq \sum a_i^1 < m$ we have $\sum x_i a_i^1 = \sum y_i a_i^1$. Contradiction. $\square$

A SURVEY OF THE HISTORY OF THE CRYPTOGRAPHIC KNAPSACK:

We will mainly survey (see Section "The trials to avoid weaknesses and attacks for the class of usual knapsacks") the additive knapsack public key systems protecting privacy and using the same encryption function as the Merkle–Hellman one. We will abbreviate this as *the usual knapsack* cryptographic schemes. (Our survey will not be exhaustive.) Then we will shortly discuss similar schemes but using different encryption functions (see Section "The case of usual knapsacks with other encryption functions"). We very briefly discuss the history of: the multiplicative knapsack schemes (see Section "The multiplicative knapsack scheme and its history"), and the use of trapdoor knapsacks in signatures (see Section "The trapdoor knapsack schemes to protect signatures and authenticity").

The Trials to Avoid Weaknesses and Attacks for the Class of Usual Knapsacks

In 1979 Shamir found that a knapsack system with a very high *density* can (probabalistically) easily be cryptanalyzed. The density of a knapsack system with public key $\mathbf{a}$ is equal to the cardinality of the image of the encryption function (see Equation (2)) divided by $\sum a_i$. This result is independent of the trapdoor used to construct the public key.

To construct public keys in *Graham–Shamir* [48] and *Shamir–Zippel* schemes one starts from other easy sequences than the superincreasing ones. Then one applies Merkle–Hellman transformations to obtain the public key. The case that only one transformation is used is called the *basic Graham–Shamir* and *basic Shamir–Zippel* scheme. For example in the Graham-Shamir scheme $\mathbf{a}^1$ is not superincreasing but can be written as:

$$\mathbf{a}^1 = \mathbf{a}' + 2^q \mathbf{a}'', \qquad \text{with} \quad a_n' < 2^{q-1}$$
$$\text{and} \quad \mathbf{a}' \text{ superincreasing.}$$

It is trivial to understand that such a sequence is easy.

In the beginning of 1981, Lenstra [34] found a polynomial time algorithm to solve the integer linear programming problem, when the number of unknowns is fixed. The complexity of the algorithm grows exponentially if the size of the number of unknowns increases. A part of Lenstra's algorithm uses a lattice reduction algorithm (more details are given in Section "The LLL algorithm"). The importance of Lenstra's work on the security of knapsack cryptosystems will be explained later.

When in 1981 Henry [27] found a method to speed up the decryption in knapsack schemes, Bell Laboratories started designing a VLSI chip for knapsack cryptosystems, boosting the perceived importance of cryptographic knapsack schemes.

In 1982 Desmedt, Vandewalle and Govaerts [16, 17] and independently Eier and Lagger [23] demonstrated that any public key which is obtained from a superincreasing sequence using the Merkle–Hellman transformation, has infinitely many decryption keys. In general, if some public key is obtained using a Merkle–Hellman transformation, then there exist infinitely many other parameters, which would result in the same public key when used to construct it. This has been called "a key observation that led eventually to the complete demise of these knapsack systems" [10].

Desmedt et al. [17] also proposed a different way to decrypt and build public keys. In previous schemes one can find *all* plaintext bits x_i from the same S^1. In their approach, the size of the knapsack, n, *grows* during the construction of the public key. Each transformation only allows to recover some bit(s) x_i of the plaintext. Let us briefly explain the other type of partially easy sequence,

called *ED* (where ED indicates that the property to find one bit x_i is *Easy* based on a Divisibility property). If d divides all a_i^j, except a_r^j, then if $S^j = \sum_{i=1}^n x_i a_i^j$, it is easy to find x_r, by checking if d divides S^j or not. The method discussed here to construct the public key, together with the partially easy sequence already discussed will be called the *Desmedt–Vandewalle–Govaerts knapsack*.

In the beginning of 1982 Lenstra, Lenstra and Lovasz found some algorithm for factoring polynomials with rational coefficients [35]. A part of this algorithm is an improvement of the lattice reduction algorithm (described in [34]). This improvement is known in the cryptographic world as the *LLL* algorithm (see shortest vector problem). Note that the LLL algorithm speeds up the algorithm to solve the integer linear programming (with the number of variables fixed) [36]. Another application of it is that it allows to find some simultaneous Diophantine approximations [35].

In April 1982 Shamir broke the basic Merkle–Hellman scheme [49, 51]. His attack uses the integer linear programming problem. Shamir was able to dramatically reduce the number of unknowns (in almost all cases) in the integer linear programming problem. In fact, the cryptanalyst first guesses the correct subsequence of the public key corresponding with the smallest superincreasing elements. The number of elements in the subsequence is small. Because the Lenstra algorithm (to solve the integer linear programming problem) is feasible if the number of unknowns is small, Shamir was able to break the basic Merkle–Hellman scheme.

A few months later Brickell et al. [3] found that by a careful construction of the public key (using the basic Merkle–Hellman scheme) the designer could avoid Shamir's attack. This work clearly demonstrated that one has to be careful with attacks, which break systems in *almost all cases*. However, as a consequence of further research, this work has made a technical mark.

About the same time Davio came up with a new and easy sequence [15]. This easy sequence is based on ED, but it allows to find all x_i at once. The construction is similar to the proof of the Chinese Remainder Theorem.

Adleman broke the basic Graham–Shamir scheme [1]. The main idea of Adleman was to treat the cryptanalytic method as a lattice reduction problem and not as a linear integer programming problem. *This idea was one of the most influential in the area of breaking cryptographic knapsack algorithms*. To solve the lattice problem he used the LLL algorithm [35]. The choice of a good lattice plays a key role in his paper.

In August 1982 Shamir presented a new knapsack scheme, known as *Shamir's ultimate knapsack scheme* [50]. The main idea is that instead of applying k Merkle–Hellman transformations, Alice uses "exactly" $n-1$ of such transformations when constructing her public key. "Exactly" means here, that after each transformation (e.g. jth) one checks if $\mathbf{a}^j$ is linearly independent of $(\mathbf{a}^1, \ldots, \mathbf{a}^{j-1})$, if not, one drops $\mathbf{a}^j$, makes a new one and tries again. The final result $\mathbf{a}^n$ is the public key. To decrypt S, Alice applies her $n-1$ reverse secret transformations. She starts with $S^n = S$ and by calculating the other S^j, similar as in the Merkle–Hellman case (see Section "The decryption"). So she obtains a set of linear equations:

$$\begin{pmatrix} S^1 \\ \vdots \\ S^{n-1} \\ S^n \end{pmatrix} = \begin{pmatrix} a_1^1 & \cdots & a_n^1 \\ \vdots & \ddots & \vdots \\ a_1^{n-1} & \cdots & a_n^{n-1} \\ a_1^n & \cdots & a_n^n \end{pmatrix} \cdot \begin{pmatrix} x_1 \\ \vdots \\ x_{n-1} \\ x_n \end{pmatrix}. \quad (8)$$

After the discussed transformations to find $\mathbf{x}$, Alice only has to solve a set of linear equations. It is important to observe that the obtained public key is one-to-one, even if $\mathbf{a}^1$ is *not* an easy sequence, or even if no partially easy sequences are used. This follows from the nonsingularity of the matrix in Equation (8).

Other research continued, trying to obtain other easy (or partially easy) knapsack sequences. Petit's [42] defined *lexicographic knapsacks*. Let $w(\cdot)$ be the Hamming weight function. $\mathbf{a}$ is called lexicographic, if and only if, $\mathbf{a}^T\mathbf{x} < \mathbf{a}^T\mathbf{y}$ for all binary $\mathbf{x}$ and $\mathbf{y}$, with $\mathbf{x} \neq \mathbf{y}$ and one of the two cases (i) $w(\mathbf{x}) < w(\mathbf{y})$ or (ii) $w(\mathbf{x}) = w(\mathbf{y})$ and $\mathbf{x}$ and $\mathbf{y}$ satisfy together $x_k \overline{y}_k = 1$ and $x_i \oplus y_i = 0$ for all $i < k$, with $\oplus$ the exclusive or. The construction of the public key is as in the Merkle–Hellman case, using Merkle–Hellman transformations.

Willett [53] also came up with another easy sequence and a partially easy sequence, which were then used similar as in the Merkle–Hellman and in the Desmedt–Vandewalle–Govaerts knapsack. We will only discuss the easy sequence. It is not to difficult to figure out how it works in the case of the partially easy sequence. The ith row of the matrix in Equation (9) corresponds with the binary representation of a_i^1.

$$(T_n \quad C_n \quad O_{n-1} \quad T_{n-1} \quad C_{n-1} \cdots$$
$$O_2 \quad T_2 \quad C_2 \quad O_1 \quad T_1 \quad C_1). \quad (9)$$

In Equation (9) the T_i are randomly chosen binary matrices, the C_i are $n \times 1$ binary column vectors such that they (C_i) are linearly independent modulo 2, and the O_i are $n \times l_i$ zero binary matrices, where $l_i \geq \log_2 n$. Let us call the locations of the C_i

t_i. To find $\mathbf{x}$ out of S_1, we first represent S_1 binary, and we call these bits s_h. As a consequence of the choice of l_i, the bits s_{t_i} are not influenced by T_{i-1} and C_{i-1}. To find $\mathbf{x}$ we have to solve modulo 2:

$$\begin{pmatrix} s_{t_1} \\ \vdots \\ s_{t_{n-1}} \\ s_{t_n} \end{pmatrix} = \begin{pmatrix} c_1^1 & \cdots & c_n^1 \\ \vdots & \ddots & \vdots \\ c_1^{n-1} & \cdots & c_n^{n-1} \\ c_1^n & \cdots & c_n^n \end{pmatrix} \cdot \begin{pmatrix} x_1 \\ \vdots \\ x_{n-1} \\ x_n \end{pmatrix} \bmod 2,$$

where the c_i^j are coefficients of C_i.

McAuley and Goodman [37] proposed in December 1982 a very similar knapsack scheme as the one proposed by Davio (see higher). The differences are that no Merkle–Hellman transformations are used and that the $\mathbf{x}$ can have more values than binary (they have to be smaller than a given value and larger than or equal to zero). The trapdoor information consists only in the secrecy of the primes which were used in the construction.

By the end of 1982 and the beginning of 1983 Desmedt, Vandewalle and Govaerts [19] generalized Shamir's ultimate knapsack scheme by generalizing the Merkle–Hellman transformation, calling it the *general knapsack scheme*. All previously discussed knapsack systems are special cases of this one [20]. In Shamir's scheme one can only choose one vector and start the transformation, while here n choices of vectors are necessary (or are done implicitly).

Around the same time Brickell cryptanalyzed *low density knapsacks* [4, 5]. A similar attack was independently found by Lagarias and Odlyzko [30]. To perform his attack, Brickell first generalized the Merkle–Hellman dominance condition. The integers he used may also be negative. Brickell called a *modular mapping* $*w \bmod m$ from $\mathbf{a}$ into $\mathbf{c}$ to have the *small sum* property if $c_i \equiv a_i w \bmod m$, and $m > \sum |c_i|$. He called mappings satisfying this property SSMM. Given $\sum x_i a_i$ one can easily calculate $\sum x_i c_i$. This is done exactly as in the reverse Merkle–Hellman case. If the result is greater than $\sum_{c_i > 0} c_i$ M is subtracted from it. He tries to find $n-1$ such transformations all starting from the public key $\mathbf{a}$. He can then solve a set of equations similar as in the ultimate scheme of Shamir (remark the difference in obtaining the matrix). To obtain such transformations in order to break, he uses the LLL algorithm choosing a special lattice. If all the *reduced lattice* basis vectors are short enough, he will succeed. This happens probably when the density is less than $1/\log_2 n$. In the other cases he uses some trick to transform the problem into one satisfying the previous condition. Arguments were given that this will succeed almost always when the density is less than 0.39. The low density attack proposed by Lagarias and Odlyzko is expected to work when the density of the knapsack is less than 0.645. These attacks break the ultimate scheme of Shamir, because the density of the public key is small as a consequence of construction method of the public key.

Lagarias found a good foundation for the attacks on the knapsack system, by discussing what he called *unusually good* simultaneous Diophantine approximations [31]. Lagarias used similar ideas [32] to analyze Shamir's attack on the basic Merkle–Hellman scheme. The main result is that Shamir overlooked some problems, but nevertheless his attack almost always works.

Brickell, Lagarias and Odlyzko performed an evaluation [6] of the Adleman's attack on multiple iterated Merkle–Hellman and Graham–Shamir schemes. They concluded that his attack on the basic Graham–Shamir scheme works, but that the version to break iterated Merkle–Hellman or Graham–Shamir scheme failed. The main reason for it was that the LLL algorithm found so-called undesired vectors, which could not be used to cryptanalyze the cited systems. Even in the case that only two transformations were applied (to construct the public key) his attack fails.

In 1983 Karnin proposed an improved time-memory-processor tradeoff [29] for the knapsack problem. The idea is related to <u>exhaustive key search</u> [21] and Hellman's <u>time-memory trade-off</u> [26], in which an exhaustive key search is used to break the system using straightforward or more advanced ideas. The result has only theoretical value if the dimension of the knapsack system n is large.

In 1984 Goodman and McAuley proposed a small modification [25] to their previous system [37]. In the new version some modulo transformation was applied.

In the same year Brickell proposed how to cryptanalyze [8] the iterated Merkle–Hellman and Graham–Shamir scheme. As usual no proof is provided that the breaking algorithm works; arguments for the heuristics are described in [8]. Several public keys were generated by the Merkle–Hellman and Graham–Shamir scheme and then turned out to be breakable by Brickell's attack. Again the LLL algorithm is the driving part of the attack. First the cryptanalyst picks out a subset of the sequence corresponding with the public key. These elements are entered in a special way in the LLL algorithm. A reduced basis for that lattice is obtained. Then one calculates the linear relation between the old and new basis for

the lattice. This last information will allow to decide whether the selected subset is "good." If it was not, one restarts at the beginning. If it was a good set, one can calculate the number of iterations that were used by the designer during the construction of the public key. Some calculation of determinants will then return an almost superincreasing sequence. Proceeding with almost superincreasing sequences was already discussed by Karnin and Hellman [28] (remarkable is the contradiction in the conclusion of their paper and its use by Brickell!).

In October 1984, Odlyzko found an effective method to cryptanalyze the McAuley–Goodman and the Goodman–McAuley scheme, using mainly <u>gcd</u>'s [41].

Later on Brickell [9] was able to break, with a similar idea as in [8], a lot of other knapsack schemes, e.g., the Desmedt–Vandewalle–Govaerts, the Davio, the Willett, the Petit and the Goodman–McAuley. The attack affects also the security of the so-called general knapsack scheme.

At Eurocrypt 85 Di Porto [22] presented two new knapsack schemes, which are very close to the Goodman–McAuley one. However they were broken during the same conference by Odlyzko.

The Case of Usual Knapsacks with other Encryption Functions

In 1979 Arazi proposed a new knapsack based additive knapsack algorithm to protect the privacy of the message [2]. The main difference with the Merkle–Hellman encryption is that random noise is used in the encryption function. The parameters which are chosen during the construction of the public key have to satisfy some properties (see [2]).

In 1983 Brickell also presented a new knapsack system [7], which is similar to the Arazi one.

One year later Brickell declared his own new scheme insecure, as a consequence of his attack on iterated knapsacks [8].

Chor and Rivest proposed in 1984 another knapsack based system [13]. The encryption process is very close to the one in the Merkle–Hellman scheme. The main difference in the encryption is that $\sum x_i \leq h$ for some given h. The trapdoor technique does not use a modular multiplication (as do almost all other knapsack schemes). The trapdoor uses the <u>discrete logarithm problem</u> [40, 43] (see also Section "The multiplicative knapsack scheme and its history"). A study of possible attacks was done, but it turned out that by a good choice of parameters all attacks known at that point of time could be avoided. New attacks were set up by the authors [13] but this did not change the above conclusion.

In 1985 Brickell broke the Arazi knapsack system [9]. In 1985 Cooper and Patterson [14] also proposed some new trapdoor knapsack algorithm, which can however be cryptanalyzed by Brickell [9]. The same attack of Brickell can break this knapsack as well as the Lagger knapsack [33].

Since 1985 interest in public key knapsacks almost vanished completely. In 1998 the Chor–Rivest scheme was finally cryptanalyzed [52].

The Multiplicative Knapsack Scheme and its History

The so called *multiplicative* knapsack here uses exactly the same encryption function as the Merkle–Hellman additive knapsack scheme. However the trapdoor is completely different in nature, because it is mainly based on a transformation from an additive knapsack problem into a *multiplicative* one. It was presented by Merkle and Hellman in their original paper [38].

Let us first explain the construction of the public key. One chooses n relative prime positive numbers $(p_1, p_2, \ldots, p_n)$, some underline{prime number} q, such that $q - 1$ has only small primes and such that

$$q > \prod_{i=1}^{n} p_i \tag{10}$$

and some primitive root b modulo q (see underline{modular arithmetic}). One then finds integers a_i, where $1 \le a_i \le q - 1$, such that $p_i \equiv b^{a_i} \bmod q$. So, a_i is the discrete logarithms of p_i base b modulo q. This explains why $q - 1$ was chosen as the product of small primes, since the Pohlig–Hellman algorithm (see underline{discrete logarithm problem}) can easily calculate these discrete logarithms in that case [43].

To decrypt one calculates $S' = b^S \bmod q$, because $b^S = b^{\sum x_i \cdot a_i} = \prod b^{x_i \cdot a_i} = \prod p_i^{x_i} \bmod q$. The last equality is a consequence of the condition in Equation (10). One can easily find the corresponding $\mathbf{x}$ starting from S', using the fact that the numbers p_i are relative prime. This last point is important, because in the general case the subset product problem is NP-complete [24].

This scheme can be cryptanalyzed by a low density attack [5, 30]. However the disadvantage is that it requires a separate run of the lattice reduction algorithm (which takes at least on the order of n^4 operations) to attack each n bit message. To overcome that problem, Odlyzko tried another attack [39]. Herein he starts from the assumption that some of the p_i are known. He then tries to find q and b. He also assumes that b, q and the a_i consist of approximately m bits. His attack will take a polynomial time if $m = O(n \log n)$. Also in this attack the LLL algorithm is the driving force. A special choice [39] of the lattice is used to attack the system. Once the b and q are found the cryptanalyst can easily cryptanalyze ciphertexts as the receiver can decrypt them.

The Trapdoor Knapsack Schemes to Protect Signatures and Authenticity

To make a underline{digital signature} the sender applies the decryption function on the plaintext. From this point of view it is easy to understand that the higher discussed knapsack schemes are not well suited for this purpose. Indeed if the decryption function is not "enough" (pseudo) invertible the sender has to perform other trials in order to generate a signature. Such a scheme was presented in the original Merkle–Hellman paper [38]. Shamir suggested a more practical one [46] in 1978.

In 1982 Schöbi and Massey proposed another version of [45] as a fast signature scheme.

In 1982–1983 Odlyzko broke [39] the Shamir's fast signature and the Schöbi-Massey one. Here also the LLL algorithm plays an important role.

SOME DETAILS: A small encyclopedia is required to discuss all schemes, weaknesses and attacks in details. Only three issues are discussed in more depth, these being: (i) why the Merkle–Hellman transformation leads to the possibility of more than one decryption key to break, (ii) the LLL algorithm and (iii) its use in the low density attack of Brickell.

The Existence of Infinitely Many Decryption Keys

Let us focus on the basic Merkle–Hellman scheme. Suppose w^{-1} and m correspond with the reverse Merkle–Hellman transformation and that a_i' was the used superincreasing sequence. We will demonstrate that other values allow to break (call these V, M, and a_i''). In order to analyze for which V and M Equations (3)–(5) (see also Equation (6)) holds let us reformulate the Merkle–Hellman transformation in terms of *linear inequalities*. $a_i'' \equiv a_i \cdot V \bmod M$ and $0 < a_i'' < M$ can be reformulated into:

$$0 < a_i'' = (a_i \cdot V - s_i \cdot M) < M, \quad s_i \text{ integer.} \tag{11}$$

Note that s_i is equal to $\lfloor (a_i \cdot V)/M \rfloor$ with $\lfloor \cdot \rfloor$ the floor function.

Using Equation (11) the conditions in Equations (3)–(5) (see also Equation (6)) and the condition of superincreasing of $\mathbf{a}''$ can be expressed as *linear inequalities* on V/M:

Equation (11) gives: $\quad \dfrac{s_i}{a_i} < \dfrac{V}{M} < \dfrac{1+s_i}{a_i} \le 1$

$$(12)$$

Equation (3) gives: $\quad \dfrac{V}{M} < \dfrac{1+\sum_{i=1}^{n} s_i}{\sum_{i=1}^{n} a_i} \quad (13)$

the condition requiring $\mathbf{a}''$ be superincreasing gives for all j, with $2 \le j \le n$:

$$\text{if } a_j < \sum_{i=1}^{j-1} a_i: \quad \frac{V}{M} < \frac{s_j - \sum_{i=1}^{j-1} s_i}{a_j - \sum_{i=1}^{j-1} a_i} \quad (14)$$

$$\text{if } a_j > \sum_{i=1}^{j-1} a_i: \quad \frac{V}{M} > \frac{s_j - \sum_{i=1}^{j-1} s_i}{a_j - \sum_{i=1}^{j-1} a_i}. \quad (15)$$

Observe that the condition in Equation (4) does not impose an extra condition on the ratio V/M. Indeed, for any V/M which satisfies the conditions in Equations (12)–(14) one can take coprime V, M in order to satisfy Equation (4).

THEOREM 1. *For each encryption key $(a_1, a_2, \ldots, a_n)$ constructed using Equations (3)–(5) from a superincreasing sequence $(a_1', a_2', \ldots, a_n')$, there exist infinitely many superincreasing sequences satisfying the conditions in Equations (3)–(5).*

PROOF. The conditions Equations (3) and (5) and superincreasing reformulated as Equations (12)–(15) can be summarized as: $L < \frac{V}{M} < U$, where L and U are rational numbers. Since there exists a superincreasing decryption key, which satisfies Equations (3)–(5) there exists an L and U such that $L < U$. So, infinitely many (V, M) satisfy the bound conditions and the condition that $\gcd(V, M) = 1$. $\qquad\square$

It is easy to generalize Theorem 1 to knapsack sequences $\mathbf{a}$ obtained by multiple Merkle–Hellman transformations [16, 17, 20, 23].

The LLL Algorithm

First we define a lattice (in the geometrical sense of the word).

DEFINITION 3. *Let $(\mathbf{v}^1, \ldots, \mathbf{v}^n)$ be a linearly independent set of real vectors in a n-dimensional real Euclidean space. The set $\{u_1\mathbf{v}^1 + \cdots + u_n\mathbf{v}^n \mid u_1, \ldots, u_n \in Z\}$, is called the lattice with basis $(\mathbf{v}^1, \ldots, \mathbf{v}^n)$.*

THEOREM 2. *Let $(\mathbf{v}^1, \ldots, \mathbf{v}^n)$ be a basis of a lattice L and let $\mathbf{v}'^i$ be the points*

$$\mathbf{v}'^i = \sum_j z_j^i \mathbf{v}^j, \quad \text{for } 1 \le i \le n$$

$$\text{and} \quad 1 \le j \le n,$$

where z_j^i are integers, then the set $(\mathbf{v}'^1, \ldots, \mathbf{v}'^n)$ is also a base for the same lattice L, if and only if $\det(z_j^i) = \pm 1$. An integer matrix Z with $\det(z_j^i) = \pm 1$ is called an unimodular matrix.

PROOF. See [12]. $\qquad\square$

As a consequence of Theorem 2 $|\det(\mathbf{v}^1, \ldots, \mathbf{v}^n)|$ is independent of a particular basis for the lattice.

For a lattice L there does not necessarily exist a set of n vectors that form an orthogonal basis for the lattice. The Lenstra Lenstra Lovasz (LLL, see shortest vector problem or [35, pp. 515–525]) algorithm finds in polynomial time a basis for a lattice L, which is nearly orthogonal with respect to a certain measure of non-orthogonality. The LLL algorithm does however not find in general the most orthogonal set of n independent vectors. As a consequence of Theorem 2 it finds short (probably not the shortest) vectors. A basis is called *reduced* if it contains relatively short vectors.

Let us briefly describe LLL. Let $\mathbf{v}_1, \mathbf{v}_2, \ldots, \mathbf{v}_n$ belong to the n-dimensional real vector space. To initialize the algorithm an orthogonal real basis $\mathbf{v}_i'$ is calculated, together with μ_j^i ($1 \le j < i \le n$), such that

$$\mathbf{v}_i' = \mathbf{v}_i - \sum_{j=1}^{i-1} \mu_j^i \mathbf{v}_j' \quad (16)$$

$$\mu_j^i = \frac{(\mathbf{v}_i, \mathbf{v}_j')}{(\mathbf{v}_j', \mathbf{v}_j')}, \quad (17)$$

where $(\cdot, \cdot)$ denotes the ordinary inner (scalar) product. In the course of the algorithm the vectors $\mathbf{v}_1, \mathbf{v}_2, \ldots, \mathbf{v}_n$ will be changed several times, but will always remain a basis for L. After every *change* the $\mathbf{v}_i'$ and μ_j^i are *updated* using Equations (16) and (17). A current subscript k is used during the algorithm. LLL starts with $k = 2$. If $k = n + 1$ it terminates. Suppose now $k \le n$, then we first check that $|\mu_{k-1}^k| \le 1/2$ if $k > 1$. If this does not hold, let r be the integer nearest to μ_{k-1}^k, and *replace* $\mathbf{v}_k$ by $\mathbf{v}_k - r\mathbf{v}_{k-1}$, (do not forget the update). Next we distinguish *two cases*. Suppose that $k \ge 2$ and $|\mathbf{v}_k' + \mu_{k-1}^k \mathbf{v}_{k-1}'|^2 < (3/4)|\mathbf{v}_{k-1}'|^2$, then we *interchange* $\mathbf{v}_{k-1}$ and $\mathbf{v}_k$, (do not forget the update), afterwards replace k by $k - 1$ and restart. In the *other case* we want to achieve that

$$|\mu_j^k| \le \tfrac{1}{2}, \quad \text{for } 1 \le j \le k - 1. \tag{18}$$

If the condition in Equation (18) does not hold, then let l be the largest index $< k$ with $\mu_l^k > 1/2$, let r be the nearest to μ_l^k and *replace* $\mathbf{b}_k$ by $\mathbf{b}_k - r\mathbf{b}_l$ (do not forget the update), repeat until the conditions Equation (18) hold, afterwards replace k by $k + 1$ and restart. Remark that if the case $k = 1$ appears one replaces it by $k = 2$.

The Use of the LLL Algorithm in Brickell's Low Dense Attack

In Section "The trials to avoid weaknesses and attacks for the class of usual knapsacks" we briefly discussed *Brickell's low dense attack*. We introduced the concept of *SSMM* and have given a sketch of Brickell's low density attack. Remember also that if the density is not low enough ($> 1/\log_2 n$) it has to be artificially lowered. We will only discuss the case that it is indeed low enough. This last part is always used as the main technique of the breaking algorithm.

The breaking is based on Theorem 3. Hereto we first have to define *short enough vector*.

DEFINITION 4. *A vector* $\mathbf{c}$ *in a lattice* L *is called short enough related to* a_1 *if*

$$\sum_{i=2}^{n} |c_i'| < a_1,$$

where $c_1' = 0$ *and* $c_i' = c_i/n$ *for* $2 \le i \le n$.

THEOREM 3. *If all vectors in the reduced basis for the lattice, with basis vectors* $\mathbf{t}^i$ *defined in Equation* (19), *are short enough related to* a_1, *then we can find* $n - 1$ *independent SSMM for* $a_1, \ldots, a_n$.

$$
\begin{aligned}
\mathbf{t}^1 &= (1 & na_2 & & na_3 & & na_4 & & \ldots & & na_n) \\
\mathbf{t}^2 &= (0 & na_1 & & 0 & & 0 & & \ldots & & 0) \\
\mathbf{t}^3 &= (0 & 0 & & na_1 & & 0 & & \ldots & & 0) \\
\mathbf{t}^4 &= (0 & 0 & & 0 & & na_1 & & \ldots & & 0) \\
&\;\;\vdots & \vdots & & \vdots & & \vdots & & \ddots & & \vdots \\
\mathbf{t}^n &= (0 & 0 & & 0 & & 0 & & \ldots & & na_1).
\end{aligned}
\tag{19}
$$

PROOF. Call the vectors of the reduced basis $\mathbf{v}^1, \mathbf{v}^2, \ldots, \mathbf{v}^n$. We will first prove that a modular mapping by $v_1^j \bmod a_1$ has the small sum property (see Section "The trials to avoid weaknesses and attacks for the class of usual knapsacks"). Since $\mathbf{v}^j$ is an integral linear combination of the vectors in Equation (19), there exist inte-

gers $(y_1^j, \ldots, y_n^j)$ such that $v_1^j = y_1^j$ and $v_i^j = y_i^j na_1 + y_1^j na_i$ for $2 \le i \le n$. Since n divides v_i^j let $u_i^j = v_i^j/n$ for $2 \le i \le n$. This implies evidently that $0 \equiv a_1 y_1^j$ and $u_i^j \equiv a_i y_1^j$ for $2 \le i \le n$. As a consequence of the short enough property we have indeed the small sum property. The independence of the $n - 1$ vectors so obtained with SSMM, is then easy to prove. $\square$

Arguments are given in [5] that the condition in Theorem 3 are almost satisfied if the density is low enough.

CONCLUSION: The encryption in the Merkle–Hellman knapsack is based on NP-completeness, however, its trapdoor was not. In secure public key cryptosystems the encryption process must be hard to invert but it must also be hard to find the original trapdoor *or another trapdoor*. Nonpublic key use of knapsack was investigated, e.g. in [18, 44].

For further details on the research on public key knapsack before 1992, consult [11].

Yvo Desmedt

References

[1] Adleman, L.M. (1983). "On breaking the iterated Merkle–Hellman public-key cryptosystem." *Advances in Cryptology—CRYPTO'82*, Lecture Notes in Computer Science, eds. D. Chaum, R.L. Rivest, and A.T. Sherman, Santa Barbara, CA, August 23–25. 1982, Plenum, New York, 303–308. More details appeared in "On breaking generalized knapsack public key cryptosystems." TR-83-207, Computer Science Dept., University of Southern California, Los Angeles, March 1983.

[2] Arazi, B. (1980). "A trapdoor multiple mapping." *IEEE Trans. Inform. Theory*, 26 (1), 100–102.

[3] Brickell, E.F., J.A. Davis, and G.J. Simmons (1983). "A preliminary report on the cryptanalysis of the Merkle–Hellman knapsack cryptosystems." *Advances in Cryptology—CRYPTO'82*, Lecture Notes in Computer Science, eds. D. Chaum, R.L. Rivest, and A.T. Sherman, Santa Barbara, CA, August 23–25, 1982. Plenum, New York, 289–301.

[4] Brickell, E.F. (1983). "Solving low density knapsacks in polynomial time." *IEEE Intern. Symp. Inform. Theory*, St. Jovite, Quebec, Canada, September 26–30, 1983. Abstract of papers, 129–130.

[5] Brickell, E.F. (1984). "Solving low density knapsacks." *Advances in Cryptology—CRYPTO'83*, Lecture Notes in Computer Science, ed. D. Chaum,

Santa Barbara, CA, August 21–24, 1983. Plenum, New York, 25–37.

[6] Brickell, E.F., J.C. Lagarias, and A.M. Odlyzko (1984). "Evaluation of the Adleman attack on multiple iterated Knapsack cryptosystems." *Advances in Cryptology—CRYPTO'83*, Lecture Notes in Computer Science, ed. D. Chaum, Santa Barbara, CA, August 21–24, 1983. Plenum, New York, 39–42.

[7] Brickell, E.F. (1983). "A new knapsack based cryptosystem." Presented at *CRYPTO'83*, Santa Barbara, CA, August 21–24, 1983.

[8] Brickell, E.F. (1985). "Breaking iterated knapsacks." *Advances in Cryptology—CRYPTO'84*, Lecture Notes in Computer Science, vol. 196, eds. G.R. Blakley and D. Chaum, Santa Barbara, CA, August 19–22, 1984. Springer-Verlag, Berlin, 342–358.

[9] Brickell, E.F. (1985). "Attacks on generalized knapsack schemes." Presented at *EUROCRYPT'85*, Linz, Austria, April 9–11, 1985.

[10] Brickell, E. and A.M. Odlyzko (1988). "Cryptanalysis: A survey of recent results." *Proc. IEEE*, 76 (5), 578–593.

[11] Brickell, E.F. and A.M. Odlyzko (1992). "Cryptanalysis: A survey of recent results." *Contemporary Cryptology*, ed. G.J. Simmons. IEEE Press, New York, 501–540.

[12] Cassels, J.W.S. (1971). *An Introduction to the Geometry of Numbers*. Springer-Verlag, Berlin.

[13] Chor, B. and R.L. Rivest (1985). "A knapsack type public key cryptosystem based on arithmetic in finite fields." *Advances in Cryptology—CRYPTO'84*, Lecture Notes in Computer Science, vol. 196, eds. G.R. Blakley and D. Chaum, Santa Barbara, CA, August 19–22, 1984. Springer-Verlag, Berlin, 54–65.

[14] Cooper, R.H. and W. Patterson (1985). "Eliminating data expansion in the Chor-Rivest algorithm." Presented at *EUROCRYPT'85*, Linz, Austria, April 9–11, 1985.

[15] Davio, M. (1983). "Knapsack trapdoor functions: An introduction." Proceedings of *CISM Summer School on: Secure Digital Communications*, ed. J.P. Longo, CISM Udine, Italy, June 7–11, 1982. Springer-Verlag, Berlin, 41–51.

[16] Desmedt, Y., J. Vandewalle, and R. Govaerts (1982). "The use of knapsacks in cryptography public key systems (Critical analysis of the security of Knapsack Public Key Algorithms)." Presented at Groupe de Contact Recherche Operationelle du F.N.R. S., Mons, Belgium, February 26, 1982. Appeared in *Fonds National de la Rechereche Scientifique, Groupes de Contact, Sciences Mathématiques*.

[17] Desmedt, Y.G., J.P. Vandewalle, and R.J.M. Govaerts (1984). "A critical analysis of the security of knapsack public key algorithms." *IEEE Trans. Inform. Theory*, IT-30 (4), 601–611. Also presented at *IEEE Intern. Symp. Inform. Theory*,

Les Arcs, France, June 1982, Abstract of papers, 115–116.

[18] Desmedt, Y., J. Vandewalle, and R. Govaerts (1982). "A highly secure cryptographic algorithm for high speed transmission." *GLOBECOM'82*, IEEE, Miami, FL, November 29–December 2, 1982, 180–184.

[19] Desmedt, Y., J. Vandewalle, and R. Govaerts (1983). "Linear algebra and extended mappings generalise public key cryptographic knapsack algorithms." *Electronics Letters*, 19 (10), 379–381.

[20] Desmedt, Y. (1984). "Analysis of the security and new algorithms for modern industrial cryptography." *Doctoral Dissertation*, Katholieke Universiteit Leuven, Belgium.

[21] Diffie, W. and M.E. Hellman (1977). "Exhaustive cryptanalysis of the NBS data encryption standard." *Computer*, 10 (6), 74–84.

[22] Di Porto, A. (1985). "A public key cryptosystem based on a generalization of the knapsack problem." Presented at *EUROCRYPT'85*, Linz, Austria, April 9–11, 1985.

[23] Eier, R. and H. Lagger (1983). "Trapdoors in knapsack cryptosystems." *Cryptography. Proc. Burg Feuerstein 1982*, Lecture Notes in Computer Science, vol. 149, ed. T. Beth. Springer-Verlag, Berlin, 316–322.

[24] Garey, M.R. and D.S. Johnson (1979). *Computers and Intractability: A Guide to the Theory of NP—Completeness*. W.H. Freeman, San Francisco, CA.

[25] Goodman, R.M. and A.J. McAuley (1985). "A new trapdoor knapsack public key cryptosystem." *Advances in Cryptology—EUROCRYPT'84*, Lecture Notes in Computer Science, vol. 209, eds. T. Beth, N. Cot and I. Ingemarsson, Paris, France, April 9–11, 1984. Springer-Verlag, Berlin, 150–158.

[26] Hellman, M.E. (1980). "A cryptanalytic time-memory trade-off." *IEEE Trans. Inform. Theory*, IT-26 (4), 401–406.

[27] Henry, P.S. (1981). "Fast decryption algorithm for the knapsack cryptographic system." *Bell Syst. Tech. J.*, 60 (5), 767–773.

[28] Karnin, E.D. and M.E. Hellman (1983). "The largest super-increasing subset of a random set." *IEEE Trans. Inform. Theory*, IT-29 (1), 146–148. Also presented at *IEEE Intern. Symp. Inform. Theory*, Les Arcs, France, June 1982, Abstract of papers, 113.

[29] Karnin, E.D. (1984). "A parallel algorithm for the knapsack problem." *IEEE Trans. on Computers*, C-33 (5), 404–408. Also presented at *IEEE Intern. Symp. Inform. Theory*, St. Jovite, Quebec, Canada, September 26–30, 1983, Abstract of papers, 130–131.

[30] Lagarias, J.C. and A.M. Odlyzko (1983). "Solving low density subset sum problems." *Proc. 24th Annual IEEE Symposium on Foundations of Computer Science*, 1–10.

[31] Lagarias, J.C. (1984). "Knapsack public key cryptosystems and diophantine approximation." *Advances in Cryptology—CRYPTO'83*, Lecture Notes in Computer Science, ed. D. Chaum, Santa Barbara, CA, August 21–24, 1983. Plenum, New York, 3–23.

[32] Lagarias, J.C. (1984). "Performance analysis of Shamir's attack on the basic Merkle–Hellman knapsack cryptosystem." *Proc. 11th Intern. Colloquium on Automata, Languages and Programming (ICALP)*, Lecture Notes in Computer Science, vol. 172, ed. J. Paredaens. Antwerp, Belgium, July 16–20, 1984. Springer-Verlag, Berlin.

[33] Lagger, H. "Public key algorithm based on knapsack systems." *Dissertation*, Technical University Vienna, Austria (in German).

[34] Lenstra, Jr., H.W., (1981). "Integer programming with a fixed number of variables." Technical Report, University of Amsterdam, Dept. of Mathematics, 81–03.

[35] Lenstra, A.K., H.W. Lenstra, Jr., and L. Lovasz (1982). "Factoring polynomials with rational coefficients." *Mathematische Annalen*, 261, 515–534.

[36] Lenstra, Jr., H.W., (1983). "Integer programming with a fixed number of variables." *Math. Oper. Res.*, 8 (4), 538–548.

[37] McAuley, A.J. and R.M. Goodman (1983). "Modifications to the Trapdoor–Knapsack public key cryptosystem." *IEEE Intern. Symp. Inform. Theory*, St. Jovite, Quebec, Canada, September 26–30, 1983. Abstract of papers, 130.

[38] Merkle, R.C. and M.E. Hellman (1978). "Hiding information and signatures in trapdoor knapsacks." *IEEE Trans. Inform. Theory*, 24 (5), 525–530.

[39] Odlyzko, A.M. (1984). "Cryptanalytic attacks on the multiplicative knapsack cryptosystem and on Shamir's fast signature system." *IEEE Trans. Inform. Theory*, IT-30 (4), 594–601. Also presented at *IEEE Intern. Symp. Inform. Theory*, St. Jovite, Quebec, Canada, September 26–30, 1983. Abstract of papers, 129.

[40] Odlyzko, A.M. (1985). "Discrete logarithms in finite fields and their cryptographic significance." *Advances in Cryptology—EUROCRYPT'84*, Lecture Notes in Computer Science, vol. 209, eds. T. Beth, N. Cot and I. Ingemarsson, Paris, France, April 9–11, 1984. Springer-Verlag, Berlin, 225–314.

[41] Odlyzko, A.M. Personal communication.

[42] Petit, M. "Etude mathématique de certains systèmes de chiffrement: les sacs à dos" (Mathematical study of some enciphering systems: The knapsack, in French). *Doctorates Thesis*, Université de Rennes, France.

[43] Pohlig, S.C. and M.E. Hellman (1978). "An improved algorithm for computing logarithms over GF(p) and its cryptographic significance." *IEEE Trans. Inform. Theory*, 24 (1), 106–110.

[44] Schaumuller-Bichl, I. (1982). "On the design and analysis of new cipher systems related to the DES." *IEEE Intern. Symp. Inform. Theory*, Les Arcs, France, 115.

[45] Schöbi, P. and J.L. Massey (1983). "Fast authentication in a trapdoor knapsack public key cryptosystem." *Cryptography, Proc. Burg Feuerstein 1982*, Lecture Notes in Computer Science, vol. 149, ed. T. Beth. Springer-Verlag, Berlin, 289–306. See also *Proc. Int. Symp. Inform. Theory*, Les Arcs, June 1982, 116.

[46] Shamir, A. (1978). "A fast signature scheme." *Internal Report*. MIT, Laboratory for Computer Science Report RM-107, Cambridge, MA.

[47] Shamir, A. (1979). "On the cryptocomplexity of knapsack systems." *Proc. Stoc 11 ACM*, 118–129.

[48] Shamir, A. and R. Zippel (1980). "On the security of the Merkle–Hellman cryptographic scheme." *IEEE Trans. Inform. Theory*, 26 (3), 339–340.

[49] Shamir, A. (1983). "A polynomial time algorithm for breaking the basic Merkle–Hellman cryptosystem." *Advances in Cryptology—CRYPTO'82*, Lecture Notes in Computer Science, eds. D. Chaum, R.L. Rivest, and A.T. Sherman, Santa Barbara, CA, August 23–25, 1982. Plenum, New York, 279–288.

[50] Shamir, A. (1982). "The strongest knapsack-based cryptosystem." Presented at *CRYPTO'82*, Santa Barbara, CA, August 23–25, 1982.

[51] Shamir, A. (1984). "A polynomial time algorithm for breaking the basic Merkle–Hellman cryptosystem." *IEEE Trans. Inform. Theory*, IT-30 (5), 699–704.

[52] Vaudenay, S. (1998). "Cryptanalysis of the chorrivest cryptosystem." *Advances in Cryptology—CRYPTO'98*, Lecture Notes in Computer Science, vol. 1462, ed. II. Krawczyk, Santa Barbara, CA, August 23–27, 1998. Springer-Verlag, Berlin, 243–256.

[53] Willett, M. (1983). "Trapdoor knapsacks without superincreasing structure." *Inform. Process. Letters*, 17, 7–11.

KNOWN PLAINTEXT ATTACK

Known plaintext attack is a scenario in which the attacker has access to pairs $(P_i, C_i), i = 1, \ldots, N$ of known plaintexts and their corresponding ciphertexts. This attack is considered to be highly practical, especially if the amount of pairs N is not too large. This attack scenario is more practical than the *chosen* plaintext attack. *Probable word* method which is a popular technique for solving classical *simple substitution* or *transposition* ciphers is an example of a known-plaintext attack. Another example is the cryptanalysis of the German Enigma cipher (see cryptomachines or [1]) using the so called *bombs*. It relied heavily on properly guessed opening words of the cryptograms (which were at the time called *cribs*). One of the most popular cribs was "Nothing to report". In modern cryptography linear cryptanalysis is a typical example of a known plaintext attack.

Alex Biryukov

Reference

[1] Deavours, C.A. and L. Kruh (1985). *Machine Cryptography and Modern Cryptanalysis*. Artech House, Boston, MA.

L

LATTICE

In mathematics, the term *lattice* is used for two very different kinds of mathematical objects, arising respectively in order theory and number theory. Here, by lattice, we always mean number-theoretical lattices. Lattice theory [3, 7] is called *Geometry of numbers*, a name due to its founder Hermann Minkowski [5].

A lattice can be defined in many equivalent ways. Informally speaking, a lattice is a regular arrangement of points in n-dimensional space. To be more formal, we need to recall a few definitions. Let $\mathbf{x}, \mathbf{y} \in \mathbb{R}^n$ denote two vectors $(x_1, \ldots, x_n)$ and $(y_1, \ldots, y_n)$, where the x_is and y_is are real numbers. Let $\langle \mathbf{x}, \mathbf{y} \rangle$ denote the Euclidean inner product of $\mathbf{x}$ with $\mathbf{y}$: $\langle \mathbf{x}, \mathbf{y} \rangle = \sum_{i=1}^{n} x_i y_i$. Let $\|\mathbf{x}\|$ denote the Euclidean norm of $\mathbf{x}$: $\|\mathbf{x}\| = \langle \mathbf{x}, \mathbf{x} \rangle^{1/2}$. A set of vectors $\{\mathbf{b}_1, \ldots, \mathbf{b}_d\}$ are said to be $\mathbb{R}$-linearly independent if and only if any equality of the form $\mu_1 \mathbf{b}_1 + \cdots + \mu_d \mathbf{b}_d = 0$, where the μ_is are real numbers, implies that the μ_is are all zero. Then the two most usual definitions of a lattice are the following ones:

- A lattice is a discrete (additive) <u>subgroup</u> of $\mathbb{R}^n$, that is, a non-empty subset $L \subseteq \mathbb{R}^n$ such that $\mathbf{x} - \mathbf{y} \in L$ whenever $(\mathbf{x}, \mathbf{y}) \in L^2$ (this is the group axiom), and where there exists a real $\rho > 0$ such that the simultaneous conditions $\mathbf{x} \in L$ and $\|\mathbf{x}\| \leq \rho$ imply that $\mathbf{x}$ be zero. With this definition, it is obvious that $\mathbb{Z}^n$ is a lattice (the <u>group</u> axiom is satisfied, and $\rho = 1/2$ works), and that any subgroup of a lattice is a lattice.
- A lattice is the set of all integer linear combinations of some set of $\mathbb{R}$-linearly independent vectors of $\mathbb{R}^n$, that is: if $\mathbf{b}_1, \ldots, \mathbf{b}_d$ are linearly independent, then $L = \left\{ \sum_{i=1}^{d} n_i \mathbf{b}_i \mid n_i \in \mathbb{Z} \right\}$ is a lattice, and $[\mathbf{b}_1, \ldots, \mathbf{b}_d]$ is said to be a *basis* of L. With this definition, it is still obvious that $\mathbb{Z}^n$ is a lattice, but it is not clear that a subgroup of a lattice is still a lattice.

It is not difficult to prove that the above definitions are in fact equivalent (see [7]). To decide at first sight whether or not a given subset L of $\mathbb{R}^n$ is a lattice, the second definition is useful only when one already knows a potential basis, which is not necessary with the first definition.

Both definitions suggest that lattices are discrete analogues of vector spaces: as a result, lattice theory bears much resemblance to linear algebra.

Lattice bases are not unique, but they all have the same number of elements, called the *dimension* or the *rank* of the lattice. Any lattice L of rank ≥ 2 has infinitely many bases. Indeed, one can see that to transform a lattice basis into another lattice basis, it is necessary and sufficient to apply a unimodular transformation, that is, a linear transformation represented by an integer matrix with determinant ± 1. This implies that the d-dimensional volume of the parallelepiped spanned by a lattice basis only depends on the lattice, and not on the choice of the basis: it is called the *volume* or *determinant* of the lattice, denoted by $\text{vol}(L)$ or $\det(L)$. By definition, it is equal to the square root of the following $d \times d$ determinant, where $(\mathbf{b}_1, \ldots, \mathbf{b}_d)$ is an arbitrary basis of L:

$$
\det(\langle \mathbf{b}_i, \mathbf{b}_j \rangle)_{1 \leq i, j \leq d}
$$
$$
= \begin{pmatrix} \langle \mathbf{b}_1, \mathbf{b}_1 \rangle & \langle \mathbf{b}_1, \mathbf{b}_2 \rangle & \ldots & \langle \mathbf{b}_1, \mathbf{b}_d \rangle \\ \langle \mathbf{b}_2, \mathbf{b}_1 \rangle & \langle \mathbf{b}_2, \mathbf{b}_2 \rangle & \ldots & \langle \mathbf{b}_2, \mathbf{b}_d \rangle \\ \vdots & & \ddots & \vdots \\ \langle \mathbf{b}_d, \mathbf{b}_1 \rangle & \langle \mathbf{b}_d, \mathbf{b}_2 \rangle & \ldots & \langle \mathbf{b}_d, \mathbf{b}_d \rangle \end{pmatrix}.
$$

The volume is useful to estimate the norm of lattice short vectors. See the entries <u>lattice reduction</u> and <u>Shortest Vector Problem</u>.

Lattices have many applications in computer science (see [2]), notably in cryptology (see [6]). See also <u>lattice based cryptography</u>, <u>lattice reduction</u>, <u>Closest Vector Problem</u>, <u>Shortest Vector Problem</u>. A classical mathematical reference about lattices is the book [1]. For an introduction to lattices, specifically from a computational point of view, the reader is referred to [4], and the references therein.

Phong Q. Nguyen

References

[1] Cassels, J.W.S. (1971). *An Introduction to the Geometry of Numbers*. Springer-Verlag, Berlin.

[2] Grötschel, M., L. Lovász, and A. Schrijver (1993). *Geometric Algorithms and Combinatorial Optimization*. Springer-Verlag, Berlin.

[3] Gruber, M. and C.G. Lekkerkerker (1987). *Geometry of Numbers*. North-Holland, Amsterdam.

[4] Micciancio, D. and S. Goldwasser (2002). *Complexity of Lattice Problems: A Cryptographic Perspective*, The Kluwer International Series in Engineering and Computer Science, vol. 671, ed. Robert Gallager. Kluwer Academic Publishers, Boston.

[5] Minkowski, H. (1896). *Geometrie der Zahlen.* Teubner-Verlag, Leipzig.

[6] Nguyen, P.Q. and J. Stern (2001). The two faces of lattices in cryptology. *Cryptography and Lattices— Proceedings of CALC 2001*, Lecture Notes in Computer Science, vol. 2146, ed. J.H. Silverman. Springer-Verlag, Berlin, 146–180.

[7] Siegel, C.L. (1989). *Lectures on the Geometry of Numbers.* Springer-Verlag, Berlin.

LATTICE REDUCTION

Among all the bases of a <u>lattice</u>, some are more useful than others. The goal of *lattice reduction* (also known as *lattice basis reduction*) is to find interesting bases. From a mathematical point of view, one is interested in proving the existence of at least one basis (in an arbitrary lattice) satisfying strong properties. From a computational point of view, one is rather interested in computing such bases in a reasonable time, given an arbitrary basis. In practice, one often has to settle for a tradeoff between the quality of the basis and the running time.

Interesting lattice bases are called *reduced*, but there are many different notions of reduction, such as those of *Minkowski, Hermite–Korkine–Zolotarev, Lenstra–Lenstra–Lovász*, etc. Typically, a reduced basis is made of vectors which are in some sense short, and which are somehow orthogonal. To explain what we mean by short, we need to introduce the so-called *successive minima* of a lattice.

The intersection of a d-dimensional lattice $L \subseteq \mathbb{R}^n$ with any bounded subset of $\mathbb{R}^n$ is always finite. It follows that there is a shortest nonzero vector in L, that is, there is $\mathbf{v} \in L \setminus \{0\}$ such that $\|\mathbf{u}\| \geq \|\mathbf{v}\|$ for all $\mathbf{u} \in L \setminus \{0\}$. Such a vector is not unique (for instance, $-\mathbf{v}$ satisfies the same property), but all such vectors must have the same norm. The first minimum of L is thus defined as $\lambda_1(L) = \|\mathbf{v}\|$. One might wonder how one could define a second-to-shortest vector: If $\mathbf{v}$ is a shortest vector, $-\mathbf{v}$ is also short but is not very interesting. To avoid such problems, one defines the successive minima as follows. For any integer k such that $1 \leq k \leq d$, the kth successive minimum of L, denoted by $\lambda_k(L)$, is the radius of the smallest hyperball centered at the origin and containing at least k linearly independent vectors of L. The successive minima can be defined with respect to any norm, but the Euclidean norm is the most common.

One can show that there are linearly independent vectors $\mathbf{v}_1, \ldots, \mathbf{v}_d$ in L such that $\|\mathbf{v}_i\| = \lambda_i(L)$

for all $1 \leq i \leq d$. Surprisingly, as soon as $d \geq 4$, such vectors may not form a basis of L: the integer linear combinations of the $\mathbf{v}_i$s may span a strict subset of L. Furthermore, as soon as $d \geq 5$, there may not exist a basis reaching all the minima: there exist d-dimensional lattices such that for all bases $(\mathbf{b}_1, \ldots, \mathbf{b}_d)$, $\|\mathbf{b}_i\| \neq \lambda_i(L)$ for at least some i. This is one of the reasons why there is no definition of reduction which is obviously better than all the others: for instance, one basis may minimize the maximum of the vector norms, while another minimizes the product of the norms, and it is not clear if one is better than the other. When we say that a reduced basis should have relatively short vectors, we mean that the ith vector of the vector is not far away from the ith minimum $\lambda_i(L)$. The orthogonality of a basis is often measured by the product of the norms of the vectors divided by the volume of the lattice: this ratio is always ≥ 1, with equality if and only if the basis is orthogonal.

Minkowski's theorem states that for all $1 \leq k \leq d$, the geometric mean of the first k minima $(\prod_{i=1}^{k} \lambda_i(L))^{1/k}$ is at most $\gamma_d \mathrm{vol}(L)^{1/d}$, where γ_d is an absolute constant (approximately equal to $\sqrt{d}$) that depends only on the dimension d, and $\mathrm{vol}(L)$ is the volume of the lattice (see the entry <u>lattice</u> for a definition). In particular, $\lambda_1(L) \leq \gamma_d \mathrm{vol}(L)^{1/d}$, but this upper bound does not hold in general for the other minima (one can easily construct lattices such that the first minimum is arbitrarily small, while the other minima are large). In a typical lattice though, we expect all the minima to be very roughly equal to $\gamma_d \mathrm{vol}(L)^{1/d}$.

Minkowski, Hermite, Korkine and Zolotarev introduced strong notions of reduction: the corresponding reduced bases have very good properties but are very difficult to obtain. For instance, bases reduced in the sense of Minkowski or of Hermite–Korkine–Zolotarev both include a shortest lattice vector, therefore finding such bases is already an NP-hard problem under randomized reductions as the lattice dimension increases (see <u>shortest vector problem</u>). Lenstra et al. [5] introduced the first notion of reduction to be interesting from both a mathematical and a computational point of view, in the sense that such reduced bases are provably made of relatively short vectors (although maybe not as short as, say, a Minkowski-reduced basis) and can be computed efficiently. More precisely, the celebrated LLL or L^3 algorithm, given as input an arbitrary basis of a d-dimensional lattice L in $\mathbb{Q}^n$, outputs (in time polynomial in the size of the basis—see <u>polynomial time</u>) a lattice basis $(\mathbf{b}_1, \ldots, \mathbf{b}_d)$ such that $\|\mathbf{b}_i\| = O((2/\sqrt{3})^d)\lambda_i(L)$ for all i. This approximation factor is exponential in the dimension d (see underline{exponential time}). Smaller approximation factors (slightly subexponential in the dimension d—see underline{subexponential time} for a definition) can be achieved in polynomial time in the size of the basis using more complex algorithms like Schnorr's *Block Korkine–Zolotarev* reduction [10].

Lattice reduction algorithms are useful because they enable to solve various lattice problems: approximating the underline{Shortest Vector Problem} and the underline{Closest Vector Problem} (see [1, 3, 6]), finding many short lattice vectors. This has proved invaluable in many areas in computer science (see [3]), notably in cryptology (see the survey [9]). In cryptanalysis, lattice reduction algorithms have been used to break various public-key cryptosystems, including many knapsack and lattice cryptosystems, and more recently certain settings of discrete-log signature schemes (see [8]). Interestingly, they are also used in the most sophisticated attacks known against the underline{RSA public key encryption} and the underline{RSA digital signature scheme}: RSA with a small private exponent, chosen-message attacks on RSA signatures with peculiar paddings, certain settings of RSA encryption with a small public exponent, etc. In particular, Coppersmith opened in [2] a new avenue for cryptanalytic applications of lattice reduction when he revisited the connection between lattices and small solutions of polynomial equations. For instance, it can be shown using the LLL algorithm that, given an integer polynomial $f(X) \in \mathbb{Z}[X]$ of degree d such that the gcd of all the coefficients of f is coprime with an integer N, then one can find in time polynomial in $(d, \log N)$ all the integers x_0 such that $f(x_0) \equiv 0 \pmod{N}$ and $|x_0| \le N^{1/d}$.

It should be emphasized that lattice reduction algorithms typically perform much better than their worst-case theoretical bounds would suggest. For instance, the LLL algorithm often outputs bases which are quite close to the successive minima (by a factor much smaller than the theoretical exponential factor): in low dimension, the LLL algorithm often outputs a shortest nonzero vector. This phenomenon has yet to be explained: the average-case behaviour of lattice reduction algorithms is mostly unknown, and it is hard to predict beforehand how good a resultant basis will be. The effectiveness of lattice reduction algorithm is another reason why lattice reduction has been so popular in cryptanalysis.

Lattice reduction is a very active research area: a lot of work is required to deepen our understanding of lattice reduction algorithms and to invent new lattice reduction algorithms.

See also underline{lattice based cryptography}, underline{lattice reduction}, underline{Closest Vector Problem}, and underline{Shortest Vector Problem}.

Phong Q. Nguyen

References

[1] Babai, L. (1986). "On Lovász lattice reduction and the nearest lattice point problem." *Combinatorica*, 6, 1–13.
[2] Coppersmith, D. (1997). "Small solutions to polynomial equations, and low exponent RSA vulnerabilities." *Journal of Cryptology*, 10 (4), 233–260.
[3] Grötschel, M., L. Lovász, and A. Schrijver (1993). *Geometric Algorithms and Combinatorial Optimization*. Springer-Verlag, Berlin.
[4] Gruber, M. and C.G. Lekkerkerker (1987). *Geometry of Numbers*. North-Holland, Amsterdam.
[5] Lenstra, A.K., H.W. Lenstra, Jr., and L. Lovász (1982). Factoring polynomials with rational coefficients. *Mathematische Ann.*, 261, 513–534.
[6] Micciancio, D. and S. Goldwasser (2002). *Complexity of Lattice Problems: A Cryptographic Perspective*, The Kluwer International Series in Engineering and Computer Science, vol. 671. Kluwer Academic Publishers, Boston.
[7] Minkowski, H. (1896). *Geometrie der Zahlen*. Teubner-Verlag, Leipzig.
[8] Nguyen, P.Q. and I.E. Shparlinski (2002). The insecurity of the Digital Signature Algorithm with partially known nonces. *Journal of Cryptology*, 15 (3), 151–176.
[9] Nguyen, P.Q. and J. Stern (2001). "The two faces of lattices in cryptology." *Cryptography and Lattices—Proceedings of CALC 2001*, Lecture Notes in Computer Science, vol. 2146, ed. J.H. Silverman. Springer-Verlag, Berlin, 146–180.
[10] Schnorr, C.P. (1987). "A hierarchy of polynomial lattice basis reduction algorithms." *Theoretical Computer Science*, 53, 201–224.

LATTICE BASED CRYPTOGRAPHY

Cryptographic applications of underline{lattices} include both cryptanalysis and the design of (provably secure) cryptographic functions. Cryptanalysis applications are usually based on underline{lattice reduction} techniques. The name "lattice based cryptography" typically refers to the second kind of applications: using lattices as a source of computational hardness in the construction of cryptographic functions which are at least as hard to break as solving some underlying lattice problem.

The study of lattice based underline{public key cryptography} has been largely stimulated by Ajtai's discovery in 1996 [1] that certain variants of the knapsack problem (see knapsack cryptosystem) are at least as hard to break on the average as the worst-case instance of certain lattice problems, e.g., solving the Shortest Vector Problem (SVP) approximately within polynomial factors. Specifically, assuming that there is no efficient algorithm to approximate SVP within a factor $g = d^c$ in the worst case (where d is the dimension of the lattice, and c is an arbitrary constant independent of d), one can build knapsack-like cryptographic one-way functions that are almost certainly hard to break (when the key is chosen at random). In 1997 Ajtai and Dwork [2] also proposed a cryptosystem with similar average-case/worst-case connection properties: decrypting random challenges is at least as hard as solving the worst-case instance of a variant of the shortest vector problem, called the *unique* SVP (uSVP). Since finding approximate solutions to lattice problems gets easier and easier as the approximation factor grows, an important question in the area is to determine the smallest factor g such that one can build cryptographic functions that are as hard to break as finding g-approximate solutions to lattice problems. At the time of this writing, the strongest known results are Micciancio's collision resistant hash functions [5] based on the inapproximability of the covering radius problem within $g \approx d^{2.5}$, and Regev's cryptosystem [7] based on the inapproximability of uSVP within $g \approx d^{1.5}$.

All these results about lattice based cryptography are mostly interesting from a theoretical point of view, as they are the only known cryptographic constructions that can be proved secure based on a worst-case complexity assumption (see the entry computational complexity for further discussion on complexity assumptions). In practice, most lattice based cryptographic functions need extremely large keys in order to avoid known cryptanalytic attacks. The reason is that the storage typically required by a d-dimensional lattice grows at least as the square of the security parameter d. So, even if the security parameter is set to a few hundreds, the resulting key can easily exceed hundreds of thousands of bits. However, practical lattice based crytographic constructions may be possible. The main practical lattice based cryptosystem is NTRU, proposed by Hoffstein et al. in 1996 [3]. NTRU achieves small key size using a special class of lattices (convolutional modular lattices) that allow for very compact representation. Unfortunately, NTRU is not known to be provably secure based on any worst-case complexity assumptions, so the security of this system relies on traditional cryptanalytic methods.

Lattice based cryptography is still a young and very active research area, and work is being done toward the design of cryptosystems that are both very efficient, and have provable security guarantees similar to Ajtai's original construction. A recent result pointing in this direction is the discovery by Micciancio [4] that certain generalized compact knapsacks are at least as hard to break as solving the worst-case instance of some lattice problems. These compact knapsacks achieve very small key size using a class of different lattices, but related to those used by NTRU. However, they do not provide an encryption scheme because they lack the trapdoor (secret key) necessary for decryption (see trapdoor one-way function). Other recent results, that may lead to provably secure practical constructions, are the lattice based identification schemes of Micciancio and Vadhan [6], where the same lattice can be shared by all users, and the public key consists of a single vector (with secret key given by the lattice point closest to the public vector).

For a detailed study of lattice based cryptographic functions with provable security properties and an expository overview of the main lattice based cryptosystems, the reader is referred to [5]. See also lattice reduction for information about the use of lattices in cryptanalysis.

Daniele Micciancio

References

[1] Ajtai, M. (1996). "Generating hard instances of lattice problems (extended abstract)." *Proceedings of the Twenty-Eighth Annual ACM Symposium on the Theory of Computing—STOC'96, Philadelphia, PA, USA, May 1996.* ACM, New York, 99–108.

[2] Ajtai, M. and C. Dwork (1997). "A public-key cryptosystem with worst-case/average-case equivalence." *Proceedings of the Twenty-Ninth Annual ACM Symposium on Theory of Computing—STOC'97, El Paso, TX, USA, May 1997.* ACM, New York, 284–293.

[3] Hoffstein, J., J. Pipher, and J.H. Silverman (1998). "NTRU: A ring based public key cryptosystem." *Algorithmic Number Theory: Third International Symposium—ANTS-III,* Lecture Notes in Computer Science, vol. 1423, ed. J. Buhler, Portland, OR, USA, June 1998. Springer, Berlin, 267–288.

[4] Micciancio, D. (2000). "Generalized compact knapsaks, cyclic lattices, and efficient one-way functions from worst-case complexity assumptions." *Proceedings of the 43rd Annual Symposium on Foundations of Computer Science—FOCS 2002,* Vancouver, BC, Canada, November 2002. IEEE Press, Piscataway, NJ, 356–365.

[5] Micciancio, D. and S. Goldwasser (2002). *Complexity of Lattice Problems: A Cryptographic Perspective,*

The Kluwer International Series in Engineering and Computer Science, vol. 671. Kluwer Academic Publishers, Boston, MA.

[6] Micciancio, D. and S. Vadhan (2003). "Statistical zero-knowledge proofs with efficient provers: Lattice problems and more." *Proceedings of the 23rd Annual International Cryptology Conference, Santa Barbara, CA, USA, August 2003. Advances in Cryptology—CRYPTO 2003*, Lecture Notes in Computer Science, vol. 2729, ed. D. Boneh. Springer-Verlag, Berlin.

[7] Regev, O. (2003). "New lattice based cryptographic constructions." *Proceedings of the Thirty-Fifth Annual ACM Symposium on Theory of Computing—STOC 2003*, San Diego, CA, USA, June 2003. ACM, New York, 407–426.

LEAST COMMON MULTIPLE

The least common multiple (lcm) of a set of positive integers $\{a_1, \ldots, a_k\}$ is the smallest positive integer that is an integer multiple of every element of the set. This is denoted by $\text{lcm}(a_1, \ldots, a_k)$. For example, $\text{lcm}(21, 91) = 273$, because 273 is a multiple of both 21 and 91, and no positive integer smaller than 273 has this property.

(Square brackets, i.e., $[a_1, \ldots, a_k]$, are also sometimes used to denote the lcm of integers $\{a_1, \ldots, a_k\}$.)

Scott Contini

LEGENDRE SYMBOL

Let p be an odd underline{prime number} and let x be an integer. If x is a quadratic residue, i.e., if x is relatively prime to p and the equation (see modular arithmetic)

$$x \equiv y^2 \pmod{p}$$

has an integer solution y, then the Jacobi symbol of x modulo p, written as (x/p) or $(\frac{x}{p})$, is $+1$. If x is a quadratic nonresidue—i.e., relatively prime to p and no square roots—then its Legendre symbol is -1. If x is not relatively prime to p then $(\frac{x}{p}) = 0$.

The Legendre symbol may be efficiently computed by modular exponentiation (see exponentiation algorithms) as

$$\left(\frac{x}{p}\right) = x^{(p-1)/2} \pmod{p}.$$

See also Jacobi symbol.

Burt Kaliski

LINEAR COMPLEXITY

The *linear complexity* of a semi-infinite sequence $\mathbf{s} = (s_t)_{t \geq 0}$ of elements of the finite field $\mathbf{F}_q$, $\Lambda(\mathbf{s})$, is the smallest integer Λ such that $\mathbf{s}$ can be generated by a linear feedback shift register (LFSR) of length Λ over $\mathbf{F}_q$, and is ∞ if no such LFSR exists. By way of convention, the linear complexity of the all-zero sequence is equal to 0. The linear complexity of a linear recurring sequence corresponds to the degree of its minimal polynomial.

The linear complexity $\Lambda(\mathbf{s^n})$ of a finite sequence $\mathbf{s^n} = s_0 s_1 \ldots s_{n-1}$ of n elements of $\mathbf{F}_q$ is the length of the shortest LFSR which produces $\mathbf{s^n}$ as its first n output terms for some initial state. The linear complexity of any finite sequence can be determined by the Berlekamp–Massey algorithm. An important result due to Massey [1] is that, for any finite sequence $\mathbf{s^n}$ of length n, the LFSR of length $\Lambda(\mathbf{s^n})$ which generates $\mathbf{s^n}$ is unique if and only if $n \geq 2\Lambda(\mathbf{s^n})$.

The linear complexity of an infinite linear recurring sequence $\mathbf{s}$ and the linear complexity of the finite sequence $\mathbf{s^n}$ composed of the first n digits of $\mathbf{s}$ are related by the following property: if $\mathbf{s}$ is an infinite linear recurring sequence with linear complexity Λ, then the finite sequence $\mathbf{s^n}$ has linear complexity Λ for any $n \geq 2\Lambda$. Moreover, the unique LFSR of length Λ that generates $\mathbf{s}$ is the unique LFSR of length Λ that generates $\mathbf{s^n}$ for every $n \geq 2\Lambda$.

For a sequence $\mathbf{s} = s_0 s_1 \cdots$, the sequence of the linear complexities $(\Lambda(\mathbf{s^n}))_{n \geq 1}$ of all subsequences $\mathbf{s^n} = s_0 \cdots s_{n-1}$ composed of the first n terms of $\mathbf{s}$ is called the *linear complexity profile* of $\mathbf{s}$.

The expected linear complexity of a binary sequence $\mathbf{s^n} = s_0 \cdots s_{n-1}$ of n independent and uniformly distributed binary random variables is

$$E[\Lambda(\mathbf{s^n})] = \frac{n}{2} + \frac{4 + \varepsilon(n)}{18} + 2^{-n}\left(\frac{n}{3} + \frac{2}{9}\right),$$

where $\varepsilon(n) = n \bmod 2$.

If $\mathbf{s}$ is an infinite binary sequence of period 2^n which is obtained by repeating a sequence $s_0 \cdots s_{2^n-1}$ of 2^n independent and uniformly distributed binary random variables, its expected linear complexity is

$$E[\Lambda(\mathbf{s})] = 2^n - 1 + 2^{-2^n}.$$

Further results on the linear complexity and on the linear complexity profile of random sequences can be found in [2].

Anne Canteaut

References

[1] Massey, J.L. (1969). "Shift-register synthesis and BCH decoding." *IEEE Transactions on Information Theory*, 15, 122–127.

[2] Rueppel, R.A. (1986). *Analysis and Design of Stream Ciphers.* Springer-Verlag, Berlin.

LINEAR CONGRUENTIAL GENERATOR

A linear congruential generator is a pseudorandom generator that produces a sequence of numbers $x_1, x_2, x_3, \ldots$ according to the following linear recurrence:

$$x_t = a x_{t-1} + b \ (\text{mod } n)$$

for $t \geq 1$ (see also modular arithmetic); integers a, b and n characterize entirely the generator, and the seed is x_0.

EXAMPLE. Let us take, for example, $a = 3$, $b = 5$, $n = 17$, and $x_0 = 2$; the sequence produced by the linear congruential generator will then be $11, 4, 0, 5, 3, 14, 13, 10, 1, 8, 12, 7, 9, 15, 16, \ldots$

Such a generator is easy to implement, and pass the following statistical tests: Golomb's randomness postulates, frequency test, serial test, poker test, runs test, autocorrelation test, and Maurer's universal statistical test. Hence, it can be considered as a good candidate for generating strong pseudorandom sequences. However, there is an important drawback: the sequence is *predictable*: given a piece of the sequence, it is easy to reconstruct all the rest of it, even if the attacker does not know the exact values of a, b and n [5, 6]. So, it would be very dangerous to use it in a cryptographic purpose.

Some variants have been considered, using either several terms in the linear recurrence equation,

$$x_t = a_1 x_{t-1} + a_2 x_{t-2} + \cdots + a_\ell x_{t-\ell} + b \ (\text{mod } n),$$

or a quadratic recurrence relation,

$$x_t = a x_{t-1}^2 + b x_{t-1} + c \ (\text{mod } n).$$

In both cases, it can be shown that the sequence remains predictable [1, 4]. Another variant has also been studied, considering that some least significant bits of the produced integers are discarded; but such sequences are still predictable [1, 3, 7]. A more precise state-of-the-art about cryptanalytic attacks of such generators can be found in [2].

Caroline Fontaine

References

[1] Boyar, J. (1989). "Inferring sequences produced by a linear congruential generator missing low-order bits." *Journal of Cryptology*, 1, 177–184.

[2] Brickell, E.F. and A.M. Odlyzko (1992). "Cryptanalysis: A survey of recent results." *Contemporary Cryptology: The Science of Information Integrity*, 501–540, IEEE Press, Piscataway, NJ.

[3] Frieze, A.M., J. Hastad, R. Kannan, J.C. Lagarias, and A. Shamir (1988). "Reconstructing truncated integer variables satisfying linear congruence." *SIAM Journal on Computing*, 17, 262–280.

[4] Krawczyk, H. (1992). "How to predict congruential generators." *Journal of Algorithms*, 13, 527–545.

[5] Plumstead, J.B. (1982). "Inferring a sequence generated by a linear congruence." *Proceedings of the IEEE 23rd Annual Symposium on Foundations of Computer Science.* IEEE Press, New York, 153–159.

[6] Plumstead, J.B. (1983). "Inferring a sequence produced by a linear congruence." *Advances in Cryptology—CRYPTO'82*, Lecture Notes in Computer Science, eds. D. Chaum, R.L. Rivest, and A.T. Sherman. Plenum Press, New York, 317–319.

[7] Stern, J. (1987). "Secret linear congruential generators are not cryptographically secure." *Proceedings of the IEEE 28th Annual Symposium on Foundations of Computer Science.* IEEE Press, New York, 421–426.

LINEAR CONSISTENCY ATTACK

The *linear consistency attack* is a divide-and-conquer technique which provides a known plaintext attack on stream ciphers. It was introduced by Zeng et al. in 1989. It has been applied to various keystream generators, like the *Jenning generator* [2], the *stop-and-go generator* of Beth and Piper [2], and the E0 cipher used in Bluetooth [1].

The linear consistency attack applies as soon as it is possible to single out a portion K_1 of the secret key and to form a system $Ax = b$ of linear equations, where the matrix A only depends on K_1 and the right-side vector b is determined by the known keystream bits. Then, an exhaustive key search on K_1 can be performed. The correct value of K_1 can be distinguished from a wrong one by checking whether the linear system is consistent or not. Once K_1 has been recovered, the solution x of the system may provide some additional bits of the secret key.

Anne Canteaut

References

[1] Fluhrer, S.R. and S. Lucks (2001). "Analysis of the E0 encryption system." *Selected Areas in Cryptography—SAC 2001*, Lecture Notes in Computer Science, vol. 2259, eds. S. Vaudenay and A.M. Youssef. Springer-Verlag, Berlin, 38–48.

[2] Zeng, K., C.H. Yang, and T.R.N. Rao (1989). "On the linear consistency test (LCT) in cryptanalysis with applications." *Advances in Cryptology—CRYPTO'89*, Lecture Notes in Computer Science, vol. 435, ed. G. Brassard. Springer-Verlag, Berlin, 164–174.

LINEAR CRYPTANALYSIS FOR BLOCK CIPHERS

Linear cryptanalysis is a powerful method of cryptanalysis of block ciphers introduced by Matsui in 1993 [1]. The attack in its current form was first applied to the Data Encryption Standard (DES), but an early variant of linear cryptanalysis, developed by Matsui and Yamagishi, was already successfully used to attack FEAL in 1992 [12]. Linear cryptanalysis is a known plaintext attack in which the attacker studies probabilistic linear relations (called *linear approximations*) between parity bits of the plaintext, the ciphertext, and the secret key. Given an approximation with high probability, the attacker obtains an estimate for the parity bit of the secret key by analyzing the parity bits of the known plaintexts and ciphertexts. Using auxiliary techniques he can usually extend the attack to find more bits of the secret key.

The next section provides some more details about the attack algorithm. Sections "Piling-up Lemma," to "Provable security against linear cryptanalysis" discuss a number of practical and theoretical aspects which play a role in linear cryptanalysis. Section "Comparison with differential cryptanalysis" points out analogies between linear and differential cryptanalysis, and Section "Extensions" concludes with some extended variants of linear cryptanalysis.

OUTLINE OF A LINEAR ATTACK: Following Matsui's notation, we denote by $A[i]$ the ith bit of A and by $A[i_1, i_2, \ldots, i_k]$ the parity bit $A[i_1] \oplus A[i_2] \oplus \cdots \oplus A[i_k]$. The first task of the attacker is to find a suitable linear approximation.

For simple linear operations such as an XOR with the key or a permutation of bits, very simple linear expressions can be written which hold with probability one. For nonlinear elements of a cipher such as S-boxes one tries to find linear approximations with probability p that maximizes $|p - \frac{1}{2}|$. Approximations for single operations inside a cipher are then further combined into approximations that hold for a single round of a cipher. By appropiate concatenation of one-round approximations, the attacker eventually obtains an approximation for the whole cipher of the type:

$$P[i_1, i_2, \ldots, i_a] \oplus C[j_1, j_2, \ldots, j_b]$$
$$= K[k_1, k_2, \ldots, k_c], \qquad (1)$$

where $i_1, i_2, \ldots, i_a$, $j_1, j_2, \ldots, j_b$ and $k_1, k_2, \ldots, k_c$ denote fixed bit locations. Note that Such approximation is interesting only if it holds with a probability $p \neq \frac{1}{2}$ (how this probability is calculated is explained in the next section). For DES Matsui found such an approximation with probability $\frac{1}{2} + 2^{-24}$. Using this approximation, a simple algorithm based on the maximum likelihood method can be used to find one parity bit $K[k_1, k_2, \ldots, k_c]$ of the key:

Given a pool of N random known plaintexts, let T be the number of plaintexts such that the left side of the Equation (1) is 0.

if $(T - N/2) \cdot (p - 1/2) > 0$ **then**
  $K[k_1, \ldots, k_c] = 0$
else
  $K[k_1, \ldots, k_c] = 1$
end if

In order for the parity bit $K[k_1, k_2, \ldots, k_c]$ to be recovered correctly with a reasonable probability, Matsui demonstrated that the amount of plaintext N needs to be in the order of $|p - \frac{1}{2}|^{-2}$. More efficient algorithms for linear cryptanalysis, which find more key bits, are described in [12].

PILING-UP LEMMA: The first stage in linear cryptanalysis consists in finding useful approximations for a given cipher (or in demonstrating that no useful approximations exist, which is usually much more difficult). Although the most biased linear approximation can easily be found in an exhaustive way for a simple component such as an S-box, a number of practical problems arise when trying to extrapolate this method to full-size ciphers. The first problem concerns the computation of the probability of a linear approximation. In principle, this would require the cryptanalyst to run through all possible combinations of plaintexts and keys, which is clearly infeasible for any practical cipher. The solution to this problem is to make a number of assumptions and to approximate the probability using the so-called *Piling-up Lemma*.

LEMMA 1. *Given n independent random variables* $X_1, X_2, \ldots, X_n$ *taking on values from* $\{0, 1\}$, *then the bias* $\epsilon = p - 1/2$ *of the sum* $X = X_1 \oplus X_2 \oplus \cdots \oplus X_n$ *is given by:*

$$\epsilon = 2^{n-1} \prod_{j=1}^{n} \epsilon_j, \tag{2}$$

where $\epsilon_1, \epsilon_2, \ldots, \epsilon_n$ *are the biases of the terms* $X_1, X_2, \ldots, X_n$.

Notice that the lemma can be further simplified by defining $c = 2\epsilon$, known as the imbalance or the correlation of an expression. With this notation, Equation (2) reduces to $c = \prod_{j=1}^{n} c_j$.

In order to estimate the probability of a linear approximation using the Piling-up Lemma, the approximation is written as a chain of connected linear approximations, each spanning a small part of the cipher. Such a chain is called a *linear characteristic*. Assuming that the biases of these partial approximations are statistically independent and easy to compute, then the total bias can be computed using Equation (2).

Although the Piling-up Lemma produces very good estimations in many practical cases, even when the approximations are not strictly independent, it should be stressed that unexpected effects can occur when the independence assumption is not fulfilled. In general, the actual bias in these cases can be both much smaller and much larger than predicted by the lemma.

MATSUI'S SEARCH FOR THE BEST APPROXIMATIONS: The Piling-up Lemma in the previous section provides a useful tool to estimate the strength of a given approximation, but the problem remains how to find the strongest approximations for a given cipher. For DES, this open problem was solved by Matsui in 1994 [15]. In his second paper, he proposes a practical search algorithm based on a recursive reasoning. Given the probabilities of the best i-round characteristic with $1 \leq i \leq n - 1$, the algorithm efficiently derives the best characteristic for n rounds. This is done by traversing a tree where branches are cut as soon as it is clear that the probability of a partially constructed approximation cannot possibly exceed some initial estimation of the best n-round characteristic.

Matsui's algorithm can be applied to many other block ciphers, but its efficiency varies. In the first place, the running time strongly depends on the accuracy of the initial estimation. Small estimations increase the size of the search tree. On the other hand, if the estimation is too large, the algorithm will not return any characteristic at all. For DES, good estimations can be easily obtained by first performing a restricted search over all characteristics which only cross a single S-box in each round. This does not work as nicely for other ciphers, however. The specific properties of the S-boxes also affect the efficiency of the algorithm. In particular, if the maximum bias of the S-box is attained by many different approximations (as opposed to the distinct peaks in the DES S-boxes), this will slow down the algorithm.

LINEAR HULLS: Estimating the bias of approximations by constructing linear characteristics is very convenient, but in some cases, the value derived in this way diverges significantly from the actual bias. The most important cause for this difference is the so-called *linear hull* effect, first described by Nyberg in 1994 [16]. The effect takes place when the correlation between plaintext and ciphertext bits, described by a specific linear approximation, can be explained by multiple linear characteristics, each with a non-negligible bias, and each involving a different set of key bits. Such a set of linear characteristics with identical input and output masks is called a *linear hull*. Depending on the value of the key, the different characteristics will interfere constructively or destructively, or even cancel out completely. If the sets of keys used in the different linear characteristics are independent, then this effect might considerably reduce the average bias of expression (1), and thus the success rate of the simple attack described above. Nyberg's paper shows, however, that the more efficient attacks described in [12], which only use the linear approximations as a distinguisher, will typically benefit from the linear hull effect.

PROVABLE SECURITY AGAINST LINEAR CRYPTANALYSIS: The existence of a single sufficiently biased linear characteristic suffices for a successful linear attack against a block cipher. A designer's first objective is therefore to ensure that such characteristic cannot possibly exist. This is usually done by choosing highly nonlinear S-boxes and then arguing that the diffusion in the cipher forces all characteristics to cross a sufficiently high minimal number of "active" S-boxes.

The above approach provides good heuristic arguments for the strength of a cipher, but in order to rigorously prove the security against linear cryptanalysis, the designer also needs to take into account more complex phenomena such as the linear hull effect. For DES-like ciphers, such security proofs were studied by Knudsen and Nyberg, first with respect to differential cryptanalysis [17], and then also applied to linear cryptanalysis [16]. The results inspired the design of a number of practical block ciphers such as MISTY (or its variant KASUMI; see KASUMI/MISTYI), Rijndael/AES, Camellia and others. Later, similar proofs were formulated for ciphers based on SP-networks [5, 8].

A somewhat more general theory for provable security against a class of attacks, including basic linear cryptanalysis, is based on the notion of decorrelation, introduced by Vaudenay [23]. The theory suggests constructions were a so-called *Decorrelation Module* effectively blocks the propagation of all traditional linear and differential characteristics.

An important remark with respect to the previous notions of provable security, however, is that ciphers which are provably optimal against some restricted class of attacks often tend to be weak when subject to other types of attacks [21, 24].

COMPARISON WITH DIFFERENTIAL CRYPTANALYSIS: Linear cryptanalysis has many methodological similarities with underlined differential cryptanalysis as is noted in [1]. *Differential characteristics* correspond to *linear approximations*. *Difference distribution tables* are replaced by *linear approximation tables*. Concatenation rule for differential characteristics: "match the differences, multiply the probabilities" corresponds to concatenation rule for linear approximations (the piling-up lemma): "match the masks, multiply the imbalances". The algorithms that search for the best characteristic or the best linear approximation are essentially the same. The notion of *differentials* has a corresponding notion of *linear hulls*. Together with striking methodological similarity between the two techniques, there is also *duality* [15] of operations: "XOR branch" and "three-forked branch" are mutually dual regarding their action on differences and masks, respectively. An important distinction between the two methods is that differential cryptanalysis works with blocks of bits, while linear cryptanalysis typically works with a single bit. The bias of the linear approximation has a sign. Thus given two approximations with the same input and output masks and equal probability but opposite signs, the resulting approximation will have zero bias, due to the cancellation of the two approximations by each other.

EXTENSIONS: The linear cryptanalysis technique has received much attention since its invention and has enjoyed several extensions. One technique is a combined differential–linear approach proposed by Langford and Hellman. Other extensions include *key-ranking* which allows for a tradeoff between data and time of analysis [6, 14, 19]; *partitioning cryptanalysis* [4] which studies correlation between partitions of the plaintext and ciphertext spaces (no practical cipher has been broken via this technique so far); $\mathcal{X}^2$ *cryptanalysis* [10, 22] has been applied successfully against several ciphers, including round-reduced versions of RC6; the use of nonlinear approximations was suggested [12, 20], but so far it provided only small improvements over the linear cryptanalysis. A full nonlinear generalization still remains evasive. The idea to use multiple approximations has been proposed in [6] though the problem of estimating the attacker's gain as well as information extraction from such approximations largely remained opened. In [7] by using a maximal likelihood framework explicit gain formulas have been derived. Define *capacity* $\bar{c}^2$ of a system of m approximations as $\bar{c}^2 = 4 \cdot \sum_{j=1}^{m} \epsilon_j^2$, where ϵ_j—are the biases of individual approximations. For a fixed attacker's gain over the exhaustive search the data complexity N of the multiple-linear attack is proportional to $\bar{c}^2$—the *capacity* of the "information channel" provided by multiple approximations. The paper also describes several algorithms which provide such gains. A conversion of a known plaintext linear attack to a chosen plaintext linear attack has been proposed in [9]. Finally note that similar techniques have been applied to stream ciphers (see Linear Cryptanalysis for Stream Ciphers).

Alex Biryukov
Christophe De Cannière

References

[1] Biham, E. (1995). "On Matsui's linear cryptanalysis." *Advances in Cryptology—EUROCRYRT'94*, Lecture Notes in Computer Science, vol. 950, ed. A. De Santis. Springer-Verlag, Berlin, 341–355.

[2] Biryukov, A., C. De Cannière, M. Quisquater (2004). "On Multiple Linear Approximations", *Advances in Cryptology, proceedings of CRYPTO 2004*, Lecture Notes in Computer Science 3152, ed. M. Franklin. Springer-Verlag, 1–22.

[3] Desmedt, Y. ed. (1994). *Advances in Cryptology—CRYPTO'94*, Lecture Notes in Computer Science, vol. 839, ed. Y.G. Desmedt. Springer-Verlag, Berlin.

[4] Harpes, C. and J.L. Massey (1997). "Partitioning cryptanalysis." *Fast Software Encryption, FSE'97*, Lecture Notes in Computer Science, vol. 1267, ed. E. Biham. Springer-Verlag, Berlin, 13–27.

[5] Hong, S., S. Lee, J. Lim, J. Sung, D. Cheon, and I. Cho (2000). "Provable security against differential and linear cryptanalysis for the SPN structure." *Proceedings of Fast Software Encryption—FSE 2000*, Lecture Notes in Computer Science, vol. 1978, ed. B. Schneier. Springer-Verlag, Berlin, 273–283.

[6] Junod, P. and S. Vaudenay (2003). "Optimal key ranking procedures in a statistical cryptanalysis." *Fast Software Encryption, FSE 2003*, Lecture Notes in Computer Science, vol. 2887, ed. T. Johansson. Springer-Verlag, Berlin, 1–15.

[7] Kaliski, B.S. and M.J. Robshaw (1994). "Linear cryptanalysis using multiple approximations." *Advances in Cryptography—CRYPTO'94*, Lecture Notes in Computer Science, vol. 839, ed. Y. Desmedt. Springer-Verlag, Berlin, 26–39.

[8] Keliher, L., H. Meijer, and S.E. Tavares (2001). "New method for upper bounding the maximum average linear hull probability for SPNs." *EUROCRYPT 2001*, Lecture Notes in Computer Science, vol. 2045, ed. B. Pfitzmann. Springer-Verlag, Berlin, 420–436.

[9] Knudsen, L.R. and J.E. Mathiassen (2001). "A chosen-plaintext linear attack on DES." *Fast Software Encryption, FSE 2000*, Lecture Notes in Computer Science, vol. 1978, ed. B. Schneier. Springer-Verlag, Berlin, 262–272.

[10] Knudsen, L.R. and W. Meier (2000). "Correlations in RC6 with a reduced number of rounds." *Proceedings of Fast Software Encryption—FSE 2000*, Lecture Notes in Computer Science, vol. 1978, ed. B. Schneier. Springer-Verlag, Berlin, 94–108.

[11] Knudsen, L.R. and M.J.B. Robshaw (1996). "Non-linear approximations in linear cryptanalysis." *Advances in Cryptology—EUROCRYPT'96*, Lecture Notes in Computer Science, vol. 1070, ed. U. Maurer. Springer-Verlag, Berlin, 224–236.

[12] Matsui, M. (1993). "Linear cryptanalysis method for DES cipher." *Advances in Cryptology—EUROCRYPT'93*, Lecture Notes in Computer Science, vol. 765, ed. T. Helleseth. Springer-Verlag, Berlin, 386–397.

[13] Matsui, M. and A. Yamagishi (1993). "A new method for known plaintext attack of FEAL cipher." *Advances in Cryptography—EUROCRYPT'92*, Lecture Notes in Computer Science, vol. 658, ed. R.A. Rueppel. Springer-Verlag, Berlin, 81–91.

[14] Matsui, M. (1994). "The first experimental cryptanalysis of the data encryption standard." *Advances in Cryptography—CRYPTO'94*, Lecture Notes in Computer Science, vol. 839, ed. Y.G. Desmedt. Springer-Verlag, Berlin, 1–11.

[15] Matsui, M. "On correlation between the order of S-boxes and the strength of DES." *Advances in Cryptology—EUROCRYPT'94*, Lecture Notes in Computer Science, vol. 950, ed. A. De Santis. Springer-Verlag, Berlin, 366–375.

[16] Nyberg, K. (1994). "Linear approximations of block ciphers." *Advances in Cryptography—EUROCRYPT'94*, Lecture Notes in Computer Science, vol. 950, ed. A. De Santis. Springer-Verleg, Berlin, 439–444.

[17] Nyberg, K. and L.R. Knudsen (1995). "Provable security against a differential attack," *Journal of Cryptology*, 8 (1), 27–38.

[18] Santis, A.D., ed. (1995). *Advances in Cryptology—EUROCRYPT'94*, Lecture Notes in Computer Science, vol. 950, ed. A. De Santis. Springer-Verlag, Berlin.

[19] Selcuk, A.A. (2002). "On probability of success in differential and linear cryptanalysis." Technical Report, Network Systems Lab, Department of Computer Science, Purdue University, 2002. Previously published at SCN 2002.

[20] Shimoyama, T. and T. Kaneko (1998). "Quadratic relation of S-box and its application to the linear attack of full round des." *Advances in Cryptology—CRYPTO'98*, Lecture Notes in Computer Science, vol. 1462, ed. H. Krawczyk. Springer-Verlag, Berlin, 200–211.

[21] Shimoyama, T., S. Moriai, T. Kaneko, and S. Tsujii (1999). "Improved higher order differential attack and its application to Nyberg-Knudsen's designed block cipher." *IEICE Transactions on Fundamentals*, E82-A (9), 1971–1980. http://search.ieice.or.jp/1999/files/e000a09 .htm#e82-a,9,1971

[22] Vaudenay, S. (1996). "On the weak keys of blowfish." *Fast Software Encryption, FSE'96*, Lecture Notes in Computer Science, vol. 1039, ed. D. Gollmann. Springer-Verlag, Berlin, 27–32.

[23] Vaudenay, S. (2003). "Decorrelation: A theory for block cipher security." *Journal of Cryptology*, 16 (4), 249–286.

[24] Wagner, D. (1999). "The boomerang attack." In *Fast Software Encryption, FSE'99*, Lecture Notes in Computer Science, vol. 1636, ed. L.R. Knudsen. Springer-Verlag, Berlin, 156–170.

LINEAR CRYPTANALYSIS FOR STREAM CIPHERS

A *linear cryptanalysis technique for* stream ciphers was presented by Golić in 1994. It relies on the same basic principles as the linear cryptanalysis for block ciphers introduced by Matsui. The linear cryptanalysis provides a known plaintext attack on various synchronous stream ciphers, which enables to distinguish the keystream from a truly random sequence. Such a *distinguishing attack* can be used for reducing the uncertainty of unknown plaintexts, or for recovering the unknown structure of the keystream generator. It may also lead to a divide-and-conquer attack when the structure of the keystream generator depends on a portion of the secret-key.

The linear cryptanalysis consists in finding some linear functions of the keystream bits which are not *balanced*, that is, which are not uniformly distributed. Such linear correlations are used for distinguishing the keystream sequence from a random sequence by a classical statistical test. Efficient techniques for finding biased linear relations among the keystream bits are presented in [2, 3].

The linear cryptanalysis leads to distinguishing attacks on several types of stream ciphers [2, 3] and to a reconstruction attack on combination generators [1].

Anne Canteaut

produce sequences with good statistical properties. LFSR refers to a feedback shift register with a linear feedback function (see Nonlinear Feedback Shift Register).

An LFSR of length L over $\mathbf{F}_q$ (see finite field) is a finite state automaton which produces a semi-infinite sequence of elements of $\mathbf{F}_q$, $\mathbf{s} = (s_t)_{t \geq 0} = s_0 s_1 \ldots$, satisfying a linear recurrence relation of degree L over $\mathbf{F}_q$

$$s_{t+L} = \sum_{i=1}^{L} c_i s_{t+L-i}, \qquad \forall t \geq 0.$$

The L coefficients $c_1, \ldots, c_L$ are elements of $\mathbf{F}_q$. They are called the *feedback coefficients* of the LFSR.

An LFSR of length L over $\mathbf{F}_q$ has the following form:

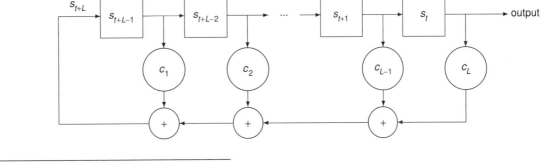

References

[1] Canteaut, A. and E. Filiol (2001). "Ciphertext only reconstruction of stream ciphers based on combination generators." *Fast Software Encryption 2000*, Lecture Notes in Computer Science, vol. 1978, ed. B. Schneier. Springer-Verlag, Berlin, 165–180.

[2] Coppersmith, D., S. Halevi, and C. Jutla (2002). "Cryptanalysis of stream ciphers with linear masking." *Advances in Cryptology—CRYPTO 2002*, Lecture Notes in Computer Science, vol. 2442, ed. M. Yung. Springer-Verlag, Berlin, 515–532.

[3] Golić, J.Dj. (1994). "Linear cryptanalysis of stream ciphers." *Fast Software Encryption 1994*, Lecture Notes in Computer Science, vol. 1008, ed. B. Preneel. Springer-Verlag, Berlin, 154–169.

LINEAR FEEDBACK SHIFT REGISTER

Linear Feedback Shift Registers (LFSRs) are the basic components of many running-key generators for stream cipher applications, because they are appropriate to hardware implementation and they

The register consists of L delay cells, called *stages*, each containing an element of $\mathbf{F}_q$. The contents of the L stages, $s_t, \ldots, s_{t+L-1}$, form the *state* of the LFSR. The L stages are initially loaded with L elements, $s_0, \ldots, s_{L-1}$, which can be arbitrary chosen in $\mathbf{F}_q$; they form the *initial state* of the register.

The shift register is controlled by an external clock. At each time unit, each digit is shifted one stage to the right. The content of the rightmost stage s_t is output. The new content of the leftmost stage is the *feedback bit*, s_{t+L}. It is obtained by a linear combination of the contents of the register stages, where the coefficients of the linear combination are given by the feedback coefficients of the LFSR:

$$s_{t+L} = \sum_{i=1}^{L} c_i s_{t+L-i}.$$

Therefore, the LFSR implements the linear recurrence relation of degree L:

$$s_{t+L} = \sum_{i=1}^{L} c_i s_{t+L-i}, \qquad \forall t \geq 0.$$

Table 1. Successive states of the LFSR with feedback coefficients $(c_1, c_2, c_3, c_4) = (0, 0, 1, 1)$ and with initial state $(s_0, s_1, s_2, s_3) = (1, 0, 1, 1)$

(t)	0	1	2	3	4	5	6	7	8	9	10	11	12	13	14	15
(s_t)	1	0	1	1	1	1	0	0	0	1	0	0	1	1	0	1
(s_{t+1})	0	1	1	1	1	0	0	0	1	0	0	1	1	0	1	0
(s_{t+2})	1	1	1	1	0	0	0	1	0	0	1	1	0	1	0	1
(s_{t+3})	1	1	1	0	0	0	1	0	0	1	1	0	1	0	1	1

EXAMPLE. Table 1 gives the successive states of the binary LFSR of length 4 with feedback coefficients $c_1 = c_2 = 0$, $c_3 = c_4 = 1$ and with initial state $(s_0, s_1, s_2, s_3) = (1, 0, 1, 1)$. This LFSR is depicted in Figure 1. It corresponds to the linear recurrence relation

$$s_{t+4} = s_{t+1} + s_t \bmod 2.$$

The output sequence $s_0 s_1 \ldots$ generated by this LFSR is $1011100\ldots$.

FEEDBACK POLYNOMIAL AND CHARACTERISTIC POLYNOMIAL:

The output sequence of an LFSR is uniquely determined by its feedback coefficients and its initial state. The feedback coefficients $c_1, \ldots, c_L$ of an LFSR of length L are usually represented by the LFSR *feedback polynomial* (or *connection polynomial*) defined by

$$P(X) = 1 - \sum_{i=1}^{L} c_i X^i.$$

Alternatively, one can use the *characteristic polynomial*, which is the reciprocal polynomial of the feedback polynomial:

$$P^*(X) = X^L P\left(\frac{1}{X}\right) = X^L - \sum_{i=1}^{L} c_i X^{L-i}.$$

For instance, the feedback polynomial of the binary LFSR shown in Figure 1 is $P(X) = 1 + X^3 + X^4$ and its characteristic polynomial is $P^* = 1 + X + X^4$.

An LFSR is said to be *non-singular* if the degree of its feedback polynomial is equal to the LFSR length (i.e., if the feedback coefficient c_L differs from 0). Any sequence generated by a non-singular LFSR of length L is periodic, and its pe-

riod does not exceed $q^L - 1$. Indeed, the LFSR has at most q^L different states and the all-zero state is always followed by the all-zero state. Moreover, if the LFSR is, singular, all generated sequences are *ultimately periodic*, that is, the sequences obtained by ignoring a certain number of elements at the beginning are periodic.

CHARACTERIZATION OF LFSR OUTPUT SEQUENCES:

A given LFSR of length L over $\mathbf{F}_q$ can generate q^L different sequences corresponding to the q^L different initial states and these sequences form a vector space over $\mathbf{F}_q$. The set of all sequences generated by an LFSR with feedback polynomial P is characterized by the following property: a sequence $(s_t)_{t \geq 0}$ is generated by an LFSR of length L over $\mathbf{F}_q$ with feedback polynomial P if and only if there exists a polynomial $Q \in \mathbf{F}_q[X]$ with $\deg(Q) < L$ such that the generating function of $(s_t)_{t \geq 0}$ satisfies

$$\sum_{t \geq 0} s_t X^t = \frac{Q(X)}{P(X)}.$$

Moreover, the polynomial Q is completely determined by the coefficients of P and by the initial state of the LFSR:

$$Q(X) = -\sum_{i=0}^{L-1} X^i \left(\sum_{j=0}^{i} c_{i-j} s_j\right),$$

where $P(X) = \sum_{i=0}^{L} c_i X^i$. This result, which is called the fundamental identity of formal power series of linear recurring sequences, means that there is a one-to-one correspondence between the sequences generated by an LFSR of length L with feedback polynomial P and the fractions $Q(X)/P(X)$ with $\deg(Q) < L$. It has two major consequences. On the one hand, any sequence generated by an LFSR with feedback polynomial P is also generated by any LFSR whose feedback polynomial is a multiple of P. This property is used in some attacks on keystream generators based on LFSRs (see Fast Correlation attack). On the other hand, a sequence generated by an LFSR with feedback polynomial P is also generated by a shorter LFSR with feedback polynomial P' if

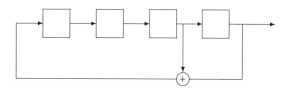

Fig. 1. Binary LFSR with feedback coefficients $(c_1, c_2, c_3, c_4) = (0, 0, 1, 1)$

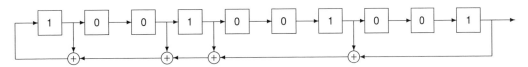

Fig. 2. Example of an LFSR of length 10

the corresponding fraction $Q(X)/P(X)$ is such that $\gcd(P, Q) \neq 1$. Thus, among all sequences generated by the LFSR with feedback polynomial P, there is one which can be generated by a shorter LFSR if and only if P is not <u>irreducible</u> over $\mathbf{F}_q$.

Moreover, for any linear recurring sequence $(s_t)_{t \geq 0}$, there exists a unique polynomial P_0 with constant term equal to 1, such that the generating function of $(s_t)_{t \geq 0}$ is given by $Q_0(X)/P_0(X)$, where P_0 and Q_0 are relatively prime. Then, the shortest LFSR which generates $(s_t)_{t \geq 0}$ has length $L = \max(\deg(P_0), \deg(Q_0) + 1)$, and its feedback polynomial is equal to P_0. The reciprocal polynomial of P_0, $X^L P_0(1/X)$, is the characteristic polynomial of the shortest LFSR which generates $(s_t)_{t \geq 0}$; it is called the <u>minimal polynomial</u> of the sequence. It determines the linear recurrence relation of least degree satisfied by the sequence. The degree of the minimal polynomial of a linear recurring sequence is the <u>linear complexity</u> of the sequence. It corresponds to the length of the shortest LFSR which generates it. The minimal polynomial of a sequence $\mathbf{s} = (s_t)_{t \geq 0}$ of linear complexity $\Lambda(\mathbf{s})$ can be determined from the knowledge of at least $2\Lambda(\mathbf{s})$ consecutive bits of $\mathbf{s}$ by the <u>Berlekamp–Massey algorithm</u>.

EXAMPLE. The binary LFSR of length 10 depicted in Figure 2 has feedback polynomial

$$P(X) = 1 + X + X^3 + X^4 + X^7 + X^{10},$$

and its initial state $s_0, \ldots, s_9$ is 1001001001.

The generating function of the sequence produced by this LFSR is given by

$$\sum_{t \geq 0} s_t X^t = \frac{Q(X)}{P(X)},$$

where Q is deduced from the coefficients of P and from the initial state:

$$Q(X) = 1 + X + X^7.$$

Therefore, we have

$$\sum_{t \geq 0} s_t X^t = \frac{1 + X + X^7}{1 + X + X^3 + X^4 + X^7 + X^{10}}$$

$$= \frac{1}{1 + X^3},$$

since $\quad 1 + X + X^3 + X^4 + X^7 + X^{10} = (1 + X + X^7)(1 + X^3)$ in $\mathbf{F}_2[X]$. This implies that $(s_t)_{t \geq 0}$ is also generated by the LFSR with feedback polynomial $P_0(X) = 1 + X^3$ depicted in Figure 3. The minimal polynomial of the sequence is then $1 + X^3$ and its linear complexity is equal to 3.

PERIOD OF AN LFSR SEQUENCE: The minimal polynomial of a linear recurring sequence plays a major role since it completely determines the linear complexity and the least period of the sequence. Actually, the least period of a linear recurring sequence is equal to the period of its minimal polynomial. The *period* (also called the *order*) of a polynomial P in $\mathbf{F}_q[X]$, where $P(0) \neq 0$, is the least positive integer e for which $P(X)$ divides $X^e - 1$. Then, $\mathbf{s}$ has maximal period $q^{\Lambda(s)} - 1$ if and only if its minimal polynomial is a *primitive polynomial* (i.e., if the period of its minimal polynomial is maximal). For instance, the sequence generated by the LFSR shown in Figure 3 has period 3 because its minimal polynomial $1 + X^3$ has period 3. This sequence is $100100100\ldots$ On the other hand, any non-zero sequence generated by the LFSR of length 4 depicted in Figure 1 has period $2^4 - 1 = 15$. Actually, the minimal polynomial of any such sequence corresponds to its characteristic polynomial $P^*(X) = 1 + X + X^4$, because P^* is irreducible. Moreover, P^* is a primitive polynomial. Any sequence $\mathbf{s} = (s_t)_{t \geq 0}$, generated by an LFSR of length L which has a primitive feedback polynomial, has the highest possible linear complexity $\Lambda(\mathbf{s}) = L$ and the highest possible period $q^L - 1$. Such sequences are called <u>maximal-length linear sequences</u> (*m-sequences*). Because of the previous optimal properties, the linear recurring sequences used in cryptography are always chosen to be *m*-sequences. Moreover, they possess good statistical properties (see <u>maximal-length linear sequences</u> for further details). In other terms, the feedback polynomial of an LFSR should always be chosen to be a primitive polynomial.

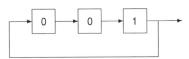

Fig. 3. LFSR of length 3 which generates the same sequence as the LFSR of Figure 2

KEYSTREAM GENERATORS BASED ON LFSRs: However, it is clear that an LFSR should never be used by itself as a keystream generator. If the feedback coefficients of the LFSR are public, the entire keystream can obviously be recovered from the knowledge of any Λ consecutive bits of the keystream, where Λ is the linear complexity of the running-key (which does not exceed the LFSR length). If the feedback coefficients are kept secret, the entire keystream can be recovered from any 2Λ consecutive bits of the keystream by the Berlekamp–Massey algorithm. Therefore, a commonly used technique to produce a pseudo-random sequence which can be used as a running-key is to combine several LFSRs in different ways in order to generate a linear recurring sequence which has a high linear complexity (see, e.g., combination generator, filter generator, ...).

Anne Canteaut

References

[1] Golomb, S.W. (1982). *Shift Register Sequences.* Aegean Park Press, revised edition.

[2] Lidl, R. and H. Niederreiter (1983). *Finite Fields.* Cambridge University Press, Cambridge.

[3] Rueppel, R.A. (1986). *Analysis and Design of Stream Ciphers.* Springer-Verlag, Berlin.

LINEAR SYNDROME ATTACK

The *linear syndrome attack* is an attack on linear feedback shift register-based keystream generators which was presented by Zeng and Huang in 1988 [2]. It is a weak version of the fast correlation attack which was independently proposed by Meier and Staffelbach [1].

Anne Canteaut

References

[1] Meier, W. and O. Staffelbach (1988). "Fast correlation attacks on stream ciphers." *Advances in Cryptology—EUROCRYPT'88*, Lecture Notes in Computer Science, vol. 330, ed. C.G. Günther. Springer-Verlag, Berlin, 301–314.

[2] Zeng, K. and M. Huang (1988). "On the linear syndrome method in cryptanalysis." *Advances in Cryptology—CRYPTO'88*, Lecture Notes in Computer Science, vol. 403, ed. S. Goldwasser. Springer-Verlag, Berlin, 469–478.

[3] Zeng, K., C.H. Yang, and T.R.N. Rao (1990). "An improved linear syndrome algorithm in cryptanalysis with applications." *Advances in Cryptology—CRYPTO'90*, Lecture Notes in Computer Science, vol. 537, eds. A.J. Menezes and S.A. Vanstone. Springer-Verlag, Berlin, 34–47.

L-NOTATION

For $t, \gamma \in \mathbf{R}$ with $0 \le t \le 1$, the notation $L_x[t, \gamma]$ is used for any function of x that equals

$$e^{(\gamma+o(1))(\log x)^t (\log \log x)^{1-t}}, \qquad \text{for } x \to \infty,$$

where logarithms are natural and where $o(1)$ denotes any function of x that goes to 0 as $x \to \infty$ (see O-notation). This function has the following properties:

- $L_x[t, \gamma] + L_x[t, \delta] = L_x[t, \max(\gamma, \delta)]$,
- $L_x[t, \gamma] L_x[t, \delta] = L_x[t, \gamma + \delta]$,
- $L_x[t, \gamma] L_x[s, \delta] = L_x[t, \gamma]$, if $t > s$,
- for any fixed k:
  - $L_x[t, \gamma]^k = L_x[t, k\gamma]$,
  - if $\gamma > 0$ then $(\log x)^k L_x[t, \gamma] = L_x[t, \gamma]$.
- $\pi(L_x[t, \gamma]) = L_x[t, \gamma]$ where $\pi(y)$ is the number of primes $\le y$.

When used to indicate runtimes and for γ fixed, $L_x[t, \gamma]$ for t ranging from 0 to 1 ranges from polynomial time to exponential time in $\log(x)$:

- runtime

$$L_x[0, \gamma] = e^{(\gamma+o(1))\log \log x} = (\log x)^{\gamma+o(1)}$$

is polynomial in $\log(x)$,

- runtimes $L_x[t, \gamma]$ with $0 < t < 1$ are examples of runtimes that are subexponential time in $\log(x)$, i.e., asymptotically greater than polynomial and less than exponential,

- runtime

$$L_x[1, \gamma] = e^{(\gamma+o(1))\log x} = x^{\gamma+o(1)}$$

is exponential in $\log(x)$.

Arjen K. Lenstra

LUBY–RACKOFF CIPHERS

In their celebrated paper [5], Luby and Rackoff showed how to construct $2n$-bit pseudorandom permutations from n-bit random Boolean functions. The constructions (see substitutions and permutations) use three and four rounds in Feistel networks [3] with randomly chosen functions in the round functions (see also block ciphers). Let L and R be respectively the left and the right n-bit halves of a $2n$-bit input. Then one round of a

Feistel network is defined as follows:

$$F(L, R) = (R, L \oplus f(R)),$$

where $f: \{0, 1\}^n \to \{0, 1\}^n$ is a randomly chosen function. In order to make the encryption and decryption routines similar, it is custom to swap the halves of the output of the last round in an r-round Feistel network. The entry on <u>Feistel ciphers</u> provides an overview of practical designs.

Luby and Rackoff's result says that in order to be able to distinguish the three-round construction from a randomly chosen $2n$-bit function with probability close to 1, an attacker needs at least $2^{n/2}$ chosen plaintexts and their corresponding ciphertexts. Such a permutation is called *pseudorandom* [5]. However, if an attacker can mount a chosen plaintext and a chosen ciphertext attack, he is able to distinguish the construction from a randomly chosen $2n$-bit function using two chosen plaintexts and one chosen ciphertext. To see this, choose two plaintexts with left halves L_1 and L_2, where $L_1 \neq L_2$ and with equal right halves R. From the corresponding ciphertexts (T_1, S_1) and (T_2, S_2), compute the ciphertext $(T_1 \oplus L_1 \oplus L_2, S_1)$ and get the corresponding plaintext. Then the right half of this plaintext equals $R \oplus S_1 \oplus S_2$, whereas this would be the case only with probability 2^{-n} in the random case. Luby and Rackoff also showed that in a combined <u>chosen plaintext and chosen ciphertext attack</u> for the four-round construction, an attacker will need roughly $2^{n/2}$ chosen texts to win with probability close to 1. Such a permutation is called *super pseudorandom*.

With q chosen plaintexts one can distinguish the three-round construction from a random function with probability

$$p = 1 - e^{-q(q-1)/2^{n+1}},$$

which is close to 1 for $q \simeq 2^{n/2}$ [1, 8]. Choose plaintexts (L_i, R) for $i = 1, \ldots, q$, where the L_is are (pairwise) distinct and R is a fixed, arbitrary value. The corresponding ciphertexts are denote by (T_i, S_i). Then for the three-round construction with probability p, one finds at least one pair (i, j) for which $i \neq j$, $L_i \oplus L_j = T_i \oplus T_j$ and $S_i = S_j$. For a random $2n$-bit function and with $q \simeq 2^{n/2}$, this happens with only very small probability. Also, with roughly $2^{n/2}$ chosen plaintexts, one can distinguish the four-round construction from a random function. Choose plaintexts (L_i, R) for $i = 1, \ldots, c2^{n/2}$, where c is an integer, the L_is are (pairwise) distinct and R is a fixed, arbitrary value. The corresponding ciphertexts are denote by (T_i, S_i). Then for the four-round construction one expects to find c pairs of plaintexts for which $L_i \oplus L_j = S_i \oplus S_j$, whereas for a random $2n$-bit function one expects to find only $c/2$ such pairs [8]. These results show that the inequalities by Luby and Rackoff are tight, that is, to distinguish the three-round and four-round constructions from a randomly chosen function with probability close to 1, an attacker needs at least but not much more than $2^{n/2}$ chosen plaintexts and their corresponding ciphertexts.

The Luby–Rackoff result has spawned a lot of research in this area, and many different constructions have been proposed, of which only a few are mentioned here. In [9, 11] it was shown that four-round super pseudorandom permutations can be constructed from only one or two (pseudo)random n-bit functions. In the four-round construction, the first and fourth functions can be replaced by simpler "combinatorial" constructions achieving the same level of security as the original construction as shown in [7], which is also a good reference for a survey of this area.

Coppersmith [2] analyzed the four-round construction. It was shown that with $n2^n$ chosen plaintexts, the round functions can be identified up to symmetry. With 8×2^n texts 99.9% of the functions are identified.

There is a trivial upper bound of $O(2^n)$ (see <u>O-notation</u>) for distinguishing constructions with r rounds for any r from a randomly chosen $2n$-bit function. This follows from the fact that the Luby–Rackoff constructions are permutations and with 2^n chosen distinct plaintexts, the resulting ciphertexts will all be distinct, whereas a collision is likely to occur for a truly random function [1, 8]. It has been studied how to distinguish the Luby–Rackoff constructions from randomly chosen $2n$-bit permutations (bijective mappings). However, in the cases using $O(2^{n/2})$ inputs this does not make much of a difference, since in these cases the probability to distinguish a $2n$-bit randomly chosen permutation from a $2n$-bit randomly chosen function is small. Also, for a fixed number of rounds, r, it has been shown that there is an upper bound of $O(2^n)$ for distinguishing the r-round construction from a randomly chosen $2n$-bit permutation [8]. More recent results [4, 6, 10] indicate that with a larger number of rounds, the lower bound for the security of the Luby–Rackoff constructions approaches 2^n.

Lars R. Knudsen

References

[1] Aiello, W. and R. Venkatesan (1996). "Foiling birthday attacks in length-doubling transformations." *Advances in Cryptology—EUROCRYPT'96,*

Lecture Notes in Computer Science, vol. 1070, ed. U. Maurer. Springer-Verlag, Berlin, 307–320.

[2] Coppersmith, D. (1996). "Luby-Rackoff: Four rounds is not enough." Technical Report RC 20674, IBM, December 1996.

[3] Feistel, H., W.A. Notz, and J.L. Smith (1975). "Some cryptographic techniques for machine-to-machine data communications." *Proceedings of IEEE*, 63 (11), 1545–1554.

[4] Knudsen, L.R. and H. Raddum (2003). "Distinguishing attack on five-round Feistel networks." *Electronics Letters*, 39 (16), 1175–1177.

[5] Luby, M. and C. Rackoff (1988). "How to construct pseudorandom permutations from pseudorandom functions." *SIAM Journal of Computing*, 17 (2), 373–386.

[6] Maurer, U. and K. Pietrzak (2003). "The security of Many-Round Luby–Rackoff Pseudo-Random Permutations." *Advances in Cryptology—EUROCRYPT 2003*, Lecture Notes in Computer Science, vol. 2656, ed. E. Biham. Springer-Verlag, Berlin, 544–561.

[7] Naor, M. and O. Reingold (1999). "On the construction of pseudorandom permutations: Luby-Rackoff revisited." *The Journal of Cryptology*, 12 (1), 29–66.

[8] Patarin, J. (1992). "New results on pseudorandom permutations generators based on the DES scheme." *Advances in Cryptology—CRYPTO'91*, Lecture Notes in Computer Science, vol. 576, ed. J. Feigenbaum. Springer-Verlag, Berlin, 301–312.

[9] Patarin, J. (1993). "How to construct pseudorandom and super pseudorandom permutations from one single pseudorandom function." *Advances in Cryptology—EUROCRYPT'92*, Lecture Notes in Computer Science, vol. 658, ed. R.A. Rueppel. Springer-Verlag, Berlin, 256–266.

[10] Patarin, J. (2003). "7 rounds are Enough for $2^{n(1-\epsilon)}$ Security." *Advances in Cryptology—CRYPTO 2003*, Lecture Notes in Computer Science, vol. 2729, ed. D. Boneh. Springer-Verlag, Berlin, 513–529.

[11] Sadeghiyan, B. and J. Pieprzyk (1993). "A construction for super pseudorandom permutations from a single pseudorandom function." *Advances in Cryptology—EUROCRYPT'92*, Lecture Notes in Computer Science, vol. 658, ed. R.A. Rueppel. Springer-Verlag, Berlin, 267–284.

M

MAA

The Message Authentication Algorithm (MAA; see also MAC algorithms) was published in 1983 by Davies and Clayden in response to a request of the UK Bankers Automated Clearing Services (BACS) [1, 2]. In 1987 it became a part of the ISO 8731 banking standard [3], which was revised in 1992. The algorithm is software oriented and has a 32-bit result; as MAA was designed for mainframes in the 1980s, its performance on 32-bit PCs is excellent (about five times faster than the Data Encryption Standard (DES)).

Several undesirable properties of MAA have been identified by Preneel et al. [4]; all these attacks exploit *internal collisions*. A forgery attack requires 2^{17} messages of 256 kbytes or 2^{24} messages of 1 kbyte; the latter circumvents the special MAA mode for long messages defined in the ISO standard. A key recovery attack on MAA requires 2^{32} chosen texts consisting of a single message block. The number of off-line 32-bit multiplications for this attack varies from 2^{44} for one key in 1000 to about 2^{51} for one key in 50. This represents a significant reduction w.r.t. exhaustive key search, which requires 3×2^{65} multiplications. Finally it is shown that MAA has 2^{33} weak keys for which it is rather easy to create a large cluster of collisions. These keys can be detected and recovered with 2^{27} chosen texts.

None of these attacks offer an immediate threat to banking applications, in which a single chosen text is often sufficient to perform a serious attack. Therefore the MAA has not been withdrawn from ISO 8731 [3] after publication of these weaknesses. Nevertheless, it would be advisable to check for the known classes of weak keys [4], and to change the key frequently. The main concern for the next years is that exhaustive key search for a 64-bit key may no longer offer an adequate security level.

B. Preneel

References

[1] Davies, D. (1985). "A message authenticator algorithm suitable for a mainframe computer." *Advances in Cryptology—CRYPTO'84*, Lecture Notes in Computer Science, vol. 196, eds. G.R. Blakely and D. Chaum. Springer-Verlag, Berlin, 393–400.

[2] Davies, D. and W.L. Price (1989). *Security for Computer Networks: An Introduction to Data Security in Teleprocessing and Electronic Funds Transfer* (2nd ed.). Wiley & Sons, New York.

[3] ISO 8731 (1992). *Banking—Approved Algorithms for Message Authentication, Part 2: Message Authentication Algorithm (MAA)*.

[4] Preneel, B., V. Rijmen, and P.C. van Oorschot (1997). "A security analysis of the message authenticator algorithm (MAA)." *European Transactions on Telecommunications*, 8 (5), 455–470.

[5] Preneel, B. and P.C. van Oorschot (1996). "On the security of two MAC algorithms." *Advances in Cruyptology—EUROCRYPT'96*, Lecture Notes in Computer Science, vol. 1070, ed. U. Maurer. Springer-Verlag, Berlin, 19–32.

MAC ALGORITHMS

INTRODUCTION: Electronic information stored and processed in computers and transferred over communication networks is vulnerable to both passive and active attacks. In a passive attack, the opponent tries to obtain information on the data sent; in an active attack, the opponent will attempt to modify the information itself, the claimed sender, and/or intended recipient.

Cryptographic techniques for information authentication focus on the origin of the data (*data origin authentication*) and on the *integrity* of the data, that is, the fact that the data has not been modified. Other aspects that can be important are the timeliness, the sequence with respect to other messages, and the intended recipient(s). Information authentication is particularly relevant in the context of financial transactions and electronic commerce. Other applications where information authentication plays an important role are alarm systems, satellite control systems, distributed control systems, and systems for access control. One can anticipate that authentication of voice and video will become increasingly important.

One can distinguish between three mechanisms for information authentication: MAC algorithms (here MAC is the abbreviation of Message Authentication Code), authentication *codes*, and digital signatures. The first two mechanisms are based on a secret key shared between sender and recipient. This means that if the sender authenticates a message with a MAC algorithm or an authentication code and later denies this, there is no way one can prove that she has authenticated the message (as the recipient could have done this as well). In technical terms, MAC algorithms and authentication codes cannot provide non-repudiation of origin. MAC algorithms are computationally secure: a necessary (but not sufficient) condition for their security is that the computing power of the opponent is limited. Authentication codes are combinatorial objects; one can compute the probability of the success of an attack exactly; this probability is independent on the computing power of the attacker. Digital signatures, introduced in 1976 by W. Diffie and M. Hellman, allow us to establish in an irrefutable way the origin and content of digital information. As they are an asymmetric cryptographic technique, they can resolve disputes between the communicating parties.

MAC algorithms have been used for a long time in the banking community and are thus older than the open research in cryptology that started in the mid seventies. However, MAC algorithms with good cryptographic properties were only introduced after the start of open research in the field. The first reference to a MAC is a 1972 patent application by Simmons et al. (reference 10 in [31]). Financial applications in which MACs have been introduced include electronic purses (such as Proton, CEPS (Common European Purse Specification), and Mondex) and credit/debit applications (e.g., the EMV-standard). MACs are also being deployed for securing the Internet (e.g., IP security, see Ipsec and transport layer security, see Transport Layer Security (TLS)). For all these applications MACs are preferred over digital signatures because they are two to three orders of magnitude faster, and MAC results are $4\ldots16$ bytes long compared to $40\ldots128$ bytes for signatures. On present day computers, software implementations of MACs can achieve speeds from $8\ldots50$ cycles/byte, and MAC algorithms require very little resources on inexpensive 8-bit smart cards and on the currently deployed Point of Sale (POS) terminals. The disadvantage is that they rely on shared symmetric keys, the management of which is more costly and harder to scale than that of asymmetric key pairs.

DEFINITION: A MAC algorithm MAC() consists of three components:
- A key generation algorithm; for most MAC algorithms the secret key K is a random bit-string of length k – typical values for k are $56\ldots256$.
- A MAC generation algorithm; this algorithm computes from the text input x and the secret key K a MAC value $\mathrm{MAC}_K(x)$, which is bit-string of fixed length m—typical values for m

are $24\ldots96$. A MAC generation algorithm can be randomized; in this case, a random string is added to the set of inputs and outputs of this algorithm. A MAC generation algorithm can be stateful; in this case the MAC generation algorithm keeps an internal state, which influences the result (e.g., a counter which is incremented after every use).
- A MAC verification algorithm; on input the text x, the MAC value $\mathrm{MAC}_K(x)$, the key K (and possibly a random string), the algorithm verifies whether the MAC value is correct or not; in practice this verification consists of a computation of the MAC value on the text and a check whether the result is identical to the MAC value provided.

Note that it is common to abuse terminology by abbreviating both the "MAC value" and the "MAC algorithm" as the "MAC."

In a communication context, sender and receiver will agree on a secret key (using a key agreement protocol. The sender will compute a MAC value for every message and append this to the message; on receipt of the message, the receiver will apply the MAC verification algorithm, which typically corresponds to recomputing the MAC value (see Figure 1).

The main security requirement for a MAC algorithm is that it should be hard to forge a MAC value on a new text, that is, to compute a MAC for a new text. The resistance of a MAC algorithm against forgeries is also known as computation resistance. The next section investigates in more detail the security of MAC algorithms.

SECURITY OF MAC ALGORITHMS: Attacks on MAC algorithms can be classified according to the type of control an adversary has over the device computing or verifying the MAC value. In a *chosen text attack*, an adversary may request and receive MACs corresponding to a number of texts of his choice, before completing his attack. In an *adaptive* chosen-text attack, requests may depend on the outcome of previous requests. In a *MAC-verification attack*, the opponent can submit text-MAC pairs of his choice to the verification device.

An opponent who tries to deceive the receiver, knows the description of the MAC algorithm, but he does not know the secret key. Attacks can be further distinguished based on their goals:

Forgery attack: This attack consists of predicting the value of $\mathrm{MAC}_K(x)$ for a text x without initial knowledge of K. If the adversary can do this for a single text, he is said to be capable of existential forgery. If the adversary is able to determine the MAC for a text of his choice,

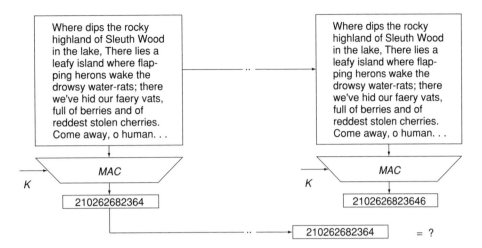

Fig. 1. Using a MAC algorithm for data authentication

he is said to be capable of <u>selective forgery</u>. Ideally, existential forgery is computationally infeasible; a less demanding requirement is that only selective forgery is so. Practical attacks often require that a <u>forgery</u> is *verifiable*, i.e., that the forged MAC is known to be correct on beforehand with probability near 1. The text on which a MAC is forged shall be new, which means that it should not be one of the texts used in the MAC generation or verification queries (as this would allow for a trivial attack).

Key recovery attack: This attack consists of finding the key K itself from a number of text/MAC pairs. Such an attack is more powerful than forgery, since it allows for arbitrary selective forgeries. One distinguishes between exhaustive search and shortcut key recovery attacks; ideally no shortcut key recovery attacks should exist.

We can now informally state the security requirement for a MAC algorithm: it should be computationally infeasible to generate an existential forgery under an adaptive chosen text attack (which also includes MAC verification queries). The success probability of an attacker is often computed as a function of m (the bit-length of the MAC) and the number q of queries.

Note that in certain environments, such as in wholesale banking applications, a chosen text attack is not a very realistic assumption: if an opponent can choose a single text and obtain the corresponding MAC, he can already make a substantial profit. Moreover, texts that are relevant may have a specific structure, which implies that an existential forgery may not pose a threat at all. However, it is better to be on the safe side, and to require resistance against chosen text attacks.

Below four attacks on MAC algorithms are considered: brute force key search; guessing of the

MAC; a generic forgery attack based on internal collisions; and attacks based on cryptanalytical weaknesses.

Brute Force Key Search

If k denotes the bit-length of the key K, one can always try all 2^k key values and check which one is correct. If m is the size of the MAC and if one assumes that $\mathrm{MAC}_K(x)$ is a random function from the key to the MAC, then verification of such an attack requires about $\lceil k/m \rceil$ text-MAC pairs. To see this, note that the expected number of keys which will take the text x to a certain given MAC value is 2^{k-m}. Extending this argument, the expected number of keys which will take $\lceil k/m \rceil$ texts to certain given MAC values is $2^{k-(m\lceil k/m \rceil)} \leq 1$. For most MAC algorithms, the value of k/m lies between 1 and 4; it is reasonable to assume that such a small number of text-MAC pairs is available. The only exceptions are certain banking systems which use one key per transaction; in this case one exploits the combinatorial rather than the cryptographic properties of the MAC algorithm; this corresponds to the use of an <u>authentication</u> *code*.

Note that unlike for confidentiality protection, the opponent can only make use of the key if it is recovered within its active lifetime (which can be reasonably short). On the other hand, a single success during the lifetime of the system might be sufficient. This depends on a cost/benefit analysis, i.e., how much one loses as a consequence of a forgery.

The only way to preclude a key search is to choose a sufficiently large key. In 2004, the protection offered by a 56-bit key is clearly insufficient. One also has to take into account what is known as a variant of 'Moore's Law': the computing power for a given cost is multiplied by four every three years. This implies that if a system is deployed with an intended lifetime of 15 years, an extra security margin of about 10 bits is recommended. Keys of 80–90 bits are adequate for medium term security (10–15 years), and long term protection (50 years or more) is offered by keys of 128 bits. The entry on exhaustive key search provides more details.

MAC Guessing Attack

A second very simple attack is to choose an arbitrary (fraudulent) text, and to append a randomly chosen MAC value. An alternative strategy is to guess the key value and compute the corresponding MAC value. Ideally, the probability that this MAC value is correct is equal to $\max(1/2^m, 1/2^k)$, where m is the number of bits of the MAC value and k is the number of bits in the key. This value should be multiplied with the expected profit corresponding to a fraudulent text, which results in the expected value of one trial. Repeated trials can increase this expected value, but note that in a good implementation, repeated MAC verification errors will result in a security alarm (the forgery is not verifiable). For most applications $k > m$ and $m = 32 \ldots 64$ is sufficient to make this attack uneconomical.

Internal Collision Attack on Iterated MAC Algorithms

Most common message authentication algorithms today are iterated MAC algorithms. The MAC input x is padded to a multiple of the block size, and is then divided into t blocks denoted x_1 through x_t. The MAC involves an initial value $H_0 = IV$, a *compression function* f, an *output transformation* g, and an n-bit ($n \geq m$) *chaining variable* H_i between stage $i - 1$ and stage i:

$$H_i = f(H_{i-1}, x_i), \qquad 1 \leq i \leq t$$
$$\text{MAC}_K(x) = g(H_t).$$

The secret key may be employed in f and/or in g.

Next we describe a general forgery attack by Preneel and van Oorschot [27, 29] that applies to all iterated MACs. Its feasibility depends on the bit sizes n of the chaining variable and m of the MAC result, the nature of the output transformation g, and the number s of common trailing blocks of the known texts ($s \geq 0$). The basic attack requires several known texts, but only a single chosen text. However, under certain conditions restrictions are imposed on the known texts; for example, if the input length itself is an input

to the output transformation, all inputs must have an equal length.

First some terminology is introduced. For an input pair (x, x') with $\text{MAC}_K(x) = g(H_t)$ and $\text{MAC}_K(x') = g(H'_t)$, a collision is said to occur if $\text{MAC}_K(x) = \text{MAC}_K(x')$. This collision is called an *internal* collision if $H_t = H'_t$, and an *external* collision if $H_t \neq H'_t$ but $g(H_t) = g(H'_t)$.

The attack starts with the following simple observation:

LEMMA 1. *An internal Collision for an iterated MAC algorithm allows a verifiable MAC forgery, through a chosen-text attack requiring a single chosen text.*

This follows since for an internal collision (x, x'), $\text{MAC}_K(x \| y) = \text{MAC}_K(x' \| y)$ for any single block y; thus a requested MAC on the chosen text $x \| y$ provides a forged MAC (the same) for $x' \| y$ (here $\|$ denotes concatenation). Note this assumes that the MAC algorithm is deterministic. Also, the forged text is of a special form, which may limit the practical impact of the attack.

The following propositions indicate the complexity to find an internal collision. They are based on the birthday paradox and extensions thereof.

PROPOSITION 1. *Let MAC() be an iterated MAC algorithm with n-bit chaining variable and m-bit result, and an output transformation g that is a permutation. An internal collision for MAC can be found using an expected number of $u = \sqrt{2} \cdot 2^{n/2}$ known text-MAC pairs of at least $t = 2$ blocks each.*

PROPOSITION 2. *Let MAC() be an iterated MAC algorithm with n-bit chaining variable and m-bit result, and output transformation g, which is a random function. An internal collision for h can be found using u known text-MAC pairs of at least $t = 2$ blocks each and v chosen texts of at least three blocks. The expected values for u and v are as follows: $u = \sqrt{2} \cdot 2^{n/2}$ and $v \approx \min(2^{n/2}, 2^{n-m})$.*

PROPOSITION 3. *Let MAC() be an iterated MAC with n-bit chaining variable, m-bit result, a compression function f which behaves like a random function (for fixed x_i), and output transformation g. An internal collision for MAC can be found using u known text-MAC pairs, where each text has the same substring of $s \geq 0$ trailing blocks, and v chosen texts. The expected values for u and v are: $u = \sqrt{2/(s+1)} \cdot 2^{n/2}$; $v = 0$ if g is a permutation or $s + 1 \geq 2^{n-m+6}$, and otherwise $v \approx 2^{n-m}/(s+1)$.*

Weaknesses of the Algorithm

The above attacks assume that no shortcuts exist to break the MAC algorithm (either for forgery or for key recovery). The security of existing MAC algorithms relies on unproven assumptions: even if the security of the MAC algorithm is reduced in a provable way to the pseudo-randomness properties of a block cipher or of the compression function of a hash function, these properties themselves cannot be proved. Therefore, it is recommended to use only well established algorithms which have been subjected to an independent evaluation and a regular review of the algorithm based on progress in cryptanalysis is recommended. A typical example of a weak construction for a MAC algorithm is one which consists of inserting a secret key into the input of a hash function.

PRACTICAL MAC ALGORITHMS: Compared to the number of block ciphers and hash functions, relatively few dedicated MAC algorithms have been proposed [12]. The main reason is that MACs have been derived from other primitives (initially from block ciphers, but also from hash functions), which reduces the need for dedicated proposals. The following section lists the most important constructions.

Based on Block Ciphers

The oldest and most popular MAC algorithm is certainly CBC-MAC, which is based on a block cipher and which has been widely standardized. For a more detailed discussion see CBC-MAC and variants. CBC-MAC is an iterated MAC algorithm. The most common padding method consists of appending a one bit followed by between 0 and $n - 1$ zero bits such that the resulting string has an input length that is a multiple of n [17]. Denote the resulting string as $x_1, x_2, \ldots, x_t$. The compression function for CBC-MAC has the following form:

$$H_i = E_K(H_{i-1} \oplus x_i), \qquad 1 \le i \le t.$$

Here $E_K(x)$ denotes the encryption of x using the k-bit key K with an n-bit block cipher E and $H_0 = IV$ is a fixed initial value, which is set equal to the all zero string. The MAC is then computed as $MAC_K(x) = g(H_t)$, where g is the output transformation.

Bellare et al. [5] have provided a security proof for this scheme with g the identity mapping; their proof is based on the pseudo-randomness of the block cipher and requires that the inputs are of fixed length. They show that CBC-MAC is a pseudo-random function, which is in fact a stronger requirement than being a secure MAC. Most of these variants are vulnerable to an internal collision attack, which requires a single chosen text and about $2^{n/2}$ known texts with n the block length of the block cipher; for a 64-bit block cipher such as DES (see the Data Encryption Standard) this corresponds to 2^{32} known texts. For $m < n$, an additional 2^{m-n} chosen texts are required, which makes the attack less realistic. It is important to note that for most schemes the gap between the lower bound (security proof) and upper bound (best known attack) is quite small. For several of these schemes shortcut key recovery attacks exist as well; lower bounds for the security against these attacks are not known for these schemes.

In practice one needs security for inputs of variable length, which can be achieved by using a different mapping g. These variants and attacks on them are discussed in more detail under CBC-MAC and variants.

The most popular choice for g was the selection of the leftmost $m < n$ bits, $m = 32$ being a very popular variant for CBC-MAC based on DES. However, Knudsen has shown that a forgery attack on this scheme requires $2 \cdot 2^{(n-m)/2}$ chosen texts and two known texts [20].

A better solution for g is the encryption of the last block with a different key, which is known as EMAC. This solution was proposed by the RIPE Consortium in [30]; Petrank and Rackoff have provided a security proof in [24].

$$g(H_t) = E_{K'}(H_t) = E_{K'}(E_K(x_t \oplus H_{t-1})),$$

where K' is a key derived from K. Further optimizations which reduce the overhead due to padding are known as XCBC (three-key MAC) [7] and OMAC [18]. EMAC and OMAC are recommended for use with the Rijndael/AES.

An alternative for g which is popular for use with *DES* consists of replacing the processing of the last block by a two-key triple encryption (with keys $K_1 = K$ and K_2); this is commonly known as the ANSI retail MAC, since it first appeared in [1]:

$$g(H_t) = E_{K_1}(D_{K_2}(H_t)).$$

Here $D_K()$ denotes decryption with key K. This construction increases the strength against exhaustive key search, but it is not without its weaknesses [29]. A better alternative is MacDES [21].

XOR-MAC by Bellare et al. [4] is a randomized construction that allows for a full parallel evaluation; a fixed number of bits in every block (e.g., 32 bits) is used for a counter, which reduces the performance. It has the advantage that it is incremental: small modifications to the input (and to the MAC) can be made at very low cost. Improved variants of XOR-MAC are XECB [14] and PMAC [8].

RMAC increases the security level of EMAC against an internal collision attack by modifying the key in the last encryption with a randomizer [19]. It has the disadvantage that its security proof requires resistance of the underlying block cipher against related key attacks. It was included in NIST's draft special publication [23] which has been withdrawn.

3GPP-MAC and RIPE-MAC are discussed in the item CBC-MAC and variants.

Based on Cryptographic Hash Functions

The availability of very fast dedicated hash functions (such as MD4 and MD5) has resulted in several proposals for MAC algorithms based on these functions. As it became clear that these hash functions are weaker than intended, they were replaced by RIPEMD-160 and by SHA-1.

The first proposed constructions were the secret prefix and secret suffix methods which can be described as follows: $MAC_K(x) = h(K \| x)$, $MAC_K(x) = h(x \| K)$. However, these schemes have some security problems [29]. Consider the secret prefix method and assume that the hash function is an iterated hash function, wherein each iteration n bits of the text (or the key) is processed. Then if one has the MAC of a text x such that the length of $K \| x$ (including padding) is a multiple of n, then it is trivial to compute the value of $MAC_K(x \| y)$ from $MAC_K(x)$ (this assumes that the output transformation is the identity function). Moreover, if the inputs x and x' have the same MAC value and if this is the result of an internal collision, the inputs $x \| y$ and $x' \| y$ will have the same MAC values.

Consider the secret suffix method and assume that the hash function is an iterated hash function. Here an attacker can try to find an internal collision for two texts x and x' ignoring the secret key K. Then if an attacker succeeds the MACs of x and x' will be identical, regardless of the value of K. It is important to notice here that the attacker can perform the computations off-line, that is, one needs no access to the MAC generation device during the first step.

A better proposal is the secret envelope method, or *envelope MAC* which can be described as $MAC_K(x) = h(K_1 \| x \| K_2)$. For this method, Bellare et al. provide a security proof in [2]. This method has been shown to be secure based on the assumption that the compression function of the hash function is a pseudo-random function (when

keyed through the chaining variable). While this is an interesting result, it should be pointed out that the compression function of most hash functions has not been evaluated with respect to this property. For the particular envelope method of Internet RFC 1828 [22], it was shown by Preneel and van Oorschot [28] that an internal collision attack (which requires about $2^{n/2}$ known texts) does not only allow for a forgery but also a key recovery attack. This attack exploits the standard padding algorithms in modern hash functions and illustrates that one has to be very careful when transforming a hash function into a MAC algorithm.

To account for such pitfalls the MDx-MAC has been proposed which extends the envelope method by also introducing secret key material into every iteration [27]. This makes the pseudo-randomness assumption more plausible. Moreover, it precludes the key recovery attack by extending the keys to complete blocks. MDx-MAC has been included in the ISO standard [17].

HMAC is the most popular hash function variant, which uses a nested construction (with padded keys):

$$MAC_K(x) = h(K_2 \| h(K_1 \| x)).$$

The security of HMAC is guaranteed if the hash function is collision resistant for a secret value H_0, and if the compression function itself is a secure MAC for one block (with the secret key in the H_i input and the text in the x_i input) [3]. While these assumptions are weaker than for the secret envelope method, it still requires further validation for existing hash functions. HMAC is used for providing message authentication in the Internet Protocol Ipsec, TLS (see Transport Layer Security) and has been included in an ISO [17] and FIPS standard [13].

Two-Track-MAC is another construction which exploits the presence of two parallel internal trails in RIPEMD-160 [11].

Dedicated MAC Algorithms

The Message Authentication Algorithm (MAA) was designed in 1983 by Davies and Clayden [9, 10]. In 1987 it became a part of the ISO 8731 banking standard [16]. Several weaknesses of MAA have been identified by Preneel et al. [26], but none of these form an immediate threat to existing applications. However, it would be advisable to check for the known classes of weak keys [26] and to change the key frequently.

A cryptanalysis of an early MAC algorithm can be found in [25]. A few MAC algorithms in use have not been published, such as the S.W.I.F.T. authenticator and the Swedish algorithm Data

Seal. Several proprietary MAC algorithms that can process only short messages algorithm have been leaked: this includes the Sky Videocrypt system of British Sky Broadcasting (which was leaked out in 1993 and replaced), the COMP128 algorithm which was used by certain GSM operators as A3/A8 algorithm (a fix for its weaknesses is proposed in [15]; it has been upgraded to COMP128-2 and COMP128-3) and the function used in the SecureID token (for which an analysis can be found in [6]). Proprietary algorithms which have not been leaked include Telepass 1 (from Bull) and the DECT Standard Authentication Algorithm (DSAA) for cordless telephony.

B. Preneel

References

[1] ANSI X9.19 (1986). *Financial Institution Retail Message Authentication*. American Bankers Association.

[2] Bellare, M., R. Canetti, and H. Krawczyk (1996). "Pseudorandom functions revisited: The cascade construction and its concrete security." *Proc. 37th Annual Symposium on the Foundations of Computer Science*. IEEE, 514–523. Full version http://www.cs.ucsd.edu/users/mihir/papers/cascade.html

[3] Bellare, M., R. Canetti, and H. Krawczyk (1996). "Keying hash functions for message authentication." *Advances in Cryptology—CRYPTO'96*, Lecture Notes in Computer Science, vol. 1109, ed. N. Koblitz. Springer-Verlag, Berlin, 1–15. Full version http://www.cs.ucsd.edu/users/mihir/papers/hmac.html

[4] Bellare, M., R. Guérin, and P. Rogaway (1995). "XOR MACs: New methods for message authentication using block ciphers." *Advances in Cryptology—CRYPTO'95*, Lecture Notes in Computer Science, vol. 963, ed. D. Coppersmith. Springer-Verlag, Berlin, 15–28.

[5] Bellare, M., J. Kilian, and P. Rogaway (2000)."The security of cipher block chaining." *Journal of Computer and System Sciences*, 61 (3), 362–399. Earlier version in *Advances in Cryptology—CRYPTO'94*, Lecture Notes in Computer Science, vol. 839, ed. Y. Desmedt. Springer-Verlag, Berlin, 1994, 341–358.

[6] Biryukov, A., J. Lano, and B. Preneel (2004). "Cryptanalysis of the alleged SecurID hash function." *Selected Areas in Cryptography 2003*, Lecture Notes in Computer Science, vol. 3006, eds. M. Matsui and R. J. Zuccheratop. Springer-Verlag, Berlin, 130–144.

[7] Black, J. and P. Rogaway (2000). "CBC-MACs for arbitrary length messages." *Advances in Cryptology—CRYPTO 2000*, Lecture Notes in Computer Science, vol. 1880, ed. M. Bellare. Springer-Verlag, Berlin, 197–215.

[8] Black, J. and P. Rogaway (2002). "A block-cipher mode of operation for parallelizable message authentication." *Advances in Cryptology—EUROCRYPT 2002*, Lecture Notes in Computer Science, vol. 2332, ed. L. Knudsen. Springer-Verlag, Berlin, 384–397.

[9] Davies, D. (1985). "A message authenticator algorithm suitable for a mainframe computer." *Advances in Cryptology—CRYPTO'84*, Lecture Notes in Computer Science, vol. 196, eds. G.R. Blakley and D. Chaum. Springer-Verlag, Berlin, 393–400.

[10] Davies, D. and W.L. Price (1989). *Security for Computer Networks: An Introduction to Data Security in Teleprocessing and Electronic Funds Transfer* (2nd ed.). Wiley & Sons, New York.

[11] den Boer, B., B. Van Rompay, B. Preneel, and J. Vandewalle (2001). "New (Two-Track-)MAC based on the two trails of RIPEMD." *Selected Areas in Cryptography 2001*, Lecture Notes in Computer Science, vol. 2259, eds. S. Vaudenay and A.M. Youssef. Springer-Verlag, Berlin, 314–324.

[12] FIPS 180-1 (1995). "Secure hash standard." NIST, US Department of Commerce, Washington, DC.

[13] FIPS 198 (2002). "The keyed-hash message authentication code (HMAC)." NIST, US Department of Commerce, Washington, DC.

[14] Gligor, V. and P. Donescu (2002). "Fast encryption and authentication: XCBC encryption and ECB authentication modes." *Fast Software Encryption*, Lecture Notes in Computer Science, vol. 2355, ed. M. Matsui. Springer-Verlag, Berlin, 92–108.

[15] Handschuh, H. and P. Paillier (2000). "Reducing the collision probability of alleged Comp128." *Smart Card Research and Applications*, Lecture Notes in Computer Science, vol. 1820, eds. J.-J. Quisquater and B. Schneier. Springer-Verlag, Berlin, 380–385.

[16] ISO 8731 (1992). "Banking—approved algorithms for message authentication, Part 1: DEA, 1987, Part 2: Message Authentication Algorithm (MAA)."

[17] ISO/IEC 9797 (2002). "Information technology—security techniques—Message Authentication Codes (MACs)." Part 1: Mechanisms using a block cipher, 1999. Part 2: Mechanisms using a dedicated hash-function, 2002.

[18] Iwata, T. and K. Kurosawa (2003). "OMAC: One key CBC MAC." *Fast Software Encryption*, Lecture Notes in Computer Science, vol. 2887, ed. T. Johansson. Springer-Verlag, Berlin, 129–153.

[19] Jaulmes, E., A. Joux, and F. Valette (2002). "On the security of randomized CBC-MAC beyond the birthday paradox limit: A new construction." *Fast Software Encryption*, Lecture Notes in Computer Science, vol. 2365, eds. J. Daemen and V. Rijmen. Springer-Verlag, Berlin, 237–251.

[20] Knudsen, L. (1997). "Chosen-text attack on CBC-MAC." *Electronics Letters*, 33 (1), 48–49.

[21] Knudsen, L. and B. Preneel (1998). "MacDES: MAC algorithm based on DES." *Electronics Letters*, 34 (9), 871–873.

[22] Metzger, P. and W. Simpson (1995). "IP authentication using keyed MD5." Internet Request for Comments 1828.

[23] NIST Special Publication 800-38B (2002). *Draft Recommendation for Block Cipher Modes of Operation: The RMAC Authentication Mode.*

[24] Petrank E. and C. Rackoff (2000). "CBC MAC for real-time data sources." *Journal of Cryptology*, 13 (3), 315–338.

[25] Preneel, B., A. Bosselaers, R. Govaerts, and J. Vandewalle (1990). "Cryptanalysis of a fast cryptographic checksum algorithm." *Computers & Security*, 9 (3), 257–262.

[26] Preneel, B., V. Rijmen, and P.C. van Oorschot (1997). "A security analysis of the message authenticator algorithm (MAA)." *European Transactions on Telecommunications*, 8 (5), 455–470.

[27] Preneel, B. and P.C. van Oorschot (1995). "MDx-MAC and building fast MACs from hash functions." *Advances in Cryptology—CRYPTO'95*, Lecture Notes in Computer Science, vol. 963, ed. D. Coppersmith. Springer-Verlag, Berlin, 1–14.

[28] Preneel, B. and P.C. van Oorschot (1996). "On the security of two MAC algorithms." *Advances in Cryptology—EUROCRYPT'96*, Lecture Notes in Computer Science, vol. 1070, ed. U. Maurer. Springer-Verlag, Berlin, 19–32.

[29] Preneel, B. and P.C. van Oorschot (1999). "On the security of iterated message authentication codes." *IEEE Trans. on Information Theory*, IT-45 (1), 188–199.

[30] RIPE (1995). "Integrity primitives for secure information systems." Final Report of RACE Integrity Primitives Evaluation (RIPE-RACE1040). Lecture Notes in Computer Science, vol. 1007, eds. A. Bosselaers and B. Preneel. Springer-Verlag, Berlin.

[31] Simmons, G.J. (1991). "How to insure that data acquired to verify treaty compliance are trustworthy." *Contemporary Cryptology: The Science of Information Integrity*, ed. G.J. Simmons. IEEE Press, Piscataway, NJ, 615–630.

MAN-IN-THE-MIDDLE ATTACK

The man-in-the-middle attack is a very old attack that has been used against a wide range of protocols, going from login protocols, entity authentication protocols, etc.

To illustrate, consider Secure Socket Layer (SSL), used to protect the privacy and authenticity of WWW traffic. Current Public Key Infrastructures are either nonexistent or have very poor security, if any (for an incident example, see [4]). This implies that a man-in-the-middle can be launched as following. Suppose Alice wants to have a secure WWW connection to Bob's WWW page. When Eve is between Alice and Bob, Eve will pretend that her made up public key is the one of Bob. So, when Alice accepts the fake certificate, she is in fact sending information to Eve. Eve can then start an SSL connection with the real WWW page of Bob. Even though encryption and authentication (see also MAC algorithms and digital signature schemes) is used, once Eve has convinced Alice that her made up key is the public key of Bob, Eve can be an active eavesdropper.

Man-in-the-middle attacks can also be launched against entity authentication schemes [1], allowing a third party, let us say Eve, to pretend to be Alice. For possible solutions consult [1–3].

Yvo Desmedt

References

[1] Bengio, S., G. Brassard, Y.G. Desmedt, C. Goutier, and J.-J. Quisquater (1991). "Secure implementations of identification systems." *Journal of Cryptology*, 4 (3), 175–183.

[2] Beth T. and Y. Desmedt (1991). "Identification tokens—or: Solving the chess grandmaster problem." *Advances in Cryptology—CRYPTO'90*, Lecture Notes in Computer Science, vol. 537, eds. A.J. Menezes and S.A. Vanstone. Springer-Verlag, Berlin, 169–176.

[3] Brands S. and D. Chaum (1994). "Distance-bounding protocols." *Advances in Cryptology—EUROCRYPT'93*, Lecture Notes in Computer Science, vol. 765, ed. T. Helleseth. Springer-Verlag, Berlin, 344–359.

[4] Erroneous VeriSign-issued digital certificates pose spoofing hazard. Updated: June 23, 2003, Microsoft Security Bulletin MS01-017, http://www.microsoft.com/technet/treeview/default.asp?url=/technet/security/bulletin/MS01-017.asp, March 22, 2001.

MARS

MARS [2] is a block cipher designed by IBM. It was proposed as an *Advanced Encryption Standard (Rijndael/AES)* candidate in 1998 and was one of the five finalists in the AES selection process.

MARS has a block size of 128 bits and accepts secret keys of variable lengths, ranging from 128 to 1248 bits. It is a word-oriented cipher, i.e., all operations are performed on 32-bit words. The main distinguishing feature of MARS is its heterogeneous structure, inspired by the theoretical work of Naor and Reingold. The cipher consists of three stages:

Forward mixing: First, four 32-bit subkeys are added to the data entering the cipher. The resulting block of four 32-bit words is then passed through 8 rounds of a "type-3" Feistel network. In each round, one data word is used to modify

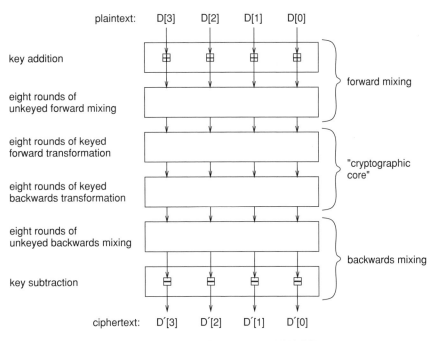

Fig. 1. High-level structure of MARS

the three other words. The Feistel network uses two fixed 8×32-bit S-boxes S_0 and S_1, and does not depend on the secret key in any way.

Cryptographic core: The core of the encryption algorithm consists of a type-3 Feistel network of 2×8 rounds. In each round, the data is modified using an *E*-function which takes as input one data word and two key words, and produces three output words. The *E*-function itself uses many different components: a 9×32-bit S-box S, a 32-bit multiplication, fixed and data-dependent rotations, an addition, and XORs. After eight "forward" rounds, eight slightly different "backwards" rounds are applied.

Backwards mixing: This layer is essentially the inverse of the forward mixing layer. Notice, however, that different subkeys are used.

At present, only a few attacks on reduced-round versions of MARS have been presented. Note that, due to its heterogeneous structure, MARS can be downscaled in many different ways. A first approach is to concentrate on the core rounds. In [1], Biham and Furman have shown impossible differentials over 8 out of 16 core rounds. An attack breaking 11 rounds using amplified boomerang techniques is presented by Kelsey, Kohno, and Schneier [3, 4]. The same authors also proposed a straight forward meet-in-the-middle attack on a MARS version with only five core rounds, but with full forward and backwards mixing.

Christophe De Cannière

References

[1] Biham, E. and V. Furman (2000). "Impossible differential on 8-round MARS core." *Proceedings of the Third AES Candidate Conference*, 186–194.

[2] Burwick, C., D. Coppersmith, E.D. Avignon, R. Gennaro, S. Halevi, C. Jutla, S.M. Matyas Jr., L. O'Connor, M. Peyravian, D. Safford, and N. Zunic (1998). "MARS—a candidate cipher for AES." *Proceedings of the First AES Candidate Conference*. National Institute of Standard and Technology.

[3] Kelsey, J., T. Kohno, and B. Schneier (2001). "Amplified boomerang attacks against reduced-round MARS and Seprepent." *Fast Software Encryption, FSE 2000*, Lecture Notes in Computer Science, vol. 1978, ed. B. Schneier. Springer-Verlag, Berlin, 75–93.

[4] Kelsey, J. and B. Schneier (2000). "MARS attacks! Preliminary cryptanalysis of reduced-round MARS variants." *Proceedings of the Third AES Candidate Conference*, 169–185.

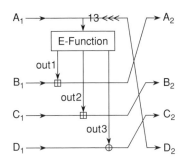

Fig. 2. One round of the type-3 Feistel network of the core (forward mode)

MASH HASH FUNCTIONS (MODULAR ARITHMETIC SECURE HASH)

MASH-1 and MASH-2 are constructions for hash functions based on modular arithmetic. A hash function is a cryptographic algorithm that takes input strings of arbitrary (or very large) length, and maps these to short fixed length output strings. MASH-1 and MASH-2 are unkeyed cryptographic hash functions that have been designed to have the following properties: preimage resistance, second preimage resistance and collision resistance.

In the following, N denotes an *RSA* modulus (see RSA public key encryption), that is, the product of two large primes and m denotes its length in bits or $m = \lceil \log_2 N \rceil$. The length n of the chaining variables in bits is then equal to the largest multiple of 16 strictly smaller than m. The length of the message blocks is equal to $n/2$ bits. The specification also needs a prime number p with $\lceil \log_2 p \rceil \leq m/2$; the prime p shall not be a divisor of N and the three most significant bits of p shall be equal to 1. The operation $\|$ denotes the concatenation of strings.

MASH-1 is defined for input strings of length $<2^{n/2}$ bits. If necessary, the string X is right-padded with '0' bits to obtain a string with bit-length a multiple of $n/2$ and the resulting string is divided into t $n/2$-bit blocks denoted with X_1, $X_2, \ldots, X_t$. Next a block X_{t+1} is added which contains the binary representation of the input string X in bits, left-padded with '0' bits. Subsequently each block X_i is expanded to an n-bit block $\tilde{X}_i$ as follows: insert four 1 bits before every 4-bit substring of X_i.

The MASH-1 compression function, which maps $3n/2$ bits to n bits, is defined as follows (see modular arithmetic):

$$H_i = \left(\left(\left(\tilde{X}_i \oplus H_{i-1} \right) \vee A \right)^2 (\text{mod } N) \right) \sim n \, \oplus H_{i-1}.$$

Here $A = \texttt{0xF00...00}$ and $\sim n$ denotes that the rightmost n bits of the m-bit result are kept. The iteration starts with the all '0' string or $H_0 = 0^n$ and runs for $1 \leq i \leq t+1$.

At the end, a rather complex output transformation is applied to H_{t+1}. First H_{t+1} is divided into four $n/4$-bit blocks defined as follows: $H_{t+1} = Y_0\|Y_1\|Y_1\|Y_3$. Define 12 $n/4$-bit blocks $Y_i = Y_{i-1} \oplus Y_{i-4}$, $4 \leq i \leq 15$. Combine the Y_i to eight additional $n/2$-bit blocks $X_{t+1+i} = Y_{2i-2}\|Y_{2i-1}$, $1 \leq i \leq 8$, transform the X_{t+1+i} blocks to $\tilde{X}_{t+1+i}$, and perform eight additional iterations of the compression function with these blocks. Finally the hash result is computed as $H_{t+1+8} \mod p$.

MASH-2 is obtained by replacing in MASH-1 the exponent 2 by the exponent $257 = 2^8 + 1$.

The redundancy in the block $\tilde{X}_i$ (four '1' bits in every byte) and the additional operations ($\vee A$ and $\sim n$) intend to preclude a correcting block attack. The complex output transformation destroys some of the mathematical structure of the modular exponentiation.

When the factorization of the modulus is not known, the best known (see integer factoring) (2nd) preimage and collision attacks on MASH-1 require $2^{n/2}$ and $2^{n/4}$ operations [2]; they are thus no better than brute force attacks. While to date no efficient attacks are known that exploit the factorization of the modulus, knowledge of the factorization may reduce the security level. Therefore it is strongly recommended that the modulus N is generated by a trusted third party (who deletes the factors after generation) or by using multiparty computation (e.g. Boneh and Franklin [1] and Frankel et al. [3]).

Both MASH-1 and MASH-2 have been included in Part 4 of ISO/IEC 10118 [4].

Bart Preneel

References

[1] Boneh, D. and M. Franklin (1997). "Effcient generation of shared RSA keys." *Advances in Cryptology—CRYPTO'97*, Lecture Notes in Computer Science, vol. 1294, ed. B. Kaliski. Springer-Verlag, Berlin, 425–439.

[2] Coppersmith, D. and B. Preneel (1995). "Comments on MASH-1 and MASH-2." ISO/IEC JTC1/SC27/N1055, February 21.

[3] Frankel, Y., P.D. MacKenzie, and M. Yung (1998). "Robust efficient distributed RSA-key generation." *Proceedings of 30th ACM Symposium on the Theory of Computing*, 663–672.

[4] ISO/IEC 10118. "Information technology—security techniques—hash-functions." Part 1: General, 2000. Part 2: Hash-functions using an n-bit block cipher algorithm, 2000. Part 3: Dedicated hash-functions, 2003. Part 4: Hash-functions using modular arithmetic, 1998.

MASTER KEY

A *master key* is a cryptographic key (typically a symmetric key (see symmetric cryptosystem)) whose sole purpose is to protect other keys, such as session keys, while those keys are in storage, in use, or in transit. This protection may take one of two forms: the master key may be used to encrypt the other keys; or the master key may be used to generate the other keys (for example, if the master key is k_0, session key k_1 may be formed by hashing (see hash function) the concatenation of k_0 and the digit "1", session key k_2 may be formed by hashing the concatenation of k_0 and the digit "2", and so on).

Master keys are usually not themselves cryptographically protected; rather, they are distributed manually or initially installed in a system and protected by procedural controls and/or by physical or electronic isolation.

Carlisle Adams

References

[1] Menezes, A., P. van Oorschot, and S. Vanstone (1997). *Handbook of Applied Cryptography*. CRC Press, Boca Raton, FL.

[2] Schneier, B. (1996). *Applied Cryptography: Protocols, Algorithms, and Source Code in C* (2nd ed.). John Wiley & Sons, New York.

MAURER'S METHOD

Maurer's method generates provably prime numbers, which are *nearly* random. The method is described in [2].

In Maurer's method, a certificate of primality for a number n is a triplet of numbers, (R, F, a), plus the prime factorization of F, where $2RF + 1 = n$, and such that (see modular arithmetic)
1. $a^{n-1} \equiv 1 \bmod n$ and
2. $a^{(n-1)/q_j} - 1$ is relatively prime to n for all $1 \leq j \leq r$, where $F = q_1^{\beta_1}, \ldots, q_r^{\beta_r}$ is the prime factorization of F.

This triplet of numbers guarantees that all prime factors of n are of the form $mF + 1$ for some positive integer m (the proof of this lemma can be found in [2] but is too complicated to be included here). In particular, if $F \geq \sqrt{n}$ then n must be prime as the product of any two primes of the form $mF + 1$ is at least $F^2 + 2F + 1 > F^2 \geq n$.

Maurer's algorithm generates a prime at random by generating R and F at random with the prime factorization of F known, and testing to see if a random a makes a certificate of primality with R and F for $n = 2RF + 1$. In order to generate F at random with known factorization, we pick sizes for the primes of F at random according to a properly constructed distribution, and then generate primes of those sizes recursively. (As a base case, random selection and trial division or some other simple test is used to generate sufficiently small

primes.) The manner in which these sizes are generated is rather complicated and involved number-theoretically, but is essentially along the lines of Bach's algorithm [1]. As any certificate actually proves that n is prime, none will ever be found for a composite number. (In fact, most will fail the single Fermat test included in the certificate. See Fermat primality test.) Furthermore, the probability that a random base a does form a certificate for $n = 2RF + 1$ when n is prime is approximately $\phi(F)/F$, where $\phi(n)$ is Euler's totient number, the number of positive values less than or equal to n which are relatively prime to n. This ratio is high (approximately $1 - \sum_{j=1}^r 1/q_j$), so the probability that a prime number will be recognized is nearly 1.

As is, Maurer's method generates prime numbers close to uniformly, so they are "nearly" random. More efficient variants are also possible at the cost of a less uniform distribution.

Moses Liskov

References

[1] Bach, E. (1988). "How to generate factored random numbers." *SIAM Journal on Computing*, 17 (4), 173–193.

[2] Maurer, Ueli M. (1995). "Fast generation of prime numbers and secure public-key cryptographic parameters." *Journal of Crypology*, 8 (3), 123–155.

MAXIMS

Here the most often quoted cryptological security maxims are listed [1].

Maxim Number One: "One should not underrate the adversary."

Della Porta's maxim: Only a cryptanalyst, if anybody, can judge the security of a cryptosystem (Auguste Kerckhoffs, [5] formulating the knowledge of the 16th century cryptologist Giambattista Della Porta [6]). To this, David Kahn remarked: "Nearly every inventor of a cipher system has been convinced of the unsolvability of his brainschild."

Kerckhoffs' maxim: "No inconvenience should occur if the cryptosystem falls into the hands of the enemy" [5].

Givierge's maxim: "Superficial complications can be illusory, for they can provide the cryptographer with a false sense of security" (Marcel Givierge, French cryptanalyst in WWI [2,3].

Rohrbach's maxim: "In judging the encryption security of a class of methods, cryptographic faults and other infringements of security discipline are to be taken into account." To this, Otto Horak remarked: "Security of a weak cipher method is not increased by trying to keep it [the method] secret." Thus, among other recommendations, the <u>key</u> has to be changed frequently, a periodic key is dangerous and, in the ideal case, a random one-time key is to be used [7].

Shannon's maxim: "The enemy knows the general system being used" [8].

Kahn's maxim: "Cryptographic errors, blunders, and faults, can significantly simplify unauthorized decryption. To this, David Kahn [4] remarked "A cryptographer's error is the cryptranalysts only hope.""

Friedrich L. Bauer

References

[1] Bauer, F.L. (2002). "Decrypted secrets." *Methods and Maxims of Cryptology*. Springer-Verlag, Berlin.

[2] Givierge, Marcel (1925). *Questions de Chiffre, Revue Militaire Française*, vol. 94 (June 1924), 398–417 (July 1924) 59–78, Paris.

[3] Givierge, Marcel (1925). *Cours de Cryptographie*. Berger-Levrault, Paris.

[4] Kahn, David (1967). *The Codebreakers*. Macmillan, New York.

[5] Kerckhoffs (1883). "Auguste, La Cryptographie militaire." *Journal des Sciences Militaires*, 9 (January) 5–38, (February) 161–191. Available on http://www.cl.ac.uk/usweatfapp2/kerckhoffs/

[6] Porta, Giambattista Della (1563). *De Furtivis Literarum Notis*, Naples.

[7] Rohrbach, Hans (1939–1946). Mathematische und Maschinelle Methoden beim Chiffrieren und Dechiffrieren. *FIAT Review of German Science*, 1939–1946: *Applied Mathematics*, 3 (I), 233–257, Wiesbaden: Office of Military Government for Germany, Field Information Agencies, 1948.

[8] Shannon, Claude E. (1949). "Communication theory of secrecy systems." *Bell Systems Technical Journal*, 28, 656–715.

MAXIMAL-LENGTH LINEAR SEQUENCE

Among the most popular *sequences* for applications are the maximal-length <u>linear feedback shift register</u> sequences (*m*-sequences). The balance, run distribution and <u>auto-correlation</u> properties of these sequences resemble properties one expects to find in random sequences. The *m*-sequences are the main ingredients in many important sequence families used in communication and in many stream cipher systems. We give a descrip-

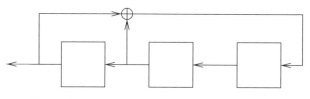

Fig. 1. Shift register for $s_{t+3} = s_{t+1} + s_t \pmod{2}$

tion of *m*-sequences over the binary alphabet $GF(2) = \{0, 1\}$, even though the sequences can be defined with symbols from any finite field with q elements, where q is a prime power.

A simple and efficient method to generate a sequence is using a linear recursion. For example the recurrence relation (see <u>modular arithmetic</u>)

$$s_{t+3} = s_{t+1} + s_t \pmod{2}$$

with initial state $(s_0, s_1, s_2) = (001)$ generates a periodic sequence

0010111 0010111 0010111 0010111 ...

of period $e = 7$. Different initial states lead to different sequences which can be cyclic shifts of each other.

Binary sequences can easily be generated in hardware using a linear shift register. One example of a shift register is shown in Figure 1. The register consists of "flip-flops" each containing a 0 or a 1. The shift register is controlled by an external clock (not shown in the figure). Each time unit shifts each bit one step to the left and replaces the right most bit by the sum (mod 2) of the two leftmost bits. The register implements the recursion $s_{t+3} = s_{t+1} + s_t \pmod{2}$, which with initial state $(s_0, s_1, s_2) = (001)$ gives the periodic sequence 0010111... above. Table 1 shows the content of the shift register at each time unit.

A linear recursion of degree n is given by

$$\sum_{i=0}^{n} f_i\, s_{t+i} = 0,$$

Table 1. Shift register content in Figure 1 generating an *m*-sequence

t	$S0$	$S1$	$S2$
0	0	0	1
1	0	1	0
2	1	0	1
3	0	1	1
4	1	1	1
5	1	1	0
6	1	0	0
7	0	0	1
.	.	.	.

where coefficients $f_i \in GF(2)$ for $0 < i < n$ and $f_0 = f_n = 1$. The *characteristic polynomial* of the recursion is defined by

$$f(x) = \sum_{i=0}^{n} f_i x^i.$$

The initial state and the given recurrence relation uniquely determine the generated sequence. A linear shift register with a characteristic polynomial of degree n generates 2^n different sequences corresponding to the 2^n different initial states and these form a vector space over $GF(2)$.

An n-bit linear shift register has at most 2^n different states. Since the zero state is always followed by the zero state, all sequences generated by a linear shift register have period at most $2^n - 1$. A *maximal length shift register sequence* (*m-sequence*) is a periodic sequence of maximal period $2^n - 1$ generated by a linear shift register of degree n. *The period of a polynomial $f(x)$ is defined as the smallest positive integer e such that $f(x)|x^e - 1$.* Let $f(x)$ be an <u>irreducible polynomial</u> of degree n and period $e = 2^n - 1$. Such a polynomial is called a *primitive polynomial* (see <u>primitive element</u>). The corresponding shift register generates an m-sequence when the initial state is nonzero. Any m-sequences has a primitive characteristic polynomial.

Binary m-sequences is perhaps the best known family of sequences. Table 2 shows some m-sequences and the corresponding characteristic polynomials. Figure 1 shows a shift register having $f(x) = x^3 + x + 1$ as characteristic polynomial and that generates an m-sequence of period $e = 7$ when the initial state is nonzero. Important properties for a binary m-sequence $\{s_t\}$ of period $2^n - 1$ are:

- (Balance property) In a period of the m-sequence there are 2^{n-1} ones and $2^{n-1} - 1$ zeros.
- (Multigram property) When t runs through $0, 1, \ldots, 2^n - 2$, the n-tuple

$$(s_t, s_{t+1}, \ldots, s_{t+n-1})$$

runs through all binary n-tuples except for the n-tuple $(0, 0, \ldots, 0)$ which do not occur.

Table 2. Characteristic polynomials and m-sequences

Degree	$f(x)$	m-sequence	Period
2	$x^2 + x + 1$	011	3
3	$x^3 + x + 1$	0010111	7
3	$x^3 + x^2 + 1$	0011101	7
4	$x^4 + x + 1$	000100110101111	15
4	$x^4 + x^3 + 1$	000111101011001	15

- (Shift-and-add property) For any τ, $0 < \tau \le 2^n - 2$, there exists a δ, depending on τ, such at

$$s_{t+\tau} + s_t = s_{t+\delta}$$

for all $t = 0, 1, 2, \ldots$.

- (Invariance under decimating by 2) There exists a shift τ of the m-sequence such that the shifted sequence $\{u_t\} = \{s_{t+\tau}\}$ is invariant under decimation with two (when every second term of the sequence is selected), i.e., $\{u_t\} = \{u_{2t}\}$.

- (Run property) Let a <u>run</u> denote a consecutive set of zeros or ones in the sequence. In a period of the m-sequence half of the runs have length 1, one-fourth have length 2, one-eight have length 3, etc., as long as the number of runs exceeds 1. Moreover, for each of these lengths, there are equally many 0-runs (<u>gaps</u>) and 1-runs (*blocks*).

EXAMPLE 1. Consider the sequence with characteristic polynomial $x^4 + x + 1$. This is a primitive polynomial and generates the m-sequence $\{s_t\} = 000100110101111$. The sequence has the properties above, it is balanced and contains 7 zeros and 8 ones. Each 4-tuple except the all zero 4-tuple occurs exactly once during a period of the sequence. The shift-and-add property is illustrated by the example

$$\begin{aligned}
s_{t+3} + s_t &= 100110101111000 + 000100110101111 \\
&= 100010011010111 \\
&= s_{t+11}.
\end{aligned}$$

The sequence $\{s_t\}$ is invariant by decimating by 2, i.e., in this case $\tau = 0$, since $\{s_{2t}\} = \{s_t\}$. Further, there are 4 runs of length 1, 2 runs of length 2, 1 run of length 3 and 1 run of length 4. The number of 0-runs and 1-runs of length <3 are the same.

Given a sequence $\{s_t\}$ of period e. In many applications it is important to compare the sequence with its cyclic shifts. The <u>autocorrelation</u> of the binary sequence $\{s_t\}$, at shift τ, is defined as

$$A(\tau) = \sum_{t=0}^{e-1} (-1)^{s_{t+\tau} - s_t}.$$

In particular $A(\tau)$ gives the number of agreements minus the number of disagreements between $\{s_{t+\tau}\}$ and $\{s_t\}$. In most applications it is desirable that a shift of the sequence looks like a "random" sequence compared to itself, i.e., that $|A(\tau)|$ is small for all $\tau \not\equiv 0 \pmod{e}$. A very important property for an m-sequence is its two-level out-of-phase auto-correlation when $\tau \not\equiv 0 \pmod{e}$.

The autocorrelation of an m-sequence $\{s_t\}$ of period $e = 2^n - 1$ is given by:

$$A(\tau) = \begin{cases} -1, & \text{for } \tau \not\equiv 0 \pmod{2^n - 1}, \\ 2^n - 1, & \text{for } \tau \equiv 0 \pmod{2^n - 1}. \end{cases}$$

To prove this, define $u_t = s_{t+\tau} - s_t$ and observe that $\{u_t\}$ obeys the same linear recursion as $\{s_t\}$. This implies that $\{u_t\}$ is an m-sequence when $\tau \not\equiv 0 \pmod{2^n - 1}$ and the balance property of m-sequences gives

$$A(\tau) = \sum_{t=0}^{e-1} (-1)^{s_{t+\tau} - s_t} = \sum_{t=0}^{e-1} (-1)^{u_t} = -1.$$

Balance, multigram and autocorrelation properties of m-sequences are properties one can expect in random binary sequences. The m-sequences obey the three Golomb's randomness postulates R1, R2 and R3. R1 is the balance property, R2 the run property and R3 the two-level autocorrelation property. Actually, these properties of m-sequences were the models for his postulates.

The pseudorandom properties of m-sequences have made them a popular building block in many communication systems and has lead to numerous practical applications, including synchronization, positioning systems, random number generation, stream cipher systems tall and multiple-access-communication.

In order to describe the m-sequence it is useful to introduce the *trace function Tr* which is a mapping from the finite field $GF(2^n)$ to the subfield $GF(2)$ given by:

$$\text{Tr}(x) = \sum_{i=0}^{n-1} x^{2^i}.$$

The trace function satisfies the following:
  (i) $\text{Tr}(x + y) = \text{Tr}(x) + \text{Tr}(y)$, for all $x, y \in GF(2^n)$.
  (ii) $\text{Tr}(x^2) = \text{Tr}(x)$, for all $x \in GF(2^n)$.
  (iii) $|\{x \in GF(2^n) \mid \text{Tr}(x) = b\}| = 2^{n-1}$ for all $b \in GF(2)$.
  (iv) Let $a \in GF(2^n)$. If $\text{Tr}(ax) = 0$ for all $x \in GF 2^n$, then $a = 0$.

Let $f(x)$ be the characteristic polynomial of the binary m-sequence $\{s_t\}$. It is well-known that the zeros of $f(x)$ belong to the finite field $GF(2^n)$. The zeros are α^{2^i} for $i = 0, 1, \ldots, n-1$ where α is a primitive element of $GF(2^n)$, i.e., an element of order $2^n - 1$. The m-sequence can be written simply in terms of the trace representation as

$$s_t = \text{Tr}(a\alpha^t), \qquad a \in GF(2^n)^*,$$

where $GF(2^n)^* = GF(2^n) \backslash \{0\}$. This follows from the properties of the trace function. First observe that $\{s_t\}$ has $f(x)$ as its characteristic polynomial

since,

$$\sum_{i=0}^{n} f_i s_{t+i} = \sum_{i=0}^{n} f_i \text{Tr}(a\alpha^{t+i})$$
$$= \text{Tr}\left(a\alpha^t \sum_{i=0}^{n} f_i \alpha^i\right)$$
$$= \text{Tr}(a\alpha^t f(\alpha^i))$$
$$= 0.$$

The $2^n - 1$ different nonzero values of $a = \alpha^\tau$, $0 \le \tau \le 2^n - 2$ correspond to all the cyclic shifts of the m-sequence. The case $a = 1$ gives the sequence with the property that $\{s_{2t}\} = \{s_t\}$.

Given an m-sequence $\{s_t\}$ of period $2^n - 1$ and let d be relatively prime to $2^n - 1$. The sequence $\{s_{dt}\}$ defined by selecting every dth term in $\{s(t)\}$ is also an m-sequence and all m-sequences of the same period can be obtained in this way. It follows from the trace representation that the characteristic polynomial of $\{s_{dt}\}$ is the primitive polynomial whose zeros are dth powers of the zeros of $f(x)$. The properties of the trace function implies that different m-sequences of the same period generated by distinct primitive characteristic polynomials are cyclically distinct. The number of binary primitive polynomials of degree n is $\phi(2^n - 1)/n$ where $\phi(x)$ is the Euler totient function, the number of positive integers less than x that are relatively prime to x (see modular arithmetic). Thus, there are $\phi(2^n - 1)/n$ cyclically distinct distinct m-sequences of period $2^n - 1$. The example below shows the two $\phi(7)/3 = 2$ cyclically distinct m-sequences of period $e = 7$.

EXAMPLE 2. The primitive polynomial $f_1(x) = x^3 + x + 1$, generates the m-sequence $\{a_t\} = 0010111\ldots$ of period $e = 2^3 - 1 = 7$. The primitive polynomial $f_2(x) = x^3 + x^2 + 1$ generates the m-sequence $\{b_t\} = 0011101\ldots$ of period $e = 2^3 - 1 = 7$. Note that $\{b_t\} = \{a_{3t}\}$ is the sequence obtained by selecting every third element of $\{a_t\}$, where indices are calculated modulo e.

A well-studied problem is to compare two different m-sequences of the same period. Let $C_d(\tau)$ denote the cross-correlation function between the m-sequence $\{s_t\}$ and its decimation $\{s_{dt}\}$. By definition, we have

$$C_d(\tau) = \sum_{i=0}^{2^n-2} (-1)^{s_{t+\tau} - s_{dt}}.$$

If $d \notin \{1, 2, \ldots, 2^{n-1}\}$, (i.e., when the two m-sequences are cyclically distinct), then $C_d(\tau)$ takes on at least three distinct values as τ varies over the set $\{0, 1, \ldots, 2^n - 2\}$. It is therefore of special interest to study the cases when exactly three values occur. The following six decimations give three-valued cross-correlation:

(1) $d = 2^k + 1$, $n/\gcd(n, k)$ odd.

(2) $d = 2^{2k} - 2^k + 1$, $n/\gcd(n, k)$ odd.

(3) $d = 2^{\frac{n}{2}} + 2^{\frac{n+2}{4}} + 1$, $n \equiv 2 \pmod 4$.

(4) $d = 2^{\frac{n+2}{2}} + 3$, $n \equiv 2 \pmod 4$.

(5) $d = 2^{\frac{n-1}{2}} + 3$, n odd.

(6) $d = \begin{cases} 2^{\frac{n-1}{2}} + 2^{\frac{n-1}{4}} - 1, & \text{if } n \equiv 1 \pmod 4 \\ 2^{\frac{n-1}{2}} + 2^{\frac{3n-1}{4}} - 1, & \text{if } n \equiv 3 \pmod 4. \end{cases}$

The crosscorrelation function of m-sequences have many applications. Gold sequences, that are based on adding m-sequences that differ by the decimation (1) above, have found extensive practical applications during several decades. The crosscorrelation of m-sequences has also several close connections to almost bent functions as well as to Almost Perfect Nonlinear (APN) functions which are very important in studying S-boxes in cryptography (see cyclic codes). In the *Advanced Encryption Standard* (see Rijndael/AES), properties of the S-boxes come from properties of the crosscorrelation of an m-sequence and its reversed m-sequence. Recently the crosscorrelation of binary and nonbinary m-sequences has been important in constructing new families of sequences with two-level autocorrelation.

Tor Helleseth

References

[1] Chabaud, F. and S. Vaudenay (1995). "Links between differential and linear cryptanalysis." *Advances in Cryptology—EUROCRYPT'94*, Lecture Notes in Computer Science, vol. 950, ed. A. De Santis. Springer, New York, 356–365.

[2] Golomb, S.W. (1982). *Shift Register Sequences.* Aegean Park Press, Laguna Hills, CA.

[3] Helleseth, T. and P.V. Kumar (1999). "Pseudonoise sequences." *The Mobile Communications Handbook* (2nd ed.), ed. J. Gibson. CRC Press and IEEE Press, Boca Raton.

[4] Helleseth, T. and P.V. Kumar (1998). "Sequences with low correlation." *Handbook of Coding Theory*, eds. V.S. Pless and W.C. Huffman. Elsevier, Amsterdam.

[5] Nyberg, K. (1994). "Differentially uniform mappings for cryptography." *Advances in Cryptology—EUROCRYPT'93*, Lecture Notes in Computer Science, vol. 765, ed. T. Helleseth. Springer, New York, 55–64.

McELIECE PUBLIC KEY CRYPTOSYSTEM

THE CRYPTOSYSTEM: This system was introduced by McEliece in 1978 [7] and is among the oldest public-key cryptography schemes. It's security is related to hard algorithmic problems of algebraic coding theory whereas for most other public-key systems it is connected to algorithmic number theory (see RSA public key encryption, Elliptic Curve Cryptography, etc.). Its main advantages are very efficient encryption and decryption procedures and a good practical and theoretical security. On the other hand, its main drawbacks are a public key of large size and a ciphertext which is larger than the cleartext.

General Idea

The cleartext of k binary digits is encoded into a codeword of $n > k$ binary digits by means of some public encoder of a *linear code* of length n and dimension k (for the standard terminology of coding theory, we refer the reader to cyclic codes). Then the ciphertext is obtained by flipping t randomly chosen bits in this codeword.

If t is less than half the *minimum Hamming distance* of the linear code, only one cleartext will correspond to a given ciphertext. If n, k, and t are large enough, computing the cleartext from the ciphertext is intractable, unless some side information on the structure of the code is known.

Description

Let $\mathcal{F}$ denote a family of binary linear codes of length n, dimension k for which computationally effective t-error correcting procedure is known.

Key generation: The legal user picks randomly and uniformly a code C from the family $\mathcal{F}$. Let G_0 be a generator matrix of C. The public key is equal to $G = SG_0P$ where S is a random non-singular binary $k \times k$-matrix and P is a random permutation $n \times n$-matrix. All calculations are done in the finite field $\mathbf{F}_2$ of two elements.

Encryption: The cleartext x is a k-bit word of $\mathbf{F}_2^k$. The ciphertext y is an n-bit word of $\mathbf{F}_2^n$ equal to $xG + e$ where e is randomly chosen with *Hamming weight t*.

Decryption: The cleartext is recovered as zS^{-1} where z is the result of applying the t-error correcting procedure of C to yP^{-1}.

In practice: McEliece proposed the use of binary Goppa codes (see Appendix) with $m = 10$, $n = 1024$ and $t = 50$. The dimension of this code is then $k = 524$. To keep up with 25 years of progress in algorithms and computers, larger codes are required and we now need $m = 11$, $n = 2048$ and $30 \le t \le 120$. Using another family of linear codes is possible but must be done with great care since it has a significant impact on the security. For instance using concatenated codes

[11] or (generalized) Reed–Solomon codes [16] is unsafe.

Related Public Key Cryptosystems

A dual encryption technique, equivalent in terms of security [6], was proposed in 1986 by Niederreiter [9] (see Niederreiter encryption scheme). A variant of the latter using Reed-Muller code was proposed a few years later [15]. A digital signature scheme was proposed in 2001 [4]. It is also possible to use the *rank metric* instead of the Hamming distance to build other encryption schemes [5] in a similar manner.

PRACTICE AND SECURITY: In this section, we assume that the family $\mathcal{F}$ is the family $\mathcal{G}_{m,n,t}$ of binary Goppa codes (see Appendix). We denote $k = n - mt$, the (designed) dimension and $r = mt$, the codimension, where $n \leq 2^m$.

Implementation Aspects

Key Size. The size of the public key is probably the worst drawback of this system. Its size is kn where k is the information block size and n the encrypted block size. For instance, with McEliece's original parameters, each encrypted block has a size of 1024 bits (for 524 bits of information) and the public-key size is 536,576 bits. If the cleartexts are uniformly distributed, the public key can be chosen in systematic form (see [13]), that is $G = (I \mid U)$ where I is the $k \times k$ identity matrix and U is $k \times (n - k)$. This reduces the key size to $k(n - k)$ bits (only the rightmost part U of G needs to be stored). This is still roughly quadratic in the block length, with the original parameters this gives 262,000 bits.

Encryption. We consider that the public key is in systematic form. Encryption consists of multiplying the information x, a binary vector of length k, by the $k \times (n - k)$ binary matrix U. The cryptogram is obtained by flipping t distinct random bits in the binary vector (x, xU) of length n. The cost of this procedure is the cost of the vector/matrix multiplication[1], that is, on average, $k(n - k)/2$ binary operations.

Decryption. Again we assume that the public key is in systematic form. Decryption consists in recovering x from $y = xG + e$. The legal user will have to apply the t-error correcting procedure for a Goppa

code (see Appendix). The actual algorithmic complexity is implementation dependent, but the cost of one decoding is roughly $\lambda t n$ operations in the finite field $\mathbf{F}_{2^m}$, where λ can vary between 2 and 5. Assuming m binary operations for each field operation, we get $\lambda t m n = \lambda(n - k)n$ binary operations for one decryption.

Security

Two aspects of security will now be discussed. The first, theoretical, gives no number and is of no help to choose the parameters; it states that any significant progress in the cryptanalysis would imply some kind of breakthrough in algorithmic coding theory. The second, practical, describes what can be done and at what cost today on a computer to break the system, and also how these costs evolve with the parameters.

Theoretical Security. The security of the system can be reduced to two well-identified problems. Both these problems address old coding theory problems and are conjectured difficult [14] on an average.

Problem 1 (Designed Distance Bounded Decoding—DDBD)

Instance: *An $r \times n$ binary matrix H and a words of $\mathbf{F}_2^r$.*

Question: *Is there a word e in $\mathbf{F}_2^n$ of weight $\leq r / \log_2 n$ such that $He^T = s$?*

Problem 2 (Goppa Code Distinguishing—GD)[2]

Instance: *An $r \times n$ binary matrix H.*

Question: *Is Ker(H) a binary Goppa code?*

Problem 1 is a specialized version of the well-known NP-complete syndrome decoding problem [2] in which the expected minimum weight in the coset is explicitly related to the code parameters. Problem 2 is a structural problem: can we somehow make a difference between a parity check matrix of a Goppa code and a random matrix?

Practical Security: Cryptanalysis. For a given set of parameters n, k and t, the two approaches for the ryptanalysis are:

- the *decoding attack*: decode t error in a known binary linear code of length n and dimension k,
- the *structural attack*: deduce from the public key G an efficient t-error correcting procedure.

The decoding attack is related to the syndrome decoding problem [2] and rapidly becomes intractable when the parameters grow. In practice, for a fixed rate ($R = k/n$) the best known

---

[1] The cost for flipping t bits is negligible, it requires a random number generator though.

[2] Ker(H) denotes the linear code of parity check matrix H.

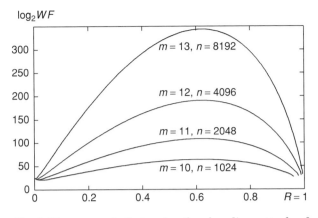

Fig. 1. Binary work factor for the decoding attack of McEliece cryptosystem

algorithms and implementations [1, 3] have a computation cost growing exponentially with t. The binary work factor, that is the average number of binary operations, required to correct t errors in a linear code of length n and transmission rate R is

$$WF(n, R, t) = P(n)2^{t \log_2 \frac{1}{1-R}},$$

where $P(n)$ roughly behaves as a polynomial in n of small degree (0–3 depending on the implementation). Therefore for an arbitrary code with the same parameters as a Goppa code (i.e., $t = (n-k)/m = n(1-R)/m$) the work factor can be written in the following way

$$WF(n, R, t) = P(n)2^{\frac{n}{m}(1-R) \log_2 \frac{1}{1-R}}.$$

The expected cost of algorithm [3] can be obtained by a Markov chain computation. An estimate for the binary work factor for decoding in various Goppa codes (whose structure is hidden) is given in Figure 1.

For the structural attack (with Goppa codes), nothing significantly better is known than enumerating all possible generator polynomial for a given support until one is found which is equal, up to a permutation, to the code generated by the

public-key (this can be done by using the support splitting algorithm [12]). The cost grows exponentially with tm and is always higher than the cost of the decoding attack.

Choosing the Parameters

It is usually agreed that, with the technology available at the start of the third millennium, a binary work factor of 2^{85} is required to insure security. Today we will thus need Goppa codes of length $n = 2048$ with a generator of degree $t \geq 30$. If we follow Moore's law (a factor 2 gain in time and memory every 18 months) and if there is no significant algorithmic improvement, a binary work factor of 2^{108} (maximum value for $m = 11$) should be enough in the 2030s. Table 1 presents the main features of the system with some parameters choice. It is interesting to note that one can simultaneously increase the security and the speed (per information bit) of the encryption/decryption procedures. Other tradeoffs between public key size, information rate and encryption/decryption speed are possible.

Appendix

Binary Goppa Code

Let m be a positive integer, and let n and t be two positive integers such that $n \leq 2^m$ and $t < n/m$. A binary Goppa code $\Gamma(L, g)$ is defined by an ordered subset $L = (\alpha_1, \ldots, \alpha_n)$ of $\mathbf{F}_{2^m}$ of cardinality n, called *support*, and a square-free monic polynomial $g(z) \in \mathbf{F}_{2^m}[z]$ of degree t without roots in L, called *generator*. It consists of all words $a = (a_1, \ldots, a_n) \in \mathbf{F}_2^n$ such that

$$R_a(z) \stackrel{\Delta}{=} \sum_{j=1}^{n} \frac{a_j}{z - \alpha_j} \equiv 0 \bmod g(z).$$

Table 1. Some parameters for the McEliece system

	$n = 1024$ $m = 10$ $t = 50$	$n = 2048$ $m = 11$ $t = 30$	$n = 4096$ $m = 12$ $t = 20$	$n = 2048$ $m = 11$ $t = 70$
Ciphertext size (in bits)	1024	2048	4096	2048
Message size (in bits)	524	1718	3856	1278
Information rate	0.51	0.84	0.94	0.62
Public key size (in KB)	32	69	113	120
Security exponent[a]	62.1	86.4	86.1	108.4
Encryption cost[b]	1	0.66	0.48	1.54
Decryption cost[b]	1	0.40	0.26	1.26

[a] Logarithm in base two of the binary work factor
[b] Per message bit, relatively to the original parameters

This code is linear, has dimension[3] $k \geq n - tm$ and minimum distance $2t + 1$ at least. We denote $\mathcal{G}_{m,n,t}$ the set of all binary Goppa codes with a support of cardinality n in $\mathbf{F}_{2^m}$ and an irreducible generator of degree t over $\mathbf{F}_{2^m}$.

Algebraic Decoding of Goppa Codes

Let a be a codeword of $\Gamma(L, g)$ and let $b = a + e$ with $e \in \mathbf{F}_2^n$ of Hamming weight t or less. Because of the distance properties of the code, given b, the words a and e are unique. There exists a unique monic polynomial $\sigma(z)$ in $\mathbf{F}_{2^m}[z]$ of degree at most t which verifies

$$\sigma(z)R_b(z) \equiv \omega(z) \bmod g(z) \qquad (1)$$

with $\omega(z)$ in $\mathbf{F}_{2^m}[z]$ of degree at most $t - 1$. The polynomial $\sigma(z)$ has exactly t roots in L which correspond to the non zero positions of the error vector e. Starting from vector b, the decoding procedure will require

- the computation of $R_b(z) \bmod g(z)$,
- the resolution of the key equation (1),
- the computation of the roots of $\sigma(z)$.

There are several variants for solving (1), including the extended <u>Euclidean algorithm</u> [8, 10] .

Nicolas Sendrier

References

[1] Barg. A. (1998). "Complexity issues in coding theory." *Handbook of Coding Theory*, vol. 1, chapter 7, eds. V.S. Pless and W.C. Huffman. North-Holland, Amsterdam, 649–754.

[2] Berlekamp, E.R., R.J. McEliece, and H.C. van Tilborg (1978). "On the inherent intractability of certain coding problems." *IEEE Transactions on Information Theory*, 24 (3), 384–386.

[3] Canteaut A. and F. Chabaud (1998). "A new algorithm for finding minimum-weight words in a linear code: Application to McEliece's cryptosystem and to narrow-sense BCH codes of length 511." *IEEE Transactions on Information Theory*, 44 (1), 367–378.

[4] Courtois. N., M. Finiasz, and N. Sendrier (2001). "How to achieve a McEliece-based digital signature scheme." *Advances in Cryptology—ASIACRYPT 2001*, Lecture Notes in Computer Science, vol. 2248, ed. C. Boyd. Springer-Verlag, Berlin, 157–174.

[5] Gabidulin, E., A. Paramonov, and O. Tretjakov (1991). "Ideals over a non-commutative ring and their application to cryptology." *Advances in Cryptology—EUROCRYPT'91*, Lecture Notes

in Computer Science, vol. 547, ed. D.W. Davies. Springer-Verlag, Berlin, 482–489.

[6] Li, Y.X., R.H. Deng, and X.M. Wang (1994). "On the equivalence of McEliece's and Niederreiter's public-key cryptosystems." *IEEE Transactions on Information Theory*, 40 (1), 271–273.

[7] McEliece, R.J. (1978). "A public-key cryptosystem based on algebraic coding theory." DSN Prog. Rep., Jet Prop. Lab., California Inst. Technol., Pasadena, CA, 114–116.

[8] MacWilliams, F.J. and N.J.A. Sloane. (1977). *The Theory of Error-Correcting Codes*, chapter 12. Alternant, Goppa and other generalized BCH codes. North-Holland, Amsterdam.

[9] Niederreiter, H. (1986). "Knapsack-type crytosystems and algebraic coding theory." *Prob. Contr. Inform. Theory*, 15 (2), 157–166.

[10] Patterson, N.J. (1975). "The algebraic decoding of Goppa codes." *IEEE Transactions on Information Theory*, 21 (2), 203–207.

[11] Sendrier, N. (1998). "On the concatenated structure of a linear code." *AAECC*, 9 (3), 221–242.

[12] Sendrier, N. (2000). "Finding the permutation between equivalent codes: The support splitting algorithm." *IEEE Transactions on Information Theory*, 46 (4), 1193–1203.

[13] Sendrier, N. (2002). *Cryptosystemes cl publique bass sur les codes correcteurs d'erreurs*. Mmoire d'habilitation diriger des recherches, Universit Paris 6.

[14] Sendrier, N. (2002). "On the security of the McEliece public-key cryptosystem." *Information, Coding and Mathematics*, eds. M. Blaum, P.G. Farrell, and H. van Tilborg. Kluwer, 141–163. Proceedings of Workshop honoring Prof. Bob McEliece on his 60th birthday.

[15] Sidel'nikov, V.M. (1994). "A public-key cryptosystem based on Reed-Muller codes." *Discrete Mathematics and Applications*, 4 (3), 191–207.

[16] Sidel'nikov, V.M. and S.O. Shestakov (1992). "On cryptosystem based on generalized Reed–Solomon codes." *Discrete Mathematics*, 4 (3), 57–63 (in Russian).

MD4-MD5

MD4 and MD5 are the initial members of the MD4 type <u>hash functions</u>. Both were designed by Rivest [1, 2]. They take variable length input messages and hash them to fixed-length outputs. Both operate on 512-bit message blocks divided into 32-bit words and produce a message digest of 128 bits. First, the message is padded according to the so-called Merkle-Damgård strengthening technique (see <u>hash functions</u> for more details). Next, the message is processed block by block by the underlying compression function. This function

---

[3] For parameters suitable with the McEliece system, the equality always holds.

initializes four 32-bit chaining variables to a fixed value prior to hashing the first message block, and to the current hash value for the following message blocks. Each step of the compression function updates in turn one of the chaining variables according to one message word. Both compression functions are organised into rounds of 16 steps each. MD4 has three such rounds, while MD5 consists of 4 rounds. In each round every message word is used just once in updating one of the chaining variables. The order in which the message words are used is different for each round. MD4 and MD5 differ in the functions used in each step, in the order in which the message words are used in different rounds, and in the number of rounds. There also exists an extended version of MD4 [1], which consists of concatenating the result of two loosely coupled instances of MD4 into a 256-bit message digest.

SECURITY CONSIDERATIONS: MD4 and MD5 have been designed to provide underline{collision resistance}. Following early collision attacks on reduced versions of MD4 (2 out of 3 rounds) by Merkle and den Boer and Bosselaers [3], Rivest designed the strengthened version MD5. It was however shown by den Boer and Bosselaers [4] that the compression function of MD5 is not collision resistant, although the collisions are of a form that is not immediately usable in practice. Late in 1995 Dobbertin found collisions for MD4. In his attack he combines algebraic techniques and optimization techniques such as genetic algorithms [5]. It can be extended in such a way that even collisions on meaningful messages are obtained: except for a few dozen bytes the complete message is under control of the attacker. His attack also applies to the compression function of extended MD4. Later Dobbertin showed that a reduced version of MD4 (2 rounds out of 3) does not offer preimage resistance [6]. Dobbertin also extended his attack on MD4 to yield collisions for the compression function of MD5. These collisions are of a different and more practical nature than those by den Boer and Bosselaers, but up to now the attack has not been extended to collisions for MD5 itself (i.e., also taking into account the initial value). Of independent interest for both MD4 and MD5 are brute force collision search attacks. In [7], van Oorschot and Wiener estimate that with a 10 million US\$ machine collisions of MD5 could be found in 21 days in 1994, which corresponds to 4 hours in 2004. To counter such collision search attacks, message digests of at least 160 bits are required.

Antoon Bosselaers

References

[1] Rivest, R.L. (1991). "The MD4 message digest algorithm." *Advances in Cryptology—CRYPTO'90*, Lecture Notes in Computer Science, vol. 537, ed. S. Vanstone. Springer-Verlag, Berlin, 303–311.

[2] Rivest, R.L. (1992). "The MD5 message-digest algorithm." Request for Comments (RFC) 1321, Internet Activities Board, Internet Privacy Task Force.

[3] den Boer, B. and A. Bosselaers (1992). "An attack on the last two rounds of MD4." *Advances in Cryptology—CRYPTO'91*, Lecture Notes in Computer Science, vol. 576, ed. J. Feigenbaum. Springer-Verlag, Berlin, 194–203.

[4] den Boer, B. and A. Bosselaers (1994). "Collisions for the compression function of MD5." *Advances in Cryptology—EUROCRYPT'93*, Lecture Notes in Computer Science, vol. 765, ed. T. Helleseth. Springer-Verlag, Berlin, 293–304.

[5] Dobbertin, H. (1998). "Cryptanalysis of MD4." *Journal of Cryptology*, 11 (4), 253–271. See also *Fast Software Encryption*, Lecture Notes in Computer Science, vol. 1039, ed. D. Gollmann. Springer-Verlag, Berlin, 1996, 53–69.

[6] Dobbertin, H. (1996). "The status of MD5 after a recent attack." *Cryptobytes*, 2 (2), 1–6.

[7] van Oorschot P.C. and M. Wiener (1999). "Parallel collision search with cryptanalytic applications." *Journal of Cryptology*, 12 (1), 1–28.

MDC-2 AND MDC-4

MDC-2 and MDC-4 are constructions for underline{hash functions} based on a underline{block cipher}, where the length in bits of the hash result is twice the block length of the block cipher. A hash function is a cryptographic algorithm that takes input strings of arbitrary (or very large) length and maps these to short fixed length output strings. MDC-2 and MDC-4 are unkeyed cryptographic hash functions which may have the following properties: underline{preimage resistance}, underline{second preimage resistance}, and underline{collision resistance}; these properties may or may not be achieved depending on the properties of the underlying block cipher. MDC-2 and MDC-4 have been designed by Brachtl et al. [1]; they are also known as the Meyer–Schilling hash functions after the authors of the first paper describing these schemes [4].

In the following, the block length and key length of the block cipher will be denoted with n and k, respectively. The encryption with the block cipher E using the key K will be denoted with $E_K(\cdot)$ and $\parallel$ denotes the concatenation of strings.

MDC-2 is an iterated hash function with a compression function that maps $2k + n$ bits to $2n$ bits.

It requires two encryptions to hash an n-bit block, hence its rate is 1/2.

$$T_i^1 = E_{u(H_{i-1}^1)}(X_i) \oplus X_i = LT_i^1 \parallel RT_i^1$$
$$T_i^2 = E_{v(H_{i-1}^2)}(X_i) \oplus X_i = LT_i^2 \parallel RT_i^2$$
$$H_i^1 = LT_i^1 \parallel RT_i^2$$
$$H_i^2 = LT_i^2 \parallel RT_i^1 \,.$$

The variables H_0^1 and H_0^2 are initialized with the values IV_1 and IV_2 respectively, and the hash result is equal to the concatenation of H_t^1 and H_t^2. The functions u, v map the ciphertext space to the key space and need to satisfy the condition $u(IV^1) \neq v(IV^2)$. For the Data Encryption Standard (DES), these mappings from 64 to 56 bits drop the parity bits in every byte and fix the second and third key bits to 01 and 10 respectively (to preclude attacks based on the complementation property and based on weak keys and semi-weak keys). By iterating this compression function in combination with MD-strengthening (see hash functions) one can construct a hash function based on this compression function.

For $k = n$, the best known preimage and collision attacks on MDC-2 require $2^{3n/2}$ and 2^n operations respectively [5]. A collision attack on MDC-2 based on *DES* ($n = 64$, $k = 56$) requires at most 2^{55} encryptions, which is not an acceptable security level in 2004. Note that the compression function of MDC-2 is rather weak: preimage and collision attacks on the compression function require at most 2^n and $2^{n/2}$ encryptions. As a consequence, one cannot obtain a meaningful lower bound for the strength of MDC-2 with a security proof based on the strength of its compression function.

The compression function of MDC-4 consists of the concatenation of two MDC-2 steps, where the plaintexts in the second step are equal to H_{i-1}^2 and H_{i-1}^1. MDC-4 requires four encryptions to hash an n-bit block, hence its rate is 1/4. For $k = n$, the best known preimage attack for MDC-4 requires $2^{7n/4}$ operations. This shows that MDC-4 is probably more secure than MDC-2 against preimage attacks. However, finding a collision for MDC-4 itself requires only 2^{n+2} encryptions. The best known attacks on the compression function of MDC-4 require $2^{3n/2}$ encryptions for a (2nd) preimage and $2^{3n/4}$ encryptions for a collision [3, 6]. Again, one cannot obtain a meaningful lower bound for the collision resistance of MDC-4 with a security proof based on the strength of its compression function.

A first observation is that no security proofs are known for MDC-2 and MDC-4. However, it is conjectured that both MDC-2 and MDC-4 achieve an acceptable security level (in 2004) against (2nd) preimage attacks for block ciphers with a block length and key length of 64 bits or more (e.g., CAST-128, IDEA, KASUMI/MISTY1). It is also conjectured that both functions achieve an acceptable security level (in 2004) against collision attacks for block ciphers with a block length and key length of 128 bits or more (e.g., Rijndael/AES, Camellia, CAST-256, MARS, RC6, TWOFISH, and SERPENT).

It is also important to note that a block cipher may have properties which pose no problem at all when they are used only for encryption, but which may result in MDC-2 and/or MDC-4 to be insecure [6, 7]. Any deviation from 'random behavior' of the encryption steps or of the key schedule could result in security weaknesses (for example, it would not be advisable to use DES-X due to the absence of a key schedule for part of the key).

MDC-2 has been included in ISO/IEC 10118-2 [2].

Bart Preneel

References

[1] Brachtl, B.O., D. Coppersmith, M.M. Hyden, S.M. Matyas, C.H. Meyer, J. Oseas, S. Pilpel, and M. Schilling (1990). "Data authentication using modification detection codes based on a public one way encryption function." U.S. Patent Number 4,908,861.

[2] ISO/IEC 10118 (1998). "Information technology—security techniques—hash-functions, Part 1: General", 2000. "Part 2: Hash-functions using an n-bit block cipher algorithm," 2000. "Part 3: Dedicated hash-functions," 2003. "Part 4: Hash-functions using modular arithmetic."

[3] Knudsen, L. and B. Preneel (2002). "Enhancing the security of hash functions using non-binary error correcting codes." *IEEE Transactions on Information Theory*, 48 (9), 2524–2539.

[4] Meyer, C.H. and M. Schilling (1988). "Secure program load with manipulation detection code." *Proc. Securicom*, 88, 111–130.

[5] Lai, X. and J.L. Massey (1993). "Hash functions based on block ciphers." *Advances in Cryptology—EUROCRYPTO'92*, Lecture Notes in Computer Science, vol. 658, ed. R.A. Rueppel. Springer-Verlag, Berlin, 55–70.

[6] Preneel, B. (1993). "Analysis and Design of Cryptographic Hash Functions." *Doctoral Dissertation*, Katholieke Universiteit Leuven.

[7] Preneel, B., R. Govaerts, and J. Vandewalle (1994). "Hash functions based on block ciphers: A synthetic approach." *Advances in Cryptology—CRYPTO'93*, Lecture Notes in Computer Science, vol. 773, ed. D. Stinson. Springer-Verlag, Berlin, 368–378.

MEET-IN-THE-MIDDLE ATTACK

Meet-in-the-middle is a classical technique of cryptanalysis which applies to many constructions. The idea is that the attacker constructs patterns that propagate from both ends to the middle of the cipher, in some cases by partial key-guessing. If the events do not match in the middle, the key-guess was wrong and may be discarded. Such attack has been applied to 7-round DES (see the Data Encryption Standard) [1], and to structural cryptanalysis of *multiple-encryption* (for example, two-key triple encryption [2,3]. A recent miss-in-the-middle attack may also be seen as a variant of this technique in which the events in the middle should *not* match, and the keys that suggest a match in the middle are filtered as wrong keys.

Alex Biryukov

References

[1] Chaum, D. and J.-H. Evertse (1986). "Cryptanalysis of DES with a reduced number of rounds; sequence of linear factors in block ciphers." *Advances in Cryptology—CRYPTO'85*, Lecture Notes in Computer Science, vol. 218, ed. H.C. Williams. Springer-Verlag, Berlin, 192–211.

[2] Markle, R.C. and M.E. Hellman (1981). "On the security of multiple encryption." *Communications of the ACM*, 24, 465–467.

[3] van Oorschot, P.C. and M.J. Wiener (1990). "A known plaintext attack, on two-key triple encryption." *Advances in Cryptology—EUROCRYPT'90*, Lecture Notes in Computer Science, vol. 473, ed. I. Dawgard. Springer-Verlag, Berlin, 318–325.

MERSENNE PRIME

A *Mersenne number* is a number of the form $M_n = 2^n - 1$ where n is a positive integer. If M_n is also prime, then it is said to be a *Mersenne prime*. The number M_n can only be prime if n is prime. Some authors reserve the term "Mersenne number" for the numbers M_p for p prime. The Mersenne primes M_p for $80 \leq p \leq 700$ occur for $p = 89, 107, 127, 521,$ and 607.

Mersenne primes have been a topic of interest in number theory since ancient times. They appear in the theory of linear feedback shift registers (LFSR). If M_p is prime, then any binary LFSR of length p with irreducible feedback polynomial and nonzero initial state generates a maximal-length shift register sequence. Equivalently, every nonzero element of the finite field of 2^p elements is a generator for the entire group of nonzero elements.

Mersenne primes are also of interest in public-key cryptography. In public-key settings such as elliptic curve cryptography, the prime modulus p (see modular arithmetic) can be chosen to optimize the implementation of the arithmetic operations in the implementation of the cryptography. Mersenne primes provide a particularly good choice because modular reduction can be performed very quickly. One typically wishes to reduce modulo M_p a 2p-bit integer n (e.g., as a step in modular multiplication). In general, modular reduction requires an integer division. However, in the special case of reduction modulo M_p, one has

$$2^p \equiv 1 \pmod{M_p}.$$

Thus one can reduce n by writing

$$n = a \cdot 2^p + b,$$

where a and b are each positive and less than M_p. Then

$$n \equiv a + b \pmod{M_p},$$

so that

$$n \bmod M_p = \begin{cases} a + b & \text{if } a + b < M_p, \\ a + b + 1 - 2^p & \text{otherwise.} \end{cases}$$

Thus reduction modulo a Mersenne prime requires an integer addition, as opposed to an integer division for modular reduction in the general case. There are two drawbacks to this method of modular reduction:

- Finding the integers a and b is easiest when p is a multiple of word size of the machine, since then there is no actual shifting of bits needed to align a and b for the modular addition. But word sizes are in practice powers of two, whereas p must be an odd prime.
- The Mersenne primes are so rare, with none between M_{127} and M_{521}, that usually there will be none of the desired magnitude.

For these reasons, cryptographers tend not to use Mersenne primes, preferring similar moduli such as pseudo-Mersenne primes and generalized Mersenne primes.

Jerome Solinas

MILLER–RABIN PROBABILISTIC PRIMALITY TEST

The Miller–Rabin probabilistic primality test is a probabilistic algorithm for testing prime numbers using modular exponentiation (see exponentiation algorithms) and the Chinese Remainder Theorem.

One property of primes is that any number whose square is congruent to 1 modulo a prime p must itself be congruent to 1 or -1. This is not true of composite numbers. If a number n is the product of k distinct prime powers, then there will be 2^k distinct "square roots" of 1 modulo n. For example, there are four square roots of 1modulo 77. The roots must be either 1 or -1 modulo 7, and either 1 or -1 modulo 11, since 7 and 11 divide 77. In order to solve these square roots, one must solve sets of equations like these:

$$a \equiv 1 \bmod 7, \quad a \equiv 1 \bmod 11$$
$$b \equiv -1 \bmod 7, \; b \equiv 1 \bmod 11$$
$$c \equiv 1 \bmod 7, \quad c \equiv -1 \bmod 11$$
$$d \equiv -1 \bmod 7, \, d \equiv -1 \bmod 11.$$

The solutions in this case are $a \equiv 1 \bmod 77$, $b \equiv 34 \bmod 77$, $c \equiv 43 \bmod 77$, and $d \equiv -1 \equiv 76 \bmod 77$. (This is a simple application of the Chinese Remainder Theorem.)

The Miller–Rabin test uses this fact about composite numbers to test them. A single round of Miller–Rabin tests whether a given base a is a "witness" to the compositeness of n, by computing a^{n-1} modulo n. This is the same as the Fermat primality test, but that test fails if n is a Carmichael number. However, to compute $a^{n-1} \bmod n$ we perform a series of squarings. The Miller–Rabin test also checks after each squaring to see if we have found a square root of 1 other than 1 or $n - 1 \bmod n$. If so, the number is composite; this additional check also catches the Carmichael numbers.

The procedure is the following, where s, the number of rounds, is a security parameter:
1. On input p, find k such that $p - 1 = q\,2^k$, where q is odd.
2. For $j = 1$ to s, do:
   (a) Generate a random base a between 2 and $p - 2$.
   (b) Compute $b = a^q \bmod p$.
   (c) For $i = 1$ to k, compute $b' = b^2 \bmod p$. If $b' \equiv 1 \bmod p$ and $b \not\equiv \pm 1 \bmod p$, output COMPOSITE. Set $b = b'$.
   (d) If $b \not\equiv 1 \bmod p$, output COMPOSITE.
3. Output PROBABLY PRIME.
(Note that the output COMPOSITE is certain; the output PROBABLY PRIME is not.)

For any odd composite n, at least $3(n - 1)/4$ of the bases a are Miller–Rabin witnesses that n is composite (a good presentation of a proof for the simpler bound $(n - 1)/2$ is in [1]). Each round of the Miller–Rabin test thus gives at least a 3/4 probability of finding a witness, if n is composite. These probabilities are independent, so if we run a 50-round Miller–Rabin test, then the probability that n is composite and we never find a compositeness witness is at most 2^{-100}. For a random prime, the probability of not finding a witness is much smaller. The number of rounds is a parameter to the primality test that can be easily changed depending on the application.

The Miller–Rabin test was described initially by Miller [2]; Rabin [3] provided further analysis, hence the name.

Moses Liskov

References

[1] Cormen, Thomas H., Charles E. Leiserson, Ronald L. Rivest, and Clifford Stein (2001). *Introduction to Algorithms*. MIT Press, Cambridge, 890–896.
[2] Miller, G.L. (1976). "Riemann's hypothesis and tests for primality." *Journal of Computer and Systems Sciences*, 13, 300–317.
[3] Rabin, M.O. (1980). "Probabilistic algorithm for testing primality." *Journal of Number Theory*, 12, 128–138.

MINIMAL POLYNOMIAL

The *minimal polynomial* of a linear recurring sequence $\mathbf{S} = (s_t)_{t \geq 0}$ of elements of $\mathbf{F}_q$ is the monic polynomial P in $\mathbf{F}_q[X]$ of lowest degree such that $(s_t)_{t \geq 0}$ is generated by the linear feedback shift register (LFSR) with characteristic polynomial P. In other terms, $P = \sum_{i=0}^{L-1} p_i X^i + X^L$ is the characteristic polynomial of the linear recurrence relation of least degree satisfied by the sequence:

$$s_{t+L} + \sum_{i=0}^{L-1} p_i s_{t+i} = 0, \qquad t \geq 0.$$

The minimal polynomial of a linear recurring sequence $\mathbf{s}$ is monic and unique; it divides the characteristic polynomial of any LFSR which generates $\mathbf{s}$. The degree of the minimal polynomial of $\mathbf{s}$ is called its linear complexity. The period of the minimal polynomial of $\mathbf{s}$ is equal to the least period of $\mathbf{s}$ (see linear feedback shift register for further details).

The minimal polynomial of a linear recurring sequence with linear complexity Λ can be recovered from any 2Λ consecutive terms of the sequence by the Berlekamp–Massey algorithm.

Anne Canteaut

MIPS-YEAR

A *MIPS-year* is perhaps the "standard" measure of computational effort in cryptography: It refers to the amount of work performed, in one year, by a computer operating at the rate of one million operations per second (1 MIPS). The actual type of operation is undefined but assumed to be a "typical" computer operation. A MIPS-year is thus approximately 2^{45} operations. The difficulty of solutions to the the RSA Factoring Challenge as well as challenges involving the Data Encryption Standard is usually given in MIPS-years; for instance, the RSA-512 benchmark took about 8000 MIPS-years, or approximately 2^{58} operations, distributed across a large number of computers. (This is somewhat less than the number of operations to search for a 56-bit DES key, as multiple operations are required to test each DES key.)

The MIPS-year is a convenient measurement, but not a perfect one. In practice, there is no "typical" computer operation, and the actual difficulty of an effort must also consider other factors such as the cost of hardware and the amount of memory required. (See Silverman [2] for discussion of some of these issues.) Recent research into the difficulty of factoring 1024-bit RSA moduli has taken a more precise approach, by giving a specific hardware design and estimating both the cost of the hardware involved and the number of operations (see factoring circuits, TWIRL). The MIPS-year nevertheless remains a helpful guideline.

The term "MIPS" is fairly old in the computer industry; Digital Equipment Corporation's VAX-11/780 is often considered the benchmark of a 1-MIPS machine. An early use of the term "MIPS-year" in cryptography may be found in a 1991 letter from Rivest to NIST regarding the security of the then-proposed Digital Signature Standard, a revised version of which is published as part of [1].

Burt Kaliski

References

[1] Rivest, Ronald L., Martin E. Hellman, John C. Anderson, and John W. Lyons (1992). "Responses to NIST's proposal." *Communications of the ACM*, 35 (7), 41–54.

[2] Silverman, R. (1999). "Exposing the mythical MIPS-year." *IEEE Computer*, 32 (8), 22–26.

MISS-IN-THE-MIDDLE ATTACK

Following the idea behind the meet-in-the-middle approach, the miss-in-the-middle attack is one of the techniques to construct *distinguishers* for the impossible differential attack. The idea is that one finds two events that propagate half way through the cipher top and bottom with certainty, but which do not match in the middle. This results in an event which is impossible for the full cipher, i.e., has zero probability. A typical tool for constructing such events would be *truncated differentials*. Note that it is sufficient that events contradict each other in a single bit in the middle. This technique was first introduced in the papers by Biham et al. [1, 2] to cryptanalyse round-reduced versions of Skipjack, IDEA and Khufu.

Alex Biryukov

References

[1] Biham, E., A. Biryukov, and A. Shamir (1999). "Cryptanalysis of Skipjack reduced to 31 rounds using impossible differentials." *Advances in Cryptology—EUROCRYPT'99*, Lecture Notes in Computer Science, vol. 1593, ed. J. Stern. Springer-Verlag, Berlin, 12–23.

[2] Biham, E., A. Biryukov, and A. Shamir (1999). "Miss in the middle attacks on IDEA and Khufu." *Fast Software Encryption, FSE'99*, Lecture Notes in Computer Science, vol. 1636, ed. L.R. Knudsen. Springer-Verlag, Berlin, 124–138.

MIX NETWORKS

It is not difficult to imagine scenarios in which message secrecy in communication is desirable. In such scenarios, encryption is a crucial tool. Unfortunately, encryption of message contents by itself may not be sufficient. Even if the contents of sensitive messages are protected, much can be inferred merely by the fact that one party is sending a message to another party. If an authoritarian regime already suspects that one party to a communication is a dissident, then the other parties to the communication become suspect as well. Accessing a crisis hotline or a patent database is a strong clue about the intentions of the user, even if the exact wording of the query remains secret. In settings where encryption is rare, the mere fact that certain messages are encrypted may cause increased scrutiny.

In 1981, Chaum [3] published a beautifully simple and elegant method to protect the identities of communicating parties: "mix networks." The basic functionality of a mix network is to provide *sender anonymity*, i.e., the identity of the originator of a message is difficult or impossible to discern for any given message delivered to any given recipient. A mix network can also provide *receiver anonymity*, i.e., the identity of the intended recipient of a message is difficult or impossible to discern for any given message originating from any given sender.

We describe Chaum's idea using the simplest mix network, which consists of a single mix. If Alice wants to send a message M to Bob, she first encrypts it using Bob's public key (see public key cryptography): $C_1 = E(M, PK_{Bob})$. Then she re-encrypts this ciphertext (together with Bob's name!) using the mix's public key: $C_2 = E(C_1 \cdot$ "Bob", $PK_{Mix})$. Then the following steps occur:

1. Alice sends C_2 to the mix.
2. The mix decrypts C_2 to recover C_1 and "Bob".
3. The mix sends C_1 to Bob.
4. Bob decrypts C_1 to recover the message M.

The identity of the originator (Alice) is protected if many senders are using the same mix at the same time. The identity of the recipient (Bob) is protected if many receivers have messages routed to them through the same mix at the same time. Of course, the mix knows who is communicating to whom. That is, there is perfect linkability by the mix of originator to recipient for every message that passes through the mix.

A general mix network is constructed similarly, except that Alice re-encrypts several times using a "chain" (or "cascade") of mixes. For example, a chain of three mixes would have Alice compute:

$$
\begin{aligned}
C_1 &= E(M, PK_{Bob}), \\
C_2 &= E(C_1 \cdot \text{"Bob"}, PK_{Mix1}), \\
C_3 &= E(C_2 \cdot \text{"Mix1"}, PK_{Mix2}), \\
C_4 &= E(C_3 \cdot \text{"Mix2"}, PK_{Mix3}).
\end{aligned}
\tag{1}
$$

Then Alice would send C_4 to Mix3, and Mix3 would send C_3 to Mix2, and Mix2 would send C_2 to Mix1, and Mix1 would send C_1 to Bob. Notice that now the mixes would have to collude to know who is communicating to whom.

Of course, this is just the high-level description of a mix network. In practice, a variety of attacks on anonymity are possible, and various design countermeasures would have to be incorporated (see also [1]):

Timing attacks: Careful observation of the timing of inputs and outputs of a single mix could enable an eavesdropper to link specific inbound and outbound messages. One defense against this is for a mix to delay the forwarding of messages until a certain number of them can be sent at the same time ("batching").

Message Length Attacks: Encryption by itself does not necessarily conceal the size of the plaintext message. If certain messages passing through the mix network differ in size, then these might be distinguishable by an outside attacker. Defenses include splitting large messages ("fragmentation") and lengthening short messages ("padding").

Absence of Communication Attacks: Certain suspects can be eliminated from consideration as potential senders of a message simply because they were idle during some relevant time period, i.e., sent no encrypted messages to a particular mix. One defense against this is for senders to continue to send null ("dummy") messages when they have nothing to communicate.

Abundance of Communication Attacks: If an adversary can inject a lot of messages into the mix network, then only a small amount of legitimate messages can be routed through the mixes at the same time. The adversary can choose his messages so that they are easily recognized as they pass through the network, and thus he may be able to infer quite a lot about the origin and destination of the few remaining messages. This is closely related to a denial of service attack, and similar defenses are possible. For example, one could limit and balance the rate at which any given sender can route messages through any given mix ("fair allocation").

Chaum's original work included several other interesting extensions. There was a technique for the receiver to reply to an anonymously transmitted message. He also showed how a mix network can be combined with pseudonyms to achieve a "general purpose" untraceable mail system. Furthermore, there was a method for the sender to specify a different chain of mixes for each message.

Chaum's mix network ideas have been implemented numerous times. Notable instances include Freedom [2], Onion Routing [5], Babel [6], and the Cypherpunk remailer system. Serjantov et al. [11] provide a good taxonomy of mix implementations and their properties.

The Crowds system of Reiter and Rubin [9] is a kind of mix network with on-the-fly randomized decisions for how many mixes a message should pass through. In fact, parties in their scheme act as both message initiators and the mixes themselves. In the Crowds protocol, some set of parties form a group called a "jondo." When a sender initiates a message, it is mixed by passing it along a path of other parties in the jondo. The length and constituency of this path is chosen randomly. Specifically, each party in the path makes a randomized decision whether to end the path. If the path is not to be ended, a second randomized decision chooses another jondo member to receive the message next. Perfect concealment is not possible for the Crowds system, but a level of "probable innocence" can be achieved.

Crowds is also vulnerable to a "predecessor attack" [9, 12]. An attacker joins a jondo, and keeps track of how often any other jondo member immediately precedes the attacker in a path. Suppose that a single sender is involved in a number of related transmissions, e.g., if a jondo member is using Crowds to anonymously surf the web, and often returns to the same website. Then the attacker will be included in some of the paths for these related transmissions. The attacker's immediate predecessor in each of these paths could be any member of the jondo, but the most frequent occupant of this position will be the original sender.

Mix networks can simplify the design of a cryptographic protocol for conducting a secret ballot election [4]. The basic idea is that each eligible voter encrypts his ballot using the public key of the tallying authority. All of the encrypted ballots from eligible voters pass through a mix network to the tallying authority. The tallying authority decrypts all of the received messages, throws away the ones that are not well-formed ballots, and then determines the outcome of the election. The privacy of each individual voter's choices is ensured by the proper functioning of the anonymizing mixes.

For applications such as a secret ballot election, it is reasonable to assume that all of the mix network inputs are available for processing at the same time. This means that the mix network design can be "synchronous." By contrast, Chaum's mix network design is asynchronous, since this is a more reasonable assumption for his motivating application of hiding traffic patterns in ongoing communication. The "re-encryption mix net" [8] is a synchronous design that makes use of more sophisticated cryptographic tools to achieve robustness despite the failure of some of the mix servers. Some of the fastest mix-based election schemes are based on this design.

Verifiable mix protocols [10] are a variant of the basic mix protocol in which correct functioning of any given mix can be publicly verified by any external observer. The primary motivation of verifiable mix protocols is to enhance the security of mix-based election protocols. Typically, the mix generates a short "proof" of correctness that can be checked against the encrypted inputs and encrypted outputs of the mix. Another recent approach to verifiable mixes is based on a "cut-and-choose" (challenge/response) approach [7].

Matthew K. Franklin

References

[1] Back, A., U. Moller, and A. Stiglic (2001). "Traffic analysis and trade-offs in anonymity providing systems." *Proceedings of the 4th Information Hiding Workshop*, Lecture Notes in Computer Science, vol. 2137, ed. I.S. Moskowitz. Springer-Verlag, Berlin, 243–254.

[2] Boucher, P., A. Shostack, and I. Goldberg (2000). Freedom 2.0 System Architecture. Zero Knowledge Systems (white paper).

[3] Chaum, D. (1981). "Untraceable electronic mail, return addresses, and digital pseudonyms." *CACM*, 24 (2), 84–88.

[4] Fujioka, A., T. Okamoto, and K. Ohta (1992). "A practical secret voting scheme for large scale elections." *Advances in Cryptology—AUSCRYPT'92*, Lecture Notes in Computer Science, vol. 718, eds. J. Seberry and Y. Zheng, Gold Coast, Queensland, Australia. Springer-Verlag, Berlin, 244–251.

[5] Goldschlag, D., M. Reed, and P. Syverson (1999). "Onion routing." *CACM*, 42 (2), 39–41.

[6] Gulcu, C. and G. Tsudik (1996). "Mixing e-mail with BABEL." *Proceedings of Symposium on Network and Distributed System Security*. IEEE Computer Society.

[7] Jakobsson, M., A. Juels, and R. Rivest (2002). "Making mix nets robust for electronic voting by partial checking." *Proc. USENIX Security*.

[8] Park, C., K. Itoh, and K. Kurosawa (1993). "Efficient anonymous channel and all/nothing election scheme." *Advances in Cryptology—EUROCRYPT'93*, Lecture Notes in Computer Science, vol. 765, ed. T. Helleseth. Springer-Verlag, Berlin, 248–259.

[9] Reiter, M. and A. Rubin (1998). "Crowds: Anonymity for web transactions." *ACM Transactions on Information and System Security*, 1 (1), 66–92.

[10] Sako, K. and J. Kilian (1995). "Receipt-free mix-type voting scheme—a practical solution to the implementation of a voting booth." *Advances in Cryptology—EUROCRYPT'95*, Lecture Notes in Computer Science, vol. 021, eds. L. Guillou and J.-J. Quisquater. Springer-Verlag, Berlin, 393–403.

[11] Serjantov, A., R. Dingledine, and P. Syverson (2002). "From a trickle to a flood: Active attacks on several mix types." *Proceedings of Information Hiding Workshop*, Lecture Notes in Computer

Science, vol. 2578, ed. F.A.P. Petitcolas. Springer-Verlag, Berlin, 36–52.

[12] Wright, M., M. Adler, B. Levine, and C. Schields (2002). "An analysis of the degradation of anonymous protocols." *Proceedings of Symposium on Network and Distributed System Security (NDSS)*.

MODES OF OPERATION OF A BLOCK CIPHER

A n-bit <u>block cipher</u> with a k-bit <u>key</u> is a set of 2^k bijections on n-bit strings. A block cipher is a flexible building block; it can be used for encryption and <u>authenticated encryption</u>, to construct <u>MAC</u> algorithms and <u>hash functions</u> [2].

When a block cipher is used for confidentiality protection, the security goal is to prevent a passive <u>eavesdropper</u> with limited computational power to learn any information on the plaintext (except for maybe its length). This eavesdropper can apply the following attacks: <u>known plaintext attacks</u>, <u>chosen plaintext attacks</u> and <u>chosen ciphertext attacks</u>.

Applications need to protect the confidentiality of strings of arbitrary length. A mode of operation of a block cipher is an algorithm which specifies how one has to apply an n-bit block cipher to achieve this. One approach is to pad the data with a padding algorithm such that the bit-length of the padded string is a multiple t of n bits, and to define a mode which works on t n-bit blocks. For example, one always appends a '1'-bit followed by as many '0' bits as necessary to make the length of the resulting string a multiple of n. An alternative is to define a mode of operation that can process data in blocks of $j \leq n$ bits.

We first discuss the five modes of operation which have been defined in the FIPS [12] (see also [22]) and ISO/IEC [16] standards: the ECB mode, the CBC mode, the OFB mode, the CTR mode, and the CFB mode. Next we discuss t some alternative modes that have been defined for <u>triple-DES</u> and modes which allow to encrypt values from finite sets.

We use the following notation: $E_K(p_i)$ denotes the encryption with a block cipher of the n-bit plaintext block p_i with the key K; similarly $D_K(c_i)$ denotes the decryption of the iphertext c_i. The operation rchop$_j(s)$ returns the rightmost j bits of the string s, and the operation lchop$_j(s)$ returns the leftmost j bits. The symbol $\|$ denotes concatenation of strings and $\oplus$ denotes addition modulo 2 (exor).

THE ELECTRONIC CODE BOOK (ECB) MODE:
The simplest mode is the ECB (Electronic Code-

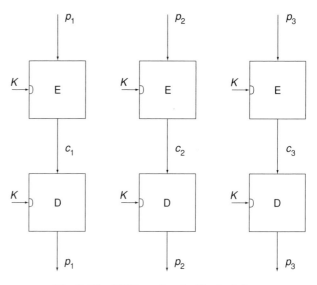

Fig. 1. The ECB mode of a block cipher

Book) mode. After padding, the plaintext p is divided into t n-bit blocks p_i and the block cipher is applied to each block; the decryption also operates on individual blocks (see Figure 1):

$$c_i = E_K(p_i) \quad \text{and} \quad p_i = D_K(c_i), \quad i = 1, \ldots, t.$$

Errors in the ciphertext do not propagate beyond the block boundaries (as long as these can be recovered). However, the ECB mode is the only mode covered in this article which does not hide patterns (such as repetitions) in the plaintext. Usage of this mode should be strongly discouraged. In the past the ECB mode was sometimes recommended for the encryption of keys; however, <u>authenticated encryption</u> would be much better for this application (or the <u>Rijndael/AES</u> key wrapping algorithm proposed by NIST).

THE CIPHER BLOCK CHAINING (CBC) MODE:
The most popular mode of operation of a block cipher is the CBC (Cipher Block Chaining) mode. The plaintext p is divided into t n-bit blocks p_i. This mode adds (modulo 2) to a plaintext block the previous ciphertext block and applies the block cipher to this result (see Figure 2):

$$c_i = E_K(p_i \oplus c_{i-1}),$$
$$p_i = D_K(c_i) \oplus c_{i-1}, \qquad i = 1, \ldots, t.$$

Note that in the CBC mode, the value c_{i-1} is used to randomize the plaintext; this couples the blocks and hides patterns and repetitions. To enable the encryption of the first plaintext block ($i = 1$), one defines c_0 as the initial value IV, which should be randomly chosen and transmitted securely to the recipient. By varying this IV, one can ensure that the same plaintext is encrypted into a different ciphertext under the same key, which is

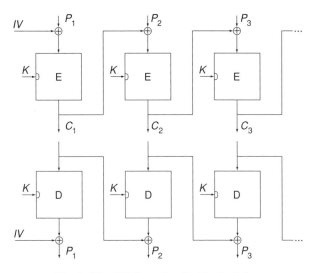

Fig. 2. The CBC mode of a block cipher

essential for secure encryption. The IV plays a similar role in the OFB, CTR and CFB modes.

The CBC decryption has a limited error propagation: errors in the ith ciphertext block will garble the ith plaintext block completely, and will be copied into the next plaintext block. The CBC decryption allows for parallelism and random access: if necessary, one can decrypt only a small part of the ciphertext. However, the encryption mode is a serial operation. To overcome this restriction, ISO/IEC 10116 [16] has defined a variant of the CBC mode which divides the plaintext into r parallel streams and applies the CBC mode to each of these streams. This requires, however, r different IV values.

A security proof of the CBC mode (with random and secret IV) against an adversary who has access to chosen plaintexts has been provided by Bellare et al. [3]; it shows that if the block cipher is secure in the sense that it is hard to distinguish it from a random permutation, the CBC mode offers secure encryption in the sense that the ciphertext is random (which implies that it does not provide the opponent additional information on the plaintext). The security result breaks down if the opponent can obtain approximately $q = 2^{n/2}$ plaintext/ciphertext pairs due to a matching ciphertext attack [18]. This can be seen as follows. Note that the ciphertext blocks c_i are random n-bit strings. After observing q n-bit ciphertext blocks, one expects to find approximately $q^2/2^{n+1}$ pairs of matching ciphertexts that is, indices (v, w) with $c_v = c_w$ (see also the birthday paradox). As a block cipher is a permutation, this implies that the corresponding plaintexts are equal, or $p_v \oplus c_{v-1} = p_w \oplus c_{w-1}$ which can be rewritten as $p_v \oplus p_w = c_{v-1} \oplus c_{w-1}$. Hence, each pair of matching ciphertexts leaks the sum of two

plaintext blocks. To preclude such a leakage, one needs to impose that $q \ll 2^{(n+1)/2}$ or $q = \alpha \cdot 2^{n/2}$ where α is a small constant (say 10^{-3}, which leads to a collision probability of 1 in 2 million). If this limit is reached, one needs to change the key. Note that the proof only considers security against chosen plaintext attacks; the CBC mode is not secure if chosen ciphertext attacks are allowed. The security against these attacks can be obtained by using authenticated encryption.

For some applications, the ciphertext should have exactly the same length as the plaintext, hence padding methods cannot be used. Two heuristic constructions have been proposed to address this problem; they are not without problems (both leak information in a chosen plaintext setting). A first solution encrypts the last incomplete block p_t (of $j < n$ bits) in OFB mode (cf. Section "The output mode"):

$$c_t = p_t \oplus \mathrm{rchop}_j(E_K(c_{t-1})).$$

A second solution is known as *ciphertext stealing* [21]: one appends the rightmost $n - j$ bits of c_{t-1} to the last block of j bits p_t, to obtain a new n-bit block:

$$c_{t-1} = E_K(p_{t-1} \oplus c_{t-2}),$$
$$c_t = E_K(p_t \parallel \mathrm{rchop}_{n-j}(c_{t-1})).$$

For the last two blocks of the ciphertext, one keeps only the leftmost j bits of c_{t-1} and n bits of c_t. This variant has the disadvantage that the last block needs to be decrypted before the one but last block.

It turns out that the common padding methods are vulnerable to side channel attacks that require chosen ciphertexts: an attacker who can submit ciphertexts of her choice to a decryption oracle can obtain information on the plaintext by noting whether or not an error message is returned stating that the padding is incorrect. This was first pointed out for symmetric encryption by Vaudenay in [24]; further results on concrete padding schemes can be found in [8, 9, 23]. The specific choice of the padding rule makes a difference: for example, the simple padding rule described in the introduction seems less vulnerable. Moreover, the implementation can to some extent preclude these attacks, for example by interrupting the session after a few padding errors. However, the preferred solution is the use of authenticated encryption.

THE OUTPUT FEEDBACK (OFB) MODE: The OFB mode transforms a block cipher into a synchronous stream cipher. This mode uses only the encryption operation of the block cipher. It consists of a finite state machine, which is initialized

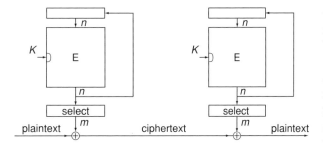

Fig. 3. The m-bit OFB mode of an n-bit block cipher

with an n-bit initial value or $s_0 = IV$. The state is encrypted and the encryption result is used as key stream and fed back to the state (see also Figure 3):

$$s_i = E_K(s_{i-1}) \quad \text{and} \quad c_i = p_i \oplus s_i, \quad i = 1, 2, \ldots$$

Treating an incomplete last block in the OFB mode is very simple: one selects the leftmost m bits of the last key word. The OFB mode can also be applied when the strings p_i and c_i consist of $m < n$ bits; in that case one uses only the m leftmost bits of each key word s_i. This results in a performance penalty with a factor n/m.

It is essential for the security of the OFB mode that the key stream does not repeat. It can be shown that the average period equals $n \cdot 2^{n-1}$ bits [14] and that the probability that an n-bit state lies on a cycle of length $< c$ is equal to $c/2^n$. This implies that after $2^{n/2}$ n-bit blocks one can distinguish the output of the OFB mode from a random string (in a random string one expects to see repetitions of n-bit blocks after $2^{n/2}$ blocks as a consequence of the <u>birthday paradox</u>, but it is highly unlikely that such repetitions occur in an OFB key stream). This suggests that one should rekey the OFB mode after $\alpha \cdot 2^{n/2}$ n-bit blocks for a small constant α. A repetition could also be induced in a different way: if IV is chosen uniformly at random for every message, the <u>birthday paradox</u> implies that IV values will repeat with high probability after approximately $2^{n/2}$ messages. The impact of such a repetition is dramatic, since it will leak the sum of all the plaintext blocks of the two messages encrypted with this IV value (for simplicity it is assumed here that all messages have equal length).

The main advantage of the OFB mode is that it has no error propagation: errors in the ith ciphertext bit will only affect the ith plaintext bit. The OFB mode does not allow for parallelism or random access.

It can be shown that the OFB mode is secure against <u>chosen plaintext attacks</u> if the block cipher is secure in the sense that it is hard to dis-

tinguish it from a random permutation. The proof requires that one changes the key after $\alpha \cdot 2^{n/2}$ n-bit blocks for small α (say 10^{-3}).

Note that an early draft of [12] included a variant of the OFB mode were only $m < n$ bits were fed back to the state, which acted as a shift register. However, this variant of the OFB mode has an average period of about $n \cdot 2^{n/2}$ bits [11]. This variant was removed because of this weakness.

THE COUNTER (CTR) MODE: The CTR mode is another way to transform a block cipher into a <u>synchronous stream cipher</u>. As the OFB mode, this mode only uses the encryption operation of the <u>block cipher</u>. It consists of a finite state machine, which is initialized with an n-bit integer IV. The state is encrypted to obtain the key stream; the state is updated as a counter mod 2^n (see also Figure 4):

$$c_i = p_i \oplus E_K(< (IV + i) \bmod 2^n) >), \quad i = 1, 2, \ldots$$

The mapping $< \cdot >$ converts an n-bit integer to an n-bit string. The processing of an incomplete final block or of shorter blocks is the same as for the OFB mode.

The period of the key stream is exactly $n \cdot 2^n$ bits. This implies that after $2^{n/2}$ n-bit blocks one can distinguish the output of the CTR mode from a random string (as for the OFB mode). This suggests that one should rekey the CTR mode after $\alpha \cdot 2^{n/2}$ n-bit blocks for a small constant α. A repeating value of IV has the same risks as for the OFB mode.

As the OFB mode, the CTR mode has no error propagation. Moreover the CTR mode allows for parallelism and for random access in both encryption and decryption.

It can be shown that the CTR mode is secure against <u>chosen plaintext attacks</u> if the block cipher is secure in the sense that it is hard to distinguish it from a random permutation [3]. Again it is recommended to change the key after $\alpha \cdot 2^{n/2}$ n-bit blocks for small α (say 10^{-3}).

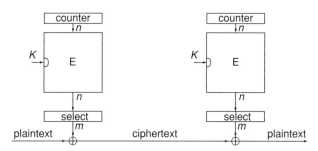

Fig. 4. The m-bit CTR mode of an n-bit block cipher

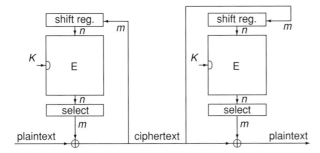

Fig. 5. The m-bit CFB mode of an n-bit block cipher

THE CIPHER FEEDBACK (CFB) MODE: The CFB mode transforms a block cipher into a <u>self-synchronizing stream cipher</u>. As the OFB and CTR mode, this mode only uses the encryption operation of the block cipher. It consists of a finite state machine, which is initialized with an n-bit initial value $s_0 = IV$. The state is encrypted and the leftmost m bits of the result are added to the m-bit plaintext block; the resulting ciphertext is fed back to the state (see also Figure 5):

$$c_i = p_i \oplus \mathrm{lchop}_m(E_K(s_{i-1})),$$
$$s_i = \mathrm{lchop}_{n-m}(s_{i-1}) \| c_i, \quad i = 1, 2, \dots$$

Treating an incomplete last block in the CFB mode is very simple: one selects the required number of bits from the output of the block cipher. The CFB mode is a factor n/m times slower than the CBC mode, since only m bits are used per encryption operation. In practice one often uses $m = 1$ and $m = 8$; this results in a significant speed penalty.

It can be shown that the CFB mode is secure against chosen plaintext attacks if the block cipher is secure in the sense that it is hard to distinguish it from a random permutation. A matching ciphertext attack also applies to the CFB mode (cf. Section "The cipher ... mode") [19]; the analysis is more complex since one can now consider n-bit ciphertext blocks which are shifted over m positions. To preclude leakage of information on the plaintexts one needs to impose that the number q of m-bit ciphertext blocks to which an opponent has access satisfies $q \ll 2^{(n+1)/2}$ or $q = \alpha \cdot 2^{n/2}$ where α is a small constant (say 10^{-3}). If this limit is reached, one needs to change the key.

The CFB decryption has a limited error propagation: errors in the ith ciphertext block will be copied into the ith plaintext block; about n subsequent plaintext bits will be completely garbled, since the error will stay for n/m steps in the state register s. From then on the decryption will recover. Moreover, if a multiple of m bits of the ciphertext are lost, synchronization will return as soon as n consecutive correct ciphertext bits have

been received. Particularly when $m = 1$, this is very attractive, since this allows for a recovery after loss of an arbitrary number of bits. The CFB decryption allows for random access and parallel processing, but the encryption process is serial.

ISO/IEC 10116 [16] specifies two extensions of the CFB mode: a first extension allows to encrypt plaintext blocks of length $m' < m; m - m'$ '1' bits are then prepended to the ciphertext c_i before feeding it back to the state. This mode offers a better speed, but increases the risk of a matching ciphertext attack. For example, if $n = 64, m = 8$, and $m' = 7$, on expects repetitions of the state after 2^{28} blocks, since the 64-bit state always contains eight '1' bits. A second extension allows for a larger state s (for example of $r \cdot n$ bits). This allows for parallel processing (with r processors) in the CFB encryption, at the cost of r IVs, a delayed error propagation and a slower synchronization.

Yet another variant of the CFB mode [1] improves the efficiency by using all the bits of $E_K(s_{i-1})$. A new encryption is only calculated if all bits of the n-bit block have been used or if a specific pattern of fixed length is observed in the ciphertext. The latter property allows resynchronization: the shorter the pattern, the faster the resynchronization, but the slower the performance.

OTHER MODES OF OPERATION: In the early 1990s, modes for <u>multiple encryption</u> of DES (see <u>Data Encryption Standard</u>) were analyzed. The simplest solution is to replace DES by <u>triple-DES</u> and to use triple-DES in one of the five modes discussed above. For triple-DES, these solutions are known as the *outer modes* [17]. However, their disadvantage is that one can only encrypt $\alpha \cdot 2^{n/2}$ blocks with a single key for small α (for example due to matching ciphertext attacks on CBC and CFB mode). This motivated research on *inner modes*, also known as interleaved or combined modes, where the modes themselves are considered as primitives (e.g., inner-CBC for triple-DES consists of three layers of single-DES in CBC mode). Biham has analyzed all the 36 double and 216 triple interleaved modes [4, 5], where each layer consists of ECB, OFB, CBC, CFB and the inverses of CBC and CFB. His goal is to recover the secret key (total break). He notes that by allowing chosen plaintext and chosen ciphertext attacks, *"all triple modes of operation are theoretically not much more secure than a single encryption."* The most secure schemes in this class require for DES 2^{67} chosen plaintexts or ciphertexts, 2^{75} encryptions, and 2^{66} storage. Biham also proposes a small set of triple modes, where a single key stream is generated in OFB mode and exored before every encryption and after the last encryption, and a few quadruple modes [4] with a higher conjectured security. However, Wagner has shown that if one allows chosen ciphertext/chosen *IV* attacks, the security of all but two of these improved modes with DES can be reduced to 2^{56} encryptions and between 2 and 2^{32} chosen chosen-*IV* texts [25]. A further analysis of the influence of the constraints on the *IV*s has been provided by Handschuh and Preneel [15]. The ANSI X9.52 standard [2] has opted for the outer modes of triple-DES. Coppersmith et al. propose the CBCM mode [10], which is a quadruple mode; this mode has also been included in ANSI X9.52. Biham and Knudsen present a certificational attack on this mode with DES requiring 2^{65} chosen ciphertexts and memory that requires 2^{58} encryptions [6]. In conclusion, one can state that it seems possible to improve significantly over the matching ciphertext attacks. However, the security results strongly depend on the model, security proofs have not been found so far and the resulting schemes are rather slow. It seems more appropriate to upgrade DES to Rijndael/AES [13].

A second area of research is on how to encrypt plaintexts from finite sets, which are not necessarily of size 2^n; this problem is partially addressed by Davies and Price in [11]; a formal treatment has been developed by Black and Rogaway in [7].

B. Preneel

References

[1] Alkassar, A., A. Geraldy, B. Pfitzmann, and A.-R. Sadeghi (2002). "Optimized self-synchronizing mode of operation." *Fast Software Encryption*, Lecture Notes in Computer Science, vol. 2355 ed. M. Mastsui. Springer-Verlag, Berlin, 78–91.

[2] ANSI X9.52 (1998). "Triple data encryption algorithm modes of operation."

[3] Bellare, M., A. Desai, E. Jokipii, and P. Rogaway (1997). "A concrete security treatment of symmetric encryption." *Proceedings 38th Annual Symposium on Foundations of Computer Science, FOCS'97*. IEEE Computer Society, 394–403.

[4] Biham, E. (1996). "Cryptanalysis of triple-modes of operation." Technion Technical Report CS0885.

[5] Biham, E. (1999). "Cryptanalysis of triple modes of operation." *Journal of Cryptology*, 12 (3), 161–184.

[6] Biham, E. and L.R. Knudsen (2002). "Cryptanalysis of the ANSI X9.52 CBCM mode." *Journal of Cryptology*, 15 (1), 47–59.

[7] Black, J. and P. Rogaway (2002). "Ciphers with arbitrary finite domains." *Topics in Cryptology CT—RSA 2002*, Lecture Notes in Computer Science, vol. 2271, ed. B. Preneel. Springer-Verlag, Berlin, 114–130.

[8] Black, J. and H. Urtubia (2002). "Sidechannel attacks on symmetric encryption schemes: The case

for authenticated encryption." *Proceedings of the 11th USENIX Security Symposium*, 327–338.

[9] Canvel, B., A.P. Hiltgen, S. Vaudenay, and M. Vuagnoux (2003). "Password interception in a SSL/TLS channel." *Advances in Cryptology—CRYPTO 2003*, Lecture Notes in Computer Science, vol. 2729, ed. D. Boneh. Springer-Verlag, Berlin, 583–599.

[10] Coppersmith, D., D.B. Johnson, and S.M. Matyas (1996). "A proposed mode for triple-DES encryption." *IBM Journal of Research and Development*, 40 (2), 253–262.

[11] Davies, D.W. and W.L. Price (1989). *Security for Computer Networks. An Introduction to Data Security in Teleprocessing and Electronic Funds Transfer* (2nd ed.). Wiley, New York.

[12] FIPS 81 (1980). "DES modes of operation." Federal Information Processing Standard (FIPS), Publication 81. National Bureau of Standards, US Department of Commerce, Washington, DC.

[13] FIPS 197 (2001). "Advanced encryption standard (AES)." Federal Information Processing Standard (FIPS), Publication 197. National Institute of Standards and Technology, US Department of Commerce, Washington, DC.

[14] Flajolet, P. and A.M. Odlyzko (1999). "Random mapping statistics." *Advances in Cryptlogy—EUROCRYPT'99*, Lecture Notes in Computer Science, vol. 1592, ed. J. Stern. Springer-Verlag, Berlin, 329–354.

[15] Handschuh, H., B. Preneel (1999). "On the security of double and 2-key triple modes of operation." *Fast Software Encryption*, Lecture Notes in Computer Science, vol. 1636, ed. L.R. Knudsen. Springer-Verlag, Berlin, 215–230.

[16] ISO/IEC 10116 (1991). "Information technology–security techniques—modes of operation of an *n*-bit block cipher algorithm." IS 10116.

[17] Kaliski, B.S. and M.J.B. Robshaw (1996). "Multiple encryption: Weighing security and performance." *Dr. Dobb's Journal*, 243, 123–127.

[18] Knudsen, L. (1994). "Block ciphers—analysis, design and applications." *PhD Thesis*, Aarhus University, Denmark.

[19] Maurer, U.M. (1991). "New approaches to the design of self-synchronizing stream ciphers." *Advances in Cryptology—EUROCRYPT'91*, Lecture Notes in Computer Science, vol. 547, ed. D.W. Davies. Springer-Verlag, Berlin, 458–471.

[20] Menezes, A., P.C. van Oorschot, and S. Vanstone (1998). *Handbook of Applied Cryptography*. CRC Press, Boca Raton, FL.

[21] Meyer, C.H., S.M. Matyas (1982). *Cryptography: A New Dimension in Data Security*. Wiley & Sons, New York.

[22] NIST (2001). *SP 800-38A Recommendation for Block Cipher Modes of Operation—Methods and Techniques.*

[23] Paterson, K.G. and A. Yau (2004). "Padding oracle attacks on the ISO CBC mode encryption standard." *Topics in Cryptology—The Cryptographers' Track at the RSA Conference*, Lecture Notes

in Computer Science, vol. 2964, ed. T. Okamoto. Springer-Verlag, Berlin, 305–323.

[24] Vaudenay, S. (2002). "Security flaws induced by CBC padding—applications to SSL, IPSEC, WTLS....." *Advances in Cryptology—EUROCRYPT 2002*, Lecture Notes in Computer Science, vol. 2332, ed. L. Knudsen. Springer-Verlag, Berlin, 534–546.

[25] Wagner, D. (1998). "Cryptanalysis of some recently-proposed multiple modes of operation." *Fast Software Encryption*, Lecture Notes in Computer Science, vol. 1372, ed. S. Vaudenay. Springer-Verlag, Berlin, 254–269.

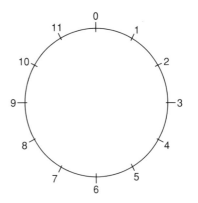

Fig. 1. Geometric view of arithmetic modulo 12

MODULAR ARITHMETIC

INTRODUCTION: Modular arithmetic is a key ingredient of many public key crypto-systems. It provides finite structures (called "rings") which have all the usual arithmetic operations of the integers and which can be implemented without difficulty using existing computer hardware. The finiteness of the underlying sets means that they appear to be randomly permuted by operations such as exponentiation, but the permutation is easily reversed by another exponentiation. For suitably chosen cases, these operations perform encryption and decryption or signature and verification. Direct applications include RSA public-key encryption and RSA digital signature scheme [17]; ElGamal public key encryption and the ElGamal digital signature scheme [3]; the Fiat–Shamir identification protocol [4]; the Schnorr Identification Protocol [18]; and the Diffie–Hellman key agreement [2].

Modular arithmetic is also used to construct finite fields and in tests during prime generation [9] (see also probabilistic primality test). Several copies of the modular structures form higher dimensional objects in which lines, planes, and curves can be constructed. These can be used to perform elliptic curve cryptography (ECC) [8, 12] and to construct *threshold schemes* threshold cryptography [19].

There are many examples of modular arithmetic in everyday life. It is applicable to almost any measurement of a repeated, circular, or cyclic process. Clock time is a typical example: seconds range from 0 to 59 and just keep repeating; hours run from 0 to 11 (or 23) and also keep repeating; days run from Sunday (0, say) to Saturday (6, say). These are examples of arithmetic modulo 60, 12 (or 24) and 7, respectively. Measuring angles in degrees uses arithmetic modulo 360.

To understand arithmetic in modulus N, imagine a line of length N units, where the whole number points $0, \ldots, N-1$ are labelled. Now connect the two end points of the line so that it forms a circle of circumference N. Performing modular arithmetic with respect to modulus N is equivalent to arithmetic with the marked units on this circle.

An example for $N = 12$ is shown in Figure 1. If one starts at number 0 and moves 14 units forward, the number 2 is reached. This is written $14 = 2 \pmod{12}$, and clearly 12 divides their difference $14 - 2$. Every 12 is thrown away in this arithmetic. Similarly, one can walk backwards 15 units from 0 and end up at 9. Hence, $-15 = 9 \pmod{12}$.

Modular addition is the same as addition of units on this circle. For example, if $N = 12$ and the numbers 10 and 4 are added on this circle, the result is 2. This is because if one starts at position 10 and moves ahead 4 units, position 2 is reached. So 4 hours after 10 o'clock is 2 o'clock. This is written $10 + 4 = 2 \pmod{12}$. We just keep the remainder (or "residue") after division by 12, i.e., $10 + 4 = 14$ becomes $14 - 12$, namely 2.

The notation for modular arithmetic is almost identical to that for ordinary (integer) arithmetic. The main difference is that most expressions and equations specify the modulus. Thus,

$$14 = 2 \pmod{12}$$

states that 14 and 2 represent the same element in a set which is called the *ring of residues mod 12*. When the modulus is clear, it may be omitted. Then we write

$$14 \equiv 2.$$

The different symbol $\equiv$ is needed because 14 and 2 are not equal as integers. The equation (or "congruence") is read as "14 is congruent to 2". All the integers in the set $\{\ldots, -22, -10, 2, 14, 26, \ldots\}$ represent the same *residue class* (or *congruence class*) modulo 12 because they all give the same remainder on division by 12, i.e., the difference between any two of them is a multiple of 12. In general, the numbers $A, A+N, A+2N, A+3N, \ldots$ and $A-N, A-2N, A-3N, \ldots$ are all *equivalent* modulo

N. Normally we work with the least non-negative representative of a class, 2 in this case, because of the convenience of the unique choice when equality is tested, and because it takes up the least space. (Note that some programming languages incorrectly implement the modular reduction of negative numbers by failing to take proper account of the sign. The Microsoft Windows calculator correctly reduces negatives, but gives the greatest non-positive value, namely -10 in our example.)

MODULAR ARITHMETIC OPERATIONS: Addition, subtraction and multiplication are performed in exactly the same way as for integer arithmetic. Strictly speaking, the arithmetic is performed on the residue classes but, in practice, we just pick integers from the respective classes and work with them instead. Thus,

$$7 \times 11 + 3 = 80 = 8 \,(\text{mod } 12).$$

In the expression on the left, we have selected the least non-negative residues to work with. The result, 80, then requires a modular reduction to obtain a least non-negative residue. Any representatives could be selected to perform the arithmetic. The answer would always differ by at most a multiple of the modulus, and so it would always reduce to the same value.

Hardware usually performs such reductions as frequently as possible in order to stop results from overflowing. Optimising integer arithmetic to perform modular arithmetic is the subject of much research. Modular multiplication is one of the most important areas of value to those implementing cryptographic functions. Montgomery [13] and Barrett [1] have created the most widely used methods for modular multiplication (see also Montgomery arithmetic). Another cryptographically important area is modular exponentiation (see exponentiation algorithms) Such operations make data-dependent use of power. This makes their use in embedded cryptosystems (e.g., smart cards) susceptible to attack through timing variations [6], compromising emanations [15], and differential power analysis [7] (see also timing attack, Radio Frequency attack, electromagnetic attack, side channel attacks, side channel analysis, and smartcard tamper resistance). Secure implementation of modular arithmetic is therefore at least as important as efficiency in such systems.

Addition, subtraction, and multiplication behave in the same way for residues as for integer arithmetic. The usual identity, commutative, and distributive laws hold, so that the set of residue classes form a "ring" in the mathematical sense. Thus,

- $N \equiv 0 \,(\text{mod } N)$.
- $A + 0 \equiv A \,(\text{mod } N)$.
- $1 \times A \equiv A \,(\text{mod } N)$.
- if $A \equiv B \,(\text{mod } N)$, then $B \equiv A \,(\text{mod } N)$.
- if $A \equiv B \,(\text{mod } N)$ and $B \equiv C \,(\text{mod } N)$, then $A \equiv C \,(\text{mod } N)$.
- if $A \equiv B \,(\text{mod } N)$ and $C \equiv D \,(\text{mod } N)$, then $A + C \equiv B + D \,(\text{mod } N)$.
- if $A \equiv B \,(\text{mod } N)$ and $C \equiv D \,(\text{mod } N)$, then $A \times C \equiv B \times D \,(\text{mod } N)$.
- $A + B \equiv B + A \,(\text{mod } N)$.
- $A \times B \equiv B \times A \,(\text{mod } N)$.
- $A + (B + C) \equiv (A + B) + C \,(\text{mod } N)$.
- $A \times (B \times C) \equiv (A \times B) \times C \,(\text{mod } N)$.
- $A \times (B + C) \equiv (A \times B) + (A \times C) \,(\text{mod } N)$.

However, division is generally a problem unless the modulus is a prime. Since

$$10 = 2 \times 5 = 2 \times 11 \,(\text{mod } 12),$$

it is clear that division by 2 (mod 12) can produce more than one answer; it is not uniquely defined. In fact, division by 2 (mod 12) is not possible in some cases: $2x \,(\text{mod } 12)$ always gives an even residue, so 3 (mod 12) cannot be divided by 2. It is easy to show that division by $A \,(\text{mod } N)$ is always well-defined precisely when A and N share no common factor. Thus division by 7 is possible in modulo 12, but not division by 2 or 3.

If 1 is divided by 7 (mod 12), the result is the *multiplicative inverse* of 7. Since $7 \times 7 = 1 \,(\text{mod } 12)$, 7 is its own inverse. Following the usual notation of real numbers, we write 7^{-1} for this inverse. For large numbers, the extended Euclidean algorithm [5] is used to compute multiplicative inverses (see Inversion in finite fields and rings).

Modular exponentiation (see exponentiation algorithms) is the main process in most of the cryptographic applications of this arithmetic. The notation is identical to that for integers and real numbers. $C^D \,(\text{mod } N)$ is D copies of C all multiplied together and reduced modulo N. As mentioned, the multiplicative inverse is denoted by an exponent $^{-1}$. Then the usual power laws, such as $x^A \times x^B = x^{A+B} \,(\text{mod } N)$, hold in the expected way.

When a composite modulus is involved, say N, it is often easier to work modulo its factors. Usually a set of co-prime factors of N is chosen such that the product is N. Solutions to the problem for each of these factors can then be pieced together into a solution modulo N using the Chinese Remainder Theorem (CRT) [14]. RSA cryptosystems which store the private key can use CRT to reduce the workload of decryption by a factor of 4.

MULTIPLICATIVE GROUPS AND EULER'S ϕ FUNCTION: The numbers which are prime to the modulus N have multiplicative inverses, as we noted above. This means that they form a group under multiplication. Consequently, each number X which is prime to N has an *order* mod N which is the smallest positive integer n such that $X^n = 1 \pmod N$. The Euler phi function ϕ gives the number of elements in this group, and it is a multiple of the order of each element. So $X^{\phi(N)} = 1 \pmod N$ for X prime to N, and, indeed, $X^{k\phi(N)+1} = X \pmod N$ for such X and any k. This last is essentially what is known as Euler's Theorem. As an example, $\{1, 5, 7, 11\}$ is the set of residues prime to 12. So these form a multiplicative group of order $\phi(12) = 4$ and $1^4 = 5^4 = 7^4 = 11^4 = 1 \pmod{12}$. A special case of this result is Fermat's "little" theorem which states that $X^{P-1} = 1 \pmod P$ for a prime P and integer X which is not divisible by P. These are really the main properties that are used in reducing the cost of exponentiation in cryptosystems and in probabilistic primality testing (see also Miller–Rabin probabilistic primality test) [11, 16].

When $N = PQ$ is the product of two distinct primes P and Q, $\phi(N) = (P-1)(Q-1)$. RSA encryption on plaintext M is performed with a public exponent E to give ciphertext C defined by $C = M^E \pmod N$. Illustrating this with $N = 35$, $M = 17$ and $E = 5$, we have $C \equiv 17^5 \equiv (17^2)^2 \times 17 \equiv 289^2 \times 17 \equiv 9^2 \times 17 \equiv 81 \times 17 \equiv 11 \times 17 \equiv 187 \equiv 12 \pmod{35}$. The private decryption exponent D must have the property that $M = C^D \pmod N$, i.e., $M^{DE} = M \pmod N$. From the above, we need a D which satisfies $DE = k\phi(N)+1$ for some k, i.e., D is a solution to $DE \equiv 1 \bmod (P-1)(Q-1)$. A solution is obtained using the Euclidean algorithm [5]. For the example, $D = 5$ since $\phi(35) = 24$ and $DE \equiv 5 \times 5 \equiv 1 \pmod{24}$. So $M \equiv 12^5 \equiv (12^2)^2 \times 12 \equiv 144^2 \times 12 \equiv 4^2 \times 12 \equiv 192 \equiv 17 \pmod{35}$, as expected. RSA (see RSA public key encryption) chooses moduli which are products of two primes so that decryption works also for texts which are not prime to the modulus. A good exercise for the reader is to prove that this is really true. CRT is useful in the proof.

PRIME FIELDS: When the modulus is a prime number P, every residue except 0 is prime to the modulus. Hence every nonzero number has a multiplicative inverse. So residues mod P form a *field* with P elements. These *prime* fields are examples of finite fields [10]. Because every nonzero has an inverse, the arithmetic of these fields is very similar to that of the complex numbers and it is possible to perform similar geometric constructions. They already form a very rich source for cryptography, such as Diffie–Hellman key agreement [2] and elliptic curve cryptography [8, 12], and will undoubtedly form the basis

for many more cryptographic primitives in the future.

S. Contini
Ç.K. Koç
C.D. Walter

References

[1] Barrett, P. (1986). "Implementing the Rivest Shamir and Adleman public-key encryption algorithm on a standard digital signal processor." *Advances in Cryptology—CRYPTO'86,* Lecture Notes in Computer Science, vol. 263, ed. A.M. Odlyzko. Springer-Verlag, Berlin, 311–323.

[2] Diffie, W. and M.E. Hellman (1976). "New directions in cryptography." *IEEE Transactions on Information Theory,* 22, 644–654.

[3] ElGamal, T. (1985). "A public key cryptosystem and a signature scheme based on discrete logarithms." *IEEE Transactions on Information Theory,* 31 (4), 469–472.

[4] Fiat, A. and A. Shamir (1986). "How to prove yourself: Practical solutions to identification and signature problems." *Advances in Cryptology—CRYPTO'86,* Lecture Notes in Computer Science, vol. 263, ed. A.M. Odlyzko. Springer-Verlag, Berlin, 186–194.

[5] Knuth, D.E. (1998). *The Art of Computer Programming, Volume 2, Semi-numerical Algorithms* (3rd ed.). Addison-Wesley, Reading, MA.

[6] Kocher, P. (1996). "Timing attack on implementations of Diffie–Hellman, RSA, DSS, and other systems." *Advances in Cryptology—CRYPTO'96,* Lecture Notes in Computer Science, vol. 1109, ed. N. Koblitz. Springer-Verlag, Berlin, 104–113.

[7] Kocher, P., J. Jaffe, and B. Jun (1999). "Differential power analysis." *Advances in Cryptology—CRYPTO'99,* Lecture Notes in Computer Science, vol. 1666, ed. M. Wiener. Springer-Verlag, Berlin, 388–397.

[8] Koblitz, N. (1987). "Elliptic curve cryptosystems." *Mathematics of Computation,* 48 (177), 203–209.

[9] Koblitz, N. (1994). *A Course in Number Theory and Cryptography* (2nd ed.). Springer-Verlag, Berlin.

[10] Lidl, R. and H. Niederreiter (1994). *Introduction to Finite Fields and Their Applications.* Cambridge University Press, Cambridge.

[11] Miller, G.L. (1976). "Riemann's hypothesis and tests for primality." *J. Computer and System Sciences,* 13, 300–317.

[12] Miller, V. (1985). "Uses of elliptic curves in cryptography." *Advances in Cryptology—CRYPTO'85,* Lecture Notes in Computer Science, vol. 218, ed. H.C. Williams. Springer-Verlag, Berlin, 417–426.

[13] Montgomery, P.L. (1985). "Modular multiplication without trial division." *Mathematics of Computation,* 44 (170), 519–521.

[14] Quisquater, J.-J. and C. Couvreur (1982). "Fast deciphering alorithm for RSA public-key cryptosystem." *Electronics Letters,* 18 (21), 905–907.

[15] Quisquater, J.-J. and D. Samyde (2001). "Electromagnetic analysis (EMA): Measures and

counter-measures for smart cards." *Smart Card Programming and Security (e-Smart 2001)*, Lecture Notes in Computer Science, vol. 2140, eds. I. Attali and T. Jensen. Springer-Verlag, Berlin, 200–210.

[16] Rabin, M.O. (1980). "Probabilistic algorithm for testing primality." *Journal of Number Theory*, 12, 128–138.

[17] Rivest, R.L., A. Shamir, and L. Adleman (1978). "A method for obtaining digital signatures and public-key cryptosystems." *Communications of the ACM*, 21 (2), 120–126.

[18] Schnorr, C.P. (1991). "Efficient signature generation by smart cards." *Journal of Cryptology*, 4, 161–174.

[19] Stinson, D.R. (2002). *Cryptography: Theory and Practice* (2nd ed.). CRC Press, Boca Raton, FL.

MODULAR ROOT

In the congruence $x^e \equiv y \pmod{n}$ (see modular arithmetic), x is said to be an eth modular root of y with respect to modulus n. For the cases that are of interest to cryptography we shall assume $\gcd(x, n) = \gcd(y, n) = 1$.

Computing modular roots is no more difficult than finding the order of the multiplicative group modulo n. In number theoretic terminology, this value is known as Euler's totient function, $\phi(n)$, which is defined to be the number of integers in $\{1, 2, \ldots, n-1\}$ that are relatively prime to n. If $\gcd(e, \phi(n)) = 1$, then there is either one or zero solutions, depending upon whether y is in the multiplicative subgroup generated by x. Assuming it is, the solution is obtained by raising both sides of the congruence to the power $e^{-1} \pmod{\phi(n)}$. If the gcd condition is not 1, then there may be more than one solution. For example, consider the special case of $e = 2$ and n an odd integer larger than 1. The congruence can have solutions only if y is a quadratic residue modulo n. Furthermore, if x is one solution, then $-x$ is another, implying that there are at least two distinct solutions.

Computing modular roots is easy when n is prime since $\phi(n) = n - 1$. The more interesting case is when n is composite, where it is known as the RSA problem. Since determining $\phi(n)$ is provably as difficult as factoring n, an important open question is whether a method exists to compute modular roots faster than factoring.

Scott Contini

MODULUS

The operand that the mod operation is computed with respect to. For instance, in the congruence $a \equiv b \bmod n$, the value n is the modulus. In RSA public-key encryption and the RSA digital signature scheme, the modulus is the integer that is the product of two large primes.

See modular arithmetic.

Scott Contini

MONDEX

Mondex is an electronic cash solution now owned by Mastercard. Developed in the second half of the 90s, today it is available as a Multos (card operationg system) application. The main feature offered is that a Mondex smart card stores digital cash in various currencies and may be used to pay for goods at merchants equipped with a Mondex retailer terminal, which transfers the payment from the customer Mondex card to the merchant Mondex card.

The chip on the Mondex card contains a program called the Mondex purse application, or purse, that stores the electronic cash and performs other Mondex operations. The purse contains one or more pockets, each storing the cash value for one individual currency.

In addition, a Mondex wallet, a pocket sized unit to store higher amounts of digital cash than the card does, is part of the system. The wallet allows a user to transfer between cards, and the a Mondex balance reader, a small device, checks the current balance on the Mondex Card. A Mondex ATM (Automated Teller Machine) is used to recharge cards or transfer money back into the account, etc.

The Mondex card logs all transactions. It stores a unique customer ID registered at the bank under which the personal information on the customer is stored. In particular, all transactions can be traced. The Mondex solution uses digital signature schemes for mutual authentication, but the underlying protocols and techniques are proprietary and not publicly available. On a worldwide scale, the use is still (at the time of writing) quite limited.

Peter Landrock

MONTGOMERY ARITHMETIC

INTRODUCTION: In 1985, P.L. Montgomery introduced an efficient algorithm [6] for computing $u = a \cdot b \pmod{n}$ where a, b, and n are k-bit binary numbers (see modular arithmetic). The algorithm is particularly suitable for implementation on general-purpose computers (signal processors or microprocessors) which are capable of performing fast arithmetic modulo a power of 2. The Montgomery reduction algorithm computes the resulting k-bit number u without performing a division by the modulus n. Via an ingenious representation of the residue class modulo n, this algorithm replaces division by n with division by a power of 2. The latter operation is easily accomplished on a computer since the numbers are represented in binary form. Assuming the modulus n is a k-bit number, i.e., $2^{k-1} \leq n < 2^k$, let r be 2^k. The Montgomery reduction algorithm requires that r and n be relatively prime, i.e., $\gcd(r, n) = \gcd(2^k, n) = 1$. This requirement is satisfied if n is odd. In the following, we summarize the basic idea behind the Montgomery reduction algorithm.

Given an integer $a < n$, we define its *n-residue* or *Montgomery representation* with respect to r as

$$\bar{a} = a \cdot r \pmod{n}.$$

Clearly, the sum or difference of the Montgomery representations of two numbers is the Montgomery representation of their sum or difference respectively:

$$\bar{a} + \bar{b} = \overline{a + b} \pmod{n}$$

and

$$\bar{a} - \bar{b} = \overline{a - b} \pmod{n}$$

It is straightforward to show that the set

$$\{i \cdot r \pmod{n} \,|\, 0 \leq i \leq n - 1\}$$

is a complete residue system, i.e., it contains all numbers between 0 and $n - 1$. Thus, there is a one-to-one correspondence between the numbers in the range 0 and $n-1$ and the numbers in the above set. The Montgomery reduction algorithm exploits this property by introducing a much faster multiplication routine which computes the n-residue of the product of the two integers whose n-residues are given. Given two n-residues $\bar{a}$ and $\bar{b}$, the *Montgomery product* is defined as the scaled product

$$\bar{u} = \bar{a} \cdot \bar{b} \cdot r^{-1} \pmod{n},$$

where r^{-1} is the (multiplicative) inverse of r modulo n (see <u>modular arithmetic</u> and <u>inversion finite fields and rings</u>), i.e., it is the number with the property.

$$r^{-1} \cdot r = 1 \pmod{n}.$$

As the notation implies, the resulting number $\bar{u}$ is indeed the n-residue of the product

$$u = a \cdot b \pmod{n}$$

since

$$\bar{u} = \bar{a} \cdot \bar{b} \cdot r^{-1} \pmod{n}$$
$$= (a \cdot r) \cdot (b \cdot r) \cdot r^{-1} \pmod{n}$$
$$= (a \cdot b) \cdot r \pmod{n}.$$

In order to describe the Montgomery reduction algorithm, we need an additional quantity, n', which is the integer with the property

$$r \cdot r^{-1} - n \cdot n' = 1.$$

The integers r^{-1} and n' can both be computed by the extended <u>Euclidean algorithm</u> [2]. The Montgomery product algorithm, which computes

$$\bar{u} = \bar{a} \cdot \bar{b} \cdot r^{-1} \pmod{n}$$

given $\bar{a}$ and $\bar{b}$, is given below:

---
function MonPro($\bar{a}, \bar{b}$)

Step 1. $t := \bar{a} \cdot \bar{b}$
Step 2. $m := t \cdot n' \pmod{r}$
Step 3. $\bar{u} := (t + m \cdot n)/r$
Step 4. **if** $\bar{u} \geq n$ **then return** $\bar{u} - n$
                     else return $\bar{u}$

The most important feature of the Montgomery product algorithm is that the operations involved are multiplications modulo r and divisions by r, both of which are intrinsically fast operations since r is a power 2. The MonPro algorithm can be used to compute the (normal) product u of a and b modulo n, provided that n is odd:

---
function ModMul(a, b, n) {n is an odd number}

Step 1. Compute n' using the extended Euclidean
         algorithm.
Step 2. $\bar{a} := a \cdot r \pmod{n}$
Step 3. $\bar{b} := b \cdot r \pmod{n}$
Step 4. $\bar{u} := \text{MonPro}(\bar{a}, \bar{b})$
Step 5. $u := \text{MonPro}(\bar{u}, 1)$
Step 6. **return** u

A better algorithm can be given by observing the property

$$\text{MonPro}(\bar{a}, b) = (a \cdot r) \cdot b \cdot r^{-1} = a \cdot b \pmod{n},$$

which modifies the above algorithm to:

---
function ModMul(a, b, n) {n is an odd number}

Step 1. Compute n' using the extended Euclidean
         algorithm.
Step 2. $\bar{a} := a \cdot r \pmod{n}$
Step 3. $u := \text{MonPro}(\bar{a}, b)$
Step 4. **return** u

However, the preprocessing operations, namely Steps 1 and 2, are rather time-consuming, especially the first. Since r is a power of 2, the second step can be done using k repeated shift and subtract operations. Thus, it is not a good idea to use the Montgomery product computation algorithm when a single modular multiplication is to be performed.

MONTGOMERY EXPONENTIATION: The Montgomery product algorithm is more suitable when several modular multiplications are needed with respect to the same modulus. Such is the case when one needs to compute a modular exponentiation, i.e., the computation of $M^e \pmod{n}$. Algorithms for modular exponentiation (see exponentiation algorithms) decompose the operation into a sequence of squarings and multiplications using a common modulus n. This is where the Montgomery product operation MonPro finds its best use. In the following, we exemplify modular exponentiation using the standard "square-and-multiply" method, i.e., the left-to-right binary exponentiation method, with e_i being the bit of index i in the k-bit exponent e:

---

function ModExp(M, e, n) {n is an odd number}

Step 1. Compute n' using the extended Euclidean
    algorithm.
Step 2. $\bar{M} := M \cdot r \pmod{n}$
Step 3. $\bar{x} := 1 \cdot r \pmod{n}$
Step 4. **for** $i = k - 1$ **down to** 0 **do**
Step 5. $\bar{x} := \text{MonPro}(\bar{x}, \bar{x})$
Step 6.   **if** $e_i = 1$ **then** $\bar{x} := \text{MonPro}(\bar{M}, \bar{x})$
Step 7. $x := \text{MonPro}(\bar{x}, 1)$
Step 8. **return** x

Thus, we start with the ordinary residue M and obtain its n-residue $\bar{M}$ and the n-residue $\bar{1}$ of 1 using division-like operations, as described above. However, once this preprocessing has been completed, the inner loop of the binary exponentiation method uses the Montgomery product operation, which performs only multiplications modulo 2^k and divisions by 2^k. When the loop terminates, we obtain the n-residue $\bar{x}$ of the quantity $x = M^e \pmod{n}$. The ordinary residue number x is obtained from the n-residue by executing the MonPro function with arguments $\bar{x}$ and 1. This is easily shown to be correct since

$$\bar{x} = x \cdot r \pmod{n}$$

immediately implies that

$$x = \bar{x} \cdot r^{-1} \pmod{n} = \bar{x} \cdot 1 \cdot r^{-1} \pmod{n}$$
$$= \text{MonPro}(\bar{x}, 1).$$

The resulting algorithm is quite fast, as was demonstrated by many researchers and engineers who have implemented it; for example, see [1, 5]. However, this algorithm can be refined and made more efficient, particularly when the numbers involved are multiprecision integers. For example, Dussé and Kaliski [1] gave improved algorithms, including a simple and efficient method for computing n'. In fact, any exponentiation algorithm can be modified in the same way to make use of MonPro: simply append the illustrated pre- and post-processing (Steps 1–3 and 7) and replace the normal modular multiplication operations in the iterative loop with applications of MonPro to the corresponding n-residues (Steps 4–6) in the above.

Here, as an example, we show how to compute $x = 7^{10} \pmod{13}$ using the Montgomery binary exponentiation algorithm.

- Since $n = 13$, we take $r = 2^4 = 16 > n$.
- Step 1 of the ModExp routine: Computation of n':
  Using the extended Euclidean algorithm, we determine that $16 \cdot 9 - 13 \cdot 11 = 1$, thus, $r^{-1} = 9$ and $n' = 11$.
- Step 2: Computation of $\bar{M}$:
  Since $M = 7$, we have $\bar{M} = M \cdot r \pmod{n} = 7 \cdot 16 \pmod{13} = 8$.
- Step 3: Computation of $\bar{x}$ for $x = 1$:
  We have $\bar{x} = x \cdot r \pmod{n} = 1 \cdot 16 \pmod{13} = 3$.
- Step 4: The loop of ModExp:

e_i	Step 5	Step 6
1	MonPro(3, 3) = 3	MonPro(8, 3) = 8
0	MonPro(8, 8) = 4	
1	MonPro(4, 4) = 1	MonPro(8, 1) = 7
0	MonPro(7, 7) = 12	

Below we show one instance of the loop computation (Steps 5 and 6). Other instances are computed similarly:

- Step 5: Computation of MonPro(3, 3) = 3:
  $t := 3 \cdot 3 = 9$
  $m := 9 \cdot 11 \pmod{16} = 3$
  $u := (9 + 3 \cdot 13)/16 = 48/16 = 3$
- Step 6: Computation of MonPro(8, 3) = 8:
  $t := 8 \cdot 3 = 24$
  $m := 24 \cdot 11 \pmod{16} = 8$
  $u := (24 + 8 \cdot 13)/16 = 128/16 = 8$
- Step 7 of the ModExp routine: $x = \text{MonPro}(12, 1) = 4$
  $t := 12 \cdot 1 = 12$
  $m := 12 \cdot 11 \pmod{16} = 4$
  $u := (12 + 4 \cdot 13)/16 = 64/16 = 4$

Thus, we obtain $x = 4$ as the result of the operation $7^{10} \pmod{13}$.

EFFICIENT MONTGOMERY MULTIPLICATION:
The previous algorithm for Montgomery multiplication is not efficient on a general purpose processor in its stated form, and so perhaps only has didactic value. Since we know that Montgomery multiplication algorithm computes

$$\text{MonPro}(a, b) = abr^{-1} \pmod{n}$$

and $r = 2^k$, we can give a more efficient bit-level algorithm which computes exactly the same value

$$\text{MonPro}(a, b) = ab2^{-k} \pmod{n}$$

as follows:

---

function MonPro (a, b) $\{n$ is odd and $a, b, n < 2^k\}$

Step 1. $u := 0$
Step 2. **for** $i = k - 1$ **downto** 0
Step 3. $\quad u := u + a_i b$
Step 4. $\quad u := u + u_0 n$
Step 5. $\quad u := u/2$
Step 6. **if** $u \geq n$ **then return** $u - n$
$\qquad\qquad$ **else return** u

---

where u_0 is the least significant bit of u and a_i is the bit with index i in the binary representation of a. The oddness of n guarantees that the division in Step 5 is exact. This algorithm avoids the computation of n' since it proceeds bit-by-bit: it needs only the least significant bit of n', which is always 1 since n' is odd because n is odd.

The equivalent word-level algorithm only needs the least significant word n'_0 (w bits) of n', which can also be easily computed since

$$2^k \cdot 2^{-k} - n \cdot n' = 1$$

implies

$$-n_0 \cdot n'_0 = 1 \pmod{2^w}.$$

Therefore, we conclude that n'_0 is equal to $-n_0^{-1}$ $(\text{mod } 2^w)$ and it can be quickly computed by the extended Euclidean algorithm or table look-up since it is only w bits (1 word) long. For the words (digits) a_i of a with index i and $k = sw$, the word-level Montgomery algorithm is as follows:

---

function MonPro (a, b) $\{n$ is odd and $a, b, n < 2^{sw}\}$

Step 1. $u := 0$
Step 2. **for** $i = s - 1$ **down to** 0
Step 3. $\quad u := u + a_i b$
Step 4. $\quad u := u + (-n_0^{-1}) \cdot u_0 \cdot n$
Step 5. $\quad u := u/2^w$
Step 6. **if** $u \geq n$ **then return** $u - n$
$\qquad\qquad$ **else return** u

This version of Montgomery multiplication is the algorithm of choice for systolic array modular multipliers [7] because, unlike classical modular multiplication, completion of the carry propagation required in Step 3 does not prevent the start of Step 4, which needs u_0 from Step 3. Such systolic arrays are extremely useful for fast SSL/TLS (see Secure Socket Layer and Transport Layer Security) servers, on the other hand, general-purpose software implementations of Montgomery multiplication rely on algorithms [4]. In particular, the Coarsely Integrated Operand Scanning (CIOS) algorithm is the preferred one for single-thread processors.

APPLICATION TO FINITE FIELDS: Since the integers modulo p form the finite field $GF(p)$, these algorithms are directly applicable for performing multiplication in $GF(p)$ by taking $n = p$. Similar algorithms are also applicable for multiplication in $GF(2^k)$, which is the finite field of polynomials with coefficients in $GF(2)$ modulo an irreducible polynomial of degree k [3].

Montgomery squaring (required for exponentiation) just uses MonPro with the arguments a and b being the same. However, in fields of characteristic 2 this is rather inefficient: all the bit products $a_i a_j$ for $i \neq j$ cancel, leaving just the terms a_i^2 to deal with. Then it may be appropriate to implement a modular operation ab^2 for use in exponentiation.

SECURE MONTGOMERY MULTIPLICATION: As a result of the data-dependent conditional subtraction in the last step of MonPro, embedded cryptosystems which make use of the above algorithms can be subject to a timing attack which reveals the secret key [10]. In the context of modular exponentiation, the final subtraction of each MonPro should then be avoided [8]. With this step omitted, all I/O to/from MonPro simply becomes bounded by $2n$ instead of n, but an extra loop iteration may be required on account of the larger arguments [9].

Ç.K. Koç
C.D. Walter

References

[1] Dussé, S.R. and B.S. Kaliski, Jr. (1990). "A cryptographic library for the Motorola DSP56000." *Advances in Cryptology—EUROCRYPT'90*, Lecture Notes in Computer Science, vol. 473, ed I.B. Damgård. Springer-Verlag, Berlin, 230–244.

[2] Knuth, D.E. (1998). *The Art of Computer Programming, Volume 2, Semi-numerical Algorithms* (3rd ed.). Addison-Wesley, Reading, MA.

[3] Koç, Ç.K. and T. Acar (1998). "Montgomery multiplication in GF(2^k)." *Designs, Codes and Cryptography*, 14 (1), 57–69.

[4] Koç, Ç. K., T. Acar, and B. S. Kaliski Jr. (1996). Analyzing and comparing Montgomery multiplication algorithms, *IEEE Micro*, 16 (3), 26–33.

[5] Laurichesse, D. and L. Blain (1991). "Optimized implementation of RSA cryptosystem." *Computers & Security*, 10 (3), 263–267.

[6] Montgomery, P.L. (1985). "Modular multiplication without trial division." *Mathematics of Computation*, 44 (170), 519–521.

[7] Walter, C.D. (1993). "Systolic modular multiplication." *IEEE Transactions on Computers*, 42 (3), 376–378.

[8] Walter, C.D. (1999). "Montgomery exponentiation needs no final subtractions." *Electronics Letters*, 35 (21), 1831–2.

[9] Walter, C.D. (2002). "Precise bounds for montgomery modular multiplication and some potentially insecure RSA moduli." *Topics in Cryptology—CT-RSA 2002*, Lecture Notes in Computer Science, vol. 2271, ed. B. Preneel. Springer-Verlag, Berlin, 30–39.

[10] Walter C.D. and S. Thompson (2001). "Distinguishing exponent digits by observing modular subtractions." *Topics in Cryptology—CT-RSA 2001*, Lecture Notes in Computer Science, vol. 2020, ed. D. Naccache. Springer-Verlag, Berlin, 192–207.

puting is one motivation for the large key sizes currently being proposed for many cryptosystems, such as the *Advanced Encryption Standard* (see Rijndael/AES), which has a 128-bit symmetric key.

The benefit of Moore's Law to users of cryptography is much greater than the benefit to opponents, because even a modest increase in computing power has a much greater impact on the key sizes that can be used, than on the key sizes that can be broken. This is a consequence of the fact that the methods available for using cryptosystems are generally polynomial time, while the fastest methods known for breaking cryptosystems are exponential time. This contrast may well be inherent to classical computing (see also quantum computers). Quantum computing, on the other hand, poses a more substantial potential threat in the future, because methods have been discovered for breaking many public-key cryptosystems in polynomial time on a quantum computer [2]. Quantum computers themselves are still in the research phase, and it is not clear if and when a sufficiently large quantum computer could be built. But if one were built (perhaps sometime in the next 30 years?) the impact on cryptography and security would be even more dramatic than the one Moore's Law has had so far.

Burt Kaliski

References

[1] Moore, Gordon (1965). "Cramming more components onto integrated circuits." *Electronics*, 38 (8).

[2] Shor, P.W. (1994). "Algorithms for quantum computation: discrete logarithms and factoring." *Proceedings of the 35th Annual IEEE Symposium on the Foundations of Computer Science*, 124–134.

MOORE'S LAW

The phenomenal rise in computing power over the past half century—which has driven the increasing need for cryptography and security as covered in this work—is due to an intense research and development effort that has produced an essentially exponential increase in the number of transistors than can fit on a chip, while maintaining a constant chip cost.

Roughly speaking, the amount of computing power available for a given cost has increased, and continues to increase by a factor of 2 every 18 months to 2 years, a pattern called *Moore's Law* after Gordon Moore of Intel, who in 1965 [1] first articulated this exponential model.

The implications to cryptography are two-fold. First, the resources available to users are continually growing, so that users can readily employ stronger and more complex cryptography. Second, the resources available to opponents are also growing. Effectively, the strength of any cryptosystem decreases by the equivalent of one symmetric-key bit every 18 months—or 8 bits every 12 years—posing a challenge to long-term security. This long-term perspective on advances in (classical) com-

MULTIPARTY COMPUTATION

Let f denote a given n-ary function, and suppose parties $P_1, \ldots, P_n$ each hold an input value $x_1, \ldots, x_n$, respectively. A **secure multiparty computation** for f is a joint protocol between parties $P_1, \ldots, P_n$ for computing $y = f(x_1, \ldots, x_n)$ *securely*. That is, even when a certain fraction of the parties is corrupted, (i) each party obtains the correct output value y and (ii) no information leaks on the input values of the honest parties beyond what is implied logically by the value of y and the values of the inputs of the corrupted parties.

Conceptually, a secure multiparty computation for function f can be viewed as an implementation

of a <u>trusted third party</u> T, which, upon receipt of the input values $x_1, \ldots, x_n$ from parties $P_1, \ldots, P_n$, respectively, produces the output value $y = f(x_1, \ldots, x_n)$. Party T is trusted for (i) providing the correct value for y and (ii) ot revealing any further information to parties $P_1, \ldots, P_n$.

A classical example is Yao's millionaires problem ($n = 2$). Parties P_1 and P_2 are two millionaires who want to see who is the richer one: writing x_1, x_2 for their respective wealths, they want to evaluate the function $f(x_1, x_2) = x_1 > x_2$. They could simply do so by telling each other the values of x_1 and x_2 but obviously this way much more information than the value of $x_1 > x_2$ is revealed. A secure two-party protocol allows them to compute the value of $x_1 > x_2$ without leaking any further information on x_1 and x_2. <u>Electronic voting</u> is another example of a secure multiparty computation, where $f(x_1, \ldots, x_n) = x_1 + \cdots + x_n$ and $x_i \in \{0, 1\}$ represent each party P_i's yes–no vote.

The theory of secure multiparty computation shows that a protocol for evaluating a given function f securely can be found, as long as f is a *computable* function, while imposing certain restrictions on the power of the corrupted parties, who are collectively called the *adversary*. A first distinction is whether the adversary is assumed to be computationally restricted, or not. In the *cryptographic model*, the adversary is assumed to be *polynomially* restricted (that is, the adversary is viewed as a probabilistic <u>polynomial-time</u> Turing machine). In the *information-theoretic model* no such restriction is assumed for the adversary. For the cryptographic model it suffices to assume *authentic* channels for each pair of parties: the messages exchanged over authentic channels cannot be changed by other (corrupted) parties; using <u>encryption</u> it is possible to hide the content of the messages. For the information-theoretic model one needs to assume a *private* (or, *secure*) channel is available to each pair of parties: the messages exchanged over private channels cannot be seen at all by other (corrupted) parties.

The adversary is called *passive* (or, *honest-but-curious*, or *semi-honest*) if it only tries to deduce information on the inputs of the honest parties by inspecting all the information available to the corrupted parties; the adversary is called *active* (or, *malicious*) if it is also allowed to let the corrupted parties deviate from the protocol in arbitrary ways. A further distinction is whether the adversary is allowed to choose which parties to corrupt adaptively. A *static* adversary must decide at the start of the protocol which parties it chooses to corrupt. An *adaptive* (or, *dynamic*) adversary may decide during the protocol which parties it chooses

to corrupt; once corrupted, however, a party remains so for the entire duration of the protocol.

A threshold (see <u>threshold scheme</u>) parameter t, $1 \leq t \leq n$, is used to indicate the maximum number of corrupted parties tolerated by a protocol for secure multiparty computation. As long as the number of corrupted parties does not exceed t, the protocol protects the interests of the honest parties. In terms of t, the main results for secure multiparty computation are as follows, where in each case the adversary may be adaptive. For the cryptographic model, any $t < n/2$ is achievable for an active adversary; for a passive adversary this can be improved to $t < n$. For the information-theoretic model, any $t < n/3$ is achievable for an active adversary; for a passive adversary this can be improved to $t < n/2$.

It is important to note that in case of an active adversary, condition (ii) above saying that each party obtains the correct output value y, can be split into two further conditions. Namely, the condition that each party actually receives an output value and the condition that, if an output value is received, then the output value is correct. If a protocol guarantees that each party receives an output value, the protocol is said to be *robust*.

The results above show that an honest majority is required to deal with an active adversary (if robustness is required). Therefore these results are not useful for the special case of two-party computation: for two parties, an honest majority comprises *all* of the parties ($t = 0$, $n = 2$). For secure two-party computation the property of robustness is thus replaced by the property of *fairness*. A protocol for two-party computation is said to be fair, if neither party can gain an advantage over the other party by quitting the protocol prematurely. For instance, a two-party protocol in which party P_1 learns the result first, and needs to send it to party P_2 is not fair, as P_1 may simply skip sending the result to P_2. Solutions to resolve this problem typically use a form of gradual release of a secret value, where the output value is released bit by bit and quitting by either party gives an advantage of at most one bit over the other party.

There exists a vast body of literature on secure multiparty computation. The paper by Yao [11] (and also [12]) and subsequent papers [1–3, 5, 6, 8, 10] build foundations for general secure multiparty computation, yielding the results mentioned above for various settings. (See also <u>verifiable secret sharing</u>). The strength of <u>oblivious transfer</u> is stressed by the result of [9]. In a similar direction, the results of [4, 7] show that threshold homomorphic cryptosystems provide a basis for efficient general secure multiparty computation as well.

A quick way to see why secure multiparty computation is possible at all, runs as follows, following [4, 7]. The basic primitive is a (probabilistic) threshold <u>homomorphic</u> cryptosystems E, where the private key is shared among parties $P_1, \ldots, P_n$. Such a public key cryptosystem is homomorphic in the sense that the product of two ciphertexts $E(x)$ and $E(y)$ results in a ciphertext $E(x + y)$, containing the sum of the values x and y. It is a threshold cryptosystem in the sense that decryption of a ciphertext $E(x)$ is done by a joint protocol between $P_1, \ldots, P_n$, resulting in the value of x, and as long as a majority of the parties is honest, ciphertexts will only be decrypted if a majority of parties agrees to do so.

Suppose function f is to be evaluated at $x_1, \ldots, x_n$, where x_i is the private input supplied by party P_i, for $i = 1, \ldots, n$. We may assume that function f is represented as an arithmetic circuit consisting of addition gates and multiplication gates (where additions and multiplications are defined over $\mathbb{Z}_N$, the integers <u>modulo</u> N, for a fixed integer $N \geq 2$). The protocol for secure computation of f then proceeds as follows. First, each party encrypts its private input value, yielding ciphertexts $E(x_1), \ldots, E(x_n)$. The circuit is then evaluated gate by gate, as described below, ultimately producing $E(f(x_1, \ldots, x_n))$ as encrypted output, from which the value $f(x_1, \ldots, x_n)$ is obtained, using threshold decryption.

The gates are evaluated as follows. An addition gate takes as input two ciphertexts $E(x)$ and $E(y)$ and produces as output a ciphertext $E(x + y)$, simply using the homomorphic property of E. A multiplication gate also takes as input two ciphertexts $E(x)$ and $E(y)$, but this time a protocol is required to produce $E(xy)$ as output value. We describe the protocol for the passive (semi-honest) case, omitting the <u>zero-knowledge</u> proofs to stop active adversaries:

1. Each party P_i, $1 \leq i \leq n$, picks a random value d_i and broadcasts ciphertexts $E(d_i)$ as well as $E(d_i y)$, where $E(d_i y)$ can be computed easily from d_i and $E(y)$.

2. Let $d = \sum_{i=1}^{n} d_i$. Using the homomorphic property of E, the parties compute ciphertext $E(x + d) = E(x) \prod_{i=1}^{n} E(d_i)$, from which they subsequently determine $x + d$, using threshold decryption. From $x + d$ and $E(y)$ one may then compute $E((x + d)y)$. Finally, using $E(dy) = \prod_{i=1}^{n} E(d_i y)$, one obtains $E(xy) = E((x + d)y)/E(dy)$, which is the desired output.

Note that all computations on x, y, and d_i's are done modulo N. Intuitively, the protocol is secure because the only values ever decrypted—apart from the output value $f(x_1, \ldots, x_n)$—are values $x + d$, where d is distributed uniformly at random and chosen jointly by $P_1, \ldots, P_n$.

Berry Schoenmakers

References

[1] Ben-Or, M., S. Goldwasser, and A. Wigderson. (1988). "Completeness theorems for noncryptographic fault-tolerant distributed computation." *Proceedings of 20th Symposium on Theory of Computing (STOC'88)*. ACM Press, New York, 1–10.

[2] Beaver, D. and S. Haber (1992). "Cryptographic protocols provably secure against dynamic adversaries." *Advances in Cryptology—EUROCRYPT'92*, Lecture Notes in Computer Science, vol. 658, ed. R.A. Rueppel. Springer-Verlag, Berlin, 307–323.

[3] Chaum, D., C. Crépeau, and I. Damgård (1988). "Multiparty unconditionally secure protocols." *Proceedings of 20th Symposium on Theory of Computing (STOC'88)*. ACM Press, New York, 11–19.

[4] Cramer, R., I. Damgård, and J.B. Nielsen (2001). "Multiparty computation from threshold homomorphic encryption." *Advances in Cryptology—EUROCRYPT 2001*, Lecture Notes in Computer Science, vol. 2045, ed. B. Pfitzmann. Springer-Verlag, Berlin, 280–300.

[5] Canetti, R., U. Feige, O. Goldreich, and M. Naor (1996). "Adaptively secure multi-party computation." *Proceedings of 28th Symposium on Theory of Computing (STOC'96)*. ACM Press, New York, 639–648.

[6] Canetti, R., Y. Lindell, R. Ostrovsky, and A. Sahai (2002). "Adaptively secure multi-party computation." *Proceedings of 34th Symposium on Theory of Computing (STOC 2002)*. ACM Press, New York, 494–503.

[7] Damgård, I. and J.B. Nielsen (2003). "Universally composable efficient multiparty computation from threshold homomorphic encryption." *Advances in Cryptology—CRYPTO 2003*, Lecture Notes in Computer Science, vol. 2729, ed. D. Boneh. Springer-Verlag, Berlin, 247–264.

[8] Goldreich, O., S. Micali, and A. Wigderson (1987). "How to play any mental game—or—a completeness theorem for protocols with honest majority." *Proceedings of 19th Symposium on Theory of Computing (STOC'87)*. ACM Press, New York, 218–229.

[9] Kilian, J. (1988). "Basing crpytography on oblivious transfer." *Proceedings of 20th Symposium on Theory of Computing (STOC'88)*. ACM Press, New York, 20–31.

[10] Rabin, T. and M. Ben-Or (1989). "Verifiable secret sharing and multiparty protocols with honest majority." *Proceedings of 21st Symposium on Theory of Computing (STOC'89)*. ACM Press, New York, 73–85.

[11] Yao, A. (1982). "Protocols for secure computations." *Proceedings of 23rd IEEE Symposium on*

Foundations of Computer Science (FOCS'82). IEEE Computer Society, 160–164.

[12] Yao, A. (1986). "How to generate and exchange secrets." *Proceedings of 27th IEEE Symposium on Foundations of Computer Science (FOCS'86)*. IEEE Computer Society, 162–167.

MULTIPLE ENCRYPTION

Composition of several ciphers is called *multiple encryption* or *cascade cipher*. See also product cipher.

Alex Biryukov

MULTIPRECISION MULTIPLICATION

The integer multiplication operation lies at the very heart of many cryptographic algorithms [5]. Naturally, tremendous effort went into developing efficient multiplication algorithms. The simplest of such algorithms is the classical "grammar school" multiplication method given as follows:

Multiprecision Multiplication Algorithm
Input: positive integers $u = (u_{m-1}u_{m-2} \ldots u_1u_0)_B$
       and $v = (v_{n-1}v_{n-2} \ldots v_1v_0)_B$
Output: The integer product $t = u \cdot v$
     For $i = 0$ to $m + n + 1$ do $t_i \leftarrow 0$;
     $c \leftarrow 0$;
     For $i = 0$ to m do
       For $j = 0$ to n do
         $(cs)_B \leftarrow t_{i+j} + u_i \cdot v_j + c$;
         $t_{i+j} \leftarrow s$;
       End For
       $t_{i+n+1} \leftarrow c$;
     End For
     Return (t)

The algorithm proceeds in a *row-wise* manner. That is, it takes one digit of one operand and multiplies it with the digits of the other operand, in turn appropriately shifting and accumulating the product. The two digit intermediary result $(cs)_B$ holds the value of the digit t_{i+j} and the carry digit that will be propagated to the digit t_{i+j+1} in the next iteration of the inner loop. The subscript B indicates that the digits are represented in radix-B. As can be easily seen the number of digit multiplications performed in the overall execution of the algorithm is $m \cdot n$. Hence, the time complexity of the classical multiplication algorithm

grows with $O(n^2)$, where n denotes the size of the operands.

The splitting technique introduced by Karatsuba [3] (see Karatsuba algorithm) reduces the number of multiplications in exchange of extra additions. The algorithm works by splitting the two operands u and v into halves

$$u = u_1B + u_0 \quad \text{and} \quad v = v_1B + v_0,$$

where $B = 2^{n/2}$ and n denotes the length of the operands. First the following three multiplications are computed:

$$d_0 = u_0v_0$$
$$d_1 = (u_1 + u_0)(v_1 + v_0)$$
$$d_2 = u_1v_1.$$

Note that these multiplications are performed with operands of half length. In comparison, the grammar school method would have required the computation of four multiplications. The product $u \cdot v$ is formed by appropriately assembling the three products together as follows

$$u \cdot v = d_0 + (d_1 - d_0 - d_2)B + d_2B^2.$$

In total, three multiplications, two additions, and two subtractions are needed compared to the four multiplications and one addition required by the classical multiplication method. Hence, the Karatsuba technique is preferable over the classical method whenever the cost of one multiplication exceeds the cost of three additions (counting additions and subtraction as same). The true power of the Karatsuba method is realized when recursively applied to all partial product computations in a multiprecision multiplication operation. For the classical Karatsuba algorithm, the length of the operands must be powers of two. An approximate analysis presented in [4] shows that a multiplication operation of two n bit numbers may be accomplished by $O(n^{\log_2 3})$ bit operations by the recursive application of the Karatsuba algorithm. More details and a generalization of the classical Karatsuba method can be found in the entry Karatsuba algorithm.

The subquadratic complexity of the Karatsuba algorithm makes it more attractive than the classical multiplication algorithm. However, in its implementation there are inconveniences one will encounter. For instance, in the computation of d_1, the operands $u_1 + u_0$ and $v_1 + v_0$ may both have an extra bit due to a carry-out in the addition. Another issue is the memory requirement for keeping temporary results during the recursion. In practice, a hybrid approach in which first the Karatsuba algorithm is applied recursively for several levels followed by the application of the classical multiplication algorithm might alleviate these problems and still provide an efficient implementation.

There is a wealth of multiplication algorithms beyond the grammar school and the Karatsuba multiplication algorithms. The key is to realize the close connection between the linear convolution and multiplication operations. Consider the multiplication of two polynomials. When two sequences are constructed using the polynomial coefficients and when their linear convolution is computed, the elements of the resulting sequence give the coefficients of the product of the two polynomials. Hence, any linear convolution algorithm may easily be adapted to compute polynomial multiplications. Furthermore, evaluating the operand and the product polynomials at $x = B = 2^w$, where w is the digit size, yields an algorithm for integer multiplication.

For instance, the elements of a 3-point convolution of the sequences $\{u_0, u_1, u_2\}$ and $\{v_0, v_1, v_2\}$ are given as

$$w_0 = u_0 v_0$$
$$w_1 = u_1 v_0 + u_0 v_1$$
$$w_2 = u_2 v_0 + u_1 v_1 + u_0 v_2$$
$$w_3 = u_2 v_1 + u_1 v_2$$
$$w_4 = u_2 v_2.$$

These expressions are exactly in the form of the coefficients of the product of the two polynomials $U(x) = u_0 + u_1 x + u_2 x^2$ and $V(x) = v_0 + v_1 x + v_2 x^2$. Evaluating the product polynomial for a particular *radix* size $B = 2^w$ gives an algorithm for integer multiplication where polynomial coefficients represent the digits of the two integer operands. Note that, the coefficients w_i may not exactly fit into a digit. Therefore, a digit carry-over operation needs to be performed through the entire length of the product in the final integer conversion step.

A well known convolution algorithm was introduced by Toom and Cook [1]. The Toom–Cook algorithm works by treating the two operands as polynomials $U(x)$ and $V(x)$ of maximum degree $k - 1$. This is done by partitioning the integer representations into k digits. Both polynomials are evaluated at $2k - 1$ points and multiplied together.

$$W(x_i) = U(x_i) \cdot V(x_i), \quad i = 0, 1, \ldots, 2k - 2.$$

This gives the evaluation of the product $U(x)V(x)$ at $2k - 1$ points which are used to form $2k - 1$ equations with the coefficients of $W(x)$ as unknowns. Solving the linear system of equations gives the coefficients of the product $W(x) = U(x) \cdot$

$V(x)$. Finally, $W(x)$ is evaluated at $B = 2^w$ and the product is obtained in integer form.

EXAMPLE 1 (Toom–Cook multiplication). We derive a 3-point Toom–Cook multiplication algorithm. Let

$$U(x) = u_0 + u_1 x + u_2 x^2$$

and

$$V(x) = v_0 + v_1 x + v_2 x^2.$$

We arbitrarily pick a sequence S of $2k - 1 = 5$ points. Let $S = \{0, 1, 2, -1, -2\}$. We then evaluate $U(x)$ and $V(x)$ for each element in S and compute their products as follows:

$$W(0) = U(0)V(0) = u_0 v_0$$
$$W(1) = U(1)V(1) = (u_0 + u_1 + u_2)(v_0 + v_1 + v_2)$$
$$W(2) = U(2)V(2) = (u_0 + 2u_1 + 4u_2)$$
$$(v_0 + 2v_1 + 4v_2)$$
$$W(-1) = U(-1)V(-1) = (u_0 - u_1 + u_2)$$
$$(v_0 - v_1 + v_2)$$
$$W(-2) = U(-2)V(-2) = (u_0 - 2u_1 + 4u_2)$$
$$(v_0 - 2v_1 + 4v_2).$$

These give us the evaluations of the product polynomial

$$W(x) = w_0 + w_1 x + w_2 x^2 + w_3 x^3 + w_4 x^4.$$

Note that when all computations are done symbolically, the evaluations of $W(x)$ are products of linear expressions of the coefficients of $U(x)$ and $V(x)$. In the final step we relate these products to the coefficients of $W(x)$ by forming the following system of equations:

$$W(0) = w_0$$
$$W(1) = w_0 + w_1 + w_2 + w_3 + w_4$$
$$W(2) = w_0 + 2w_1 + 4w_2 + 8w_3 + 16w_4$$
$$W(-1) = w_0 - w_1 + w_2 - w_3 + w_4$$
$$W(-2) = w_0 - 2w_1 + 4w_2 - 8w_3 + 16w_4.$$

By solving the equations the coefficients are obtained as follows:

$$w_0 = W(0)$$
$$w_1 = \tfrac{2}{3}W(1) - \tfrac{1}{12}W(2) - \tfrac{2}{3}W(-1) + \tfrac{1}{12}W(-2)$$
$$w_2 = -\tfrac{5}{4}W(0) + \tfrac{2}{3}W(1) - \tfrac{1}{24}W(2) + \tfrac{2}{3}W(-1)$$
$$\quad - \tfrac{1}{24}W(-2)$$
$$w_3 = -\tfrac{1}{6}W(1) + \tfrac{1}{12}W(2) + \tfrac{1}{6}W(-1) - \tfrac{1}{12}W(-2)$$
$$w_4 = \tfrac{1}{4}W(0) - \tfrac{1}{6}W(1) + \tfrac{1}{24}W(2) - \tfrac{1}{6}W(-1)$$
$$\quad + \tfrac{1}{24}W(-2).$$

With the equations relating the coefficients of the

product $W(x) = U(x)V(x)$ to the coefficients of the input operands, the multiplication algorithm is obtained.

As seen in the example above only in the second step of Toom–Cook's algorithm multiplications are computed. The number of multiplications is fixed as $2n - 1$. In the initial polynomial evaluation step multiplications with small integers are performed which are ignored here, since they may be implemented with inexpensive shifts and additions. In the final step where the coefficients of $W(x)$ are computed divisions by small integers are required. In fact as the length of the convolution grows the fractions grow radically. Considering a recursive implementation of the Toom–Cook algorithm the complexity can be shown to be $O(n^{\log_k(2k-1)})$ [4]. By choosing appropriately large k, the complexity can be brought close to $O(n^{1+\epsilon})$ for any $\epsilon > 0$ value. It should be noted, however, that this complexity figure ignores the additions as well multiplications and divisions with small constant integers. The number of such operations becomes more serious as k grows (and ϵ decreases).

As noted before, the multiplication operation is equivalent to linear convolution. This immediately suggests a Fourier Transform based approach for multiplication. One advantage of this technique over the Toom–Cook method and other direct convolution methods is in the lower number of additions and constant multiplications which were ignored in the complexity figure. But more importantly, it allows one to utilize Fast Fourier Transform techniques and achieve $O(n \log n)$ speed.

The Discrete Fourier Transform of a sequence is defined as follows:

DEFINITION 1 (Discrete Fourier Transform (DFT)). *Let s be a sequence of length d consisting of elements from an algebraic domain. Let g be a primitive d-th root of unity in that domain, i.e., $g^d = 1$ and let d be invertible in the domain. Then the Discrete Fourier Transform of s is defined as the sequence* **S** *whose elements are given as*

$$\mathbf{S}_k = \sum_{i=0}^{d-1} s_i g^{ik}.$$

The inverse transform is defined as

$$s_k = \frac{1}{d} \sum_{i=0}^{d-1} \mathbf{S}_i g^{-ik}.$$

There are many choices for the domain of the transformation such as a complex field, a *finite field*, or an integer ring. If the domain is a complex field, floating point operations may be needed and special attention must be given to handle rounding errors. If the domain is chosen as a finite field, than modular reductions become necessary. The third choice, an integer ring, gives more flexibility in choosing the modulus. In practice, special moduli of form $2^k \pm 1$ may be chosen to eliminate costly reductions.

After a domain is chosen and a DFT is set up as defined above, the following outlines an integer multiplication algorithm:
1. Partition both integer operands into equal sized blocks treating them as sequence elements.
2. Compute the DFT of both sequences.
3. Compute the componentwise product of the DFT of the two sequences.
4. Compute the inverse DFT of the product sequence.
5. Treat the sequence elements as the digits of the integer product.

The asssociated algorithm is given below:

DFT Based Integer Multiplication Algorithm
Input: positive integers $u = (u_{m-1}u_{m-2} \ldots u_1 u_0)_B$
      and $v = (v_{n-1}v_{n-2} \ldots v_1 v_0)_B$
Output: The integer product $w = u \cdot v$
    For $k = 0$ to $d - 1$ do
      $\mathbf{U}_k \leftarrow \sum_{i=0}^{d-1} u_i g^{ik}$;
      $\mathbf{V}_k \leftarrow \sum_{i=0}^{d-1} v_i g^{ik}$;
    End For
    For $k = 0$ to $d - 1$ do
      $\mathbf{W}_k \leftarrow \mathbf{U}_k \cdot \mathbf{V}_k$;
    End For
    For $k = 0$ to $d - 1$ do
      $w_k = \frac{1}{d} \sum_{i=0}^{d-1} \mathbf{W}_i g^{-ik}$;
    End For
    Return (w)

While the overall method is quite simple, an efficient algorithm is obtained only if the parameters are carefully chosen and a particular Fast Fourier Transform can be applied for computing the two forward transforms and the final inverse transform.

In an earlier work Shönhage and Strassen [6] introduced a DFT-based integer multiplication method that achieves an exciting asymptotic complexity of $O(n \log n \log \log n)$. The Shönhage and Strassen method is based on the Fermat number transform where the domain of the DFT is the integer ring Z_{2^m+1}. In this method the DFT is recursively turned into shorter DFT's of the same kind. Due to the special <u>pseudo-Mersenne</u> structure of the modulus $2^m + 1$, the method requires no multiplications for implementing the reductions. There are many other methods derived from special DFT and convolution algorithms. For an excellent survey on DFT based multiplication algorithms the reader is referred to [2].

Despite the tremendous improvement in the asymptotic complexity, the majority of DFT based algorithms have a large computational overhead associated with the forward and inverse transformations. Therefore they only become effective when the operands are longer than several thousand bits.

Finally, it is worth recognizing the relationship between the multiplication and squaring operations. Although highly redundant, a multiplication algorithm may be used in a trivial manner to accomplish a squaring. On the other hand, an integer multiplication may be achieved via two squarings by using the following simple trick:

$$u \cdot v = \frac{1}{4}[(u+v)^2 - (u-v)^2].$$

This identity may be useful when a fast squaring algorithm is available. See also Multiprecision Squaring.

Berk Sunar

References

[1] Cook, S.A. (1966). "On the Minimum Computation Time of Functions". *Master's Thesis*, Harvard University, Boston, MA, USA.

[2] Crandall, R. and C. Pomerance (2001). *Prime Numbers: A Computational Perspective*. Springer-Verlag, New York.

[3] Karatsuba, A. and Y. Ofman (1963). "Multiplication of multidigit numbers on automata". *Sov. Phys. Dokl. (English translation)*, 7 (7), 595–596.

[4] Knuth, D.E. (1997). *The Art of Computer Programming. Volume 2: Seminumerical Algorithms* (3rd ed.). Addison-Wesley, Reading, MA.

[5] Menezes, A.J., P.C. van Oorschot, and S.A. Vanstone (1997). *Handbook of Applied Cryptography*. CRC Press, Boca Raton, FL.

[6] Schönhage, A. and V. Strassen (1971). "Schnelle Multiplikation großer Zahlen." *Computing*, 7, 281–292.

MULTIPRECISION SQUARING

The integer squaring operation is a crucial operation for many cryptographic primitives. Squaring plays a central role in public-key cryptography

where operations such as exponentiation (see exponentiation methods) require a large number of integer squarings to be computed. The squaring operation can be thought of as a special case of multiplication, where both operands are the same. In this case, the multiplication algorithm has symmetries which are exploited. Consider a "grammar-school" multiplication performed with 4-digit operands $u = (u_3 u_2 u_1 u_0)_B$ and $v = (v_3 v_2 v_1 v_0)_B$ in radix $B = 2^w$ notation. For $u = v$ and $v_i = u_i$ the partial product array is as follows:

$$
\begin{array}{rccccc}
 & & u_3 & u_2 & u_1 & u_0 \\
\times & & u_3 & u_2 & u_1 & u_0 \\
\hline
 & & u_3 u_0 & u_2 u_0 & u_1 u_0 & u_0^2 \\
 & u_3 u_1 & u_2 u_1 & u_1^2 & u_0 u_1 & \\
u_3 u_2 & u_2^2 & u_1 u_2 & u_0 u_2 & & \\
u_3^2 & u_2 u_3 & u_1 u_3 & u_0 u_3 & &
\end{array}
$$

We observe that the array is symmetric across the diagonal, and odd numbered columns have a squared middle term in their diagonal. Hence, in a column only about half of the multiplication needs to be computed. The result is multiplied by two, and if the column is odd numbered, the squared term is added. In the following we present a multiprecision squaring algorithm that is based on the algorithms given in [1, 2]

Multiprecision Squaring Algorithm
Input: positive integer $u = (u_{m-1} u_{m-2} \ldots u_1 u_0)_B$
Output: The integer $t = u^2$

    For $i = 0$ to $2m - 1$ do $t_i \rightarrow 0$;
    For $i = 0$ to $m - 1$ do
      $(cs)_B \leftarrow t_{2i} + u_i^2$;
      $t_{2i} \leftarrow s$;
      For $j = i + 1$ to $m - 1$ do
        $(cs)_B \leftarrow t_{i+j} + 2u_i \cdot u_j + c$;
        $t_{i+j} \leftarrow s$;
      End For
      $t_{i+m} \leftarrow s$;
    End For
    $(cs)_B \leftarrow t_{2m-2} + u_{m-1}^2$;
    $t_{2m-2} \leftarrow s$; $t_{2m-1} \leftarrow c$;
    Return (t)

Note that product sum operation performed in the inner loop may may not fit into a double digit $(cs)_B$ and may require an extra bit due to the term $2u_i u_j$. The total number of multiplications performed in the algorithm is easily seen to be $(m^2 + m)/2$. The multiplication by the constant 2 in the inner loop of the algorithm may be implemented by simple shift and hence is not counted as a multiplication. See also Multiprecision Multiplication.

Berk Sunar

References

[1] Knuth, D.E. (1997). *The Art of Computer Programming. Volume 2: Seminumerical Algorithms* (3rd ed.). Addison-Wesley, Reading, MA.

[2] Menezes, A.J., P.C. van Oorschot, and S.A. Vanstone (1997). *Handbook of Applied Cryptography*. CRC Press, Boca Raton, FL.

MULTISET ATTACK

Multiset attack is a generic class of attacks which covers several recently designed (typically chosen plaintext attacks), which appeared in the literature under three different names: the *Square attack* [1], *the saturation attack* [4], the *integral cryptanalysis* [3]. The first such attack was discovered by Knudsen during analysis of the cipher Square [1] and was thus called "Square attack". A similar attack was used by Lucks [4] against the cipher Rijndael/AES and called "saturation" attack. Later Biryukov and Shamir have shown an attack of similar type breaking arbitrary 3 round SPN (see also substitution–permutation (SP) network) with secret components (the so-called *SASAS* scheme, which consists of five layers of substitutions and affine transforms). Gilbert–Minier's "collision" attack [2] on 7-rounds of Rijndael as well as Knudsen–Wagner's [3] "integral" cryptanalysis of 5-rounds of MISTY1 also fall into the same class.

The main feature behind these attacks is that unlike a differential attack in which the attacker studies the behavior of pairs of encryptions, in a multiset attack the attacker looks at a larger, carefully chosen set of encryptions, in which parts of the input text forms a multiset. A multiset is different from a regular notion of a set, since it allows the same element to appear multiple times. The element of a multiset is thus a pair (*value, multiplicity*), where *value* is the value of the element and *multiplicity* counts the number of times this value appears in the multiset. The attacker then studies the propagation of multisets through the cipher. The effect of the cipher on a multiset is in the changing of values of the elements but preserving some of the multiset properties like: multiplicity; or "integral" (i.e., sum of all the components); or causing a reduced set of values which would increase the probability of birthday-like events inside the cipher.

Multiset attacks are currently the best known attacks for the AES (see Rijndael/AES) due to its byte-wise structure. This new type of attack is a promising direction for future research.

Alex Biryukov

References

[1] Daemaen, J, L.R. Knudsen, and V. Rijmen (1997). "The block cipher Squar." *Proceedings of Fast Software Encryption—FSE'97*, Lecture Notes in Computer Science, vol. 1267, ed. E. Biham. Springer-Verlag, Berlin, 149–165.

[2] Gilbert, H. and M. Minier (2000). "A collision attack on seven rounds of Rijndael." *Proceedings of the Third AES Candidate Conference*, 230–241.

[3] Knudsen, L.R. and D. Wagner (2002). "Integral cryptanalysis (extended abstract)." *Fast Software Encryption, FSE 2002*, Lecture Notes in Computer Science, vol. 2365, eds. J. Daemen and V. Rijmen. Springer-Verlag, Berlin, 112–127.

[4] Lucks, S. (2000). "Attacking seven rounds of Rijndael under 192-bit and 256-bit keys." *Proceedings of the Third AES Candidate Conference*, 215–229.

N

NEEDHAM–SCHROEDER PROTOCOLS

In 1978, Needham and Schroeder [2] proposed symmetric and public key based protocols for key establishment (more specifically, key transport) (see symmetric cryptosystem, public key cryptography, or [1]). Their protocols satisfy a number of properties, including mutual identification of the participants, key authentication, and the establishment of a shared key. Of historical importance, the Needham–Schroeder symmetric key protocol forms the basis for the Kerberos authentication protocol [1].

The symmetric key based version (see symmetric cryptosystem) employs a trusted server T (see Trusted Third Party) that is online, or active during the key establishment. In this protocol, user A wishes to establish a key with user B, and initiates a protocol with T for this purpose. Both A and B (as well as any other user that will interact with T) respectively share symmetric keys K_A and K_B with T. At a high level, this protocol involves A securely interacting with T to establish a shared key K, followed by A securely sharing K with B.

In more detail, the symmetric key based version of Needham–Schroeder proceeds as follows:

1. A sends T a message containing its identifier I_A, an identifier for A's intended communicant B, I_B, and a nonce N_A (see Challenge–Response Identification) whose purpose is explained below.
2. T returns to A a message symmetrically encrypted using K_A. This encrypted message contains N_A, I_B, K (the symmetric key that A will use to securely communicate with B), and a message M_B specifically encrypted for B, containing the key K and I_A.
3. A decrypts the received message from T, ensures that N_A and I_B match those originally sent to T, and then forwards the message M_B to B.
4. Upon receipt, B decrypts M_B and ensures it contains I_A. Using K, B selects a nonce N_B and encrypts and sends the encrypted nonce to A.
5. A decrypts, subtracts 1 from the nonce value, and returns the result to B, encrypted with K.
6. B ensures that A was able to decrypt and subtract 1 from the nonce before using K for further secure communications with A.

Steps 1–3 establish K for both A and B. A includes a nonce N_A in step 1 to ensure that an attacker does not repeat (from an earlier session) the response received from T in step 2 (see Replay Attack). An explicitly includes the identifier I_B in step 1 to ensure that an attacker does not cause T to encrypt K for the attacker, rather than B. Steps 4 and 5 serve to provide mutual key authentication. A shortcoming of this protocol is that B is not able to determine the *freshness* of K. In particular, an attacker with knowledge of K from a previous session can replay step 3 to provide B with the old key, and then impersonate A at steps 4 and 5. Kerberos uses time stamps to avoid this shortcoming.

The public key based version of Needham–Schroeder (see public key cryptography) does not require an actively involved trusted third party T, although it does require some process to ensure the authenticity of the shared public keys. Assuming A and B have access to and are able to validate each others' public keys, the protocol proceeds as follows.

1. A selects a random symmetric key K_A, encrypts this value using the public key of B, and sends the result to B.
2. Upon receipt, B recovers K_A, similarly selects a symmetric key K_B, encrypts the concatenation of both symmetric keys using the public key of A, and sends the result to A.
3. A recovers both K_A and K_B, ensures that K_A matches the previously sent value, and returns K_B to B, encrypted using the public key of B.

At the end of this exchange, both A and B have copies of K_A and K_B. They can then compute a shared session key K as a function of both K_A and K_B.

It should be noted that the protocols above do not necessarily correspond with those originally presented by Needham and Schroeder [2], though they represent the most commonly accepted descriptions.

Mike Just

References

[1] Menezes, A., P. van Oorschot, and S. Vanstone (1997). *Handbook of Applied Cryptography*. CRC Press, Boca Raton, FL.
[2] Needham, R.M. and M.D. Schroeder (1978). "Using encryption for authentication in large networks of computers." *Communications of the ACM*, 21, 993–999.

NESSIE PROJECT

INTRODUCTION: NESSIE (New European Schemes for Signature, Integrity and Encryption) [19] was a research project within the Information Societies Technology (IST) Programme of the European Commission (IST-1999-12324). The seven NESSIE participants were: Katholieke Universiteit Leuven (Belgium), coordinator; Ecole Normale Supérieure (France); Royal Holloway, University of London (UK); Siemens Aktiengesellschaft (Germany); Technion—Israel Institute of Technology (Israel); Université Catholique de Louvain (Belgium); and Universitetet i Bergen (Norway).

NESSIE was a 40 month project, which started in January 2000. The goal of the NESSIE project was to put forward a portfolio of strong cryptographic algorithms that has been obtained after an open call and been evaluated using a transparent and open evaluation process. NESSIE has also developed a software toolbox to support the security and performance evaluation.

In February 2000, the NESSIE project has published an open call for a broad set of algorithms providing confidentiality, data integrity, and authentication. These algorithms include block ciphers (not restricted to 128-bit block ciphers), synchronous and self-synchronizing stream ciphers, hash functions, MAC algorithms, digital signature schemes, public-key encryption schemes, and identification schemes. In September 2000, more than 40 algorithms were received from major players in response to the NESSIE call. Two-thirds of the submissions came from industry, and there was some industry involvement in every five out of six algorithms. During 12 months, a first security and performance evaluation phase took place, which was supported by contributions from more than 50 external researchers. In September 2001, the selection of a subset of 25 algorithms for the second phase was announced. In the second phase of the project, the remaining algorithms were subjected to a thorough security evaluation, combined with a performance evaluation that will produce realistic performance estimates of optimized implementations. The selection of the portfolio of 17 recommended algorithms was announced in February 2003. This article discusses the NESSIE process and its results.

THE NESSIE CALL: In the first year of the project, an open call for the submission of cryptographic algorithms, as well as for evaluation methodologies for these algorithms has been launched. The scope of this call has been defined together with the project industry board (consisting of more than 25 companies) and was published in February 2000. The deadline for submissions was September 29, 2000. In response to this call NESSIE received 40 submissions, all of which met the submission requirements.

The NESSIE call includes a request for a broad set of cryptographic algorithms. While key management protocols are also very important, it was felt that they should be excluded from the call, as the scope of the call is already rather wide. The scope of the NESSIE call is much wider than that of the Rijndael/AES call launched by NIST [20], which was restricted to 128-bit block ciphers. It is comparable to that of the RACE Project RIPE (Race Integrity Primitives Evaluation, 1988–1992) [24] (confidentiality algorithms were excluded from RIPE for political reasons) and that of the Japanese CRYPTREC project [6] (which also includes key establishment protocols and pseudorandom number generation). Another difference is that both the Rijndael/AES competition and CRYPTREC intend to produce algorithms for government standards. The results of NESSIE will *not* be adopted by any government or by the European commission and will *not* become NESSIE Standards. However, the intention is that relevant standardization bodies [8, 9] will adopt these results.

The call also specified the main selection criteria which will be used to evaluate the proposals. These criteria are long-term security, market requirements, efficiency, and flexibility. Submissions could be targeted toward a specific environment (such as 8-bit smartcards or high-end 64-bit processors), but it is clearly an advantage to offer a wide flexibility of use. Security is put forward as the most important criterion, as security of a cryptographic algorithm is essential to achieve confidence and to build consensus.

For the *security requirements* of symmetric algorithms, two main security levels are specified, named *normal* and *high*. The minimal requirements for a symmetric algorithm to attain either the normal or high security level depend on the key length, internal memory, or output length of the algorithm. For block ciphers a third security level, *normal-legacy*, is specified, with a block size of 64 bits compared to 128 bits for the normal and high security levels. The motivation for this request is applications such as UMTS/3GPP, which intend to use 64-bit block ciphers for the next 10–15 years (to reduce hardware costs). For the asymmetric algorithms, a varying security level is accepted, with a minimum of about 2^{80} Triple-DES encryptions.

If selected by NESSIE, the algorithm should preferably be available royalty-free. If this is not possible, then access should be nondiscriminatory. The submitter should state the position concerning intellectual property and should update it when necessary.

The submission requirements were much less stringent than for the AES competition, particularly in terms of the requirement for software implementations (only "portable C" is mandatory).

THE EVALUATION PROCESS: The NESSIE evaluation process was an open process: it consisted of both internal evaluations by the project partners and external evaluations by submitters, project industry board members, and outsiders. All information w.r.t. the evaluation has been made public. Information was exchanged through a Web site (with a discussion forum); four open workshops were held to stimulate interactions and to disseminate the results.

The evaluation process was divided into two phases: after one year, the most promising submissions were selected and subjected to a more in-depth evaluation during the second phase. At the end of the first phase, minor modifications could be made to the algorithms provided that they did not invalidate the previous analysis.

The evaluation consisted of a security evaluation and a performance evaluation. Software have been developed to support this evaluation process. An additional criterion, which has only been taken into account at the very end, was the intellectual property status of the submission.

Security Evaluation

Each algorithm has been subjected to a thorough evaluation, according to the state of the art of cryptanalysis in the specific domain. An algorithm could not be selected if it failed to meet the security level required in the call or if it failed to meet a security claim made by the designer. Every weakness which has been identified has been verified independently. Moreover, if available, security proofs of algorithms have been checked in full detail. This has resulted in an extensive security report of more than 300 pages in [23]. The security of the algorithm was clearly the most important aspect of the evaluation.

Performance Evaluation

The performance has been evaluated for a broad range of platforms, taking into account both software and hardware implementations. First of all

a theoretical approach has been established. Each algorithm was dissected into three parts: setup (independent of key and data), precomputations (independent of data, e.g., key schedule), and the algorithm itself (that must be repeated for every use). Next a set of test platforms has been defined on which each candidate may be tested: 32-bit and 64-bit machines, 8-bit smartcards, and Field Programmable Gate Arrays (FPGAs). Then rules have been established which specify how performance should be measured on these platforms; this includes the definition of a custom designed API for each algorithm. The limited resources of the project did not allow for the evaluation of dedicated hardware implementations (ASICs).

For 32-bit and 64-bit machines, extensive tests have been run on 30 platforms (including Intel Pentium with MS Windows and Linux, Compaq Alpha, Sun Sparc, Apple PowerPC). Optimized C code has been developed and compared to Assembly language results provided by the submitters and by external parties. For smart cards, efforts have been concentrated on legacy block ciphers (with 64-bit blocks), digital signature schemes, and the identification scheme (note that only one identification scheme has been submitted). For FPGAs, benchmarking efforts have been focused on legacy block ciphers and digital signature schemes.

The project has attempted to evaluate to which extent implementations could be made more resistant against side channel attacks (such as timing attacks [14], power attacks [15], and fault analysis [3, 4]).

Tools

The NESSIE project has developed two types of tools. The general tools are not specific for the algorithms to be analyzed. Special tools, which are specific for the analysis of one algorithm, have been developed when a special tool was required for the cryptanalysis of an algorithm.

For the evaluation of the symmetric submissions, a comprehensive set of general tools was available within the project. These tools are in part based on an improved version of the statistical tools developed by the RIPE (RACE Integrity Primitives Evaluation) project [24]. As expected, all the submitted algorithms passed these tests.

In September 2000, the US NIST published a suite of statistical tests for the evaluation of sequences of random or pseudo-random bits; this document has been revised in December 2000 [21]. A careful comparison has been made between the RIPE and NIST test suites.

The NESSIE project has also developed a new generic tool to analyze block ciphers with differential [2] and linear cryptanalysis [17]. This tool is based on a general description language for block ciphers.

THE NESSIE SUBMISSIONS

Response to the NESSIE Call

The cryptographic community has responded very enthusiastically to the call. Thirty-nine algorithms have been received (43 when counting variants) along with one proposal for a testing methodology. After an interaction process, which took about 1 month, all submissions complied with the requirements of the call. The 26 symmetric algorithms include 17 block ciphers, which is probably not a surprise given the increased attention to block cipher design and evaluation as a consequence of the AES competition organized by NIST:

- Six 64-bit block ciphers: CS-Cipher, Hierocrypt-L1, IDEA, Khazad, MISTY1 and Nimbus;
- Seven 128-bit block ciphers: Anubis, Camellia, Grand Cru, Hierocrypt-3, Noekeon, Q and SC2000;
- One 160-bit block cipher: SHACAL;
- Three block ciphers with a variable block length: NUSH (64, 128 and 256 bits), RC6 (at least 128 bits), and SAFER++ (64 and 128 bits).
- Six synchronous stream ciphers: BMGL, Leviathan, LILI-128, SNOW, SOBER-t16, and SOBER-t32.
- Two MAC algorithms: Two-Track-MAC and UMAC; and
- One collision-resistant hash function: Whirlpool.

The 13 (or 17 counting variants) asymmetric algorithms are:

- Five asymmetric encryption schemes: ACE Encrypt, ECIES, EPOC, PSEC, and RSA-OAEP (both EPOC and PSEC have three variants);
- Seven digital signature algorithms: ACE Sign, ECDSA, ESIGN, FLASH, QUARTZ, RSA-PSS, and SFLASH; and
- One identification scheme: GPS.

The submissions came from four continents; the majority of submissions originated within industry; seven came from academia, and six were the result of a joint effort between industry and academia.

Selection for Phase II

On September 24, 2001, the NESSIE project announced the selection of candidates for the second phase of the project. The most important reason for rejection was a security weakness. Another reason to eliminate an algorithm could be that a similar algorithm existed with better security (for comparable performance) or with significantly better performance (for comparable security). However, very few algorithms were eliminated because of performance reasons. It should also be noted that the selection was more competitive in the area of block ciphers, where many strong contenders were considered.

Designers of submitted algorithms were allowed to make small alterations to their algorithms before entering phase II; the main criterion to accept these alterations was that they should improve the algorithm and not substantially invalidate the existing security analysis. Altered algorithms are indicated with a * in the list below.

The following symmetric algorithms have been selected for phase II (17 algorithms, but 14 if variants are only counted once):

- 64-bit block ciphers: IDEA, Khazad*, MISTY1, SAFER++64;
- 128-bit block cipher: Camellia, RC-6, SAFER ++128;
- 160-bit block cipher: SHACAL;
- 256-bit block cipher: SHACAL-2 and RC-6;
- Synchronous stream ciphers: BMGL*, SNOW*, SOBER-t16, and SOBER-t32;
- MAC algorithms: Two-Track-MAC and UMAC; and
- Collision-resistant hash function: Whirlpool*.

The NESSIE project intended to benchmark the solutions against well-established standards. The 128-bit block ciphers have been compared to AES/Rijndael [7, 10]. A 256-bit version of SHACAL has been introduced based on the new member of the SHA family SHA-256 [9, 13]. The MAC algorithms have been compared to HMAC [1, 12] and EMAC (cf. CBC-MAC and variants) [12, 22]. The hash function Whirlpool has been compared to the following members of the SHA family: SHA-256, SHA-384, and SHA-512 [9, 13].

The 11 asymmetric algorithms are:

- Public-key encryption algorithms: ACE-KEM* (derived from ACE Encrypt), EPOC-2*, PSEC-KEM* (derived from PSEC-2), ECIES*, RSA-OAEP*;
- Digital signature algorithms: ECDSA, ESIGN*, RSA-PSS, SFLASH*, QUARTZ*; and
- Identification scheme: GPS*.

Many of the asymmetric algorithms have been updated at the beginning of phase II. For the asymmetric encryption schemes, these changes were driven in part by developments during the first phase of the NESSIE project [11, 16, 25]. A second reason for these changes was the progress of the draft standard ISO/IEC 180033-2 within ISO/IEC JTC1/SC27 [26]. The standard defines a hybrid encryption scheme, consisting of two components: a KEM (Key Encapsulation Mechanism),where the asymmetric encryption is used to encrypt a symmetric key, and a DEM (Data Encapsulation Mechanism), which protects both secrecy and integrity of the bulk data with symmetric techniques (a "digital envelope"). This approach is slightly more complicated for encryption of a short plaintext, but it offers a more general solution with clear advantages. Two of the five NESSIE algorithms (ACE Encrypt and PSEC-2) have been modified to take into account this development. In addition, the NESSIE project has decided to evaluate RSA-KEM and ECIES-KEM, two variants defined in the working draft of ISO/IEC 18033-2 [26]. At the same time some other improvements have been introduced; as an example, ACE-KEM can be based on any abstract group, which was not the case for the original submission ACE Encrypt.

For the digital signature schemes, three out of five schemes (ESIGN, QUARTZ, and SFLASH) have been altered. In this case, there are particular reasons for each algorithm (correction for the security proof to apply, improve performance, or preclude a new attack). GPS has been modified to fix a small weakness.

The NESSIE Portfolio

Seventeen algorithms have been selected for the final NESSIE portfolio; twelve of these are NESSIE submissions, and the remaining five originate from well-established standards.

- 64-bit block ciphers: MISTY1;
- 128-bit block cipher: AES and Camellia;
- 256-bit block cipher: SHACAL-2;
- Synchronous stream ciphers: none;
- MAC algorithms: Two-Track-MAC, UMAC, EMAC, and HMAC;
- Collision-resistant hash function: Whirlpool, SHA-256, SHA-384, and SHA-512 (these three are counted as one family of algorithms);
- Public-key encryption algorithms: ACE-KEM, PSEC-KEM, and RSA-KEM;
- Digital signature algorithms: ECDSA, RSA-PSS, and SFLASH; and
- Identification scheme: GPS.

A full description of these algorithms as well as a detailed motivation for their selection and of the settings for which they are recommended is provided in [23].

One year after the announcement of the final portfolio, two developments can be reported:

an unintended property has been discovered in Whirlpool, which has been corrected in ISO/IEC 10118-3 [13]. The digital signature algorithm SFLASHv2, which was not recommended for general use but only for low-cost implementations, has been upgraded by its designers to SFLASHv3 [5] due to a new weakness identified in SFLASHv2. SFLASHv2 is no longer recommended by its designers. It is fair to state that if the NESSIE project would have been aware of this attack, it would not have recommended SFLASHv2: it seems that the area of multivariate cryptology is not completely mature yet, but one can hope that progress will be made in the coming years.

While it would be ideal for users of the NESSIE results that all algorithms recommended by NESSIE would be in the public domain, it is clear that this is not a realistic requirement. In one case, problems concerning the intellectual property status of a submission have resulted in it not being recommended (at the request of the submitter). An unexpected benefit from the NESSIE process is that several submitters have made it easier to use their algorithms. Twelve algorithms are in the public domain. For PSEC-KEM, the conditions are very favorable (license fees have to be paid only under exceptional circumstances). For ACE-ECRYPT, ECDSA, GPS, and SFLASH, licenses are available on a nondiscriminatory basis, with some additional conditions (users should offer reciprocate licenses if applicable). The reader is referred to [19, 23] for more detailed information.

CONCLUSION: The NESSIE project has been able to attract high quality submissions received from key players in the community; moreover, the broader cryptographic community has contributed to the evaluation process. Open competitions (such as AES, CRYPTREC [6], NESSIE, and RIPE [24]) can bring a clear benefit to the cryptographic research community and to the users and implementors of cryptographic algorithms. By asking cryptographers to design concrete and fully specified schemes, they are forced to make choices, to think about real life optimizations, and to consider all the practical implications of their research. While leaving many options and variants in a construction may be very desirable in a research paper, it is often confusing for a practitioner. Implementors and users can clearly benefit from the availability of a set of well defined algorithms that are described in a standardized way.

The developments in the last years have also shown that this approach can result in a better understanding of the security of cryptographic algorithms. We have also learned that concrete security proofs are an essential tool to build confidence, particularly for public key cryptography (where constructions can be reduced to mathematical problems believed to be hard) and for constructions that reduce the security of a scheme to other cryptographic algorithms. At the same time, we have learned that it is essential to study proofs for their correctness and to evaluate the efficiency of such reductions. Most asymmetric algorithms needed small modifications and corrections between the first and second phases, which shows that many subtle issues are involved in the specification of these algorithms.

For most symmetric cryptographic algorithms, confidence in existing standards has been strengthened and a few new algorithms have been put forward. The submitted stream ciphers were designed by experienced researchers and offer a very good performance, but none of them seems to meet the very stringent security requirements. While some of these algorithms may be adequate for many applications, this negative result shows that in the coming years, further work is necessary to investigate whether we can develop new stream ciphers that offer a very high security level (comparable to AES) while being substantially faster.

Finally, it is important to stress that the NESSIE portfolio is not a list of NESSIE standards, and the results of NESSIE will not be adopted "by default" by any government or by the European commission. However, one can anticipate that these results will be integrated into the existing cryptographic standards.

B. Preneel

References

[1] Bellare, M., R. Canetti, and H. Krawczyk (1996). "Keying hash functions for message authentication." *Advances in Cryptology—CRYPTO'96*, Lecture Notes in Computer Science, vol. 1109, ed. N. Koblitz. Springer-Verlag, Berlin, 1–15. Full version http://www.cs.ucsd.edu/users/mihir/papers/hmac.html

[2] Biham, E. and A. Shamir (1993). *Differential Cryptanalysis of the Data Encryption Standard.* Springer-Verlag, Berlin.

[3] Biham, E. and A. Shamir (1997). "Differential fault analysis of secret key cryptosystems." *Advances in Cryptology—CRYPTO'97*, Lecture Notes in Computer Science, vol. 1294, ed. B. Kaliski. Springer-Verlag, Berlin, 513–525.

[4] Boneh, D., R.A. DeMillo, and R.J. Lipton (1997). "On the importance of checking cryptographic protocols for faults." *Advances in Cryptology—EUROCRYPT'97*, Lecture Notes in Computer Science, vol. 1233, ed. W. Fumy. Springer-Verlag, Berlin, 37–51.

[5] Courtois, N.T., L. Goubin, and J. Patarin (2003). "SFLASHv3, a fast asymmetric signature scheme." Cryptology ePrint Archive: Report 2003/211.

[6] CRYPTREC project (2003). http://www.ipa.go.jp/security/enc/CRYPTREC/index-e.html

[7] Daemen, J. and V. Rijmen (2001). *The Design of Rijndael. AES—The Advanced Encryption Standard.* Springer-Verlag, Berlin.

[8] FIPS 180-1 (1995). "Secure hash standard." Federal Information Processing Standard (FIPS), Publication 180-1, National Institute of Standards and Technology, US Department of Commerce, Washington, DC.

[9] FIPS 180-2 (2002). "Secure hash standard." Federal Information Processing Standard (FIPS), Publication 180-2, National Institute of Standards and Technology, US Department of Commerce, Washington, DC (Change notice 1 published on December 1, 2003.)

[10] FIPS 197 (2001). *Advanced Encryption Standard (AES).* NIST, US Department of Commerce, Washington, DC.

[11] Fujisaki, E., T. Okamoto, D. Pointcheval, and J. Stern (2001). "RSA-OAEP is secure under the RSA assumption." *Advances in Cryptology—CRYPTO 2001*, Lecture Notes in Computer Science, vol. 2139, ed. J. Kilian. Springer-Verlag, Berlin, 260–274.

[12] ISO/IEC 9797. "Information technology—security techniques—message authentication codes (MACs)." Part 1: Mechanisms using a block cipher, 1999, Part 2: Mechanisms using a dedicated hash-function, 2002.

[13] ISO/IEC 10118. "Information technology—security techniques—hash-functions." Part 1: General, 2000, Part 2: Hash-functions using an n-bit block cipher algorithm, 2000, Part 3: Dedicated hash-functions, 2003. Part 4: Hash-functions using modular arithmetic, 1998.

[14] Kocher, P. (1996). "Timing attacks on implementations of Diffie–Hellman, RSA, DSS, and other systems." *Advances in Cryptology—CRYPTO'96*, Lecture Notes in Computer Science, vol. 1109, ed. N. Koblitz. Springer-Verlag, Berlin, 104–113.

[15] Kocher, P., J. Jaffe, and B. Jun (1999). "Differential power analysis." *Advances in Cryptology—CRYPTO'99*, Lecture Notes in Computer Science, vol. 1666, ed. M.J. Wiener. Springer-Verlag, Berlin, 388–397.

[16] Manger, J. (2001). "A chosen ciphertext attack on RSA optimal asymmetric encryption padding (OAEP) as standardized in PKCS #1 v2.0." *Advances in Cryptology—CRYPTO 2001*, Lecture Notes in Computer Science, vol. 2139, ed. J. Kilian. Springer-Verlag, Berlin, 230–238.

[17] Matsui, M. (1994). "The first experimental cryptanalysis of the Data Encryption Standard."

Advances in Cryptology—CRYPTO'94, Lecture Notes in Computer Science, vol. 839, ed. Y. Desmedt. Springer-Verlag, Berlin, 1–11.

[18] Menezes, A.J., P.C. van Oorschot, and S.A. Vanstone (1997). *Handbook of Applied Cryptography*. CRC Press, Boca Raton, FL.

[19] NESSIE. http://www.cryptonessie.org

[20] NIST. AES Initiative, http://www.nist.gov/aes

[21] NIST (2001). *A Statistical Test Suite for Random and Pseudorandom Number Generators for Cryptographic Applications*. NIST Special Publication 800-22. National Institute of Standards and Technology, US Department of Commerce, Washington, DC.

[22] Petrank, E. and C. Rackoff (2000). "CBC MAC for real-time data sources." *Journal of Cryptology*, 13 (3), 315–338.

[23] Preneel, B., A. Biryukov, C. De Cannière, S.B. Ors, E. Oswald, B. Van Rompay, L. Granboulan, E. Dottax, G. Martinet, S. Murphy, A. Dent, R. Shipsey, C. Swart, J. White, M. Dichtl, S. Pyka, M. Schafheutle, P. Serf, E. Biham, E. Barkan, Y. Braziler, O. Dunkelman, V. Furman, D. Kenigsberg, J. Stolin, J.J. Quisquater, M. Ciet, F. Sica, H. Raddum, L. Knudsen, and M. Parker. *Final Report of NESSIE, New European Schemes for Signatures, Integrity, and Encryption*, Lecture Notes in Computer Science, vol. 2274, eds. D. Naccache and P. Paillier. Springer-Verlag, Berlin, 297–309.

[24] RIPE (1995). "Integrity primitives for secure information systems." Final Report of RACE Integrity Primitives Evaluation (RIPE-RACE 1040), Lecture Notes in Computer Science, vol. 1007, eds. A. Bosselaers and B. Preneel. Springer-Verlag, Berlin.

[25] Shoup, V. (2001). "OAEP reconsidered." *Advances in Cryptology—CRYPTO 2001*, Lecture Notes in Computer Science, vol. 2139, ed. J. Kilian. Springer-Verlag, Berlin, 239–259.

[26] Shoup, V. (2001). *A Proposal for an ISO Standard for Public key Encryption*, Version 2.0. Available from http://www.shoup.net

NIEDERREITER ENCRYPTION SCHEME

THE CRYPTOSYSTEM: This public-key cryptosystem was introduced by H. Niederreiter in 1986 [2]. Its security is, as for the McEliece public key cryptosystem, related to difficult algorithmic problems of algebraic coding theory. For the standard terminology of coding theory, we refer the reader to the entry cyclic codes.

It has the same advantages (efficient encryption and decryption) and drawbacks (public-key size, transmission rate) as the McEliece system. The block size however is smaller.

General Idea

Consider an $r \times n$ matrix H, with $r < n$, over a finite field Fq. Let C denote the linear code of parity check matrix H.

The cryptogram is a linear combination of any t columns of H. For a given cryptogram, this linear combination is unique if and only if the minimum distance of C is at least $2t + 1$. Recovering the linear combination from the cryptogram is difficult, unless a t-error correcting procedure is known for C.

Description

Let $\mathcal{F}$ be a family of q-ary linear codes of length n and dimension k $(r = n - k)$ for which computationally effective t-error correcting procedure is known.

Key generation: The legal user picks randomly and uniformly a code C in the family $\mathcal{F}$. Let H_0 be a parity check matrix of C. The public key is equal to $H = SH_0P$ where S is a random non-singular binary $r \times r$ matrix and P a random permutation $n \times n$ matrix.

Encryption: The cleartext is a word x of F_q^n of *Hamming weight* t. The *ciphertext is a word of F_q^r equal to* Hx^T.

Decryption: The cleartext is recovered by finding the (unique) linear combination of t columns (or less) of H producing the cyphertext.[1]

PRACTICE AND SECURITY: We consider from now on that the family $\mathcal{F}$ in which the key was chosen is, as for the McEliece public key cryptosystem, a family $\mathcal{G}_{m,n,t}$ of *binary Goppa codes* (see Appendix in McEliece public key cryptosystem). In that case both systems are globally equally secure [1]. Many features of the systems differ however. In particular

- The public key can be chosen in systematic form without restriction,
- The block length (cleartext and cyphertext) is shorter,
- The information rate is (slightly) lower,
- Encryption and decryption costs (per information bit) are comparable but encryption is a little bit faster and decryption a little bit slower.

Implementation Aspects

Encoding constant weight words. There is a one-to-one correspondence between words of weight t and

---

[1] This is the same problem as correcting t errors occurring in a codeword of C.

Table 1. Some parameters for the Niederreiter system with binary Goppa codes

	$n = 1024$ $m = 10$ $t = 50$	$n = 2048$ $m = 11$ $t = 30$	$n = 4096$ $m = 12$ $t = 20$	$n = 2048$ $m = 11$ $t = 70$
Ciphertext size (in bits)	500	330	240	770
Message size (in bits)	284.0	222.0	178.8	435.8
Information rate	0.57	0.67	0.74	0.57
Public key size (in kB)	32	69	113	120
Security exponent[a]	62.1	86.4	86.1	108.4
encryption cost[b]	1	0.51	0.30	1.40
Decryption cost[b]	1	0.61	0.44	1.70

[a] Logarithm in base two of the binary work factor.
[b] Per message bit, relatively to the original parameters.

length n with the integers in the interval $|1, \binom{n}{t}|$. Computing this correspondence exactly is relatively expensive (quadratic in the block length n). However, there exist efficient approximate solutions with a cost proportional to the block length n (see [3]).

Encryption. Encryption will consist in adding t columns of the public key, that is $t^2 m$ binary operations.

Decryption. Decryption is roughly like decoding t errors in a binary Goppa code. The algorithmic complexity is proportional to $t^2 m^3$ binary operations.

Choosing the parameters. For given Goppa codes parameters, the security of the system against decoding and structural attack is the same as for the McEliece public key cryptosystem. Table 1 presents the main features of the system for various values of the parameters.

The cost for encoding/decoding words of length n and weight t is not included in the encryption/decryption cost. In theory this cost is negligible (with the approximate solution). In practice, however, it is significant in a software implementation, particularly for encryption.

Nicolas Sendrier

References

[1] Li, Y.X., R.H. Deng, and X.M. Wang (1994). "On the equivalence of McEliece's and Niederreiter's public-key cryptosystems." *IEEE Transactions on Information Theory*, 40 (1), 271–273.
[2] Niederreiter, H. (1986). "Knapsack-type cryptosystems and algebraic coding theory." *Prob. Contr. Inform. Theory*, 15 (2), 157–166.
[3] Sendrier, N. (2002). *Cryptosystmes cl publique bass sur les codes correcteurs d'erreurs.* Mmoire d'habilitation diriger des recherches, Universit Paris 6.

NONINTERACTIVE PROOF

A *noninteractive proof* may be viewed as a variation of an interactive proof [3], with two differences. Firstly, the interaction between the prover and the verifier is limited to the exchange of a *single* message, which is sent by the prover to the verifier. Secondly, both the prover and verifier are given access to a uniformly random bit string, referred to as the *common reference string*. Intuitively, the common reference string provides a way to challenge the prover (even though the verifier is not allowed to send any message to the prover).

A noninteractive zero-knowledge proof, as introduced in [1, 2], is a noninteractive proof satisfying the zero-knowledge property. In this case, the zero-knowledge property holds if it is possible to efficiently generate *both* the common reference string and a (valid) noninteractive zero-knowledge proof.

Berry Schoenmakers

References

[1] Blum, M., A. De Santis, S. Micali, and G. Persiano (1991). "Non-interactive zero-knowledge proof systems." *SIAM Journal on Computing*, 20 (6), 1084–1118.
[2] Blum, M., P. Feldman, and S. Micali (1988). "Non-interactive zero-knowledge and its applications." *Proceedings of the 20th ACM Symposium on the Theory of Computing*, 103–112.
[3] Goldreich, O. (2001). *Foundations of Cryptography—Basic Tools.* Cambridge University Press, Cambridge

NONLINEAR FEEDBACK SHIFT REGISTER

Feedback shift registers (FSRs) are the basic components of many keystream generators. The most commonly used are linear feedback shift registers (LFSRs). Here, we will focus on general FSRs that are also called nonlinear feedback shift registers in order to distinguish the linear case from the nonlinear one. Hence, we are interested in finite state automata of the form:

The feedback bit is computed as $s_{t+1} = f(s_t, \ldots, s_{t-L+1})$, where f can be any function in L variables. We will focus here on the binary case, since it is the most useful for implementation purposes (so f is a Boolean function). During each unit of time, the internal bit sare shifted and s_{t-L+1} is being output; the new left bit is s_{t+1}.

can be produced by such a register (this is due to the fact that it is an automaton with a finite set of states—the states being given by the internal values of the register).

Sequences achieving this period are de Bruijn sequences and more precise information can be found on them in their dedicated entry.

These general registers have not been studied as much as LFSRs, and many open questions remain. Some basic information can be found in [1], where it is stated that if we want the output sequence not to have a *preperiod* (so it starts to cycle right from the start), f must be of the form:

$$f(s_t, \ldots, s_{t-L+1}) = g(s_t, \ldots, s_{t-L+2}) + s_{t-L+1}.$$

But we almost have no information on the properties that g must fulfill in order to obtain a good

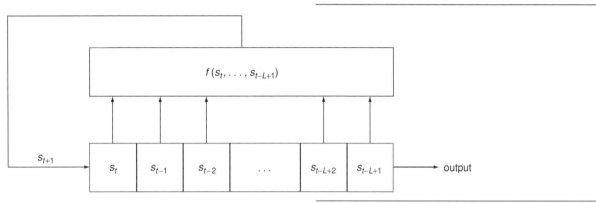

EXAMPLE. Let us consider $L = 3$ and let the *feedback function* $f(s_t, s_{t-1}, s_{t-2}) = s_t \cdot s_{t-1} + s_t + s_{t-2}$ (in its algebraic normal form, with binary operations "$\cdot$"-AND and "$+$"-XOR) be given by the following truth table:

s_t	s_{t-1}	s_{t-2}	$f(s_t, s_{t-1}, s_{t-2})$
0	0	0	0
0	0	1	1
0	1	0	0
0	1	1	1
1	0	0	1
1	0	1	0
1	1	0	0
1	1	1	1

If we assume that the first three bits of the sequence (also the initial state of the NLFSR) are $s_0 = 0, s_1 = 1, s_2 = 0$, then the sequence produced by the NLFSR is $010011010011\ldots$.

It is well known that any binary sequence produced by such a register has a period that can be at most equal to 2^L, and that any periodic sequence

keystream as output. Some studies have been done on the construction of some kind of de Bruijn sequences, but not for all.

An important point is also to study the linear complexity of the output sequence. It is easy to understand that for a fixed length L, a (NL)FSR can produce sequences with much higher linear complexity than LFSRs. The linear complexity of a sequence produced by a (NL)FSR of length L is upper bounded by 2^L (instead of L for a LFSR of the same length). It is also of importance to see how we can reconstruct the shortest (NL)FSR generating a given sequence. The best known algorithm for doing this is described in [2, 3]; its complexity is linear in space and memory. Since there is no constraint on the degree of f in this algorithm, Jansen calls *maximum order complexity of the sequence* the length of the shortest FSR that generates it. At the same time the notion of *complexity spectrum* of a sequence was introduced, that corresponds to the sets of the lengths, indexed by k, of the shortest FSRs that generate the sequence and with feedback functions of degree at most k. It contains the linear complexity ($k = 1$), the quadratic complexity ($k = 2$), and so on up to the maximum order complexity. Almost nothing is known about these quantities: how they are related, etc. A study about the quadratic case can be found in [4].

Caroline Fontaine

References

[1] Golomb, S.W. (1982). *Shift Register Sequences* (revised edition). Aegean Park Press, Laguna Hills, CA.

[2] Jansen, CEES J.A. (1989). "Investigations on nonlinear stream cipher systems: Construction and evaluation methods." *PhD Thesis*, Technical University of Delft.

[3] Jansen, CEES J.A. and D.E. Boekee (1990). "The shortest feedback shift register that can generate a given sequence." *Advances in Cryptology—CRYPTO'89*, Lecture Notes in Computer Science, vol. 435, ed. G. Brassard. Springer-Verlag, Berlin, 90–99.

[4] Youssef, A.M. and G. Gong (2000). "On the quadratic span of binary sequences." Technical Report, University of Waterloo, CORR 2000–20.

NONLINEARITY OF BOOLEAN FUNCTIONS

The *Hamming distance* between two n-variable Boolean functions $f: F_2^n \mapsto F_2$ and $g: F_2^n \mapsto F_2$ equals the size of the set $\{x \in F_2^n / f(x) \neq g(x)\}$. The *nonlinearity* of an n-variable Boolean function f is the minimum Hamming distance between f and all affine functions $a \cdot x \oplus a_0 = a_1 x_1 \oplus \cdots \oplus a_n x_n \oplus a_0$ (with $a \in F_2^n$; $a_0 \in F_2$). The nonlinearity is an affine invariant in the sense that if two functions f and g are such that there exists an affine automorphism A of F_2^n such that $g = f \circ A$, then they have same nonlinearity. The nonlinearity of cryptographic functions must be high, since the existence of affine approximations of the Boolean functions involved in a cryptosystem (a stream cipher or a block cipher) permits to build attacks on this system (see Linear cryptanalysis for block ciphers and Linear cryptanalysis for stream ciphers). The nonlinearity criterion can be quantified through the *Walsh transform* $\widehat{f}$ (see Boolean functions):

$$\mathcal{NL}(f) = 2^{n-1} - \frac{1}{2} \max_{a \in F_2^n} |\widehat{f}(a)|. \qquad (1)$$

Parseval's relation (see Boolean functions) then implies

$$\mathcal{NL}(f) \leq 2^{n-1} - 2^{(n/2)-1}. \qquad (2)$$

This bound, valid for every Boolean function, is called the *universal nonlinearity bound*.

Asymptotically, almost all Boolean functions on F_2^n have nonlinearities greater than $2^{n-1} - \sqrt{n} \, 2^{(n-1)/2}$.

Equality occurs in (2) if and only if $|\widehat{f}(a)|$ equals $2^{n/2}$ for every vector a. The corresponding functions, with nonlinearity $2^{n-1} - 2^{(n/2)-1}$, are called *bent*. They can exist only for even values of n, because $2^{n-1} - 2^{(n/2)-1}$ must be an integer, and they do exist for every n even, see [3]. An n-variable Boolean function f is bent if and only if, for any nonzero vector $a \in F_2^n$, the Boolean function $D_a f(x) = f(x) \oplus f(x + a)$ is balanced. Thus, the n-variable bent functions are those Boolean functions satisfying the propagation criterion $PC(n)$ (see Propagation characteristics of Boolean functions). For every even integer n greater than or equal to 4, the algebraic degree (see Boolean functions) of any bent function on F_2^n is at most $n/2$.

For n odd, inequality (2) cannot be tight. The maximum nonlinearity of n-variable Boolean functions then lies between $2^{n-1} - 2^{(n-1)/2}$ (which can always be achieved by functions of algebraic degree 2) and $2^{n-1} - 2^{(n/2)-1}$. It has been shown that it equals $2^{n-1} - 2^{(n-1)/2}$ when $n = 1$, 3, 5, 7 and is greater than $2^{n-1} - 2^{(n-1)/2}$ for every odd $n \geq 15$.

Bent functions being not balanced (i.e., their values being not uniformly distributed), they are improper for use in cryptosystems. For this reason, it is necessary to study the maximum nonlinearity of balanced functions. But this value is unknown for any $n \geq 8$.

The universal nonlinearity bound can be improved when we restrict ourselves to subclasses of functions. There exists a nontrivial upper bound on the nonlinearity of resilient functions: their nonlinearity is upper bounded by $2^{n-1} - 2^{m+1}$ (see Correlation immune and resilient Boolean functions for more details).

A notion of nonlinearity of mappings from F_2^n to F_2^n also exists: the nonlinearity of such function F is the minimum nonlinearity of all the nonzero linear combinations of the coordinate functions of F. It is upper bounded by $2^{n-1} - 2^{(n-1)/2}$ for every function F (cf. [1, 3]). This bound is tight for every odd n; the functions which achieve it are called *almost bent*.

Claude Carlet

References

[1] Chabaud, F. and S. Vaudenay (1995). "Links between differential and linear cryptanalysis." *Advances in Cryptology—EUROCRYPT'94*, Lecture Notes in Computer Science, vol. 950, ed. A. De Santis. Springer-Verlag, Berlin, 356–365.

[2] Rothaus, O.S. (1976). "On bent functions". *J. Comb. Theory*, 20A, 300–305.

[3] Xiao, Guo-Zhen, C. Ding, and W. Shan (1991). *The Stability Theory of Stream Ciphers*, Lecture Notes in Computer Science, vol. 561, eds. C. Ding, G. Xiao, and W. Shan. Springer-Verlag, Berlin.

NONMALLEABILITY

NONMALLEABILITY BY EXAMPLE: In describing the security offered by a cryptographic primitive, one must specify both what it means to break the primitive and what is the power of the adversary. *Nonmalleability* lies in the first category. For concreteness, let us focus on the problem of public-key encryption against a passive eavesdropper.

Goldwasser and Micali [10] laid the theoretical groundwork for modern cryptography, both giving a clean definition of secrecy in a complexity-theoretic framework, and providing a candidate public-key cryptosystem generator that, under prevailing computational assumptions, satisfies that definition. For the case of a passive eavesdropper listening to an encrypted conversation, their notion, called semantic security, is still the "gold standard" of security. One simple and intuitive description of semantic security is

For all relations R, seeing $\alpha \in E(x)$ does not help one to find y such that $R(x, y)$ holds.

(By $\alpha \in E(x)$ we mean that α is an element of the set of legal encryptions of x.) In other words, the adversary "learns nothing" about the plaintext from the ciphertext (it is assumed that the adversary is restricted to probabilistic polynomial time).

More than 20 years later, cryptography is now routinely used as a tool among *participants* in a protocol, not just to hide information from outside eavesdroppers but also from malicious participants in the protocol. In this case, the adversary is no longer passive, but active—and semantic security no longer suffices.

To see this, consider the following toy coin-flipping protocol. There are three players: Alice, Bob, and a judge. The goal is for Alice and Bob to produce an unbiased bit with the help of the judge (who for some reason cannot just flip a coin and announce the result). The judge chooses a (public key, private key) pair using a public-key cryptosystem generator, and publishes E, the encryption algorithm. First Alice, and then Bob, chooses a random bit, encrypts it using E, and sends the result to the judge. The judge decrypts and announces the two bits, and the result is the exclusive-OR of these bits.

The (erroneous) intuition is that, since Bob does not know Alice's bit, any ciphertext he produces is the encryption of a bit that is *independent* of Alice's bit, and so the exclusive-OR is unbiased. But suppose Bob could ensure that his bit is the same as Alice's; then the outcome will always be 0, *even if Bob has no idea what he is saying*. Similarly, if Bob could ensure that his bit is always the complement of Alice's, he can force an outcome of 1. It is a straightforward exercise for Bob to achieve either of these goals if the cryptosystem used is the one proposed in [10], despite the semantic security of the scheme.

As the above example shows, something stronger than semantic security is required; we call this property *non-malleability*:

For all polynomial time computable relations R, seeing $\alpha \in E(x)$ does not help one to find $\beta \in E(y)$, where $\beta \neq \alpha$, such that $R(x, y)$ holds.

In other words, seeing the ciphertext does not help a polynomial time bounded adversary to generate an *encryption* of a plaintext value related to x.

Just as semantic security specifies one kind of break of a cryptosystem (a break is a violation of semantic security), nonmalleability, or its opposite, malleability, specifies an alternate notion of what it means to break a cryptosystem.

We now turn to the means of attack available to the adversary. For public-key cryptography, three kinds of attack are defined in the literature: chosen-plaintext (this is the weakest form of attack that makes sense against a public-key cryptosystem, since an eavesdropper can encrypt plaintexts of her choice using the public encryption key), chosen ciphertext attacks in the preprocessing mode, and chosen ciphertext attacks in the postprocessing mode. In the preprocessing mode, the adversary may make polynomially many queries of a decryption oracle before being presented with a challenge ciphertext (for which it tries to find a violation of semantic security or nonmalleability). In the postprocessing mode, the adversary carries out a preprocessing attack, is presented with a challenge ciphertext α, and is then permitted the postprocessing attack: it may query the decryption oracle with any polynomial number of ciphertexts *other than the exact string α* before trying to *maul* (violate the nonmalleability of) the ciphertext.

Given two types of requirements (semantic security and nonmalleability) and three types of attacks (eavesdropping, cca-pre, cca-post), there are six possible combinations, of which five turn out to be distinct for public-key cryptosystems. Interestingly, under the strongest attack, chosen ciphertext in the postprocessing mode, nonmalleability, and semantic security are equivalent. This is not so surprising: in the public-key case any violation of semantic security is clearly a violation of nonmalleability (having found a y related to x, in the public-key case it is trivial to obtain $E(y)$ where y is related to x: just encrypt y using the public key). For the other direction, if it were possible to maul an encryption $\alpha \in E(x)$ to obtain an encryption $\beta \in E(y)$, for y related x (and where $\beta \neq \alpha$), then by feeding β to the decryption oracle one would obtain a *plaintext* y related to x.

The concept of nonmalleability has application beyond encryption. For instance, in the toy coin-flipping protocol described above, one role of the encryption step is to *commit* to the bits before opening them and computing their exclusive-OR. In the toy protocol, the opening of the <u>commitment</u> is made trivial by the presence of the judge; in a less centralized protocol the commitment would be done by secret sharing. But the problem of malleability is the same: *secrecy* of Alice's committed value does not ensure *independence* from the value committed to by Bob. If Bob can maul Alice's commitment messages so as to commit to the same or opposite value at will, the fact that Alice's value is secret is irrelevant.

A natural response is to require Alice and Bob to prove *knowledge* of their committed values: intuitively, if Bob *knows* what he is saying (this can be made precise), then his bit must be independent of Alice's bit—otherwise we would have a violation of semantic security. But again, there is a malleability problem: perhaps the <u>proof of knowledge</u> itself is malleable, even if it is a <u>zero-knowledge</u> proof of knowledge.

ONE SOLUTION: The approach taken in [8] for achieving nm-cca-post encryption based on any *semantically secure* public-key cryptosystem generator $\mathcal{G}$ (as defined in [10]) is as follows.

1. The public key consists of n *pairs* of public keys, $(E_0^1, E_1^1), \ldots, (E_0^n, E_1^n)$, drawn according to $\mathcal{G}$, together with a reference string σ for <u>noninteractive</u> <u>zero-knowledge</u> proofs (NIZKs).
2. To encrypt a plaintext message m:
   - Choose an instance of a <u>digital signature</u> <u>scheme</u>;
   - View the public verification key of the signature scheme as a sequence of bits selecting one key (under which to encrypt the plaintext) from each pair of keys in the public key—a total of n keys.
   - Encrypt the plaintext m under each of the n selected keys.

   - Provide a NIZK of consistency of encryptions using the reference string σ—that is, provide a noninteractive zero-knowledge proof that all ciphertexts are valid encryptions, under the selected keys, of the same plaintext.
3. The ciphertext consists of the public verification key for the signature scheme, the n ciphertexts, and the proof of consistency of encryptions.

The intuition for nonmalleability is straightforward. Assume the attacker is given a ciphertext α that it wishes to maul. That is, the attacker is trying to construct a valid ciphertext $\beta \neq \alpha$ encrypting a message m' that is related to m. If the attacker uses for β the same instance of the signature scheme as was used in generating α, then it will not know the secret signing key corresponding to the public verification key, and so will be unable to generate a signature on any content other than the content already signed in α, preventing it from generating a different valid ciphertext. If, instead, the attacker changes the signature scheme, then there will be at least one pair of encryption keys so that, without loss of generality, α contains an encryption $E_0^i(m)$, and the adversary must generate $E_1^i(m')$ for some m' related to m (possibly $m' = m$); intuitively, since the two keys are chosen completely independently, the attacker has been given no clue how to generate an encryption of a related m' under E_1^i (even if it could easily do so under E_0^i because it has seen $E_0^i(m)$ and encryptions under any individual key may be malleable).

FORMAL DEFINITION: A completely general definition of nonmalleability, suitable to all the settings we have described, is beyond the scope of this encyclopedia entry. We therefore restrict our attention to nonmalleability of a public-key cryptosystem under a chosen ciphertext attack in the postprocessing mode.

We first define precisely the power of the cca-post adversary, denoted $\mathcal{A}$. Let R be a polynomial-time computable relation. Let n be the security parameter for the *nonmalleable under cca-post* public-key cryptosystem generator $\mathcal{G}'$. $\mathcal{A}$ receives the public key $E \in_R \mathcal{G}'(n)$ and can adaptively choose a sequence of ciphertexts $c_1, c_2, \ldots$. On each of them, $\mathcal{A}$ receives the corresponding plaintext (or "invalid" message, if the ciphertext is invalid). It then produces a distribution $\mathcal{M}$ on messages of length $\ell(n)$, for some polynomial ℓ, by producing a description of the polynomial-time machine that can generate this distribution.

Next m is selected according to $\mathcal{M}$, that is, $m \in_R \mathcal{M}$, and $\mathcal{A}$ receives as a challenge a ciphertext $c \in_R E(m)$, together with some "side-information" about m in the form of $h(m)$. (The function h is polynomially computable; $h(\cdot)$ is sometimes called the *history variable*, and it can be thought of as encapsulating prior knowledge that $\mathcal{A}$ may have concerning the plaintext.) $\mathcal{A}$ then engages in a second phase of adaptively choosing ciphertexts $c'_1, c'_2, \ldots$ and learning their decryptions. The only restriction is that $c \notin \{c'_1, c'_2, \ldots\}$.

At the end of the process, $\mathcal{A}$ produces a polynomially bounded length vector of ciphertexts $(f_1, f_2, \ldots)$ not containing the challenge ciphertext c, with each $f_i \in E(\beta_i)$, and a *cleartext* string σ which we assume contains a description of $\mathcal{M}$. Let $\beta = (\beta_1 \beta_2 \ldots)$. $\mathcal{A}$ is considered to have succeeded with respect to R if $R\ (m, \beta, \sigma)$ holds. (We seperate β from σ because the goal of the adversary is to produce encryptions of the elements in β.) Let $\pi(\mathcal{A}, R)$ be the probability that $\mathcal{A}$ succeeds, where the probability is over the coin flips of the key generator, $\mathcal{A}$, $\mathcal{M}$ and the encryption of m.

Let $\mathcal{A}'$ be an *adversary simulator* that does not have access to the encryptions or to the decryptions but can pick a distribution $\mathcal{M}'$. On output E, $\mathcal{A}'$ produces $\mathcal{M}'$, from which $m \in_R \mathcal{M}'$ is chosen. $\mathcal{A}'$ receives $h(m)$ and, without the benefit of a chosen ciphertexts attack, produces a vector of ciphertexts $(f_1 f_2, \ldots)$, where each $f_i \in E(\beta_i)$, and a string σ containing $\mathcal{M}'$. Let $\beta = (\beta_1, \beta_2, \ldots)$. As above $\mathcal{A}'$ is considered to have succeeded with respect to R if $R(m, \beta, \sigma)$. Let $\pi'(\mathcal{A}', R)$ be the probability that $\mathcal{A}'$ succeeds where the probability is over the coin-flips of the key generator, $\mathcal{A}'$ and $\mathcal{M}'$.

Note that $\mathcal{A}'$ has much less power than $\mathcal{A}$: not only does it not have access to the ciphertext $c \in E(m)$, but it cannot perform *any* type of chosen ciphertext attack, even in choosing the distribution $\mathcal{M}'$. Note also that the fact that $\mathcal{M}'$ is given to R prevents $\mathcal{A}'$ from choosing trivial distributions.

DEFINITION 1. *A public-key cryptosystem generator is nonmalleable with respect to chosen ciphertext attacks in the postprocessing mode if for all probabilistic polynomial-time adversaries $\mathcal{A}$ as above, there exists a probabilistic polynomial-time adversary simulator $\mathcal{A}'$ such that for all relations $R(\alpha, \beta, \sigma)$ computable in probabilistic polynomial time, $|\pi(\mathcal{A}, R) - \pi'(\mathcal{A}', R)|$ is subpolynomial in n.*

HISTORY: Nonmalleability was first defined and explored by Dolev et al. [8], who obtained non-malleable public-key encryption under all three forms of attack, as well as nonmalleable commitment schemes and nonmalleable zero-knowledge interactive proof systems, all under general assumptions, and also treated nonmalleability of shared-key cryptosystems. Additional constructions under general assumptions appear in [5, 11, 12]. The first practical nm-cca-post cryptosystem is due to Cramer and Shoup [3], and is based on the Decisional Diffie–Hellman assumption (see [4] for schemes based on other assumptions). Canetti and Goldwasser constructed a threshold variation of the Cramer–Shoup public key system.

Barak obtained the first constant-round nonmalleable commitment schemes and zero-knowledge proofs [1]. Relaxed versions of nonmalleable commitment, some in a model in which the sender and receiver have access to a shared guaranteed-random string or other so-called *public parameters*, have also been explored (see, e.g., [6, 7, 9]).

Cynthia Dwork

References

[1] Barak, B. (2002). "Constant-round coin-tossing with a man in the middle or realizing the shared random string model." *Proc. FOCS 2002*, 345–355.

[2] Canetti, R. and S. Goldwasser (1999). "An efficient *threshold* public key cryptosystem secure against adaptive chosen ciphertext attack." *Advances in Cryptology—EUROCRYPT'99*, Lecture Notes in Computer Science, vol. 1592, ed. J. Stren. Springer-Verlag, Berlin, 90–106.

[3] Cramer, R. and V. Shoup (1998). "A practical public key cryptosystem secure against adaptive chosen ciphertext attacks." *Advances in Cryptology—CRYPTO'98*, Lecture Notes in Computer Science, vol. 1462, ed. H. Krawczyk. Springer-Verlag, Berlin.

[4] Cramer, R. and V. Shoup (2002). "Universal hash proofs and a paradigm for adaptive chosen ciphertext secure public-key encryption." *Advances in Cryptology—EUROCRYPT 2002*, Lecture Notes in Computer Science, vol. 2332, ed. L. Knudsen. Springer-Verlag, Berlin, 45–64.

[5] De Santis, A., G. Di Crescenzo, R. Ostrovsky, G. Persiano, and A. Sahai (2001). "Robust non-interactive zero-knowledge." *Advances in Cryptology—CRYPTO 2001*, Lecture Notes in Computer Science, vol. 2139, ed. J. Kilian. Springer-Verlag, Berlin, 566–598.

[6] Di Crescenzo, G., Y. Ishai, and R. Ostrovsky (1998). "Non-interactive and non-malleable commitment." *Proceedings STOC 1998*, 141–150.

[7] Di Crescenzo, G., J. Katz, R. Ostrovsky, and A. Smith (2001). "Efficient and non-interactive non-malleable commitment." *Advances in Cryptology—EUROCRYPT 2001*, Lecture Notes in Computer Science, vol. 2045, ed. B. Pfitzmann. Springer-Verlag, Berlin, 40–59.

[8] Dolev, D., C. Dwork, and M. Naor (2000). "Non-malleable cryptography." *SIAM J. Comp*, 30 (2), 391–437.

[9] Fischlin, M. and R. Fischlin (2000). "Efficient non-malleable commitment schemes." *Advances in Cryptology—CRYPTO 2000*, Lecture Notes in Computer Science, vol. 1880, ed. M. Bellare. Springer-Verlag, Berlin, 413–431.

[10] Goldwasser, S. and S. Micali (1984). "Probabilistic encryption." *J. Comput. Syst. Sci.*, 28, 270–299.

[11] Lindell, Y. (2003). "A simpler construction of CCA2-secure public-key encryption under general assumptions." *Advances in Cryptology—EUROCRYPT 2003*, Lecture Notes in Computer Science, vol. 2656, ed. E. Biham. Springer-Verlag, Berlin, 241–254.

[12] Sahai, A. (1999). "Non-malleable non-interactive zero knowledge and adaptive chosen-ciphertext security". *Proceedings FOCS 1999*, 543–553.

NONREPUDIATION

INTRODUCTION

"A word is not a crystal, transparent and unchanged; it is the skin of a living thought, and may vary greatly in color and content according to the circumstances and the time in which it is used."

Oliver Wendell Holmes
Towne v. Eisner (1918).

"Abuse of words has been the great instrument of sophistry and chicanery, of party, faction, and division of society."

John Adams (1819).

In the area of security technology and protocols, it is entirely possible that no other word has caused as much disagreement, confusion, and anxiety as the word *nonrepudiation*. First coined in the 1980s, the concept of nonrepudiation continues to perplex and cause arguments well into the 21st century. As one commentator observed in 2000, "over 100 messages were exchanged without anyone really being able to uncontestably [sic] define what [the] digitalSignature and nonRepudiation [bits] really signified . . . [and] no one can agree on what the term 'nonRepudiation' actually means, exemplified by a ∼200-message debate in mid-1999 which couldn't reach any useful conclusion" [1]. Even in more recent times, the meanings of nonrepudiation and the nonRepudiation bit continue to confound. During an exchange in early 2002, well over 100 messages regarding nonrepudiation were exchanged on the PKIX listserver. In late 2002, almost 200 more messages were exchanged after an X.509 meeting during which over a dozen PKI experts spent almost 2 days discussing the language for describing the nonRepudiation bit.

REPUDIATION, NONREPUDIATION, AND ELECTRONIC COMMERCE: The notion of repudiation has existed for hundreds of years.[1] Generally, to repudiate means "[t]o cast off; to disavow; to have nothing to do with; to renounce; to reject, . . . , to refuse, to acknowledge or to pay; to disclaim."[2] In legal terms, "repudiation of a contract means refusal to perform a duty or obligation owed to another party."[3] Traditionally, repudiation occurs when a party to a contract unequivocally asserts that he or she will not perform a future obligation as required by the contract. Known as "anticipatory repudiation,"[4] this will result in a contract being breached. In the context of information security and electronic commerce, one goal is to reduce as much as possible (or, as some might erroneously assert, eliminate) the ability of a party to repudiate an electronic transaction. Given the perception of the Internet as the next "Wild Wild West" and the understandable hesitation to conduct transactions (particularly ones of significant value) with faceless strangers hundreds or thousands of miles away, it is no wonder that the very first papers on public key technology declared that the use of public key cryptography would "protect against the threat of dispute" [2] and that a recipient of a message would be able to "convince a judge that . . . she has proof that [the signer] signed the document" [3].

Historically, the term nonrepudiation most likely evolved from the discussion of repudiation in the New Directions in Cryptography paper by Whit Diffie and Marty Hellman. The authors stated that "[u]nforgeable digital signatures are needed" to protect against a message being "later repudiated by either the transmitter or sender."

As public key technology progressed in the late 1970s and into the 1980s, nonrepudiation became one of four main features touted by proponents of the technology. Along with confidentiality (provided by encryption), authentication, and data integrity, nonrepudiation via digital signature schemes was hailed as the enabling technology for electronic commerce. Some commentators declared nonrepudiation to be "the elusive holy grail" of Internet security [4].

The main problem with the use of the term nonrepudiation was that, from a technical standpoint,

---

[1] The etymology of the word repudiate reveals that *repudiare*, the Latin root for repudiate, has origins dating back to the mid-1500s
[2] Webster's Revised Unabridged Dictionary (1913).
[3] Black's Law Dictionary, Sixth Ed., 1303 (1991).
[4] Under US law, anticipatory repudiation in the context of the sale of goods is covered in §2-610 (Anticipatory Repudiation) of the Uniform Commercial Code (UCC).

absolute prevention of repudiation via a single digital signature was thought to be an achievable goal. From a legal standpoint, however, repudiation can always occur, for any number of reasons both legitimate and dishonest (thus relinquishing non repudiation to, at best, a less than absolute status). As a result, or perhaps in spite of this, numerous definitions for the term developed. The meanings span a continuum ranging from the complete inability to deny sending or receiving a message to simply providing evidence of a message. As will be seen shortly, the most appropriate definition tends toward the latter end of the spectrum.

Thus, although digital signatures were often advertised as "providing nonrepudiation," a digital signature by itself on a message or transaction cannot *prevent* that message or transaction from being repudiated. Even with the nonRepudiation bit being set in a corresponding digital certificate, the most that a digital signature can provide is *support* or *evidence* for the recipient of the digital signature to be used *at a later time* to prove to a *third party* that the message or transaction took place. At a 1999 joint meeting between the Information Security Committee (ISC) of the Electronic Commerce Division of the Section of Science and Technology Law within the American Bar Association (see ABA digital signature guidelines) (consisting of a number of individuals from the legal, technical, and business communities) and the Internet Engineering Task Force (IETF), the conclusion was reached that the setting of the nonRepudiation bit is a "necessary but not sufficient" condition for providing a nonrepudiation service. As discussed elsewhere, however, even the setting of the nonRepudiation bit is not an absolutely necessary condition for an entity to successfully refute a repudiation by another party.

APPEARANCE IN STANDARDS: In the late 1980s, the concept of nonrepudiation began making its way into technical standards. One of the first appearances in the technical literature occurred in ISO 7498-2, the "Information Processing Systems—Open Systems Interconnection—Basic Reference Model—Part 2 : Security Architecture" (1989). This standard recognized nonrepudiation as a service that "may take one or both of two forms." The two forms comprised nonrepudiation with proof of origin, which was said to "protect against any attempt by the sender to falsely deny sending the data or its contents," and nonrepudiation with proof of delivery, which was supposed to "protect against any subsequent attempt by the recipient to falsely deny receiving the data or its contents."

In 1993, the ITU released the X.509 standard, which was titled "Information Technology—Open Systems Interconnection—The Directory: Authentication framework." This standard, which has become the *de facto* standard for PKI (see Public Key Infrastructure) certificates, further complicated the landscape by defining a nonRepudiation bit. The complications arose not from the fact that the bit was provided, but from how its use was to be interpreted. While some people recognized that this bit (which would have been asserted by the Certification Authority (CA)) could only be used to show that *at the time the certificate was issued* that the *certificate* was intended to provide support for the service of nonrepudiation, others felt that if the bit was set in a certificate used to digitally sign *a particular message or transaction*, that message or transaction could never be repudiated.

In 1996, the ABA Information Security Committee released the Digital Signature Guidelines ("DSG").[5] As defined in the DSG, nonrepudiation provides "strong and substantial evidence of the identity of the signer of a message and of message integrity, sufficient to prevent a party from successfully denying the origin, submission, or delivery of the message and the integrity of its contents." The DSG went on to note that:

> [These] Guidelines define nonrepudiation not as an automatic result of technical mechanisms, but as a property which can ultimately only be determined after recourse to available dispute resolution mechanisms such as a court or arbitrator . . . Nonrepudiation as defined in this Guideline 1.20 is intended to express a legal conclusion something less than a final determination by a court of last resort, but something more than a naked rebuttable presumption as is now provided by simple e-mail.

Thus, the DSG was possibly the first articulation of a rebuttable presumption regarding digital signatures.

In 1997, ISO published 10181-4, "Information Technology—Open Systems Interconnection—Security Frameworks for Open Systems: Nonrepudiation framework", which was the first detailed technical treatment of the nonrepudiation issue. The standard, titled The Non-repudiation Framework, provided the first full technical treatment of the topic of nonrepudiation. The standard focuses on the notion of nonrepudiation being an evidence generating process. The process described ISO 10181-4 involved a

---

[5] See http://www.abanet.org/scitech/ec/isc/digital_signature.html (last visited December 10, 2002).

number of different entities and several distinct technical processes. One particularly interesting point about ISO 10181-4 is that it stresses the time-based nature of a nonrepudiation service. In particular, it contemplates a process that produces evidence at one point in time that can be used by a party to a transaction for a message at a later point in time. The standard provides an excellent foundation upon which a definition of nonrepudiation can be based. The nonrepudiation framework in ISO 10181-4 is also consistent with the X.509 standard and is consistent with the American Bar Association digital signature guidelines. Another important thing to note is that the current (as of November 2002) proposed changes to the X.509 standard reflect the notion of evidence that is described in ISO 10181-4 and the digital signature guidelines.

Building on the Digital Signature Guidelines, the ABA ISC released the PKI Assessment Guidelines ("PAG")[6] in 2001. The PAG further elaborated on the ideas of nonrepudiation first proposed in the DSG. Specifically, the PAG stated that:

> the digital signature does not by itself result in legal "nonrepudiation." When a subscriber attempts to repudiate a transaction or communication, there may be factual and legal questions and disputes that, if not settled, will need to be resolved in litigation, arbitration, or other alternative dispute resolution mechanism, in order to determine whether the attempted repudiation is ultimately successful. The unique value of PKI is its technological ability to provide robust factual inferences of nonrepudiation, through cryptography, that will serve to provide credible evidence sufficiently strong to persuade a disinterested third party (the ultimate dispute resolution authority), that a subscriber originated a particular transaction or communication. Once the legal proceedings produced a final judgment to that effect, then legal nonrepudiation has occurred.

The PAG, in an attempt to discern between the absolute ability for a decision to be made based upon a technical mechanism (e.g., the assertion of the nonRepudiation bit) and the less than absolute nature of the overall nonrepudiation service, described the notion of "legal nonrepudiation" as part of the information security lexicon. Although both the DSG and the PAG recognized the distinction between an "automatic result of technical mechanisms" and "a property which can ultimately only

be determined after recourse to available dispute resolution mechanisms," the introduction of the term legal nonrepudiation seems to have further complicated the issue. Some people in the community view the use of the term "legal nonrepudiation" as further breaking down into separate stages the technical and legal aspects leading to an actual determination of nonrepudiation. In actuality, these two stages are inextricably linked.

PERSISTENT PROBLEMS: Despite the valiant attempts in the above partial list of standards and the numerous other writings that have occurred on the subject, a number of problems persist regarding nonrepudiation. As already mentioned above, nonrepudiation is not an absolute—a party can always repudiate a message. The best example is if the person is under duress (e.g., if another persons forces the signature to be applied against the intentions of the signer). Other valid reasons for repudiation include forgery, unconscionable conduct, fraud, and undue influence [5]. The definition of nonrepudiation in ISO 7498-2 (see preceding section) provides one example of how this quality of absoluteness may have seeped into the general understanding of nonrepudiation. Although the definition in ISO 7498-2 is consistent with the conclusions reached below, the subtlety of the phrase "*falsely* deny" to capture the notion that the nonrepudiation service is not absolute seems to have been lost on many readers. Read in the alternative, this means that the nonrepudiation service is *not* intended to protect against the true or valid denial of sending or receiving a message (i.e., for the reasons stated earlier in this paragraph).

A second related problem involves sufficiency. Many view a <u>digital signature</u> (particularly one whose corresponding digital certificate has the nonRepudiation bit asserted) as always being sufficient to provide nonrepudiation. Although this might be possible in certain very limited circumstances, it will not be true universally. Many other pieces of evidence may be utilized to further prove *or* further deny the validity of a particular message. For example, a person may utilize a digital certificate (without the nonRepudiation being set) to sign a message constituting an electronic contract. By itself, this might be easily repudiated, particularly if the applicable policy requires the nonRepudiation bit to be set for a transaction to be nonrepudiable. If, however, the signer/sender includes an email that says, "I know that policy says that I cannot be bound, but by this email and the signed paper copy that I am sending forthwith, I agree to be bound to this transaction," the

──────────

[6] See http://www.abanet.org/scitech/ec/(2000).isc/pag/pag.html (last visited December 12, 2002).

ability to repudiate would be significantly diminished. This notion of extrinsic evidence contributing to the overall service of nonrepudiation must be well understood in order for software to be properly written that will be taking certain actions based on whether the nonRepudiation bit is set or not.

A third issue involves the notion of risk. Despite all attempts to reduce a transactional process to purely technological means, such transactions will invariably involve some amount of business risk. No current technology can make a transaction absolutely trusted. The concept of non-repudiation as an evidence producing technological mechanism, however, provides an additional element in the overall risk assessment process.

TECHNOLOGY ISSUES: As if the topic weren't already complicated enough, additional consideration must be given to the dynamics of the technology necessary for providing a nonrepudiation service. Ideally, only a completely trusted software process should be able to produce a digital signature.[7] Otherwise, some doubt could be cast on whether the signer's intentions were accurately captured or whether the signer was even aware of the digital signature being applied using the signer's private key. Unfortunately, commercial deployments of such systems remain difficult due to the costs of producing a trusted system.

As discussed in the preceding section, some risk will always exist that a digital signature could be applied in a surreptitious manner since perfect security is currently not possible. Even without the human element, technical problems exist that could cast doubt on the validity of a particular digital signature (e.g., malware or flawed software). Each of these contributes to the overall evidentiary calculus that make up the nonrepudiation service.

WHAT'S AHEAD?: At an ISO meeting in September 2002 to work through the next version of X.509, the participants devoted almost 2 days to the discussion of the nonrepudiation bit. Recognizing the interpretational problems that had developed, the group exchanged ideas and shared perspectives in an attempt to better define the bit. Two initial considerations involved either deprecating the bit altogether or renaming the bit (both of which, while solving some problems, would likely raise other problems). Ultimately, the group decided to tackle the difficult job of refining the definition of nonrepudiation.

---

[7] *Id.*, under the section entitled "Trusted Computing Systems."

After much deliberation and impassioned advocacy, the group agreed on the following language in Draft Corrigendum 6 to describe the role of the nonrepudiation bit in an X.509 digital certificate:

> nonRepudiation [bit]: for verifying digital signatures which are intended to be used as evidence if a subsequent dispute arises, to prevent a signer from falsely denying involvement in a transaction. This bit does not, in itself, provide this assurance, but can be used together with other tools, such as an assertion of intent by the signer, an assertion from a third party notary to the transaction, a binding contract, policy statements etc, to assist in determining whether a signer's denial of involvement is a true or false claim

The important things to note here are: (a) the emphasis on the use of the bit as evidence in a determination of the validity of a transaction (consistent with the ISO 10181-4 nonrepudiation framework) and (b) the notion that the NR bit, *by itself*, does not provide assurance of nonrepudiation.

In November 2002 the Information Security Committee within the American Bar Association responded to the draft set of changes to the X.509 standard related to the nonrepudiation bits. In their response, the American Bar Association laid out six points to try to further explain the legal and technical impact of the nonrepudiation bit on a nonrepudiation service. The first point that was made stated that the setting of the nonrepudiation bits could provide at least some quantum of evidence in a dispute resolution involving a digitally signed message. The committee noted however that the quantum of evidence that would be provided as a result of only the nonrepudiation bit being set (i.e., without any extrinsic evidence) was not significant. The second point raised by the committee was that the facts that the nonrepudiation bit being set in a certificate is not conclusive or dispositive evidence that a signer intends to be bound by a specific digitally signed communication or transaction. Correspondingly, the third point was that the fact that the nonrepudiation bit was cleared is also not conclusive or dispositive evidence that a signer does not intend to be bound. The fourth point flows naturally, which states that the totality of the circumstances surrounding a digitally signed communication is *critical* to understanding and evaluating the effect of the non-repudiation bit on the evidence gathering and evidence generating process of a nonrepudiation service. A fifth point elaborates further by stating that a certificate policy or certification practice statement can help control the significance of a nonRepudiation bit within a particular PKI and more specifically on the evidence generating process of the nonrepudiation service within that PKI. The final point raised by the committee was that the notion of technical nonrepudiation does not necessarily translate into legal nonrepudiation. As discussed above, the notion of a distinction between technical nonrepudiation and legal nonrepudiation may actually cause more confusion than clarification since legal nonrepudiation is somewhat of a misnomer.

CONCLUSION: In conclusion, despite the confusion and problems that the concept of nonrepudiation has caused within the information security community, the concept of nonrepudiation remains an important and useful concept and should be maintained as a description of one of the security services provided by digital signatures. It must be made clear, however, that the use of a nonRepudiation bit within a certificate has *no* bearing on whether a particular signature in a single transaction can or cannot be repudiated. A host of other evidentiary issues would need to be considered by a fact finder in order to fully understand whether, *for that particular transaction*, a repudiated signature will hold up or not. Therefore, the overall concept of nonrepudiation does need further definition and explanation (including decisions from a competent court). The interaction and knowledge transfer that has been occurring between the legal, business, and technical communities in these various areas will allow the overall security community to develop a better understanding of the nonrepudiation service amongst all entities. The recent proposed changes to the X.509 standard evidence this trend toward evolving the concept of nonrepudiation.

Randy V. Sabett

References

[1] Gutmann, Peter (2000). "X.509 style guide." Found at http://www.cs.auckland.ac.nz/~pgut001/pubs/x409guide.txt (last visit November 23, 2002).

[2] Diffie, Whitfield and Martin E. Hellman (1976). "New directions in cryptography." *IEEE Trans. Info. Theory*, 22 (6), 644–654.

[3] Rivest, R.L., A. Shamir, and L. Adleman (1978). "A method for obtaining digital signatures and public-key cryptosystems." *Communications of the ACM*, 21 (2), 120–126.

[4] Merrill, Charles R. (1999). "Time is of the essence: Establishing the critical elements of electronic commerce." Found at http://www.pkilaw.com/surety_time_long_3.htm, (last visited December 1, 2002).

[5] McCullagh, Adrian and William Caelli (2000). "Non-repudiation in the digital environment." *First Monday*, 5 (8). Found at http://www.firstmonday.org/issues/issues5_8/mccullagh/index.html (last visited November 18, 2002).

NONSECRET ENCRYPTION

Modern cryptology was sparked by the groundbreaking work of Martin Hellman, a professor at Stanford University, and two graduate students who worked with him in 1974 and 1975, Ralph Merkle and Whitfield Diffie (see Figure 1). In April 1976, they started to publish their work in three papers [3, 4, 9], which revealed the concept of public key cryptography to the public. The work of Diffie and Hellman, which is now referred to as Diffie–Hellman key agreement, allows two people, or more properly their computers, to agree upon a secret key if they have only exchanged authenticated public keys of each other in the first place. Before Public Key Cryptography, it was unanimously agreed in the cryptography community that the only way for two parties to establish secure communications was to first exchange a secret key of some kind. This seemed to be simple common sense: if the recipients didn't have a secret to give them some leverage, how could they be in a better position to decrypt the message than an eavesdropper? Practically speaking, this meant that one of the parties first had to send a trusted person to the second party with a secret key (which typically took a fair amount of time), or send the key through an existing encryption channel that couldn't be completely trusted (if it was broken, all of the keys transmitted over that channel were also broken).

Fig. 1. Ralph C. Merkle, Martin Hellman, and Whitfield Diffie (from left to right)

In order to set up a group of n people to encrypt messages to each other, one had to exchange $N = 1/2n(n - 1)$ secret keys among them, one between each pair of group members. The discovery of public key cryptography reduced the burden of key generation and exchange from a number N, quadratic in n, to simply n because each group member only needed to publish its own public key in an authenticated fashion. Based on these public keys, any two group members could agree dynamically on a new secret key every time they wished to set up a communication session with each other. There is no way known for an eavesdropper to pick up the secret key by listening in. Without public key cryptography the problem of key management for a large group of participants in a communication network was not practical to handle.

After reading the Diffie–Hellman paper, three researchers at the Massachusetts Institute of Technology (MIT) named Ronald Rivest, Adi Shamir, and Leonard Adleman (RSA) began searching for a practical mathematical function to implement a complete PKC approach. After working on more than 40 candidates, they finally discovered an elegant algorithm based on the product of two prime numbers, which went on to become one of the dominant solutions used on the Internet. Their algorithm, named RSA after the initials of its inventors, was not only an elegant implementation of public key encryption (see RSA public key encryption) but also of digital signatures (see RSA digital signature scheme). Although the US intelligence tried to stop further publication, the news was spreading fast within and outside of the research community: Martin Gardner introduced the concept of public key cryptography in his widely read column Mathematical Games in the August 1977 edition of Scientific American. Then in February 1978, Rivest et al. published a more detailed paper on their work [10], and the popular press followed soon after. These works opened the door to a totally new area of cryptology and sparked such a huge amount of research in

the following decades that public key cryptography is held as the most important development in cryptology since the invention of polyalphabetic ciphers during the Renaissance.

As the British Labour Party strived for more openness of the British Government Communications Head Quarters (GCHQ) organization, the GCHQ revealed in December 1997 that researchers at GCHQ did some work in the early 1970s in the field of "nonsecret encryption," which is related to public-key cryptography, but without inclusion of the concept of digital signatures. Although the claims of GCHQ are not verifiable since the work was not published, and there are no evidentiary artifacts available such as original copies of the papers (David Kahn does not mention nonsecret encryption in his voluminous history of cryptography [7] either), there is still evidence for their accurateness. In 1979, US National Security Agency (NSA) chief Bobby Inman publicly stated that, all the noise about Diffie–Hellman and RSA aside, the intelligence establishment had known about public key cryptography for some time [8]. Whitfield Diffie pursued this track and in the early 1980s, he pried out of an NSA source the names of James Ellis and Clifford Cocks at GCHQ Communications-Electronics Security Group (see Figure 2). The work of the GCHQ researchers is described below according to James Ellis in his paper "The History of Non-Secret Encryption" [6]. Ellis began thinking about the shared secret key problem in the late 1960s when he discovered an old Bell Labs paper dated October 1944 titled "Final Report on Project C43" [1], describing a clever method of secure telephone conversation between two parties without any prearrangement. If John calls Mary, then Mary can add a random amount of noise to the phone line to drown out John's message in case any eavesdroppers are listening. However, at the same time Mary can also record the telephone call, then later play it back and subtract the noise she had added, thereby leaving

Fig. 2. James H. Ellis, Clifford C. Cocks, Malcolm J. Williamson (from left to right)

John's original message for only her to hear. While there were practical disadvantages to this method, it suggested that the logical possibility existed: there might be methods of establishing secure communications without first exchanging a shared secret key.

Ellis thought about this seemingly paradoxical idea, and while lying in bed one night developed an existence proof that the concept was possible with mathematical encryption, which he recorded in a secret CESG report titled The Possibility of Non-Secret Encryption in January 1970 [5]. This showed logically that there could be an encryption method that could work without prior pre-arrangement, and the quest in GCHQ then turned to finding a practical method to embody the theoretical concept.

The first workable mathematical formula for nonsecret encryption was discovered by Clifford Cocks, which he recorded in 1973 in a secret CESG report titled A Note on Non-Secret Encryption [2]. This work describes a special case of RSA public key encryption where the enciphering exponent e is set equal to the public modulus $n = pq$, without any mention of the application to digital signatures. A few months later in 1974, Malcolm Williamson (see Figure 2) discovered a mathematical expression based on the commutativity of exponentiation that he recorded in a secret report titled Non-Secret Encryption Using A Finite Field [11], and which describes a key exchange method similar to that discovered by Diffie, Hellman, and Merkle.

As Steven Levy writes in his article "The Open Secret" [8]: "So concerned was the GCHQ with this possibility that it not only looked at the schemes internally—finding no inherent flaws—but also took the unusual step in 1974 of going to a renowned outsider, Professor R.F. Churchhouse of the University of Wales, presenting him with the mathematical foundation of Cocks's idea, and asking whether it was secure. Churchhouse concluded that as long as no one figured out a fast way of factoring huge primes—something that no mathematician had ever come close to—the scheme was sound."

Even Williamson believed that the whole venture was too risky. When he finally wrote up a revised version of his key scheme, he cited these reservations as the reason for the two-year delay. "I find myself in an embarrassing position," he wrote. "I have come to doubt the whole theory of nonsecret encryption. The trouble is that I have no proof that the method ... is genuinely secure." He conceded he could not find anything wrong with the system, though, "and would be grateful if any-

one else can." No one did. But by then the GCHQ had tacitly concluded it wasn't worth the effort to implement a public key cryptosystem.

Looking back at the discovery of nonsecret encryption at the GCHQ after more than 25 years of research in public key cryptography, it appears as if going public was the only way to establish public key cryptography as a technology mature enough to be used in commercial systems, banking, telecommunications, health care, etc. As we know today, almost all public key based cryptographic systems rely on unproven complexity theoretic assumptions, e.g., the assumption that integer factorization is hard, or that taking discrete logarithms is hard, or the Strong RSA Assumption, and many more. Some of them have been investigated by mathematicians for hundreds of years, yet remain neither proven nor falsified. There is a chance that someone might come up some day with a smart algorithm proving that some of these assumptions are false. At least the assumptions of integer factorization and discrete logarithms are generally believed not to be invalid any time soon, and this belief is based on more than 25 years of public research, and tremendous efforts of finding more efficient algorithms, which is partly stimulated by challenges and significant rewards for solving them. Nevertheless, all algorithms found so far are of at least subexponential complexity, impractical to be performed as long as the keys are chosen of appropriate length. The case of GCHQ shows that for a new method such as nonsecret encryption, which was so far out of the ordinary at its time, and which after 25 years is still based on a number of unproven complexity theoretic assumptions, only a moderate level of confidence can be achieved within the walls of an intelligence agency. The concept of public key cryptography needed a large open research community to explore all possible weaknesses and risks it might have. If public key cryptography had not been discovered outside of the intelligence community, it would most likely not be used today, not even in the military and intelligence communities because the spies and spooks had hardly gained hardly enough confidence in the new technique and would be too paranoid of potential trap doors.

Gerrit Bleumer

References

[1] Bell Telephone Laboratory (1944). Final Report of Project C43. 23.

[2] Cocks, Clifford C. (1973). "A note on non-secret encryption." CESG Report. Available from the CESG Portal at http://www.cesg.gov.uk/indexNS.cfm

[3] Diffie, Whitfield and Martin E. Hellman (1976). "New directions in cryptography." *IEEE Transactions on Information Theory*, 22 (6), 644–654.

[4] Diffie, Whitfield and Martin E. Hellman (1976). "Multiuser cryptographic techniques." *AFIPS Conference Proceedings*, vol. 45, *1976 National Computer Conference, June 7–10, New York City, New York*, 109–112.

[5] Ellis, James H. (1970). "The possibility of secure non-secret digital encryption." CESG Report. Available from the CESG Portal at http://www.cesg.gov.uk/indexNS.cfm

[6] Ellis, James H. (1987). "The history of non-secret encryption." CESG Report. Available from the CESG portal at http://www.cesg.gov.uk/indexNS.cfm

[7] Kahn, David (1967). *The Codebreakers: The Story of Secret Writing*. Macmillan Publishing Co., New York. Also Kahn, David (1978). *IEEE Transactions on Information Theory*, IT-24, 525.

[8] Levy, Steven (1999). "The open secret." *Wired*, 108–115. http://www.wired.com/wired/archive/7.04/crypto_pr.html

[9] Merkle, Ralph C. and Martin E. Hellman (1977). "Hiding information and receipts in trap door knapsacks." Presented at the *Internal Symposium on Information Theory, Cornell University, Ithaca, NY, October 1977*.

[10] Rivest, Ronald L., Adi Shamir, and Leonard Adleman (1978). "A Method for obtaining digital signatures and public-key cryptosystems." *Communications of the ACM*, 21 (2), 120–126. Reprinted: 26 (1) (1983) 96–99.

[11] Williamson, Malcolm J. (1974). "Non-secret encryption using a finite field." CESG Report.

NTRU

THE NTRU CONVOLUTION RING AND THE NTRU LATTICE: The NTRUENCRYPT cryptosystem, first proposed in 1996, is a public-key cryptosystem based on polynomials in the *polynomial convolution ring* $R = \mathbb{Z}[X]/(X^N - 1)$. Addition of polynomials in R is simple polynomial addition. Multiplication is polynomial multiplication followed by reduction mod $X^N - 1$ (also known as taking the *convolution product* of the coefficient vectors). The ring is determined by the integer N. Here, $*$ denotes multiplication in R. The coefficients of polynomials in this ring will frequently be reduced modulo some quantity v, where v is an integer or polynomial of small degree. This reduction will always be denoted explicitly.

To show how operations in R are related to lattice operations, we consider the ring homomorphism between polynomials in the ring R and the $N \times N$ matrices known as *circulant matrices*

with integer coefficients. Under this homomorphism, a polynomial $f(X) = \sum_i f_i X^i$ maps to the matrix

$$\begin{pmatrix} f_0 & f_{N-1} & f_{N-2} & \cdots & f_1 \\ f_1 & f_0 & f_{N-1} & \cdots & f_2 \\ \vdots & \vdots & \vdots & \ddots & \vdots \\ f_{N-1} & f_{N-2} & f_{N-3} & \cdots & f_0 \end{pmatrix}.$$

It is easy to verify that this mapping respects addition and multiplication, i.e., that it is a ring homomorphism as stated.

In NTRU operations, the coefficients of polynomials in R are frequently reduced mod q, $q \in \mathbb{Z}$. The *convolution modular lattice* defined by a given polynomial h and integer q is the $2N$-dimensional lattice defined by the row vectors of a $2N \times 2N$ matrix of the form

$$\begin{pmatrix} I & H \\ 0 & qI \end{pmatrix},$$

where the submatrices are all $N \times N$, and H is the circulant matrix defined by h. If h has been chosen in such a way that it is known that there is a particularly short vector in this lattice, the lattice is known as an NTRU *lattice*. The hard problems presented to an attacker by the NTRU systems—recovering the private key from a public key, recovering a plaintext from a ciphertext, and signing a message representative—can all be considered instances of solving a Closest Vector Problem (CVP) or a Shortest Vector Problem (SVP) in an NTRU lattice.

DESCRIPTION OF THE NTRUENCRYPT SYSTEM: NTRUENCRYPT [6] is a public-key cryptosystem whose difficulty is based on CVP or SVP in the NTRU lattice. In addition to the ring dimension N and the "big modulus" $q \in \mathbb{Z}\mathbb{Z}$, NTRUENCRYPT uses a "little modulus" p, which has typically been taken to be the integer 2 or 3 or the polynomial $2 + X$ (operations mod $2 + X$ are discussed in [7]). The quantities p, q, and $X^N - 1$ must be relatively prime (generate the unit ideal) in the ring $\mathbb{Z}[X]$. Additional parameters include the distributions $\mathcal{D}_f, \mathcal{D}_g, \mathcal{D}_m, \mathcal{D}_r$, which are described below.

- NTRUENCRYPT **key generation** selects "small" polynomials f and g. We will make much use of the concept of small polynomials, which are defined for our purposes as polynomials of degree $N - 1$ whose coefficients are small relative to those of random mod q polynomials. The polynomials f and g are drawn from distributions $\mathcal{D}_f$ and $\mathcal{D}_g$, respectively. These must be defined as part of the parameter set. For example, a parameter set might define $\mathcal{D}_f$ and $\mathcal{D}_g$ to be the sets of binary polynomials with d_f and d_g ones, and $N - d_f$ and $N - d_g$ zeros, respectively. For a given definition of $\mathcal{D}_f, \mathcal{D}_g$, key generation proceeds as follows.

1. Randomly generate "small" polynomials f and g in $\mathcal{D}_f, \mathcal{D}_g$ respectively.
2. Invert $f \bmod q$ in R to obtain f_q. Also invert $f \bmod p$ in R to obtain f_p. These calculations are straightforward using an analog of the extended Euclidean algorithm. The probability that a random polynomial has an inverse is discussed in [17].
3. Calculate the public key h as $h \equiv p \cdot g \cdot f_q \bmod q$. In other words, h satisfies $f \cdot h \equiv p \cdot g \pmod{q}$. The private key is the pair (f, f_p).

- The NtruEncrypt **encryption primitive** uses polynomials m and r which are drawn from distributions $\mathcal{D}_m$ and $\mathcal{D}_r$, respectively. The definitions of $\mathcal{D}_m$ and $\mathcal{D}_r$ are part of the parameter set. Polynomials in $\mathcal{D}_m, \mathcal{D}_r$ are small relative to random mod q polynomials: for example, a parameter set might define $\mathcal{D}_r$ to be the set of binary polynomials with d_r ones, and $\mathcal{D}_m$ to be the set of all binary polynomials. For a given definition of $\mathcal{D}_r, \mathcal{D}_m$, encryption proceeds as follows.
  1. Encode the message M as a polynomial in $\mathcal{D}_m$.
  2. Randomly select a polynomial r in $\mathcal{D}_r$.
  3. Calculate the ciphertext e as $e \equiv r \cdot h + m \pmod{q}$.

- The NtruEncrypt **decryption primitive** consists of the following operations:
  1. Calculate $a \equiv f \cdot e \pmod{q}$, performing the mod q reduction into the interval $[A, A + q - 1]$, where the choice of A is discussed later.
  2. Recover m by calculating $m \equiv f_p \cdot a \pmod{p}$.

To see why decryption works, substitute $h \equiv p \cdot g \cdot f_q$ and $e \equiv r \cdot h + m$ into the above equations. This gives

$$a \equiv p \cdot r \cdot g + f \cdot m \bmod q.$$

If this were an equality over $\mathbb{Z}$, rather than simply $\pmod{q}$, it is clear that step 2 would recover m. Recommended NtruEncrypt parameter sets $N, p, q, \mathcal{D}_f, \mathcal{D}_g, \mathcal{D}_m, \mathcal{D}_r$ give a high probability that the *width* of $p \cdot r \cdot g + f \cdot m$ (its largest coefficient minus its smallest) will be less than q. Therefore, a range $[A, A + q - 1]$ can be found such that the equation above is an equality over $\mathbb{Z}$, and by reducing into this range the decrypter can recover m. A will depend on the expected value of the average coefficient of $p \cdot r \cdot g + f \cdot m$; in other words, it will depend on the values of $f(X = 1), r(X = 1), g(X = 1)$, and m $(X = 1)$. The first three of these values are fixed for a particular parameter set, and the final value can be obtained from $e(X = 1)$;

therefore, the decrypter can choose an appropriate value for A and recover m.

OPERATING CHARACTERISTICS OF NTRUEN-CRYPT: The running time of NtruEncrypt increases with the security parameter N at a rate between $\Theta(N \ln(N))$ and $\Theta(N^2)$ (see O-notation). For comparison purposes, ignoring fast multiplication techniques and precomputation where relevant, RSA encryption (with small exponent) is $\Theta(N^2)$ and RSA decryption is $\Theta(N^3)$, while encryption and decryption for both elliptic curve cryptography and cryptosystems based on the discrete logarithm problem are $\Theta(N^3)$. While the actual value of the security parameter N obviously differs in practice among these various forms of public-key cryptography, NtruEncrypt's "subquadratic" running time is nonetheless notable.

NtruEncrypt parameter choices can be optimized for speed in various ways [7, 8], including: (1) taking f to be of the form $1 + p \cdot F$ with F small and thus $f_p = 1$; (2) taking f to have the form $f_1 \cdot f_2 + f_3$ with f_1, f_2, f_3 small (and similarly for r); and (3) taking q to be a power of 2. Different parameter choices lead to slightly different instances of the underlying hard problem.

SECURITY OF THE NTRUENCRYPT SYSTEM: Ntru keys and ciphertexts can be attacked by lattice reduction applied to the $2N$-dimensional Ntru lattice defined by h (or, in the formulation given above, defined by $p^{-1}h \bmod q$) [1, 6, 18]. The breaking time appears experimentally to depend on the structure of f and g, or r and m, in that the greater the norm of these small polynomials (relative to q), the longer the reduction takes. "Zero-forcing attacks" [13] use the rotational symmetry of the problem to let an attacker reduce the problem to a somewhat lower dimensional lattice. If N is not prime, say N is divisible by n, then an attacker can apply reduction methods in a lattice of dimension $2n$ [3]. Thus N should always be chosen to be prime.

Since f, g, r, and m are drawn from a relatively small space of polynomials, they may be vulnerable to an exhaustive search. Known meet-in-the-middle (i.e., collision) type methods take roughly the square root of the time required for an exhaustive search [16]. The key and message spaces must therefore be chosen to give an adequate level of combinatorial security.

Raw NtruEncrypt ciphertexts are malleable in many ways and thus leak some information about m. For example, if a ciphertext e decrypts to m, then $X \cdot e$ will decrypt to $X \cdot m$; $e + X^i$ will, with reasonable probability, decrypt to $m + X^i$; the sum of the coefficients of the ciphertext, i.e., the value $e(X = 1)$, reveals the sum of the coefficients of the message $m(1) \mod q$. Adaptive chosen ciphertext attacks based on these observations are described in [9, 12]. Therefore, NTRUENCRYPT should only be used with a padding scheme such that there is a known proof of security relative to certain assumptions on the difficulty of a certain hard problem [15].

In addition, NTRUENCRYPT is vulnerable to attacks based on decryption failures [9, 10, 14]. In these attacks, an attacker attempts to create validly encrypted messages $e = r + h \cdot m$ such that the width of $p \cdot r \cdot g + f \cdot m$ is greater than q, so that the decryption primitive described above will not be able to decrypt them. Decryption failures leak information and can allow the recovery of the private key f.

In general, if a security level of k bits is desired, then the chance that the width of $p \cdot r \cdot g + f \cdot m$ is greater than q for random r, g, f, m should be at most $1/2^k$. The padding scheme should ensure that r and m are drawn uniformly at random from some set, and that they cannot be directly selected by the sender [11]. Since small values of q increase lattice security and reduce bandwidth, q should be chosen as small as possible, consistent with ensuring an acceptably low decryption failure probability.

Recommended parameter sets for 80-bit security are as follows [16, 18, 19]. In these sets, the polynomials F, g, r are binary with d_F, d_g, d_r ones respectively.

$$N = 251, \quad p = 2, \quad q = 239, \quad f = 1 + p \cdot F,$$
$$d_F = d_g = d_r = 72.$$

Public keys and ciphertexts for such parameters can be specified in 2004 bits, while private keys in 251 bits. A full description of how to implement NTRUENCRYPT using the recommended encryption scheme is contained in the EESS#1 standard [2].

NTRUSIGN: NTRUSIGN, a digital signature scheme based on solving CVP in the NTRU lattice, was first presented in 2001 and published in 2003 [5]. NTRUSIGN is not computationally zero-knowledge and information from a transcript of signatures eventually leaks the private key [4]. An extension to the algorithm, based on "perturbing" the message before signing it, is claimed to reduce the information leakage to such an extent that a transcript-based attack needs an impractically large number of signatures to succeed [5]. NTRUSIGN is fully described, with parameters, in the EESS#1 standard [2].

DERIVATION OF NTRU: The origin of the acronym "NTRU" is obscure, with suggested candidates including "Number Theory Research Unit" and the more whimsical "Number Theorists 'R' Us".

William Whyte

References

[1] Coppersmith, D. and A. Shamir (1997). "Lattice attacks on NTRU." *Advances in Cryptology—EUROCRYPT'97*, Lecture Notes in Computer Science, vol. 1233, ed. W. Fumy. Springer-Verlag, Berlin, 52–61.

[2] Efficient Embedded Security Standard #1, Version 2. (2003). http://www.ceesstandards.org

[3] Gentry, C. (2001). "Key recovery and message attacks on NTRU-Composite." *Advances in Cryptology—EUROCRYPT 2001*, Lecture Notes in Computer Science, vol. 2045, ed. B. Pfitzmann. Springer-Verlag, Berlin.

[4] Gentry, C. and M. Szydlo (2002). "Cryptanalysis of the revised NTRU signature scheme." *Advances in Cryptology—EUROCRYPT 2002*, Lecture Notes in Computer Science, vol. 2332, ed. L. Knudsen. Springer-Verlag, Berlin, 299–320.

[5] Hoffstein, J., N. Howgrave-Graham, J. Pipher, J. Silverman, and W. Whyte (2003). "NTRUSign: Digital signatures in the NTRU lattice." *CT-RSA 2003*. Springer-Verlag, Berlin, 122–140.

[6] Hoffstein, J., J. Pipher, and J. Silverman (1998). "NTRU: A ring based public key cryptosystem." *Algorithmic Number Theory (ANTS III), Portland, OR, June 1998*, Lecture Notes in Computer Science, vol. 1423, ed. J.P. Buhler. Springer-Verlag, Berlin, 267–288.

[7] Hoffstein, J. and J. Silverman (2001). *Optimizations for NTRU, Public Key Cryptography and Computational Number Theory (Warsaw, September 11–15, 2000)*. Walter de Gruyter, Berlin, 77–88.

[8] Hoffstein, J. and J. Silverman (2003). "Random small Hamming weight products with applications to cryptography." *Com2MaC Workshop on Cryptography Pohang, Korea, June 2000. Discrete Applied Mathematics*, 130, 37–49.

[9] Hong, J., J.W. Han, D. Kwon, and D. Han (2003). "Key recovery attacks on NTRU without ciphertext validation routine." *ACISP 2003*, Lecture Notes in Computer Science, vol. 2727, eds. R. Safari-Naini and J. Seberry. Springer–Verlag, Berlin.

[10] Howgrave-Graham, N., P. Nguyen, D. Pointcheval, J. Proos, J.H. Silverman, A. Singer, and W. Whyte (2003). "The impact of decryption failures on the security of NTRU encryption." *Advances in Cryptology—CRYPTO 2003*. Lecture Notes in Computer Science, vol. 2729, ed. D. Boneh. Springer-Verlag, Berlin.

[11] Howgrave-Graham, N., J. Silverman, A. Singer, and W. Whyte (2003). NAEP: Provable security

in the presence of decryption failures. http://www.ntru.com/cryptolab/articles.htm

[12] Jaulmes, E. and A. Joux (2000). "A chosen-ciphertext attack against NTRU." *Advances in Cryptology—CRYPTO 2000*, Lecture Notes in Computer Science, vol. 1880, ed. M. Bellare. Springer-Verlag, Berlin.

[13] May, A. and J.H. Silverman (2001). "Dimension reduction methods for convolution modular lattices." *Conference on Lattices and Cryptography (CaLC 2001)*, Lecture Notes in Computer Science, vol. 2146, ed. J.H. Silverman. Springer-Verlag, Berlin, 111–127.

[14] Meskanen, T. and A. Renvall (2003). "A wrap error attack against NTRUEncrypt." University of Turku Technical Report TUCS 507, presented at *WCC 2003*. http://www.tucs.fi/Research/Series/techreports/

[15] Nguyen, P. and D. Pointcheval (2002). "Analysis and improvements of NTRU encryption paddings." *Advances in Cryptology—CRYPTO 2002*, Lecture Notes in Computer Science, vol. 2442, ed. M. Yung. Springer-Verlag, Berlin.

[16] NTRU Cryptosystems Technical Report #004. A meet-in-the-middle attack on an NTRU private key. www.ntru.com/cryptolab/tech_notes.htm

[17] NTRU Cryptosystems Technical Report #009. Invertibility in truncated polynomial rings. www.ntru.com/cryptolab/tech_notes.htm

[18] NTRU Cryptosystems Technical Report #012. Estimated breaking times for NTRU lattices. www.ntru.com/cryptolab/tech_notes.htm

[19] NTRU Cryptosystems Technical Report #018. Estimating decryption failure probabilities for NTRUEncrypt. www.ntru.com/cryptolab/tech_notes.htm

NUMBER FIELD

A *number field* $Q(\alpha)$, also called an *algebraic number field*, is an extension field of the rational numbers Q, generated from a root α of a polynomial

$$f(x) = f_d x^d + f_{d-1} x^{d-1} + \cdots + f_1 x + f_0,$$

where the coefficients $f_0, \ldots, f_d$ are rational numbers and d is the *degree*. A number field $Q(\alpha)$ is thus to the rational numbers Q as a finite field F_{q^d} is to the underlying subfield F_q. In particular, every element of a number field $Q(\alpha)$ has the form

$$a_{d-1}\alpha^{d-1} + a_{d-2}\alpha^{d-2} + \cdots + a_1\alpha + a_0,$$

where $a_0, \ldots, a_{d-1}$ are rational numbers, and every element of $Q(\alpha)$ can be generated from field operations on elements of Q and α. Since Q is infinite, a number field is likewise infinite.

As an example, the number field $Q(i)$ where $i = \sqrt{-1}$, also called an *imaginary quadratic field*, consists of the values of the form

$$a + bi$$

where a, b are rational numbers. Here, i is a root of the polynomial

$$x^2 + 1 = 0.$$

An element of a number field $Q(\alpha)$ has a *norm*, which is a rational number with certain multiplicative properties (e.g., the norm of a product of two elements is the product of their norms); the norm is a way of expressing the "size" of an element.

Number fields are employed in cryptography primarily in solutions to integer factoring and the discrete logarithm problem. See the entry Number Field Sieve in particular for a detailed discussion of how specially constructed number fields are employed in integer factoring, where one searches for elements of the number field whose norm is a smooth number (see smoothness).

Burt Kaliski

NUMBER FIELD SIEVE

The Number Field Sieve, or *NFS*, is one of a family of algorithms which attempt to factor an integer N by finding integers x and y such that $x^2 \equiv y^2$ mod N (see modular arithmetic), in which case gcd $(x + y, N)$ is then a (possibly nontrivial) factor of N. It is the asymptotically fastest known algorithm for integer factoring when the factors have no exploitable special characteristics (compare, for example, the Elliptic Curve Method which is particularly effective at discovering relatively small factors). There are two forms of the NFS distinguished by whether N itself has an especially simple form (the Special NFS or *SNFS*) or is a general integer (the General NFS or *GNFS*). With optimal choice of parameters, the NFS factors a composite integer N in heuristic time $L_N[1/3, c]$ (see L-notation) where $c = (\frac{32}{9})^{1/3} \approx 1.526$ for the SNFS and $c = (\frac{64}{9})^{1/3} \approx 1.923$ for the GNFS. Coppersmith's variant of the GNFS reduces c to $(1/3)(92 + 26\sqrt{13})^{1/3} \approx 1.902$ but his version is not competitive until N is much larger than values which can be factored at present.

A version of the NFS may also be used to solve discrete logarithm problems; we do not describe that application here.

HISTORY OF THE NFS: The idea of representing N as the difference of two squares $x^2 - y^2$ with factorization $(x - y)(x + y)$ goes back to Fermat in

1643. Some 50 years later Legendre observed that a pair (x, y) such that

$$x^2 \equiv y^2 \bmod N, \quad 0 < x \neq y < N, \quad x + y \neq N \quad (1)$$

suffices to factor N. Kraïtchik in the 1920s proposed looking for values of a such that $a \equiv x^2 \bmod N$ (i.e., a is a quadratic residue modulo N) and to piece together solutions of this congruence to obtain solutions of (1).

In 1970, Morrison and Brillhart introduced two key ideas: generating small quadratic residues (in their CFRAC algorithm by evaluating the continued fraction expansion of $\sqrt{N}$) and requiring that they be *smooth* which is defined as factoring entirely into a limited number of primes contained within a factor base. The value a and its prime factorization are called a relation. CFRAC starts by generating a large number of relations and then uses linear algebra to piece together a solution to (1). CFRAC was the first algorithm with heuristic subexponential running time. It should be noted that in this approach the problem of factoring a large integer N has been replaced in part by the problem of factoring a large number of much smaller integers. A major limitation of CFRAC is that it uses trial division to identify and factor small quadratic residues and division is a slow operation on most computers.

In 1982, Pomerance found a way (the Quadratic Sieve) of using sieves to speed the finding of smooth quadratic residues. The speed of the Quadratic Sieve is limited by the size of the quadratic residues, which are approximately $\sqrt{N}$. If the size of these numbers could be reduced the sieve would run faster. The NFS exploits this observation.

In 1988, Pollard suggested a factoring method that was very well suited to numbers that are close to a high power of an integer. His algorithm was a form of the SNFS. Before long this method had been extended to the GNFS so that it could be used to factor general composites. Reference [5] describes the initial development of the NFS. By 1990 the SNFS had been developed to the point where Lenstra et al. used it to factor $2^{512} + 1$, a number with 155 digits [6]. By December 2002 the SNFS record stood at 233 digits [11]. The GNFS first set a factoring record in 1996 when it was applied to a 130-digit RSA Factoring Challenge integer [3] and again in 1999 when it factored a 155-digit challenge [1]. A 174-digit challenge was factored in December 2003 [4].

STRUCTURE OF THE ALGORITHM: All versions of the NFS consist of four distinct phases: polynomial selection, sieving, linear algebra, and extraction of square roots in an algebraic number field. At least two polynomials are chosen—more than two may be used but in the following we restrict our description to two for simplicity. With the aid of a sieve and two factor bases, one per polynomial, we find a large number of pairs of integers which generate smooth values (see smoothness) of both polynomials simultaneously. These pairs are also called relations. A polynomial value is the norm of an element in the number field generated by a root of the polynomial (incidentally, the origin of the name Number Field Sieve). As in other algorithms which use factor bases, linear algebra techniques are used to select and combine relations which form pairs of squares in the number fields. The square roots in the number fields are taken and, via homomorphisms, are converted into rational integers x and y such that $x^2 \equiv y^2 \bmod N$.

POLYNOMIAL SELECTION: We need to find two polynomials over Z,

$$f_1(x) = c_{1,d_1} x^{d_1} + c_{1,d_1-1} x^{d_1-1} + \cdots + c_{1,0},$$
$$f_2(x) = c_{2,d_2} x^{d_2} + c_{2,d_2-1} x^{d_2-1} + \cdots + c_{2,0},$$

with $f_1 \neq \pm f_2$, both irreducible over Z, and having content = $\gcd(c_{id_i}, \ldots, c_{i,0})$ equal to 1 (see greatest common divisor), together with an integer m that is a common root modulo N of f_1 and f_2. The value of $d_1 + d_2$ is normally in the range 4–8 when factoring N with 100–250 digits or so. A fairly simple approach to finding f_i and m that has been used successfully, and is the basis of many later and more sophisticated methods, is first to choose an integer m near to $N^{1/(d_1+1)}$ and then determine f_1 from the radix—m expansion of N. Many trial values of m are selected and the corresponding polynomials ranked by absolute size of coefficients and, perhaps, other criteria. Very frequently f_2 is chosen to be the linear polynomial $x - m$ once f_1 and its root m have been chosen. The coefficients $c_{i,j}$ are generally chosen to be as small as possible, as this speeds up the overall algorithm. The only difference between the SNFS and the GNFS is that the special form of N usually dictates the polynomials used in the SNFS and that the coefficients are of size $o(1)$—i.e., bounded by a (possibly large) constant—whereas in the GNFS the coefficients grow as $O(N^{1/(d+1)})$. For example, the factorization of $N = 2^{512} + 1$ used the polynomials $x^5 + 8$ and $x - m$ which have a common root $m = 2^{103} \bmod N$. The GNFS factors N with no special form and searching for polynomials with small coefficients and a known root is generally a time-consuming task. Other properties of the polynomials, e.g., the behavior of their roots modulo small primes, are also important and the development of methods for finding good polynomials is currently an active area of research [9].

Each polynomial f_i defines an algebraic number field. Let α_i be a complex root of f_i, that is $f(\alpha_i) = 0$. We already know that $f_i(m) \equiv 0 \bmod N$ so there is a natural homomorphism $\phi_i : \alpha_i \mapsto m$. This homomorphism is exploited in the square root phase of the NFS. For integers a and b, the norms $N(a - b\alpha_i)$ equal $f_i(a/b)b^{d_i}$.

SIEVING: With each polynomial we associate a *factor base* consisting of all underline{prime numbers} p up to a *factor base bound* for which the polynomial has at least one root modulo p. The factor base bounds may be the same for each polynomial, but need not be. We also set a *large prime bound* for each polynomal; again this value may be the same or different for each polynomial and it is typically about 100 times the factor base bound. The siever looks for coprime integer pairs (a, b) for which each norm $N(a - b\alpha_i)$ factors into primes contained in the respective factor base while allowing a small number of *large primes* which are between the factor base bound and the large prime bound. Allowing two large primes for each polynomial is very common.

The simplest form of sieving, *line sieving*, proceeds as in the following sketch. First a value is fixed for b. Then a block of memory, representing each value of a in a fixed range, is initialized to zero. The roots of f_i modulo a prime p in the ith factor base fall into one or more sets; the members of each set are equispaced within the range of a under consideration (they are p apart), so locating any one permits all the others in the set to be found very rapidly. One member of each set of roots is located for the first polynomial and then all the sieve locations corresponding to the roots of $f_1 \bmod p$ are incremented by an approximation to $\log p$. This process is then repeated for all the other primes in the factor base associated with f_1. Upon completion of this process, those memory locations which hold values close to $\log N(a - b\alpha_1)$ indicate norms which are very likely to factor into primes from the first factor base together with at most a few large primes. These "promising" locations are noted. Next, the memory is reinitialized to zero and the same sieving procedure carried out with f_2 and the second factor base. Those locations which appear to be promising in both sieves simultaneously indicate the values of (a, b) which may give the norms we are looking for. These norms are

investigated further, by trial division perhaps, and if both norms are smooth a relation (a, b) is output, together with the prime factorization of the norms. In this respect, the NFS differs from the QS in that two values are required to be smooth (see underline{smoothness}) simultaneously. This entire procedure is then repeated, perhaps with a different range of a values with the same b, or with a different value for b. Note that the sieving procedure is very easily distributed over many machines working concurrently: each machine sieves over distinct values of b and ranges of a values.

An alternative sieving technique, *lattice sieving* [10], sieves over a subset of (a, b) pairs for which one of the norms is chosen to be divisible by a particular large prime, the so-called special-q. That norm, when divided by the special-q, will be smaller and so more likely to be smooth. Only a small subset of the possible norms will be divisible by any one special-q but the sieve may be repeated many times, each with a different special-q. The relative performance of the line and lattice sieving algorithms depends somewhat on implementation details, with the lattice siever tending to be the more efficient.

LINEAR ALGEBRA: The output from the siever is a very large collection of relations each of which contains the prime factorization of a pair of norms of polynomials, together with the corresponding values of a and b. The linear algebra stage attempts to find sets of relations such that in their product each prime in each factorization appears an even number of times. Consider a matrix in which each column represents a relation and each row is a prime from the factor bases; elements of the matrix are set to 1 if that prime occurs an odd number of times in that relation and 0 otherwise. Finding the required sets of relations corresponds to finding linear dependencies mod 2 in the matrix.

It would be possible to use standard techniques from linear algebra such as Gaussian elimination or Lanczos [9] or Wiedemann algorithms directly on this matrix, but it would be very large and very expensive to process. For example, when factoring a 150-digit integer with the GNFS, the siever may produce a matrix with around 80 million rows and columns, albeit an extremely sparse one. Accordingly, a filtering and merging process is almost universally used to reduce the matrix size before proceeding with the matrix stage proper. First all singletons (relations in which a prime appears only once in the entire collection) are removed because they cannot possibly form part of a useful linear combination. Subsequent merging phases seek for sets of two or more relations which share a common large prime and create new sets in which the large prime occurs twice in each set. The new sets may then be treated as relations which no longer contain the large prime (because the linear algebra is performed over $\mathbf{F_2}$) and so the matrix size has been reduced, though at a probable increase in its density.

SQUARE ROOT: The output of the linear algebra stage is a set of linear dependencies, modulo 2, within the exponent vectors of the prime factorization of a number of norms of two polynomials. That is, we have a set S of relations (a, b) such that the products $\prod_{(a,b)\in S}(a - b\alpha_i)$ are squares of algebraic integers, say γ_1^2 and γ_2^2, in the number fields $\mathbf{Q}(\alpha_i)$, where α_i is a root of f_i for $i = 1, 2$. By applying the homomorphisms $\phi_1 : \alpha_1 \mapsto m$ and $\phi_2 : \alpha_2 \mapsto m$ we find a congruence of two squares modulo N, namely

$$(\phi_1(\gamma_1))^2 \equiv (\phi_1(\gamma_1^2))$$
$$\equiv \prod_{(a,b)\in S} (a - bm)$$
$$\equiv (\phi_2(\gamma_2^2)) \equiv (\phi_2(\gamma_2))^2.$$

For a nonlinear polynomial, this means that we need to extract a square root of the product of a large number of terms in the respective number field. Couveignes [2] first developed an algorithm to perform this root extraction, but Montgomery's [7] subsequent and faster algorithm has now replaced it in general use. Their methods are rather complex and details are beyond the scope of this encyclopedia entry.

Paul Leyland

References

[1] Cavallar, S., B. Dodson, A.K. Lenstra, W. Lioen, P.L. Montgomery, B. Murphy, H. te Riele, K. Aardal, J. Gilchrist, G. Guillerm, P. Leyland, J. Marchand, F. Morain, A. Muffett, C. Putnam, and P. Zimmermann (2000). "Factorization of a 512-bit RSA modulus." *Advances in Cryptology—EUROCRYPT 2000*, Lecture Notes in Computer Science, vol. 1807, ed. B. Preneel. Springer-Verlag, Berlin, 1–18.

[2] Couveignes, J.-M. (1993). *Computing a Square Root for the Number Field Sieve*, Lecture Notes in Mathematics, vol. 1554. Springer-Verlag, Berlin.

[3] Cowie, J., B. Dodson, R.-M., Elkenbracht-Huizing, A.K. Lenstra, P.L. Montgomery, and J. Zayer (1994). "A world wide number field sieve factoring record: On to 512 bits." *Advances in Cryptography—ASIACRYPT'96*, Lecture Notes in Computer Science, vol. 1163, eds. K. Kin and T. Matsumoto. Springer-Verlag, Berlin, 382–394.

[4] Franke, J. et al. (2003). RSA-576. http://www.loria.fr/zimmerma/records/rsa576, http://www.rsasecurity.com/rsalabs/challenges/factoring/rsa576.html

[5] Lenstra, A.K. and H.W. Lenstra (eds.) (1993). *The Development of the Number Field Sieve*, Lecture Notes in Mathematics, vol. 1554. Springer-Verlag, Berlin.

[6] Lenstra, A.K., H.W. Lenstra, M.S. Manasse, and J.M. Pollard (1993). "The factorization of the ninth Fermat number." *Mathematics of Computation*, 61, 319–349.

[7] Montgomery, P.L. (1993). "Square roots of products of algebraic numbers." *Proceedings of Symposia in Applied Mathematics, Mathematics of Computation, 1943–1993*.

[8] Montgomery, P.L. (1995). "A block Lanczos algorithm for finding dependencies over GF(2)." *Advances in Cryptology—EUROCRYPT'95*, Lecture Notes in Computer Science, vol. 921, eds. L.C. Guillou and J.-J. Quisquater. Springer-Verlag, Berlin, 106–120.

[9] Montgomery, P.L. and B. Murphy (1999). "Improved polynomial selection for the number field sieve." Extended Abstract for the *Conference on the Mathematics of Public-Key Cryptography, The Fields Institute, Ontario, Canada*.

[10] Pollard, J.M. (1993). *The Lattice Sieve*, in: The development of the number field sieve, Lecture Notes in Math, vol. 1554, eds. A.K. Lenstra and H.W. Lenstra, Jr. Springer-Verlag, Berlin, 43–49.

[11] te Riele, H. (2000). "233-digit SNFS factorization." http://listserv.nodak.edu/scripts/wa.exe?A2=ind0011&L=nmbrthry&F=&S=&P=1207

NUMBER THEORY

Number theory is the study of natural numbers. It includes topics such as <u>prime numbers</u>, <u>integer factoring</u>, diophantine equations (for example, finding integer solutions to equations of the form $x^2 + y^2 = z^2$), continued fractions (example: $\sqrt{2} = 1 + \frac{1}{2+\frac{1}{2+\cdots}}$), properties of algebraic numbers (for example, numbers like $3 + 4 \cdot i$ where $i^2 = -1$), and a broad range of other mathematical problems.

Despite having evolved over at least 3000 years, number theory has had relatively few applications prior to the age of the digital computer. It was studied mainly for the reason of mathematical curiosity, without a lot of thought about how it could be applied to real world problems. In fact, a leading number theorist of the first half of the 20th century, G.H. Hardy, was quite proud that he spent his whole life working on mathematics that had no practical value.

Nowadays, things are very different. Our everyday lives involve many activities that rely on cryptography based upon number theory. A few examples include e-commerce (specifically, transactions that require sending credit card information over the Internet), mobile phones, cable and satellite television, online banking, and stock trading through hand-held devices. Had the application of number theory never been discovered, it is quite likely that at least some of these examples would not be possible today, while others would be a lot more complex and more vulnerable to fraud.

What got all of this started can be traced back to a research publication in the mid 1970s. The paper, entitled "New Directions in Cryptography" [19] by Stanford University researchers Whit Diffie and Martin Hellman, is perhaps the most important paper in the subject's history. One of the problems that they solved was the key exchange problem, which asks how two people (say, Alice and Bob) in separate locations can agree upon a secret encryption key, while any eavesdroppers would not be able to figure it out. This was an essential piece of the puzzle that was missing in order for cryptography to be used as it is today. For instance, when one sends credit card information over the Internet to some Web site one has never visited before, it can only be encrypted after the client browser and the server agree upon a secret encryption key.

Diffie and Hellman's solution was based upon modular arithmetic—one of the most basic topics in number theory. The idea was that the two people first agree upon a very large prime, p, and a multiplicative generator, g, modulo that prime, which were not kept secret—any eavesdropper would know these values but it would not help him in figuring out the secret key that they will compute. Alice and Bob would then choose secret integer values a and b, respectively, which they would not tell to anybody else—not even each other. Alice would compute g^a mod p while Bob computes g^b mod p, and then they exchange these values. When Alice receives g^b mod p, she raises it to the power a modulo p, and likewise Bob raises Alice's value to the power b modulo p. The result is that they both computed $g^{a \cdot b}$ mod p, which can be used as the secret key. Yet, any eavesdropper would not be able to compute the same values without knowing a or b, or so we think!

Note that the eavesdropper has all values except a and b. When he sees g^a mod p, can he figure out a? This is known as the discrete logarithm problem, and is the basis for the security of the key

exchange method, which is known as the Diffie–Hellman key agreement, or DH for short. At the time Diffie and Hellman presented their protocol, there were no methods for solving discrete logarithms efficiently. Since then, many new number theoretic algorithms have been developed to attack DH, but none of them is fast enough to break it. The protocol is still considered very secure today, and is widely used.

A second important idea that Diffie and Hellman introduced in their research paper was the concept of public-key cryptography. In traditional cryptography, there was always one secret key that was used to both encrypt and decrypt. Diffie and Hellman demonstrated that it should be possible to design methods of encryption where two keys are used: a *public key* that is used for encryption, and a *private key* that is used for decryption. The public key would be made public, so that anybody who wants to send you an encrypted message can do so by using it. The private key would be kept only to yourself, so that you alone can decrypt the scrambled messages. For this to work, it must be the case that nobody can figure out somebody's private key from his or her public key, except for the person who created the key pair in the first place!

Diffie and Hellman then showed that public key cryptography can be used for more than just encryption. In particular, they introduced the concept of a digital signature scheme. Suppose Alice wanted to sign her name on a digital document. She would do so by "encrypting" it with her private key. Bob, or anybody else could verify that she signed it by "decrypting" it with her public key. This proves that Alice signed the document: only Alice was able to compute the function with her private key since only she knows her private key. Yet anybody can verify that she signed it because anybody and everybody can look up her public key. Digital signatures are reversing the roles of the public and private keys in order to demonstrate to the world that somebody wanted his or her name at the bottom of some electronic document. This was a monumental concept, from which the science of cryptography began taking on more of a role than just encryption!

Diffie and Hellman had not yet discovered a public key encryption method by the time of their publication. But soon after, three researchers from M.I.T. came up with one whose security was based upon the difficulty of factoring integers. The method is known as the RSA cryptosystem (see entries RSA public-key encryption and RSA digital signature scheme), which comes from the names of the inventors: Ron Rivest, Adi Shamir, and Len Adleman. RSA is based upon the idea that anybody can find really large prime numbers easily and multiply them together, but there is no fast way known to recover those primes back when given only the product. To the beginner, this is not obvious: factoring and primality testing may seem to have about the same complexity. However, we shall see that there are tricks from number theory that allow us to determine primality much faster than factoring.

The RSA algorithm works similar to DH: they both involve modular exponentiation (see underline{exponentiation algorithms}). However, in RSA the modulus for the exponentiation is a composite number that is a product of two large primes, and each person has his or her own separate modulus. The public and private key pairs are derived from the modulus using knowledge of the secret prime divisors. Only the person who created the modulus can figure out public and private key pairs that work with it, since only that person knows the prime divisors.

Along with the publication of the RSA cryptosystem, the authors presented a challenge. They encrypted a secret message using a 129-digit modulus that was the product of a 64-digit and a 65-digit prime, and offered $100 to anybody who could decrypt the message. At that time, they estimated that the amount of time to factor the number, known as RSA-129, would take more than 40 quadrillion years using the best factoring algorithm known. They even stated that "with such a huge modulus the message will never be recovered." They also started the company RSA Data Security, which would sell the new, patented encryption technology and offer cash prizes for factoring numbers of increasing sizes, starting at 100-digits (see underline{RSA Factoring Challenge}). The number RSA-129 was eventually factored [3] by a new factoring method, but the RSA cryptosystem still remains secure for large enough moduli, and is ubiquitous in electronic communications today.

It is hard to emphasize enough how much of an impact the DH and RSA cryptosystems influenced research in number theory, especially the computational aspect. Prior to these cryptosystems, the study of factorization and primality testing was done by a small group of individuals who shared their results among each other, but did not always publish their work [12]. Their field of interest soon became a gold mine, and unfortunately for them, some of their research was reinvented. On the other hand, the reinventions were occasionally accompanied by new, advanced mathematical analyses that revealed useful and surprising results. For those interested in some of the fascinating computational number theory research prior to public key cryptography, one nice source is [13].

In the remainder of this article we talk more specifically about problems in number theory that are of interest to cryptography, with an emphasis on why they are important and how the two fields have shaped each other. The reader should be aware that this article is only scratching the surface: to write a comprehensive essay on number theory in cryptography is enough to fill an encyclopedia.

MODULAR ARITHMETIC: it is difficult to do much number theory or cryptography at all without underline{modular arithmetic}. Two numbers a and b are said to be congruent modulo n if their difference $a - b$ is divisible by n. This is written $a \equiv b \bmod n$. The congruence relation, $\equiv$, behaves in many ways like the very familiar equality relation, $=$.

Both the DH and RSA cryptosystems involve heavy use of modular arithmetic, and there are many other cryptographic primitives that do as well. When the modulus n is prime, arithmetic with the set of integers modulo n is a special case of a more general set of algebraic objects known as underline{finite fields}. Finite fields are used everywhere in cryptography, and are one of the most important concepts an apprentice in the subject must master.

The literature in cryptography abounds with research on fast modular and finite field arithmetic. Like with automobiles, some people can never have enough speed. For example, consider an e-commerce server which must handle numerous purchases per second. It is critical that the cryptography does not heavily delay the transactions, or else people will just start shopping elsewhere. A good beginning survey on fast modular arithmetic is given in [21]. Included in the survey is an explanation of Montgomery's very clever method of computing the mod operation without actually doing division (see underline{Montgomery arithmetic}). See also the entry underline{exponentiation algorithms}.

There is one modular arithmetic tool that deserves special mention: the underline{Chinese Remainder Theorem} or CRT. CRT is a method for quickly transforming a solution of a congruence modulo a composite integer to a solution modulo its prime divisors, and vice versa. It is so named because the 4th century AD Chinese mathematician Sun Tsu Suan-Ching seems to be the first to consider problems of this nature. CRT is often included in RSA implementations to speed up decryption and digital signature operations.

PRIME NUMBERS: There are two important facts that we glossed over when we talked about DH and RSA:

1. For security reasons, we need to be sure that there are enough large <u>prime numbers</u>. It should not be possible that somebody is able to make a list of all primes up to a certain size and then use it to factor somebody's RSA modulus, and hence break RSA.
2. It is essential that we are able to quickly determine whether very large numbers are prime. It is not obvious how to do this.

One of the most basic and most elegant theorems in number theory is that there are infinitely many prime numbers, which was proved in Euclid's *Elements* (approximately 300 BC). But that in itself does not tell us how many there are of some size. Originally conjectured by Legendre but first proved by de la Vallée Poussin and independently by Hadamard, the *Prime Number Theorem* answers the latter question. The theorem states that asymptotically, the number of primes up to some integer x is on the order of $x/\log x$. Using this theorem, we can estimate the number of 100-digit primes to be on the order of 4×10^{97}. This is far larger than the conjectured number of atoms in the universe (on the order of 10^{78}), so it is certainly not possible to make a list of all such primes to break RSA.

Addressing the second point, suppose n is a number that we want to test for primality. How do we do it? The first method that we learn is to trial divide by all numbers up to $\sqrt{n}$, and if none other than 1 divides n, then n is prime. This takes $O(n^{0.5})$ steps. To the untrained eye, this may seem fast. But in fact, it is not. This is actually <u>exponential time</u> in the *size* of the number which is $\log n$. An equivalent way of writing the running time is $O(e^{0.5 \cdot \log n})$ (remark: in number theory literature, it is customary to use "log" to mean "natural log"), which makes its inefficiency more clear. In contrast, a polynomial time algorithm is of the form $O(\log^c n)$ for some constant integer c. So, for cryptographic applications, we must have better methods.

The most important concept in modern primality testing applied to cryptography goes back to a 17th century French lawyer named Pierre de Fermat, who spent much of his free time playing with mathematics. Known as <u>Fermat's Little Theorem</u>, he claimed every odd prime number, p, has the property that $a^{p-1} \bmod p \equiv 1$ for all integers a not divisible by p. On the other hand, it is quite rare to find composite integers that have this property. Those that do are called <u>pseudoprimes</u>, or more specifically, base-a pseudoprimes.

Fermat's Little Theorem gives an efficient method to determine whether an odd number n is prime or not, with some chance of error (see <u>Fermat primality test</u>): simply choose an integer a that is not divisible by n, compute $a^{n-1} \bmod n$, and call it prime if the result is 1 or composite otherwise. The test never lies when it declares a number to be composite, but sometimes, though very rarely, lies when it says "prime."

Since there is some chance of error, it is tempting to reduce or eliminate it by repeating the test with other bases. Unfortunately, this strategy has limits due to a disruptive class of numbers that are pseudoprimes to nearly every base. They are called *Carmichael numbers*. The smallest is 561.

Carmichael numbers present a bit of an obstacle to Fermat's primality testing method. For cryptographic applications, we need to be sure to an exceedingly high degree of confidence that the number we think is prime really is. Otherwise, the cryptography may not be as secure as we think. For instance, the RSA algorithm will still work if we accidently choose Carmichael numbers instead of primes for our modulus, but the modulus becomes much easier to factor, hence making it insecure. This brings up several problems, such as:

- Can we detect with very high probability whether our number really is prime and not a Carmichael number?
- How much time does it take to be confident enough that our number is really prime?
- How many Carmichael numbers are there? Is there a simple formula that estimates how many Carmichael numbers there are between 1 and x?

The first two questions have been answered based upon a modified version of Fermat's test which is stronger (more accurate) and very slightly faster. The modified version is due to a few different people who independently arrived at more or less the same test (Artjuhov, Selfridge, and Miller/Rabin; see [16] for details). Today it is often called the <u>Miller–Rabin probabilistic primality test</u> or sometimes the *strong pseudoprime test*. In Rabin's publication, he showed that for any odd composite number larger than 9, at most one fourth of the bases will falsely fool the test into believing that the number is prime. This implies that if a large odd number passes the test for 50 random bases, then the probability is no more than 2^{-100} that it is composite, which is sufficiently small for cryptographic applications. But, note that this is a worst case analysis: if the number was chosen at random, then with high probability it is not a pseudoprime to any base in the first place, so fewer tests should suffice. The average case analysis was done in [18].

We conclude with a few brief remarks about research in primality testing. In [2], it is shown that there are infinitely many Carmichael numbers, which was a longstanding open conjecture. They also show that an asymptotic formula for the number of Carmichael numbers up to x is at least $x^{2/7}$, but a tight bound on the asymptotics is still not known. Bach has shown that the strong pseudoprime test can be turned into a primality proving algorithm[1] under the assumption of the widely believed extended Riemann hypothesis [4]. However, this is mainly of theoretical interest, since such a proof would take much longer to complete than the tests we do today, which are good enough for cryptographic purposes. A new primality test which generalizes *many previous primality tests* was developed by Grantham [20]. Known as the *quadratic Frobenius test*, it achieves a better worst case analysis per unit of time than the Miller–Rabin test. See [17] for the average case analysis and [27] for an improved variant of the test for numbers congruent to 1 mod 4. Recently, [1] gave the first deterministic polynomial-time primality proving algorithm under no assumptions. So far it is impractical, but future research may some day change this.

See the entry primality test for a more complete discussion of these issues.

DISCRETE LOGARITHMS AND INTEGER FACTORIZATION:

Before talking about discrete logarithms and factoring, it is best to first ask whether the security of DH and RSA really rely on the hardness of solving these problems. It is certainly true that if one can compute discrete logarithms or if one can factor, then one can break DH or RSA. But what about the other way around? It is unknown whether that is the case. This is a very important research area, since quite often an attacker will find another way around breaking a system rather than the "hard problem" that the designer thinks it depends upon. In fact, RSA really relies on the difficulty of computing roots modulo a composite number, which is not known to be as difficult as factoring, except when the exponent is even. Even exponents are not used for RSA, but they are used in a variant of RSA known as the Rabin cryptosystem [30]. Consequently, the Rabin variant has been proved to be as difficult to break as

factoring the modulus, whereas RSA itself has not. Further, the research of [10] argues that breaking RSA may not be as difficult as factoring. These examples motivate the important topic of *provable security* (see "reductions" in computational complexity), which means proving that breaking some system is equivalent to solving some problem that we hope and think is very difficult to do.

Despite the lack of provability, the best methods known for breaking RSA or DH when implemented properly[2] are factoring and solving discrete logarithms respectively (see the entries Diffie–Hellman problem and RSA problem for further discussion). So let's now discuss algorithms for solving the mathematical problems, starting with factoring. Factoring algorithms are categorized either as *general purpose* or *special purpose*. General purpose algorithms have running times that depend upon only the length of the number that is being factored, while special purpose algorithms depend upon properties of the number. Most often, "properties of the number" has some relation to the size of the prime divisors, but not always. Special purpose algorithms are very useful for many reasons, but it is the general purpose ones that seem to have the most potential for attacking public key cryptosystems such as RSA.

Suppose n is an integer that we want to factor. Rather surprisingly, many of the best factoring algorithms have running times described by the formula (see L-notation)

$$L_n[t, \gamma] = e^{(\gamma + o(1)) \cdot (\log n)^t \cdot (\log \log n)^{1-t}} \qquad (1)$$

for some constant numbers $\gamma > 0$ and $t \in [0, 1]$. Let's take a closer look at this formula.

First, if the reader does not feel comfortable with the $o(1)$, he or she can ignore it for the purpose of this discussion. Plugging in the values $t = 1$ and $\gamma = 0.5$ gives $e^{(0.5 + o(1)) \cdot \log n}$. By ignoring the $o(1)$ and simplifying, this comes to be $n^{0.5}$ which is exponential time. The reader should recognize this as the worst case run time of trial division (the first algorithm everybody learns for factoring). Setting $t = 0$ and keeping $\gamma = 0.5$ gives $e^{(0.5 + o(1)) \cdot \log \log n}$ which (again ignoring the $o(1)$) simplifies to $(\log n)^{0.5}$. This is polynomial time. In fact, for any constant value of γ, the running time is polynomial when t is 0. The significance of these observations is that the t value is more important in terms of the asymptotic running time: smaller values result in better algorithms, at least in theory.

---

[1] The difference between a *primality test* and a *primality proving algorithm* is that a proof is never wrong, while a test has a small chance of error.

[2] Proper implementation is a subject in its own right. It is not trivial to do everything right, and the slightest mistake could entirely defeat the security.

Pomerance's <u>quadratic sieve</u>[3] from the early 1980s is a general purpose factoring algorithm that achieves $t = 0.5$ and $\gamma = 1$. Silverman demonstrated the practical value of the algorithm by distributing it on many workstations [36] in order to factor hundreds of numbers for the book [13]. A. Lenstra, and the numerous people he worked with, took it further. Not only did they develop important practical improvements, but they also distributed their code on the Internet so volunteers could help them factor large numbers [24]. This is a trend that many people would follow for many different research areas. Their work on the quadratic sieve peaked when they factored the RSA-129 challenge number [3], and decoded the secret message.

There is a special purpose factoring algorithm that has running time that is asymptotically at least as good as the quadratic sieve[4]. It is H. Lenstra's <u>Elliptic Curve Method for factoring (ECM)</u> [25]. In spite of what theory suggests, the quadratic sieve tends to perform much better in practice when attempting to factor the difficult numbers. On the other hand, ECM is very fast at finding small prime factors and uses much less memory than the quadratic sieve, which are major benefits.

The overall best general purpose factorization algorithm is the <u>Number Field Sieve</u> (NFS), which has $t = \frac{1}{3}$ and $\gamma = (64/9)^{1/3}$. The original idea of NFS came from Pollard in the early 1990s, but the ideas were further developed and refined by many people (for details, see <u>integer factoring</u>, <u>Number Field Sieve</u>, and [22]). The number field sieve has been used to factor numbers of 512-bits (155-digits) and more. Such factorizations are very significant, since originally 512-bits was the industrywide defacto standard. Thanks to mathematical analysis and real factorization results, it was clear well in advance that 512-bits was not enough. The number field sieve has many fine details and uses advanced mathematics in the area of algebraic number theory. One should not attempt to learn it until he or she first has a solid understanding of the the quadratic sieve.

Many of the best general purpose factoring algorithms have been translated into similar algorithms for solving discrete logarithms (for details, see <u>discrete logarithm problem</u>). The discrete logarithm methods of Coppersmith et al. [15] for a prime p attain the same asymptotic running time as the quadratic sieve applied to a composite n. There also exists a number field sieve algorithm for discrete logarithms [33, 34]. The general strategy that these algorithms use when applied to discrete logarithms is labeled <u>index calculus</u>.

As the running time formulas of these algorithms might suggest, their analyses are not exactly trivial. But, underlying the $L_n[t, \gamma]$ formula is one common concept. In order to factor a particular integer or to solve a particular discrete logarithm, just about all of the most powerful algorithms use or depend upon factorizations of smaller, randomly distributed integers (or in some cases small polynomials). The integers that are sought are those that factor into small primes, which are known as *smooth numbers*. (See the entry <u>smoothness</u> for details.) Let $\psi(x, y)$ be the number of integers up to x that factor into primes up to y. These values are called y-smooth. If y is not too small, then $\psi(x, y)$ is asymptotically $x \cdot u^{-u(1+o(1))}$ where $u = \log x / \log y$ [16]. Analysis of this function was first done by Dickman in the 1930s, but the research of [14] filled in the gaps to make it applicable to the analysis of the algorithms mentioned in this section. Smoothness has additional applications in areas of cryptography other than factoring and discrete logarithms.

ELLIPTIC CURVES: We mentioned that the Diffie Hellman protocol involves arithmetic modulo a prime p, which is a special case of finite field arithmetic. Finite fields have two operations: multiplication and addition. DH makes use of only the multiplication operation, so it is not using all of the properties of what makes it a finite field. The properties that it does use are classified under a simpler algebraic object, known as an algebraic *group*. Algebraic groups are usually taught in an undergraduate level modern algebra or abstract algebra class.

It turns out that any algebraic group could potentially be used to create a variant of the Diffie Hellman protocol, as well as some other public key schemes (such as <u>ElGamal public key encryption</u> or the Digital Signature Algorithm (see <u>Digital Signature Standard</u>)). The high-level structure of the protocol remains the same, but the modular arithmetic is replaced with arithmetic from the new group. One simple example is if the normal Diffie Hellman protocol uses addition instead of

---

[3] The quadratic sieve evolved from Schroeppel's linear sieve, and Kraitchik had independently invented a similar algorithm many years earlier. See <u>integer factoring</u> and the <u>quadratic sieve</u> for details on each person's contribution.

[4] If the number is a product of two primes that are equal in size, then it is asymptotically the same as the quadratic sieve in run time. Otherwise, it is better.

multiplication as the primary operation. The protocol still works, and in fact is much faster. There is only one problem: it is easy to break!

There are many examples of groups, but one that has become particularly valuable to cryptography today is the group of points on an underlined elliptic curve. The mathematics of elliptic curves is too involved to explain here, so we just highlight the benefits. Unlike RSA or regular DH, there has been almost no progress in attacking the underlying mathematical problem that is the basis of elliptic curve cryptography, except in a few special cases that are not a threat to elliptic curves in general. Because of this, people are able to use much smaller key sizes for elliptic curves than they would for other public key cryptosystems. This is extremely important in constrained environments such as smart cards, where memory can be very limited—especially nonvolatile memory. There are also some speed advantages to using elliptic curves in certain applications, though RSA is faster in others.

Elliptic curves are a hot topic of modern mathematics research and the application to cryptography has proliferated new research. Sample topics include analysis of the security, fast algorithms for counting points on an elliptic curve, and fast elliptic curve arithmetic (one nice example that uses "complex multiplication" is given in [37]). Moreover, the rich algebraic structure of elliptic curves is providing new tools which are turning out to be of high cryptographic value [5, 8] (see identity based cryptosystems).

GENERAL CRYPTANALYSIS: Cryptography has numerous pitfalls [35]. It is very easy to make subtle errors that can lead to very large security breaches. In order to reduce this risk, cryptographic standards such as PKCS [32] are carefully written to specify exactly how everything should be implemented. Regardless, people quite frequently do things their own way without having much of an understanding of how cryptosystems get broken. Common errors range from the simple, such as not checking for special cases (for example, a surprisingly large number of amateur public key cryptosystems can be broken if the attacker sends in an integer like 0 or 1) to the rather complex, such as being susceptible to attacks like [6] or [11]. As a rule, if one does not have a very good understanding of how to break cryptosystems, then one should not be making them.

In general, knowledge of cryptanalytic techniques will prove to be invaluable to not only researchers, but to system architects and security engineers as well. Such techniques tend to use numerous tools from many areas of number theory. The more an information security specialist learns about number theory, the more valuable his or her knowledge will be. Further reading about design and cryptanalysis of number theoretic ciphers can be found in [38].

MISCELLANEOUS: This article has focused mainly on public key cryptography. Number theory is also an important tool in many other areas of cryptography, including pseudorandom number generators, zero-knowledge based protocols, and threshold cryptography (see [9] for a nice example of shared RSA key generation that has direct application to threshold cryptography).

Scott Contini

References

[1] Agarwal, M., N. Saxena, and N. Kayal. "PRIMES is in P." Available from http://www.cse.iitk.ac.in/news/primality.html

[2] Alford, W., A. Granville, and C. Pomerance (1994). "There are infinitely many Carmichael numbers." *Ann. of Math.*, 139, 703–722.

[3] Atkins, D., M. Graff, A., K. Lenstra, and P.C. Leyland (1995). "The magic words are squeamish ossifrage." *Advances in Cryptography—ASIACRYPT'94*, Lecture Notes in Computer Science, vol. 917, eds. J. Pieprzyk and R. Saavi-Naini. Springer-Verlag, Berlin, 263–277.

[4] Bach, E. (1985). "Analytic methods in the analysis and design of number-theoretic algorithms." *ACM Distinguished Dissertation*. MIT Press, Cambridge, 1985.

[5] Barreto, P. (2002). "The pairing-based crypto lounge." http://planeta.terra.com.br/informatica/paulobarreto/pblounge.html

[6] Bleichenbacher, D. (1998). "Chosen ciphertext attacks against protocols based on the RSA encryption standard PKCS #1." *Advances in Cryptology—CRYPTO'98*, Lecture Notes in Computer Science, vol. 1462, ed. H. Krawczyk. Springer-Verlag, Berlin, 1–12.

[7] Boneh, D. (1999). "Twenty years of attacks on the RSA cryptosystem." *AMS*, 46 (2), 203–213.

[8] Boneh, D. and M. Franklin (1998). "Identity based encryption from the Weil pairing." *Advances in Cryptology—CRYPTO 2001*, Lecture Notes in Computer Science, vol. 2139, ed. J. Kilian. Springer-Verlag, Berlin, 213–229.

[9] Boneh, D. and M. Franklin (2001). "Efficient generation of shared RSA keys." *J. ACM*, 48 (4), 702–722.

[10] Boneh, D. and R. Venkatesan (1998). "Breaking RSA may not be equivalent to factoring." *Advances in Cryptography—ASIACRYPT'98*, Lecture Notes in Computer Science, vol. 1514, eds. K. Ohta and D. Pie. Springer-Verlag, Berlin, 25–34.

[11] Brier, E., C. Clavier, J.-S. Coron, and D. Naccache (2001). "Cryptanalysis of RSA signatures with fixed-pattern padding." *Advances in Cryptology—CRYPTO 2001*, Lecture Notes in Computer Science, vol. 2139, ed. J. Kilian. Springer-Verlag, Berlin, 433–439.

[12] Brillhart, J. (1966). Private communication.

[13] Brillhart, J., D.H. Lehmer, J.L. Selfridge, B. Tuckerman, and S.S. Wagstaff, Jr. (1998). "Factorizations of $b^n \pm 1$, $b = 2, 3, 5, 6, 7, 10, 11, 12$ up to high powers." *Contemporary Mathematics* (2nd ed.), vol. 22. AMS, Providence.

[14] Canfield, E., P. Erdös, and C. Pomerance (1983). "On a problem of Oppenheim concerning 'factorisatio numerorum'." *J. Number Theory*, 17, 1–28.

[15] Coppersmith, D., A.M. Odlyzko, and R. Schroeppel (1986). "Discrete logarithms in GF(p)." *Algorithmica*, 1, 1–15.

[16] Crandall, R. and C. Pomerance (2001). *Prime Numbers, a Computational Perspective*. Springer, Berlin, 237.

[17] Damgård, I. and G. Frandsen, "An extended quadratic Frobenius primality test with average case error estimates." Unpublished document.

[18] Damgård, I., P. Landrock, and C. Pomerance (1993). "Average case error estimates for the strong probable prime test." *Math. Comput.*, 61, 177–194.

[19] Diffie, W. and M. Hellman (1976). "New directions in cryptography." *IEEE Trans. Inf. Theory*, 22, 644–654.

[20] Grantham, J. (1998). "A probable prime test with high confidence." *J. Number Theory*, 72, 32–47.

[21] Koç, Ç.K. (1994). "High-speed RSA implementation." RSA Laboratories Technical Report #TR 201.

[22] Lenstra, A.K. and H. Lenstra Jr. (eds.) (1993). *The Development of the Number Field Sieve*, Lecture Notes in Mathematics, vol. 162. Springer-Verlag, Berlin.

[23] Lenstra, A.K., H. Lenstra Jr., M.S. Manasse, and J.M. Pollard (1993). "The factorization of the ninth Fermat number." *Math. Comput*, 61, 319–349.

[24] Lenstra, A.K. and M.S. Manasse (1990). "Factoring by electronic mail." *Advances in Cryptology—EUROCRYPT'89*, Lecture Notes in Computer Science, vol. 434, eds. J.-J. Quisquater and J. Vandewalle. Springer-Verlag, Berlin, 355–371.

[25] Lenstra Jr., H. (1987). "Factoring integers with elliptic curves." *Ann. of Math.* 2, 649–673.

[26] Montgomery, P. (1985). "Modular multiplication without trial division." *Math., Comp.*, 44, 519–521.

[27] Müller, S. (2001). "A probable prime test with very high confidence for $n \equiv 1$ mod 4." *Advances in Cryptography—ASIACRYPT 2001*, Lecture Notes in Computer Science, vol. 2248, ed. C. Boyd. Springer-Verlag, Berlin, 87–106.

[28] Odlyzko, A.M. (1985). "Discrete logarithms in finite fields and their cryptographic significance." *Advances in Cryptology—EUROCRYPT'84*, Lecture Notes in Computer Science, vol. 209, eds. T. Beth, N. Cot and I. Ingemarsson. Springer-Verlag, Berlin, 224–313.

[29] Pomerance, C. (1985). "The quadratic sieve factoring algorithm." *Advances in Cryptology—EUROCRYPT'84*, Lecture Notes in Computer Science, vol. 209, eds. T. Beth, N. Cot, and I. Ingemarsson. Springer-Verlag, Berlin, 169–182.

[30] Rabin, M.O. (1979). "Digitalized signatures and public-key functions as intractable as factorization." MIT/LCS/TR-212, MIT Laboratory for Computer Science.

[31] Rivest, R., A. Shamir, and L. Adleman (1978). "A method for obtaining digital signatures and public-key cryptosystems." *Commun. ACM*, 21, 120–126.

[32] Laboratories, RSA (2004). "Public-key cryptography standards." http://www.rsasecurity.com/ rsalabs/pkcs/

[33] Schirokauer, O. (2000). "Using number fields to compute logarithms in finite fields." *Math. Comp.*, 69, 1267–1283.

[34] Schirokauer, O., D. Weber, and T.F. Denny (1996). "Discrete logarithms: The effectiveness of the index calculus method." *ANTS*, Lecture Notes in Computer Science, vol. 1122, ed. H. Cohen. Springer, Berlin, 337–361.

[35] Schneier, B. "Security pitfalls in cryptography." http://www.schneier.com/ essay-pitfalls.html.

[36] Silverman, R.D. (1987). "The multiple polynomial quadratic sieve." *Math. Comput.*, 48, 329–339.

[37] Solinas, J.A. (1997). "An improved algorithm for arithmetic on a family of elliptic curves." *Advances in Cryptology—CRYPTO'97*, Lecture Notes in Computer Science, vol. 1294, ed. B.S. Kaliski. Springer-Verlag, Berlin, 357–371.

[38] Wagstaff, Jr., S.S. (2002). *Cryptanalysis of Number Theoretic Ciphers*. CRC Press, Boca Raton, FL.

NYBERG–RUEPPEL SIGNATURE SCHEME

Nyberg–Rueppel proposed a signature scheme [1] which is of the message recovery type (see digital signature scheme) and based on the discrete logarithm problem. The following gives a typical interpretation of the Nyberg–Rueppel signature scheme.

Key generation: a prime number p, a prime factor q of $p - 1$, an element g of order q in the group of integers modulo p, a secret key $x(0 < x < q)$. The public key consists of p, q, g, and $y = g^x \bmod p$ (see also modular arithmetic).

Signing: for message m, compute $r = m \cdot g^k \bmod p$, $\acute{r} = r \bmod q$, $s = -k - r' \cdot x \bmod q$, and output (r, s). Verification: verify $s < q$, compute $r' = r \bmod q$, and check that $g^s \cdot y^{r'} \cdot r = m$.

However, it is advised to apply some redundant function R to a message m and use $R(m)$ instead of m. The reason is as follows: if a valid signature (r, s) for m is given, then $(r, s + t \bmod q)$ is a valid signature for $m \cdot g^t \bmod q$. Therefore by using $R(m)$ instead of m, one can neglect the possibility of $R(m) \cdot g^t$ being in the image space of R.

Kazue Sako

Reference

[1] Nyberg, K. and R.A. Rueppel (1995). "Message recovery for signature schemes based on the discrete logarithm problem." *Advances in Cryptology—EUROCRYPT'94*, Lecture Notes in Computer Science, vol. 950, ed. A. De Santis. Springer, Berlin, 182–193.

O

OAEP: OPTIMAL ASYMMETRIC ENCRYPTION PADDING

It has been noticed that the plain RSA public key encryption cannot be used directly for practical purpose, paddings are required, in order to rule out basic attacks.

THE RSA–PKCS #1 V1.5 ENCRYPTION: A widely deployed padding for RSA-based encryption is defined in the PKCS #1 v1.5 standard: for any modulus $2^{8(k-1)} \le n < 2^{8k}$, in order to encrypt a message m, one defines the k-byte long string $M = 02 \parallel r \parallel 0 \parallel m$, where r is a string of randomly chosen non-zero bytes (at least 8). This block is thereafter encrypted with the RSA permutation, $C = M^e \bmod n$ (see modular arithmetic). When decrypting a ciphertext C, the decryptor applies RSA inversion by computing $M = C^d \bmod n$ and then checks that the result M matches the expected' format. If so, the decryptor outputs the last part as the plaintext. Otherwise, the ciphertext is rejected. Intuitively, this padding seems sufficient to rule out all the well-known weaknesses of the plain RSA system, but without any formal proof or guarantee. Surprisingly, in 1998, Bleichenbacher [3] showed that a simple active attack can completely break RSA–PKCS #1. This attack applies to real systems such as a Web server using SSL v3.0.

THE OPTIMAL ASYMMETRIC ENCRYPTION PADDING: For some time, people have tried to provide security proofs for cryptographic protocols in the "reductionist" sense [4]. To do so, one presents an algorithm that uses an effective adversary as a sub-program to break some underlying hardness assumption (such as the RSA assumption, or the intractability of the integer factorization).

The Random Oracle Model

A few years ago, a new line of research started with the goal of combining provable security with efficiency, still in the "reductionist" sense. To achieve this goal, Bellare and Rogaway [1] formalized a heuristic suggested by Fiat and Shamir [7]. This heuristic consisted in making an idealized assumption about some objects, such as hash functions, according to which they were assumed to behave like truly random functions. This assumption is known as the random oracle model. We stress that security proofs in this model are not strong proofs. However, one can also consider random-oracle-based proofs under the assumption that the adversary is generic, whatever may be the actual implementation of the hash function. In other words, we may assume that the adversary does/cannot use any specific weakness of the hash functions used in practice.

Description of OAEP

At the time Bleichenbacher published his attack on RSA–PKCS #1 v1.5, the only efficient and "provably secure" encryption scheme based on RSA was the Optimal Asymmetric Encryption Padding (OAEP) proposed by Bellare and Rogaway [2]. OAEP can be used with any trapdoor one-way permutation f (see trapdoor one-way function and substitutions and permutations). To encrypt a message m using the encryption scheme f-OAEP, first apply the OAEP procedure described in Figure 1. Here r is a random string and G, H are hash functions. The resulting values $[s \parallel t]$ are then encrypted using f, namely $C = f(s, t)$.

OAEP and Provable Security

Bellare and Rogaway proved that OAEP padding used with any trapdoor one-way permutation f provides a semantically secure encryption scheme. By adding some redundancy (the constant value 0^{k_1} at the end of the message, as shown in Figure 1), they furthermore proved it to be *weakly*

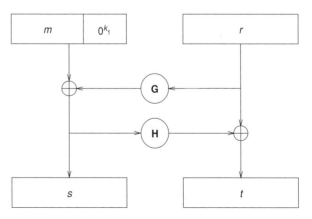

Fig. 1. OAEP padding

plaintext-aware. Plaintext-awareness is a property of encryption schemes in the random oracle model which informally means that in order to build a valid ciphertext, one needs to know the corresponding plaintext. The *weak* part in the original definition was that the awareness of the plaintext of any valid ciphertext built by the adversary hold, while the adversary had not received any valid ciphertext from any source. For such a scheme, a decryption oracle access on such a ciphertext does not provide any information to the adversary, and thus until he has received the challenge ciphertext: a valid ciphertext. Unfortunately, the adaptive chosen ciphertext attack model gives the adversary a full-time access to a decryption oracle, even after receiving the challenge ciphertext. Therefore, semantic security together with *weak* plaintext-awareness only implies the semantic security against non-adaptive chosen-ciphertext attacks (a.k.a. lunchtime attacks [11]—IND–CCA1).

However, even if the semantic security against adaptive chosen-ciphertext attacks (IND–CCA2) had never been proven, it was widely admitted until Shoup's counter-example [15]: he indeed showed that if there exists a trapdoor one-way permutation g for which it is easy to compute $g(x \oplus a)$ from $g(x)$ and a, then OAEP cannot be IND–CCA2 secure for an arbitrary trapdoor one-way permutation f.

Fixing the OAEP Proof of Security

However, from a carefull analysis of this counter-example, one can see that for the attack to work, the adversary has to be able to partially invert the permutation f. Therefore, let us define the *partial-domain one-wayness* of a permutation f to be the intractability of deducing s from $C = f(s, t)$. Fujisaki et al. [8] formally proved this fact: If f is partial-domain one-way, then f-OAEP is IND–CCA2 secure. We note that partial-domain one-wayness is a stronger property than one-wayness: a function might be one-way but still not partial-domain one-way.

Fortunately, the homomorphic properties of RSA makes that the RSA permutation is partial-domain one-way if and only if RSA is one-way. Altogether, this proves the widely believed IND–CCA2 security of RSA–OAEP assuming that RSA is a trapdoor one-way permutation, and thus under the widely admitted RSA assumption.

OAEP ALTERNATIVES: Shoup also proposed a formal security proof of RSA–OAEP, but in the particular case where the encryption exponent e is equal to 3 only. However, many people believe that the RSA trapdoor one-way permutation with exponent 3 may be weaker than with greater exponents. Therefore, he also proposed a slightly modified version of OAEP, called OAEP+ (see Figure 2), which can be proved secure under the one-wayness of the permutation. It uses the variable redundancy $R(m, r)$ instead of the constant 0^{k_1}. It is thus a bit more intricate than the original OAEP. Boneh [5] also proposed a new padding scheme, SAEP+, to be used with the Rabin primitive [12] or RSA. It is simpler than OAEP+, hence the name Simplified Asymmetric Encryption Padding: whereas OAEP+ is a two-round Feistel network, SAEP+ is a single-round. But as OAEP+, it is provably secure, whatever the exponent is.

David Pointcheval

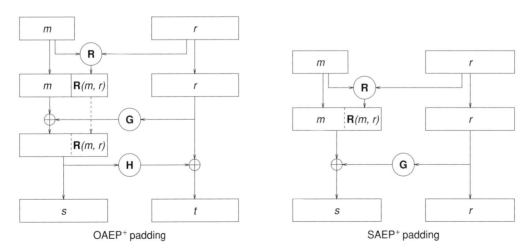

OAEP+ padding   SAEP+ padding

Fig. 2. OAEP+ and SAEP+ paddings

References

[1] Bellare, M. and P. Rogaway (1993). "Random oracles are practical: A paradigm for designing efficient protocols." *Proceedings of the 1st CCS*. ACM Press, New York, 62–73.

[2] Bellare, M. and P. Rogaway (1995). "Optimal asymmetric encryption—how to encrypt with RSA." *Advances in Cryptology—EUROCRYPT'94*, Lecture Notes in Computer Science, vol. 950, ed. A. De Santi. Springer-Verlag, Berlin, 92–111.

[3] Bleichenbacher, D. (1998). "A chosen ciphertext attack against protocols based on the RSA encryption standard PKCS #1." *Advances in Cryptology—CRYPTO'98*, Lecture Notes in Computer Science, vol. 1462, ed. H. Krawazy. Springer-Verlag, Berlin, 1–12.

[4] Blum, M. and S. Micali (1984). How to generate cryptographically strong sequences of pseudorandom bits. *SIAM Journal on Computing*, 13, 850–864.

[5] Boneh, D. (2001). "Simplified OAEP for the RSA and rabin functions." *Advances in Cryptology—CRYPTO 2001*, Lecture Notes in Computer Science, vol. 2139, ed. J. Kilian. Springer-Verlag, Berlin, 275–291.

[6] Boneh, D. and G. Durfee (2000). "Cryptanalysis of RSA with private key d less than $N^{0.292}$." *IEEE Transactions on Information Theory*, 46 (4), 1339–1349.

[7] Fiat, A. and A. Shamir (1987). "How to prove yourself: Practical solutions of identification and signature problems." *Advances in Cryptology—CRYPTO'86*, Lecture Notes in Computer Science, vol. 263, ed. A. Odlyzko. Springer-Verlag, Berlin, 186–194.

[8] Fujisaki, E., T. Okamoto, D. Pointcheval, and J. Stern (2001). "RSA–OAEP is secure under the RSA assumption." *Advances in Cryptology—CRYPTO 2001*, Lecture Notes in Computer Science, vol. 2139, ed. J. Kilian. Springer-Verlag, Berlin, 260–274.

[9] Håstad, J. (1988). "Solving simultaneous modular equations of low degree." *SIAM Journal of Computing*, 17, 336–341.

[10] Miller, G. (1976). "Riemann's hypothesis and tests for primality." *Journal of Computer and System Sciences*, 13, 300–317.

[11] Naor, M. and M. Yung (1989). "Universal one-way hash functions and their cryptographic applications." *Proceedings of the 21st STOC*. ACM Press, New York, 33–43.

[12] Rabin, M.O. (1978). "Digitalized signatures." *Foundations of Secure Computation*, eds. R. Lipton and R. De Millo. Academic Press, New York, 155–166.

[13] Rackoff, C. and D.R. Simon (1992). "Non-interactive zero-knowledge proof of knowledge and chosen ciphertext attack." *Advances in Cryptology—CRYPTO'91*, Lecture Notes in Computer Science, vol. 576, ed. J. Feigenbaum. Springer-Verlag, Berlin, 433–444.

[14] Rivest, R., A. Shamir, and L. Adleman (1978). "A method for obtaining digital signatures and public key cryptosystems." *Communications of the ACM*, 21 (2), 120–126.

[15] Shoup, V. (2001). "OAEP reconsidered." *Advances in Cryptology—CRYPTO 2001*, Lecture Notes in Computer Science, vol. 2139, ed. J. Kilian. Springer-Verlag, Berlin, 239–259.

[16] Wiener, M. (1990). "Cryptanalysis of short RSA secret exponents." *IEEE Transactions on Information Theory*, 36 (3), 553–558.

OBLIVIOUS TRANSFER

Oblivious transfer (OT) is a two-party protocol between a sender and a receiver, by which the sender transfers some information to the receiver, the sender remaining oblivious, however, to what information the receiver actually obtains. The most basic form of oblivious transfer, as introduced by Rabin [4], is a protocol achieving the following functionality. The sender uses one bit b as its private input to the protocol; the receiver does not provide any private input to the protocol. At the completion of the protocol, the receiver either gets the bit b or an undefined value $\perp$. Both cases occur with probability 50%, and the receiver knows whether it gets b or $\perp$. However, the sender does not know whether bit b was transferred successfully or not.

Despite its somewhat strange functionality, OT turns out to be sufficiently powerful to construct a *secure* multiparty computation for *any* computable function, as follows from the completeness result proved by Kilian [3]. In many cases, a slightly more advanced form of oblivious transfer is used, known as "chosen one-out-of-two" OT, denoted by $\binom{2}{1}$-OT. In a $\binom{2}{1}$-OT, the sender uses two private input bits b_0, b_1 and the receiver uses one private input bit s. At the completion of the protocol, the receiver gets the bit x_s, whereas the sender does not get any information on the value of s, i.e., the sender does not know which bit was selected by the receiver. These two basic types of oblivious transfer are equivalent in the sense that either type can be constructed from the other one, using a polynomial time transformation (see [2]: Rabin's OT can be achieved using a single $\binom{2}{1}$-OT; for the other direction, however, one requires $O(k)$ instances of Rabin's OT to construct a single $\binom{2}{1}$-OT, where k is a security parameter).

Many oblivious protocols have been proposed over the years. As a simple example, consider the following $\binom{2}{1}$-OT protocol, proposed by [1]. Let g denotes a generator of a cyclic group G of order p, where p is a large underline{prime}. Let h denotes a random element of G, $h \neq 1$, such that the underline{discrete logarithm} of h with respect to g is not known to any party.

The protocol runs as follows. The receiver picks a random value $x_s \in \mathbb{Z}_p$, and sets $y_s = g^{x_s}$ and $y_{1-s} = h/g^{x_s}$, where s denotes the receiver's private input bit. The receiver sends y_0 to the sender. Upon receipt of y_0, the sender sets $y_1 = h/y_0$. The sender then computes two underline{ElGamal public key encryptions}: E_0 contains message g^{b_0} encrypted under public key y_0, and E_1 contains message g^{b_1} encrypted under public key y_1, where b_0, b_1 denote the private input bits of the sender. The sender sends the ordered pair (E_0, E_1) to the receiver. Finally, the receiver decrypts E_s, using the private key x_s, to obtain b_s.

Since the receiver cannot know both $\log_g y_0$ and $\log_g y_1$ (as this implies knowledge of $\log_g h$), the receiver cannot decrypt the other bit b_{1-s}. On the other hand, the value of y_0 is clearly independent of s, hence the sender does not learn which bit the receiver chooses.

Berry Schoenmakers

References

[1] Bellare, M. and S. Micali (1989). "Non-interactive oblivious transfer and application." *Advances in Cryptology—CRYPTO'89*, Lecture Notes in Computer Science, vol. 435, ed. G. Brassand. Springer-Verlag, Berlin.

[2] Crépeau, C. (1987). "Equivalence between two flavours of oblivious transfer." *Advances in Cryptology—CRYPTO'87*, Lecture Notes in Computer Science, vol. 293, ed. C. Pomerance. Springer-Verlag, Berlin, 350–354.

[3] Kilian, J. (1988). "Basing crpytography on oblivious transfer". *Proceedings of 20th Symposium on Theory of Computing (STOC'88)*. ACM Press, New York, 20–31.

[4] Rabin, M. (1981). "How to exchange secrets by oblivious transfer." Technical Memo TR-81, Aiken Computation Laboratory, Harvard University.

ONE-TIME PASSWORD

A *one-time* underline{password} is a password that is used only once, in an underline{authentication} session, and then thrown away and never used again. Because of the single use of this password, it is impervious to underline{eavesdropping} (i.e., it is safe from passive adversaries who listen to an authentication session and then later use the overheard information to attempt impersonation). In practice, the user and the system typically share a list of one-time passwords and then cross an entry off the list after it is used (a new list must be generated and shared once all entries have been crossed off).

Variations on the basic one-time password concept include mechanisms for generating a sequence of passwords from a single shared password (e.g., using a underline{one-way function}) and mechanisms for generating a new password as a function of the current time. Such variations seek to eliminate the need for list maintenance at both the user and the server.

Carlisle Adams

References

[1] Kaufman, C., R. Perlman, and M. Speciner (1995). *Network Security: Private Communication in a Public World*. Prentice Hall, Englewood Cliffs, NJ.

[2] Menezes, A., P. van Oorschot, and S. Vanstone (1997). *Handbook of Applied Cryptography*. CRC Press. Boca Raton, FL.

ONE-WAY FUNCTION

Informally, a one-way function is a function for which computation in one direction is straightforward, while computation in the reverse direction is far more difficult. This is typically described in a more formal, though still not rigorous, way [3–5] as a function f with domain X and range (codomain) Y where $f(x)$ is 'easy' to compute for all $x \in X$; but for 'virtually all' elements $y \in Y$, it is 'computationally infeasible' to find an x such that $f(x) = y$. The function f is a *one-way permutation* when f is a bijective one-way function and $X = Y$ (see also underline{substitutions and permutations}).

The seminal paper of Diffie and Hellman [1] was the first to set down the potential of one-way functions in the development of underline{public-key cryptography}. The interesting, and important, feature of the one-way function is the asymmetry in computational effort required to perform a function evaluation and its reverse. Diffie and Hellman provided a familiar example of such asymmetry in the difficulty of undoing the action of a sophisticated compiler that translates an easily understood program written in a high-level language into almost unintelligible machine code. A more mathematical example was provided by the contrast between exponentiation and taking discrete logarithms in some underline{finite field}. If we consider the integers modulo a large underline{prime number} p together with some (primitive) base element g, then given *a* it is straightforward to compute $g^a \bmod p$ (see modular arithmetic). However given $g^b \bmod p$, it is computationally difficult to recover *b* (for most *b*). This problem, with the additional stipulation that $p-1$ have a large prime factor [7], provides the security of Diffie–Hellman key agreement as well as many other cryptographic mechanisms that have been developed since 1976. See the discrete logarithm problem.

It is notable that although Diffie and Hellman introduced and explored the role of one-way functions in the development of public-key cryptography, the earlier use of one-way functions for secure password validation is attributed to Needham and Schroeder [6]. The concept of a one-way function is also to be found when considering the properties of a cryptographic hash function.

Since their introduction, considerable research has been conducted into providing a more rigorous foundation to the intuitive description given by Diffie and Hellman. Providing a suitable framework within which to analyze and formalize the concept of a one-way function is not straightforward and builds on an interchange between fields as diverse as computational complexity [2], information theory [8], and number theory [1]. Nevertheless, the fundamental practical attributes of one-way functions and their companion trapdoor one-way functions have been vitally important in the development of cryptography and public-key cryptography, in particular.

Matt Robshaw

References

[1] Diffie, W. and M.E. Hellman (1976). "New directions in cryptography." *IEEE Transactions on Information Theory*, 22, 644–654.
[2] Garey, M.R. and D.S. Johnson (1979). *Computers and Intractability: A Guide to the Theory of NP-Completeness*. W. Freeman, San Francisco.
[3] Goldreich, O. (1999). *Modern Cryptography, Probabilistic Proofs and Pseudorandomness*. Springer-Verlag, Berlin.
[4] Massey, J.L. (1992). "Contemporary cryptology: An introduction." *Contemporary Cryptology, The Science of Information Integrity*. ed. G.J. Simmons. IEEE Press.
[5] Menezes, A.J., P.C. van Oorschot, and S.A. Vanstone (1997). *Handbook of Applied Cryptography*. CRC Press, Boca Raton, FL.
[6] Needham, R.M. and M.D. Schroeder (1978). "Using encryption for authentication in large networks of computers." *Communications of the ACM*, 21, 993–999.
[7] Pohlig, S.C. and M.E. Hellman (1978). "An improved algorithm for computing logarithms over *GF(p)* and its cryptographic significance." *IEEE Transactions on Information Theory*, 24, 106–110.
[8] Yao, A.C. (1982). "Theory and applications of trapdoor functions." *Proceedings of the IEEE 23rd Annual Symposium on Foundations of Computer Science*, 80–91.

O-NOTATION

O-notation is a convenient way of expressing the relationship between positive-valued functions as their inputs grow. The notation $T(x) = O(U(x))$ means, informally, that the function $T(x)$ is at most the same "order" as $U(x)$. More precisely, this means that there exists a constant $c > 0$ such that for all sufficiently large *x*,

$$T(x) \le cU(x).$$

The value of $T(x)$ for small *x* is not necessarily constrained; the notation only indicates the asymptotic behavior.

This notation is often employed as a shorthand for algorithm running times, as it conceals implementation-specific details and focuses instead on the rate of growth. For instance, a typical algorithm for modular multiplication takes $O(x^2)$ time, where *x* is the length of the operands (or more typically in cryptography, the key size or security parameter). The actual time will be some implementation-specific constant times x^2, and may also involve a linear or constant "overhead". The notation $O(x^2)$ focuses on the highest-order effects and is helpful in assessing the growth of the running time as the size of the operands increase.

Four related notations have also been defined:

- "big-Ω" notation: $T(x) = \Omega(U(x))$ means that $T(x) \ge cU(x)$ for some constant $c > 0$, for sufficiently large *x*;
- "Θ" notation: $T(x) = \Theta(U(x))$ means that $T(x) = O(U(x))$ and $T(x) = \Omega(U(x))$;
- "little-o" notation: $T(x) = o(U(x))$ means that for every constant *c*, $T(x) < cU(x)$ for all sufficiently large *x*;
- "little-ω" notation: $T(x) = \omega(U(x))$ means that for every constant *c*, $T(x) > cU(x)$ for all sufficiently large *x*.

The notations are analogous to the usual arithmetic comparison operators $>, \ge, =, \le,$ and $<$:

Notation	Operator
ω	$>$
Ω	$\ge$
Θ	$=$
O	$\le$
o	$<$

For instance, if $T(x) = O(U(x))$ and $U(x) = O(V(x))$, then $T(x) = O(V(x))$; if $T(x) = o(U(x))$ then $U(x) = \omega(T(x))$, and so on.

The notations $O(1)$ and $o(1)$ deserve special explanation. $O(1)$ denotes a function of x that is bounded by a constant as $x \to \infty$, since one has that

$$T(x) \leq c \times 1 = c$$

for some constant c and all sufficiently large x. Likewise, $o(1)$ denotes a function of x that tends toward 0 as $x \to \infty$.

The various notations can also be employed within more complex mathematical expressions. For instance, in the expression

$$T(x) = e^{(\gamma + o(1))(\log x)^t (\log \log x)^{1-t}}, \quad \text{for } x \to \infty,$$

the first term in the exponent tends toward γ as $x \to \infty$.

Technically, notation such as $O(U(x))$ denotes the *set* of all functions which asymptotically grow more slowly than $U(x)$. Thus, formally one might write $T(x) \in O(U(x))$ to denote membership in this set. Also, in mathematics, one is sometimes concerned with the relationship between functions as the input approaches some finite value, rather than as it tends toward infinity. However, in cryptography, the notation $T(x) = O(U(x))$ is standard, and the limit $x \to \infty$ is assumed.

Burt Kaliski

OPTIMAL EXTENSION FIELDS (OEFs)

Optimal extension fields (OEFs) are a family of underline{finite fields} with special properties. They were designed in a way that leads to efficient field arithmetic if implemented in software. OEFs were introduced first in [3] and independently in [7]. They are defined as follows:

DEFINITION 1. *An Optimal Extension Field is a finite field $GF(p^m)$ such that:*
1. *p is a* prime number *of the form $2^n \pm c$, $\log_2 c \leq \lfloor \frac{1}{2}n \rfloor$ (such primes are also referred to as pseudo-Mersenne prime),*
2. *An irreducible binomial $P(x) = x^m - \omega$ exists over $GF(p)$.*

An example of an OEF is the field $GF(p^6)$ with the prime $p = 2^{32} - 387$ and the irreducible polynomial $x^6 - 2$. Note that the cardinality of this OEF is *roughly* $(2^{32} - 387)^6 \approx 2^{192}$.

The main motivation for OEFs is that the field parameters can be chosen such that they are a good match for the processor on which the field arithmetic is to be implemented. In particular, it is often an advantage to choose an OEF $GF(p^m)$ such that the prime p can be represented within one register of the target processor. For instance, in the OEF example given above, $GF((2^{32} - 387)^6)$, the prime $2^{32} - 387$ fits nicely in the registers of a 32-bit CPU. In this situation, field arithmetic can be implemented rather efficiently. The following theorem from [5] describes the cases when an irreducible binomial over $GF(p)$ exists:

THEOREM 1. *Let $m \geq 2$ be an integer and $\omega \in GF(p)^*$. Then the binomial $x^m - \omega$ is irreducible in $GF(p)[x]$ if and only if the following two conditions are satisfied: (i) each prime factor of m divides the order e of ω over $GF(p)$, but not $(p-1)/e$; (ii) $p \equiv 1 \bmod 4$ if $m \equiv 0 \bmod 4$.*

An important corollary is given in [4]:

COROLLARY 1. *Let ω be a primitive element for $GF(p)$ and let m be a divisor of $p-1$. Then $x^m - \omega$ is an irreducible polynomial.*

A brief outline of the arithmetic algorithms of OEFs follows, where we distinguish between arithmetic in the subfield $GF(p)$ and arithmetic in the extension field $GF(p^m)$. Extension field arithmetic requires subfield calculations as a "subroutine".

Subfield addition and subtraction. If p can be represented in one register, all elements of $GF(p) = \{0, 1, \ldots, p-1\}$ can be represented as simple one-word integers. Addition is straightforward and very efficient: One performs a regular integer addition and, if the sum is larger than p, the modulus p is subtracted from the sum (see underline{modular arithmetic}). Subtraction can be done analogously.

Subfield multiplication. Due to the fact that p is a underline{pseudo-Mersenne prime}, subfield multiplication is also efficient. In fact, the overall performance of OEFs greatly relies on the fact that subfield multiplication is fast. In a typical implementation, in a first step the two operands $a, b \in GF(p)$ are multiplied yielding the integer product $d = a \times b$. This is done with one integer multiplication. Note that in the general case, d has about twice the bit length of p if we assume that a and b both have about the same bit lengths as the modulus p. Due to the special form of p, the following algorithm allows an efficient reduction $d \bmod p$ without performing an explicit integer division. We present a form of such a modular reduction algorithm, adapted from [6]. The operator $\gg$ is taken to mean "right shift".

ALGORITHM 1. *Fast Subfield Modular Reduction*

Require: $p = 2^n - c, \log_2 c \leq \lfloor \frac{1}{2} n \rfloor, d < p^2$ is
the integer to reduce
Ensure: $r \equiv d \bmod p$
$q_0 \leftarrow d \gg n$
$r_0 \leftarrow d - q_0 2^n$
$r \leftarrow r_0$
$i \leftarrow 0$
while $q_i > 0$ **do**
    $q_{i+1} \leftarrow q_i c \gg n$
    $r_{i+1} \leftarrow q_i c - (q_{i+1} \gg n)$
    $i \leftarrow i + 1$
    $r \leftarrow r + r_i$
end while
while $r \geq p$ **do**
    $r \leftarrow r - p$
end while

This reduction algorithm requires two integer multiplications and some shifts and additions. If we ignore the latter operations, the main costs for one OEF subfield multiplication are $1 + 2 = 3$ integer multiplications.

An important special case is OEFs where the prime has the form $p = 2^n \pm 1$. In this case, the modulo reduction itself can be performed with one addition or subtraction, and the main costs for an entire subfield multiplication are one integer multiplication. The reduction method for primes $2^n - 1$ is described in the entry <u>Mersenne prime.</u> OEFs with primes $2^n \pm 1$ are sometimes referred to as Type I OEFs [3].

Extension field addition and subtraction. Addition of two field elements is simply an addition of the corresponding coefficients of the two elements. The coefficient additions follow $GF(p)$ arithmetic rules. Subtraction is done analogously.

Extension field multiplication. It is usually advantageous to represent elements of OEFs in a standard (or polynomial) basis. Field multiplication can be performed in two stages. First, we perform an ordinary polynomial multiplication of two field elements $A(x)$ and $B(x)$, resulting in an intermediate product $C'(x)$ of degree less

than or equal to $2m - 2$:

$$
\begin{aligned}
C'(x) &= A(x) \times B(x) \\
&= c'_{2m-2} x^{2m-2} + \cdots + c'_1 x + c'_0, \\
&\quad c'_i \in GF(p),
\end{aligned} \tag{1}
$$

The schoolbook method to calculate the coefficients $c'_i, i = 0, 1, \ldots, 2m - 2$, requires m^2 multiplications and $(m - 1)^2$ additions in the subfield $GF(p)$. Optionally, the <u>Karatsuba algorithm</u> can be applied here to reduce the number of coefficient multiplications. For instance, for fields $GF(p^6)$, the polynomial multiplication can be performed with 18 subfield multiplications (as opposed to 36 with the schoolbook method) when applying the Karatsuba algorithm recursively [2].

In the second stage of the OEF multiplication, the intermediate result $C'(x)$ has to be reduced modulo the irreducible polynomial $P(x) = x^m - \omega$. We note that the following congruences hold: $x^m \equiv \omega \bmod P(x)$, $x^{m+1} \equiv \omega x \bmod P(x), \ldots, x^{2m-2} \equiv \omega x^{m-2} \bmod P(x)$. Hence, the terms $c'_m x^m, \ldots, c'_{2m-2} x^{2m-2}$ can each be reduced with one multiplication by ω and one addition in the subfield. Thus, the entire modulo reduction requires at most $m - 1$ multiplications by ω and $m - 1$ additions, where both of these operations are performed in $GF(p)$.

Extension field inversion. The two most useful methods for inversion are the Euclidean algorithm and reduction of the extension field inversion to subfield inversion via the Itoh–Tsujii algorithm. See <u>Inversion in Finite Fields and Rings</u> for a detailed description of those two inversion methods.

Extension field exponentiation. One can use either one of the standard exponentiation techniques, such as the <u>sliding window method</u>. A particularly fast method for OEFs is the one described in [1] which is based on the fact that the Frobenius automorphism can be computed efficiently in OEFs.

A generalization of OEFs, that is fields $GF(p^m)$, $p > 2$, with p not necessarily a pseudo-Mersenne prime and the field polynomial not necessarily a binomial, is discussed in [1]. Tables with OEFs and more details about the arithmetic of OEFs are described in [2].

OEFs are applicable as underlying algebraic structure for cryptosystems that rely on finite fields. In particular, they appear useful for <u>elliptic curve cryptosystems</u> and schemes using the <u>discrete logarithm in finite fields</u>. At the time of writing, it remains an open question whether elliptic curve cryptosystems and discrete logarithm schemes which use OEFs (rather than *prime fields* or fields $GF(2^m)$) have cryptographic weaknesses.

Christof Paar

References

[1] Avanzi, R.A. and P. Mihăilescu (2003). "Generic efficient arithmetic algorithms for PAFFs (Processor Adequate Finite Fields) and related algebraic structures." *Workshop in Selected Areas in Cryptography (SAC)*, Lecture Notes in Computer Science, vol. 3006, eds. M. Matsui and R. J. Zuccheratop. Springer-Verlag, Berlin, Germany, 130–144.

[2] Bailey, D.V. and C. Paar (2001). "Efficient arithmetic in finite field extensions with application in elliptic curve cryptography." *Journal of Cryptology*, 14 (3), 153–176.

[3] Bailey, D.V. and C. Paar (1998). "Optimal extension fields for fast arithmetic in public-key algorithms." *Advances in Cryptology—CRYPTO'98*, Lecture Notes in Computer Science, vol. 1462, ed. H. Krawczyk. Springer-Verlag, Berlin, Germany, 472–485.

[4] Jungnickel, D. (1993). *Finite Fields*. B.I.-Wissenschaftsverlag.

[5] Lidl, R. and H. Niederreiter (1983). "Finite fields." *Encyclopedia of Mathematics and its Applications*, vol. 20. Addison-Wesley, Reading, MA, USA.

[6] Menezes, A.J., P.C. van Oorschot, and S.A. Vanstone (1997). *Handbook of Applied Cryptography*. CRC Press, Boca Raton, FL, USA.

[7] Mihăilescu, P. (1997). "Optimal galois field bases which are not normal." *Recent Result Session, Fast Software Encryption*.

ORDER

The *order* of a group $G = (S, \circ)$ is the number of elements in the set S, and likewise the order of a ring or field $(S, +, \times)$ is the number of elements in S.

In a group, the order of an element $g \in S$ is the least positive integer k such that $g^k = 1$ (where the group law is written multiplicatively and 1 denotes the identity element).

In a ring or field, the order of an element typically refers to *multiplicative order*, that is, the order of the element under the multiplication operation in the multiplicative group.

The order of each element in a group divides the order of the group.

Burt Kaliski

OVERSPENDER DETECTION

In electronic payment schemes and electronic cash that use prepaid electronic coins, there is a natural fraud scenario where a customer tries to use an already spent electronic coin a second or third time. An important security requirement in such schemes is thus overspender detection, i.e., an effective way of determining the culprit who has spent an electronic coin twice. A related and equally important security requirement is to detect coins that are overspent or even better to prevent coins from being overspent (see overspending prevention).

If electronic coins bear the identities of customers who have withdrawn them, then overspender detection is easy to accomplish and appears an almost trivial security requirement. If electronic coins can be spent anonymously (see anonymity) though, then each coin alone carries too little information to identify the spender. It is possible, however, to design systems where two or more transactions of spending the same coin reveal enough information to the verifier (bank) to recover the identity of the overspender.

Gerrit Bleumer

OVERSPENDING PREVENTION

In electronic payment schemes and electronic cash that use prepaid electronic coins, there is a natural fraud scenario where a customer tries to use an already spent electronic coin a second or third time. An important security requirement in such schemes is thus to detect coins that are overspent or even better to prevent coins from being overspent. A related and equally important security requirement is to detect culprits who have overspent one or more of their electronic coins (see overspender detection).

If electronic coins bear the identities of customers who have withdrawn them, overspending detection and prevention are easy to accomplish and appear an almost trivial security requirement. The verifiers of coins only need to check in a database whether the electronic coin at hand has been spent before, and if so, who the respective double spender is. If electronic coins can be spent anonymously (see anonymity) though, then each coin alone carries too little information to identify the spender. Double spending of anonymous coins can therefore only be prevented by using hardware security devices that cannot be manipulated by their holders (under reasonable assumptions). It is possible, however, to design anonymous electronic coin schemes where two or more spending transactions of the same coin reveal enough information to the verifier (bank) to detect whether a coin has been spent before, to reject the second attempt to pay with such a coin, and to recover the identity of the overspender.

Gerrit Bleumer

P

PAILLIER ENCRYPTION AND SIGNATURE SCHEMES

In the spirit of earlier encryption schemes suggested by Goldwasser–Micali, Benaloh, Naccache–Stern, and Okamoto–Uchiyama, Paillier proposed in 1999 a public-key cryptosystem [4] (see public-key cryptograhy) based on the properties of nth powers modulo n^2 where n is an RSA modulus (see modular arithmetic and RSA public key encryption). The original observation is that the function $\mathcal{E}(x, y) = g^x y^n \bmod n^2$ is a one-way trapdoor permutation (see trapdoor one-way function) over the group $\mathbb{Z}_n \times \mathbb{Z}_n^* \simeq \mathbb{Z}_{n^2}^*$ where the trapdoor information is the factorization of n. The group $\mathbb{Z}_{n^2}^*$ is of order $n\phi$ where $\phi = \phi(n)$ is Euler's totient function of n and the base $g \in \mathbb{Z}_{n^2}^*$ is an element of order $\alpha \cdot n$ for some divisor α of ϕ (for instance $n + 1$ for which $\alpha = 1$). Noting $L(u) = (u - 1)/n$ when $u = 1 \bmod n$, x is recovered from $w = \mathcal{E}(x, y)$ as $x = L(w^\phi \bmod n^2)/L(g^\phi \bmod n^2) \bmod n$ and y is then $(wg^{-x})^{1/n} \bmod n$. When the factorization ϕ of n is unknown, however, recovering x from $\mathcal{E}(x, y)$ and even deciding if $x = 0$ are believed to be hard problems (see computational complexity). This is referred to as the [Decisional] Composite Residuosity assumption ([D]CR for short).

Given the public key (g, n), the Paillier encryption of $m \in [0, n - 1]$ is $c = \mathcal{E}(m, r)$ where r is chosen at random in $[0, n - 1]$. Encryption is therefore probabilistic (see probabilistic public-key encryption) and features homomorphic properties as multiplying $\mathcal{E}(m_1, r_1)$ by $\mathcal{E}(m_2, r_2)$ modulo n^2 provides an encryption of $m_1 + m_2 \bmod n$. This property makes this factoring-based cryptosystem particularly attractive for many cryptographic applications. Paillier encryption is semantically secure (resp. one-way) against chosen-plaintext attacks under the DCR (resp. CR) assumption and was shown to hide $\mathcal{O}(\log n)$ plaintext bits under a slightly stronger assumption [1]. The cryptosystem was extended in several directions. Damgård–Jurik suggested an extension modulo n^s for $s \geq 2$. Variations of different flavors and distributed versions [2,5] of the scheme were introduced (see threshold cryptography). Galbraith showed an embodiment of the scheme on elliptic curves over rings [3].

To sign $m \in 0, n^2 - 1$ under the private key ϕ, one inverts $\mathcal{E}$ to get s_1 and s_2 such that $\mathcal{E}(s_1, s_2) = m$ and the signature is (s_1, s_2). The verification consists in checking if $\mathcal{E}(s_1, s_2) = m$. In virtue of the homomorphic property of $\mathcal{E}$, this provides a blind signature) scheme. As for RSA, a hash function or a padding can be applied before signing to ensure strong security against existential forgery.

Pascal Paillier

References

[1] Catalano, Dario, Rosario Gennaro, and Nick Howgrave-Grahlam (2001). "The bit security of paillier's encryption schemes and its applications." *Advances in Cryptology—EUROCRYPT 2001,* Lecutre Notes in Computer Science, vol. 2045, ed. B. Pfitzmann, Springer-Verlag, Berlin, 229–243.

[2] Catalano, Dario, Rosario Gennaro, Nick Howgrave-Graham, and Phong Q. Nguyen (2001). "Paillier's cryptosystem revisited." *Proceedings of the 8th ACM conference on Computer and Communications Security* ACM Press, New York, 206–214.

[3] Galbraith, Steven D. (2002). "Elliptic curve paillier schemes." *Journal of Cryptology*, 15 (2), 129–138.

[4] Paillier, Pascal (1999). "Public-key cryptosystems based on composite-degree residuosity classes." *Advances in Cryptology—EUROCRYPT'99,* Lecture Notes in Computer Science, vol. 1592, ed. J. Stern. Springer-Verlag, Berlin, 223–238.

[5] Damgård, Ivan and Mads Jurik (2001). "A generalization, a simplification and some applications of paillier's probabilistic public-key system." *Public Key Cryptography—PKC 2001*, Lecture Notes in Computer Science, vol. 1992, ed. K. Kim. Springer-Verlag, Berlin, 119–136.

PASSWORD

A password is a secret that is presented to a verifier to prove a claim, typically during user authentication. A verifier can determine that the claimant knows the secret by comparing the password with a value he has on store. This can be the secret itself, a unique value computed from the secret using, e.g., a one-way function, or a known document encrypted using the password as a key. Passwords are a simple and convenient authentication mechanism but can only provide limited security. More secure authentication based on cryptographic protocols and hardware is possible, but these options are typically less convenient for users and more expensive to operate.

As a secret, passwords must be protected. Storing or transmitting a password in plain text is a risk, as is writing it down on a sticky note and attaching it to a screen. While these observations are obvious, they are often disregarded in practice. Passwords are also vulnerable to interception during the actual authentication process if the connection between the terminal and the verifier software is not secure, or if the terminal itself can be manipulated, e.g., to record keystrokes. To prevent attacks on the login process that intercept passwords on the path between the terminal and the verifier, the *Orange Book* security criteria (cf. Security Evaluation Criteria) describe a *Trusted Path*, with which systems can provide secure communication so that users can directly communicate with the secure parts of the system.

A characteristic of passwords is that they are directly entered by users, e.g., at a terminal login, or when accessing protected Web sites. Consequently, passwords can only be composed of characters that can be typed on a keyboard. Typical password mechanisms usually only consider the first n characters of a password, so the number of possible passwords is 95^n (with 95 printable characters). For a standard UNIX system, which considers a maximum of eight characters, this amounts to 6.6×10^{15} different values for passwords that are at least eight characters long. A Windows 2000 (TM) password, e.g., can be up to 127 characters long, which gives maximum of 1.4×10^{251} different combinations. While it is typically not feasible for a casual attacker to try all possible combinations of passwords, such an exhaustive key search is still a possible threat when considering more dedicated and resourceful attackers. With current processing and storage technology, it is possible to precompute and store all possible password values for a sufficiently limited search space.

However, the actual search space for typical passwords is considerably smaller than the theoretical limit because passwords need to be remembered by human users. Because arbitrary combinations of characters with no apparent meaning are hard to remember, requiring the use of such passwords would lead to users writing down their passwords, which induces the obvious risk of exposure. Users therefore typically choose shorter passwords using some form of mnemonics or simply familiar terms. This means that the entropy (cf. information theory) of passwords is low, so passwords are vulnerable to guessing.

A particular kind of password guessing attacks are dictionary attacks. A dictionary attack is carried out by trying candidate passwords from a large dictionary with popular words and terms, such as movie actors, characters from cartoons or literature, animal names, computer science, astrological terms, etc. Because attacks like these are simple and have proven to be very effective [1, 2], passwords are sometimes encrypted based on an additional bit string, a salt, which requires attackers using precomputed dictionaries with encrypted passwords to include 2^n variations of each encrypted password, n being the number of bits used for the salt.

If an attacker can install and run programs on the target host system, password security may also be attacked using a Trojan Horse login program. This program would masquerade as a regular system login screen on a computer terminal and capture and store the passwords entered by unsuspecting users. The program may then print a rejection message to the terminal before terminating and starting the regular login program. The existence of the Trojan Horse is unnoticed by users, who believe they simply mistyped their password during the first login attempt. This attack will not work on systems that provide a Trusted Path because communications via such a path are by definition initiated exclusively by the user, i.e., a login prompt would only appear as a system reaction to a user action, e.g., a keyboard interrupt, as in Windows NT (TM).

"A good password is one that is easily remembered, yet difficult to guess [3]." Because of the importance of choosing "good" passwords for maintaining password security, users should be educated and given guidelines for choosing passwords. The most important ones are summarized as rules of thumb below:

1. Choose long passwords to enlarge the search space for an attacker.
2. Do not choose words that are likely to appear in a dictionary, not even with variations.
3. Do not base passwords on any public information about yourself (birthdate, hobby, children's names), because it may help attackers in guessing.
4. Use the initial letters of the words in a sentence that you can remember, e.g., turn a sentence like "my daughter prefers muffins over cheese cake most of the time" into a password "mdpmoccmott". Words created in this manner are unlikely to be found in any dictionary.
5. Insert punctuation characters and digits freely, and capitalize some characters, e.g., turn "mdpmoccmott" into "mD4pMo,cCmott" to further enlarge the search space.

Three main lines of defense against attacks on password security can be identified:

1. *Reactive*—carry out internal dictionary attacks to check for weak passwords. If a weak password is found the respective account should be blocked until the user chooses a new password.

2. *Proactive*—educate users about the importance of choosing good passwords. Prevent them from choosing weak passwords, i.e., reject passwords that are too short, that are found within a dictionary, or are otherwise considered easy to guess. Proactive password checks can be integrated into the system programs that modify a password so that any new password is checked to be reasonably secure before it is accepted.

3. *Secretive*—protect password files so that not even the encrypted passwords can be obtained, which could otherwise be targets of off-line dictionary attacks. For this reason it is standard practice in modern systems to "shadow" out passwords from the password file, i.e., maintain a separate file that can only be read by system administrators.

In summary, passwords are a compromise between convenience, user acceptance, and cost on the one hand and security on the other. In highly sensitive environments, alternative authentication mechanisms should be considered, e.g., mechanisms based on biometric identification, on external devices (cf. Hardware Security Modules) that, e.g., generate one-time passwords, or on combinations of these mechanisms.

Gerald Brose

References

[1] Feldmeier, D.C. and P.R. Karn (1990). "Unix password security—ten years later." *Advances in Cryptology—CRYPTO'89*, Lecture Notes in Computer Science, vol. 435, ed. G. Brassard. Springer-Verlag, Berlin, 44–63.

[2] Klein, D.V. (1990). "'Forcing the cracker:' A survey of and improvements to password security." *Proc. 2nd USENIX Security Symposium*, 5–14.

[3] Morris, R. and K. Thompson (1979). "Password Security: A case history." *Communications of the ACM*, 22 (11), 594–597.

PAYMENT CARD

A payment card is a magnetic stripe or chip card used in a payment scheme, such as debit and credit schemes, as well as electronic cash schemes like Mondex or Proton. Electronic cash schemes require a chip card, and debit and credit schemes are migrating toward chip card technology.

Peter Landrock

PEM, PRIVACY ENHANCED MAIL

PEM provides a number of security mechanisms for protecting electronic mail transferred over Internet by defining a number of protocol extensions and processing procedures for mail messages following RFC 822 (Request For Comment). The security mechanisms include encryption of mails such that only the intended recipient(s) can read the contents of the mail. Other security mechanisms supported by PEM provide message authentication and integrity as well as digital signatures. Using the latter, nonrepudiation of origin may be achieved.

The security mechanisms provided can be based on symmetric cryptography (see key) or public key cryptography. In the latter case X.509 certificates are used to provide the public keys of the sender and recipient as needed (when mails are signed the public key of the originator is required, and when mails are encrypted the public key of the recipient is required).

PEM is defined in detail in RFC 1421 through 1424. RFC 1421 [2] defines the extensions to RFC 822 mail messages and RFC 1423 [4] specifies the cryptographic algorithms (including formats, parameters, and modes of use) to be used within PEM. RFC 1422 [3] and RfC1424 [5] deal with the key management issues—in particular the use of X.509 certificates.

Torben Pedersen

References

[1] http://www.ietf.org/proceedings/94mar/charters/pem-charter.html

[2] RfC1421: Privacy Enhancement for Internet Electronic Mail: Part I: Message Encryption and Authentication Procedures. See http://www.rfc-editor.org/rfc.html

[3] RfC1422: Privacy Enhancement for Internet Electronic Mail: Part II: Certificate-Based Key Management (RFC 1422). See http://www.rfc-editor.org/rfc.html

[4] RfC1423: Privacy Enhancement for Internet Electronic Mail: Part III: Algorithms, Modes, and

Identifiers (RFC 1423). See http://www.rfc-editor
.org/rfc.html

[5] RfC1424: Privacy Enhancement for Internet Electronic Mail: Part IV: Key Certification and Related Services (RFC 1424). See http://www.rfc-editor.org/rfc.html

PENETRATION TESTING

Penetration testing is part of a security assessment (e.g., Audit) or certification process (e.g., Common Criteria) with as objective to locate and eliminate security vulnerabilities that could be exploited to gain access to the *security target* (system, device or module) by a potential attacker. In this context, the objective of an attacker is to gain access to the security target by breaking or circumventing its security measures. In other words, an attacker needs only to find one vulnerability to successfully penetrate the security target, while the penetrations testers' objective task is to identify all vulnerabilities. Penetration test constraints as well as constraints for an attacker are time, money, amount of effort and resources. Given these constraints (e.g., resources and time), exploiting a vulnerability and successfully penetrating a security target might be practically infeasible even though it is theoretically possible. Hence, finding a security vulnerability does not imply that it always can be exploited easily. Therefore, the penetration tester's objective and task is much broader and labor intensive than that of an attacker.

When a successful penetration in the security target is possible, then this provides evidence that the current security measures are inadequate and need to be strengthened. In addition, when the penetration went unnoticed by the security target, it shows that the intrusion detection mechanisms do not provide an adequate level of security assurance. On the other hand, when a penetration test is not successful, it does not provide evidence that the security target is secure. In the latter case, the penetration test provides evidence that under the given test conditions, the security target did not show obvious exploitable security flaws.

A penetration test assumes the presence of a *security boundary*, as the test is aimed at penetrating it. The security boundary separates the internal components of the security target from the external components (e.g., outside world), and this separation is enforced by various security measures. In general, a security target is composed of hardware, software and human components. Depending on the human component,

security assumptions are made and a penetration test might also include human vulnerabilities (e.g., social engineering).

In general, a penetration test assesses the state of the security target at a given point in time, in contrast to penetration tests based on monitoring techniques (e.g., intrusion detection). Hence, any change in the internal and external environments might lead to new vulnerabilities which would require the execution of a new penetration test. Therefore, in practice certain types of penetration tests based on monitoring techniques analyze security target events on a regular interval or even in real-time, and can search for anomalies that might indicate exploitable vulnerabilities. It should be noted that these penetration tests profiles need to be updated on a regular basis.

A penetration test can be passive (observing) and active (interact). The passive penetration test infers with the existence of vulnerability of the security target in a non-intrusive manner. The *passive penetration test* probes the target boundary and scans for security weaknesses in the target environment, without aiming at gaining access and without adversely affecting its normal operation. *Active penetration testing*, on the other hand, includes tests that are more intrusive by nature. The active penetration test (see also physical attacks) exploits security flaws inherently present in the technology of the target environment, design faults or flaws in the object's parameter configuration.

There are two extreme kinds of penetration tests *Zero-Knowledge Penetration Test (ZKPT)* and *Full-Knowledge Penetration Test (FKPT)*. A ZKPT is based on a black box approach, which assumes zero knowledge or assumptions about the security target. The objective of a ZKPT is to identify how much information about the security target can be gained (i.e., is leaked), how the security target can be modeled, and what vulnerabilities can be identified that could be exploited by potential attackers. A FKPT is based on Kerckhoff's principle [1] (see Maxims), and assumes full knowledge of the internal and external components except from the critical security parameters. Based on this knowledge, the security target is tested for security vulnerabilities that could be exploited by potential attackers. In contrast to the FKPT, ZKPT has the advantage that the tester is not tempted to be biased by the security target design specifications and documentation, and will also explore areas beyond its design. Therefore, ZKPT is often the first phase of a penetration testing followed by an FKPT. Between ZKPT and FKPT several hybrid forms exist, which are optimized based on the security target and the testing constraints.

Tom Caddy

Reference

[1] Kerckhoff, Auguste (1883). "La cryptographie militaire." *Journal des Sciences Militarires*, 9, 5–38. http://www.cl.cam.ac.uk/users/fapp2/kerckhoffs/

PERFECT FORWARD SECRECY

Perfect forward secrecy (PFS for short) refers to the property of key-exchange protocols in which the exposure of long-term keying material, used in the protocol to negotiate session keys, does not compromise the secrecy of session keys established before the exposure. The most common way to achieve PFS in a key-exchange protocol is by using the Diffie–Hellman key agreement with ephemeral exponents to establish the value of a session key, while confining the use of the long-term keys (such as private signature keys) to the purpose of authenticating the exchange (see authentication). In this case, once a session key is no longer used and is erased from memory then there is no way for the attacker to find this key except by cryptanalyzing the Diffie–Hellman exchange (or other applications that used the session key). In particular, finding the long-term authentication key is of no use in learning the session-key value. One essential element for achieving PFS with the Diffie–Hellman exchange is the use of *ephemeral* exponents which are erased from memory as soon as the exchange is complete. (This should include the erasure of any other information from which the value of these exponents can be derived such as the state of a pseudo-random generator used to compute these exponents.)

The PFS property of authenticated Diffie–Hellman exchanges can be highlighted by contrasting it with other forms of key exchange, such as key-transport protocols, where session keys are transmitted between the peers in the exchange encrypted under long-term public keys (see public key cryptography). In this case, the exposure of the long term secret decryption key will compromise the secrecy of *all* session keys (including those erased from memory) that were exchanged under the corresponding public encryption key. This is a major security threat which, in particular, makes the long-term key extremely attractive for attack.

The property of perfect forward secrecy is especially relevant to scenarios in which the exchanged session keys require secrecy protection beyond their lifetime, such as in the case of session keys used for data encryption. In contrast, applications that use the shared keys only for the sake of authentication may not need PFS in their key-exchange protocol (in most cases, finding an expired shared authentication key is of no value for the attacker).

In principle, any public key encryption scheme can be used to build a key exchange with PFS by using the encryption scheme with ephemeral public and private keys. From a practical point of view this requires that the key generation for the encryption scheme be fast enough. For most applications this disqualifies, for example, the use of ephemeral RSA public key encryption for achieving PFS, since the latter requires the generation of two long prime numbers for each exchange, a relatively costly operation. Currently, most systems that provide PFS are based on the Diffie–Hellman key agreement. These protocols can be found in practice (e.g., in the IKE key-exchange protocol [4] for the IPsec standard) and they have been widely studied in many research papers. Some of the well-known key-exchange protocols that provide PFS are STS [2], ISO-9798 [5], EKE [1], SKEME [6], MQVC [8], and SIGMA [7]. The term "perfect forward secrecy" was first introduced in [3].

Hugo Krawczyk

References

[1] Bellovin, S.M. and M. Merritt (1992). "Encrypted key exchange: Password-based protocols secure against dictionary attacks." *Proceedings of the IEEE Computer Society Symposium on Research in Security and Privacy*, May, 72–84.

[2] Diffie, W., P.C. van Oorschot, and M. Wiener (1992). "Authentication and authenticated key exchanges." *Designs, Codes and Cryptography*, 2, 107–125.

[3] Günther, C.G. (1990). "An identity-based key-exchange protocol." *Advances in Cryptology—Eurocrypt'89*, Lecture Notes in Computer Science, vol. 434, eds. J.-J. Quisquater and J. Vandewalle. Springer-Verlag, Berlin, 29–37.

[4] Harkins, D., and D. Carrel (ed.) (1998). "The Internet Key Exchange (IKE)." *RFC 2409*, November.

[5] ISO/IEC IS 9798-3 (1993). "Entity authentication mechanisms—Part 3: Entity authentication using asymmetric techniques."

[6] Krawczyk, H. (1996). "SKEME: A versatile secure key exchange mechanism for internet." *Proceedings of the 1996 Internet Society Symposium on Network and Distributed System Security*, February, 114–127.

[7] Krawczyk, H. (2003). "SIGMA: The 'SIGn-and-MAc' approach to authenticated Diffie–Hellman and its use in the IKE protocols." *Advances in Cryptology—CRYPTO 2003*, Lecture Notes in Computer Science, vol. 2729, ed. D. Boneh. Springer-Verlag, Berlin, 399–424.

[8] Law, L., A. Menezes, M. Qu, J. Solinas, and S. Vanstone (2003). "An efficient protocol for authenticated key agreement." *Designs, Codes and Cryptography*, 28, 211–223.

PERSONAL IDENTIFICATION NUMBER (PIN)

A *Personal Identification Number (PIN)* is a relatively short (typically 4–8 digits) numeric string that is used as a password to authenticate a user to a device such as a smart card, an Automated Teller Machine (ATM), or a mobile phone. Standards addressing the management and security of PINs include ANSI X9.8 and ISO 9564.

Carlisle Adams

Reference

[1] Menezes, A., P.C. van Oorschot, and S. Vanstone (1997). *Handbook of Applied Cryptography*. CRC Press, Boca Raton, FL.

PHYSICAL ATTACKS

The term "physical attacks" has two quite different meanings in the field of IT security. The first one describes mechanisms to physically penetrating a rather large perimeter, e.g., overcoming an access control system for a server room. The related techniques, for instance picking locks and bridging fences, are outside the scope of the Encyclopedia. Technologies used for perimeter security involve, for instance, intrusion detection sensors and alarm systems.

In the context of cryptographic implementations, "physical attack" is understood as a term which encompasses all attacks based on physical means against cryptographic devices. Physical attacks are of relevance if an adversary gains physical access to the cryptographic device or its near-by environment, e.g., a smart card.

There are two different objectives which have to be regarded in order to counter physical attacks. The first one aims to prevent the disclosure and/or modification of the internal data (e.g., cryptographic keys and application data). For its realization, tamper resistant and tamper response measures are implemented. Another—weaker—approach focuses on the question of whether or not a cryptographic module has been tampered with. For this, tamper detection characteristics are needed. Note that tamper evidence neither prevent the breaking into the cryptographic boundary nor the disclosure of internal data of the cryptographic module.

The term "cryptographic boundary" [2] defines the physical bounds of the cryptographic device that encloses all relevant security components (hardware and software). There are typically external interfaces to the cryptographic boundary, for instance, lines for data communication and power supply. These lines are generally untrusted as they are controlled externally.

In [1], five attack scenarios which indicate the main areas of physical attacks are defined: penetration, monitoring, manipulation, modification, and substitution.

Penetration: Penetration is an active, invasive attack against the cryptographic module. This includes a breaking into the cryptographic boundary of the module. The aim is to intercept data at the internal communication lines or to read out the memory in order to determine the secret keys stored inside the security module.

Monitoring: Monitoring is a passive, noninvasive attack that leaves the cryptographic boundary intact. This class of attacks makes use of the inherent leakage of the cryptographic module, e.g., by measuring the electromagnetic emanation. TEMPEST investigations and side channel analysis are prominent passive attacks based on monitoring.

Manipulation: Manipulation is a noninvasive attack that leaves the cryptographic boundary intact. The attacks aim to obtain a service in an unintended manner [1], mainly at the logical interface. Manipulating attacks may also include changed environmental conditions. For instance, the cryptographic module might be operated under extreme operating conditions which includes the power supply and the environmental temperature. Noninvasive Fault Attacks belong to this category.

Modification: Modification is an active, invasive attack that includes breaking into the cryptographic boundary of the module. Unlike penetration attacks, the aim is to modify internal connections or the internal memories used.

Substitution: Substitution includes the removal of the cryptographic module which is then substituted by an emulating device with a modified implementation of security functions. The cryptographic boundary is not of primary interest in this attack. Note that the removed module can be used for a comprehensive analysis of the internal construction.

Examples for security requirements include smart card IC protection profiles used by Common Criteria evaluations [3, 4] and the FIPS 140 security requirements [2]. The latter provides information about the implementation of secure computer systems for the use in unprotected areas.

Kerstin Lemke
Christof Paar

References

[1] ISO 13491-1 (1998). *Banking—Secure cryptographic devices (retail), Part 1: Concepts, requirements and evaluation methods* (1st ed.).
[2] FIPS PUB 140-2 (2002). "Security requirements for cryptographic modules." National Institut of Standards and Technology, available at http://csrc.nist.gov/cryptval/
[3] BSI-PP-0002. "Smartcard IC platform protection profile, v1.0." Available at http://www.bsi.bund.de/cc/pplist/ssvgpp01.pdf
[4] PP/9806. "Smartcard integrated circuit protection profile v2.0." Available at http://www.ssi.gouv.fr/site_documents/PP/PP9806.pdf

PKCS

PKCS, the Public-Key Cryptography Standards, are specifications produced by RSA Laboratories in cooperation with interested developers worldwide in order to promote PKI solutions. The PKCS series are referenced in many formal and de facto standards, including ANSI X9, PKIX, SET, S/MIME, and SSL.

The PKCS series is under constant development. Currently, the list comprises:

PKCS #1:   RSA Cryptography Standard (includes former #2 and 4)
PKCS #3:   Diffie–Hellman Key Agreement Standard
PKCS #5:   Password-Based Cryptography Standard
PKCS #6:   Extended-Certificate Syntax Standard
PKCS #7:   Cryptographic Message Syntax Standard
PKCS #8:   Private-Key Information Syntax Standard
PKCS #9:   Selected Attribute Types
PKCS #10: Certification Request Syntax Standard
PKCS #11: Cryptographic Token Interface Standard
PKCS #12: Personal Information Exchange Syntax Standard
PKCS #13: Elliptic Curve Cryptography Standard
PKCS #15: Cryptographic Token Information Format Standard

In addition, there are guidelines for contributions to the PKCS series.

Peter Landrock

PKIX—PUBLIC KEY INFRASTRUCTURE (X.509)

PKIX is a working group under IETF, Internet Engineering Task Force, established in 1995 [1]. The charter of the PKIX working group defines the scope and goals of the working group: to develop Internet standards needed to support an X.509-based PKI. This includes profiling ITU PKI standards and developing new standards related to the use of X.509 certificates on Internet. Most results of the working group are published as Request For Comments (RFC).

X.509 version 3 certificates and version 2 of certificate revocation lists as defined by ITU permit a number of standard extensions as well as privately defined extensions. RFC2459 and RFC3280 profile such certificates and certificate revocation lists by recommending usage of standard extensions and also defines a few new extensions. Other standards profile the use of attribute certificates and qualified certificates, while a schema to support PKIX in LDAP is defined in RFC2559.

An informational standard (RFC2527) describes a framework for the definition of certificate policies and *certificate practice statements*.

A number of protocols related to the use of X.509 certificates on Internet have been defined. Central to the management of certificates is RFC2511, which defines a format for certificate requests. Most notably these requests include the possibility to prove possession of the private key corresponding to the public key to be certified and they provide means for authenticating the requester. These certificate requests are used in RFC2510 and RFC2797, which define management protocols for a number of PKI services including issuing, updating and revoking certificates. The protocols in RFC2510 have a number of options and the standard gives a profile for conforming implementations. RFC2797 gives alternative management protocols based on certificate management syntax (and hence much the same security syntax as in S/MIME).

The Online Certificate Status Protocol (OCSP) defined in RFC2560 provides a mechanism for getting timely information about the status of a certificate. This may not be possible using regularly published certificate revocation lists, as a newly revoked certificate may not be listed in the revocation list held by the relying party. In OCSP a request for the status of a number of certificates is answered with a signed message containing the status of these certificates.

RFC3161 defines a protocol for requesting and getting a secure timestamp. Such timestamps may be used as part of a non-repudiation mechanism by timestamping a signed message, but they can also be used in many other applications—including some that are not related to the use of X.509 certificates. A request for a timestamp does not identify the requester and a timestamp on a particular message is obtained without revealing the contents of the message—only the hash value of the message (see Collision-Resistant Hash Function) is revealed to the time stamping authority. Basically, the timestamp is a signed structure containing the hash value of the message and the time. The timestamp also contains policy information indicating the applicability of the timestamp. If necessary more information can be placed in timestamps using the extension mechanism also used in X.509 certificates.

Torben Pedersen

Reference

[1] http://www.ietf.org/

PLAYFAIR CIPHER

This is digraphic, bipartite *substitution* (see substitutions and permutations) $V_{25} \times V_{25} \to V_{25} \times V_{25}$, using a 5×5 Polybios square and a 'crossing step' $(\alpha\beta)(\gamma\delta) \mapsto (\alpha\delta)(\gamma\beta)$. In the example below, the key Palmerstone has been used to make the Polybios square (see mixed, alphabet).

	1	2	3	4	5
1	P	A	L	M	E
2	R	S	T	O	N
3	B	C	D	F	G
4	H	I	K	Q	U
5	V	W	X	Y	Z

ag $\mapsto$ (12)(35) $\mapsto$ (15)(32)

$\mapsto$ EC

If a bigram is found in the same row (or the same column), the 'crossing step' degenerates: it takes the cyclically right neighbor: am $\mapsto$ LE; ae $\mapsto$ LP; aa $\mapsto$ LL (or the neighbor cyclically below: dl $\mapsto$ KT;

dx $\mapsto$ KL). Modified rules are common, especially the one concerning the doubles: perhaps one letter of the pair will be omitted or replaced with a null; in cases like 'less seven' this will lead to encrypt 'le s& s& se ve n&', where & is a null.

A modified Playfair uses two Polybios squares, one for the first and one for the second character of the bigrams, e.g.,

```
A Y K I H     Y X U H A
L B M N P     T R K B I
Q R C O G     P M C G S     bu ↦ KY
Z X V D S     F D L Q V
F W U T E     E N O W Z
```

In case a bigram is found in the same row or column, rules similar to the ones above are to be applied.

Friedrich L. Bauer

Reference

[1] Bauer, F.L. (1997). "Decrypted secrets." *Methods and Maxims of Cryptology*. Springer-Verlag, Berlin.

PMAC

PMAC is a MAC algorithm designed by Black and Rogaway [1] in 2002. A MAC algorithm is a cryptographic algorithm that computes a complex function of a data string and a secret key; the resulting MAC value is typically appended to the string to protect its authenticity. PMAC is a deterministic MAC algorithm based on a block cipher. In contrast to CBC-MAC, PMAC is fully parallelizable, which means that up to t processors can be used in parallel, where t is the number of n-bit blocks of the message. For serial computations, PMAC is about 8% slower than CBC-MAC. The computation of PMAC does not require the knowledge in advance of the message length.

In the following, the block length and key length of the block cipher will be denoted with n and k, respectively. The encryption with the block cipher E using the key K will be denoted with $E_K(\cdot)$. An n-bit string consisting of zeroes will be denoted with 0^n. The length of a string x in bits is denoted with $|x|$. If $i \geq 1$ is an integer, $\text{ntz}(i)$ is the number of trailing 0-bits in the binary representation of i. For a string x with $|x| < n$, $\text{pad}_n(x)$ is the string of length n obtained by appending to the right a '1' bit followed by $n - |x| - 1$ '0' bits. Considered the finite field $GF(2^n)$ defined using the irreducible polynomial $p_n(x)$; here $p_n(x)$ is the lexicographically first polynomial, chosen among the irreducible polynomials of degree n that have a minimum number of non-zero coefficients. For $n = 128$, $p_n(x) = x^{128} + x^7 + x^2 + x + 1$. Denote the string consisting of the rightmost n coefficients (corresponding to x^{n-1} through the constant term) of $p_n(x)$ with $\tilde{p}_n$; for example $\tilde{p}_{128} = 0^{120}10000111$. The operation multx (s) on an n-bit string considers s as an element in the finite field $GF(2^n)$ and multiplies it by the element corresponding to the monomial x in $GF(2^n)$. It can be computed as follows, where s_{n-1} denotes the leftmost bit of s, and $\ll$ denotes a left shift operation.

$$\text{multx}(s) = \begin{cases} s \ll 1 & \text{if } s_{n-1} = 0 \\ (s \ll 1) \oplus \tilde{p}_n & \text{if } s_{n-1} = 1. \end{cases}$$

Similarly, the operation invx(s) multiplies the string s by the element corresponding to the monomial x^{-1} in $GF(2^n)$. It can be computed as follows, where $\gg$ denotes a right shift operation:

$$\text{invx}(s) = \begin{cases} s \gg 1 & \text{if } s_{n-1} = 0 \\ (s \gg 1) \oplus \tilde{p}_n & \text{if } s_{n-1} = 1. \end{cases}$$

The first step of a PMAC computation precomputes some key-dependent constants. Define $\mu \leftarrow \lceil \log_2 t_{\max} \rceil$ where $t_{\max}$ is the maximum number of n-bit blocks in the message. For a key K, one computes $L \leftarrow E_K(0^n)$ and sets $L(0) \leftarrow L$. Next one defines $L(i) \leftarrow \text{multx}(L(i-1))$ for $1 \leq i \leq \mu$ and $L(-1) \leftarrow \text{invx}(L)$.

PMAC can now be defined as follows. Set $t \leftarrow \lceil |x|/n \rceil$; in the special case that t equals 0 set $t \leftarrow 1$. Split the input x (of maximum length $n2^n$ bits) into t n-bit blocks $x_1, x_2, \ldots, x_t$, where the last block may be shorter than n bits. Set $\Delta \leftarrow 0^n$ and $\Sigma \leftarrow 0^n$. Now compute for $1 \leq i \leq t-1$:

$$\Delta \leftarrow \Delta \oplus L(\text{ntz}(i))$$
$$H_i \leftarrow E_K(x_i \oplus \Delta)$$
$$\Sigma \leftarrow \Sigma \oplus H_i.$$

As a final step, set $\Sigma \leftarrow \Sigma \oplus \text{pad}_n(x_t)$. Finally

$$H_t = \begin{cases} E_K(\Sigma) & \text{if } |x_t| < n \\ E_K(\Sigma \oplus L(-1)) & \text{if } |x_t| = n. \end{cases}$$

The PMAC value consists of the leftmost m bits of H_t.

The designers of PMAC have proved that PMAC is a secure MAC algorithm if the underlying block cipher is a pseudo-random permutation; the security bounds are meaningful if the total number of input blocks is significantly smaller than $2^{n/2}$. The best attack known on PMAC is a forgery attack based on internal collisions [2]. In early 2004, PMAC has not yet been included in any standards.

B. Preneel

References

[1] Black, J. and P. Rogaway (2000). "CBC-MACs for arbitrary length messages." *Advances in Cryptology—CRYPTO 2000*, Lecture Notes in Computer Science, vol. 1880, ed. M. Bellare. Spring-Verlag, Berlin, 197–215.

[2] Preneel, B. and P.C. van Oorschot (1995). "MDx-MAC and building fast MACs from hash functions." *Advances in Cryptology—CRYPTO'95*, Lecture Notes in Computer Science, vol. 963, ed. D. Coppersmith. Spring-Verlag, Berlin, 1–14.

POLICY

Policy describes how environments dictate the behavior of applications and services. Security policies specify how relevant conditions mandate how, when, and/or to whom access to controlled resources is given. For example, the access rights defined by a UNIX file-system state a policy: read, write, and execute bits define the kinds of operations a user is permitted perform on the files or directories. The policy is *evaluated* by determining whether the user has sufficient permissions to perform the operation at the point at which access is attempted. The policy is *enforced* by the file-system by allowing or preventing the operation.

Security policy has historically been divided into two broad classes: provisioning policy and authorization (or access control) policy. Provisioning policies define how software is configured to meet the requirements of the local environment. As is illustrated by the UNIX file-system policy, authorization policies map entities and resources onto allowable action. The following considers these broad classes in serial.

One can view any kind of software configuration as provisioning policy. That is, any aspect of software behavior configured at run-time is policy. This is relevant to the current discussion where configuration affects how security is provided. For example, one of the central goals of the ssh remote access utility [1] is to provide confidentiality over the session. The ssh policy states which cryptographic algorithm (e.g., Triple DES, Blowfish) should be used to encrypt session traffic (e.g., as specified by host-local configuration file). The policy (i.e., encryption algorithm) dictates how the goal (i.e., confidentiality) is achieved. In general, the degree to which a user or administrator can influence security is largely dictated by the scope of the system's provisioning policies.

General-purpose policy management services support the creation, storage, and enforcement of policy. Note that while these services can be used

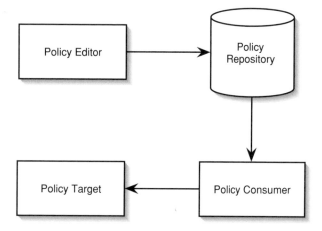

Fig. 1. IETF Policy Framework Working Group architecture—architecture supporting the creation, distribution and enforcement of network management policies

to manage <u>authorization policy</u>, they have historically been targeted to provisioning policy. The IETF Policy Framework Working Group (PWG) has developed a widely adopted reference architecture and lexicon for policy management [2]. This framework defines a collection of components and operations used to manage the network (commonly referred to as *policy-based networking*). Depicted in Figure 1, the architecture defines four logical policy components; *policy editors*, *policy repositories*, *policy consumers*, and *policy targets*.

A policy editor provides interfaces for specifying and validating policy specifications. The policy editor is responsible for detecting (and potentially resolving) inconsistencies in the specification. For example, an ssh policy that mandates both *DES* (see <u>Data Encryption Standard</u>) and <u>Triple DES</u> be used for confidentiality is erroneous (only one algorithm should be specified). Such a policy would be flagged as errored, and where available, corrected using a resolution algorithm. The policy editor delivers validated policies to policy repositories as dictated by the environment. A policy repository stores the policies to be used by an administrative domains. The policy repository does not act on or interpret policy.

A policy consumer translates policy into action. The consumer acquires and evaluates the policy relevant to the current environment. The resulting action is communicated to the set of policy targets. Targets enforce policy by performing the actions that implement the defined semantics. For example, again consider the ssh policy. A consumer would interpret policy to determine which algorithm is to be used for confidentiality. ssh would then act as a policy target by configuring the ssh client and subsequently using it to encrypt the traffic (and thus enforce confidentiality).

Authorization policy describes to whom and under what circumstances access to resources is granted. These policies are further defined by an <u>authentication</u> policy and an <u>*access control*</u> policy. The authentication policy states how the identity of the requesting entity must be established. For example, an authentication policy for a UNIX system is the password: the user must provide the appropriate password at the login prompt to be allowed access to the system. How *authentication* is performed is largely defined by environment needs, and outside the scope of this section.

An access control policy maps an identity established during authentication and other information to a set of rights. Rights, often called permissions, defines the types of operations that can be granted, e.g., read, write, and execute on a UNIX file system. The structure and meaning of these policies are defined by their *access control model*. (There are many models, we choose to focus on one.)

One popular model is the *role based* access control model. In this model, the policy defines collections of permissions called *roles* [3]. Users assume roles as they perform different tasks within the system. Hence, the set of rights is strictly defined by the set of rights allowed to the roles they have assumed.

The following example illustrates the use of role-based access control policy. Consider a simplified sealed bid online auction application. Three entities participate in this application: an auctioneer, a bidder, and a seller. The bidder *bids* for goods *sold* by the seller. Once all bids are placed, the auctioneer *clears* the auction by opening the bids and declaring the highest bidder the winner. Once a winner is declared, interested parties can *view* the result. The access control policy for the online auction defines four permissions (described above), bid, sell, clear, and view, and three roles, seller, buyer, and auctioneer. Table 1 describes a policy that assigns the permissions to roles.

The auction policy is enforced at run-time by evaluating the permissions associated with the roles that they have assumed. In any role-based

Table 1. Example role-based policy for online auction application

	Permissions			
Role	Sell	Bid	Clear	View
Seller	Yes	No	No	Yes
Buyer	No	Yes	No	Yes
Auctioneer	No	No	Yes	Yes

system, a user assumes roles through a explicit or implicit software process. The example online auction maps user accounts to roles. For example, the *sally* account is mapped to the seller role. When a user logs into the application as *sally*, she automatically (implicitly) assumes the *seller* role. From that point on, she is free to perform any action allowed by that role. Other accounts are similarly mapped to the *buyer* and *auctioneer* roles, and governed by the associated policy.

One might ask why we do not just assign a user the union of all rights assigned to each role they may assume. Firstly, there may be legitimate reasons that a user be explicitly prohibited from simultaneously assuming more than one role. For example, in the above example, it may be important that a user be prevented from assuming both auctioneer and bidder roles. If a user were to assume both roles, it could "cheat" by viewing all bids before placing its own.

The second reason role-based models are useful is convenience. Roles provide a powerful abstraction for dealing with the potentially many rights that must be managed. In separating the access control from the user, one simplifies the process of evaluating. Moreover, this eliminates the need to alter access control policy each time a entities position changes.

Many access control models have been defined, studied, and ultimately used in real environments. Each model defines a unique way of viewing the relations between protected artifacts, actions, and the entities that manipulate them. Because applications and environments view these concepts in vastly different ways, it is unlikely that any one model will be universally applicable. Hence, like the use policy itself, the selection of a model is a function of taste and system requirements.

Trust Management (TM) systems blur the lines between authorization and provisioning policy. TM policies, frequently called <u>credentials</u>, define statements of trust. More specifically, they state that an entity is trusted to access a particular resource under a specified set of conditions. An entity supplies the appropriate set of credentials when accessing a protected resource.[1] If the supplied credentials are authentic and consistent with a *local policy*, the entity is permitted access.

Note that the TM policy alters the traditional policy flow: each entity accessing the resource supplies those policies that are needed to perform an action. This eliminates the need for the policy

---

[1] Policies can define access to actions, rather than resource. However, the evaluation of such policies relating to action is identical to resource-oriented policy.

```
KeyNote-Version: "2"
Authorizer: "DSA : 4401ff92"    # the Alice CA
Licensees: "DSA : abc991"       # Bob DSA key
Conditions: ((app_domain == "ssh") && (crypt == "3DES"));
Signature: "DSA-SHA1 : 8912aa"
```

Fig. 2. KeyNote credential—specifies a policy stating that Alice allows Bob to connect via SSH if 3DES encryption is used

consumer to discover and acquire policy. It is incumbent upon the user to supply the set of credentials that allow access. Hence, TM systems enable policy in creating widely distributed systems.

Provisioning policy is specified in TM systems through the access criteria. Hence, TM systems do not specify provisioning directly, but mandate how the environment provisioning must be provisioned to allow access. This model of policy again departs from traditional uses: the policy infrastructure passively assesses whether the environment is correctly provisioned, rather than actively provisioning it.

The KeyNote system [4] provides a standardized language for expressing trust management policies. KeyNote credentials have three cryptographic components: an *authorizer*, a *licensee*, and a *signature*. The authorizer identifies the authority issuing the policy. KeyNote authorities are cryptographic keys. The signature creates a cryptographically strong (and verifiable) link between the policy and the authorizer. The licensee is to the entity to which the policy refers, e.g., the entity to be allowed access.

KeyNote credentials express a *says* relation: an authority says some entity some aspect has some rights under a set of conditions. To illustrate, the KeyNote credential defined in Figure 2 expresses a policy governing `ssh` access. In this credential, the authorizer authority states that an licensee entity has the right to access the host only if <u>Triple DES</u> is used to implement confidentiality. Note again that both the authorizer and licensee are not really entities, but keys. Hence, any entity that has access to the private key of the licensee, can use this credential to gain access to the host. Moreover, any entity with access to the private key of the authorizer can create and sign such credentials.

On first viewing, the <u>credential</u> in Figure 2 may appear to be ambiguous. The credential does specify which hosts are to be governed. This ambiguity is the source of much of the power of trust management. This credential is only accepted by those systems which have a *local policy* that (directly or indirectly) states that the authorizer has dominion over `ssh` access.

Note that the local policy need not specifically identify the authority in the above credential. The local policy can simply state the authorities it trusts to make policy decisions. Further credentials (typically supplied by the user) that express the delegation by the trusted authorities over `ssh` access are used to construct a delegation chain. Hence, access is granted where a chain of credentials beginning at the local policy and terminating at the access granting credential can be found.

Computing environments are becoming more fluid. Increasing requirements for software systems to be more open, and at the same time, more secure place unique demands on the supporting infrastructure. New environments and changing threats mandate that the kinds of security provided be reexamined and reconfigured constantly. Consequently, systems need to be more flexible and adaptive. Policy is increasingly used as the means by which correct behavior is defined.

Policy is not a panacea. While significant strides have been made in the construction, distribution and use of policy, many areas require further exploration. For example, we know little about the security implications of enforcing several policies (i.e., composing policies) within the same domain. Future policy systems need to find techniques that identify and mitigate interactions between policies. A more systemic area of investigation is scale: how do we deploy policy systems that are feasible in networks the size of the Internet. It is the answers to these questions, rather than the specifics of policy construction, that will determine the degree to which policy systems will be adopted in the future.

Patrick McDaniel

References

[1] Ylonen, Tatu (1996). "SSH—secure login connections over the internet." *Proceedings of 6th USENIX UNIX Security Symposium*. USENIX Association, San Jose, CA, 37–42.
[2] Internet Engineering Task Force (2002). Policy Framework Working Group. http://www.ietf.org/html.charters/policy-charter.html
[3] Sandhu, Ravi S., Edward J. Coyne, Hal L. Feinstein, and Charles E. Youman (1996). "Role-based access control models." *IEEE Computer*, 29 (2), 38–47.
[4] Blaze, M., J. Feignbaum, J. Ioannidis, and A. Keromytis (1999). "The KeyNote trust management system—version 2." *Internet Engineering Task Force*, RFC 2704.

POLYBIOS SQUARE ENCRYPTION

This is a monographic, bipartite *substitution* (see <u>substitutions and permutations</u>) $V_{25} \longrightarrow W_5^2$, known in antiquity (Polybios) for the Greek alphabet. In a modern form, a standard alphabet Z_{25} (left) or a mixed <u>alphabet</u> V_{25} (right) is inscribed into a 5×5 checkerboard (Polybios square), like

	1	2	3	4	5			1	2	3	4	5
1	a	b	c	d	e		1	p	a	l	m	e
2	f	g	h	i	k	or	2	r	s	t	o	n
3	l	m	n	o	p		3	b	c	d	f	g
4	q	r	s	t	u		4	h	i	k	q	u
5	v	w	x	y	z		5	v	w	x	y	z

An encryption example is given by: heaven $\mapsto$ 23151151121533 resp. heaven $\mapsto$ 411512511525

Friedrich L. Bauer

Reference

[1] Bauer, F.L. (1997). "Decrypted secrets." *Methods and Maxims of Cryptology*. Springer-Verlag, Berlin.

POLYNOMIAL TIME

A *polynomial-time* algorithm is one whose running time grows as a polynomial function of the size of its input. Let x denote the length of the input to the algorithm (typically in bits, but other measures are sometimes used). Let $T(x)$ denote the running time of the algorithm on inputs of length x. Then the algorithm is polynomial-time if the running time satisfies

$$T(x) \leq c_d x^d + c_{d-1} x^{d-1} + \cdots + c_1 x + c_0$$

for all sufficiently large x, where the degree $d \geq 0$ and the coefficients $c_d, \ldots, c_0$ are constants. It is easy to see that the running time will also satisfy the simpler bound

$$T(x) \leq c_d' x^d$$

for a (possibly) larger constant c_d', and (possibly) larger x. In <u>O-notation</u>, this would be written $T(x) = O(x^d)$.

The term "polynomial" comes from the fact that the bound has same general form as a polynomial in x. The definition of polynomial-time is more general, however; in particular, it does not require that d be an integer.

The definition just given is somewhat informal; for instance, it assumes that the algorithm takes the same time for all inputs of a given size. If the time varies among inputs of a given size, then $T(x)$ should be an upper bound on the running time for that size. Moreover, if the algorithm involves randomization, then $T(x)$ should be an upper bound on the expected running time of the algorithm (given random choices of the algorithm) for inputs of size x.

As noted above, x refers to the length of the input. However, sometimes it is convenient just to let x be the input itself, as for example when expressing the time it takes to find a prime of length x bits, where x itself is the input to the prime-finding algorithm. (Alternatively, as is standard in complexity theory, one could redefine the input to the prime-finding algorithm as a string of x ones.)

The distinction between polynomial-time algorithms and those with higher complexity such as subexponential time and exponential time is a major concern in computational complexity theory, but in practical cryptography, the primary issue is how fast an algorithm runs for practical-size inputs. For instance, while a polynomial-time algorithm will be faster for all sufficiently large inputs than one that is not polynomial-time, but the crossover-point may be well beyond practical sizes. Likewise, the existence of a polynomial-time algorithm to solve some supposedly hard problem (e.g., integer factoring) would not necessarily compromise security, if the polynomial-time algorithm turned out to have a very large degree d. The search for polynomial-time algorithms is nevertheless an important one in cryptography, as it generally provides for scalability in a system design, since key sizes can be increased with only a moderate effect on running time. Likewise, the lack of polynomial-time algorithms for hard problems provides assurance that a moderate increase in key sizes can have a significant effect on security, which helps to stay ahead of improvements in computing power (see Moore's Law).

A polynomial-time algorithm technically belongs to the higher complexity classes, since each complexity class only expresses an upper bound on the running time, not a lower bound. However, in typical discussions on cryptography (see, e.g., L-notation), the term "subexponential time" excludes algorithms that are polynomial time, and likewise "exponential time" excludes subexponential time and polynomial time.

In addition to running time, the "polynomial" metric is sometimes applied to other measures of complexity in cryptography, such as the memory requirements of an algorithm or the ratio of difficulty between two problems. In general, a *polynomial function*, in this context, is one that is bounded asymptotically by a constant power of its input. In these cases, one may also consider degrees $d < 1$ (which would does not make sense for an algorithm, since it takes $O(x)$ time just to *read* the input). A function that is $O(\sqrt{x})$, or that is $O(1)$—a constant—is technically polynomial in x. If $d < 1$ then the function of x is sometimes called a *small polynomial*. If the function grows more slowly than x^d for every degree $d > 0$, then the function is called *subpolynomial*. An example of a subpolynomial function is $O(\log x)$.

Burt Kaliski

PORTA ENCRYPTION

This particular encryption method works on an alphabet of $N = 2 \cdot \nu$ letters. A Porta encryption step (Giambattista della Porta, 1563) is a simple *substitution* (see substitutions and permutations) consisting of ν swaps (cycles of length 2). Typically, there are ν such swaps, each one designated by *two* key letters in a polyphonic way. A Porta encryption is self-reciprocal. For an example of a PORTA encryption see the section "autokey" in the entry key.

A *Porta table* for Z_{20} (G.B. della Porta and M. Argenti, 1589) is given by:

A B	(am) (bn) (co) (dp) (eq) (fr) (gs) (ht) (iu) (lx)
C D	(ax) (bm) (cn) (do) (ep) (fq) (gr) (hs) (it) (lu)
E F	(au) (bx) (cm) (dn) (eo) (fp) (gq) (hr) (is) (lt)
G H	(at) (bu) (cx) (dm) (en) (fo) (gp) (hq) (ir) (ls)
I L	(as) (bt) (cu) (dx) (em) (fn) (go) (hp) (iq) (lr)
M N	(ar) (bs) (ct) (du) (ex) (fm) (gn) (ho) (ip) (lq)
O P	(aq) (br) (cs) (dt) (eu) (fx) (gm) (hn) (io) (lp)
Q R	(ap) (bq) (cr) (ds) (et) (fu) (gx) (hm) (in) (lo)
S T	(ao) (bp) (cq) (dr) (es) (ft) (gu) (hx) (im) (ln)
U X	(an) (bo) (cp) (dq) (er) (fs) (gt) (hu) (ix) (lm)

Friedrich L. Bauer

Reference

[1] Bauer, F.L. (1997). "Decrypted secrets." *Methods and Maxims of Cryptology*. Springer-Verlag, Berlin.

PREIMAGE RESISTANCE

Preimage resistance is the property of a hash function that it is hard to invert, that is, given an element in the range of a hash function, it should be computationally infeasible to find an input that maps to that element. This property corresponds to one-wayness, which is typically used for functions with input and output domain of similar size (see one-way function). A minimal requirement for a hash function to be preimage resistant is that the length of its result should be at least 80 bits (in 2004). Preimage resistance needs to be distinguished from two other properties of hash functions: second preimage resistance and collision resistance. A hash function is said to be a one-way hash function (OWHF) if it is both preimage resistant and second preimage resistant. A natural question is to investigate how these concepts are related; it turns out that the answer is rather subtle and requires a formalization [8]. A simplification is that under certain conditions, preimage resistance is implied by the two other properties.

An informal definition is not fully satisfactory, as one can always apply the function to a small set of inputs and store the result; for this set of hash results, it is easy to find a preimage. In order to write a formal definition, one needs to specify according to which distribution the element in the range is selected (e.g., one could choose an element uniformly in the domain and apply the hash function, or one could choose an element uniformly in the range), and one needs to express the probability of finding a preimage for this element. Moreover, one often introduces a class of functions indexed by a public parameter, which is called a key. One could then distinguish between three cases: the probability can be taken over the random choice of elements in the range, over the random choice of the parameter, or over both simultaneously. As most practical hash functions have a fixed specification, the first approach is more relevant to applications.

The definition of a one-way function was given in 1976 by Diffie and Hellman [2]. Preimage resistance of hash functions has been studied in [1, 3–7, 9, 10]. For a complete formalization and a discussion of the relation between the variants and between hash functions properties, the reader is referred to Rogaway and Shrimpton [8].

B. Preneel

References

[1] Damgård, I.B. (1990). "A design principle for hash functions." *Advances in Cryptology—CRYPTO'89*, Lecture Notes in Computer Science, vol. 435, ed. G. Brassard. Springer-Verlag, Berlin, 416–427.

[2] Diffie, W. and M.E. Hellman (1976). "New directions in cryptography." *IEEE Trans. on Information Theory*, IT-22 (6), 644–654.

[3] Gibson, J.K. (1990). "Some comments on Damgård's hashing principle." *Electronics Letters*, 26 (15), 1178–1179.

[4] Merkle, R. (1979). *"Secrecy, Authentication, and Public Key Systems."* UMI Research Press.

[5] Preneel, B. (1993). "Analysis and Design of Cryptographic Hash Functions." *Doctoral Dissertation*, Katholieke Universiteit Leuven.

[6] Preneel, B. (1999). "The state of cryptographic hash functions." *Lectures on Data Security*, Lecture Notes in Computer Science, vol. 1561, ed. I. Damgård. Springer-Verlag, Berlin, 158–182.

[7] Rabin, M.O. (1978). "Digitalized signatures." *Foundations of Secure Computation*, eds. R. Lipton, R. DeMillo. Academic Press, New York, 155–166.

[8] Rogaway, P. and T. Shrimpton (2004). "Cryptographic hash function basics: Definitions, implications, and separations for preimage resistance, second-preimage resistance, and collision resistance." *Fast Software Encryption*, Lecture Notes in Computer Science, vol. 3017, eds. B.K. Roy and W. Meier. Springer-Verlag, Berlin.

[9] Stinson, D. (2001). "Some observations on the theory of cryptographic hash functions." Technical Report 2001/020, University of Waterloo.

[10] Zheng, Y., T. Matsumoto, and H. Imai (1990). "Connections between several versions of one-way hash functions." *Proceedings of SCIS90, The 1990 Symposium on Cryptography and Information Security*, Nihondaira, Japan, January 31–February 2.

PRETTY GOOD PRIVACY (PGP)

Pretty Good Privacy (PGP) is the most widely used software package for email and file protection. It was primarily developed by Philip R. Zimmermann in the early 1990s and allows to encrypt and digitally sign email messages, individual files or protect complete file systems. Over a decade, PGP advanced from a niche market product of the cryptography community to a mainstream application with easy-to-use user interface. Today, it is available on all major operating systems and integrates as a plug-in to email systems, such as Lotus Notes, Microsoft Outlook, Novell GroupWise and the Eudora mail system. PGP brings its own key management applications and key server, but also interoperates with X.509-compliant public key infrastructures (PKI) and LDAP-based directory services. The current FIPS 140-1 compliant products support Smart-Cards and PKCS#11-compliant cryptographic devices. Beside the commercial PGP, which is available in a variety of product editions with customized features, a culture of compliant, free software packages has evolved. Partly they stem

Fig. 1. Phil Zimmermann

from the original PGP source code, published as non-commercial, international versions. Others, like GNU Privacy Guard (GnuPG), have been founded on the OpenPGP standard [14] and were developed independently.

The creator of PGP, Philip Zimmermann, describes himself as a privacy and civil rights activist. He intended to create a public-key encryption software that would support every PC user to privately exchange electronic mail. The actual development of PGP was motivated by the increasing power and efficiency with which U.S. law-enforcement and other governmental agencies could analyze huge amounts of electronic communication, and the emerging U.S. Senate Bill 266, 1991, which would have forced manufacturers to add back doors to secure communication products, permitting governmental authorities access to all conveyed information in decoded form. Senate Bill 266 was later discarded after vigorous protests of libertarian and industry groups [11, 12].

HISTORY: The first version of PGP for Microsoft DOS was released in June 1991 as an Internet give-away through a friend of Zimmermann. It spread rapidly world-wide [6], but faced several problems: Firstly, the novel symmetric cipher algorithm *Bass-O-Matic* turned out to be fundamentally flawed. Moreover, the patented RSA algorithm was implemented without a valid U.S. licence. Earlier, in 1991, Zimmermann had tried to obtain a free licence from RSA Data Security, Inc., but without success, probably because PGP resembled too closely their own product 'Mailsafe'. He decided not to buy a commercial license in order to keep a door open to later distribute PGP as shareware. After all, because the software utilized strong cryptography, PGP fell under the U.S. jurisdiction of the International Traffic in Arms

Regulations (ITAR) and Arms Export Control Act (AECA). Therefore, it was illegal to export it from the U.S. to other countries, for instance, through the Internet or Bulletin Board Systems (BBS).

The U.S. Customs Bureau interviewed Zimmermann and others involved in the distribution, caused by complaint from RSA Data Security; documents relating to PGP were subpoenaed from Zimmermann. In addition, a grand jury started gathering evidence, whether to indict Zimmermann on unlawful arms-export. The case raised a controversial public debate on privacy and government-controlled cryptography politics. Zimmermann became a symbol of the civil rights and privacy movement, while facing an up to 5 years imprisonment sentence and huge legal expenses. The investigations lasted from 1993 to 1996. Eventually, the prosecution was declined, at least partially impelled by the public opposition [3, 4].

Version 2.0 released in September 1992, replaced Bass-O-Matic by the Swiss International Data Encryption Algorithm (IDEA) as its principal block cipher, which was licence free for non-commercial products. This PGP version was developed by an informal team of international programmers including Phil Zimmermann, Branko Lankester (Netherlands) and Peter Gutmann (New Zealand) and published its results outside the United States.

Zimmermann sold the exclusive rights for commercial versions to ViaCrypt, which released ViaCrypt PGP 2.4 shortly after the purchase in November 1993 [6]. ViaCrypt resolved the RSA and IDEA patent infringements. Several major and minor ViaCrypt versions leaded up to PGP 4.5.

In spring 1994, the Massachusetts Institute of Technology (MIT) issued PGP version 2.6 as freeware. This was possible because the MIT negotiated an agreement with RSA Data Security to incorporate the RSAREF toolkit, the freely redistributable reference source code of the RSA algorithm (see RSA public key encryption). As part of the deal, the software contained a signature format incompatibility toward earlier versions, which infringed upon RSA patents. Therefore, users were forced to upgrade to 2.6 or be locked out.

In June 1997, PGP 5.0 was published. The source code was completely re-written for the first time since v1.0. New features included a graphical user interface for Microsoft Windows variants and the Apple Macintosh. The collection of cryptographic algorithms was extended by the block ciphers CAST and triple DES (for session keys), the *Secure Hash Algorithm* (SHA family) and <u>RIPEMD-160</u> as additional hash functions and, maybe most importantly, by introducing <u>Diffie–Hellmann</u> <u>Digital Signature Standard</u> (DH/DSS) keys, whose patents had recently expired. One new feature was introduced as Corporate Message Recovery (CMR) or Additional Decryption Key (ADK); it permitted an additional, implicit encryption of message session keys under a corporate escrow key, firmly bound to the users certificate. Numerous critics publicly disapproved the feature as a means of antagonistic surveillance, arguing that the available group addressing feature would solve a company's legitimate interest in message recovery already. Furthermore, a serious implementation flaw was revealed (see below), which demonstrated vulnerabilities against subsequent certificate modifications that would attach functioning ADKs, but remain undetected by the software [13].

1997 ViaCrypt merged with Zimmermanns company Phils Pretty Good Software (PPGS) into PGP Inc. In December of the same year, Network Associates, Inc. (formerly McAfee Associates) acquired the PGP Inc.

Beginning with v5.0i, an official international PGP branch was established outside the US by Ståle Schumacher Ytteborg from Norway. Although cryptographic software in electronic form was still strictly export regulated under US laws, no such restrictions applied to printed material. Therefore, the full PGP source code was printed in an OCR font to a 600-page hard cover by MIT Press and shipped from the US to Europe perfectly legal. Until the US export controls loosened in 1999, all major versions up to v6.5.1i were convcycd in printed form, then scanned electronically to recover the original source code and republished internationally [10].

The product releases from v5.0 up to v7.1 added various utilities and functionality such as disk encryption, a graphical key administration tool and key server, a firewall and a virtual private network (VPN) solution and several adaptations to third-party communication software. Up through v6.5.8, NAI disclosed the program source code for public peer review. PGP Corp. continued this practice beginning with v8.0.

After Zimmermann left NAI in February 2001, NAI declared the PGP business segment for sale in July 2001. In order to rescue the PGP product suite, the PGP Corporation was founded with Zimmermann and the cryptographer Bruce Schneier as members of the technical advisory board. It took over most of the product units and copyrights relating to PGP about a year later in June 2002 and announces the shipping of PGP v8.0 for autumn of that year.

OTHER PRODUCTS/PLATFORMS: An independent implementation based on the OpenPGP standard [14], GNU Privacy Guard or GnuPG, was released in version 1.0 in September 1999 by Werner Koch (Germany). Created as a free UNIX command-line utility under the GNU Public License (GPL) in the first place, the software and a large set of auxiliary projects were ported to all major operating systems. GnuPG is functionally complete and OpenPGP-compatible, but avoids patent-protected algorithms, particularly <u>IDEA</u>, in its core distribution [5]. GnuPG was funded in part by the German Federal Ministry of Economics and Technology (BMWi) in an effort to improve and spread secure email exchange. The international PGP [10] and OpenPGP Alliance [9] web sites provide information on further PGP-enabled products.

STANDARDS: Zimmermann compiled an initial specification describing data structures and methods employed in his implementations. Later the PGP2 format was derived and became an Internet Engineering Task Force (IETF) "work in progress" Internet draft, RFC1991, in August 1996 [7]. Since 1998, the specification is further developed as OpenPGP Message Format by the OpenPGP Working Group of the IETF [14]. RFC2440 gives a complete specification of formats and algorithmic methods for secure, electronic communication and file storage. The document intends to provide a framework for interoperable OpenPGP application and defines, in addition to the description of cryptographic techniques, a variety of packet types for message and key items.

WEB OF TRUST: PGP established a decentralized trust model, where each party acts as a user and as a <u>certification authority</u> (CA); all users can be introducers to the Web Of Trust, generate their key pairs, distribute their own public keys, and certify those of other users. As opposed to a centralized, hierarchical, <u>X.509</u>-compliant <u>public key infrastructure</u> (PKI), no standardized or company-specific security policies are applied and no additional information is certified except the mere public key binding to the user's name and email address. PGP users maintain their personal trust relationships locally and rate them by discrete trust values, which mirror the keys validity in the program interaction or user interface. A general framework for trust management structures is presented by Blaze [2].

SECURITY ISSUES: In August 2000, the implementation of the aforementioned CMR/ADK feature turned out to be flawed. Subpackets of type CMR/ADK are intended to be contained inside an area of the public key packet that is validated with the key pairs self-signature; when located outside, such subpackets need to be ignored for message encryption (and their existence reported as warning to the user). Ralf Senderek inspected several PGP versions up to 6.5.1i and found this condition not satisfied for keys stored under the v4 packet format that was newly introduced with PGP v5.0 [13]. He placed unprotected recovery key information into both *RSA* and *DH/DSS* public key packets and showed that the respective program versions still encrypted messages under both keys. PGP commercial and freeware versions 6.5.8 and later had this bug fixed; later PGP versions offer an option to generate legacy RSA keys, i.e., compliant to the immune v3 format. GnuPG does not support an ADK/CMR feature.

In March 2001, two Czech cryptologists, Klima and Rosa, discovered a weakness in the OpenPGP private keyring specification that may lead to the disclosure of signature keys [8]. Both common packet formats, v3 for *RSA* and v4 for *RSA* and *DH/DSS* keys, are attacked through similar steps, although the respective algorithm parameters are encrypted under a strong, passphrase-derived, symmetric cipher. The stored parameters of the key packets are suitably modified, allowing to recalculate the original private keys, if a message, signed under the tampered key, can be obtained by an attacker. For instance, v3 key packets (only the RSA algorithm is supported) hold the private key parameters enciphered, but their length fields and a 16-bit checksum are stored in the clear. v4 key packets encrypt the length fields and checksum, too. Nevertheless, Klima and Rosa give a prospect of about a half to successfully exploit the CFB mode properties on the last plaintext block of a 1024-bit RSA key packet using the Rijndael/AES block cipher. Regarding DH/DSS v4 key packets, the attack takes advantage of the fact that the public key data fields are not protected, but are nevertheless used during signature generation. Replacing the subgroup generator and a 1024-bit prime parameter with smaller fabricated values reduces the discrete logarithm computation to a feasible problem. In a practical scenario, an attacker needs to gain full access to the victim's private keyring file twice. A first time, so that the keyring can be tampered, and a second time after the victims signature key was recovered, to restore the original keyring to cover up the attack. PGP subjects RSA keys to a built-in integrity check during the signing process, so that altered key parameters are detected, but other products should undergo a comprehensive audit. However, DH/DSS keys of PGP 5.0–7.04 are affected, later versions are fixed, as well as GnuPG beginning with v1.0.7. The Klima/Rosa vulnerability is not highly critical, because the attacker needs to have physical access to the victims computer. Assuming the keyring files cannot be obtained through a network link, which certainly is good practice, it might be more efficient, for example, to replace PGP executables on the target computer, once its location was penetrated, in order to recover the users passphrase directly by monitoring key strokes. On the other side, the format specification does not define a proper, inherent keyring protection.

An implementation flaw in PGP versions 5.0–7.04 allows the execution of malicious code after opening an ASCII armoured file under Microsoft Windows operating systems. Chris Anley (United States) from the security company @stake showed that PGP creates a temporary file of arbitrary names and arbitrary contents on the target machine during the process of parsing the ASCII armour [1]. Assuming the file name was chosen to be an appropriate dynamic link library (DLL) and contains executable code with valid library entry functions, then, due to the Windows DLL loading strategy, the manipulated code might be executed by a subsequent application started. The vulnerability occurs because temporary files are generated in directories such as the current directory, which are also included in the DLL loaders search path and because PGP does not properly remove the temporary file if the parsing fails. NAI supplied patches for PGP 7.03 and 7.04; later releases are not affected anymore.

Clemens Heinrich

References

[1] Anley, Chris (2001). "Windows PGP (Pretty Good Privacy) ASCII armor parser vulnerability." @state security advisory. http://www.atstake.com/research/advisories/2001/a040901-1.txt, April 2001.

[2] Blaze, Matt, Joan Feigenbaum, and Jack Lacy (1996). "Decentralized trust management." *AT&T Research*. Originally published in *Proc. IEEE Conference on Security and Privacy*.

[3] Diffie, Whitfield and Susan Landau (1998). *Privacy on the Line*. MIT Press, Cambridge, MA.

[4] Electronic Frontier Foundation legal cases archive: archive, http://www.eff.org

[5] GnuPG, GNU Privacy Guard web site: http://www.gnupg.org

[6] Garfinkel, Simson (1995). *Pretty Good Privacy*. OReilly & Associates, Inc.

[7] Internet Engineering Task Force, IETF: http://www.ietf.org

[8] Klima, Vlastimil and Tomá Rosa (2001). "Attack on private signature keys of the OpenPGP format." PGPTM programs and other applications compatible with OpenPGP, 2001, http://eprint.iacr.org/2002/076.pdf

[9] OpenPGP Alliance: http://www.openpgp.org

[10] International PGP web site: http://www.pgpi.org

[11] Zimmermann, Philip R. (1996). Testimony of Philip R. Zimmermann to the Subcommittee on Science, Technology and Space, of the U.S. Senate Committee on Commerce, Science, and transportation; http://www.philzimmermann.com/testimony.shtm, 26 June 1996.

[12] Zimmermann, Philip R. (1999). "Why I Wrote PGP", http://www.philzimmermann.com

[13] Senderek, Ralf (2000). *Key Experiments—How PGP deals with Manipulated Keys*, http://senderek.de/security/key-experiments.html, August 2000.

[14] Callas, Jon, Lutz Donnerhacke, and Hal Finney (2002). OpenPGP Message Format. IETF internet draft RFC2440, August 2002.

PRIMALITY PROVING ALGORITHM

A primality test algorithm which, contrary to a probabilistic primality test, always outputs a correct result. See prime number for further discussion and examples of such algorithms.

Anton Stiglic

PRIMALITY TEST

A *primality test* is a criterion that can be used to test whether an integer is prime. The term is also used to designate an algorithm that tests whether a given integer is prime based on some criteria. See also prime number, primality proving algorithm, and probabilistic primality test.

Anton Stiglic

PRIME GENERATION

Different methods can be used to generate primes. The most common use is a primality test. Typically, a random candidate is chosen and tested for primality, if the candidate is found to be composite, another candidate is chosen at random. The process is repeated until a prime is found.

One can also test a sequence of integers derived from one randomly chosen candidate. For example, chose a random odd integer n and apply a *sieving procedure* (see sieving) to the sequence $n, n + 2, n + 4, \ldots, n + 2k$, for some k. (With high probability, a prime will be found within $O(\ln n)$ steps.) Then test the sequence incrementaly until the first prime if found. *Probable primes* are generated in this way by using a probabilistic primality test, while a so-called *primality proving algorithm* can be used to generate integers that are guaranteed to be prime. Some primality proving algorithms generate a certificate of primality, which can be used to independently verify the primality of an integer. Another method consists in directly constructing integers that are guaranteed to be prime. These prime generation algorithms are typically recursive; Maurer's method is an example. Primes with certain properties, such as safe primes and strong primes, can also be generated using specialized techniques. For a more detailed discussion on prime generation, and references, see prime number.

Anton Stiglic

PRIME NUMBER

A *prime number* is an integer, greater than 1, whose only divisors (positive integer factors) are 1 and itself. For example, the prime divisors of 10 are 2 and 5. The first 7 primes are 2, 3, 5, 7, 11, 13, and 17. A whole number greater than 1 that is not prime is called a *composite*.

Prime numbers play a central role in number theory (also known as *higher arithmetic*), as can be observed by the *fundamental theorem of arithmetic*, which can be stated as follows: Any positive integer (other than 1) can be written as the product of prime numbers in a unique way (disregarding the order of the factors). Thus, prime numbers can be viewed as the multiplicative building blocks from which all whole numbers are constructed. This is a significant theorem with an analogy to chemistry, saying that just as any natural compound can be broken into basic elements from the periodic table, in a unique way, so can any integer be broken into a product of primes. It is interesting to note that if 1 were considered to be prime, the uniqueness property of the fundamental theorem of arithmetic would not hold. For example, if 1 were a prime, 6 could be written as 2×3, or $1 \times 2 \times 3$. Thus, rather than considering 1 as a prime and losing the uniqueness property, 1 is considered to be neither a prime nor a composite, instead it is called a *unit*.

Prime numbers play a major role in cryptography, especially in public key cryptography. For instance, in their landmark 1976 paper "New Directions in Cryptography", Diffie and Hellman present a secure key agreement protocol that can be carried out over public communication channels (see Diffie–Hellman key agreement). Their protocol gave birth to the notion of public-key cryptography. The operations in the protocol they describe are based on arithmetic modulo a large prime number (more precisely, the arithmetic was done in a Galois Field of a large prime order, see modular arithmetic and finite field). Since then, many other cryptographic systems that were proposed were based on the use of large prime numbers and mathematical results surrounding these. For example, large prime numbers are used in public-key encryption schemes, key agreement, digital signature schemes, and pseudo-random number generators.

One basic question that can be asked about prime integers is—how may of them exist? The answer is that there are infinitely many of them! The earliest proof of the infinitude of primes is due to the Greek mathematician Euclid (300 B.C.) and can be found in his work, the *Elements*. A modified version of Euclid's proof is: Suppose that there is a finite set of prime integers, say $\{p_1, p_2, \ldots, p_k\}$ for some fixed value k. Now, consider $N = (p_1 \cdot p_2 \cdots p_k) + 1$ (the product can be computed since there are only a finite number of terms). N is not equal to 1 and is greater than any prime in our hypothetical set of finite primes, thus N is a composite and should be divisible by a prime. However, for any prime p_i in our finite set or primes, p_i does not divide N since p_i divides $p_1 \cdot p_2 \cdots p_n$ so it cannot divide $p_1 \cdot p_2 \cdots p_i + 1$ (this is due to a basic theorem in number theory that states that if an integer greater than 1 divides an integer x, it cannot possibly divide $x + 1$ as well). Thus, we have a contradiction! Since the only unfounded assumption we made is about the finite size of the set of primes, we conclude that this assumption is erroneous, and that the number of primes must be infinite.

Not only do we know that there is an infinity of primes, we also know that in any sufficiently large interval there are a good number of them and we can approximate the quantity. This is important, since even though we proved that there is an infinity of primes, we might wonder what is the chance of a large, randomly chosen number being prime?

Let $\pi(x)$ denote the number of primes smaller or equal to x. It can be shown that for any $x \geq 114$,

$$\frac{x}{\ln(x)} < \pi(x) < \frac{5}{4} \cdot \frac{x}{\ln(x)}.$$

This result is, in fact, a refinement of the so-called *Prime Number Theorem*, which states that $\pi(x)$ is *asymptotic* to $x/\ln(x)$. These results allow us to conclude, among other things, that the chance of a random integer x being prime is about $1/\ln(x)$.

Since prime numbers are used in many different algorithms and protocols, the following questions may also be asked: How does one generate a prime integer? And, given an integer, how does one verify whether or not it is prime?

Two techniques exist that answer these questions for small numbers, the *Sieve of Eratosthenes* and *Trial Division*. The sieve of Eratosthenes allows one to generate all the primes less than or equal to a given positive integer n. The technique consists of initially listing all the numbers from 2 up to n in order. For example, to generate all the primes up to 20 we would start with the sequence

$$2, 3, 4, 5, 6, 7, 8, 9, 10, 11, 12, 13, 14, 15, 16, 17,$$
$$18, 19, 20.$$

Starting from 2 (the first prime in the sequence), delete all the multiples of 2 with the exception of 2 itself. In our example, this gives

$$2, 3, 5, 7, 9, 11, 13, 15, 17, 19.$$

Next, starting from 3 (the next prime in the sequence), delete all the multiples of 3 (except 3 itself). Our example gives

$$2, 3, 5, 7, 11, 13, 17, 19.$$

In general, if the sequence at the tth state is such that p is the tth prime, then the next step is to keep p and delete all the other multiples of p. In our example, the last sequence we came up with is the final sequence, containing all the primes up to 20. This technique is only efficient for small numbers.

Trial division, on the other hand, is a technique that allows us to test the primality of an integer. It is perhaps the simplest primality test one can think of. Given an odd integer n (the only even prime is 2), try to divide it by all odd integers less than n. The number is prime if no divisors are found. It suffices to try dividing it by all odd integers up to $\sqrt{n}$, since if n is composite, at least one prime factor of n must be less than or equal to the square root of n. This primality test is very inefficient, however, since to determine if an integer n is prime, one needs to execute about $\sqrt{n}$ divisions. This is exponential in the size of n and thus not considered to be efficient for large numbers, since the number of steps is not bounded by a polynomial in the size of n (see polynomial time).

For large integers (such as the ones used in public-key cryptography), we need techniques that are much more efficient for generating and testing primes. A method used in practice to generate large prime numbers is based on the use of an efficient primality test, and can be described as follows. To generate a large prime, execute the following steps:

1. Generate as a candidate an odd random integer n of appropriate size.
2. Test n for primality.
3. If n is determined to be a composite, start again at step 1.

Results regarding the density of prime numbers assure us that we will find a prime number in a reasonable amount of time. As we stated before, the chance of a random integer x being prime is about $1/\ln(x)$, and this chance becomes at least $2/\ln(x)$ if we only test odd integers, since we know that all of the even integers greater than 2 are composite. Thus, if the candidates that are chosen are 2048-bit odd integers, then one can expect to test at most about $\ln(2^{2048})/2$ of them, or about 710 odd integers. Instead of choosing a new random candidate in each iteration of the loop, we could start from a random odd integer candidate n and test the sequence $n, n+2, n+4, \ldots$ until we find a prime. This procedure is called an incremental search. With high probability, a prime will be found within $O(\ln n)$ steps (see O-notation), although there are currently no proofs guaranteeing that a prime will be found quickly for all starting points. Using specific search sequences may allow us to increase the chance that a candidate is prime. For example, to speed up the process, one can take the proposed sequence and apply a *sieving procedure* to remove composites with small prime factors, then test the integers in the resulting sequence. A sieving procedure allows us to eliminate most composites in the sequence rapidly, see sieving. The efficiency of generating large random prime numbers in this way is based on the efficiency of the primality test that is used. A lot of work has been done in primality testing throughout the years. The subject is a very important one, as demonstrated by the following quote, well known among mathematicians, from article 329 of *Disquisitiones Arithmeticae* (1801) by C.F. Gauss:

The problem of distinguishing prime numbers from composite numbers, and of resolving the latter into their prime factors is known to be one of the most important and useful in arithmetic. It has engaged the industry and wisdom of ancient and modern geometers to such an extent that it would be superfluous to discuss the problem at length... Further, the dignity of the science itself seems to require that every possible means be explored for the solution of a problem so elegant and so celebrated.

There are two types of primality tests: true and probabilistic. Given a candidate integer, a true primality test will *prove* whether the candidate is prime without any probability of error, while a probabilistic primality test can declare that the candidate is *probably* prime with a certain probability of error. The former can be used to generate so-called *provable primes*, the latter can be used to generate *probable primes*. Algorithms implementing true primality tests that prove the primality of an integer are also called *primality proving algorithms*. Most probabilistic primality tests are correct when they declare a candidate to be composite, but may mistakenly declare a composite to be prime. Typically, when using a probabilistic primality test, step 2 of the prime number generation algorithm, described earlier, is executed multiple times in order to decrease the probability of it falsely declaring a composite to be prime. In what follows, we discuss various probabilistic and true primality testing algorithms.

In the 17th century, Fermat came up with a theorem referred to as Fermat's Little Theorem. The contrapositive of this theorem says that given an integer n, if there exists an integer a, $1 \leq a \leq n - 1$ such that $a^{n-1} \not\equiv 1 \bmod n$, then n is guaranteed to be a composite (see modular arithmetic for a discussion of this type of arithmetic). This is the basis of the Fermat primality test. To generate a random prime using Fermat's test with the method we described earlier, do the following:

1. Generate a random integer n of appropriate size.
2. Generate a random integer a, $1 \leq a \leq n - 1$ and verify whether or not $a^{n-1} \equiv 1 \bmod n$.
3. If the above equality is true, declare that the number is probably prime, else start again at step 1.

If the chosen candidate is a composite, there is a good chance that the integer a, in step 2, will render the equivalence false, thus proving from Fermat's Little Theorem that n is composite. In step 2, an integer a for which the equivalence is satisfied for a composite n is called a *Fermat liar* (to primality) for n, while such an n is called a *pseudoprime to the base a* (see pseudoprime). For a given candidate, step 2 can be executed multiple times, with different values of a, while the equivalence remains true, in order to increase the confidence that n is prime. However, the test is not a guarantee of primality, since there exist integers n for which all integers a are Fermat liars. Such n are called *Carmichael numbers*.

In 1976, Miller [24] described a true primality test that determines in polynomial time, assuming the *Extended Riemann Hypothesis* (ERH), whether a given number is prime. ERH is widely believed to be true, but mathematicians have been trying (and failing!) to prove it for over 100 years.

A year later, Solovay and Strassen discovered a randomized algorithm for testing primality based on a criterion due to Euler ([4,31]). The algorithm has a probability of error that can be made arbitrarily small for all inputs. In other words, when the algorithm declares an integer to be prime, there is a small probability that in fact the integer is a composite. This probability of error can be made to be as small as 2^{-100}, or even smaller, while the algorithm still remains efficient. A random value a for which a composite n passes the Euler criteria is called an *Euler liar* (to primality) for n, while such an n is called an *Euler pseudoprime to the base a*. The test resembles Fermat's test, but it doesn't have the drawback of having composites for which every base is a liar, which is one reason why the probability of error can be made arbitrarily small for any input. Furthermore, the result is *unconditional*, meaning that it is not based on any unproven hypothesis (such as ERH).

Rabin [27, 28] later modified Miller's algorithm to present it as an *unconditional*, but randomized <u>polynomial-time</u> algorithm, a form that is comparable to the Solovay and Strassen algorithm, with a probability of error that can be made arbitrarily small as well.

This last algorithm (with some optimizations from Knuth [20]) is commonly referred to as the <u>Miller–Rabin probabilistic primality test</u>, or the *strong pseudoprime test*. A random value a for which a composite n passes the primality criteria of the Miller–Rabin test is called a *strong liar* (to primality) for n, while such an n is called a *strong pseudoprime to the base a*. The Miller–Rabin primality test is more efficient then the Solovay–Strassen test and is always correct at least as often (the set of strong liars is a strict subset of the set of Euler liars, which in turn is a strict subset of Fermat liars). The Miller–Rabin test is the most commonly used test for generating primes in practice. One iteration of the Miller–Rabin test will err in declaring a composite integer to be a prime with probability less than $\frac{1}{4}$, while t iterations will err with probability less than $(\frac{1}{4})^t$. That is to say that $\mathrm{prob}[Y_t \mid X] \leq (\frac{1}{4})^t$, where X stands for *n is composite* and Y_t stands for *RepeatRabin(n,t)*

returned "prime", RepeatRabin(n, t) being the algorithm that executes t iterations of Miller–Rabin's test and outputs "composite" as soon as any iteration declares n to be prime, else returns "prime" if each iteration passed. The algorithm is always correct when it declares an integer to be composite. A more interesting result stating that $\mathrm{prob}[X \mid Y_t] \leq (\frac{1}{4})^t$ can also be proven using elementary probability theory and the fact that $\mathrm{prob}[Y_1 \mid X]$ is actually much smaller than $\frac{1}{4}$ [9]. In fact, the error probabilities given are *worst case estimates*. Damgård et al. [16] gave numerical upper bounds for these probabilities which are much smaller when choosing candidates independently from a uniform distribution. So, for example, when looking for a 1000-bit prime, to get a probability of error that is less than 2^{-100}, one can simply choose independently, and from a uniform distribution, 1000-bit numbers and subject each candidate to up to three independent iterations of the Miller–Rabin test, until one is found that passes three consecutive tests. There are also low upper bounds for the probability of error when using an incremental search with Miller–Rabin, these are given by Brandt and Damgård [12].

Other probabilistic algorithms exist that are efficient and have a small probability of error on each round when declaring an integer to be a prime, there are also mixes of tests that seem to be very good.

For example, one can use two iterations of the Miller–Rabin test followed by a single iteration of the Lucas probable prime test to generate 1000-bit primes with probability of error smaller than 2^{-100} [3, 10, 13]. The advantage of this last test is that while composite integers that fool multiple rounds of Miller–Rabin are well known, there is yet no known composite integer that passes even a single Miller–Rabin test to the base 2 followed by a single Lucas probable prime test [26].

The Frobenius–Grantham test [19], on the other hand, is slightly more complicated to implement than the other probabilistic tests mentioned. Furthermore, each iteration takes about three times as long as one iteration of the Miller–Rabin test, although the *worst case* probability of error on each iteration is considerably smaller than that of Miller–Rabin. To achieve an error probability that is less than 2^{-100} for 768-bit candidates or greater, one need only run two iterations of the Frobenius–Grantham test.

Several primality proving algorithms also exist. In 1983, Adleman et al. [2] presented a deterministic primality proving algorithm, whose running time is bounded, for sufficiently large n, by $k \cdot (\ln n)^{c \cdot (\ln \ln \ln n)}$, for some constants k and c, making it almost polynomial in the size of the input. Cohen and Lenstra [14, 15] simplified the algorithm both theoretically and algorithmically. Further improvements were made by Bosma and van der Hulst [11]. See also [23] for a generalization of the theory used in these tests. The version of this algorithm used in practice is randomized. The algorithm is referred to by different names, such as the Adleman–Pomerance–Rumely or the Cohen–Lenstra–Bosma algorithm, the Jacobi Sum Test, and the Cyclotomy Method. Although efficient in practice (the primality of numbers that are several hundred decimal digits long can be proved in just a few minutes on fast workstations), the algorithm is not easy to program and the possibility of undetected bugs is very likely.

Later, Goldwasser and Kilian [17] proposed a randomized primality proving algorithm, based on elliptic curves, running in *expected* polynomial-time on almost all inputs. This algorithm is inefficient in practice. Adleman and Huang [1] extended these results and designed a primality proving algorithm whose expected running time is polynomial for all inputs. This established, without the use of any unproven assumptions, that the problem of determining whether or not an integer n is prime can be solved, without any probability of error, by a randomized algorithm that runs in time polynomial in the size of the input of the problem. Atkin [5–7] developed a similar algorithm known as Elliptic Curves for Primality Proving (ECPP), which is efficient in practice. The interesting feature of these algorithms is that they produce a certificate of primality, which is a small set of numbers associated with an integer that can be used to prove more efficiently that the integer is prime. So even though the algorithm is also difficult to implement, the results (more specifically the primality certificate) can be verified by an independent implementation, allowing bugs in the code to be detected. ECPP has been used to prove the primality of numbers having more than 1000 decimal digits.

Finally, in 2002, Agrawal et al. [8] discovered a primality testing algorithm that runs in polynomial time in the size of the candidate, for all candidates, without any randomization, no probability of error and not based on any unproven assumptions (such as ERH). This demonstrates that the problem of determining whether or not a given integer is prime is in the complexity class P (see computational complexity). Even though the algorithm is in P, it is far from being as efficient as the probabilistic tests previously mentioned (e.g., Miller–Rabin). In practice, the probabilistic algorithms are preferred for generating prime numbers because of their efficiency and since the likelihood of an error can be made acceptably small, thus conferring no practical advantage to primality proving algorithms for the generation of primes. For instance, the probability that the Miller–Rabin test erroneously declares a composite integer to be prime can efficiently be made smaller than the probability that a computer running a primality proving algorithm would have an undetectable hardware error, leading to an erroneous result.

Other primality proving algorithms exist that are efficient in practice for candidates n having special forms, such as *Mersenne numbers*, or for example when the factorization of $n - 1$ is known. A Mersenne number is an integer of the form $2^m - 1$, for $m \geq 2$. A Mersenne number that is a prime is called a *Mersenne prime*. The *Lucas–Lehmer primality test for Mersenne numbers* has been used to prove the primality of integers of over 4 million digits. (Notably, $2^{13466917} - 1$ was proved to be prime using this algorithm.)

The following table summarizes the algorithms that have been discussed. We use the following notation: *Rand* stands for a randomized algorithm, while *Det* stands for a deterministic algorithm. *PT* is for polynomial-time. *Prob* stands for a probabilistic primality test, while *PProv* designates a primality proving algorithm and *PProvCert* a PProv that also produces a certificate of primality. Note that a deterministic polynomial-time primality testing algorithm implicitly provides a trivial certificate of primality consisting simply of the number determined to be prime.

Other techniques do not generate prime numbers by applying a primality test to randomly chosen candidates, rather they construct, in a special way, integers that are guaranteed to be primes. Examples of constructive prime generation algorithms are Shawe–Taylor's algorithm [30] and Maurer's method [21, 22], which are both recursive randomized algorithms that return guaranteed primes. The former will reach roughly 10% of all primes of a specified size, while the latter generates primes that are almost uniformly distributed over the set of all primes of a specified size.

Various standards describe algorithms for generating prime numbers. For example, the sole purpose of the ANSI standard X9.80—2001 [3] is to describe primality tests and prime number generation techniques for public-key cryptography. The standard is intended to be the normative reference in this topic for other ANSI X9 standards. Three probabilistic methods for testing integers for primality are described (Miller–Rabin, Lucas and Frobenius–Grantham), four deterministic

Algorithm name	Type	Based on	Year	Reference
Solovay–Strassen	Prob. Rand, PT	Euler criteria	1977	[4, 31]
Miller–Rabin (strong pseudoprime test)	Prob. Rand, PT	Fermat's Little Theorem and the fact that ± 1 are the only square roots of 1 modulo an odd prime	1976–1980	[20, 24, 27, 28]
Adleman–Pomerance–Rumely (Cohen–Lenstra–Bosma, Jacobi Sum Test, Cyclotomic Method)	PProv. Det, almost PT	Set of congruences which are analogues of Fermat's theorem in certain cyclotomic rings	1983	[2, 11, 14, 15]
Goldwasser–Kilian	PProvCert. Expected PT	Elliptic curves	1986–1992	[1, 17]
Elliptic Curve Primality (ECPP, Atkin, Atkin–Morain)	PProvCert. Expected PT	Elliptic curves	1986–1993	[5–7]
Frobenius–Grantham	Prob. Rand, PT	Quadratic polynomials and Frobenius automorphism	1998	[19]
Agrawal–Kayal–Saxena	PProvCert. Det, PT	Variation on Fermat's Little Theorem	2002	[8]

methods (including ECPP and trial division, for sufficiently small primes), along with methods for generating prime numbers that use these primality tests, as well as a description of a sieving procedure. The standard also describes two methods for direct construction of prime numbers (Maurer's algorithm and the Shawe–Taylor algorithm). There are also descriptions of methods for generating primes with additional properties, such as safe primes and strong primes. NIST's FIPS 186-2 [25] describes the Miller–Rabin primality test and an algorithm for generating primes with certain properties for the Digital Signature Standard. The algorithm uses the Miller–Rabin test or any other primality test where the probability of a non-prime number passing the test can efficiently be made to be at most 2^{-80}. The algorithm generates two sufficiently large primes, q and p, in such a way that q divides $p - 1$. Furthermore, the algorithm provides a *seed* that can be used to demonstrate, under reasonable assumptions, that the primes were not intentionally constructed to be "weak" in a way that the entity who constructed them could subsequently exploit their structure to recover other entities' cryptographic private keys.

Anton Stiglic

References

[1] Adleman, L.M. and M.-D.A. Huang (1992). "Primality Testing and Abelian Varieties over Finite Fields." Lecture Notes in Mathematics, vol. 1512, eds. E. Gimenez and C. Paulin-Mohring. Springer-Verlag, Berlin, Heidelberg, New York.

[2] Adleman, L.M, C. Pomerance, and R.S. Rumely, (1983). "On distinguishing prime numbers from composite numbers." *Annals of Mathematics*, 117, 173–206.

[3] American National Standards for Financial Services (2001). Prime number generation, primality testing, and primality certificates: X9.80.

[4] Atkin, A.O.L. and R.G. Larson (1982). "On a primality test of Solovay and Strassen." *SIAM Journal on Computing*, 11 (4), 789–791.

[5] Atkin, A.O.L. (1986). *Lecture Notes of a Conference*, Boulder, Colorado, August 1986.

[6] Atkin, A.O.L. and F. Morain, (1993). "Elliptic curves and primality proving." *Mathematics of Computation,* 61 (203), 29–68.

[7] Atkin, A.O.L. and F. Morain, (1992). "Finding suitable curves for the elliptic curve method of factorization." *Mathematics of Computation*, 60 (201), 399–405.

[8] Agrawal, M., N. Kayal, and N. Saxena (2002). *PRIMES is in P*.

[9] Beauchemin, P., C. Brassard, C. Crépeau, C. Goutier, C. Pomerance (1998). "The generation of random numbers that are probably prime." *Journal of Cryptology*, 1, 53–64.

[10] Baillie, R.J. and S.S. Wagstaff, Jr. (1980). "Lucas pseudoprimes." *Mathematics of Computation*, 35, 1391–1417.

[11] Bosma, W. and M.-P. van der Hulst (1990). "Faster primality testing." *Advances in Cryptology—EUROCRYPT'89*, Lecture Notes in Computer Science, vol. 434, eds. J.-J Quisquater and J. Vandewalle. Springer-Verlag, Berlin, 10–13, 652–656.

[12] Brandt, J. and I. Damgård (1993). "On generation of probable primes by incremental search."

Advances in Cryptology—CRYPTO'92, Lecture Notes in Computer Science, vol. 740, ed. E.F. Brickell. Springer Verlag, Berlin, 358–370.

[13] Pomerance, C., J.L., Selfridge, and S.S. Wagstaff, Jr. (1980). The pseudoprimes to 25×10^9. *Mathematics of Computation*, 35 (151), 1003–1026.

[14] Cohen, H. and A.K. Lenstra (1986). "Implementation of a new primality test." *Mathematics of Computation*, 48 (177), 103–121, S1-S4.

[15] Cohen, H. and H.W. Lenstra, Jr. (1984). "Primality testing and Jacobi sums." *Mathematics of Computation*, 42 (165), 297–330.

[16] Damgård, I., P., Landrock, and C. Pomerance (1993). "Average case error estimates for the strong probable prime test." *Mathematics of Computation*, 61 (203), 177–194.

[17] Goldwasser, S. and J. Kilian (1986). "Almost all primes can be quickly certified." *Proceedings of the 18th Annual ACM Symposium on Theory of Computing*, 316–329.

[18] Gordon, J. (1985). "Strong primes are easy to find." *Advances in Cryptology—EUROCRYPT'84*, Lecture Notes in Computer Science, vol. 209, eds. T. Beth, N. Cot and I. Ingemarsson. Springer-Verlag, Berlin, 216–223.

[19] Grantham, J. (1998). "A probable prime test with high confidence." *Journal of Number Theory*, 72, 32–47.

[20] Knuth, D. (1981). *The Art of Computer Programming—Seminumerical Algorithms*, vol. 2 (2nd. ed.). Addison-Wesley, Reading, MA.

[21] Maurer, U.M. (1995). "Fast generation of prime numbers and secure public-key cryptographic parameters." *Journal of Cryptology*, 8 (3), 123–155.

[22] Maurer, U.M. Fast generation of secure RSA-moduli with almost maximal diversity. *Advances in Cryptology—EUROCRYPT'89*, Lecture Notes in Computer Science, vol. 434, eds. J.-J. Quisquater and J. Vandewalle. Springer-Verlag, Berlin, 636–647.

[23] Mihăilescu, P. (1997). "Cyclotomy of Rings and Primality Testing." *PhD Thesis*, Swiss Federal Institute of Technology Zürich, Diss. ETH No. 12278.

[24] Miller, G.L. (1976). "Reimann's hypothesis and tests for primality." *Journal of Computer and System Sciences*, 13, 300–317.

[25] NIST (1993). "Digital signature standard." FIPS PUB 186, February 1993.

[26] Pomerance, C. (1984). *Are There Counter-Examples to the Baillie-PSW Primality Test?*

[27] Rabin, M.O. (1976). "Probabilistic algorithms." *Algorithms and Complexity: New Directions and Recent Results*, ed. J.F. Traub. Academic Press, New York, 21–39.

[28] Rabin, M.O. (1980). "A probabilistic algorithm for testing primality." *Journal of Number Theory*, 12, 128–138.

[29] Rivest, R.L., and R.D. Silverman (1998). "Are 'strong' primes needed for (RSA)?" Technical Report. RSA Data Security, Inc., Redwood City, CA, USA.

[30] Shawe-Taylor, J. (1986). Generating strong primes. *Electronics Letters*, 22, 875–877.

[31] Solovay, R. and V. Strassen (1977). "A fast Monte-Carlo test for primality." *SIAM Journal on Computing*, 6 (1), 84–85.

PRIMITIVE ELEMENT

A primitive element of a <u>finite field</u> is a root of the <u>field polynomial</u> that is also a <u>generator</u> of the multiplicative <u>group</u> of the field.

Let $f(x)$ be the field polynomial of degree n for an <u>extension field</u> $\mathbf{F}_{q^d}$ constructed over a subfield $\mathbf{F}_q$, and let α be a root of $f(x)$. If α is also a generator of $\mathbf{F}_{q^d}^*$, i.e., if the set of elements

$$\alpha, \alpha^2, \alpha^3, \ldots$$

traverses all elements in $\mathbf{F}_{q^d}^*$, then α is a *primitive element* and *f(x)* is called a *primitive polynomial*.

Not all field polynomials are primitive, but it is often convenient for implementation to use a primitive polynomial. Every finite field $\mathbf{F}_{q^d}$ has $\phi(q^d - 1)$ primitive elements, and $\phi(q^d - 1)/d$ primitive polynomials, where ϕ is <u>Euler's totient function</u>.

Burt Kaliski

PRIVACY

A "name", or *identity*, is a set of information that distinguishes an entity from every other within a given environment. In some environments, the "name" may just be a given name; in other environments, it will be a given name and a family name; in still others, it may include additional data such as a street address, or may be some other form entirely (e.g., an employee number). In all cases, however, the identity depends upon the environment: the size and characteristics of the environment determine the amount of information required for uniqueness.

Regardless of the information tied together to form the identity, however, this "name", or "nym", will be one of three possible types: an *anonym* ("no name"); a <u>pseudonym</u> ("false name"); or a *veronym* ("true name"). These three categories of nyms are defined by the amount of linkage that is possible across transactions. Anonyms are meaningless or null names for one-time use, and so no linkage can be made (1) to the actual entity involved in any given transaction and (2) between transactions. Pseudonyms are either meaningless or apparently meaningful names for multiple uses that bear no relation to the true name. Thus, no linkage can be made to the actual entity involved in any given transaction, but it is clear that the same (unknown) entity is involved in different transactions. Veronyms are, or very readily disclose, the name of the physical entity in the real world, and so linkage can be made both to the entity involved and across different transactions. Note that linkage to a real-world entity (such as a machine, device, software process, or human person) is central to the notion of *privacy*, but may or may not be relevant to any given application—for many applications, anonyms, pseudonyms, and veronyms are all valid identities for the entities with which they deal.

PRIVACY CONCEPTS: *Privacy* may be defined as an entity's ability to control how, when, and to what extent personal information about it is communicated to others (see Brands [2], p. 20). In order to understand privacy, then it is important to understand what "personal information" is and to understand the ways that personal information can be controlled.

Personal Information

Personal information can include intrinsic physical data such as name, birth date, and gender, but can also include location data (such as home address and telephone number), financial data (such as salary, bank account balance, or credit card number), user-created data (such as confidential correspondence or a list of personal preferences), and data assigned by other entities (such as a bank account number or a social security number). Despite this wide range of types of personal information, it is sometimes useful to distinguish the subset of personal information that reveals the identity of the entity ("identifying information") from all the other types because identifying information can be sensitive on its own, whereas other types of personal information (e.g., salary) are typically sensitive only when revealed in conjunction with identifying information.

Communication of identifying information to unintended parties may be referred to as "exposure": the identity of the entity is exposed to others. Communication of other types of personal information to unintended parties may be referred to as "disclosure": this information has been disclosed to others. Both exposure and disclosure may be direct or indirect. Direct exposure/disclosure is the determination of user identity or other personal information by an observer or by another participant in the exchange from the explicit contents of a single transaction or message. Indirect exposure/disclosure is the determination of user identity or other personal information by inference or from the correlation of the contents of several transactions or messages.

A primary goal of any privacy infrastructure, then, is to provide confidentiality of identifying information and other personal information when desired. That is, the infrastructure must enable an entity to limit the release of this information both directly and indirectly in accordance with the entity's wishes.

Techniques for Control

There are a number of ways in which confidentiality of identifying information may be enabled by a privacy infrastructure. For example, anonymous Web or e-mail services may be provided to entities, or protocols accepting anonymous authentication (proof of ownership of certain attributes without the need to reveal identifying data) may be supported. For environments in which multiple transactions will take place over time or across many servers, the use of pseudonymous identifiers may be supported—either one at a time per entity, or an unlimited number at a time per entity, depending upon the requirements of the environment (see the Trusted Computing Platform Alliance architecture [5] for an example of multiple pseudonym support in a computing environment).

There are two primary techniques for providing confidentiality of non-identifying personal information: encryption and access control. Personal information may be encrypted so that it may only be seen by intended recipients while it is in transit (e.g., as it travels over a Secure Sockets Layer, SSL, channel) or while it is in storage. In storage, it may be encrypted at the application layer for specific entities ("Alice", "Bob", "Charlene") or for particular roles ("Faculty", "Senior Managers", "Hospital Staff"). Encrypting personal information ensures that—at least until it is decrypted—no unintended recipients will be able to read the data.

Access control limits who may do what with a given resource (see authorization architecture). When the resource is personal information, access control allows an entity to explicitly specify which other entities may (or may not) read this information. The effect is similar to encryption (only a restricted set of entities may see the data), but the data itself may not be rendered unintelligible; rather, a Policy Enforcement Point (PEP) enforces decisions rendered in accordance with access policies created by the subject/owner of the personal information. Any entity not meeting the conditions stipulated by the access policy will be prevented by the PEP from looking at the protected data.

More generally, an entity may wish to control other aspects of its personal data than simply who can read it. In particular, the entity may want to control such things as collection, use, access, dissemination, and retention of its personal information. Access policies and reliable decision and enforcement engines are again the technique for this control. In some environments, however, there are two components of the access policy that must be taken into account. If personal data will be stored on a server or site remote from the entity, then the owner or administrator of that site may create a policy stating what its practices are with respect to privacy (e.g., "We collect and use your name and address information for the purpose of shipping goods to you; we retain this data for a maximum of six months; and we do not sell or in any other way communicate this data to any other party for any reason"). This server policy is one component of the complete access policy. The other component is the policy created by the entity itself, specifying use, retention, and so on, of its data. When access to this data is requested by any entity, both components of the access policy must be consulted and followed by the decision engine. Any conflicts between the two components (e.g., the server says it will retain personal data for 6 months, but the owner/subject of the data specifies a retention period of no longer than 3 months) must be resolved in an acceptable fashion (e.g., the owner/subject policy takes precedence over the site's policy).

Another critical aspect in controlling the communication of personal information to other entities is known as "individual access". This means that the owner/subject of personal data is given read/write access to at least a subset of this data in order to ensure correctness and completeness of the information. This includes access to data created by the entity itself (e.g., address and credit card information), access to data created about the user (e.g., medical examination results), and access to audit information regarding data use and dissemination.

PRIVACY PRINCIPLES: A number of organizations, regulatory bodies, and government agencies have taken a keen interest in privacy issues in recent times. This has resulted in a relatively large collection of privacy guidelines and privacy principles that have been developed in Europe,

North America, and around the world. Examples of such guidelines and principles include Bill C-6, the CSA Model Code, the EU Directive on Data Protection, the Health Insurance Portability and Accountability Act (HIPAA), the Gramm–Leach–Bliley Act (GLBA), the OECD Privacy Principles, and the Fair Information Practice Principles.

Although not identical, many of these initiatives do have significant overlap in the issues that they cover. These include topics such as the following: accountability, identifying purposes, user consent, limiting collection of data, limiting use/disclosure/retention of data, accuracy, safeguards, openness, individual access, and compliance. Notions such as user notice and user choice also figure prominently in much of this work.

It may be noted, however, that virtually all of these initiatives are focused on privacy principles that are of interest to the corporate or server entity that holds personal information on behalf of users. There is very little comparable work that is focused on principles that are of specific interest to the users themselves. Although these views will coincide in many ways, the change in perspective can lead to subtle differences or changes in emphasis. User-focused privacy principles deal with the issues that are most important to the owner/subject of personal data and will include the four areas discussed in the previous section: confidentiality of identifying information; confidentiality of other personal information; access to personal data for the purpose of ensuring correctness; and control over the use, retention, dissemination, and so on, of personal information.

PRIVACY TECHNOLOGIES: A number of technologies have been developed over the years to enable and preserve the privacy of entities participating in transactions on electronic networks such as the Internet. Collectively, these have come to be known as Privacy Enhancing Technologies (PETs). This section provides examples of some PETs that are intended to address different portions of the privacy spectrum.

Onion routing (see also <u>MIX networks</u>) and subsequent technologies built on that framework such as *Crowds*, addresses the requirement for anonymity—confidentiality of identifying information—in Web browsing and in messaging (primarily e-mail) environments. The basic idea is that the message from the originator is bounced around a relatively large set of participants in a fairly random fashion before being sent to the eventual destination. The destination, as well as the other participants, may know the immediately preceding hop, but do not know whether there were any hops prior to that one. Thus, the identity of the originator (including machine identity, IP address, and so on) is kept hidden from all observers. Some technologies allow the possibility of pseudonymous identifiers for participants so that transactions may be linked over time to a single (unknown) entity for continuity, context building, personalization, or similar purposes.

Encryption is commonly used as a technology to provide confidentiality of data, including personal data. As noted above, encrypting data for specific individuals or roles can ensure that unintended parties are not able to read the data while it is in transit or in storage. Symmetric encryption algorithms (see symmetric cryptosystem) are used to scramble the data, with key management/key establishment supported by out-of-band means or by technologies such as Kerberos, Public Key Infrastructure (PKI), or Secure Sockets Layer. Encrypting personal data for a role can be an attractive option when the individuals that need to see that data may vary over time, but it does require a supporting infrastructure in which, at any given time, one entity can reliably determine whether or not another entity is a valid member of that role. A PKI that uses attribute certificates, Security Assertion Markup Language (SAML) assertions, or similar technology for its attribute management (see authorization architecture) is one way to build this supporting infrastructure [1].

Access control is an important technology for restricting the dissemination of personal information. Rules and policies can be established regarding who can do what with this data, and trusted components can make and enforce access decisions on the basis of these policies. Technologies such as SAML and eXtensible Access Control Markup Language (XACML) provide the data structures, protocols, and policy syntax to allow a relatively comprehensive access control infrastructure to be built in the service of privacy. Other technologies such as CORBA and DCE can also provide much of the required functionality in this area.

Finally, technologies that address the control (by both server and entity) of personal information include Platform for Privacy Preferences Project (P3P) and A P3P Preference Exchange Language (APPEL). P3P [3] is a mechanism for Web sites to advertise their privacy policies and practices in a machine-readable way so that a user browsing to a site may automatically be warned if the site's policy does not match his/her own preferences. P3P policies can include statements about the types of personal data that an owner/subject can access, the ways in which the owner/subject can resolve disputes about the site's privacy practices, the purpose(s) for which personal data will be used by the site, other recipients with whom personal data may be shared, and the period for which personal data will be retained by the site.

APPEL [4] is a technology to give users some control over the use of their personal information. With P3P, a user can browse to a site, download its privacy policy, compare this policy with his/her own preferences, and make an informed decision as to whether to continue dealing with this site (e.g., submit personal information or make a purchase). APPEL allows the possibility of some negotiation with the site: after downloading the site's P3P policy and examining it, the owner/subject can respond with his/her own list of preferences (a narrowed-down set of values of various elements in the policy). The site can then choose to treat this user's personal data in accordance with this more restricted P3P policy, or can reject this modified policy. Depending upon the site's response, the user can again decide whether to continue dealing with this site. APPEL gives users a bit more control over how their data is used than P3P alone, which simply informs the user of a site's current practices.

Carlisle Adams

References

[1] Adams, C. and S. Lloyd (2003). *Understanding PKI: Concepts, Standards, and Deployment Considerations* (2nd ed.). Addison-Wesley, Reading, MA.

[2] Brands, S. (2000). *Rethinking Public Key Infrastructures and Digital Certificates: Building in Privacy*. MIT Press, Cambridge, MA.

[3] Cranor, L., M. Langheinrich, M. Marchiori, M. Presler-Marshall, and J. Reagle (2002). "The platform for privacy preferences 1.0 specification." W3C Recommendation, 16 April 2002; see http:/www.w3.org/TR/P3P

[4] Cranor, L., M. Langheinrich, and M. Marchiori (2002). "A P3P preference exchange language 1.0." W3C Working Draft, 15 April 2002; see http:/www.w3.org/TR/P3P-preferences for current status.

[5] The Trusted Computing Platform Alliance; see http:/www.trustedpc.org/home/home.htm for details.

PRIVILEGE MANAGEMENT

Privilege management is a subset of general "authorization data" management (see authorization architecture) in which the data being managed are *privileges* granted or bestowed upon entities in an environment. A *privilege* may be defined as follows [1]: "a right or immunity granted as a peculiar benefit, advantage, or favor".

Carlisle Adams

Reference

[1] Merriam-Webster OnLine, http://www.m-w.com/cgi-bin/dictionary

PROBABILISTIC PRIMALITY TEST

A *probabilistic primality test* is a primality test that has a probability of error. Such tests typically pick a random number called a *witness* and verify some criteria involving the witness and the candidate being tested. Most probabilistic primality tests used in practice will not err when declaring an integer to be composite, but have a probability of error when declaring an integer to be prime: they either declare a candidate to be *definitely composite* or *probably prime*. A candidate that passes such a test—whether prime or composite—is called a *probable prime* for that test. A composite number that erroneously passes such a test is called a pseudoprime.

See prime number for further discussion and examples of such tests.

Anton Stiglic

PROBABILISTIC PUBLIC-KEY ENCRYPTION

Probabilistic public-key encryption is a public-key encryption scheme (see public key cryptography) where the ciphertext of the same message under the same public key differs on every run of a probabilistic Turing machine. That is, a random coin toss of the Turing machine is used in the encryption algorithm. The notion was proposed in contrast to the RSA public-key encryption scheme, which is deterministic in the sense that the ciphertext is always fixed given a public key and a plaintext.

The early date examples of probabilistic public key encryption schemes are the Goldwasser Micali encryption *scheme* and the ElGamal public-key encryption scheme, and many others followed. It is known that an encryption scheme that satisfies provable security such as semantic security must be probabilistic. Since the original RSA encryption scheme was not probabilistic, several padding techniques such as OAEP are considered so that the padded RSA scheme becomes probabilistic and satisfies security properties, such as semantic security.

Kazue Sako

Reference

[1] Goldwasser, S. and S. Micali (1984). "Probabilistic encryption." *Journal Computer and System Sciences*, 28, 270–299.

PRODUCT CIPHER, SUPERENCRYPTION

Product ciphers are ciphers that are built as a composition of several different functions. In a special case when all the functions are the same, the cipher is called *iterative cipher* and the functions are called *rounds*. The intuition behind such constructions is inspired by the analogy with mixing transformations studied in the theory of Dynamical Systems, which was first noted by Shannon [3]. The IBM team followed this approach in the design of *Lucifer* and the Data Encryption Standard (DES). Rounds of iterative ciphers are typically keyed with different *subkeys* or at least involve different round constants to break self-similarity, which otherwise would be vulnerable to slide attacks. Note that the individual round transformation might be cryptographically weak. The strength of the whole construction relies on the number of iterations. The choice of the proper *number of rounds* is a difficult task, which is performed via cryptanalysis of the cipher. Most of the modern block-ciphers have an iterative structure for the reasons of compact hardware, software implementation and in order to facilitate the analysis. The typical number of rounds in modern ciphers ranges between 8 and 32. By their structure, iterative block ciphers may be divided into two large classes: Feistel ciphers and substitution–permutation networks (SPN).

If one takes the composition of several ciphers one gets what is called multiple encryption or *cascade cipher* or *superencryption* (this term is often used in communications when some information which was encrypted off-line is transmitted via an encrypted communication link). Multiple encryption is often used to enhance security of a single encryption function, though at the expense of speed. Note that multiple encryption does not necessarily improve security; consider for example *simple substitution ciphers*, where the product of substitutions is another substitution. One may prove that multiple encryption is at least as secure as the first component [1, 2] if the keys are chosen independently. A typical example of multiple encryption is <u>Triple-DES</u>.

Alex Biryukov

References

[1] Even, S. and O. Goldreich (1985). "On the power of cascade ciphers." *ACM Transactions on Computer Systems*, 3, 108–116.
[2] Maurer, U.M. and J.L. Massey (1993). "Cascade ciphers: The importance of being first." *Journal of Cryptology*, 6 (1), 55–61.
[3] Shannon, C.E. (1949). Communication theory of secrecy systems." *Bell System Technical Journal*, 28, 656–715.

PROOF OF KNOWLEDGE VS. PROOF OF MEMBERSHIP

A basic <u>interactive proof</u> for a statement of the form $x \in L$, where L is some formal language is often called a *proof of membership*. For instance, if L_H is the language of graphs containing a Hamiltonian cycle, a proof for the statement $x \in L_H$ shows that x is a Hamiltonian graph. However, the proof does not necessarily show that the prover actually *knows* a Hamiltonian cycle for the graph.

A *proof of knowledge* is similar to an <u>interactive proof</u>, except that the soundness condition is strengthened in the following way. Referring to the example statement $x \in L_H$, the verifier accepts the proof only if the prover actually knows a Hamiltonian cycle for x, where "knowing" is characterized as follows. Generally speaking, if one is given access to a successful prover for the statement $x \in L_H$, then one should be able to extract a Hamiltonian cycle for x. Such an efficient algorithm for extracting a Hamiltonian cycle (in this example) from a prover is in general called a *knowledge extractor*.

The formal notion of a proof of knowledge developed from the results in [1–5]. Many zero-knowledge proofs of membership are in fact zero-knowledge proofs of knowledge. The <u>Fiat–Shamir identification protocol</u> is an early example of a zero-knowledge proof of knowledge. The <u>Schnorr identification protocol</u> is another well-known example.

Berry Schoenmakers

References

[1] Bellare, M. and O. Goldreich (1992). "On defining proofs of knowledge." *Advances in Cryptology—CRYPTO'92*, Lecture Notes in Computer Science, vol. 740 ed. E.F. Brickell. Springer-Verlag, Berlin, 390–420.
[2] Feige, U., A. Fiat, and A. Shamir (1988). "Zero-knowledge proofs of identity." *Journal of Cryptology*, 1 (2), 77–94.
[3] Feige, U. and A. Shamir (1990). "Witness indistinguishable and witness hiding protocols." *Proceedings of 22nd Symposium on Theory of Computing (STOC'90)*. ACM Press, New York, 416–426.
[4] Goldreich, O. (2001). *Foundations of Cryptography—Basic Tools*. Cambridge University Press, Cambridge.
[5] Goldwasser, S., S. Micali, and C. Rackoff (1989). "The knowledge complexity of interactive proof systems." *SIAM Journal on Computing*, 18, 186–208. Preliminary version in *17th ACM Symposium on the Theory of Computing*, 1982.

PROPAGATION CHARACTERISTICS OF BOOLEAN FUNCTIONS

Let n and l be positive integers such that $l \le n$. Let f be a Boolean function on F_2^n and let $a \in F_2^n$. We denote by $D_a f$ the *derivative* of f with respect to a, that is: $D_a f(x) = f(x) \oplus f(a + x)$. This notion is related to the differential attack (see <u>Differential cryptanalysis</u> for <u>block ciphers</u>).

The function f satisfies the *propagation criterion PC(l)* of degree l if, for all $a \in E$ of weight at most l, the function $D_a f$ is balanced (that is, takes the values 0 and 1 equally often) [3]. In other words, f satisfies $PC(l)$ if the autocorrelation coefficient $\sum_{x \in F_2^n} (-1)^{f(x) \oplus f(x+a)}$ is null for every $a \in F_2^n$ such that $1 \le w_H(a) \le l$. The *strict avalanche criterion SAC* corresponds to $PC(1)$.

The functions satisfying $PC(n)$ are the bent functions (see <u>Nonlinearity of Boolean functions</u>).

If n is even, then $PC(n-2)$ implies $PC(n)$. Since bent functions are never balanced, there exist balanced n-variable $PC(l)$ functions for n even only if $l \le n - 3$. For odd $n \ge 3$, the functions satisfying $PC(n-1)$ are those functions of the form $g(x_1 \oplus x_n, \ldots, x_{n-1} \oplus x_n) \oplus \ell(x)$, where g is bent and ℓ is affine.

The only known upper bound on the *algebraic degree* (see <u>Boolean functions</u>) of a $PC(l)$ function is $n - 1$. A lower bound on the *nonlinearity* (see <u>Boolean functions</u>) of $PC(l)$ functions exists: it is lower bounded by $2^{n-1} - 2^{n-\frac{1}{2}l-1}$. Equality can occur only if $l = n - 1$ (n odd) and $l = n$ (n even).

Constructions of $PC(l)$ functions have been given in [2], for instance.

It is needed, for some cryptographic applications, to have Boolean functions that still satisfy $PC(l)$ when we keep constant a certain number k of their coordinates (whatever are these coordinates and whatever are the constant values chosen for them). We say that such functions satisfy the *propagation criterion $PC(l)$ of order k*. This notion, introduced in [3], is a generalization of the strict avalanche criterion of order k, $SAC(k)$, introduced in [1]. $SAC(k)$ functions, that is, $PC(1)$ of order k functions, have algebraic degrees at most $n - k - 1$ (see [3]). A construction of $PC(l)$ of order k functions based on Maiorana-McFarland's method is given in [2].

There exists another notion, similar to $PC(l)$ of order k, but stronger [3]: a Boolean function satisfies $EPC(l)$ of order k if every derivative $D_a f$ corresponding to a direction $a \neq 0$ of weight at most l is k-resilient.

Claude Carlet

References

[1] Forré, R. (1989). "The strict avalanche criterion: spectral properties of Boolean functions and an extended definition." *Advances in Cryptology—CRYPTO'88*, Lecture Notes in Computer Science, vol. 403, ed. S. Goldwasser. Springer-Verlag, Berlin, 450–468.

[2] Kurosawa, K. and T. Satoh (1997). "Design of $SAC/PC(\ell)$ of order k Boolean functions and three other cryptographic criteria." *Advances in Cryptology—EUROCRYPT'97*, Lecture Notes in Computer Science, vol. 1233, ed. W. Fumy. Springer-Verlag, Berlin, 434–449.

[3] Preneel, B., W. Van Leekwijck, L. Van Linden, R. Govaerts, and J. Vandevalle (1991). "Propagation characteristics of Boolean functions." *Advances in Cryptology—EUROCRYPT'90*, Lecture Notes in Computer Sciences, vol. 473, ed. I.B. Damgård. Springer-Verlag, Berlin, 161–173.

PROTOCOL

In the context of cryptography, "protocol" is a shorthand for "cryptographic protocol". A cryptographic protocol is a distributed algorithm describing precisely the interactions of two or more entities to achieve certain security objectives. The entities interact with each other by exchanging messages over private and/or public communication channels.

Important classes of protocols are: key exchange protocols or key establishment protocols, challenge–response protocols, identification verification protocols, and zero-knowledge protocols. A more practical example is the Secure Socket Layer (SSL) protocol for establishing a secure communication link between two entities, a client and a server.

Cryptographic protocols are often used as building blocks for constructing cryptographic schemes (or systems). In general, a cryptographic scheme may be composed of several cryptographic algorithms and/or cryptographic protocols.

Berry Schoenmakers

Reference

[1] Menezes, A.J., P.C. van Oorschot, and S.A. Vanstone (1997). *Handbook of Applied Cryptography*. CRC Press, Boca Raton, FL.

PROTON

Proton is a proprietary e-cash smart card solution developed by Banksys in Belgium, which is CEPS and EMV compliant. The security is based on symmetric cryptography, which basically means that the security relies on the protection of one key which has been used to encrypt data on the cards.

The card may be uploaded at an ATM or similar device. The cardholder keys in his *PIN* (see Personal Identification Number) and selects the account from which the amount should be debited.

The ATM or load device then links with the smart card chip and loads the amount onto the chip. The card can be used at stores, payphones, vending machines, pay-per-view TV set-top box, and Internet low-value payments, and the solution has been employed in a number of countries.

Transactions are fully accountable, with centrally available audit trail, but only small amounts are allowed, and payments are provided without the use of PIN-codes for cost limitation and user friendliness.

Peter Landrock

PSEUDO-MERSENNE PRIME

A prime of the form

$$p = 2^m - k,$$

where k is an integer for which

$$0 < |k| < 2^{\lfloor m/2 \rfloor}.$$

If $k = 1$, then p is a Mersenne prime (and m must necessarily be a prime). If $k = -1$, then p is called a *Fermat prime* (and m must necessarily be a power of two).

Pseudo-Mersenne primes are useful in public-key cryptography because they admit fast modular reduction (see modular arithmetic) similar to Mer- senne primes. If n is a positive integer less than p^2, then n can be written

$$n = u \cdot 2^{2m} + a \cdot 2^m + b,$$

where $u = 0$ or 1 and a and b are nonnegative integers less than 2^m. (It is only rarely true that $u = 1$, and never true if $k > 0$.) Then

$$n \equiv u \cdot k^2 + a \cdot k + b \pmod{p}.$$

Repeating this substitution a few times will yield n modulo p. This method of modular reduction requires a small number of additions and subtractions rather than the usual integer division step.

This method works best when m is a multiple of the word size of the machine being used and k is expressible in one machine word. Thus the first recorded use of pseudo-Mersenne primes in elliptic curve cryptography [1] specifies that $|k| < 2^{32}$.

More recently, pseudo-Mersenne primes have been used in the construction of optimal extension fields. These are fields of the form $\mathbb{F}_p[\alpha]$ where p is a pseudo-Mersenne prime of single word size (e.g., $2^{32} - 5$) and α is a root of an irreducible binomial equation $x^n - b$ over $\mathbb{F}_p$. Optimal extension fields are efficient fields for elliptic curve cryptography because the algebraic operations can be carried out using single-precision arithmetic modulo a pseudo-Mersenne prime (see [2]).

Jerome Solinas

References

[1] Crandall, Richard E. (1992). "Method and apparatus for public key exchange in a cryptographic system." U.S. Patent # 5,159,632, October 27, 1992.

[2] Bailey, Daniel and Christof Paar (1998). "Optimal extension fields for fast arithmetic in public-key algorithms." *Advances in Cryptology—CRYPTO'98*, Lecture Notes in Computer Science, vol. 1462, ed. H. Krawczyk. Springer-Verlag, Berlin, 472–485.

PSEUDO-NOISE SEQUENCE (PN-SEQUENCE)

Sequences which obey the three randomness postulates of Golomb are usually called pseudo-noise sequences (PN-sequences). The primary examples of such sequences are the maximum-length linear sequences, i.e., the m-sequences (see also Golomb's randomness postulates).

Tor Helleseth

PSEUDONYMS

In the paper based world, a pseudonym is a fictitious name for an individual. Examples are pen names, aliases, legalized names, or working names. In the realm of electronic communication, a pseudonym (or sometimes just 'nym') can be any identifier that an individual or object uses in a particular context. The purpose of using pseudonym is to identify an individual or object in the respective context while not identifying it in other (unintended) contexts. In effect, one keeps transactions in different contexts separate such that observers who have access to both contexts cannot link the transactions of the same individual or those related to the same object (see anonymity and unlinkability).

In oppressive political environments, people use pseudonyms to make public political statements and to hide the identities of their correspondents. In more tolerant environments people use pseudonyms to avoid embarrassment, harassment, or even loss of their jobs. Examples are discussing alcoholism, depression, sexual preferences, etc. Several servers for managing e-mail pseudonyms are available on the internet [4]. Pseudonyms are a dual use technology with many very beneficial uses and a lot of serious misuses. See Froomkin [3].

The following types of pseudonyms can be distinguished:

Person pseudonym: A pseudonym that is used over an individual's life-time or remaining lifetime. Examples are author names, working names, social security numbers, driver's licence numbers, photo ID (to a certain extent), fingerprint, genetic print, etc.

Member (role) pseudonym: A pseudonym associated to an individual for as long as the individual acts as a member, employee, volunteer, or under a contract. Examples are bank account numbers, credit card numbers, e-mail addresses, *PGP* public keys (see <u>Pretty Good Privacy</u>), customer numbers, cell phone numbers, member IDs for affiliations in universities, colleges, or other educational programs, membership in sports clubs, etc.

Relationship pseudonym: A pseudonym used by one person *A* in all communications with another person *B*. If *B* is a bank and an insurance and *A* is a customer of *B* and holds an insurance police with *B* then *A* might want to use the same relationship pseudonym with *A*. For example, because *B* can be expected to find out anyway that the account holder and the insurance police holder are the same person.

Session pseudonym: Usually a more technical kind of pseudonym because it is used only for the duration of a communication session. For example, a session ID in a TCP/IP protocol stack that is kept for the duration of one session.

Transaction pseudonym: A technical kind of pseudonym that is used only for a single transaction. For example, electronic coins can be withdrawn for one transaction pseudonym and be spent for another transaction pseudonym.

According to Pfitzmann and Köhntopp [5], these types of pseudonyms can be related in order of the increasing anonymity and unlinkability they can achieve (see Figure 1). Pseudonyms are a helpful—but usually not sufficient—aid to achieve <u>privacy</u> including <u>anonymity</u> and <u>unlinkability</u>. They were introduced as a technical concept in cryptography by David Chaum [1,2]. Privacy is about protecting identities of indivduals against third party interests. An individual acts anonymously if it leaves no information that traces back to the individual's identity, i.e., the registercd name, address, photo ID, e-mail address, etc. One way to act anonymously is to use a pseudonym that is chosen independently of the individual's identity, preferably a randomly chosen number. More privacy can be achieved by using different pseudonyms in different contexts, thereby achieving unlinkability between transactions in different contexts. For example, a patient could use a medical ID and a health insurance ID that have no commonalities, like an encoded birth date. Then if a treating physician and an employee of the health insurance collaborate without both having "more identifying material" about the patient, they could hardly figure out that they are actually talking about the same patient. As Rao and Rohatgi [6] have emphasized, however, the use of pseudonyms alone is hardly ever enough to achieve effective anonymity or unlinkability, because in real life transactions carry a lot of context information that can be analyzed and linked to the context information of other transactions.

In cryptography, pseudonyms have important applications in untraceable <u>electronic cash</u>, unlinkable <u>credentials</u>, <u>electronic voting schemes</u>, secure auction systems, and <u>blind signatures</u>.

Gerrit Bleumer

References

[1] Chaum, David (1981). "Untraceable electronic mail, return addresses, and digital pseudonyms." *Communications of the ACM*, 24 (2), 84–88.

[2] Chaum, David (1986). "Showing credentials without identification—signatures transferred between unconditionally unlinkable pseudonyms." *Advances in Cryptology—EUROCRYPT'85*, Lecture Notes in Computer Science, vol. 219, ed. F. Pichler. Springer-Verlag, Berlin, 241–244.

[3] Froomkin, Michael (1996). "Flood control on the information ocean: Living with anonymity, digital cash, and distributed databases." *University of Pittsburgh Journal of Law and Commerce*, 395 (15).

[4] Mazieres, David, Frans Kaashoek (1998). "The desgin, implementation and operation of an email pseudonym server." *5th ACM Conference on Computer and Communications Security*. ACM Press, New York, 27–36.

[5] Pfitzmann, Andreas, Marit Köhntopp (2001). "Anonymity, unobservability, and pseudonymity—a proposal for terminology." *Designing Privacy Enhancing Technologies*, Lecture in Computer Science, vol. 2009, ed. H. Frederrath. Springer-Verlag, Berlin, 1–9.

[6] Rao, Josyula R., Pankaj Rohatgi (2000). "Can pseudonyms really guarantee privacy?" *9th Usenix Symposium*, August.

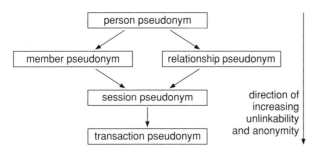

Fig. 1. Types of pseudonyms

PSEUDOPRIME

A *pseudoprime* (without any other qualification) is a *composite* that passes the *Fermat primality test*. More precisely, *n* is a *pseudoprime* to the base a (noted as a-PSP) if n is *composite* and $a^{n-1} \equiv \mod n$ (see modular arithmetic). Similarly, a *strong pseudoprime* (SPSP) is a composite that passes the Miller–Rabin probabilistic primality test (also known as the *strong primality test*). Other qualifications for the term *pseudoprime* exist for composites that are declared to be prime by other probabilistic primality tests. The term *probable prime* is related but is not restricted to composite numbers.

Anton Stiglic

PSEUDORANDOM FUNCTION

A *pseudorandom function* is a deterministic function of a key and an input that is indistinguishable from a truly random function of the input. More precisely, let s be a security parameter, let K be a key of length s bits, and let $f(K, x)$ be a function on keys K and inputs x. Then f is a pseuodorandom function if:

- f can be computed in polynomial time in s; and
- if K is random, then f cannot be distinguished from a random function in polynomial time.

In this context, "distinguishability" refers to the ability of an algorithm to tell whether a function is not truly random. Let g be a truly random function of x with the same output length as f. Suppose a polynomial-time algorithm A is given access to a "oracle" which, on input x, either consistently returns $f(K, x)$, or consistently returns $g(x)$. After some (polynomial) number of accesses to the oracle, the algorithm outputs a guess, b, as to whether the oracle is f or g. Let ϵ be A's *advantage*, i.e., the difference in probabilities

$$\epsilon = \big| \Pr[b = \text{``}f\text{''} \mid \text{oracle is } f] \\ - \Pr[b = \text{``}f\text{''} \mid \text{oracle is } g] \big|.$$

If the inverse $1/\epsilon$ grows faster than any polynomial in s for all polynomial-time algorithms A, then the function f is said to be indistinguishable from a random function.

Pseudorandomness is a stronger requirement than being a *one-way function*, since a one-way function only needs to be hard to invert; a pseudorandom function also needs to be hard to *guess* when the key is unknown.

Pseudorandom functions have many applications in cryptography, as they provide a way to turn an input into a value that is effectively random. This is helpful for computing MAC algorithms, deriving keys from other keys, and more generally for replacing random number generators in an application with a deterministic function, when a secret key is available. Goldreich, Goldwasser and Micali [1] showed how to construct a pseudorandom function from a pseudorandom number generator; in practice, a pseudorandom function is often constructed from a hash function, as, for example, in the popular Secure Sockets Layer (SSL) protocol.

Burt Kaliski

Reference

[1] Goldreich, O., S. Goldwasser, and S. Micali (1986). "How to construct random functions." *Journal of the ACM*, 33 (4), 210–217.

PSEUDO-RANDOM NUMBER GENERATOR

Many cryptographic primitives require random numbers, to be used as keys, challenges, unique identifiers, etc. However, generating random values on a computer is in fact a very difficult task. One possible method for such generation is to use a pseudo-random numbers generator.

A pseudo-random number generator (PRNG) is a function that, once initialized with some random value (called the *seed*), outputs a sequence that appears random, in the sense that an observer who does not know the value of the seed cannot distinguish the output from that of a (true) random bit generator.

It is important to note that a PRNG is a deterministic process: put back in the same state, it will reproduce the same sequence, as will two PRNGs initialized with the same seed. This property makes PRNGs suitable for use as stream ciphers.

Pseudo-random generators aimed for cryptographic applications must not be confused with those used for most other purposes (simulations, probabilistic algorithms, . . .). Whereas the latter must basically have good properties in terms of the statistical distribution of their output, the former must also resist against active adversaries disposing of an extended description of the generator's structure, observing a large quantity of output, and using advanced cryptanalysis methods in order to predict future output. Good statistical distribution is a necessary, but not sufficient condition for this. Classical PRNGs do not resist such attacks, and specific (namely cryptographically secure) PRNGs must be used for security purposes.

An important notion related to random numbers generation is that of *entropy* (see information theory), that basically measures the amount of information carried by a sequence (or, in other words, the amount of *a priori* uncertainty about a sequence). Since the sequence generated by a PRNG is entirely determined by its seed, the entropy cannot be greater than that of the seed, no matter how long the generated sequence is. This distinguishes PRNGs from a (true) random bit generator. However, provided that the seed is long enough to make exhaustive search beyond the reach of today's computing power, this limitation does not rule out the use of PRNGs in some cryptographic applications.

PRNG STRUCTURE: A PRNG is typically composed of a seed repository, an output function producing some random-looking bits from the seed, and a feedback function that iteratively transforms the seed.

It can be shown that a PRNG is necessarily periodic: the sequence it produces will repeat itself after a (possibly extremely long) period. Care must be taken that the generator is not used to produce more data than a full cycle.

PRNG EXAMPLES: This section reviews some well-known PRNGs constructs, although the list is in no way deemed to be exhaustive.

Linear Feedback Shift Registers

A very simple and efficient (but insecure) construct is given by linear feedback shift registers (LFSRs). A LFSR is composed of a register, which is an array of memory cells, each capable of storing one binary value, and a feedback function, which consists in the XOR operator applied to selected cells of the register (Figure 1). For each new unit of time, the following operations are performed:
1. The content of the last memory cell is output.
2. The register is processed through the feedback function. In other words, selected memory cells are XORed together to produce one bit of feedback.

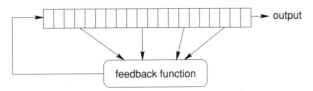

Fig. 1. Linear feedback shift register

3. Each element of the register advances one position, the last element being discarded, and the first one receiving the result of the feedback function.

LFSRs can be implemented very efficiently in hardware, can produce sequences with large periods and good statistical distribution, and are theoretically well understood. However, they are not cryptographically secure: efficient techniques are known to reconstruct the content of the register, and hence the full LFSR's output, based on the observation of a short output sequence.

Nevertheless, LFSRs can be used as building blocks to construct a secure PRNG. Common combination techniques consist in:
- using a nonlinear function to combine the output of several LFSRs;
- using the output of one (or a combination of) LFSR(s) to clock one (or a combination of) other LFSR(s).

However, great care has to be taken in the design of such a combination, which must be carefully analyzed against known cryptanalytic techniques. Many designs prove much less secure than they first appear, as is for example witnessed by the Geffe generator [2], that succumbs to correlation attacks [5].

Blum–Blum–Shub Generator

The Blum–Blum–Shub PRNG is a strong generator, in the sense that it has been proved secure under the assumption that integer factorization is intractable [8]. However, its slowness limits its usefulness to very specific cases.

The generator works by repeatedly squaring a (secretly-initialized) value modulo n (see modular arithmetic), and outputting its least significant bit (Algorithm 1). A modular squaring operation is thus required for each output bit. More efficient versions, outputting more than one bit per squaring, have also been proved secure [5].

ALGORITHM 1. Blum–Blum–Shub pseudo-random generator

Generate two large secret primes p and q, such that $p \equiv 3 \bmod 4$ and $q \equiv 3 \bmod 4$.
Compute $n = pq$
Generate a random integer s such that
$\quad \gcd(s, n) = 1$
$x_0 \leftarrow s^2 \bmod n$
for $i \leftarrow 1$ to l **do**
$\quad x_i \leftarrow x_{i-1}^2 \bmod n$
$\quad$ Output the least significant bit of x_i
end for

Other Constructions

Another frequently used structure for PRNGs consists in processing a register through a one-way function, using (all or part off) the result as output. The register is then updated, either using a counter method (e.g., increase register by one), or by copying part of the one-way function's output into it.

This is for example the case of the ANSI X9.17 generator (based on the Data Encryption Standard, or of the FIPS 186 generator (based on the Data Encryption Standard or the hash function SHA). A detailed description of these algorithms can be found in [5].

STATISTICAL TESTS: Although it does not constitute a sufficient condition for a PRNG to be cryptographically secure, passing statistical tests is certainly necessary, since any statistical defect can be used by an attacker to gain information about the PRNG's future output. Statistical tests thus provide a first indication on the quality of a generator, although it has to be completed by a careful analysis of the generator's structure and its resistance to cryptanalysis.

Several test suites have been proposed. Among the most well-known, we can cite:

- Golomb's randomness postulates were one of the first attempts to establish necessary conditions for randomness. Nowadays, they are of mere historical interest.
- **Knuth** [3] proposes a set of simple randomness tests to apply to a sequence.
- **FIPS 140-1** [6] is a standard test series defined by the U.S. National Institute of Standards and Technology (NIST), inspired by Knuth's basic tests. This standard was superseded by FIPS 140-2.
- **FIPS 140-2** [7] (see FIPS 140) defines a more comprehensive battery of tests, including Maurer's Universal Statistical test.
- **Maurer's Universal Statistical test** [4] is a test capable of detecting a large class of defects. In addition, this test provides an estimate of the entropy of the source. A better estimate can be obtained using a modified version of the test [1].

PRNG INITIALIZATION: Since the output is entirely determined by the value of the seed, it is very important for the seed to be initialized to some truly random, unpredictable value (see random bit generator). A PRNG can be viewed as a tool expanding a small random value into a much longer sequence.

This random initialization can be performed only once (e.g. during on-factory personalization phase). Other applications require occasional PRNG re-seeding.

Generating this truly random seed is usually outside the scope of PRNG definitions that rarely address the issue. However, its importance must not be underestimated.

François Koeune

References

[1] Coron, J.-S. (1999). "On the security of random sources." *Proceedings of PKC'99*, Lecture Notes in Computer Science, vol. 1560, eds. H. Imai and Y. Zheng. Springer-Verlag, Berlin.

[2] Geffe, P. (1973). "How to protect data with ciphers that are really hard to break." *Electronics*, 46, 99–101.

[3] Knuth, D.E. (1997). *The Art of Computer Programming* (3rd. ed.). *Computer Science and Information Processing, Vol. 2. Seminumerical Algorithms*. Addison-Wesley, Reading, MA.

[4] Maurer, U.M. (1992). "A universal statistical test for random bit generators." *Journal of Cryptology*, 5 (2), 89–105.

[5] Menezes, A.J., P.C. van Oorschot, and S.A. Vanstone, (1997). *Handbook of Applied Cryptography*. CRC Press, Boca Raton, FL.

[6] US Department of Commerce—National Institute of Standards and Technology (1994). *FIPS 140-1: Security requirements for Cryptographic Modules*, available at http://csrc.nist.gov/publications/fips/fips1401.pdf

[7] US Department of Commerce (1999). *FIPS 140-2: Security requirements for Cryptographic Modules (Draft)*, available at http://csrc.nist.gov/publications/fips/dfips140-2.pdf

[8] Vazirani, U.V. and V.V. Vazirani (1985). "Efficient and secure pseudo-random number generation." *Advances in Cryptology—CRYPTO'84*, Lecture Notes in Computer Science, vol. 196, eds. G.R. Blakley and D. Chaum. Springer, Berlin, 193–202.

PUBLIC KEY CRYPTOGRAPHY

Public key cryptography is a method to encrypt messages using a nonsecret key. The term public key cryptography also includes various others cryptographic methods using a nonsecret key, such as authentication, digital signature schemes, and key agreement. Here we describe public key encryption schemes.

Conventional cryptography, also known as symmetric cryptosystem, uses a secret key to encrypt messages; the same key is required to decrypt these messages. In a public key encryption scheme however, knowledge of the key used to encrypt messages (which we call *encryption key* in the sequel) does not allow one to derive the key to decrypt the messages. Therefore an encryption key can be made public without endangering the security of the decryption key. A pair of encryption key and decryption key is generated for each receiver, and all the encryption keys are published. When sending a secret message to a receiver, the sender picks the public encryption key of the receiver and encrypts the message with it. The encrypted message can be recovered by using the corresponding secret decryption key, which only the intended receiver has.

Public key cryptography solves the key agreement problem and key management problem of conventional, symmetric, key cryptosystems, where it is the concern of a sender to agree with the receiver on a secret encryption key. Moreover, they both have to store these keys secretly while the number of the keys grows linearly in the number of possible receivers. The disadvantage of public key cryptography is in its processing speed. Therefore it is often used to for key agreement purposes only. A secret key that is used to encrypt a message via symmetric key cryptography is encrypted using public key cryptography and is attached to the encrypted message. The receiver first decrypts the secret key using his own decryption key and then decrypts the message using the recovered secret key.

The design of a public key cryptosystem can be based on a trapdoor one-way function. Some examples of public key encryption schemes are: RSA public key encryption, ElGamal public key encryption, Rabin cryptosystem, Goldwasser–Micali encryption scheme, Blum–Goldwasser public-key encryption scheme, and Paillier encryption and signature schemes.

A public key encryption scheme is comprised of three algorithms: a *key generation algorithm*, an *encryption algorithm* and a *decryption algorithm*. The security of a public key encryption scheme is evaluated through security measures called semantic security. If an encryption scheme is semantically secure, then an adversary, who is given an encryption of either one of the two plaintexts, cannot guess which one is encrypted with a probability more than $1/2$.

In practice, it is important that the published public key is indeed the key of the intended receiver, i.e., that the key is certified (see certificate).

Otherwise, a sender could be encrypting the classified message using the wrong key, and the message could be obtained by the adversary. In order to certify keys, the notion of a public key infrastructure has been developed.

Kazue Sako

PUBLIC KEY INFRASTRUCTURE

This term, or PKI, is used as a label for any security architecture where end-users have digital signature capability and possibly encryption as well based on public key cryptography. It is a general misconception that such solutions always require the use of certificates. Certificates are only required if end-users have to communicate with more than a few entities, e.g., with each other. Examples of PKI solutions without certificates are electronic banking solutions where customers may generate signatures but only communicate with the bank. But is possible to build complex PKIs as well without certificates, where instead an online service can provide the status of a public key at any point in time (so-called instant certificates). In more conventional solutions, the infrastructure is based on some general standard, such as X.509 or EMV, where each user has a certificate on each of his public key(s). If certificates are required, it is often necessary to provide means for verifying whether a certificate—which on the face of it appears to be valid—has not been revoked for some reason. This is handled by means of revocation lists or on-line inquiry protocols regarding the status of a certificate (such as *OCSP*, the On-line Certificate Status Protocol).

Peter Landrock

PUBLIC KEY PROXY ENCRYPTION

In a public key proxy encryption scheme as introduced by Blaze and Strauss [1, 2] there are a collection of recipients who generate their public key pairs of public encryption keys and private decryption keys (see public key cryptography) and one or more proxy principals who are provided with proxy keys. A proxy key $\pi_{A \to B}$ with respect to recipients A and B allows a proxy principal to convert ciphertext that can be decrypted by A into ciphertext that can be decrypted by B giving the same plain text. The conversion of ciphertexts works such that the proxy principal learns no information about the underlying plaintext, which implies that the proxy principal does not learn enough information about the deciphering keys of A and B.

Note that proxy encryption is different from passing the decryption key of A and the encryption key of B to the proxy principal, such that the proxy principal first decrypts an incoming message using the decryption key of A and then re-encrypts the plain text to B. In this case, the proxy principal would learn all plain texts underlying the cipher texts being transformed. In addition, proxy encryption is usually a faster operation than first decrypting and afterwards re-encrypting again.

Compared to a conventional public key encryption scheme, the confidentiality requirement of a public key proxy encryption scheme holds not only against underlined{eavesdroppers} and unintended recipients, but also against distrusted proxy principals. The main security requirement on public key proxy encryption schemes is:

Confidentiality: An attacker who has access to public encryption keys, proxy keys, and cipher texts, cannot figure information about the corresponding plain texts, which implies that he cannot recover information about the corresponding decryption keys.

Historically, the term *proxy cryptosystem* was introduced by Mambo and Okamoto [4], but their trust model is the same as that of conventional public key encryption. They only consider senders and recipients, but not separate proxy principals that are distrusted by the recipients $A, B, \ldots$.

Public key proxy encryption schemes have been developed by Blaze and Strauss [1, 2] into a separate class of cryptographic schemes. They have categorized public key proxy encryption schemes as follows.

Trust model between recipients: Clearly, if Alice empowers a proxy to divert her cipher texts to another recipient Bob, then she trusts Bob. But Bob does not need to trust Alice. More precisely a public key proxy encryption scheme is called *symmetric* if Alice can figure Bob's private decryption key if she learns the proxy key $\pi_{A \to B}$. Otherwise, it is called *asymmetric*.

Input to proxy key computation: An asymmetric public key proxy encryption scheme is called *active* if it is feasible to compute a proxy key $\pi_{A \to B}$ only if Bob provides his private decryption key as an input. If it is feasible to compute a proxy key $\pi_{A \to B}$ without the input of Bob's private decryption key (but only his public encryption key), then the scheme is called *passive*.

Anonymity of proxy keys: Public key proxy encryption schemes can be categorized according to how much information they reveal about the public encryption keys of the two recipients who introduced the proxy key. A public key proxy encryption scheme is called *transparent* if a proxy key $\pi_{A \to B}$ reveals the public keys of Alice and Bob. A scheme is called *translucent* if an attacker who guesses the public encryption keys of Alice and Bob can verify his guess by using the proxy key $\pi_{A \to B}$. A scheme is *opaque*, if a proxy key $\pi_{A \to B}$ reveals no information about the public encryption keys of Alice and Bob.

To illustrate the concept of proxy encryption, we consider a simple proxy encryption scheme proposed by Blaze and Strauss, which is based on a variant of the underlined{ElGamal encryption scheme}. Consider p a large safe underlined{prime}, q a large prime factor of $p - 1$, and g be an element of the underlined{group} $\mathbb{Z}_p^*$ such that g is of order q, i.e., q is the smallest positive integer such that $g^q = 1 \pmod{p}$ (see underlined{modular arithmetic}). Alice and Bob be two recipients with respective public key pairs (x_i, y_i) $(i \in \{A, B\})$, where the decryption keys x_i are chosen independently and uniformly at random from $\mathbb{Z}_q^*$ and the encryption keys are computed such that $y_i = g^{x_i} \pmod{p}$. Alice can encrypt a message m for Bob by computing the ciphertext $c = (c_1, c_2) = (mg^k, y_B^k)$ consisting of two components, and Bob can decrypt the ciphertext by computing $m' = c_1 / c_2^{a^{-1}} \pmod{p}$.

In order to install a proxy between them, Alice and Bob compute the proxy key $\pi_{A \to B} = x_B / x_A \pmod{q}$. If neither Alice nor Bob trusts the other one, they could use a two party computation underlined{protocol} (see underlined{multiparty computation}) whose output is given to the proxy. This way, neither Alice nor Bob learns the private decryption key of the other party and the proxy learns none of the private decryption keys.

If $c_A = (mg^k, y_A^k)$ is an encrypted message m to Alice, then the proxy can transform it into an encrypted message m to Bob as follows:

$$c_B = \left(c_{A1}, c_{A2}^{\pi_{A \to B}} \right) = (mg^k, g^{x_A k x_B / x_A}) = (mg^k, y_B^k).$$

This scheme is shown by Blaze and Strauss [1] to achieve confidentiality in the above sense under the computational Diffie–Hellman assumption and the underlined{discrete logarithm problem}. It is symmetric because Alice can easily figure Bob's private decryption key if she learns the proxy key $\pi_{A \to B} = x_B / x_A \pmod{q}$ because she knows her own private decryption key x_A. It is active because computing the proxy key requires Bob's private decryption key x_B as input.

No asymmetric public key proxy encryption scheme has been proposed to date.

Blaze and Strauss [1] propose proxy encryption as a useful mechanism in restricted computing environments such as smart cards. Girard proposes to use proxy encryption in the personalization process of smart cards [3]. Dually related to the notion of public key proxy encryption is that of public key proxy signatures.

Gerrit Bleumer

References

[1] Blaze, Matt and Martin Strauss (1998). "Atomic proxy cryptography." Technical Report 98.5.1, AT&T Labs Research, http://www.research.att.com/library/trs

[2] Blaze, Matt, Gerrit Bleumer, and Martin Strauss (1998). "Divertible protocols and atomic proxy cryptography." *Advances in Cryptology—EUROCRYPT'98*, Lecture Notes in Computer Science, vol. 1403, ed. K. Nyberg. Springer-Verlag, Berlin, 127–144.

[3] Girard, Pierre (2000). "Secure personalization using proxy cryptography." *CARDIS'98*, Lecture Notes in Computer Science, vol. 1820, eds. J.-J. Quisquater, B. Schneier. Springer-Verlag, Berlin, 326–335.

[4] Mambo, Masahiro and Eiji Okamoto (1997). "Proxy cryptosystem: Delegation of the power to decrypt ciphertexts." *IEICE Transaction*, E80-A (1), 54–63. http://search.ieice.org/1997/pdf/a010054.pdf

PUBLIC KEY PROXY SIGNATURES

In a public key proxy signature scheme as introduced by Blaze and Strauss [1, 2] there are a collection of signers who generate their public key pairs of public verification keys and private signing keys (see public key cryptography) and one or more proxy principals who are provided with proxy keys. A proxy key $\pi_{A \to B}$ with respect to signers A and B allows a proxy principal to convert a digital signature valid with respect to A's verifying key into a signature for the same message that is valid with respect to B's verifying key. The conversion of signatures works such that the proxy principal learns no information about the signing keys of either A or B.

Compared to a conventional public key signature scheme, the unforgeability requirement of a public key proxy signature scheme holds not only for general active attackers without access to the signers private key, but also for distrusted proxy principals. The main security requirement on public key proxy signature schemes is:

Unforgeability: An attacker who has access to public verifying keys, proxy Keys, and pairs of messages with valid signatures (or can adaptively choose the messages for which he gets valid signatures) cannot produce signatures for any new messages, which implies that he cannot gather information about the corresponding private signing keys.

Historically, the term *proxy signatures* was introduced by Mambo and Okamoto [3], but their trust model is the same as that of conventional digital signatures. They only consider signers and verifiers, but not separate proxy principals that are distrusted by the signers $A, B, \ldots$.

Proxy signature schemes have been developed by Blaze and Strauss [1, 2] into a separate class of cryptographic schemes. They have categorized proxy signature schemes as follows.

Trust model between recipients: Clearly, if Bob empowers a proxy to transform Alice's signatures into Bob's own signatures, then he clearly trusts Alice. But Alice does not need to trust Bob. More precisely a proxy signature scheme is called *symmetric* if Bob can figure out Alice's private signing key if he learns the proxy key $\pi_{A \to B}$. Otherwise, it is called *asymmetric*.

Input to proxy key computation: An asymmetric proxy signature scheme is called *active* if it is feasible to compute a proxy key $\pi_{A \to B}$ only if Alice provides her private signing key as an input. If it is feasible to compute a proxy key $\pi_{A \to B}$ without the input of Alice's private signing key (but only her public verifying key), then the scheme is called *passive*.

Anonymity of proxy keys: Proxy signature schemes can be categorized according to how much information they reveal about the public verifying keys of the two recipients who introduced the proxy key. A proxy signature scheme is called *transparent* if a proxy key $\pi_{A \to B}$ reveals the public verifying keys of Alice and Bob. A scheme is called *translucent* if an attacker who guesses the public verifying keys of Alice and Bob can verify his guess by using the proxy key $\pi_{A \to B}$. A scheme is *opaque*, if a proxy key $\pi_{A \to B}$ reveals no information about the public verifying keys of Alice and Bob.

To illustrate the concept of proxy signatures, we consider a simple proxy signature scheme, which is based on a variant of the ElGamal signature scheme. Consider p a large safe prime number, q a large prime factor of $p - 1$, and g be an element of the group $\mathbb{Z}_p^*$ such that g is of order q, i.e., q is the smallest positive integer such that $g^q = 1 \pmod{p}$. Let Alice and Bob be two signers with respective public key pairs (x_i, y_i) ($i \in \{A, B\}$), where the private signing keys x_i are chosen independently and uniformly at random from $\mathbb{Z}_q^*$ and the public verification keys are computed such that $y_i = g^{x_i} \pmod{p}$. $h(m)$ denotes a collision resistant hash function that takes any binary string $m \in \{0, 1\}^+$ as input and returns a value in $\mathbb{Z}_q$. Alice can sign a message m by choosing some $k \in_R \mathbb{Z}_q$ and computing the signature $\sigma = (r, s) = (g^k, x_A r + k h(m))$ consisting of two components, and anyone with access to Alice's verifying key y_A can verify that σ is a valid signature for m by checking that $g^s = y_A^r r^{h(m)} \pmod{p}$.

In order to install a proxy between them, Alice and Bob compute the proxy key $\pi_{A \to B} = x_B - x_A \pmod{q}$. If neither Alice nor Bob trusts the other one, they could use a two party computation protocol whose output is given to the proxy. This way, neither Alice nor Bob learns the private signing key of the other party and the proxy learns none of the private signing keys.

If $\sigma_A = (g^k, x_A r + k h(m))$ is a signature by Alice for message m, then the proxy can transform it into a signature σ_B by Bob for message m as follows:

$$
\begin{aligned}
\sigma_B &= (\sigma_{A1}, \sigma_{A2} + \pi_{A \to B} r) \\
&= (g^k, x_A r + k h(m) + (x_B - x_A) r) \\
&= (g^k, x_B r + k h(m)).
\end{aligned}
$$

This scheme is symmetric because Bob can easily figure Alice's private signing key x_A if he learns the proxy key $\pi_{A \to B} = x_B - x_A \pmod{q}$ because he knows his own private signing key x_B. It is active because computing the proxy key requires Alice's private signing key x_A as input.

No asymmetric proxy signature scheme has been proposed to date.

Dually related to the notion of proxy signatures is that of public key proxy encryption.

Gerrit Bleumer

References

[1] Blaze, Matt and Martin Strauss (1998). "Atomic proxy cryptography." Technical Report 98.5.1, AT&T Labs Research, http://www.research.att.com/library/trs

[2] Blaze, Matt, Gerrit Bleumer, and Martin Strauss (1998). "Divertible protocols and atomic proxy cryptography." *Advances in Cryptology—EUROCRYPT'98*, Lecture Notes in Computer Science, vol. 1403, ed. K. Nyberg. Springer-Verlag, Berlin, 127–144.

[3] Mambo, Masahiro, Keisuke Usuda and Eiji Okamoto (1996). "Proxy signatures: Delegation of the power to sign messages." *IEICE Transactions Fundamentals*, E79-A (9), 1338–1354. http://search.ieice.org/1996/pdf/e79-a_9_1338.pdf

QUADRATIC RESIDUE

Let n be an odd, positive integer and let x be an integer that is relatively prime to n (see underline{modular arithmetic}). The integer x is a *quadratic residue* modulo n if the equation

$$x \equiv y^2 \,(\mathrm{mod})\, n$$

has an integer solution y. In other words, the integer x is a square modulo n. The integer x is a *quadratic non-residue* otherwise.

If n is an odd underline{prime number}, then exactly half of all integers x relatively prime to n are quadratic residues. If n is the product of two distinct odd primes p and q, then the fraction is one-quarter.

See also underline{Jacobi symbol}, underline{Legendre symbol}, and underline{Quadratic Residuosity Problem}.

Burt Kaliski

QUADRATIC RESIDUOSITY PROBLEM

Let n be the product of two distinct odd underline{prime numbers} p, q, and let x be an integer such that the underline{Jacobi symbol} $(x/n) = +1$. The *Quadratic Residuosity Problem* (QRP) is to determine, given x and n, whether x is a underline{quadratic residue} modulo n (see underline{modular arithmetic}). (All quadratic residues have Jacobi symbol $+1$, but not necessarily the reverse.) This problem is easy to solve given the factors p and q, but is believed to be difficult given only x and n. However, it is not known whether the problem is equivalent to factoring the modulus n.

The QRP is the basis for several cryptosystems, including the underline{Goldwasser–Micali encryption scheme} and Cocks' identity-based encryption scheme [1] (see underline{identity-based cryptosystems}).

Burt Kaliski

Reference

[1] Cocks, Clifford (2001). "An identity based encryption scheme based on quadratic residues." *Cryptography and Coding*, Lecture Notes in Computer Science, vol. 2260, ed. B. Honary. Springer, Berlin, 360–363.

QUADRATIC SIEVE

INTRODUCTION: The Quadratic Sieve (QS) and its variants are the first of the modern underline{integer factoring} algorithms to be able to routinely factor abitrary integers in the 60+ digit range on just a single PC. They are the successor to the earlier Continued Fraction Method of Morrison and Brillhart [4] (see underline{integer factoring}) and the predecessor to the underline{Number Field Sieve} [2]. All three of these algorithms share features in common. They are based upon the observation that if $A^2 \equiv B^2 \bmod N$ and $A \not\equiv \pm B \bmod N$ (see underline{modular arithmetic}), then $GCD\,(A+B, N)$ and $GCD\,(A-B, N)$ are proper factors of N. The Quadratic Sieve will factor an arbitrary integer N, in heuristic time $L_N[1/2, 1] = \exp((1 + o(1))\,(\log N)^{1/2}\,(\log\log N)^{1/2})$ (see underline{L-notation}). The entries underline{integer factoring} and underline{Number Field Sieve} discuss the history of these algorithms and we do not repeat that discussion here. For those interested in implementing QS, reference [7] gives all of the necessary details. The version of QS that is detailed here is known as the Multiple Polynomial Quadratic Sieve (MPQS). Other variants are known.

KEY IDEAS: The Quadratic Sieve, like other sieving algorithms (and the underline{Elliptic Curve Method for factoring}), is based on the idea of underline{smooth numbers} (see also underline{smoothness}). A number x is said to be y—*smooth* if all of its underline{prime} factors are less than y. QS generates many *relations* of the form

$$S^2 = r \bmod N,$$

where the pair (S, r) is generated from a quadratic polynomial in such a way that r is small compared to N. The algorithm then *attempts* to factor r using a fixed set of primes called a underline{factor base}. The largest prime in the factor base is then the smoothness bound. Most such values will not factor. However, a sieve can be used to attempt to factor many such rs simultaneously. It is the speed of sieving on modern computers that makes QS and NFS very effective methods. The sieve works by observing that if a prime p divides the value of a polynomial $Q(x)$, then it divides $Q(x + kp)$ for all k.

Once a sufficient number of smooth relations have been found, a subset is then extracted such that the product of the rs in the subset is a perfect square. From that, $A^2 \equiv B^2 \bmod N$ is easily computed.

The subset is found by solving a system of equations mod 2. This is referred to as the *linear algebra* phase of the algorithm.

GENERATION OF RELATIONS: The algorithm starts by computing a factor base $FB = \{p_i | (\frac{N}{p_i}) = 1,$ p prime, $i = 1, \ldots, F\}$ for some appropriate value of F and $p_0 = -1$ for the sign. $(\frac{N}{p_i})$ is the <u>Legendre symbol</u> and indicates that p_i must be a <u>quadratic residue</u> of N.

The following is then repeated until enough smooth relations have been collected:

- Construct a quadratic polynomial $Q(x) = Ax^2 + Bx + C$ and solve $Q(x) \equiv 0 \bmod p_i$ for all i. There will be two roots r_{1i} and r_{2i} for each prime.
- Initialize a sieve array to zero over the interval $[-M, M]$ for an appropriate value of M.
- For all $p_i \in FB$ add the value $\lfloor \log(p_i) \rfloor$ to the locations $r_1, r_1 \pm p_i, r_1 \pm 2p_i, \ldots$ and $r_2, r_2 \pm p_i, r_2 \pm 2p_i, \ldots$
- The value of $Q(x)$ will be approximately $M\sqrt{N}$ over $[-M, M]$ so compare each sieve location with $T = \log(M) + \log(N)/2$. Fully factored values will have their corresponding sieve value close to T. For these, construct the exact factorization by division. It is also possible (and usually quicker) to find the factorization by re-sieving. See the section on optimization for how to do this. We then have

$$S^2 \equiv Q(x) \equiv \prod_{i=0}^{F} p_i^{\alpha_i} \quad \bmod N.$$

The value of S is easily computed from the value of x because of the special way the coefficients of Q are computed. Let

$$\vec{v} = \{\alpha_1, \alpha_2, \ldots, \alpha_F\} \bmod 2.$$

Collect a total of at least $F + 1$ factored relations. One then finds a set whose product is a square by finding a linear dependency over $GF(2)$ from the matrix formed by letting each $\vec{v}$ be a row in the matrix.

COMPUTATION OF POLYNOMIAL COEFFI-CIENTS: The coefficients for each polynomial are derived from a prime number $D = 3 \bmod 4$ with $(\frac{D}{N}) = 1$. Each prime D yields a different polynomial. This makes parallel implementation of this algorithm easy. Simply give different sets of Ds to different machines and let each one run independently.

To compute the coefficients, we start by letting $A = D^2$ with $D \approx (N/2)^{1/4}/\sqrt{M}$. The value of D is chosen this way to minimize the maximum value of $Q(x)$ over the interval $[-M, M]$. The computation of B, C and S depends on A and is detailed in [7].

OPTIMIZATION AND PARAMETER SELECTION: It is often useful, rather than to factor just N to factor kN for some small value of k. This can have the effect of allowing more small quadratic residues in the factor base. In this case, replace N with kN in all of the computations outlined above. It may also be *necessary* in order for the algorithm to work. Since $N = B^2 - 4AC$ and the right-hand side is 0 or 1 mod 4, if $N \equiv 3 \bmod 4$, then we must multiply N by k so that $kN \equiv 1 \bmod 4$. However, this requirement may be avoided by taking $2B$, rather than just B as the middle coefficient for Q. The Knuth–Schroeppel function may be used to evaluate the effectiveness of different values of k. See [7] for details.

Rather than demand that r be fully factored over the factor base, it is very useful to allow a small number of somewhat larger primes to appear in the factorization of r. This is known as the *large prime variation*. Let

$$r = \prod_{i=0}^{F} p_i^{\alpha_i} P_1 P_2 \ldots,$$

where the P_i are allowed to be somewhat larger primes than those in the factor base. The <u>Birthday Paradox</u> now becomes useful here. The set of fully factored rs will now be quite large and we can expect many of the large primes P_i to appear more than once. When it does, we may multiply the corresponding relations together and obtain P_i^2 on the right-hand side of each relation. For N up to about 85 digits, using one large prime is quite effective. For N greater than about 85 digits, two primes are effective. Limited experience suggests that for N above 120 digits, three primes may be effective. Once the factor base primes are removed, P_1 and P_2 may then be split via Pollard's rho (see <u>integer factoring</u>) or SQUFOF algorithms (see [6] for a definition of SQUFOF). Both are effective.

The smallest primes p_i take the longest time to sieve and their logarithms contribute the least to the accumulated sum at each sieve location. It is worthwhile to replace the smallest primes (up to about 30) with small powers of those primes. This has the effect of greatly speeding sieve time while losing only a tiny fraction of factored relations. This is known as the *small prime variation*.

It is also worthwhile to partition the sieve interval $[-M, M]$ into pieces that will fit in cache while sieving. This too can greatly improve the speed of sieving, and it cuts down on memory requirements.

The cost of changing polynomials is dominated by the cost of computing $(2A)^{-1}$ mod p_i for each p_i. A method for greatly reducing this cost is known as the self-initializing Quadratic Sieve (SIQS). Details may be found in [5].

A fair bit of time is taken by the reconstruction of the actual factorization of r. The sieving process merely identifies which r are smooth. It does not produce the factorization. This is readily accomplished by trial division of r by the factor base primes. However, as N, and hence the size of F increases, trial division starts taking a larger and larger percentage of the run time. This may be alleviated by finding the factorizations by resieving. Now, however, instead of accumulating $\log p_i$, one simply stores the p_i that hit the identified smooth locations. It is now a simple matter to produce the actual factorization.

Suggested values for the parameters M and F as well as additional coding and computational considerations may be found in [7].

Robert D. Silverman

References

[1] Caron, T. and R.D. Silverman (1988). "Parallel implementation of the quadratic sieve." *Journal of Supercomputing*, 1, 273–290.

[2] Lenstra, A. and H.W. Lenstra, Jr. (eds.) (1992). *The Development of the Number Field Sieve*, Lecture Notes in Mathematics, vol. 1554. Springer-Verlag, Berlin.

[3] Lenstra, A. and M. Manasse (1994). "Factoring with two large primes." *Math. Comp.*, 63, 785–798.

[4] Morrison, M. and J. Brillhart (1975). "A method of factoring and the factorization of F_7." *Math. Comp.*, 29, 183–205.

[5] Pomerance, C., J.W. Smith, and R. Tuler (1988). "A pipeline architecture for factoring large integers with the quadratic sieve." *SIAM J. Comp.*, 17 (2), 387–403.

[6] Riesel, H. (1987). *Prime Numbers and Computer Methods for Factorization*. Volume 57 of *Progress in Mathematics*. Birkhäuser.

[7] Silverman, R.D. (1987). "The multiple polynomial quadratic sieve." *Math. Comp.*, 48, 329–339.

QUANTUM CRYPTOGRAPHY

QUANTUM CRYPTOGRAPHY [A]: Quantum Cryptography was born in the early 1970s when Stephen Wiesner wrote "Conjugate Coding," which unfortunately took more than 10 years to see the light of print [50]. In the mean time, Charles H. Bennett (who knew of Wiesner's idea) and Gilles Brassard picked up the subject and brought it to fruition in a series of papers that culminated with the demonstration of an experimental prototype that established the technological feasibility of the concept [5]. Quantum cryptographic systems take advantage of Heisenberg's uncertainty relations, according to which measuring a quantum system, in general, disturbs it and yields incomplete information about its state before the measurement. Eavesdropping on a quantum communication channel therefore causes an unavoidable disturbance, alerting the legitimate users. This yields a cryptographic system for the distribution of a secret random key between two parties initially sharing no secret information (however they must be able to authenticate messages) that is secure against an eavesdropper having at her disposal unlimited computing power. Once this secret key is established, it can be used together with classical cryptographic techniques such as the Vernam cipher (one-time pad) to allow the parties to communicate meaningful information in absolute secrecy.

Quantum cryptography is best known for key distribution [7]. A short summary of this so-called *BB84 protocol* is provided in the Section "Qnantum Key Distribution." A remarkable surge of interest in the international scientific and industrial communities has propelled quantum cryptography into mainstream computer science and physics. Furthermore, quantum cryptography is becoming increasingly practical at a fast pace. The first quantum key distribution prototype, built in 1989, worked over a distance of 32 cm [5], [11]. Since then, many additional experimental demonstrations have been set up, covering distances of tens of kilometers. Consult [46] or [42] for popular accounts of the state of the art in experimental quantum cryptography.

The Various Uses of Quantum Physics for Cryptography

In addition to key distribution, quantum techniques may also assist in the achievement of more subtle cryptographic goals, important in the post-cold war world, such as protecting private information while it is being used to reach public decisions. Such techniques, pioneered by Crépeau [10], [15], allow two people to compute an agreed-upon function $f(x, y)$ on private inputs x and y when one person knows x, the other knows y, and neither is willing to disclose anything about his private input to the other, except for what follows logically from one's private input and the function's output. The classic example of such discreet decision making is the "dating problem," in which two people seek a way of making a date if and only if each likes the other, without disclosing any further information. For example, if Alice likes Bob but Bob does not like Alice, the date should be called off without Bob finding out that Alice likes him. On the other hand, it is logically unavoidable for Alice to learn that Bob does not like her, because if he did the date would be on.

Indeed, two applications of quantum physics to cryptography were discovered well before quantum key distribution: quantum bank notes that are impossible to counterfeit and quantum multiplexing that allows one party to send two messages to another party in a way that the receiver can obtain either message at his choice, but reading one destroys the other irreversibly [50]. (The notion of multiplexing was reinvented 10 years later by Michael Rabin in the context of classical cryptography under the name of oblivious transfer [43], [28].) Unfortunately, even its author, Stephen Wiesner, knew from the start that the quantum multiplexing protocol could be defeated with arbitrary measurements performed by the receiver of the strings. Thus, a more elaborate quantum oblivious transfer protocol was designed subsequently [10] under the assumption of the existence of a bit commitment scheme [19], a result unlikely to be possible classically as argued by Impagliazzo and Rudich [34]. Another quantum cryptographic task that has been studied extensively is indeed bit commitment [15]. Unfortunately it turned out that early claims of security of certain quantum protocols for this task were after all insecure as showed by Mayers [39] and independently by Lo and Chau [37]. This no-go theorem was later extended to *any* Quantum Bit Commitment scheme consistent with quantum physics [40], [38].

On a closely related topic, various Quantum Coin Tossing protocols have been also introduced [7] as well as a lower bound of $1/\sqrt{2}$ on the bias of such a protocol in a very general quantum mechanical framework [1].

Quantum Key Distribution

The purpose of quantum key distribution is to enable two honest parties, Alice and Bob, to agree on a random cryptographic key in a situation where eavesdropping is possible. By transmitting one of four possible nonorthogonal quantum states, Alice may send to Bob a random bit-stream that she knows exactly and of which Bob will randomly select a constant fraction. These four possible states may be the $0°$, $45°$, $90°$, and $135°$ polarizations of a photon. According to quantum mechanics, orthogonally polarized photons (($0°, 90°$) or ($45°, 135°$)) are perfectly distinguishable whereas nonorthogonal photons (($0°, 45°$), ($45°, 90°$), etc.) are not. When Alice sends Bob a random state from these four, he may choose to measure whether it is ($0°, 90°$) or ($45°, 135°$). If he makes the correct measurement then he detects perfectly the original state. If he makes the wrong measurement then he detects a random state among the two he was trying to distinguish. When Alice later tells him which was the correct measurement, he keeps the correctly measured states and discards the others. Thus, in a perfect world, Bob would receive 50% of Alice's photons in their exact original state and discard the other 50% of the photons. If we assign binary value 0 to $0°$ and $45°$ and value 1 to $90°$ and $135°$, then their agreed bit-stream is obtained by the correctly measured 50% of the photons.

However, the world is not perfect. Therefore, a fraction of the correctly measured photons will be detected incorrectly. Also, if an eavesdropper (Eve) tries to measure Alice's photons before they reach Bob, errors will be induced by the fact that she is measuring information about the photons' polarizations. Moreover, these two situations are indistinguishable from each other: natural noise or induced noise looks the same. (Indeed, part of the "natural" noise is produced by "nature" eavesdropping on Alice and Bob!) The beauty of quantum cryptography is that an estimate on the noise level leads to an estimate of the information obtained by Eve. Consequently, a three-phase classical protocol allows Alice and Bob to extract an agreed upon, smaller secret cryptographic key from their noisy, partly eavesdropped bit-stream. These three phases are called "error estimation," "information reconciliation," and "privacy amplification."

Error Estimation. Error estimation is performed by having one of Alice or Bob pick at random a certain number t of bits previously transmitted according to the correct measurement and announce them to the other party. The latter compares these bits with his/her own copy and announces the number of errors e. For large enough samples, the ratio e/t should be a reasonable estimate of the fraction of errors left in the undisclosed bits.

Information Reconciliation. Although interactive error correction such as [16] was first encouraged in [5], Crépeau pointed out that traditional error-correcting codes may be used here as well [10]. In both cases, this process will disclose some extra information about the remaining (corrected) bits to any potential eavesdropper. This extra information must be taken into account in the last privacy amplification phase.

Privacy Amplification. Assume an eavesdropper is left with only ℓ bits of Rényi (collision) entropy about the bit-stream W of size n resulting from the information reconciliation phase. If Alice and Bob can estimate ℓ from error estimation and error correction, they may produce a new smaller bit-stream K of size nearly ℓ from W. Let H be a uniformly selected hash function from a Strongly Universal Set [49] mapping n bits to $\ell - s$ bits. Then we obtain a tight bound on the uncertainty $\mathbf{H}(H(W) \mid H, E) \leq 2^{-s}$ (see information theory for definitions) where E is the eavesdropping information (including error correction). This means that if one of Alice or Bob picks a random hash function h and announces it publicly to the other, they are able to use it to replace their longer string W by $K = h(W)$ that is almost perfectly secret to the eavesdropper [9] with nearly probability 1.

Eavesdropping. The key distribution protocol described above has been proven secure regardless of the eavesdropper's strategy and computing power. The first proof of this theorem is due to Mayers [41]. However, the very technical nature of that proof encouraged many alternate proofs to be developed such as those of Biham, et al. [14], Shor and Preskill [45], Gottesman and Lo [31], etc. A more powerful security proof in the universal composability framework was recently demonstrated by Ben-Or, et al. [13].

Alternative Quantum Key Distribution Protocols and Implementations

The original quantum key distribution protocol uses four different polarization states of single photons as carrier of quantum information [7], but other approaches have been put forward. Early variations were to use only two nonorthogonal states rather than four [4], and to use phase modulation rather than polarization [26], [48]. A more fundamental variation, due to Ekert [25], was to make use of Einstein–Podolsky–Rosen entangled pairs [24], which allows the key to remain protected by quantum mechanics even in storage, rather than merely in transit. More interestingly, Ekert's scheme can benefit from powerful quantum techniques that were discovered only years later, such as entanglement distillation [12], [23]. Prototypes of entanglement-based quantum cryptography, working over kilometers of fiber, came soon after the theoretical invention [26] as well as much more recently [27].

The past decade has seen an explosion in experimental demonstrations of quantum cryptography, with distances ever increasing, sometimes at the expense of giving up the Holy Grail of unconditional security. We can mention only a few examples here. A plug-and-play device built in Switzerland was tested successfully using 67 km of optical fiber laid out between Geneva and Lausanne [47]. More recent experiments achieve even further distances such as 150 km of optical fiber [35]. The notion of *quantum repeaters* has been discussed in order to achieve even greater distances [18]. Free-space prototypes have shown the possibility of line-of-sight quantum cryptography over distances of tens of kilometers [33], [36], making it legitimate to dream of a quantum-cryptographic satellite-based global network [44]. A thorough survey of quantum cryptography, with an emphasis on technological issues, can be found in [29]. A living roadmap of the work ahead can be obtained in [6].

Finally, we point out that quantum key distribution is now available as a commercial product. Information about quantum-cryptographic products can be found at the Web sites of the Swiss company *id Quantique* (www.idquantique.com) and the American corporation *MagiQ Technologies* (magiqtech.com).

Cryptography on Quantum Data

The last component of quantum cryptography is the *cryptography on quantum data* where cryptographic tools are developed for information imbedded in quantum systems. A first example is known as the *one-time quantum pad* where the sender Alice and receiver Bob share *a priori* a pair of maximally entangled particles and use them to teleport [8] an arbitrary qubit (quantum bit), for example, from Alice to Bob. The only public transmission of this scheme is a pair of classical random bits from the sender to receiver, allowing him to reconstruct the original state she wanted to communicate.

A second example is the *Quantum Vernam Cipher* [2] where a classical key of four possible values is used by Alice who applies one of four unitary (Pauli) operators to an arbitrary system of a single qubit that may then be transmitted to Bob. Bob decrypts the state by applying the inverse unitary operator. The quantum description of the state transmitted is the same regardless of the original state to be transferred as long as the key is uniformly distributed and a secret to an eavesdropper. An interesting difference between the quantum and classical scenarios is that two key bits are required to encrypt a general qubit in the quantum setting [2], but this may be reduced to nearly one key bit to encrypt a qubit almost perfectly if we tolerate arbitrarily small errors, as long as it is not entangled with Eve [32]. It was also demonstrated recently how the secret key used for Quantum Vernam Cipher may be re-used [22] when the plaintexts are classical.

Quantum error-correcting codes have led to the notion of *Quantum Message Authentication* [3] that allows Alice to send Bob a message in such a way that any tampering of the transmitted message will either result in detection of the tampering or actual correction of the tampering back to the original message. Surprisingly, quantum authentication requires quantum encryption, whereas classically these two tasks are fairly independent of each other. A very interesting notion of *Uncloneable Encryption*, linking Quantum Encryption and Quantum Authentication, was later introduced by Gottesman [30].

We conclude with a short list of recent quantum cryptographic applications: *Quantum Secret Sharing* [17] where the secret to be shared is a quantum state, *Verifiable Quantum Secret Sharing* [20] offers the extra guarantee that if enough honest people are involved the secret may be uniquely reconstructed, *Multi-Party Quantum Computation* [20] allows multiparty evaluation of a quantum circuit in which each party secretly provides some of the input quantum states, and *Quantum Zero-Knowledge* [21] that generalizes the classical notion although "rewinding" a quantum computer is impossible.

Gilles Brassard
Claude Crépeau

References

[1] Ambainis, A. (2004). "A new protocol and lower bounds for quantum coin flipping." *J. Comput. Syst. Sci.*, 68 (2), 398–416.

[2] Ambainis, A., M. Mosca, A. Tapp, and R. de Wolf (2000). "Private quantum channels." In *FOCS'00: Proceedings of the 41st IEEE Symposium on Foundations of Computer Science*, 547–553.

[3] Barnum, H., C. Crépeau, D. Gottesman, A. Smith, and A. Tapp (2002). "Authentication of quantum messages." In *FOCS'02: Proceedings of the 43rd IEEE Symposium on Foundations of Computer Science*, 449–458.

[4] Bennett, C.H. (1992). "Quantum cryptography using any two nonorthogonal states." *Phys. Rev. Lett.*, 68, 3121–3124.

[5] Bennett, C.H., F. Bessette, G. Brassard, L. Salvail, and J. Smolin (1992). "Experimental quantum cryptography." *J. Cryptol.*, 5 (1), 3–28.

[6] Bennett, C.H., D. Bethune, G. Brassard, N. Donnangelo, A.K. Ekert, C. Elliott, J. Franson, C. Fuchs, M. Goodman, R. Hughes (Chair), P. Kwiat, A. Migdall, S.-W. Nam, J. Nordholt, J. Preskill, and J. Rarity (2004). "A quantum information science and technology roadmap, Part 2: Quantum cryptography, Version 1.0." *Advanced Research and Development Activity (ARDA)*, July 2004. Available at http://qist.lanl.gov/qcrypt_map.shtml.

[7] Bennett, C.H., and G. Brassard (1984). "Quantum cryptography: Public key distribution and coin tossing." In *Proceedings of IEEE International Conference on Computers, Systems and Signal Processing, Bangalore, India*, 175–179.

[8] Bennett, C.H., G. Brassard, C. Crépeau, R. Jozsa, A. Peres, and W.K. Wootters (1993). "Teleporting an unknown quantum state via dual classical and Einstein–Podolsky–Rosen channels." *Phys. Rev. Lett.*, 70, 1895–1899.

[9] Bennett, C.H., G. Brassard, C. Crépeau, and U. Maurer (1995). "Generalized privacy amplification." *IEEE Transaction on Information Theory*, 41 (6), 1915–1923.

[10] Bennett, C.H., G. Brassard, C. Crépeau, and M.-H. Skubiszewska (1991). "Practical quantum oblivious transfer." In *Advances in Cryptology: Proceedings of Crypto'91*, 351–366.

[11] Bennett, C.H., G. Brassard, and A.K. Ekert (1992). "Quantum cryptography." *Scientific American*, 267 (4), 50–57.

[12] Bennett, C.H., G. Brassard, S. Popescu, B. Schumacher, J.A. Smolin, and William K. Wootters (1996). "Purification of noisy entanglement and faithful teleportation via noisy channels." *Physical Review Letters*, 76, 722–725.

[13] Ben-Or, M., M. Horodecki, D.W. Leung, D. Mayers, and J. Oppenheim (2005). "The universal composable security of quantum key distribution." In *Proceedings of Second Theory of Cryptography Conference: TCC 2005, Cambridge, MA, USA, February 10–12, 2005*. http://arXiv.org/abs/quant-ph/0409078.

[14] Biham, E., M. Boyer, P.O. Boykin, T. Mor, and V. Roychowdhury (2000). "A proof of the security of quantum key distribution (extended abstract)." In *STOC'00: Proceedings of the 32nd Annual ACM Symposium on Theory of Computing*, 715–724.

[15] Brassard, G., C. Crépeau, R. Jozsa, and D. Langlois (1993). A quantum bit commitment scheme provably unbreakable by both parties. In *FOCS'93: Proceedings of the 34th IEEE Symposium on Foundations of Computer Science*, 362–371.

[16] Brassard, G., and L. Salvail (1993). "Secret-key reconciliation by public discussion." In *Advances in Cryptology: Proceedings of Eurocrypt'93*, 410–423.

[17] Cleve, R., D. Gottesman, and H.-K. Lo (1999). "How to share a quantum secret." *Phys. Rev. Lett.*, 83 (3), 648–651.

[18] Collins, D., N. Gisin, and H. De Riedmatten (2005). "Quantum relays for long distance quantum cryptography." *J. Mod. Optics*, 52 (5), 735.

[19] Crépeau, C., P. Dumais, D. Mayers, and L. Salvail (2004). "Computational collapse of quantum state with application to oblivious transfer." In *Proceedings of First Theory of Cryptography Conference: TCC 2004, Cambridge, MA, USA, February 19–21*, 374–393.

[20] Crépeau, C., D. Gottesman, and A. Smith (2002). "Secure multi-party quantum computation." In *STOC'02: Proceedings of the 34th Annual ACM Symposium on Theory of Computing*, 643–652.

[21] Damgård, I., S. Fehr, and L. Salvail (2004). "Zero-knowledge proofs and string commitments withstanding quantum attacks." In *Advances in Cryptology: Proceedings of Crypto'04*, 254–272.

[22] Damgård, I., T. Pedersen, and L. Salvail (2004). "On the key-uncertainty of quantum ciphers and the computational security of one-way quantum transmission." In *Advances in Cryptology: Proceedings of Eurocrypt'04*, 91–108.

[23] Deutsch, D., A.K. Ekert, R. Jozsa, C. Macchiavello, S. Popescu, and A. Sanpera (1996). "Quantum privacy amplification and the security of quantum cryptography over noisy channels." *Physical Review Letters*, 77, 2818–2821. Erratum (1998). *Physical Review Letters*, 80, 2022.

[24] Einstein, A., B. Podolsky, and N. Rosen (1935). "Can quantum-mechanical description of physical reality be considered complete?" *Physical Review*, 47, 777–780.

[25] Ekert, A.K. (1991). "Quantum cryptography based on Bell's theorem." *Phys. Rev. Lett.*, 67, 661–663.

[26] Ekert, A.K., J. Rarity, P. Tapster, and G. Palma (1992). "Practical quantum cryptography based on two-photon interferometry." *Physical Review Letters*, 69, 1293–1295.

[27] Enzer, D.G., P.G. Hadley, R.J. Hughes, C.G. Peterson, and P.G. Kwiat (2002). "Entangled-photon six-state quantum cryptography." *New Journal of Physics*, 4, 45.1–45.8.

[28] Even, S., O. Goldreich, and A. Lempel (1985). "A randomized protocol for signing contracts." *Commun. ACM*, 28 (6), 637–647.

[29] Gisin, N., G. Ribordy, W. Tittel, and H. Zbinden (2002). "Quantum cryptography." *Reviews of Modern Physics*, 74, 145–195.

[30] Gottesman, D. (2003). "Uncloneable encryption." *Quantum Information and Computation*, 3, 581–602. http://arXiv.org/abs/quant-ph/0210062.

[31] Gottesman, D., and H.-K. Lo (2003). "Proof of security of quantum key distribution with two-way classical communications." *IEEE Trans. Inf. Theory*, 49, 457–475.

[32] Hayden, P., D. Leung, P.W. Shor, and A. Winter (2004). Randomizing quantum states: Constructions and applications. *Commun. Math. Phys.*, 250, (2), 371–391.

[33] Hughes, R.J., J.E. Nordholt, D. Derkacs, and C.G. Peterson (2002). "Practical free-space quantum key distribution over 10 km in daylight and at night." *New Journal of Physics*, 4, 43.1–43.14.

[34] Impagliazzo, R., and S. Rudich (1989). "Limits on the provable consequences of one-way permutations." In *STOC'89: Proceedings of the 21st Annual ACM Symposium on Theory of Computing*, 44–61.

[35] Kimura, T., Y. Nambu, T. Hatanaka, A. Tomita, H. Kosaka, and K. Nakamura (2004). "Single-photon interference over 150-km transmission using silica-based integrated-optic interferometers for quantum cryptography." *Electronics Letters*, 43 (9), L1217–L1219.

[36] C. Kurtsiefer, P. Zarda, M. Halder, H. Weinfurter, P.M. Gorman, P.R. Tapster, and J.G. Rarity (2002). "Quantum cryptography: A step towards global key distribution." *Nature*, 419, 450.

[37] Lo, H.-K., and H.F. Chau (1997). "Is quantum bit commitment really possible?" *Physical Review Letters*, 78 (17), 3410–3413. Originally http://arXiv.org/abs/quant-ph/9603004.

[38] Lo, H.-K., and H.F. Chau (1998). "Why quantum bit commitment and ideal quantum coin tossing are impossible." *Physica D*, 120, 177–187.

[39] Mayers, D. (1995). "The trouble with quantum bit commitment." http://arXiv.org/abs/quant-ph/9603015. The author first discussed the result in Montréal at a workshop on quantum information theory held in October 1995.

[40] Mayers, D. (1997). "Unconditionally secure quantum bit commitment is impossible." *Physical Review Letters*, 78 (17), 3414–3417.

[41] Mayers, D. (2001). "Unconditional security in quantum cryptography." *J. ACM*, 48 (3), 351–406.

[42] Ouellette, J. (2005). "Quantum key distribution." *The Industrial Physicist*, 10 (6), 22–25.

[43] Rabin, M. (1981). "How to exchange secrets by oblivious transfer." Technical Report TR-81, Harvard Aiken Computation Laboratory.

[44] Rarity, J.G., P.R. Tapster, P.M. Gorman, and P. Knight (2002). "Ground to satellite secure key exchange using quantum cryptography." *New Journal of Physics*, 4, 82.1–82.21.

[45] Shor, P.W., and J. Preskill (2000). "Simple proof of security of BB84 quantum key distribution protocol." *Phys. Rev. Lett.*, 85, 441–444.

[46] Stix, G. (2005). "Best-kept secrets—quantum cryptography has marched from theory to laboratory

to real products." *Scientific American*, 280 (1), 78–83.

[47] Stucki, D., N. Gisin, O. Guinnard, G. Ribordy, and H. Zbinden (2002). "Quantum key distribution over 67 km with a plug & play system." *New Journal of Physics*, 4, 41.1–41.8.

[48] Townsend, P., J. Rarity, and P. Tapster (1993). "Single photon interference in 10 km long optical fibre interferometer." *Electronics Letters*, 29, 1291–1293.

[49] Wegman, M., and J. Carter (1981). "New hash functions and their use in authentication and set equality." *J. Comput. Syst. Scz.*, 22, 265–279.

[50] Wiesner, S. (1983). "Conjugate coding." *Sigact News*, 15 (1), 78–88. Original manuscript written circa 1970.

R

RABIN CRYPTOSYSTEM

The smaller the public exponent in the RSA public key encryption or RSA digital signature schemes, the more efficient the encryption process is. Michael O. Rabin thus suggested to use $e = 2$ into an encryption scheme [12]. But things are not as simple as for RSA.

MODULAR SQUARING: Thanks to the Euler's theorem, one can easily extract modular eth roots, until e is co-prime to $\varphi(n)$ (see Euler's Totient function) and the latter value is known: $d = e^{-1}$ mod $\varphi(n)$ helps to get it. Unfortunately, $e = 2$ is not co-prime to $\varphi(n)$, moreover squaring is not a bijection in the group $\mathbb{Z}_n^*$, for $n = pq$ (see also modular arithmetic), and even in $\mathbb{Z}_p^*$ for a prime number p: if x is a square root of y in $\mathbb{Z}_p^*$, then $-x$ is also a square root of y. More formally, the function $f : x \mapsto x^2$ mod p from $\mathbb{Z}_p^*$ into $\mathbb{Z}_p^*$ is a morphism, whose kernel is $\{-1, +1\}$. As a consequence, the cardinality of the image of f is exactly $(p - 1)/2$: an element in $\mathbb{Z}_p^*$ is either a square with two square roots, or a non-square without any square root (see also quadratic residue).

Once again, the Chinese Remainder Theorem helps to know more about squares in $\mathbb{Z}_n^*$ for a composite $n = pq$. Indeed, y is a square in $\mathbb{Z}_n^*$ if and only if it is a square in both $\mathbb{Z}_p^*$ and $\mathbb{Z}_q^*$: there are only $\varphi(n)/4$ squares which admit four distinct square roots: $x, -x, z$ and $-z$. Since they are distinct, $x + z \neq 0$ mod n and $x - z \neq 0$ mod n. However, $x^2 = z^2 = y$ mod n. Then,

$$x^2 - z^2 = (x - z)(x + z) = 0 \text{ mod } n.$$

As a consequence, $\gcd(x - z, n) \in \{p, q\}$: the ability to compute modular square roots helps to factor the modulus.

In this other direction, Euler's theorem does not help any more, since 2 is not co-prime to $\varphi(n)$ (whatever n is, either a prime or a composite integer). But first, for prime moduli, methods are known to compute square roots. Particularly, for *Blum primes* p, which satisfy $p = 3$ mod 4 (see also Blum integer), if y is a square in $\mathbb{Z}_p^*$, then the square roots are $\pm y^{(p+1)/4}$ mod p. Then, for computing square roots in $\mathbb{Z}_n^*$, one can simply use the Chinese Remainder Theorem: from $y \in \mathbb{Z}_n^*$, one uses the isomorphism from $\mathbb{Z}_n^*$ onto $\mathbb{Z}_p^* \times \mathbb{Z}_q^*$. One then computes the square roots in $\mathbb{Z}_p^*$ and $\mathbb{Z}_q^*$. The inverse isomorphism on the four possible pairs leads to the four possible square roots

of y. Therefore, the square root problem in $\mathbb{Z}_n^*$, with $n = pq$, is equivalent to the factorization of n, which is a stronger formal result than for RSA.

THE RABIN PRIMITIVE: Granted the equivalence between the modular square root problem and the factorization of the modulus, it is natural to try to use it for cryptographic applications: Rabin suggested a public-key cryptosystem [12].

- Key generation: randomly choose two large Blum primes p and q, and compute $n = pq$. The public key is thus the modulus n, while the private consists of its factorization (p, q).
- Encryption: in order to encrypt a message $m \in \mathbb{Z}_n^*$, one computes $c = m^2$ mod n.
- Decryption: given the ciphertext c, granted the factorization of n, one can extract the square roots.

Unfortunately, a problem arises here because of the non-injectivity of the square function: four plaintexts are possible. Redundancy in the plaintext is thus required to help the recipient to make a choice. Furthermore, the algebraic structure allows several kinds of attacks, as RSA suffers, and thus paddings are required to solve the two problems. The SAEP$^+$ padding which contains redundancy can be applied to the Rabin primitive. It would then lead to an efficient encryption scheme, provably IND–CCA2 secure under the intractabilty of integer factoring, in the random oracle model.

David Pointcheval

References

[1] Bellare, M. and P. Rogaway (1993). "Random oracles are practical: A paradigm for designing efficient protocols." *Proceedings of the 1st CCS*. ACM Press, New York, 62–73.

[2] Bellare, M. and P. Rogaway (1995). "Optimal asymmetric encryption—how to encrypt with RSA." *Advances in Cryptography—EUROCRYPT'94*, Lecture Notes in Computer Science, vol. 950, ed. A. De Santis. Springer-Verlag, Berlin, 92–111.

[3] Bleichenbacher, D. (1998). "A chosen ciphertext attack against protocols based on the RSA encryption standard PKCS #1." *Advances in Cryptography—CRYPTO'98*, Lecture Notes in Computer Science, vol. 1462, ed. H. Krawczyk. Springer-Verlag, Berlin, 1–12.

[4] Blum, M. and S. Micali (1984). "How to generate cryptographically strong sequences of

pseudorandom bits." *SIAM Journal on Computing*, 13, 850–864.

[5] Boneh, D. (2001). "Simplified OAEP for the RSA and Rabin functions." *Advances in Cryptography— CRYPTO 2001*, Lecture Notes in Computer Science, vol. 2139, ed. J. Kilian. Springer-Verlag, Berlin, 275–291.

[6] Boneh, D. and G. Durfee (2000). "Cryptanalysis of RSA with private key d less than $N^{0.292}$." *IEEE Transactions on Information Theory*, 46 (4), 1339–1349.

[7] Fiat, A. and A. Shamir (1987). "How to prove yourself: Practical solutions of identification and signature problems." *Advances in Cryptography— CRYPTO'86*, Lecture Notes in Computer Science, vol. 263, ed. A.M. Odlyzko. Springer-Verlag, Berlin, 186–194.

[8] Fujisaki, E., T. Okamoto, D. Pointcheval, and J. Stern (2001). "RSA–OAEP is secure under the RSA assumption." *Advances in Cryptography— CRYPTO 2001*, Lecture Notes in Computer Science, vol. 2139, ed. J. Kilian. Springer-Verlag, Berlin, 260–274.

[9] Håstad, J. (1988). "Solving simultaneous modular equations of low degree." *SIAM Journal of Computing*, 17, 336–341.

[10] Miller, G. (1976). "Riemann's hypothesis and tests for primality." *Journal of Computer and System Sciences*, 13, 300–317.

[11] Naor, M. and M. Yung (1989). "Universal one-way hash functions and their cryptographic applications." *Proceedings of the 21st STOC*. ACM Press, New York, 33–43.

[12] Rabin, M.O. (1978). "Digitalized signatures." *Foundations of Secure Computation*, eds. R. Lipton and R. De Millo. Academic Press, New York, 155–166.

[13] Rackoff, C. and D.R. Simon (1992). "Non-interactive zero-knowledge proof of knowledge and chosen ciphertext attack." *Advances in Cryptology—CRYPTO'91*, Lecture Notes in Computer Science, vol. 576, ed. J. Feigenbaum. Springer-Verlag, Berlin, 433–444.

[14] Rivest, R., A. Shamir, and L. Adleman (1978). "A method for obtaining digital signatures and public key cryptosystems." *Communications of the ACM*, 21 (2), 120–126.

[15] Shoup, V. (2001). "OAEP reconsidered." *Advances in Cryptology—CRYPTO 2001*, Lecture Notes in Computer Science, vol. 2139, ed. J. Kilian. Springer-Verlag, Berlin, 239–259.

[16] Wiener, M. (1990). "Cryptanalysis of short RSA secret exponents." *IEEE Transasctions on Information Theory*, 36 (3), 553–558.

RABIN DIGITAL SIGNATURE SCHEME

In [5] Rabin described a <u>trapdoor one-way function</u> that can be used for <u>digital signatures</u> and for public-key encryption (see <u>public key cryptography</u>). Here we focus on Rabin's digital signature system.

Rabin's trapdoor function makes use of <u>modular arithmetic</u> and is defined as follows: (i) let $N = pq$ be a product of two distinct equal size <u>prime numbers</u>, (ii) define the function $F : \mathbb{Z}_N^* \to \mathbb{Z}_N^*$ as $F(x) = x^2 \in \mathbb{Z}_N^*$. Since N is a product of two distinct primes, the function F is a 4-to-1 map on $\mathbb{Z}_N^*$ (every element in the image of F has exactly four pre-images). Rabin shows that inverting this function is as hard as factoring the modulus N. However, given the factorization of N it is easy to find all four pre-images for a given element in the image of F. Hence, the factorization of N serves as a trapdoor for this function.

In the <u>random oracle model</u> there are several generic methods for building a secure signature scheme from a 4-to-1 trapdoor function. Here we describe Rabin signatures using the Full Domain Hash method [2].

Key Generation. Given a security parameter $\tau \in \mathbb{Z}$ as input do the following:

1. Generate two random τ-bit primes p, q where $p = q = 3 \bmod 4$. Set $N = pq$.
2. Pick an element $w \in \mathbb{Z}_N^*$ such that the <u>Jacobi symbol</u> of w over N is equal to -1. In other words, w is a <u>quadratic residue</u> modulo exactly one of p or q.
3. Let H be a <u>hash function</u> $H : \{0, 1\}^* \to \mathbb{Z}_N$.
4. Output the public key (N, w, H) and the private key (N, p, q, w, H).

Signing. To sign a message $m \in \{0, 1\}^*$ using the private key (N, p, q, w, H) do:

1. Compute $x = H(m) \in \mathbb{Z}_N$. If x is not in $\mathbb{Z}_N^*$ output 'fail' and abort. This is extremely unlikely to happen.
2. One can show that exactly one of $\pm x, \pm xw \in \mathbb{Z}_N^*$ must be a quadratic residue. Let $y \in \{\pm x, \pm xw\}$ be that value. To find y, find the unique element in $\{\pm x, \pm xw\}$ for which the <u>Legendre symbol</u> is equal to 1 over both p and q.
3. Let $s \in \mathbb{Z}_N^*$ be the square root of y in $\mathbb{Z}_N^*$. Output s as the signature on m.

Verifying. To verify a message/signature pair $(m, s) \in \{0, 1\}^* \times \mathbb{Z}_N$ using the public key (N, p, q, w, H) do:

1. Compute $x = H(m) \in \mathbb{Z}_N$.
2. Check if $s^2 \in \{\pm x, \pm xw\}$. If so, accept the signature. Otherwise, reject.

Note that signature verification is fast requiring a single modular squaring. When the hash function $H : \{0, 1\}^* \to \mathbb{Z}_N$ is modeled as a random oracle, one can show that the signature scheme is <u>existentially</u> unforgeable under a chosen message attack assuming that factoring random <u>Blum</u>

integers is intractable [2, 4]. To do so, one shows that a forging algorithm can be used to factor the modulus N in the public key.

We note that Rabin signatures can be shortened by a factor of 2 using a cute trick due to Bleichenbacher [1]. The basic idea is to output only half the bits of s (the most significant ones). Let $\hat{s}$ be the resulting signature. Its length is τ-bits as opposed to 2τ-bits. During verification, the least significant bits of the signature can be recovered using Coppermith's algorithm [3]. Indeed, given $x, \hat{s} \in \mathbb{Z}$, Coppersmith's algorithm can test whether there exists a $0 \leq \Delta < 2^{\tau}$ such that $(\hat{s} \cdot 2^{\tau} + \Delta)^2 = x \bmod N$. If Δ exists the algorithm will find it, thus recovering the missing bits of the signature. For a more efficient method see [1].

Dan Boneh

References

[1] Bleichenbacher, Daniel (2004). "Compressing Rabin signatures." *Proceedings CT-RSA, Topics in Cryptology: CT-RSA 2004*, Lecture Notes in Computer Science, vol. 2964, ed. Okamoto. Springer-Verlag, 126–128.

[2] Mihir Bellare and Phillip Rogaway (1996). "The exact security of digital signatures: How to sign with RSA and Rabin." *Advances in Cryptology—EUROCRYPT'96*, Lecture Notes in Computer Science, vol. 1070, ed. U. Maurer. Springer-Verlag, Berlin, 399–416.

[3] Coppersmith, D. (1997). "Small solutions to polynomial equations, and low exponent RSA vulnerabilities." *Journal of Cryptology*, 10, 233–260.

[4] Coron Jean-Sébastien (2000). "On the exact security of full domain hash." *Advances in Cryptology—CRYPTO 2000*, Lecture Notes in Computer Science, vol. 1880, ed. M. Bellare. Springer-Verlag, Berlin, 229–35.

[5] Rabin, M. (1979). "Digital signatures and public key functions as intractable as factorization." Technical Report MIT/LCS/TR-212, MIT Laboratory for Computer Science.

RADIO FREQUENCY ATTACKS

SUMMARY: Conquest without fight is the basic idea of Sun Tzu in *The art of war* written already 2500 years ago. Force an enemy to retreat immediately, strike with high precision without leaving of the origin of the blow, shut down an adversary's communications networks, disrupt its power supplies, yet still leaving buildings intact are the properties conferred to an attack that uses electromagnetic waves at the right frequencies. Most types of matter are transparent to microwaves, and waves coming from an electromagnetic blast are difficult to stop in an appropriate manner. A mastered generation of microwaves may not only disrupt or damage electronic equipment, but may also create faults and even completely destroy it. Solar storms constitute a good illustration of the kind of disruption that an electronic equipment might be submitted to when exposed to electromagnetic disruptions (satellite communications, . . .).

Today's computers and other electronics devices are sensitive to computer attacks such as worms, viruses or logical bombs (see Trojan horses, computer viruses and worms). Electromagnetic radiation leakage is well known and remains the subject of studies for some devices [1, 7, 12, 15]. The electronic components that make up common devices may, however, be disrupted by intense electromagnetic variations in their near surroundings. The sensitiveness of the equipment has thus to be taken into account. Electromagnetic waves that are able to destroy electronics (Compton effect) can be obtained by different means. They may be used for specific purposes by determined will (e-bomb Mk-84, Argus Project on the 27th of August 1958), but they may also have accidental results (Blackout of Hawaii and disturbances in radio-navigations in 1958). The imagination of scenarists and the movie world made that the general public became aware, not only of the existence but also of the power, of ElectroMagnetic Pulse (EMP), through recent Hollywood films like Golden Eye (1995), Broken Arrow (1996), Matrix (1999) or Ocean's Eleven (2001). This type of attack is, however, not yet very common and seems to be part of advanced research projects [5]. This may be due to the particular constraints they involve like mastering impulse modes, high currents and high temperature supraconductivity. Anyway, the technology will however become more available and the number of incidents will increase. Has a new Pandora's box been opened [8]?

INTRODUCTION AND DEFINITIONS: A sufficiently intense and correctly directed electromagnetic field may disrupt the functioning of an electric circuit to different degrees. The circuit may become completely useless. According to Carlo Kopp [10], electromagnetic weapons can nowadays be realized by nations with limited powers, close to those at the disposal of the big nations at the beginning of World War II. More than 20 countries have development programs for some kind of radiofrequency weapons. These weapons do not leave any trace behind, damage buildings, may strike without anybody noticing it a priori and may be created by ill-intentioned people without yielding any big benefit, but generating enough power to cause delicate situations. The idea of using fields for destruction purposes is rather old and today it has become possible for an amateur to construct an e-bomb in his garage. The Internet also contributes to the dissemination of technical data related to this technology [6, 7], and details the materials and open literature or reference material to a high number of persons.

There are two types of RF attacks, namely high and low power attacks. The high power attacks are the most devastating ones, requiring mastering techniques like explosively pumped flux generator (Field Compress Generator), homopolar generators, plastic explosives, high performance detonation and magnetohydrodynamics (MHD). This is why they are almost inaccessible to non-governmental entities, but almost exclusively reserved to government funded, specialized research units (or to government supported terrorists). Moreover, their final bulk and form factor are rather imposing and close to the size of more conventional bombs. Attacks using low power weapons, on the contrary, are considerably easier, less expensive (really very low cost) and are accessible to many people, provided that they dispose of the required level of expertise. An experimented technician or an engineer can design, fabricate and experiment such a device. Very old antennas and military amplifiers can, for instance, be bought from military surplus and can easily be modified to serve these purposes. It is not even necessary to look for components that have this high potential, capable of yielding this kind of power.

Ultra-wideband and narrowband weapons are the basis of radiofrequency attacks. Ultra-wideband devices emit over a large frequency range, nanoseconds long burst of low energy (about 10 J/pulse) are radiated hundred of times within a second. Their destructive power is directly linked to the strength of the source and to the distance from the target. On the opposite, narrowband weapons emit a very reduced spectrum or a unique sinus wave (one frequency) at very high power which can reach thousand kilojoules per pulse, and this hundreds of times a second.

Electromagnetic waves, like acoustic waves, may however interact with human and other living beings. Radiotherapy treatment of cancers is a proof thereof, without even mentioning food cooking in a microwave oven. In the same way

as audio waves with frequencies of a few Hertz, microwaves may have a disturbing effect on the human organism (VMADS and crowd control). But this is not the subject of this article. It is important not to confuse instruments allowing conducting an RF attack with jamming devices that are used to temporarily disrupt the electromagnetic spectrum within some frequency span and in a given region. This is not the subject of this article either.

Principles and Description

VLSI stands for Very Large Scale Integration, and the integration density of modern chips is ever growing (system on chip, ...). Modern VLSI chips are extremely sensitive to voltage surges, and could be burned out by even small current leakages resulting from an EMP. The higher the circuit's density, the more vulnerable it is of course. It is also important to notice that, as the power voltage of chips and computers is going down and down, their susceptibility is increasing.

ElectroMagnetic Pulse

Arthur Compton discovered the Compton effect in 1923. When a beam of X-rays of well-defined frequency is scattered through an angle by sending the radiation through a metallic foil, the frequency of the scattered radiation is different from the original one. So if a nuclear blast occurs at high altitudes, the gamma rays following the detonation are the source of HEMP. When the rays encounter the upper regions of the atmosphere, the molecules interact with the rays depending on the atmospheric density and burst conditions. So the energy of the gamma rays is transferred to the electron of an air molecule. The EMP effect may, in fact, decompose into three periods: early time (0–1 μs), intermediate time (1 μs to a tenth of a second) and the last but not the least late time which involves magnetohydrodynamic properties. At the first order, gamma radiations cause bursts of electrons from the photoelectric Compton effect. Contrary to high altitude detonation, a surface burst EMP (SBEMP) can be produced by a nuclear burst close to the earth's surface. In such a case, particular attention must be given to surge protection to dissipate the high currents. Ground connection can enable in certain cases to reduce or send back the currents that are created by the EMP. An Air-burst EMP is the third possibility between low and high altitudes (0.5–30 km). The effects are a melt of both previous cases.

System generated EMP (SGEMP) and Internal EMP (IEMP) are of course obtained from the direct interaction of nuclear gamma rays and X-rays from a system. The first case is very important outside the atmosphere, typically for satellites. The emission of electrons from internal EMP can create current generation and electromagnetic fields within cavities create the second one.

Coupling

Energy distribution lines or telecommunication wires are omnipresent, both at the level of buildings as within electrical circuits. This is why they are privileged targets for radio frequency attacks. As the emitted wave rapidly attenuates, it is more interesting to create a surge in a wire and let it propagate, rather than to increase indefinitely the emission power (which implies very complex problems). But therefore one should be able to optimize the coupling between the radiated wave and the circuit that will receive it. Water supply circuits or metal pieces in a building, without electrical function, may also help to make EMP induced currents to transit (dangerous in case of cross talk).

It is sometimes necessary to call upon theoretical physics and electronics to compute the equivalent circuit to determine the EMP induced voltage. Electronic structures are often approximated and it is not always possible to consider that the radiation is a plane wave. So after the experiment, the empirical data should be compared with the experimental data coming from complex analytical computations. Transient effects also have to be taken into account and therefore typical coupling models and shielded cables coupling models are used. These models consider the transfer impedance and the conductivity, when they can be calculated.

Transmission line theory is nevertheless the most commonly used method to determine the effects of an EMP on aerial and buried conductors. The length of the conductor has then to be compared with the length of the radiated wave, which is one of the principal characteristics together with the characteristic impedance and the charge impedance. Computers allow then finding an analytical solution.

All electronic circuits contain resistors, capacitors or inductances, which provide them with one or several resonance frequencies. In this way, the electromagnetic pulse can make the circuit resonating at several dominant frequencies. According to the electrical resonance, the oscillation can

be long live, but this highly depends on the shielding and grounding characteristics.

It is also important to notice the cross talk effects. A cable that conveys current can, depending on its distance to other surrounding cables (water pipe, twisted pair, . . .), create by radiation a current in another cable that was initially not affected by the problem.

Damage

The equipment's susceptibility is an important parameter to know if it can be avoided, affected, disrupted or destroyed by a radio frequency attack. Because of Ohm's law, high impedance and low voltage signals are most susceptible to interfere. The obtained currents are sometimes even sufficient to fuse the silicon on the chip. We can, however, distinguish several entering points for a device under test. The first one is called the front door and concerns elements for which the nature or structure favors the effects of the waves. Computer or other electronic devices have attached antennas that constitute a nice front door for an electromagnetic pulse to penetrate into a device [13]. The back door, on the contrary, may be composed of an unshielded wire.

Damages may thus be of different natures, going from disruptions to lasting destruction, including thermal related failure, metallization burn out and the avalanche effects in active components [9, 19]. Passive components are destroyed by voltage breakdown or induced thermal overstress. This is true for resistors (overheating) and capacitors (dielectric breakdown) [18].

Even for cases where the result would not be directly exploitable, the real problem in the use of an RF attack by non-expert persons would be the Denial of Services (DOS). Without necessary reaching the needed destruction level, it could be possible to damage an unprotected and important structure. The replacement or the repairing time may then be prohibitive with respect to the function of the machine.

Energy

Maxwell's law explains that the effect of an electromagnetic wave follows an inverse square law with increasing distance, so the strength of the wave dissipates quickly as it moves away from the initial point of the blast. It is thus necessary to get substantial energies (GigaWatts) in very short times to create electromagnetic disturbances such as the ones described in this article. The sources of the energies put at stake may be twofold: nuclear or electrical energy but even sometimes converted mechanical energy. Obviously, the results are better when using nuclear energy.

Ultrawideband bombs create en electromagnetic pulse like a nuclear detonation, but a conventional or chemical explosive replaces the nuclear part. The microwave source relies on an extremely fast switching device. But narrow band bombs are based on magnetron or vircator (Virtual Cathode Oscillator).

Non-nuclear Techniques

Forty years ago, Andréï Sakharov (Nobel Peace Price 1975) has elaborated, together with Altshuler, Voitenko and Bichenkov, and at the same time as Clarence Marx, the first explosively pumped flux generator (FCG) [14]. He obtained an intense field by discharging a capacitor inside a solenoid and managed to crush the field lines with a peripheral explosive. Then, also by using the energy of an explosive, he constructed the MK-2. The locking up of the explosive within a copper tube containing an inductance, made it to deform the tube by short-circuiting the spirals of the solenoid one after another. As the speed is very high, the sudden decrease of the inductance and the conservation of the flux occasion a brutal increase of the intensity. The very big FCG have already reached some tens of GigaWatts, and as for magnetrons they can be cascaded, the output of the first one supplying the entry of the second.

The microwave source used is an extremely fast switching device. Narrowband e-bombs use a virtual cathode oscillator tube or a variant of a magnetron. The idea behind the Vircator is accelerating a high current electron beam against a mesh anode. Many electrons will pass through

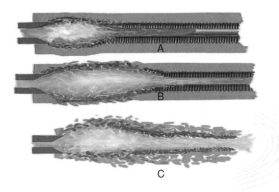

Courtesy of John Batchelor © http://www.jbatchelor.com

the anode, forming a bubble of space charge behind the mesh anode. Under the proper conditions, this space charge region will oscillate at very high frequencies. If the space charge region is placed into a resonant cavity which is appropriately tuned, very high peak powers may be obtained.

NUCLEAR TECHNIQUES AND TED: Schamiloglu's team and the U.S. Air Force's Shiva Star is a pulsed-power system used to simulate the effects of nuclear weapons. TED are very simple devices at low cost. TED does not generate a pure sine wave, it operates completely differently than narrow band devices. Instead of generating a burst of smooth sinus it generates a single spike of energy.

Protection and Countermeasures

The first countermeasure consists in performing a real measure of the susceptibility of the involved equipment. Constructing an analytical model of the disruptions that can be generated on a device is very difficult; this is the reason why tests are carried out. The Federal Communication Commission (FCC) establishes the testing method for emission certification of commercial products. A lot of FCC measures are indeed carried out nowadays and in the best cases without adding all possible options to the equipment. For commercial purposes, the great majority of present machines are constructed in such a way as to respect the electromagnetic compatibility standards. Each year, the aircraft direction notices, however, cases where portable computers or electronic gadgets manage to disrupt the aircraft's electronics. The same applies to machines that are used in hospitals and their interaction with portable telephones. So, the required level of protection is to be evaluated before starting to devise and apply a whole series of countermeasures.

Computers are very sensitive to RF attacks, because these can be propagated through their power cable, network interface, the wire mouse, the keyboard, Input/Output slots and cables, and all kinds of apertures (peripherals, buttons, ventilation, ...). Telecommunications Electronics Material Protected from Emanating Spurious Transmissions (TEMPEST) are better protected than ordinary computers, because the reduction of their radiation level also reduces their sensitiveness. Machines abiding by the military "Milspec" norms, or "ruggedised" machines, already exist, but their power and cost discourages possible buyers. Solutions exist nonetheless, but they often involve high costs. The use of optical fibers would, for instance, highly reduce the sensitiveness of machines. Nowadays nothing forbids the use of a keyboard or a mouse that uses optical communication from one side to the other, like micro-spies already do by modulating the light signal of an optical fiber. The construction of a comprehensive shield and the use of ferrite beads, together with a non-electrical coupling of the connectors would already reduce the risks. Another advantage would be the reduction of mass loops, which often disrupt the connection of peripherals. A commercial computer, armored against RF attacks, does not need the same level of protection as a military machine, but it requires to continue working in the presence of disruptions on, for example, the power supply lines.

PC switch-mode supplies are particularly sensitive to electromagnetic disruptions, because they all contain components that are ideally sensitive to an attack. It is, however, possible to lower their fragility by modifying slightly the structures of their power supplies. A perfect power supply would recreate its energy in a Faraday cage. One should therefore use a dynamo that is mechanically coupled to a motor, situated outside the room.

Network components are also hardly protected and cables, scattered around entire buildings, allow conveying the induced signal. Commutation electronics that are used in network devices are, of course, very sensitive to RF attacks. The use of optical fibers, which convey light but do not offer electrical conduction, would allow reducing the risks. Optical fibers are immune against any kind of electromagnetic attack.

These attacks disrupt or damage logic devices, but there are some components that they are not in power to act upon. Vacuum tubes were used during the conquest of space for their weak sensitiveness to solar radiation disruptions. These tubes do not use silicon as a basic component, so the Compton effect does not create any noticeable disruption on their functioning. It should be noticed that some fight aircraft have still some of their sensitive parts equipped with tubes in order to avoid electromagnetic disruptions. These tubes suffer, however, from some other drawbacks, such as their mechanical resistance.

Perfect protection against RF attacks consists in a complete isolation of the sources of disruption. Therefore, one should use a Faraday cage. The Faraday cage allows protecting sensitive equipment by using a metallic and ferromagnetic box that closes hermetically. External disruptions cannot enter the cage. A good compromise to perform an opening is to make a hole with sufficiently small diameter into the Faraday cage and adding a tube with a length of eight times the diameter of the hole. Otherwise, the classical RF traps or the ferrites grommets should be used. If the inside of the cage is to be ventilated, the air arrival should be extremely well controlled in order to prevent undesirable radiations from entering. In the same way, the communication with the outside should be performed through optical fibers. The envelope of Faraday cages is in general connected to the ground, except for those that are situated at some height because of the ground effect that they would engender. The risks involved in the cold war have made these techniques to be already mastered since a long time.

The price of a Faraday cage is very high and it is not always easy to put them in place. So the protection level has sometimes to be degraded, contenting oneself with electrostatic shielding. Minimal protection consists in rounding the cables with a flexible metal sheath. Simply connecting these loops to the ground is not enough. The electromagnetic compatibility has to be taken into account (but this is not the subject here).

Targets and Properties

Potential targets of RF attacks are manifold. Financial systems, telecommunications, medical centers, critical infrastructures as airport radars, transportation means (aircrafts, electronic ignition cars) are all aimed at.

Properties

Radio-frequency weapons have, compared to their drawbacks (Hugh power consumption), numerous properties. First of all, these weapons are tunable (which is nice when a target seems to be invulnerable to particular frequencies) and some are even reusable. But they can be triggered off and fired from miles away at any time and in every circumstance. Contrary to weapons that need ballistic computations, an RF attack is gravity independent and hardly detectable. These are low cost weapons and the materials needed could easily be acquired in a large city, without conventional counter-terrorist agencies being able to trace them easily. But above all, these weapons allow multiple target acquisition and do not concentrate on a single target, they are instantaneous and non-lethal to humans, which makes them very adapted to denial of service attacks.

Further Reading

We will present here our work based on the perturbation of a smart card or a crypto processor using intense electromagnetic fields, our aim being to create a fault in order to apply cryptanalytic methods [4] such as Differential Fault Analysis [3]. Our method also permits destroying a chip, but there are very few advantages in doing so. Attackers generally want to create transient faults in a cryptosystem, and permanent faults are quite rare (security sensor destruction ...).

Traditional electromagnetic analysis is based on a sensing coil located in the near field of the chip. Measurement is thus passive. In some circumstances, this gives similar information to that obtained by measuring the chip's power consumption (i.e., in power analysis). It has, however, been shown that electromagnetic analysis gives strictly more information, as the coil can pick up the magnetic fields generated by local signals that are not present outside the chip. This was highly significant in itself; it also turned out to be important for later analysis and protection work. An alternating current in a coil near a conducting surface creates an electromagnetic field. Here we send a high current in a coil very close to the chip, so it is an active measurement. An active measurement can interfere with the local or total activity of the processor. Depackaging the chip is not necessary to apply this attack, but the ability to see the surface of the chip improves the precision. The application of the traditional techniques of depackaging with concentrated nitric acid and acetone is still an easy way to open many chips [Anderson].

We developed a variant of this technique using their electromagnetic probing tools. By placing a small coil next to a target component in a smartcard chip and passing a current pulse through the coil, they found that they could induce a sufficiently large eddy current in the chip to cause a targeted malfunction.

Our sensor is composed of a touch point coming from a microscope and a wire. The wire is wound on the test probe. The current injected in the coil creates an electromagnetic field. The test probe concentrates the lines of the field. So the field obtained at the end of the needle is relatively intense. The current injected into the coil can be obtained by using a simple camera flash gun. The intense magnetic field allows moving charges, so we create a movement of charges through the grid oxide of the transistors. Charges are then stocked in the grid oxide by a tunnel effect with high energy.

At present silicon manufacturers shrink transistors to increase their density on a chip. But if the reduction in the thickness of the grid oxide is too high, the transistor will cease functioning. Thus we imagine that our attack will be easier to use in the future although its precision will be reduced. In order to quantify the number of charges brought to or withdrawn from the grid oxide, we decided to heat the components. The increase in the electronic shocks resulting from the increase in temperature could make the quantity of loads present in the grid oxide evolve. For a static RAM, commercially available from a big silicon manufacturer, 95.81% of the remaining faults were maintained after 100 hours at $420°$ K. It seems not trivial to stop this attack, and designers should continue to use hardware and software countermeasures.

Conclusion

Electromagnetic attacks find their theoretical foundations in the basic physics introduced by Faraday and in the work of Compton, Einstein, Oppenheimer, Sakharov, Marx, and many others that have contributed to finalizing these weapons. The world's principal armies already have these devices at their disposal and others are trying to obtain them. The security of existing systems is not always conceived in a way to resist this type of attacks. By highly reducing the order of magnitude, it is also possible to use the basic principles of these attacks for inserting faults or disrupting the functioning of a cryptographic machine. Other attacks such as, for instance, glitch insertion on power supplies, clock disruption (metastability) or optical attacks [17] are however also possible.

Intense electromagnetic fields can easily be created and may allow disrupting the good functioning of electronic components. The required components are common used and can be easily obtained. It is possible to disrupt a cryptographic component by acting close to the device or to disrupt an entire building at a distance of some hundred meters, which of course depends on the level of energy. Using fields leaves behind few traces, and finding back their source is a very difficult task. Unfortunately, many devices offer a non-zero coupling and are thus a very easy victim for electromagnetic disruptions. The damage can then be very important and may require the replacement of the electronics. Countermeasures exist, but are not often used.

The security domain has to take into account the possibilities of pirating by RF disruptions. The attacks can spread from a simple deny of service to the destruction of a machine, passing by disruptions that are more difficult to detect. The interaction with the cryptographic world is without any doubt to be situated in the very local use of intense fields, used to influence the functioning of cryptoprocessors or ciphering machines.

Jean-Jacques Quisquater
Samyde David

References

[1] Agrawal, D., B. Archambeault, J.R. Rao, and P. Rohatgi (2002). "The EM side-channel(s)," *Proceeding of the Cryptographic Hardware and Embedded Systems CHES 2002*, Lecture Notes in Computer Science, vol. 2523, eds. B.S. Kaliski Jr., Ç.K. Koç, and C. Paar. Springer-Verlag, Redwood City, USA, 29–45. Also available at http://ece.gmu.edu/crypto/ches02/talks.htm

[2] Anderson, R., and M. Kuhn (1996). "Tamper Resistance—a cautionary note." *Proceedings of the Second Usenix Workshop on Electronic Commerce*, 1–11. Also available at http://www.cl.cam.ac.uk/~mgk25/tamper.pdf

[3] Biham, E. and A. Shamir (1997). "Differential fault analysis of secret key cryptosystems." *Advances in Cryptology—CRYPTO'97*, Lecture Notes in Computer Science, vol. 1294, ed. B. Kaliski. Springer-Verlag, Santa Barbara, CA, 513–525. Also available at http://citeseer.nj.nec.com/biham97differential.html

[4] Boneh, D., R.A. Demillo, and R.J. Lipton (1997). "On the importance of checking cryptographic protocols for faults." *Advances in Cryptology—EUROCRYPT'97*, Lecture Notes in Computer Science, vol. 1233, ed. W. Fumy. Springer-Verlag, Santa Barbara, CA, 37–51. Also available at http://citeseer.nj.nec.com/boneh97importance.html

[5] Bresselin, S. (2003). "Une avancée franco-allemande vers la bombe "E", *Air & Cosmos*, No. 1877, 14th February.

[6] http://www.cryptome.org

[7] Gandolfi, K., C. Mourtel, and F. Olivier (2001). "Electromagnetic attacks: Concrete results." *Proceedings of the Cryptographic Hardware and Embedded Systems CHES 2001*, Lecture Notes in Computer Science, vol. 2162, eds. Ç.K. Koç, D. Naccache, and C. Paar, Springer-Verlag, Paris, France, 251–261. Also available at http://www.gemplus.com/smart/r_d/publications/pdf/GMO01ema.pdf

[8] Abrams, M. (2003). "Dawn of the E-Bomb." *IEEE Spectrum 2003*. Also available at http://www.spectrum.ieee.org/WEBONLY/publicfeature/nov03/1103ebom.html

[9] Jenkins, C.R. and D.L. Durgin (1975). "EMP susceptibility of integrated circuits." *IEEE Transactions on Nuclear Science*, NS-22 (6).

[10] Kopp, C. (1997). "Hardening your computing assets." Asia/Pacific Open Systems Review. Computer Magazine Group, NSW under the title of "Information Warfare—Part 2," Australia. Also available at http://www.globalsecurity.org/military/library/report/1997/harden.pdf

[11] Kopp, C. (1996). The electromagnetic bomb—a weapon of electrical mass destruction. USAF CADRE Air Chronicles, Melbourne, Australia. Also available at http://www.airpower.maxwell.af.mil/airchronicles/kopp/apjemp.html

[12] Kuhn, Markus G. and Ross J. Anderson (1998). "Soft tempest: Hidden data transmission using electromagnetic emanations." *Proceeding of Information Hiding, Second International Workshop, IH'98*, Portland, OR, USA, 124–142. Also available at http://www.cl.cam.ac.uk/~mgk25/ih98-tempest.pdf

[13] Lee, K.S.H., T.K. Liu, and L. Morins (1978). "EMP Response of aircraft antennas." *IEEE Transactions on Antenna Properties*. IEEE AP-26, 1.

[14] Petit, J-P. (2003). "Armes secrètes américaines - L'extraordinaire témoignage d'un scientifique." Collection Aux marches de la science, Albin Michel, ISBN: 2226136169.

[15] Quisquater, J.-J. and D. Samyde (2001). "Electro-Magnetic Analysis (EMA): Measures and countermeasures for smart cards." *Proceedings of the International Conference on Research in Smart Cards E-Smart*, Lecture Notes in Computer Science, vol. 2140, eds. I. Attali and T. Jensen. Springer-Verlag, Cannes, France, 200–210.

[16] http://www.voltsamps.com

[17] Skorobogatov S. and R. Anderson (2002). "Optical fault induction attacks." *Proceedings of the Cryptographic Hardware and Embedded Systems CHES 2002*, Lecture Notes in Computer Science, vol. 2523, eds. B.S. Kaliski Jr., Ç.K. Koç and C. Paar. Springer-Verlag, Redwood City, USA. 2–12. Also available at http://ece.gmu.edu/crypto/ches02/talks.htm

[18] Tasca, D.M., D.C. Wunsch, and H. Domingos (1975). "Device degradation by high amplitude currents and response characteristics of discrete resistors." *IEEE Transactions on Nuclear Science*, NS-22 (6).

[19] Wunsch, D.C. and R.R. Bell (1968). "Determination of threshold failure levels of semiconductor diodes and transistors due to pulse voltages." *IEEE Transactions on Nuclear Science*, NS-15 (6).

RANDOM BIT GENERATOR

DEFINITION: A random bit generator is a system whose output consists of fully unpredictable (i.e., statistically independent and unbiased) bits. In security applications, the unpredictability of the output implies that the generator must be also not observable and not manipulable by any attacker.

A random bit generator basically differs with respect to a pseudorandom number generator, because the complete knowledge of the generator structure and of whatever previously generated

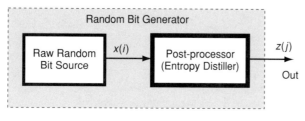

Fig. 1. Random bit generator

sequence does not result in any knowledge of any other bit. This means that the entropy of a sequence of n output bits should be ideally equal to n. On the contrary, the entropy of a <u>sequence</u> of n output bits of a pseudorandom generator cannot be greater than its seed, whatever n is.

Since pseudorandom generators are suitable in those applications where just a flat statistic is needed, random generators are suitable in applications where unpredictability is also needed.

DESCRIPTION: A true random bit generator has necessarily to be based on some kind of non-deterministic phenomena that could implement the source of the system randomness. The random sources commonly used can present several statistic defects, due to physic limitations, implementation issues, or to extern attacks aimed to manipulation. For this reason, usually, the scheme of a random bit generator consists of a raw random stream source and of a post-processor as shown in Figure 1.

The post-processor compresses the sequence produced by the source, so that it distills the entropy (see <u>information theory</u>) in the outgoing sequence. The amount of required compression depends both on the effective entropy of the source and on the efficiency of the post-processing algorithm. Of course, since the post-processor cannot oppose every malfunction or attack attempt, the random bit generator should be provided with alarm sensors capable of revealing those anomalies for which the post-processing cannot compensate.

It is worthwhile to note that even if the post-processing uses cryptographic functions similar to those used by pseudorandom generators, it works in an opposite way. In fact, a pseudorandom generator expands its input (i.e., its seed); the post-processing, on the contrary, compresses it.

RAW RANDOM BIT SOURCES: In the applications, the random source can be constructed of dedicated hardware devices; otherwise; the random source can use software procedures to extract random processes from the platform on which the generator is implemented.

Generators of the first type are commonly called hardware-based (HW); generators of the second type are software-based (SW).

SW-Based Generators

Generally, SW-based generators are implemented on computer systems and the values typically exploited as raw stream sources are obtained from:
- event timings:
  - mouse movements and clicks;
  - keystrokes;
  - disk and network accesses;
- data depending on the history of the system and/or on a large amount of events:
  - system clock;
  - I/O buffers;
  - load or network statistics.

It is easy to understand that these types of sources are far from ideal. Entropy is mostly low and difficult to evaluate as well as the actual robustness with respect to observation and manipulation. SW-based generators should use more than one source, in order to be protected from the possibility that one or more sources could be compromised. Missing statistic evaluation, the post-processing should be planned assuming that the source entropy is very low; it should therefore execute a drastic compression and use a robust hash algorithm.

HW-Based Generators

HW-based generators present the advantage of having a clearer model of the source and of its possible interaction with the outside. Generally, HW-based generators can be implemented using the common integrated technologies, and can be inserted in tamper resistant protections, in order to be protected from observation and manipulation. Normally, the nature of these sources (faster, higher in quality and more protected) lowers the need of prost-processing permitting to obtain a much higher throughput than the one obtainable from SW-based generators.

Typically the raw random stream source is a system that generates <u>sequence</u> X by sampling and quantizing an analog non-deterministic value S (Figure 2). We can consider two possible quantization modes which represent two very common cases:
- a sign mode $x(i) = \text{sign}(s(i))$;
- a mod 2 mode $x(i) = \lfloor s(i) \rfloor \bmod 2$.

The quantization mod 2 (see <u>modular arithmetic</u>) presents the advantage of limiting or making negligible the effects of deterministic

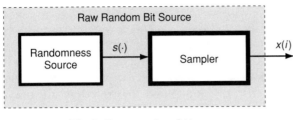

Fig. 2. Raw random bit source

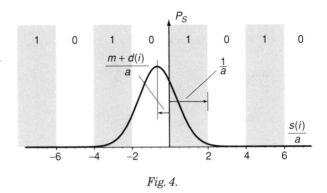

Fig. 4.

components of S. Let

$$S = a \cdot R + m + D$$

where R represents a normalized random process, a is an amplitude factor, m is an offset and D represents a possible deterministic process. We can observe that, no matter what is the amplitude of D and m, their effect becomes negligible when the amplitude of $a \cdot R$ is big enough.

As an example, Figures 3 and 4 represent the case in which R has a normalized Gaussian distribution, $a = 0.5$ and $(m + d(i))/a = 0.6$. The x axis is divided into 0 and 1 zones to show how the R distribution is partitioned between 0 and 1 value samples. The mod 2 quantization mode results in 0 and 1 bands as shown in Fig. 4. Of course, in actual implementations, these bands can be asymmetric and this issue must be taken into account as possible cause of offset on the sequence $x(i)$.

Randomness Sources in HW-Based Generators

Random sources commonly used by HW-based generators use phenomena such as:
- electronic noises (thermal, shot, avalanche);
- phase noises; and
- flip-flop metastabilities.

All these types of sources can be implemented by means of standard electronic devices.

Electronic Noise Sources. Among electronic noises, both the thermal noise in a resistor and the junc-tion shot noise offer the advantage of being white noises with a Gaussian distribution whose intensity depends on physical values easy to keep under control. In fact, in the case of thermal noise, the power density of the noise depends merely on the resistor value and the absolute temperature, while in the case of the shot noise, the power density of the noise depends only on the current that flows through the junction. These phenomena make possible the realization of sources endowed of a simply analytic statistical model, instead of based on an empiric model depending on several technological and implementation factors. The deriving advantages are the following:
- the possibility to evaluate a priori source entropy and defects, and consequently the possibility to design a suitable post-processing;
- availability of a statistical model useful for verifying that the source works correctly.

In Figure 5 we see the general scheme of a raw random bit source based on direct amplification technique. The noise's source consists of the resistor; the clocked comparator performs the sampling and the quantization of the amplified noise and the low pass feedback loop compensates the offsets due to the amplifier and the comparator.

This kind of source can have a very high throughput. The only limitation is the bandwidth of the noise amplifier, given that as the sampling frequency increases, so does the correlation among samples. In fact, the sampling frequency can reach the amplifier high pass cut off frequency since in any case the post-processing must be designed to

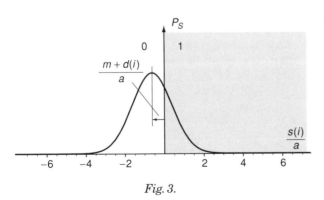

Fig. 3.

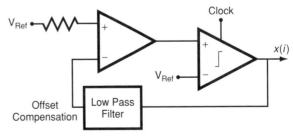

Fig. 5.

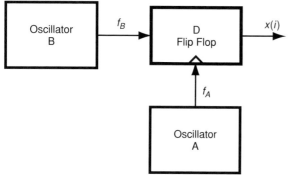

Fig. 6.

be able to remove the stronger correlation that can result from an attack.

On the other hand, this design requires an accurate implementation. In fact the high-gain amplification needed for the thermal noise involves a critical offset compensation and a high sensitivity to internal or external interfering signals. Since the sign mode quantisation is used, both an excess in the offset m or in an interfering signal D can block or force the source (see Figure 3).

Phase Noise Sources. Generally, sources based on phase noise have a simple and robust implementation. In Figure 6 a basic and typical scheme is depicted: a slow oscillator A samples a fast oscillator B. The D type flip-flop performs the mod 2 sampling of the phase difference between the two oscillators. Width and symmetry of the quantization bands depend respectively on the frequency and duty cycle of the oscillator B.

In fact this scheme has an intrinsic periodic behavior due to the phase shifting that always occurs if f_B/f_A is not an integer. This effect is negligible if the phase noise is large with respect to half the period of the sampled oscillator (i.e., if in Figure 4, noise distribution overflows the width of quantisation bands).

More generally, since the mod 2 quantization is used, this device is robust with respect to any phase process that could be superimposed on the phase noise.

Basically the obtainable throughput depends on the f_B value and on the intrinsic noise of the oscillators. In fact, once f_B is maximized (i.e., quantization bands are made as narrow as possible), the sampling period $1/f_A$ must be long enough in order to accumulate a sufficient phase noise between two subsequent samples. Hence, noisier oscillators allow a higher sampling frequency.

A more efficient phase noise exploitation can be obtained by means of a phase control that allows sampling the output of oscillator B on its edges. In this way, even a little amount of phase noise is sufficient to get a random output. Basically this solution implies a phase control that plays the same role than the offset compensation in Figure 5 scheme.

Generally, the statistical model of phases noises is not known a priori, being determined by several technological and implementation factors. Anyhow, in some solutions, an electronic noise source can be used to cause phases noises in an oscillator. In that way, it is possible to obtain a source whose phases noise characterization is known a priori since it is directly derived from an electronic noise.

Flip-Flop Metastability Sources. Actually, the random sources based on flip-flop metastability also exploit electronic and phase noises. Flip-flop metastability occurs when input signals are very close to the threshold and/or when data and clock signals switch very close in time to one another. In this condition, a small variation in levels or in phases results in a different output value.

The main implementation issue is the control of level or phase that is needed to give rise to metastability. Basically these are the same offset and phase control functions that are involved in the implementation of sources based on electronic and phase noises.

POST-PROCESSING: In practice, every kind of raw random stream source can present defects as offset, auto-correlation or cross-correlation with other phenomena. Auto-correlation can occur because of the limitation in frequency bandwidth that is intrinsic in any physical randomness source. Other defects can be done to implementation issues as:

- electrical or timing offsets (e.g., amplifier offset in Figure 5 design, unbalanced duty cycle of the sampled oscillator in Figure 6 design);
- intrinsic design behaviors (e.g., in Figure 5 design, the offset compensation suppresses long sequences of equal symbols; in Figure 6 design, a lack of phase noise results in a periodic behavior);
- interference with internal or external signals (e.g., power supply fluctuation, clock signals, switching signals coupled via the silicon substrate or via power supply, etc.).

However, the most important problems can be caused by manipulation attempts. For instance, an attacker trying to inject a signal D, wanting to force the exit $x(i)$.

The design of random sources which are intrinsically very stable and robust with respect to these problems, results inevitably in a complex design and in drastic penalties on the performance. Moreover, it seems quite impossible, especially in consumer applications, to design a source that could resist a well-equipped attacker. A more effective approach seems to be the use of reasonably good random sources together with a suitable post-processing. According to the quality and robustness of the source, the post-processing can be based on simple scrambling and mixing techniques, or even on an actual hash algorithm.

Note the fact that an effective post-processing unavoidably hides the defects of the source, even when they are so massive to completely compromise the entropy of the generator. Practically, the more the entropy of the source tends to zero, the more the post-processor tends to act as if it were a pseudorandom generator.

This behavior makes it impossible to check the malfunction of the source through statistical tests applied to the output of the generator. On the contrary, most of the possible source anomalies can be revealed by means of statistical tests applied directly at the source, that is before that the behavior of the source is masked by post-processing and quantization. Obviously, the source must have a statistical characterization and a behavior that allow distinguishing normal defects and variances with respect to faults malfunction and attacks.

Depending on the kind of source, testing can even be very simple. As an example, in the device of Figure 5, defects such as decreasing of the noise frequency bandwidth and/or of the amplification as well as increasing of the offset, all can be revealed by the decrease in the number of transition of $x(i)$.

A simple test performed on the amplitude of S (i.e., before sampling and quantization) can also reveal the attempt to force S by means of the superimposition of a known signal D. In fact this will inevitably result in an increase of the S amplitude. It must be noticed that the attacker can use a signal D such that the statistic of $x(i)$ is not changed. This means that, after quantization, since the amplitude information is removed, no statistical test can reveal this kind of attack.

Due to their simplicity, tests on the source can be executed even continually during generator operation. This allows to perform real time check of malfunctions or even to dynamically tune the amount of post-processing compression depending on the estimated source quality.

Marco Bucci

References

[1] Bagini, V. and M. Bucci (1999). "A design of reliable true random number generator for cryptographic applications." *Proceedings of Workshop Cryptographic Hardware and Embedded Systems, CHES '99*, Lecture Notes in Computer Science, vol. 1717, eds. Ç.K. Koç and C. Paar. 204–218.

[2] Bendat, J.S. (1958). *Principles and Applications of Random Noise Theory*. Wiley, New York.

[3] Davis, D., R. Ihaka, and Philip Fenstermacher (1994). "Cryptographic randomness from air turbulence in disk drives." *Advances in Cryptology—CRYPTO'94*, Lecture Notes in Computer Science, vol. 839, ed. Y. Desmedt. Springer-Verlag, Heidelberg, Germany, 114–120.

[4] Dichtl, M. and N. Janssen (2000). "A high quality physical random number generator." *Proceedings of Sophia Antipolis Forum Microelectronics (SAME 2000)*, 48–53.

[5] FIPS 140-1. (1994). "Security requirements for cryptographic modules." Nat'l Institute of Standards and Technology, GPO, Washington, DC.

[6] Gude, M. (1985). "Concepts for a high performance random number generator based on physical random phenomena." *Frequenz*, 39 (7–8), 187–190.

[7] Holman, W.T., J.A. Connelly, and A.B. Downlatabadi (1997). "An integrated analog/digital random noise source." *IEEE Transactions on Circuits and Systems I*, 44 (6), 521–528.

[8] Jun, B. and P. Kocher (1999). "The intel random number generator." Cryptography Research Inc., white paper prepared for Inter Corp., http://www.cryptography.com/resources/whitepapers/IntelRNG.pdf

[9] Knuth, D.E. (1981). *The Art of Computer Programming* (2nd ed.). Addison-Wesley, Reading, MA.

[10] Maddocks, R.S., S. Matthews, E.W. Walker, and C.H. Vincent (1972). "A compact and accurate generator for truly random binary digits." *Journal of Physics*, E5 (8), 542–544.

[11] Memezes, A.J., P.C. Oorschot, and S.A. Vanstone (2001). *Handbook of Applied Cryptology*. CRC Press, Boca Raton, FL.

[12] Murry, H.F. (1970). "A general approach for generating natural random variables." *IEEE Transactions on Computers*, C-19, 1210–1213.

[13] Papoulis, A. (1965). *Probability, Random Variables and Stochastic Processes*. McGraw-Hill, New York.

[14] Petrie, C.A. (1997). "An integrated random bit generator for applications in cryptography." *PhD Thesis*, Georgia Institute of Technology.

[15] Petrie C.S. and J.A. Connelly (1996). "Modeling and simulation of oscillator-based random number generators." *Proceedings of IEEE Int'l Symposium on Circuits and Systems, ISCAS'96*, vol. 4, 324–327.

[16] Petrie, C.S. and J.A. Connelly (2000). "A noise-based IC random number gnerator for aplications in cyptography." *IEEE Transactions on Circuits and Systems I*, 47 (5), 615–621.

[17] Schneier, B. (1996). *Applied Cryptography* (2nd ed.). John Wiley & Sons, New York.

[18] Trichina, E., M. Bucci, D. De Seta, and R. Luzzi (2001). "Supplementary cryptographic hardware for smart cards." *IEEE Micro*, 21 (6), 26–35.

[19] Vincent, C.H. "The generation of truly random binary numbers." *Journal of Physics*, E3 (8), 594–598.

[20] Vincent, C.H. (1971). "Precautions for the accuracy in the generation of truly random binary numbers." *Journal of Physics*, E4 (11), 825–828.

RANDOM ORACLE MODEL

The random oracle model was introduced by Bellare and Rogaway [2]. The idea is a simple one: namely provide all parties of a protocol—good and bad alike—with access to a (public) function h and then prove the protocol to be correct assuming that h maps each input to a truly random output, i.e. it behaves like a truly random oracle. Later, in practice, one sets h to some specific function derived in some way from a standard cryptographic hash function like SHA-1 [5], MD5 [6], RIPEMD-160 [4], or others. It is clear though that any specific function will not be random because it is deterministic, i.e., it returns the same value when given the same input. (Also see Bellare's overview of the random oracle model in [1].)

The random oracle model buys efficiency and, as Rogaway claims, security guarantees, which, although not at the same level as those provided by the standard "provable security approach," are arguably superior to those provided by a totally ad hoc protocol design.

The overly skeptical might say that a security proof in the random oracle model gains nothing because the function h that one actually uses in the final protocol is not random. Here is another way to look at it. In practice, attacks on schemes involving a from SHA-1 derived h and number theory will often themselves treat h as random. Bellare and Rogaway call such attacks *generic*. In other words, cryptanalysis of these "mixed" protocols is usually done by assuming h to be random. But then proofs in the random oracle model apply, and indeed show that such generic attacks will fail

unless the underlying number-theoretic problems are easy to solve. In other words, the analysis at least provably excludes a certain common class of attacks, namely generic ones.

It is important to choose carefully the instantiating function h. The intuition stated by Bellare and Rogaway in [2] is that the resulting protocol is secure as long as the protocol and the hash function are sufficiently "independent," which means the protocol does not itself refer to the hash function in some way. This is a fuzzy guideline that needs more work in the future.

An important step in better understanding the random oracle model was taken by Canetti *et al.* [3]. They show there exist protocols secure in the random oracle model but insecure under any instantiations in which we substitute a function from a small family of efficiently computable functions. Their examples however are somewhat contrived, and this kind of situation does not arise with any of the "real" cryptographic mechanisms in the literature.

In comparison with a totally ad hoc design, a proof in the random oracle model has the benefit of judging the protocol under a strong, formal notion of security, even if this assumes some underlying primitive to be very strong. This is better than not formally modeling the security of the protocol at all. This explains why the random oracle model is viewed as a "bridge" between theory and practice [2].

Gerrit Bleumer

References

[1] Bellare, Mihir (1999). "Practice-oriented provable security." *Lectures on Data Security*, Lecture Notes in Computer Science, vol. 1561. Springer-Verlag, Berlin.

[2] Bellare, Mihir and Philip Rogaway (1993). "Random oracles are practical: A paradigm for designing efficient protocols." *1st ACM Conference on Computer and Communications Security, Proceedings, Fairfax, November 1993*. ACM Press, New York, 62–73.

[3] Canetti, Ran, Oded Goldreich, and Shai Halevi (1998). "The random oracle methodology, revisited." *30th Symposium on Theory of Computing (STOC) 1998*. ACM Press, New York, 209–218.

[4] Dobbertin, Hans, Antoon Bosselaers, and Bart Preneel (1996). "RIPEMD-160: A strengthened version of RIPEMD." *Fast Software Encryption, Third International Workshop, Cambridge, UK, February 21–23, 1996, Proceedings*, Lecture Notes in Computer Science, vol. 1039, ed. D. Gollman. Springer-Verlag, Berlin, 71–82.

[5] National Institute of Standards and Technology (NIST) (1995). Secure Hash Standard. Federal Information Processing Standards Publication (FIPS PUB 180-1).

[6] Rivest, Ronald (1992). The MD5 Message-Digest Algorithm; RFC1321. http://www.faqs.org/rfcs/rfc1321.html

RC4

RC2 and RC4 are ciphers developed by R. Rivest for RSA Data Security, Inc. These are proprietary, and their details have not been officially-published. RC2 is a variable-key-size <u>block cipher</u>, and RC4 a variable-key-size <u>stream cipher</u> (the key size may vary between 1 up to 2048 bits). RC4 is used a lot today; you can find it, for example, in <u>secure socket layer/SSL</u>, in Wi-Fi security protocols, etc.

Nevertheless, some reverse engineering has been done, and an algorithm was accessible on the Internet in 1994, that gave the same output as RC4. Several studies and attacks have been published since 1997. A good starting point on the subject, with a description of RC4, and a good state of the art is [5]. The published attacks are [1–4, 6–8]. But in practice, RC4 remains secure, if well used (see [9] for an example of bad usage in Wi-Fi environment).

Caroline Fontaine

References

[1] Fluhrer, S.R. and D.A. McGrew (2000). "Statistical analysis of the alleged RC4 key stream generator." FSE'00, Lecture Notes in Computer Science, vol. 1978, ed. B. Schneier. Springer-Verlag, Berlin, 19-ff.

[2] Fluhrer, S., I. Mantin and A. Shamir (2001). "Weaknesses in the key Scheduling algorithm of RC4." SAC'01, Lecture Notes in Computer Science, vol. 2259, eds. S. Vaudenay and A.M. Youssef. Springer-Verlag, Berlin, 1-ff.

[3] Golic, J. (1997). "Linear statistical weakness of alleged RC4 key stream generator." Advances in Cryptology—EUROCRYPT'97, Lecture Notes in Computer Science, vol. 1233, ed. W. Fumy. Springer-Verlag, Berlin, 226–238.

[4] Knudsen, L.R., W. Meier, B. Preneel, V. Rijmen, and S. Verdoolaege (1998). "Analysis methods for (alleged) RC4." Advances in Cryptology—ASIACRYPT'98, Lecture Notes in Computer Science, vol. 1514, eds. K. Ohta and D. Pei. Springer-Verlag, Berlin, 327-ff.

[5] Mantin, I. (2001). "Analysis of the Stream Cipher RC4." Master's Thesis, Weizmann Institute of Science.

[6] Mantin, I. and A. Shamir (2001). "A practical attack on broadcast RC4." FSE'01, Lecture Notes in Computer Science, vol. 2355, ed. M. Matsui. Springer-Verlag, Berlin, 152-ff.

[7] Mironov, I. (2002). "(Not so) random shuffles of RC4." Advances in Cryptology—CRYPTO 2002, Lecture Notes in Computer Science, vol. 2442, ed. M. Yung. Springer-Verlag, Berlin, 304-ff.

[8] Mister, S. and S.E. Tavares (1998). "Cryptanalysis of RC4-like ciphers." SAC'98, Lecture Notes in Computer Science, vol. 1556, eds. S. Tavares and H. Meijer. Springer-Verlag, Berlin, 131-ff.

[9] Stubblefield, Loannidis, and Rubin (2001). "Using the Fluhrer, Mantin, and Shamir attack to break WEP." AT&T Labs Technical Report.

RC5

RC5 is an iterative secret-key <u>block cipher</u> designed by Rivest [5] in 1995. It has variable parameters such as the <u>key</u> size, the block size, and the number of rounds. A particular (parameterized) RC5 encryption algorithm is designated as RC5-$w/r/b$, where w is the word size (one block is made of two words), r is the number of rounds ($r = 2h$), and b is the number of bytes for the secret key. The "nominal" choice for the algorithm, RC5-32/12/16, has a 64-bit block size, 12 rounds and a 128-bit key. The secret key is first expanded into a table of $2h + 2$ secret words S_i of w bits according to the key schedule. Let (L_0, R_0) denote the left and right halves of the plaintext. Note that a w-bit word is equivalently viewed as an integer modulo 2^w (see <u>modular arithmetic</u>). Then the encryption algorithm is given by:

$$
\begin{aligned}
&L_1 \leftarrow L_0 + S_0 \bmod 2^w \\
&R_1 \leftarrow R_0 + S_1 \bmod 2^w \\
&\text{for } i = 1 \text{ to } 2h \text{ do} \\
&\quad L_{i+1} \leftarrow R_i \\
&\quad R_{i+1} \leftarrow ((L_i \oplus R_i) \lll R_i) + S_{i+1} \bmod 2^w
\end{aligned}
$$

where "$\oplus$" represents bit-wise exclusive-or, and "$X \lll Y$" is the (data-dependent) rotation of X to the left by the $\log_2 w$ least significant bits of Y. The ciphertext is (L_{2h+1}, R_{2h+1}) and each half-round i involves exactly one subkey S_i. Kaliski and Yin [3] studied both <u>differential</u> and <u>linear attacks</u> on nominal RC5. Knudsen and Meier [4] further improved over their attacks and also showed that RC5 has <u>weak keys</u>. The best differential (<u>chosen plaintext</u>) and linear (<u>known plaintext</u>) attacks on RC5 today are respectively the one by Biryukov and Kushilevitz [1] which breaks 12 rounds of RC5-32/12/16 with 2^{44} chosen plaintexts using partial differentials, and the one by Borst et al. [2] which breaks RC5 up to 10 rounds using multiple linear approximations. Both results suggest that RC5 should be used with at least 16 rounds.

Helena Handschuh

References

[1] Biryukov, A. and E. Kushilevitz (1998). "Improved cryptanalysis of RC5." *Advances in Cryptology— EUROCRYPT'98*, Lecture Notes in Computer Science, vol. 1403, ed. K. Nyberg. Springer-Verlag, Berlin, 85–99.

[2] Borst, J., B. Preneel, and J. Vandewalle (1999). "Linear cryptanalysis of RC5 and RC6." *Fast Software Encryption—Sixth International Workshop*, Lecture Notes in Computer Science, vol. 1636, ed. L.R. Knudsen. Springer-Verlag, Berlin, 16–30.

[3] Kaliski B.S. and Y.L. Yin (1995). "On differential and linear cryptanalysis of the RC5 encryption algorithm." *Advances in Cryptology—CRYPTO'95*, Lecture Notes in Computer Science, vol. 963, ed. D. Coppersmith. Springer-Verlag, Berlin, 171–184.

[4] Knudsen, L.R. and W. Meier (1996). "Improved differential attacks on RC5." *Advances in Cryptology— CRYPTO'96*, Lecture Notes in Computer Science, vol. 1109, ed. N. Koblitz. Springer-Verlag, Berlin, 216–228.

[5] Rivest, R. (1995). "The RC5 encryption algorithm." *Fast Software Encryption—Second International Workshop*, Lecture Notes in Computer Science, vol. 1008, ed. B. Preneel. Springer-Verlag, Berlin, 86–96.

RC6

RC6 is an iterative secret-key block cipher designed by Rivest et al. [5] in 1998. It has variable parameters such as the key size, the block size, and the number of rounds. A particular (parameterized) RC6 encryption algorithm is designated as RC6 (w, r, b), where w is the word size (one block is made of four words), r is the number of rounds, and b is the number of bytes for the secret key. The three "nominal" choices for the algorithm as submitted to the American Advanced Encryption Standard (Rijndael/AES) contest and to the European NESSIE contest are RC6 (32, 20, 16), RC6 (32, 20, 24) and RC6 (32, 20, 32). All three versions have a 128-bit block size, 20 rounds and only differ in the key-size which is respectively 128, 196 and 256 bits long. The secret key is first expanded into an array of $2r + 4$ secret w-bit words S_i according to the key scheduling algorithm. Let (A, B, C, D) denote the four w-bit words of the plaintext. Note that a w-bit word

is equivalently viewed as an integer modulo 2^w (see modular arithmetic). Then the encryption algorithm is given by:

$$
\begin{aligned}
&B \leftarrow B + S_0 \bmod 2^w \\
&D \leftarrow D + S_1 \bmod 2^w \\
&\text{for } i = 1 \text{ to } r \text{ do} \\
&\quad A \leftarrow ((A \oplus f(B)) \lll f(D)) + S_{2i} \bmod 2^w \\
&\quad C \leftarrow ((C \oplus f(D)) \lll f(B)) + S_{2i+1} \bmod 2^w \\
&\quad (A, B, C, D) \leftarrow (B, C, D, A) \\
&A \leftarrow A + S_{2r+2} \bmod 2^w \\
&C \leftarrow C + S_{2r+3} \bmod 2^w
\end{aligned}
$$

where "$\oplus$" represents bit-wise exclusive-or, "$X \lll Y$" is the (data-dependent) rotation of X to the left by the $\log_2 w$ least significant bits of Y, and the function f plays the role of a pseudo-random generator defined by:

$$ f(x) = (x(2x + 1) \bmod 2^w) \lll \log_2 w. $$

The ciphertext is (A, B, C, D) and each round r involves exactly two subkeys S_i. RC6 overcomes certain weaknesses of its predecessor RC5 by introducing fixed rotations as well as a quadratic function to determine the data-dependent rotations. Contini et al. [1] on one hand, and Iwata and Kurosawa [3] on the other hand, both give a comprehensive analysis of the contribution of these features to the security of RC6. The best statistical attack on RC6 by Gilbert et al. [2] breaks RC6 (32, 14, 16) and the best correlation attack by Knudsen and Meier [4] further enables a distinguisher and a key-recovery attack on RC6 (32, 15, 32). For a fraction of the keys called weak keys, RC6 is vulnerable to a multiple linear attack up to 18 rounds [6].

Helena Handschuh

References

[1] Contini, S., R.L. Rivest, M.J.B. Robshaw, and Y.L. Yin (1999). "Improved analysis of some simplified variants of RC6." *Fast Software Encryption— Seventh International Workshop*, Lecture Notes in Computer Science, vol. 1636, ed. L.R. Knudsen. Springer-Verlag, Berlin, 1–15.

[2] Gilbert, H., H. Handschuh, A. Joux, and S. Vaudenay (2000). "A statistical attack on RC6." *Fast Software Encryption—Seventh International Workshop*, Lecture Notes in Computer Science, vol. 1978, ed. B. Schneier. Springer-Verlag, Berlin, 64–74.

[3] Iwata, T. and K. Kurosawa (2000). "On the pseudo-randomness of the AES finalists—RC6 and serpent." *Fast Software Encryption—Seventh International Workshop*, Lecture Notes in Computer Science, vol. 1978, ed. B. Schneier. Springer-Verlag, Berlin, 231–243.

[4] Knudsen, L.R. and W. Meier (2000). "Correlations in RC6 with a reduced number of rounds." *Fast Software Encryption—Seventh International Workshop*, Lecture Notes in Computer Science, vol. 1978, ed. B. Schneier. Springer-Verlag, Berlin, 94–108.

[5] Rivest, R.L. M.J.B. Robshaw, R. Sidney, and Y.L. Yin (1998). "The RC6 block cipher." *AES—The First Advanced Encryption Standard Candidate Conference, Conference Proceedings*.

[6] Shimoyama, T. M. Takenaka, and T. Koshiba (2002). "Multiple linear cryptanalysis of a reduced round RC6." *Fast Software Encryption—Ninth International Workshop*, Lecture Notes in Computer Science, vol. 2365, eds. J. Dalmen and V. Rijmen. Springer-Verlag, Berlin, 76–88.

RECIPIENT ANONYMITY

Recipient anonymity is achieved in a messaging system if an eavesdropper picks up messages from the communication line of a sender can—after some time of monitoring the network—not tell with better probability than pure guessing who has eventually received the messages. During the attack, the eavesdropper may listen on all communication lines of the network including those that connect the potential senders to the network, he may send and receive his own messages. It is clear that all messages in such a network must be encrypted to the same length in order to keep the attacker from distinguishing different messages by their content or length. The anonymity set for any particular message attacked by the eavesdropper is the set of all network participants that will have received a message within a certain time window after the attacked message was sent. This time window of course depends on latency characteristics and node configurations of the network itself.

Recipient anonymity against computationally unrestricted attackers can be achieved by broadcast, e.g., DC-Network [1, 4], by Mix-Network [4], or by anonymous information retrieval [2]. Note that recipient anonymity is weaker than *recipient unobservability*, where the attacker cannot even determine whether or not a participant has received a (meaningful) message. Recipient unobservability can be achieved with either of the above techniques by adding dummy traffic.

Gerrit Bleumer

References

[1] Chaum, David (1985). "Security without identification: Transaction systems to make big brother obsolet." *Communications of the ACM*, 28 (10), 1030–1044.

[2] Chaum, David (1981). "Untraceable Electronic Mail, Return Addresses, and Digital Pseudonyms." *Communications of the ACM*, 24 (2), 84–88.

[3] Cooper, David A. and Kenneth P. Birman (1995). "The design and implementation of a private message service for mobile computers." *Wireless Networks*, 1, 297–309.

[4] Waidner, Michael (1990). "Unconditional sender and recipient untraceability in spite of active attacks." *Advances in Cryptography—EUROCRYPT'89*, Lecture Notes in Computer Science, vol. 434, eds. J.-J. Quisquater and J. Vandewalle. Springer-Verlag, Berlin, 302–319.

REED–MULLER CODES

It is well-known that any property of Reed–Muller codes is a property of Boolean functions. Reed Muller codes provide a natural way to quantify the *degree*, the nonlinearity, the correlation-immunity or the propagation characteristics of a Boolean function [1]. On the other hand, Reed Muller codes are an important class of error-correcting codes, in particular they can be viewed as extended cyclic codes. They play a crucial role in the study of important families of cryptographic mappings, such as permutations on finite fields.

Here, we present the multivariable definition of Reed-Muller codes. More on Reed Muller codes can be found in [2].

DEFINITION 1. *Define* $\{\mathbf{F}_2^m, +\}$ *as an ordered vector space:*

$$\mathbf{F}_2^m = \{\mathbf{v}_0, \mathbf{v}_1, \ldots, \mathbf{v}_{2^m-1}\}, \tag{1}$$

where $\mathbf{v}_i$ *is an m-dimensional binary vector (often onetakes* $\mathbf{v}_i$ *as the binary representation of integer i). The Reed-Muller code of length* 2^m *and order r,* $0 \leq r \leq m$, *denoted by* $\mathcal{R}(r, m)$, *is the binary code of length* 2^m *consisting of all codewords* $(f(\mathbf{v}_0), f(\mathbf{v}_1), \ldots, f(\mathbf{v}_{2^m-1}))$ *where f is any Boolean function of m variables whose algebraic degree is less than or equal to r.*

Note that $\mathcal{R}(0, m)$ corresponds to the set of constant functions and that $\mathcal{R}(1, m)$ corresponds to the set of functions which are affine or constant.

The *r*th order Reed–Muller code is usually constructed by using as basis the set of *monomials* of degree at most *r*. One can interpret $\mathbf{F}_2^m$ as an $m \times 2^m$ binary matrix, the *i*th row of which is the sequence of values of variable $\mathbf{x}_i$ evaluated for the successive vectors $\mathbf{v}_i$. Thus, $\mathbf{x}_i$ is in fact a Boolean function over $\mathbf{F}_2^m$: for each *j*, $\mathbf{x}_i(\mathbf{v}_j)$ equals the *i*th symbol of vector $\mathbf{v}_j$. This function produces a unique codeword whose symbols are labelled with respect to (1). For the sake of simplicity, we denote by $\mathbf{x}_i(\mathbf{v}_j)$ by $x_{i,j}$.

Now, any monomial $\prod_{i \in I} \mathbf{x}_i$ is, in the same way, a Boolean function g satisfying for $0 \le j \le 2^m - 1$

$$g(\mathbf{v}_j) = \begin{cases} 1, & \text{if } x_{i,j} = 1, \text{ for all } i \in I, \\ 0, & \text{otherwise.} \end{cases}$$

In this way, a basis for $R(r,m)$ can be obtained. An explicit construction for $m = 4$ is presented in the example below.

PROPOSITION 1. *The set of monomials*

$$\mathbf{x}_1^{e_1} \mathbf{x}_2^{e_2} \cdots \mathbf{x}_m^{e_m}, \quad e_i \in \{0, 1\}, \quad \sum_{i=1}^{m} e_i \le r,$$

is a basis of code $\mathcal{R}(r, m)$.

EXAMPLE. Construction of a generator matrix $\mathcal{G}_r$ for each code $\mathcal{R}(r, 4)$, $0 \le r \le 4$.

Matrix $\mathcal{G}_r$ is obtained by computing the codewords produced by the monomials of degree less than or equal to r, using Proposition 1. Observe that $\mathcal{G}_1$ is a 5×16 matrix, $\mathcal{G}_2$ is a 11×16 matrix, etc. Note also that the lines of $\mathcal{G}_1$ are the all-one vector and the lines of matrix $\mathbf{F}_2^4$ (with the given fixed order).

1	1 1 1 1 1 1 1 1 1 1 1 1 1 1 1 1	$\mathcal{G}_0$
$\mathbf{x}_4$	0 0 0 0 0 0 0 0 1 1 1 1 1 1 1 1	
$\mathbf{x}_3$	0 0 0 0 1 1 1 1 0 0 0 0 1 1 1 1	
$\mathbf{x}_2$	0 0 1 1 0 0 1 1 0 0 1 1 0 0 1 1	
$\mathbf{x}_1$	0 1 0 1 0 1 0 1 0 1 0 1 0 1 0 1	$\mathcal{G}_1$
$\mathbf{x}_3\mathbf{x}_4$	0 0 0 0 0 0 0 0 0 0 0 0 1 1 1 1	
$\mathbf{x}_2\mathbf{x}_4$	0 0 0 0 0 0 0 0 0 0 1 1 0 0 1 1	
$\mathbf{x}_1\mathbf{x}_4$	0 0 0 0 0 0 0 0 0 1 0 1 0 1 0 1	
$\mathbf{x}_2\mathbf{x}_3$	0 0 0 0 0 0 1 1 0 0 0 0 0 0 1 1	
$\mathbf{x}_1\mathbf{x}_3$	0 0 0 0 0 1 0 1 0 0 0 0 0 1 0 1	
$\mathbf{x}_1\mathbf{x}_2$	0 0 0 1 0 0 0 1 0 0 0 1 0 0 0 1	$\mathcal{G}_2$
$\mathbf{x}_2\mathbf{x}_3\mathbf{x}_4$	0 0 0 0 0 0 0 0 0 0 0 0 0 0 1 1	
$\mathbf{x}_1\mathbf{x}_3\mathbf{x}_4$	0 0 0 0 0 0 0 0 0 0 0 0 0 1 0 1	
$\mathbf{x}_1\mathbf{x}_2\mathbf{x}_4$	0 0 0 0 0 0 0 0 0 0 0 1 0 0 0 1	
$\mathbf{x}_1\mathbf{x}_2\mathbf{x}_3$	0 0 0 0 0 0 0 1 0 0 0 0 0 0 0 1	$\mathcal{G}_3$
$\mathbf{x}_1\mathbf{x}_2\mathbf{x}_3\mathbf{x}_4$	0 0 0 0 0 0 0 0 0 0 0 0 0 0 0 1	$\mathcal{G}_4$

Clearly, $\mathcal{R}(r, m)$ is a linear subspace of the 2^m-dimensional binary vector space. One can easily verify that its dimension is $1 + m + \cdots + \binom{m}{r}$. It can also be shown (see [2]) that different codewords in $\mathcal{R}(r, m)$ differ in at least 2^{m-r} coordinates. One says that $\mathcal{R}(r, m)$ has *minimum distance* 2^{m-r} (see Boolean functions). This proves:

PROPOSITION 2. *The rth order Reed–Muller code* $\mathcal{R}(r, m)$, $0 \le r \le m$, *has length* $n = 2^m$, *dimension* $1 + m + \cdots + \binom{m}{r}$ *and minimum distance* 2^{m-r}.

Pascale Charpin

References

[1] Massey, J.L. (1995). "Some applications of coding theory in cryptography." *Codes and Ciphers: Cryptography and Coding IV*, ed. P.G. Farell. Formara Ltd., Essex, England, 33–47.

[2] Pless, V.S., W.C. Huffman, and R.A. Brualdi (1998). "An introduction to algebraic codes." *Handbook of Coding Theory, Part 1: Algebraic Coding*, Chapter 1. Elsevier, Amsterdam, The Netherlands.

REGISTRATION AUTHORITY

In Public Key Infrastructure (PKI) solutions, the most important procedural step is the identification and registration of users in the system. This is normally handled by so-called (Local) Registration Authorities ((L)RA). There may be a number of (L)RAs attached to a Certification Authority (CA), and a typical role is to provide the CA with the credentials of the user and possibly his public key (see public key cryptography) through an authenticated channel.

Peter Landrock

RELATED KEY ATTACK

The first attacks of this type were developed independently by Biham [1] and Knudsen [4] and the notion of a *related key attack* was defined by Biham [1]. The idea of the attack is that the attacker knows (or chooses) a relation between several keys and is given access to encryption functions with such related keys. The goal of the attacker is to find the keys themselves. If the relation is known but cannot be changed by the attacker, the attack is called a *known related key*; and if the attacker may choose the relation, it is called a *chosen related key attack*. The scenario of the attack is very powerful in terms of the attacker's capabilities and thus quite unrealistic in practice. Still these attacks may be seen as important certificational weaknesses for the key-schedule of a cipher. A line of ciphers have been shown to have weaknesses in this attack scenario [2, 3], namely: IDEA, GOST, G-DES, SAFER, Triple-DES, 3-WAY, Biham–Biryukov-DES, CAST, DES-X, NewDES, RC2, and TEA. Recently a new type of cryptanalytic attack called slide attack has been developed. It can be viewed as a variant of a related key attack, in which a relation of the key with itself is exploited. Slide attacks are known plaintext or chosen plaintext attacks and thus are more practical than related key attacks since they do not require the attacker to know relations between different encryption keys.

Alex Biryukov

References

[1] Biham, E. (1994). "New types of cryptanalytic attacks using related keys." *Journal of Cryptology*, 4, 229–246. An earlier version appeared in the proceedings of *EUROCRYPT'93*, Lecture Notes in Computer Science, vol. 765, ed. T. Helleseth.

[2] Kelsey, J., B. Schneier, and D. Wagner (1996). "Key-schedule cryptanalysis of IDEA, G-DES, GOST, SAFER, and Triple-DES." *Advances in Cryptology— CRYPTO'96*, Lecture Notes in Computer Science, vol. 1109, ed. N. Koblitz. Springer-Verlag, Berlin, 237–251.

[3] Kelsey, J., B. Schneier, and D. Wagner (1997). "Related-key cryptanalysis of 3-WAY, Biham-DES, CAST, DES-X, NewDES, RC2, TEA." *International Conference on Information and Communications Security, ICICS'97*, Lecture Notes in Computer Science, vol. 1334, eds. Y. Han, T. Okamato, and S. Qing. Springer-Verlag, Berlin.

[4] Knudsen, L.R. (1993). "Cryptanalysis of LOKI91." *Advances in Cryptography—ASIACRYPT'92*, Lecture Notes in Computer Science, vol. 718, eds. J. Seberry and Y. Zheng. Springer-Verlag, Berlin, 22–35.

RELATIVELY PRIME

Two integers n, x are *relatively prime* if they have no common integer factors other than $+1$ and -1, i.e., their greatest common divisor is 1.

Burt Kaliski

RELAY ATTACK

In a relay attack, a party, say Eve, will use the resource of a second party in an unauthorized way. A typical example is related to e-mail. Suppose Eve wants to send spam e-mail to lots of users, but does not have the resources (e.g., bandwidth). She will try to use Alice's machine to have it send all this e-mail. This may result in a denial-of-service against Alice's machine. Old mail servers allow mail to be relayed.

Yvo Desmedt

REPLAY ATTACK

A *replay attack* is an attack in which the adversary records a commumication session and replays the entire session, or some portion of the session, at a later point in time. The replayed message(s) may be sent to the same verifier as the one that participated in the original session, or to a different verfier. The goal of the replay attack may be impersonation (see impersonation attack), or it may be some other deception (e.g., a successful protocol exchange to transfer money from A's account to B's account may be replayed by B in an attempt to transfer more money than A had intended).

Carlisle Adams

References

[1] Menezes, A., P. van Oorschot, and S. Vanstone (1997). *Handbook of Applied Cryptography*. CRC Press, Boca Raton, FL.

[2] Schneier B. (1996). *Applied Cryptography: Protcols, Algorithms, and Source Code in C* (2nd ed.). John Wiley & Sons.

RESPONSE

In a cryptographic identification scheme, a *response* is the answer by a claimant to a question posed by a verifier in a challenge–response protocol. More generally, a response is any answer to a question or statement made by another party.

Carlisle Adams

RESYNCHRONIZATION ATTACK

Synchronous stream ciphers require some procedure for resynchronizing in the case of synchronization loss. This opens doors to new attack scenarios. A typical stream cipher encrypts the stream in fixed data blocks, called *frames* (or packets) by keeping the same secret key for all the frames but mixing the new *initial value (IV)* or the *frame-counter* for each frame (see, e.g., the A5/1 cipher). This allows for easy synchronization as well as for the late entry mechanism in the case of multi-party communication. On the one hand, such mode of operation produces only short streams for any fixed state which reduces the chances of some attacks, but on the other hand, it may open doors to new analysis techniques which will attack the resynchronization mechanism itself. Depending on the way IV and the key are loaded and mixed into the *state* of the stream cipher the scheme may be susceptible to differential, linear, slide or other attacks. A typical resynchronization attack on stream ciphers is given in [3]. For more recent results on the subject see [1, 2, 4, 5, 6].

Alex Biryukov

References

[1] Biryukov, A. and D. Wagner (1999). "Slide attacks." *Proceedings of Fast Software Encryption—FSE'99*, Lecture Notes in Computer Science, vol. 1636, ed. L.R. Knudsen. Springer-Verlag, Berlin, 245–259.

[2] Borissov, Y., S. Nikova, B. Preneel, and J. Vandewalle (2002). "On a resynchronization weakness in a class of combiners with memory." *Third International Conference, SCN 2002*, Amalfi, Italy, September 11–13. Revised Papers, Lecture Notes in Computer Science, vol. 2576, eds. S. Cimato, C. Galdi, and G. Persiano. Springer-Verlag, Berlin, 164–173.

[3] Daemen, J., R. Govaerts, and J. Vandewalle (1993). "Resynchronization weaknesses in synchronous stream ciphers." *Advances in Cryptology—EUROCRYPT'93*, Lecture Notes in Computer Science, vol. 765, ed. T. Helleseth. Springer-Verlag, Berlin, 159–167.

[4] Ekdahl, P. and T. Johansson (2003). "Another attack on A5/1." *IEEE Transactions on Information Theory*, 49, 1–7.

[5] Frederik Armknecht, Joseph Lano, Bart Preneel (2004) "Extending the Resynchronization Attack." *Proceedings of Selected Areas in CRYPTOGRAPHY 2004*, Lecture Notes in Computer Science, vol. 3357, Springer, Berlin, 19–38.

[6] Golic, J.D. and G. Morgari (2003). "On resynchronization attack." *Fast Software Encryption, FSE 2003*, Lecture Notes in Computer Science, vol. 2887, ed. T. Johansson. Springer-Verlag, Berlin, 100–110.

RIGHTS MANAGEMENT

Rights management is a subset of general "authorization data" management (see authorization architecture) in which the data being managed are *rights* associated with entities in an environment. A *right* may be defined as follows [1]: "something to which one has a just claim; the power or privilege to which one is justly entitled; the property interest possessed under law or custom and agreement in an intangible thing especially of a literary and artistic nature".

A Digital Rights Management (DRM) architecture is an authorization architecture in which the data being managed are *rights*, and the components that enforce access control decisions (the Policy Enforcement Points) are distributed throughout the environment. In particular, a PEP resides in each piece of local client equipment to ensure that a client can only watch the movies, play the games, listen to the music, and so on, for which s/he has the appropriate rights. Further information regarding DRM can be found at [2] and [3].

Carlisle Adams

References

[1] Merriam-Webster OnLine, http://www.m-w.com/cgi-bin/dictionary

[2] Wikipedia, http://en.wikipedia.org/wiki/Digital_rights_management

[3] Cover Pages, http://xml.coverpages.org/drm.html

RIGHT-TO-LEFT EXPONENTIATION

Many exponentiation methods have two variants: one that examines exponents starting at the most significant digit and going down to the least significant one, i.e., *in left-to-right* direction (assuming big-endian notation); and a related one that examines exponents in the opposite direction, i.e., *right-to-left*. For specific methods, see the entries on binary exponentiation, 2^k-ary exponentiation, and sliding window exponentiation. There is a general duality between left-to-right and right-to-left exponentiation (this is explained in [1], answer to exercise 4.6.3-39] by representing addition chains (see fixed-exponent exponentiation) as directed multi-graphs such that reversing all arcs turns one into another).

Bodo Möller

Reference

[1] Knuth, D.E. (1998). *The Art of Computer Programming—Vol. 2: Seminumerical Algorithms* (3rd ed.). Addison-Wesley, Reading, MA.

RIJNDAEL/AES

After explaining the relation between Rijndael and AES, we describe the features of Rijndael. This is followed by the description of the Rijndael cipher structure, the round transformation and its steps, and the key schedule.

RIJNDAEL AND AES: On October 2, 2000, the US federal agency National Institute of Standards and Technology (NIST) officially announced that Rijndael would become the Advanced Encryption Standard (AES). NIST chose Rijndael from a set of 15 candidates after a 3-year public and fully open selection and evaluation process. The choice was motivated in an excellent 116-page report in which they summarize all contributions and motivate the choice [4].

Both Rijndael and AES are block ciphers that provide a mapping from *plaintext blocks* to *ciphertext blocks* and vice versa under a *cipher key*. Rijndael supports all combinations of block lengths and key lengths that are a multiple of 32 bits with a minimum of 128 bits and a maximum of 256 bits. The Rijndael reference specification can be found in [1, Appendix E]. AES is equal to Rijndael limited to a block length of 128 bits and supports for key length of 128, 192 or 256 bits. AES is specified in [3].

FEATURES OF RIJNDAEL: The most important feature of Rijndael is its consistent good performance among a wide range of platforms.

On smartcards, Rijndael can be implemented using less than 1 kbyte of code (tables included), and using 36 bytes of memory. Since the text input and the key take both 16 bytes, only 4 bytes extra are required for temporary variables. On the other hand, on high-end processors, Rijndael can exploit cache and parallelism to achieve a significant speedup.

In order to allow fast key setup times, Rijndael has a lightweight key schedule. Fast key setup times are important in systems that switch keys often such as financial authorisation schemes or IPsec security.

Along with its performance, Rijndael has shown to have a high security margin with respect to all types of cryptanalysis. Moreover, its structure and choice of operations facilitates implementations that are resistant against side channel attacks.

These features are a result of the application of the following design principles:

Keep it simple: No complexity is added unless there is a demonstrated need for it. The aim is to have a design that is secure against known attacks, without introducing new vulnerabilities. There is no reason to go beyond that, i.e., to add an excessive amount of extra layers of complexity in the hope that this will provide extra security.

Modularity: The design is composed of different building blocks, or steps, each with their own functionality. Building blocks are selected according to specific quantitative selection criteria.

Symmetry and parallelism: All steps can be parallelized and act in a symmetrical way on the input. The large degree of parallelism allows to trade-off area for speed in a flexible way in hardware implementations.

Choice of operations: All steps are defined with operations in the finite field $GF(2^8)$ and can be implemented using XOR and table lookup instructions only. The fact that no arithmetic operations are used, saves space on hardware platforms. The use of operations gives the programmer a lot of flexibility for implementing platform-dependent optimizations on a wide range of processors.

An important factor in the design of Rijndael is the wide trail strategy. This strategy defines diffusion and nonlinearity criteria for the building blocks of the cipher to provide high resistance against differential and linear cryptanalysis in an efficient way. For a detailed treatment, we refer to [1, Chapter 9].

BLOCK CIPHER STRUCTURE: When encrypting, the bytes of a plaintext block are mapped onto the elements of a *state*, the state undergoes a transformation and the elements of the state are mapped onto the bytes of a ciphertext block. Rijndael is a *key-iterated* block cipher: the transformation from plaintext to ciphertext can be seen as the repeated application of an invertible round transformation, alternated with the addition of round keys. The number of rounds is denoted by N_r. An encryption consists of an initial key addition, denoted by AddRoundKey, followed by $N_r - 1$ applications of the transformation Round, and finally one application of FinalRound. The initial key addition and every round take as input the State and a *round key*. The round key for round i is denoted by ExpandedKey[i], and ExpandedKey[0] denotes the input of the initial key addition. The derivation of ExpandedKey from the CipherKey is denoted by KeyExpansion. A high-level description of Rijndael in pseudo-C notation is shown in List 1.

The number of rounds depends on the block length and the key length. Let N_b be the block length divided by 32 and N_k the length of the cipher key divided by 32. Then we have:

$$N_r = \max(N_k, N_b) + 6. \qquad (1)$$

The number of rounds was determined by considering the known types of cryptanalysis and adding

```
Rijndael (State, CipherKey)
{
KeyExpansion(CipherKey, ExpandedKey);
AddRoundKey (State, ExpandedKey[0]);
for (i = 1; i < Nr; i ++) Round (State,
    ExpandedKey[i]);
FinalRound (State, ExpandedKey[ Nr ]);
}
```

List 1. High-level algorithm for encryption with Rijndael.

a security margin. For example, for a key length and block length of 128 bits, the best shortcut attack was on 6 rounds. We added 4 rounds resulting in $N_r = 10$. For a detailed treatment, we refer to [1, Section 3.5]. The encryption and decryption algorithms of Rijndael are not the same, but do have the same structure. For a treatment of these aspects, we refer to [1, Section 3.7].

THE ROUND TRANSFORMATION: The round transformation is denoted `Round`, and is a sequence of four invertible transformations, called *steps*. This is shown in List 2. The final round of the cipher is slightly different: with respect to the round transformation, the `MixColumns` step has been removed. It is denoted `FinalRound` and also shown in List 2. The steps are specified in the following subsections.

Representation

The state is a rectangular array of elements of $GF(2^8)$ of four rows and N_b columns.

In the specification of Rijndael, a byte with bits $b_7b_6b_5b_4b_3b_2b_1b_0$ maps to an element in $GF(2^8)$ given by the following polynomial:

$$b(x) = b_7x^7 + b_6x^6 + b_5x^5 + b_4x^4 + b_3x^3$$
$$+ b_2x^2 + b_1x + b_0. \qquad (2)$$

```
Round(State, ExpandedKey[i])
{
SubBytes(State);
ShiftRows(State);
MixColumns(State);
AddRoundKey(State,ExpandedKey[i]);
}
FinalRound(State,ExpandedKey[Nr])
{
SubBytes(State);
ShiftRows(State);
AddRoundKey(State,ExpandedKey[Nr]);
}
```

List 2. The Rijndael round transformation.

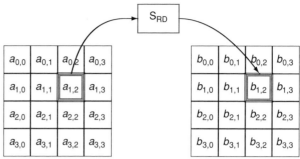

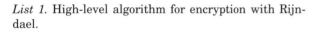

Fig. 1. SubBytes acts on the individual bytes of the state

where the coefficients b_i are elements of $GF(2)$ (i.e., bits). In this representation, addition consists of addition of polynomials and multiplication corresponds with multiplication of polynomials modulo the following underline{irreducible polynomial}:

$$m(x) = x^8 + x^4 + x^3 + x + 1. \qquad (3)$$

In the following of this chapter, we denote constants in $GF(2^8)$ by the hexadecimal notation of the corresponding byte value. For example, 57 corresponds with bit string 01010111 and hence with the polynomial $x^6 + x^4 + x^2 + x + 1$.

The SubBytes Step

The `SubBytes` step is the only non-linear transformation of the round. `SubBytes` is a bricklayer permutation consisting of an invertible S-box applied to the elements of the state. Figure 1 illustrates the effect of the `SubBytes` step on the state.

The same S-box is used for all byte positions. This is a design choice motivated by concerns of simplicity and implementation cost.

The S-box is defined by the following function in $GF(2^8)$:

$$f_{RD}(x) = 05 \cdot x^{254} \quad + 09 \cdot x^{253} + F9 \cdot x^{251}$$
$$+ 25 \cdot x^{247} + F4 \cdot x^{239} + 01 \cdot x^{223} \qquad (4)$$
$$+ B5 \cdot x^{191} + 8F \cdot x^{127} + 63.$$

The S-box was constructed as the composition of two invertible mappings:
- Multiplicative inverse: to have the desired nonlinearity properties as required by the wide trail strategy. This choice was inspired by Nyberg [2].
- Affine mapping: to complicate the algebraic expression without affecting the nonlinearity properties. This was inspired by algebraic attacks such as interpolation attacks.

In hardware and software the S-box can be implemented as a look-up table with 256 entries. In

Table 1. ShiftRows: shift offsets for different block lengths

Block length	C_0	C_1	C_2	C_3
128	0	1	2	3
160	0	1	2	3
192	0	1	2	3
224	0	1	2	4
256	0	1	3	4

hardware the area taken by an S-box can be reduced by exploiting the internal structure of the S-box.

The ShiftRows Step

The ShiftRows step is a transposition that cyclically shifts the rows of the state, each over a different offset, as imposed by the wide trail strategy.

Row 0 is shifted over C_0 bytes, row 1 over C_1 bytes, row 2 over C_2 bytes and row 3 over C_3 bytes. The shift offsets C_2 and C_3 depend on the block length. The different values are specified in Table 1. Figure 2 illustrates the effect of the ShiftRows step on the state.

The MixColumns Step

The MixColumns step is a bricklayer permutation operating on the state column by column.

The columns of the state are considered as polynomials over $GF(2^8)$ and multiplied modulo $x^4 + 1$ with a fixed polynomial $c(x)$:

$$c(x) = 03 \cdot x^3 + 01 \cdot x^2 + 01 \cdot x + 02. \tag{5}$$

This polynomial has been selected as one of the simplest polynomial that has a *branch number* equal to 5. The branch number is a measure that expresses the diffusion power of a mapping in the context of the wide trail strategy. As illustrated in Figure 3, the modular multiplication with a fixed polynomial can be written as a matrix multiplication.

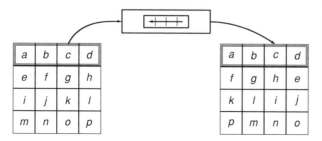

Fig. 2. ShiftRows operates on the rows of the state

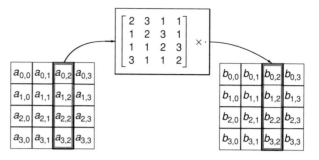

Fig. 3. MixColumns operates on the columns of the state

The polynomial $c(x)$ is coprime to $x^4 + 1$ and therefore has an inverse modulo $x^4 + 1$. The inverse polynomial $d(x)$ is defined by

$$(03 \cdot x^3 + 01 \cdot x^2 + 01 \cdot x + 02) \cdot d(x) \equiv 01$$
$$(\bmod x^4 + 1). \tag{6}$$

This yields:

$$d(x) = 0B \cdot x^3 + 0D \cdot x^2 + 09 \cdot x + 0E. \tag{7}$$

In hardware implementations, these linear maps can be efficiently hardwired. In software implementations, table-lookups can be used to efficiently exploit a wide range of processors. On 32-bit processors, the sequence of the steps SubBytes, ShiftRows and MixColumns can be implemented by a single sequence of table lookups.

The Key Addition

The key addition is denoted AddRoundKey. In this transformation, the state is modified by adding a round key to it. The addition in $GF(2^8)$ corresponds with the bitwise XOR operation. The round key length is equal to the block length.

KEY SCHEDULE: The key schedule consists of two components: the key expansion and the round key selection. The key expansion specifies how ExpandedKey is derived from the cipher key.

The expanded key can be seen as a rectangular array with four rows of elements in $GF(2^8)$. The key expansion function depends on the key length: there is a version for keys up to 224 bits and a version for keys longer than 224 bits. For their detailed specification, we refer to [1, Section 3.6]. In both versions of the key expansion, the first N_k columns of the expanded key are filled with the cipher key. The following columns are computed recursively in terms of previously defined columns. The recursion uses the elements of the previous column, the bytes of the column

N_k positions earlier, and round constants RC[j]. The round constants are independent of N_k and defined by RC[j] = x^{j-1}. The recursive key expansion allows on-the-fly computation of round keys on memory-constrained platforms.

Joan Daemen
Vincent Rijmen

References

[1] Daemen, J. and V. Rijmen (2002). "The design of Rijndael." *AES—Advanced Encryption Standard. Information Security and Cryptography*. Springer-Verlag, Berlin, Heidelberg, New York.

[2] Nyberg, K. (1994). "Differentially uniform mappings for cryptography." *Advances in Cryptology— EUROCRYPT'93*, Lecture Notes in Computer Science, vol. 765, ed. T. Helleseth. Springer-Verlag, Berlin, 55–64.

[3] Advanced Encryption Standard (AES), Federal Information Processing Standard (FIPS), Publication 197, National Bureau of Standards, U.S. Department of Commerce, Washington, DC, November 2001.

[4] Nechvatal, James, Elaine Barker, Lawrence Bassham, William Burr, Morris Dworkin, James Foti, Edward Roback, *Report on the Development of the Advanced Encryption Standard (AES)* Computer Security Division, Information Technology Laboratory, NIST, Technology Administration, U.S. Department of Commerce, Washington, DC.

RING

A *ring* $R = (S, +, \times)$ is the extension of a <u>group</u> $(S, +)$ with an additional operation $\times$, subject to the following additional axioms:

- **Commutativity of $+$:** For all $x, y \in S$, $x + y = y + x$.
- **Closure of $\times$:** For all $x, y \in S$, $x \times y \in S$.
- **Associativity of $\times$:** For all $x, y, z \in S$, $(x \times y) \times z = x \times (y \times z)$.
- **Distributivity of $\times$ over $+$:** For all $x, y, z \in S$, $x \times (y + z) = (x \times y) + (x \times z)$ and $(x + y) \times z = (x \times z) + (y \times z)$.

In other words, a ring can be viewed as the extension of a commutative additive group with a multiplication operation. The rings of interest in cryptography generally also have an identity element:

- **Identity of $\times$:** There exists a multiplicative *identity element*, denoted 1, such that for all $x \in S, x \times 1 = 1 \times x = 1$.

Let S^* denote the elements that have a multiplicative inverse; these are sometimes called the *units*

of the ring. (The additive identity 0 does not have a multiplicative inverse. Nonzero elements that do not have multiplicative inverses are called the *zero divisors* of the ring.) Then $(S^*, \times)$ is a group with respect to the multiplication operation; it is called the *multiplicative group* of the ring, or sometimes the *group of units*.

A ring is *commutative* if the multiplicative group is also commutative, i.e., for all $x, y \in S$, $x \times y = y \times x$. (The additive group is always commutative as noted above.)

The most common ring in public-key cryptography is the ring of integers modulo a composite number n. Here, the ring, denoted Z_n, consists of the set of integers (i.e., residue classes) modulo n and the ring operations are modular addition and multiplication (see <u>modular arithmetic</u>). The multiplicative group, denoted Z_n^*, consists of the integers relatively prime to the modulus n, and its <u>order</u> is $\phi(n)$, where ϕ is <u>Euler's totient function</u>. For instance, if $n = pq$ where p and q are distinct primes, then $\phi(n) = (p-1)(q-1)$. It is easy to determine $\phi(n)$ given the primes p and q and difficult without them; this fact is one basis for the security of the <u>RSA problem</u>.

See also <u>field</u>, <u>finite field</u>.

Burt Kaliski

RIPEMD FAMILY

The RIPEMD Family designates a family of five different <u>hash functions</u>: RIPEMD, RIPEMD-128, RIPEMD-160, RIPEMD-256, and RIPEMD-320 [1, 2]. They take variable length input messages and hash them to fixed-length outputs. They all operate on 512-bit message blocks divided into sixteen 32-bit words. RIPEMD (later replaced by RIPEMD-128/160) and RIPEMD-128 produce a hash value of 128 bits, RIPEMD-160, RIPEMD-256, and RIPEMD-320 have a hash result of 160, 256, and 320 bits, respectively. All the five functions start by padding the message according to the so-called Merkle–Damgård strengthening technique (see <u>hash functions</u> for more details). Next, the message is processed block by block by the underlying compression function. This function initializes an appropriate number of 32-bit chaining variables to a fixed value to hash the first message block, and to the intermediate hash value for the following message blocks.

In RIPEMD, RIPEMD-128, and RIPEMD-160, two copies are made from the chaining variables, and both these sets of line variables are processed independently by two parallel lines. Each step of such a parallel line updates in turn one of the line variables using a different message word W_x. After 16 steps all message words have been used once, and are reused in the next 16 steps, but in a different order. This is repeated 3, 4 or 5 times depending on the algorithm. In the last step, the initial values of the chaining variables are combined with both sets of line variables to form the intermediate hash value. When all consecutive message blocks have been hashed, the last intermediate hash value is the hash value for the entire message. RIPEMD-256 and RIPEMD-320 are derived from RIPEMD-128 and RIPEMD-160, respectively, by turning the line variables into chaining variables and by replacing the combination of line variables at the end by a simple feedforward of the initial values of the chaining variables. In addition the contents of two chaining variables belonging to different lines is exchanged after every 16 steps. The following provides an overview of RIPEMD-160, RIPEMD-128, and their twins RIPEMD-320 and RIPEMD-256.

PADDING: The message is appended with a binary one and right-padded with a variable number of zeros followed by the length of the original message (modulo 2^{64}) coded over two binary words. The total padded message length must be a multiple of the message block size.

INITIAL VALUES: The RIPEMD Family uses up to ten 32-bit initial values defined as follows:

$$IV_0 = 67452301_x \quad IV_5 = 76543210_x$$
$$IV_1 = \text{EFCDAB89}_x \quad IV_6 = \text{FEDCBA98}_x$$
$$IV_2 = \text{98BADCFE}_x \quad IV_7 = \text{89ABCDEF}_x$$
$$IV_3 = 10325476_x \quad IV_8 = 01234567_x$$
$$IV_4 = \text{C3D2E1F0}_x \quad IV_9 = \text{3C2D1E0F}_x$$

RIPEMD-160 COMPRESSION FUNCTION: Five 32-bit chaining variables h_0, h_1, h_2, h_3, h_4 are either initialized to the fixed values IV_0 through IV_4 for the first 512-bit message block or to the intermediate hash value for the following message blocks. Let "$X^{\lll n}$" represent the cyclic rotation of X to the left by n bits, and let "+" represents addition

modulo 2^{32}. Then the compression function works as follows:

$$A \leftarrow h_0, A' \leftarrow h_0$$
$$B \leftarrow h_1, B' \leftarrow h_1$$
$$C \leftarrow h_2, C' \leftarrow h_2$$
$$D \leftarrow h_3, D' \leftarrow h_3$$
$$E \leftarrow h_4, E' \leftarrow h_4$$
for i = 0 to 79 do
$$T \leftarrow (A + f_i(B,C,D) + W_{r(i)} + K_i)^{\lll s_i(r(i))} + E$$
$$T' \leftarrow (A' + f_{79-i}(B',C',D') + W_{r'(i)}$$
$$\quad + K_i')^{\lll s_i(r'(i))} + E'$$
$$A \leftarrow E, A' \leftarrow E'$$
$$E \leftarrow D, E' \leftarrow D'$$
$$D \leftarrow C^{\lll 10}, D' \leftarrow C'^{\lll 10}$$
$$C \leftarrow B, C' \leftarrow B'$$
$$B \leftarrow T, B' \leftarrow T'$$
$$T \leftarrow h_1 + C + D'$$
$$h_1 \leftarrow h_2 + D + E'$$
$$h_2 \leftarrow h_3 + E + A'$$
$$h_3 \leftarrow h_4 + A + B'$$
$$h_4 \leftarrow h_0 + B + C'$$
$$h_0 \leftarrow T$$

where the ordering of message words $r(i)$ and $r'(i)$, the non-linear functions f_i, the shifts s_i, and the constants K_i and K_i' are defined as:

1. **Ordering of the message words $r(i)$ and $r'(i)$.** Take the following permutation ρ:

j	0	1	2	3	4	5	6	7	8	9	10	11	12	13	14	15
$\rho(j)$	7	4	13	1	10	6	15	3	12	0	9	5	2	14	11	8

Further define the permutation π by setting $\pi(j) = (9j + 5) \bmod 16$. The ordering of the message words $r(i)$ and $r'(i)$ is then given by the following table:

	$0 \le i \le 15$	$16 \le i \le 31$	$32 \le i \le 47$	$48 \le i \le 63$	$64 \le i \le 79$
$r(i)$	i	$\rho(i-16)$	$\rho^2(i-32)$	$\rho^3(i-48)$	$\rho^4(i-64)$
$r'(i)$	$\pi(i)$	$\rho\pi(i-16)$	$\rho^2\pi(i-32)$	$\rho^3\pi(i-48)$	$\rho^4\pi(i-64)$

2. **Non-linear functions f_i.** Let "$\oplus$", "$\vee$", "$\wedge$", and "$\neg$" represent, respectively, bit-wise exclusive-or, bit-wise or, bit-wise and, and bit-wise complement:

$$f_i(x,y,z) = x \oplus y \oplus z, \qquad 0 \le i \le 15$$
$$f_i(x,y,z) = (x \wedge y) \vee (\neg x \wedge z), \quad 16 \le i \le 31$$
$$f_i(x,y,z) = (x \vee \neg y) \oplus z, \qquad 32 \le i \le 47$$
$$f_i(x,y,z) = (x \wedge z) \vee (y \wedge \neg z), \quad 48 \le i \le 63$$
$$f_i(x,y,z) = x \oplus (y \vee \neg z), \qquad 64 \le i \le 79$$

3. Shifts $s_i(j)$

j	0	1	2	3	4	5	6	7	8	9	10	11	12	13	14	15
$0 \leq i \leq 15$	11	14	15	12	5	8	7	9	11	13	14	15	6	7	9	8
$16 \leq i \leq 31$	12	13	11	15	6	9	9	7	12	15	11	13	7	8	7	7
$32 \leq i \leq 47$	13	15	14	11	7	7	6	8	13	14	13	12	5	5	6	9
$48 \leq i \leq 63$	14	11	12	14	8	6	5	5	15	12	15	14	9	9	8	6
$64 \leq i \leq 79$	15	12	13	13	9	5	8	6	14	11	12	11	8	6	5	5

4. Constants K_i and K_i'.

	$0 \leq i \leq 15$	$16 \leq i \leq 31$	$32 \leq i \leq 47$	$48 \leq i \leq 63$	$64 \leq i \leq 79$
K_i	00000000_x	$5A827999_x$	$6ED9EBA1_x$	$8F1BBCDC_x$	$A953FD4E_x$
K_i'	$50A28BE6_x$	$5C4DD124_x$	$6D703EF3_x$	$7A6D76E9_x$	00000000_x

RIPEMD-128 COMPRESSION FUNCTION: The main difference with RIPEMD 160 is that a hash result and chaining variable of 128 bits (four 32-bit words) is used and that there are only 64 steps. Four 32-bit chaining variables h_0, h_1, h_2, h_3 are either initialized to the fixed values IV_0 through IV_3 for the first 512-bit message block or to the intermediate hash value for the following message blocks. Then the compression function works as follows:

$$
\begin{aligned}
&A \leftarrow h_0, A' \leftarrow h_0 \\
&B \leftarrow h_1, B' \leftarrow h_1 \\
&C \leftarrow h_2, C' \leftarrow h_2 \\
&D \leftarrow h_3, D' \leftarrow h_3 \\
&\text{for } i = 0 \text{ to } 63 \text{ do} \\
&\quad T \leftarrow (A + f_i(B, C, D) + W_{r(i)} + K_i)^{\lll s_i(r(i))} \\
&\quad T' \leftarrow (A' + f_{63-i}(B', C', D') + W_{r'(i)} \\
&\qquad + K_i')^{\lll s_i(r'(i))} \\
&\quad A \leftarrow D, A' \leftarrow D' \\
&\quad D \leftarrow C, D' \leftarrow C' \\
&\quad C \leftarrow B, C' \leftarrow B' \\
&\quad B \leftarrow T, B' \leftarrow T' \\
&T \leftarrow h_1 + C + D' \\
&h_1 \leftarrow h_2 + D + A' \\
&h_2 \leftarrow h_3 + A + B' \\
&h_3 \leftarrow h_0 + B + C' \\
&h_0 \leftarrow T
\end{aligned}
$$

where the ordering of message words $r(i)$ and $r'(i)$, the non-linear functions f_i, the shifts s_i, and the constants K_i and K_i' are defined as in RIPEMD-160, except that $K_i' = 00000000_x$ for $48 \leq i \leq 63$.

RIPEMD-320 COMPRESSION FUNCTION: Ten 32-bit chaining variables $h_0, h_1, h_2, h_3, h_4, h_5, h_6, h_7, h_8, h_9$ are either initialized to the fixed values IV_0 through IV_9 for the first 512-bit message block or to the intermediate hash value for the following message blocks. Then the compression

function works as follows:

$$
\begin{aligned}
&A \leftarrow h_0, A' \leftarrow h_5 \\
&B \leftarrow h_1, B' \leftarrow h_6 \\
&C \leftarrow h_2, C' \leftarrow h_7 \\
&D \leftarrow h_3, D' \leftarrow h_8 \\
&E \leftarrow h_4, E' \leftarrow h_9 \\
&\text{for } i = 0 \text{ to } 79 \text{ do} \\
&\quad T \leftarrow (A + f_i(B, C, D) + W_{r(i)} + K_i)^{\lll s_i(r(i))} \\
&\qquad E \\
&\quad T' \leftarrow (A' + f_{79-i}(B', C', D') + W_{r'(i)} \\
&\qquad + K_i')^{\lll s_i(r'(i))} + E' \\
&\quad A \leftarrow E, A' \leftarrow E' \\
&\quad E \leftarrow D, E' \leftarrow D' \\
&\quad D \leftarrow C^{\lll 10}, D' \leftarrow C'^{\lll 10} \\
&\quad C \leftarrow B, C' \leftarrow B' \\
&\quad B \leftarrow T, B' \leftarrow T' \\
&\quad \text{if } i = 15 \text{ then} \\
&\qquad T \leftarrow B, B \leftarrow B', B' \leftarrow T \\
&\quad \text{else if } i = 31 \text{ then} \\
&\qquad T \leftarrow D, D \leftarrow D', D' \leftarrow T \\
&\quad \text{else if } i = 47 \text{ then} \\
&\qquad T \leftarrow A, A \leftarrow A', A' \leftarrow T \\
&\quad \text{else if } i = 63 \text{ then} \\
&\qquad T \leftarrow C, C \leftarrow C', C' \leftarrow T \\
&\quad \text{else if } i = 79 \text{ then} \\
&\qquad T \leftarrow E, E \leftarrow E', E' \leftarrow T \\
&h_0 \leftarrow h_0 + A, h_5 \leftarrow h_5 + A' \\
&h_1 \leftarrow h_1 + B, h_6 \leftarrow h_6 + B' \\
&h_2 \leftarrow h_2 + C, h_7 \leftarrow h_7 + C' \\
&h_3 \leftarrow h_3 + D, h_8 \leftarrow h_8 + D' \\
&h_4 \leftarrow h_4 + E, h_9 \leftarrow h_9 + E'
\end{aligned}
$$

where the ordering of message words $r(i)$ and $r'(i)$, the non-linear functions f_i, the shifts s_i, and the constants K_i and K_i' are defined as in RIPEMD-160.

RIPEMD-256 COMPRESSION FUNCTION: The main difference with RIPEMD 320 is that a hash result and chaining variable of 256 bits (eight 32-bit words) is used and that there are only 64 steps. Eight 32-bit chaining variables h_0, h_1, h_2, h_3, h_4, h_5, h_6, h_7 are either initialized to the fixed values IV_0 through IV_3 and IV_5 through IV_8 for the first 512-bit message block or to the intermediate hash value for the following message blocks. Then the compression function works as follows:

$$
\begin{aligned}
&A \leftarrow h_0, A' \leftarrow h_4 \\
&B \leftarrow h_1, B' \leftarrow h_5 \\
&C \leftarrow h_2, C' \leftarrow h_6 \\
&D \leftarrow h_3, D' \leftarrow h_7 \\
&\text{for } i = 0 \text{ to } 63 \text{ do} \\
&\quad T \leftarrow (A + f_i(B, C, D) + W_{r(i)} + K_i)^{\lll s_i(r(i))} \\
&\quad T' \leftarrow (A' + f_{63-i}(B', C', D') + W_{r'(i)} \\
&\qquad + K'_i)^{\lll s_i(r'(i))} \\
&\quad A \leftarrow D, A' \leftarrow D' \\
&\quad D \leftarrow C, D' \leftarrow C' \\
&\quad C \leftarrow B, C' \leftarrow B' \\
&\quad B \leftarrow T, B' \leftarrow T' \\
&\quad \text{if } i = 15 \text{ then} \\
&\qquad T \leftarrow A, A \leftarrow A', A' \leftarrow T \\
&\quad \text{else if } i = 31 \text{ then} \\
&\qquad T \leftarrow B, B \leftarrow B', B' \leftarrow T \\
&\quad \text{else if } i = 47 \text{ then} \\
&\qquad T \leftarrow C, C \leftarrow C', C' \leftarrow T \\
&\quad \text{else if } i = 63 \text{ then} \\
&\qquad T \leftarrow D, D \leftarrow D', D' \leftarrow T \\
&h_0 \leftarrow h_0 + A, h_4 \leftarrow h_4 + A' \\
&h_1 \leftarrow h_1 + B, h_5 \leftarrow h_5 + B' \\
&h_2 \leftarrow h_2 + C, h_6 \leftarrow h_6 + C' \\
&h_3 \leftarrow h_3 + D, h_7 \leftarrow h_7 + D'
\end{aligned}
$$

where the ordering of message words $r(i)$ and $r'(i)$, the non-linear functions f_i, the shifts s_i, and the constants K_i and K'_i are defined as in RIPEMD-160, except that $K'_i = 00000000_x$ for $48 \leq i \leq 63$.

RIPEMD: The original RIPEMD consists of essentially two parallel versions of MD4, with some improvements to the shifts and the order of the message words; the two parallel instances differ only in the round constants. At the end of the compression function, the words of left and right halves are added to each other and to the initial values of the chaining variable.

SECURITY CONSIDERATIONS: The RIPEMD Family has been designed to provide collision resistance. RIPEMD was developed in 1992 in the framework of the EC-RACE project RIPE [1]. In 1995, Dobbertin found collisions for reduced versions of RIPEMD [3]. Due to these partial attacks, RIPEMD was upgraded in 1996 by Dobbertin

et al. to RIPEMD-128 (as plug-in substitute for RIPEMD) and RIPEMD-160 [2]. At the same time, the variants RIPEMD-256 and RIPEMD-320 were introduced as well. An additional reason for the introduction of RIPEMD-160 are brute force collision search attacks. In [4], van Oorschot and Wiener estimate that with a 10 million US$ machine collisions of MD5 can be found in 21 days in 1994, which corresponds to 4 hours in 2004. To counter such collision search attacks, hash values of at least 160 bits are required. RIPEMD-128 and RIPEMD-160 are included in ISO/IEC 10118-3 [5].

Antoon Bosselaers

References

[1] RIPE (1995). "Integrity primitives for secure information systems." *Final Report of RACE Integrity Primitives Evaluation (RIPE-RACE 1040)*, Lecture Notes in Computer Science, vol. 1007, eds. A. Bosselaers, and B. Preneel. Springer-Verlag, Berlin.

[2] Dobbertin, H., A. Bosselaers, and B. Preneel (1996). "RIPEMD-160: A strengthened version of RIPEMD." *Fast Software Encryption*, Lecture Notes in Computer Science, vol. 1039, ed. D. Gollmann, Springer-Verlag, Berlin, 71–82. Final version available at http://www.esat.kuleuven.ac.be/~cosicart/pdf/AB-9601/. More information on all aspects of RIPEMD-xxx can be found at http://www.esat.kuleuven.ac.be/~bosselae/ripemd160/

[3] Dobbertin, H. (1992). "RIPEMD with two-round compress function is not collisionfree." *Journal of Cryptology*, 10 (1), 51–69.

[4] van Oorschot, P.C. and M. Wiener (1999). "Parallel collision search with cryptanalytic applications." *Journal of cryptology*, 12 (1), 1–28.

[5] ISO/IEC 10118-3 (2003). "Information technology—security techniques—hash-functions—Part 3: Dedicated hash-functions."

RSA DIGITAL SIGNATURE SCHEME

In the basic formula for the RSA cryptosystem [30] (see also RSA Problem, RSA public-key encryption), a digital signature s is computed on a message m according to the equation (see modular arithmetic)

$$s = m^d \bmod n, \tag{1}$$

where (n, d) is the signer's RSA private key. The signature is verified by recovering the message m with the signer's RSA public key (n, e):

$$m = s^e \bmod n. \tag{2}$$

Though the meaning of the value m that is signed with this formula has changed over the years, the basic formula has remained the same since it was introduced in 1977. The purpose of this entry is to survey the main approaches based on that formula, each of which can be referred to as an *RSA digital signature scheme* (see also digital signature schemes for general definitions).

In the definition just given, the value m is itself the message. This is helpful for illustration, but introduces a few challenges:

1. The set of messages that can be signed with the basic formula is limited to the set of integers in the range $[0, n-1]$. A larger value of m could be signed, but the value $m \bmod n$ is all that would be protected.

2. Some messages are quite easy to sign: for instance, $m = 0$ always has the signature 0 regardless of the signer's private key, and similarly for $m = 1$ and $m = n - 1$.

3. Every signature value s corresponds to some message m by Equation (2). It is easy therefore to construct valid message-signature pairs given only the signer's public key by starting with the signature value s (though the messages so obtained might not be anything meaningful).

4. Finally, the signer's willingness to provide signatures on some messages can be exploited by an opponent to obtain signatures on additional messages. As one example, if the opponent wants a signature on a message m but does not want the signer to see m, the opponent can instead ask to have the message $m' = mr^e \bmod n$ signed, where r is a random value. Let s' be the signature on m', and let $s = s'r^{-1} \bmod n$. Then s is the signature on m. Though this property has the benefit of enabling blind signatures, it also presents an avenue for attack. More generally, RSA signatures have a multiplicative property that can be an advantage or a disadvantage depending on the situation: given signatures on messages m_1 and m_2, it is straightforward to determine the signatures on any combination of products of the form $m_1^i m_2^j$.

Due to the various concerns just described, RSA digital signature schemes in practice are typically constructed so that the value m is not the message itself, but rather a *message representative* derived from the message. This approach employs four design principles that address the concerns just noted:

- *Large message space*: The set of messages that can be signed should be as large as possible. For this reason, the value m is typically derived in some way from the result of applying a hash function to the actual message to be signed.

- *Nontrivial message representative*: Message representatives such as 0 and 1 should be avoided, or should be very unlikely. In some schemes, this is accomplished by giving m a "random" appearance; in others, just by some padding.

- *Sparse message representative space*: Only a small fraction of values of m should be valid message representatives; this makes it unlikely for that a random signature s will correspond to a valid m. For this reason, the value m generally has some structure that can be verified, and which is unlikely to occur at random. Alternatively, if many values of m are valid, then it should be difficult to find a message with a given message representative.

- *Non-multiplicativity*: Multiplicative relationships between signatures should be avoided. Randomness or padding, and structure, both help in this regard.

One of the earliest and simplest improvements over the basic formula is to define the message representative m as the hash of the message M being signed, $m = Hash(M)$. This meets the first three design goals: messages of any length can be signed; hash values such as 0 and 1 are very unlikely; and, assuming the hash values are sufficiently shorter than the modulus n, only a small fraction of values of m will be valid hash values.

However, multiplicative relationships are still a potential problem. In particular, if the hash output is significantly shorter than the modulus (e.g., 160 bits vs. 1024 bits with today's parameter sizes), then it is possible to attack the signature scheme by methods from index calculus. Although the opponent cannot factor the modulus, the attacker can readily factor the much shorter hash values. From a sufficiently large set of signatures, the attacker can thereby solve for the signatures on *all* values m in a factor base by index calculus—and from those construct the signature on *any* message M for which $Hash(M)$ is smooth with respect to that factor base (see smoothness). (This approach, observed in the design of the PKCS #1 specification [14] discussed below, is an extension of an early attack on the RSA cryptosystem by Desmedt and Odlyzko [7].)

On the other hand, if the hash value is as long as the modulus n, then factoring the hash value is as hard as factoring the modulus, so the attack just described is not a concern. This is the basis for the *Full Domain Hash* (FDH) scheme of Bellare and Rogaway [2]. In fact, FDH turns out to have an additional security benefit, which is that it is possible in the <u>random oracle model</u> to obtain a proof that the signature scheme is as difficult to break, as the RSA Problem is to solve. (Recall that the RSA Problem is to solve for x such that $y = x^e$ mod n, given y, n and e, where the target value y is random.) It is instructive to explore that proof briefly, as it is a good example of the modern design of cryptographic schemes that has resulted from the insights of Bellare and Rogaway and other contemporary researchers.

In the proof, the attacker is assumed to have the ability to do the following:

- Obtain signatures from the actual signer on some number of chosen messages M.
- Evaluate the hash function on some number of chosen messages M.

The underlying hash function is modeled as a random oracle, meaning that the attacker's probability of success is taken over a *random* choice of hash function. Accordingly, it is not enough that the attack works well for *some* hash function; rather, it must work well, on average, for *any* (theoretical) hash function. This is a strong assumption on the attack, but a reasonable starting point for analysis.

The attacker's goal is to produce a new signature s' on a new message M', and to be able to do so in a reasonable amount of time with high probability.

Suppose now that there is such an attacker, represented by an algorithm A. In the proof, this algorithm A is transformed to a second algorithm B that breaks the RSA Problem in a similar amount of time and probability. The "reduction" (see <u>computational complexity</u>) from B to A takes advantage of the fact that the hash function is a black box, so the attacker cannot look inside. Accordingly, it is possible to "simulate" both the signer and the hash function so that they appear to be actual ones to the algorithm A, but such that when algorithm A forges the signature, it will in fact be solving an instance of the RSA Problem.

The reduction employs a table that has a signature-hash entry for each message, and goes something like this:

1. When the attacker asks to obtain a signature on a message M, the simulated signer first looks in the table to see if there's an entry for the message. If so, the signature part of that entry is returned. If not, see Step 3 below.

2. When the attacker asks to evaluate the hash function on a message M, the "simulated" hash function likewise first looks and returns the hash part of an entry, if there is one. If not, see Step 3 next.

3. In either case, if there isn't an entry for the message, then a new one is created. First, a random

signature s is generated. Second, a message representative $m = s^e$ mod n is computed. Third, the pair (m, s) is entered in the table as the hash-signature pair for the message M. Finally, the hash or signature is returned according to the attacker's request.

This simulation looks just like an actual signer and random <u>hash function</u>, because the hash value for a given message is random (since s is random), and the hash-signature pair for each message is internally consistent. Thus, an attacker will be just as successful when interacting with this simulation, as in the actual environment. But notice that the simulation doesn't need the actual signer's private key. So the simulator, combined with the algorithm A, produces another algorithm B, also independent of the actual signer's private key, which can then be applied to solve the RSA Problem.

In order to produce a forged signature on some message other than by accident, the attacker has to ask for the hash of the message. Otherwise, since the hash value is random, the probability that the signature will match is $1/n$, i.e., essentially 0. Since this signature is a forgery and FDH is deterministic, though, the attacker can't also ask for a signature on the same message. This means that there will be at least one message that the attacker asks to be hashed, but doesn't ask to be signed.

The one extra message gives the simulator an opportunity to embed an instance of the RSA Problem into the hash value. This is done with a small change to Step 2 above. In particular, at one, randomly selected time during the interaction with the attacker, instead of constructing a new entry, the step returns, as the hash value on that message, the target value y for the RSA Problem to be solved. Now, if the attacker also asks for a signature on the same message, the simulation will fail. But if the attacker produces a new signature on that message—a forgery—then the signature will be the solution to the RSA Problem on y!

Algorithm B is thus able to solve the RSA Problem whenever algorithm A produces a forgery, provided that the hash value involved in the forgery is the one selected by the simulator (and Algorithm A uses the hash value, rather than guessing the signature). Suppose algorithm A succeeds with probability ϵ_A in time T_A, and asks for at most Q messages to be hashed. Then algorithm B succeeds with probability at least $(\epsilon_A - 1/n)/Q$ in time just slightly more than T_A. Conversely, this means that if there is no algorithm for solving the RSA Problem that succeeds with probability greater than ϵ_B in time T_B, then there is no algorithm for forging signatures with probability greater than $\epsilon_B Q + 1/n$ in time T_B.

Since in practice an attacker can run a hash function a very large number of times, the ratio between the probabilities ϵ_A and ϵ_B may be quite large. Thus, the security proof itself may not imply as great a minimum difficulty as one might desire for forging FDH signatures, at least for typical parameter sizes. For instance, if one believes that the probability of breaking the RSA problem in a certain amount of time is at most 2^{-80}, and that an attacker can run the hash function 2^{64} times, then the proof only shows that the probability of forgery is at most about 2^{-16}. This doesn't mean that there's an algorithm that is this successful; it just means that such an algorithm can't be ruled out by the proof itself. Still, the line of reasoning is better than for "ad hoc" designs, where there is no clear connection between the difficulty of forgery and the RSA Problem.

The reason that the reduction for FDH is so "loose" (i.e., the ratio is so high) is that the simulator is able to embed the RSA Problem into only one hash value. Another scheme by the same authors, the *Probabilistic Signature Scheme* (PSS) [3], overcomes this limitation by introducing a random value (called a *seed*, which plays a role similar to _salt_) into each signature operation. As a result, each signature is independent of previous hash operations, so each hash value can embed a separate instance of the RSA Problem.

In PSS, the probability of success for breaking the RSA Problem is about the same as the probability for forging signatures, which is the best security reduction one can achieve.

(Some improvement in the security reduction for FDH can be obtained by a better proof technique, as Coron has shown [4]. Coron also gives a very careful analysis of the effect of the size of the salt on the reduction in PSS.)

In addition to the tight security proof, PSS has a second advantage: It has a simple variant, called PSS-R, that provides *message recovery*. In PSS-R, part or all of the message can be carried in the message representative in addition to the hash value. This is a return to the goal of the original formula, but with stronger security properties.

For the long term, PSS and PSS-R may well prevail as the most common RSA signature schemes, and they are found in newer standards such as PKCS #1 v2.1 [15], ISO/IEC 9796-2 [12] and the forthcoming IEEE P1363a [10]. However, in the short term, other signature schemes are better established. The most common today in practice is the PKCS #1 v1.5 scheme, introduced in 1991 in the first set of Public-Key Cryptography

Standards from RSA Laboratories [14]. The scheme has an ad hoc design where the message representative is constructed from a hash value with simple padding. On the one hand, no practical attack has been developed on this scheme, though some specially constructed cases have been shown to have weaknesses (see [6]). On the other hand, no security proof is available either, and it seems unlikely that one would be developed. Thus, like primitives from symmetric cryptography, the security of the scheme depends on resistance against specific attacks rather than a security reduction from a hard problem.

Another scheme today, found more in standards documents than in practice, is the scheme in ANSI X9.31 [1], which is also in ISO/IEC 14888-3 [13]. (A variant with message recovery is in ISO/IEC 9796-2 [12].) This scheme also has an ad hoc design with similar security properties to PKCS #1 v1.5.

An early scheme with message recovery can be found in ISO/IEC 9796-1 [11]. The scheme is particularly attractive for implementation since no hash function is involved. However, the scheme has turned out to be vulnerable to attack in some cases, as shown by Grieu [9] and in drafts circulated by Coppersmith, Halevi and Jutla. The standard has since been withdrawn. (Not all implementations of the standard are affected by the attacks; in particular, implementations where the message being signed and "recovered" is itself a hash value are not affected.)

Some of the RSA signature schemes mentioned also have variants based on the Rabin cryptosystem; this is true of the schemes in ANSI X9.31 and the various ISO/IEC documents, as well as PSS and PSS-R, though not the PKCS #1 v1.5 scheme. The variants based on the Rabin cryptosystem have the advantage that they allow the public exponent $e = 2$, so that signature verification is very efficient; but on the other hand some extra steps are required due to the fact that not every message representative m may have a square root modulo n. (See Rabin digital signature scheme for further discussion.)

A complementary approach to the schemes just described, which is primarily of research interest so far, is to derive the public exponent e itself from the message, where the value m is fixed within the public key. The advantage of this approach, described by Gennaro et al. [8], is that it is possible to obtain a tight security proof in the standard model, where the hash function is only assumed to have a certain "division-intractability" property—it does not need to be modeled as a random oracle. The difficulty of forging a signature can be shown to be closely related to the Strong RSA Assumption (see again RSA Problem). (Note though that the initial analysis needed some improvements [5].) A related approach is presented by Vanstone and Qu [17]; in their approach, both *e* and *m* may be derived from the message.

Burt Kaliski

References

[1] American National Standard X9.31-1998. *Digital Signatures Using Reversible Public Key Cryptography for the Financial Services Industry (rDSA)*.

[2] Bellare, Mihir and Phillip Rogaway (1993). "Random oracles are practical. A paradigm for designing efficient protocols." *Proceedings of 1st ACM Conference on Computer and Communications Security*. ACM Press, 62–73.

[3] Bellare, Mihir and Phillip Rogaway (1996). "The exact security of digital signatures—how to sign with RSA and Rabin." *Advances in Cryptology—EUROCRYPT'96*, Lecture Notes in Computer Science, vol. 1070, ed. U. Maurer. Springer, Berlin, 399–416.

[4] Coron, Jean-Sébastien (2002). "Optimal security proofs for PSS and other signature schemes." *Advances in Cryptology—EUROCRYPT 2002*, Lecture Notes in Computer Science, vol. 2332, ed. L.R. Knudsen. Springer, Berlin, 272–287.

[5] Coron, Jean-Sébastien and David Naccache (2000). "Security analysis of the Gennaro–Halevi–Rabin signature scheme." *Advances in Cryptology—EUROCRYPT 2000*, Lecture Notes in Computer Science, vol. 1807, ed. B. Preneel. Springer, Berlin, 91–101.

[6] Coron, Jean-Sébastien, David Naccache, and Julien P. Stern (1999). "On the security of RSA padding." *Advances in Cryptology—CRYPTO'99*, Lecture Notes in Computer Science, vol. 1666, ed. M. Wiener. Springer, Berlin, 1–18.

[7] Desmedt, Yvo and Andrew M. Odlyzko (1986). "A chosen text attack on the RSA cryptosystem and some discrete logarithm schemes." *Advances in Cryptology—CRYPTO'85*, Lecture Notes in Computer Science, vol. 218, ed. H.C. Williams. Springer, Berlin, 516–522.

[8] Gennaro, Rosario, Shai Halevi and Tal Rabin (1999). "Secure hash-and-sign signatures without the random oracle." *Advances in Cryptology—EUROCRYPT'99*, Lecture Notes in Computer Science, vol. 1592, ed. J. Stern. Springer, Berlin, 123–139.

[9] Grieu, François (2000). "A chosen messages attack on the ISO/IEC 9796-1 signature scheme." *Advances in Cryptology—EUROCRYPT 2000*, Lecture Notes in Computer Science, vol. 1807, ed. B. Preneel. Springer, Berlin, 70–80.

[10] IEEE P1363a Draft D12 (2003). *Standard Specifications for Public Key Cryptography: Amendment 1—Additional Techniques*. IEEE P1363 Working Group.

[11] ISO/IEC 9796-1:1999 (withdrawn). *Information Technology—Security Techniques Digital Signatures Giving Message Recovery—Part 1: Mechanisms Using Redundancy*.

[12] ISO/IEC 9796-2:2002. *Information Technology—Security Techniques—Digital Signatures Giving Message Recovery—Part 2: Integer Factorization Based Mechanisms*.

[13] ISO/IEC 14888-3:1998. *Information Technology—Security Techniques—Digital Signatures with Appendix—Part 3: Certificate-Based Mechanisms*.

[14] PKCS #1 v1.5 (1993). *RSA Encryption Standard*. RSA Laboratories.

[15] PKCS #1 v2.1 (2002). *RSA Cryptography Standard*. RSA Laboratories.

[16] Rivest, Ronald L., Adi Shamir, and Leonard M. Adleman (1978). "A method for obtaining digital signatures and public-key cryptosystems." *Communications of the ACM*, 21 (2), 120–126.

[17] Vanstone, Scott A and Minghua Qu. *Digital Signature Protocol with Reduced Bandwidth*. U.S. Patent No. 6,097,813, August 1, 2000 (Filed May 15, 1997).

RSA FACTORING CHALLENGE

Starting in 1991, RSA Data Security offered a set of "challenges" intended to measure the difficulty of integer factoring. The challenges consisted of a list of 41 *RSA Numbers*, each the product of two primes of approximately equal length, and another, larger list of *Partition Numbers* generated according to a recurrence.

The first five of the RSA Numbers, ranging from 100 to 140 decimal digits (330–463 bits), were factored successfully by 1999 (see [2] for details on the largest of these). An additional 512-bit (155-digit) challenge number was later added in view of the popularity of that key size in practice; it was also factored in 1999 [3].

In addition to the formal challenge numbers, an old challenge number first published in August 1977, renamed 'RSA-129', was factored in 1994 [1].

The Quadratic Sieve was employed for the numbers up to RSA-129, and the Number Field Sieve for the rest. The work factor in MIPS-years, summarized in Table 1, was roughly in line with expectations for these methods as techniques steadily improved. It is noteworthy that the effort for RSA-130, taking advantage of the Number Field Sieve, was less than that for RSA-129.

Cash prizes of more than US $50,000 have been awarded to the winners over the duration of the contest.

Table 1. Results of the RSA factoring challenge (QS = quadratic sieve; NFS = number field sieve); adapted from [4]

Number	Bits	Year	Method	MIPS-Years
RSA-100	330	1991	QS	7
RSA-110	364	1992	QS	75
RSA-120	397	1993	QS	830
RSA-129	426	1994	QS	5000
RSA-130	430	1996	NFS	1000
RSA-140	463	1999	NFS	2000
RSA-155	512	1999	NFS	8000
RSA-160	530	2003	NFS	Not given
RSA-576	576	2003	NFS	Not given

The current RSA Factoring Challenge has numbers from 576 to 2048 bits. As of this writing only the first challenge number, RSA-576, has been factored. (The numbers in the current challenge are designated by their length in bits rather than decimal digits.) See http://www.rsasecurity.com/rsalabs/challenges/ for more information.

Burt Kaliski

References

[1] Atkins, D., M. Graff, A.K. Lenstra, and P.C. Leyland (1995). "The magic words are SQUEAMISH OSSIFRAGE." *Advances in Cryptology—ASIACRYPT'94*, Lecture Notes in Computer Science, vol. 917, eds. J. Pieprzyk and R. Safavi-Naini. Springer, Berlin, 263–277.

[2] Cavallar, Stefania, Bruce Dodson, Arjen Lenstra, Paul Leyland, Walter Lioen, Peter L. Montgomery, Brian Murphy, Herman te Riele, and Paul Zimmermann (1999). "Factorization of RSA-140 using the number field sieve." *Advances in Cryptology—ASIACRYPT'99*, Lecture Notes in Computer Science, vol. 1716, eds. K.Y. Lam, E. Okamoto, and C. Xing. Springer, Berlin, 195–207.

[3] Cavallar, Stefania, Bruce Dodson, Arjen Lenstra, Paul Leyland, Walter Lioen, Peter L. Montgomery, Brian Murphy, Herman te Riele, et al. (2000). "Factorization of a 512-bit RSA modulus." *Advances in Cryptology—EUROCRYPT 2000*, Lecture Notes in Computer Science, vol. 1807, ed. B. Preneel. Springer, Berlin, 1–18.

[4] Contini, Scott (1999). "The Factorization of RSA-140." *RSA Laboratories' Bulletin* 10. March 8, 1999. Available from http://www.rsasecurity.com/rsalabs/bulletins/

RSA PROBLEM

INTRODUCTION: In RSA public-key encryption [30], Alice encrypts a plaintext M for Bob using Bob's public key (n, e) by computing the ciphertext

$$C = M^e \pmod{n}, \tag{1}$$

where n, the *modulus*, is the product of two or more large primes, and e, the *public exponent*, is an (odd) integer $e \geq 3$ that is relatively prime to $\phi(n)$, the order of the multiplicative group $\mathbf{Z}_n^*$. (See also Euler's totient function, modular arithmetic for background on these concepts.)

Bob, who knows the corresponding RSA private key (n, d), can easily decrypt, since $de = 1 \pmod{\phi(n)}$ implies that

$$M = C^d \pmod{n}. \tag{2}$$

An adversary may learn C by eavesdropping, and may very well also know Bob's public key; nonetheless such an adversary should not be able to compute the corresponding plaintext M.

One may formalize the task faced by this adversary as the *RSA Problem*:

The RSA Problem: Given an RSA public key (n, e) and a ciphertext $C = M^e \pmod{n}$, to compute M.

To solve the RSA Problem an adversary, who doesn't know the private key, must nonetheless invert the RSA function.

The *RSA Assumption* is that the RSA Problem is hard to solve when the modulus n is sufficiently large and randomly generated, and the plaintext M (and hence the ciphertext C) is a random integer between 0 and $n - 1$. The assumption is the same as saying that the RSA function is a trapdoor one-way function (the private key is the trapdoor).

The randomness of the plaintext M over the range $[0, n - 1]$ is important in the assumption. If M is known to be from a small space, for instance, then an adversary can solve for M by trying all possible values for M.

The RSA Problem is the basis for the security of RSA public-key encryption as well as RSA digital signature schemes.

See also surveys by Boneh [10] and Katzenbeisser [24].

RELATIONSHIP TO INTEGER FACTORING: The RSA Problem is clearly no harder than integer factoring, since an adversary who can factor the modulus n can compute the private key (n, d) from the public key (n, e).

However, it is not clear whether the converse is true, that is, whether an algorithm for integer factoring can be efficiently constructed from an algorithm for solving the RSA Problem.

Boneh and Venkatesan [9] have given evidence that such a construction is unlikely when the public exponent is very small, such as $e = 3$ or 17. Their result means that the RSA Problem for very small exponents could be easier than integer factoring, but it doesn't imply that the RSA Problem is actually easier, i.e., efficient algorithms are still not known. For larger public exponents, the question of equivalence with integer factoring still open as of this writing.

RECOVERING THE PRIVATE KEY: Clearly, if the adversary could compute Bob's private key (n, d) from his public key (n, e), then the adversary could decrypt C using Equation (2).

However, de Laurentis [15] and Miller [27] have shown that computing an RSA private key (n, d) from the corresponding RSA encryption key (n, e) is as hard as factoring the modulus n into its prime factors p and q. As already noted, given the factors p and q, it is easy to compute d from e, and conversely there is a probabilisitic polynomial-time algorithm which takes as input n, e, and d, and which factors n into p and q. (See also Fact 1 in Boneh [10].)

If the modulus n was chosen as the product of two "sufficiently large" randomly-chosen prime numbers p and q, then the problem of factoring n appears to be intractable. Thus, the private exponent d is protected from disclosure by the difficulty of factoring the modulus n.

An adversary might also try to compute d using some method of solving the discrete logarithm problem. For example, an adversary could compute the discrete logarithm of M to the base M^e (mod n). If d is too small (say, less than 160 bits), then an adversary might be able to recover it by the baby step-giant step method.

Even if d is too large to be recovered by discrete logarithm methods, however, it may still be at risk.

For example, Wiener [33] has shown that if the secret exponent is less than $n^{1/4}/3$, an adversary can efficiently compute d given n and e. An improved bound of $n^{0.292}$ has been presented by Boneh and Durfee [8] (see Wiener, Doneh-Durfee and May attacks on the RSA public key cryptosystem).

However, it does appear to be the case that if the RSA parameters were chosen large enough, then the adversary cannot solve the RSA Problem by computing the private RSA exponent of the recipient.

SELF-REDUCIBILITY: It is conceivable that someone could devise a clever procedure for solving the RSA Problem without factoring the modulus n or determining the private key d. An adversary might, for example, have a procedure that decrypts a small fraction of "weak" ciphertexts. However, the RSA procedure enjoys a certain kind of "self-reducibility", since it is multiplicative:

$$(MR)^e = M^e R^e \pmod{n}.$$

An adversary can transform a given ciphertext M^e into another one $(MR)^e$ by multiplying it by the encryption R^e of a randomly chosen element R of $\mathbf{Z}_n^*$. Since the result has a chance of being a "weak" ciphertext, it follows that if there is an adversarial procedure A that can decrypt a fraction ϵ of ciphertexts, then there is another (randomized) adversarial procedure A' that can decrypt all ciphertexts in expected running time that is polynomial in the running time of A, in $1/\epsilon$, and in $\log n$ (see polynomial time). (See Motwani and Raghavan [28, Section 14.4].)

Self-reducibility is a double-edged sword. On the one hand, it provides assurance that "all" random ciphertexts are equally hard to invert. This property has been helpful in the security proofs for several public-key encryption and signature schemes based on the RSA Problem. On the other hand, self-reducibility provides an avenue for an adversary to gain information about the decryption of one ciphertext from the decryption of other ciphertexts (see "chosen ciphertext attacks") below.

LOW PUBLIC EXPONENT RSA: A user of the RSA cryptosystem may reasonably wish to use a public exponent e that is relatively short: common choices are $e = 3$ or $e = 2^{16} + 1 = 65537$. Using a short public exponent results in faster public-key encryption and faster public-key signature verification. Does this weaken RSA?

If the public exponent is small and the plaintext M is very short, then the RSA function may be easy to invert: in particular, if $M < \sqrt[e]{N}$, then $C = M^e$ over the integers, so M can be recovered as $M = \sqrt[e]{C}$.

Håstad [22] shows that small public exponents can be dangerous when the same plaintext is sent to many different recipients, even if the plaintext is "padded" in various (simple) ways beforehand.

Coppersmith et al. [12] give a powerful "related messages" attack, which is effective when the public exponent is small, based on the LLL algorithm [25] for lattice reduction.

Because of these concerns, small public exponents are sometimes avoided in industry standards and in practice. However, the concerns can also be addressed with appropriate padding schemes (see "chosen ciphertext attacks" below), provided they are correctly implemented. For digital signature schemes, small public exponents are generally not an issue.

STRONG RSA ASSUMPTION: The *Strong RSA Assumption* was introduced by Barić and Pfitzmann [3] and by Fujisaki and Okamoto [18] (see also [13]).

This assumption differs from the RSA Assumption in that the adversary can select the public exponent e. The adversary's task is to compute, given a modulus n and a ciphertext C, *any* plaintext M and (odd) public exponent $e \geq 3$ such that $C = M^e$ (mod n). This may well be easier than solving the RSA Problem, so the assumption that it is hard is a stronger assumption than the RSA Assumption. The Strong RSA Assumption is the basis for a variety of cryptographic constructions.

BIT-SECURITY OF RSA ENCRYPTION: It is conceivable that RSA could be "secure" in the sense that the RSA Assumption holds (i.e., RSA is hard to invert), yet that RSA "leaks" information in that certain plaintext bits are easy to predict from the ciphertext. Does RSA provide security to individual bits of plaintext?

Goldwasser et al. [21] first studied the bit-security of RSA, showing that an adversary who could reliably extract from a ciphertext the least signficant bit (lsb) of the plaintext would in fact be able to decrypt RSA efficiently (i.e., obtain the entire plaintext efficiently).

This line of research was pursued by other researchers. For example, Vazirani and Vazirani [32]) showed that the adversary could still decrypt even with an lsb procedure that was only $0.732 + \epsilon$ accurate. They also showed that the low-order $\log(\log(n))$ bits of plaintext are $3/4 + \epsilon$ secure.

Chor and Goldreich [11] improved this result to show that the least-significant bit of RSA plaintext cannot be predicted with probability better than $1/2 + 1/\text{poly}(\log(n))$ (under the RSA Assumption). Alexi et al. [1, 2] completed this result to show that the least-significant $\log(\log(n))$ bits are secure in the same sense. (Fischlin and Schnorr [17] provide a simpler and tighter proof of this result.)

Håstad and Näslund [23] have shown that *all* of the plaintext bits are well-protected by RSA, in the sense that having a nontrivial advantage for predicting any one plaintext bit would enable the adversary to invert RSA completely.

The results about bit-security of RSA generally involve a *reduction* technique (see computational complexity), where an algorithm for solving the RSA Problem is constructed from an algorithm for predicting one (or more) plaintext bits. Like self-reducibility, bit-security is a double-edged sword. This is because the security reductions also provide an avenue of attack on a "leaky" implementation. If an implementation of an RSA decryption operation leaks some bits of the plaintext, then an adversary can potentially solve the RSA Problem for *any* ciphertext just by observing the implementation's behavior on some number of other ciphertexts. Such attacks have been described by Bleichenbacher [7] and by Manger [26].

CHOSEN CIPHERTEXT ATTACKS: An adversary may be able to decrypt an RSA ciphertext C if he can obtain decryptions (e.g., from the legitimate recipient) of other ciphertexts $C_1, C_2, \ldots, C_k$ (which may or may not be related to C). Such attacks are known as chosen ciphertext attacks (CCA1 and CCA2, depending on whether the C_is are allowed to depend upon C (of course they can't be equal to C)); see Bellare et al. [4] for details.

(The attacks related to bit-security are a special case of chosen-ciphertext attacks in which the adversary only obtains partial information about the decryption, not the full plaintext.)

Davida [14] first studied chosen ciphertext attacks for RSA, utilizing the multiplicative property of RSA.

Desmedt and Odlyzko [16] provided another chosen ciphertext attack, based on obtaining the decryption of many small primes.

To defeat chosen ciphertext attacks, researchers have turned to (possibly randomized) "padding" schemes that (reversibly) transform a plaintext before encryption.

One such proposal is Optimal Asymmetric Encryption Padding (OAEP) [5] which has been proven secure for chosen ciphertext attacks by Fujisaki et al. [19] under the RSA assumption. Other proposals that also avoid chosen ciphertext attacks have better security properties [29, 31]. See also RSA public-key encryption for related discussion.

Chosen-ciphertext attacks on digital signature schemes are the analogue to chosen ciphertext attacks on public-key encryption, and various padding shemes have been developed to defeat them as well, such as the Probabilistic Signature Scheme (PSS) of Bellare and Rogaway [6] and the scheme of Gennaro et al. [20]. See also RSA digital signature scheme.

CONCLUSIONS: The RSA Problem is now over a quarter century old. The elegant simplicity of the problem has led to numerous observations over the years, some yielding attacks, others avoiding them. Public-key encryption schemes and digital signature schemes have been developed whose strength is derived fully from the RSA Problem. The remaining open question, still, is how closely the security of the RSA Problem depends on integer factoring, and as with any hard problem in cryptography, whether any methods faster than those currently available for solving the problem will ever be discovered.

Ronald L. Rivest
Burt Kaliski

References

[1] Alexi, W.B., B. Chor, O. Goldreich, and C.P. Schnorr (1984). "RSA/Rabin bits are $1/2 + 1/poly(\log(N))$ secure." *Proceedings of FOCS'84*, Singer Island, IEEE, 449–457.

[2] Alexi, W.B., B. Chor, O. Goldreich, and C.P. Schnorr (1988). "RSA and Rabin functions: Certain parts are as hard as the whole." *SIAM Journal of Computing*, 17 (2), 194–209.

[3] Barić, Niko and Birgit Pfitzmann (1997). "Collision-free accumulators and fail-stop signature schemes without trees." *Advances in Cryptology—EUROCRYPT'97*, Lecture Notes in Computer Science, vol. 1233, ed. W. Fumy. Springer-Verlag, Berlin, 480–494.

[4] Bellare, M., A. Desai, D. Pointcheval, and P. Rogaway (1998). "Relations among notions of security for public-key encryption." *Advances in Cryptology—CRYPTO'98*, Lecture Notes in Computer Science, vol. 1462, ed. H. Krawczyk. Springer-Verlag, Berlin, 26–45.

[5] Bellare, Mihir and Phillip Rogaway (1996). "Optimal asymmetric encryption—how to encrypt with RSA." *Advances in Cryptology—EUROCRYPT'94*, Lecture Notes in Computer Science, vol. 950, ed. A. DeSantis. Springer-Verlag, Berlin, 92–111.

[6] Bellare, Mihir and Phillip Rogaway (1996). "The exact security of digital signatures—how to sign with RSA and Rabin." *Advances in Cryptology—EUROCRYPT'96*, Lecture Notes in Computer Science, vol. 1070, ed. U. Maurer. Springer-Verlag, Berlin, 399–416.

[7] Bleichenbacher. D. (1988). "Chosen ciphertext attacks against protocols based on the RSA encryption standard PKCS #1." *Advances in Cryptology—CRYPTO'98*, Lecture Notes in Computer Science, vol. 1462, ed. H. Krawczyk. Springer, Berlin, 1–12.

[8] Boneh, D. and G. Durfee (2000). "Cryptanalysis of RSA with private key d less than $N^{0.292}$." *IEEE Transactions on Information Theory*, 46 (4), 1339–1349.

[9] Boneh, D. and R. Venkatesan (1988). "Breaking RSA may not be equivalent to factoring." *Advances in Cryptology—EUROCRYPT'98*, Lecture Notes in Computer Science, vol. 1403, ed. K. Nyberg. Springer, Berlin, 59–71.

[10] Boneh, Dan (1999). "Twenty years of attacks on the RSA cryptosystem." *Notices of the AMS*, 46 (2), 203–213.

[11] Benny, Chor and Oded Goldreich (1985). "RSA/Rabin least significant bits are $\frac{1}{2} + 1/poly(\log n)$ secure." *Advances in Cryptology—CRYPTO'84*, Lecture Notes in Computer Science,

vol. 196. eds. G.R. Blakley and D.C. Chaum. Springer, Berlin, 303–313.

[12] Coppersmith, D., M. Franklin, J. Patarin, and M. Reiter (1996). "Low-exponent RSA with related messages." *Advances in Cryptography—EUROCRYPT'96*, Lecture Notes in Computer Science, vol. 1070, ed. V. Maurer. Springer-Verlag. Berlin, 1–9.

[13] Cramer, Ronald and Victor Shoup (2000). "Signature schemes based on the strong RSA assumption." *ACM Transactions on Information and System Security*, 3 (3), 161–185.

[14] Davida, G. (1982). "Chosen signature cryptanalysis of the RSA (MIT) public key cryptosystem." *Technical Report Tech Report TR-CS-82-2*, Deptartment of EECS, University of Wisconsin, Milwaukee.

[15] DeLaurentis, J.M. (1984). "A further weakness in the common modulus protocol for the RSA cryptoalgorithm." *Cryptologia*, 8, 253–259.

[16] Desmedt, Y. and A. M. Odlyzko (1986). "A chosen text attack on the RSA cryptosystem and some discrete logarithm schemes." *Advances in Cryptology—CRYPTO'85*, Lecture Notes in Computer Science, vol. 218, ed. H.C. Williams. Springer, Berlin, 516–522.

[17] Fischlin, Roger and Claus-Peter Schnorr (2000). "Stronger security proofs for RSA and Rabin bits." *Journal of Cryptology*, 13 (2), 221–244.

[18] Fujisaki, Eiichiro and Tatsuaki Okamoto (1997). "Statistical zero knowledge protocols to prove modular polynomial relations." *Advances in Cryptology—CRYPTO'97*, Lecture Notes in Computer Science, vol. 1294, ed. Burton S. Kaliski Jr. Springer-Verlag, Berlin, 16–30.

[19] Fujisaki, Eiichiro, Tatsuaki Okamoto, David Pointcheval, and Jacques Stern (2004). "RSA-OAEP is secure under the RSA assumption." *Journal of Cryptology*, 17 (2), 81–104.

[20] Gennaro, Rosario, Shai Halevi, and Tal Rabin. (1999). "Secure hash-and-sign signatures without the random oracle." *Advances in Cryptography—EUROCRYPT'99*, Lecture Notes in Computer Science, vol. 1592, ed. J. Stern. Springer-Verlag, Berlin, 123–139.

[21] Goldwasser, S., S. Micali, and P. Tong (1982). "Why and how to establish a private code on a public network." *Proc. FOCS'82*, IEEE. Chicago, 134–144.

[22] Håstad, J. (1988). "Solving simultaneous modular equations of low degree." *SIAM Journal of Computing*, 17, 336–341.

[23] Johan Håstad and Mats Näslund (1998). "The security of individual RSA bits." *IEEE Symposium on Foundations of Computer Science*, 510–521.

[24] Stefan Katzenbeisser (2001). *Recent Advances in RSA Cryptography*. Kluwer Academic Publishers.

[25] Lenstra, A.K., H.W. Lenstra, Jr., and L. Lovász (1982). "Factoring polynomials with rational coefficients." *Mathematische Ann.*, 261, 513–534.

[26] Manger, J. (2001). "A chosen ciphertext attack on RSA Optimal Asymmetric Encryption Padding (OAEP) as standardized in PKCS #1 v2.0." *Advances in Cryptology—CRYPTO 2001*, Lecture

Notes in Computer Science, vol. 2139, ed. J. Kilian. Springer, Berlin, 260–274.

[27] Gary, L. Miller (1976). "Riemann's hypothesis and tests for primality." *Journal of Computer and Systems Sciences*, 13 (3), 300–317.

[28] Motwani Rajeev and Prabhakar Raghavan (1995). *Randomized Algorithms*. Cambridge University Press, Cambridge.

[29] Okamoto, T. and D. Pointcheval (2001). "REACT: Rapid enhanced-security asymmetric cryptosystem transform." *Proceedings Cryptographers' Track RSA Conference (CT-RSA) 2001*, Lecture Notes in Computer Science, vol. 2020, ed. D. Naccache. Springer, Berlin, 159–175.

[30] Ronald, L. Rivest, Adi Shamir, and Leonard M. Adleman (1978). "A method for obtaining digital signatures and public-key cryptosystems." *Communications of the ACM*, 21 (2), 120–126.

[31] Shoup. V. (2001). *A Proposal for an ISO Standard for Public Key Encryption (Version 2.1)*. Manuscript, December 20. Available from http://shoup.net/papers/

[32] Vazirani Umesh and Vijay Vazirani (1984). "RSA bits are .732 + ε secure." *Proceedings CRYPTO'83*, ed. D. Chaum. Plenum Press, New York, 369–375.

[33] Wiener, M. (1990). "Cryptanalysis of short RSA secret exponents." *IEEE Transactions on Information Theory*, 36 (3), 553–558.

RSA PUBLIC-KEY ENCRYPTION

TRAPDOOR ONE-WAY PERMUTATIONS: A one-way function is a function f that anyone can compute efficiently, however inverting f is hard. Such a primitive is the basis of modern cryptography, and relies on the open problem $\mathcal{P}$ vs. $\mathcal{NP}$ (see computational complexity). As a consequence, any $\mathcal{NP}$-complete problem should lead to such a one-way function candidate. Unfortunately, $\mathcal{NP}$-complete problems are not so convenient for cryptographic applications, because either they are hard to solve for very large instances only, or very few instances are hard but the problem is easy on average. Furthermore, such a primitive is not enough for public-key encryption.

A trapdoor one-way permutation primitive (see also substitutions and permutations) is a permutation f onto a set X that anyone can compute efficiently; however inverting f is hard unless one is also given some "trapdoor" information. Given the trapdoor information, computing g the inverse of f becomes easy. Naively, a trapdoor one-way permutation defines a simple public-key encryption scheme (see public key cryptography): the description of f is the public key and the trapdoor, or equivalently the inverse permutation g, is the

secret key. As a consequence, in order to send a message $m \in X$ to the owner of the public key f, one computes $c = f(m)$. The recipient is the only one to know the trapdoor, and thus the only one able to compute $m = g(c)$.

NUMBER THEORY: A first simple candidate that may come to mind as a one-way function, except $\mathcal{NP}$-complete problems, is integer multiplication: while it is easy to multiply two prime integers p and q to get the product $n = p \cdot q$, there is no easy way to get back p and q from n. Indeed, the product of two integers p and q of similar bit-size k just requires a quadratic amount of time in k. However, the factorization of any integer n, which consists of either writing n as a product of prime numbers $n = \Pi p_i^{v_i}$—which decomposition is unique up to a permutation of the factors—or just extracting one factor, is much more intricate. Factorization is indeed believed to be a quite difficult problem (see integer factorization), especially for products of two primes of similar sizes larger than 384 bits each.

Unfortunately, integer multiplication is just one-way. And no trapdoor can make inversion easier. However, some algebraic structures are based on the factorization of an integer n, where some computations are difficult without the factorization of n, but easy with it: in the finite quotient ring $\mathbb{Z}_n = \mathbb{Z}/n\mathbb{Z}$ (see modular arithmetic), one can easily compute basic operations (equality test, addition, subtraction or multiplication). About inversion, Bézout's theorem gives a theoretical result, while the extended Euclidean algorithm gives the constructive version, since it explicitly computes u and v:

THEOREM 1. (Bézout). *Let a and n be two integers, then there exist u, $v \in \mathbb{Z}$ such that $au + nv = 1$ if and only if $\gcd(a, n) = 1$.*

As a consequence, for any $a \in \mathbb{Z}_n$, a is invertible if and only if a is coprime to n and, the extended Euclidean algorithm efficiently provides the inverse u. Furthermore, the following corollary comes from the fact that a prime number is co-prime with any positive integer, less than itself:

COROLLARY 1. *The integer p is a prime $\Leftrightarrow$ the ring $\mathbb{Z}_p$ is a field.*

Therefore, the multiplicative group $\mathbb{Z}_p^*$ of the inverses modulo the prime p contains all the non-zero elements. When n is not a prime, $\mathbb{Z}_n$ is not a field, but the Chinese Remainder Theorem provides the structure of $\mathbb{Z}_n$, with an explicit isomorphism of rings:

THEOREM 2. (Chinese Remainder Theorem). *Let* $n = m_1 m_2$ *be a composite integer, where* $\gcd(m_1, m_2) = 1$. *Then the ring* $\mathbb{Z}_n$ *is isomorphic to the product ring* $\mathbb{Z}_{m_1} \times \mathbb{Z}_{m_2}$.

About the multiplicative group $\mathbb{Z}_n^*$, one gets the following corollary:

COROLLARY 2. *Let* $n = m_1 m_2$ *be a composite integer, where* $\gcd(m_1, m_2) = 1$.

$$(\mathbb{Z}_n^*, \times) \simeq (\mathbb{Z}_{m_1}^*, \times) \times (\mathbb{Z}_{m_2}^*, \times).$$

The well-known <u>Euler's Totient Function</u> $\varphi(n)$ is defined by the cardinality of the multiplicative group $\mathbb{Z}_n^*$. Thanks to the Chinese Remainder Theorem, and in the above corollary, this function is weakly multiplicative, which means:

$$\gcd(m_1, m_2) = 1 \Rightarrow \varphi(m_1, m_2) = \varphi(m_1)\varphi(m_2).$$

Since $\varphi(p^v) = p^v - p^{v-1}$ for any prime p and any valuation $v \geq 1$, one can deduce that for any integer n

$$n = \prod_1^\ell p_i^{v_i} \Rightarrow \varphi(n) = n \times \prod_1^\ell \left(1 - \frac{1}{p_i}\right).$$

THEOREM 3. *The computation of* $\varphi(n)$ *is polynomially equivalent to the factorization of* n.

PROOF. It is clear that the factorization of n easily leads to the value of $\varphi(n)$ with the above formula. Furthermore, Miller's algorithm [10] outputs the factorization of any n, given a multiple of $\varphi(n)$ $\quad\square$

MODULAR POWERS AND ROOTS: In any group (denoted multiplicatively), the power to a scalar c can be performed with a linear complexity in the size k of this scalar, using the *square-and-multiply* technique (see also <u>exponentiation algorithms</u>):

$$x^c = x^{\sum_{i=0}^{i=k} c_i 2^i} = \prod_{i=0}^{i=k} x^{c_i \times 2^i} = \prod_{i=0}^{i=k} x_i^{c_i},$$

where $x_0 = x$ and $x_i = x_{i-1}^2,$

where $c_0, c_1, \ldots, c_k$ denotes the binary expansion of c. On the other hand, root-extraction (see <u>modular root</u>) does not admit any generic algorithm, unless the order of the group is known. Indeed, the classical Lagrange's theorem provides a solution:

THEOREM 4. (Lagrange's Theorem). *Let G be any group, and c its cardinality, for any element* $x \in G, x^c = 1$.

This theorem applied to the particular situation of the multiplicative group $\mathbb{Z}_n^*$ becomes:

THEOREM 5. (Euler's Theorem). *Let n be any integer, for any element* $x \in \mathbb{Z}_n^*, x^{\varphi(n)} = 1 \bmod n$.

Therefore, for any integer e relatively prime to $\varphi(n)$, and any $x \in \mathbb{Z}_n^*$, if one takes $d = e^{-1} \bmod \varphi(n)$ which means that there exists an integer $k \in \mathbb{Z}$ such that $ed + k\varphi(n) = 1$ and the values of d (and k) can be easily computed with the extended Euclidean algorithm, then

$$(x^e)^d = x^{ed} = x^{1 - k\varphi(n)} = x \cdot (x^{\varphi(n)})^{-k} = x \bmod n.$$

As a consequence, $y = x^d \bmod n$ is the eth root of x in $\mathbb{Z}_n^*$.

As previously seen, the eth power can be easily computed using the *square-and-multiply* method. The above relation allows to easily compute eth roots, by computing dth powers, where $ed = 1 \bmod \varphi(n)$. However, to compute eth roots, one requires to know an integer d such that $ed = 1 \bmod \varphi(n)$. And therefore, $ed - 1$ is a multiple of $\varphi(n)$ which is equivalent to the knowledge of the factorization of n. This provides a trapdoor one-way permutation $f_{n,e}$ whose inverse $g_{n,d}$ requires the knowledge of d, or equivalently the factorization of n:

$$f_{n,e} : x \mapsto x^e \bmod n, \qquad g_{n,d} : y \mapsto y^d \bmod n.$$

THE RSA PRIMITIVE

The RSA Problem

In 1978, Rivest et al. [14] defined the following problem.

DEFINITION 1. (The RSA Problem). *Let* $n = pq$ *be the product of two large primes and e an integer relatively prime to* $\varphi(n)$. *For a given* $y \in \mathbb{Z}_n^*$, *compute* $x \in \mathbb{Z}_n^*$ *such that* $x^e = y \bmod n$.

We have seen above that with the factorization of n (the trapdoor), this problem can be easily solved. However, nobody knows whether the factorization is required, but nobody knows how to do without it either, hence the RSA assumption.

DEFINITION 2. (The RSA Assumption). *For any product of two large primes,* $n = pq$, *the RSA problem is intractable* (*presumably as hard as the factorization of* n).

The Plain RSA Cryptosystem

In the RSA cryptosystem, the setup consists of choosing two large prime numbers p and q, and computing the RSA modulus $n = pq$. The public key is n together with an exponent e (relatively prime to $\varphi(n) = (p-1)(q-1)$). The secret key d is defined to be the inverse of e modulo $\varphi(n)$. Encryption and decryption are defined as follows:

$$\mathcal{E}_{n,e}(m) = m^e \bmod n, \qquad \mathcal{D}_{n,d}(c) = c^d \bmod n.$$

Security Weaknesses

Unfortunately, encryption in this naive public-key scheme is not secure: the accepted security requirement for an encryption scheme is the so-called <u>semantic security</u> (indistinguishability) against an <u>adaptive chosen-ciphertext attack</u> [13] or IND–CCA2 for short. But the RSA primitive does not provide by itself an IND–CCA2 secure encryption scheme. Under a slightly stronger assumption than the intractability of the integer factorization, it gives a cryptosystem that is only one-way under <u>chosen-plaintext attack</u>—a very weak level of security. Semantic security fails because encryption is deterministic: one can indeed easily note that a deterministic encryption algorithm can never achieve semantic security, since one can check a plaintext candidate by simply re-encrypting it. Even worse, under a CCA2 attack, the attacker can fully decrypt a challenge ciphertext $c = m^e \bmod n$ using the <u>homomorphic</u> property of RSA:

$$\mathcal{E}_{n,e}(m_1) \cdot \mathcal{E}_{n,e}(m_2) = \mathcal{E}_{n,e}(m_1 m_2 \bmod n) \bmod n.$$

To decrypt $c = m^e \bmod n$ using a CCA2 attack, one first computes $c' = c \cdot 2^e \bmod n$, then asks for the decryption of $c' \neq c$ and gets $2m \bmod n$, finally one can deduce m.

But these are not the only weaknesses of the plain RSA cryptosystem. More subtle attacks have been found.

Multicast Encryption. Håstad [9] showed how to use the Chinese Remainder Theorem to recover the plaintext sent using the plain RSA encryption to several recipients which all have a common and small public exponent. Since the encryption cost is directly related to the size of the public exponent e, it is natural that people want to use the smallest value, that is $e = 3$.

Let us assume that Alice, Bob and Carole have three distinct RSA moduli n_a, n_b and n_c, but a common public exponent $e_a = e_b = e_c = e = 3$. If Daniel sends the same message m to each of them, Eve can intercept

$$c_a = m^3 \bmod n_a, \quad c_b = m^3 \bmod n_b,$$
$$c_c = m^3 \bmod n_c,$$

and then, granted the <u>Chinese Remainder Theorem</u>, she can compute $c \in \mathbb{Z}_{n_a n_b n_c}$ which satisfies

$$c = c_a \bmod n_a, \quad c = c_b \bmod n_b, \quad c = c_c \bmod n_c,$$

This system is also satisfied by m^3, but the isomorphism between $\mathbb{Z}_{n_a} \times \mathbb{Z}_{n_b} \times \mathbb{Z}_{n_c}$ and $\mathbb{Z}_{n_a n_b n_c}$ says that there should be a unique solution, thus $c = m^3 \bmod n_a n_b n_c$, which equality holds in $\mathbb{Z}$ since $m^3 < n_a n_b n_c$. An easy third root in $\mathbb{Z}$ leads back to m.

Small Exponents. As already noticed, people may be interested in using small exponents, either a small encryption exponent to speed up the encryption process, or a small decryption exponent to speed up the decryption process.

A short encryption exponent may be dangerous, as already remarked above, but even in a single-user environment if the message to be encrypted is short too: let us assume that one uses a 2048-bit modulus n for a quite secure application (such as electronic transactions), with public exponent $e = 3$. Then, one uses this cryptosystem to encrypt a short message m (a transaction, a credit card number, etc.) over less than 64 characters (that is less than 512 bits). The ciphertext is $c = m^3 \bmod n$, but even in $\mathbb{Z}$ since m^3 is over 1576 bits only which is less than n: the modular reduction does not apply—it may be faster! But on the other hand, to recover m from c, one just has to compute a third root in $\mathbb{Z}$, which does not require any factorization.

One may think about a short decryption exponent, since the decryption process is often performed by a low-power device (i.e., smart card). But let us assume that $n = pq$, with $p < q < 2p, e$ is the encryption exponent such that the decryption exponent d is less than $n^{1/4}/3$, and $ed = 1 + k\varphi(n)$:

$$\left| \frac{e}{n} - \frac{k}{d} \right| = \left| \frac{ed - kn}{nd} \right| = \left| \frac{1 - k(p+q-1)}{nd} \right|$$
$$\leq \frac{k(p+q)}{nd} \leq \frac{3k}{d\sqrt{n}}.$$

Since $e \leq \varphi(n)$ and $d \leq n^{1/4}/3$, necessarily $3k \leq n^{1/4}$, and $1/n^{1/4} \leq 1/3d$. As a consequence, the above difference is upper-bounded by $1/2d^2$. Using the continued fractions result, this bound says that k/d is one of the *convergents* of the *continued fraction* (see <u>integer factoring</u>) expansion of e/n: in polynomial time, one gets d [6, 16].

CONCLUSION: The RSA function is a quite nice primitive for cryptographic purposes: it provides not only a public-key encryption scheme, but also signature (see <u>RSA digital signature scheme</u>). However, the basic primitive cannot be used directly because of the very strong algebraic structure. As a consequence, practical RSA-based cryptosystems randomly pad the plaintext prior to encryption. This randomizes the ciphertext and eliminates the homomorphic property. Paddings have been proposed that provably rule out any attack, under the sole RSA assumption (see OAEP).

David Pointcheval

References

[1] Bellare, M. and Rogaway (1993). "Random oracles are practical: A paradigm for designing efficient protocols." *Proceedings of the 1st CCS*. ACM Press, New York, 62–73.

[2] Bellare, M. and P. Rogaway (1995). "Optimal asymmetric encryption—how to encrypt with RSA." *Advances in Cryptology—EUROCRYPT'94*, Lecture Notes in Computer Science, vol. 950, ed. A. De Santis. Springer-Verlag, Berlin, 92–111.

[3] Bleichenbacher, D. (1998). "A chosen ciphertext attack against protocols based on the RSA encryption standard PKCS #1." *Advances in Cryptology—CRYPTO'98*, Lecture Notes in Computer Science, vol. 1462, ed. H. Krawczyk. Springer-Verlag, Berlin, 1–12.

[4] Blum, M. and S. Micali (1984). "How to generate cryptographically strong sequences of pseudorandom bits." *SIAM Journal on Computing*, 13, 850–864.

[5] Boneh, D. (2001). Simplified OAEP for the RSA and Rabin Functions. *Advances in Cryptology—CRYPTO 2001*, Lecture Notes in Computer Science, vol. 2139, ed. J. Kilian. Springer-Verlag, Berlin, 275–291.

[6] Boneh, D. and G. Durfee (2000). "Cryptanalysis of RSA with private key d less than $N^{0.292}$." *IEEE Transactions on Information Theory*, 46 (4), 1339–1349.

[7] Fiat, A. and A. Shamir (1987). "How to prove yourself: Practical solutions of identification and signature problems." *Advances in Cryptology—CRYPTO'86*, Lecture Notes in Computer Science, vol. 263, ed. A.M. Odlyzko. Springer-Verlag, Berlin, 186–194.

[8] Fujisaki, E., T. Okamoto, D. Pointcheval, and J. Stern (2001). "RSA–OAEP is secure under the RSA assumption." *Advances in Cryptography—CRYPTO 2001*, Lecture Notes in Computer Science, vol. 2139, ed. J. Kilian. Springer-Verlag, Berlin, 260–274.

[9] Håstad, J. (1988). "Solving simultaneous modular equations of low degree." *SIAM Journal of Computing*, 17, 336–341.

[10] Miller, G. (1976). "Riemann's hypothesis and tests for primality." *Journal of Computer and System Sciences*, 13, 300–317.

[11] Naor, M. and M. Yung (1989). "Universal one-way hash functions and their cryptographic applications." *Proceedings of the 21st STOC*. ACM Press, New York, 33–43.

[12] Rabin, M.O. (1978). "Digitalized signatures." *Foundations of Secure Computation*, eds. R. Lipton and R. De Millo. Academic Press, New York, 155–166.

[13] Rackoff, C. and D.R. Simon (1992). "Noninteractive zero-knowledge proof of knowledge and chosen ciphertext attack." *Advances in Cryptology—CRYPTO'91*, Lecture Notes in Computer Science, vol. 575, ed. J. Feigenbaum. Springer-Verlag, Berlin, 433–444.

[14] Rivest, R., A. Shamir, and L. Adleman (1978). "A method for obtaining digital signatures and public key cryptosystems." *Communications of the ACM*, 21 (2), 120–126.

[15] Shoup, V. (2001). "OAEP reconsidered." *Advances in Cryptography—CRYPTO 2001*, Lecture Notes in Computer Science, vol. 2139, ed. J. Kilian. Springer-Verlag, Berlin, 239–259.

[16] Wiener, M. (1990). "Cryptanalysis of short RSA secret exponents." *IEEE Transactions on Information Theory*, 36 (3), 553–558.

RUN

A run in a binary sequence is a set of consecutive 0s or 1s. A run of 0s is ofter denoted a 0-run or a gap and a run of 1s is often denoted a 1-run or a *block*. A gap of length k is a set of k consecutive 0s flanked by 1s. A block of length k is a set of k consecutive 1s flanked by 0s. A run of length k is a gap of length k or a block of length k.

Tor Helleseth

Reference

[1] Golomb, S.W. (1982). *Shift Register Sequences*. Aegean Park Press, Laguna Hills, CA.

RUNNING-KEY

In a stream cipher, the *running-key,* also called the *keystream*, is the sequence which is combined, digit by digit, with the plaintext sequence for obtaining the ciphertext sequence. The running key is generated by a finite state automaton called the *running-key generator* or the *keystream generator* (see stream cipher).

Anne Canteaut

S

SAFE PRIME

A *safe prime* is a prime number p of the form $p = 2q + 1$ where q is also prime. In such a case, q is called a *Sophie Germain prime*. Safe primes are used in some implementations of the Diffie–Hellman key exchange protocol, for example, to protect against certain types of attacks.

Anton Stiglic

SALT

A *salt* is a t-bit random string that may be prepended or appended to a user's password prior to application of a one-way function in order to make dictionary attacks less effective. Both the salt and the hash (or encryption) of the augmented password are stored in the password file on the system. When the user subsequently enters a password, the system looks up the salt associated with that user, augments the password with the salt, applies the one-way function to the augmented password, and compares the result with the stored value.

It is important to note that the work factor for finding a particular user's password is unchanged by salting because the salt is stored in cleartext in the password file. However, it can substantially increase the work factor for generating random passwords and comparing them with the entire password file, since each possible password could be augmented with any possible salt. The effort required to find the password associated with an entry in the password file is multiplied by the smaller of {the number of passwords, 2^t} compared with a password file containing hashes (or encryptions) of unsalted passwords.

Another benefit of salting is that two users who choose the same password will have different entries in the system password file; therefore, simply reading the file will not reveal that the passwords are the same.

Carlisle Adams

References

[1] Denning, D. (1982). *Cryptography and Data Security*. Addison-Wesley, Reading, MA.

[2] Kaufman, C., R. Perlman, and M. Speciner (1995). *Network Security: Private Communication in a Public World*. Prentice-Hall, Englewood Cliffs, NJ.

[3] Menezes, A., P. van Oorschot, and S. Vanstone (1997). *Handbook of Applied Cryptography*. CRC Press, Boca Raton, FL.

SCHNORR DIGITAL SIGNATURE SCHEME

The Schnorr signature scheme [6] is derived from Schnorr's identification protocol using the Fiat–Shamir heuristic [2]. The resulting digital signature scheme is related to the Digital Signature Standard (DSS). As in DSS, the system works in a subgroup of the group $\mathbb{Z}_p^*$ for some prime number p. The resulting signatures have the same length as DSS signatures. The signature scheme works as follows:

Key Generation. Same as in the DSS system. Given two security parameters $\tau, \lambda \in \mathbb{Z}$ $(\tau > \lambda)$ as input do the following:
1. Generate a random λ-bit prime q.
2. Generate a random τ-bit prime prime p such that q divides $p - 1$.
3. Pick an element $g \in \mathbb{Z}_p^*$ of order q.
4. Pick a random integer $\alpha \in [1, q]$ and compute $y = g^\alpha \in \mathbb{Z}_p^*$.
5. Let H be a hash function $H : \{0, 1\}^* \to \mathbb{Z}_q$.

The resulting public key is (p, q, g, y, H). The private key is (p, q, g, α, H).

Signing. To sign a message $m \in \{0, 1\}^*$ using the private key (p, q, g, α, H) do:
1. Pick a random $k \in \mathbb{Z}_p^*$.
2. Compute $r = g^k \in \mathbb{Z}_p^*$. Set $c = H(m\|r) \in \mathbb{Z}_q$ and $s = \alpha c + k \in \mathbb{Z}_q$.
3. Output the pair $(s, c) \in \mathbb{Z}_q^2$ as the signature on m.

Verifying. To verify a message/signature pair $(m, (s, c))$ using the public key (p, q, g, y, H) do:
1. Compute $v = g^s y^{-c} \in \mathbb{Z}p$.
2. Accept the signature if $c = H(m\|v)$. Otherwise, reject.

We first check that the verification algorithm accepts all valid message/signature pairs. For a valid message/signature pair we have

$$v = g^s y^{-c} = g^{\alpha c + k} y^{-c} = (y^c g^k) y^{-c} = g^k \in \mathbb{Z}p$$

and therefore $H(m\|v) = H(m\|g^k) = c$. It follows that a valid message/signature is always accepted.

The signature can be shown to be existentially unforgeable (see existential forgery) under a chosen message attack in the random oracle model, assuming the discrete logarithm problem in the group generated by g is intractable. This proof of security is a special case of a general result that shows how to convert a public-coin authentication protocol (a protocol in which the verifier only contributes randomness) into a secure signature scheme in the random oracle model [1, 5]. In the proof of security, the function H is assumed to be a random oracle. In practice, one derives H from some cryptographic hash function such as SHA-1.

To discuss signature length we fix concrete security parameters. At the present time the discrete-log problem in the cyclic group $\mathbb{Z}_p^*$ where p is a 1024-bit prime is considered intractable [3] except for a very well funded organization. Schnorr signatures use a subgroup of order q of $\mathbb{Z}_p^*$. When q is a 160-bit prime, the discrete log problem in this subgroup is believed to be as hard as discrete-log in all of $\mathbb{Z}_p^*$, although proving this is currently an open problem. Hence, for the present discussion we assume p is a 1024-bit prime and q is a 160-bit prime. Since a Schnorr signature contains two elements in $\mathbb{Z}_q$ we see that, with these parameters, its length is 320-bits.

Schnorr signatures are efficient and practical. The time to compute a signature is dominated by one exponentiation and this exponentiation can be done offline, i.e. before the message is given. Verifying a signature is dominated by the time to compute a multi-exponentiation of the form $g^a h^b$ for some $g, h \in \mathbb{Z}_p$ and $a, b \in \mathbb{Z}_q$. Multi-exponentiations of this type can be done at approximately the cost of a single exponentiation [4, p. 617].

Dan Boneh

References

[1] Abdalla, M., J. An, M. Bellare, and C. Namprempre (2002). "From identification to signatures via the fiat-shamir transform: Minimizing assumptions for security and forward-security." *Advances in Cryptology—EUROCRYPT 2004*, Lecture Notes in Computer Science, vol. 2332, ed. Lars Knudsen. Springer-Verlag, Berlin, 418–33.

[2] Fiat, Amos and Adi Shamir (1986). "How to prove yourself: Practical solutions to identification and signature problems." *Advances in Cryptology—CRYPTO'86*, Lecture Notes in Computer Science, vol. 263, ed. Andrew M. Odlyzko. Springer-Verlag, Berlin, 186–194.

[3] Lenstra, Arjen and Eric Verheul (2001). "Selecting cryptographic key sizes." *Journal of Cryptology*, 14 (4), 255–293.

[4] Menezes, Alfred J., Paul C. van Oorschot, and Scott A. Vanstone (1997). *Handbook of Applied Cryptography*. CRC Press, Boca Raton, FL.

[5] Ohta, Kazuo and Tatsuaki Okamoto (1998). "On concrete security treatment of signatures derived from identification." *Advances in Cryptology—CRYPTO'98*, Lecture Notes in Computer Science, vol. 1462, ed. H. Krawczyk. Springer-Verlag, Berlin, 354–369.

[6] Schnorr, C. (1991). "Efficient signature generation by smart cards." *Journal of Cryptology*, 4 (3), 161–174.

SCHNORR INDENTIFICATION SCHEME

In its simplest form, an identification protocol involves the presentation or submission of some information (a "secret value") from a claimant to a verifier (see Identification). Challenge-response identification is an extension in which the information submitted by the claimant is the function of both a secret value known to the claimant (sometimes called a "prover"), and a challenge value received from the verifier (or "challenger").

Such a challenge-response protocol proceeds as follows. A verifier V generates and sends a challenge value c to the claimant C. Using his/her secret value s and appropriate function $f()$, C computes the response value $v = f(c, s)$, and returns v to V. V verifies the response value v, and if successful, the claim is accepted. Choices for the challenge value c, and additionally options for the function $f()$ and secret s are discussed below.

Challenge-response identification is an improvement over simpler identification because it offers protection against replay attacks. This is achieved by using a challenge value that is time-varying. Referring to the above protocol, there are three general types of challenge values that might be used. The property of each is that the challenge value is not repeatedly sent to multiple claimants. Such a value is sometimes referred to as a *nonce*, since it is a value that is "*n*ot used more than *once*." The challenge value could be a randomly generated value (see Random bit generation), in which case V would send a random value c to C. Alternatively, the challenge value might be a sequence number, in which case the verifier V would maintain a sequence value corresponding to each challenger. At each challenge, the stored sequence number would be increased by (at least) one before sending to the claimant. Finally, the challenge value might be a function of the current time. In this case, a challenge value need not be sent from V to C, but could be sent by C, along with the computed verifier. As long as the time chosen was within an accepted threshold, V would accept.

There are three general classes of functions and secret values that might be used as part of a challenge-response protocol. The first is symmetric-key based in which the claimant C and verifier V *a priori* share a *secret* key K. The function $f()$ is a symmetric encryption function (see Symmetric Cryptosystem), a hash function, or a *Message Authentication Code* (MAC algorithms). Both Kerberos (see Kerberos authentication protocol) and the Needham–Schroeder protocol are examples of symmetric-key based challenge-response identification. In addition, the protocols of ISO/IEC 9798-2 perform identification using symmetric key techniques.

Alternatively, a public key based solution may be used. In this case, the claimant C has the private key in a public key cryptosystem (see Public Key Cryptography). The verifier V possesses a public key that allows validation of the public key corresponding to C's private key. In general, C uses public key techniques (generally based on number-theoretic security problems) to produce a value v, using knowledge of his/her private key. For example, V might encrypt a challenge value and send the encrypted text. C would decrypt the encrypted text and return the value (i.e., the recovered plaintext) to V (note that in this case it would only be secure to use a random challenge, and not a sequence number or time-based value). Alternatively, V might send a challenge value to C and ask C to digitally sign and return the challenge (see Digital Signature Schemes). The Schnorr identification protocol is another example of public key based challenge-response identification.

Finally, a zero-knowledge protocol can be used. In this case, the challenger demonstrates knowledge of his/her secret value without revealing any information (in an information theoretic sense— see "information theoretic security" in glossary) about this value. Such protocols typically require a number of "rounds" (each with its own challenge value) to be executed before a claimant may be successfully verified.

Mike Just

References

[1] Menezes, A., P. van Oorschot, and S. Vanstone (1997). *Handbook of Applied Cryptography*. CRC Press, Boca Raton, FL.

[2] Stinson, D.R. (1995). *Cryptography: Theory and Practice*. CRC Press, Boca Raton, FL.

SEAL

SEAL stands for Software-optimized Encryption ALgorithm. It is a binary additive stream cipher (see the entry concerning synchronous stream ciphers). It has been proposed in 1993, and several versions have been published: SEAL 1.0, SEAL 2.0 and SEAL 3.0 [3,4]. Some attacks have been published that show how SEAL 1.0, SEAL 2.0 [2] and later SEAL 3.0 [1] can be distinguished from a true random function. But there is no really practical attack for the moment.

SEAL has been designed to be really efficient in its software implementation, mainly for 32-bit processors. It is a length-increasing pseudorandom function that maps a 32-bit sequence number n to an L-bit keystream, under control of a 160-bit secret key. a more precise description can be found in the original papers.

Caroline Fontaine

References

[1] Fluhrer, S.R. (2001). "Cryptanalysis of the SEAL 3.0 pseudorandom function family." *FSE'01*, Lecture Notes in Computer Science, vol. 2355, ed. M. Matsui. Springer-Verlag, Berlin, 135-ff.

[2] Handschuh, H. and H. Gilbert (1997). "χ^2-cryptanalysis of the SEAL encryption algorithm." *FSE'97*, Lecture Notes in Computer Science, vol. 1687, eds. O. Nierstrasz and M. Lemoine. Springer-Verlag, Berlin, 1–12.

[3] Rogaway, P. and D. Coppersmith (1994). "A software-optimized encryption algorithm." *FSE'94*, Lecture Notes in Computer Science, vol. 809, ed. R.J. Anderson. Springer-Verlag, Berlin, 56–63.

[4] Rogaway, P. and D. Coppersmith (1998). "A software-optimized encryption algorithm." *Journal of Cryptology*, 11 (4), 273–287.

SECOND PREIMAGE RESISTANCE

Second preimage resistance is the property of a hash function that it is computationally infeasible to find any second input that has the same output as a given input. This property is related to preimage resistance and one-wayness; however, the later concept is typically used for functions with input and output domain of similar size (see one-way function). Second preimage resistance is also known as weak collision resistance. A minimal requirement for a hash function to be second preimage resistant is that the length of its result should be at least 80 bits (in 2004). A hash function is said to be a one-way hash function (OWHF) if it is both preimage resistant and second preimage resistant. The relation between collision resistance, second preimage resistance and preimage resistance is rather subtle, and it depends on the formalization of the definition: it is shown in [6] that under certain conditions, collision resistance implies second preimage resistance and second preimage resistance implies preimage resistance.

In order to formalize the definition, one needs to specify according to which distribution the first element in the domain is selected and one needs to express the probability of finding a second preimage for this element. Moreover, one often introduces a class of functions indexed by a public parameter, which is called a key. One could then distinguish between three cases: the probability can be taken over the random choice of elements in the range, over the random choice of the parameter, or over both simultaneously. As most practical hash functions have a fixed specification, the first approach is more relevant to applications. The second case is known as a Universal One-Way Hash Function or UOWHF.

The definition of a one-way function was given in 1976 by Diffie and Hellman [1]. Second preimage resistance of hash functions has been introduced by Rabin in [5]; further work on this topic can be found in [2–4, 7, 8]. For a complete formalization and a discussion of the relation between the variants and between hash functions properties, the reader is referred to Rogaway and Shrimpton [6].

B. Preneel

References

[1] Diffie, W. and M.E. Hellman (1976). "New directions in cryptography." *IEEE Transactions on Information Theory*, IT-22 (6), 644–654.

[2] Merkle, R. (1979). *Secrecy, Authentication, and Public Key Systems*. UMI Research Press.

[3] Preneel, B. (1993). "Analysis and Design of Cryptographic Hash Functions." *Doctoral Dissertation*, Katholieke Universiteit Leuven.

[4] Preneel, B. (1999). "The state of cryptographic hash functions." *Lectures on Data Security*, Lecture Notes in Computer Science, vol. 1561, ed. I. Damgård. Springer-Verlag, Berlin, 158–182.

[5] Rabin, M.O. (1978). "Digitalized signatures." *Foundations of Secure Computation*, eds. R. Lipton and R. DeMillo. Academic Press, New York, 155–166.

[6] Rogaway, P. and T. Shrimpton (2004). "Cryptographic hash function basics: Definitions, implications, and separations for preimage resistance, second-preimage resistance, and collision resistance." *Fast Software Encryption*, Lecture Notes in Computer Science, vol. 3017, eds. B.K. Roy and W. Meier. Springer-Verlag, Berlin, 371–388.

[7] Stinson, D. (2001). "Some observations on the theory of cryptographic hash functions." Technical Report 2001/020, University of Waterloo.

[8] Zheng, Y., T. Matsumoto, and H. Imai (1990). "Connections between several versions of one-way hash functions." *Proceedings SCIS90, The 1990 Symposium on Cryptography and Information Security, Nihondaira, Japan, January 31–February 2*.

SECRET SHARING SCHEMES

Informally speaking, a secret sharing scheme (SSS, for short) allows one to share a secret among n participants in a such a way that some sets of participants called *allowed* coalitions can recover the secret exactly, while any other sets of participants (*non-allowed* coalitions) cannot get any additional (i.e., a posteriori) information about the possible value of the secret. the SSS with the last property is called *perfect*. The set Γ of all allowed coalitions is called an access structure.

The history of SSS began in 1979 when this problem was introduced and partially solved by Blakley [1] and Shamir [2] for the case of (n, k)-threshold schemes where the access structure consists of all sets of k or more participants. Consider the simplest example of (n, n)-threshold scheme. There is a dealer who wants to distribute a secret s_0 among n participants. Let s_0 be an element of some finite additive group G. For instance, G is the group of binary strings of length m with addition by modulo 2, i.e., $G = GF(2)^m$ (see finite field). The dealer generates a random sequence $s_1, \ldots, s_n$ such that $\sum_{i=1}^{n} s_i = s_0$ (for instance, by generating independently elements $s_1, \ldots, s_{n-1} \in G$ and then putting $s_n := s_0 - \sum_{i=1}^{n-1} s_i$). Then the dealer sends *privately* to each ith participant the elements s_i called share, i.e., other participants have no information about the value of s_i. It is easy to see that any coalition of less then n participants has no information except of a priori information about s_0 and all participants together recover the value of the secret as $\sum_{i=1}^{n} s_i$. These simple schemes appear to be enough for the realization of arbitrary monotone (i.e., if $A \in \Gamma$ and $A \subset B$ then $B \in \Gamma$) access structure Γ. Namely, for any allowed coalition $A \in \Gamma$ let the above realize (independently) an $(|A|, |A|)$-threshold scheme, i.e., send to the ith, participant as many shares s_i^A) as the number of allowed coalitions to which this participant belongs it is enough to consider only maximum allowed coalitions).

The probabilistic model of an SSS for the general case is the following (see [4, 5]). There are $n + 1$ sets $S_0, S_1, \ldots, S_n$ and the probability distribution P on their Cartesian product $S = S_0 \times \cdots \times S_n$. A pair (P, S) is called a perfect probabilistic SSS realizing the access structure Γ if the following two properties hold:

- participants of an allowed set A (i.e., $A \in \Gamma$) together can recover the secret exactly (formally, $P(S_0 = c_0 \mid S_i = c_i, i \in A) \in \{0, 1\}$ if $A \in \Gamma$);
- participants forming a non-allowed set A ($A \notin \Gamma$) cannot get additional information beyond their *a priori* information about s_0, i.e., $P(S_0 = c_0 \mid S_i = c_i, i \in A) = P(S_0 = c_o)$ if $A \notin \Gamma$. These conditions can be reformulated in the language of entropy (see <u>information theory</u>) as $H(S_i, i \in A \cup 0) = H(S_i, i \in A) + \delta_\Gamma(A)H(S_0)$, where $\delta_\Gamma(A) = 0$ if $A \in \Gamma$, and $\delta_\Gamma A) = 1$, otherwise.

There are also combinatorial models of SSSs. An arbitrary set $V \subset S$ is called the "code" of the combinatorial SSS, and its codewords called "sharing rules". The simplest combinatorial model demands that at first, for every set $A \in \Gamma$ the 0th coordinate of any codeword from V is uniquely determined by the values of the coordinates from the set A and, secondly, for every set $A \notin \Gamma$ and for any given selection of values of the coordinates from the set A the number of codewords with given value of 0th coordinate does not depend on this value. This model is a particular case of the probabilistic model, namely, when all nonzero values of P are equal. The most general definition of combinatorial models, which contain probabilistic ones as a particular case, was given in [6, 7].

For both types of models the "size" of share, provided to any participant, cannot be smaller than the "size" of the secret, where "size" is defined as $log|S_i|$ or $H(S_i)$ respectively for combinatorial and probabilistic statements of the problem. Special attention has been paid to so-called *ideal* SSSs, where the size of any share coincides with the size of the secret. For example, any (n, k)-threshold scheme can be realized as an ideal perfect SSS (see <u>threshold cryptography</u>). It was shown in a chain of papers [6–9] that the access structures of ideal perfect SSS correspond to a special class of matroids [10]. On the other hand, any access structure can be realized as a perfect SSS but probably not very efficient (ideal). At least the above given realization demands for some access structures to distribute shares which size is exponentially (in n) larger than the secret's size. An infinite family of <u>access structures</u> was constructed such that for any perfect realization the size of the shares is at least $n/\ln n$ times larger than the size of the secret [11].

To generate <u>shares</u> the dealer of an SSS has to use some source of randomness, say $r \in X$, where X is in some probabilistic space, and any share s_i is a function of s_0 and r, i.e., $s_i = f_i(s_0, r)$. A linear realization of SSS (or, *linear* SSS) means that all functions $f_i(\cdot)$ are linear. To make it formal: let $s_0, \ldots, s_n$ be elements in m_i-dimensional vector spaces $(i = 0, 1, \ldots, n)$ over some <u>finite field</u> $GF(q)$ of q elements, let r be an element of the l-dimensional vector space over the same field. Then a linear SSS is generated by some $(m_0 + l) \times m$ matrix G according to the formula $s = (s_0, \ldots, s_n) = xG$, where $m = \sum_{i=0}^n m_i$ and vector x is the concatenation of vectors s_0 and r. Consider vector spaces $V_0, \ldots, V_n$, where V_i is the linear subspace generated by the columns of G that correspond to s_i, i.e., by columns g_j, where $j = m_0 + \cdots + m_{i-1} + 1, \ldots, m_0 + \cdots + m_i$. Then matrix G realizes the access structure Γ perfectly if and only if [12,13]:

- for any set $A \in \Gamma$ the linear span of subspaces $\{V_a : a \in A\}$ (i.e., the minimal vector subspace containing all these subspaces V_a) contains the subspace V_0;
- for any set $A \notin \Gamma$ the linear span of subspaces $\{V_a : a \in A\}$ intersects with the linear subspace V_0 only by the vector $\mathbf{0}$.

All aforementioned examples of SSS are linear. Note that if all dimensions m_i are equal to 1 then the matrix G can be considered as a generator matrix of a linear error-correcting code (see <u>cyclic codes</u>). In particular, it gives another description of <u>Shamir's threshold schemes</u> via Reed–Solomon codes [14]. This "coding theory approach" was further developed and generalized to the case of arbitrary linear codes and their minimal words as only possible access structures [16]. Surely, a linear SSS with all dimensions $m_i = 1$ is ideal, but multidimensional linear SSSs with all $m_i = m_0$ give a larger class of ideal SSS [15]. It is an open question if any ideal SSS can be realized as a (multidimensional) linear ideal SSS?

Modifications of assumptions of the secret sharing problem's statement such as perfectness of the scheme, honesty of the dealer and participants, sending shares via secure, private channels and so on lead to many variations of secret sharing problem. Among them: *ramp schemes*, publicly <u>verifiable secret schemes</u>, SSS with cheaters, SSS with public reconstruction [17], and <u>visual secret sharing schemes</u>.

Robert Blakley
Gregory Kabatiansky

References

[1] Blakley, R. (1979). "Safeguarding cryptographic keys." *Proceedings of AFIPS 1979 National Computer Conference*, vol. 48, Va. Arlington, *New York*, 313–317.

[2] Shamir, A. (1979). "How to share a secret." *Communications of the ACM*, 22 (1), 612–613.

[3] Ito, M., A. Saito, and T. Nishizeki (1993). "Multiple assignment scheme for sharing secret." *Journal of Cryptology*, 6, 15–20.

[4] Karnin, E.D., J.W. Greene, and M.E. Hellman (1983). "On secret sharing systems." *IEEE Transactions on Informatiom Theory*, 29 (1), 231–241.

[5] Capocelli, R.M., A. De Santis, L. Gargano, and U. Vaccaro (1993). "On the size of shares for secret sharing schemes." *Journal of Cryptology*, 6, 157–167.

[6] Brickell, E.F. and D.M. Davenport (1991). "On the classification of ideal secret sharing schemes." *Journal of Cryptology*, 4, 123–134.

[7] Blakley, G.R. and G.A. Kabatianski (1997). "Generalized ideal secret sharing schemes and matroids." *Problems of Information Transmission*, 33 (3), 102–110.

[8] Kurosawa, K., K. Okada, K. Sakano, W. Ogata, and S. Tsujii (1993). "Nonperfect secret sharing schemes and matroids." *Advances in Cryptology—EUROCRYPT'93*, Lecture Notes in Computer Science, vol. 765, ed. T. Helleseth. Springer-Verlag, Berlin, 126–141.

[9] Jackson, W.-A. and K.M. Martin (1998). "Combinatorial models for perfect secret sharing schemes." *Journal of Combinatorial Mathematics and Combinatorial Computing*, 28, 249–265.

[10] Welsh, D.J.A. (1976). *Matroid Theory*. Academic Press, New York.

[11] Csirmaz, L. (1995). "The size of a share must be large." *Advances in Cryptology—EUROCRYPT'94*, Lecture Notes in Computer Science, vol. 950, ed. A. De Santis. Springer-Verlag, Berlin, 13–22.

[12] Blakley, G.R. and G.A. Kabatianski (1994). "Linear algebra approach to secret sharing schemes." *Error Control, Cryptology and Speech Compression*, Lecture Notes in Computer Science, vol. 829, eds. A. Chmora, and S.B. Wicker. Springer, Berlin, 33–40.

[13] van Dijk, M. (1995). "A linear construction of perfect secret sharing schemes." *Advances in Cryptology—EUROCRYPT'94*, Lecture Notes in Computer Science, vol. 950, ed. A. De Santis. Springer-Verlag, Berlin, 23–34.

[14] McEliece, R.J. and D.V. Sarwate (1981). "On secret sharing and Reed–Solomon codes." *Communications of the ACM*, 24, 583–584.

[15] Ashihmin, A. and J. Simonis (1998). "Almost affine codes." *Designs, Codes and Cryptography*, 14 (2), 179–197.

[16] Massey, J. (1993). "Minilal codewords and secret sharing." *Proceedings of Sixth Joint Swedish– Russian Workshop on Information Theory, Molle, Szweden*, 246–249.

[17] Beimel, A. and B. Chor (1998). "Secret sharing with public reconstruction." *IEEE Transactions on Informatiom Theory*, 44 (5), 1887–1896.

SECURE SIGNATURES FROM THE "STRONG RSA" ASSUMPTION

In the late 1990's it was realized that by making a somewhat stronger intractability assumption than RSA (see RSA problem), it is possible to devise digital signature schemes that are fairly efficient, and at the same time have a rigorous proof of security (without resorting to the random-oracle heuristic). The intractability assumption states that given a modulus n (see modular arithmetic) of unknown factorization and an element x in the ring Z_n^*, it is hard to come up with an exponent $e \geq 2$ and an element y in Z_n^*, such that $y^e = x$ (mod n). This assumption, first used by Barić and Pfitzmann in the context of fail-stop signatures [1], is called the *strong RSA assumption* (or the *flexible RSA assumption*).

A simple way of using this assumption for signatures was described by Gennaro et al. [7]. In their construction, the public key (see public key cryptography) is a triple (n, x, H), where n is product of two "quasi-safe primes", x is a random element in Z_n^*, and H is a hash function, mapping strings to odd integers. The secret key consists of the factorization of n. To sign a message m, the signer picks a random string r, computes $e \leftarrow H(m, r)$, and then, using the factorization of n, finds an element $y \in Z_n^*$ such that $y^e = x$ (mod n). To verify a signature (r, y) on message m, the verifier checks that $y^{H(m,r)} = x$ (mod n).

Gennaro et al. proved that this scheme is secur—in the sense of existential unforgeability (see also existential forgery) under an adaptive chosen message attack (EU-CMA)—under some conditions on the function H. The first condition is that H is *division intractable*. This means that it is hard to come up with a list of pairs, $(m_i, r_i), i = 1, \ldots, t$, where the integer $H(m_t, r_t)$ divides the product $\prod_{i=1}^{t=1} H(m_i, r_i)$. The other condition on H means, informally, that one cannot reduce breaking the strong RSA assumption to "breaking the hash function H". It is shown in [7] that hash functions satisfying these conditions exist if the strong RSA assumption holds. (It is also shown in [7] that if H is modeled as a random oracle, then it satisfies these conditions, provided that its output is sufficiently long. However, Coron and Naccache showed in [2] that the output of H in this case should be at least as long as the modulus n.)

Cramer and Shoup [4], followed by Fischlin [6], proposed more efficient signatures based on the strong RSA assumption. In the Cramer–Shoup public key scheme, the public key is a 5-tuple (n, h, x, e', H), where n is a product of two "safe primes", h, x are random quadratic residues in Z_n^*, e' is an odd prime, and H is a hash function, mapping strings to integers. If ℓ denotes the output length of H, then the prime e' must be at least $(\ell + 1)$-bit long. The secret key consists of the factorization of n. To sign a message m, the signer picks a new $(\ell + 1)$-bit prime $e \neq e'$ and a random quadratic residue $y \in Z_n^*$, sets $x' \leftarrow (y')^{e'} h^{-H(m)} \pmod{n}$, and then, using the factorization of n, finds an element $y \in Z_n^*$ such that $y^e = x h^{H(x')} \pmod{n}$. To verify a signature (e, y, y') on message m, the verifier checks that e is an odd $(\ell + 1)$-bit number, $e \neq e'$, sets $x' \leftarrow (y')^{e'} h^{-H(m)} \pmod{n}$, and checks that $y^e = x h^{H(x')} \pmod{n}$.

In the scheme of Fischlin, the public key is a 5-tuple (n, g, h, x, H), where n, h, x, H are as in the Cramer–Shoup scheme, and g is yet another random quadratic residue in Z_n^*. Again, the secret key is the factorization of n, and we use ℓ to denote the output length of H. To sign a message m, the signer picks a new $(\ell + 1)$-bit prime e and a random ℓ-bit string α, and then, using the factorization of n, finds an element $y \in Z_n^*$ such that $y^e = x g^\alpha h^{\alpha \oplus H(m)} \pmod{n}$. To verify a signature (e, α, y) on message m, the verifier checks that e is an odd $(\ell + 1)$-bit number, that α is an ℓ-bit string, and that $y^e = x g^\alpha h^{\alpha \oplus H(m)} \pmod{n}$. Fischlin also proposed other variations of this scheme, where the prime e can be chosen even shorter than $\ell + 1$ bits (and the computation made slightly more efficient), at the price of a longer public key.

For all of these schemes, it is proved that they are secure (in the sense of EU-CMA), assuming the strong RSA assumption and the collision-intractability of the hash function H, and as long as the signer never uses the same prime e for two different signatures. The Cramer-Shoup signature scheme was generalized by Damgård and Koprowski to any group where extracting roots is hard [5]. The same generalization can be applied to the schemes described by Fischlin.

It is interesting to note that the Cramer-Shoup signature scheme can be viewed as a simple variation on an earlier scheme by Cramer and Damgård [3]. The Cramer-Damgård scheme is based on a tree construction, and one of its parameters is the depth of that tree. For a tree of depth one, the scheme maintains in the public key a list of t odd primes $e_1, \ldots, e_t$, and can be used to generate upto t signatures, using a different prime each time. The Cramer-Shoup scheme is obtained essentially by letting the signer choose the odd primes "on the fly" instead of committing to them ahead of time in the public key. One pays for this flexibility by having to make a stronger intractability assumption. Since the attacker can now choose the exponent e relative to which it issues the forgery, one has to assume the strong RSA assumption (whereas the standard RSA assumption suffices for the Cramer–Damgård scheme.)

A curious feature of the Cramer-Shoup and Fischlin schemes (as well as some instances of the Gennaro–Halevi–Rabin scheme) is that when there is an a-priori polynomial bound on the total number of signatures to be made, the signer can generate its public/private key pair even without knowing the factorization of the modulus n. (This is done using the same strategy as the simulator in the security proof of these schemes.) That makes it possible in some cases to have many users in a system, all sharing the same modulus n.

Dan Boneh

References

[1] Barić, N. and B. Pfitzmann (1997). "Collision-free accumulators and fail-stop signature schemes without trees." *Advances in Cryptology—EUROCRYPT'97*, Lecture Notes in Computer Science, vol. 1233, ed. W. Fumy. Springer-Verlag, Berlin, 480–494.

[2] Coron, J.S. and D. Naccache (2000). "Security analysis of the Gennaro–Halevi–Rabin signature scheme." *Advances in Cryptology—EUROCRYPT 2000*, Lecture Notes in Computer Science, vol. 1807, ed. B. Preneel. Springer-Verlag, Berlin, 91–101.

[3] Cramer, R. and I.B. Damgård (1996). "New generations of secure and practical RSA-based signatures." *Advances in Cryptology—CRYPTO'96*, Lecture Notes in Computer Science, vol. 1109, ed. N. Koblitz. Springer-Verlag, Berlin, 173–186.

[4] Cramer, R. and V. Shoup (2000). "Signature schemes based on the strong RSA assumption." *ACM Transactions on Information System Security (ACM-TISSEC)*, 3 (3), 161–185.

[5] Damgård, I.B. and M. Koprowski (2002). "Generic lower bounds for root extraction and signature schemes in general groups." *Advances in Cryptology—EUROCRYPT 2002*, Lecture Notes in Computer Science, vol. 2332, ed. L. Knudsen. Springer-Verlag, Berlin, 256–271.

[6] Fischlin, M. (2003). "The Cramer–Shoup strong-RSA signature scheme revisited." *Public Key Cryptography—PKC 2003*, Lecture Notes in Computer Science, vol. 2567, ed. Y.G. Desmedt. Springer-Verlag, Berlin, 116–129.

[7] Gennaro, R., S. Halevi, and T. Rabin (1999). "Secure hash-and-sign signatures without the random oracle." *Advances in Cryptology—EUROCRYPT'99*, Lecture Notes in Computer Science, vol. 1592, ed. J. Stem. Springer-Verlag, Berlin, 123–139.

SECURE SOCKET LAYER (SSL)

GENERAL: *Secure Socket Layer* (SSL) denotes the predominant security protocol of the Internet for World Wide Web (WWW) services relating to electronic commerce or home banking. The majority of web servers and browsers support SSL as the defacto standard for secure client-server communication. The Secure Socket Layer protocol builds up point-to-point connections that allow private and unimpaired message exchange between strongly authenticated parties.

CLASSIFICATION: In the ISO/OSI reference model [8], SSL resides in the session layer between the transport layer (4) and the application layer (7); with respect to the Internet family of protocols this corresponds to the range between TCP/IP and application protocols such as HTTP, FTP, Telnet, etc. SSL provides no intrinsic synchronization mechanism; it relies on the data link layer below.

Netscape developed the first specification of SSL in 1994, but only publicly released and deployed the next version, SSLv2, in the same year [6]. With respect to public key cryptography, it relies mainly on RSA encryption (RSA public key cryptosystem) and X.509-compliant certificates. Block ciphers, such as DES (see Data Encryption Standard), Triple DES (3DES), and RC4, along with hash functions like MD5 and SHA, complement the suite of algorithms. SSLv3 followed in 1995, adding cryptographic methods such as Diffie–Hellman key agreement (DH), support for the FORTEZZA key token, and the Digital Signature Standard (DSS) scheme [5].

This article focuses on SSL version 3.0 and its designated successor protocol Transport Layer Security (TLS) 1.0, which the *Internet Engineering Task Force* (IETF) published for the first time in 1999 [3]. The IETF published the most recent Internet-Draft for TLS 1.1 in October 2002 [4].

LAYER STRUCTURE: SSL splits into distinct layers and message types (see Figure 1). The *handshake* message sequence initiates the communication, establishes a set of common parameters

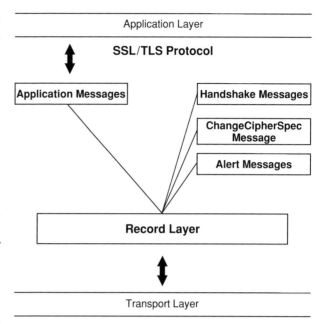

Fig. 1. SSL layer and message structure

like the protocol version, applicable cryptographic algorithms (*cipher suites*), and assures the validity of the message sequence. During the handshake, the participants accomplish the negotiated authentication and derive the session key material.

The *record layer* fragments the full data stream into records with a maximum size of 2^{14} bytes and envelopes them cryptographically under the current session keys. Records contain a keyed message authentication code (HMAC). The initial handshake presupposes a NULL cipher suite applying no encryption and no HMAC. The record layer fully provides the use of compression. However, for patent reasons the core specifications name no method explicitly, except for the mandatory NULL algorithm, which practically makes compression an incompatible, implementation-dependent feature.

Additional *alert* messages inform on exceptional protocol conditions or one participant's demand to end the communication (*closure* alert).

BASIC PROTOCOL SEQUENCE: The SSL handshake accomplishes three goals. First, both parties agree on a *cipher suite*, i.e. the set of cryptographic algorithms that they intend to use for application data protection. Second, they establish a common *master_secret* in order to derive their session key material. Third, the participant's identities are authenticated. Although the SSL specification permits anonymous, server-only and mutual authentication, it is customary to only assert the server's identity.

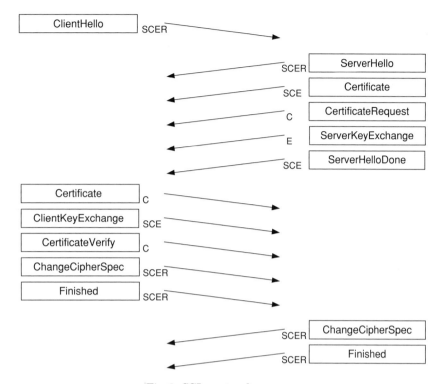

Fig. 2. SSL protocol sequence

Figure 2 gives an overview of the SSL protocol variants. It comprises four different handshake sequences, each identified by a capital letter:

- **S** denotes the server-authenticated message flow.
- **C** marks the sequence with additional client authentication.
- **E** shows the handshake variant with ephemeral Diffie–Hellman key agreement.
- **R** stands for the handshake of resumed sessions. Note, that the message pairs *ChangeCipherSpec/Finished* of client and server are drawn in reverse order; the server message pair follows *ServerHello* immediately.

The client opens the connection with a *ClientHello* message, which contains its selection of acceptable cipher suites and a random number.

The server chooses a supported cipher suite and adds another random number, which together builds the *ServerHello* response. Later, these two random numbers become part of the session's *master_secret*.

SERVER AUTHENTICATION: The server appends a *Certificate* message, which holds a X.509 certificate bearing its identity and public key. Most often RSA keys (see RSA digital signature scheme) are used. DSS signed certificates usually carry a long term DH public key. If multiple levels of the public key hierarchy separate the server certificate from the root authority certificate present in the client browser, then the server is required to deliver an ordered list of all dependent certificates. The empty *ServerHelloDone* message finishes this sequence.

The client confirms the validity of the certificate chain up to one of its built-in root certificates. It generates another random number and encrypts it under the server's RSA public key. This encrypted *pre_master_secret* forms the *ClientKeyExchange* message.

DH/DSS cipher suites might demand the client to create (ephemeral) DH keys matching the server's domain parameters for the *ClientKeyExchange* message. Note, that if both parties own certificates with group-compatible, fixed DH public keys, every connection generates the identical *pre_master_secret*.

Both sides derive the shared session key material independently in two steps. First, the *key derivation function* (KDF) transforms the client's *pre_master_secret* and both exchanged random numbers into the *master_secret*. Afterwards, the KDF is re-applied to the *master_secret* and both random values to compute the final key block. With respect to the chosen cipher suite, the key block is broken up into a maximum of six segments used as directional encryption keys (with initialization vectors) and directional HMAC keys.

CLIENT AUTHENTICATION: SSL servers can demand client authentication through a *Certificate-Request* message. It contains the permitted certificate types, i.e. signature algorithms, and a list of trusted certification authorities identified by their respective distinguished name.

The client answers with a *Certificate* message requiring either a single certificate or the full certification chain. In addition, it creates a *Certificate-Verify* message that contains a digest of the previous handshake messages, signed by the private key corresponding to its certificate.

EPHEMERAL DIFFIE–HELLMAN: The ephemeral Diffie–Hellman key agreement (DH) embeds into the *ServerKeyExchange* and the *Client-KeyExchange* messages. Both sides send their DH public keys and, together with their own DH private keys, calculate a shared *pre_master_secret*. Note, that anonymous DH cipher suites are susceptible to man-in-the-middle attacks and protect only against passive eavesdropping (eavesdropper).

Both sides end the handshake sequence with two further messages: *ChangeCipherSpec* indicates the shift to the newly negotiated cipher parameters. *Finished* is the first message encrypted under the new keys and declares the handshake sequence complete. It holds an HMAC digest over the whole handshake sequence and the negotiated *master_secret* in order to ensure that no message tampering remains undetected.

The cryptographic state is established and confirmed on both sides and the record layer now encrypts and authenticates application data under its new session keys.

The alert message *CloseAlert* indicates the protocol end, followed by the TCP layer's closing FIN packet.

PROTOCOL RESUMPTION: SSL permits the resumption of formerly established sessions, in order to shortcut the handshake, and preserve CPU cycles by avoiding, for instance, the computationally expensive *pre_master_secret* decryption.

Depending on whether the server supports this feature, it sends a session identification string enclosed in the *ServerHello* message. After establishing a valid cryptographic state, this ID refers to the stored *master_secret*. Subsequent client connections indicate their intention to resume a session by specifying the ID in the *ClientHello* message.

Resumed sessions possess unique key blocks, because the key generation process recombines the stored *master_secret* with fresh random nonce out of both *Hello* messages.

ADDITIONAL INFORMATION: SSL permits the re-negotiation of its cipher suites during the course of the application protocol through a simple repetition of the handshake sequence.

The Internet Assigned Numbers Authority (IANA) assigns unique TCP port numbers to SSL/TLS-enabled protocols, which are marked with the appended letter s; for example port 443 for HTTPS or 989/990 for FTPS [7].

SECURITY ANALYSIS, BUGS: Several authors have analysed the SSL protocol suite, stating in consensus that, beginning with v3.0, it is mature and without major design flaws [11, 12, 14].

Wagner and Schneier conclude in their analysis that *"In general SSL 3.0 provides excellent security against eavesdropping and other passive attacks. Although export-weakened modes offer only minimal confidentiality protection, there is nothing SSL can do to improve that fact."* [14]

Some minor attacks are known, however, cautious implementation seems to prevent their applicability [13].

The man-in-the-middle *version rollback* attack tries to establish a SSL v2.0 handshake protocol between SSLv3/TLS-capable parties in compatibility mode. Due to a serious flaw in SSLv2, an active adversary is capable to enforce an export weakened cipher suite, and *brute-force attack* (see cryptanalysis) to session keys directly. The SSLv2 attack is called *cipher suite rollback*. Reference [3] gives recommendations on how to detect downgrade attempts by embedding a short, well defined pattern into the PKCS#1 padding data (PKCS) of RSA encryptions, instead of using of purely random bytes. If an SSLv3/TLS-capable server finds the pattern, it will recognize that the other party operates in backwards compatibility mode although a later protocol version is supported.

Bleichenbacher published an attack against PKCS#1 (version 1) formatted RSA encryptions known as the *million message attack* [1]. Probing an SSL/TLS server with chosen *ClientKeyExchange* messages might reveal the encrypted *pre_master_secret* after about 2^{20} attempts, if the server issues error alerts that allow to distinguish between correctly and incorrectly formatted messages (chosen ciphertext attack). The TLS specification recommends as a countermeasure to treat PKCS#1 formatting errors by simply continuing the handshake protocol with a randomly generated *pre_master_secret*. In this case, the server behaves in the same way, whether or not the formatting is correct [4].

The general method of timing cryptanalysis [10] is applicable against SSL/TLS servers, if a large number of unbiased measurements of private key operations is available. Practical attacks were shown for example by Brumley and Boneh and Klime, et al. Countermeasures against timing cryptanalysis usually degrade performance, for instance by randomly delaying cryptographic operations or by holding a constant response time without dependence of the algorithms' execution paths.

Clemens Heinrich

References

[1] Bleichenbacher, Daniel (1998). "Chosen ciphertext attacks against protocols based on RSA encryption standard PKCS#1." *Advances in Cryptology—CRYPTO'98*, Lecture Notes in Computer Science, vol. 1462, ed. H. Krawczyk. Springer-Verlag, Berlin.

[2] Brumley, David and Dan Boneh (2003). "Remote Timing Attacks are Practical." Proceedings of the 12th USENIX Security Symposium, Washington, D.C.

[3] Dierks, Tim and Christopher Allen (1999). The TLS Protocol Version 1.0, IETF Internet-Draft RFC 2246, January (expired), http://www.ietf.org/rfc/rfc2246.txt

[4] Dierks, Tim and Eric Rescorla (2002). The TLS Protocol Version 1.1; IETF Internet-Draft, http://www.ietf.org/

[5] Freier, Alan, Philip Karlton, and Paul Kocher (1996). The SSL 3.0 Protocol; Netscape Communications Corp.

[6] Hickman, Kipp (1995). The SSL Protocol; Netscape Communications Corp.

[7] Internet Assigned Numbers Authority (IANA). http://www.iana.org

[8] ISO 7498-2. "Information processing systems—open systems interconnection—basic reference model—Part 2: Security Architecture." ISO International Standard 7498-2; First edition 1989-02-15.

[9] Klima, Vlastimil, Ondrej Pokorny, and Tomas Rosa (2003). "Attacking RSA-based sessions in SSL/TLS." http://eprint.iacr.org/2003/053.

[10] Kocher, Paul (1997). *Timing Attacks on Implementations of Diffie–Hellman, RSA, DSS, and Other Systems. Advances in Cryptology—CRYPTO'97*, Lecture Notes in Computer Science, vol. 1109, ed. B.S. Kaliski Jr. Springer, Berlin, 104–113.

[11] Mitchell, John C., Vitaly Shmatikov, and Ulrich Stern (1998). "Finite state analysis of SSL 3.0." *Proceedings of the 7th USENIX Security Symposium, San Antonio, Texas*.

[12] Paulson, Lawrence C. (1999). "Inductive analysis of the Internet protocol TLS." *Security Protocols: 6th International Workshop 1998*, Lecture Notes in Computer Science, vol. 1550, eds. Bruce Christianson, Bruno Crispo, William S. Harbison, and Michael Roe. Springer, Berlin.

[13] Rescorla, Eric (2000). *SSL and TLS: Designing and Building Secure Systems*. Addison-Wesley, Reading, MA.

[14] Wagner, David and Bruce Schneier (1997). Analysis of the SSL 3.0 Protocol (revised).

SECURITY

The security of underlined encryption against unauthorized decryption, unauthorized changing of the data, etc. Security should depend completely on the key. One distinguishes between the following two types of security:

Computational security: quantitative security against unauthorized decryption, based on particular (usually mathematical) assumptions like the inherent difficulty of factoring sufficiently long numbers. Often a *security parameter* denotes the computational level of the security.

Unconditional security: security against unauthorized decryption assuming that the cryptanalyst has unlimited computing facilities (so, security in an information theoretic sense).

Friedrich L. Bauer

Reference

[1] Menezes, A.J., P.C. van Oorschot, and S.A. Vanstone (1997). *Handbook of Applied Cryptography*. CRC Press, Boca Raton, FL.

SECURITY ARCHITECTURE

Given any system characterized by a number of devices and/or users communicating with specific communication protocols, the (related) security architecture refers to the enhancing security solution based on cryptographic techniques, protocols, and secure storage, as well as protection of keys and messages.

Examples of security architectures based on Public key techniques include X.509 and EMV. As part of the security architecture, a number of Trusted Third Parties may be defined, such as Registration Authorities, Certification Authorities, and Time Stamping Authorities. These are entities that are not part of the original system as such, but are introduced as part of the security architecture.

Other examples are <u>Kerberos</u>, based on conventional cryptography and bespoke key management architectures e.g. to handle online PIN-code (see <u>Personal Identification Number</u>) verification, which is characterized by key hierarchies, starting with session keys, or data keys at the bottom, which are protected or exchanged by key encryption keys, perhaps comprising several layers, and the top layer consisting of the so-called master keys.

Peter Landrock

SECURITY EVALUATION CRITERIA

Security Evaluation Criteria are usually presented as a set of parameter thresholds that must be met for a system to be evaluated and deemed acceptable. These criteria are established based on a Threat Assessment to establish the extent of the data sensitivity, the security policy, and the system characteristics. The system is evaluated, the evaluation is measured against the criteria, and then an assessment is made of whether or not the system security characteristics meet the requirements as specified by the Security Evaluation Criteria. The criteria is typically unique to each system, the environment it is in and how it is used.

Important past frameworks of security evaluation criteria have been the following:

TCSEC by US Department of Defense (1985): The Trusted Computer System Evaluation Criteria (TCSEC) is a collection of criteria that was previously used to grade or rate the security offered by a computer system product. No new evaluations are being conducted using the TCSEC although there are some still ongoing at this time. The TCSEC is sometimes referred to as the *"Orange Book"* because of its orange cover. The current version is dated 1985 (DOD 5200.28-STD, Library No. S225,711). The TCSEC, its interpretations, and guidelines all have different color covers and are sometimes known as the *"Rainbow Series"* [1]. It is available at http://www.radium.ncsc.mil/tpep/library/rainbow/5200.28-STD.html.

***ITSEC* by the European Commission (1991):** The Information Technology Security Evaluation Criteria (ITSEC) is a European-developed criteria filling a role roughly equivalent to the TCSEC. Although the ITSEC and TCSEC have many similar requirements, there are some important distinctions. The ITSEC places increased emphasis on integrity and availability, and attempts to provide a uniform approach to the evaluation of both products and systems. The ITSEC also introduces a distinction between doing the right job (effectiveness) and doing the job right (correctness). In so doing, the ITSEC allows less restricted collections of requirements for a system at the expense of more complex and less comparable ratings and the need for effectiveness analysis of the features claimed for the evaluation.

***CTCPEC* by CSE Canada (1993):** The Canadian Trusted Computer Product Evaluation Criteria is the Canadian equivalent of the TCSEC. It is somewhat more flexible than the TCSEC (along the lines of the ITSEC) while maintaining fairly close compatibility with individual TCSEC requirements.

Common Criteria ISO 15408 (2001): In 1990, the Organization for Standardization (ISO) sought to develop a set of international standard evaluation criteria for general use. The CC project was started in 1993 in order to bring all these (and other) efforts together into a single international standard for IT security evaluation. The new criteria was to be responsive to the need for mutual recognition of standardized security evaluation results in a global IT market. The common criteria combine the best aspects of TCSEC and ITSEC and aims to supersede both of them [2].

Tom Caddy
Gerrit Bleumer

References

[1] Trusted Product Evaluation Program: Computer Security Evaluation FAQ. http://www.radium.ncsc.mil/tpep/process/faq.html

[2] www.commoncriteria.org, http://csrc.nist.gov/cc

SECURITY STANDARDS ACTIVITIES

This article describes a number of highly visible security standards activities. It cannot be exhaustive, but it does include many standards bodies that are influencing the security industry and product development. Many of the standards are interrelated; for example, <u>X.509</u> public key certificates have been profiled for use in the Internet by the PKIX working group of the Internet Engineering Task Force (IETF), and that profile has been augmented for Qualified Certificates, which are used to identify human beings involved in electronic commerce.

X.509: ITU-T Recommendation X.509 defines public key <u>certificates</u> and <u>attribute certificates</u>. ITU-T was previously known as CCITT, which has been developing telecommunications standards for decades. X.509 [40, 41] is part of a joint effort between ITU-T and the International Organization for Standardization (called ISO), which developed the X.500 series of standards. The documents have numbers assigned by both standards bodies, but the numbers assigned by ITU-T are traditionally used to refer to the documents. Within this series, X.509 defines the public key certificate to provide authentication in a ubiquitous global directory environment. While the envisioned directory deployment has never materialized, the certificate format has been used in small, closed, networks as well as large, open, deployments. The public key certificate enables secure communication between entities that were unknown to each other prior to the communication. Deployments of these certificates are known as <u>public key infrastructure</u> (PKI). To bring PKI to large multinational corporations or to millions of Internet users, a common certificate format is necessary. X.509 defines a general, flexible certificate format. The widespread adoption of X.509 is due to two factors. First, X.509 is technically suitable for many environments. Second, it was developed at an important time. It became an international standard at a time when a number of vendors were ready to begin implementing certificate-based products.

X.509 includes a powerful extension mechanism. It was defined for Version 3 certificates and Version 2 CRLs (<u>Certificate Revocation</u> Lists). Arbitrary extensions can be defined and incorporated into a certificate or CRL, and a criticality flag indicates whether or not a certificate using system needs to understand and examine this extension as part of the verification process. Thus, certificate and CRL contents can readily be tailored to specific environments. The inclusion of a particular critical extension can restrict use of the certificate to a particular application environment.

X.509 also specifies the format of the attribute certificate. The attribute certificate is used in conjunction with a public key certificate to provide additional information about the named entity. Attribute certificates are most often used to express authorization information.

Although X.509 is an international standard, the ITU-T continues to maintain the document and develop enhancements. Most of the maintenance takes the form of clarifying text, and most of the enhancements take the forms of new standard extensions. Any problems found through operational experience are addressed in the standard through a formal defect reporting and resolution process.

PKIX: The Internet Engineering Task Force (IETF) is responsible for creating standards for the Internet. For the most part, the IETF develops protocols. This work is carried out by a number of working groups, which are organized into Areas. Within the Security Area, the <u>PKIX (Public Key Infrastructure using X.509)</u> working group was formed at the end of 1995 with the explicit intention of tailoring the X.509 public key certificate to the Internet environment. Essentially, the group set out to define an Internet PKI. Quickly, the group realized that defining an Internet PKI was more extensive than profiling the X.509 certificate. Thus, the PKIX charter was written to encompass four major activities:

1. X.509 certificate and certificate revocation list (CRL) profile;
2. Certificate management protocols;
3. Operational protocols; and
4. Certificate Policy (CP) and Certification Practice Statement (CPS) framework.

The first activity was the original motivating task. The profile includes detailed specification of the mandatory, optional, critical, and non-critical extensions in a *PKIX-compliant* certificate or CRL. The profile was published in January 1999 [15], and updated in April 2002 [37]. The profile is likely to be refined to provide guidance on the use of international character sets [39]. Also, the qualified certificate profile was developed in January 2001 [34], and the attribute certificate profile was developed in April 2002 [38].

The second activity was to specify the protocols for management operations required in the Internet PKI, including certification of entities and their key pairs, <u>certificate revocation</u>, key backup and recovery (see <u>key management</u>), <u>Certification Authority</u> (CA) key rollover, and crosscertification. Two competing protocols were developed: CMP [16] and CMC [29].

The third activity, operational protocols, was to specify the protocols for day-to-day Internet PKI operation, such as certificate retrieval, CRL retrieval, and on-line certificate status checking. The results, to date, include several important specifications. It tells how to use FTP and HTTP to access repositories [20]. Others tell how to use an LDAPv2 directory as a repository [18,21]. Another specification defines the Online Certificate Status Protocol (OCSP) [19]. And, others are being developed to address the use of LDAPv3 directories.

Finally, the fourth activity, guidance to CP and CPS authors, provides topics and formats for these documents. The guidance was originally published in March 1999 [17]. Since that time, the American Bar Association's Information Security Committee has reviewed it. With the assistance of these lawyers, an update is in progress.

PKIX has played an essential role in bringing PKI concepts to the Internet. The protocols and functions it has defined make a PKI possible, even in the diverse Internet, because their flexibility and generality can satisfy the requirements of greatly differing environments. The PKIX work continues to evolve, and the charter was expanded in 1999 to include additional work items, including time-stamping protocols. The <u>Time-Stamp</u> Protocol (TSP) [36] was published in August 2001.

LDAP: The Lightweight Directory Access Protocol (LDAP) [2] was originally conceived as a simple to describe and simple to implement subset of the capability of the X.500 Directory Access Protocol (DAP). Over time, the subset of functions and features has expanded. Today, it is used as the access protocol for many repositories, some of which are based on X.500 directories, but many are not. As part of this evolution, the "lightweight" aspect of the protocol has diminished. Nevertheless, many vendors worldwide use LDAPv2 [6] and LDAPv3 [8]. The IETF LDAPext Working Group has been formed to specify useful extensions for LDAPv3, such as an authentication and access control mechanism.

An LDAPv2 schema [21] has been specified for LDAP-compliant repositories that contain certificate and CRL information. This facilitates interoperability between PKI products from different vendors in an LDAP environment. In a joint effort between the LDAPext and PKIX working groups, a similar schema is being developed for LDAPv3.

S/MIME: In 1995, a consortium of industry vendors led by RSA Data Security, Inc., developed a companion security solution to the Multipurpose Internet Mail Extensions (MIME) specifications, which are the basis for any email message that goes beyond simple text. For example, an email message that includes bold text or includes an attachment makes use of MIME. Secure MIME (S/MIME) specifies encryption and digital signatures for MIME messages. While a formal standards body did not develop the original S/MIME specifications, many important product vendors embraced S/MIME. To build on and expand this success, the consortium released change control of the S/MIME Version 2 documents [9, 10] to the IETF in 1997.

The IETF S/MIME Working Group developed significant enhancements, resulting in S/MIME Version 3 [22–26]. The primary focus of the S/MIME Working Group was to develop an algorithm independent protocol and incorporate a number of new security features into the specifications, while preserving compatibility with the earlier specification whenever possible. In particular, the S/MIME Version 3 specifications include support for sending encrypted messages to large mail lists, security labels on messages (for example, "company proprietary," "secret," or "top secret"), and signed message receipts. These signed receipts provide proof that the intended recipient received a signed message that contained a request for a receipt.

The S/MIME Version 3 specifications include discussion of PKI concepts such as certificate format, certificate processing, and CRLs. These specifications are compatible with the X.509 profile developed by the PKIX Working Group, and they provide additional details for the use of X.509 certificates in the email environment. Further, provision is made in the message envelope to carry an arbitrary numbers of certificates and CRLs to assist the recipient with the task of path construction and certificate validation.

IPSEC: IPsec is designed to provide interoperable, high quality, cryptographically-based security the Internet Protocol (both Version 4 (IPv4) and Version 6 (IPv6)). The security services offered include access control, connectionless integrity, data origin authentication, protection against replays, confidentiality, and limited traffic flow confidentiality. The services are provided by the use of two traffic security protocols, the Authentication Header (AH) [11] and the Encapsulating Security Payload (ESP) [12], and through the use of cryptographic key management procedures and protocols.

The Internet Key Exchange (IKE) protocol [13] is used to establish the symmetric keying material needed by AH and ESP. IKE provides for strong, X.509-certificate-based authentication of the IP layer entities, and it is compatible with the certificate profile developed by the PKIX Working Group. However, a companion document is being developed to describe details of certificate usage in the IPsec environment.

The IPsec Working Group is working on the second version of IKE. The primary goal of the update is to simplify the protocol. The simplification is targeted at increased interoperability.

TLS: The <u>Transport Layer Security (TLS)</u> specification [7] is the IETF standards-track version of the Secure Sockets Layer Version 3.0 (SSLv3.0) protocol found in millions of Web browsers and Web servers. The history has many parallels to S/MIME. The original specification was developed outside of any standards body, and then it was released to the IETF, who took over configuration control and made enhancements.

TLS creates a secure channel between two transport layer entities, providing certificate-based authentication, information integrity, and data confidentiality. The TLS specification discusses X.509 certificates, and it is mostly compatible with the profile developed by the PKIX Working Group. The few conflicts are associated with compatibility with SSLv3.0 (and earlier) implementations that were developed well in advance of the PKIX profile. The PKIX X.509 certificate profile appears to meet the goals of the Internet community. Interestingly, non-IETF standards groups are also using the PKIX certificate profile.

AAA: In 1997, the IETF created the first standard Authentication, Authorization, and Accounting (AAA) protocol, called RADIUS (Remote Access Dial-In User Service) [30–32]. As the name implies, RADIUS is designed for use with Dial-In Access Servers. RADIUS has been a big success, displacing many proprietary protocols. RADIUS is widely implemented as Network Access Servers (NASs) serving analog and digital dial-in Point-to-Point Protocol (PPP) service, and it is the prevalent Internet Service Provider (ISP) access model. RADIUS has been adapted for use with DSL (using PPPOE) and cable access (using DOCSIS). RADIUS has been successful because it offers a simple and flexible model for client-server exchanges. However, this simple model does not have sufficient security for some new applications, and it also lacks support for server-initiated control.

The IETF AAA Working Group is responsible for building a more secure and capable AAA protocol. A number of proposals were evaluated in June 2000, and the working group selected the Diameter protocol [35]. Diameter is designed to be upwards compatible with RADIUS, but many of the messaging underpinnings have been upgraded to be more secure. Security is provided using CMS and IPsec. For better response time, the SCTP (Stream Control Transmission Protocol) transport is supported as an alternative.

Diameter explicitly supports server-to-client requests and message forwarding. These capabilities have previously been forced into RADIUS [33]. Diameter also includes explicit support for application suite additions. Application designs have been drafted for Mobile IP authentication and

third generation wireless telecommunications [1] sessions.

SPKI: The IETF formed the Simple Public Key Infrastructure (SPKI) Working Group in 1996. In many ways, it is an alternative to PKIX. One fundamental premise of SPKI is that X.509 is a complicated and bulky certificate format that, by explicitly binding a key pair to an identity, rests upon an inherently flawed foundation. SPKI proponents argue that the concept of a globally unique identity will never be realized. Instead, they advocate the use of the public key as an identity. Where necessary and meaningful, a name or other identifying information may be associated with a public key, but this is optional and, it is only intended to have local significance.

The SPKI specifications [27,28] discuss the concepts and philosophy behind this approach to an Internet PKI. A detailed certificate format and processing rules are included. SPKI explicitly encompasses authorization as well as authentication. The sophisticated certificate format makes it possible to express, in a general way, the permitted uses of the certified public key. This capability (not surprisingly) diminishes the intended simplicity of the Simple Public Key Infrastructure. Although SPKI embodies a number of interesting ideas and research contributions, it has not gained widespread support.

OPENPGP: As with the S/MIME and TLS Working Groups, the IETF OpenPGP Working Group was formed to develop a standard based on a protocol that was developed outside of any standards body. The popular Pretty Good Privacy (PGP) email security package was brought to the IETF so that interoperable implementations from different vendors could be developed. OpenPGP [14] defines email message protection and the PGP certificate format (an alternative to both X.509 and SPKI). Despite a loyal installed base, OpenPGP has not seen corporate or government adoption. OpenPGP is viewed as an individual-to-individual solution. The user-centric trust model cannot easily be centrally controlled by an organization.

XML SECURITY: Prominent standards bodies are actively developing XML (eXtensible Markup Language) security specifications, including the World Wide Web Consortium (W3C) and the Organization for the Advancement of Structured Information Standards (OASIS).

The W3C is developing specifications for the XML syntax with respect to encryption (XML

Encryption) and digital signature (XML Signature), as well as XML protocols for key management (XML Key Management Specification) that allow a client to obtain key information (including values, certificates, management, or trust data) from a Web service.

The OASIS Security Services Technical Committee is developing the Security Assertion Markup Language (SAML), an XML framework for exchanging authentication and authorization information. The underlying authentication mechanism may be PKI-based, but SAML encompasses a number of other authentication technologies as well. A number of other OASIS technical committees are likely to build upon SAML, as well as the W3C specifications mentioned above, to provide security; such committees include Business Transaction Processing (BTP), electronic business XML (ebXML), Provisioning Services Markup Language (PSML), eXtensible Access Control Markup Language (XACML), Web Services Security (WSS), and Digital Signature Services (DSS).

IEEE P802: Local Area Network (LAN) and Metropolitan Area Network (MAN) standards encompass a number of data communications technologies and the applications of these technologies. There is no single technology that is applicable to all applications. Correspondingly, no single local or metropolitan area network standard is adequate for all applications. As a result, the Institute of Electrical and Electronics Engineers (IEEE) Standards Association sponsors several working groups and technical advisory groups within Project 802. Security is the focus of IEEE 802.10, which has seen little market adoption. However, other working groups have also developed security relevant standards.

IEEE 802.1X specifies port-based access controls. It provides a means of authenticating and authorizing devices attached to a LAN port, preventing access when authentication and authorization fails.

IEEE 802.11 includes the ability to encrypt wireless LAN traffic using the Wired Equivalent Privacy (WEP) protocol. Unfortunately, WEP has many flaws. IEEE 802.11 is presently working on a short-term and a long-term replacement for WEP, called TKIP and CCMP, respectively. The Temporal Key Integrity Protocol (TKIP) is intended to replace WEP on current hardware, and it is implemented by firmware and driver software upgrades. The Counter and CBC-MAC Protocol (CCMP) is intended for future generations of product. Future product generations will likely implement both

TKIP and CCMP for compatibility with currently fielded devices.

IEEE 802.15 is developing security solutions for personal area networks, and IEEE 802.3 is developing security solutions for some uses of Ethernet. Clearly, more customers are demanding security solutions. Other working groups are likely to have security initiatives in the near future.

IEEE P1363: IEEE Project 1363 is developing standard specifications for public key cryptography, which includes mathematical primitives for key derivation, public-key encryption, and digital signatures. P1363 has been adopted as an IEEE standard, although work continues on a companion document, called IEEE P1363a, which will specify additional techniques. A study group is investigating newer schemes and protocols not considered in P1363 and P1363a; such specifications will appear over time as P1363–1, P1363–2, and so on.

ANSI X9F: The American National Standards Institute (ANSI) committee X9 (Financial Services) develops and publishes standards for the financial services industry. These standards facilitate delivery of financial products and services. Subcommittee X9F is responsible for security, and it includes working groups that focus on cryptographic tools (X9F1), security protocols (X9F3), and digital signature and certification policy (X9F5), among others. X9F has published many standards (the X9 on-line catalog can be found at http://www.x9.org), and many of its standards become international standards through a close working relationship with ISO TC68.

INFLUENTIAL ACTIVITIES: Some activities that are not part of any formal security standards body are influencing security standards development, the security industry, and product development. Again, this discussion cannot be exhaustive, but a number of the highly visible security standards influencing activities are discussed.

U.S. FPKI

The U.S. Federal Public-Key Infrastructure (FPKI) is an initiative by the U.S. Government to define a PKI suitable for its own use. One focus is the production of an acceptable profile for X.509 certificates and CRLs, where there is significant harmonization with the PKIX certificate profile, but the ultimate goal is a full PKI specification. This specification will encompasses all relevant PKI entities, including end entities, registration authorities (RAs), certification authorities (CAs), and Bridge CAs. It also includes the security-relevant protocols between these entities, as well as the operational policies and procedures required for the PKI.

The U.S. FPKI specifications impose compliance requirements on vendors wanting to sell PKI products to the U.S. Government. To the greatest extent possible, commercial standards have been referenced and profiled. The hope is that the FPKI is sufficiently similar to PKIs for other environments that compliance will not unduly restrict vendors.

The Minimum Interoperability Specifications for PKI Components (MISPC) [5] is one component of the full U.S. FPKI vision. The goal in MISPC is to understand and to specify the minimum functionality required of PKI entities that will still enable them to interoperate usefully with other PKI entities. Thus, for example, the certificate and CRL profile portion of MISPC identifies which of the many optional fields in the X.509 and PKIX specifications must be implemented. Interestingly, MISPC is more than a detailed specification; a CD containing a complete reference implementation compliant with the specification is also available. Thus, vendors have a straightforward way of testing whether their products are MISPC compliant.

GOC PKI

The Government of Canada Public-Key Infrastructure (GOCPKI) is the first large-scale governmental PKI initiative in the world. Its goal similar to the U.S. FPKI, but it defines a PKI suitable for Canadian federal government use. It is a full PKI specification, including certificate and CRL profiles, entity functionality and characteristics, communications protocols, and operational policies and procedures. The GOC PKI will impose compliance requirements on vendors, but it is hoped that this will not preclude Commercial Off-the-Shelf (COTS) products.

JCP

The Java Community Process (JCP) is an open organization of international Java developers and licensees whose charter is to develop and revise Java technology specifications, reference implementations, and technology compatibility kits. This group publishes Java Specification Requests (JSRs), and several are related to security and PKI. For example, JSR 55 discusses certification path creation, building, and verification; JSR 74 discusses many of the Public Key Cryptography Standards (PKCS) published by RSA Laborato-

ries; JSR 104 discusses XML trust services; JSR 105 discusses XML Digital Signature services; JSR 106 discusses XML Digital Encryption services; and JSR 155 discusses Web Services Security Assertions based on the OASIS SAML specification. These and related efforts are expected to eventually be included in future versions of the Java 2 Micro Edition (J2ME), Java 2 Standard Edition (J2SE), and Java 2 Enterprise Edition (J2EE) platforms. Further information can be found at [4].

ICE-CAR

The Interworking Certification Infrastructure for Commerce, Administration and Research (ICE-CAR) project, a successor to the ICE-TEL project, began in January 1999. The objective of this project is to provide all of the technology components that are needed to support the secure use of the Internet for commercial and administrative applications in Europe. These applications include e-commerce, intra-organizational communication, health-care applications, and research. An additional goal was to promote the availability of technically compatible and interconnectable PKIs, which guarantee the authenticity and validity of public keys used in these environments. The project has produced numerous technical reports that are available for download from the *Deliverables* section of the main Web site; see [3] for further details.

Russ Housley

References

[1] See http://www.3gpp2.org/
[2] Howes, T. and M. Smith (1997). *LDAP: Programming Directory-Enabled Applications with Lightweight Directory Access Protocol*. Macmillan Technical Publishing, Indianapolis.
[3] See http://ice-car.darmstadt.gmd.de/ice-car-home.html
[4] See http://www.jcp.org/
[5] National Institute of Standards and Technology. Minimum Interperability Specification for PKI Components, Version 1, June 1997.
[6] Yeong, W., T. Howes, and S. Kille (1995). Lightweight Directory Access Protocol. RFC 1777.
[7] Dierks, T. and C. Allen (1999). The TLS Protocol Version 1.0. RFC 2246.
[8] Wahl, M., T. Howes, and S. Kille (1997). Lightweight Directory Access Protocol (v3). RFC 2251.
[9] Dusse, S., P. Hoffman, B. Ramsdell, L. Lundblade, and L. Repka (1998). S/MIME Version 2 Message Specification. RFC 2311.

[10] Dusse, S., P. Hoffman, B. Ramsdell, and J. Weinstein (1998). S/MIME Version 2 Certificate Handling. RFC 2312.

[11] Kent, S. and R. Atkinson (1998). IP Authentication Header. RFC 2402.

[12] Kent, S. and R. Atkinson (1998). IP Encapsulating Security Payload (ESP). RFC 2406.

[13] Harkins, D. and D. Carrel (1998). The Internet Key Exchange (IKE). RFC 2409.

[14] Callas, J., L. Donnerhacke, H. Finney, and R. Thayer (1998). OpenPGP Message Format. RFC 2440.

[15] Housley, R., W. Ford, W. Polk, and D. Solo (1999). Internet X.509 Public Key Infrastructure Certificate and CRL Profile. RFC 2459.

[16] Adams, C. and S. Farrell (1999). Internet X.509. Public Key Infrastructure Certificate Management Protocols. RFC 2510.

[17] Chokhani, S. and W. Ford (1999). Internet X.509 Public Key Infrastructure Certificate Policy and Certification Practices Framework. RFC 2527.

[18] Boeyen, S., T. Howes, and P. Richard (1999). Internet X.509 Public Key Infrastructure Operational Protocols—LDAPv2. RFC 2559.

[19] Myers, M., R. Ankney, A. Malpani, S. Galperin, and C. Adams (1999). X.509 Internet Public Key Infrastructure Online Certificate Status Protocol—OCSP. RFC 2560.

[20] Housley, R. and P. Hoffman (1999). Internet X.509 Public Key Infrastructure Operational Protocols: FTP and HTTP. RFC 2585.

[21] Boeyen, S., T. Howes, and P. Richard (1999). Internet X.509 Public Key Infrastructure LDAPv2 Schema. RFC 2587.

[22] Housley, R. (1999). Cryptographic Message Syntax. RFC 2630.

[23] Rescorla, E. (1999). Diffie-Hellman Key Agreement Method. RFC 2631.

[24] Ramsdell, B. (ed.). (1999). S/MIME Version 3 Certificate Handling. RFC 2632.

[25] Ramsdell, B. (ed.). (1999). S/MIME Version 3 Message Specification. RFC 2633.

[26] Hoffman, P. (ed.). (1999). Enhanced Security Services for S/MIME. RFC 2634.

[27] Ellison, C. (1999). SPKI Requirements. RFC 2692.

[28] Ellison, C., B. Frantz, B. Lampson, R. Rivest, B. Thomas, and T. Ylonen (1999). SPKI Certificate Theory. RFC 2693.

[29] Myers, M., X. Liu, J. Schaad, and J. Weinstein (2000). Certificate Management Messages over CMS. RFC 2797.

[30] Rigney, C., S. Willens, A. Rubens, and W. Simpson (2000). Remote Authentication Dial in User Service (RADIUS). RFC 2865.

[31] Rigney, C. (2000). RADIUS Accounting. RFC 2866.

[32] Rigney, C., W. Willats, and P. Calhoun (2000). RADIUS Extensions. RFC 2869.

[33] Mitton, D. (2000). Network Access Servers Requirements: Extended RADIUS Practices. RFC 2882.

[34] Santesson, S., W. Polk, P. Barzin, and M. Nystrom (2001). Internet X.509 Public Key Infrastructure Qualified Certificates Profile. RFC 3039.

[35] Mitton, D., M. St.Johns, S. Barkley, D. Nelson, B. Patil, M. Stevens, and B. Wolff (2001). Authentication, Authorization, and Accounting: Protocol Evaluation. RFC 3127.

[36] Adams, C., P. Cain, D. Pinkas, and R. Zuccherato (2001). Internet X.509 Public Key Infrastructure Time-Stamp Protocol (TSP). RFC 3161.

[37] Housley, R., W. Polk, W. Ford, and D. Solo (2002). Internet X.509 Public Key Infrastructure Certificate and Certificate Revocation List (CRL) Profile. RFC 3280.

[38] Farrell, S. and R. Housley (2002). An Internet Attribute Certificate Profile for Authorization. RFC 3281.

[39] Yergeau, F. (1998). UTF-8, a Transformation Format of ISO 10646. RFC 2279.

[40] ITU-T. (1997). Recommendation X.509: The Directory—Authentication Framework.

[41] ITU-T. (2000). Recommendation X.509: The Directory—Public Key and Attribute Certificate Frameworks.

SELECTIVE FORGERY

Selective forgery is a message related forgery against a cryptographic digital signature scheme. Given a victim's verifying key, a selective forgery is successful if the attacker finds a signature s for a message m selected by the attacker prior to the attack, such that the signature s is valid for m with respect to the victim's verifying key.

Gerrit Bleumer

SELF-SHRINKING GENERATOR

The self-shrinking generator is a clock-controlled generator that has been proposed in [1]; it is strongly related to the shrinking generator, but uses only one Linear Feedback Shift Register (LFSR) R, producing a maximum-length linear sequence.

Its principle is really easy to get: the output sequence of the LFSR is partitioned into pairs of bits. According to the value of the pair, one bit is added to the keystream, and then the pair is discarded and we go to the next one. More precisely:

Pair	Bit added
10	0
11	1
01	no bit added
00	no bit added

EXAMPLE. Let us consider that R has length four, and that its feedback is given by $s_{t+1} = s_t + s_{t-3}$. If the initial state is $s_0 s_1 s_2 s_3 = 1010$, then the output of the LFSR is 10101100100011110101100100011101011001000-1111010110010001111... This gives the following output for the whole scheme: 00101101001011...

A recent survey on the possible attacks is [2].

Caroline Fontaine

References

[1] Meier, W. and O. Staffelbach (1995). "The self-shrinking generator." *Advances in Cryptology—EUROCRYPT'94*, Lecture Notes in Computer Science, vol. 950, ed. A. De Santis. Springer-Verlag, Berlin, 205–214.

[2] Zenner, E., M. Krause, and S. Lucks (2001). "Improved cryptanalysis of the self-shrinking generator." ACIPS 2001, Lecture Notes in Computer Science, vol. 2119, eds. V. Varadharajan and Y. Mu. Springer-Verlag, Berlin, 21–35.

SELF-SYNCHRONIZING STREAM CIPHER

In a self-synchronizing, or asynchronous, stream cipher, the keystream depends on the secret key of the scheme, but also of a fixed number, say t, of ciphertext digits (that have already been produced, or read; this distinguishes it from a synchronous stream cipher). It can be viewed as follows:

According to its design, such a scheme is able to resynchronize the keystream with the message with just a few correct bits of ciphertext. This means that if some bits are inserted or deleted in the ciphertext, just a small part of the plaintext will not be obtained correctly; the next set of t consecutive correct bits in the ciphertext will be sufficient to resynchronize the keystream and produce the following bits of the plaintext correctly.

Let us now consider that one bit of the ciphertext has been altered during the transmission. This will induce some errors in the decryption of the next t bits; after this, decryption will go on correctly.

What can an active attacker do with such a scheme? According to the propagation of each error in a ciphertext on about t bits of plaintext, it is more difficult for an attacker to forge a plaintext of its choice than in a synchronous stream cipher. Moreover, it is also more difficult for him to desynchronize the keystream, since the scheme is able to resynchronize it by itself. If the attacker wants to desynchronize all the keystream, he has to do a lot of modifications on the ciphertext. Nevertheless, some complementary mechanisms, that can ensure authentication or integrity of the ciphertext are welcome to help the receiver check that all is going well.

At last, since each plaintext digit influences the whole ciphertext (through the feedback of the ciphertext on the keystream generation), the statistical properties of the plaintext are dispersed in the ciphertext, and such a scheme may be more resistant against attacks based on plaintext redundancy, than synchronous stream ciphers.

Good references are [1] and [2].

Caroline Fontaine

References

[1] Maurer, U.M. (1991). "New approaches to the design of self-synchronizing stream ciphers." *Advances in Cryptology—EUROCRYPT'91*, Lecture Notes in Computer Science, vol. 547, ed. D.W. Davies. Springer-Verlag, Berlin, 458–471.

[2] Rueppel, R.A. (1986). *Analysis and Design of Stream Ciphers*. Springer-Verlag, Berlin.

SEMANTIC SECURITY

Semantic security is a notion to describe the security of an encryption scheme.

An adversary is allowed to choose between two plaintexts, m_0 and m_1, and he receives an encryption of either one of the plaintexts. An encryption scheme is semantically secure, if an adversary cannot guess with better probability than 1/2 whether the given ciphertext is an encryption of message m_0 or m_1. The notion is also referred to as *indistinguishability of encryptions* and noted as IND. Historically the word "semantic" came from the definition that the encryption reveals no information no matter what kind of semantics are embedded in the encryption. It has been proven that the definition describing this requirement is equivalent to the indistinguishability of encryptions. The notion of semantic security can be further distinguished by the power of adversary. More specifically, a powerful adversary may have access to an encryption *oracle* and/or decryption oracle at various stages of the guessing game. Here, an encryption oracle is an oracle that provides an encryption of a queried plaintext, and a decryption oracle provides the decryption of a queried ciphertext (see also random oracle model).

The notion of semantic security can be applied to both symmetric cryptosystems and public key cryptosystems. But since the concrete security analysis of a public key encryption scheme is more tractable, the term is more frequently used to discuss the security of public key encryption schemes.

In a public key encryption scheme, the adversary can always access the encryption oracle, because he can encrypt by himself. Therefore the semantic security must be achieved against such an adversary. Such security is called "semantically secure against chosen plaintext attack" and written IND-CPA. The threat of adversary who has access to decryption oracle is called chosen ciphertext attack (CCA). If a public-key scheme is semantically secure against an adversary who has access to a decryption oracle before determining the pair of plaintexts m_0 and m_1, it is called IND–CCA1. If a public-key scheme is semantically secure against an adversary who has access to a decryption oracle not only before receiving a target ciphertext but also during the guessing stage, then it is defined as IND–CCA2. It is regarded that this type of adversary is the most powerful. Therefore the scheme achieving IND–CCA2 is considered most secure. (There is a restriction on this type of adversary, namely that he cannot receive an answer of the target ciphertext from decryption oracle.)

Besides semantic security, there are related notions such as non-malleability and *plaintext awareness*.

Kazue Sako

Reference

[1] Bellare, M., A. Desai, D. Pointcheval, and P. Rogaway (1998). "Relations among notions of security for public-key encryption schemes." *Advances in Cryptography—CRYPTO'98*, Lecture Notes in Computer Science, vol. 1462, ed. H. Krwawczyk. Springer-Verlag, Berlin, 26–45.

SENDER ANONYMITY

Sender anonymity is achieved in a messaging system if an eavesdropper who picks up messages from the communication line of a recipient cannot tell with better probability than pure guessing who sent the messages. During the attack, the eavesdropper may also listen on all communication lines of the network including those that connect the potential senders to the network and he may send his own messages. It is clear that all messages in such network must be encrypted to the same length in order to keep the attacker from distinguishing different messages by their content or length. The *anonymity* set for any particular message attacked by the eavesdropper is the set of all network participants that have sent message within a certain time window before the attacked message was received. This time window ofcourse depends on latency characteristics and node configurations of the network itself.

Sender anonymity can be achieved against computationally restricted eavesdroppers by MIX networks [1] and against computationally unrestricted eavesdroppers by DC networks [2,3].

Note that sender anonymity is weaker than sender unobservability, where the attacker cannot even determine whether or not a participant sends a message. Sender unobservability can be achieved with MIX networks and DC networks by adding dummy traffic.

Gerrit Bleumer

References

[1] Chaum, David (1981). "Untraceable electronic mail, return addresses, and digital pseudonyms." *Communications of the ACM*, 24 (2), 84–88.
[2] Chaum, David (1985). "Security without identification: Transaction systems to make big brother obsolete." *Communications of the ACM*, 28 (10), 1030–1044.
[3] Chaum, David (1988). "The dining cryptographers problem: Unconditional sender and recipient untraceability." *Journal of Cryptology*, 1 (1), 65–75.

SEQUENCES

Sequences have many applications in modern communication systems, including signal synchronization, navigation, radar ranging, Code-Division Multiple-Access (CDMA) systems, random number generation, spread-spectrum communications, cryptography, in particular in stream cipher systems.

In stream cipher systems it is essential to construct sequences with good random properties, long periods, and large linear complexity. To achieve many of these goals one often generates sequences using linear recurrence relations. The *period* of a sequence $\{s_t\}$ is the smallest integer ε such that $s_{t+\varepsilon} = s_t$ for all t. We will explain how the period of a generated sequence is completely determined by the characteristic polynomial of the sequence. For linear sequences the period of the sequences generated can easily be controlled, which makes them good building blocks in stream cipher systems.

A linear recursion of degree n with binary coefficients is given by

$$\sum_{i=0}^{n} f_i \, s_{t+i} = 0,$$

where $f_i \in GF(2) = \{0, 1\}$ for $0 < i < n$ and $f_0 = f_n = 1$. The *characteristic polynomial* of the recursion is defined by

$$f(x) = \sum_{i=0}^{n} f_i x^i.$$

The initial state $(s_0, s_1, \ldots, s_{n-1})$ and the given recursion uniquely determines the generated sequence. A linear shift register with a characteristic polynomial $f(x)$ of degree n generates 2^n different sequences corresponding to the 2^n different initial states and these form a vector space over $GF(2)$ which is denoted $\Omega(f)$.

The maximum period of a sequence generated by a linear shift register is at most $2^n - 1$. This follows since a sequence is completely determined by n-successive bits in the sequence and period 2^n is impossible since n successive zeros implies the all zero sequence. Sequences with the maximal period $2^n - 1$ are called *m*-sequences. For example, with initial state $(s_0, s_1, s_2) = (001)$, then $f(x) = x^3 + x + 1$ generates the *m*-sequence 0010111.

EXAMPLE 1. Let the recursion be

$$s_{t+4} + s_{t+3} + s_{t+2} + s_{t+1} + s_t = 0 \pmod 2$$

with characteristic polynomial $f(x) = x^4 + x^3 + x^2 + x + 1$. The sequences in $\Omega(f)$ consists of the $2^4 = 16$ sequences corresponding to the sequences $\{(0), (00011), (00101), (01111)\}$ and their cyclic shifts.

To analyze properties of linear sequences, we associate a generating function $G(x)$ to the sequence $\{s_t\}$, and let

$$G(x) = \sum_{t=0}^{\infty} s_t x^t.$$

Let $f^*(x) = \sum_{i=0}^{n} f_{n-i} x^i$ be the *reciprocal polynomial* of the characteristic polynomial of $f(x)$ of the

sequence. Then, we can compute the product

$$
\begin{aligned}
G(x) f^*(x) &= (s_0 + s_1 x + s_2 x^2 + \cdots) \\
&\quad \times (1 + f_{n-1} x + \cdots + f_1 x^{n-1} + x^n) \\
&= \sum_{t=0}^{\infty} c_t x^t.
\end{aligned}
$$

The coefficient c_{t+n} of x^{t+n} for any $t \geq 0$ becomes

$$c_{t+n} = \sum_{i=0}^{n} f_i s_{t+i} = 0$$

as a consequence of the recurrence relation. Hence,

$$G(x) f^*(x) = \phi^*(x)$$

for some polynomial $\phi^*(x)$ of degree at most $n - 1$. Its reciprocal polynomial $\phi(x)$ is given by

$$
\begin{aligned}
\phi(x) &= s_0 x^{n-1} + (s_1 + f_{n-1} s_0) x^{n-2} + \cdots \\
&\quad + (s_{n-1} + f_{n-1} s_{n-2} + \cdots + f_1 s_0) \\
&= \sum_{i=0}^{n-1} \left(\sum_{j=0}^{n-1-i} f_{i+j+1} s_j \right) x^i.
\end{aligned}
$$

There is a one-to-one correspondence between any sequence $\{s_t\}$ in $\Omega(f)$ and any polynomial $\phi^*(x)$ of degree $\leq n - 1$. All sequences generated by $f(x)$ can therefore be described by

$$\Omega(f) = \left\{ \frac{\phi^*(x)}{f^*(x)} \; \middle| \; deg(\phi^*(x)) < deg(f) = n \right\}.$$

For example, the *m*-sequence 0010111 in $\Omega(x^3 + x + 1)$ can be written

$$
\begin{aligned}
\frac{x^2}{1 + x^2 + x^3} &= x^2 + x^4 + x^5 + x^6 + x^9 + x^{11} + x^{12} + \cdots \\
&= (x^2 + x^4 + x^5 + x^6)(1 + x^7 + x^{14} + \cdots
\end{aligned}
$$

In particular, a simple consequence of the above description of $\Omega(f)$ is that $\Omega(f) \subset \Omega(g)$ if and only if $f(x)$ divides $g(x)$.

The generating function $G(x)$ for a periodic sequence of period ε can be written as

$$
\begin{aligned}
G(x) &= \left(s_0 + s_1 x + \cdots + s_{\varepsilon-1} x^{\varepsilon-1}\right) \\
&\quad \times (1 + x^\varepsilon + x^{2\varepsilon} + \cdots) \\
&= \frac{s_0 + s_1 x + \cdots + s_{\varepsilon-1} x^\varepsilon}{1 - x^\varepsilon}.
\end{aligned}
$$

Combining the two expressions for $G(x)$, we obtain the identity

$$(x^\varepsilon - 1)\phi(x) = \sigma(x) f(x),$$

where $\sigma(x) = s_0 x^{\varepsilon-1} + s_1 x^{\varepsilon-2} + \cdots + s_{\varepsilon-1}$, contains all the information of a period of the sequence.

The *period of the polynomial* $f(x)$ is the smallest positive integer e such that $f(x)$ divides $x^e - 1$. The importance of the period e of $f(x)$, is that in order to find the period of all the sequences in $\Omega(f)$, it is enough to find the period of $f(x)$.

Since $f(x)$ divides $x^e - 1$ it follows that $\Omega(f) \subset \Omega(x^e - 1)$, the set of sequences where $s_{t+e} = s_t$, i.e., of period dividing e. Hence, all sequences generated by $f(x)$ has period dividing e. Let the sequence $\{s_t\}$ correspond to the polynomial $\phi(x)$. If $gcd(f(x), \phi(x)) = 1$ then as a consequence of the identity $(x^\varepsilon - 1)\phi(x) = \sigma(x)f(x)$, it follows that $\{s_t\}$ has *smallest* period e, since in this case $f(x)$ must divide $x^\varepsilon - 1$ and thus $\varepsilon \geq e$ which implies that $\varepsilon = e$.

In particular, when $f(x)$ is an <u>irreducible polynomial</u> of period e, then all the nonzero sequences in $\Omega(f)$ have period e. For example the polynomial $f(x) = x^4 + x^3 + x^2 + x + 1$ in Example 1 is irreducible and divides $x^5 + 1$ and has period 5, and therefore all nonzero sequences in $\Omega(f)$ have period 5.

To determine the cycle structure of $\Omega(f)$ for an arbitrary polynomial $f(x)$ that can be factored as $f(x) = \prod f_i(x)^{k_i}$, $f_i(x)$ irreducible, one first needs to determine the cycle structure of $\Omega(f_i^{k_i})$ and then the cycle structure for $\Omega(gh)$ when $gcd(g, h) = 1$.

Cycle structure of $\Omega(f^r)$

Let $f(x)$ be an irreducible polynomial of period e. Let k be defined such that $2^k < r \leq 2^{k+1}$. Then $\Omega(f^r) \setminus \Omega(f)$ contains

$$2^{n2^j} - 2^{n2^{j-1}}$$

sequences of period $e2^j$ for $j = 1, 2, \ldots, k$ and

$$2^{nr} - 2^{n2^k}$$

sequences of period $e2^{k+1}$.

EXAMPLE 2. Let $f(x) = x^3 + x + 1$ be the characteristic polynomial with $e = 7$, that generates an m-sequence. The number of sequences of each period in $\Omega(f^3)$ is therefore

Number 1 7 56 448
Period 1 7 14 28

Cycle structure of $\Omega(gh)$ when $gcd(g, h) = 1$.

In this case it can be shown that each sequence $\{s_t\}$ in $\Omega(gh)$ can be written uniquely

$$\{s_t\} = \{u_t\} + \{v_t\}$$

where $\{u_t\} \in \Omega(g)$ and $\{v_t\} \in \Omega(h)$. Further, the period of the sum $\{u_t\} + \{v_t\}$ is equal to the <u>least common multiple</u> of the period of the two sequences, i.e.,

$$per(s_t) = lcm(per(u_t), per(v_t)).$$

To find the cycle structure of $\Omega(gh)$, suppose $\Omega(g)$ contains d_1 cycles of length λ_1 and $\Omega(h)$ contain d_2 cycles of length λ_2. Add in all possible ways the corresponding $d_1\lambda_1$ sequences from $\Omega(g)$ and the $d_2\lambda_2$ sequences from $\Omega(h)$. This gives $d_1\lambda_1 d_2\lambda_2$ distinct sequences all of period $lcm(\lambda_1, \lambda_2)$. Formally we can write this as $[d_1(\lambda_1)][d_2(\lambda_2)] = [d(\lambda)]$ where $d = d_1 d_2 gcd(\lambda_1, \lambda_2)$ and $\lambda = lcm(\lambda_1, \lambda_2)$.

EXAMPLE 3. Let $f_1(x) = x^3 + x + 1$ and $f_2(x) = x^4 + x^3 + x^2 + x + 1$. The cycle structure of $\Omega(f_1)$ can be written $[1(1) + 1(7)]$, and similarly for $\Omega(f_2)$ as $[1(1) + 3(5)]$. Combining the cycle structure as described above, gives the cycle structure $[1(1) + 1(7) + 3(5) + 3(35)]$.

The discussion above shows that the period of all sequences in $\Omega(f)$ is completely determined from the periods of the divisors of $f(x)$. This way of controlling the periods is one of the main reasons for using linear recursions as building blocks in stream ciphers.

The sequence $\{s_t\}$ can be expressed in terms of the zeros of its characteristic polynomial $f(x)$ of degree n. In the case when the zeros of $f(x)$ are simple, which is the case when the sequence has odd period, then $\{s_t\}$ has a unique expansion in the form

$$s_t = \sum_{i=1}^{n} a_i \alpha_i^t$$

for some constants a_i and where α_i, $1 \leq i \leq n$ are the zeros of $f(x)$.

The main problem with linear recursions in cryptography is that it is easy to reconstruct a sequence $\{s_t\}$ generated by a characteristic polynomial $f(x)$ of degree n from the knowledge of $2n$ consecutive bits in $\{s_t\}$, since this gives a system of n equations for determining the unknown coefficients of $f(x)$. The <u>Berlekamp–Massey</u> algorithm is an efficient method for finding $f(x)$ in this way. Several methods exist to increase the linear span, i.e, the smallest degree of the linear recursion that generates the sequence. We just mention a few simple ones obtained by multiplying sequences.

Let $\{u_t\}$ and $\{v_t\}$ be two sequences of odd period. Then, $u_t = \sum a_i \alpha_i^t$ where α_i, $1 \leq i \leq n$, are the zeros of the characteristic polynomial of $\{u_t\}$ and $v_t = \sum b_j \beta_j^t$ where β_j, $1 \leq j \leq m$, are the zeros of the characteristic polynomial of $\{v_t\}$. Then the sequence $\{w_t\} = \{u_t v_t\}$ can be written as

$$w_t = \sum a_i b_j (\alpha_i \beta_j)^t.$$

This shows that $\{w_t\}$ is generated by the polynomial with the, at most, nm different zeros $\alpha_i \beta_j$ for

$1 \le i \le n,\ 1 \le j \le n$. If $gcd(per(u_t), per(v_t)) = 1$, then it can be shown that $per(w_t) = per(u_t)per(v_t)$. However, the number of zeros and ones in sequence $\{w_t\}$ will in general not be *balanced* even if this is the case for $\{u_t\}$ and $\{v_t\}$. This follows since $w_t = 1$ if and only if $u_t = v_t = 1$, and thus only 1/4 of the elements in $\{w_t\}$ will be 1's when $\{u_t\}$ and $\{v_t\}$ are balanced.

Often one considers sequences of the form

$$w_t = s_{t+\tau_1}s_{t+\tau_2}\cdots s_{t+\tau_k},$$

where s_t is an m-sequence. A closer study of the zeros of the characteristic polynomial of $\{w_t\}$ shows that the linear span is at most $\sum_{i=1}^{k}\binom{n}{i}$ and frequently the equality holds.

Every Boolean function in n variables, $f(x_1, x_2, \ldots, x_n)$, can be written uniquely as the sum

$$f(x_1, x_2, \ldots, x_n) = u_0 + \sum_{i=1}^{n} u_i x_i + \sum_{i=1}^{n}\sum_{j=1}^{n} u_{ij}x_i x_j$$
$$+ \cdots + u_{12\ldots n}x_1 x_2 \cdots x_n$$

with binary coefficients (see *algebraic normal* form in Boolean functions).

One can determine the linear span obtained by combining n different m-sequences $\{a_t\}, \{b_t\}, \ldots, \{c_t\}$ with characteristic polynomials of pair-wise relative prime degrees $e_1, e_2, \ldots, e_n$ using a Boolean combining function. From the Boolean function $f(x_1, x_2, \ldots, x_n)$ we construct a sequence $w_t = f(a_t, b_t, \ldots, c_t)$. Then the linear span of the combined sequence is equal to $f(e_1, e_2, \ldots, e_n)$, evaluated over the integers.

It is important in applications of sequences in communication systems as well as in stream cipher systems to generate sequences with good auto- and cross-correlation properties.

Let $\{u(t)\}$ and $\{v(t)\}$ be two binary sequences of period e. The cross-correlation of the sequences $\{u(t)\}$ and $\{v(t)\}$ at shift τ is defined as

$$C_{u,v}(\tau) = \sum_{t=0}^{e-1}(-1)^{u_{t+\tau}-v_t}$$

where the sum $t + \tau$ is computed modulo e. In the case when the two sequences are the same, we denote this by the auto-correlation at shift τ.

For synchronization purposes one prefers sequences with low absolute values of the maximal out-of-phase auto-correlation, i.e., $|C_{u,u}(\tau)|$ should be small for all values of $\tau \neq 0 \pmod{e}$.

Let $\mathcal{F}$ be a family consisting of M sequences

$$\mathcal{F} = \{s_i(t) : i = 1, 2, \cdots, M\},$$

where each sequence $\{s_i(t)\}$ has period e.

The cross-correlation between two sequences $\{s_i(t)\}$ and $\{s_j(t)\}$ at shift τ is denoted by $C_{i,j}(\tau)$.

In Code-Division Multiple-Access (CDMA) applications it is desirable to have a family of sequences with certain properties. To facilitate synchronization, it is desirable that all the out-of-phase auto-correlation values ($i = j, \tau \neq 0$) are small. To minimize the interference due to the other users in a multiple access situation, the cross-correlation values ($i \neq j$) must also be kept small. For this reason the family of sequences should be designed to minimize

$$C_{max} = max\{|C_{i,j}| : 1 \le i, j \le M,$$
$$\text{and either } i \neq j \text{ or } \tau \neq 0\}.$$

For practical applications in communication systems one needs a family $\mathcal{F}$ of sequences of period e, such that the number of users $M = |\mathcal{F}|$ is large and simultaneously C_{max} is small. Also in stream ciphers it is of importance that the generated sequences have good auto-correlation and cross-correlation properties.

Tor Helleseth

References

[1] Golomb, S.W. (1982). *Shift Register Sequences*. Aegean Park Press, Laguna Hills, CA.

[2] Helleseth, T. and P.V. Kumar (1999). "Pseudonoise sequences." *The Mobile Communications Handbook*, ed. J.D. Gibson. Chapter 8. CRC/IEEE Press, Boca Raton, FL/Piscataway, NJ.

[3] Helleseth, T. and P.V. Kumar (1998). "Sequences with low correlation." *Handbook of Coding Theory*, eds. V.S. Pless and W.C. Huffman. Chapter 21. Elsevier, Amsterdam.

[4] Selmer, E.S. (1966). *Linear Recurrence Relations over Finite Fields*. Lecture Notes, Department of Mathematics, University of Bergen, Norway.

[5] Zierler, N. (1959). "Linear recurring sequences." *J. Soc. Indust. Appl. Math.*, 7, 31–48.

SERPENT

Serpent is a 128-bit block cipher designed by Anderson et al. and first published in 1998 [1]. Later that year the cipher was slightly modified [2] and proposed as a candidate for the *Advanced Encryption Standard* (Rijndael/AES). In 1999 it was selected as one of the five finalists of the AES competition.

Serpent is a 32-round substitution–permutation (SP) network operating on 128-bit blocks. Each round consists of a key mixing operation, a layer of 32 copies of a 4×4-bit S-box, and (except in the last round) a linear transformation. The replicated

S-box differs from round to round and is selected from a set of eight different S-boxes. The last (incomplete) round is followed by a final key mixing operation. An additional bit permutation before the first round and after the last key mixing layer is applied to all data entering and leaving the SP network. The 128-bit subkeys mixed with the data in each round are generated by linearly expanding a 128-bit, 192-bit, or 256-bit secret key, and passing the result through a layer of S-boxes.

The initial and final permutations, the S-boxes, and the linear transformation have all been designed in order to allow an optimized implementation in software using the "bitslice" technique [3]. The idea is to construct a complete description of the cipher using only logical bit-operations (as in hardware) and then execute 32 (or 64) operations in parallel on a 32-bit (or 64-bit) processor.

Serpent is considered to have a rather high security margin. The best attacks published so far break about 1/3 of the rounds. Kelsey, Kohno, and Schneier [7] presented a first attack breaking 9 rounds with a time complexity slightly faster than exhaustive key search. This amplified boomerang attack was improved and extended by one round by Biham et al. [5]. The best attacks so far are the linear and the differential-linear attacks presented in [4] and [6]. Both break 11 rounds out of 32.

Christophe De Cannière

References

[1] Anderson, R.J., E. Biham, and L.R. Knudsen (1998). "Serpent: A new block cipher proposal." *Fast Software Encryption, FSE'98*, Lecture Notes in Computer Science, vol. 1372, ed. S. Vaudenay. Springer-Verlag, Berlin, 222–238.

[2] Anderson, R.J., E. Biham, and L.R. Knudsen (1998). "Serpent: A proposal for the advanced encryption standard." *Proceedings of the First AES Candidate Conference. National Institute of Standards and Technology, August.*

[3] Biham, E. (1997). "A fast new DES implementation in software." *Fast Software Encryption, FSE'97*, Lecture Notes in Computer Science, vol. 1267, ed. E. Biham. Springer-Verlag, Berlin, 260–272.

[4] Biham, E., O. Dunkelman, and N. Keller (2002). "Linear cryptanalysis of reduced round Serpent." *Fast Software Encryption, FSE 2001*, Lecture Notes in Computer Science, vol. 2355, ed. M. Matsui. Springer-Verlag, Berlin, 16–27.

[5] Biham, E., O. Dunkelman, and N. Keller (2002). "New results on boomerang and rectangle attacks." *Fast Software Encryption, FSE 2002*, Lecture Notes in Computer Science, vol. 2365, eds. J. Daemen and V. Rijmen. Springer-Verlag, Berlin, 1–16.

[6] Biham, E., O. Dunkelman, and N. Keller (2003). "Differential-linear cryptanalysis of Serpent." *Fast Software Encryption, FSE 2003*, Lecture Notes in Computer Science, vol. 2887, ed. T. Johansson. Springer-Verlag, Berlin, 9–21.

[7] Kelsey, J., T. Kohno, and B. Schneier (2001). "Amplified boomerang attacks against reduced-round MARS and Serpent." *Fast Software Encryption, FSE 2000*, Lecture Notes in Computer Science, vol. 1978, ed. B. Schneier. Springer-Verlag, Berlin, 75–93.

SET

SET (Secure Electronic Transactions) is a standard for a payment protocol for credit card payments over the Internet and was developed in 1996–97 as a joint initiative of MasterCard, VISA, IBM, Microsoft, Netscape and others as a more secure alternative to Secure Socket Layer SSL, which never really caught on.

SET assumes the existence of appropriate infrastructure within the card organisation, and entails the communication between the registered Payer (cardholder), Payee (merchant) and the Payment Gateway Provider, i.e. the Acquirer or a Payment Service Provider. The main purpose of the protocol is to secure this communication in such a way that neither the Payee, nor the Payment Gateway Provider can access all purchase transaction details. Thus the Payee has access to the order information only and not the credit card details, while the Payment Gateway Provider has access to the payment information only.

SET is a PKI-solution (see public key infrastructure). The Certificate Authority (CA) hierarchy consists of a Root CA that signs the certificates of each of the credit card brand CA's. These CA's sign certificates for the Cardholder CA (the Card Issuer), Merchant CA (the Customer Acquirer) and the Payment CA. These CA's then in turn sign the certificates for the cardholder, merchant, and payment gateway provider, respectively, using the X.509 v3 format. Neither cardholder's name, nor card number are shown in the certificates. Rather a number is used that has been computed from the credit card number and other input by the Issuer.

In short, the protocol works as follows: the cardholder indicates that he wants to initiate payment for his order. The merchant then identifies himself with his certificate and provides the cardholder with the public key of the payment gateway provider. The cardholder encrypts the payment information using this key, thus ensuring the merchant cannot access this information and signs the payment instruction. The merchant forwards the payment information to the payment gateway provider in an authorization request, and the payment gateway provider verifies the content and authorizes accordingly.

Peter Landrock

SHA FAMILY (SECURE HASH ALGORITHM)

The SHA (Secure Hash Algorithm) Family designates a family of six different hash functions: SHA-0, SHA-1, SHA-224, SHA-256, SHA-384, and SHA-512 [7, 8]. They take variable length input messages and hash them to fixed-length outputs. The first four operate on 512-bit message blocks divided into 32-bit words and the last two on 1024-bit blocks divided into 64-bit words. SHA-0 (the first version of SHA since replaced by SHA-1) and SHA-1 produce a message digest of 160 bits, SHA-224 of 224 bits, SHA-256 of 256 bits, SHA-384 of 384 bits and SHA-512 of 512 bits respectively. All six functions start by padding the message according to the so-called Merkle-Damgård strengthening technique. Next, the message is processed block by block by the underlying compression function. This function initializes an appropriate number of chaining variables to a fixed value to hash the first message block, and to the current hash value for the following message blocks. Each step i of the compression function updates in turn one of the chaining variables according to one message word W_i. As there are more steps in the compression function than words in a message block, an additional message schedule is applied to expand the message block. In the last step, the initial value of the chaining variable is added to each variable to form the current hash value (or the final one if no more message blocks are available). The following provides an overview of SHA-1, SHA-256, and SHA-512. SHA-0 is almost identical to SHA-1, SHA-224 to SHA-256 and SHA-384 to SHA-512.

PADDING: The message is appended with a binary one and right-padded with a variable number of zeros followed by the length of the original message coded over two binary words. The total padded message length must be a multiple of the message block size.

SHA-1 Compression Function

Five 32-bit chaining variables A, B, C, D, E are either initialized to

$$A \leftarrow IV_1 = 67452301_x$$
$$B \leftarrow IV_2 = \text{EFCDAB89}_x$$
$$C \leftarrow IV_3 = 98\text{BADCFE}_x$$
$$D \leftarrow IV_4 = 10325476_x$$
$$E \leftarrow IV_5 = \text{C3D2E1F0}_x$$

for the first 512-bit message block or to the current hash value for the following message blocks. The first sixteen words of the message schedule are initialized to input message words. The following 64 message schedule words W_i are computed as

$$W_i \leftarrow (W_{i-3} \oplus W_{i-8} \oplus W_{i-14} \oplus W_{i-16}) \lll 1,$$
$$16 \le i \le 79$$

where "$\oplus$" represents bit-wise exclusive-or, and "$X \lll n$" is the cyclic rotation of X to the left by n bits. Then the compression function works as follows:

$$
\begin{aligned}
&\text{for } i = 0 \text{ to } 79 \text{ do} \\
&\quad T \leftarrow W_i + A \lll 5 + f_i(B, C, D) + E + K_i \\
&\quad\quad \mod 2^{32} \\
&\quad B \leftarrow A \\
&\quad C \leftarrow B \lll 30 \\
&\quad D \leftarrow C \\
&\quad E \leftarrow D \\
&\quad A \leftarrow T
\end{aligned}
$$

where the nonlinear functions f_i are defined by

$$f_{if}(X, Y, Z) =$$
$$(X \land Y)|(\neg X \land Z), \quad 0 \le i \le 19$$
$$f_{xor}(X, Y, Z) =$$
$$(X \oplus Y \oplus Z), \quad 20 \le i \le 39, \quad 60 \le i \le 79$$
$$f_{maj}(X, Y, Z) =$$
$$((X \land Y)|(X \land Z)|(Y \land Z), \quad 40 \le i \le 59$$

and the constants K_i by

$$K_i \leftarrow 5\text{A}827999_x, 0 \le i \le 19$$
$$K_i \leftarrow 6\text{ED9EBA1}_x, 20 \le i \le 39$$
$$K_i \leftarrow 8\text{F1BBCDC}_x, 40 \le i \le 59$$
$$K_i \leftarrow \text{CA62C1D6}_x, 60 \le i \le 79.$$

After 80 steps, the output value of each chaining variable is added to the previous intermediate hash value according to the Davies–Meyer construction to give the new intermediate hash value. When all consecutive message blocks have been hashed, the last intermediate hash value is the final overall hash value.

SHA-0

The only difference between SHA-1 and SHA-0 is the fact that there is no left rotation by one bit in the message schedule of SHA-0. In other words, the 64 message schedule words W_i for SHA-0 are computed as

$$W_i \leftarrow W_{i-3} \oplus W_{i-8} \oplus W_{i-14} \oplus W_{i-16}, \ 16 \le i \le 79.$$

SHA-256 and SHA-512 Compression Functions

Eight chaining variables A, B, C, D, E, F, G, H are initialized to fixed values H_0 to H_7 for the first message block, and to the current intermediate hash value for the following blocks. The first sixteen w-bit words (where $w = 32$ for SHA-256 and $w = 64$ for SHA-512) of the message schedule are initialized to the input message words. The following $r - 16$ (where $r = 64$ for SHA-256 and $r = 80$ for SHA-512) message schedule words W_i are computed as

$$W_i = \sigma_1(W_{i-2}) + W_{i-7} + \sigma_0(W_{i-15})$$
$$+ W_{i-16} \bmod 2^w, \qquad 16 \le i \le r - 1$$

where σ_0 and σ_1 represent linear combinations of three rotated values of the input variable. Then the compression function works as follows:

```
for i = 0 to r do
    T_1  ←  H + Σ_1(E) + f_if(E, F, G) + K_i
            + W_i mod 2^w
    T_2  ←  Σ_0 (A) + f_maj(A, B, C)  mod 2^w
    H ← G
    G ← F
    F ← E
    E ← D + T_1  mod 2^w
    D ← C
    C ← B
    B ← A
    A ← T_1 + T_2  mod 2^w
```

where Σ_0 and Σ_1 again represent linear combinations of three rotated values of the input variable and K_i is a different w-bit constant for each step i. Finally, the output value of each chaining variable is added to the previous intermediate hash value according to the Davies–Meyer construction to give the new intermediate hash value. When all consecutive message blocks have been hashed, the last intermediate hash value is the final overall hash value.

SHA-224

The SHA-224 hash computations are exactly the same as those of SHA-256, up to the following two differences: the constants H_0 to H_7 used in SHA-224 are not the same as those used in SHA-256, and the SHA-224 output is obtained by truncating the final overall hash value to its 224 leftmost bits.

SHA-384

The SHA-384 hash computations are exactly the same as those of SHA-512, up to the following two differences: the constants H_0 to H_7 used in SHA-384 are not the same as those used in SHA-512, and the SHA-384 output is obtained by truncating the final overall hash value to its 6 leftmost words.

SECURITY CONSIDERATION: All six SHA functions belong to the MD4 type hash functions and were introduced by the American National Institute for Standards and Technology (NIST). SHA was published as a Federal Information Processing Standard (FIPS) in 1993. This early version is known as SHA-0. In 1994, a minor change to SHA-0 was made, and published as SHA-1 [7]. SHA-1 was subsequently standardized by ISO [5]. The following generation of SHA functions with much larger message digest sizes, namely 256, 384, and 512 bits, was introduced in 2000 and adopted as a FIPS standard in 2002 [8] as well as an ISO standard in 2003 [5]. The latest member of the family, namely SHA-224, was adopted in a Change Notice to FIPS 180-2 in 2004. This latter generation of hash functions provides theoretical security levels against collision search attacks which are consistent with the security levels expected from the three standard key sizes of the *Advanced Encryption Standard* (see Rijndael/AES) (128, 192, and 256 bits). The first attack known on SHA-0 is by Chabaud and Joux [2]. They show that in about 2^{61} evaluations of the compression function it is possible to find two messages hashing to the same value whereas a brute-force attack exploiting the birthday paradox requires about 2^{80} evaluations in theory. In 2004, Biham and Chen introduce the neutral bit technique and find near-collisions on the compression function of SHA-0 [1] as well as collisions on reduced-round versions of SHA-1. In August 2004, Joux, Carribault, Jalby and Lemuet [6] first provide a full collision on SHA-0 using two four-block messages and requiring a complexity of 2^{51} compression function computations. In February 2005, Wang, Yin and Yu [10] announce full collisions on SHA-0 in 2^{39} hash operations and report that collisions on SHA-1 can be obtained in less than 2^{69} hash operations. Saarinen [9] addresses the existence of slid pairs in SHA-1. The first security analysis on SHA-256, SHA-384 and SHA-512 in 2003 is by Gilbert and Handschuh [3]. They show that collisions can be found with a reduced work factor for weakened variants of these functions. Subsequently, Hawkes and Rose show that second pre-image attacks are far easier than expected on SHA-256 [4]. However these observations do not lead to actual attacks in 2004.

Helena Handschuh

References

[1] Biham, E. and R. Chen (2004). "Near-collisions of SHA-0." *Advances in Cryptology—CRYPTO 2004*, Lecture Notes in Computer Science, vol. 3152, ed. M. Franklin. Springer-Verlag, Berlin, 290–305.

[2] Chabaud, F. and A. Joux (1998). "Differential collisions in SHA-0." *Advances in Cryptology—CRYPTO'98*, Lecture Notes in Computer Science, vol. 1462, ed. H. Krawczyk. Springer-Verlag, Berlin, 56–71.

[3] Gilbert, H. and H. Handschuh (2004). "Security analysis of SHA-256 and sisters." *Selected Areas in Cryptography—SAC 2003*, Lecture Notes in Computer Science, vol. 3006, eds. M. Matsui and R. Zuccherato. Springer-Verlag, Berlin, 175–193.

[4] Hawkes P. and G. Rose (2004). On Corrective Patterns for the SHA-2 Family. http://eprint.iacr.org/2004/207.

[5] ISO/IEC 10118-3 (2003). "Information technology—security techniques—hash-functions—Part 3: Dedicated hash-functions."

[6] Joux, A., P. Carribault, W. Jalby, and C. Lemuet (2004). "Collisions in SHA-0." Presented at the rump session of *CRYPTO 2004*, August.

[7] National Institute of Standards and Technology (NIST). (1995). FIPS Publication 180-1: Secure Hash Standard.

[8] National Institute of Standards and Technology (NIST). (2002). FIPS Publication 180-2: Secure Hash Standard.

[9] Saarinen, M.-J.O. (2003). "Cryptanalysis of block ciphers based on SHA-1 and MD5." *Fast Software Encryption—FSE 2003*, Lecture Notes in Computer Science, vol. 2887, ed. Thomas Johansson. Springer-Verlag, Berlin, 36–44.

[10] Wang, X., Y.-L. Yin, and H. Yu (2005). Collision Search Attacks on SHA-1. Unpublished manuscript.

SHAMIR'S THRESHOLD SCHEME

In [1], Shamir proposed an elegant "polynomial" construction of a perfect threshold schemes (see threshold cryptography). An (n, k)-threshold scheme is a particular case of secret sharing scheme when any set of k or more participants can recover the secret exactly while any set of less than k participants gains no additional, i.e. a posteriori, information about the secret. Such threshold schemes are called *perfect* and they were constructed in [2] and [1]. Shamir's construction is the following.

Assume that the set S_0 of secrets is some finite field $GF(q)$ of q elements (q should be prime power) and that the number of participants of SSS $n < q$. The dealer chooses n different nonzero elements (points) $x_1, \ldots, x_n \in GF(q)$, which are publicly known. To distribute a secret s_0 the dealer generates randomly coefficients $g_1, \ldots, g_{k-1} \in GF(q)$, forms the polynomial $g(x) = s_0 + g_1 x + \cdots + g_{k-1} x^{k-1}$ of degree less than k and sends to the i-th articipant the share $s_i = g(x_i)$. Clearly any k participants can recover the whole polynomial $g(x)$ and, in particular, its zero coefficient (or $g(0)$), since any polynomial of degree l is uniquely determined by its values in $l + 1$ points and Lagrange interpolation formula shows how to determine it. On the other hand, the point 0 can be considered as an "evaluation point" x_0, corresponding to the dealer, since $s_0 = g(0)$. Then the above consideration shows that for any given shares $s_1 = g(x_1), \ldots, s_{k-1} = g(x_i)$ all possible values of s_0 are equally probable, hence the scheme is perfect.

For some applications it is convenient to have the maximal possible number n of participants equal to q, especially for $q = 2^m$. For Shamir's scheme $n < q$ but the following simple modification allows to have $n = q$. Namely, the dealer generates a random polynomial of the form $f(x) = f_0 + f_1 x + \cdots + f_{k-2} x^{k-2} + s_0 x^{k-1}$ and distribute shares $s_i = f(x_i)$, where the x_i are different but not necessary nonzero elements of $GF(q)$. The perfectness of this scheme can be proved either directly (along the line of the above proof by considering the polynomial $h(x) = f(x) - s_0 x^{k-1}$ of degree at most $k - 2$), or as an application of established in [3] the relationship between perfect (n, k)-threshold schemes and $(n + 1, k)$ Reed–Solomon codes (see cyclic codes), since the above construction is equivalent to so-called 2-lengthening of Reed–Solomon codes.

Robert Blakley
Gregory Kabatiansky

References

[1] Shamir, A. (1979). "How to share a secret." *Communications of the ACM*, 22 (1), 612–613.

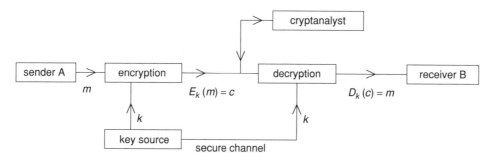

Fig. 1. The conventional cryptosystem

[2] Blakley, R. (1979). "Safeguarding cryptographic keys." *Proceedings of AFIPS 1979 National Computer Conference*, 48, 313–317.

[3] McEliece, R.J. and D.V. Sarwate (1981). "On secret haring and Reed–Solomon codes." *Communications of the ACM*, 24, 583–584.

SHANNON'S MODEL

Although <u>symmetric cryptosystems</u> have been around for at least two thousand years (see for instance <u>Caesar cipher</u>), it was only in 1949 that Claude Shannon gave a formal mathematical description of these systems [1].

In his description, a sender A (often called Alice) wants to send a message m to a receiver B (who is called Bob). The message is called a *plaintext* and is taken from a finite set, called plaintext space $\mathcal{M}$. Of course, Alice may send more messages.

Since the transmission channel is insecure (a person called Eve is also connected to the channel), Alice applies a mapping E_k to m. The result c is called the *ciphertext* and is an element of a set $\mathcal{C}$, the ciphertext space. The mapping E_k is called the <u>encryption</u> function. It is c that Alice sends to Bob and so it will be c that is intercepted by Eve.

Clearly, the encryption function E_k must be a one-to-one mapping, since Bob must be able to retrieve the plaintext/message m from the ciphertext c by means of the *decryption* function D_k. In formula: $D_k(c) = m$.

Since more people may want to use the same cryptosystem and since Alice and Bob do not want to use the same mapping too long for security reasons, their function is taken from a large set $\mathcal{E}$ of one-to-one mappings from $\mathcal{M}$ to $\mathcal{C}$. It is for this reason that the encryption and decryption functions carry a label k. This k is called the <u>key</u> and is taken from the so-called *key-space* $\mathcal{K}$. It is the set $\mathcal{E} = \{E_k \mid k \in \mathcal{K}\}$ that describes the cryptosystem. Quite clearly Alice and Bob must use the same key k. To this end, they use a *secure channel*, a communication line without any eavesdroppers. A possibility is that they agreed beforehand on the key, another possibility is that one has sent the key by means of a courier to the other. Nowadays <u>public key cryptography</u> is often used for this purpose.

Normally, the same cryptosystem $\mathcal{E}$ will be used for a long time and by many people, so it is reasonable to assume that $\mathcal{E}$ is also known to the cryptanalyst. It is the frequent changing of the key that has to provide the security of the data. This principle was already clearly stated by the Dutchman Auguste Kerckhoff (see <u>maxims</u>) in the 19th century.

Often $\mathcal{M} = \mathcal{C}$ in which case one wants the number of plaintexts that are mapped to a particular ciphertext (under different keys) to be the same. In that case the ciphertext does not give any information about the plaintext (see <u>information theory</u>).

The cryptanalyst who is connected to the transmission line can be:

Passive (eavesdropping): The cryptanalyst tries to find m (or even better k) from c.

Active (tampering): The cryptanalyst tries to actively manipulate the data that are being transmitted. For instance, she alters a transmitted ciphertext.

Henk van Tilborg

Reference

[1] Shannon, C.E. (1949). "Communication theory and secrecy systems." *Bell Systems Technical Journal*, 28, 656–715.

SHARE

Share is a portion of information distributed by a <u>secret sharing scheme</u> (SSS) to a given user. In the standard definition of SSS, shares are distributed via secure, private channels in such a way that each participant only knows his own share [1, 2]. We note that it is also possible to organize SSS in case of public channels [3].

Robert Blakley
Gregory Kabatiansky

References

[1] Shamir, A. (1979). "How to share a secret." *Communications of the ACM*, 22 (1), 612–613.

[2] Blakley, R. (1979). "Safeguarding cryptographic keys." *Proceedings of AFIPS 1979 National Computer Conference*, 48, 313–317.

[3] Beimel, A. and B. Chor (1998). "Secret sharing with public reconstruction." *IEEE Transactions on Informatiom Theory*, 44 (5), 1887–1896.

SHORTEST VECTOR PROBLEM

The Shortest Vector Problem (SVP) is the most famous and widely studied computational problem on lattices. Given a lattice $\mathcal{L}$ (typically represented by a basis), SVP asks to find the shortest nonzero vector in $\mathcal{L}$. The problem can be defined with respect to any norm, but the Euclidean norm is the most common (see the entry lattice for a definition). A variant of SVP (commonly studied in computational complexity theory) only asks to compute the length (denoted $\lambda(\mathcal{L})$) of the shortest nonzero vector in $\mathcal{L}$, without necessarily finding the vector.

SVP has been studied by mathematicians (in the equivalent language of quadratic forms) since the 19th century because of its connection to many problems in the number theory. One of the earliest references to SVP in the computer science literature is [7], where the problem is conjectured to be NP-hard.

A cornerstone result about SVP is *Minkowski's first theorem*, which states that the shortest nonzero vector in any n-dimensional lattice has length at most $\gamma_n \det(L)^{1/n}$, where γ_n is an absolute constant (approximately equal to $\sqrt{n}$) that depends only of the dimension n, and $\det(L)$ is the determinant of the lattice (see the entry lattice for a definition).

The upper bound provided by Minkowski's theorem is tight, i.e., there are lattices such that the shortest nonzero vector has length $\gamma_n \det(L)^{1/n}$. However, general lattices may contain vectors much shorter than that. Moreover, Minkowski's theorem only proves that short vectors exist, i.e., it does not give an efficient algorithmic procedure to find such vectors. An algorithm to find the shortest nonzero vector in two-dimensional lattices was already known to Gauss in the 19th century, but no general methods to efficiently find (approximately) shortest vectors in n-dimensional lattices were known until the early 1980s. A g-approximation algorithm for SVP is an algorithm that on input a lattice $\mathcal{L}$, outputs a nonzero lattice vector of length at most g times the length of the shortest vector in the lattice. The LLL lattice reduction algorithm ([4], see lattice reduction) can be used to approximate SVP within a factor $g = O((2/\sqrt{3})^n)$ where n is the dimension of the lattice. Smaller approximation factors (slightly subexponential in n—see subexponential time for a definition) can be achieved in polynomial time using more complex algorithms like Schnorr's *Block Korkine–Zolotarev* reduction [6].

No efficient (polynomial time) algorithm to compute the length of the shortest vector in a lattice is known to date (leave alone actually finding the shortest vector). The NP-hardness of SVP (in the Euclidean norm) was conjectured by van Emde Boas in 1981 [7]. The conjecture remained wide open until 1997, when Ajtai proved that SVP is NP-hard to solve exactly under randomized reductions [1]. The strongest NP-hardness result for SVP known to date is due to Micciancio [5], who showed that SVP is NP-hard even to approximate within any factor less than $\sqrt{2}$. Stronger (but still subpolynomial) inapproximability results are known for SVP in the ℓ_∞ norm [2]. On the other hand, Goldreich and Goldwasser [3] have shown that (under standard complexity assumptions) SVP cannot be NP-hard to approximate within small polynomial factors $g = O(\sqrt{n/\log n})$.

As is the case with the related Closest Vector Problem, finding a good approximation algorithm (i.e., a polynomial-time algorithm with polynomial approximation factors) is one of the most important open questions in the area. Indeed, the hardness of approximating SVP within certain polynomial factors can be used as the basis for the construction of provably secure cryptographic functions (see lattice based cryptography).

Daniele Micciancio

References

[1] Ajtai, M. (1998). "The shortest vector problem in L_2 is NP-hard for randomized reductions (extended abstract)." *Proceedings of the Thirtieth Annual ACM Symposium on Theory of Computing—STOC'98, Dallas, TX*. ACM Press, New York, 10–19.

[2] Dinur, I. (2000). "Approximating SVP_∞ to within almost-polynomial factors is NP-hard." *Proceedings of the 4th Italian Conference on Algorithms and Complexity—CIAC 2000, Rome*, Lecture Notes in Computer Science, vol. 1767, eds. G. Bongiovanni, G. Gambosi, and R. Petreschi. Springer, Berlin, 263–276.

[3] Goldreich, O. and S. Goldwasser (2000). "On the limits of nonapproximability of lattice problems." *Journal of Computer and System Sciences*, 60 (3), 540–563. Preliminary version in *STOC'98*.

[4] Lenstra, A.K., H.W. Lenstra, Jr., and L. Lovász (1982). "Factoring polynomials with rational coefficients." *Mathematische Annalen*, 261, 513–534.

[5] Micciancio, D. "The shortest vector problem is NP-hard to approximate to within some constant." *SIAM Journal on Computing*, 30 (6), 2008–2035. Preliminary version in *FOCS'98*.

[6] Schnorr, C.-P. (1987). "A hierarchy of polynomial time lattice basis reduction algorithms." *Theoretical Computer Science*, 53 (2–3), 201–224.

[7] van Emde Boas, P. (1981). "Another NP-complete problem and the complexity of computing short vectors in a lattice." Technical Report 81-04, Mathematische Instituut, University of Amsterdam, Available on-line at URL http://turing.wins.uva.nl/peter/

SHRINKING GENERATOR

The shrinking generator is a <u>clock-controlled generator</u> that has been proposed in 1993 [1]. It is based on two <u>Linear Feedback Shift Registers (LFSRs)</u>, say R_1 and R_2. The idea is that R_1's output will *decimate* R_2's output. At each step, both are clocked; if R_1 output a 1, then R_2's output bit is included in the keystream, else (if R_1 outputs a 0) R_2's output bit is discarded.

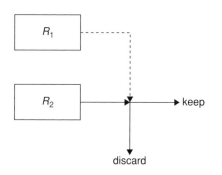

EXAMPLE. Let us consider R_1 of length three, with the feedback relation $s_{t+1} = s_t + s_{t-2}$, and R_2 of lenth four, with the feedback relation $s_{t+1} = s_t + s_{t-3}$. Then the following happens (the first row concerns only the initialization; the internal states are of the form $s_t s_{t-1} s_{t-2}$ or $s_t s_{t-1} s_{t-2} s_{t-3}$):

R_1		R_2		
State	Output	State	Output	Output
010		0101		
001	0	1010	1	
100	1	1101	0	0
110	0	0110	1	
111	0	0011	0	
011	1	1001	1	1
101	1	0100	1	1
010	1	0010	0	0
001	0	0001	0	
100	1	1000	1	1
110	0	1100	0	
111	0	1110	0	
011	1	1111	0	0
101	1	0111	1	1
⋮	⋮	⋮	⋮	⋮

The inventors discussed some security points in their paper. More recent results have been given in [2, 5]. A discussion about the implementation and the use of a buffer (in order to avoid the irregular rate of the output) is presented in [3] and [4].

Caroline Fontaine

References

[1] Coppersmith, D., H. Krawczyk, and Y. Mansour (1994). "The shrinking generator." *Advances in Cryptology—CRYPTO'93*, Lecture Notes in Computer Science, vol. 773, ed. D.R. Stinson. Springer-Verlag, Berlin, 22–39.

[2] Golic, J. (1995). "Intrinsic statistical weakness of keystream generators". *Advances in Cryptology—ASIACRYPT'94*, Lecture Notes in Computer Science, vol. 917, eds. J. Pieprzyk and R. Safari-Naini. Springer-Verlag, Berlin, 91–103.

[3] Kessler, I. and H. Krawczyk (1995). "Minimum buffer length and clock rate for the shrinking generator cryptosystem." IBM Research Report RC19938, IBM T.J. Watson Research Center, Yorktown Heights, NY, USA.

[4] Krawczyk, H. (1994). "The shrinking generator: Some practical considerations." *Fast Software Encryption*, Lecture Notes in Computer Science, vol. 809, ed. R.J. Anderson. Springer-Verlag, Berlin, 45–46.

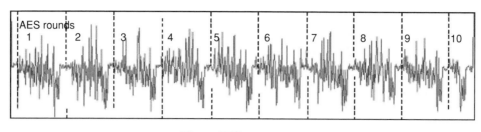

Fig. 1. AES power trace

[5] Johansson, T. (1998). "Reduced complexity correlation attack on two clock-controlled generators." *Advances in Crptology—ASIACRYPT'98*, Lecture Notes in Computer Science, vol. 1514, eds. K. Ohta and D. Pei. Springer-Verlag, Berlin, 342–356.

SIDE-CHANNEL ANALYSIS

INTRODUCTION: Electronic devices have to comply with consumption constraints especially on autonomous equipments, like mobile phones. Power analysis has been included into most certification processes regarding products dealing with information security such as smart cards.

The electrical consumption of any electronic device can be measured with a resistor inserted between the ground or Vcc pins and the actualground in order to transform the supplied current into a voltage easily monitored with an oscilloscope.

Within a micro-controller the peripherals consume differently. For instance writing into nonvolatile memory requires more energy than reading. Certain chips for smart cards enclose a crypto-processor, i.e., a particular device dedicated to specific cryptographic operations, which generally entails a consumption increase. The consumption trace of a program running inside a micro-controller or a microprocessor is full of information. The signal analysis may disclose lots of things about the used resources or about the process itself. This illustrates the notion of *side channel* as a source of additional information.

Basically a power consumption trace exhibits large scale patterns most often related to the structure of the executed code. The picture below (Figure 1) shows the power trace of a smart-card chip ciphering a message with the Advanced Encryption Standard (AES). The ten rounds are easily recognised with nine almost regular patterns first followed by a shorter one.

Zooming into a power signal exhibits a local behaviour in close relationship with the silicon technology. At the cycle scale, the consumption curve looks roughly like a capacitive charge and discharge response.

A careful study of several traces of a same code with various input data shows certain locations where power trace patterns have different heights. The concerned cycles indicate some data dependence also called *information leakage*. They may be magnified by a variance analysis over a large number of executions with random data. For instance, by ciphering many random plaintexts with a secret-key algorithm, it is possible to distinguish the areas sensitive to input messages from the constant areas that correspond to the key schedule.

INFORMATION LEAKAGE MODEL: The characterisation of data leakage (namely, finding the relationships between the data and the variability of consumption) has been investigated by several researchers. The most common model consists in correlating these variations to the Hamming weight of the handled data, i.e., the number of nonzero bits. Such a model is valid for a large number of

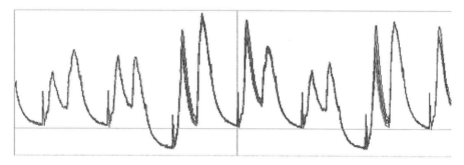

Fig. 2. Information leakage

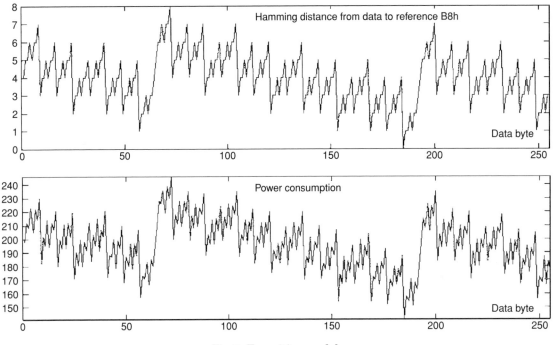

Fig. 3. Transition model

devices. However it can be considered as a special case of the transition model which assumes that the energy is consumed according to the number of bits switched for going from one state to the next one. This behaviour is represented by the Hamming distance between the data and some *a priori* unknown constant, i.e., the Hamming weight of the data XOR-ed with this constant.

As shown in the next picture (Figure 3), for an 8-bit micro-controller, the transition model may seem rough but it suffices to explain many situations, provided that the reference constant state is known. In most microprocessors this state is either an address or an operating code. Each of them has a specific binary representation and therefore a different impact in the power consumption: this is why each cycle pattern is most often different from its neighbours.

Some technologies systematically go through a clear "all-zeros" state that explains the simpler Hamming-weight model.

STATISTICAL ANALYSES: With information leakage models in mind, it is possible to designstatistical methods in order to analyse the data leakage. They require a large amount of power traces assigned to many executions of the same-code with varying data, generally at random, and make use of statistical estimators such as averages, variances and correlations. The most famous method is due to Paul Kocher et al. and is called <u>Differential Power Analysis</u> (DPA).

Basically the purpose of DPA is to magnify the effect of a single bit inside a machine word. Suppose that a random word in a Ω-bit processor is known and uniformly distributed. Suppose further that the associated power consumption obeys the Hamming-weight model. On average the Hamming weight of this word is $\Omega/2$. Given N words, two populations can be distinguished according to an arbitrary selection bit: the first population, $\mathcal{S}_0$, is the set of t words whose selection bit is 0 and the second population, $\mathcal{S}_1$, is the set

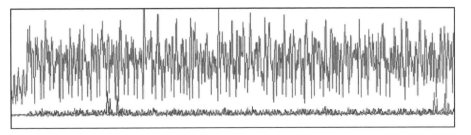

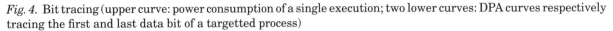

Fig. 4. Bit tracing (upper curve: power consumption of a single execution; two lower curves: DPA curves respectively tracing the first and last data bit of a targetted process)

of $N\!-\!t$ words whose selection bit is 1. On average, the words of set S_0 will have a Hamming weight of $(\Omega - 1)/2$ whereas the words of set S_1 will have a Hamming weight of $(\Omega + 1)/2$. The same bias can be seen through the corresponding power consumption traces since it is supposed to be correlated with the Hamming weight of the data. Let C_0 and C_1 respectively denote the averaged power consumption traces of the blue curvesets S_0 and S_1. The *DPA trace* is defined as the difference $C_0 - C_1$.

The resulting DPA curve has the property of erecting bias peaks at moments when the selection bit is handled. It looks like noise everywhere else: indeed, the constant components of the signal are cancelled by the subtraction whereas dynamic ones are faded by averaging, because they are not coherent with the selection bit.

This approach is very generic and applies to many situations. It works similarly with the transition model. Of course the weight of a single selection bit is relatively more important in processors with short words like 8-bit chips. If the machine word is larger, the same DPA bias can be obtained by increasing the number of trials.

A first application of DPA is called *bit tracing*. It is a useful reverse engineering tool for monitoring a predictable bit during the course of a process. In principle a DPA peak rises up each time it is processed. This brings a lot of information about an algorithm implementation. To achieve the same goal Paul Fahn and Peter Pearson proposed another statistical approach called *Inferential Power Analysis* (IPA). The bits are inferred from the deviation between a single trace and an average trace possibly resulting from the same execution: for instance the average trace of a DES round (see Data Encryption Standard) can be computed over its sixteen instances taken from a single execution. IPA does not require the knowledge of the random data to make a prediction on a bit value. But as counterpart it is less easy to implement and the interpretation is less obvious.

After Paul Kocher, Thomas Messerges et al. have proposed to extend DPA by considering multiple selection bits in order to increase the signal to noise ratio (SNR). If the whole machine word is taken into account, a global approach consists in considering the transition model as suggested by Jean-Sébastien Coron et al.

FROM POWER ANALYSIS TO POWER ATTACKS: Obviously, if the power consumption is sensitive to the executed code or handled data, critical information may leak through power analysis. This

```
k  ← bitsize(d)
y  ← x
for i = k − 2 downto 0 do
    y ← y² (mod n)
    if (bit i of d is 1) then y ← y · x (mod n)
endfor
return y
```

Fig. 5. Square-and-multiply exponentiation algorithm

section explains how to turn a side-channel analysis into an attack.

SPA-Type Attacks

A first type of power attacks is based on *Simple Power Analysis* (SPA). For example, when applied to an *unprotected* implementation of an RSA public key encryption scheme, such an attack may recover the whole private key (i.e., signing or decryption key) from a single power trace.

Suppose that a private RSA exponentiation, $y = x^d \bmod n$ (see modular arithmetic), is carried out with the square-and-multiply algorithm (see also exponentiation algorithms). This algorithm processes the exponent bits from left to right. At each step there is a squaring, and when the processed bit is 1 there is an additional multiplication. A straightforward (i.e., unprotected) implementation of the square-and-multiply algorithm is given in Figure 5.

The corresponding power curve exhibits a sequence of consumption patterns among which some have a low level and some have a high level. These calculation units are assigned to a cryptoprocessor handling n-bit arithmetic. Knowing that a low level corresponds to a squaring and that a high level corresponds to a multiplication, it is fairly easy to read the exponent value from the power trace:

- a low-level pattern followed by another low-level pattern indicates that the exponent bit is 0, and
- a low-level pattern followed by a high-level pattern indicates that the exponent bit is 1.

This previous picture also illustrates why the Hamming weight of exponent d can be disclosed by a timing measurement.

DPA-Type Attacks

Historically, DPA-type attacks—that is, power attacks based on *Differential Power Analysis* (DPA)—were presented as a means to retrieve the bits of a DES key.

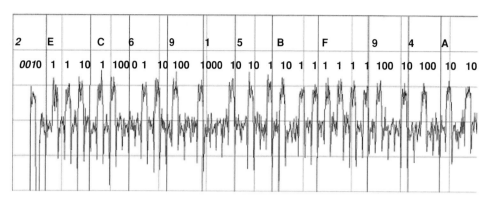

Fig. 6. SPA trace of the basic square-and-multiply algorithm

At the first round of DES, the output nibble of the ith S-box ($1 \leq i \leq 8$) can be written as $S_i(M \oplus K)$ where

- M is made of 6 bits constructed from the input message after IP- and E-permutations: it has to be chosen at random but is perfectly known and predictable, and
- K is a 6-bit sub-key derived from the key scheduling.

Rising up a DPA bias would require the knowledge of the output nibble. As K is unknown to the adversary this is not possible. But sub-key K can be easily exhausted as it can take only $2^6 = 64$ possible values. Therefore the procedure consists in reiterating the following process for $0 \leq \widehat{K} \leq 63$:

1. form sets $S_0 = \{M \mid g(\text{S-box}_i(M \oplus \widehat{K})) = 0\}$ and $S_1 = \{M \mid g(\text{S-box}_i(M \oplus \widehat{K})) = 1\}$ where selection function g returns the value of a given bit in the output nibble; and
2. compute the corresponding DPA curve.

In principle the bias peak should be maximised when the guess $\widehat{K}$ is equal to the real sub-key K. Then inverting the key schedule permutation leads to the value of 6 key bits. In other words the DPA operator is used to validate sub-key hypotheses. The same procedure applies to the 7 other S-boxes of the DES. Therefore the whole procedure yields $8 \times 6 = 48$ key bits. The 8 remaining key bits can be recovered either by exhaustion or by conducting a similar attack on the second round.

The main feature of a DPA-type attack resides in its genericity. Indeed it can be adapted to many cryptographic routines as soon as varying and known data are combined with secret data through logical or arithmetic operations.

A similar attack can be mounted against the first round of Rijndael/AES; the difference being that there are 16 byte-wise bijective substitutions and therefore 256 guesses for each. Finally, we note that DPA-type attacks are not limited to symmetric algorithms, they also apply to certain (implementations of) asymmetric algorithms, albeit in a less direct manner.

Other Attacks

Amongst the other statistical attacks, IPA is more difficult and less efficient. Its purpose is to retrieve key bits without knowing the processed data. It proceeds by comparing the power trace of a DES round with an average power trace computed for instance over the 16 rounds. In principle, key bits could be inferred this way because the differential curve should magnify the bits deviation where they are manipulated.

Dictionary (or template) attacks can be considered as a generalisation of IPA to very comfortable but realistic situations. They have been widely studied in the field of smart cards where information on secret key or personal identification numbers (PIN) could potentially be extracted. They consist in building a complete dictionary of all possible secret values together with the corresponding side-channel behaviour (e.g., power trace) when processed by the device (e.g., for authentication purpose). Then a secret value embedded in a twin device taken from the field can be retrieved by comparing its trace and the entries of the dictionary.

In practice, things do not happen that easily for statistical reasons and application restrictions. Only part of the secret is disclosed and the information leakage remains difficult to exploit fully.

Finally, in addition to power consumption, other side channels can be considered; possible sources of information leakage include running time or electro-magnetic radiation.

COUNTERMEASURES: The aforementioned attacks have all been published during the second half of the 1990s. In view of this new threat the manufacturers of cryptographic tokens have

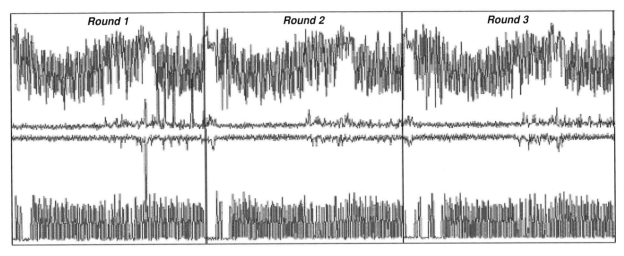

Fig. 7. DPA trace of the three first rounds of DES (two upper (respectively lower) curves: power consumption curve of maxima (respectively minima) of a single execution and DPA curve of maxima (respectively minima))

designed a large set of dedicated countermeasures especially to thwart statistical attacks like DPA. All the related research activity has now resulted in <u>tamper resistant</u> devices widely available in the market. It has given rise to the new concept of "secure implementation" which states that information leakage is not only due to the specification of an application (cryptographic processing or whatever), but also to the way it is implemented.

If information leaks through a physical side-channel there are two defensive strategies. The first consists in decorrelating the secret data from the side-channel. The second consists in decorrelating the side-channel from the secret data. The borderline between both is sometimes fuzzy but roughly speaking, the former is rather software oriented and intends to mask the data since they have to leak anyway, whereas the latter is more hardware oriented and intends to shut the side-channel physically in order to make the device tamper-resistant.

Chip manufacturers have introduced into their hardware designs many security features against power attacks. They are stricto sensu countermeasures since they aim at impeding the power measurement and make the recorded signal unworkable.

- Some countermeasures consist in blurring the signal using smoothing techniques, additive noise or desynchronisation effects. These countermeasures are poorly efficient against SPA working on broad scale traces. They are rather designed against statistical attacks. They may require some complementary circuits to generate parasitic components into the consumed current. Desynchronisation aims at misaligning a set of power traces by the means of unstable clocking or the insertion of dummy cycles at random, making the statistical combination of several curves ineffective.

- Other countermeasures rather intend to decrease or cancel the signal at the source. Reduction is a natural consequence of the shrinking trend in the silicon industry that diminishes the power consumption of each elementary gate. More interesting (and expensive) is the emerging technology called "precharged dual rail logic" where each bit is represented by a double circuitry. At a given time a logical 0 is represented physically by a 01, and a logical 1 by 10. The transition to the next time unit goes through a physical 00 or 11 state so that the same amount of switching occurs whatever the subsequent state is. Consequently if both rails are perfectly balanced, the overall consumption of a microprocessor does not depend on the data anymore.

Software countermeasures enclose a large variety of techniques going from the application level to the most specific algorithmic tricks. One can classify them into three categories: application constraints, timing counter-measures and data masking.

- Application constraints represent an obvious but often forgotten means to thwart statistical analyses. For instance DPA requires known data with a high variability. An application wherein an input challenge (or an output cryptogram) would be strictly formatted, partially visible and constrained to vary within hard limits (like a counter) would resist DPA fairly well.

- Timing countermeasures mean the usage of empirical programming tricks in order to tune the time progress of a process. A critical instruction may have its execution instant randomised by software: if it never occurs at the same time, statistical analysis becomes more difficult. Conversely other situations require the code to be executed in a constant time, in order to protect it from SPA or timing analysis. For instance a conditional branch may be compensated with a piece of fake code with similar duration and electrical appearance.

- Data masking (also known as whitening or randomization), covers a large set of numerical techniques designed by cryptographers and declined in various manners according to the algorithm they apply to. Their purpose is to prevent the data from being handled in clear and to disable any prediction regarding their behavior when seen through the side channel. For example, the modular exponentiation $y = x^d \bmod n$ (as used in the <u>RSA public key cryptosystem</u>) can be evaluated as:

$$y = \{(x + r_1 n)^{d + r_2 \varphi(n)} \bmod r_3 n\} \bmod n$$

for randoms r_i and where ϕ denotes <u>Euler totient function</u>.

To illustrate how fuzzy the borderline between hardware and software countermeasures can be, we have mentioned that for instance desynchronisation can be implemented by hardware or software means. The same remark applies to data masking for which some manufacturers have designed dedicated hardware tokens or mechanisms such as bus encryption or wired fast implementations of symmetric algorithms.

The experience shows that combined countermeasures act in synergy and increase the complexity in a much larger proportion than the sum of both. For instance the simple combination of desynchronisation tricks and data masking makes DPA (or more sophisticated variants thereof) quite harmless. In the same way, new hardware designs resist the most state-of-the-art and best equipped experts.

Marc Joye
Francis Olivier

References

[1] Chari, Suresh, Charanjit S. Jutla, Josyula R. Rao, and Pankaj Rohatgi (1999). "Towards sound approaches to counteract power-analysis attacks." *Advances in Cryptology—CRYPTO'99*, Lecture Notes in Computer Science, vol. 1666, ed. M. Wiener. Springer-Verlag, Berlin, 398–412.

[2] Coron, Jean-Sébastien, Paul Kocher, and David Naccache (2001). "Statistics and secret leakage." *Financial Cryptography (FC 2000)*, Lecture Notes in Computer Science, vol. 1962, ed. Y. Frankel. Springer-Verlag, Berlin, 157–173.

[3] Fahn, Paul N. and Peter K. Pearson (1999). "IPA: A new class of power attacks." *Cryptographic Hardware and Embedded Systems (CHES'99)*, Lecture Notes in Computer Science, vol. 1717, eds. Ç.K. Koç and C. Paar. Springer-Verlag, Berlin, 173–186.

[4] Gandolfi, Karine, Christophe Mourtel, and Francis Olivier (2001). "Electromagnetic analysis: Concrete results." *Cryptographic Hardware and Embedded Systems—CHES 2001*, Lecture Notes in Computer Science, vol. 2162, eds. Ç.K. Koç, D. Naccache, and C. Paar. Springer-Verlag, Berlin, 251–261.

[5] Kocher, Paul (1996). "Timing attacks on implementations of Diffie–Hellman, RSA, DSS, and other systems." *Advances in Cryptology—CRYPTO'96*, Lecture Notes in Computer Science, vol. 1109, ed. N. Koblitz. Springer-Verlag, Berlin, 104–113.

[6] Kocher, Paul, Joshua Jaffe, and Benjamin Jun (1999). "Differential power analysis." *Advances in Cryptology—CRYPTO'99*, Lecture Notes in Computer Science, vol. 1666, ed. M. Wiener. Springer-Verlag, Berlin, 388–397.

[7] Messerges, Thomas S. (2000). "Using second-order power analysis to attack DPA resistant software." *Cryptographic Hardware and Embedded Systems—CHES 2000*, Lecture Notes in Computer Science, vol. 1965, eds. Ç.K. Koç and C. Paar. Springer-Verlag, Berlin, 238–251.

[8] Messerges, Thomas S., Ezzy A. Dabbish, and Robert H. Sloan (2002). "Examining smart-card security under the threat of power analysis attacks." *IEEE Transactions on Computers*, 51 (5), 541–552.

SIDE-CHANNEL ATTACKS

Side-Channel Attacks or Environmental Attacks of cryptographic modules exploit characteristic information extracted from the implementation of the cryptographic primitives and protocols. This characteristic information can be extracted from timing, power consumption, or electromagnetic radiation features (see <u>tempest</u>). Other forms of side-channel information can be a result of hardware or software faults, computational errors, and changes in frequency or temperature. Side-channel attacks make use of the characteristics of the hardware and software elements as well as the implementation structure of the cryptographic primitive. Therefore, in contrast to analyzing the mathematical structure and properties of the cryptographic primitives only, <u>side-channel analysis</u> also includes the implementation. Some implementations are more vulnerable to specific side-channel attacks than others. Examples of attacks based on side-channel analysis are Differential Power Attacks examining power traces (see <u>Differential Power Analysis</u>), Timing Attacks measuring the amount of time used to complete cryptographic operations (see Timing Attack), and Fault Induction Attacks exploiting errors in the computation process of cryptographic primitives (see Fault Attacks).

Tom Caddy

SIEVING

Sieving refers to a process for selecting candidates for further processing among a set of elements. The "sieve" is the test that an element must pass to be considered further.

In many cases, by employing arithmetic progressions, it is possible to identify multiple candidates from the set more efficiently than if each element were tested separately. (Indeed, sometimes the term "sieving" refers only to this speedup.) For instance, in the *Sieve of Eratosthenes* (see prime number), candidate primes are selected from a range of integers by crossing off elements divisible by small primes $2, 3, 5, 7, 11, 13, \ldots$. Crossing off every second, third, fifth, seventh element and so on is generally faster than testing each element separately for divisibility by small primes.

Sieving is the first and major phase of the fastest general algorithms for integer factoring and for solving the discrete logarithm problem. Here, the candidates sought are those that are divisible only by small primes or their equivalent (see smoothness and factor base). Specific examples of sieving are described further in the entries Number Field Sieve, Quadratic Sieve, sieving in function fields, and index calculus. See also TWIRL for a recent design for efficient sieving in hardware.

Burt Kaliski

SIEVING IN FUNCTION FIELDS

Function fields are analogous constructions to number fields, where the role of the integers is replaced by polynomials. The coefficients of these polynomials are elements of finite fields for all cryptographically relevant applications. But in contrast to number fields, function fields over finite fields (so they are called) have interesting properties, notably concerning smoothness of elements, an important topic for sieving.

Notably, there exists a provable bound for the necessary size of a *factor base*, a set of elements generating a larger, targeted set of elements to be factored. This is due to the fact that the analog of

the Riemann hypothesis has been proven in the function field case. The most important application to cryptography (although a theoretical result) is the existence of a provable subexponential-time algorithm by Adleman et al. in 1992 [1] for solving the discrete logarithm problem in the Jacobian of a hyperelliptic curve (in short called hyperelliptic cryptosystems), a generalization of the group of points of an elliptic curve, and an analog to the ideal class group of a quadratic number field in function fields. The result is mostly of a theoretical nature, since hyperelliptic cryptosystems, as proposed by Koblitz in 1989 [8], are considered to be not practical enough because of their complicated arithmetic. Yet there exist some implementations in the group around Frey showing this performance is not as bad as expected, especially because the size of the elements is considerably smaller than for elliptic curves which might make them even more suitable for small computing devices such as smart cards.

The first implementation actually solving hyperelliptic cryptosystems has been done by R. Flassenberg and the author in 1997 [4]. They applied a sieving technique to accelerate a variant of the Hafner–McCurley algorithm [6] (known to solve discrete logarithms (see discrete logarithm problem) in ideal class groups of quadratic number fields). The basic idea of sieving in function fields is to find a good representation of polynomials by integers, which allows one to "jump" from one polynomial to another, and to increment the exponent in the cell of a three dimensional matrix. All the other optimizations known from the number field sieve could then be applied. Later, Smart [9] compared the Hafner-McCurley variant with the original, theoretically faster, Adleman–De Marrais–Huang variant which allows one to construct sparse systems of linear equations, therefore being better suited for curves of larger genus (the genus basically being the size of the discriminant of the function field). It turned out that the size of the cryptosystems N.P. Smart experienced with was still too small. Later on, N.P. Smart implemented the sieving technique for superelliptic cryptosystems (where the degree of the corresponding function field is at least 3) based on a joint work with Galbraith and the author [5]. None of these implementations was ever even close to the size real cryptosystems would use, but no one ever started a massively parallel project as for the number field sieve for these types of cryptosystems. As a consequence, one cannot sincerely decide about the practical usefulness of hyperelliptic cryptosystems.

This is very different to the other application of sieving in function fields: namely, to compute the discrete logarithm in a finite field. This can be done by constructing for a given finite field a function field with the following property: there is an embedding of subgroups of the multiplicative group of the finite field into the Jacobian of a curve corresponding to the field. This mapping has been used for applying the Adleman-De Marrais-Huang result to finite fields with small characteristic and high degree, resulting in a subexponential algorithm for discrete logarithms in this type of field [2]. Since solving discrete logarithms in finite fields is of general interest, especially for finite fields with characteristic 2, there are a few implementations of these algorithms.

An important point is how the function field is constructed. Whereas Adleman and Huang [2] were looking for the most simple representation for an optimal performance of the necessary function field arithmetic, Joux and Lercier in 2001 [7] generalized this approach to get better asympotic running times. They proved their theoretical result by solving a discrete logarithm in the finite field of size 2^{521} in approximately one month on one machine. Moreover, they showed that the specialized algorithm of Coppersmith [3], which holds the actual discrete logarithm record (in a field of size 2^{607}), is a special case of their algorithm in the case of characteristic 2. But since the record computation, done by Thome in 2001 [10], was performed using massively parallel computations for collecting relations in the sieving part of the algorithm, there is still room for practical improvements of the computation of discrete logarithms by using Joux' and Lercier's ideas. Especially, there are no known practically relevant results for characteristics different from 2 as of this writing.

Sachar Paulus

References

[1] Adleman, L.M., J. De Marrais, and M.-D. Huang (1994). "A subexponential algorithm for discrete logarithms over the rational subgroup of the Jacobian of large genus hyperelliptic curves over finite fields." *ANTS-1: Algorithmic Number Theory*, Lecture Notes in Computer Science, vol. 877, eds. L.M. Adleman and M.-D. Huang. Springer-Verlag, Berlin, 28–40.

[2] Adleman, L.M. and M.-D. Huang (1999). "Function field sieve method for discrete logarithms over finite fields." *Information and Computation*, 151, 5–16.

[3] Coppersmith, Don (1984). "Fast evaluation of Logarithms in Fields of Characteristic Two, IEEE. Transaction on Information Theory, vol. IT-30 (4), 587–594.

[4] Flassenberg, R. and S. Paulus (1999). "Sieving in function fields." *Experimental Mathematics*, 8, 339–349.

[5] Galbraith, S.D., S. Paulus, and N.P. Smart (2001). "Arithmetic on superelliptic curves." *Mathematics of Computation*, 71, 393–405.

[6] Hafner, James L. and Kevin S. McCurley (1991). "Asymptotically fast triangularization of matrices over rings." *SIAM Journal of Computer*, 20 (6), 1068–1083.

[7] Joux, A. and R. Lercier (2002). "The function field sieve is quite special." *Proceedings of ANTS-V*, Lecture Notes in Computer Science, vol. 2369, eds. C. Fieker and D.R. Kohel. Springer-Verlag, Berlin, 431–445.

[8] Koblitz, N. (1989). "Hyperelliptic cryptosystems." *Journal of Cryptology*, 1 (3), 139–150.

[9] Smart, N.P. (1999). "On the performance of hyperelliptic cryptosystems." *Advances in Cryptology—EUROCRYPT'99*, Lecture Notes in Computer Science, vol. 1592. Springer-Verlag, Berlin, 165–175.

[10] Thome, E. (2001). "Computation of discrete logarithms in GF(2^{607})." *Advances in Cryptology—ASIACRYPT 2001*, Lecture Notes in Computer Science, vol. 2248, ed. C. Boyd. Springer-Verlag, Berlin, 107–124.

SIGNCRYPTION

INTRODUCTION: Encryption and digital signature schemes are fundamental cryptographic tools for providing privacy and authenticity, respectively, in the public-key setting. Traditionally, these two important building-blocks of public-key cryptography have been considered as *distinct* entities that may be *composed* in various ways to ensure *simultaneous* message privacy and authentication. However, in the last few years a new, separate primitive—called *signcryption* [14]—has emerged to model a process simultaneously achieving privacy and authenticity. This emergence was caused by many related reasons. The obvious one is the fact that given that *both* privacy and authenticity are simultaneously needed in so many applications, it makes a lot of sense to invest special effort into designing a tailored, more efficient solution than a mere composition of signature and encryption. Another reason is that viewing authenticated encryption as a separate *primitive* may conceptually simplify the design of complex protocols which require both privacy and authenticity, as signcryption could now be viewed as an "indivisible" atomic operation. Perhaps most importantly, it was noticed by [2,3] (following some previous work in the symmetric-key setting [4,10]) that proper modeling of signcryption is not so obvious. For example, a straightforward composition of signature and encryption might not always work; at least, unless some special care is applied [2]. The main reason for such difficulties is the fact that signcryption is a complex *multi-user* primitive, which opens a possibility for some subtle attacks (discussed below), not present in the settings of stand-alone signature and encryption.

Defining Signcryption

Syntactically, a signcryption scheme consists of the three efficient algorithms (Gen, SC, DSC). The key generation algorithm $\text{Gen}(1^\lambda)$ generates the key-pair $(\text{SDK}_U, \text{VEK}_U)$ for user U, where λ is the security parameter, SDK_U is the signing/decryption key that is kept private, and VEK_U is the verification/encryption key that is made public. The randomized signcryption algorithm SC for user U implicitly takes as input the user's secret key SDK_U, and explicitly takes as input the message m and the identity of the recipient ID_R, in order to compute and output the *signcryptext* on Π. For simplicity, we consider this identity ID_R, to be a public key VEK_R of the recipient R, although ID's could generally include more convoluted information (as long as users can easily obtain VEK from ID). Thus, we write $\text{SC}_{\text{SDK}_U}(M, \text{ID}_R)$ as $\text{SC}_{\text{SDK}_U}(m, \text{VEK}_R)$, or simply $\text{SC}_U(m, \text{VEK}_R)$. Similarly, user U's deterministic designcryption algorithm DSC implicitly takes the user's private SDK_U and explicitly takes as input the signcryptext $\tilde{\Pi}$ and the senders' identity ID_S. Again, we assume $\text{ID}_S = \text{VEK}_R$ and write $\text{DSC}_{\text{SDK}_U}(\Pi, \text{VEK}_S)$, or simply $\text{DSC}_U(\Pi, \text{VEK}_S)$. The algorithm outputs some message $\tilde{m}$, or $\perp$ if the signcryption does not verify or decrypt successfully. Correctness of property ensures that for any users S, R, and message m, we have $\text{DSC}_R(\text{SC}_S(m, \text{VEK}_R), \text{VEK}_S) = m$.

We also remark that it is often useful to add another optional parameter to both SC and DSC algorithms: a *label L* (also termed *associated data* [11]). This label can be viewed as a public identifier which is "inseparably bound" to the message m inside the signcryptext. Intuitively, designcrypting the signcryptext Π of m with the wrong label should be impossible, as well as changing Π into a valid signcryptext $\tilde{\Pi}$ of the same m under a different label.

Security of Signcryption

Security of signcryption consists of two distinct components: one ensuring privacy, and the other—authenticity. On a high level, privacy is defined somewhat analogously to the privacy of an ordinary encryption, while authenticity—to that of an ordinary digital signature. For example, one can talk about indistinguishability of signcryptexts under chosen ciphertext attack, or existential unforgeability of signcryptexts under chosen message attack, among others. For concreteness, we concentrate on the above two forms of security too, since they are the strongest.

However, several new issues come up due to the fact that signcryption/designcryption take as an extra argument the identity of the sender/recipient. Below, we semiformally introduce some of those issues (see [2] for in-depth technical discussion, as well as formal definitions of signcryption).

- *Simultaneous Attacks.* Since the user U utilizes its secret key SDK_U to both send and receive the data, it is reasonable to allow the adversary $\mathcal{A}$ oracle access to both the signcryption and the designcryption oracle for user U, irrespective of whether $\mathcal{A}$ is attacking privacy or authenticity of U.

- *Two- vs. Multi-user Setting.* In the simplistic two-user setting, where there are only two users S and R in the network, the explicit identities become redundant. This considerably simplifies the design of secure signcryption schemes (see below), while providing a very useful intermediate step towards general, multi-user constructions (which are often obtained by adding a simple twist to the basic two-user construction). Intuitively, the security in the two-user model already ensures that there are no weaknesses in the way the message is encapsulated inside the signcryptext, but does not ensure that the message is bound to the identities of the sender and/or recipient. In particular, it might still allow the adversary a large class of so called *identity fraud* attacks, where the adversary can "mess up" correct user identities without affecting the hidden message.

- *Public* NonRepudiation? In a regular digital signature scheme, anybody can verify the validity of the signature, and unforgeability of the signature ensures that a signer S indeed certified the message. Thus, we say that a signcryption scheme provides *nonrepudiation* if the recipient can extract a regular (publicly verifiable) digital signature from the corresponding signcryptext. In general, however, it is a-priori only clear that the *recipient R* is sure that S sent the message. Indeed, without R's secret key SDK_R others might not be able to verify the authenticity of the message, and it might not be possible for R to extract a regular signature of m. Thus, signcryption does not necessarily provide nonrepudiation. In fact, for some applications we might explicitly want *not* to have nonrepudiation. For example, S might be willing to send some confidential information to R only under the condition that R cannot convince others of this fact. To summarize, nonrepudiation is an optional feature which some schemes support, others do not, and others explicitly avoid!

- *Insider vs. Outsider Security.* In fact, even with R's secret key SDK_R it might be unclear to an observer whether S indeed sent the message m to R, as opposed to R "making it up" with the help of SDK_R. This forms the main basis for distinction between *insider-* and *outsider-secure* signcryption. Intuitively, in an outsider-secure scheme the adversary must compromise communication between two honest users (whose keys he does not know). Insider-secure signcryption protects a given user U even if his partner might be malicious. For example, without U's key, one cannot forge signcryptext from U to any other user R, even with R's secret key. Similarly, if honest S sent $\Pi = SC_S(m, VEK_U)$ to U and later exposed his key SDK_S to the adversary, the latter still cannot decrypt Π. Clearly, insider-security is stronger than outsider-security, but might not be needed in a given application. In fact, for applications supporting message repudiation, one typically does not want to have insider-security.

Supporting Long Inputs

Sometimes, it is easier to design natural signcryption schemes supporting short inputs. Below we give a general method how to create signcryption SC′ supporting arbitrarily long inputs from SC which only supports fixed-length (and much shorter) inputs. The method was suggested by [8] and uses a new primitive called *concealment*. A concealment is a publicly known randomized transformation, which, on input m, outputs a *hider* h and a *binder* b. Together, h and b allow one to recover m, but separately, (1) the hider h reveals "no information" about m, while (2) the binder b can be "meaningfully opened" by at most one hider h. Further, we require $|b| \ll |m|$ (otherwise, one could trivially set $b = m$, $h = \emptyset$). Now, we let $SC'(m) = \langle SC(b), h \rangle$ (and DSC' is similar). It was shown in [8] that the above method yields a secure signcryption SC′. Further, a simple construction of concealment was given: set $h = E_\tau(m)$, $b = \langle \tau, H(h) \rangle$, where E is a <u>symmetric-key</u> one-time secure encryption (with short key τ) and H is a collision-resistant <u>hash function</u> (with short output).

CURRENT SIGNCRYPTION SCHEMES: We now survey several signcryption schemes achieving various levels of provable security.

Generic Composition Schemes

The two natural composition paradigms are "encrypt-then-sign" ($\mathcal{E}t\mathcal{S}$) and "sign-then-encrypt" ($\mathcal{S}t\mathcal{E}$). More specifically, assume Enc is a semantically secure encryption against chosen ciphertext attack, and Sig is an existentially unforgeable signature (with message recovery) against chosen message attack. Each user U has a key for for Sig and Enc. Then the "basic" $\mathcal{E}t\mathcal{S}$ from S to R outputs $\mathsf{Sig}_S(\mathsf{Enc}_R(m))$, while $\mathcal{S}t\mathcal{E}$—$\mathsf{Enc}_R(\mathsf{Sig}_S(m))$. Additionally, [2] introduced a novel generic composition paradigm for *parallel* signcryption. Namely, assume we have a secure <u>commitment</u> scheme, which on input m, outputs a commitment c and a decommitment d (where c is both hiding and binding). Then "commit-then-encrypt-and-sign" ($\mathcal{C}t\mathcal{E}\&\mathcal{S}$) outputs a pair $\langle \mathsf{Enc}_R(d), \mathsf{Sig}_S(c) \rangle$. Intuitively, the scheme is private as public c reveals no information about m (while d is encrypted), and authentic since c binds one to m. The advantage of the above scheme over the sequential $\mathcal{E}t\mathcal{S}$ and $\mathcal{S}t\mathcal{E}$ variants is the fact that expensive signature and encryption operations are performed in parallel. In fact, by using trapdoor commitments in place or regular commitments, most computation in $\mathcal{C}t\mathcal{E}\&\mathcal{S}$—including the expensive computation of both public-key signature and encryption—can be done off-line, even before the message m is known!

It was shown by [2] that all three basic composition paradigms yield an insider-secure signcryption in the two-user model. Moreover, $\mathcal{E}t\mathcal{S}$ is outsider-secure even if Enc is secure only against the chosen plaintext attack, and $\mathcal{S}t\mathcal{E}$ is outsider-secure even if Sig is only secure against no message attack. Clearly, all three paradigms are insecure in the multiuser model, since no effort is made to bind the message m to the identities of the sender/recipient. For example, intercepting a signcryptext of the form $\mathsf{Sig}_S(e)$ from S to R, an adversary $\mathcal{A}$ can produce $\mathsf{Sig}_\mathcal{A}(e)$, which is a valid signcryptext from $\mathcal{A}$ to R of the same message m, even though m is *unknown* to $\mathcal{A}$. [2] suggest a simple solution: when encrypting, always append the identity of the sender to the message, and when signing, of the recipient. For example, a multi-user secure variant of $\mathcal{E}t\mathcal{S}$ is $\mathsf{Sig}_S(\mathsf{Enc}_R(m, VEK_S), VEK_R)$. Notice, if Enc and/or Sig support labels, these identities can be part of the label rather than the message.

Finally, we remark that $\mathcal{S}t\mathcal{E}$ and $\mathcal{C}t\mathcal{E}\&\mathcal{S}$ always support nonrepudiation, while $\mathcal{S}t\mathcal{E}$ might or might not.

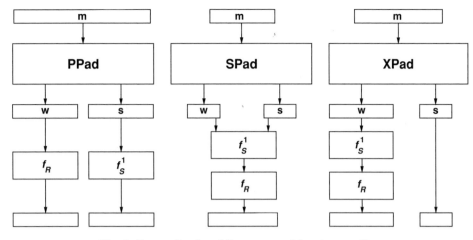

Fig. 1. Generalized paddings as used by signcryption

Schemes from Trapdoor Permutations

The generic schemes above validate the fact that signcryption can be built from ordinary signature and encryption, but will be inefficient unless the latter are efficiently implemented. In practice, efficient signature and encryption schemes, such as <u>OAEP</u> [5], OAEPP+ [13], PSS-R [6], are built from trapdoor permutations, such as RSA, and are analyzed in the <u>random oracle model</u>. Even with these efficient implementations, however, the generic schemes will have several drawbacks. For example, users have to store two independent keys, the message bandwidth is suboptimal and the "exact security" of the scheme is not as good as one might expect. Thus, given that practical schemes are anyway built from trapdoor permutations, it is natural to have highly optimized *direct* signcryption constructions from trapdoor permutations (in the random oracle model).

This is the approach of [9]. In their model, each user U independently picks a trapdoor permutation f_U (together with its trapdoor, denoted f_U^{-1}) and publishes f_U as its public key (see also <u>trapdoor one-way function</u> and <u>substitutions and permutations</u>). (Notice, only a *single* key is chosen, unlike what is needed for the generic schemes.) Then, [9] considers the following three paradigms termed **P**-Pad, **S**-Pad and **P**-Pad. Each paradigm proceeds by constructing a padding scheme produces $\pi(m) = w|s$, and then composing it with the corresponding permutations of the sender and the

recipient as shown in Figure 1. Table 1 also shows how the corresponding approaches could be used for plain signature and encryption as well.

The convenience of each padding scheme depends on the application for which it is used. As was shown in [9], **P**-Pad signcryption provides parallel application of "signing" f_S^{-1} and "encrypting" f_R, which can result in efficiency improvements on parallel machines. However, the minimum ciphertext length is twice as large as compared to **S**-Pad, yet the exact security offered by **S**-Pad is not as tight as that of **P**-Pad. Finally, **X**-Pad regains the optimal exact security of **P**-Pad, while maintaining ciphertext length nearly equal to the length of the trapdoor permutation (by achieving quite short s).

It remains to describe secure padding schemes π for **P**-Pad, **S**-Pad and **X**-Pad. All constructions offered by [9] are quite similar. One starts with any *extractable commitment* (c, d), where c is the <u>commitment</u> and d is the decommitment. Such schemes are very easy to construct in the <u>random oracle model</u>. For example, if $|m| = n$, for any $0 \leq a \leq n$, the following scheme is an extractable commitment: split $m = m_1|m_2$, where $|m_1| = a$, $|m_2| = n - a$, and set

$$c = G(r) \oplus m_1|H(m_2|r)$$
$$d = m_2|r$$

where G and H are random oracles (with appropriate input/output lengths) and r is a random salt.

Table 1. Signcryption Schemes Based on Trapdoor Permutations.

Padding Type	Encryption	Signature	Signcryption
P-Pad (**P**arallel Padding)	$f_R(w)|s$	$w|f_S^{-1}(s)$	$f_R(w)|f_S^{-1}(s)$
S-Pad (**S**equential Padding)	$f_R(w|s)$	$f_S^{-1}(w|s)$	$f_R(f_S^{-1}(w|s))$
X-Pad (e**X**tended sequential Padding)	$f_R(w)|s$	$f_S^{-1}(w)|s$	$f_R(f_S^{-1}(w))|s$

To get a secure padding scheme for the **P**-Pad paradigm, one should then apply the <u>Feistel</u> Transform to the resulting pair (d, c), with yet another random oracle F as the round function. Namely, set $w = c$, $s = F(c) \oplus d$. For example, using the extractable commitment above with $a = n$, we get nothing else but the OAEP padding, while $a = 0$ would give the PSSR padding! For arbitrary a, [9] call the resulting hybrid between PSSR and OAEP *Probabilistic Signature-Encryption Padding* (PSEP).

To get the padding π sufficient for either **S**-Pad or **P**-Pad, one only needs to perform one more Feistel round to the construction above: $w' = s$, $s' = F'(s) \oplus w$, and set $\pi(m) = w'|s'$. Coincidentally, the resulting π also gives a very general construction of the so called *universal padding schemes* [7].

As described, the paddings π_1 and π_3 above would only give insider security in the two-user setting. To get multi-user security, all one needs to do is to prepend the pair $(\mathsf{VEK}_S, \mathsf{VEK}_R)$ to all the inputs to the random oracles F and F': namely, create effectively independent F and F' for every sender-recipient pairing! More generally, the paddings above also provide label support, if one sticks the label L as part of the inputs to F and F'.

Finally, we remark that **P**-Pad, **X**-Pad and **X**-Pad always support non-repudiation.

Schemes Based on Gap Diffie–Hellman

Finally, we present two very specific, but efficient schemes based on the so called *Gap Diffie–Hellman* assumption. Given a cyclic <u>group</u> G of prime order q, and a generator g of G, the assumption states that the computational <u>Diffie–Hellman</u> <u>problem</u> (CDH) is computationally hard, even if one is given oracle access to the <u>decisional Diffie–Hellman</u> (DDH) oracle. Specifically, it is hard to compute g^{ab} from g^a and g^b, even if one can test whether a tuple $\langle g^x, g^y, g^z \rangle$ satisfies $z = xy \bmod q$.

In both schemes, the user U chooses a random $x_U \in \mathbb{Z}_q$ as its secret key VEK_U, and sets its public key $\mathsf{SDK}_U = y_U = g^{x_U}$. The scheme of [1] is based on the following noninteractive key agreement between users S and R. Namely, both S and R can compute the quantity $Q_{SR} = g^{x_R x_S} = y_S^{x_R} = y_R^{x_S}$. They then set the key $K_{SR} = H(Q_{SR})$, where H is a random oracle, and then always use K_{SR} to perform symmetric-key authenticated encryption of the message m. For the latter, they can use any secure symmetric-key scheme, like "encrypt-then-mac" [4] or OCB [12]. The resulting signcryption scheme can be shown to be outsider-secure for both privacy and authenticity, in the multi-user setting. Clearly, it is not insider-secure, since both S and

R know the key K_{SR}. In fact, the scheme is perfectly repudiable, since all the signcryptexts from S could have been easily faked by R.

To get insider-security for authenticity under the same assumption, one can instead consider the following scheme, originally due to [14], but formally analyzed by [3]. Below G and H are random oracles with appropriate domains, and E is a one-time secure symmetric-key encryption (e.g., one-time pad will do). To signcrypt a message from S to R, S chooses a random $x \in \mathbb{Z}_q$, computes $Q = y_R^x$, makes a symmetric key $K = H(Q)$, sets $c \leftarrow E_K(m)$, computes the "validation tag" $r = G(m, y_A, y_B, Q)$ and finally $t = x(r + x_S)^{-1} \bmod q$. Then S outputs $\langle c, r, t \rangle$ as the signcryption of m. To designcrypt $\langle c, r, t \rangle$, R first recovers g^x via $w = (y_S g^r)^t$, then recovers the Diffie–Hellman key $Q = w^{x_R}$, the encryption key $K = H(Q)$ and the message $m = D_K(c)$. Before outputting m, however, it double checks if $r = G(m, y_A, y_B, Q)$. While this scheme is insider-secure for authenticity, it is still not insider-secure.

We also mention that the scheme supports public <u>nonrepudiation</u>. All that R has to do is to reveal Q, m and a proof that $Q = w^{x_R}$ (which can be done noninteractively using the <u>Fiat-Shamir</u> heuristics, applied to the three-move proof that $\langle g, y_R, w, Q \rangle$ form a DDH-tuple).

Yevgeniy Dodis

References

[1] An, Jee Hea (2001). "Authenticated encryption in the public-key setting: Security notions and analyses." *Cryptology ePrint Archive*, Report 2001/079.

[2] An, Jee Hea, Yevgeniy Dodis, and Tal Rabin (2002). "On the security of joint signature and encrytion." *Advances in Cryptology—EUROCRYPT 2002, April 28–May 2*, Lecture Notes in Computer Science, vol. 2332, ed. L. Knudsen. Springer-Verlag, Berlin. Available from http://eprint.iacr.org/2002/046/

[3] Baek, Joonsang, Ron Steinfeld, and Yuliang Zheng (2002). "Formal proofs for the security of signcryption." *5th International Workshop on Practice and Theory in Public Key Cryptosystems—PKC 2002, February*, Lecture Notes in Computer Science, vol. 2274, eds. D. Naccache and P. Paillier. Springer-Verlag, Berlin, 80–98.

[4] Bellare, Mihir and Chanathip Namprempre (2000). "Authenticated encryption: Relations among notions and analysis of the generic composition paradigm." *Advances in Cryptology—ASIACRYPT 2000, Kyoto, Japan, December 3–7*, Lecture Notes in Computer Science, vol. 1976, ed. T. Okamoto. Springer-Verlag, Berlin, 531–545.

[5] Bellare, Mihir and Phillip Rogaway (1995). "Optimal asymmetric encryption." *Advances in Cryptology—EUROCRYPT'94, May 9–12*, Lecture

Notes in Computer Science, vol. 950, ed. A. De Santis. Springer-Verlag, Berlin, 92–111. Revised version available from http://www-cse.ucsd.edu/users/mihir/

[6] Bellare, Mihir and Phillip Rogaway (1996). "The exact security of digital signatures: How to sign with RSA and Rabin." *Advances in Cryptology—EUROCRYPT'96, May 12–16*, Lecture Notes in Computer Science, vol. 1070, ed. U. Maurer. Springer-Verlag, Berlin, 399–416. Revised version appears in http://www-cse.ucsd.edu/users/mihir/papers/crypto-papers.html

[7] Coron, Jean-Sébastian, Marc Joye, David Naccache, and Pascal Pailler (2002). "Universal padding schemes for RSA." *Advances in Cryptology—CRYPTO 2002, August 18–22*, Lecture Notes in Computer Science, vol. 2442, ed. M. Yung. Springer-Verlag, Berlin. Available from http://eprint.iacr.org/2002/115/

[8] Dodis, Yevgeniy and Jee Hea An (2003). "Concealment and its applications to authenticated encryption." *Advances in Cryptology—EUROCRYPT 2003, May 4–8*, Lecture Notes in Computer Science, vol. 2656, ed. E. Biham. Springer-Verlag, Berlin, 312–329.

[9] Dodis, Yevgeniy, Michael J. Freedman, Stanislaw Jarecki, and Shabsi Walfish (2004). Versatile Padding Schemes for Joint Signature and Encryption. In Birgit Pfitzmann, editor, Eleventh ACM Conference on Computer and Communication Security pages 196–205. ACM, October 25–29, 2004.

[10] Krawczyk, Hugo (2001). "The order of encryption and authentication for protecting communications (or: How secure is ssl?)." *Advances in Cryptology—CRYPTO 2001, August 19–23*, Lecture Notes in Computer Science, vol. 2139, ed. J. Kilian. Springer-Verlag, Berlin, 310–331.

[11] Rogaway, Phillip (2002). "Authenticated-encryption with associated-data." *Ninth ACM Conference on Computer and Communication Security, November 17–21*, ed. Ravi Sandhu. ACM Press, New York, 98–107.

[12] Rogaway, Phillip, Mihir Bellare, John Black, and Ted Krovetz (2001). "OCB: A block-cipher mode of operation for efficient authenticated encryption." *Eighth ACM Conference on Computer and Communication Security, November 5–8*, ed. Pierangela Samarati. ACM Press, New York, 196–205. Full version available from http://www.cs.ucsdavis.edu/~rogaway

[13] Shoup, Victor (2001). "OAEP reconsidered." *Advances in Cryptology—CRYPTO 2001, August 19–23*, Lecture Notes in Computer Science, vol. 2139, ed. J. Kilian. Springer-Verlag, Berlin, 240–259.

[14] Zheng, Y. (1997). "Digital signcryption or how to achieve cost(signature & encryption) ≪ cost(signature) + cost(encryption)." *Advances in Cryptology—CRYPTO'97, August 17–21*, Lecture Notes in Computer Science, vol. 1294, ed. B.S. Kaliski Jr. Springer-Verlag, Berlin, 65–179.

SIGNED DIGIT EXPONENTIATION

Signed digit exponentiation is an approach for computing powers in any group in which the inverse A^{-1} of any group element A can be computed quickly (such as the groups of points on an elliptic curve employed in elliptic curve cryptography). It is related to sliding window exponentiation: while in sliding window exponentiation each window corresponds to a positive digit value, signed digit exponentiation additionally makes use of the corresponding negative digit values, and the ease of inversion makes these extra digits available almost for free. This often makes signed digit exponentiation faster when using the same amount of memory for storing group elements, and allows it to reach approximately the same speed with less memory.

Let $B_k = \{\pm 1, \pm 3, \ldots, \pm(2^k - 1)\}$ where k is a positive integer; and let a base-two representation of an exponent e be given using the digit set $\{0\} \cup B_k$, i.e.

$$e = \sum_{i=0}^{l-1} e_i 2^i, \quad e_i \in \{0\} \cup B_k.$$

Assuming that l is chosen such that $e_{l-1} \neq 0$, the *left-to-right signed digit exponentiation method* computes g^e as follows where g is any group element; cf. the left-to-right sliding window exponentiation method.

$$G_1 \leftarrow g$$
$$A \leftarrow g \circ g$$
for $d = 3$ to $2^k - 1$ step 2 **do**
$\quad G_d \leftarrow G_{d-2} \circ A$

if $e_{l-1} > 0$ **then**
$\quad A \leftarrow G_{e_{l-1}}$
else
$\quad A \leftarrow G_{-e_{l-1}}^{-1}$
for $i = l - 2$ down to 0 **do**
$\quad A \leftarrow A \circ A$
$\quad$ **if** $e_i \neq 0$ then
$\quad\quad$ **if** $e_i > 0$ then
$\quad\quad\quad A \leftarrow A \circ G_{e_i}$
$\quad\quad$ **else**
$\quad\quad\quad A \leftarrow A \circ G_{-e_i}^{-1}$
return A

The *right-to-left signed digit exponentiation method* computes g^e as follows; cf. the right-to-left sliding window exponentiation method. Note that the algorithm as written can be optimized similarly to the right-to-left 2^k-ary exponentiation or sliding window exponentiation methods to avoid (at least) 2^{k-1} applications of the group operation.

> **for** $d = 1$ to $2^k - 1$ step 2 **do**
> > $B_d \leftarrow$ identity element
>
> $A \leftarrow g$
>
> **for** $i = 0$ to $l - 1$ **do**
> > **if** $e_i \neq 0$ **then**
> > > **if** $e_i > 0$ **then**
> > > > $B_{e_i} \leftarrow B_{e_i} \circ A$
> > >
> > > **else**
> > > > $B_{-e_i} \leftarrow B_{-e_i} \circ A^{-1}$
> > >
> > > **if** $i < l - 1$ **then**
> > > > $A \leftarrow A \circ A$
>
> {Now $g^e = \Pi_{d \in \{1,3,\ldots,2^k-1\}} B_d^d.$}
>
> **for** $d = 2^k - 1$ to 3 step -2 **do**
> > $B_{d-2} \leftarrow B_{d-2} \circ B_d$
> >
> > $B_1 \leftarrow B_1 \circ (B_d \circ B_d)$
>
> **return** B_1

For both the left-to-right and the right-to-left variant, it remains to be considered how signed digit representations of exponents e using the digit set $\{0\} \cup B_k$ with $B_k = \{\pm 1, \pm 3, \ldots, \pm(2^k - 1)\}$ can be obtained. An algorithm for the simplest case $k = 1$ is due to Reitwiesner [1]; the representation obtained by it (using digits $\{-1, 0, 1\}$) is known as the *nonadjacent form (NAF)* of e. The generalization for an arbitrary parameter k was simultaneously suggested by multiple researchers; the following algorithm is from [2]:

> $c \leftarrow e$
>
> $i \leftarrow 0$
>
> **while** $c > 0$ **do**
> > **if** c is odd **then**
> > > $d \leftarrow c \bmod 2^{k+1}$
> > >
> > > **if** $d > 2^k$ **then**
> > > > $d \leftarrow d - 2^{k+1}$
> > >
> > > $c \leftarrow c - d$
> >
> > **else**
> > > $d \leftarrow 0$
> >
> > $e_i \leftarrow d; i \leftarrow i + 1$
> >
> > $c \leftarrow c/2$
>
> **return** $e_{i-1}, \ldots, e_0$

This algorithm is a variant of right-to-left scanning as used in sliding window exponentiation with an effective window size of $k + 1$. For efficiency considerations, if the cost of inverting groups elements and the additional cost for obtaining the appropriate representation of e can be neglected, signed digit exponentiation differs from sliding window exponentiation with the same parameter k in that the expected number of nonzero digits in the representation is approximately $l/(k + 2)$ instead of approximately $l/(k + 1)$ (but the maximum possible length of the signed digit representation is longer: while l cannot exceed the length of the binary representation of e for sliding window exponentiation, it can be said length plus 1 for signed digit exponentiation).

Bodo Möller

References

[1] Reitwiesner, G.W. (1960). "Binary arithmetic." *Advances in Computers*, 1, 231–308.
[2] Solinas, J.A. (2000). "Efficient arithmetic on Koblitz curves." *Designs, Codes and Cryptography*, 19, 195–249.

SIMULTANEOUS EXPONENTIATION

Various schemes for public-key cryptography involve computing power products in some commutative group (or commutative semigroup). A straightforward way to compute a power product

$$\prod_{j=1}^{n} g_j^{e_j}$$

is to compute the individual powers $g_j^{e_j}$ using binary exponentiation or some other exponentiation method, and perform $n - 1$ applications of the group operation to multiply these partial results. However, specialized algorithms for computing power products are often faster. The task of computing a power product is sometimes called *multi-exponentiation*, and performing a multiexponentiation by a procedure that does not involve computing the partial results $g_j^{e_j}$ is known as *simultaneous exponentiation*. Two methds for multiexponentiation that both generalize left-to-right sliding window exponentiation are *simultaneous sliding window exponentiation*, which is due to Yen, Laih. and Lenstra [3] (based on the simultaneous 2^k-ary exponentiation method from Straus [2]), and *interleaved sliding window exponentiation* [1]. Like the sliding window method for single exponentiations, these methods use the binary representation of exponents, on which nonoverlapping windows are placed such that every nonzero bit is covered by one of the windows. Simultaneous sliding window exponentiation and interleaved sliding window exponentiation use different approaches for placing windows; sometimes the former is faster, sometimes the latter.

Simultaneous sliding window exponentiation uses windows up to some maximum width k that span across all n exponents; e.g., for exponents e_1, e_2, e_3 with binary representations 1011010, 0011001, and 1001011 and $k = 2$:

e_1	1	0	1	1	0	1	0
e_2	0	0	1	1	0	0	1
e_3	1	0	0	1	0	1	1

Such windows can be found by left-to-right scanning: look at the binary representations of the exponents simultaneously, going from left to right, starting a new window whenever a nonzero bit is encountered, choosing the maximum width up to k for this particular window such that one of the rightmost bits is also nonzero. The result of collapsing the window values into the right-most row of each window can be considered a base-two representation

$$(e_1, \ldots, e_n) = \sum_{i=0}^{l-1} (e_{1,i}, \ldots, e_{n,i}) 2^i$$

of the vector of exponents, e.g.

e_1	1 0 0 3 0 0 2
e_2	0 0 0 3 0 0 1
e_3	1 0 0 1 0 0 3

for the above example. Assume we have such a representation with l chosen minimal, i.e. $(e_{1,l}, \ldots, e_{n,l}) \neq (0, \ldots, 0)$. To perform a simultaneous sliding window exponentiation, first products

$$G_{(d_1, \ldots, d_n)} = \prod_{j=1}^{n} g_j^{d_j}$$

of small powers are computed and stored for all possible window values, namely for the tuples $(d_1, \ldots, d_n)$ with $d_j \in \{0, 1, \ldots, 2^k - 1\}$ for $j = 1, \ldots, n$ such that at least one of the d_j is odd. There are $2^{nk} - 2^{n(k-1)}$ such tuples, and computing the table of those products can be done with

$$2^{nk} - 2^{n(k-1)}$$

applications of the group operation, n of which are squarings ($g_1, \ldots, g_n$ appear in the table and are

available without any computation; once $g_j \circ g_j$ for $j = 1, \ldots, n$ have been computed as temporary values, each of the $2^{nk} - 2^{n(k-1)} - n$ remaining table values can be obtained by using the group operation once). The multi-exponentiation result then is computed using the table of small powers:

$A \leftarrow G_{(e_{1,l-1}, \ldots, e_{n,l-1})}$
for $i = l - 2$ down to 0 **do**
  $A \leftarrow A \circ A$
  if $(e_{1,l}, \ldots, e_{n,l}) \neq (0, \ldots, 0)$ **then**
    $A \leftarrow A \circ G_{(e_{1,i}, \ldots, e_{n,i})}$
return A

For random b-bit exponents, this requires at most another $b - 1$ squaring operations and on average approximately another

$$b \cdot \frac{1}{k + \frac{1}{2^n - 1}}$$

general group operations. Note that in practice it is not necessary to completely derive the representation

$$(e_{1,l-1}, \ldots, e_{n,l-1}), \ldots, (e_{1,0}, \ldots, e_{n,0})$$

before starting the exponentiation; instead, left-to-right scanning can be used to determine it window by window when it is needed.

In *interleaved sliding window exponentiation*, each single exponent has independent windows up to some maximum width k; e.g., for exponents e_1, e_2, e_3 with binary representations 1011010, 0011001, and 1001011 and $k = 3$:

e_1	1 0 1	1 0 1	0
e_2	0 0	1 1 0 0	1
e_3	1 0 0	1 0 1	1

For each exponent, such windows can be found by left-to-right scanning: look at the binary representation of the respective exponent, going from left to right, starting a new window whenever a nonzero bit is encountered, choosing the maximum width up to k for this particular window such that the rightmost bit is also nonzero. To perform an interleaved sliding window exponentiation, first for each g_j, the powers for odd exponents 1 up to $2^k - 1$ are computed and stored:

for $j = 1$ to n **do**
  $G_{j,1} \leftarrow g$
  $A \leftarrow g \circ g$
  for $d = 3$ to $2^k - 1$ step 2 **do**
    $G_{j,d} \leftarrow G_{j,d-2} \circ A$

Rule A	Rule B
$w_1^{k+1} = G^k(w_1^k) \oplus w_4^k \oplus counter^k$	$w_1^{k+1} = w_4^k$
$w_2^{k+1} = G^k(w_1^k)$	$w_2^{k+1} = G^k(w_1^k)$
$w_3^{k+1} = w_2^k$	$w_3^{k+1} = w_1^k \oplus w_2^k \oplus counter^k$
$w_4^{k+1} = w_3^k$	$w_4^{k+1} = w_3^k$

Fig. 1. Rule A and Rule B

Then the multi-exponentiation result is computed using that table of powers. The following algorithm shows how this computation can be implemented including left-to-right scanning of exponents up to b bits. The algorithm accesses the bits $e_j[i]$ of the binary representations

$$e_j = \sum_{i=0}^{b-1} e_j[i]2^i, \quad e_j[i] \in \{0, 1\}$$

of the exponents; the notation $e_j[i \ldots h]$ is shorthand for $\Sigma_{v=h}^i e_j[v]2^{v-h}$.

> $A \leftarrow$ identity element
> **for** $j = 1$ to n **do**
> $\quad window\_position_j \leftarrow -1$
> **for** $i = b - 1$ down to 0 **do**
> $\quad A \leftarrow A \circ A$
> $\quad$ **for** $j = 1$ to n **do**
> $\quad\quad$ **if** $window\_position_j = -1$
> $\quad\quad$ and $e_j[i] = 1$ **then**
> $\quad\quad\quad h \leftarrow i - k + 1$
> $\quad\quad\quad$ **if** $h < 0$ **then**
> $\quad\quad\quad\quad h \leftarrow 0$
> $\quad\quad\quad$ **while** $e_j[h] = 0$ **do**
> $\quad\quad\quad\quad h \leftarrow h + 1$
> $\quad\quad\quad window\_position_j \leftarrow h$
> $\quad\quad\quad E_j \leftarrow e_j[i \ldots h]$
> $\quad\quad$ **if** $window\_position_j = i$ **then**
> $\quad\quad\quad A \leftarrow A \circ G_{j,E_j}$
> $\quad\quad\quad window\_position_i \leftarrow -1$
> **return** A

The algorithm as written can be improved by a simple optimization: while A still has its initial value, omit the statement $A \leftarrow A \circ A$, and use a direct assignment $A \leftarrow G_{j,E_j}$ instead of the first assignment $A \leftarrow A \circ G_{j,E_j}$. With this optimization, an interleaved sliding window exponentiation takes up to $n + b - 1$ squarings and on average about

$$n \cdot \left(2^{k-1} - 1 + \frac{b-1}{k+1}\right)$$

general group operations.

Interleaved sliding window exponentiation essentially interleaves the operations of n single exponentiations using left-to-right sliding window exponentiation, saving many of the squarings. In groups where computing inverses of elements is possible very quickly, it is possible to similarly interleave the operations of n single exponentiations using left-to-right signed digit exponentiation for faster multi-exponentiation.

Bodo Möller

References

[1] Moller, B. (2001). "Algorithms for multi-exponentiation." *Selected Area in Cryptography— SAC 2001*, Lecture Notes in Computer Science, vol. 2259, eds. S. Vaudenay and A.M. Youssef. Springer-Verlag, Berlin, 165–180.

[2] Straus, E.G. (1964). "Problems and solutions: Addition chains of vectors." *American Mathematical Monthly*, 71, 806–808.

[3] Yen, S.-M., C.-S. Laih, and A.K. Lenstra (1994). "Multi-exponentiation." *IEE Proceedings—Computers and Digital Techiques*, 141, 325–326.

SKIPJACK

Skipjack [6] is the secret key encryption algorithm (see symmetric cryptosystem) developed by the NSA for the Clipper chip initiative (including the Capstone chip and the Fortezza PC card). It was implemented in tamper-resistant hardware and its structure was kept secret since its introduction in 1993.

On June 24, 1998, Skipjack was declassified, and its description was made public on the web site of NIST [6]. It is an *iterative* block cipher with 64-bit block, 80-bit key and 32 rounds. It has two types of rounds, called Rule A and Rule B. Each round is described in the form of a linear feedback shift register with an additional nonlinear keyed G permutation. Rule B is basically the inverse of Rule A with minor positioning differences. Skipjack applies eight rounds of Rule A, followed by eight rounds of Rule B, followed by another eight rounds of Rule A, followed by another eight rounds of Rule B. The original definitions of Rule A and

Rule B are given in Figure 1, where *counter* is the round number (in the range 1 to 32), G is a four-round Feistel permutation whose F function is defined as an 8×8-bit S box, called *F Table*, and each round of G is keyed by eight bits of the key.

The key schedule of Skipjack takes a 10-byte key, and uses four of them at a time to key each G permutation. The first four bytes are used to key the first G permutation, and each additional G permutation is keyed by the next four bytes cyclically, with a cycle of five rounds.

Skipjack has been subject to intensive analysis [2–5]. For example, Skipjack reduced to (the first) 16 rounds can be attacked with 2^{17} chosen plaintexts and 2^{34} time of analysis [5], which may be reduced to 2^{14} texts and 2^{16} steps using the *yoyo-game* approach [1]. Attacking the middle 16 rounds of Skipjack requires only 3 chosen plaintexts and 2^{30} time of analysis. The currently most successfull attack against the cipher is the imposible differential attack which breaks 31 rounds out of 32, marginally faster than exhaustive search.

In addition, it is worth noting that Skipjack can be attacked by a generic time-memory tradeoff approach requiring 2^{80} steps of precomputation and 2^{54} 80-bit words (i.e., 2^{60} bits) of memory, but then each search for a key requires only 2^{54} steps of computation.

Alex Biryukov

References

[1] Biham, E., A. Biryukov, O. Dunkelman, E. Richardson, and A. Shamir (1999). "Initial observations on skipjack: Cryptanalysis of Skipjack-3XOR." *Selected Areas in Cryptography, SAC'98*, Lecture Notes in Computer Science, vol. 1556, eds. S.E. Tavares and H. Meijer. Springer-Verlag, Berlin, 362–376.

[2] Biham, E., A. Biryukov, and A. Shamir (1999). "Cryptanalysis of Skipjack reduced to 31 rounds using impossible differentials." *Advances in Cryptology—EUROCRYPT'99*, Lecture Notes in Computer Science, vol. 1592, ed. J. Stern. Springer-Verlag, Berlin, 12–23.

[3] Granboulan, L. (2001). "Flaws in differential cryptanalysis of Skipjack." *Proceedings of Fast Software Encryption—FSE 2001*, Lecture Notes in Computer Science, vol. 2335, ed. M. Matsui. Springer-Verlag, Berlin, 328–335.

[4] Hwang, K., W. Lee, S. Lee, S. Lee, and J. Kim (2002). "Saturation attacks on round Skipjack." *Fast Software Encryption—FSE 2002*, Lecture Notes in Computer Science, vol. 2365, eds. J. Daemen and V. Rijmen. Springer-Verlag, Berlin, 100–111.

[5] Knudsen, L.R., M. Robshaw, and D. Wagner (1999). "Truncated differentials and Skipjack." *Advances in Cryptology—CRYPTO'99*, Lecture Notes in Computer Science, vol. 1666, ed. M. Wiener. Springer-Verlag, Berlin, 165–180.

[6] NIST (1998). "SKIPJACK and KEA algorithm specification." Technical Report, http://csrc.nist.gov/ CryptoToolkit/skipjack/skipjack-kea.htm. Version 2.0.

SLIDE ATTACK

Slide attack is generic attack designed by Biryukov and Wagner [1, 2]. It can be applied in both known plaintext or chosen plaintext scenarios. It can be viewed as a variant of a related key attack, in which a relation of the key with itself is exploited. The main feature of this attack is that it realizes a dream of cryptanalysts: if the cipher is vulnerable to such an attack, the complexity of the attack is independent of the number of rounds of the ciphel. A typical slide of one encryption against another by one round (under the same key) is shown in Figure 1. If the equation $F_1(P_0, K_1) = P_1$ holds, the pair is called a *slid pair*. The attacker would then obtain two equations:

$$F_1(P_0, K_1) = P_1, \qquad F_r(C_0, K_r) = C_1,$$

where the second equation would hold for free due to sliding. These equations involve only a single round function, and thus could be solved by the attacker for the secret subkeys K_1, K_r of these rounds. The attacker may create properly *slid pairs* (P_0, P_1) by birthday paradox or by careful construction. For an arbitrary cipher the attack has complexity of $2^{n/2}$ known-plaintexts, where n is the blocksize. For a Feistel cipher complexity is reduced to $2^{n/4}$ chosen plaintexts.

Several ciphers or slight modifications of existing ciphers have been shown vulnerable to such attacks: for example the Brown-Seberry variant of the Data Encryption Standard (DES) [3] (rotations in key-schedule are by seven positions, instead of varying 1, 2 rotations as in the original DES), DES-X, the *Even-Mansour scheme* [4], arbitrary Feistel ciphers with 4-round periodic key-schedule as well as round-reduced versions of GOST. The basic attack has been extended into a *slide-with a twist*, a technique where encryption is slid against decryption and *complementary slide*

$$P_0 \rightarrow F_1 \; F_2 \; F_3 \; \dots \; F_r \rightarrow C_1$$
$$P_1 \rightarrow F_1 \; F_2 \; F_3 \; \dots \; F_r \rightarrow C_2.$$

Fig. 1. A typical slide attack

technique [2], where the inputs to the rounds do not have to be identical but may have the difference which is canceled out by a difference in the keys. In the same paper another generalization of the technique for the case of a composition of strong round functions is given.

It is clear that slide-attack would apply to any *iterative* construction which has enough *self-similarity* in its rounds. It could be applied to block-ciphers as described above, to stream-ciphers (see for example resynchronization attack on *WAKE-ROFB* [1]) or to MAC and hash-functions (see for example a recent *slid pair* discovery for SHA-1 by Saarinen [5].

In practice the attack seems easy to avoid by breaking the similarity of the round transforms by applying round counters (as is done for example in Skipjack) or different random constants in each round (as in Rijndael/AES, SHA-256 and many other constructions). Whether such simple changes are indeed sufficient is a matter of further research.

Alex Biryukov

References

[1] Biryukov, A. and D. Wagner (1999). "Slide attacks." *Proceedings of Fast Software Encryption—FSE'99*, Lecture Notes in Computer Science, vol. 1636, ed. L.R. Knudsen. Springer-Verlag, Berlin, 245–259.

[2] Biryukov, A. and D. Wagner (2000). "Advanced slide attacks." *Advances in Cryptology—EUROCRYPT 2000*, Lecture Notes in Computer Science, vol. 1807, ed. B. Preneel. Springer-Verlag, Berlin, 589–606.

[3] Brown, L. and J. Seberry (1990). "Key scheduling in DES type cryptosystems." *Advances in Cryptology—AUSCRYPT'90*, Lecture Notes in Computer Science, vol. 453, eds. J. Seberry and J. Pieprzyk. Springer-Verlag, Berlin, 221–228.

[4] Even, S. and Y. Mansour (1997). "A construction of a cipher from a single pseudorandom permutation." *Journal of Cryptology*, 10 (3), 151–162.

[5] Saarinen, M.-J.O. (2003). "Cryptanalysis of block ciphers based on SHA-1 and MD5." *Proceedings of Fast Software Encryption—FSE 2003*, Lecture Notes in Computer Science, vol. 2887, ed. T. Johansson. Springer-Verlag, Berlin, 36–44.

SLIDING WINDOW EXPONENTIATION

Sliding window exponentiation is an approach for computing powers in any group (or semigroup).

Like 2^k-ary exponentiation, it generalizes binary exponentiation and is parameterized by a positive integer k, where the case $k = 1$ is the same as binary exponentiation. Sliding window exponentiation can be considered an improved variant of 2^k-ary exponentiation: with identical $k \geq 2$, sliding window exponentiation needs storage for fewer group elements and usually performs less applications of the group operation than 2^k-ary exponentiation. However, the algorithms for sliding window exponentiation are slightly more complicated. 2^k-ary exponentiation uses the 2^k-ary representation of exponents, which can be considered as looking at the binary representation through fixed windows of width k:

$$\boxed{0\ 0\ 1}\boxed{1\ 1\ 0}\boxed{1\ 0\ 0}\boxed{0\ 1\ 1}\boxed{0\ 0\ 1}\boxed{0\ 1\ 0}$$

The sliding window method is based on the observation that fewer windows of width up to k can suffice to cover all nonzero exponent bits if one allows the windows to take arbitrary positions. Also, one can arrange for all windows to be odd-valued (i.e., have a 1 as the rightmost bit). Then the bits covered by each single window correspond to a value in the set $B_k = \{1, 3, \ldots, 2^k - 1\}$, and the number of possible window values is less than with the 2^k-ary exponentiation method. Covering the binary representation of the exponent by such windows yields a base-two representation of the exponent that uses the digit set $\{0\} \cup B_k$. One possible way to determine windows for a given exponent is to look at the binary representation of the exponent from left to right, starting a new window whenever a nonzero bit is encountered, choosing the maximum width up to k for this particular window such that the rightmost bit is also nonzero:

$$0\ 0\ \boxed{1\ 1\ 1}\ 0\ \boxed{1}\ 0\ 0\ 0\ \boxed{1\ 1}\ 0\ 0\ \boxed{1\ 0\ 1}\ 0$$
$$\Rightarrow\qquad 7\ 0\ 1\ 0\ 0\ 0\ 0\ 3\ 0\ 0\ 0\ 0\ 5\ 0$$

Another possibility is to look at the binary representation of the exponent from right to left, starting a new width-k window whenever a nonzero bit is encountered:

$$0\ \boxed{0\ 1\ 1}\ \boxed{1\ 0\ 1}\ 0\ 0\ \boxed{0\ 1\ 1}\ 0\ 0\ \boxed{1\ 0\ 1}\ 0$$
$$\Rightarrow\qquad 3\ 0\ 0\ 5\ 0\ 0\ 0\ 0\ 3\ 0\ 0\ 0\ 0\ 5\ 0$$

Such left-to-right scanning or right-to-left scanning yields a representation

$$e = \sum_{i=0}^{l-1} e_i 2^i, \quad e_i \in \{0\} \cup B_k.$$

In the following we assume that we have such a representation of some positive integer e with l chosen minimal; thus, $e_{l-1} \neq 0$.

The *left-to-right sliding window exponentiation method* computes g^e, where g is an element of the group (or semigroup), as follows. First the powers for odd exponents 1 up to $2^k - 1$ are computed and stored:

$G_1 \leftarrow g$

$A \leftarrow g \circ g$

for $d = 3$ to $2^k - 1$ step 2 **do**

$\quad G_d \leftarrow G_{d-2} \circ A$

Then g^e is computed using the tables of powers $G_1 = g, G_3 = g^3, \ldots, G_{2^K - 1} = g^{2^K - 1}$:

$A \leftarrow G_{e_{l-1}}$

for $i = l - 2$ down to 0 **do**

$\quad A \leftarrow A \circ A$

$\quad$ **if** $e_i \neq 0$ **then**

$\quad\quad A \leftarrow A \circ G_{e_i}$

return A

Note that in practice it is not necessary to completely derive the representation $e_{l-1}, \ldots, e_0$ before starting the exponentiation; instead, left-to-right scanning can be used to determine it digit by digit when it is needed without storing it completely. Left-to-right sliding window exponentiation is a slight modification of the method described in [2, proof of Theorem 3]; the idea to use variable windows is from [2, p. 912])

Like binary exponentiation and 2^k-ary exponentiation, sliding window exponentiation has a variant that performs a right-to-left exponentiation:

for $d = 1$ to $2^k - 1$ step 2 **do**

$\quad B_d \leftarrow$ identity element

$A \leftarrow g$

for $i = 0$ to $l - 1$ **do**

$\quad$ **if** $e_i \neq 0$ **then**

$\quad\quad B_{e_i} \leftarrow B_{e_i} \circ A$

$\quad$ **if** $i < l - 1$ **then**

$\quad\quad A \leftarrow A \circ A$

{Now $g^e = \prod_{d \in \{1,3,\ldots,2^k - 1\}} B_d^d$; this can be computed as follow :}

for $d = 2^k - 1$ to 3 step -2 **do**

$\quad B_{d-2} \leftarrow B_{d-2} \circ B_d$

$\quad B_1 \leftarrow B_1 \circ (B_d \circ B_d)$

return B_1

Again, in practice it is not necessary to completely derive the representation $e_{l-1}, \ldots, e_0$ before

starting the exponentiation; here, right-to-left scanning can be used to determine it digit by digit when it is needed. The algorithm as written can be optimized similarly to the right-to-left 2^k-ary exponentiation method to avoid (at least) 2^{k-1} applications of the group operation. The idea used to perform sliding window exponentiation in right-to-left fashion is due to Yao [3]; the sub-algorithm shown above for computing $\prod_{d \in \{1,3,\ldots,2^k - 1\}} B_d^d$ is due to Knuth [1, answer to exercise 4.6.3–9].

The number of group operations performed during a sliding window exponentiation with maximum window width k depends on the length l of the sliding window representation and on the number of digits in the representation $e_{l-1}, \ldots, e_0$ that are non-zero. For any b-bit exponent ($2^{b-1} \leq e \leq 2^b$), the length l is bounded by $b - k < l \leq b$. Assume that left-to-right or right-to-left scanning is performed on a sequence of independently and uniformly random bits; then a new window will be started on average every $k + 1$ bits. For b-bit exponents, one bit is necessarily nonzero, and both scanning techniques will usually have an unused part in the final window when the end of the exponent is reached. The expected number of nonzero values among $e_{l-1}, \ldots, e_0$ for random b-bit exponents lies between $b/(k + 1)$ and $1 + (b - 1)/(k + 1)$.

Using the upper bounds to derive estimates for average performance that are on the safe side (i.e. slightly pessimistic) gives b squaring operations (one time $g \circ g$ and $b - 1$ times $A \circ A$) and

$$2^{k-1} - 1 + \frac{b - 1}{k + 1}$$

general group operations for left-to-right sliding window exponentiation, or

$$2^{k-1} - 2 + b$$

squaring operations ($b - 1$ times $A \circ A$ and $2^{k-1} - 1$ times $B_d \circ B_d$) and

$$\underbrace{1 + \frac{b - 1}{k + 1}}_{\text{loop over } i} + \underbrace{2 \cdot (2^{k-1} - 1)}_{\text{loop over } d} - \underbrace{2^{k-1}}_{\text{optimization}} = 2^{k-1} - 1 + \frac{b - 1}{k + 1}$$

general group operations for right-to-left sliding window exponentiation with the optimization explained above.

In some groups, such as those employed in elliptic curve cryptography, computing inverses of elements is a very fast operation. For such groups, better performance than with ordinary sliding window exponentiation can often be obtained by using signed digit exponentiation instead.

Bodo Möller

References

[1] Knuth, D.E. (1998). *The Art of Computer Programming—vol. 2: Seminumerical Algorithms* (3rd ed.). Addison-Wesley, Reading, MA.

[2] Thurber, E.G. (1973). "On addition chains $l(mn) \leq l(n) - b$ and lower bounds for $c(r)$." *Duke Mathematical Journal*, 40, 907–913.

[3] Yao, A.C.-C. (1976). "On the evaluation of powers." *SIAM Journal on Computing*, 5, 100–103.

SMARTCARD TAMPER RESISTANCE

Tamper-resistant cryptographic modules are devices intended for applications that need to protect stored cryptographic keys and intermediate results of algorithms against unauthorized access. The most popular portable form is the smartcard, which has the form of a banking plastic card with embedded microcontroller. The typical interfaces are either five visible electrical contacts (for ground, power supply, reset, clock, and a bi-directional serial port) or an induction loop. Typical smartcard processors are 8-bit microcontrollers with a few hundred bytes of RAM and 4–64 kilobytes of ROM or non-volatile writable memory (NVRAM). Battery-like small steel cans ("crypto buttons"), CardBus/PCMCIA modules, and various PCI plug-in cards for non-portable applications are other popular form factors for tamper-resistant modules.

Smartcards are used in applications with both tamper-resistance and tamper-evidence requirements. Tamper resistance means that stored information must remain protected, even when the attacker can work on several samples of the module undisturbed for weeks in a well-equipped laboratory. Tamper evidence is a weaker requirement in which the regular holder of the module must merely be protected against unnoticed access to information stored in the module.

One common application for tamper-resistant smartcards are pay-TV conditional-access systems, where operators hand out millions of cards to customers, each of which contains the key necessary to descramble some subscription TV service. Pirates who manage to extract the key from one single issued card can use it to produce and sell illicit clone cards. Most proposed forms of digital rights management (DRM) mechanisms are based on some form of tamper-resistant element in the user system.

Examples for smartcard applications where operators can rely more on just a tamper-evidence requirement are digital signature identity cards, banking cards, and GSM subscriber identity modules. Here, stored secrets are specific to a single card or cardholder and can be revoked, should the module get stolen.

There are four broad categories of attacks against tamper-resistant modules:

- *Software attacks* use the normal communication interface of the processor and exploit security vulnerabilities found in protocols, cryptographic algorithms, or the software implementation. Countermeasures involve very careful design and in-depth implementation reviews, possibly augmented by formal techniques.

- *Microprobing* techniques access the chip surface directly, such that the attacker is able to observe, manipulate, and interfere with the integrated circuit. This has been the dominant form of attack against pay-TV conditional-access cards since about 1993. Chemical depackaging (e.g., with fuming nitric acid) is used to dissolve conventional packaging materials without damaging the silicon chip. Microscopes with micromanipulators are then used to place fine tungsten hairs onto micrometer-wide on-chip bus lines, in order to establish an electrical contact between the chip circuits and recording equipment such as digital oscilloscopes. The glass passivation layer that covers the metal interconnects can be broken mechanically or removed with UV laser pulses. The content of the main memory can then be reconstructed from observed on-chip bus traffic, a process that can be simplified by damaging the instruction decoder to prevent the execution of jump commands. Attackers have also succeeded in accessing the memory with the help of circuitry placed on the chip by the manufacturer for post-production testing. Modern chips with smaller feature sizes require the use of focused ion-beam workstations. With these, the surface of a de-packaged chip can be modified inside a vacuum chamber. A beam of accelerated gallium ions and various added processing gases remove chip material or deposit either conducting and insulating substances with a resolution of tens of nanometers. This not only allows attackers to modify the metal connections between the transistors, effectively to edit the processor design, but also helps in establishing larger probing pads for the connection of recording equipment. Countermeasures involve more difficult-to-remove packaging materials (e.g., silicon, silicon carbide), obfuscated circuits, additional top-layer metal sensor meshes, the careful destruction of test circuitry before the chip is delivered to customers, and the design of instruction decoders that frustrate modifications aimed at simplifying access to all memory locations [1].

- *Fault generation* techniques or fault attacks use abnormal environmental conditions to generate malfunctions in the processor aimed at providing additional access. A simple example would be a deliberately caused and carefully timed glitch that disrupts the correct execution of a single security-critical machine instruction, such as the conditional branch at the end of a password comparison. Carefully placed, a single glitch can help to bypass many layers of cryptographic protection. Such glitches have been generated by increasing the provided clock frequency for a single cycle, by brief supply voltage fluctuations, by applying light flashes to the entire chip or single gates, and with the help of electromagnetic pulses. Another class of fault generation attacks attempts to reduce the entropy generated by hardware random-bit generators. For example, where multiple noisy oscillators are used to generate randomness, externally applied electromagnetic fields with carefully selected frequencies can result in a phase lock and more predictable output. Countermeasures against fault generation include adding filters into supply lines, regular statistical checks of random-bit generators, redundant consistency checks in the software, and new logic design techniques that lead to inherently glitch-resistant circuits.

- Eavesdropping or side-channel analysis techniques monitor with high time resolution the characteristics of all supply and interface connections and any other electromagnetic radiation produced by a processor. A simple example is the determination of the length of the correct prefix of an entered password from the runtime of the string-compare routine that rejects it. This can significantly reduce the average number of guesses needed to find the correct string. The nature of the executed instruction, as well as parts of the processed data, are evident in the power-supply current of a CPU. A conditional branch that takes effect can easily be distinguished from one that passes through by examining with an oscilloscope the voltage drop over a 10Ω resistor inserted into the processor's ground connection line. The current consumed by the write operation into memory cells is often proportional to the number of bits that change their value. Even status register

flags and Hamming weights of data processed in arithmetic units can show up in power consumption curves. The technique of differential power analysis determines secret-key bits by correlating measured current curves with externally simulated intermediate results of a symmetric cipher. It has been demonstrated as a practical attack technique, even in situations where there has not been a microprobing attack first to disassemble the software in the targeted smartcard. Countermeasures include the addition of filters and shields against compromising emanations, circuitry and routines for adding random noise and delays, new balanced or dual-rail logic design techniques that lead to inherently less information in the power signal, and algorithmic techniques for reducing the number of intermediate results useful for eavesdroppers.

Microprobing requires time and careful preparation in a laboratory environment and is therefore primarily a challenge of tamper resistance. The other three attack classes are noninvasive and can, with suitable preparation, be performed in just a few seconds with attack equipment that could be disguised as a regular smartcard reader. The holder of the card might not notice such an attack, and then even the tamper evidence would be lost.

Markus Kuhn

References

[1] Kömmerling, Oliver and Markus G. Kuhn (1999). "Design principles for tamper-resistant smartcard processors." *Proceedings of the USENIX Workshop on Smartcard Technology (Smartcard '99), Chicago, IL, USA, May 10–11*, Berkeley, CA, US. USENIX Association, 9–20, ISBN 1-880446-34-0.

[2] Weingart, Steve H. (1965). "Physical security devices for computer subsystems: A survey of attacks and defenses." *Workshop on Cryptographic Hardware and Embedded Systems (CHES 2000)*, Lecture Notes in Computer Science, vol. 1965. Springer-Verlag, Berlin, 302–317.

S/MIME

S/MIME (see also Security Standards Activities) is the IETF Internet Security Syntax for MIME (Multipurpose Internet Mail Extensions), currently available in version 3, under constant development and communicated in a range of RFCs (abbreviation for "Request for Comments"). It basically specifies the syntax for the integration of various cryptographic mechanisms and algorithms within the MIME format scope.

The Cryptographic Message Syntax (CMS) (see RFC 3369) is cryptographic algorithm independent, but, typically, applying an actual algorithm is not entirely defined uniquely and requires some attendance and care for seamless interoperability.

As part of the specification update, a new suite of "mandatory to implement" algorithms are constantly being selected, reflected in updates to Certificate Handling (RFC 2632), and S/MIME v3 Message Specification (RFC 2633).

Building on the CMS Compressed Data content type specified in RFC 3274, the update to RFC specifies conventions for message compression as well as to message signature and encryption. Few are used in reality.

To aid implementers, documentation containing example output for CMS is made available, some of which for example, include structures and signed attributes defined in the Enhanced Security Services (ESS) (RFC 2634) document.

CMS, and thus S/MIME version 3 and later, permit the use of previously distributed symmetric key-encryption keys, and the underlying Public Key Infrastructure (PKI) is based on the PKIX standard, e.g. for certificates and CRLs (see certificate revocation), whilst the underlying syntax for cryptographic mechanisms rely on the PKCS standards.

Peter Landrock

Reference

[1] See http://www.networksorcery.com/enp/ for all RFCs.

SMOOTHNESS

A natural number n is called B-smooth if its factorization does not contain any prime factors larger than B, i.e.

$$n = \prod_{p \leq B} p^{n_p}.$$

Analogously, we can also consider elements of the polynomial ring $\mathbb{F}_p[x]$. For a polynomial f of degree $\deg(f)$ define the norm of f to be $p^{\deg(f)}$. An element of $\mathbb{F}_p[x]$ is called B-smooth if its factorisation does not contain any irreducible polynomials of norm greater than B.

For $t, c \in \mathbb{R}$ such that $0 \leq t \leq 1$ the complexity-theoretic L-notation is defined by

$$L_x[t, \gamma] = e^{(\gamma + o(1))(\log x)^t (\log \log x)^{1-t}},$$

where $x \to \infty$. Note that for $t = 0$ this equals $(\log x)^\gamma$, while for $t = 1$ we obtain x^γ (neglecting the $o(1)$ term). Hence we see that for values of t between 0 and 1 the function L interpolates between polynomial time and exponential time behaviour. For these values we say that L is subexponential in x (see subexponential time).

The main observation about the distribution of smooth numbers in an interval $[0, a]$ is that if the smoothness bound B is chosen subexponentially in x, then the probability that a random integer in this interval is B–smooth (or more precisely the inverse of that probability) is also subexponential.

More precisely, set $a = L_x[r, \alpha]$ and $B = L_x[s, \beta]$, where $r, s, \alpha, \beta \in \mathbb{R}_{>0}$ and $s < r \leq 1$, then the probability that a random number in $[0, a]$ is B–smooth is given (see smoothness probability) by

$$L_x[r - s, -\alpha(r - s)/\beta] \tag{1}$$

where $x \to \infty$.

A similar result holds for the polynomial case: Assume $r, s, \alpha, \beta \in \mathbb{R}_{>0}$ such that $r \leq 1$ and essentially $s < r$. Then the probability that a random element of $\mathbb{F}_p[x]$ of norm bounded by $L_x[r, \alpha]$ is $L_x[s, \beta]$–smooth is given exactly by expression 1 (see [3] for details).

Smooth numbers or polynomials are used in the most effective methods to factor natural numbers (see integer factoring) and compute discrete logarithms in finite fields (see discrete logarithm problem). The overall subexponential complexity of these methods is a direct consequence of the fact that the number of smooth elements in a given interval grows subexponentially if the smoothness bound is chosen subexponential as well.

For further background, please see [1, 2, 4].

Kim Nguyen

References

[1] Canfield, E.R., P. Erdös, and C. Pomerance (1983). "On a problem of Oppenheim concerning 'Factorisatio Numenerorum.'" *Journal of Number Theory*, 17, 1–28.

[2] De Bruijn, N.G. (1966). "On the number of positive integers $\leq x$ and free of prime factors $> y$." *Indag. Math.*, 38, 239–247.

[3] Odlyzko, A.M. (1985). "Discrete logarithms in finite fields and their cryptographic significance." *Advances in Cryptology—EUROCRYPT'84*, Lecture Notes in Computer Science, vol. 209, eds. T. Beth, N. Cot, and I. Ingemarsson. Springer-Verlag, Berlin.

[4] Ramaswami, V. (1949). "The number of positive integers $\leq x$ and free of prime divisors $< x^c$, and a problem of S.S. Pillai." *Duke Math. J.*, 16, 99–109.

SMOOTHNESS PROBABILITY

Let $\alpha, \beta, r, s \in \mathbf{R}_{>0}$ with $s < r \leq 1$. With L_x as in L-notation, it follows from [1, 2] that a random positive integer $\leq L_x[r, \alpha]$ is $L_x[s, \beta]$-smooth (see smoothness) with probability

$$L_x[r - s, -\alpha(r - s)/\beta], \quad \text{for } x \to \infty.$$

Arjen K. Lenstra

References

[1] Canfield, E.R., P. Erdös, and C. Pomerance (1983). "On a problem of Oppenheim concerning 'Factorisatio Numerorum.'" *Journal of Number Theory*, 17, 1–28.
[2] De Bruijn, N.G. (1966). "On the number of positive integers $\leq x$ and free of prime factors $> y$, II. *Indag. Math.*, 38, 239–247.

SOLITAIRE

Solitaire is a stream cipher designed to be implemented using a deck of cards. It was invented by Bruce Schneier for use in the novel *Cryptonomicon*, by Neal Stephenson [1], where it was called Pontifex. Solitaire gets its security from the inherent randomness in a shuffled deck of cards. By manipulating this deck, a communicant can create a string of "random" letters which he then combines with his message. Solitaire can be simulated on a computer, but it is designed to be used by hand.

Manual ciphers are intended to be used by spies in the field who do not want to be caught carrying evidence that they send and receive encrypted messages. In David Kahn's book *Kahn on Codes* [2], he describes a real pencil-and-paper cipher used by a Soviet spy. Both the Soviet algorithm and Solitaire take about the same amount of time to encrypt a message: most of an evening.

Solitaire, as described in the appendix to *Cryptonomicon*, has a cryptographic weakness. While this weakness does not affect the security of short messages, Solitaire is not recommended for actual use.[1]

Bruce Schneier

---

[1] See http://www.schneier.com/solitaire.html

References

[1] Stephenson, Neal (2002). *Cryptonomicom*. Avon Eos Books, Avon.
[2] Kahn, David (1984). *Kahn on Codes: Secrets of the New Cryptology*. Macmillan Publishing Co., London.

SPKI/SDSI

SPKI (Simple Public Key Infrastructure) [2, 1] was developed starting in 1995 to remedy shortcomings [3] in the existing ID certificate definitions: X.509 and PGP (see Pretty Good Privacy). It provided the first authorization certificate definition [4, 5]. Originally, SPKI used no names for keyholders but, after the merger with SDSI (Simple Distributed Security Infrastructure), now includes both named keyholders and named groups or roles—specifying authorization grants to names and definitions of names (membership in named groups).

In public-key security protocols, the remote party (the *prover*) in a transaction is authenticated via public key cryptography. Upon completion of that authentication, the *verifier* has established that the prover has control over a particular private key—the key that corresponds to the public key the verifier used. This public key is itself a good identifier for the prover. It is a byte string that is globally unique. It also has the advantages of not requiring a central ID creator or distributor and of being directly usable for authentication. However, since anyone can create a key pair at any time, a raw public key has no security value. It is the purpose of a certificate to give value or meaning to this public key.

ID certificate systems bind names to public keys. This is an attempt to directly answer the question "who is that other party?". The shortcomings of ID certificates that SPKI addresses are:

1. Because there is no single, global name source, names are not globally unique. Therefore mapping from public key to name can introduce nonuniqueness. In SPKI, the real identifier is a public key or its cryptographic hash—each of which is globally unambiguous.

2. Names have no special value to a computer, but are strongly preferred by people over raw keys or hash values. However, people have a limited ability to distinguish from among large numbers of names, so the use of names can introduce scaling problems. The original SPKI did not use names, but SDSI names are defined by

Table 1. Certificate name sources

Type	Source of names
X.509	Certificate Authority (CA)
PGP	End Entity (EE)
SDSI	Relying Party (RP)

the *Relying Party (RP)* and presumably limited to the set that the RP can distinguish.

3. Name assignments are made by some Certificate Authority *(CA)* and the introduction of that additional component reduces overall system security. In SPKI/SDSI there is no CA, in the X.509 sense.

4. The real job is to make a security decision and a name by itself does not give enough information to make that decision. SPKI carries authorization information.

There are certain characteristics of SPKI/SDSI that set it apart from other certificate systems: SDSI names, authorization algebra, threshold subjects, canonical S-expressions and certificate revocation (see authorization architecture, authorization management, and authorization policy).

SDSI NAMES: Keys, and by implication their keyholders, need to be identified. SPKI uses the public key itself or its cryptographic hash as the ID of the key and the keyholder. This ID is globally unique and requires no issuer, therefore no expense or added insecurity of an ID issuer. For computers and the protocols between them, this ID is nearly perfect: globally unique and directly authenticable. The hash of the key has the added advantage of being fixed length.

For humans, such IDs fail miserably. They have no mnemonic value. SPKI uses SDSI names for human interfaces. Each human in a system using SPKI/SDSI maintains his or her own dictionary mapping between that human's preferred name for a keyholder and the public key or hash. The human operator can see friendly and meaningful names displayed via a UI, while the underlying system uses the key or its hash as an ID.

Source of Names

There is sometimes confusion among *X.509*, *PGP*, and SDSI—all of which build name certificates. The best way to distinguish them is via the source of the names used (see Table 1).

X.509 started out planning to use globally unique assigned names from the one global X.500 directory. That single directory has never been created and is unlikely ever to be. This leaves X.509

names to be chosen by the CA that issues a certificate. PGP leaves choice of name up to the person generating the key. SDSI gives choice of name to the person who will need to use that name.

Advantage of SDSI Names. When a name is used by a human, the correctness of that use depends on whether the human calls the correct person, thing, or group, to mind on seeing the name. When SDSI names are used, the one who chose that name is the same person who must correctly understand the linkage between the name and the person, thing, or group.

For example, the RP might choose the SDSI name "John Smith", if the RP knows only one John Smith—but a global naming authority would form "John Smith 3751" or jsmith39@localisp.net and require the RP to somehow deduce from that name which John Smith was intended. If the RP has an offline channel to John Smith and can ask him what his global ID is, then the RP can keep a local mapping from his preferred "John Smith" to the global name—but that is exactly what happens with SDSI (the global name being the hash of a key). If the RP does not have off-line contact with this John Smith, then the RP is forced to guess which John Smith is behind the name—and that guess is a source of security error [6].

Group Names

Both X.509 and PGP assume that the name is of an individual. SDSI names are of groups or roles. A named individual is a group of one.

Globally Unique SDSI Names

There are times when a SDSI name needs to be included in a certificate: when rights are assigned to the name or the name is added to some other named group. Since SDSI names are inherently local, a global form must be constructed. For example:

```
(name (hash sha1
  #14dc6cb49900bdd6d67f03f91741cfefa2d26fa2#)
  Leanna)
```

stands for the name Leanna in the local dictionary of the keyholder of the key that hashes via SHA1 to 14dc6cb49900bdd6d67f03f91741cfefa2d26fa2.

This is an advantage of SPKI/SDSI over other ID certificate forms. SDSI knew that names were local and had to do something to make them globally unique while X.509 and PGP assumed names were global, even when they were not.

AUTHORIZATION ALGEBRA: SPKI carries authorization information in its certificates and (Access Control List) ACL entries. This authorization is constrained to be in a language defined by SPKI so that the SPKI library can perform set intersections over authorizations. Each authorization is a set of specific permissions. That set is expressed as an enumeration (a literal set) or in a closed form (e.g., ranges of strings or numbers). The language is defined by intersection rules [2] that were designed to permit the intersection of two authorization sets to be expressed in the same closed form (see also authorization architecture, authorization management, and authorization policy).

By contrast, X.509v3 certificate extensions can be used to carry permission information, but because the extension is completely free-form, custom code must be written to process each different extension type.

FORMAT: An SPKI certificate has five fields:
1. Issuer: the key of the certificate issuer.
2. Subject: a key, hash, SDSI name or threshold subject construct.
3. Delegation: a Boolean, indicating whether the Subject is allowed to delegate some or all of the rights granted here.
4. Tag: a canonical S-expression listing a set of rights granted by the issuer to the subject.
5. Validity: limits on validity: not-before or not-after dates, requirements to check online status or to get a revocation list, etc.

An SPKI ACL entry has fields 2.5 of the above since the authority (issuer) of an ACL entry is the machine that holds it.

A name membership certificate has four fields:
1. Issuer
2. Name being defined
3. Subject (key, hash or name)
4. Validity
There is one certificate for each member of a name.

A *threshold subject* is a list of N subjects (possibly including a subordinate threshold subject) and a parameter K. Only when K of the N subjects agree to delegate some rights or sign some document is that certificate or ACL entry considered valid. (The keys used by these subjects need not be in the same algorithm so, among other things, a threshold subject might tolerate the catastrophic break of one algorithm.)

Canonical S-expressions (CSEXP)

SPKI/SDSI certificates are expressed and communicated as canonical S-expressions. An S-expression is of power equivalent to XML (Extensible Markup Language). Canonical S-expressions are binary forms with only one possible encoding. S-expressions in SPKI/SDSI are constrained to have each list start with an atom (the equivalent of an XML element name). Atoms are binary strings, with an explicit length stated, so CSEXP creation is trivial. CSEXP parsing requires under 10KB of code, in the open-source implementation. If element names are kept small, CSEXP binary forms are smaller than equivalent ASN.1 forms.

CERTIFICATE REVOCATION: At the time SPKI was designed, X.509 used *Certificate Revocation Lists (CRL)* that were optional and were not dated. A new CRL could be issued at any time and would override any prior CRL. In SPKI, revocation is deterministic. Each certificate that could be subject to revocation includes the revocation/validation agent's key and URL and all validity instruments (CRLs, etc.) have contiguous, non-overlapping date ranges.

IMPLEMENTATIONS: SPKI certificates are used in HP's eSpeak and several prototype systems. It is available in open source code in two sourceforge.net projects: CDSA and JSDSI. SPKI's spiritual descendent XrML V.2 [5] is in use in Microsoft's Rights Management Services (RMS).

Carl Ellison

References

[1] Ellison, Carl. SPKI/SDSI Certificates; http://theworld.com/~cme/html/spki.html

[2] Ellison, Carl, Bill Frantz, Butler Lampson, Ronald Rivest, Brian Thomas, and Tatu Ylönen (1999). SPKI Certificate Theory, IETF RFC2693, September 1999, ftp://ftp.isi.edu/in-notes/rfc2693.txt

[3] Ellison, Carl (2002). "Improvements on conventional PKI wisdom." *1st Annual PKI Research Workshop*, April 2002, http://www.cs.dartmouth.edu/~pki02/Ellison/

[4] Blaze, Matt. KeyNote; http://www.crypto.com/trustmgt/kn.html

[5] ISO/IEC JTC1/SC29/WG11/N5231: XrML V.2 (MPEG-21 Rights Expression Language) http://www.chiariglione.org/mpeg/standards/mpeg-21/mpeg-21.htm#_Toc23297977

[6] Dohrmann, Steve and Carl Ellison (2002). "Public-key support for collaborative groups." *1st Annual PKI Research Workshop, April 2002*, 139–148. http://www.cs.dartmouth.edu/~pki02/Dohrmann/

SSH

Secure Shell, a product from SSH Communications Security, allows the user to log into another machine over a network, to execute commands in a remote machine, and to move files from one machine to another. For sometime, it was developed as a standard under IEFT.

SSH basically provides strong authentication and secure communications over insecure channels. It was originally intended as a replacement for various UNIX commands such as telnet, rlogin, rsh, and rcp. For SSH2, there was in addition a replacement for FTP, namely sftp.

When the standardisation was terminated, there were two versions of Secure Shell available: SSH1 and SSH2, which unfortunately are quite different and incompatible. As for the use of cryptographic algorithms, SSH1 supported DES (the Data Encryption Standard) Triple-DES, IDEA, and Blowfish, for encryption, while SSH supports 3DES, Blowfish, Twofish, and a few others. For authentication, SSH1 supported RSA digital signature scheme, while SSH2 supported the Digital Signature Standard.

Peter Landrock

The protocol, in which Alice (A) and Bob (B) want to exchange a key, works as following:

Step 1. A sends B $\alpha := g^{r_A}$ computed in $\langle g \rangle$, where r_A is chosen uniformly random in Z_q.

Step 2. B chooses r_B uniformly random in Z_q and computes $\beta := g^{r_B}$ in $\langle g \rangle$, $k_B := \alpha^{r_B}$ in $\langle g \rangle$ and $\gamma_B := E_{k_B}(sign_B(\alpha, \beta))$, where α and β are concatenated. B sends A: β, γ_B.

Step 3. A computes $k_A := \beta^{r_A}$ and verifies whether the string $D_{k_A}(\gamma_B)$ is the digital signature of (α, β), signed by B. If so, she sends B: $\gamma_A := E_{k_A}(sign_A(\alpha, \beta))$ and views k_A as the authenticated key exchanged with B.

Step 4. B verifies whether the string $D_{k_B}(\gamma_A)$ is the digital signature of (α, β) signed by A. If so, B regards k_B as the authenticated key exchanged with A.

As in the Diffie–Hellman key agreement scheme, if there are no dishonest parties, Alice and Bob will exchange the same key, i.e. $k_A = k_B$.

Yvo Desmedt

Reference

[1] Diffie, W., P.C. van Oorschot, and M.J. Wiener (1992). "Authentication and authenticated key exchanges." *Designs, Codes and Cryptography*, 2, 107–125.

STATION-TO-STATION PROTOCOL

In a two-party *authenticated key exchange* the legitimate parties can compute a secret key, while at the same time being certain about the authenticity of the parties with whom they exchange a key. The scheme must, in particular, be secure against a man-in-the-middle attack.

A popular authenticated version of the Diffie–Hellman key exchange protocol is the *Station-to-Station protocol*. It was proposed by Diffie-van Oorschot-Wiener [1].

Let $\langle g \rangle$ be a suitable finite cyclic group of large enough order in which the computational Diffie–Hellman problem is (assumed to be) hard. We assume that q (not necessarily prime) is a multiple of the order of g and publicly known. Let $sign_A(m)$ indicate the digital signature of the bitstring m by party A. So, $sign_A(m)$ can be verified using the public key of A. Let $E_k(m)$ be a conventional encryption of the bitstring m using the conventional key k. If k is too long, one assumes it is hashed (see hash function). The corresponding decryption is written as $D_k(\cdot)$.

STREAM CIPHER

A *stream cipher* is a symmetric cryptosystem (see key) which operates with a time-varying transformation on individual plaintext digits. By contrast, block ciphers operate with a fixed transformation on large blocks of plaintext digits. More precisely, in a stream cipher a sequence of plaintext digits, $m_0 m_1 \ldots$, is encrypted into a sequence of ciphertext digits $c_0 c_1 \ldots$ as follows: a pseudorandom sequence $s_0 s_1 \ldots$, called the running-key or the *keystream*, is produced by a finite state automaton whose initial state is determined by a secret key. The ith keystream digit only depends on the secret key and on the $(i - 1)$ previous plaintext digits. Then, the ith ciphertext digit is obtained by combining the ith plaintext digit with the ith keystream digit.

Stream ciphers are classified into two types: synchronous stream ciphers and *asynchronous stream ciphers*. The most famous stream cipher is the Vernam cipher, also called *one-time pad*, that leads to perfect secrecy (the ciphertext gives no information about the plaintext).

Stream ciphers have several advantages which make them suitable for some applications. Most notably, they are usually faster and have a lower hardware complexity than block ciphers. They are also appropriate when buffering is limited, since the digits are individually encrypted and decrypted. Moreover, synchronous stream ciphers are not affected by errorpropagation (see also non-linear feedback shift register).

Anne Canteaut

References

[1] Rueppel, R.A. (1986). *Analysis and Design of Stream Ciphers*. Springer-Verlag, Berlin.

[2] Vernam, G.S. (1926). "Cipher printing telegraph systems for secret wire and radio telegraphic communications." *Journal of the American Institute of Electrical Engineers*, 55, 109–115.

STRONG PRIME

A strong prime [1] is an integer p such that

- p is a large prime.
- $p-1$ has a large prime number factor, denoted r.
- $p+1$ has a large prime factor.
- $r-1$ has a large prime factor.

The precise qualification of "large" depends on specific attacks the strong prime is intended to protect against. For a long time, strong primes were believed to be necessary in the cryptosystems based on the RSA problem in order to guard against two types of attacks: factoring of the RSA modulus by the $p+1$ and Pollard $p-1$ factoring methods, and "cycling" attacks. Rivest and Silverman [2] published a paper in 1999 arguing that strong primes are unnecessary in the RSA public key encryption system. There are two points in their argument. First, that the use of strong primes provides no additional protection against factoring attacks, because the Elliptic Curve Method for factoring is about as effective as the $p+1$ and the $p-1$ methods (though none is particularly likely to succeed for random, large primes) and is not prevented by the strong prime conditions. Furthermore, the Number Field Sieve can factor RSA modulus with near certainty in less time than these methods. (See integer factoring for a discussion on factoring methods.) Secondly, they argue that cycling attacks are extremely unlikely to be effective, as long as the primes used are large. This has recently been formally proven in [3]. Thus, in the current state of knowledge,

there is no rationale for requiring strong primes in RSA. A new factoring method might once again make strong primes desirable for RSA, or on the contrary exploit the properties of strong primes in order to factor more efficiently and thus make strong primes appear to be dangerous.

Anton Stiglic

References

[1] Gordon, J. (1985). "Strong primes are easy to find." *Advances in Cryptology—EUROCRYPT'84*, Lecture Notes in Computer Science, vol. 209, eds. T. Beth, N. Cot, and I. Ingemarson. Springer-Verlag, Berlin, 216–223.

[2] Rivest, R.L. and R.D. Silverman (1998). "Are 'strong' primes needed for RSA?" Technical Report, RSA Data Security, Inc., Redwood City, CA, USA.

[3] Shparlinski, I., J.B. Friedlander, and C. Pomerance (2001). Period of the power generator and small values of Carmichael's function. Math. Comp., vol. 70, 1591–1605.

STRONG RSA ASSUMPTION

Let $1 < \tau \in \mathbb{Z}$ be a security parameter. Let $N = pq$ be a product of two random τ-bit primes and let s be an element of the group $\mathbb{Z}_N^*$ (see also modular arithmetic). The strong-RSA problem is defined as follows:

given (N, s) as input, output a pair $a, b \in \mathbb{Z}$ such that $a^b = s \bmod N$ and $b \neq \pm 1$.

Loosely speaking, the Strong-RSA assumption states that for a sufficiently large τ the strong RSA problem is intractable.

The Strong-RSA assumption was introduced by Baric and Pfitzman [2]. The assumption is used to construct efficient signature schemes that are existentially unforgeable under a chosen message attack *without* the random oracle model. One such system is described in [4] and another in [3]. The Strong-RSA assumption is also the basis of several efficient group signature schemes [1].

Dan Boneh

References

[1] Ateniese, Giuseppe, Jan Camenisch, Marc Joye, and Gene Tsudik (2000). "A practical and provably secure coalition-resistant group signature scheme." *Advances in Cryptology—CRYPTO 2000, August*, Lecture Notes in Computer Science, vol. 1880, ed. M. Bellare. Springer-Verlag, Berlin, 255–70.

[2] Baric, N. and B. Pfitzman (1997). "Collision-free accumulators and fail-stop signature schemes without trees." *Proceedings of Eurocrypt*, Lecture Notes in Computer Science, vol. 1233, ed. W. Fumy. Springer-Verlag, Berlin, 480–494.

[3] Cramer, Ronald and Victor Shoup (2000). "Signature schemes based on the strong RSA assumption." *ACM* Transactions on Information and System Security (ACM TISSEC), 3 (3), 161–185, extended abstract in *Proc. 6th ACM Conf. on Computer and Communications Security, 1999*.

[4] Gennaro, Rosario, Shai Halevi, and Tal Rabin (1999). "Secure hash-and-sign signatures without the random oracle." *Advances in Cryptology—EUROCRYPT'99*, Lecture Notes in Computer Science, vol. 1592, ed. J. Stern. Springer-Verlag, Berlin, 123–139.

STRUCTURAL CRYPTANALYSIS

Structural Cryptanalysis is a branch of *Cryptanalysis* which studies the security of cryptosystems described by generic block diagrams. It analyses the syntactic interaction between the various blocks, but ignores their semantic definition as particular functions. Typical examples include meet-in-the-middle attacks on multiple encryptions, the study of various chaining structures used in modes of operation, and the properties of Feistel structures or substitution–permutation networks with a small number of rounds.

Structural attacks are often weaker than actual attacks on given cryptosystems, since they cannot exploit particular weaknesses (such as bad differential cryptanalysis properties or weak *avalanche effect*) of concrete functions. The positive side of this is that they are applicable to large classes of cryptosystems, including those in which some of the internal functions are unknown or key dependent. Structural attacks often lead to deeper theoretical understanding of fundamental constructions, and thus they are very useful in establishing general design rules for strong cryptosystems.

Alex Biryukov

SUBEXPONENTIAL TIME

A *subexponential-time* algorithm is one whose running time as a function of the size k of its input grows more slowly than b^x for *every* base $b > 1$. That is, for every constant base $b > 1$, the running time $T(x)$ satisfies

$$T(x) < b^x$$

for all sufficiently large x. In *O-notation*, this would be written $T(x) = 2^{o(x)}$ or $e^{o(x)}$.

(In computational complexity, subexponential security sometimes refers to the related notion that for all $\epsilon > 0$, $T(x) < 2^{x^\epsilon}$, for all sufficiently large x.)

Subexponential-time algorithms occur in cryptography in connection with the discrete logarithm problem and integer factoring. The fastest algorithms known for those problems (i.e., the ones that grow most slowly as a function of input size) typically have running times of the form

$$e^{(\gamma + o(1))(\log x)^t (\log\log x)^{1-t}}, \qquad \text{for } x \to \infty,$$

for some constants $\gamma > 0$ and $0 < t < 1$, where x is the order of the finite field in which discrete logarithms are being computed, or the modulus to be factored. The size of the input to these algorithms is proportional to the length in bits of x, so the running time, being subexponential in $\log x$, is subexponential in the input size as well (see L-notation).

For further discussion, see exponential time and polynomial time.

Burt Kaliski

SUBGROUP

A subset of elements of a group that is itself a group, i.e., that follows the group axioms (closure, associativity, identity, inverse). For example, if $G = (S, \times)$ is a group, then for any $g \in S$, the set of elements

$$g, g^2, g^3, \ldots$$

(together with the multiplication operation) is a subgroup of G. The order of any subgroup of a group G divides the order of the group G itself; this is known as *Lagrange's theorem*.

Burt Kaliski

SUBGROUP CRYPTOSYSTEMS

In cryptographic applications it is often advantageous to replace a generator of the multiplicative group $\mathbf{F}_{p^t}^*$ of a finite field $\mathbf{F}_{p^t}$ of characteristic p by a generator g of a subgroup of $\mathbf{F}_{p^t}^*$, as originally suggested by Schnorr [2]. The subgroup

$\langle g \rangle$ generated by g must be chosen in such a way that solving the <u>discrete logarithm problem</u> in $\langle g \rangle$ is not easier than computing discrete logarithms in $\mathbf{F}_{p^t}^*$.

Because of the Pohlig–Hellman algorithm (see <u>discrete logarithm problem</u>), the order of g must be chosen in such a way that it contains a sufficiently large prime factor. Usually, g is chosen in such a way that its order q is prime. Because q divides the order $p^t - 1$ of $\mathbf{F}_{p^t}^*$ and because $p^t - 1 = \prod_{s \text{ dividing } t} \Phi_s(p)$, where $\Phi_s(X)$ is the tth cyclotomic polynomial (as defined in the generalization of Pollard's $p - 1$ method—see <u>integer factoring</u>), the prime order q of g divides $\Phi_s(p)$ for one of the s dividing t. However, if q divides $\Phi_s(p)$ for some $s < t$, then $\langle g \rangle$ can effectively be embedded in the proper subfield $\mathbf{F}_{p^s}$ of $\mathbf{F}_{p^t}$. This has the undesirable consequence that the discrete logarithm problem in $\langle g \rangle$ can be solved in the multiplicative group $\mathbf{F}_{p^s}^*$ of the substantially smaller field $\mathbf{F}_{p^s}$, which is easier than solving it in $\mathbf{F}_{p^t}^*$. Thus, in order not to affect the hardness of the discrete logarithm problem in $\langle g \rangle$, the order q of g must be chosen as a sufficiently large prime divisor of $\Phi_t(p)$. Given q, a proper g can be found as $g = h^{(p^t-1)/q}$ for any $h \in \mathbf{F}_{p^t}^*$ such that $g \neq 1$.

If $t = 1$, this implies that p must be chosen so that $\Phi_1(p) = p - 1$ has a large enough prime factor q. If $t = 2$, however, q must be a large prime factor of $\Phi_2(p) = p + 1$ and if $t = 6$ of $\Phi_6(p) = p^2 - p + 1$. The case $t = 1$ corresponds to the traditional and conceptually easiest choice of using the prime field $\mathbf{F}_{p^t} = \mathbf{F}_p$: for 1024-bit security, representation of elements of the subgroup $\langle g \rangle$ requires about 1024 bits. The latter two cases, $t = 2$ and $t = 6$ (or, more generally, t divisible by 2 or 6, respectively) are of interest because they allow a more efficient representation of the subgroup elements when LUC or XTR are used (where LUC [3] refers to 'Lucas' because of LUC's use of Lucas sequences, and XTR [1] is an abbreviation of ECSTR which stands for *efficient compact subgroup trace representation*). For 1024-bit security $1024/2 = 512$ bits suffice for even t when using LUC and $1024/3 \approx 342$ bits suffice for t divisible by 6 when using XTR. Let f be the factor indicating the improvement in representation size: $f = 2$ for LUC and $f = 3$ for XTR.

For any finite field $\mathbf{F}_u$, <u>extension field</u> $\mathbf{F}_{u^v}$, and $w \in \mathbf{F}_{u^v}$ the *trace* $Tr(w)$ of w over $\mathbf{F}_u$ is defined as the sum of the v conjugates of w over $\mathbf{F}_u$: $Tr(w) = \sum_{0 \leq i < v} w^{u^i} \in \mathbf{F}_u$ (the inclusion in $\mathbf{F}_u$ because $Tr(w)^u = Tr(w)$). LUC and XTR work by representing elements of $\langle g \rangle$ by their trace over the subfield $\mathbf{F}_{p^{t/f}}$. The resulting representation advantage of a factor f compared to the traditional representation applies in principle to any element of $\mathbf{F}_{p^t}$. When applied to the order-$\Phi_t(p)$ subgroup G of $\mathbf{F}_{p^t}^*$ with t as above, however, the trace representation has other important advantages: given $Tr(w) \in \mathbf{F}_{p^{t/f}}$ for any $w \in G$, it determines w and its conjugates uniquely and the trace of any power of w can be computed very efficiently. Since g was chosen in such a way that $\langle g \rangle \subset G$, this fast 'exponentiation' applies to the subgroup $\langle g \rangle$ as well. LUC with $t = 2$ and XTR with $t = 6$ allow very efficient methods to find proper p and q of cryptographically relevant sizes. For large choices of t parameter selection becomes more cumbersome. For details of the exponentiation and parameter selection methods, see [3] for LUC and [1] and [4] for XTR.

The fact that the distinction between subgroup elements and their $p^{t/f}$-th powers (i.e., their conjugates over $\mathbf{F}_{p^{t/f}}$) is lost, has been shown (see [1]) to have no negative impact on the security of LUC and XTR. A potential disadvantage of the trace-based systems is that they complicate ordinary multiplication of subgroup elements (represented by their traces).

Arjen K. Lenstra

References

[1] Lenstra, A.K. and E. Verheul (2000). "The XTR public key system." *Advances in Cryptology—CRYPTO 2000*, Lecture Notes in Computer Science, vol. 1880, ed. M. Bellare. Springer-Verlag, Berlin, 1–19.

[2] Schnorr, C.P. (1991). "Efficient signature generation by smart cards." *Journal of Cryptology*, 4, 161–174.

[3] Smith, P. and C. Skinner (1995). "A public-key cryptosystem and a digital signature system based on the Lucas function analogues to discrete logarithms." *Proceedings ASIACRYPT'94*, Lecture Notes in Computer Science, vol. 917, eds. J. Pieprzyk and R. Safari-Naini. Springer-Verlag, Berlin, 357–364.

[4] Stam, M. and A.K. Lenstra (2001). "Speeding up XTR." *Proceedings ASIACRYPT 2001*, Lecture Notes in Computer Science, vol. 2248, ed. C. Boyd. Springer-Verlag, Berlin, 125–143.

SUBSTITUTIONS AND PERMUTATIONS

A substitution cipher is usually described by a sequence or list of single substitutions, each of which is commonly denoted by an arrow, like $\mathrm{p} \longmapsto \pi$.

Example: The Russian-English ISO transliteration (using diacritical marks) is a substitution.

```
A B V G D E · Z I F̃ K L M N O P R S T U F H C Q X W _ Y ^ / Yu "
↓ ↓ ↓ ↓ ↓ ↓ ↓ ↓ ↓ ↓ ↓ ↓ ↓ ↓ ↓ ↓ ↓ ↓ ↓ ↓ ↓ ↓ ↓ ↓ ↓ ↓ ↓ ↓ ↓ ↓ ↓
A B V G D E Ž Z I Ĭ K L M N O P R S T U F H C Č Š Šč ' Y " Ė Ju Ja
```

A substitution may have homophones (see encryption).

A *permutation* is a one-to-one mapping from an alphabet to itself.

A substitution may be described by two lines: the first one being the standard alphabet, the second one being a mixed alphabet (see alphabet). An example is a given below:

```
a b c d e f g h i j k l m n o p q r s t u v w x y z
↓ ↓ ↓ ↓ ↓ ↓ ↓ ↓ ↓ ↓ ↓ ↓ ↓ ↓ ↓ ↓ ↓ ↓ ↓ ↓ ↓ ↓ ↓ ↓ ↓ ↓
B E K P I R C H S Y T M O N F U A G J D X Q W Z L V
```

Note that we have used small letters for the plaintext and capital letters for the ciphertext.

In mathematics, there is a commonly used, simplified notation with two lines bracketed together:

$$\downarrow \begin{pmatrix} a\ b\ c\ d\ e\ f\ g\ h\ i\ j\ k\ l\ m\ n\ o\ p\ q\ r\ s\ t\ u\ v\ w\ x\ y\ z \\ B\ E\ K\ P\ I\ R\ C\ H\ S\ Y\ T\ M\ O\ N\ F\ U\ A\ G\ J\ D\ X\ Q\ W\ Z\ L\ V \end{pmatrix}$$

This is convenient for encryption. For decryption, it is worth while to rearrange the list:

$$\uparrow \begin{pmatrix} q\ a\ g\ t\ b\ o\ r\ h\ e\ s\ c\ y\ l\ n\ m\ d\ v\ f\ i\ k\ p\ z\ w\ u\ j\ x \\ A\ B\ C\ D\ E\ F\ G\ H\ I\ J\ K\ L\ M\ N\ O\ P\ Q\ R\ S\ T\ U\ V\ W\ X\ Y\ Z \end{pmatrix}$$

or

$$\downarrow \begin{pmatrix} A\ B\ C\ D\ E\ F\ G\ H\ I\ J\ K\ L\ M\ N\ O\ P\ Q\ R\ S\ T\ U\ V\ W\ X\ Y\ Z \\ q\ a\ g\ t\ b\ o\ r\ h\ e\ s\ c\ y\ l\ n\ m\ d\ v\ f\ i\ k\ p\ z\ w\ u\ j\ x \end{pmatrix}$$

There is also the cycle notation which is shorter

(a b e i s j y l m o f r g c k t d p u x z v q) (h) (n) (w)

but this notation is inconvenient both for encryption and decryption. The cycle is generated by iterating the substitution on a arbitrarily chosen starting letter; whenever a cycle is closed, a new starting letter is chosen until all letters are exhausted.

Self-reciprocal permutations are permutations that, when applied twice, restore the original. Put equivalently, they are their own inverse. Their cycle notation shows a decomposition in 2-cycles and 1-cycle, for example:

(a n) (b x) (d s) (e i) (f v) (g h) (k u) (l c) (m q) (o w)

(p y) (j) (r) (t) (z)

If a self-reciprocal permutation has no 1-cycle (so n is even) there is also the following notation

$$\updownarrow \begin{pmatrix} a\ b\ c\ d\ e\ f\ g\ h\ i\ j\ k\ l\ m \\ n\ o\ p\ q\ r\ s\ t\ u\ v\ w\ x\ y\ z \end{pmatrix}$$

The Enigma machine of the German *Wehrmacht* used a (properly) selfreciprocal permutation. This was thought to be particularly practical since the same machine could be used for encryption and decryption, disregarding the fact that this opened ways for a cryptanalytic attack (see noncoincidence exhaustion in Cryptanalysis).

A *substitution cipher* in general replaces certain groups of characters by certain other groups of characters. This may be described by a list, e.g.

for $\mathbb{Z}_2^3 \to \mathbb{Z}_2^3$:

$$(000) \longmapsto (001),\ (001) \longmapsto (010),\ (010) \longmapsto (011),\ (011) \longmapsto (100),$$
$$(100) \longmapsto (101),\ (101) \longmapsto (110),\ (110) \longmapsto (111),\ (111) \longmapsto (000).$$

We shall give some more terms that one may see in this context. A *monographic* substitution is a substitution *of* single characters, while a *unipartite* substitution is a substitution *by* single characters.

A *simple* substitution is a substitution of single characters by single characters, so it is a monographic, unipartite substitution.

A *digraphic* substitution is a substitution *of bigrams* (ordered pairs of characters). A *bipartite* substitution is a substitution *by* bigrams. Finally, a *bigram* substitution is a substitution of bigrams by bigrams, so a digraphic, bipartite substitution.

In general, an *n-graphic* substitution is a substitution *of* n-tuples of characters (n-grams) and an *n-partite* substitution is a substitution *by* n-tuples of characters. Similarly, a *polygraphic* substitution is an n-graphic substitution, $n \geq 2$, and a *multipartite* substitution is an n-partite substitution, $n \geq 2$.

A *linear substitution* is a block encryption $\mathbb{Z}_N^n \to \mathbb{Z}_N^m$ that is the composition of a translation t and an homogenous part φ which is additive with respect to addition modulo N (for all $x, y \in \mathbb{Z}_N^n$: $\varphi(x + y) = \varphi(x) + \varphi(y)$).

A *null* is meaningless ciphertext character, the encryption image of the empty plaintext word. It is used, e.g., for swamping the plaintext statistics or masking the occurrence of idle times.

A *straddling encryption* or *straddling cipher* is a substitution with encryption steps $V^{(l)} \to W^{(m)}$, where $Z^{(k)}$ denotes the set of all sequences of at most k characters from Z, in formula $\{\varepsilon\} \cup Z \cup Z^2 \cup Z^3 \ldots \cup Z^k$, where Z^n is the set of all words of length n over the alphabet Z, and ε denotes the empty word.

Example: $Z_{20}^{(3)} \to \mathbb{Z}^{(2)}$ with the homophonic substitution

$$\downarrow \begin{pmatrix} \text{che} & \text{con} & \text{non} & \text{et} & \text{a} & \text{b} & \text{c} & \text{d} & \text{e} & \text{f} & \text{g} & \text{h} & \text{i} \\ 44 & 64 & 00 & 08 & 1 & 86 & 02 & 20 & 62 & 22 & 06 & 60 & 3 \\ & & & & 82 & & & & & & & & \end{pmatrix}$$

$$\downarrow \begin{pmatrix} \text{l} & \text{m} & \text{n} & \text{o} & \text{p} & \text{q} & \text{r} & \text{s} & \text{t} & \text{v} & \text{z} & \epsilon \\ 24 & 26 & 84 & 9 & 66 & 68 & 28 & 42 & 80 & 04 & 88 & 5 \\ & & & 40 & & & & & & & & 7 \end{pmatrix}$$

Both 5 and 7 are in this example (Matteo Argenti, 1590) nulls. Other elements of $Z_{20}^{(3)}$ have no image, except by composition of their individual letters.

Let a *block* be a text of predetermined length. Then a block cipher or *block encryption* is a substitution with encryption steps $V^n \to W^m$, i.e.

without straddling. The block length is usually rather high (for instance, the Data Encryption Standard has a block length of $m = n = 64$, and the *Advanced Encryption Standard* (see Rijndael/AES) has a block length of $m = n = 128$, 192, or 256 bits). The same block encryption step with its key is repeated on and on, thus, each bit of ciphertext in a given block normally depends on the complete corresponding plaintext block, with as consequence the possibility of error propagation over the full block.

A stream cipher (also called *stream encryption*) is a substitution $(V^n)^* \to (W^m)^*$ between infinite series of blocks, controlled by a key generating algorithm. The generated key may have a finite period. Autokey or other cipher feedback is excluded.

A *transposition cipher* or *tranposition* does not substitute the characters of a message, but permutes their position: it may be considered as a special case of a polygraphic substitution $V^n \to V^n$ of the kind

$$(x_1, x_2, \ldots, x_n) \longmapsto (x_{\pi(1)}, x_{\pi(2)}, \ldots, x_{\pi(n)}),$$

where π is a permutation of the *subscripts* $\{1, 2, \ldots, n\}$. It can be performed by multiplication of $(x_1, x_2, \ldots, x_n)$ with a *permutation matrix*, i.e., an $n \times n$ $\{0, 1\}$-matrix such that in every row and in every column, one occurs just once. This extreme property makes cryptanalysis of transposition ciphers very different from cryptanalysis of normal substitution ciphers and explains why alternating composition of substitutions and transpositions (see "pastry dough mixing" below) is so effective.

A *grille* is a tool, usually in the form of punch cards, that can be rotated to perform a transposition of the letters.

Pastry dough mixing stands for a composition of alternating substitutions and transpositions. It was already recommended by Shannon in 1949 and used, e.g., in the DES cryptosystem. The expression 'pastry dough mixing' was introduced by Eberhard Hopf in the mathematical theory of compact spaces.

Friedrich L. Bauer

Reference

[1] Bauer, F.L. (1997). "Decrypted secrets." *Methods and Maxims of Cryptology*. Springer-Verlag, Berlin.

SUBSTITUTION– PERMUTATION (SP) NETWORK

Shannon [1] suggested to use several mixing layers interleaving substitutions and permutations to build strong <u>block ciphers</u>. Such design is called a *substitution–permutation sandwich* or a substitution-permutation network (SPN). Although weak on its own, a line of *substitutions* followed by a *permutation* has good "mixing" properties: substitutions add to local *confusion* and permutation "glues" them together and spreads (*diffuses*) the local confusion to the more distant sub-blocks (see also <u>substitutions and permutations</u>). If one considers flipping a single bit at the input of such a network, it effects the m output bits of particular S-box which in turn are sent to different S-boxes by a permutation. Thus inputs/outputs of up to m S-boxes would be effected by the *avalanche* of change. These are again permuted into different S-boxes, covering almost all the S-boxes of the network. On the output of such network about half of the bits are effected by change and are flipped and about half of the bits are not flipped. This makes an outcome of a single bit change at the input hard to predict, especially if secret key bits are mixed into the block between the layers of encryption. Without a secret key the SPN performs a complex but fully deterministic function of its inputs. Modern ciphers tend to use linear or affine mappings instead of permutations, which allows them to achieve better diffusion in fewer iterations. Such networks are called substitution-linear (SLN) or substitution-affine networks (SAN). The current block encryption standard <u>Rijndael/AES</u> is a SLN cipher.

Alex Biryukov

Reference

[1] Shannon, C.E. (1949). "Communication theory of secrecy system." *Bell System Technical Journal*, 28, 656–715.

SUMMATION GENERATOR

The summation generator is based on a *combination* of n <u>Linear Feedback Shift Registers (LFSRs)</u> and was first proposed in [5, 6]. The combining function is an addition over the set of integers. From a binary point of view, it is a *nonlinear function*, with maximum <u>correlation immunity</u>. The output bit is the least significant bit of the integer sum.

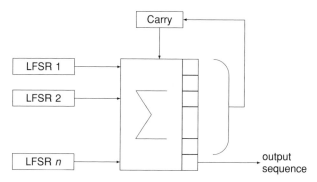

Powerful attacks exist in the case $n = 2$ [1, 4]. Hence, it is better to use several LFSRs, with moderate lengths, than just a few large ones. But it has also been shown that this scheme is vulnerable if all the LFSRs are short [3]. A <u>Fast Correlation Attack</u> has recently been presented in [2]. (See also <u>combination generator</u>.)

Caroline Fontaine

References

[1] Dawson, E. (1993). "Cryptanalysis of summation generator." *Advances in Cryptology—ASIACRYPT'92*, Lecture Notes in Computer Science, vol. 718, eds. J. Seberry and Y. Zheng. Springer-Verlag, Berlin, 209–215.

[2] Golic, J., M. Salmasizadeh, and E. Dawson (2000). "Fast correlation attacks on the summation generator." *Journal of Cryptology*, 13, 245–262.

[3] Klapper, A. and M. Goresky (1995). "Cryptanalysis based on 2-adic rational approximation." *Advances in Cryptology—CRYPTO'95*, Lecture Notes in Computer Science, vol. 963, ed. D. Coppersmith. Springer-Verlag, Berlin, 262–273.

[4] Meier, W. and O. Staffelbach (1992). "Correlation properties of combiners with memory in stream ciphers." *Journal of Cryptology*, 5, 67–86.

[5] Rueppel, R.A. (1986). *Analysis and Design of Stream Ciphers*. Springer-Verlag, Berlin.

[6] Rueppel, R.A. (1986). "Correlation immunity and the summation generator." *Advances in Cryptology—CRYPTO'85*, Lecture Notes in Computer Science, vol. 218, ed. H.C. Williams. Springer-Verlag, Berlin, 260–272.

SYMMETRIC CRYPTOSYSTEM

The type of cryptography in which the same <u>key</u> is employed for each of the operations in the cryptosystem (e.g., encryption and decryption), and thus that same key, typically a secret, must be shared by the parties performing the various operations. See also underline{block cipher}, underline{stream cipher}, underline{MAC algorithms}, and (for the contrasting notion) underline{asymmetric cryptosystem}.

Equivalent names are conventional cryptosystem, secret key cryptosystem, classical cryptosystem, and private key cryptosystem.

Burt Kaliski

SYNCHRONOUS STREAM CIPHER

A synchronous underline{stream cipher} consists of a cipher, in which the keystream is generated independently of the plaintext and of the ciphertext. It can be depicted as follows:

The keystream is usually produced by a pseudo-random generator, parameterized by a key, which is the secret key of the whole scheme.

This means that it is impossible to dynamically check the synchronization between the keystream and the message. The keystreams generated by the sender (encryption), and by the receiver (decryption) must be perfectly synchronized. If synchronization is lost, then decryption fails immediately. If we want to be able to resynchronize both signals, we need some additional techniques (through reinitialization, or by putting some marks in the message, . . .).

Nevertheless, there is an advantage, in terms of errors of transmission. If the ciphertext is altered by some errors, then this will only affect the decryption of the wrong bits, but this will have no effect on the others.

These two properties (perfect synchronization needed, no propagation of errors) lead to some active attacks: the first one could be to modify the ciphertext in order to desynchronize the message and the keystream during decryption (this can easily be achieved by deleting or inserting some bits, for example); the second one consists in modifying the values of some bits, in order to modify the plaintext obtained after decryption (this can be powerful if the attacker knows sufficient information about the message in order to choose the meaning of the modified plaintext). This implies that it is important to use, at the same time as encryption, some integrity/authentication techniques in order to avoid such attacks.

Most of the underline{stream ciphers} used nowadays (see for example underline{E0} and underline{SEAL}) are *binary additive stream ciphers*; they are synchronous stream ciphers, in which all the data (plaintext, keystream, and ciphertext) are binary, and that simply add (through the XOR function) the message (plaintext/ciphertext) to the keystream.

A good reference on the topic is [1].

Caroline Fontaine

Reference

[1] Rueppel, R.A. (1986). *Analysis and Design of Stream Ciphers*. Springer-Verlag, Berlin.

T

TAMPER DETECTION

Tamper detection is the ability of a device to sense that it is under physical attack and initiate defensive actions through tamper response. The tamper detection design can be implemented to sense different types, techniques, and sophistication of tampering. Examples of techniques used to detect tampering may include any or all of the following; switches to detect the opening of doors or access covers, sensors to detect changes in light or pressure within the device, or a barrier or matrix to detect drilling or penetrating the device boundary. To be effective, the detection mechanism must be active regardless of the current logical state of the module.

Tom Caddy

TAMPER RESISTANCE

Tamper resistance is the ability of a device to defend against a threat. The device is often a cryptographic module, but this characteristic could also be important for noncryptographic devices that need to protect design intellectual property or provide evidence of the device integrity for trust or warranty purposes. A device that is designed to have relatively more tamper resistance will require a higher level of attack skills, added work effort, and more sophisticated equipment to perform the tamper actions. Added tamper resistance reduces the probability of compromising the device without damaging the device or triggering a tamper event that will erase keys or disable the device. Often a goal of tamper resistance is to cause the attacker to inflict clear visible evidence on the device, so that the operator and/or manufacturer is aware of the tamper actions and can take appropriate precautions and actions.

Tom Caddy

TAMPER RESPONSE

Tamper Response is the action a device (cryptographic module) performs in order to prevent misuse of the cryptographic module or disclosure of Critical Security Parameters (CSPs) that are generated or stored within the device. The response mechanism is typically triggered by either a signal from a sensor designed to detect (see tamper detection) that the module is in a threat condition or by an explicit command from an operator. The objective of the tamper response is to zeroize (erase) all memory locations that contain cryptographic keys, passwords, PINs (see Personal Identification Number), or other critical security parameters that need to be protected from disclosure to hostile entities. To be effective the response also needs to prevent the module from being misused while in the threat condition, by inhibiting authentication, key management, and cryptographic services from being initiated. The response action needs to be completed quickly enough to prevent the threat from compromising the integrity and trust of the device and CSPs before the response actions have been completed.

Tom Caddy

TEMPEST

The term TEMPEST was initially a US military codeword for a secret research project that originated in the late 1950s and studied compromising emanations. Later, it became the name of the resulting family of (still classified) US and NATO protection standards and test procedures for equipment and facilities that are specially shielded against the unwanted broadcast of sensitive information via spurious electromagnetic, acoustic and other emissions from information processing equipment. The term is sometimes also used as a synonym for compromising emanations and related protective measures in general.

Markus Kuhn

References

[1] Russel, D. and G.T. Gangemi (1991). *Computer Security Basics*. O'Reilly & Associates, Sebastopol, CA, Chapter 10: TEMPEST. ISBN 0-937175-71-4.
[2] McNamara, Joel (1996–2002). "The complete, unofficial TEMPEST information page." Internet Web page, http://www.eskimo.com/~joelm/tempest.html

[3] NACSIM 5000. (1982). "TEMPEST fundamentals." National Security Agency, Fort George G. Meade, MD, Partially declassified transcript: http://cryptome.org/nacsim-5000.htm

THRESHOLD CRYPTOGRAPHY

INTRODUCTION: In modern cryptography most schemes have been developed for a scenario with one sender and one receiver. However, there are scenarios in which many receivers (or many senders) need to share the power to use a cryptosystem. The main motivation for threshold cryptography was to develop techniques to deal with the multi-sender/multi-receiver scenarios.

To illustrate the aforementioned scenarios we first discuss several particular cases of threshold cryptography to clarify its importance. To motivate *threshold decryption*, take the setting of key escrow [4, p. 210]. In Micali's approach [33] as well as the NIST proposal Clipper Chip proposal [7], a threshold scheme is used. Key Escrow agents have shares of each user's secret key. When a court order is received, the law enforcement receives these shares from the Key Escrow agents. This permits recovering the user's secret key. A major disadvantage of these schemes is that once these shares of a user have been provided, the law enforcement receives the technical means (the key) to decrypt any ciphertext received (or sent) by this user. This includes even those ciphertexts sent before the court order was issued. A solution that reduces this risk is one in which a threshold of shareholders decrypt specific messages *without leaking during this process anything about other plaintexts or the secret key*. Achieving the above is precisely the goal of threshold decryption. So, while in threshold schemes the shareholders (called the Key Escrow Agencies in the Clipper proposal) *reveal* their shares to the law enforcement agency, in threshold decryption they *use* their shares; only the plaintext (see cryptosystem) is revealed to the law enforcement agency.

Now consider threshold signatures. Often documents do not originate from a single source but from multiple. Indeed, consider a democracy, e.g. the parliament, in which no member has the power himself/herself to make a law. There, a majority is required. So, the original concept of digital signature needs to be adapted. A seemingly trivial approach would be to have the members have shares of the secret key (corresponding to the parliament public key), and when a message needs to be signed, a member voting for the proposal, just reveals his/her share to the speaker (chair). The speaker can then compute the secret key and sign the message. However, this approach has a major security problem. Indeed, the speaker obtaining the secret key could fraudulently (now or in the future) authenticate other messages alone. The real solution to prevent this type of fraud is to have the members calculate the digital signature together *without divulging, to each other or to others, anything more about digital signatures for different messages, or about the secret key used in the process*. Achieving this is the goal of threshold signatures.

In general, any traditional cryptosystem has a threshold variant. Consider a scheme that has some security requirement S against an adversary. In such a system usually a party has some secret key. To achieve the threshold variant, first, this secret key is shared among l parties, often called shareholders. To achieve a threshold cryptography variant one needs the following conditions:

Reliability: The new cryptographic scheme, when executed by t of the shareholders, satisfies the security condition S.

Threshold Security: Any attempt by $t - 1$ shareholders to help a passive (active) adversary break the original security condition S will fail.

A scheme satisfying these conditions is called a t-out-of-l threshold cryptosystem (for a more detailed description among these lines, see [17]).

So, one can speak, for example, about:

- **threshold pseudorandomness** in which a threshold of parties generate jointly a pseudorandom string.

- **threshold zero-knowledge interactive proof of knowledge** in which a threshold of parties prove to know jointly a secret, without divulging to each others and outsiders anything additionally about this secret.

- **threshold authentication codes** in which a threshold of parties is required to generate a MAC (see MAC algorithms).

We will first focus on the case insiders try to help a passive adversary. Note that in the above description of threshold cryptography, the threshold condition can be replaced by a general access structure, as already addressed by [20]. From now on when we speak about threshold cryptography, we include the case the access structure is a general access one.

Since threshold cryptography is a very active research area, it is impossible in a few pages to give a complete survey of all the work done and all of the subtle issues.

DIFFERENT APPROACHES: We distinguish two main approaches and a variant.

A General Approach

For the case of the general approach [10], let us express the underlying cryptographic scheme as $f_k(input)$, where k is a secret key and, in the case of, for example:

threshold decryption: the *input* is a ciphertext,
threshold signatures: the *input* is a message to sign.

In the general approach, one defines a parameterized function g such that $g_{input}(k) = f_k(input)$ for each k and input. The general approach now works when the function g_{input} is an <u>homomorphism</u>, i.e.

$$g_{input}(k_1) * g_{input}(k_2) = g_{input}(k_1 + k_2), \quad (1)$$

where $k_1, k_2 \in K(+)$, an Abelian <u>group</u>.

Note that many cryptographic schemes satisfy this condition. Examples are the one-time pad (see <u>key</u>) encryption/decryption and the <u>RSA digital signature scheme</u>. A part of the decryption in the <u>ElGamal public key encryption</u> scheme also satisfies this condition. Indeed when receiving (c_1, c_2), to decrypt the receiver computes $m' := c_2 \cdot (c_1^a)^{-1}$. The exponentiation is the only part that uses the secret key a. This exponentiation trivially satisfies Equation (1).

We now explain how one can use such a cryptosystem to make a threshold cryptographic variant. To facilitate the reading, we start with a simple case. In general one uses a multiplicative secret sharing scheme, which is discussed in Section "Using a multiplicative secret sharing scheme."

A 2-out-of-2 Case. Equation (1) can be used immediately to obtain a 2-out-of-2 threshold cryptographic variant of the cryptographic scheme based on g. Indeed, assume $k \in K(+)$ is the secret key, where K is a <u>group</u> (a quasigroup [1] is sufficient). Now assume that k_1 was chosen uniformly random in K and that $k_2 = -k_1 + k$. Give the first participant k_1 and the second k_2 as shares of k.

Each time now that the cryptographic scheme must be used on some input a, participant i computes the *partial result* $\alpha_i := g_a(k_i) = f_{k_i}(a)$, i.e. the underlying cryptographic scheme is used as if the key is replaced by the share k_i. (If we talk about

threshold signatures the partial result is called *partial signature*, etc.)

Now, from α_1 and α_2 anybody can compute what the output of the original cryptographic scheme would have been using Equation (1). If this result is transmitted outside some organization, the party that performs this computation is called a *combiner*.

The proof of security is rather easy in this case. Since slightly different arguments are needed dependent whether the security condition is an unconditional or condition one, we give a sketch. First, the reliability condition is satisfied due to the <u>homomorphic</u> property of g and the fact that $k = k_1 + k_2$. The threshold security property follows from the following argument. Suppose one participant could help an adversary in beating the security condition S, then the adversary does not need that participant (informally, the original scheme does not satisfy the condition S). We demonstrate that the adversary can just simulate that participant. Indeed k_1 occurs uniformly random, and due to the one-time pad property [39] so does k_2. Suppose the inside participant is P_1, then the adversary chooses a uniformly random k_1 and computes $g_a(k_1)$. Since $g_a(k) = g_a(k_1) \cdot g_a(k_2)$, the task for the adversary to compute $g_a(k)$ is as hard as the computation of $g_a(k_2)$.

The more general case of a t-out-of-l threshold cryptographic scheme (or the use of a general access structure) is only a little more complex.

Using a Multiplicative Secret Sharing Scheme. First we define what a *multiplicative secret sharing* scheme is.

DEFINITION 1. Let $\mathcal{K}(*)$ be a <u>group</u> from which the secrets are chosen and $\mathcal{S}_i$ be a set from which the <u>shares</u> of the ith participant are chosen. Let $\mathcal{A} = \{P_1, \ldots, P_l\}$ be a set of l participants, and $\Gamma_\mathcal{A}$ be an <u>access structure</u> over $\mathcal{A}$. A multiplicative sharing scheme for $\mathcal{G}$ is a secret sharing scheme in which the dealer distributes the shares in such a way that if P_i gets s_i, $i = 1, \ldots, n$, then for each $\mathcal{B} = \{P_{i_1}, \ldots, P_{i_{|B|}}\} \in \Gamma_\mathcal{A}$, any secret $k \in \mathcal{K}$ can be expressed as

$$k = \psi_{i_1, \mathcal{B}}(s_{i_1}) \cdot \psi_{i_2, \mathcal{B}}(s_{i_2}) \cdot \cdots \cdot \psi_{i_{|B|}, \mathcal{B}}(s_{i_{|B|}}), \quad (2)$$

where $\psi_{j, \mathcal{B}}$, $j \in \mathcal{B}$, are functions from $\mathcal{S}_j$ to $\mathcal{K}$ which any P_j can compute.

If the participants P_j are polynomially bounded, then one should assume that $\psi_{i, \mathcal{B}}(s_j)$, $j \in \mathcal{B}$, can be computed in <u>polynomial time</u>.

Note that when K is an Abelian group, and we use additive notation, then Equation (2) becomes:

$$k = \sum_{i \in \mathcal{B}} \psi_{i,\mathcal{B}}(s_i). \tag{3}$$

We now explain how to achieve the threshold cryptographic scheme. Let the participants have shares s_i of the secret key k. To execute the cryptographic scheme on input a, when the set $\mathcal{B}$ is known, each participant P_i in $\mathcal{B}$ computes the partial result $g_a(\psi_{i,\mathcal{B}}(s_i)) = f_{\psi_{i,\mathcal{B}}(s_i)}(a)$. These partial results are sent to the combiner, who basically performs a multiplication. Indeed due to the homomorphic property of g, we have that

$$g_a(k) = \prod_{i \in \mathcal{B}} g_a(\psi_{i,\mathcal{B}}(s_i)). \tag{4}$$

The case that $\mathcal{B}$ is not known in advance is briefly discussed in Section "Unknown subsets of participants".

We now list some examples of multiplicative secret sharing schemes. If K is a finite field, then <u>Shamir's threshold scheme</u> is multiplicative. This allows to explain the threshold <u>ElGamal public key encryption</u> scheme [12]. In the case of the <u>RSA digital signature scheme</u>, $K = Z_{\phi(n)}$ (see <u>modular arithmetic</u> and <u>Euler's totient function</u>) and so Shamir's threshold scheme cannot be used. An additional problem is that making $\phi(n)$ public would allow an enemy to break RSA. To address this issue *black box secret sharing* schemes were studied. These work, regardless what K is, provided one can perform the group operation in the Abelian group K efficiently. For details of such schemes see [8, 15].

(For more details about multiplicative secret sharing, see, e.g. [9, 11, 13] and also [20, 24].)

Unknown Subset of Participants. An interesting case occurs when the set of participants is unknown before the final output is computed by the combiner, and one wants to avoid interaction between the shareholders and the combiner.

This issue can be resolved in the case the functions $\psi_{i,\mathcal{B}}$ correspond with multiplications with the scalar $a_{i,\mathcal{B}}$. This can be addressed using module theory [28]. Still these scalars depend on $\mathcal{B}$. So, this multiplication cannot be executed by the participant. Using a property of modules [29], the combiner can instead perform an equivalent operation. For more details consult, e.g. [10].

Particular Approaches

The general approach suffers from two main problems, being:

- not all cryptographic schemes satisfy the homomorphic property of Equation (1). In its generality no practical solution has been proposed to address this problem. For some algorithms, such as the <u>Digital Signature Standard</u> (DSS), a practical approach was studied in [26, 32].
- Using the most general approach does not necessarily result in the most efficient scheme. The use of a black box secret sharing scheme, when K is a <u>finite field</u> is just one example. Shoup [40] proposed a scheme specifically designed to achieve threshold RSA signatures.

A Variant of the General Approach

It seems that the last approach should outperform the first in efficiency. However, King [30] (see also [31]) demonstrated that when the set of shareholders that will jointly execute RSA is known in advance, that the scheme in [19] outperforms Shoup's scheme.

So, the idea to start from a general scheme and then to (sub-)optimize it to a particular underlying cryptographic scheme, is not necessarily a bad approach.

ENHANCEMENTS: The aforementioned schemes can be viewed as basic schemes. Indeed, until now we assumed that insiders would only help a passive adversary. When protecting against insiders trying to help an active adversary, several issues need to be addressed. These are now surveyed and references are given.

Robustness

To analyze the reliability aspect, let us focus on Equation (4). It is clear that if one (or more) shareholder sends one wrong partial result, $g_a(\psi_{i,\mathcal{B}}(s_i))$ the result will (likely) be wrong. If a public key system is used one can, using the public key, verify that the result is wrong. When the numbers of wrong partial results is small, and a public key system is used, an exhaustive search will evidently find out the subset of participants that sent the wrong partial result [38]. One can then recompute the result $g_a(k)$ by ignoring the wrong partial results, provided one has at least $t + e$ partial results, where e is the number of wrong ones.

A threshold scheme that addresses this issue is called robust. Several schemes that do not need above exhaustive search have been presented for variant underlying cryptographic schemes, e.g. [23, 25, 26].

To avoid this exhaustive search, in the case of an underlying public key scheme, the following can be used. If t participants each prove that what was sent is the correct partial result, then one can ignore all other partial results.

No Trusted Dealer

One can wonder who computes the secret share the participant will use. In the first schemes a single trusted dealer was often used. It is clear that such an approach can best be avoided, however this is not always that easy.

If the secret key can be chosen randomly, as is usually the case in a discrete log setting, then the following use of *homomorphic secret sharing* can be utilized towards abolishing the need for a trusted dealer. We first define the notion of homomorphic secret sharing.

DEFINITION 2. Let $(s_1, s_2, \ldots, s_l)$ be a share assignment of the key k and similarly $(s'_1, s'_2, \ldots, s'_l)$ be the shares of the key k'. Assume operations, denoted using "+", are defined on the share spaces and the key space. A secret sharing scheme is called homomorphic [3] if $((s_1 + s'_1), (s_2 + s'_2), \ldots, (s_l + s'_l))$ is a possible share assignment of the key $k + k'$.

Shamir secret sharing scheme is homomorphic. In fact, any multiplicative secret sharing scheme in which each function $\psi_{i,\mathcal{B}}$ (for $\mathcal{B} \in \Gamma_A$) is a homomorphism, forms a homomorphic secret sharing scheme.

We now survey how to avoid a dealer. The first participant chooses a uniformly random key k_1 and plays distributor of this key generating shares $(s_{1,1}, s_{1,2}, \ldots, s_{1,l})$. The first participant sends, using a secure channel, the shares $s_{1,i}$ to participant i and $1 \leq i \leq l$. Now, t participants, let say those in $\mathcal{B} \subset \mathcal{A}$, will perform similar operations (choosing the randomness independently) and create shares $s_{j,i}$ instead of $s_{1,i}$ and send those privately to participant i. A participant i can then compute the share $s_i = \sum_{j \in B} s_{j,i}$. Since the sharing scheme is homomorphic, s_i is a share of the key $k = \sum_{j \in B} k_j$. If the keys k_j belong to an Abelian group (see also [24]) and the secret sharing scheme is perfect, then $t - 1$ shareholders have no information about the secret key k.

The first use of this idea in the context of threshold cryptography was by Pedersen in [35]. Pedersen's scheme also guarantees that the distribution is <u>verifiable</u>, i.e., that the shares the shareholders received will always recompute the same secret key. Pedersen's scheme also guarantees that this secret key corresponds to the public key that is made public (see <u>public key cryptography</u>).

Note that the problem of avoiding a trusted dealer is much more complex in the context of RSA (see, e.g. [5, 36]).

Proactive Security and its Generalizations

One can wonder what should happen when a share is stolen or lost. Worse, what happens when an outsider collects more shares than the threshold? As already observed in [16], it is a bad idea to change the public key of a group, in particular when this group is well known. Those who have not updated their public key database will use the old one. Also the new public key must be certified enough times independently before it can be trusted.

The solution that has been proposed to address this problem is to get new *guaranteed correct shares* without relying on a trusted dealer and to keep the old public key as long as is reasonable possible. The old shares should be destroyed and the update should be done frequently enough, taken the power of the enemy who may collect shares into account. This solution is called *proactive* secret sharing.

We briefly explain how the use of homomorphic secret sharing can contribute towards achieving proactive threshold cryptography. We assume that the secret sharing scheme is homomorphic.

If $(s_1, s_2, \ldots, s_l)$ is a share assignment for the key k and $(s'_1, s'_2, \ldots, s'_l)$ is a uniformly random share assignment for the "key" 0, then $(s''_1, s''_2, \ldots, s''_l) = (s_1 + s'_1, s_2 + s'_2, \ldots, s_l + s'_l)$ is a new share assignment for the same key k. Assume that one trusts t shareholders. Then t participants, denoted by j, can each contribute their own random $(s'_{j,1}, s'_{j,2}, \ldots, s'_{j,l})$. This is done in a similar way as in Section "No Trusted Dealer," however the shares correspond with the "keys" 0. When working in an Abelian group (see also [24]) and when the secret sharing scheme is perfect, the resulting share $s''_i = s_i + \sum_{j \in B} s'_{j,i}$ will be guaranteed independent of the original share s_i, due to the properties of the one-time-pad [39]. Both s_i and s''_i are shares of the same key k.

One should note that the schemes are more complex since each contributing shareholder needs to prove that his contribution $(s'_1, s'_2, \ldots, s'_l)$ consists of shares of 0. Also, achieving proactive threshold RSA is more complex. The concept of proactive secret sharing is based on [34] and its combination with threshold cryptography has been studied in [22, 27, 37].

The following is a more general problem. How, given shares for authorized subsets of the participants in A, specified by an access structure Γ_A, can the participants in A, without the use of a trusted dealer, distribute new shares for an access structure $\Gamma'_{A'}$, where A' is the new set of participants. If $\Gamma_A \nsubseteq \Gamma'_{A'}$ it is clear that some shareholders must destroy their shares. The security requirement is that no unauthorized set of parties learns anything new about the secret key. For more details see [2, 6, 13, 21].

OTHER SURVEYS: Other surveys of threshold cryptography can for example be found in [14, 18].

Yvo Desmedt

References

[1] Albert, A.A. (1943). "Quasigroups I." *Transactions of the American Mathematical Society*, 54, 507–519.

[2] Bao, F., R. Deng, Y. Han, and A. Jeng (1997). "Design and analysis of two basic protocols for use in TTP-based key escrow." *Information Security and Privacy, Second Australian Conference, ACISP'97, Sydney, NSW, Australia, July 7–9*, Lecture Notes in Computer Science, vol. 1270, eds. V. Varadharajan, J. Pieprzyk, and Y. Mu. Springer-Verlag, Berlin, 261–270.

[3] Benaloh, J.C. (1987). "Secret sharing homomorphisms: Keeping shares of a secret secret." *Advances in Cryptology—CRYPTO'86, Santa Barbara, CA, August 11–15*, Lecture Notes in Computer Science, vol. 263, ed. A. Odlyzko. Springer-Verlag, Berlin, 251–260.

[4] Beth, T. (1990). "Zur Sicherheit der Informationstechnik." *Informatik-Spektrum*, 13, 204–215 (in German).

[5] Boneh, D. and M. Franklin (1997). "Efficient generation of shared RSA keys." *Advances in Cryptology—CRYPTO'97, Santa Barbara, CA, August 17–21*, Lecture Notes in Computer Science, vol. 1294, ed. B.S. Kaliski. Springer-Verlag, Berlin, 425–439.

[6] Chen, L., D. Gollmann, and C. Mitchell (1997). "Key escrow in mutually mistrusting domains." *Security Protocols, Cambridge, UK, April 10–12*, Lecture Notes in Computer Science, vol. 1189, ed. M. Lomas. Springer-Verlag, Berlin, 139–153.

[7] A proposed federal information processing standard for an escrowed encryption standard (EES). Federal Register, July 30, 1993.

[8] Cramer, R. and S. Fehr (2002). "Optimal blackbox secret sharing over arbitrary abelian groups." *Advances in Cryptology—CRYPTO 2002, Santa Barbara, CA, August 18–22*, Lecture Notes in Computer Science, vol. 2442, ed. M. Yung. Springer-Verlag, Berlin, 272–287.

[9] Di Crescenzo, G. and Y. Frankel (1999). "Existence of multiplicative secret sharing schemes with polynomial share expansion." *Proceedings of the Tenth Annual ACM-SIAM Symposium on Discrete Algorithms, January, 17–19, Baltimore, MD.*

[10] De Santis, A., Y. Desmedt, Y. Frankel, and M. Yung (1994). "How to share a function securely." *Proceedings of the Twenty-Sixth Annual ACM Symposium*. Theory of Computing (STOC), *May 23–25, Montréal, Québec, Canada*, ACM Press, 522–533.

[11] Desmedt, Y., G. Di Crescenzo, and M. Burmester (1995). "Multiplicative non-abelian sharing schemes and their application to threshold cryptography." *Advances in Cryptology—Asiacrypt'94, Wollongong, November–December 1994*, Lecture Notes in Computer Science, vol. 917, eds. J. Pieprzyk and R. Safavi-Naini. Springer-Verlag, Berlin, 21–32.

[12] Desmedt, Y. and Y. Frankel (1990). "Threshold cryptosystems." *Advances in Cryptology—CRYPTO'89, Santa Barbara, CA, August 20–24*, Lecture Notes in Computer Science, vol. 435, ed. G. Brassard. Springer-Verlag, Berlin, 307–315.

[13] Desmedt, Y. and S. Jajodia (1997). "Redistributing secret shares to new access structures and its applications." Technical Report ISSE-TR-97-01, George Mason University, July, ftp://isse.gmu.edu/pub/techrep/97_01.jajodia.ps.gz

[14] Desmedt, Y.G. (1994). "Threshold cryptography." *European Trans. on Elecommunications*, 5 (4), 449–457 (Invited paper).

[15] Desmedt, Y.G. and Y. Frankel (1994). "Homomorphic zero-knowledge threshold schemes over any finite abelian group." *SIAM Journal on Discrete Mathematics*, 7 (4), 667–679.

[16] Desmedt, Y. (1988). "Society and group oriented cryptography: A new concept." *Advances in Cryptology—CRYPTO'87, Santa Barbara, CA, August 16–20*, Lecture Notes in Computer Science, vol. 293, ed. C. Pomerance. Springer-Verlag, Berlin, 120–127.

[17] Desmedt, Y. (1993). "Threshold cryptosystems." *Advances in Cryptology—ASIACRIPT'92, Old Coast, Queensland, December*, Lecture Notes in Computer Science, vol. 718, eds. J. Seberry and Y. Zheng. Springer-Verlag, Berlin, 3–14 (Invited paper).

[18] Desmedt, Y. (1997). "Some recent research aspects of threshold cryptography." *Information Security, proceedings, September 17–19, 1997, Tatsunokuchi*, Ishikawa, Japan. Lecture Notes in Computer Science, vol. 1396, eds. E. Okamoto, G. Davida, and M. Mambo. Springer-Verlag, Berlin, 158–173. (Invited Lecture).

[19] Frankel, Y. and Y. Desmedt (1992). "Parallel reliable threshold multisignature." Tech. Report TR-92-04-02, Dept. of EE & CS, University of Wisconsin–Milwaukee, ftp://ftp.cs.uwm.edu/pub/tech_reports/desmedt-rsa-threshold_92.ps

[20] Frankel, Y. and Y. Desmedt (1992). "Classification of ideal homomorphic threshold schemes over finite Abelian groups." *Advances in Cryptology—EUROCRYPT'92, Balatonfüred, Hungary*, Lecture

Notes in Computer Science, vol. 658, ed. R.A. Rueppel. Springer-Verlag, Berlin, 25–34.

[21] Frankel, Y., Y.P. Gemmell, P.D. MacKenzie, and M. Yung (1997). "Optimal resilience proactive public key cryptosystems." *38th Annual Symp. on Foundations of Computer Science (FOCS)*, October 20–22, Miami Beach, FL, USA. IEEE Computer Society Press, Los Abumitos, CA.

[22] Frankel, Y., P. Gemmell, P.D. MacKenzie, and M. Yung (1997). "Proactive RSA." *Advances in Cryptology—CRYPTO'97, Santa Barbara, CA, August 17–21,* Lecture Notes in Computer Science, vol. 1294, ed. B. S. Kaliski. Springer-Verlag, Berlin, 440–454.

[23] Frankel, Y., P. Gemmell, and M. Yung (1996). "Witness-based cryptographic program checking and robust function sharing." *Proceedings of the Twenty-Eighth Annual ACM Symposium on Theory of Computing*, ACM Press. *May 22–24,* 499–508.

[24] Frankel, Y., Y. Desmedt, and M. Burmester (1993). "Non-existence of homomorphic general sharing schemes for some key spaces." *Advances in Cryptology—CRYPTO'92, Santa Barbara, CA, August 16–20,* Lecture Notes in Computer Science, vol. 740, ed. E. F. Brickell. Springer-Verlag, Berlin, 549–557.

[25] Gennaro, R., S. Jarecki, H. Krawczyk, and T. Rabin (1996). "Robust and efficient sharing of RSA functions." *Advances in Cryptology—CRYPTO'96, Santa Barbara, CA, August 18–22,* Lecture Notes in Computer Science, vol. 1109, ed. N. Koblitz. Springer-Verlag, Berlin, 157–172.

[26] Gennaro, R., S. Jarecki, H. Krawczyk, and T. Rabin (1996). "Robust threshold DSS signatures." *Advances in Cryptology—EUROCRYPT'96, Zaragoza, Spain, May 12–16,* Lecture Notes in Computer Science, vol. 1070, ed. U. Maurer. Springer-Verlag, Berlin, 354–371.

[27] Herzberg, A., S. Jarecki, H. Krawczyk, and M. Yung, (1995). "Proactive secret sharing." *Advances in Cryptology—CRYPTO'95, Santa Barbara, CA, August 27–31,* Lecture Notes in Computer Science, vol. 963, ed. D. Coppersmith. Springer-Verlag, Berlin, 339–352.

[28] Jacobson, N. (1985). *Basic Algebra I*. W. H. Freeman and Company, New York.

[29] Jacobson, N. (1989). *Basic Algebra II*. W. H. Freeman and Company, New York.

[30] King, B. (1976). "Improved methods to perform threshold RSA." *Advances in Cryptology—ASIACRYPT 2000, December 2000, Kyoto, Japan,* Lecture Notes in Computer Science, vol. 1976, ed. T. Okamoto. Springer-Verlag, Berlin, 359–372.

[31] King, B. (2000). "Algorithms to speed up computations in threshold RSA." *Information Security and Privacy, 5th Australian Conference, ACISP2000, Brisbane, Australia, July 10–12,* Lecture Notes in Computer Science, vol. 1841, eds. E. Dawson, A. Clark, and C. Boyd. Springer-Verlag, Berlin, 443–456.

[32] Langford, S.K. (1995). "Threshold DSS signatures without a trusted party." *Advances in Cryptology—CRYPTO'95, Santa Barbara, CA, August 27–31,* Lecture Notes in Computer Science, vol. 963, ed. D. Coppersmith. Springer-Verlag, Berlin, 397–409.

[33] Micali, S. (1993). "Fair public-key cryptosystems." *Advances in Cryptology—CRYPTO'92, Santa Barbara, CA, August 16–20,* Lecture Notes in Computer Science, vol. 740, ed. E.F. Brickell. Springer-Verlag, Berlin, 113–138.

[34] Ostrovsky, R. and M. Yung (1991). "How to withstand mobile virus attacks." *Proceedings of the 10-th Annual ACM Symp. on Principles of Distributed Computing, August 19–21, Montreal, Quebec, Canada,* ACM Press 51–60.

[35] Pedersen, T.P. (1991). "A threshold cryptosystem without a trusted party." *Advances in Cryptology—EUROCRYPT'91, April 1991, Brighton, UK,* Lecture Notes in Computer Science, vol. 547, ed. D.W. Davies. Springer-Verlag, Berlin, 522–526.

[36] Poupard, G. and J. Stern (1998). "Generation of shared RSA keys by two parties." *Advances in Cryptology—ASIACRYPT'98, Beijing, China, Ocotober,* Lecture Notes in Computer Science, vol. 1514, eds. K. Ohta and D. Pei. Springer-Verlag, Berlin, 11–24.

[37] Rabin, T. (1998). "A simplified approach to threshold and proactive RSA." *Advances in Cryptology—CRYPT'98,* Lecture Notes in Computer Science, vol. 1462, ed. H. Krawczyk. Springer, Berlin, 89–104.

[38] Reiter, M.K. and K.P. Birman (1994). "How to securely replicate services." *ACM Transactions on Programming Languages and Systems,* 16 (3), 986–1009.

[39] Shannon, C.E. (1949). "Communication theory of secrecy systems." *Bell System Techn. Jour.,* 28, 656–715.

[40] Shoup, V. (2000). "Practical threshold signatures." *Advances in Cryptology—EUROCRYPT 2000, Bruges, Belgium, May 14–18,* Lecture Notes in Computer Science, vol. 1807, ed. B. Preneel. Springer-Verlag, Berlin, 207–220.

THRESHOLD SIGNATURE

Threshold signatures are underline{digital signatures} where signers can establish groups such that only certain subsets of the group can produce signatures on behalf of the group. The collection of subsets that are authorized to produce signatures is called the underline{access structure} of a threshold scheme. More particularly, a (t, n)-threshold signature scheme is a digital signature scheme where any t or more signers of a group of n signers can produce signatures on behalf of the group. In general, a threshold signature does not reveal the actual group members that have cooperated to produce it.

Multisignatures are threshold signatures with the additional feature that they reveal the identities of the group members who produced them [2, 12]. In multisignatures, the signing members are not anonymous at all. The special case of a $(1, 1)$-threshold signature scheme is an ordinary underline{digital signature} scheme. The goal of a threshold signature scheme is to enforce dual control over the signing capability (choose $t > 1$) or to eliminate single points of failure (choose $n > 1$), or both. Each group of signers can be managed by a *trusted group authority*, which oversees joining and leaving the group. Obviously, many groups can choose to be managed by the same trusted group authority, or a group can choose to fully distribute the group management among its members such that every member is involved in all management transactions. Comprehensive overviews of threshold signatures are given by Desmedt [4, 5].

In a threshold signature scheme, each signing member of a group has an individual signing key pair (see also public key cryptography). If individuals generate their key pairs without having to agree on common domain parameters, the threshold signature scheme is called separable [3]. An individual is registered for a group by presenting a suitable ID certificate to the respective trusted group authority and submitting her or his public verifying key. The trusted group authority constructs a group key pair, which consists of a private group key and a public group key, and publishes the public group key through one or more authentic channels such as a public key infrastructure (PKI). A member leaves a group by revoking her or his public verifying key from the trusted group authority. It is the responsibility of the trusted group authority to keep track of who belongs to the group at any point of time.

Any subset of $\geq t$-out-of-n members of a group G can produce a signature. To do so, each member contributes a *partial signature* to a *designated combiner*, and the combiner derives the intended threshold signature from the partial signatures. Everyone who has access to the public group key of group G can verify the threshold signature. The designated combiner can be a real entity such as the trusted group authority, or it can be a virtual entity whose operations are computed in a distributed fashion among all group members [8]. A threshold signature scheme is *robust*, if the designated confirmer can verify the validity of each partial signature before accepting it as an input to a threshold signature [8, 13]. Many of the early threshold signatures were not robust [2, 6, 7].

The most important security property of threshold signatures is security against existential forgery. In ordinary digital signature schemes, one considers adaptive chosen message attacks, i.e., an attacker who has repeated oracle access to signers in order to get messages signed, but without access to the signer's private signing key. In threshold signatures, one considers stronger attackers, who can take complete control over some of the group members at some time (including access to their private signing keys). Such an adversary is called *static* if he chooses the group members to be controlled before the attack begins. Otherwise, the adversary is called *adaptive*. Note that this kind of adaptivity is different from the one introduced by Goldwasser, Micali and Rackoff [10] for attackers against ordinary digital signature schemes. *Resilience* is the number or fraction of cheating members a group can tolerate during group key generation or update while it still maintains unforgeability of the resulting threshold signatures (also see secret sharing schemes). The adversary is called *erasure-free* if it does not require that group members erase all information from their computer systems once this information is no longer needed, e.g., replaced private signing keys [11]. Further characteristics of adversaries are discussed in [11].

Key management is a particularly important issue in threshold signatures. For example, if the public group key changes whenever members join or leave the group or update their private signing keys, then it becomes a burden for the trusted group authority to publish the public group keys for all recent time-intervals in a timely fashion. Moreover, if a private signing key of a group member is compromised, then the attacker can freely produce signatures on behalf of the group until the respective public verifying key is revoked. Therefore, all the signatures of the victimized group members must be regarded invalid if there is no way of distinguishing the signatures produced by the honest group members from those produced by the attackers. These problems are addressed by an approach called *proactive security*. Proactive threshold signature schemes allow individual group members to join or leave a group or update their private signing keys without affecting the respective public group key. By dividing the lifetime of all individual private signing keys into discrete time intervals, and by tying all signatures to the time interval when they are produced, group members who are revoked in time interval i have their signing capability effectively stripped away in time interval $i + 1$, while all their signatures produced in time interval i or before (and of course the signatures of all other group members) remain verifiable and valid [9, 13]. Proactive security in threshold signature schemes is similar to *forward security* in group signature schemes.

Of particular interest are those proactive threshold schemes, where the effort of signing and verification and the length of each signature is independent of the number of signers who actually produce a signature and of the group size. An example of such threshold signature scheme is proposed by Rabin [13]. Multisignatures with these properties are proposed in [1]. They are based on the *Gap Diffie–Hellman Groups*, i.e., groups in which the *computational* Diffie–Hellman (CDH) problem is hard, while the decisional Diffie–Hellman (DDH) problem is easy to solve.

A threshold signature scheme has the following operations: (i) An operation for generating pairs, a private signing key, and a public verifying key for an individual, (ii) an operation for generating pairs of a private group key and a public group key for a trusted group authority, (iii) operations for group management such as joining and revoking group members and updating their individual key pairs, (iv) an operation for signing messages, (v) an operation for verifying signatures against a public group key.

The characteristic security requirements of a threshold signature scheme are:

Unforgeability: Resistance against existential forgery under chosen message attacks by computationally restricted attackers.

Threshold: In general, any of the authorized subsets of group members can produce a threshold signature, but no other coalitions. In (t, n)-threshold signature schemes, any $\geq t$-out-of-n group members can produce a threshold signature, but no less than t group members can do so.

Robustness: If all partial signatures contributing to one threshold signature are valid for a message m with respect to the private verifying keys of their respective group members, then the resulting threshold signature will be valid for m with respect to the public group key.

Traceability (optional): Any coalition of cheating signers cannot produce a valid threshold signature that will not reveal all of their identities to a verifier.

Proactive Security: The members of a group can update their individual keys without changing the public group key. The updating can tolerate a certain fraction of cheating group members.

In general, computing threshold signatures can be regarded as a multiparty computation; a problem that has practical solutions [14]. The point of specific constructions [14] of threshold signatures is in either being more efficient than general multi-

party computation, or in having additional features such as robustness, proactive security, non-erasure, etc.

The early construction (see [4]) employed a (verifiable) secret sharing scheme in order to share a private group key among all of its members. The group members then used their private shares in order to produce partial signatures, which then had to be combined by the trusted group center to arrive at a threshold signature that is verifiable against the public verifying key of the group.

Gerrit Bleumer

References

[1] Boldyreva, Alexandra (2003). "Efficient threshold signature, multisignature and blind signature schemes based on the Gap-Diffie–Hellman-Group signature scheme." *Public Key Cryptography (PKC) 2003*, Lecture Notes in Computer Science, vol. 2567, ed. Y. G. Desmedt. Springer-Verlag, Berlin.

[2] Boyd, Colin (1989). "Digital multisignatures." *1st IMA Symposium on Cryptography and Coding, Cirencester 1986*. Clarendon Press, Oxford, 241–246.

[3] Camenisch, Jan and Markus Michels (1999). "Separability and efficiency of generic group signatures." *Advances in Cryptology—CRYPTO'99*, Lecture Notes in Computer Science, vol. 1666, ed. J. Wiener. Springer-Verlag, Berlin, 413–430.

[4] Desmedt, Yvo (1993). "Threshold cryptography." *Advances in Cryptology—ASIACRYPT'92*, Lecture Notes in Computer Science, vol. 718, eds. J. Seberny and Y. Zheng. Springer-Verlag, Berlin, 3–14.

[5] Desmedt, Yvo (1997). "Some recent research aspects of threshold cryptography." *Information Security, First International Workshop ISW'97*, Lecture Notes in Computer Science, vol. 1196, eds. Lubin G. Vulkov, Jerzy Wasniewski, and Plamen Y. Yalamov. Springer-Verlag, Berlin, 158–173.

[6] Desmedt, Yvo G. and Yair Frankel (1990). "Threshold cryptosystems." *Advances in Cryptology—CRYPTO'89*, Lecture Notes in Computer Science, vol. 435, ed. G. Brassard. Springer-Verlag, Berlin, 307–315.

[7] Desmedt, Yvo and Yair Frankel (1992). "Shared generation of authenticators and signatures." *Advances in Cryptology—CRYPTO'91*, Lecture Notes in Computer Science, vol. 576, ed. J. Feigenbaum. Springer-Verlag, Berlin, 457–469.

[8] Gennaro, Rosario, Stanislaw Jarecki, Hugo Krawczyk, and Tal Rabin (1996). "Robust threshold DSS signatures." *Advances in Cryptology—EUROCRYPT'96*, Lecture Notes in Computer Science, vol. 1070, ed. U. Maurer. Springer-Verlag, Berlin, 354–371.

[9] Herzberg, Amir, Stanislaw Jarecki, Hugo Krawczyk, and Moti Yung (1995). "Proactive

secret sharing, or: How to deal with perpetual leakeage." *Advances in Cryptology—CRYPTO'95,* Lecture Notes in Computer Science, vol. 963, ed. D. Coppersmith. Springer-Verlag, Berlin, 339–352.

[10] Goldwasser, Shafi, Silvio Micali, and Ronald L. Rivest (1988). "A digital signature scheme secure against adaptive chosen-message attacks." *SIAM Journal on Computing,* 17 (2), 281–308.

[11] Jarecki, Stanislaw and Anna Lysyanskaya (2000). "Adaptively secure threshold cryptography. *Introducing Concurrency, Removing Erasures.*" *Advances in Cryptology—EUROCRYPT 2000,* Lecture Notes in Computer Science, vol. 1807, ed. B. Preneel. Springer-Verlag, Berlin, 221–242.

[12] Micali, Silvio, Kazuo Ohta, and Leonid Reyzin (2001). "Accountable subgroup multisignatures." *8th ACM Conference on Computer and Communication Security (CCS-8).* ACM Press, New York, 245–254.

[13] Rabin, Tal (1998). "A simplified approach to threshold and proactive RSA." *Advances in Cryptology—CRYPTO'98,* Lecture Notes in Computer Science, vol. 1462, ed. H. Krawczyk. Springer-Verlag, Berlin, 89–104.

[14] Shoup, Victor (2000). "Practical threshold signatures." *Advances in Cryptology—EUROCRYPT 2000,* Lecture Notes in Computer Science, vol. 1807, ed. B. Preneel. Springer-Verlag, Berlin, 207–220.

TIME-MEMORY TRADEOFFS

Many searching problems allow time-memory tradeoffs. That is, if there are K possible solutions to search over, the time-memory tradeoff allows the solution to be found in T operations (time) with M words of memory, provided the time-memory product $T \times M$ equals K. Cryptanalytic attacks based on exhaustive key search are the typical context where time-memory tradeoffs are applicable.

Due to large key sizes, exhaustive key search usually needs unrealistic computing powers and corresponds to a situation where $T = K$ and $M = 1$. However, if the same attack has to be carried out numerous times, it may be possible to execute the exhaustive search in advance and store all the results in a memory. Once this precomputation is done, the attack could be performed almost instantaneously, although in practice, the method is not realistic because of the huge amount of memory needed: $T = 1, M = K$. The aim of a time-memory tradeoff is to mount an attack that has a lower online processing complexity than exhaustive key search, and lower memory complexity than a table lookup. The method can be used to invert any

one-way function and was originally presented by Hellman in [1].

THE ORIGINAL METHOD: Let $E_K(X) : 2^n \times 2^k \to 2^n$ denote an encryption function of a n-bit plaintext X under a k-bit secret key K. The time-memory tradeoff method needs to define function g that maps ciphertexts to keys: $g : 2^n \to 2^k$. If $n > k, g$ it is a simple reduction function that drops some bits from the ciphertexts (e.g., in the Data Encryption Standard (DES), $n = 64, k = 56$). If $n < k, g$ adds some constant bits. Then we define

$$f(K) = g(E_K(P)), \tag{1}$$

where P is a fixed chosen plaintext. Computing $f(K)$ is almost as simple as enciphering, but computing K from $f(K)$ is equivalent to cryptanalysis. The time-memory tradeoff method is composed of a precomputation task and an online attack that we describe as follows.

Precomputation task: The cryptanalyst first chooses m different start points: $SP_1, SP_2, \ldots, SP_m$ from the key space. Then he computes encryption chains where $X_{i,0} = SP_i$ and $X_{i,j+1} = f(X_{i,j})$, for $1 \le j \le t$:

$$X_{0,0} \xrightarrow{f} X_{0,1} \xrightarrow{f} X_{0,2} \xrightarrow{f} \ldots \ldots \xrightarrow{f} X_{0,t}$$
$$X_{1,0} \xrightarrow{f} X_{1,1} \xrightarrow{f} X_{1,2} \xrightarrow{f} \ldots \ldots \xrightarrow{f} X_{1,t}$$
$$X_{2,0} \xrightarrow{f} X_{2,1} \xrightarrow{f} X_{2,2} \xrightarrow{f} \ldots \ldots \xrightarrow{f} X_{2,t}$$
$$\ldots \ldots$$
$$X_{m,0} \xrightarrow{f} X_{m,1} \xrightarrow{f} X_{m,2} \xrightarrow{f} \ldots \ldots \xrightarrow{f} X_{m,t} \tag{2}$$

To reduce the memory requirements, the cryptanalyst only stores start and end points ($SP_i = X_{i,0}, EP_i = X_{i,t}$) and sorts the $\{SP_i, EP_i\}_{i=1}^{m}$ on the end points. The sorted table is stored as the result of this precomputation.

Online attack: Now we assume that someone has chosen a key K and the cryptanalyst intercepts or is provided with $C = E_K(P)$. Then he can apply the function g to obtain $Y = g(C) = f(K)$ and follows the algorithm:

ALGORITHM 1. Online attack

1. If $Y = EP_i$, then either $K = X_{i,t-1}$ or EP_i has more than one inverse image. We refer to this latter event as a false alarm. If $Y = EP_i$, the cryptanalyst therefore computes $X_{i,t-1}$ and checks if it is the key, for example by seeing if it deciphers C into P

2. If Y is not an end point or a false alarm occurred, the cryptanalyst computes $Y = f(Y)$ and restarts step 1.

Note that the cryptanalyst needs to access the table lookup every time a new Y is computed. If all $m \times t$ elements are different, the probability of success PS would be $\frac{m \times t}{2^k}$. The actual probability of success depends on how the precomputed chains cover the key space. Unfortunately, there is a chance that chains starting at different keys collide and merge. The larger is a table, the higher is the probability that a new chain merges with a previous one. Each merge reduces the number of distinct keys that are actually covered by the table. If f is a random function, then the probability of success is bounded by:

$$PS_{table} \geq \frac{1}{N} \sum_{i=1}^{m} \sum_{j=0}^{t-1} \left(1 - \frac{it}{N} \right)^{j+1}. \qquad (3)$$

Equation 3 indicates that, for a fixed value of N, there is not much to be gained by increasing m or t beyond the point at which $mt^2 = N$. To obtain a high probability of success, a more effcient method is to generate multiple tables using a different function g for each table. The probability of success with r tables is:

$$PS_{tot} \geq 1 - (1 - PS_{table})^r. \qquad (4)$$

Chains of different tables can collide, but not merge, since the function g is different for every table.

DISTINGUISHED POINTS AND RAINBOW TABLES: The idea of using distinguished points (DPs) in time-memory tradeoffs refers to Rivest in [2]. If $\{0, 1\}^k$ is the key space, a DP property of order d is usually defined as an easily checked property that holds for 2^{k-d} different elements of $\{0, 1\}^k$, e.g. having d bits of the key locked to zero. In a time-memory tradeoff using DPs, the start and end points of the precomputed chains fulfill a DP property. As a consequence, the chains have variable length but detectable extreme points. This greatly reduces the number of table lookups during the online attack from t to 1.

A remarkable property of the DP method is that mergers can be easily detected and therefore, can possibly be rejected during the precomputation in order to build perfect tables [3]. The major drawback of DPs is that they introduce variable chain lengths and they are more difficult to analyze [4].

An alternative solution to reduce the number of table lookups is to use the rainbow tables presented in [5]. That is to use a different function g for each point in a chain:

$$X_{0,0} \xrightarrow{f_1} X_{0,1} \xrightarrow{f_2} X_{0,2} \xrightarrow{f_3} \dots \dots \xrightarrow{f_t} X_{0,t}$$
$$X_{1,0} \xrightarrow{f_1} X_{1,1} \xrightarrow{f_2} X_{1,2} \xrightarrow{f_3} \dots \dots \xrightarrow{f_t} X_{1,t}$$
$$X_{2,0} \xrightarrow{f_1} X_{2,1} \xrightarrow{f_2} X_{2,2} \xrightarrow{f_3} \dots \dots \xrightarrow{f_t} X_{2,t}$$
$$\dots \dots$$
$$X_{m,0} \xrightarrow{f_1} X_{m,1} \xrightarrow{f_2} X_{m,2} \xrightarrow{f_3} \dots \dots \xrightarrow{f_t} X_{m,t}. \qquad (5)$$

Two rainbow chains can only merge if they collide at the same position. Other collisions do not provoke a merge. The method is extremely easy to analyze and one rainbow table may contain t times more chains than an original table. This reduces the number of table lookups from t to 1.

As a consequence, rainbow tables are the easiest and most effcient way to perform a time-memory tradeoff. DP methods have a more theoretical interest but may also be used to detect collisions (e.g., of hash function) as suggested in [6,7]

Jean-Jacques Quisquater
François-Xavier Standaert

References

[1] Hellman, M. (1980). "A cryptanalytic time-memory tradeoff." *IEEE Transactions on Information Theory*, 26, 401–406.

[2] Denning, D. (1982). *Cryptography and Data Security*. Addison-Wesley, Reading, MA, 100.

[3] Borst, J. (2001). "Block ciphers: Design, Analysis and Side-Channel Analysis." *PhD Thesis*, Departement of Electrical Engineering, Katholieke Universiteit Leuven.

[4] Standaert, F.X., G. Rouvroy, J.-J. Quisquater, and J.D. Legat (2002). "A time-memory tradeoff using distinguished points: New analysis and FPGA results." *Proceedings of CHES 2002*, Lecture Notes in Computer Science, vol. 2523, eds. B.S. Kaliski Jr., Ç.K. Koç, and C. Paar. Springer-Verlag, Berlin, 593–609.

[5] Oechslin, P. (2003). "Making a faster cryptanalytic time-memory trade-off." *Advances in Cryptology—CRYPTO 2003*, Lecture Notes in Computer Science, vol. 2729, ed. D. Boneh. Springer-Verlag, Berlin, 617–630.

[6] Quisquater, J.J. and J.P. Delescaille (1990). "How easy is collision search?" Application to DES. *Advances in Cryptology—EUROCRYPT'89*, Lecture Notes in Computer Science, vol. 434, eds. J.-J. Quisquater and J. Vandewalle. Springer-Verlag, Berlin, 429–434.

[7] Van Oorschot, P.C. and M.J. Wiener (1999). "Parallel collision search with cryptanalytic applications." *Journal of Cryptology*, 12 (1), 1–28.

TIME-STAMPING

INTRODUCTION: As more and more of the world's data is created, stored, and manipulated online, it is sometimes crucial to verify when a digital document or record was created or last modified. Ideally, any *time-stamping* procedures to do this should depend only on the data in the document in question, and should not be susceptible to tampering—either with the data, or with the time and date.

To be precise, and to fix our vocabulary, a time-stamping system consists of two procedures: a *certification* procedure, which produces for any digital record a *time-stamp certificate* attesting to the time of certification; and a *validation* procedure, which checks whether or not a given record–certificate pair is valid. Naturally, the aim is to ensure that the only pairs that will pass the validation test are those consisting of a record and a correctly computed time-stamp certificate for that record.

In many situations, a time-stamp certificate for a document will be more important as an attestation to its integrity than it is in indicating the precise time of its creation.

A particularly significant application of digital time-stamping is to enable later verification of the validity of a long-lived <u>digital signature</u> for a document, as explained in Section "Long-lived digital signatures".

TECHNIQUES: Perhaps, the most straightforward solution to the time-stamping problem would make use of a trusted server, as follows. Certification of a record would consist of sending a copy of it to the server, which would save it as a dated entry in a secure database, and return a receipt which would serve as the time-stamp certificate. One would validate this certificate by querying the server. This is clearly an unsatisfactory solution to the problem, except in special circumstances (e.g., when the owner of the database is the only party that will ever need to verify certificates).

The first step towards improving this naive solution is to make use of a *one-way* <u>hash function</u> (see also <u>one-way function</u>). In practice, every digital time-stamping scheme begins its certification procedure by hashing the record which is its input. Since it is infeasible to compute any other bit-string input that produces the same *hash value* output, the resulting hash value can serve as a proxy for the record itself in all further processing; the validation procedure includes a recomputation of the record's hash value.

In addition, the hash value is short, whatever the length of the original record, and preserves the confidentiality of the record.

The task now is to bind this hash value to a date or time in a verifiable manner. This can be done in two essentially different ways, leading to two different families of time-stamping algorithms, those using <u>digital signatures</u> and those based entirely on one-way hashing. We describe each of these in turn.

Hash-and-Sign

One can achieve an improvement in the naive solution above by replacing the transmission of a copy of the record with the transmission of its hash value. But this is still unsatisfactory in its continued reliance on the server's securely maintaining its database. One can avoid this burden by the use of digital signatures.

In a *hash-and-sign* time-stamping scheme, the hash value of the record to be certified is sent to a trusted entity, a time-stamping server, which uses a digital signature algorithm to sign a statement that amounts to the claim that "I saw *this* hash value at *this* time." The resulting signature is the time-stamp certificate for the record.

According to what is desired, the "trusted" time-stamping server can be local or remote, implemented in software or hardware, with greater or lesser protection for its private signing key, and with varying levels of precision and of reliability for the time value it uses.

The procedure to validate a certificate consists mainly of validating the digital signature; as with any digital-signature system, this requires:
1. checking that the digital signature is correctly computed, using the server's public key; and
2. checking the validity of the key used in Step 1.
In the first step, the verifier checks the mathematical correctness of the signature. In the second step, however, the validity that is checked is not mathematical but rather semantic, corresponding to these questions: Do I trust that *this* public key was deemed to be "valid" for time-stamping purposes by *this* time-stamping server at *this* time? If so, do I trust that the key is still valid now (at the time of validation)? If it is no longer valid, either because it was later compromised or revoked or because its validity period has expired, do I have an independent reason to believe that the signature was computed at the time asserted? The procedures sketched in Section "Long-lived Digital Signatures" provide two examples of ways to supply such an independent reason.

In general, the meaning of "validity" here may depend on several factors, including for example the design of the system, the management of the time-stamping server, and the relationships among the owner of the record whose time-stamp certificate is in question, the verifier of the time-stamp certificate, and the time-stamping server. Especially, if the certificate is not a recent one, determining validity may require access to trustworthy archived data recording the history of key-validity status in the system. This point is further discussed in Section "Long-lived digital signatures." To be charitable, not many currently deployed PKI systems (see public-key infrastructure) make a serious effort to build this component with an eye towards long-lasting secure use of the system. In fact, some of them completely ignore it.

Linking

In a time-stamping scheme based on *hash-linking*, the hash value of the record to be certified is combined—by means of further hash computations—with other hash values, and the resulting *witness hash value* is stored, along with the time of storage, in a *secure repository*. A server managing the repository returns a time-stamp certificate consisting of:

1. the time of storage;
2. a pointer to the location in the repository of the witness; and
3. the list of hash values that can be used to compute (or recompute) the witness hash value from the hash value of the record.

(Optionally, the certificate may be digitally signed by the repository server.) Validation of a record–certificate pair consists of computing the hash value of the record, computing a tentative value for the witness hash, and comparing this with the value stored in the repository. (If the certificate is accompanied by a digital signature, then validation will also include a validation of the signature, as in a pure hash-and-sign scheme.)

This basic approach to "linking" of hash values can be executed in several ways. The simplest method to describe would simply store a record's hash value directly in the repository.

To lessen the load on the repository, a time-stamping server can compute a hash-linked chain as follows: keep a running current summary hash value v_i, and with each new request compute a new value v_{i+1} as the hash of the concatenation of v_i and the incoming request. At appropriate intervals, the latest summary hash value is stored in the repository as a witness value. The time-stamp certificate for each time-stamp request in the interval is the list of hash values needed to recompute the witness value at the end of the chain.

In a more elaborate variation, certification requests—i.e., hash values of records to be time-stamped—are sent to a coordinating server. At regular intervals the server builds a Merkle binary tree out of all the requests received during the interval: the leaves are the requests, and each internal node is the hash of the concatenation of its two children [13]. The root of this tree is hashed together with the previous interval's witness hash to produce the current witness hash. The linking information that is returned in each requester's time-stamp certificate is the list of sibling hash values along the path from the requester's leaf up to the witness hash (each one accompanied by a bit—denoted below as R and L, respectively—indicating whether it is the right or the left sibling).

For example, Figure 1 illustrates this process for an interval during which the requests $y_1, \ldots, y_8$ were received. In this diagram, H_{12} is the hash of the concatenation of y_1 and y_2, H_{14} is the hash of the concatenation of H_{12} and H_{34}, and similarly for the other nodes, and W_i and W_{i-1} are the

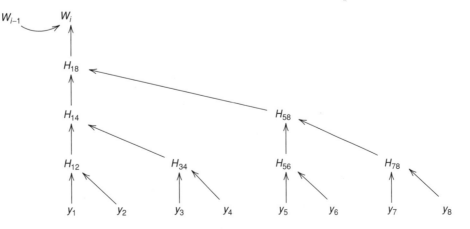

Fig. 1. Linked Merkle trees

respective witness hash values for the current and the previous intervals. The time-stamp certificate for the third request (the one containing hash value y_3), for example, is

$$[t_i; (y_4, R), (H_{12}, L), (H_{58}, R), (W_{i-1}, L)],$$

where t_i is the time associated with the current round. If h denotes the one-way hash function employed by the system (and "|" denotes string concatenation), then one validates the claim that this is a correct time-stamp certificate for a digital record d by computing, in turn, $v_3 = h(d)$, $y_{34} = h(v_3 | y_4)$, $y_{14} = h(H_{12} | y_{34})$, $y_{18} = h(y_{14} | H_{58})$, and finally $z = h(W_{i-1} | y_{18})$, and then querying the repository for the witness value W_i stored at the location associated with the time t_i; the validation is deemed to be correct exactly when $z = W_i$.

Variations of this procedure have been implemented by [7,12] and by Surety, which has offered it as a commercial service since 1995 [16].

The "secure" repository can be managed in a number of different ways, according to the desired reliability and longevity of the system. For example, the repository could consist of one or more hash values published in a widely distributed newspaper. Surety backs up the sequence of interval witness hash values in its repository by, once a week, using that week's sequence of witness hashes to compute, in turn, a weekly "super-witness" hash value that is placed in a classified advertisement in the U.S. national edition of the Sunday *New York Times* [16].

Time-stamping based on hash-linking was first proposed by [8–10], and later elaborated by [3,4,7] and other researchers.

Accumulators. In the scheme just described, a time-stamp request that is included in a Merkle tree with n leaves results in a time-stamp certificate of length $\log_2 n$, and validating the certificate requires $\log_2 n$ hash computations. This certificate can be regarded as a proof of membership in the list of leaves whose "accumulated hash value" is the root of the tree.

Benaloh and de Mare proposed the use of a *one-way accumulator*, which is a hash function h with the additional property that it is "quasi-commutative": for all appropriate inputs x, y_1, y_2, we have $h(h(x, y_1), y_2) = h(h(x, y_2), y_1)$ [5]. By means of such a function, it is easy to give proofs of membership of constant size, independent of the length of the list.

Benaloh and de Mare constructed one-way accumulators as follows. Given an RSA modulus N (see RSA problem), they propose the function $h_N(x, y) = x^y \bmod N$ (which is considerably slower to compute than more traditional one-way hash functions, as long as N is reasonably large). In this scheme, anyone knowing the factorization of N can compute hash collisions at will, so it is only applicable if one is willing to trust the party that chooses N. There are techniques by which many parties can jointly generate an RSA modulus so that no party knows its factorization (see [6] and later references), but these are unlikely to be used for a widely deployed high-throughput time-stamping system.

Nyberg has described a one-way accumulator without a trapdoor (such as the factorization for the RSA construction), but this construction is neither space- nor time-efficient [14].

Researchers have further investigated the properties of one-way accumulators, proposing a number of constructions (including [2,15]), but it is still an open research problem to devise a one-way accumulator that is secure, efficient, and useable in practice.

LONG-LIVED DIGITAL SIGNATURES: The discussion in Section "Hash-and-sign" of the two distinct steps for validating the digital signature that constitutes the time-stamp certificate of a hash-and-sign time-stamping scheme is of course completely generic, and applies as well to any digital signature that is meant to be long-lived.

A secure system must allow the verifier of a digital signature s on a document d to ascertain that the public key p used to check the correctness of s was "valid" at the time when s was computed for document d. Clearly, this requires in turn that the verifier have the means to know *when* s was computed; in other words, this requires the use of a time-stamping scheme of some sort.

In a typical PKI (public-key infrastructre), the validity of a public key is checked by means of a combination of *public-key certificates* (see certificate), certificate revocation lists (CRLS), and signed statements by trusted third parties such as Online Certificate Status Protocol (OCSP) servers. In a particular PKI, let V denote the data (certificates, etc.) needed to validate the public key used for signature s on document d. Here are two different ways to integrate time-stamping securely:

- The receiver of (d, s) assembles the key-validating data V, requests a time-stamp certificate c for (d, s, V), and saves (d, s, V, c). A later verifier needs to revalidate each of s, V, and c.
- The signer of d computes a time-stamp certificate c for (d, s) and saves (d, s, c). Later verifiers of this triple must retrieve (from an appropriate service) a trustworthy archived version of V, and revalidate all the data.

By time-stamping the concatenation (d, s), and not just the signature s, one avoids the possibility that the hash function used in the computation of the signature may later be subject to devastating attack, making it possible to use s as a valid signature for another document d' that collides with d (cf. Section "Renewability").

Naturally, other choices are possible for dividing up the responsibilities. However it is done, secure time-stamping is crucial.

RENEWABILITY: As computational resources increase, and as researchers make advances in cryptanalytic algorithms, the cryptographic security of any time-stamping system may be called into question. This raises the question [3]: can time-stamp certificates be renewed?

Suppose that an implementation of a particular time-stamping system is in place, and consider the pair (d, c_1), where c_1 is a valid time-stamp certificate (in this implementation) for the digital document d. Now suppose that some time later an improved time-stamping system is implemented and deployed—by replacing the hash function used in the original system with a new hash function, or even perhaps after the invention of a completely new algorithm.

Is there any way to use the new time-stamping system to buttress the guarantee of integrity supplied by the certificate c_1 in the face of potential later attacks on the old system?

One could simply submit d as a request to the new time-stamping system; but this would lose the connection to the original time of certification. Another possibility is to submit c_1 as a request to the the new time-stamping system. But that would be vulnerable to the later existence of a devastating attack on the hash function used in the computation of c_1, as follows: If an adversary could find another document d' that collides with d under that hash function, then he could use this renewal system to back-date d' to the original time.

Suppose instead that the pair (d, c_1) is time-stamped by the new system, resulting in a new certificate c_2, and that some time after this is done (i.e., at a definite later date), the original method is compromised. The certificate c_2 provides evidence not only that the document d existed prior to the time of the new certificate, but also that it existed at the time stated in the original certificate, c_1; prior to the compromise of the old implementation, the only way to create a valid time-stamp certificate was by legitimate means.

Observe that the supposition that the first system will be compromised at a definite time after the second system is launched is not an unreasonable one. Advances in cryptanalytic attacks on hash functions typically proceed incrementally, and well before a hash function is completely broken, fielded systems can swap in a new hash function.

STANDARDS: There are international standards in place for time-stamp mechanisms and formats, including those of the IETF for hash-and-sign time-stamping [1], and those of the ISO for both hash-and-sign and hash-linking [11].

Stuart Haber
Henri Massias

References

[1] Adams, C., P. Cain, D. Pinkas, and R. Zuccherato (2001). "Internet X.509 public key infrastructure time stamp protocols (TSP)." Internet Engineering Task Force, IETF RFC 3161, http://www.ietf.org/

[2] Baric, N. and B. Pfitzmann (1997). Collision-free accumulators and fail-stop signature schemes without trees." *Advances in Cryptology—EUROCRYPT'97*, Lecture Notes in Computer Science, vol. 1233, ed. W. Fumy. Springer-Verlag, Berlin, 480–494.

[3] Bayer, D., S. Haber, and W.S. Stornetta (1993). "Improving the efficiency and reliability of digital time-stamping." *Sequences II: Methods in Communication, Security, and Computer Science*, eds. R.M. Capocelli, A. De Santis, and U. Vaccaro. Springer-Verlag, Berlin, 329–334. (Proceedings of the *Sequences Workshop*, Positano, Italy, 1991.)

[4] Benaloh, J. and M. de Mare (1991). "Efficient broadcast time-stamping." Technical Report TR-MCS-91-1. Clarkson University Department of Mathematics and Computer Science.

[5] Benaloh, J. and M. de Mare (1993). "One-way accumulators: A decentralized alternative to digital signatures." *Advances in Cryptology—EUROCRYPT'93*, Lecture Notes in Computer Science, vol. 765, ed. Tor Helleseth. Springer-Verlag, Berlin, 274–285.

[6] Boneh, D. and M. Franklin (2001). "Efficient generation of shared RSA keys." *Journal of the ACM*, 48 (4), 702–722.

[7] Buldas, A., P. Laud, H. Lipmaa, and J. Villemson (1998). "Time-stamping with binary linking schemes." *Advances in Cryptology—CRYPTO'98*, Lecture Notes in Computer Science, vol. 1462, ed. Hugo Krawczyk. Springer-Verlag, Berlin, 486–501.

[8] Haber, S. and W.S. Stornetta (1991). "How to time-stamp a digital document." *Journal of Cryptology*, 3 (2), 99–111.

[9] Haber, S. and W. S. Stornetta (1992). "Method for secure time-stamping of digital documents." U.S. Patent 5,136,647, issued August 1992; Re. 34,954, reissued May 1995.

[10] Haber, S. and W.S. Stornetta (1992). "Digital document time-stamping with catenate certificate." U.S. Patent 5,136,646, August 1992.

[11] International Organization for Standardization (2002–2003). Information technology—Security techniques—Time-stamping services. ISO/IEC 18014 (parts 1–3), http://www.iso.ch/

[12] Massias, H., X. Serret Avila, and J.-J. Quisquater (1999). "Design of a secure timestamping service with minimal trust requirements." *Twentieth Symposium on Information Theory in the Benelux*, eds. P. Vanroose, A. Barbé, E.C. van der Meulen. May 1999, 79–86.

[13] Merkle, R. (1980). "Protocols for public key cryptosystems." *Proceedings of the 1980 Symposium on Security and Privacy*. IEEE Computer Society Press, Los Alamitos, CA, 122–133.

[14] Nyberg, K. (1996). "Commutativity in cryptography." *Proceedings of the First International Workshop on Functional Analysis at Trier University*, ed. Walter de Gruyter. 331–342.

[15] Sander, T. (1999). "Efficient accumulators without trapdoor." *Proceedings of the Second International Conference on Information and Communication Security*, Lecture Notes in Computer Science, vol. 1726, eds. Vijay Varadharajan and Yi Mu. Springer-Verlag, Berlin, 252–262.

[16] Surety. http://www.surety.com

TIMING ATTACK

INTRODUCTION: When the running time of a cryptographic device is not constant, this time may leak information about the secret parameters involved, so that careful timing measurement and analysis may allow to recover the system's secret key. This idea first appeared in the scientific literature in 1996 [7].

Targeting implementation specificities, timing attacks belong to the side-channel attacks family, of which they were one of the first representatives (see also side-channel analysis).

TIME MEASUREMENT: To conduct the attack, the adversary needs to collect a set (sample) of messages, together with their processing time by the cryptographic device. Running time might be obtained by measuring the question-answer delay, by monitoring the processor activity, etc.

One privileged target for timing attacks—as well as other side-channel attacks—is that of smart cards. A classical smart card (like defined by the ISO 7816 standard) is not equipped by an internal clock, but has its clock ticks provided by the terminal it is inserted in. Hence, a rogue terminal makes it straightforward to obtain a very accurate measurement of the running time.

TIMING ANALYSIS: Timing attacks use the divide-and-conquer strategy. The secret key is recovered by parts, by predicting and verifying some correlation between a partial key value and the expected running time.

For each part of the key, the attack goes as follows: First, depending on a guess for this partial key's value, define a criterion about the expected running time. Then, check whether the actual running times match the criterion. If they do, conclude the partial key guess was correct; if they do not, conclude it was wrong and repeat with another guess.

EXAMPLE: ATTACK AGAINST RSA WITH MONTGOMERY MULTIPLICATION: Consider the case of a private *RSA* exponentiation (see RSA public key encryption) where modular multiplications are performed using the square and multiply (Algorithm 1; see also binary exponentiation method) and Montgomery arithmetic. As can be seen from function MonPro() in Montgomery arithmetic, the Montgomery algorithm will behave differently depending on whether the temporary value $\overline{u}$ is less (in which case $\overline{u}$ is returned) or greater than n (in which case $\overline{u} - n$ is returned). Depending on the input value, this algorithm will have to perform an additional subtraction or not, which will impact its running time.

ALGORITHM 1. Square and multiply algorithm computing $g^e \bmod n$

$$
\begin{aligned}
&a \leftarrow g \\
&\textbf{for } i = l - 2 \text{ down to } 0 \textbf{ do} \\
&\quad a \leftarrow a.a \bmod n \\
&\quad \textbf{if } e_i = 1 \textbf{ then} \\
&\quad\quad a \leftarrow a.g \bmod n \\
&\quad \textbf{end if} \\
&\textbf{end for} \\
&\text{return } a
\end{aligned}
$$

Therefore, a simple method to recover the first bit of the secret key [3] is to assume it is equal to 1 (partial guess), and, for each message in the sample, to simulate the first multiplication by g. The sample set is then divided in two subsets, depending on whether that multiplication requires an additional subtraction or not. The criterion is "running time for the first subset is longer than for the second". Simple comparison between the average (actual) running times can then be used to validate or invalidate the partial key guess. Once the first secret bit is known, the process can be repeated against the second secret bit, without the need for additional measurements.

More powerful statistical tools greatly improve the attack's efficiency [8].

OTHER APPLICATIONS: Timing attacks were applied against other underline{asymmetric cryptosystems} (GPS [2], underline{elliptic curve-based cryptosystems} [6]) and underline{symmetric cryptosystems} (RC5 [5], underline{Rijndael/ AES} [8]), as well as against some internet underline{protocols} (SSL-TLS [1]), although the latter's efficiency quickly decreases when the number of relaying machines grow.

COUNTERMEASURES: Two types of countermeasures can be applied against timing attacks. The first one consists in eliminating timing variations whereas the second renders these variations useless for an attacker.

The only absolute way to prevent timing attacks is to make the computation strictly constant time, independent of the input. However, this would imply a very severe performance drawback, especially for asymmetric cryptosystems, since this constant time would be that of the slowest possible case (for RSA, for example, this would correspond to an exponent equal to $111\ldots1$). Such a countermeasure would therefore not be very practical.

More efficient, although less generic, countermeasures can be applied to defeat *specific* timing attacks. For example, several countermeasures against the aforementioned attack on RSA focus on the additional subtraction in Montgomery multiplication [4,9].

Adding random delays to the algorithm in order to hide time variations is an intuitive, but not very efficient countermeasure, since it is equivalent to adding white noise to a source. Such noise can easily be filtered out for an increase in sample size.

The second type of countermeasure consists in hiding the internal state, so that the attacker cannot simulate internal computations any more. For example, Kocher [7] suggests the following blinding strategy: before computing the modular exponentiation (see underline{modular arithmetic}), choose a random pair—we refer the reader to this reference for a way to generate such pairs at a reasonable cost—(v_i, v_f) such that $v_f^{-1} = v_i^e$; multiply the message by $v_i \pmod{n}$ and multiply back the output by $v_f \pmod{n}$ to obtain the searched result. These countermeasures make internal computations impossible to simulate by the attacker, therefore preventing exploitation of knowledge of the running times. Although they are not guaranteed to eliminate *all* possible timing attacks, these countermeasures are pretty efficient. In addition, blinding techniques have also proven efficient against other underline{side-channel attacks}.

François Koeune

References

[1] Canvel, B., A. Hiltgen, S. Vaudenay, and M. Vuagnoux (2003). "Password interception in a SSL/TLS channel." *Advances in Cryptology—CRYPTO 2003* Lecture Notes in Computer Science, vol. 2729, ed. D. Boneh. Springer-Verlag, Berlin.

[2] Cathalo, J., F. Koeune, and J.-J. Quisquater (2003). "A new type of timing attack: Application to GPS." *Cryptographic Hardware and Embedded Systems— CHES* 2003. Lecture Notes in Computer Science, vol. 2779, eds. Burton, S. Kaliski, Çetin K. Koç, and Christof Paar. Springer-Verlag, Berlin.

[3] Dhem, J.-F., F. Koeune, P.-A. Leroux, P. Mestré, J.-J. Quisquater, and J.-L. Willems (1998). "A practical implementation of the timing attack." *Proceedings of CARDIS 1998, Smart Card Research and Advanced Applications*. Lecture Notes in Computer Science, vol. 1820, eds. J.-J. Quisquater and B. Schneier. Springer, Berlin.

[4] Dhem, J.F. (1998). "Design of an efficient public-key cryptographic library for risc-based smart cards." *PhD Thesis*, UCL Crypto Group, Laboratoire de microélectronique (DICE), Université catholique de Louvain.

[5] Handschuh, H. and H. Heys (1998). "A timing attack on RC5." *Proceedings of SAC'98*, Lecture Notes in Computer Science, vol. 1556, eds. S.E. Tavares and H. Meijer. Springer, Berlin, 306–318.

[6] Katagi, M., I. Kitamura, T. Akishita, and T. Takagi (2002). "A timing attack on hyperelliptic curve cryptosystems." Cryptology ePrint Archive: Report 2002/203. Available at http://eprint.iacr.org

[7] Kocher, P. (1996). "Timing attacks on implementations of Diffie–Hellman, RSA, DSS, and other systems," *Advances in Cryptology—CRYPTO'96, Santa Barbara, CA*, Lecture Notes in Computer Science, vol. 1109, ed. N. Koblitz. Springer, Berlin, 104–113.

[8] Schindler, W., J.-J. Quisquater, and F. Koeune (2001). "Improving divide and conquer attacks against cryptosystems by better error detection correction strategies." *Proceedings of 8th IMA International Conference on Cryptography and Coding*, Lecture Notes in Computer Science, vol. 2260, ed. B. Honary. Springer Berlin. December 2001, 245–267.

[9] Colin, D. Walter (1999). "Montgomery's multiplication technique: How to make it smaller and faster." *Cryptographic Hardware and Embedded Systems—CHES'99, August*, Lecture Notes in Computer Science, vol. 1717, eds. Çetin K. Koç and Christof Paar. Springer-Verlag, Berlin, 80–93.

TOKEN

A token usually refers to a handheld device, used in connection with two-factor authentication. The token shares a key with a central server and provides user authentication and possibly message authentication. The token may come with or without a (restricted) *pin* pad (see Personal Identification Number), and a display. Various protocols may be deployed, time or counter dependent, or based on a challenge–response approach.

Whichever the protocol, an input (challenge), is provided to the key token either time or counter based by the token itself, or through a user input using the pin pad. A key dependent response is then calculated by the token and displayed, which is then keyed in by the user on the work station or terminal connected to the central server, where the response is verified using the same key.

This notion is not to be to confused with a *key token* which is often used to mean an encryption using a key encryption key of a key and some additional information related to the key.

Peter Landrock

TRAITOR TRACING

INTRODUCTION: Traitor tracing is a method for providing personal decryption keys for users, such that (1) there is a single encryption key corresponding to all the decryption keys, and (2) any (illegitimate) decryption key, even one that was generated by a *coalition* of corrupt users (*traitors*), identifies personal keys that were used to generate it. The concept of traitor tracing was introduced by Chor et al. [2].

Tracing the source of illegitimate keys is important if these keys enable access to sensitive data. The data can be encrypted to keep its confidentiality but at some point it must be revealed in the clear to the parties using it, who must therefore have corresponding decryption keys. In some scenarios corrupt parties (the traitors), who have legitimate access to decryption keys, wish to further distribute the decrypted data to other users. In many cases it is ineffective for the traitors to leak the decrypted data, since the economics of scale make it much more expensive for them, compared to the legitimate distributor, to distribute this data in a timely manner. This argument seems to hold for example in cases such as the distribution of encrypted pay-TV programs, access to online databases, or distribution of content in encrypted high-capacity media such as DVDs.

An alternative and cheaper approach for the traitors is to further distribute the decryption *keys* that enable the decryption of the encrypted content. (These decryption keys are typically secured in tamper-resistant software or hardware, for example in a smartcard, but such security measures are often broken by dedicated hackers, for example by using differential power analysis.) The decryption keys are much shorter than the encrypted data itself, and therefore it is much easier to distribute them. The purpose of traitor tracing is that given an illegitimate key, for example, one found in a pirate decoding device, it would be possible to trace traitors whose keys were used to generate this key. Note that a coalition of several traitors might collude to generate an illegitimate key by mixing information from the different personal keys of the coalition members.

Traitor tracing is different from fingerprinting or watermarking, whose goal is to trace illegitimate copies of the content itself. These methods have better functionality than tracing, since they enable authorized parties to trace the source of content even after its decryption. On the down side, their overhead is much higher (especially that of fingerprinting), and their security guarantees are weaker (especially in the case of watermarking).

Traitor tracing also provides different functionality than broadcast encryption. Tracing enables the identification of the source of a piracy problem, i.e., the parties whose keys are used to enable illegal usage of content. Broadcast encryption can then be used to take measures against the piracy by preventing further usage of these keys.

A different version of tracing, described below, is *dynamic traitor tracing*. Given a real-time broadcast from a pirate who is illegitimately distributing content, dynamic traitor tracing enables the identification of the source of the keys that are used for generating the illegitimate broadcast.

TRACING SCHEMES: If different parties receive the same decryption key then it would be impossible to tell which of them leaked it. Each party should therefore receive a personal key, different from the key of any other party. A simple solution is to provide each party with an independent personal key, encrypt each data block using a random key, and then separately encrypt this key using each of the different personal keys. The problem with this approach is that its overhead is linear in the number of users, i.e., given N users the system needs to distribute N additional encrypted messages. Since N might be large (in the millions), the overhead is undesirable.

Tracing schemes are usually designed to be secure against coalitions that contain a limited number of traitors. Let us denote by k an upper bound on the size of a coalition of traitors. Following is a description of the basic tracing scheme suggested by Chor, Fiat and Naor in [2], which is secure against coalitions of up to k parties. (In addition, [2] contains more efficient schemes. See also [3] for a more detailed discussion.)

- **Initialization:** The system uses a table of l rows and $2k^2$ columns. Each table entry contains an independent key. Each user is mapped to a random location in every row. The user's personal key contains the keys of the entries to which the user is mapped, a total of l keys.
- **Encryption:** The data is encrypted using a random key S. Then l random <u>shares</u> $S_1, \ldots, S_l$ are generated, subject to the constraint that $S_1 \oplus \cdots \oplus S_l = S$. Each share S_i, for $1 \leq i \leq l$, is independently encrypted using every key in row i, giving a total of $2lk^2$ encryptions for all the shares.
- **Decryption:** Each user has a key from every row i, enabling it to decrypt the share S_i. The user can then compute $S = S_1 \oplus \ldots \oplus S_l$, and decrypt the data.
- **Tracing:** The tracing procedure is given a pirate decoder that was generated by a coalition of at most k traitors. This decoder must contain a key from every row. Assume, without loss of generality, that it contains one key from every row. Then at least one traitor contributed l/k or more of the keys in the decoder. On the other hand, the personal set of keys of each other user is expected to intersect with only $l/(2k^2)$ keys of the decoder. The tracing algorithm therefore identifies the user whose personal set of keys has the largest intersection with the set of keys of the pirate decoder, and declares it to be a traitor. Setting the number of rows to be $l = k^2 \log N$ ensures that this user is a traitor with high probability.

Two major measures of the overhead of a tracing system are the size of the personal key of each user ($k^2 \log N$ keys in the scheme described above), and the total communication overhead ($2k^4 \log N$ encryptions in this example). The overhead can be substantially improved using more advanced techniques, such as mapping users into smaller subsets and running a different tracing scheme for every subset.

A further significant improvement in the overhead is achieved using *threshold tracing* [5]. The difference between this method and basic tracing

schemes is that the latter can trace the source of keys of any pirate decryption device which can decrypt content with nonnegligible probability, whereas threshold tracing is only effective against pirate devices that succeed in the decryption with probability greater than a given threshold t, (e.g., $t = 90\%$). The use of threshold tracing is quite appealing, however, since decryption devices that cannot decrypt a substantial fraction of the content are not very attractive, and threshold tracing is considerably more efficient than basic traitor tracing.

The tracing operation is based on examining a pirate decryption device. Conceptually, it is simpler to imagine that we can apply reverse engineering to the device, find out exactly which keys it is using, and trace their source. In practice, however, the reverse engineering operation might be quite complex, and it is preferable to perform *black-box tracing*, which is based on the functionality of the decryption device, rather than on obtaining its keys. Specifically, the tracing procedure operates by sending specially crafted encrypted messages to the decoder and examining how it decrypts them. The tracing schemes in [2, 3, 5] support black-box tracing.

Many tracing schemes are based on combinatorial constructions, and there is considerable research on designing codes supporting tracing (see, e.g., [7]). There are also several schemes that support both tracing and <u>broadcast encryption</u>, enabling both the identification of traitors and the disabling of their keys (see, e.g., [6]).

PUBLIC-KEY TRAITOR TRACING: The tracing methods of [2] and their like are based on combinatorial or probabilistic constructions, and can be used for either <u>symmetric-key</u> or <u>public-key</u> encryption. Boneh and Franklin [1] introduced an efficient public-key only tracing scheme which is based on an algebraic (number theoretic) construction. The security of this system is based on the <u>decisional Diffie–Hellman assumption</u>. Its overhead is linear in k, and does not depend on the number of users N. Furthermore, tracing is deterministic, and ensures that all parties who contributed their keys to the pirate devices are traced. On the downside, the system does not support full black-box tracing, except for some specific cases.

The system operates by using a fixed base of $2k$ field elements. Each user receives a private personal key which is a solution to the <u>discrete log problem</u> of representing a known value relative to the base. The paper shows that any useful pirate key must be a convex combination of private keys. However, the private personal keys are derived from a Reed–Solomon code (see cyclic codes) in such a way that any $2k$ keys are linearly independent. Therefore any convex combination of at most k of them can be efficiently traced to the keys that were used to generate it.

DYNAMIC TRAITOR TRACING: The tracing schemes described above are only effective against pirates that generate illegitimate decryption devices. These schemes are not helpful against pirates that rebroadcast content using a pirate broadcasting system. Dynamic traitor tracing was suggested by Fiat and Tassa [4]. It operates against such pirates by examining their broadcasts and dynamically changing the method used to encrypt the content, until the keys used by the pirates are identified. Dynamic tracing is different than watermarking schemes since the latter trace the source of leaking content by performing a "post mortem" examination of a pirate copy, and do not adaptively change the way the content is encrypted.

Dynamic tracing is based on the "watermarking assumption", i.e. the ability to generate different versions of the same content such that all versions have the same functionality (e.g. look the same to humans), while it is impossible for the receivers of these versions to generate a new version that hides its source and retains reasonable quality.

After a pirate rebroadcast of the content is identified, dynamic tracing generates different versions of the content on the fly. The tracing algorithm compares the pirate broadcast to the different versions, splits the set of legitimate receivers into subsets, and identifies a subset of receivers that contains a traitor. Further splitting is performed based on feedback learned from the rebroadcast of the content, until the identity of the traitor becomes known. Unlike static tracing, there is no need to decide in advance on an upper bound to the number of colluding traitors. The tracing process adapts to the number of traitors on the fly. In addition tracing can be rerun until all traitors are traced (whereas in the static case it is only assured that one of the traitors is identified).

Benny Pinkas

References

[1] Boneh, D. and M. Franklin (1999). "An efficient public key traitor tracing scheme." *Advances in Cryptology—CRYPTO'99*, Lecture Notes in Computer Science, vol. 1666, ed. J. Wiener. Springer-Verlag, Berlin, 338–353.

[2] Chor, B., A. Fiat, and M. Naor (1994). "Tracing traitors." *Advances in Cryptology—CRYPTO'94*, Lecture Notes in Computer Science, vol. 839, ed. Y.G. Desmedt. Springer-Verlag, Berlin, 480–491.

[3] Chor, B., A. Fiat, M. Naor, and B. Pinkas (2000). "Tracing traitors." *IEEE Transactions on Information Theory*, 46 (3), 893–910.

[4] Fiat, A. and T. Tassa (2001). "Dynamic traitor tracing." *Journal of Cryptology*, 14 (3), 211–223, Previous version appeared in the proceedings of CRYPTO'99.

[5] Naor, M. and B. Pinkas (1998). "Threshold traitor tracing." *Advances in Cryptology—CRYPTO'98*, Lecture Notes in Computer Science, vol. 1462, ed. H. Krawczyk. Springer-Verlag, Berlin, 502–517.

[6] Naor, M. and B. Pinkas (2001). "Efficient trace and revoke schemes." *Proceedings of Financial CRYPTO 2000*, Lecture Notes in Computer Science, vol. 1962, ed. Y. Frankel. Springer-Verlag, Berlin.

[7] Stinson, D.R. and R. Wei (1998). "Combinatorial properties and constructions of frameproof codes and traceability schemes." *SIAM Journal Discrete Mathematics*, 11, 41–53.

TRANSPORT LAYER SECURITY (TLS)

The *Transport Layer Security* (TLS) standard defines the successor of the prominent Internet security protocol SSL (see Secure Socket Layer). Almost all cryptographically protected World Wide Web (WWW) communication relies on protocols of the TLS/SSL suite, like for example e-commerce, home banking or email services. The protocol is supported by major browser products, such as the current versions of Microsoft Internet Explorer, Netscape Navigator, Opera and the Mozilla projects.

TLS secures client-server sessions through strong cryptographic methods against eavesdropping (eavesdropper), tampering, and forgery. It allows connections to be anonymous, server-only, and mutually authenticated, and transfers the application data confidentially and integrity protected.

The rough outline of the protocol is very similar to SSLv3; therefore, the descriptions given in SSL apply to TLS as well.

According to Rescorla Ref. [7], the technical differences between SSLv3 and TLS are very small; most important changes relate to the construction of message authentication codes (HMAC) and the key expansion method. The new *pseudorandom function* (PRF) transforms secret data, contributed by the client side, and both participants' random numbers into a raw key block, which is subsequently broken up into session keys. Although the overall design of the HMAC and PRF algorithms were quite similar to SSL, in detail the alterations made TLS largely incompatible to SSL.

A number of cryptographic methods were defined as mandatory for implementations, such as the Diffie–Hellman key agreement (*DH*) and authentication following the Digital Signature Standard (*DSS*). The requirements for the block cipher Triple DES (3DES) raised difficulties concerning the US export regulations on cryptographic software. Until January 2000, when these restrictions were loosened, no standard compliant TLS implementation could be exported legally.

The *Internet Engineering Task Force* (IETF) released TLS version 1.0 in 1999 as RFC2246 [9]. The current Internet-Draft is version 1.1 [10]; among other things, it addresses vulnerabilities noted by Moeller [6], relating to padding ambiguities and alert definitions of CBC cipher suites.

Several extensions to the core standard have been specified. Chown complements cipher suites for the *Advanced Encryption Standard* (Rijndael/AES) [8]; Hollenbeck [3] assigns two additional compression methods. Gupta, Blake-Wilson, Moeller and Hawk specify key exchange mechanisms based on Elliptic Curve Cryptography, especially the (ephemeral) Elliptic Curve Diffie–Hellman (ECDH) key agreement (elliptic curve protocols for key agreement schemes) and the Elliptic Curve Digital Signature Algorithm (ECDSA, elliptic curve protocols for signatures) Ref. [2].

The adaptation of TLS to other than X.509-compliant public key infrastructures (PKIX) is addressed by Mavroyanopoulos for OpenPGP (see Pretty Good Privacy) [4] and by Medvinsky and Hur for the symmetric key-based Kerberos authentication system [5].

A general guideline for generic TLS extensions is given by Blake-Wilson, Nystrom et al. [1]. Note that some of the referenced articles are in IETF-Internet-Draft status and subject to change.

Clemens Heinrich

References

[1] Simon, Blake-Wilson, Magnus Nystrom, David Hopwood, Jan Mikkelsen, and Tim Wright (2002). "Transport Layer Security (TLS) extensions." IETF Interntet-Draft, http://www.ietf.org/

[2] Gupta, Vipul, Simon Blake-Wilson, Bodo Moeller, and Chris Hawk (2002). "ECC cipher suites for TLS." IETF Internet-Draft, http://www.ietf.org/

[3] Hollenbeck, Scott (2002). "Transport Layer Security protocol compression methods." IETF Internet-Draft, http://www.ietf.org/

[4] Mavroyanopoulos, Nikos (2002). "Using OpenPGP keys for TLS authentication." IETF Internet-Draft, http://www.ietf.org/

[5] Medvinsky, Ari and Matthew Hur (1999). "Addition of Kerberos cipher suites to transport layer security (TLS)." Network Working Group RFC 2712, Internet resource: http://www.ietf.org/rfc/rfc2712.txt

[6] Moeller, Bodo (2001). "Security of CBC Ciphersuites in SSL/TLS: Problems and Countermeasures". http://www.openssl.org/~bodo/tls-cbc.txt/

[7] Rescorla, Eric (2000). SSL and TLS. *Designing and Building Secure Systems*. Addison-Wesley, Reading, MA.

[8] Chown, Pete (2002). "Advanced encryption standard (AES) ciphersuites for transport layer security (TLS)." Network Working Group RFC 3268, Internet resource: http://www.ietf.org/rfc/rfc3268.txt

[9] Dierks, Tim and Allen Christopher (1999). The TLS Protocol Version 1.0, IETF Internet-Draft RFC 2246, (expired), Internet resource: http://www.ietf.org/rfc/rfc2246.txt

[10] Dierks, Tim and Eric Rescorla (2002). "The TLS Protocol Version 1.1." IETF Interntet-Draft, http://www.ietf.org/

TRAPDOOR ONE-WAY FUNCTION

A trapdoor one-way function is a one-way function with an additional requirement. Informally, a one-way function might be described as a function for which evaluation in one direction is straightforward, while computation in the reverse direction is far more difficult. Such a function becomes a trapdoor one-way function when we add the requirement that computation in the reverse direction becomes straightforward when some additional (trapdoor) information is revealed [3].

While there are alternative descriptions [2] we might describe a trapdoor one-way function as a function f with domain X and range (codomain) Y where $f(x)$ is 'easy' to compute for all $x \in X$ but for 'virtually all' elements $y \in Y$ it is 'computationally infeasible' to find an x such that $f(x) = y$. Yet, given certain trapdoor information z, it is easy to describe an 'efficient' function g_z with domain Y and range X such that $g_z(y) = x$ and $f(x) = y$. Just as a bijective one-way function with identical domain and range is a one-way permutation, a bijective trapdoor one-way function with identical domain and range is a *trapdoor permutation* (see also substitutions and permutations).

Paralleling the theoretical development of one-way functions, trapdoor one-way functions were introduced by Diffie and Hellman [1] and their implications for the development of public-key cryptography explored in that remarkable paper. The full power of trapdoor one-way functions was elegantly captured in the proposal for public-key cryptography and the introduction of digital signature schemes.

However, unlike the case of one-way functions, no candidate trapdoor one-way function was proposed by Diffie and Hellman in their original paper [1]. Instead, arguably the most important proposal for a trapdoor one-way function (in fact a trapdoor one-way permutation) is due to Rivest, Shamir, and Adleman and is known as RSA (see RSA public key encryption) [5]. The difficulty of integer factoring is vital to the security of RSA and the trapdoor that allows the legitimate receiver to decrypt an encrypted message effectively consists of the factorization of some large number. The knapsack problem [4] (see knapsack cryptosystem provides an example of a candidate trapdoor one-way function that was compromised by subsequent cryptanalysis.

The theoretical treatment of both one-way functions and trapdoor one-way functions is extensive [6], yet in practice there remain only a few practical and trusted proposals on which much of the machinery for the electronic information revolution relies.

Matt Robshaw

References

[1] Diffie, W. and M.E. Hellman (1976). "New directions in cryptography." *IEEE Transactions on Information Theory*, 22, 644–654.

[2] Massey, J.L. (1992). "Contemporary cryptology: An introduction." *Contemporary Cryptology, The Science of Information Integrity*, ed. G.J. Simmons. IEEE Press, Piscataway, NJ.

[3] Menezes, A.J., P.C. van Oorschot, and S.A. Vanstone (1997). *Handbook of Applied Cryptography*. CRC Press, Boca Raton, FL.

[4] Merkle, R.C. and M.E. Hellman (1978). "Hiding information and signatures in trapdoor knapsacks." *IEEE Transactions on Information Theory*, 24, 525–530.

[5] Rivest, R.L., A. Shamir, and L.M. Adleman (1978). "A method for obtaining digital signatures and and public-key cryptosystems." *Communications of the ACM*, 21, 120–126.

[6] Yao, A.C. (1982). "Theory and applications of trapdoor functions." *Proceedings of the IEEE 23rd Annual Symposium on Foundations of Computer Science*, 80–91.

TRIPLE-DES

Since the introduction of the Data Encryption Standard (DES) in the mid 1970s, cryptanalysts have been increasingly concerned about the 56-bit secret key used in the cipher and its vulnerability to exhaustive key search. In 1977, Diffie and Hellman [2] estimated the cost of a machine capable of recovering a 56-bit key within a day at US\$20 million. In 1993, Wiener provided a detailed design for a machine which would reduce the average search time to 3.5 hours [10]. The design consisted of 57,000 custom chips and had an estimated cost of US\$1 million. Half a decade later, the Electronic Frontier Foundation (EFF) actually built the first search machine. The US\$250,000 machine was called "Deep Crack" [3] and ran through the complete key space in nine days.

As it became clear that DES did not provide adequate security because of its 56-bit secret key, the cipher was gradually replaced by Triple-DES (also known as TDEA). Triple-DES is a multiple encryption scheme. The idea of triple encryption was proposed by Diffie and Hellman, who noted that double encryption schemes did not improve the security much. Matyas and Merkle suggested a variant on this scheme, which was included in the ANSI X9.52 [1] standard, and replaced single DES as FIPS approved symmetric algorithm of choice in 1999 (FIPS 46-3 [7]). Triple-DES itself is now being replaced in favor of the Advanced Encryption Standard (Rijndael/AES). The transition is slow, however, and both encryption schemes are expected to coexist for many years (see for example NIST SP 800-67 [8]).

The Triple-DES encryption algorithm consists of three applications of DES. The ANSI X9.52 variant is defined as

$$C = E_{K_3}(D_{K_2}(E_{K_1}(P)))\,,$$

where P and C are 64-bit plaintext and ciphertext blocks, and $E_K(\cdot)$ and $D_K(\cdot)$ denote the DES encryption and decryption functions. The standard specifies three different ways of choosing the 56-bit keys K_1, K_2, and K_3:

- Keying Option 1: K_1, K_2, and K_3 are independent (168 secret bits);
- Keying Option 2: K_1 and K_2 are independent, $K_3 = K_1$ (112 secret bits);
- Keying Option 3: $K_1 = K_2 = K_3$ (56 secret bits).

The last keying option is equivalent to a single DES encryption and provides backward compatibility with older systems. Because of this convenient feature, the *E-D-E* scheme described above is usually preferred over the *E-E-E* variant, which consists of three consecutive encryptions.

Option 2 (two-key Triple-DES) effectively doubles the number of key bits and triples the number of rounds of DES. Option 1 (three-key Triple-DES) triples both. As a result, the algorithm is strengthened both against cryptanalytical attacks (linear and differential cryptanalysis) and against exhaustive key search. Still, the security gain obtained by doing multiple encryptions is not as high as one might expect, mainly because of meet-in-the-middle attacks. It is especially true for double encryption schemes, but the technique also applies to two-key and three-key Triple-DES. However, none of the attacks pose an immediate threat, as they currently require an impractical amount of resources.

A first meet-in-the-middle attack on two-key Triple-DES was proposed by Merkle and Hellman [6]. Their attack takes 2^{56} operations, but requires 2^{56} *chosen* plaintexts (see chosen plaintest attack) and a large amount of memory to store 2^{56} 64-bit data blocks. The attack was later improved by van Oorschot and Wiener [9], who demonstrated that an adversary could recover the 112-bit key in 2^{120-t} steps, if he is given 2^t *known* plaintexts (see known plaintext attack) and 2^t blocks of storage.

The classical meet-in-the-middle attack applied to three-key Triple-DES requires three known plaintexts, 2^{56} units of storage and in the order of 2^{112} operations. In [5], Lucks proposes an alternative attack, requiring the same amount of memory, 2^{45} known plaintexts and 2^{108} analysis steps. Additionally, Kelsey, Schneier, and Wagner [4] have demonstrated that three-key Triple-DES is vulnerable to related key attacks.

Two-key and three-key Triple-DES have considerably longer secret keys than single DES, but they still operate on 64-bit data blocks. This implies that the cipher starts leaking information about the plaintext when more than 2^{32} data blocks are encrypted without changing the key (a matching ciphertext attack as a consequence of the birthday paradox). This limitation, together with the fact that Triple-DES is rather slow, motivates the switch to a 128-bit block cipher such as the Advanced Encryption Standard.

Christophe De Cannière

References

[1] ANSI-X9.52 (1998). "Triple date encryption algorithm modes of operation." Revision 6.0.

[2] Diffie, W. and M. Hellman (1997). "Exhaustive cryptanalysis of the NBS data encryption standard." *Computer*, 10 (6), 74–84.

[3] Electronic Frontier Foundation (EFF) (1998). "DES cracker." http://www.eff.org/DEScracker/

[4] Kelsey, J., B. Schneier, and D. Wagner (1996). "Key-schedule cryptanalysis of IDEA, G-DES, GOST, SAFER, and Triple-DES." *Advances in Cryptology—CRYPTO'96*, Lecture Notes in Computer Science, vol. 1109, ed. N. Koblitz. Springer-Verlag, Berlin, 237–251.

[5] Lucks, S. (1998). "Attacking triple encryption." *Fast Software Encryption, FSE'98*, Lecture Notes in Computer Science, vol. 1372, ed. S. Vaudenay. Springer-Verlag, Berlin, 239–257.

[6] Merkle, R.C. and M.E. Hellman (1981). "On the security of multiple encryption." *Communications of the ACM*, 24, 465–467.

[7] National Institute of Standards and Technology (1979). "FIPS-46: Data Encryption Standard (DES)." Revised as FIPS 46-1:1988, FIPS 46-2:1993, FIPS 46-3:1999, available at http://csrc.nist.gov/publications/fips/fips46-3/fips46-3.pdf

[8] National Institute of Standards and Technology (2004). *Recommendation for the Triple Data Encryption Algorithm (TDEA) Block Cipher.* SP-800-67, NIST, March 2004. Draft available at http://csrc.nist.gov/publications/drafts.html

[9] van Oorschot, P.C. and M.J. Wiener (1990). "A known plaintext attack on two-key triple encryption." *Advances in Cryptology—EUROCRYPT'90*, Lecture Notes in Computer Science, vol. 473, ed. I. Damgård. Springer-Verlag, Berlin, 318–325.

[10] Wiener, M. (1996). "Efficient des key search." *Practical Cryptography for Data Internetworks*, 31–79. Presented at the rump session of *CRYPTO'93*. Reprinted in *Practical Cryptography for Data Internetworks*, ed. W. Stallings, IEEE Computer Society Press, pp. 31–79.

TROJAN HORSES, COMPUTER VIRUSES AND WORMS

A program which is different from the specified (specs) one, is said to contain a *Trojan horse*. The Trojan horse may be malicious. It is difficult to decide whether a program is free of Trojan horses.

A Trojan horse that, when executed, can modify other computer programs, e.g., by copying itself (or a part of it) into these, is called a *computer virus*. Protection mechanisms used against computer viruses are to:

use digitally signed computer programs. Provided these digitally signed programs were developed in a secure and trusted environment, then one can detect modifications to the digitally signed program [3]. (For implementation issues see [1].) If the environment was not trusted, known computer viruses may be in the digitally signed program and remain undetected unless virus scanners are used.

use virus scanners. To a known computer virus corresponds a <u>fingerprint</u> (also known as a "signature" in the computer virus literature). Before running a program (e.g., at the start-up of the computer) a virus scanner will check the program for fingerprints of known computer viruses. The disadvantage of this method is that it cannot detect new computer viruses that differ enough from old ones.

Originally, a *computer worm* was a means of performing distributed computation using segments [4]. Today it is primarily used to indicate Trojan Horses that spread, using network resources. A hybrid may have characteristics of both a worm as well as of a computer virus. A collection of scientific texts on the topic can be found in e.g. [2].

Yvo Desmedt

References

[1] Davida, G.I., Y.G. Desmedt, and B.J. Matt (1989). "Defending systems against viruses through cryptographic authentication." *Proceedings 1989 IEEE Symposium on Security and Privacy*, IEEE Computer Society, May 1989, *Oakland, CA*, 312–318.

[2] Hoffman, L.J. (ed.) (1990). *Rogue Programs: Viruses, Worms and Trojan Horses*. Van-Nostrand Reinhold, New York.

[3] Pozzo, M.M. and T.E. Gray (1987). "An approach to containing viruses." *Computers and Security*, 6 (4), 321–331.

[4] Shoch, J.F. and J. Hupp (1982). "The 'worm' programs: Early experience with a distributed computation." *Communication of ACM*, 25 (3), 172–180.

TRUNCATED DIFFERENTIALS

The notion of a truncated differential was defined by Knudsen in [2] and was applied to <u>cryptanalyse</u> the cipher *SAFER* due to its word-oriented operations [3]. Truncated differentials are an extension of the notion of *differentials*, used in <u>differential cryptanalysis</u>. The main idea is to leave part of the difference unspecified, thus clustering several differentials together. This can be done by specifying m-bit constraints on the whole block (where m is smaller than the block size n), like: $(A, -A, B, 2B)$, where A, B can take any value as was done in [2]; or by fixing part of the data block to certain value and allowing the rest to vary arbitrarily, like: $(0, *, 3, *, 255, *, *)$, where $*$ may take any value.

Such "wild-card" differentials were introduced in the cryptanalysis of the hash-function *Snefru* [1]. Truncated differentials are a powerful tool against ciphers with word-oriented structure, and play an important role in extensions of differential techniques such as <u>impossible-differentials</u> and <u>boomerang attacks</u>. Truncated differentials are often combined with a technique of packing data into *structures*, which sometimes allow to exploit truncated differentials even with probabilities lower than 2^{-n}. See also <u>differential cryptanalysis</u>.

Alex Biryukov

References

[1] Biham, E. and A. Shamir (1991). "Differential cryptanalysis of Snefru, Khafre, REDOC-II, LOKI and Lucifier." *Advances in Cryptology—CRYPTO'91*, Lecture Notes in Computer Science, vol. 576, ed. J. Feigenbaum. Springer-Verlag, Berlin, 156–171.

[2] Knudsen, L.R. (1995). "Truncated and higher order differentials." *Fast Software Encryption, FSE'94*, Lecture Notes in Computer Science, vol. 1008, ed. B. Preneel. Springer-Verlag, Berlin, 196–211.

[3] Knudesen, L.R. and T.A. Berson (1996). "Truncated differentials of SAFER." *Fast Software Encryption, FSE'96*, Lecture Notes in Computer Science, vol. 1039, ed. D. Gollmann. Springer-Verlag, Berlin, 15–26.

TRUST MODELS

INTRODUCTION: <u>Public-key infrastructure</u> (PKI) manages trust in electronic transactions. The principal elements used for maintaining that trust are the contents of the <u>certificates</u> and the security safeguards in effect in the environments of the various parties involved. These two elements are derived by a risk management procedure from the business purpose of the exchanges, as captured in the certificate <u>policy</u>.

Before discussing trust management in PKI, a definition of the word "trust" is required.

Reference [1] defines trust in the following way:

"Generally, an entity can be said to "trust" a second entity when it (the first entity) makes the assumption that the second entity will behave exactly as the first entity expects."

The first entity makes this assumption about a relevant area of the second entity's behaviour, and so the trust between them is limited to that specific area. In PKI the behaviour of interest is related to the distribution and use of public keys for electronic commerce. Different types of trust relationship are capable of conveying different types of assurance between the parties. A trust relationship based upon <u>public-key cryptography</u> technology is intended to ensure the authenticity of the second entity's identifying descriptor and the enforceability of commitments undertaken by both entities.

TRUST RELATIONSHIPS: Trust is a well-established concept, and there are many examples of conventional trust relationships, including those between a bank and its account holders, between an employer and its employees, between a government and its citizens, between the media and its subscribers, between an industry association and its members, and so on. These conventional trust relationships also play an essential role in establishing new trust relationships based on public-key technology.

In the realm of public-key technology, a necessary step towards establishing a trust relationship is for the first entity to import a public key from the second one and protect its integrity for storage or communication to other entities. The entity that imports the public key is known as the relying party, because it intends to **rely** upon the public key for protecting subsequent exchanges with the key-holder (the entity from whom the key is imported). Figure 1 illustrates a basic direct trust relationship between a key-holder and a relying party.

Any entity may act, simultaneously, as both a relying party and a key-holder. But, for the sake of simplicity, these two roles are separated throughout this discussion.

In order to avoid confusion between the two parties, the public key import operation must be performed in a manner that preserves the key's authenticity and integrity (i.e., it must be received, unmodified, from the correct key-holder) and its "clarity" (i.e., the relying party's understanding of the approved uses for the public key must be the same as the key-holder's understanding). These security properties can only be established by means of an existing trust relationship capable of conveying the necessary assurances. So, it appears to be axiomatic that a trust relationship cannot be "created" where there is no existing trust relationship. Rather, existing trust relationships can only

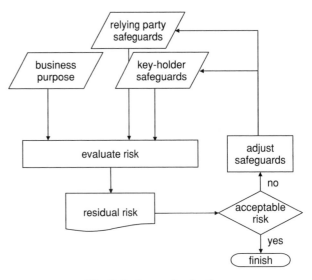

Fig. 2. Safeguard selection

be "qualified" and "combined" to form trust relationships with new characteristics. This paper discusses ways of building trust relationships with desirable characteristics including business controls, based on public-key technology, and using existing conventional trust relationships as their starting point.

One way of building on an existing trust relationship, to form a new trust relationship based on public-key techniques with integrity and clarity, is to precede the import of a public key with an exchange of a trusted <u>authentication token</u> for the key-holder to the relying party. Following that exchange, an automated transfer of the public key and its qualifying information can be conducted between the key-holder and the relying party. The trustworthiness of that exchange depends upon the information covered in the exchange of the authentication token for the key-holder.

The authentication token may take the form of a displayable string of characters (which the relying party can conveniently read and enter at a computer keyboard), or some pre-existing shared secret information, which is linked with the public key and qualifying information. Proper transfer of the authentication token relies upon the existing trust relationship. The authenticity and integrity of the public key and its qualifying information can then be protected by relying on this authentication token.

An essential component of the qualifying information is an identifying descriptor for the key-holder. The descriptor may be unique or shared, or some combination thereof. Sometimes it is the key-holder's name, but this is not necessarily the case. In many applications, the relying party's end-goal is to associate a privilege with the key-holder, and it will use the public key to authenticate the

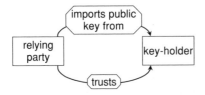

Fig. 1. Direct trust relationship

key-holder merely as an initial step in controlling the granting of that privilege. In other circumstances, the qualifying information may indicate directly that the key-holder possesses the required privilege. In self-service and inter-personal messaging applications, the key-holder's identifying descriptor is commonly sufficient.

The property of clarity may be implemented by the qualifying information in a number of different ways. It may be partially and implicitly expressed in the type of the public key, because for technical reasons not all public keys can be used for all business purposes. It may be explicitly encoded in key-usage codes and it may be included by reference in the form of certificate policy identifiers.

Risk

According to the X.509 definition of trust, the risk that the key-holder might fail to behave as expected naturally attaches to the relying party. Some examples of the elements of risk in a public-key-based trust relationship are the following:

- the identifying descriptors associated with a key are incorrect or misleading;
- the public-key holder's private key has been discovered by another entity;
- the public-key holder's implicit privilege has been withdrawn recently;
- the public-key holder has a prevailing right not to be bound by its signature in the way the relying party expects;
- the public-key holder does not adequately protect the confidentiality of the sensitive information with which it is entrusted;
- etc.

For dealings between individuals, where the relying party has a close and long-standing relationship with the community of key-holders, this allocation of risk is appropriate, because the relying party is able to evaluate its risk and decide whether or not to accept it. But, in electronic-commerce, the relying party may either be unqualified to evaluate its risk or will evaluate it and choose not to accept it. The following strategies can be used to manage risk.

- **Minimize:** to minimize risk, the risk taker attempts to reduce the probability of a loss-causing event as much as practicable. To avoid a loss-causing event, each party attempts to perform according to the other's expectations.
- **Avoid or contain consequences:** if a failure of one party's expectations occurs, the other party tries to reduce the economically significant consequences, as much as is practicable. For example, disclosure of the key-holder's pri-

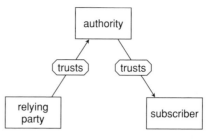

Fig. 3. Trusted third party

vate key may breach an obligation to a relying party, but if the key-holder informs the relying party before the relying party suffers any harm, then the damages due to the breach are only nominal.

- **Bear the residual risk:** it is generally not possible or cost-effective to reduce the risk of a loss-causing event to zero, so the residual risk must be borne, usually by spreading it among a large group of risk bearers, so that financially the risk amounts to a fixed, budgetable expense rather than the unpredictable possibility of a crippling loss. This spreading can often happen over time, so that the time value of money is involved, as well as over geography, societies, etc.

Suitable safeguards are chosen according to the intended business purpose of the key by means of the risk management process, as shown in Figure 3.

The safeguards applied by the relying party and the key-holder are evaluated in the context of the intended business purpose of the public key, and if the residual risk is not acceptable, then adjustments to the choice of safeguards must be made.

Independent of the question of selecting suitable safeguards, there is the question of "assurance". That is, if, when properly implemented, the chosen safeguards reduce the risk to an acceptable level for the intended business purpose, each party requires assurance that the safeguards are indeed properly implemented, both in its own environment and in those of the other parties.

In the majority of its dealings, the relying party will be able to identify a suitable trusted third party in the form of an appropriate traditional source of trust to assist it in bearing its risk; suitable sources include, but are not limited to, an employer, bank, doctor, government, etc.

The Trusted Third Party

When reliance is placed on an authority, the key-holder is commonly referred to as a subscriber, because, sometimes, the key-holder elects to subscribe to a service operated by the authority. As we place greater and greater reliance in electronic commerce systems, the authority may be required to accept a significant measure of risk from the relying party. An authority with a close and long-standing relationship with the subscriber community will be better placed to mitigate risk associated with registering and controlling the behaviour of that community. In order to restrict its risk acceptance to matters that are under its *direct* control, the authority will have to redistribute that portion of its risk that is associated with other matters to the parties, such as the subscriber and the relying party, who *do* have direct control over them.

The involvement of a <u>trusted third party</u>, or "authority", is shown in Figure 3.

The relying party trusts the authority; the authority trusts the subscriber; so, the relying party trusts the subscriber.

In general, a single relying party may rely on more than one primary source of trust for its dealings in different aspects of its life. However, for simplicity, this paper assumes a single "trust anchor."

Authentication and Certification Authorities

Relying parties tend to redistribute risk to authorities, which are "close to", or have a direct and long-standing relationship with, the subscriber community. The main reason for this is that proximity facilitates familiarity, so people close at hand have access to better information and evidence. In economic terms, they can confirm the accuracy of certified information more cheaply and easily than remote persons, whose information is more likely to be derivative, based on heuristic assumptions, etc.

Such authorities may take one of two forms:

- An <u>authentication</u> authority has only conventional trust relationships with the relying party and the subscriber, not a public-key relationship. The public-key relationship exists directly only between the relying party and the subscriber.
- A <u>certification authority</u> has a public-key relationship established between itself and the relying party and between itself and the subscriber as a precursor to the establishment of the direct public-key relationship between the relying party and the subscriber.

An *authentication* authority trust relationship is comprised of the following elements:

(a) The subscriber provides an authentication token to the authentication authority.
(b) Upon successful registration, the authority makes the authentication token, and applica-

ble qualifying information, available to the relying party.
(c) A relying party can then obtain the subscriber's public key directly from the subscriber and use the authentication token to confirm its authenticity and suitability to its business purpose.

A *certification* authority trust relationship is comprised of the following elements:

(a) The subscriber's public key is supplied to the certification authority by an authentic means.
(b) The authority's public key is supplied to the relying party by an authentic means.
(c) Subsequently, the subscriber's public key and qualifying information are supplied to the relying party either directly from the authority, or by some other communications path, with its authenticity, integrity and clarity protected by a digital signature applied by the authority. The corresponding data structure is called a certificate and the most common means for implementing this scheme is <u>X.509</u> [1]. The <u>certificate</u> can be viewed as the secure means by which the certification authority communicates trust to the relying party.

The main advantages of a certification authority over an authentication authority are:

(a) Evidence of the role of the certification authority appears in the sequence of certificates used by the relying party to validate the subscriber's public key, whereas evidence of the role of the authentication authority does not;
(b) Consequently, in the case of the certification authority, the relying party identifies the basis of its trust with the authority that introduced it to the subscriber, rather than with the subscriber itself, as is the case with the authentication authority;
(c) The certification authority can automatically revoke the trust in the subscriber, whereas the authentication authority cannot; and
(d) The mechanisms and protocols are defined for the function of the certification authority, but not for the function of the authentication authority.

The main advantages of an authentication authority over a certification authority are:

(a) The authentication authority does not have to be implemented in an automated information processing system, whereas the certification authority does; and
(b) when using an authentication authority, the certificate path contains one fewer certificate than it does when using a certification authority.

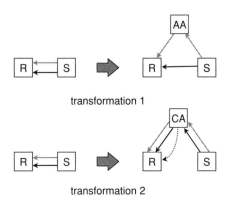

transformation 1

transformation 2

Fig. 4. Basic trust transformations

TRUST MODELS: Figure 4 illustrates the two elementary trust transformations, based upon the authentication authority and the certification authority, introduced earlier. The light arrows represent exchanges in a conventional trust relationship. The dark arrows represent exchanges in a trust relationship derived from the conventional trust relationship using the first exchange. The broken arrows represent certificates.

These transformations can be applied repeatedly to form more complex compound trust models. Five such models of particular interest are shown in Figure 5 to Figure 9. The characteristics of these models are discussed further below.

Subscriber Registration Authority. The subscriber registration authority model is shown in Figure 5.

This model results from applying *transformation 2* and then *transformation 1* to the subscriber relationship. It is useful when the CA is remote from the subscriber community. In this configuration, the authentication authority is commonly referred to as a subscriber registration authority. Although there are two authorities, there is only one certificate, and the involvement of the authentication authority is invisible to the relying party.

● The relying party has as its trust anchor the public key of its local CA.
● A subscriber certificate issued by the local CA, signed with the private key that corresponds to the public key that is the relying party's trust anchor, is the authentic means by which the

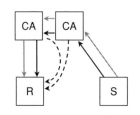

Fig. 6. Direct cross-certification

local CA provides the subscriber's public key to the relying party to be trusted.

Direct Cross-certification. The direct cross-certification model is shown in Figure 6.

This model results from applying *transformation 2* twice to the relying party relationship. Direct cross-certification is an applicable model when authorities operated by separate organizational entities enter into a direct trust relationship. In this case, there are two authorities and two certificates, and the involvement of each authority is visible to the relying party.

● The relying party has as its trust anchor the public key of its local CA.
● The cross-certificate, issued by the local CA to the remote CA and signed with the private key corresponding to the public key that is the relying party's trust anchor, is the authentic means by which the local CA provides the remote CA's public key to the relying party to be trusted.
● The subscriber certificate issued by the remote CA, signed with the private key corresponding to the public key in the previous cross-certificate, is the authentic means by which the remote CA provides the relying party with the subscriber's public key to be trusted.

Two-tier Hierarchy. The two-tier hierarchy model is shown in Figure 7.

This model results from applying *transformation 1* to the relying party relationship shown in Figure 6. In this case, the authentication authority is more commonly referred to as a certificate list manager. The two-tier hierarchy is an

Fig. 5. Subscriber authentication authority

Fig. 7. Two-tier hierarchy

applicable model when the certificate list manager and the subordinate CA are operated by separate organizational entities and their trust relationship is facilitated by a third entity, which operates the root CA. There are three authorities, but only two certificates and the involvement of the authentication authority is not recorded in the list of certificates, which, in conjunction with the business transaction, form the complete evidence.

- The certificate list manager is the means by which the relying party establishes trust in the public key of its local CA.
- A subordinate cross-certificate issued by the root CA, signed by the private key corresponding to the public key that is the relying party's trust anchor, is the authentic means by which the root CA supplies the subordinate CA's public key to the relying party to be trusted.
- The subscriber certificate issued by the subordinate CA, signed by the private key corresponding to the public key certified in the subordinate certificate, is the authentic means by which the subordinate CA provides the subscriber's public key to the relying party to be trusted.

Bridge Certification Authority. The bridge certification authority model is shown in Figure 8.

This model results from applying *transformation 2* to the relying party relationship shown in Figure 6. The bridge certification authority is an applicable model when the two spoke certification authorities are operated by separate organizational entities and their trust relationship is facilitated by a third entity, which operates the bridge CA. There are three authorities and three certificates, so the role of each authority is recorded in the list of certificates that form the evidence.

- The relying party has, as its trust anchor, the public key of its local spoke CA.
- A cross-certificate issued by its local spoke CA, signed with the private key corresponding to the public key that is the relying party's trust anchor, is the authentic means by which the local

spoke CA provides the relying party with the bridge CA's public key to be trusted.

- A cross-certificate issued by the bridge CA, signed with the private key corresponding to the public key in the cross-certificate issued by the local spoke CA, is the authentic means by which the bridge CA provides the relying party with the remote spoke CA's public key to be trusted.
- The subscriber certificate issued by the remote spoke CA, signed with the private key corresponding to the public key certified in the cross-certificate issued by the bridge CA, is the authentic means by which the remote spoke CA provides the subscriber's public key to the relying party to be trusted.

The bridge certification authority model is becoming a popular model, especially in environments where a large number of CAs need to cross-certify with one another in order to facilitate trusted transactions among the members of their respective subscriber and relying party communities. In particular, this model is popular with national governments where it enables the domain of trust to be expanded beyond the domain of a single government department or agency to the complete national government. Each agency CA cross-certifies bi-directionally with the central bridge CA. This enables a simple certification path to be built between parties within any agencies in the complete bridge community through a single common point of trust. Each agency need establish only a single cross-certification relationship, that with the bridge, and they are immediately capable of trusting certificates issued throughout the bridged community. Without a bridge CA, individual direct cross-certification relationships would need to be established between every pair of agencies/departments within that national government. A bridge CA acts as an introducer and facilitator of trust to the total environment, while enabling each spoke in the model to retain relative autonomy over their own domain.

Bridge Authentication Authority. The bridge authentication authority model is shown in Figure 9.

This model results from applying *transformation 1* to the relationship between the two authorities shown in Figure 6. The bridge authentication authority is an applicable model when the two spoke certification authorities are operated by separate organizational entities and their trust relationship is facilitated by a third entity, which operates the bridge authentication authority. This model is useful when the third entity

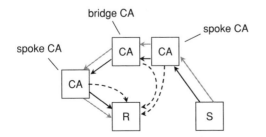

Fig. 8. Bridge certification authority

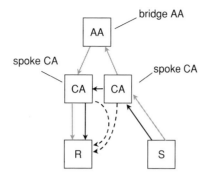

Fig. 9. Hub authentication authority

is not equipped to operate an automated information system. There are three authorities, but only two certificates. Therefore, the involvement of the bridge authentication authority is not recorded in the list of certificates that form the evidence.

The mechanism used by the relying party to establish trust in the subscriber's key is identical to that described for the bridge certification authority model above.

Summary

The trust transformations described above may be repeatedly applied to create even more elaborate trust models. However, the five shown above are the ones of most practical interest.

Different trust models are suited to different business situations. But, no matter which trust model is chosen, the relying party expectation is that its trust requirements are satisfied by the authority upon which it relies directly, and if that authority makes private arrangements to redistribute its risk to other authorities or subscribers, then this in no way diminishes its obligation to its relying parties. So, it must take whatever measures are necessary to control the behaviour of other authorities and subscribers whose keys it has certified, directly or indirectly, so that its risk remains under control.

In practical terms, the trust brand displayed to the relying party will be that of the certification authority upon which it relies directly. Where an authentication authority is involved, its role is invisible to the relying party at the time of validating the subscriber certificate. Although a significant measure of risk may be accepted by an authentication authority, the relying party appears to rely solely on the certification authority whose public key it has imported directly.

TRUST MODEL BUSINESS CONTROLS: With any of the trust models described above, the business requirements reflected in the policies of the authorities involved in the structure need to be

reflected and enforced within the trust model itself. X.509 [1] standardizes a set of tools that can be used to impose technical business controls on the trust models intended to reflect the business requirements associated with risk as described earlier. The standard tools include:

- Extensions in end-user (subscriber) certificates
- Extensions in cross-certificates
- Variable inputs to the path validation process.

The set of extensions used to reflect business controls in certificates is both comprehensive and flexible. The intent of this flexibility is to enable each authority to reflect its specific business requirements through an appropriate set of optional extensions, thereby minimizing risk. The path validation algorithm in X.509 specifies standard inputs to the process as well as the rules for processing each of the business control related certificate extensions. In addition to the inclusion of business control extensions in certificates, the ability to initialize values for some variables in the standard path validation process enables further refinement of the business controls on a per-relying-party basis. If all relying parties in the local domain have exactly the same set of business requirements, these requirements can generally be addressed solely through inclusion of appropriate extensions in cross-certificates issued by the local CA to remote CAs. If, however, different subsets of the local relying party community have different business requirements, then the business controls imposed through extensions in cross-certificates may need to be complemented with relying-party-specific business controls through the initialization of path validation variables for those subsets. Note that business controls imposed through this configuration technique cannot expand the acceptable set of remote certificates beyond that permitted by the business controls in the cross-certificates. Configuration of relying party path validation variables can only further restrict the acceptable set of remote certificates.

Certificate Extensions

The set of standard extensions related to business controls is outlined below. Many of these can be flagged as either critical or noncritical to the use of the certificate. If an extension is flagged critical, then a relying party shall not trust the certificate without understanding and processing that extension. If an extension is flagged noncritical then a relying party that does not understand that extension is permitted to ignore that extension in determining whether or not to trust the certificate. In general, if business controls are included in certificates they should be flagged critical to ensure that the business requirements are respected.

Certificate Policies Extension. In an end-user certificate this extension indicates the certificate policies under which this certificate can be used. In a cross-certificate the extension indicates the set of certificate policies for which this certificate can be used in a certification path. The standard path processing procedure defines the steps relying parties must go through to determine whether, based on policy, the certificate and path are acceptable. Processing this extension is one of those steps.

Policy Mappings Extension. This extension can only appear in cross-certificates. It enables trust to be placed in remotely issued certificates that have been issued under certificate policies that are equivalent to local policies understood by the relying party but which have different identifiers. Mapping policies has the effect that a remote policy is considered equivalent to a local policy for purposes of path processing.

Policy Constraints Extension. This extension can only appear in cross-certificates. It enables a certification authority to 'turn off' policy mappings that may appear in other certificates in the path. For purposes of path validation, the presence of this extension in a cross-certificate indicates that any path that would only be acceptable through mapping of policies as indicated in subsequent certificates is not an acceptable path. This extension can also be used by a certification authority to require to acceptable policy identifiers to be present in all certificates in the path, for the path to be acceptable.

Inhibit Any Policy Extension. X.509 defines a specific identifier that can be present in the certificate policy's extension to indicate that the certificate can be used under any policy. This identifier becomes a 'wild card' in policy matching and is considered a match to any other policy identifier. This extension, which only appears in cross-certificates, can be used by a certification authority to prevent that special identifier from being an acceptable policy match in the certification path.

Basic Constraints Extension. This extension can be present in end-user certificates as well as in cross-certificates. In an end-user certificate, its sole purpose is to indicate that the subject is an end-entity and not a certification authority. As such this certificate can only be the final certificate in a certification path and not an intermediary certificate. When this extension appears in a cross-certificate it serves two purposes. First it identifies the subject as a certification authority and therefore indicates that the certificate can be used as an intermediary certificate in a certification path. Second, it enables a certification authority to impose a restriction on the number of subsequent intermediary certificates that can appear in a certification path that includes this certificate. In terms of business controls, this enables an authority to reflect, within the certificates, its policy with respect to some of the boundaries of the remote trusted environment.

Name Constraints Extension. This extension can only appear in cross certificates. It enables a certification authority to impose further boundary restrictions on the set of remote certificates that can be trusted through this cross-certificate. Name subtrees that are permitted and names that are prohibited can both be specified.

Relying Party Configuration

The path processing procedure defined in X.509 defines the standard set of inputs to the process. Many of these have predetermined initial values while some are configurable, with default values defined in the standard. These configurable variables can be used to enhance the enforcement of business controls for the relying party community. Business controls deployed through extensions in cross-certificates, while very useful, necessarily must be generic enough to satisfy the requirements of the whole community of local relying parties. In many cases this is insufficient as different sets of relying parties perform different business functions and therefore have varying requirements in terms of business controls used to determine which set of certificates are acceptable for a given relying party set or even a given application or specific transaction. For example, assume that some relying parties in the local community are working with a purchase order system and that only certificate policy x is acceptable for that application and those relying parties, while other relying parties in the same community may be exchanging emails with remote users and that policy y is an acceptable policy for those relying parties. Cross-certificates issued by their local certification authority to other remote authorities would necessarily include both policy x and policy y as acceptable for that cross-certification relationship. In order to further restrict the specific trust placed by each distinct subset of the local relying party community, additional business controls are necessary. These further controls are realized by configuring the initial values for the relevant inputs to the path processing procedure with distinct values for each subset of the community.

Initial Policy Set. This variable, which defaults to "any policy" is used to establish a different set of acceptable policies for each distinct subset of the relying party community. In the above example relying parties that use the purchasing system would have this variable initialized to the value for policy x and the other relying parties would be initialized to the value for policy y.

Explicit Policy Indicator. This variable is used to configure some relying parties to only accept paths where acceptable policies are included in every certificate in the path, while other relying parties can be configured to the default setting for this variable, which does not require the policy to be present in each certificate. This variable is particularly useful where the relying party community has some instances where the policy is absolutely required and others where it is not. In such an environment, setting the "requires explicit policy" extension in cross certificates is not an option as it would require that all local relying parties only accept paths where each and every certificate contains an acceptable policy identifier.

Policy Mapping Inhibit Indicator. This variable is particularly valuable where the local relying party is divided with some for which policy mapping is an acceptable feature and others where it is not and must be inhibited. As with the previous variable, this situation cannot be handled by setting the policy constraints extension in cross certificates to inhibit policy mapping as this would impact all relying parties and would force no mappings to ever be acceptable to any local relying party.

Inhibit Any Policy Indicator. As with the other variables, the ability to configure this variable for relying parties is particularly valuable where the relying party community is heterogeneous, with some instances where the special identifier for "any policy" is an acceptable policy and others where it is not. The relying parties can be configured so that only the appropriate instances have this identifier prohibited. As with the other variables, the related cross-certificate extension (inhibit any policy) is not suitable unless the complete relying party community shares exactly the same business requirement for this identifier.

Permitted and Excluded Subtrees. At present, X.509 has a specific predetermined value for each of these (empty set). The only way to restrict acceptance of remote certificates, based on names, is through the name constraints extension in cross-certificates. This has been identified as a shortcoming and work is underway at the time of this writing to enhance [1] to allow initial values for these two variables to be configurable as well. Once this work is completed it would be possible for subsets of relying parties to be configured to accept remote certificates issued to users in different naming subtrees. For example, this would enable relying parties in the marketing department to be configured so that only their counterparts in the remote organization's marketing department would be acceptable, while relying parties in the local finance team could be configured to accept certificates issued to their remote finance counterparts. As with the other variables, this is particularly valuable if the relying party community is not completely homogenous with respect to their requirements. Otherwise the name constraints extension in cross-certificates can satisfy the common requirements for the total relying party community.

Combined Business Controls

Together, the set of certificate extensions and the set of configurable path validation variables provide a comprehensive toolkit for authorities to enforce business controls on the trust that local relying parties place in remotely issued certificates. Extensions placed in cross-certificates can be thought of as enabling business controls. Although they actually restrict the set of remote certificates that will be trusted locally, their purpose is to enable the complete set of local relying parties to place trust in the remote certificates that satisfy their business requirements and as such these can be considered 'enabling' business controls. The ability to configure subsets of the local relying community with different values for the path processing variables enables authorities to enforce business controls at a much finer granularity and support the overall specific business requirements of each subset of the local relying party community. The business controls established through this process have the effect of further restricting the subset of remote certificates than any relying party will trust beyond that which would be trusted if reliance was placed solely on the business controls established in the extensions in cross certificates. Therefore, these controls can be considered 'restricting' business controls.

SUMMARY: As illustrated in this paper there are a variety of trust models that can be implemented in Public Key Infrastructure. No single trust model is the best for all environments. Rather, the business environment, business requirements for electronic commerce, and policies related to minimizing risk determine the most appropriate trust model for any particular environment. Regardless of the trust model deployed, the trust established through the infrastructure must reflect and satisfy the business requirements for each of the parties involved in the electronic commerce transactions. Through a set of standardized technical tools, PKI enables those business requirements to be reflected in the certificates themselves and enforced by standard path validation engines to ensure the security associated with a transaction is in accordance with the relevant security policies that govern the transactions themselves.

Sharon Boeyen

Reference

[1] ITU-T Recommendation X.509 (2000 E). "Information Technology." Open systems interconnection—The Directory: Public-key and attribute certificate frameworks.

TRUSTED THIRD PARTY

A *Trusted Third Party (TTP)*, sometimes referred to as a Trusted Authority, is an entity within a given community that is trusted by all entities in that community to properly perform a particular service. In some architectures, the TTP must store and adequately protect long-term secrets. The compromise of such secrets will immediately render insecure all future communications in the network and may also render insecure all past communications protected by using those secrets; this situation will persist until new long-term secrets can be established. In some architectures the TTP may have the ability to read all messages. Typically, the service performed by the TTP will enhance the security of other messages or transactions between entities in the community; example services include key generation, entity authentication, time stamping, and notarization.

Examples of TTPs in well-known architectures include the Key Distribution Center (KDC) in Kerberos, the Certification Authority (CA) in a PKI (see Public Key Infrastructure), and the Naming Authority (or name server) in an organizational environment.

Carlisle Adams

References

[1] Menezes, A., P. van Oorschot, and S. Vanstone (1997). *Handbook of Applied Cryptography*. CRC Press, Boca Raton, FL.
[2] Stinson, D. (1995). *Cryptography: Theory and Practice*. CRC Press, Boca Raton, FL.

TWIRL

TWIRL (The Weizmann Institute Relation Locator) is an electronic device designed to speed up the sieving part of modern integer factoring algorithms such as the Quadratic Sieve and the Number Field Sieve. It was designed in 2003 by Shamir and Tromer [3], and improves upon the optoelectronic TWINKLE device of Shamir [2]. See also: factoring circuits.

The goal of sieving algorithms is to find all the smooth numbers in an interval of candidate integers x, where a *smooth number* (see smoothness) is x such that (almost) all prime divisors of x are small primes belonging to a factor base. For example, to factor 1024-bit composites TWIRL would test about 10^{15} numbers against a factor base consists of all primes smaller than about 10^{10}, and repeat this about 10^8 times for different collections of numbers.

The standard implementation of sieving algorithms is to use computers with large amounts of RAM and devote one memory location to each candidate x. For each prime p in the factor base, the value $\log(p)$ is added to every pth memory location. At the end of the algorithm the accumulated value in each memory location is approximately equal to the binary size of the smooth part of the corresponding x, and thus the output consists of the indices of all the memory locations whose values exceed a certain threshold.

The TWIRL device is based on three basic ideas. The first, borrowed from TWINKLE, is exchanging the roles of space and time. Instead of assigning a memory location to each candidate x and iterating over the primes p in the factor base, TWIRL assigns a circuit element on a silicon wafer to each prime p and, during the operation of the wafer, assigns a clock cycle to each candidate x. Each circuit element emits a $\log(p)$ value in those clock cycles that correspond to x which are divisible by p. These emissions are summed using an addition pipeline, and compared to a threshold.

The second idea is to speed up the operation by "compressing time", i.e., by handling thousands of consecutive candidates x in every clock cycle. Each prime p needs to be simultaneously tested against many candidates at each clock cycle, which necessitates means for scheduling, routing, and congestion control for the emitted $\log(p)$ values.

The third idea is to also "compress space". In the factor base, most p values are very large and thus contribute $\log(p)$ very seldom. Hence, thousands of primes can be bunched together and handled by shared circuitry, in which each prime is represented very compactly in DRAM-type memory which is inspected and updated cyclically by a simple dedicated processor.

The TWIRL and TWINKLE devices have attracted considerable attention, but it is not known whether any devices of this type have been built so far, and if so, what was their actual performance. Theoretical analysis indicates that TWIRL-assisted factorization of 1024-bit numbers can be done in 6 weeks on a device whose construction cost is about \$10 M using 90 nm VLSI process technology [1, 3]. For smaller key sizes, a single TWIRL wafer (which would cost a few thousand dollars) can factor a 768-bit composite within a month, or a 512-bit composite in under 3 minutes. However, in all these cases there would be a large initial R&D investment (on the order of \$20 M) due to the use of a custom VLSI design.

Adi Shamir
Eran Tromer

References

[1] Lenstra, A.K., E. Tromer, A. Shamir, B. Dodson, J. Hughes, W. Kortsmit, and P. Leyland (2003). "Factoring estimates for a 1024-bit RSA modulus." *Advances in Cryptology—ASIACRYPT 2003*, Lecture Notes in Computer Science, vol. 2894, ed. C.S. Laith. Springer-Verlag, Berlin.

[2] Shamir, A. (1999). "Factoring large numbers with the TWINKLE device." *Proceedings of CHES'99*, Lecture Notes in Computer Science, vol. 1717, eds. Ç.K. Koç and C. Paar. Springer-Verlag, Berlin.

[3] Shamir, A. and E. Tromer (2003). "Factoring large numbers with the TWIRL device." *Proceedings of*

CRYPTO 2003, Lecture Notes in Computer Science, vol. 2729, ed. D. Boneh. Springer-Verlag, Berlin.

TWO-FACTOR AUTHENTICATION

This term indicates user <u>authentication</u> where two factors are involved in the process: something the user knows (like a static <u>password</u>) and something the user possesses, like a <u>token</u> or a mobile phone used for one-time passwords.

Peter Landrock

TWOFISH

Twofish [4] is a 128-bit <u>block cipher</u> submitted to the *Advanced Encryption Standard* (<u>Rijndael/AES</u>) selection process by Bruce Schneier et al. It was one of the five AES finalists in 1999.

Twofish has a <u>Feistel</u>-type design and was inspired by the 64-bit block cipher Blowfish [3], published in 1993. Twofish has 16 rounds and accepts 128-bit, 192-bit, and 256-bit secret keys. The 64-bit F-function used in the Feistel network consists of two parallel 32-bit branches called g-functions. Each of them contains a layer of four key-dependent 8×8-bit S-boxes and a 4-byte linear transform based on a Maximum Distance Separable (MDS; see <u>cyclic codes</u>) code. The outputs of both branches are combined using a Pseudo–Hadamard Transform (PHT) and the result is mixed with two 32-bit subkeys. Two additional 128-subkeys are XORed with the data before the first and after the last round (this is called input and output whitening).

The *key schedule* (see <u>block cipher</u>) of Twofish has to provide 1280 bits of key material and 4 key-dependent S-boxes used in the g-function. The main component in the key schedule is a 32-bit h-function, which is constructed by alternating layers of fixed S-boxes and key additions, and ends with an MDS-transform. The schedule can be implemented in different ways and allows different tradeoffs between setup time and throughput.

Today, the best attack on reduced-round Twofish is the saturation attack by Lucks [1] (see <u>multiset attack</u>). The attack breaks 8 rounds out of 16 and requires 2^{127} chosen plaintexts (i.e., half of the codebook). Another interesting cryptanalytical result is an observation by Mirza and Murphy [2] concerning the key schedule.

Christophe De Cannière

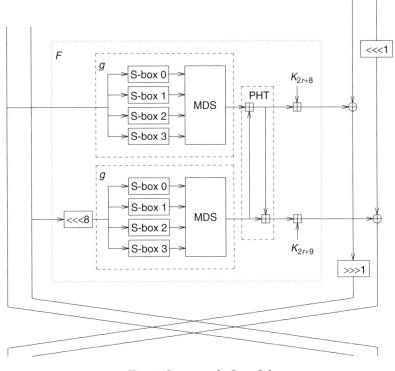

Fig. 1. One round of twofish

References

[1] Lucks, S. (2002). "The saturation attack—a bait for Twofish." *Fast Software Encryption, FSE 2001*, Lecture Notes in Computer Science, vol. 2355, ed. M. Matsui. Springer-Verlag, Berlin, 1–15.

[2] Mirza, F. and S. Murphy (1999). "An observation on the key schedule of Twofish." *Proceedings of the Second AES Candidate Conference*. National Institute of Standards and Technology, 151–154.

[3] Schneier, B. (1994). "Description of a new variable-length key, 64-bit block cipher (Blowfish)." *Fast Software Encryption, FSE'93*, Lecture Notes in Computer Science, vol. 809, ed. R. J. Anderson. Springer-Verlag, Berlin, 191.

[4] Schneier, B., J. Kelsey, D. Whiting, D. Wagner, C. Hall, and N. Ferguson (1998). "Twofish: A 128-bit block cipher." *Proceedings of the First AES Candidate Conference*, National Institute of Standards and Technology.

2^k-ARY EXPONENTIATION

2^k-ary exponentiation is an approach for exponentiation in any <u>group</u> (or semigroup). It generalizes <u>binary exponentiation</u> and is based on the 2^k-ary representation of the exponent. The positive integer parameter k determines the memory usage and the expected running time of the <u>exponentiation algorithm</u>. The case $k = 1$ of 2^k-ary exponentiation is the same as binary exponentiation.

First we describe how the *left-to-right 2^k-ary exponentiation method*, for an arbitrary integer $k \geq 1$, computes g^e where g is an element of the group (or semigroup) and e is a positive integer whose 2^k-ary representation is

$$e = \sum_{i=0}^{l-1} e_i 2^{ki}, \quad e_i \in \{0, \dots, 2^k - 1\}.$$

We assume that l is chosen minimal; thus, $e_{l-1} \neq 0$. First, the small powers for exponents 1 up to $2^k - 1$ are computed and stored; this can be done like this:

$G_1 \leftarrow g$
for $d = 2 \, \text{to} \, 2^k - 1$ **do**
$\quad G_d \leftarrow G_{d-1} \circ g$

Second, the result g^e is computed using this table of small powers $G_1 = g, \dots, G_{2^k-1} = g^{2^k-1}$:

$A \leftarrow G_{e_l} - 1$
for $i = l - 2$ down to 0 **do**
$\quad$ **for** $j = 1$ to k **do**
$\quad\quad A \leftarrow A \circ A$
$\quad$ **if** $e_i \neq 0$ **then**
$\quad\quad A \leftarrow A \circ G_{e_i}$
return A

Left-to-right 2^k-ary exponentiation is due to Brauer [1].

Like binary exponentiation, 2^k-ary exponentiation has a variant that performs right-to-left exponentiation, i.e., that starts scanning e at the least significant digit e_0 rather than at the most significant digit e_{l-1}. Whereas the left-to-right method uses a fixed table of values $G_1, \dots, G_{2^k-1}$, the right-to-left method uses a dynamically modified array.

> **for** $d = 1$ to $2^k - 1$ **do**
>> $B_d \leftarrow$ identity element
>
> $A \leftarrow g$
>
> **for** $i = 0$ to $l - 1$ **do**
>> **if** $e_i \neq 0$ **then**
>>> $B_{e_i} \leftarrow B_{e_i} \circ A$
>>
>> **if** $i < l - 1$ **then**
>>> **for** $j = 1$ to k **do**
>>>> $A \leftarrow A \circ A$
>
> {Now $g^e = \Pi_{d=1}^{2^k-1} B_d^d$; this can be computed as follows:}
>
> **for** $d = 2^k - 1$ downto2 **do**
>> $B_{d-1} \leftarrow B_{d-1} \circ B_d$
>>
>> $B_1 \leftarrow B_1 \circ B_d$
>
> **return** B_1

The algorithm as written can be optimized so that (at least) $2^k - 1$ applications of the group operation are avoided: for each of the variables B_d, an assignment of the form $B_{e_i} \leftarrow B_{e_i} \circ A$ or $B_{d-1} \leftarrow B_{d-1} \circ B_d$ can be simplified to $B_{e_i} \leftarrow A$ or $B_{d-1} \leftarrow B_d$, respectively, when the variable still contains the identity element. (In the final loop, one more application of the group operation can be saved in the case that both B_{d-1} and B_d still have their initial value. Below, we disregard this possible additional optimization for simplicity as in practice it will rarely have an effect.) An array of $2^k - 1$ flags can be used to implement this optimization (cf. the similar optimization in the pseudo-code for the right-to-left binary exponentiation method). Right-to-left 2^k-ary exponentiation is due to Yao [3]; the subalgorithm show above for computing $\Pi_{d=1}^{2^k-1} B_d^d$ is due to Knuth [2, answer to exercise 4.6.3-9].

The most significant 2^k-ary digit e_{l-1} is nonzero for all exponents with a length from $l(k - 1) + 1$ to lk bits. Assuming that all other bits are uniformly and independently random, the expected number of 2^k-ary digits among $e_0, \dots, e_{l-2}$ that will be nonzero is $(l - 1)\frac{2^k-1}{2^k}$. Thus, the left-to-right 2^k-ary exponentiation method as shown above will perform $(l - 1)k$ squaring operations ($A \circ A$) and

$$2^k - 2 + (l - 1)\frac{2^k - 1}{2^k}$$

general group operations on average. The right-to-left 2^k-ary exponentiation method will also perform $(l - 1)k$ squaring operations and on average

$$(l - 1)\frac{2^k - 1}{2^k} + 1 + 2(2^k - 2) - (2^k - 1)$$
$$= 2^k - 2 + (l - 1)\frac{2^k - 1}{2^k}$$

general group operations (this tally includes $(l - 1)\frac{2^k-1}{2^k} + 1$ for the loop over i, $2(2^k - 2)$ for the loop over d, and a reduction of $2^k - 1$ achieved by the optimization explained above, which affects both loops).

In practice it is often the case that group operations can be implemented such that computing a square $A \circ A$ is faster than a general multiplication. For slightly improved speed in such scenarios, both the algorithm for computing the table of the G_i for the left-to-right method and the algorithm for computing the product $\prod_{d=1}^{2^k-1} B_d^d$ in the right-to-left method can be modified to use $2^{k-2} - 1$ squarings instead of $2^{k-2} - 1$ of the general multiplications. (In the left-to-right method, G_d can be computed as $G_{d/2}^2$ instead of $G_{d-1} \circ g$ whenever d is even. An according modification for the right-to-left method follows from the duality between left-to-right and right-to-left exponentiation; see right-to-left exponentiation).

2^k-ary exponentiation is convenient in that the 2^k-ary representation of the exponent e can be easily obtained from its binary representation. However, a different generalization of binary exponentiation can further reduce the expected number of group operations needed for each exponentiation; see sliding window exponentiation.

Bodo Möller

References

[1] Brauer, A. (1939). "On addition chains." *Bulletin of the American Mathematical Society*, 45, 736–739.

[2] Knuth, D.E. (1981). *The Art of Computer Programming—Vol. 2: Seminumerical Algorithms* (2nd ed.). Addison-Wesley, Reading, MA.

[3] Yao, A.C.-C. (1976). "On the evaluation of powers." *SIAM Journal on Computing*, 5, 100–103.

U

UNDENIABLE SIGNATURES

Undeniable signatures are digital signatures that can be verified only by some help from the signer. These were introduced by Chaum and van Antwerpen [2]. Unlike an ordinary digital signature that can be verified by anyone who has accessed the public verifying key of the signer (*universal verifiability*), an undeniable signature can only be verified by engaging in a—usually interactive—protocol with the signer. The outcome of the protocol is an affirming or rejecting assertion telling the verifier whether the undeniable signature has originated from the alleged signer or not. The verifier cannot enforce a clarification about a signature's validity because a signer can always refuse to cooperate, but nonrepudiation is still guaranteed since a signer cannot convince a verifier that a correct signature is invalid or that an incorrect signature is valid.

Undeniable signatures are useful for signers of nonpublic sensitive information who seek to keep control over who can verify their signatures. For example, a company producing software for safety critical systems could deliver its executables with undeniable signatures. This would allow registered customers to verify the origin of the software, while software pirates could not do so. In case a significant bug is discovered later in the software, a registered customer could hold the software company liable for the bug and perhaps for its consequences.

An undeniable signature scheme has three operations: one for generating pairs of a private signing key and a public verifying key (see also public key cryptography), one for signing messages, and a *confirming operation* for proving signatures valid (confirmation) or invalid (disavowal). The confirming operation must have two defined outputs to signal confirmation or disavowal in order to distinguish three possible cases: (a) the signature in question is valid (operation returns "confirm"), (b) the signature in question is invalid (operation returns "disavow"), and (c) the alleged signer is not willing or not available to cooperate and let the verifier find out whether (a) or (b) holds (operation fails). This latter problem is addressed by designated confirmer signatures.

The characteristic security requirements of an undeniable signature scheme are:

- **Unforgeability:** Resistance against existential forgery under adaptive chosen message attacks by a computationally restricted attacker.
- **Invisibility:** A cheating verifier, given a signer's public verifying key, a message, and an undeniable signature, cannot decide with non-negligible probability better than a pure guess whether the signature is valid for the message with respect to the signer's verifying key or not.
- **Soundness:** A cheating signer cannot misuse the confirming operation in order to prove a valid signature invalid (nonrepudiation), or an invalid signature valid (false claim of origin).
- **Nontransferability:** A cheating verifier obtains no information from the confirming operation that allows him to convince a third party that the alleged signature is valid or invalid, regardless of if the signature is in fact valid or not.

The property of nontransferability was intended by the original work of Chaum and van Antwerpen [2], but Jakobsson [7] showed that their particular undeniable signature construction cannot achieve nontransferability against mutually distrusting but interacting verifiers. Jakobsson et al. [8] proposed undeniable signature constructions that satisfy non-transferability as well.

Constructions of undeniable signatures have been based on groups, in which the discrete logarithm problem is hard [2, 4, 9] and on the problem of factoring integers [5].

Undeniable signature schemes can be equipped with additional features: the confirming operation can be noninteractive according to [8]. Pedersen [10] suggested distribution of the power of confirming signatures over a set of delegates in order to increase the availability of individual signers. Harn and Yang [6] proposed the concept of undeniable threshold signatures, where certain subsets (coalitions) of signers are authorized to produce signatures on behalf of a whole set of signers. Efficient and secure constructions were proposed by Michels and Stadler [9]. Chaum et al. proposed convertible undeniable fail-stop signatures [3], where signers can convert their undeniable signatures into fail-stop signatures.

Sakurai and Yamane [11] have proposed undeniable blind signatures.

CONVERTIBLE UNDENIABLE SIGNATURES:
Convertible undeniable signatures [1] are an interesting extension to undeniable signatures. In a convertible undeniable signature scheme, a signer can convert each individual undeniable signature into an ordinary digital signature that is universally verifiable. Upon request by a verifier, the signer provides an individual receipt for a requested undeniable signature to the verifier. Henceforth, the verifier can unlock the respective undeniable signature and forward it together with the receipt to any third party, who can now verify the signature against the signer's public verifying key. Moreover, the signer can provide a universal receipt that instantly allows a recipient to universally verify all signatures of the respective signer. In effect, convertible undeniable signature schemes support signers in gradually increasing the verifiability of their signatures in a controlled fashion.

Let us reconsider the software company mentioned in the introduction and imagine it is going bankrupt. It may still have contractual liabilities to support its customers in verifying their software for a number of years. However, this service may be too costly and there may even be no need any more to further control who is verifying which software packages. In this case, the company could release a universal receipt on their Web page, which would henceforth allow anyone to verify the signatures of its software packages at any time.

A convertible undeniable signature scheme has the same three operations as an undeniable signature scheme and the following three additional operations:

- **An individual conversion operation**, which takes as input a message, an undeniable signature, a signer's private signing key, and returns an ordinary, i.e., universally verifiable signature.
- **A universal conversion operation**, which takes as input a signer's private signing key and returns a universal receipt that allows to convert all undeniable signatures valid with respect to the signer's public verifying key into ordinary, i.e., universally verifiable signatures.
- **A universal verifying operation**, which takes as input a message, a converted undeniable signature, and a signer's public verifying key and returns whether the signature is valid or not with respect to the alleged signer's public verifying key.

The characteristic security requirements of a convertible undeniable signature scheme include those of an undeniable signature scheme and additional security requirements guaranteeing that valid (invalid) signatures can only be converted into valid (invalid) signatures, etc. These additional security requirements have not yet been formalized in the open literature.

Gerrit Bleumer

References

[1] Boyar, Joan, David Chaum, Ivan Damgård, and Torben Pedersen (1991). "Convertible undeniable signatures." *Advances in Cryptology—CRYPTO'90*, Lecture Notes in Computer Science, vol. 537, eds. A.J. Menezes and S.A. Vanstone. Springer-Verlag, Berlin, 189–205.

[2] Chaum, David and Hans van Antwerpen (1990). "Undeniable signatures." *Advances in Cryptology—CRYPTO'89*, Lecture Notes in Computer Science, vol. 435, ed. G. Brassand. Springer-Verlag, Berlin, 212–216.

[3] Chaum, David, Eugène van Heijst, and Birgit Pfitzmann (1990). "Cryptographically strong undeniable signatures, unconditionally secure for the signer." *Advances in Cryptology—CRYPTO'91*, Lecture Notes in Computer Science, vol. 576, ed. J. Feigenbaum. Springer-Verlag, Berlin, 470–484.

[4] Damgård, Ivan Bjerre and Torben P. Pedersen (1996). "New convertible undeniable signature schemes." *Advances in Cryptology—EUROCRYPT'96*, Lecture Notes in Computer Science, vol. 1070, ed. U. Maurer. Springer-Verlag, Berlin, 372–386.

[5] Gennaro, Rosario, Hugo Krawczyk, and Tal Rabin (1997). "RSA-based undeniable signatures." *Advances in Cryptology—CRYPTO'97*, Lecture Notes in Computer Science, vol. 1294, ed. B.S. Kaliski. Springer-Verlag, Berlin, 132–149.

[6] Harn, Lein and Shoubao Yang (1993). "Group-oriented undeniable signature schemes without the assistance of a mutually trusted party." *Advances in Cryptology—ASIACRYPT'92*, Lecture Notes in Computer Science, vol. 718, eds. J. Sebermy and Y. Zheng. Springer-Verlag, Berlin, 133–142.

[7] Jakobsson, Markus (1995). "Blackmailing using undeniable signatures." *Advances in Cryptology—EUROCRYPT'94*, Lecture Notes in Computer Science, vol. 950, ed. A. De Santis. Springer-Verlag, Berlin, 425–427.

[8] Jakobsson, Markus, Kazue Sako, and Russell, Impagliazzo (1996). "Designated verifier proofs and their applications." *Advances in Cryptology—EUROCRYPT'96*, Lecture Notes in Computer Science, vol. 1070, ed. U. Maurer. Springer-Verlag, Berlin, 143–154.

[9] Michels, Markus and Markus Stadler (1997). "Efficient convertible undeniable signature schemes." *International Workshop on Selected Areas in Cryptography (SAC'97)*. Springer-Verlag, Berlin, 231–244.

[10] Pedersen, Pryds Torben (1991). "Distributed provers with applications to undeniable signatures (Extended abstract)." *Advances in Cryptology—EUROCRYPT'91*, Lecture Notes in Computer Science, vol. 547, ed. D.W. Davies. Springer-Verlag, Berlin, 221–242.

[11] Sakurai, Kouichi and Yoshinori Yamane (1996). "Blind decoding, blind undeniable sigatures, and their applications to privacy protection." *Information Hiding (IHW'96)*, Lecture Notes in Computer Science, vol. 1174, ed. R. Anderson. Springer-Verlag, Berlin, 257–264.

UNIVERSAL ONE-WAY HASH FUNCTIONS

A Universal One-Way Hash Function (UOWHF) is a class of hash functions indexed by a public parameter (called a key), for which finding a second preimage is hard. The main idea is that first the challenge input is selected, and subsequently the function instance (or parameter) is chosen. Only then should the opponent try to find a second input with the same output as the challenge. A UOWHF is a weaker notion than a collision resistant *hash function* (CRHF). In a CRHF, the opponent is first given the key and then he has to produce two colliding inputs. Finding collisions for a fixed parameter of a UOWHF may be rather easy, but this will not help the opponent to violate the security requirement, as the instance is chosen *after* the challenge. This also implies that the birthday paradox does not apply to a UOWHF and a hash result of 80 bits may offer adequate security (in 2004). Simon [8] has shown that there exists an oracle relative to which a UOWHF exists, but no CRHF.

The concept of UOWHF has been introduced by Naor and Yung [5]. Bellare and Rogaway [1] propose the alternative name Target Collision Resistant (TCR) hash function. Alternative definitions and generalizations were introduced by Zheng et al. [9] and Mironov [4].

UOWHFs can replace CRHFs in the construction of efficient digital signature schemes: in this case the signer needs to pick first the message m and then the key K and sign the pair $(K, h_K(m))$. Note that this has the disadvantage that a cheating signer could reverse the order: first choose K, then find a collision (m, m'), and later on claim that he has signed m' rather than m. It depends on the context whether or not this is a problem. However, this situation can be avoided by employing a CRHF.

Naor and Yung construct a UOWHF based on a strongly universal hash function and a one-way permutation [5]. Rompel [6] describes an interesting scheme to turn any one-way function in a UOWHF, which shows that one-way functions imply digital signature schemes. Impagliazzo and Naor present a construction based on the subset sum problem [2] (see also knapsack cryptographic schemes). None of these constructions is very efficient.

Naor and Yung [5] describe a composition construction. Several papers have studied the problem of constructing an efficient UOWHF based on a fixed size UOWHF which compresses n bits to m bits (with $n > m$). Bellare and Rogaway show that the Merkle–Damgård construction, which is used for CRHFs (see hash functions), does *not* work [1]. They present a linear construction, that is sequential, and a tree construction, that allows for a parallel implementation. The linear construction has been optimized by Shoup [7]: his scheme requires 2^t invocations of the fixed size UOWHF and a key of t m-bit strings to hash a message of length $2^t(m - n) + m$ bits to m bits; this has been shown to be optimal [4]. The tree construction has been optimized by Lee et al. [3]; the best scheme allows for a t-fold parallelism in exchange for a slightly larger key (see [3] for a comparison of several alternatives).

Bart Preneel

References

[1] Bellare, M. and P. Rogaway (1997). "Collision-resistant hashing: Towards making UOWHFs practical." *Advances in Cryptology—CRYPTO'97*, Lecture Notes in Computer Science, vol. 1294, ed. B. Kaliski. Springer-Verlag, Berlin, 470–484.

[2] Impagliazzo, R. and M. Naor (1996). "Efficient cryptographic schemes provably as secure as subset sum." *Journal of Cryptology*, 9 (4), 199–216.

[3] Lee, W., D. Chang, S. Lee, S. Sung, and N. Nandi (2003). "New parallel domain extenders for UOWHFs." *Advances in Cryptology—ASIACRYPT 2003*, Lecture Notes in Computer Science, vol. 2894, ed. C.S. Lai. Springer-Verlag, Berlin, 208–227.

[4] Mironov, I. (2001). "Hash functions: from Merkle-Damgård to Shoup." *Advances in Cryptology—EUROCRYPT 2001*, Lecture Notes in Computer Science, vol. 2045, ed. B. Pfitzmann. Springer-Verlag, Berlin, 166–181.

[5] Naor, M. and M. Yung (1990). "Universal one-way hash functions and their cryptographic

applications." *Proceedings of 21st ACM Symposium on the Theory of Computing*, 387–394.

[6] Rompel, J. (1990). "One-way functions are necessary and sufficient for secure signatures." *Proceedings of 22nd ACM Symposium on the Theory of Computing*, ACM, New York, 387–394.

[7] Shoup, V. (2000). "A composition theorem for universal one-way hash functions." *Advances in Cryptology—EUROCRYPT 2000*, Lecture Notes in Computer Science, vol. 1807, ed. B. Preneel. Springer-Verlag, Berlin, 445–452.

[8] Simon, D. (1998). "Finding collisions on a one-way street: Can secure hash functions be based on general assumptions?" *Advances in Cryptology—EUROCRYPT'98*, Lecture Notes in Computer Science, vol. 1403, ed. K. Nyberg. Springer-Verlag, Berlin, 334–345.

[9] Zheng, Y., T. Matsumoto, and H. Imai (1990). "Connections between several versions of one way hash functions." *Transactions on IEICE E*, E73 (7), 1092–1099.

UNLINKABILITY

Unlinkability of two events occurring during a process under observation of an attacker is the property that the two events appear to the attacker *after the process* exactly as much related—or unrelated—as they did *before the process* started (see [1]).

In order to apply this notion to a particular cryptographic scheme, the attacker model needs to be specified further. For example, whether it is a *passive attacker* such as an eavesdropper, or an *active attacker* (see cryptanalysis for this terminology). If passive, which communication lines he can observe and when. If active, how he can interact with the honest system participants (e.g., *oracle access*) and thereby stimulate certain behavior of the honest participants, or how many honest participants he can control entirely (see *resilience* in threshold signature), and whether the attacker is computationally restricted or computationally unrestricted (see computational security). Based on a precise attacker model, certain events occurring in a given cryptographic scheme can then be defined as unlinkable.

An individual who interacts with other individuals or authorities may keep its interactions unlinkable by using different pseudonyms in different transactions. As Rao and Rohatgi [3] showed, this may not be a sufficient measure to achieve unlinkability, but it is usually a necessary one. Anonymity, untraceability, and privacy are all closely related to the notion of unlinkability. In fact, many privacy oriented payment schemes,

credential schemes, electronic voting schemes, and secure auction schemes are built around the notion of unlinkability and employ transaction pseudonyms (see [2]).

Gerrit Bleumer

References

[1] Chaum, David (1981). "Untraceable electronic mail, return addresses, and digital pseudonyms." *Communications of the ACM*, 24 (2), 84–88.

[2] Chaum, David (1986). "Showing credentials without identification—signatures transferred between unconditionally unlinkable pseudonyms." *Advances in Cryptology—EUROCRYPT'85*, Lecture Notes in Computer Science, vol. 219, ed. F. Pichler. Springer-Verlag, Berlin, 241–244.

[3] Rao, Josyula R. and Pankaj Rohatgi (2000). "Can pseudonyms really guarantee privacy?" *9th Usenix Symposium, August 2000*.

UNTRACEABILITY

Untraceability of an object during a process under observation of an attacker is the property that the attacker cannot follow the trace of the object as it moves from one participant or location to another. A standard example is e-mail, which in certain anonymizing networks cannot be observed by attackers to flow from a sender to a recipient through a sequence of network nodes (see [1]). Another prominent example is an electronic coin, which in certain privacy oriented electronic cash schemes cannot be traced being spent from one participant to another.

Gerrit Bleumer

Reference

[1] Chaum, David (1981). "Untraceable electronic mail, return addresses, and digital pseudonyms." *Communications of the ACM*, 24 (2), 84–88.

USER AUTHENTICATION

User authentication is identical to entity authentication except that the term "entity" is restricted to denoting a human user (as opposed to a server, a process, a device, a computer terminal, or any other system entity). In practice, this restriction typically limits the possible authentication techniques to password-based schemes.

Carlisle Adams

V

VERIFIABLE ENCRYPTION

Verifiable encryption is an encryption scheme where one can prove some property of a message m, while the message is given in an encrypted form. When an encryption scheme is secure, the encryption $E(m)$ should reveal no information regarding m. But this property may not be suitable in cases where checking a property of the encrypted content is required before processing the encrypted data. Verifiable encryption is useful in such cases. An example of such a case is a key escrow scheme. In key escrow schemes, a sender wants to prove that a key given in an encrypted form under the escrow agent's public key is indeed the right key to decrypt the encrypted message that the sender is transmitting to the receiver. Another example is a group signature scheme where the information to identify the signer is encrypted under the public key of the *trusted group authority* (also known as *group manager*) so that the authority can trace the signer in a case of dispute.

Verifiable encryption is also used in the context of electronic voting. In a yes–no voting, a voter encrypts '1' if voting for 'yes' and '−1' if voting for 'no'. Using homomorphic encryption (see homomorphism), the addition of encryption yields the encryption of the sum of these values. If the sum is a positive number, it means the majority was for yes-vote. However, it is important that a vote is the encryption of either 1 or −1 and not 100 or 1000. Verifiable encryption can be used in this purpose.

Verifiability is usually achieved by employing a zero-knowledge protocol to prove possession of knowledge regarding the property of the encrypted message.

Kazue Sako

References

[1] Camenisch, J. and I. Damgård (2000). "Verifiable encryption, group encryption, and their applications to separable group signatures and signature sharing schemes." *Proceedings of Asiacrypt 2000*, Lecture Notes in Computer Science, vol. 1976, ed., T. Okamoto. Springer, Berlin, 331–345.

[2] Camenisch, J. and V. Shoup (2003). "Practical verifiable encryption and decryption of discrete logarithms." *Proceedings of CRYPTO 2003*, Lecture Notes in Computer Science, vol. 2729, ed. D. Boneh. Springer, Berlin, 126–144.

VERIFIABLE SECRET SHARING

A basic secret sharing scheme is defined to resist *passive attacks* only, which means that its security depends on the assumption that all parties involved run the protocols as prescribed by the scheme. After taking part in the distribution protocol, a non-qualified set of participants is not able to deduce (part of) the secret from their shares.

In many applications, however, a secret sharing scheme is also required to withstand active attacks. This is accomplished by *verifiable secret sharing* (VSS) schemes, as first introduced in 1985 [3]. Specifically, a VSS scheme is required to withstand the following two types of active attacks:

- a dealer sending inconsistent or incorrect shares to some of the participants during the distribution protocol, and
- participants submitting incorrect shares during the reconstruction protocol.

Clearly, Shamir's threshold scheme is not a VSS scheme, since it does not exclude either of these attacks.

A well-known example is Feldman's VSS scheme [4]. Informally, Feldman's scheme runs as follows for the case of (t, n)-threshold access structures, $1 \leq t \leq n$. Let g denotes a generator of a cyclic group G of order p (where p is a large prime number, $p > n$). The distribution protocol and reconstruction protocol for dealer and participants $P_1, \ldots, P_n$ are defined as follows:

- **Distribution.** Let $s \in_R \mathbb{Z}_p$ (see modular arithmetic) denotes the secret to be distributed by the dealer. The dealer chooses a random polynomial in $\mathbb{Z}_p[x]$ of the form

  $$a(x) = s + \alpha_1 x + \cdots + \alpha_{t-1} x^{t-1},$$

  subject to the condition that $\alpha_0 = s$. The dealer sends share $s_i = a(i)$ to participant P_i in private, for $i = 1, \ldots, n$. In addition, the dealer broadcasts values $B_j = g^{\alpha_j}$, $0 \leq j < t$. Upon receipt of its share s_i, each participant verifies the validity of the share by evaluating the following equation:

  $$g^{s_i} = \prod_{j=0}^{t-1} B_j^{i^j}. \tag{1}$$

- **Reconstruction.** Each share s_i contributed by participant P_i is verified using Equation (1). The secret $s = a(0)$ is then recovered as in Shamir's threshold scheme, using t valid shares.

Because of Equation (1), it is impossible for the dealer to give out inconsistent shares (to honest participants). The only way the dealer may cheat is by sending incorrect shares to some participants. Each participant receiving an incorrect share during the distribution protocol is supposed to file a complaint against the dealer. If the dealer receives t or more complaints, the distribution is said to fail. Otherwise, the dealer is required to broadcast the correct shares s_i for all participants P_i who filed a complaint. If the decisional Diffie–Hellmann problem is hard for group G, Feldman's VSS scheme is secure against cheating by the dealer and cheating by at most $t - 1$ of the participants as long as $2(t - 1) < n$.

Another well-known example is Pedersen's VSS scheme [10]. The schemes by Feldman and Pedersen are called *non-interactive* because the distribution protocol does not require any interaction between the dealer and participants, nor between participants among each other, except for the filing of complaints.

Publicly verifiable secret sharing (PVSS) schemes, as introduced by Stadler [12], remove the need for interaction entirely by ensuring that anyone (not just the participants) is able to verify the shares distributed by the dealer (see also [5, 11, 13]).

Many examples of *interactive* VSS schemes can be found in papers presenting protocols for *secure multiparty computation* (see [1, 2, 8] and later papers). Typically, interactive VSS schemes rely on the use of authentication tags, checking vectors, or similar ideas. For instance, the VSS scheme of [1] lets the dealer choose a random bivariate polynomial $p(x, y)$ of degree at most $t - 1$, such that $s = p(0, 0)$ for a secret s. The dealer sends each participant P_i two univariate polynomials of degree at most $t - 1$: $a_i(x) = p(x, i)$ and $b_i(y) = p(i, y)$. Each participant is supposed to check that $a_i(i) = b_i(i)$. Furthermore, each pair of participants P_i and P_j checks that $a_i(j) = b_j(i)$. An interactive protocol between the dealer and the participants is used to determine whether distribution was successful or not.

Apart from their use in secure multiparty computation, VSS schemes are specifically used in the construction of *distributed key generation* (DKG) protocols, which are in turn a basic tool in threshold cryptography. The object of a DKG protocol is to let a group of n parties jointly generate a key, consisting of a private key and a public key, such that the private key is shared among the n parties (see, e.g., [6, 7, 9]).

Berry Schoenmakers

References

[1] Ben-Or, M., S. Goldwasser, and A. Wigderson (1988). "Completeness theorems for noncryptographic fault-tolerant distributed computation." *Proceedings of 20th Symposium on Theory of Computing (STOC'88)*. ACM Press, New York, 1–10.

[2] Chaum, D., C. Crépeau, and I. Damgård (1988). "Multiparty unconditionally secure protocols." *Proceedings of 20th Symposium on Theory of Computing (STOC'88)*. ACM Press, New York, 11–19.

[3] Chor, B., S. Goldwasser, S. Micali, and B. Awerbuch (1985). "Verifiable secret sharing and achieving simultaneity in the presence of faults." *Proceedings of 26th IEEE Symposium on Foundations of Computer Science (FOCS'85)*. IEEE Computer Society Press, Los Alamitos, CA, 383–395.

[4] Feldman, P. (1987). "A practical scheme for non-interactive verifiable secret sharing." *Proceedings of 28th IEEE Symposium on Foundations of Computer Science (FOCS'87)*. IEEE Computer Society Press, Los Alamitos, CA, 427–437.

[5] Fujisaki, E. and T. Okamoto (1998). "A practical and provably secure scheme for publicly verifiable secret sharing and its applications." *Advances in Cryptology—EUROCRYPT'98*, Lecture Notes in Computer Science, vol. 1403, ed. K. Nyberg. Springer-Verlag, Berlin, 32–46.

[6] Gennaro, R., S. Jarecki, H. Krawczyk, and T. Rabin (1999). "Secure distributed key generation for discrete-log based cryptosystems." *Advances in Cryptology—EUROCRYPT'99*, Lecture Notes in Computer Science, vol. 1592, ed. J. Stern. Springer-Verlag, Berlin, 295–310.

[7] Gennaro, R., S. Jarecki, H. Krawczyk, and T. Rabin (2003). "Secure applications of pedersens distributed key generation protocol." *Cryptographers' Track RSA 2003*, Lecture Notes in Computer Science, vol. 2612, ed. M. Joye. Springer-Verlag, Berlin, 373–390.

[8] Goldreich, O., S. Micali, and A. Wigderson (1987). "How to play any mental game—or—a completeness theorem for protocols with honest majority." *Proceedings of 19th Symposium on Theory of Computing (STOC'87)*. ACM Press, New York, 218–229.

[9] Pedersen, T. (1991). "A threshold cryptosystem without a trusted party." *Advances in Cryptology—EUROCRYPT'91*, Lecture Notes in Computer Science, vol. 547, ed. D.W. Davies. Springer-Verlag, Berlin, 522–526.

[10] Pedersen, T.P. (1992). "Non-interactive and information-theoretic secure verifiable secret sharing." *Advances in Cryptology—CRYPTO'91*, Lecture Notes in Computer Science, vol. 576, ed. J. Feigenbaum. Springer-Verlag, Berlin, 129–140.

[11] Schoenmakers, B. (1999). "A simple publicly verifiable secret sharing scheme and its application to electronic voting." *Advances in Cryptology—CRYPTO'99*, Lecture Notes in Computer Science, vol. 1666, ed. J. Wiener. Springer-Verlag, Berlin, 148–164.

[12] Stadler, M. (1996). "Publicly verifiable secret sharing." *Advances in Cryptology—EUROCRYPT'96*, Lecture Notes in Computer Science, vol. 1070, ed. U. Maurer. Springer-Verlag, Berlin, 190–199.

[13] Young, A., and M. Yung (2001). "A PVSS as hard as discrete log and shareholder separability." *Public Key Cryptography—PKC'01*, Lecture Notes in Computer Science, vol. 1992, ed. K. Kim. Springer-Verlag, Berlin, 287–299.

VERNAM CIPHER

This is an encryption operating on groups of 5-bit words, by means of coordinate-wise addition modulo 2 (see modular arithmetic). Normally, the CCIT2-code is used. The key should be a completely random, one-time key (see key).

The *Vernam table* for $Z_{32} = Z_2^5$ (Gilbert S. Vernam, 1917) looks like:

	0	a	b	c	d	e	f	g	h	i	j	k	l	m	n	o	p	q	r	s	t	u	v	w	x	y	z	2	3	4	5	1
0	0	A	B	C	D	E	F	G	H	I	J	K	L	M	N	O	P	Q	R	S	T	U	V	W	X	Y	Z	2	3	4	5	1
A	A	0	G	F	R	5	C	B	Q	S	4	N	Z	1	K	3	Y	H	D	I	W	2	X	T	V	P	L	U	O	J	E	M
B	B	G	0	Q	T	O	H	A	F	1	L	P	J	S	Y	E	K	C	W	M	D	V	U	R	2	N	4	X	5	Z	3	I
C	C	F	Q	0	U	K	A	H	G	4	S	E	M	L	5	P	O	B	2	J	V	D	T	X	W	3	1	R	Y	I	N	Z
D	D	R	T	U	0	4	2	W	X	K	5	I	3	Y	S	Z	1	V	A	N	B	C	Q	G	H	M	O	F	L	E	J	P
E	E	5	O	K	4	0	N	3	Y	U	R	C	W	X	F	B	Q	P	J	2	2	Z	I	1	L	M	H	T	S	G	D	A	V
F	F	C	H	A	2	N	0	Q	B	J	I	5	1	Z	E	Y	3	G	U	4	X	R	W	V	T	O	M	D	P	S	K	L
G	G	B	A	H	W	3	Q	0	C	M	Z	Y	4	I	P	5	N	F	T	1	R	X	2	D	U	K	J	V	E	L	O	S
H	H	Q	F	G	X	Y	B	C	0	L	1	3	I	4	O	N	5	A	V	Z	2	W	R	U	D	E	S	T	K	M	P	J
I	I	S	1	4	K	U	J	M	L	0	F	D	H	G	R	V	T	Z	N	A	P	E	O	Y	3	W	Q	5	X	C	2	B
J	J	4	L	S	5	R	I	Z	1	F	0	2	B	Q	U	W	X	M	E	C	3	N	Y	O	P	V	G	K	T	A	D	H
K	K	N	P	E	I	C	5	Y	3	D	2	0	X	W	A	Q	B	O	S	R	1	4	Z	M	L	G	V	J	H	U	F	T
L	L	Z	J	M	3	W	1	4	I	H	B	X	0	C	V	R	2	S	O	Q	5	Y	N	E	K	U	A	P	D	G	T	F
M	M	1	S	L	Y	X	Z	I	4	G	Q	W	C	0	T	2	R	J	P	B	N	3	5	K	E	D	F	O	U	H	V	A
N	N	K	Y	5	S	F	E	P	O	R	U	A	V	T	0	H	G	3	I	D	M	J	L	1	Z	B	X	4	Q	2	C	W
O	O	3	E	P	Z	B	Y	5	N	V	W	Q	R	2	H	0	C	K	L	X	4	1	I	J	S	F	D	M	A	T	G	U
P	P	Y	K	O	1	Q	3	N	5	T	X	B	2	R	G	C	0	E	M	W	I	Z	4	S	J	A	U	L	F	V	H	D
Q	Q	H	C	B	V	P	G	F	A	Z	M	O	S	J	3	K	E	0	X	L	U	T	D	2	R	5	I	W	N	1	Y	4
R	R	D	W	2	A	J	U	T	V	N	E	S	O	P	I	L	M	X	0	K	G	F	H	B	Q	1	3	C	Z	5	4	Y
S	S	I	M	J	N	2	4	1	Z	A	C	R	Q	B	D	X	W	L	K	0	Y	5	3	P	O	T	H	E	V	F	U	G
T	T	W	D	V	B	Z	X	R	2	P	3	1	5	N	M	4	I	U	G	Y	0	Q	C	A	F	S	E	H	J	O	L	K
U	U	2	V	D	C	I	R	X	W	E	N	4	Y	3	J	1	Z	T	F	5	Q	0	B	H	G	L	P	A	M	K	S	O
V	V	X	U	T	Q	1	W	2	R	O	Y	Z	N	5	L	I	4	D	H	3	C	B	0	F	A	J	K	G	S	P	M	E
W	W	T	R	X	G	L	V	D	U	Y	O	M	E	K	1	J	S	2	B	P	A	H	F	0	C	I	5	Q	4	3	Z	N
X	X	V	2	W	H	M	T	U	D	3	P	L	K	E	Z	S	J	R	Q	O	F	G	A	C	0	4	N	B	I	Y	1	5
Y	Y	P	N	3	M	H	O	K	E	W	V	G	U	D	B	F	A	5	1	T	S	L	J	I	4	0	2	Z	C	X	Q	R
Z	Z	L	4	1	O	T	M	J	S	Q	G	V	A	F	X	D	U	I	3	H	E	P	K	5	N	2	0	Y	R	B	W	C
2	2	U	X	R	F	S	D	V	T	5	K	J	P	O	4	M	L	W	C	E	H	A	G	Q	B	Z	Y	0	1	N	I	3
3	3	O	5	Y	L	G	P	E	K	X	T	H	D	U	Q	A	F	N	Z	V	J	M	S	4	I	C	R	1	0	W	B	2
4	4	J	Z	I	E	D	S	L	M	C	A	U	G	H	2	T	V	1	5	F	O	K	P	3	Y	X	B	N	W	0	R	Q
5	5	E	3	N	J	A	K	O	P	2	D	F	T	V	C	G	H	Y	4	U	L	S	M	Z	1	Q	W	I	B	R	0	X
1	1	M	I	Z	P	V	L	S	J	B	H	T	F	A	W	U	D	4	Y	G	K	O	E	N	5	R	C	3	2	Q	X	0

Friedrich L. Bauer

Reference

[1] Bauer, F.L. (1997). "Decrypted secrets." *Methods and Maxims of Cryptology*. Springer-Verlag, Berlin.

VIGENÈRE ENCRYPTION

The Vigenère encryption is an encryption with shifted standard alphabets. It makes use of a so-called *Vigenère table*. For Z_{26} ("tabula recta" of Trithemius, 1518), this table looks like:

	a	b	c	d	e	f	g	h	i	j	k	l	m	n	o	p	q	r	s	t	u	v	w	x	y	z
A	A	B	C	D	E	F	G	H	I	J	K	L	M	N	O	P	Q	R	S	T	U	V	W	X	Y	Z
B	B	C	D	E	F	G	H	I	J	K	L	M	N	O	P	Q	R	S	T	U	V	W	X	Y	Z	A
C	C	D	E	F	G	H	I	J	K	L	M	N	O	P	Q	R	S	T	U	V	W	X	Y	Z	A	B
D	D	E	F	G	H	I	J	K	L	M	N	O	P	Q	R	S	T	U	V	W	X	Y	Z	A	B	C
E	E	F	G	H	I	J	K	L	M	N	O	P	Q	R	S	T	U	V	W	X	Y	Z	A	B	C	D
F	F	G	H	I	J	K	L	M	N	O	P	Q	R	S	T	U	V	W	X	Y	Z	A	B	C	D	E
G	G	H	I	J	K	L	M	N	O	P	Q	R	S	T	U	V	W	X	Y	Z	A	B	C	D	E	F
H	H	I	J	K	L	M	N	O	P	Q	R	S	T	U	V	W	X	Y	Z	A	B	C	D	E	F	G
I	I	J	K	L	M	N	O	P	Q	R	S	T	U	V	W	X	Y	Z	A	B	C	D	E	F	G	H
J	J	K	L	M	N	O	P	Q	R	S	T	U	V	W	X	Y	Z	A	B	C	D	E	F	G	H	I
K	K	L	M	N	O	P	Q	R	S	T	U	V	W	X	Y	Z	A	B	C	D	E	F	G	H	I	J
L	L	M	N	O	P	Q	R	S	T	U	V	W	X	Y	Z	A	B	C	D	E	F	G	H	I	J	K
N	N	O	P	Q	R	S	T	U	V	W	X	Y	Z	A	B	C	D	E	F	G	H	I	J	K	L	M
O	O	P	Q	R	S	T	U	V	W	X	Y	Z	A	B	C	D	E	F	G	H	I	J	K	L	M	N
P	P	Q	R	S	T	U	V	W	X	Y	Z	A	B	C	D	E	F	G	H	I	J	K	L	M	N	O
Q	Q	R	S	T	U	V	W	X	Y	Z	A	B	C	D	E	F	G	H	I	J	K	L	M	N	O	P
R	R	S	T	U	V	W	X	Y	Z	A	B	C	D	E	F	G	H	I	J	K	L	M	N	O	P	Q
S	S	T	U	V	W	X	Y	Z	A	B	C	D	E	F	G	H	I	J	K	L	M	N	O	P	Q	R
T	T	U	V	W	X	Y	Z	A	B	C	D	E	F	G	H	I	J	K	L	M	N	O	P	Q	R	S
U	U	V	W	X	Y	Z	A	B	C	D	E	F	G	H	I	J	K	L	M	N	O	P	Q	R	S	T
V	V	W	X	Y	Z	A	B	C	D	E	F	G	H	I	J	K	L	M	N	O	P	Q	R	S	T	U
W	W	X	Y	Z	A	B	C	D	E	F	G	H	I	J	K	L	M	N	O	P	Q	R	S	T	U	V
X	X	Y	Z	A	B	C	D	E	F	G	H	I	J	K	L	M	N	O	P	Q	R	S	T	U	V	W
Y	Y	Z	A	B	C	D	E	F	G	H	I	J	K	L	M	N	O	P	Q	R	S	T	U	V	W	X
Z	Z	A	B	C	D	E	F	G	H	I	J	K	L	M	N	O	P	Q	R	S	T	U	V	W	X	Y

An encryption example with the keytext "GOLD" of length 4 is given by:

```
plaintext  m u c h h a v e i t r a v e l l e d
keytext    G O L D G O L D G O L D G O L D G O
ciphertext S I N K N O G H O H C D B S W O K R
```

Friedrich L. Bauer

Reference

[1] Bauer, F.L. (1997). "Decrypted secrets." *Methods and Maxims of Cryptology*. Springer-Verlag, Berlin.

VIRUS PROTECTION

Computer viruses are probably the most well-known and widespread threat to computer security. Some viruses such as Chernobyl, Melissa and The Love Bug spread so rampantly that they became common household names. Microsoft Windows is the operating system most plagued with viruses though viruses are not restricted to it. Viruses can cause damage in many ways: they can delete or corrupt data on the infected host; they can use the infected host to mount attacks such as Denial of Service on other hosts; and they cause enormous wastage of computing resources. The threat of a fast-spreading malicious virus bringing down millions of computers in the matter of a few minutes looms over us. Such a virus could cost billions of dollars in losses and would be a disaster for today's computer-driven economy.

WHAT IS A VIRUS?: The first thing in a study of viruses is obviously knowing exactly what one is. Unfortunately, it is not easy to find a satisfactory definition because of several reasons. Over the years, the term virus has been overloaded with many definitions. Often a virus is mistaken for a *Worm* or a *Trojan horse* (see Trojan horses, computer viruses and worms), and vice-versa. This is sometimes unavoidable because a "good" virus (one that spreads rapidly, avoids detection and causes lot of damage) often has several characteristics of a worm and Trojan horse too.

Firstly, a virus is a computer program written in a similar way as other normal programs. Infact, anybody with even the most modest programming knowledge can write one [1, 3, 6]. The most distinguishing property of a virus is that *the virus program copies itself to other programs or documents so that the virus code is executed whenever the program is run or the document is opened.* Programs to which the virus copies or attaches itself are said to be *infected* with the virus. When the infected program is run, the virus searches for other uninfected files and tries to attach itself to them too.

Malicious action of viruses includes but is not restricted to deleting or zeroing the files, trashing the BIOS, leaving backdoors, spying private information, using the infected machine to mount Denial of Service (DoS) attacks, etc. Even if the virus does not perform any such destructive activity, it might impede the normal working of computer systems by consuming too much CPU and memory or causing too much network traffic.

Often, the virus' malicious action is triggered by a time bomb or logic bomb. These are pieces of code that get activated when a certain date or time is reached (time bomb) or when some given logic condition becomes true (logic bomb). For example, the famous CIH or Chernobyl virus was triggered to destroy files on the infected machine on 26th April, the date of the Chernobyl nuclear disaster.

INFECTION OF FILES: This section briefly describes how a virus infects a file. Traditionally, most viruses have infected executables. This is because the goal of a virus is to run on the local host and the obvious way to achieve this is by attaching itself to an executable. Recently, a new category of viruses, called macro-viruses, have surfaced that attach themselves to document files and are able to run whenever the document is opened. In this section, the two types of viruses are discussed in turn.

Traditional Executable Virus

The traditional virus targets executables and either overwrites the entire file or attaches itself to it so that the virus also runs whenever the executable is run. Viruses normally attach themselves in such a way that the virus runs first after which the program proceeds normally. This strategy ensures that the virus runs even in cases when the infected program crashes or runs forever. It also makes detection hard since the program seems to run normally to the user.

Attaching a virus to text-based executables such as shell scripts is trivial—just put the virus code in the beginning—but suffers from the obvious disadvantage of being easily detectable if anybody happens to view the code. Since most executables are binaries, attaching a virus to a binary executable is more common. It is more complicated but has the advantage that the virus is better hidden. A very general overview of how a virus attaches itself to a binary is given below.

All operating systems have a minimum unit of hard disk access called block (it is often 512 or 1024 bytes). Files on the hard disk always occupy an integer number of blocks, thus the last block of a file is only partly used. For example, if the block size is 512 bytes, and if a program needs only 998 bytes, it would occupy 2 blocks on disk, of which 998 bytes are the program itself and 26 bytes are unused. Viruses fit themselves within the unused space and thus do not require any additional blocks. Figure 1 gives a graphical picture of how this is done. It shows the structure of a binary before and after a virus attaches itself to it. The virus is able to modify the binary so as to fit in the available unused space in such a manner

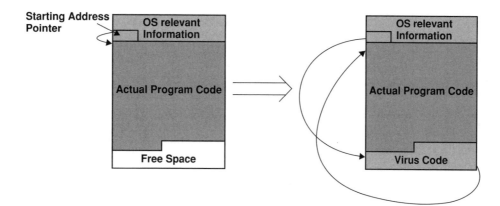

Fig. 1. How a virus attaches itself to a binary executable

that when the binary is launched, the virus runs first and then the program. The binary contains a location called the *starting address pointer* that points to the first instruction to be executed and is required by the operating system to load the binary. The virus, after fitting itself into the unused space, modifies the starting pointer to point to the virus' first instruction. At the end of the virus code, a jump instruction sends control to the program's first instruction so that it is able to run normally.

Macro-virus

Some software packages allow their associated data files to contain script like code that is executed when the file is opened. Viruses exploit this feature by attaching themselves to data files in the form of a script. The Microsoft Office software suite has been the most vulnerable to this type of viruses. Programs such as Microsoft Word and Excel files can contain macros (VB Script code) that is executed when the document is opened. The fact that Word and Excel are widely used and the documents are often shared makes this a very attractive target for viruses.

VIRUS PROPAGATION AND INITIATION: The previous section described how a virus infects a file. This section discusses how it initiates the infection. Infection requires two things: firstly, the virus needs to reach the host (propagation) and secondly, it needs to run on the host at least once in order to initiate the infection (initiation).

Any communication medium used to transfer data between computer systems can be used for propagation. Since the widespread deployment of computer networks, they have become the de facto medium. More specifically, some of the common ways viruses spread are:

1. **Email.** Email has been the most popular transport for viruses in the last few years. Melissa was the first virus to spread widely through email. Since then, Happy 99, Worm.ExploreZip, BubbleBoy, The Love Bug and many others have used it. The Love Bug sent out emails from the infected host to addresses in Microsoft Outlook's addressbook. The message had the subject "I Love You" and asked the recipient to open the accompanying attachment that contained the virus. People got pulled by the love message into opening the attachment that let the virus loose on their machine. BubbleBoy exploited a feature in Microsoft Outlook that allowed it to execute code on the local host when the email was previewed. Another common feature

viruses exploit is that some operating systems (such as Windows) decide what action to perform on a file (whether to execute it, open it with Microsoft Word, etc.) based on just the file's extension. Moreover, the extension is often hidden from the user and the operation is performed automatically without user intervention. The combination of unsuspecting user actions, social engineering, software bugs/idiosyncrasies along with email as the underlying transport has been very successful for viruses.

2. **Internet Downloads.** A very common way to get viruses on the desktop is by downloading infected files from the Internet. When the infected file is opened, it infects the local machine. This strategy is often employed by virus writers to launch their viruses by posting them on the Internet in the guise of useful programs.

3. **Floppies.** Before networks became widespread, floppies were the most common medium through which viruses moved from one machine to another. There are two ways in which this is done. The first way is to use floppies to move infected files between machines. The second way is to infect a floppy's boot sector so that whenever the floppy is used in a machine (for any purpose), the virus gets transferred to the machine.

4. **Infect Boot Sector.** A common technique viruses employ to ensure that they get activated on an infected host is to install themselves in the boot sector or partition sector of the host's disk drive. This activates the virus every time the system boots up. These viruses are significantly more difficult to remove with surety.

GUIDELINES TO PREVENT VIRUS INFECTION: As described above, viruses depend hugely on unsafe usage practices to spread fast. Curbing some of these would greatly reduce the risk of virus infection. Some guidelines for safe computer usage from the perspective of virus prevention are as follows:

1. Do not carelessly open executables or macro-supported documents downloaded from the Internet or received as email attachments. If there is any way to verify the authenticity and integrity of such files using digital signatures or cryptographic checksums, it should be done. If such techniques are not available, files should be downloaded only from reputed websites or checked with the sender of the email. Files should be passed through an anti-virus software before being executed or opened.

2. If possible, support for macros or similar scripting ability in documents should be disabled. In

particular, macro support in Microsoft Office software such as Microsoft Word and Microsoft Excel should be turned off unless absolutely required.

3. Do not allow operating systems to hide file extensions from the user or make security critical decisions (such as opening a file received as an email attachment) on its own.

4. Be extremely careful when booting systems from floppies. Floppies should not be carelessly left in drives since many systems have their BIOS configured to first try to boot from floppy. If it is necessary to use a floppy to boot a system, it should be thoroughly checked to be clean of viruses. Boot from floppy option in the BIOS should be disabled.

5. Most viruses are operating system specific. Thus, having a heterogeneous computing environment greatly helps in ensuring that not all machines get infected or compromised at the same time.

ANTI-VIRUS SOFTWARE: Anti-virus software has become more and more important over the last few years and has become a necessity on the more vulnerable operating systems. Even though users can adopt safe usage practices such as the ones mentioned above, viruses still get through. Over 50,000 viruses were known in 2000 and new ones get written everyday. Keeping track of all viruses and protecting a Windows desktop against them is not possible without anti-virus software. Unfortunately, it is impossible to build software that is always able to correctly determine whether a file is virus-infected or not [3]. However, anti-virus software is definitely a potent and effective weapon against all known viruses and, to some extent, against new viruses too.

Many anti-virus softwares are available today in the market. The basic detection technique involves matching patterns in files against a virus "signature" database. A virus signature is a code pattern that represents the virus and can identify it. This technique is successful only against known viruses whose signature is available. The most important thing with a virus signature database is to continuously update it with new virus signatures. It is safe to say that an anti-virus software is only as good as its signature database. Therefore, anti-virus software companies include update mechanisms in their software.

To detect new viruses, some virus checkers employ heuristics. Heuristics can be put into two categories: static and dynamic. Static heuristics go through a file, analyzing its structure and looking for malicious patterns. They use this information to decide whether the file contains a virus. Dynamic heuristics set up a controlled virtual environment where the suspect program is run and its behavior monitored. Based on the observations, the heuristic determines whether or not the file is infected. As both the static and dynamic schemes are heuristics, they are not always correct. The exact details of the scheme determine the trade-offs between the false positive and false negative rates as well as efficiency. In general, though static schemes have the advantage of being fast whereas dynamic schemes provide a lower false positive rate. Dynamic schemes are often susceptible to the logic and whims of viruses affecting their false negative rate. Some other heuristics use cryptographic checksums to detect viruses.

THE LATEST IN VIRUSES: Unfortunately for the security community, viruses are getting sophisticated and harder to detect. Polymorphic viruses and encrypted viruses are two of the latest viruses aimed at making detection harder. Polymorphic viruses change themselves (i.e., change the code) every now and then, allowing only a small detection window. Encrypted viruses encrypt the virus code so that it does not match any regular patterns. The encryption key can also be changed, and this results in a polymorphic encrypted virus. Not only are viruses getting harder to detect, but they are also getting faster in spreading, some techniques for which are discussed in [10].

WHERE TO LEARN MORE ABOUT VIRUSES: Viruses are a vast area of study and involve lot of low-level detail that cannot be included here. Some of the earliest ground-laying academic work on viruses is presented in [1, 3, 4, 7]. Duff [6] presents a simple virus on the UNIX operating system, and helps in demystifying the process of writing viruses. Most security books devote a chapter or a few sections to viruses. In particular, [5] is a good reference.

The CERT Coordination Center [2] is an excellent reference for practical details about viruses. It keeps an up to date list of viruses describing their symptoms, operating systems affected, safeguards, etc.

Some good on-line sources for reference are viruslist.com [12] and the websites of two major anti-virus software providers: McAfee [8, 9] and Symantec [11]. VXHavens [13] contains virus code and material on how to write a virus.

Pavan Verma

References

[1] Adleman, L. (1988). "An abstract theory of computer viruses." *Advances in Cryptology—CRYPTO'88*, Lecture Notes in Computer Science, vol. 403, ed. S. Goldwasser. New York, NY, 354–374.

[2] The CERT Coordination Center. http://www.cert.org

[3] Cohen, F. (1986). "Computer Viruses." *PhD Thesis*, University of Southern California.

[4] Cohen, F. (1989). "Practical defenses against computer viruses." *Computers and Security*, 8 (2), 149–160.

[5] Denning, P. (1990). *Computers Under Attack: Intruders, Worms and Viruses*. Addison-Wesley, Reading, MA.

[6] Duff, T. (1989). "Experience with viruses on UNIX systems." *Computing Systems*, 2 (2).

[7] Israel, H. (1987). "Computer viruses: Myth or reality?" *Tenth National Computer Security Conference Proceedings*, 238–244.

[8] McAfee Security. http://www.mcafee.com

[9] Network Associates Technology, Inc. http://www.nai.com

[10] Staniford, Stuart, Vern Paxson, and Nicholas Weaver (2002). "How to own the Internet in your spare time." *Proceedings of the 11th USENIX Security Symposium*.

[11] Symantec Corporation. http://www.symantec.com

[12] Viruslist.com. http://www.viruslist.com/eng/

[13] VX Havens. http://vx.netlux.org/index.shtml

VISUAL SECRET SHARING SCHEMES

Visual secret sharing schemes (VSSS, for shortness) were introduced in 1995 [1]. They form a particular case of secret sharing schemes with the additional restriction that *allowed* (also called *privileged* or *trusted* in access structure) coalitions can recover the secret by visual means.

More precisely, consider the case when just one bit of visual information, i.e., a black or white pixel, has to be distributed as a secret s among n participants of VSSS. To realize this, the dealer uses two sets B and W of binary (Boolean) $n \times m$ matrices in the following way. To distribute a "black" secret ($s = 1$), he randomly chooses a matrix $B = (b_{i,j}) \in B$ and sends to the ith participant a transparency consisting of m subpixels, where the jth subpixel is black if $b_{i,j} = 1$ and the jth subpixel is white if $b_{i,j} = 0$. For distributing $s = 0$, the dealer does the same but with a randomly chosen matrix $W = (w_{i,j}) \in W$. An allowed coalition recovers the secret by stacking their transparencies (called shares) and setting $s = 1$ if the stack is "sufficiently" black. The latter means calculating the number c of black subpixels in the stack, i.e., the *Hamming weight* (see cyclic codes) of the stack, and comparing it with some threshold.

Note that stacking transparencies is equivalent to evaluating the *OR* of the corresponding Boolean vectors $\mathbf{b}_i = (b_{i,1}, \ldots, b_{i,m})$ or $\mathbf{w}_i = (w_{i,1}, \ldots, w_{i,m})$. For the most interesting case of a visual (n, k) threshold scheme where any set of k or more participants can recover the secret "visually" (see Conditions 1 and 2 below), while any set of less than k participants gains no additional, i.e., a posteriori, information about the secret (see Condition 3 below) the formal definition is the following [1].

Two collections B and W of $n \times m$ binary matrices are called (n, k) VSSS if the following three conditions are met:

1. For any $B \in B$, the OR of any k of the n rows of B has Hamming weight of at least d_1.
2. For any $W \in W$, the OR of any k of the n rows of W has Hamming weight of at most d_0.
3. For any subset $\{i_1, \ldots, i_l\} \subset \{1, \ldots, n\}$ with $l < k$, the two collections obtained by restricting each matrix in B and W to the rows $i_1, \ldots, i_l$ are indistinguishable in the sense that they contain the same matrices with the same frequencies.

The value $\alpha = (d_1 - d_0)/m$ is called the *contrast* of the VSSS. Two other important parameters of VSSS are:

- m, the number of subpixels in a share, or *length* of VSSS;
- r, the number of matrices in collections B and W.

Obviously the contrast should be sufficiently large while keeping m and r as small as it possible. Construction of a VSSS turned out to be a much more complicated problem than that of an ordinary SSS, even for the case of threshold (n, k)-schemes.

For the simplest case of an (n, n) threshold scheme, the optimal visual version consists of all $2^{n-1}!$ permutations of the columns of the $n \times 2^{n-1}$ matrix W, which columns of W are all binary vectors $x = (x_1, \ldots, x_n)$ with $\sum_{i=1}^{n} x_i = 0 \pmod{2}$ (in words, W is the code book of the $(n, n-1)$ *even-weight* code defined by a single overall parity-check symbol), and also of all $2^{n-1}!$ permutations of columns of the matrix B, where, analogously, the columns of B are all binary vectors $y = (y_1, \ldots, y_n)$ such that $\sum_{i=1}^{n} y_i = 1 \pmod{2}$. This scheme has parameters $m = d_1 = 2^{n-1}$ and $\alpha = 1/m$, and is optimal since in any visual (n, n)-scheme $\alpha \le 1/2^{n-1}$ and $m \ge 2^{n-1}$ [1].

For general visual threshold (n, k)-schemes, various constructions were given in [1] yielding $\alpha = 2^{-\Omega(k)}$, which were further improved in [2, 3]. Starting with [2, 4] the technique of linear programming was applied to evaluate the maximal possible contrast of (n, k) visual threshold schemes. This led [5] to the asymptotically answer $\lim 4^n \alpha = 4$ for fixed k and growing n (see [6] for small values of $k = 3, 4, 5$). Generalizations of VSSS to the colored case have been also considered, see [3].

Robert Blakley
Gregory Kabatiansky

References

[1] Naor, M. and A. Shamir (1995). "Visual cryptography." *Advances in Cryptology—EUROCRYPT'94*, Lecture Notes in Computer Science, vol. 950, ed. A. De Santis. Springer-Verlag, Berlin, 1–12.

[2] Droste, S. (1996). "New results on visual cryptography." *Advances in Cryptology—CRYPTO'96*, Lecture Notes in Computer Science, vol. 1109, ed. N. Koblitz. Springer-Verlag, Berlin, 401–415.

[3] Verheul, E.R. H.C.A. van Tilborg (1997). "Constructions and properties of k out of n visual secret sharing schemes." *Designs, Codes and Cryptography*, 11 (2), 179–196.

[4] Hofmeister, T.M. Krause and H.U. Simon (2000). "Optimal k out of n secret sharing schemes in visual cryptography." *Theoretical Computer Science*, vol. 240, 471–485.

[5] Krause, M. and H.U. Simon (2000). "On contrast optimal secret sharing schemes in visual cryptography: determining the optimal contrast exactly." Lecture Notes in Computer Science, vol. 1776, eds. G. H. Gonnet, D. Panario, and A. Viola. Springer, Berlin, 280–291.

[6] Blundo, C.P. D'Arco, A. De Santis, and D.R. Stinson (2003). "Contrast optimal threshold visual cryptography schemes." *SIAM Journal on Discrete Mathematics*, vol. 16, 224–261.

W

WATERMARKING

Watermarking is a method in computer security by which identifiers of sources or *copyright owners* of digital or analog signals are embedded into the respective signals themselves in order to keep track of where a signal comes from or who the copyright owners are. Thus, watermarking is for copy protection of electronic content, predominantly, digital content. The signal carrying the content before a watermark is embedded is sometimes called the *cover-signal*, and the piece of data carrying the copyright ID (and other optional information) is called the *watermark*. The term "watermark" dates back to the 13th century when paper makers used traditional watermarks to differentiate their products from each other. Thus, traditional watermarks serve as authentication tags of print media, while electronic watermarks serve as copyright tags of the content itself.

The demand for watermarking comes mainly from the movie and music industry, which is trying to limit the pirating of their digital audio, video and artwork.

The characteristic security requirements on digital watermarking schemes, also called *electronic copyright marking schemes* (ECMS), are as follows:

Unobstrusiveness: Watermarks should not degrade the quality of the cover-signal in a way that annoys consumers when listening to or watching the content. The watermark may still be visible though, like, for example, the logo of a TV station that is continuously visible in one of the corners of the screen.

Robustness: Watermarks should be embedded into the content, in such a way that any signal transform of reasonable strength cannot remove the watermark. Examples of transformations are digital-to-analog conversion, compression or decompression, or the addition of small amounts of noise. Hence, pirates trying to remove a watermark will not succeed unless they degrade the quality of the cover-signal so much that the result is of little or no commercial interest any more.

Note how watermarking is different from digital steganography, which was originally introduced as *invisible communication* by Gus Simmons [7]. In steganography, the sender hides a message in a cover signal in order to conceal the *existence* of a message being transferred. The main goal is imperceptability of the hidden message. Unobstrusiveness and robustness are no primary concerns in steganography. The cover signal may be degraded to a certain extent and no precautions are taken to avoid losing the hidden message should the cover signal be manipulated. Watermarking and steganography are different topics in the area of *information hiding*.

Watermarking is also different from authentication and non-repudiation (see unforgeability). On the one hand, robustness is a stronger requirement than unforgeability, because an attacker against robustness is already successful if he destroys or removes a watermark, whereas a forger is only successful if he comes up with an authentication tag that verifies against the verification key of an existing sender. In another sense, robustness is a weaker requirement than nonrepudiation, because there are watermarking schemes where all decoders use the same secret that is used to create a watermark. Thus any decoder could be used to pick up a watermark from some cover signal *A* and embed it to another signal *B*, thus making the impression that the copyright owner of signal *A* also owns a copyright for signal *B*. There are watermarking mechanisms, though, that depend not only on the sender ID, but also on the cover-signal itself, and thus can be used for authentication purposes.

Important classes of digital watermarking schemes are the following:

- *Blind watermarking* (sometimes called *public watermarking*) means that a decoder can detect a watermark without a need to look at the cover signal. Only the sender's secret key is required in order to re-construct some random sequence that the sender used to construct the watermark in the first place. These types of schemes can be used easily in mass market electronic equipment or software. In some cases you may need extra information to help your decoder (in particular to synchronise its random sequence on the possibly distorted test signal).

- *Non-blind watermarking* (sometimes called *private watermarking*) means that a decoder requires the sender's secret key and the original cover signal in order to detect the watermark in it.

- *Asymmetric watermarking* (sometimes called *public-key watermarking*) means that the sender uses a private key to construct the watermark in a similar way as a digital signature is created by using a private key, and any decoder who has access to the corresponding public key can detect and recognize the respective watermark (see also public key cryptography). The public key would neither allow to reconstruct the sender's private key (under certain complexity theoretic assumptions), nor to forge a watermark, nor to remove an embedded watermark. No sufficiently robust and secure asymmetric watermarking schemes are known [3].

Prominent examples of electronic copyright marking schemes are the following:

DVD (*Digital Video Disk* later reissued as *Digital Versatile Disk*) [1]. This technique has been cryptanalyzed by Petitcolas et al. in [5].

SDMI (*Secure Digital Music Initiative*) [6]. This technique uses *echo hiding*, which was cryptanalyzed by Craver et al. [2].

A good introduction and papers about modern watermarking techniques is collected by Katzenbeisser and Petitcolas [4].

Gerrit Bleumer

References

[1] Bloom, Jefrey A., Ingemar J. Cox, Ton Kalker, Jean-Paul M.G. Linnartz, Matthew L. Miller, C., and Brendan S. Traw (1999). "Copy protection for DVD video." *Proceedings of the IEEE*, 87 (7), 1267–1276.

[2] Craver, Scott A., Min Wu, Bede Liu, Adam Subblefield, Ben Swartzlander, Dan S. Wallach, Drew Dean, and Edward W. Felten (2002). "Reading between the lines: Lessons from the SDMI challenge." *10th Usenix Security Symposium 2002*, 353–363; http://www.usenix.org/

[3] Hachez, Gaël and Jean-Jacques Quisquater (2002). "Which directions for asymmetric watermarking?" *Proceedings of the XI European Signal Processing Conference (EUSIPCO 2002)*, Toulouse, France, September 2002.

[4] Katzenbeisser, Stefan, and Fabien A.P. Petitcolas (eds.) (2000). *Information Hiding Techniques for Steganography and Digital Watermarking*. Artech House.

[5] Petitcolas, Fabien A.P., Ross Anderson, and Markus G. Kuhn (1998). "Attacks on copyright marking systems." *Information Hiding*. Lecture Notes in Computer Science, vol. 1525, ed. D. Aucsmith. Springer-Verlag, Berlin, 218–238.

[6] Secure Digital Music Initiative. http://www.sdmi.org

[7] Simmons, Gustavus J. (1984). "The prisoners' problem and the subliminal channel." *Advances in Cryptology—CRYPTO'83*, Lecture Notes in Computer Science, ed. D. Chaum. Plenum Press, New York, 51–67.

WEAK KEYS

The strength of the encryption function $E_K(P)$ may differ significantly for different keys K. If for some set WK of keys the encryption function is much weaker than for the others, this set is called a *class of weak keys*. The attack technique that succeeds against the keys in the class WK is called a *membership test* for the class. For example, if the test uses differential cryptanalysis, then it will be called a *differential membership test*.

Suppose the key space has k bits, so that complexity of exhaustive key search is 2^k. Suppose there exists a class of weak keys of size 2^f, with a complexity of the membership test of 2^w. If $2^w < 2^f$, exploiting weak keys is more efficient than using the exhaustive search. In other words if the choice of the key of the cryptosystem is restricted to a weak-key class, the attack succeeds if it is faster than exhaustive search over this restricted key-class.

The following attack model allows to compare the conventional attacks and the attacks using weak keys. Suppose the attacker is given an access to the block box performing encryption/decryption function. Suppose that the box has a key-reset button, which causes the key to change inside the box uniformly at random. We call the attack successful if the attacker can recover at least one of the keys of the box faster than exhaustive search (or in a relaxed scenario, is able to distinguish a box with a cipher from a box with a collection of random-permutations). The measure of complexity of such an attack in a weak key scenario is 2^{k-f+w}. This can be compared directly to the complexities of the conventional attacks, in which the attacker will try to break a "fixed" key, i.e., will not touch the key-reset button. The larger the weak-key class the faster the membership test—the better the attack would be. A typical example of a cipher with large weak-key classes is IDEA. For example, a class of 2^{63} weak keys out of total 2^{128} keys has been reported [2] for a full 8.5-round IDEA. The membership test has negligible complexity given only 20 chosen plaintexts. The measure of complexity in this case would be $2^{k-f+w} \approx 2^{128-63+4} = 2^{69}$ steps to recover one of the 128-bit keys of the black-box containing the IDEA cipher. An example of a cipher completely broken by the weak key analysis is *Lucifer* [1]. In this 128-bit block cipher half of the keys can be discovered by a differential membership test using 2^{36} chosen plaintexts and analysis steps. The attack complexity measure in this case is $2^{128-127+36} = 2^{37}$. In the case of *DES* (see Data Encryption Standard) there is a set of four keys for which the cipher is an involution, i.e., $DES_k(DES_k(m)) = m$.

Alex Biryukov

References

[1] Ben-Aroya, I. and E. Bihim (1993). "Differential crytanalysis of Lucifer." *Advances in Cryptology—CRYPT'93*, Lecture Notes in Computer Science, vol. 773, ed. D.R. Stinson. Springer-Verlag, 187–199. See also *Journal of Cryptology*, 9 (1), 21–34, 1996.

[2] Hawkes, P. (1998). "Differential-linear weak key classes of IDEA." *Advances in Cryptology—EUROCRYPT'98*, Lecture Notes in Computer Science, vol. 1403, ed. K. Nyberg. Springer-Verlag, 112–126.

WEB SECURITY

INTRODUCTION: Millions of users use the World Wide Web (web) everyday to obtain information, make purchases, move money, etc. Virtually every company uses the web to enhance its business, whether by publishing information on the company web site or by using the web site for electronic commerce. The web enabled business-to-consumer and business-to-business electronic commerce. Everyday, people and businesses publish sensitive information on their web sites and transfer sensitive data through the Internet to web servers. All of those actions involve certain security risks, which affect users, web servers, and networks that host web sites.

The risks are severe from the Webmaster's perspective. A web server opens a window into your local network that the entire Internet can peer through. Most web users will only look at the information that is provided to them and will not try to access things they are not authorized to see. Unfortunately, some users will try to circumvent access controls and gain unauthorized access to confidential information. They may even try to break into the host computer on which the web server is running and use it to gain access to other computers on the organization's local area network. The results can range from the merely embarrassing (i.e., the discovery one morning that your site's home page has been replaced by an obscene parody) to the damaging (i.e., the theft of your entire database of customer information).

The end-users aren't without risk either. Active content, such as ActiveX controls and Java applets, introduces the possibility that web browsing will introduce viruses or other malicious software into the user's system.

Both end-users and web administrators need to worry about the confidentiality of data transmitted across the web. Confidential information might be intercepted by an attacker when documents are transmitted from a web server to a browser or when the end-user sends private information back to the web server inside a fill-out form. To prevent information theft, it might be necessary to encrypt data traveling between clients and servers. The *SSL* and *TLS* protocols (see Secure Socket Layer and Transport Layer Security) were developed to enable secure transfer of confidential information between web servers and clients. These protocols encrypt the data so that potential eavesdroppers on the Internet cannot read the information.

Authentication is essential to most web applications. It is important for the clients to know that they are sending personal information, such as credit card numbers, to the right web site. It is just as important for the server to authenticate clients. Currently, servers use usernames and passwords to authenticate clients. Clients, on the other hand, use certificates to authenticate servers. See authentication *codes* for more information on different authentication schemes and possible vulnerabilities.

SSL and TLS protect data in transit between web clients and web servers. Steps must also be taken to prevent theft of that data from servers or clients themselves. The following sections describe existing threats in web server and web client security and possible solutions.

WEB BROWSER SECURITY: From the beginning of the web, developers have looked for ways to enhance static HTML pages and display dynamic and more interactive content on web pages. Sun, Netscape, and Microsoft have developed technologies to automatically download and run programs on demand. Such programs are often called *mobile code*.

Downloading and running unknown programs on a user's computer introduces serious security risks. Many programs have bugs that may cause the browser or even the computer to crash. Some bugs can also be exploited to gain access to private information on the computer. Other mobile programs are malicious by design. They might erase all of the information on your computer's disk, plant a virus, or scan your keyboard to obtain passwords, credit card numbers, etc. They might also search for confidential information stored on your computer and transmit it to some location on the Internet. Some companies have also used mobile code to learn the email addresses or browsing history of unaware users.

This section describes different mobile code technologies and related security risks, as well as examples of security breaches due to flaws in mobile code implementations.

MIME

Multipurpose Internet Mail Extensions (MIME) [4, 5] is an Internet standard that specifies how messages must be formatted so that they can be exchanged between different email systems. MIME is also used by web servers and browsers. MIME is a very flexible format, permitting one to include virtually any type of file or document in an email message. Specifically, MIME messages can contain text, images, audio, video, or other application specific data.

MIME defines a number of header fields describing a message, including *Content-Type*. The Content-Type field is used to specify the nature of the data in the body of a MIME message. For example, *Content-Type: image/jpeg* describes a jpeg image file.

Mailers and browsers use MIME to determine how to handle a particular message. In case of *image/jpeg* file, an email program or a browser can invoke a picture viewer to display the file.

MIME was designed to be flexible and extensible. It allows an "application" media type which is used to describe data to be processed by some type of external application. This could cause serious security risks if the external application is not secure or if the application is a language interpreter. For example, *application/PostScript* indicates a PostScript program. The PostScript language includes a number of dangerous features that make it risky to invoke external PostScript interpreters to process files obtained via email or downloaded from the web. Some dangerous operations in the language include "deletefile," "renamefile," etc.

Plug-ins

Plug-ins were developed to extend the functionality of a browser. Often written by a third-party, they usually add the capability to display a nonstandard media type in the browser window (animations, video clips, etc.).

Traditionally, plug-ins are manually downloaded by the web user. Modern browsers offer

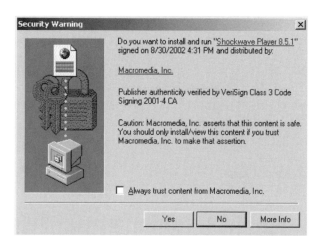

Fig. 1. Security warning displayed by Internet Explorer when trying to install a plug-in

to automatically download plug-ins for unknown media types. Plug-ins must be installed using a "setup" utility. During installation, a plug-in registers itself with the browser. If the browser downloads a file with a registered file type, the plug-in is automatically loaded into the browser's memory and started. By default, browsers display a security warning when a plug-in is about to be downloaded and installed (Figure 1). Modern web browsers allow users to turn off warnings or to disable downloading of executable files altogether. Plug-ins for Internet Explorer are usually signed using Microsoft's *Authenticode* technology. The signature clearly states who created the plug-in. A user can download the plug-in if he trusts the plug-in maker. Downloading plug-ins made by trustworthy organizations only can minimize the risks described below. Section 2.4 describes the Authenticode technology in more detail.

Plug-ins pose several risks to a user's computer:

- The plug-in might be a malicious program. Such program can do anything from stealing credit card numbers from your disk to deleting important files and dialing expensive phone numbers.
- The plug-in might be a legitimate plug-in, but the downloaded copy might have been modified by third parties. Such plug-in might change your security settings, or even contain a virus.
- The plug-in might contain a bug that can be exploited by someone who wants to steal information or damage your computer.
- The plug-in might implement a programming language that can be misused by an attacker.

To limit risk, one should only download widely used plug-ins with no known security problems. Plug-ins should be downloaded only from the original source or another reliable web site to reduce the risk of getting a modified plug-in. It is also important to keep track of updates to different plug-ins and to install security patches as they become available.

Java

Java is an object-oriented programming language that was first designed at Sun Microsystems for use in embedded devices such as television set-top boxes. With the failure of interactive cable TV trials, Java was repositioned as the language for the World Wide Web. Java was designed to run independently of the underlying hardware and operating system. Java programs can run on virtually any platform. This cross-platform compatibility makes Java an ideal candidate for *mobile code* programs.

In a typical Java implementation, Java programs are compiled into a processor-independent *bytecode*. The bytecode is loaded into a computer's memory by the *Java Class Loader*. Finally, the bytecode is executed on a *Java virtual machine* (JVM).

The Java virtual machine can run Java programs directly on an operating system such as Windows or Unix; alternately, the JVM can be embedded inside a web browser, allowing programs to be executed as they are downloaded from the web. Java bytecode can be executed directly using the interpreter, or the JVM can use a *just-in-time* compiler to convert the bytecode into native machine code. Java can also be compiled directly into machine code and run on a target machine. Compiling Java code for a particular architecture is done for performance reasons but compiled code looses its portability.

Java was designed to be a *safe* programming language, which increases reliability of programs written in it. The main way that Java achieves this reliability is by providing automatic memory management mechanisms.

Java has a built-in garbage collection system. Programmers do not have to manually use standard memory services (i.e., *malloc*, *free*, etc.) since they are automatically called by the Java environment. Programmers also do not have to worry about memory leaks due to the automatic memory allocation and deallocation. Java also has automatic bounds checking on all strings and arrays, which eliminates buffer overflows. Buffer overflow errors are a major source of bugs and security flaws in C and C++ programs. Java lacks pointers that can be arithmetically manipulated and dereferenced, which helps avoid memory leaks and bugs. Unlike C and C++, Java is a strongly

typed language and it provides a sophisticated exception mechanism.

Since most security problems are due to bugs and programming errors, programs written in the Java language are thought to be more secure than programs written in traditional languages such as C or C++.

Having safe programming language protects users from many conventional security problems but it cannot protect users from programs that are intentionally malicious. It is necessary to place limits on what downloaded programs can do.

Java Security Architecture. Java employs a variety of techniques to limit what a downloaded program can do:

- *Java sandbox*
  Java programs (applets) are prohibited from directly manipulating a computer's hardware or making calls to the operating system. Java programs run on a virtual computer inside a restricted virtual space.
- *SecurityManager class*
  Java allows programs to run with different sets of privileges, depending on where the program comes from. For example, programs downloaded from an untrusted Internet source should be more restricted than programs run directly from the user's hard disk. The *SecurityManager* class determines whether "dangerous" operations (disk writes, network access, etc.) should be allowed.
- *Class Loader*
  It is important that malicious code cannot disable security checks built into the Java system. One way to do this would be to have a malicious program disable or modify the *SecurityManager* class. Class Loader examines classes to make sure they do not violate the runtime system.
- *Bytecode Verifier*
  The Bytecode Verifier makes sure that the downloaded bytecode was created by compiling a valid Java program. For example, it ensures that the downloaded program neither forges pointers nor does it violate access restrictions or the types of objects.

Originally, Java applets downloaded from the web were restricted in their capabilities: applets could not access files on the user's disk; they could not initiate network connections to any computer other than the computer from which an applet was downloaded; they could not receive network connections, etc. With time, web browser makers (Microsoft, Netscape, etc.) were forced by web content developers to relax those restrictions and to allow downloaded Java programs to execute more

- *Disable Java*
  This setting prevents Java programs and applets from executing.
- *High safety*
  This setting allows applets to run in the Java sandbox. Applets cannot access files on the disk and can only connect to the web site from which they were downloaded.
- *Medium safety*
  Applets run in the Java sandbox. They can also access "scratch space," and access certain files as allowed by the user.
- *Low safety*
  Applets run outside the sandbox. Applets have access to all resources on the computer and can make connections to any computer on the Internet.
- *Custom*
  This setting allows the user to specify each individual permission.

Fig. 2. Default Java security policies in Internet Explorer

functions. Rather than giving all Java applets full access to a user's computer, those companies implemented user control over the additional capabilities.

Users can customize privileges given to downloaded applets through Java security policy. Internet Explorer allows you to create multiple security policies for different categories of web sites and then to assign web sites to these categories. Microsoft calls these categories *security zones*. They can be configured through the "Security" tab of the Internet Options control panel.

Microsoft Internet Explorer uses four different policies for running Java programs. The policy used for a given program depends on the current security zone for the web page that you are viewing. A user can also create custom policies. The default Java security policies are showed in Figure 2.

Java Security Problems. Unfortunately, in the short time since its release, a number of security holes have been found in Java. Most of the problems were implementation errors such as bugs in the Java runtime system that allowed specially designed Java applets to execute arbitrary programs on the user's machine. A buffer overflow bug in some Windows 95/NT versions of the Java virtual machine was discovered in September 1998. This bug could potentially cause your computer to crash. A demonstration of how the bug could compromise your computer can be found at http://www.eyeone.no/KillerApp/KillerApp.htm. A list of vulnerable browsers can also be found at that URL.

Major errors found in Java implementations include bugs in the Java virtual machine that let programs violate Java's type system. These types of bugs made it possible for applets to execute arbitrary code. Class library bugs found in a number of Java implementations allowed hostile programs to

learn "private information" about the user. A number of fundamental design errors, leading to web spoofing and unrestricted network access, were also found. Most of the problems were fixed shortly after they were reported. For a list of problems found by a security group at Princeton University, see [11].

Java is believed to be a far better alternative than the other forms of active content. Unfortunately, it is difficult to guarantee that implementations of JVM are free of bugs that could open security holes due to the complexity of the Java environment. In addition to Java security architecture described above, recent versions of Java allow cryptographic signing of programs similar to Microsoft's *Authenticode* described in the next section.

ActiveX

ActiveX is a technology developed by Microsoft Corporation for distributing software over the Internet. ActiveX "controls" can be embedded in Web pages, where they typically appear as smart interactive graphics. You can think of ActiveX controls as self-installing plug-ins.

ActiveX controls are automatically downloaded when the web browser encounters a <OBJECT> tag on a web page. Depending on the current security settings of the browser, the tags are either ignored or cause the software to be downloaded. After downloading, the ActiveX control may be run automatically, or the user may be prompted as to whether the control should be allowed to execute or not. Figure 3 shows the ActiveX security settings in Internet Explorer.

ActiveX controls can do anything a regular program can—from displaying a new file type to upgrading your computer's operating system. This amount of power comes with increased risks.

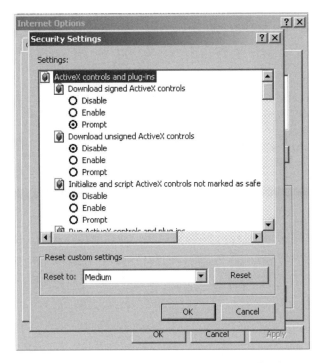

Fig. 3. ActiveX security settings in Internet Explorer

Instead of trying to develop a new system for downloading code that would be "safe" (like Java), Microsoft decided that each ActiveX control can be digitally "signed" by its author in such a way that the signature cannot be altered or repudiated using a system called "Authenticode". The digital signatures are then certified by a trusted "certifying authority", such as VeriSign, to create the equivalent of a shrink-wrapped software package. When a digital certificate is granted, the software developer pledges that the software is free from viruses and other malicious components. If you download a signed ActiveX control and it crashes your machine, you will at least know who to blame. In other words, "Authenticode" vouches for the downloaded program but does nothing to restrict its execution.

There are many potential problems with the Authenticode technology. To illustrate that Authenticode could not provide safety, a software developer Fred McLain (mclain@halcyon.com) published an ActiveX control named Exploder in 1996. This control, which has been fully signed and certified, performs a clean shutdown of any Windows 95 machine that downloads the Exploder control (using Microsoft Internet Explorer version 3.0 or higher). After learning about Exploder, Microsoft and VeriSign jointly revoked Fred McLain's certified digital signature, claiming that he had violated the agreement when the certificate was issued. Therefore, if you are running a newer

version of Internet Explorer, you'll see a message that the Exploder's software certificate is invalid.

While Exploder does not cause any data loss, a less friendly control might reformat the user's hard disk or plant a virus. For example, in February 1997, Chaos Computer Club of Hamburg, Germany demonstrated an ActiveX control that would hack into a copy of Quicken on a computer and initiate a funds transfer (without the user entering any passwords into the program).

For a detailed report on ActiveX security see the report published by CERT [1].

JavaScript

JavaScript was developed by Netscape to add forms validation, local computation, and more interactivity to web browsers. JavaScript programs reside in HTML files, usually surrounded by <SCRIPT> tags. JavaScript programs can also open new windows, fill out fields in forms, jump to new URLs, process image maps, change the HTML content of the page itself, compute mathematical results, and perform many other functions. JavaScript can also modify the appearance of web browsers, making elements of the web browser appear and disappear dynamically.

JavaScript in its basic form should be more secure than Java or other mobile code technologies since there are no JavaScript methods that directly access the network or the files on the user's computer.

Unfortunately, most JavaScript implementations have had a number of significant bugs, which opened a number of security holes. As with Java, JavaScript was developed without a formal security model. Security relied on hunches and ad-hoc methods developed by creators. This lack of a security model was the main cause of a large number of security problems discovered in JavaScript.

Some of the major problems found include a bug in the JavaScript implementation in Netscape Communicator 4.5 and 4.04–4.05, which allows a Web page to read arbitrary files from the user's machine and transmit those files across the Internet. A similar bug was also found in Internet Explorer versions 4.0–4.01 and prerelease versions of IE 5.0. A different type of error, which allows remote sites to read the user's browsing history, was found in Netscape Navigator 3.04, 4.07 and 4.5.

Many other bugs have been discovered (see [3]). Although in most cases Microsoft and Netscape quickly corrected the reported bugs, users were not quick to upload new patches.

Other JavaScript Attacks. JavaScript can be used to mount a <u>denial-of-service</u> attack against the user's computer. These attacks can reside on web pages or they can be sent to users via email. Most of those attacks result in the web browser crashing. The computer itself can also crash as a result of these attacks. A simple JavaScript denial-of-service attack is shown below:

```
<script>
while (1) {
alert("Denial of Service");
}
</script>
```

JavaScript can also be used to spoof user context. For example, the status line of a browser normally displays the URL that will be accessed if the user clicks on a link. Using JavaScript, an attacker can change the status line of a browser, tricking a user into believing that that the URL points someplace else. For example, an HTML link can display the URL http://www.shopping-mall.com/order.html when a mouse is moved over the link, but clicking on the link will jump to the web page http://www.hacker.com/gotcha.html.

WEB SERVER SECURITY: Installing a web server can be a very simple operation that has profound implications. A web server provides a place where clients, customers, and partners come to learn more about your organization, to shop, to exchange ideas, or to find entertainment. Unfortunately, the increased visibility can attract some unwelcome attention. A web site host is a natural target for attack. Some attackers might attempt to break the web server's access control in order to view confidential information, customer records, etc. Others might try to "Webjack" your site and modify its content. Webjackings are becoming more and more common and many widely known sites have been vandalized in recent years.

Breaking into a web server can also be a prelude to an attack on other computers on the local area network. Once inside the web server host, an attacker can attempt to access file servers, databases, and other confidential systems.

Web server security can be divided into three different areas: *physical security* (see also <u>physical attacks</u>), *host security, and application security.*

Physical Security

If the server hardware isn't secure, nothing is. Surprisingly, physical security is ignored by most organizations. The role of physical security is to protect computer hardware and the data stored on it. Physical security threats include theft as well as natural (and man-made) disasters. An advanced security system designed to prevent security breaches becomes useless if an attacker can simply steal hardware (including the web server or its data drives) from the organization's buildings.

Physical security deals with restricting access to the hardware on which your servers are running as well as protecting desktops and laptops and the data that may be stored on them. Part of physical security is also prevention and recovery after disasters (i.e., fire, earthquake, war, terrorism). It may be impossible to prevent most of such disasters, but organizations that are prepared will recover faster. Redundant servers at different locations and off-site backup are some of the essential techniques for disaster recovery. For more on physical security, see Chapter 14 of [8].

Host Security

The host computer, on which the web server is running, must be secure. If an attacker can gain access to the computer through security holes in the operating system or other applications, then the web server itself and the data it serves may be compromised.

Many organizations that run servers on the Internet do not secure them against an external attack. There are a number of techniques that an attacker can use to gain access and control over a target computer. Many of these techniques allow the attacker to run unrestricted code on the target. Some of these techniques include remote exploits, malicious programs, and <u>password</u> theft:

- *Remote exploits*
  Most computer systems contain vulnerabilities that allow an attacker to compromise or disable the system over the network. Many of the remote exploits are based on the *buffer overflow* technique. This technique relies on the way C and C++ programs handle input data and the fact that historically, many programs did not check if input data is larger than the input buffer.

- *Malicious programs*
  Many attackers will deliver malicious programs (often hidden inside other, non-malicious software) to users, hoping that some of them will run it. When executed, such programs often give attackers remote access to the compromised machine. These programs are called *back doors* because they allow access to the system bypassing standard security measures.

Similarly, *trojan horses* often give control of the system to the attacker.

Viruses and *worms* are self-replicating programs that travel from computer to computer via the Internet or as email attachments (see <u>virus protection</u> or <u>Trojan horse, computer viruses and worms</u>). Viruses can often exhibit malicious behavior and delete/modify files on the system. Both viruses and worms can be used to install trojan horses or back doors on the infected system.

- *Stolen usernames and passwords*
On many computer systems, it is possible for a regular user to gain root or administrative access through a number of exploits. Because of these exploits, many attackers try to gain access as a regular user (any user) and then obtain administrative privileges through the exploit.

An attacker might attempt to obtain user passwords through *dictionary attacks*, *social engineering*, or *password sniffing*. Dictionary attacks take advantage of the fact that many users use simple English words as their passwords. Social engineering is a broad term for trying to "trick" users into divulging their passwords to the attacker (i.e., an attacker may pretend to be a system administrator). Password sniffing is done by "sniffing" network packets between the user and the remote system. Many users still use remote access programs that send user name and password in the clear (i.e., telnet, ftp, etc.), allowing an attacker to easily obtain passwords from sniffed network packets.

Most organizations attempt to protect their web servers and internal networks with <u>firewalls</u>. Firewalls are sophisticated filters installed between the organization's network and the Internet. They allow system administrators to decide which protocols will pass through the firewall and which protocols will be dropped. A firewall can be used to protect a web server by restricting incoming network traffic only to HTTP traffic on port 80.

Many organizations also use *intrusion detection* systems to warn them if their network security has been compromised. Intrusion detection systems can monitor network traffic looking for patterns indicating a remote attack or a security breach. Other systems also monitor files on servers to find out when an intruder changes them (possibly leaving behind a backdoor).

See entries for <u>firewall</u>, <u>intrusion detection</u> for more information on these topics.

Security Policy. Every organization should have a comprehensive *security policy*. A security <u>policy</u>

clearly states what is allowed and what is not allowed. It can be used as a guide when configuring systems and services. It should also be considered when installing or purchasing new services. A security policy should include the following:

- Who is allowed access (see also <u>access control</u>), what are the restrictions on that access, and who authorizes that access.
- Who is responsible for security, for upgrades, for backups, and for maintenance?
- What type of information is allowed to be published on web pages?
- Which external users are allowed access to pages and data served?
- What kind of evaluation/verification is required before new pages/information are published on the web site?
- What should be done in case of security breach/incident/break-in?
- How can the security policy be updated and who is authorized to make those changes?

Even a very good security policy is useless if it is not enforced. A number of tools and methods exist that help organizations enforce their security policy. Auditing tools are very important in checking system's compliance with the security policy. Such tools can test system and security settings of different servers as well as software configurations of desktop machines. Best common practices for security also include scanning of logs generated by servers, patching software to the latest versions, etc. Another emerging tactic is external *red teams*. This method involves an external group of security consultants who evaluate the organization's security policy, test the compliance of different systems to that policy, and attempt to uncover weaknesses in the organization's security defenses.

Minimize Services. Every network service running on a host can potentially open a security hole that can be used to compromise the web server and the data it is serving. To minimize threats to the web server, it is important to minimize the number of services running on the host. It is best to run the web server as a stand-alone process on a dedicated machine. Some services that should potentially be restricted are: Domain Name Service (DNS), Mail, Telnet, FTP, etc. Bugs in DNS and Mail services (i.e., sendmail) can be used to compromise the system. These services should be run on a separate machine. Telnet and FTP <u>protocols</u> send user names and <u>passwords</u> without <u>encryption</u>, allowing potential attackers an easy way into the system. Both should be replaced by secure access utilities (i.e., <u>SSH</u> (Secure SHell) and <u>SCP</u> (Secure CoPy)).

Logging. System logs can help to determine that a server is under attack or that it has been broken into. They are also invaluable when recovering from a security incident. They can be used to determine how the attacker broke in and even help to track down the perpetrator. Attackers often try to cover their tracks by modifying the system log file. *Forensic computing* attempts to correlate data from a number of different log files (data sources) to determine whether a machine has been compromised and what parts of the system may have been modified.

Unix and Windows NT/2000 offer advanced logging options. Events can be written to a single file, multiple files, or even sent to a different computer over the network. If security is of high concern, it is a good idea to set up a separate, very secure, log server, which receives events from other servers on the organization's network.

Application Security

Web servers were designed to display static information such as brochures, FAQs, product catalogs, etc. The need for more dynamic content delivery, such as user customizable web pages and shopping cart type applications, required special extensions to web servers. Web servers must execute specialized code every time a page is requested. There are four primary techniques for creating dynamic web content:

- *CGI and servlets*
  CGI programs are executables that are independent of the main Web server binary. When a remote user requests a URL that points at a CGI script, the server executes the script and sends the results back to the requesting browser. CGI scripts can perform database queries, perform calculations, and generate other types of dynamic content. Servlets are written in Java and are similar to CGI programs. Servlets are continuously running (unlike CGI programs which are started separately for each request), waiting for requests to be passed from the web server. Most web servers have modules that support the servlet API.
- *Server plug-ins, loadable modules, etc.*
  Modern web servers often include API for extension modules. They are usually written in C or C++ and are loaded into the web server at run time. Plug-ins and modules are much faster than CGI scripts because they don't require a new process to be started for each web interaction.
- *Embedded scription languages*
  Languages such as Microsoft's ASP, PHP, server-side JavaScript, and mod_perl allow developers to place scripts into web pages. An interpreter in the web server runs the script before the resulting page is sent to the web browser.
- *Embedded web server*
  Some systems do not use an off-the-shelf web server but rather embed an HTTP server into the custom web application itself.

The above extension techniques pose serious security risks and can compromise the security of a web server or the host on which it is running. This is because potentially any program can be run through these extensions. This could include programs with security problems, programs with exploitable bugs, programs that give outsiders access to the web server host, and programs that modify or even delete critical system files.

The web administrator (webmaster) has to not only worry about the security of the web server and the host on which it is running, but also the security of every program and script written by web and content developers.

A number of common CGI security problems have emerged over the years:

1. Use of interpreters as CGI scripts
   This occurs when a powerful interpreter, such as PERL (practical extraction and report language) or TCL (Tool Command Language) interpreters, is installed in the common CGI script directory. This mistake allows anybody on the Internet to run arbitrary perl, TCL, etc. commands on the server computer.
2. Flawed memory management
   The most common problem is the failure to check the length of input data before copying it into a fixed length buffer. This leads to the buffer overflow problem which can be exploited to execute arbitrary machine code instructions on the server.
3. Passing unchecked user input to command interpreters
   One of the most common CGI scripting bugs is the failure to validate user input. This is most serious when unchecked user input is passed to command interpreter (shell), allowing an attacker to execute arbitrary shell commands on the server.

Servlets are a popular alternative to CGI scripts and are believed to be more secure. The use of the Java language eliminates many memory management bugs which have plagued CGI scripts but other security problems remain the same. Potential bugs in the servlet module could also open security holes.

To reduce the risk introduced by web applications, the programs themselves should be designed to only perform the desired functions. More importantly, they should be inspected and verified to ensure that that is the case. The programs should run in a restricted environment. If one of these programs is compromised by an attacker, the damage will be limited.

Hidden Fields, Cookies

Many web applications use hidden HTML fields and cookies to store session information. The ability to store such information on the client makes programming web applications easier. Unfortunately, many web applications fail to validate information from hidden fields and cookies. Even though users aren't meant to see or modify that information, malicious users can read and modify the contents of those fields in an attempt to find vulnerabilities in the server side of the web application. For example, the authors of [6] identified a number of ecommerce sites that fail to validate the information in cookies. As a result, small changes to the cookies can give an attacker access to unauthorized information.

A simple way to solve many problems with hidden fields and cookies is to use cryptography to make the stored information more secure. The use of the right cryptographic techniques can prevent attackers from reading the information stored in cookies and hidden fields, and allow server applications to detect unauthorized modifications to this information. The data fields should be encrypted using a symmetric encryption function (see symmetric cryptosystem) with a secret key. Then, an HMAC function of the encrypted string should be calculated to protect the data from unauthorized modifications. Because symmetric encryption functions and HMAC functions (such as MD5) are very fast, this should not introduce much overhead to the server.

Databases

Databases help create powerful and flexible web applications. Many web applications use databases that are external to the web server. They can be used to store user preferences, shopping carts, order status, etc. External databases often store confidential information such as customer records and therefore must be protected from unauthorized access.

CGI scripts implementing the web application connect to the database through local area network and issue queries. It is important to authenticate the script issuing the query to make sure it is authorized to access the information.

It is also important to filter and validate SQL queries before sending them to the database. If a query contains data from a client, it could contain special characters that could potentially allow the attacker to execute arbitrary queries and gain access to confidential information.

CONCLUSION: The World Wide Web is here to stay. Companies use it to enhance their business by providing a number of services, including business-to-business and business-to-consumer ecommerce. Users use the web to find information, shop, bank, chat, etc. In the sections above, we described a number of security issues facing both web site administrators and users.

Users have to worry about their personal information traveling across the Internet, the security of the browser itself, and the security of numerous plug-ins or applications downloaded and installed off the web. The World Wide Web presents many threats to user's privacy as well. Personal information can be stolen from servers or it can be intercepted while traveling across the Internet. User's web browsing is being tracked and logged by web servers as well as Internet service providers (ISPs). Cookies can be used to create an electronic fingerprint of a person's online activities. See privacy for more information on privacy threats and protection.

Web site administrators, on the other hand, have to worry about the security of the host on which the web server is running, the security of the web server software, and the security of all web applications running on the server.

This section offers only a small glimpse at the issues of web security. We recommend a number of additional resources for readers interested in learning more on the subject.

A number of books on web security have been published since the *Internet revolution* began [8, 10, 12]. These books cover security issues for users and webmasters in greater detail as well as present additional information related to web security. The issue of host security is its own area and has been studied for years before the web was invented [7, 9]. These two books present detailed information on securing Unix and Windows NT/2000 servers on the Internet. Other valuable security resources are the World Wide Web Security FAQ [3] and the CERT Coordination Center [2]. The CERT web site reports the most recent security vulnerabilities and fixes.

Lukasz Opyrchal

References

[1] Results of the Security in ActiveX workshop (August 2000). www.cert.org/reports/activeX_report .pdf

[2] CERT Coordination Cente. www.cert.org

[3] The World Wide Web Security FAQ. http://www.w3
.org/Security/Faq

[4] Freed, N. and N. Borenstein (1996). "Multipurpose intenet mail extensions (mime) part one: Format of internet message bodies." *RFC 2045*.

[5] Freed, N. and N. Borenstein (1996). "Multipurpose intenet mail extensions (mime) part two: Media types." *RFC 2046*.

[6] Fu, Kevin, Emil Sit, Kendra Smith, and Nick Feamster (2001). "Dos and don'ts of client authentication on the Web." *10th USNIX Security Symposium*, August 2001, Washington, DC.

[7] Garfinkel, Simson and Gene Spafford (1996). *Practical Unix and Internet Security*. O'Reilly.

[8] Garfinkel, Simson and Gene Spafford (2002). *Web Security, Privacy and Commerce*. O'Reilly.

[9] Norberg, Stefan and Deborah Russell (2000). *Security Windows NT/2000 Servers for the Internet*. O'Reilly.

[10] Rubin, Aviel, Daniel Geer, and Marcus Ranum (1997). *Web Security Sourcebook*. John Wiley and Sons, Inc.

[11] Secure Internet Programming. www.cs.princeton.edu/sip

[12] Stein, Lincoln D. (1997). *Web Security*. Addison-Wesley, Reading, MA.

WIENER, BONEH–DURFEE, AND MAY ATTACKS ON THE RSA PUBLIC KEY CRYPTOSYSTEM

These cryptanalytic results (see cryptanalysis) show that the RSA public key encryption scheme is not secure under certain conditions on the RSA parameters.

To use RSA one first chooses two large prime numbers p and q. The public modulus is $N = pq$, and p and q are kept secret (see modular arithmetic). A public exponent e and a secret (private) exponent d are chosen satisfying the RSA equation $ed = 1 \mod(N - p - q + 1)$. Typically, the secret exponent d is of approximately the same bit-length as the public modulus N, the public exponent e has a bit-length no greater than the bit-length of N, and the primes p and q are approximately the same bit-length.

Because RSA decryption and signature generation depend on the length of the secret exponent d, it is tempting to use a short d to speed up these operations. However, in 1990, Wiener [1] showed that if the length of d is less than one-quarter of the length of N, then there is an efficient attack which computes the secret exponent d from the public modulus N and public exponent e. This attack is based on the method of continued fractions (see integer factoring). One of the countermeasures suggested by Wiener is to use a public modulus e much larger than N; as e ranges from N to $N^{1.5}$, shorter d are required for this attack to succeed. Once e is greater than $N^{1.5}$, Wiener's attack provides no information about d.

In 2000, Boneh and Durfee [3] showed a heuristic attack which recovers d from e and N when the length of d is less than 0.292 times the length of N. The attack uses ideas due to Coppersmith [2] for finding solutions to polynomial equations using lattices. As in Wiener's attack, increasing the length of e decreases the effectiveness of the attack; the Boneh–Durfee attack works up to e close to $N^{1.875}$. The attack works for larger d when e is chosen to be much shorter than N. The attack also improves when the lengths of p and q are vastly different.

In 2000, Durfee and Nguyen [4] showed a similar heuristic attack that recovers d from e and N in a variant of RSA which uses atypical ratios between the lengths of p and q as well as e and N. The size of d for which the attack succeeds is a complicated function of these ratios; see [4] for details.

To enjoy the computational efficiency of a short secret exponent without exposure to the Wiener or Boneh–Durfee attacks, the typical countermeasure is to use short secrets for an RSA variant using the Chinese Remainder Theorem (CRT). In RSA–CRT, one chooses secrets d_p and d_q satisfying $ed_p = 1 \mod(p - 1)$ and $ed_q = 1 \mod(q - 1)$. RSA decryption and signature generation times depend on the lengths of d_p and d_q. If d_p and d_q are chosen to be of short bit-length, the efficiency gains due to short secrets can be realized without exposure to the Wiener or Boneh–Durfee attacks.

In 2002, May [5] showed an attack that succeeds when d_p and d_q are small, provided that p and q are chosen to be of substantially different bit lengths. The ranges for which the attack succeeds is a complicated function of these bit lengths; see [5] for details.

Glenn Durfee

References

[1] Wiener, M. (1990). "Cryptanalysis of short RSA secret exponents." *IEEE Transactions on Information Theory*, 36 (3), 553–558.

[2] Coppersmith, D. (2001). "Finding small solutions to small degree polynomials." *Proceedings of Cryptography and Lattices Conference*, Lecture Notes in

Computer Science, vol. 2146, ed. J.H. Silverman. Springer-Verlag, Berlin.

[3] Boneh, D. and G. Durfee (2000). "Cryptanalysis of RSA with private key d less than $N^{0.292}$." *IEEE Transactions on Information Theory*, 46 (4), 1339–1349.

[4] Durfee, G. and P. Nguyen (2000). "Cryptanalysis of RSA schemes with short secret exponent from Asiacrypt'99." *Proceedings of ASIACRYPT 2000*, Lecture Notes in Computer Science, vol. 1976, ed. T. Okamoto. Springer-Verlag, Berlin, 14–29.

[5] May, A. (2002). "Cryptanalysis of unbalanced RSA with small CRT-exponent." *Proceedings of CRYPTO 2002*, Lecture Notes in Computer Science, vol. 2442, ed. M. Yung. Springer-Verlag, Berlin, 242–256.

WITNESS HIDING

Witness hiding is a property attributed to an interactive proof or an interactive argument, similar in nature to the zero-knowledge property. If an interactive proof is zero-knowledge, it is also witness hiding, but the converse does not necessarily hold. For many applications, witness hiding protocols are sufficiently secure and more efficient than zero-knowledge protocols.

Loosely speaking, an interactive proof is witness hiding if an arbitrarily cheating verifier, after engaging in many executions of the protocol with an honest prover, is not able to compute a witness unless the verifier is able to compute one anyway, without interacting with the prover at all. In this context, a witness may be thought of as a private key, corresponding to a public key. If an interactive proof is witness hiding, it is not excluded that a cheating verifier learns some fraction of the bits of a witness. However, knowledge of such a fraction of the bits cannot be feasibly extended to knowing all of the bits, if the proof is witness hiding.

Witness hiding protocols were introduced in [2], followed by practical constructions in [4] and [1]. Witness hiding protocols are constructed by means of **witness-indistinguishable** protocols. An interactive proof is witness-indistinguishable if an arbitrarily cheating verifier, after engaging in many executions of the protocol with an honest prover, is not able to tell which witness the prover is using. If there is only one possible witness, then an interactive proof is trivially witness-indistinguishable. If several witnesses are possible, the views of the protocol as seen by the verifier should be indistinguishable—and this holds even when the verifier knows all possible witnesses.

Okamoto's variation of Schnorr's identification protocol is an example of a witness hiding protocol [4]. Let g denote a generator of a cyclic group G of order p, where p is a large prime number. Let h denote a random element of G, $h \neq 1$, such that the discrete logarithm of h with respect to g is not known to any party.

A key pair is generated by choosing $w, x \in \mathbb{Z}_p$ uniformly at random, and setting $y = g^w h^x$ as the public key and (w, x) as the corresponding private key (see modular arithmetic and public key encryption). The protocol runs as follows.

The prover chooses $u, v \in \mathbb{Z}_p$ uniformly at random, sets $a = g^u h^v$ and sends a to the verifier. Next, the verifier chooses a challenge $c \in \mathbb{Z}_p$ and sends c to the prover. The prover computes the responses $r = u + cw \bmod p$, $s = v + cx \bmod p$ and sends r, s to the verifier. Finally, the verifier checks that $g^r h^s = ay^c$ holds.

Note that for each public key y, there are exactly p possible private keys ("witnesses") (w', x') such that $y = g^{w'} h^{x'}$. However, it can be proved that whatever a cheating verifier does, when interacting with an honest prover, the verifier is not able to tell which witness the prover is using—and this holds even when the verifier knows the values of the possible witnesses (w', x'). Therefore, the protocol is witness-indistinguishable.

To see that the protocol is also witness hiding, one reasons at follows. Suppose, on the contrary, that a cheating verifier is able to find a witness (w', x'). Since the protocol is witness-indistinguishable, the probability that the witness found by the verifier is equal to the witness used by the prover is $1/p$. That is, with very high probability, we see that the prover and the verifier (viewing these as a combined algorithm) are able to compute two distinct witnesses (w, x) and (w', x'). From $y = g^w h^x$ and $y = g^{w'} h^{x'}$, it follows that $h = g^{(w-w')/(x'-x)}$, implying that the discrete log of h with respect to g is known. This contradicts the assumption that this discrete log is not known to any party.

Berry Schoenmakers

References

[1] Cramer, R., I. Damgård, and B. Schoenmakers (1994). "Proofs of partial knowledge and simplified design of witness hiding protocols." *Advances in Cryptology—CRYPTO'94*, Lecture Notes in Computer Science, vol. 839, ed. Y.G. DeSnedt. Springer-Verlag, Berlin, 174–187.

[2] Feigel, U. and A. Shamir (1990). "Witness indistinguishable and witness hiding protocols." *Proceedings of 22nd Symposium on Theory of*

Computing (STOC'90), ed. J.H. Silverman. ACM Press, New York, 416–426.

[3] Goldreich, O. (2001). *Foundations of Cryptography—Basic Tools*. Cambridge University Press, Cambridge.

[4] Okamoto, T. (1993). "Provably secure and practical identification schemes and corresponding signature schemes." In *Advances in Cryptology—CRYPTO'92*, Lecture Notes in Computer Science, vol. 740, ed. E.F. Brickell. Springer-Verlag, Berlin, 31–53.

X.509

The X.509 standard (ITU, the International Telecommunication Union, formerly CCITT) is the best known <u>Public Key Infrastructure</u> (PKI) standard for electronic commerce, even though originally it was developed for applications like authorized reading and writing access to records in data bases rather than electronic commerce. The standard defines <u>certification authorities</u> which issues <u>certificates</u> on users. These certificates are specified as well as based on the ASN1 syntax with a number of records, some of which are mandatory, whereas others are optional. The original specification was rather limited in its scope, and the current commonly used version is version 3. The Certification Authorities are off-line authorities, each connected to one or several <u>Registration Authorities</u>, and the certificates are made available in Directories. The original setting was X.500 distributed databases, and it was anticipated to have a global setup, where CA's would identify each other through so-called certificate paths. In this way, any two registered users would be able to access each other's public keys through an appropriate certificate path if the same CA did not certify them. The standard further specifies the use of *blacklists*, or *revocation lists*, which is one of the weaknesses of the whole setup, as these revocation lists only are updated at discrete points in time, and a user can never be sure—when he receives a valid certificate from another user—whether meanwhile the certificate has been revoked even though it appears to be valid.

Another important point often missed in realizations is that <u>non-repudiation</u> cannot be achieved unless the architecture offers independent <u>time stamping</u>. Indeed, if a signed message is accepted on the basis of a valid <u>digital signature</u> and a valid certificate, and the involved public key later is revoked by its owner, the signed message may not any longer have any value to the verifier unless he can prove that he received the valid signature and certificate before the public key was revoked, e.g., by obtaining an signature from a time stamping authority on the hash (see <u>hash function</u>) of the said message appended with a time stamp at any point in time before the revocation.

Peter Landrock

ZERO-KNOWLEDGE

Zero-knowledge is a property attributed to inter-active proofs, interactive arguments and non-interactive proofs. Since the soundness property protects the interest of the verifier, the zero-knowledge property protects the interest of the prover. By means of a zero-knowledge proof, the prover is able to convince the verifier of the validity of a given statement, *without releasing any knowledge beyond the validity of the statement.* (Note that the notion of witness hiding proofs provides an alternative to the notion of zero-knowledge proofs.)

In other words, from executing a zero-knowledge protocol with an honest prover, the verifier should learn nothing beyond the validity of the statement. This is captured by stating that whatever the verifier 'sees' when interacting with the prover by means of the zero-knowledge protocol can be *efficiently simulated* by the verifier itself. It is crucial to note that the zero-knowledge condition should be satisfied even if the verifier deviates from the protocol in arbitrary ways.

As a simple example, consider the following protocol. Let g denote a generator of a cyclic group G of order p, where p is a large prime number. A key pair is generated by choosing $x \in \mathbb{Z}_p$ uniformly at random, and setting $y = g^x$ as the public key and x as the corresponding private key (see also modular arithmetic and public key cryptography). The protocol for proving knowledge of x on common input y runs as follows.

The prover chooses $u \in \mathbb{Z}_p$ uniformly at random, sets $a = g^u$ and sends a to the verifier. Next, the verifier chooses a challenge $c \in \{0, 1\}$ uniformly at random and sends c to the prover. The prover computes the response $r = u + cx \mod p$ and sends it to the verifier. Finally, the verifier checks that $g^r = ay^c$ holds.

The zero-knowledge property follows from the fact that the outputs of the following two probabilistic polynomial-time algorithms are identically distributed, where V^* denotes an arbitrarily cheating verifier. It is assumed that V^* is given as a rewindable black-box. A triple (a, c, r) is called a *conversation*, as it consists of the messages exchanged during a run of the protocol.

Real conversations
Input: private key x
Output: conversation (a, c, r)
1. Choose random $u \in \mathbb{Z}_p$
2. Set $a = g^u$
3. Send a to V^*
4. Receive $c \in \{0, 1\}$ from V^*
5. Set $r = u + cx \mod p$
6. Output (a, c, r)

Simulated conversations
Input: public key y
Output: conversation (a, c, r)
1. Choose random $c \in \{0, 1\}, r \in \mathbb{Z}_p$
2. Set $a = g^r y^{-c}$
3. Send a to V^*
4. Receive $c' \in \{0, 1\}$ from V^*
5. If $c \neq c'$ rewind V^* to point prior to accepting a and go to step 1
6. Output (a, c, r)

At step 5 of the simulation, the probability that $c = c'$ is exactly $1/2$, since $c \in \{0, 1\}$ is chosen uniformly at random. Hence, on average two iterations are required to generate a simulated transcript (a, c, r).

The conclusion is that no matter what algorithm (or "strategy") a cheating verifier V^* follows in trying to extract useful information from the prover, the same algorithm can be used to generate identically distributed conversations without needing the cooperation of the prover. Whereas the real conversations are generated using the private key x as input, the simulated conversations are generated using only the public key y as input.

In general, the distributions of the real conversations and the simulated conversations do not need to be identical. *Perfect* zero-knowledge means that the distributions are indeed identical. *Almost-perfect* or *statistical* zero-knowledge means that the distributions are statistically indistinguishable (i.e., the statistical distance between the distributions is negligible). Similarly, *computational* zero-knowledge (see also computational complexity) means that the distributions are polynomially indistinguishable (i.e., cannot be efficiently distinguished).

By engaging in a zero-knowledge protocol multiple times, a cheating verifier may collect many valid conversations. In general, the simulation for a single run of a <u>protocol</u> can be easily extended to a simulation for multiple runs of the protocol as long as the runs are sequential, that is, the second run starts only after the first run is finished, and so on. In other words, the zero-knowledge property is preserved under *sequential composition*. However, *parallel composition*, where a prover is engaged in several runs of a protocol at the same time, in general, does not preserve the zero-knowledge property; running the above simulation k times in parallel does not result in an efficient simulation as the chances that $c = c'$ holds at step 5 for all runs at the same time will be only 2^{-k}.

The concept of zero-knowledge was introduced by Goldwasser et al. in the early eighties (journal version appeared in [6]). In [7] it was subsequently proved that a zero-knowledge interactive proof exists for every language in NP. *Non-interactive* zero-knowledge proofs were introduced in [1, 2]. There are many varieties of zero-knowledge proofs, see [8] for an overview. Examples of some advanced notions are *concurrent* zero-knowledge [4, 5] and *resettable* zero-knowledge [3].

Berry Schoenmakers

References

[1] Blum, M., P. Feldman, and S. Micali (1988). "Non-interactive zero-knowledge and its applications." *Proceedings of the 20th ACM Symposium on the Theory of Computing*, 103–112.

[2] Blum, M., A. De Santis, S. Micali, and G. Persiano (1991). "Non-interactive zero-knowledge proof systems." *SIAM Journal on Computing*, 20 (6), 1084–1118.

[3] Canetti, R., O. Goldreich, S. Goldwasser, and S. Micali (2000). "Resettable zero-knowledge." *Proceedings of the 32nd ACM Symposium on the Theory of Computing*, 235–244.

[4] Damgård, I. (2000). "Efficient concurrent zero-knowledge in the auxiliary string model." *Advances in Cryptology—EUROCRYPT 2000*, Lecture Notes in Computer Science, vol. 1807, ed. B. Preneel. Springer-Verlag, Berlin, 418–430.

[5] Dwork, C., M. Naor, and A. Sahai (1998). "Concurrent zero-knowledge." *Proceedings of the 30th ACM Symposium on the Theory of Computing*, 409–418.

[6] Goldwasser, S., S. Micali, and C. Rackoff (1989). "The knowledge complexity of interactive proof systems." *SIAM Journal on Computing*, 18, 186–208. Preliminary version in *17th ACM Symposium on the Theory of Computing*, 1982.

[7] Goldreich, O., S. Micali, and A. Wigderson (1991). "Proofs that yield nothing but their validity or all languages in NP have zero-knowledge proof systems." *Journal of the ACM*, 38 (1), 691–729. Preliminary version in *27th IEEE Symposium on Foundations of Computer Science*, 1986.

[8] Goldreich, O. (2001). *Foundations of Cryptography—Basic Tools*. Cambridge University Press, Cambridge.

Index